McGraw-Hill Ryerson

# Chemistry

## Author Team

**Dr. Frank Mustoe**
The University of Toronto Schools
Toronto, Ontario

**Michael P. Jansen**
Crescent School
Toronto, Ontario

**Dr. Michael Webb**
Michael J. Webb Consulting Inc.
Toronto, Ontario

**Ted Doram**
Bowness High School
Calgary, Alberta

**Christy Hayhoe**
Professional Writer
Toronto, Ontario

**Jim Gaylor**
Formerly with St. Michael
Catholic Secondary School Board
Stratford, Ontario

**Anita Ghazariansteja**
Birchmount Collegiate
Scarborough, Ontario

## Consultants

**Ted Doram**
Bowness High School
Thornhill Secondary School
Calgary, Alberta

**Greg Wisnicki**
Anderson Collegiate and Vocational Institute
Whitby, Ontario

**Dr. Audrey Chastko**
Science Curriculum Leader
Springbank Community High School
Calgary, Alberta

**Dr. Penny McLeod**
Former Head of Science
Thornhill Secondary School
Thornhill, Ontario

## Probeware Specialist

**Kelly Choy**
Minnedosa Collegiate
Minnedosa, Manitoba

## Technology Consultants

**Alex Annab**
Head of Science
Iona Catholic Secondary School
Mississauga, Ontario

**Wilf Kazmaier**
Multimedia Specialist
Calgary, Alberta

Toronto   Montréal   Boston   Burr Ridge, IL   Dubuque, IA   Madison, WI   New York   San Francisco
St. Louis   Bangkok   Bogotá   Caracas   Kuala Lumpur   Lisbon   London   Madrid   Mexico City
Milan   New Delhi   Santiago   Seoul   Singapore   Sydney   Taipei

## McGraw-Hill Ryerson Limited
*A Subsidiary of The McGraw-Hill Companies*

**COPIES OF THIS BOOK MAY BE OBTAINED BY CONTACTING:**
McGraw-Hill Ryerson Ltd.

**E-MAIL:**
*orders@mcgrawhill.ca*

**TOLL FREE FAX:**
1-800-463-5885

**TOLL FREE CALL:**
1-800-565-5758

**OR BY MAILING YOUR ORDER TO:**
McGraw-Hill Ryerson
Order Department,
300 Water Street,
Whitby ON, L1N 9B6

Please quote the ISBN and title when placing your order.

**McGraw-Hill Ryerson Chemistry**

Copyright © 2004, McGraw-Hill Ryerson Limited, a Subsidiary of The McGraw-Hill Companies. All rights reserved. No part of this publication may be reproduced or transmitted in any form or by any means, or stored in a data base or retrieval system, without the prior written permission of McGraw-Hill Ryerson Limited, or, in the case of photocopying or other reprographic copying, a licence from The Canadian Copyright Licensing Agency, (Access Copyright). For an Access Copyright license, visit www.accesscopyright.ca or call toll free to 1-800-893-5777.

Any request for photocopying, recording, or taping of this publication shall be directed in writing to CANCOPY.

The information and activities in this textbook have been carefully developed and reviewed by professionals to ensure safety and accuracy. However, the publishers shall not be liable for any damages resulting, in whole or in part, from the reader's use of the material. Although appropriate safety procedures are discussed in detail and highlighted throughout the textbook, safety of students remains the responsibility of the classroom teacher, the principal, and the school board/district.

0-07-097106-4

*http://www.mcgrawhill.ca*

1 2 3 4 5 6 7 8 9 0 TCP 0 9 8 7 6 5 4 3

Printed and bound in Canada

Care has been taken to trace ownership of copyright material contained in this text. The publisher will gladly take any information that will enable it to rectify any reference or credit in subsequent printings. Please note that products shown in photographs in this textbook do not reflect an endorsement by the publisher of those specific brand names.

**National Library of Canada Cataloguing in Publication Data**

McGraw-Hill Ryerson chemistry/Frank Mustoe ... [et al.].—Revision

For grade 11 and 12

Includes index.

Previously published as: McGraw-Hill Ryerson chemistry 11 and McGraw-Hill Ryerson chemistry 12.

ISBN 0-07-097106-4

1. Chemistry. I. Mustoe, Frank J., 1947-. II. Title: McGraw-Hill Ryerson chemistry 11. III. Title: McGraw-Hill Ryerson chemistry 12.

QD33. M335 2003      540      C2003-904132-8

**The Chemistry Team**
SCIENCE PUBLISHER: Jane McNulty
MANAGER OF EDITORIAL SERVICES: Linda Allison
PROJECT MANAGER: Keith Owen Richards
DEVELOPMENTAL EDITORS: Jon Bocknek, Lois Edwards,
    Sara Goodchild, Christy Hayhoe
SENIOR SUPERVISING EDITOR: Linda Allison
EDITORIAL ASSISTANT: Erin Parton
PERMISSIONS EDITOR: Pronk&Associates
SPECIAL FEATURES CO-ORDINATOR: Jill Bryant
PRODUCTION SUPERVISOR: Yolanda Pigden
PRODUCTION CO-ORDINATOR: Jennifer Wilkie
DESIGN AND ELECTRONIC PAGE MAKE-UP: Pronk&Associates
SET-UP PHOTOGRAPHY: Ian Crysler
SET-UP PHOTOGRAPHY CO-ORDINATOR: Shannon O'Rourke
TECHNICAL ILLUSTRATION: Theresa Sakno, Jun Park, Pronk&Associates
COVER IMAGE: Ken Edwards/Science Source/Photo Researchers Inc.

## Acknowledgements

The authors, editors, project manager, and publisher of *McGraw-Hill Ryerson Chemistry* would like to extend our deepest thanks, first and foremost, to academic and pedagogical reviewers Sheldon Gillam, Murray Park, and Kevin Toope. Their diligent efforts in dividing up the assessment of the complete manuscript at various stages of development, ensuring its pedagogical soundness and readability, have contributed immensely to the final product. Their ability to see how the smallest details fit into the "big picture" has aided both authors and editors in presenting the material in this text as effectively and creatively as possible.

The editors also benefited from the insights and advice provided by the other outstanding educators from Atlantic Canada, listed below, who reviewed selected chapters, checked the Glossary, and provided safety reviews for additional labs.

We extend thanks as well to Eric Grace, who contributed several excellent new features to this edition, to Dr. Frank Mustoe, who provided expert guidance during the development of the *McGraw-Hill Chemistry Teachers' Resource*, and to Jim Gaylor, who showed great care in producing the *Solutions Manual*.

Last but not least, we wish to thank the gifted designers at Pronk&Associates for their close collaboration with us on this project.

### Pedagogical and Academic Reviewers

**Sheldon Gillam**
Gill Memorial Academy
Musgrave Harbour, Newfoundland and Labrador

**Murray Park**
Science Department Head
O'Donel High School
Mount Pearl, Newfoundland and Labrador

**Kevin Toope**
Prince of Wales Collegiate
St. John's, Newfoundland and Labrador

**Peter Atkinson**
Vice-Principal
Kennebecasis Valley High School
Rothesay, New Brunswick

**Vicki Atkinson**
Kennebecasis Valley High School
Rothesay, New Brunswick

**Tammy Lawlor**
Fredericton High School
Fredericton, New Brunswick

**Walter Prokopiw**
Formerly at Horton High School
Wolfville, Nova Scotia

**Kathy Reid**
Science Department Head
St. Patrick's High School
Halifax, Nova Scotia

**Sean Ward**
Tantramar Regional High School
Sackville, New Brunswick

**Jocelyn Wells**
St. Malachy's High School
Saint John, New Brunswick

### Accuracy Reviewers

**Dr. Michael C. Baird**
Queen's University
Kingston, Ontario

**Dr. Christopher Flinn**
Memorial University of Newfoundland
St. John's, Newfoundland and Labrador

**Dr. R.J. Gillespie**
McMaster University
Hamilton, Ontario

**Ian Krouse**
Formerly with University of Calgary
Calgary, Alberta
Purdue University
West Lafayette, Indiana

### Safety Reviewers

**Dr. Margaret-Ann Armour**
University of Alberta
Edmonton, Alberta

**Ian Fogarty**
Riverview High School
Moncton, New Brunswick

**John Henry**
Science Teacher's Association of Ontario Safety Committee
Hamilton, Ontario

**Margaret Redway**
Fraser Scientific and Business Services
Delta, British Columbia

**Our cover:** The image on our cover shows a computer-generated model of a $C_{60}$ fullerene molecule, containing a molecule of methanol. Fullerenes are spherical molecules of carbon. Since their discovery in 1985, they have fascinated scientists with their perfect geometry and their range of potential applications, including superconductors, rocket fuels, and lubricants. Scientists can manipulate the properties of a fullerene by inserting an atom or a small molecule into it, as shown here, or by bonding a different chemical group to its outside surface. You will examine the connections between structure, bonding, and properties in Unit 2.

# Contents

**Safety in Your Chemistry Laboratory and Classroom**    x

## Chapter 1   Studying Matter and Its Changes    2

**1.1** Observing Matter    3
   **Chemistry Bulletin:** Air Canada Flight 143    7
   *ThoughtLab*: Mixtures, Pure Substances, and Changes    10
**1.2** Building Blocks of Matter    12
**1.3** Communicating About Matter    22
   *Investigation 1-A*: Analyzing an Industrial Process    34
   Chapter 1 Review    37

## UNIT 1   Stoichiometry    40

### Chapter 2   The Mole    42

**2.1** Isotopes and Average Atomic Mass    43
**2.2** The Avogadro Constant and the Mole    47
   *ThoughtLab*: The Magnitude of $6.02 \times 10^{23}$    50
**2.3** Molar Mass    55
   **Chemistry Bulletin:** Chemical Amounts in Vitamin Supplements    61
**2.4** Molar Volume    66
   *ThoughtLab*: Molar Volumes of Gases    69
   Chapter 2 Review    75

### Chapter 3   Chemical Proportions in Compounds    78

**3.1** Percentage Composition    79
   *ThoughtLab*: Percent by Mass and Percent by Number    83
**3.2** The Empirical Formula of a Compound    87
   *Investigation 3-A*: Determining the Empirical Formula of Magnesium Hydroxide    92
**3.3** The Molecular Formula of a Compound    95
   **Careers in Chemistry:** Analytical Chemistry    96
**3.4** Finding Empirical and Chemical Formulas by Experiment    99
   *Investigation 3-B*: Determining the Chemical Formula of a Hydrate    104
   Chapter 3 Review    107

### Chapter 4   Quantities in Chemical Reactions    110

**4.1** Introducing Stoichiometry    111
   *ExpressLab*: Mole Relationships in a Chemical Reaction    112
   **Canadians in Chemistry:** Dr. Stephen Beauchamp    123
**4.2** The Limiting Reactant    128
   *ThoughtLab*: The Limiting Item    129
   *Investigation 4-A*: Limiting and Excess Reactants    132

| | | |
|---|---|---|
| **4.3** | Percentage Yield | 137 |
| | *Investigation 4-B*: Determining the Percentage Yield of a Chemical Reaction | 142 |
| | **Careers in Chemistry:** Chemical Engineer | 144 |
| | **Chemistry Bulletin:** Nickel Mining at Voisey's Bay | 145 |
| | Chapter 4 Review | 149 |
| | Unit 1 Review | 152 |

## UNIT 2  Structure and Properties    156

### Chapter 5  Chemical Bonding    158

| | | |
|---|---|---|
| **5.1** | Elements and Compounds | 159 |
| | *ExpressLab*: A Metal and a Compound | 160 |
| | *ThoughtLab*: Ionic or Molecular | 161 |
| **5.2** | Bond Formation | 165 |
| | **Careers in Chemistry:** Metallurgist | 173 |
| **5.3** | Bonds as a Continuum | 174 |
| | Chapter 5 Review | 181 |

### Chapter 6  Structure and Properties of Substances    184

| | | |
|---|---|---|
| **6.1** | Covalent Bonds and Structures | 185 |
| | **Tools & Techniques:** AIM Theory and Electron Density Maps | 194 |
| | *Investigation 6-A*: Modelling Molecules | 197 |
| **6.2** | Intermolecular Forces | 202 |
| | *Investigation 6-B*: Investigating the Properties of Water | 207 |
| | *ThoughtLab*: Properties of Liquids | 210 |
| | **Canadians in Chemistry:** Dr. R.J. Le Roy | 213 |
| **6.3** | Structure Determines Properties | 216 |
| | **Canadians in Chemistry:** Dr. Geoffrey Ozin | 220 |
| | *Investigation 6-C*: Properties of Substances | 222 |
| | **Chemistry Bulletin:** Ionic Liquids: A Solution to the Problem of Solutions | 223 |
| | Chapter 6 Review | 225 |
| | Unit 2 Review | 230 |

## UNIT 3  Solutions and Solubility    234

### Chapter 7  Solutions and Their Concentrations    236

| | | |
|---|---|---|
| **7.1** | Types of Solutions | 237 |
| | *ThoughtLab*: Matching Solutes and Solvents | 241 |
| **7.2** | Factors That Affect Rate of Dissolving and Solubility | 243 |
| | *Investigation 7-A*: Plotting Solubility Curves | 249 |
| | *ExpressLab*: The Effect of Temperature on Soda Water | 251 |
| | **Chemistry Bulletin:** Solvents and Coffee: What's the Connection? | 253 |

Contents • MHR  v

| | | |
|---|---|---|
| **7.3** | The Concentration of Solutions | 255 |
| | **Careers in Chemistry:** Product Development Chemist | 265 |
| | *Investigation 7-B*: Determining the Concentration of a Solution | 269 |
| **7.4** | Preparing Solutions | 271 |
| | *Investigation 7-C*: Estimating Concentration of an Unknown Solution | 274 |
| | Chapter 7 Review | 277 |

### Chapter 8  Solution Stoichiometry — 280

| | | |
|---|---|---|
| **8.1** | Making Predictions About Solubility | 281 |
| | *Investigation 8-A*: The Solubility of Ionic Compounds | 283 |
| **8.2** | Reactions in Aqueous Solutions | 288 |
| | *Investigation 8-B*: Qualitative Analysis | 296 |
| **8.3** | Stoichiometry in Solution Chemistry | 299 |
| | **Chemistry Bulletin:** Water Quality | 309 |
| | Chapter 8 Review | 310 |
| | Unit 3 Review | 314 |

### UNIT 4  Organic Chemistry — 318

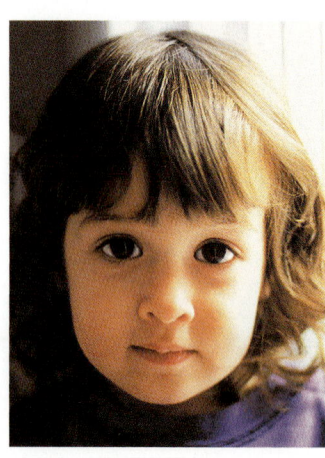

### Chapter 9  Hydrocarbons — 320

| | | |
|---|---|---|
| **9.1** | Introducing Organic Compounds | 321 |
| **9.2** | Representing Hydrocarbons | 324 |
| | *Investigation 9-A*: Modelling Organic Compounds | 328 |
| | *ExpressLab*: Molecular Shapes | 329 |
| **9.3** | Classifying Hydrocarbons | 331 |
| | *Investigation 9-B*: Comparing the Reactivity of Alkanes and Alkenes | 350 |
| | *Investigation 9-C*: Structures and Properties of Aliphatic Compounds | 359 |
| **9.4** | Refining and Using Hydrocarbons | 365 |
| | **Tools & Techniques:** Oil Refining in Newfoundland and Labrador | 369 |
| | **Canadians in Chemistry:** Dusanka Filipovic | 370 |
| | Chapter 9 Review | 371 |

### Chapter 10  Hydrocarbon Derivatives — 376

| | | |
|---|---|---|
| **10.1** | Functional Groups | 377 |
| | *ThoughtLab*: Comparing Intermolecular Forces | 380 |
| **10.2** | Single-Bonded Functional Groups | 386 |
| **10.3** | Functional Groups with the C=O Bond | 402 |
| | *Investigation 10-A*: Preparing a Carboxylic Acid Derivative | 408 |
| | *Investigation 10-B*: Comparing Physical Properties | 418 |
| | Chapter 10 Review | 421 |

| Chapter 11 | Organic Compounds on a Larger Scale | 426 |
|---|---|---|
| 11.1 | Polymer Chemistry | 427 |
| | *Investigation 11-A*: Examining Polymers | 432 |
| | **Chemistry Bulletin:** Degradable Plastics: Garbage That Takes Itself Out | 435 |
| 11.2 | Natural Polymers and Other Biomolecules | 437 |
| | **Canadians in Chemistry:** Dr. Raymond Lemieux | 443 |
| 11.3 | Organic Compounds and Everyday Life | 445 |
| | *ThoughtLab*: Risk Benefit Analyses of Organic Products | 448 |
| | *ThoughtLab*: Problem Solving with Organic Compounds | 451 |
| | **Careers in Chemistry:** Polymer Chemist | 451 |
| | Chapter 11 Review | 453 |
| | Unit 4 Review | 456 |

## UNIT 5  Kinetics and Equilibrium  460

| Chapter 12 | Rates of Chemical Reactions | 462 |
|---|---|---|
| 12.1 | Factors That Affect Reaction Rates | 463 |
| | *Investigation 12-A*: Studying Reaction Rates | 464 |
| | **Tools & Techniques:** Methods for Measuring Reaction Rates | 466 |
| 12.2 | Collision Theory and Reaction Mechanism | 469 |
| | **Canadians in Chemistry:** Dr. Maud L. Menten | 483 |
| | Chapter 12 Review | 486 |

| Chapter 13 | Reversible Reactions and Chemical Equilibrium | 488 |
|---|---|---|
| 13.1 | Recognizing Equilibrium | 489 |
| | *ExpressLab*: Modelling Equilibrium | 491 |
| 13.2 | The Equilibrium Constant | 494 |
| | *Investigation 13-A*: Measuring an Equilibrium Constant | 501 |
| 13.3 | Predicting the Direction of a Reaction | 517 |
| | *Investigation 13-B*: Perturbing Equilibrium | 521 |
| | **Chemistry Bulletin:** Le Châtelier's Principle: Beyond Chemistry | 525 |
| | **Careers in Chemistry:** Anesthesiology: A Career in Pain Management | 534 |
| | Chapter 13 Review | 535 |
| | Unit 5 Review | 538 |

## UNIT 6  Acids and Bases  542

| Chapter 14 | Properties of Acids and Bases | 544 |
|---|---|---|
| 14.1 | Defining Acids and Bases | 545 |
| | *Investigation 14-A*: Observing Properties of Acids and Bases | 545 |

| | | |
|---|---|---|
| **14.2** | Strong and Weak Acids and Bases | 560 |
| | *Investigation 14-B*: The Effect of Dilution on the pH of an Acid | 575 |
| | **Chemistry Bulletin:** The Chemistry of Oven Cleaning | 577 |
| | Chapter 14 Review | 579 |

### Chapter 15 Acid-Base Equilibria and Reactions — 582

| | | |
|---|---|---|
| **15.1** | Revisiting Acid-Base Strength | 583 |
| **15.2** | Acid-Base Reactions and Titration Curves | 599 |
| | *Investigation 15-A*: The Concentration of Acetic Acid in Vinegar | 606 |
| | *Investigation 15-B*: $K_a$ of Acetic Acid | 613 |
| | **Careers in Chemistry:** Dangerous Goods Inspection | 615 |
| | Chapter 15 Review | 617 |
| | Unit 6 Review | 620 |

### UNIT 7 Thermochemistry — 624

### Chapter 16 Theories of Energy Changes — 626

| | | |
|---|---|---|
| **16.1** | Temperature Change and Heat | 627 |
| | *ThoughtLab*: Factors in Heat Transfer | 631 |
| **16.2** | Enthalpy Changes | 639 |
| **16.3** | Heating and Cooling Curves | 651 |
| | *ExpressLab*: Construct a Heating Curve and a Cooling Curve | 653 |
| | Chapter 16 Review | 657 |

### Chapter 17 Measuring and Using Energy Changes — 660

| | | |
|---|---|---|
| **17.1** | The Technology of Heat Measurement | 661 |
| | **Tools & Techniques:** The First Ice Calorimeter | 662 |
| | *Investigation 17-A*: Determining the Enthalpy of a Neutralization Reaction | 668 |
| | *ExpressLab*: The Energy of Dissolving | 670 |
| | *Investigation 17-B*: The Enthalpy of Combustion of a Candle | 671 |
| **17.2** | Hess's Law of Heat Summation | 677 |
| | *Investigation 17-C*: Hess's Law and the Enthalpy of Combustion of Magnesium | 682 |
| **17.3** | Fuelling Society | 692 |
| | **Chemistry Bulletin:** Hot Ice | 695 |
| | **Careers in Chemistry:** Oil Spill Advisor | 697 |
| | **Chemistry Bulletin:** Lamp Oil and the Petroleum Age | 699 |
| | *ThoughtLab*: Comparing Energy Sources | 701 |
| | Chapter 17 Review | 703 |
| | Unit 7 Review | 706 |

viii MHR • Contents

## UNIT 8 Electrochemistry — 710

### Chapter 18 Oxidation-Reduction Reactions — 712

- 18.1 Defining Oxidation and Reduction — 713
  - **Chemistry Bulletin:** Aging: Is Oxidation a Factor? — 717
  - *Investigation 18-A*: Single Displacement Reactions — 718
- 18.2 Oxidation Numbers — 721
  - T*houghtLab*: Finding Rules for Oxidation Numbers — 723
- 18.3 The Half-Reaction Method for Balancing Equations — 730
  - **Tools & Techniques:** The Breathalyzer Test: A Redox Reaction — 739
  - *Investigation 18-B*: Redox Reactions and Balanced Equations — 740
- 18.4 The Oxidation Number Method for Balancing Equations — 747
  - Chapter 18 Review — 751

### Chapter 19 Cells and Batteries — 756

- 19.1 Galvanic Cells — 757
  - *Investigation 19-A*: Measuring Cell Potentials of Galvanic Cells — 762
  - **Careers in Chemistry:** Explosives Chemist — 766
- 19.2 Standard Cell Potentials
  - *ThoughtLab*: Assigning Reference Values — 774
- 19.3 Electrolytic Cells — 776
  - *Investigation 19-B*: Electrolysis of Aqueous Potassium Iodide — 784
- 19.4 Faraday's Law — 790
  - *Investigation 19-C*: Electroplating — 794
- 19.5 Issues Involving Electrochemistry — 798
  - **Canadians in Chemistry:** Dr. Viola Birss — 804
  - Chapter 19 Review — 807
  - Unit 8 Review — 810

Appendix A: Answers to Selected Numerical Chapter and Unit Review Questions — 814
Appendix B: Supplemental Practice Problems — 822
Appendix C: Alphabetical List of Elements and Periodic Table of the Elements — 835
Appendix D: Math and Chemistry — 838
Appendix E: Chemistry Data Tables — 845
Glossary — 851
Index — 861
Credits — 870

# Safety in Your Chemistry Laboratory and Classroom

**The following *Safety Precautions* symbols appear throughout *Chemistry*, whenever an investigation or ExpressLab presents possible hazards.**

appears when there is a danger to the eyes, and safety goggles, safety glasses, or a face shield should be worn

appears when substances that could burn or stain clothing are used

appears when objects that are hot or cold must be handled

appears when sharp objects are used, to warn of the danger of cuts and punctures

appears when toxic substances that can cause harm through ingestion, inhalation, or skin absorption are used

appears when corrosive substances, such as acids and bases, that can damage tissue are used

warns of caustic substances that could irritate the skin

appears when chemicals or chemical reactions that could cause dangerous fumes are used and ventilation is required

appears as a reminder to be careful when you are around open flames and when you are using easily flammable or combustible materials

warns of danger of electrical shock or burns from live electrical equipment

Actively engaging in laboratory investigations is essential to gaining a hands-on understanding of chemistry. Following safe laboratory procedures should not be seen as an inconvenience in your investigations. Instead, it should be seen as a positive way to ensure your safety and the safety of others who share a common working environment. Familiarize yourself with the following general safety rules and procedures. It is your responsibility to follow them when completing any of the investigations or ExpressLabs in this textbook, or when performing other laboratory procedures.

## General Precautions

- Always wear safety glasses and a lab coat or apron in the laboratory. Wear other protective equipment, such as gloves, as directed by your teacher or by the Safety Precautions at the beginning of each investigation.

- If you wear contact lenses, always wear safety goggles or a face shield in the laboratory. Inform your teacher that you wear contact lenses. Generally, contact lenses should not be worn in the laboratory. If possible, wear eyeglasses instead of contact lenses, but remember that eyeglasses are not a substitute for proper eye protection.

- Know the location and proper use of the nearest fire extinguisher, fire blanket, fire alarm, first aid kit, and eyewash station (if available). Find out from your teacher what type of fire-fighting equipment should be used on particular types of fires. (See "Fire Safety" on page xiii.)

- Do not wear loose clothing in the laboratory. Do not wear open-toed shoes or sandals. Accessories may get caught on equipment or present a hazard when working with a Bunsen burner. Ties, scarves, long necklaces, and dangling earrings should be removed before starting an investigation.

- Tie back long hair and any loose clothing before starting an investigation.

- Lighters and matches must not be brought into the laboratory.

- Food, drinks, and gum must not be brought into the laboratory.

- Inform your teacher if you have any allergies, medical conditions, or physical problems (including hearing impairment) that could affect your work in the laboratory.

## Before Beginning Laboratory Investigations

- Listen carefully to the instructions that your teacher gives you. Do not begin work until your teacher has finished giving instructions.

- Obtain your teacher's approval before beginning any investigation that you have designed yourself.

- Read through all of the steps in the investigation before beginning. If there are any steps that you do not understand, ask your teacher for help.

- Be sure to read and understand the Safety Precautions at the start of each investigation or Express Lab.

x   MHR • Safety in Your Chemistry Laboratory and Classroom

- Always wear appropriate protective clothing and equipment, as directed by your teacher and the Safety Precautions.
- Be sure that you understand all safety labels on materials and equipment. Familiarize yourself with the WHMIS symbols on this page.
- Make sure that your work area is clean and dry.

**During Laboratory Investigations**

- Make sure that you understand and follow the safety procedures for different types of laboratory equipment. Do not hesitate to ask your teacher for clarification if necessary.
- Never work alone in the laboratory.
- Remember that gestures or movements that may seem harmless could have dangerous consequences in the laboratory. For example, tapping people lightly on the shoulders to get their attention could startle them. If they are holding a beaker that contains an acid, for example, the results could be very serious.
- Make an effort to work slowly and steadily in the laboratory. Be sure to make room for other students.
- Organize materials and equipment neatly and logically. For example, do not place materials that you will need during an investigation on the other side of a Bunsen burner from you. Keep your bags and books off your work surface and out of the way.
- Never taste any substances in the laboratory.
- Never touch a chemical with your bare hands.
- Never draw liquids or any other substances into a pipette or a tube with your mouth.
- If you are asked to smell a substance, do not hold it directly under your nose. Keep the object at least 20 cm away, and waft the fumes toward your nostrils with your hand.
- Label all containers holding chemicals. Do not use chemicals from unlabelled containers.
- Hold containers away from your face when pouring liquids or mixing reactants.
- If any part of your body comes in contact with a potentially dangerous substance, wash the area immediately and thoroughly with water.
- If you get any material in your eyes, do not touch them. Wash your eyes immediately and continuously for 15 min, and make sure that your teacher is informed. A doctor should examine any eye injury. If you wear contact lenses, take them out immediately. Failing to do so may result in material becoming trapped behind the contact lenses. Flush your eyes with water for 15 min, as above.
- Do not touch your face or eyes while in the laboratory unless you have first washed your hands.
- Do not look directly into a test tube, flask, or the barrel of a Bunsen burner.
- If your clothing catches fire, smother it with the fire blanket or with a coat, or get under the safety shower.
- If you see any of your classmates jeopardizing their safety or the safety of others, let your teacher know.

**WHMIS (Workplace Hazardous Materials Information System) symbols are used in Canadian schools and workplaces to identify dangerous materials. Familiarize yourself with the symbols below.**

 Poisonous and Infectious Material Causing Immediate and Serious Toxic Effects

 Poisonous and Infectious Material Causing Other Toxic Effects

 Flammable and Combustible Material

 Compressed Gas

 Corrosive Material

 Oxidizing Material

 Dangerously Reactive Material

 Biohazardous Infectious Material

**Heat Source Safety**

- When heating any item, wear safety glasses, heat-resistant safety gloves, and any other safety equipment that your teacher or the Safety Precautions suggests.
- Always use heat-proof, intact containers. Check that there are no large or small cracks in beakers or flasks.
- Never point the open end of a container that is being heated at yourself or others.
- Do not allow a container to boil dry unless specifically instructed to do so.
- Handle hot objects carefully. Be especially careful with a hot plate that may look as though it has cooled down, or glassware that has recently been heated.
- Before using a Bunsen burner, make sure that you understand how to light and operate it safely. Always pick it up by the base. Never leave a Bunsen burner unattended.
- Before lighting a Bunsen burner, make sure there are no flammable solvents nearby.
- If you do receive a burn, run cold water over the burned area immediately. Make sure that your teacher is notified.
- When you are heating a test tube, always slant it. The mouth of the test tube should point away from you and from others.
- Remember that cold objects can also harm you. Wear appropriate gloves when handling an extremely cold object.

**Electrical Equipment Safety**

- Ensure that the work area, and the area of the socket, is dry.
- Make sure that your hands are dry when touching electrical cords, plugs, sockets, or equipment.
- When unplugging electrical equipment, do not pull the cord. Grasp the plug firmly at the socket and pull gently.
- Place electrical cords in places where people will not trip over them.
- Use an appropriate length of cord for your needs. Cords that are too short may be stretched in unsafe ways. Cords that are too long may tangle or trip people.
- Never use water to fight an electrical equipment fire. Severe electrical shock may result. Use a carbon dioxide or dry chemical fire extinguisher. (See "Fire Safety" on the next page.)
- Report any damaged equipment or frayed cords to your teacher.

**Glassware and Sharp Objects Safety**

- Cuts or scratches in the chemistry laboratory should receive immediate medical attention, no matter how minor they seem. Alert your teacher immediately.
- Never use your hands to pick up broken glass. Use a broom and dustpan. Dispose of broken glass as directed by your teacher. Do not put broken glassware into the garbage can.

- Cut away from yourself and others when using a knife or another sharp object.
- Always keep the pointed end of scissors and other sharp objects pointed away from yourself and others when walking.
- Do not use broken or chipped glassware. Report damaged equipment to your teacher.

**Fire Safety**

- Know the location and proper use of the nearest fire extinguisher, fire blanket, and fire alarm.
- Understand what type of fire extinguisher you have in the laboratory, and what type of fires it can be used on. (See below.) Most fire extinguishers are the ABC type.
- Notify your teacher immediately about any fires or combustible hazards.
- Water should only be used on Class A fires. Class A fires involve ordinary flammable materials, such as paper and clothing. Never use water to fight an electrical fire, a fire that involves flammable liquids (such as gasoline), or a fire that involves burning metals (such as potassium or magnesium).
- Fires that involve a flammable liquid, such as gasoline or alcohol (Class B fires) must be extinguished with a dry chemical or carbon dioxide fire extinguisher.
- Live electrical equipment fires (Class C) must be extinguished with a dry chemical or carbon dioxide fire extinguisher. Fighting electrical equipment fires with water can cause severe electric shock.
- Class D fires involve burning metals, such as potassium and magnesium. A Class D fire should be extinguished by smothering it with sand or salt. Adding water to a metal fire can cause a violent chemical reaction.
- If someone's hair or clothes catch on fire, smother the flames with a fire blanket. Do not discharge a fire extinguisher at someone's head.

**Clean-Up and Disposal in the Laboratory**

- Clean up all spills immediately. Always inform your teacher about spills.
- If you spill acid or base on your skin or clothing, wash the area immediately with a lot of cool water.
- You can neutralize small spills of acid solutions with sodium hydrogen carbonate (baking soda). You can neutralize small spills of basic solutions with sodium hydrogen sulfate or citric acid.
- Clean equipment before putting it away, as directed by your teacher.
- Dispose of materials as directed by your teacher, in accordance with your local School Board's policies. Do not dispose of materials in a sink or a drain unless your teacher directs you to do so.
- Wash your hands thoroughly after all laboratory investigations.

---

**Web LINK**

**CAUTION** **The Use and Sharing of Computer Files**

Be sure to install an up-to-date virus checker on your computer, because even a disk given to you by a close friend could contain a virus. Never open an e-mail file if you do not know the source of the message, since you could, inadvertently, infect your computer with a virus that could destroy files containing many hours of your work. If you discover that you have a virus in your computer, notify your teacher and peers immediately.

Visit the *Chemistry* web site at **www.mcgrawhill.ca/links/atlchemistry** where you will find links to many useful, reputable educational institutions and organizations—excellent sources of material for research.

# CHAPTER 1
# Studying Matter and Its Changes

**Chapter Preview**

1.1 Observing Matter
1.2 Matter and the Atom
1.3 Communicating About Matter

Imagine a chemical that
- is a key ingredient in most pesticides
- contributes to environmental hazards, such as acid rain, the greenhouse effect, and soil erosion
- helps to spread pollutants that are present in all contaminated rivers, lakes, and oceans
- is used in vast quantities by every industry on Earth
- can produce painful burns to exposed skin
- causes severe illness or death in either very low or very high concentrations in the body
- is legally discarded as waste by individuals, businesses, and industries
- has been studied extensively by scientists throughout the world

In 1996, a high school student wrote a report about this chemical for a science fair project. The student called the chemical dihydrogen monoxide. The information in the student's report was completely factual. As a result, 86% of those who read the report—43 out of 50 students—voted in favour of banning the chemical. What they did not realize was that "dihydrogen monoxide" is simply another name for water.

What if you did not know that water and dihydrogen monoxide are the same thing? What chemical knowledge and skills can help you distinguish genuine environmental issues from pranks like this one?

This chapter will reacquaint you with the science of chemistry. You will revisit important concepts and skills from previous grades. You will also prepare to extend your knowledge and skills in new directions.

What mistake in measuring matter nearly resulted in an airplane disaster in 1983? Read on to find the answer to this question later in this chapter.

# Observing Matter

## 1.1

Many people, when they hear the word "chemistry," think of scientists in white lab coats. They picture bubbling liquids, frothing and churning inside mazes of laboratory glassware.

Is this a fair portrayal of chemistry and chemists? Certainly, chemistry happens in laboratories. Laboratory chemists often do wear white lab coats, and they do use lots of glassware! Chemistry also happens everywhere around you, however. It happens in your home, your school, your community, and the environment. Chemistry is happening right now, inside every cell in your body. You are alive because of chemical changes and processes.

**Chemistry** is the study of matter and its composition. Chemistry is also the study of what happens when matter interacts with other matter. When you mix ingredients for a cake and put the batter in the oven, that is chemistry. When you pour soda water on a stain to remove it from your favourite T-shirt, that is chemistry. When a scientist puts a chunk of an ice-like solid into a beaker, causing white mist to ooze over the rim, that is chemistry, too. Figure 1.1 illustrates this interaction, as well as several other examples of chemistry in everyday life.

**Section Preview/Outcomes**

In this section you will

- **identify** chemical substances and chemical changes in everyday life
- **classify** properties of matter and changes of matter
- **communicate** your understanding of the following terms: *chemistry, matter, properties, physical property, chemical property, physical changes, chemical changes, pure substance, element, compound*

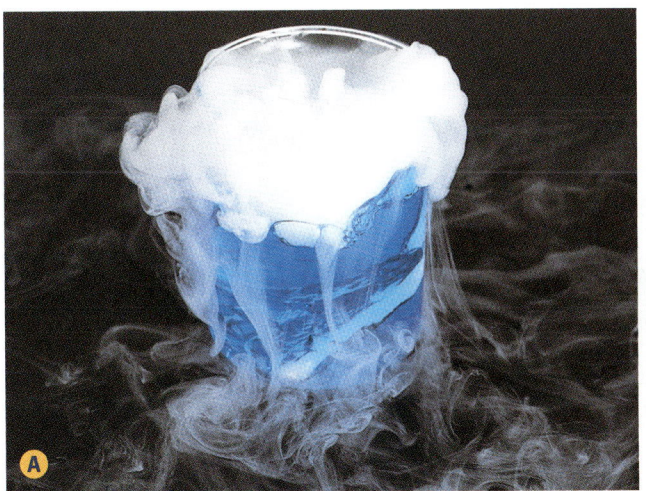

**Figure 1.1**

**A** Frozen (solid) carbon dioxide is also known as "dry ice." It changes to a gas at temperatures higher than −78°C. In this photograph, warm water has been used to speed up the process, and food colouring has been added.

**B** Green plants use a chemical process, called photosynthesis, to convert water and carbon dioxide into the food substances they need to survive. All the foods that you eat depend on this process.

**C** Your home is full of products that are manufactured by chemical industries. The products that are shown here are often used for cleaning. Some of these products, such as bleach and drain cleaner, can be dangerous if handled improperly.

Chapter 1 Studying Matter and Its Changes • MHR   3

## Chemistry: A Blend of Science and Technology

Like all scientists, chemists try to describe and explain the world. Chemists start by asking questions such as these:

- Why is natural gas such an effective fuel?
- How can we separate a mixture of crude oil and water?
- Which materials dissolve in water?
- What is rust and why does it form?

To answer these questions, chemists develop models, conduct experiments, and seek patterns. They observe various types of chemical reactions, and they perform calculations based on known data. They build continuously on the work and the discoveries of other scientists.

Long before humans developed a scientific understanding of the world, they invented chemical techniques and processes. These techniques and processes included smelting and shaping metals, growing crops, and making medicines. Early chemists invented technological instruments, such as glassware and distillation equipment.

Present-day chemical technologists continue to invent new equipment. They also invent new or better ways to provide products and services that people want. Chemical technologists ask questions such as the following:

- How can we redesign this motor to run on natural gas?
- How can we contain and clean up an oil spill?
- What methods can we use, or develop, to make water safe to drink?
- How can we prevent iron objects from rusting?

## Chemistry: A Science of Matter

Water is the most striking feature of our planet. It is visible from space, giving Earth a vivid blue colour. You can observe water above, below, and at Earth's surface. Water is a component of every living thing, from the smallest bacterium to the largest mammal and the oldest tree. You drink it, cook with it, wash with it, skate on it, and swim in it. Legends and stories involving water have been a part of every culture in human history. No other kind of matter is as essential to life as water.

> As refreshing as it may be, water straight from the tap seems rather ordinary. Try this: Describe a glass of water to someone who has never seen or experienced water before. Be as detailed as possible. See how well you can distinguish water from other kinds of matter.

In addition to water, there are millions of different kinds of matter in the universe. The dust specks suspended in the air, the air itself, your chair, this textbook, your pen, your classmates, your teacher, and you—all these are examples of matter. In the language of science, **matter** is anything that has mass and volume (takes up space). In the rest of this chapter, you will examine some key concepts related to matter. You have encountered these concepts in previous studies. Before you continue, complete the Concept Check activity to see what you recall and how well you recall it. As you proceed through this chapter, assess and modify your answers.

### CONCEPT CHECK

From memory, explain and define each of the following concepts. Use descriptions, examples, labelled sketches, graphic organizers, a computer FAQs file or Help file, or any combination of these. Return to your answers frequently during this chapter. Modify them as necessary.

- states of matter
- properties of matter
- physical properties
- chemical properties
- physical change
- chemical change
- mixture
- pure substance
- element
- compound

## Describing Matter

You must observe matter carefully to describe it well. When describing water, for example, you may have used statements such as these:

- Water is a liquid.
- It has no smell.
- Water is clear and colourless.
- It changes to ice when it freezes.
- Water freezes at 0°C.
- Sugar dissolves in water.
- Oil floats on water.

Characteristics that help you describe and identify matter are called **properties**. Figure 1.2 shows some properties of water and hydrogen peroxide.

Examples of properties include physical state, colour, odour, texture, boiling temperature, density, and flammability (combustibility). Table 1.1 lists some common properties of matter. You will gain direct experience with most of these properties as you study chemistry.

**Figure 1.2** Liquid water is clear, colourless, odourless, and transparent. Hydrogen peroxide (an antiseptic liquid that many people use to clean wounds) has the same properties. It differs from water, however, in other properties, such as boiling point, density, and reactivity with acids.

**Table 1.1** Common Properties of Matter

| Physical Properties ||  Chemical Properties |
| Qualitative | Quantitative | |
|---|---|---|
| physical state<br>colour<br>odour<br>crystal shape<br>malleability<br>ductility<br>hardness<br>brittleness | melting point<br>boiling point<br>density<br>solubility<br>electrical conductivity<br>thermal conductivity | reactivity with water<br>reactivity with air<br>reactivity with pure oxygen<br>reactivity with acids<br>reactivity with pure substances<br>combustibility (flammability)<br>toxicity<br>decomposition |

Properties may be physical or chemical. A **physical property** is a property that you can observe without changing one kind of matter into something new. For example, iron is a strong metal with a shiny surface. It is solid at room temperature, but it can be heated and formed into different shapes. These properties can all be observed without changing iron into something new.

A **chemical property** is a property that you can observe when one kind of matter is converted into a different kind of matter. For example, a chemical property of iron is that it reacts with oxygen to form a different kind of matter: rust. Rust and iron have completely different physical and chemical properties.

Figure 1.3 on the following page shows another example of a chemical property. Glucose test paper changes colour in the presence of glucose. Thus, a chemical property of glucose test paper is that it changes colour in response to glucose. Similarly, a chemical property of glucose is that it changes the colour of glucose test paper.

Recall that some properties of matter, such as colour, and flammability, are *qualitative*. You can describe them in words, but you cannot measure them or express them numerically. Other properties, such as density and boiling point, can be measured and expressed numerically. Such properties are *quantitative*.

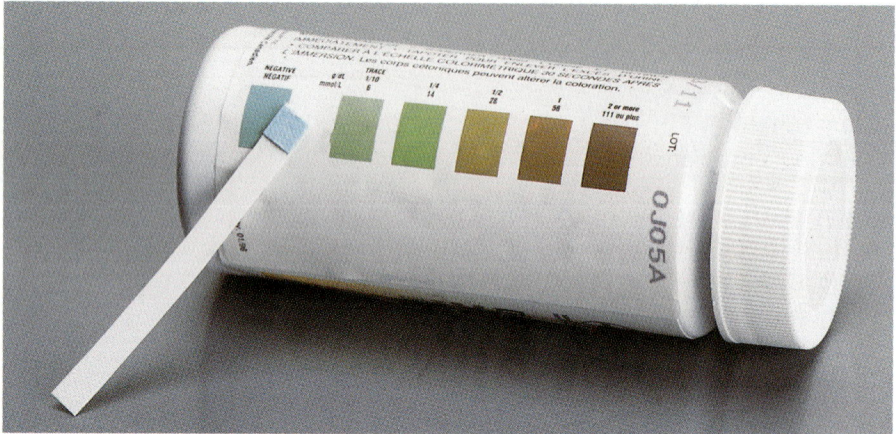

**Figure 1.3** People with diabetes rely on a chemical property to help them monitor the amount of glucose (a simple sugar) in their blood.

Try the problems below to practise distinguishing between chemical and physical properties, and between quantitative and qualitative properties. The Chemistry Bulletin that follows describes what can happen when quantitative properties are measured incorrectly.

### Practice Problems

1. State whether each of the following sentences describes a chemical property or a physical property.
    (a) Solid potassium permanganate, $KMnO_{4(s)}$, exists at room temperature as purple crystals.
    (b) Gold can be hammered into thin sheets (it is malleable) and drawn into thin wires (it is ductile).
    (c) Sodium metal reacts with water to form dissolved sodium hydroxide, $NaOH_{(aq)}$, and hydrogen gas, $H_{2(g)}$.
    (d) Solid sodium chloride, $NaCl_{(s)}$, does not conduct electricity.
    (e) Magnesium, $Mg_{(s)}$, burns in air with a bright, white flame.
    (f) Ammonia, $NH_{3(g)}$, has a strong, distinct odour.

2. Classify the following observations about the physical properties of diamond as qualitative or quantitative.
    (a) At room temperature, diamond is a clear, colourless, crystalline solid.
    (b) Diamond has no odour.
    (c) Diamond melts at 3800 K.
    (d) Diamond has a density of 3.51 g/cm$^3$.
    (e) Diamond is neither malleable nor ductile.

3. The hardness of a substance is sometimes expressed numerically according to Moh's scale. Classify each of the following statements about the hardness of diamond as qualitative or quantitative.
    (a) Diamond has a hardness of 10.0 on Moh's scale.
    (b) Diamond is one of the hardest substances known to science.

# Chemistry Bulletin

## Air Canada Flight 143

Air Canada Flight 143 was en route from Montréal to Edmonton on July 23, 1983. The airplane was one of Air Canada's first Boeing 767s, and its systems were almost completely computerized.

While on the ground in Montréal, Captain Robert Pearson found that the airplane's fuel processor was malfunctioning. As well, all three fuel gauges were not operating. Pearson believed, however, that it was safe to fly the airplane using manual fuel measurements.

Partway into the flight, as the airplane passed over Red Lake, Ontario, one of two fuel pumps in the left wing failed. Soon the other fuel pump failed and the left engine flamed out. Pearson decided to head to the closest major airport, in Winnipeg. He began the airplane's descent. At 8400 m, and more than 160 km from the Winnipeg Airport, the right engine also failed. The airplane had run out of fuel.

In Montréal, the ground crew had determined that the airplane had 7682 L of fuel in its fuel tank. Captain Pearson had calculated that the mass of fuel needed for the trip from Montréal to Edmonton was 22 300 kg. Since fuel is measured in litres, Pearson asked a mechanic how to convert litres into kilograms. He was told to multiply the amount in litres by 1.77.

By multiplying 7682 L by 1.77, Pearson calculated that the airplane had 13 597 kg of fuel on board. He subtracted this value from the total amount of fuel for the trip, 22 300 kg, and found that 8703 kg more fuel was needed.

To convert kilograms back into litres, Pearson divided the mass, 8703 kg, by 1.77. The result was 4916 L. The crew added 4916 L of fuel to the airplane's tanks.

This conversion number, 1.77, had been used in the past because the density of jet fuel is 1.77 *pounds* per litre. Unfortunately, the number that should have been used to convert litres into kilograms was 0.803. The crew should have added 20 088 L of fuel, not 4916 L.

First officer Maurice Quintal calculated their rate of descent. He determined that they would never make Winnipeg. Pearson turned north and headed toward Gimli, an abandoned Air Force base. Gimli's left runway was being used for drag-car and go-kart races. Surrounding the runway were families and campers. It was into this situation that Pearson and Quintal landed the airplane.

Tires blew upon impact. The airplane skidded down the runway as racers and spectators scrambled to get out of the way. Flight 143 finally came to rest 1200 m later, a mere 30 m from the dazed onlookers.

Miraculously no one was seriously injured. As news spread around the world, the airplane became known as "The Gimli Glider."

### Making Connections

1. You read that the airplane should have received 20 088 L of fuel. Show how this amount was calculated.

2. Use print or electronic resources to find out what caused the loss of the *Mars Climate Orbiter* spacecraft in September 1999. How is this incident related to the "Gimli Glider" story? Could a similar incident happen again? Why or why not?

## Classifying Matter and Its Changes

Matter is constantly changing. Plants grow by converting matter from the soil and air into matter they can use. Water falls from the sky, evaporates, and condenses again to form liquid water in a never-ending cycle. You can probably suggest many more examples of matter changing.

Matter changes in response to changes in energy. Adding energy to matter or removing energy from matter results in a change. Figure 1.4 shows a familiar example of a change involving matter and energy—a change of state.

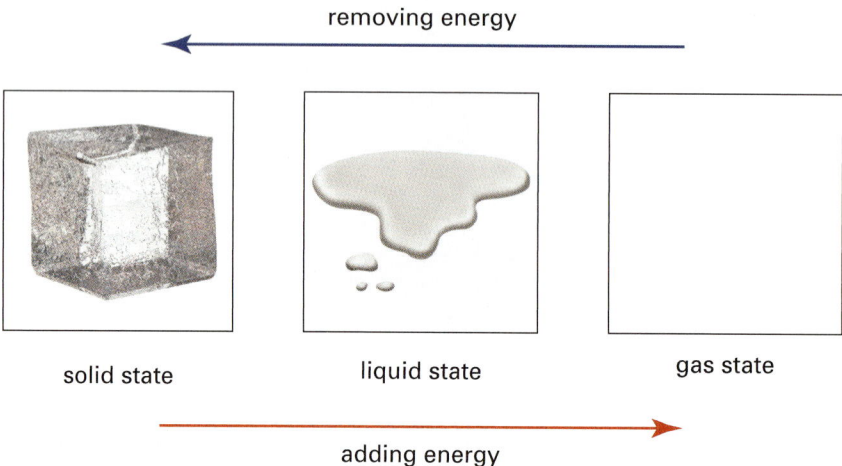

**Figure 1.4** Like all matter, water can change its state when energy is added or removed.

Review the following terms, which are used to describe changes of state.
- *Melting* (or *fusion*) is the change from a solid to a liquid.
- *Freezing* is the change from a liquid to a solid.
- *Vaporization* is the change from a liquid to a gas.
- *Condensation* is the change from a gas to a liquid.
- *Sublimation* is the word used to describe both the change from a gas directly to a solid and the change from a solid directly to a gas.

### Physical and Chemical Changes in Matter

A change of state alters the appearance of matter. The composition of matter remains the same, however, regardless of its state. For example, ice, liquid water, and water vapour are all the same kind of matter: water. Melting and boiling other kinds of matter have the same result. The appearance and some other physical properties change, but the matter retains its identity—its composition. Changes that affect the physical appearance of matter, but *not* its composition, are **physical changes**.

Figure 1.5 shows a different kind of change involving water. Electrical energy is passed through water, causing it to decompose. Two completely different kinds of matter result from this process: hydrogen gas and oxygen gas. These gases have physical and chemical properties that are different from the properties of water and from each other's properties. Therefore, decomposing water is a change that affects the composition of water. Changes that alter the composition of matter are called **chemical changes**. Iron rusting, wood burning, and bread baking are three examples of chemical changes.

You learned about physical and chemical properties earlier in this chapter. A physical change results in a change of physical properties only. A chemical change results in a change of both physical and chemical properties.

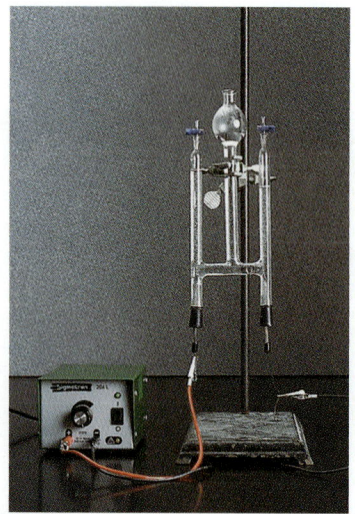

**Figure 1.5** An electrical current is used to decompose water. This process is known as electrolysis.

## Practice Problems

**4.** Classify each situation as either a physical change or a chemical change. Explain your reasoning.

(a) A rose bush grows from a seed that you have planted and nourished.

(b) A green coating forms on a copper statue when the statue is exposed to air.

(c) Your sweat evaporates to help balance your body temperature.

(d) Frost forms on the inside of a freezer.

(e) Salt is added to clear chicken broth.

(f) Your body breaks down the food you eat to provide energy for your body's cells.

(g) Juice crystals dissolve in water.

(h) An ice-cream cone melts on a hot day.

### mind STRETCH

Before adopting the metric system, Canadians measured temperature in units called Fahrenheit degrees (°F). Based on the Fahrenheit scale, water boils at 212°F and freezes at 32°F. A few countries, including the United States, still use the Fahrenheit scale. Without checking any reference materials, design a method for converting Fahrenheit temperatures to Celsius temperatures, and back again. Show your work, and explain your reasoning.

## Classifying Matter

All matter can be classified into two groups: mixtures and pure substances. A **mixture** is a physical combination of two or more kinds of matter. For example, soil is a mixture of sand, clay, silt, and decomposed leaves and animal bodies. If you look at soil under a magnifying glass, you can see these different components. Figure 1.6 shows another way to see the components of soil.

The components in a mixture can occur in different proportions (relative quantities). Each individual component retains its identity. Mixtures in which the different components are clearly visible are called *heterogeneous mixtures*. The prefix "hetero-" comes from the Greek word *heteros*, meaning "different."

Mixtures in which the components are blended together so well that the mixture looks like just one substance are called *homogeneous mixtures*. The prefix "homo- " comes from the Greek word *homos*, meaning "the same." Saltwater, clean air, and grape juice are common examples. Homogeneous mixtures are also called solutions. You will investigate solutions in Unit 3.

A **pure substance** has a definite composition, which stays the same in response to physical changes. A lump of copper is a pure substance. Water (with nothing dissolved in it) is also a pure substance. Diamond, carbon dioxide, gold, oxygen, and aluminum are also pure substances.

Pure substances are further classified into elements and compounds. An **element** is a pure substance that cannot be separated chemically into any simpler substances. Copper, zinc, hydrogen, oxygen, and carbon are examples of elements.

A **compound** is a pure substance that results when two or more elements combine chemically to form a different substance. Compounds can be broken down into elements using chemical processes. For example, carbon dioxide is a compound. It can be separated into the elements carbon and oxygen. The Concept Organizer on the next page outlines the classification of matter at a glance. The ThoughtLab reinforces your understanding of properties, mixtures, and separation of substances.

**Figure 1.6** To see the components of soil, add some soil to a glass of water. What property is responsible for separating the components?

### Language LINK

The word "pure" can be used to mean different things. In ordinary conversation, you might say that orange juice is "pure" if no other materials have been added to it. How is this meaning of pure different from the scientific meaning in the term "pure substance?"

## Concept Organizer — The Classification System for Matter.

**Matter**
- anything that has mass and volume
- found in three physical states: solid, liquid, gas

**Mixtures**
- physical combinations of matter in which each component retains its identity

**Heterogenous Mixtures (Mechanical Mixtures)**
- all components are visible

**Homogeneous Mixtures (Solutions)**
- components are blended so that it looks like a single substance

Physical Changes →

**Pure Substances**
- matter that has a definite composition

**Elements**
- matter that cannot be decomposed chemically into simpler substances

Chemical Changes ↕

**Compounds**
- matter in which two or more elements are chemically combined

## ThoughtLab  Mixtures, Pure Substances, and Changes

You frequently use your knowledge of properties to make and separate mixtures and substances. You probably do this most often in the kitchen. Even the act of sorting clean laundry, however, depends on your ability to recognize and make use of physical properties. This activity is a "thought experiment." You will use your understanding of properties to mix and separate a variety of chemicals, all on paper. Afterward, your teacher may ask you to test your ideas in the laboratory.

### Procedure

1. Consider the following chemicals: table salt, water, baking soda, sugar, iron filings, sand, vegetable oil, milk, and vinegar. Identify each chemical as a mixture or a pure substance.

2. Which of these chemicals can you mix together *without* producing a chemical change? In your notebook, record as many of these physical combinations as you can.

3. Which of these chemicals can you mix together to produce a chemical change? Record as many of these chemical combinations as you can.

4. Record a mixture that is made with four of the chemicals. Then suggest one or more techniques that you can use to separate the four chemicals from one another. Write notes and sketch labelled diagrams to show your techniques. Identify the properties that your techniques depend on.

### Analysis

1. In step 2, what properties of the chemicals did you use to determine your combinations?

2. In step 3, what properties did you use to determine your combinations?

### Application

3. Exchange your four-chemical mixture with a partner. Do not include your notes and diagrams. Challenge your partner to suggest techniques to separate the four chemicals. Then assess each other's techniques. What modifications, if any, would you make to your original techniques?

10   MHR • Chapter 1 Studying Matter and Its Changes

## Section Summary

Notice that the classification system for matter, shown in the Concept Organizer on the previous page, is based mainly on the changes that matter undergoes:

- physical changes to separate mixtures into elements or compounds
- chemical changes to convert compounds or elements into different compounds or elements

To explain how and why these chemical changes occur, you must look "deeper" into matter. You must look at its composition. This is what you will do in the next section. You will see how examining the composition of matter leads to a different classification system: the periodic table. You will also see how the periodic table allows chemists to make predictions about the properties and behaviour of matter.

## Section Review

**1** Based on your current understanding of chemistry, list five ways in which chemistry and chemical processes affect your life.

**2** Copy Figure 1.4 into your notebook.
   (a) Add the following labels in the appropriate places: evaporation, condensation, melting, freezing, solidifying. **Note:** Some labels may apply to the same places on the diagram.
   (b) You may recall that sublimation is a change of state in which a solid changes directly into a gas or a gas changes directly into a solid. Add the label "sublimation" to your diagram for question 1.
   (c) Which of the changes in the diagram absorb energy? Which changes release energy?

**3** List three mixtures that you use or encounter frequently.
   (a) Explain how you know that each is a mixture.
   (b) Classify each mixture as either heterogeneous or homogeneous.

**4** List three pure substances that you use or encounter frequently.
   (a) Explain how you know that each is a pure substance.
   (b) Try to classify each substance as an element or compound. Explain your reasoning.

**5** Carbon can exist as diamond or as graphite (the "lead" in pencils). List three physical properties that differ for diamond and graphite at room temperature. State one physical property that is the same for diamond and graphite at room temperature.

**6** You are given a mixture of wood chips, sand, coffee grounds, table salt, and water. Design a process to clean the water.

# 1.2 Matter and the Atom

**Section Preview/Outcomes**

In this section, you will

- **define** and **describe** the relationships among atomic number, mass number, and atomic mass
- **describe** elements in the periodic table in terms of energy levels and electron arrangements
- **communicate** your understanding of the following terms: *atom, atomic mass unit, atomic number, mass number, atomic symbol, ion, periodic table, periodic law, energy levels, periodic trends, valence electrons*

As you read in the previous section, elements are pure substances that cannot be broken down further by chemical means. But elements themselves are made up of tiny particles called atoms. What are atoms?

John Dalton was a British teacher and self-taught scientist. In 1809, he described atoms as solid, indestructible particles that make up all matter. In the two centuries since that time, scientists have modified and built upon Dalton's model of the atom, using the ever-evolving atomic theory to explain their observations about matter and its interactions.

Today, physicists and chemists have developed a highly complex and detailed model of the structure of the atom. Fortunately, a more simplified model of the atom explains many observations about chemicals and chemical changes.

## The Atom

An **atom** is the smallest particle of an element that still retains the identity of the element. For example, the smallest particle of the graphite in your pencil is a carbon atom. (Graphite is a form of the element carbon.)

An average atom is about $10^{-10}$ m in diameter. Such a tiny size is difficult to visualize. If an average atom were the size of a grain of sand, a strand of your hair would be about 60 m in diameter!

Atoms themselves are made up of even smaller particles. These *subatomic particles* are protons, neutrons, and electrons. Protons and neutrons cluster together to form the central core, or *nucleus*, of an atom. Fast-moving electrons occupy the space that surrounds the nucleus of the atom. As their names imply, subatomic particles are associated with electrical charges. Table 1.2 and Figure 1.7 summarize the general features and properties of an atom and its three subatomic particles.

**Table 1.2** Properties of Protons, Neutrons, and Electrons

| Subatomic particle | Charge | Symbol | Mass (in g) | Radius (in m) |
|---|---|---|---|---|
| electron | 1− | $e^-$ | $9.02 \times 10^{-28}$ | smaller than $10^{-18}$ |
| proton | 1+ | $p^+$ | $1.67 \times 10^{-24}$ | $10^{-15}$ |
| neutron | 0 | $n^0$ | $1.67 \times 10^{-24}$ | $10^{-15}$ |

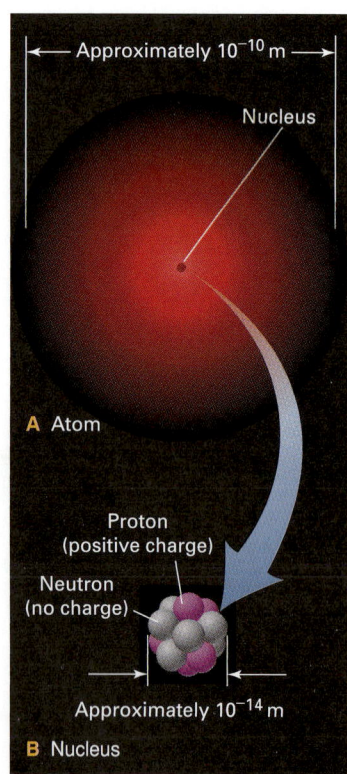

**Figure 1.7** This illustration shows a modern view of an atom. Notice that a fuzzy, cloud-like region surrounds the atomic nucleus. Electrons exist in this region, which represents most of the atom's volume.

## Expressing the Mass of Subatomic Particles

As you can see in Table 1.2, subatomic particles are incredibly small. Suppose that you could count out protons or neutrons equal to

602 000 000 000 000 000 000 000 (or $6.02 \times 10^{23}$)

and put them on a scale. They would have a mass of about 1.0 g. This means that one proton or neutron has a mass of approximately

$$\frac{1.0 \text{ g}}{6.02 \times 10^{23}} = 0.000\ 000\ 000\ 000\ 000\ 000\ 000\ 001\ 66 \text{ g}$$
$$= 1.7 \times 10^{-24} \text{ g}$$

Chemists use a unit called an **atomic mass unit** (symbol **u**) to compare the mass of subatomic particles. A proton has a mass of about 1 u, which is equal to $1.66 \times 10^{-24}$ g.

## The Nucleus of an Atom

All the atoms of a particular element have the same number of protons in their nucleus. For example, all hydrogen atoms—anywhere in the universe—have one proton. All helium atoms have two protons. All oxygen atoms have eight protons. Chemists use the term **atomic number** (symbol $Z$) to refer to the number of protons in the nucleus of each atom of an element.

As you know, the nucleus of an atom also contains neutrons. In fact, the mass of an atom is due to the combined masses of its protons and neutrons. Therefore, an element's **mass number** (symbol $A$) is the total number of protons and neutrons in the nucleus of one of its atoms. Each proton or neutron is counted as one unit of the mass number. For example, an oxygen atom that has 8 protons and 8 neutrons in its nucleus has a mass number of 16. A uranium atom that has 92 protons and 146 neutrons has a mass number of 238.

Information about an element's protons and neutrons is often summarized using the chemical notation shown in Figure 1.8. The letter X represents the **atomic symbol** for an element. (The atomic symbol is also called the *element symbol*.) Each element has a different atomic symbol. All chemists, throughout the world, use the same atomic symbols. Appendix C, at the back of this book, lists the elements in alphabetical order, along with their symbols.

**CHEM FACT**

The number $6.02 \times 10^{23}$ is called the Avogadro constant. In Chapter 2, you will learn more about the Avogadro constant.

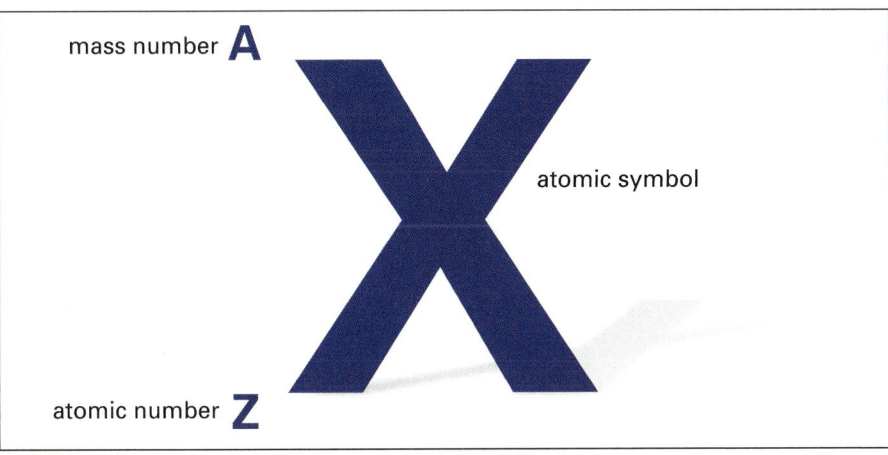

**Figure 1.8** The atomic number and atomic symbol identify the element.

Notice what the chemical notation in Figure 1.8 does, and does not, tell you about the structure of an element's atoms. For example, consider the element fluorine: $^{19}_{9}F$. The mass number (the superscript 19) indicates that fluorine has a total of 19 protons and neutrons. The atomic number (subscript 9) indicates that fluorine has 9 protons. Neither the mass number nor the atomic number tells you how many neutrons fluorine has. You can calculate this value, however, by subtracting the atomic number from the mass number.

Number of neutrons = Mass number − Atomic number
= A − Z

Thus, for fluorine,

Number of neutrons = A − Z
= 19 − 9
= 10

**CHEM FACT**

A proton is about 1837 times more massive than an electron. According to Table 1.2, the mass of an electron is $9.02 \times 10^{-28}$ g. This value is extremely small compared to the mass of a proton or a neutron. Therefore, the number of electrons do not affect the mass number.

## Math LINK

Expressing numerical data about atoms in units such as metres is like using a bulldozer to move a grain of sand. Atoms and subatomic particles are so small that they are not measured using familiar units. Instead, chemists often express the dimensions of atoms in nanometres (1 nm = 1 × 10$^{-9}$ m) and picometres (1 pm = 1 × 10$^{-12}$ m).

- Convert the diameter of a proton and a neutron into nanometres and then picometres.
- Atomic and subatomic sizes are hard to imagine. Create an analogy to help people visualize the size of an atom and its subatomic particles. (The first sentence of this feature is an example of an analogy.)

## CHEM FACT

You cannot use the atomic number to infer the number of neutrons, because atoms of the same element may have differing numbers of neutrons. Two atoms with the same number of protons but different numbers of neutrons are called *isotopes*. You need to know the mass number of an atom and its atomic number to determine the number of neutrons it has. You will learn more about isotopes in Chapter 2.

Now try a few similar calculations in the Practice Problem below.

### Practice Problems

5. Copy the table below into your notebook. Fill in the missing information. Use a periodic table, if you need help identifying the atomic symbol.

| Chemical notation | Element | Number of protons | Number of neutrons |
|---|---|---|---|
| $^{11}_{5}B$ | (a) | (b) | (c) |
| $^{208}_{82}Pb$ | (d) | (e) | (f) |
| (g) | tungsten | (h) | 110 |
| (i) | helium | (j) | 2 |
| $^{239}_{94}Pu$ | (k) | (l) | (m) |
| $^{56}_{26}$(n) | (o) | 26 | (p) |
| (q) | bismuth | (r) | 126 |
| (s) | (t) | 47 | 60 |
| $^{20}_{10}$(u) | (v) | (w) | (x) |

## Using the Atomic Number to Infer the Number of Electrons

As just mentioned, the atomic number and mass number do not give you direct information about the number of neutrons in an element. They do not give you the number of electrons, either. You can infer the number of electrons, however, from the atomic number. The atoms of each element are electrically neutral. This means that their positive charges (protons) and negative charges (electrons) must balance one another. In other words, *in the neutral atom of any element, the number of protons is equal to the number of electrons*. For example, a neutral hydrogen atom contains one proton, so it must also contain one electron. A neutral oxygen atom contains eight protons, so it must contain eight electrons.

What happens when an atom is not neutral, but has a charge? When an atom has greater or fewer electrons than it has protons, it is called an **ion**. An ion is an atom with a negative or positive charge. Ions are represented using their atomic symbol and a superscript showing their charge.

- A magnesium ion with a charge of 2+ is represented as $Mg^{2+}$. The ion $Mg^{2+}$ has 12 protons and 10 electrons. Adding the charge of the protons and electrons gives the net charge of the ion:
  +12 + (−10) = +2
- A chloride ion with a charge of 1− is represented as $Cl^{-}$ (you do not include the number "1"). The ion $Cl^{-}$ has 17 protons and 18 electrons. Adding the charge of the protons and electrons gives the net charge of the ion:
  +17 + (−18) = −1

You will learn more about ion formation in Chapter 5.

In a way, electrons are on the "front lines" of atomic interactions. The number and arrangement of the electrons in an atom determine how the atom will react, if at all, with other atoms.

14  MHR • Chapter 1 Studying Matter and Its Changes

## The Periodic Table

By the mid 1800's, there were 65 known elements. Chemists studied these elements intensively and recorded detailed information about their reactivity and the masses of their atoms. Some chemists began to recognize patterns in the properties and behaviour of many of these elements. (See Figure 1.9.)

Other sets of elements display similar trends in their properties and behaviour. For example, oxygen (O), sulfur (S), selenium (Se), and tellurium (Te) share similar properties. The same is true of fluorine (F), chlorine (Cl), bromine (Br), and iodine (I). These similarities prompted chemists to search for a fundamental property that could be used to organize all the elements.

One chemist, Dmitri Mendeleev (1834–1907), sequenced the known elements in order of increasing atomic mass. The result was a table of the elements, organized so that elements with similar properties were arranged in the same column. Because Mendeleev's arrangement highlighted periodic (repeating) patterns of properties, it was called a **periodic table**.

The modern periodic table is a modification of the arrangement first proposed by Mendeleev. Instead of organizing elements according to atomic mass, the modern periodic table organizes elements according to atomic number. According to the **periodic law**, *the chemical and physical properties of the elements repeat in a regular, periodic pattern when they are arranged according to their atomic number.*

Figures 1.10 and 1.11 on the following pages outline the key features of the modern periodic table. Take some time to review these features. Another version of the periodic table, containing additional data, appears on the inside back cover of this textbook, as well as in Appendix C.

> **Language LINK**
>
> The term *periodic* means "repeating in an identifiable pattern." For example, a calendar is periodic. It organizes the days of the months into a repeating series of weeks. What other examples of periodicity can you think of?

lithium, Li
6.94 u

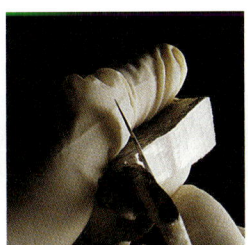
Sodium, Na
22.99 u

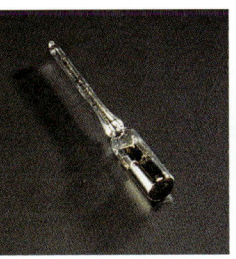
Potassium, K
39.10 u

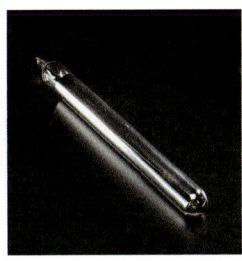
Rubidium, Rb
85.47 u

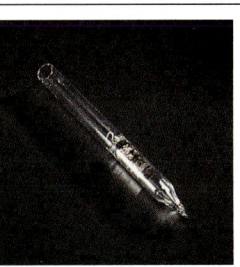
Cesium, Cs
132.91 u

**Shared Physical Properties**
- soft
- malleable
- ductile
- good conductors of electricity
- grey and shiny ("metallic"-looking)

**Shared Chemical Properties**
- are very reactive
- react vigorously with water
- combine with chlorine to form a white solid that dissolves easily in water

**Figure 1.9** These five elements share many physical and chemical properties. However, as shown, they have widely differing atomic masses.

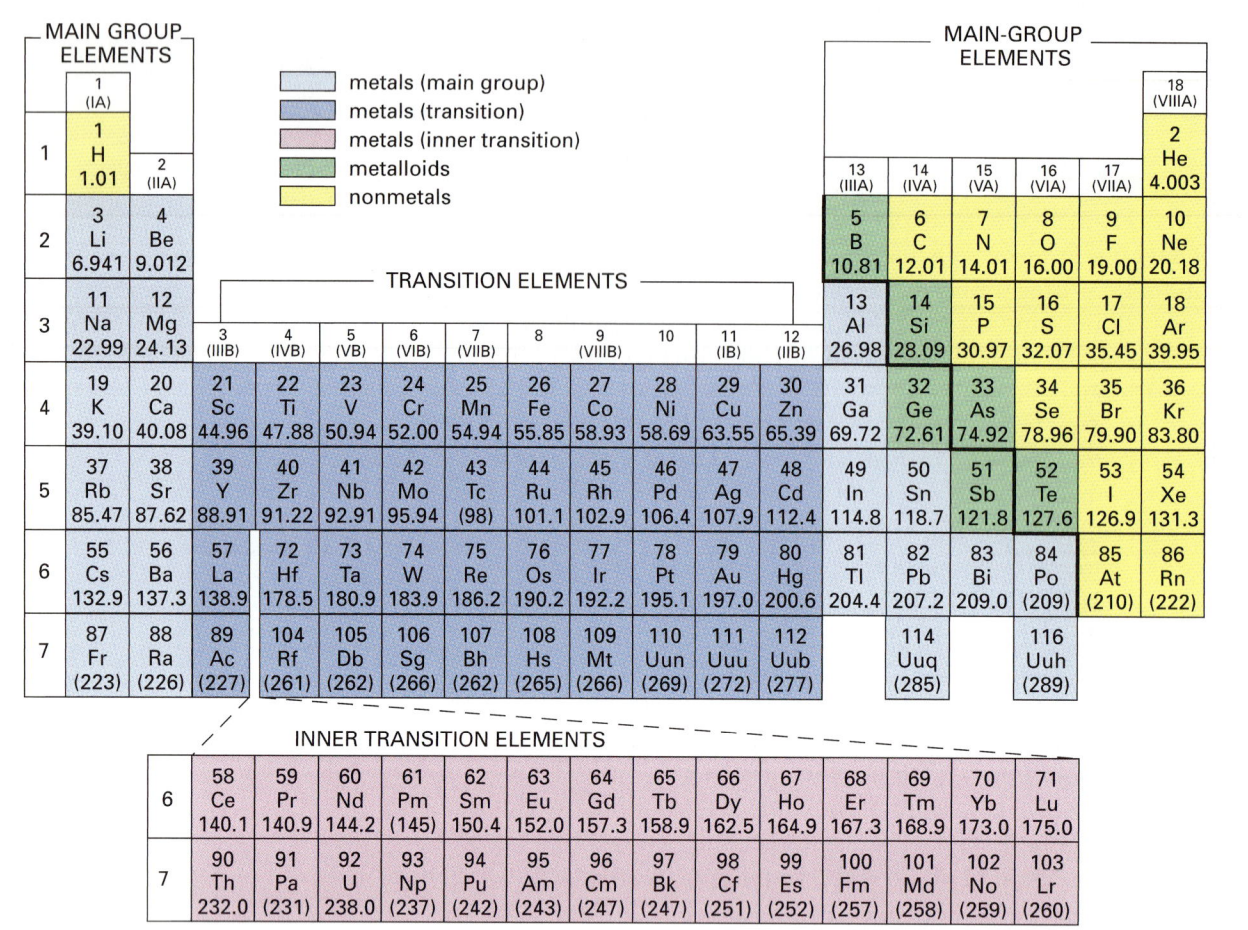

- Each element is in a separate box, with its atomic number, atomic symbol, and atomic mass. (Different versions of the periodic table provide additional data and details.)
- Elements are arranged in seven numbered periods (horizontal rows) and 18 numbered groups (vertical columns).
- Groups are numbered according to two different systems. The current system numbers the groups from 1 to 18. An older system numbers the groups from I to VIII, and separates them into two categories labelled A and B. Both of these systems are included in this textbook.
- The elements in the eight A groups are the main-group elements. They are also called the representative elements.

- The elements in the ten B groups are known as the transition elements. (In older periodic tables, Roman numerals are used to number the A and B groups.)
- Within the B group transition elements are two horizontal series of elements called inner transition elements. They usually appear below the main periodic table. Notice, however, that they fit between the elements in Group 3 (IIIB) and Group 4 (IVB).
- A bold "staircase" line runs from the top of Group 13 (IIIA) to the bottom of Group 16 (VIA). This line separates the elements into three broad classes: metals, metalloids (or semi-metals), and non-metals. (See Figure 1.11 on the next page for more information.)

- Group 1 (IA) elements are known as *alkali metals*. They react with water to form alkaline, or basic, solutions.
- Group 2 (IIA) elements are known as *alkaline earth metals*. They react with oxygen to form compounds called oxides, which react with water to form alkaline solutions. Early chemists called all metal oxides "earths."
- Group 17 (VIIA) elements are known as *halogens*, from the Greek word *hals*, meaning "salt." Elements in this group combine with other elements to form compounds called salts.
- Group 18 (VIIIA) elements are known as *noble gases*. Noble gases do not combine in nature with any other elements.

**Figure 1.10** The key features of the periodic table are summarized here. Much of your work in this course will focus on elements 1 through 20.

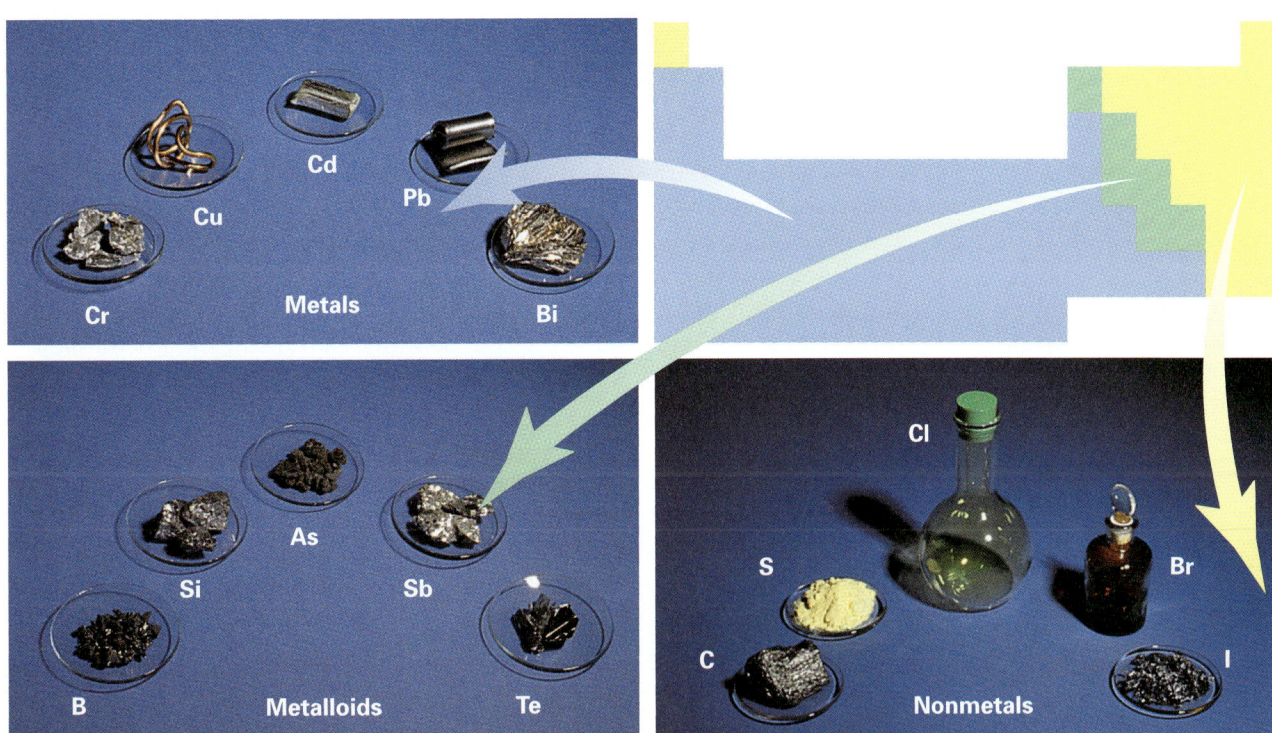

**Figure 1.11** Several examples from each of the three main classes of elements are shown here. Find where they appear in the periodic table in Figure 1.10.

### Practice Problems

6. Identify the name and symbol of the element in each of the following locations of the periodic table:
   (a) Group 14 (IVA), Period 2
   (b) Group 11 (IB), Period 4
   (c) Group 18 (VIIIA), Period 6
   (d) Group 1 (IA), Period 1
   (e) Group 12 (IIB), Period 5
   (f) Group 2 (IIA), Period 4
   (g) Group 17 (VIIA), Period 5
   (h) Group 13 (IIIA), Period 3

## Electrons and the Periodic Table

You have seen how the periodic table organizes elements so that those with similar properties are in the same group. You have also seen how the periodic table shows a clear distinction among metals, non-metals, and metalloids. Other details of the organization of the periodic table, however, may seem baffling. Why, for example, are there different numbers of elements in the periods?

The reason for this, and other details of the periodic table's organization, involves the number and arrangement of electrons in the atoms of each element. To appreciate the importance of electrons to the periodic table, it is necessary to revisit the structure of the atom.

## Electron Energy Levels

Electrons in an atom have certain allowed energies that enable the atom to remain stable. These allowed energies can be thought of as **energy levels**. Electrons are associated with specific energy levels. They can move from one permitted energy level to another, but they cannot exist between energy levels.

### History LINK

Mendeleev did not develop his periodic table in isolation. He built upon work that had been done by other chemists, in other parts of the world, over several decades. Research other ideas that were proposed for organizing the elements. Include Mendeleev's work in your research. What was it about his arrangement that convinced chemists to adopt it?

Chapter 1 Studying Matter and Its Changes • MHR  **17**

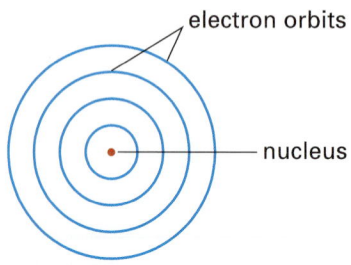

**Figure 1.12** Danish physicist Niels Bohr (1885–1962) was the first to hypothesize that electrons in an atom must have only certain allowed energies, symbolized here.

## CONCEPT CHECK

Examine the following illustration. Then answer these questions.

- Which book possesses more potential energy? Why?
- Can a book sit between shelves instead of on a shelf as shown?
- How does the potential energy of a book on a higher shelf change if it is moved to a lower shelf?
- How do you think this situation is related to electrons and the potential energy they possess when they move in different energy levels?

By absorbing a specific quantity of energy, an electron can move to a higher energy level. By emitting the same quantity of energy, an electron can move back to its original energy level. Figure 1.12 is a simplified picture of energy levels. The diagram shows the energy levels in two dimensions for simplicity. In three dimensions, they would have the shape of spherical shells.

There is a limit to the number of electrons that can occupy each energy level. For example, a maximum of two electrons can occupy the first energy level. A maximum of eight electrons can occupy the second energy level. The **periodic trends** (repeating patterns) that result from organizing the elements by their atomic number are linked to the way in which electrons occupy and fill energy levels. (See Figure 1.13.)

As shown in Figure 1.13A, a common way to show the arrangement of electrons in an atom is to draw circles around the atomic symbol. Each circle represents an energy level. Dots represent electrons that occupy each energy level. This kind of diagram is called a Bohr-Rutherford diagram. It is named after two scientists who contributed their insights to the atomic theory.

Figure 1.13B shows that the first energy level is full when two electrons occupy it. Only two elements have two or fewer electrons: hydrogen and helium. Hydrogen has one electron, and helium has two. These elements, with their electrons in the first energy level, make up Period 1 of the periodic table.

As you can see in Figure 1.13C, Period 2 elements have two occupied energy levels. The second energy level is full when eight electrons occupy it. Neon, with a total of ten electrons, has its first and second energy levels filled. Notice how the second energy level fills with electrons as you move across the period from lithium to fluorine.

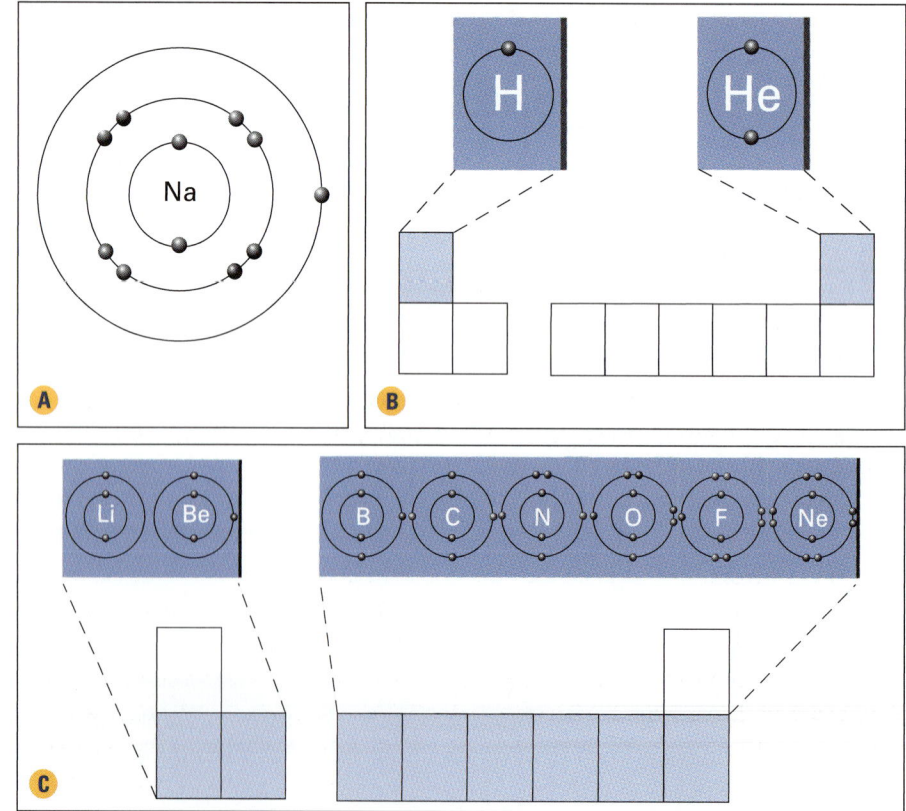

**Figure 1.13** (A) A Bohr-Rutherford diagram (B) Hydrogen and helium have a single energy level. (C) The eight Period 2 elements have two energy levels.

**18** MHR • Chapter 1 Studying Matter and Its Changes

## Patterns Based on Energy Levels and Electron Arrangements

The structure of the periodic table is closely related to energy levels and the arrangement of electrons. Two important patterns result from this relationship. One involves periods, and the other involves groups.

### The Period-Related Pattern

As you can see in Figure 1.13, elements in Period 1 have electrons in one energy level. Elements in Period 2 have electrons in two energy levels. This pattern applies to all seven periods. *An element's period number is the same as the number of energy levels that the electrons of its atoms occupy.* Thus, you could predict that Period 5 elements have electrons that occupy five energy levels. This is, in fact, true.

What about the inner transition elements—the elements that are below the periodic table? Figure 1.14 shows how this pattern applies to them. Elements 58 through 71 belong in Period 6, so their electrons occupy six energy levels. Elements 90 through 103 belong in Period 7, so their electrons occupy seven energy levels. Chemists and chemical technologists tend to use only a few of the inner transition elements (notably uranium and plutonium) on a regular basis. Thus, it is more convenient to place all the inner transition elements below the periodic table.

| | 1 IA | 2 IIA | | | | | | | | | | | | | | | 3 IIIB | 4 IVB | 5 VB | 6 VIB | 7 VIIB | 8 VIIIB | 9 | 10 | 11 IB | 12 IIB | 13 IIIA | 14 IVA | 15 VA | 16 VIA | 17 VIIA | 18 VIIIA |
|---|---|---|---|---|---|---|---|---|---|---|---|---|---|---|---|---|---|---|---|---|---|---|---|---|---|---|---|---|---|---|---|---|
| 1 | 1 H | | | | | | | | | | | | | | | | | | | | | | | | | | | | | | | 2 He |
| 2 | 3 Li | 4 Be | | | | | | | | | | | | | | | | | | | | | | | | | 5 B | 6 C | 7 N | 8 O | 9 F | 10 Ne |
| 3 | 11 Na | 12 Mg | | | | | | | | | | | | | | | | | | | | | | | | | 13 Al | 14 SI | 15 P | 16 S | 17 Cl | 18 Ar |
| 4 | 19 K | 20 Ca | | | | | | | | | | | | | | | 21 Sc | 22 Ti | 23 V | 24 Cr | 25 Mn | 26 Fe | 27 Co | 28 Ni | 29 Cu | 30 Zn | 31 Ga | 32 Ge | 33 As | 34 Se | 35 Br | 36 Kr |
| 5 | 37 Rb | 38 Sr | | | | | | | | | | | | | | | 39 Y | 40 Zr | 41 Nb | 42 Mo | 43 Tc | 44 Ru | 45 Rh | 46 Pd | 47 Ag | 48 Cd | 49 In | 50 Sn | 51 Sb | 52 Te | 53 I | 54 Xe |
| 6 | 55 Cs | 56 Ba | 57 La | 58 Ce | 59 Pr | 60 Nd | 61 Pm | 62 Sm | 63 Eu | 64 Gd | 65 Tb | 66 Dy | 67 Ho | 68 Er | 69 Tm | 70 Yb | 71 Lu | 72 Hf | 73 Ta | 74 W | 75 Re | 76 Os | 77 Ir | 78 Pt | 79 Au | 80 Hg | 81 Tl | 82 Pb | 83 Bi | 84 Po | 85 At | 86 Rn |
| 7 | 87 Fr | 88 Ra | 89 Ac | 90 Th | 91 Pa | 92 U | 93 Np | 94 Pu | 95 Am | 96 Cm | 97 BK | 98 Cf | 99 Es | 100 Fm | 101 Md | 102 No | 103 Lr | 104 Rf | 105 Db | 106 Sg | 107 Bh | 108 Hs | 109 Mt | 110 Uun | 111 Uuu | 112 Uub | | 114 Uuq | | 116 Unh | | |

**Figure 1.14** The "long form" of the periodic table includes the inner transition metals in their proper place.

### The Group-Related Pattern

The second pattern emerges when you consider the electron arrangements in the main-group elements: the elements in Groups 1 (IA), 2 (IIA), and 13 (IIIA) to 18 (VIIIA). All the elements in each main group have the same number of electrons in their outermost occupied energy level. The electrons that occupy the outer energy level are called **valence electrons**. The term "valence" comes from a Latin word that means "to be strong." "Valence electrons" is a suitable name because the outer energy level electrons are the electrons involved when atoms form compounds. In other words, valence electrons are responsible for the chemical behaviour of elements.

You can infer the number of valence electrons in any main-group element from its group number. For example, Group 1 (IA) elements have one valence electron. Group 2 (IIA) elements have two valence electrons. For elements in Groups 13 (IIIA) to 18 (VIIIA), the number of valence electrons is the same as the second digit in the current numbering system. It is the same as the only digit in the older numbering system. For example, elements in Group 15 (VA) have 5 valence electrons. The elements in Group 17 (VIIA) have 7 valence electrons.

### Practice Problems

**7.** Draw boxes to represent the first 20 elements in the periodic table. Using Figure 1.13 as a guide, sketch the electron arrangements for these elements using Bohr-Rutherford diagrams.

**8.** Based on the patterns of the periodic table, identify the number of valence electrons for each of the following elements.

(a) chlorine
(b) helium
(c) indium
(d) strontium
(e) rubidium
(f) lead
(g) antimony
(h) selenium
(i) arsenic
(j) xenon

## Section Summary

You have seen that the structure of the periodic table is directly related to energy levels and arrangements of electrons. The patterns that emerge from this relationship enable you to predict the number of valence electrons for any main group element. They also enable you to predict the number of energy levels that an element's electrons occupy. These skills will be very useful to you when you study chemical bonding in Chapter 5.

In the final section of this chapter, you will learn another set of skills that you will use throughout your study of chemistry: writing chemical formulas, chemical names, and chemical equations.

## Section Review

**1** Copy the table below into your notebook. Use a graphic organizer to show the relationship among the headings of each column. Then fill in the blanks with the appropriate information. (Assume that the atoms are neutral.)

| Element name | Atomic number | Mass number | Number of protons | Number of electrons | Number of neutrons |
|---|---|---|---|---|---|
| (a) | (b) | 108 | (c) | 47 | (d) |
| (e) | (f) | (g) | 33 | (h) | 42 |
| (i) | 35 | (j) | (k) | (l) | 45 |
| (m) | 79 | 179 | (n) | (o) | (p) |
| (q) | (r) | (s) | (t) | 50 | 69 |

**2** For each of the following ions, write the number of protons and the number of electrons.

(a) lithium ion, $Li^+$
(b) calcium ion, $Ca^{2+}$
(c) bromide ion, $Br^-$
(d) sulfide ion, $S^{2-}$

**3** Identify the group number for each of these sets of elements. Write the names and symbols of two elements from each group.

  (a) alkali metals
  (b) noble gases
  (c) halogens
  (d) alkaline earth metals

**4** Identify the element that is described by the following information. Refer to a periodic table as necessary.

  (a) It is a Group 14 metalloid in the third period.
  (b) It is a Group 15 metalloid in the fifth period.
  (c) It is the other metalloid in Group 15.
  (d) It is a halogen that exists in the liquid state at room temperature.

**5** Using writing, sketches, or both, explain to someone who has never seen the periodic table how it can be used to tell at a glance the number of valence electrons in the atoms of an element.

**6** How many valence electrons are there in an atom of each of the following elements?

  (a) neon
  (b) bromine
  (c) sulfur
  (d) strontium
  (e) sodium
  (f) chlorine
  (g) helium
  (h) tin
  (i) magnesium
  (j) silicon

**7** Using print or electronic resources, or both, find one technological application for each of the following elements:

  (a) europium
  (b) neodymium
  (c) carbon
  (d) nitrogen
  (e) silicon
  (f) mercury
  (g) ytterbium
  (h) bromine
  (i) chromium
  (j) krypton

# 1.3 Communicating About Matter

**Section Preview/Outcomes**

In this section, you will

- **write** names and formulas for binary ionic and molecular compounds
- **balance** chemical equations
- **classify** chemical equations
- **communicate** your understanding of the following terms: *systematic name, binary compounds, binary molecular compounds, binary ionic compound, zero sum rule, hydrocarbons*

**Figure 1.15** The aftermath of a chemical explosion at a Bayer plant in Germany

Mistakes in chemistry can be costly and dangerous. A Bayer chemical plant in Germany manufactures a chemical used to kill parasites. One of the reactions involved in the production of this chemical involves potassium carbonate, $K_2CO_3$. On June 8, 1999, however, potassium hydroxide, KOH, was added by mistake. When the reactants were heated, a violent reaction took place, causing an explosion and a fire. Although no one was seriously injured, 91 people, including workers and nearby residents, required hospital treatment. Damage to the chemical plant was estimated at over $100 million.

Fortunately, accidents such as the explosion at the Bayer plant are relatively rare. To minimize the chance of mistakes, chemists communicate clearly and efficiently about chemicals and chemical reactions. They use names and chemical formulas to refer to specific substances, and they use chemical formulas and symbols to represent chemical reactions.

Who decides how compounds should be named? The International Union of Pure and Applied Chemistry (IUPAC) is a global organization of scientists that is responsible for setting standards in chemistry. Committees of the IUPAC make recommendations on how to name compounds.

IUPAC assigns each compound a unique **systematic name** to distinguish it from other chemicals. A compound's systematic name gives you enough information to write its formula.

## Binary Compounds

Compounds composed of two elements are called **binary compounds**. Sodium chloride, NaCl, is a binary compound, as is nitrogen dioxide, $NO_2$. Compounds such as sodium nitrate, $NaNO_3$, that contain atoms of more than two elements, are not binary compounds. One way to identify a binary compound is to examine the end of its name. The names of binary compounds usually end with the suffix "-ide."

## Writing Names for Molecular Compounds

When atoms of two different elements group together as molecules, they form **binary molecular compounds**. *Usually the two elements in binary molecular compounds are non-metallic elements.* The following three rules will help you write the names of binary molecular compounds.

**Step 1** The first element in the name and formula is usually the one that is farther to the left on the periodic table.

For example, in a compound containing carbon and oxygen, the carbon is named first because carbon is to the left of oxygen.

**Step 2** The suffix "-ide" is attached to the name of the second element.

For example, in a compound containing carbon and oxygen, the name "oxygen" is changed to "oxide".

22 MHR • Chapter 1 Studying Matter and Its Changes

**Step 3** Use prefixes to indicate how many atoms of each type are present in one molecule of the compound. Table 1.3 lists the first ten prefixes. Note that the prefix "mono" is used only for the second element. Drop the "o" if the second element is oxygen (e.g., monoxide, not monooxide).

For example, a compound consisting of molecules with one carbon atom and two oxygen atoms, $CO_2$, is called carbon dioxide. A compound consisting of molecules with one carbon atom and one oxygen atom, CO, is called carbon monoxide.

**Table 1.3** Prefixes for Binary Molecular Compounds

| Prefix | Number of atoms |
|---|---|
| mono- | 1 |
| di- | 2 |
| tri- | 3 |
| tetra- | 4 |
| penta- | 5 |
| hexa- | 6 |
| hepta- | 7 |
| octa- | 8 |
| nona- | 9 |
| deca- | 10 |

## Writing Formulas for Molecular Compounds

The systematic names for molecular compounds tell you how to write their formulas. For example, the name nitrogen dioxide tells you that each molecule of the compound contains one atom of nitrogen and two atoms of oxygen. You write: $NO_2$. The subscript "2" indicates that there are two oxygen atoms in the molecule. You do not include the subscript "1"; it is implied when there is no subscript.

Many molecular compounds are known primarily by their trivial names. Water, $H_2O$, ammonia, $NH_3$, and sucrose, $C_{12}H_{22}O_{11}$ are three common examples. Trivial names do not provide information about how the formula of the compound is written. Therefore, you should memorize the formulas for compounds that are usually known by their trivial names. In your notebook, keep a list of compounds with trivial names and their formulas for handy reference. Add to the list as you encounter new compounds.

Note that many elements exist as polyatomic molecules. The elements $H_2$, $N_2$, $O_2$, $F_2$, $Cl_2$, $Br_2$, and $I_2$ are all diatomic. Sulfur exists as $S_8$, and white phosphorus exists as $P_4$.

Study Table 1.4 to see how the names and formulas of molecular compounds containing nitrogen and oxygen correspond. Then test your naming and formula-writing skills by answering the Practice Problems that follow.

**Table 1.4** Names and Formulas of Oxides of Nitrogen

| Formula | Systematic name | Trivial name | Comments |
|---|---|---|---|
| NO | nitrogen monoxide | nitric oxide | • helps to maintain blood pressure<br>• pollutant from vehicle exhaust |
| $N_2O$ | dinitrogen monoxide | nitrous oxide | • also known as laughing gas<br>• used as the propellant gas in cans of whipped cream |
| $NO_2$ | nitrogen dioxide | none | • brown gas<br>• used to manufacture nitric acid |
| $N_2O_3$ | dinitrogen trioxide | none | • deep blue liquid |
| $N_2O_4$ | dinitrogen tetroxide | none | • used in rocket fuels |
| $N_2O_5$ | dinitrogen pentoxide | none | • dissolves in water to form nitric acid |

### Practice Problems

9. Write names for each of the following molecular compounds.
   - (a) $CCl_4$
   - (b) $PCl_5$
   - (c) $SF_6$
   - (d) $P_2O_5$

10. Write the chemical formula for each of the following molecular compounds.
    - (a) sulfur dioxide
    - (b) carbon monoxide
    - (c) diarsenic trisulfide
    - (d) silicon tetrachloride

11. Write systematic names for each of the following oxides of sulfur.
    - (a) $SO$
    - (b) $SO_2$
    - (c) $SO_3$
    - (d) $S_2O_3$

12. Carborundum is a hard black solid, used as an abrasive. It has the chemical formula SiC. What is the systematic name for carborundum?

## Names for Binary Ionic Compounds

A **binary ionic compound** is composed of *ions of one metal element and ions of one non-metal element* grouped together in a lattice structure. The name of the compound is formed from the names of its elements, according to these rules:

1. The first element in the name is the metal. For example, use the name sodium to name the metal in NaCl. Use the name calcium to name the metal in $CaCl_2$.

2. The second element, the non-metal, is named as an ion. In other words, the suffix "-ide" is attached to the name. For example, use chloride to name the non-metal in NaCl and $CaCl_2$.

3. Put the parts of the name together. For example, NaCl is called sodium chloride, and $CaCl_2$ is named calcium chloride.

Note that the names of ionic compounds do not contain prefixes. This does not pose a problem when the compounds contain metal ions from Groups 1 or 2, because they always have the same charge. Metal ions from Group 1 always form ions with a charge of 1+. Metal ions from Group 2 always form ions with a charge of 2+.

Compounds containing transition metals, however, require that you identify which ion is involved.

## The Stock System and the Classical System

Transition metals may form two or more different ions. Two systems are available to name ionic compounds that contain transition metal ions.

- The Stock system was devised by the German chemist Alfred Stock. For the Stock system, use Roman numerals to indicate the charge of the metal cation. Place a Roman numeral in brackets after the name of the first element. Examples are copper(II) oxide, CuO, and copper(I) oxide, $Cu_2O$.

- For the classical system, use the suffix –ic to indicate the metal ion with the greater charge and the suffix –ous to indicate the metal ion with the smaller charge. The earliest discovered elements are sometimes named

**CHEM FACT**

Today, the Stock system is more widely used than the classical system. The advantage of the Stock system is that it can name elements that form more than two ions, while the classical system cannot. Unless otherwise directed, use the Stock system to name compounds containing transition metals.

using Latin names in the classical system. For example, using the classical system, FeO is named ferrous oxide, and $Fe_2O_3$ is named ferric oxide. Other common examples are stannous oxide, SnO, and mercuric nitride, $Hg_3N_2$.

## Writing Formulas for Binary Ionic Compounds

A chemical formula for ionic compounds indicates the ratio of ions that are present in the compound. When you write the chemical formula of a neutral ionic compound, the sum of the charges on the positive ions plus the sum of the charges on the negative ions must equal zero. This statement is known as the **zero sum rule**. To write a chemical formula, follow the steps below.

**Step 1** Write the unbalanced formula, placing the metal ion first.

**Step 2** Write the charge of the ion on top of the appropriate symbol. Recall that metal ions from Group 1 have a charge of 1+. Metal ions from Group 2 have a charge of 2+. Halogen ions have a charge of 1−. Oxide ions have a charge of 2−. Ion charges are listed in Appendix E.

**Step 3** Cross over the numerical value of each charge and write this number as the subscript for the other ion in the compound. Do not include negative or positive signs, and do not include the subscript 1 in the formula.

**Step 4** If necessary, divide through by any common factor.

**Step 5** Check your answer by multiplying the number of each ion by its charge. The total should add up to zero.

The following Sample Problem shows you how to use these steps to write a chemical formula for an ionic compound. Note that the steps are listed for only the first two solutions.

### Sample Problem
### Writing a Chemical Formula From the Name of a Compound

**Problem**

Write the chemical formula of each compound.

(a) magnesium fluoride
(b) zinc telluride
(c) aluminum chloride
(d) cobalt(II) chloride
(e) tin(IV) oxide
(f) iron(II) sulfide

**What Is Required?**

Use the name of each compound to write a chemical formula for the compound.

**What Is Given?**

You are given the names of the compounds. From each name, you can identify the types of atoms that are present in the compound.

*Continued...*

> **Continued...**
>
> ### Plan Your Strategy
> Use the periodic table to find the charge of each ion in the name. Then follow the steps you have just learned to write each chemical formula.
>
> ### Act on Your Strategy
>
> **(a)** The charges are +2 for Mg and −1 for F.
>
> **Step 1** Magnesium has a positive valence, so write the unbalanced formula as MgF.
>
> **Step 2** Write the valences above each element.
> $$+2 \quad -1$$
> $$Mg \quad F$$
>
> **Step 3** Cross over the numerical value of each charge.
> $$MgF_2$$
>
> **Step 4** Since there is only one Mg atom, you do not need to divide by a common factor.
>
> **(b)** The charges are +2 for Zn, and −2 for Te.
>
> **Step 1** Zinc has a positive valence, so write the unbalanced formula as ZnTe.
>
> **Step 2** Write the valences above each element.
> $$+2 \quad -2$$
> $$Zn \quad Te$$
>
> **Step 3** Cross over the numerical value of each valence.
> $$Zn_2Te_2$$
>
> **Step 4** Divide by the common factor, 2, to give the chemical formula ZnTe.
>
> **(c)** +3 −1
> Al Cl
> AlCl$_3$
>
> **(d)** +2 −1
> Co Cl
> CoCl$_2$
>
> **(e)** +4 −2
> Sn O
> Sn$_2$O$_4$ = SnO$_2$
>
> **(f)** +2 −2
> Fe S
> Fe$_2$S$_2$ = FeS
>
> ### Check Your Solution
> If you have time, use each chemical formula to name the compound. Then check that your name and the original name match. Check that the net charge of the ions in each formula is zero.

## Practice Problems

> **13.** Name each compound.
> - **(a)** KCl
> - **(b)** MgBr$_2$
> - **(c)** Li$_2$O
> - **(d)** NaF
>
> **14.** Give the Stock name of each compound.
> - **(a)** TiO$_2$
> - **(b)** CoCl$_2$
> - **(c)** NiBr$_2$
> - **(d)** HgO
> - **(e)** SnCl$_2$
>
> **15.** Write the correct chemical formula of each compound.
> - **(a)** potassium sulfide
> - **(b)** magnesium nitride
> - **(c)** silver chloride
> - **(d)** aluminum fluoride

**16.** Write the chemical formula for each compound.

(a) mercury(II) nitride

(b) mercury(II) sulfide

(c) tin(IV) chloride

(d) copper(I) bromide

## Polyatomic Ions

Many ionic compounds are not binary because one or both ions contain atoms of more than one element. These polyatomic ions consist of two or more different atoms, which are joined by covalent bonds. Some examples of polyatomic ions are shown in Table 1.5. A more complete list is provided in Appendix E.

Compounds containing polyatomic ions are classified as ionic. To name these compounds, use the name of the cation, followed by the name of the anion. For example, $NH_4NO_3$ is named ammonium nitrate. When writing formulas, you must use parentheses around the polyatomic ion when more than one is present in a formula unit. For example, aluminum nitrate must contain three nitrate ions, $NO_3^-$, for every aluminum ion, $Al^{3+}$, in order to have a net charge of zero. The formula is therefore written $Al(NO_3)_3$. The Sample Problem below gives two examples of writing names and formulas for compounds that contain polyatomic ions.

**Table 1.5** Common Polyatomic Ions

| Name | Chemical formula |
| --- | --- |
| ammonium | $NH_4^+$ |
| hydroxide | $OH^-$ |
| carbonate | $CO_3^{2-}$ |
| nitrate | $NO_3^-$ |
| sulfate | $SO_4^{2-}$ |
| hydrogen carbonate | $HCO_3^-$ |
| hydrogen sulfate | $HSO_4^-$ |
| phosphate | $PO_4^{3-}$ |

### Sample Problem

#### Writing a Chemical Formula with Polyatomic Ions

**Problem**

Write the chemical formula of each compound.

(a) ammonium sulfide

(b) copper(II) phosphate

**What Is Required?**

Use the name of each compound to write a chemical formula for the compound.

**What Is Given?**

You are given the names of the compounds. From each name, you can identify the ions that are present in the compound.

**Plan Your Strategy**

Use the periodic table and a table of ions to find the charge of each ion. Then follow the steps you learned above to write each chemical formula.

(a) The charges are +1 for ammonium and −2 for sulfide.

**Step 1** Ammonium has a positive charge, so write the unbalanced formula as $NH_4S$.

(b) The charges are +2 for copper and −3 for phosphate

**Step 1** Copper has a positive charge, so write the unbalanced formula as $CuPO_4$.

*Continued...*

**Step 2** Write the charges above each element.

$$+1 \quad -2$$
$$NH_4 \quad S$$

**Step 3** Cross over the numerical value of each charge.

$(NH_4)_2S$

**Step 4** Since there is only one chloride ion, you do not need to divide by a common factor.

**Step 5** Check to ensure that the total charge of the ions in the formula add up to zero.
$[2 \times (+1)] + (-2) = 0$

**Step 2** Write the charges above each element.

$$+2 \quad -3$$
$$Cu \quad PO_4$$

**Step 3** Cross over the numerical value of each charge.

$Cu_3(PO_4)_2$

**Step 4** The numbers 3 and 2 do not have a common factor.

**Step 5** Check to ensure that the total charge of the ions in the formula add up to zero.
$[3 \times (+2)] + [2 \times (-3)] = 0$

**Check Your Solution**

If you have time, use each chemical formula to name the compound. Then check that your name and the original name match.

### Practice Problems

**17.** Name each compound.
- (a) $CaCO_3$
- (b) $Ba(NO_3)_2$
- (c) $Na_2SO_4$
- (d) $NH_4OH$

**18.** Use the Stock system to name each compound.
- (a) $CuOH$
- (b) $Fe_2(SO_4)_3$
- (c) $Pb(NO_3)_2$
- (d) $Co(HSO_4)_2$

**19.** Write the chemical formula of each compound.
- (a) sodium hydroxide
- (b) sodium carbonate
- (c) ammonium hydrogen carbonate
- (d) magnesium phosphate

**20.** Write the chemical formula of each compound.
- (a) copper(II) hydroxide
- (b) manganese(II) hydrogen carbonate
- (c) cobalt(III) nitrate
- (d) lead(II) hydrogen sulfate

## Hydrogen Compounds

One important group of compounds seems to break the naming rules given in this section. You might think that HCl, for example, would be ionic because it contains hydrogen (found with the metals on the periodic table) and a non-metal. In fact, HCl, like other compounds whose formulas begin with hydrogen, is known to be molecular.

The names of these molecular substances, however, do not use prefixes. The correct name for the molecular compound HCl is hydrogen chloride, not hydrogen monochloride. Similarly, $H_2S$ is hydrogen sulfide, not dihydrogen monosulfide. Many compounds that contain hydrogen are acids. For example, $H_2SO_4$, hydrogen sulfate, is called sulfuric acid when it is dissolved in water. You will learn more about acids in Unit 6. Compounds containing hydrogen and carbon, such as $CH_4$ or $C_2H_6$, have another set of naming rules, which you will learn about in Unit X.

## Writing Balanced Chemical Equations

If chemical names and formulas are the words of the language of chemistry, then chemical equations are the sentences. A chemical equation shows the reactants (the starting materials) and the products (the new materials that form) involved in a chemical change. Usually, chemical equations also include subscripts beside each chemical formula to show the state of the substance (see Table 1.6).

Chemical equations are balanced to reflect the fact that atoms are conserved in a chemical reaction. In other words, in a chemical reaction, no atoms are created, and no atoms are destroyed. Figure 1.16 will help you visualize this process.

**Table 1.6** Subscripts for States

| State | Abbreviation |
|---|---|
| solid | (s) |
| liquid | (ℓ) |
| gas | (g) |
| aqueous (water) solution | (aq) |

**1**

Na + Cl₂ ⟶ NaCl

The equation is unbalanced. The mass of the reactants is greater than the mass of the products.

Add a coefficient of 2 in front of NaCl. The equation is still unbalanced. The mass of the products is greater than the mass of the reactants.

**2**

Na + Cl₂ ⟶ 2NaCl

**3**

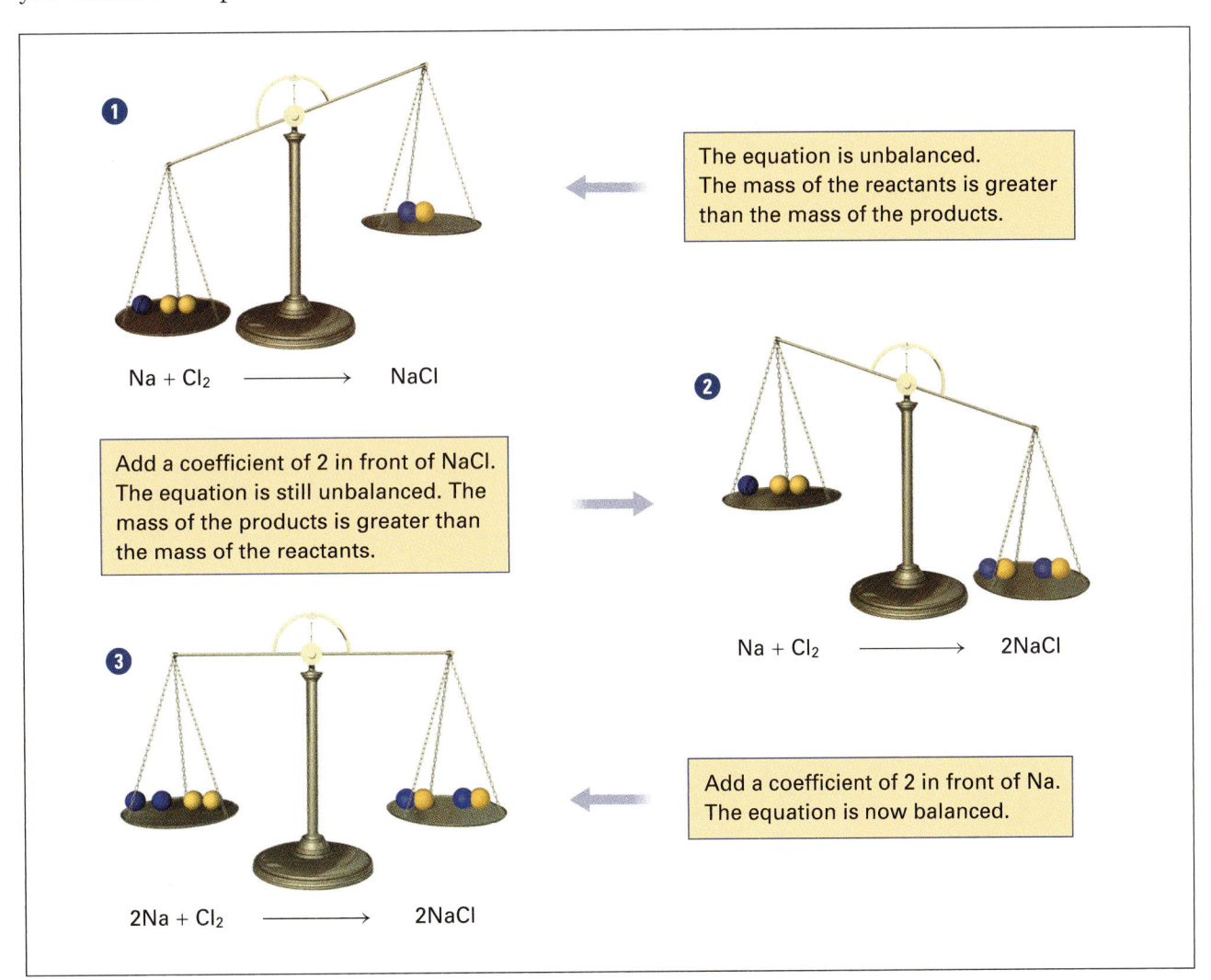

Add a coefficient of 2 in front of Na. The equation is now balanced.

2Na + Cl₂ ⟶ 2NaCl

**Figure 1.16** A balanced chemical equation must show that atoms are conserved.

As shown in Figure 1.16, the process of balancing an equation by inspection involves trial and error—going back and forth between reactants and products to find the correct balance. The systematic approach that is outlined below can be helpful.

1. Write out the unbalanced equation. Ensure that you have copied all the chemical formulas correctly.

2. Begin with the most complex substance—the substance with the largest number or greatest variety of atoms. Balance the atoms that occur in the largest numbers. Leave hydrogen, oxygen, and elements that occur in smaller numbers until later.

3. Balance any polyatomic ions that occur on both sides of the equation as one unit, rather than as separate atoms.

4. Balance any hydrogen or oxygen atoms that occur in a combined and uncombined state.

5. Balance any other element that occurs in its uncombined state.

6. Check your answer by counting the number of atoms of each element on each side of the equation.

Examine the following Sample Problem to see how to balance a chemical equation. Then try the Practice Problems to test your balancing skills.

### Sample Problem
### Balancing Chemical Equations

**Problem**

Copper(II) nitrate reacts with potassium hydroxide to form potassium nitrate and solid copper(II) hydroxide. Balance the equation.

$$Cu(NO_3)_{2(aq)} + KOH_{(aq)} \rightarrow Cu(OH)_{2(s)} + KNO_{3(aq)}$$

**What Is Required?**

The atoms of each element on the left side of the equation should equal the atoms of each element on the right side of the equation.

**Plan Your Strategy**

Balance the polyatomic ions first ($NO_3^-$, then $OH^-$). Check to see whether the equation is balanced. If not, balance the potassium and copper ions. Check your equation again.

**Act on Your Strategy**

There are two $NO_3^-$ ions on the left, so put a 2 in front of $KNO_3$.

$$Cu(NO_3)_{2(aq)} + KOH_{(aq)} \rightarrow Cu(OH)_{2(s)} + 2KNO_{3(aq)}$$

To balance the two $OH^-$ ions on the right, put a 2 in front of the KOH.

$$Cu(NO_3)_{2(aq)} + 2KOH_{(aq)} \rightarrow Cu(OH)_{2(s)} + 2KNO_{3(aq)}$$

Check to see that the copper and potassium ions are balanced. They are, so the equation above is balanced.

### Check Your Solution
Tally the number of each type of atom on each side of the equation.

$$Cu(NO_3)_{2(aq)} + 2KOH_{(aq)} \rightarrow Cu(OH)_{2(s)} + 2KNO_{3(aq)}$$

| | Left Side | Right Side |
|---|---|---|
| Cu | 1 | 1 |
| N | 2 | 2 |
| K | 2 | 2 |
| O | 8 | 8 |
| H | 2 | 2 |

### Math LINK
What does it mean when a fraction is expressed in lowest terms? The fraction $\frac{5}{10}$, expressed in lowest terms, is $\frac{1}{2}$. Similarly, the equation

$$4H_{2(g)} + 2O_{2(g)} \rightarrow 4H_2O_{(\ell)}$$

is balanced, but it can be simplified by dividing all the coefficients by two.

$$2H_{2(g)} + O_{2(g)} \rightarrow 2H_2O_{(\ell)}$$

Write the balanced equation $6KClO_{3(s)} \rightarrow 6KCl_{(g)} + 9O_{2(s)}$ so that the coefficients are the lowest possible whole numbers. Check that the equation is still balanced.

### Practice Problems

**21.** Balance each equation.
   (a) $(NH_4)_3PO_{4(aq)} + Ba(OH)_{2(aq)} \rightarrow Ba_3(PO_4)_{2(s)} + NH_4OH_{(aq)}$
   (b) $Li_{(s)} + H_2O_{(\ell)} \rightarrow LiOH_{(aq)} + H_{2(g)}$
   (c) $NH_{3(g)} + O_{2(g)} \rightarrow N_{2(g)} + H_2O_{(\ell)}$
   (d) $Al(NO_3)_{3(aq)} + H_2SO_{4(aq)} \rightarrow Al_2(SO_4)_{3(aq)} + HNO_{3(aq)}$
   (e) $Cu(NO_3)_{2(s)} \rightarrow CuO_{(s)} + NO_{2(g)} + O_{2(g)}$
   (f) $C_6H_{6(\ell)} + O_{2(g)} \rightarrow CO_{2(g)} + H_2O_{(g)}$

**22.** Aqueous copper(II) nitrate reacts with aqueous potassium hydroxide to form aqueous potassium nitrate and solid copper(II) hydroxide. Write a balanced chemical equation to represent this reaction.

## Classifying Chemical Reactions

Identifying types of chemical reactions helps chemists to predict what will happen when two substances interact. Most chemical reactions can be classified as one of four main types of reactions. These four types of reactions are classified by examining the numbers and types of reactants and products.

The names for the different types of reactions often provide a clue to what happens during a reaction. For example, a *synthesis reaction* occurs when two or more substances react to form a single, different substance. The word "synthesis" signals that a single substance is synthesized from two or more simpler substances. Ammonia, a key component of fertilizers (see Figure 1.17), can be made using a synthesis reaction. Table 1.7 on the following page shows the four main types of reactions.

**Figure 1.17** Ammonia, a component of fertilizers, is produced by the synthesis reaction between nitrogen and hydrogen.

| Reaction type | General form | Example |
|---|---|---|
| synthesis reaction | A + B → C | $2Mg_{(s)} + O_{2(g)} → 2MgO_{(s)}$<br><br>Magnesium reacts with oxygen to form magnesium oxide. The bright white flame that accompanies the reaction can seriously harm your eyes if you view it without protection. |
| decomposition reaction | AB → C + D | $NH_4NO_{3(s)} → N_2O_{(g)} + 2H_2O_{(g)}$<br><br>At high temperatures, ammonium nitrate explodes, decomposing into dinitrogen monoxide and water. |
| single displacement reaction | A + BC → B + AC<br>YZ + X → YX + Z | $Cu_{(s)} + 2AgNO_{3(aq)} → Cu(NO_3)_{2(aq)} + 2Ag_{(s)}$<br><br>A single displacement reaction occurs when copper wire combines with silver nitrate solution. |
| double displacement reaction | AX + BY → AY + BX | $Pb(NO_3)_{2(aq)} + 2KI_{(aq)} → PbI_{2(s)} + 2KNO_{3(aq)}$<br><br>Soluble lead(II) nitrate and potassium iodide react to form soluble potassium nitrate and yellow, insoluble lead(II) iodide. |

## Combustion Reactions

Many combustion reactions do not fit into any of the four categories in Table 1.7. Therefore, combustion reactions are usually classified separately. A combustion reaction occurs when a compound reacts in the presence of oxygen to form oxides and energy. For example, sulfur reacts with oxygen to produce sulfur dioxide.

$$S_{8(s)} + 8O_{2(g)} \rightarrow 8SO_{2(g)}$$

Note that the reaction above is both a synthesis reaction and a combustion reaction. More complex combustion reactions are common for compounds that are composed of carbon and hydrogen atoms. These compounds are called **hydrocarbons**.

- **Complete combustion** of a hydrocarbon occurs when the hydrocarbon reacts completely in the presence of sufficient oxygen. The complete combustion of a hydrocarbon produces only water vapour and carbon dioxide gas as products. The complete combustion of propane, $C_3H_8$, is shown below.

$$C_3H_{8(g)} + 5O_{2(g)} \rightarrow 3CO_{2(g)} + 4H_2O_{(g)}$$

- **Incomplete combustion** of a hydrocarbon occurs when there is not enough oxygen present for the hydrocarbon to react completely. The incomplete combustion of a hydrocarbon produces water, carbon dioxide, carbon monoxide, and solid carbon in varying amounts. More than one balanced equation is possible for the incomplete combustion of a hydrocarbon. One possible equation for the incomplete combustion of propane is shown below.

$$2C_3H_{8(g)} + 7O_{2(g)} \rightarrow 2CO_{2(g)} + 8H_2O_{(g)} + 2CO_{(g)} + 2C_{(s)}$$

Try the following problems to practice classifying chemical reactions. In Investigation 1-A, you will apply your classification skills.

### Practice Problems

23. Classify each of the following reactions as synthesis, decomposition, single displacement, double displacement, complete combustion, or incomplete combustion. Balance the equations if necessary.

    (a) $C_8H_{18(l)} + O_{2(g)} \rightarrow CO_{2(g)} + H_2O_{(g)}$
    (b) $Pb_{(s)} + H_2SO_{4(aq)} \rightarrow PbSO_{4(s)} + H_{2(g)}$
    (c) $Al_2(SO_4)_{3(aq)} + K_2CrO_{4(aq)} \rightarrow K_2SO_{4(aq)} + Al_2(CrO_4)_{3(s)}$
    (d) $Mg_{(s)} + N_{2(g)} \rightarrow Mg_3N_{2(s)}$
    (e) $N_2O_{4(g)} \rightarrow 2NO_{2(g)}$
    (f) $(NH_4)_2S_{(aq)} + Pb(NO_3)_{2(aq)} \rightarrow PbS_{(s)} + 2NH_4NO_{3(aq)}$
    (g) $Sn_{(s)} + AgNO_{3(aq)} \rightarrow Sn(NO_3)_{2(aq)} + Ag_{(s)}$
    (h) $Cu_{(s)} + 2AgNO_{3(aq)} \rightarrow Cu(NO_3)_{2(aq)} + 2Ag_{(s)}$
    (i) $2CH_{4(g)} + 3O_{2(g)} \rightarrow 2CO_{(g)} + 4H_2O_{(g)}$

24. Based on the reactants, predict what type of reaction will occur. Then write the balanced chemical equation. You may omit the state subscripts.

    (a) $Al + O_2 \rightarrow$
    (b) $C_3H_8 + O_2 \rightarrow$ (with sufficient oxygen)
    (c) $Mg + Fe(NO_3)_3 \rightarrow$
    (d) $HCl + KOH \rightarrow$

# Investigation 1-A

**SKILL FOCUS: Analyzing and interpreting**

## Analyzing an Industrial Process

Magnesium and its compounds have numerous applications in the world around you. If you have ever flown in a jet airplane, driven in a car, used a flash bulb to take a picture, or enjoyed a fireworks display, magnesium has touched your life. Magnesium touches your life in another vital way, as well. It is a nutrient that your body's cells need to release energy from food and perform other life functions. Compounds such as magnesium oxide are also used to make fertilizers and insulation for pipes, refine sugar, and treat waste water.

### Procedure

The diagram shows how magnesium is obtained commercially from seawater, using a technique called the Dow process. The raw materials are seawater, seashells, and hydrochloric acid. Examine the diagram, and the numbered steps, to identify the physical and chemical changes involved.

1. Seashells are mostly calcium carbonate, $CaCO_3$. They are heated to produce calcium oxide and carbon dioxide.
   (a) Write the word equation and the balanced chemical equation.
   (b) Identify the type of chemical reaction.

2. The calcium oxide is added to seawater, and a number of changes occur. First calcium oxide reacts with water to form calcium hydroxide.
   (a) Write the word equation and the balanced chemical equation.
   (b) Identify the type of chemical reaction.
   
   Next the calcium hydroxide reacts with magnesium ions, $Mg^{2+}_{(aq)}$, in the seawater to produce magnesium hydroxide and calcium ions.
   
   (c) Write the word equation and the balanced chemical equation.
   (d) Identify the type of chemical reaction.

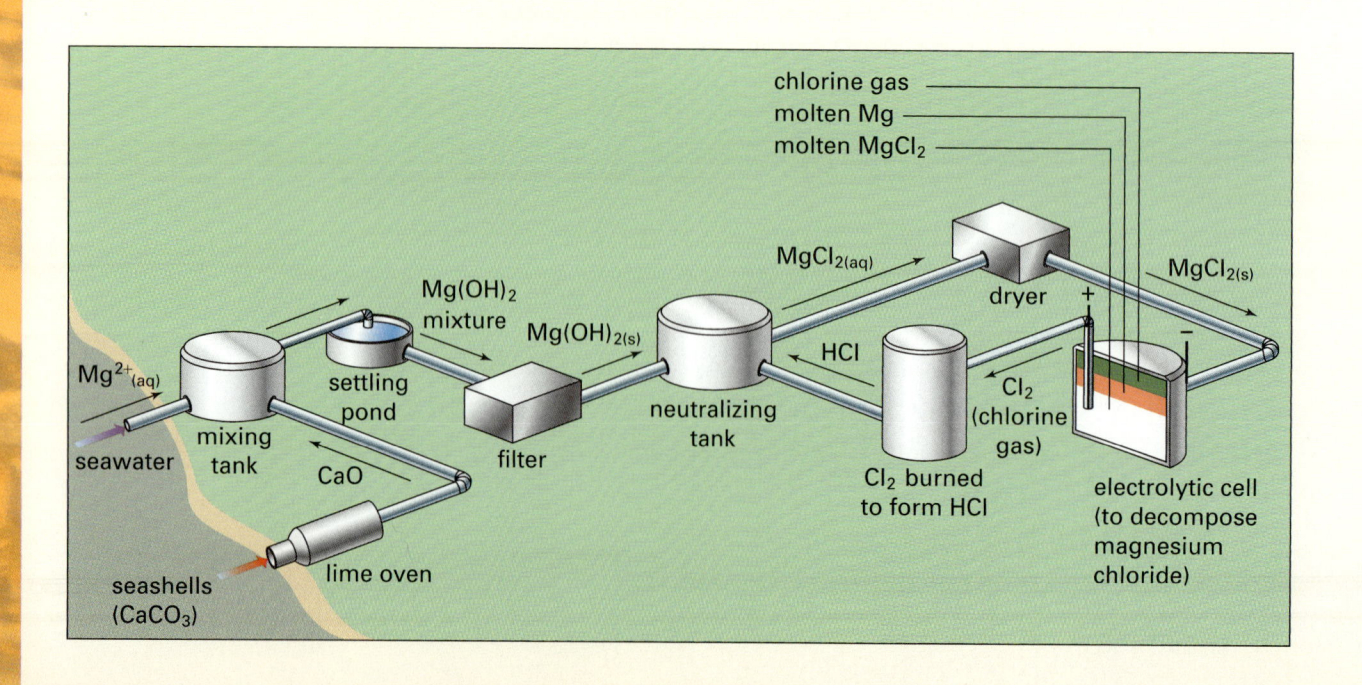

3. Magnesium hydroxide is insoluble. It is pumped as a suspension through filters. How does a filter separate $Mg(OH)_{2(s)}$ from water?

4. In the next step, the magnesium hydroxide reacts with hydrochloric acid to produce an aqueous solution of magnesium chloride and water.
   (a) Write the word equation and the balanced chemical equation.
   (b) Identify the type of chemical reaction.

5. Water is evaporated from the magnesium chloride solution. The resulting solid is melted at 700°C and decomposed by passing electric current through it.
   (a) What is the name of the physical change that occurs first in the magnesium chloride?
   (b) Name the process in which an electric current is used to decompose a compound.
   (c) Write the balanced chemical equation for this step.

6. One of the products in step 5 is chlorine gas. This can be burned with natural gas, which is mostly methane, $CH_{4(g)}$, and oxygen to provide the hydrochloric acid needed for step 4. Balance the equation for this reaction.
   $CH_{4(g)} + O_{2(g)} + Cl_{2(g)} \rightarrow HCl_{(g)} + CO_{(g)}$

## Analysis

When the space shuttle *Challenger* exploded in 1986, scientists recovered the data recorder from the ocean several weeks after the accident. They were not able to analyze the tape at first because magnesium from the recorder case had reacted with the water. The tapes were coated with insoluble magnesium hydroxide, $Mg(OH)_{2(s)}$, which had to be removed.

1. Write a balanced chemical equation for the reaction between water and magnesium to form magnesium hydroxide, and identify the type of reaction.

2. Scientists removed the magnesium hydroxide by reacting it carefully with nitric acid, $HNO_{3(aq)}$. The skeleton equation is
   $Mg(OH)_{2(s)} + HNO_{3(aq)} \rightarrow Mg(NO_3)_{2(aq)} + H_2O_{(\ell)}$
   Balance the equation, and identify the type of reaction.

3. Why was magnesium hydroxide converted to magnesium nitrate?

## Section Summary

In this section you learned how to name and write formulas for binary compounds and for compounds containing polyatomic ions. You also learned how to write balanced chemical equations and how to classify reactions based on their equations. You will use these skills, along with the skills you reviewed in sections 1.1 and 1.2, as you work through the units in this textbook. In Unit 1, for example, you will use chemical formulas and balanced chemical equations to analyze chemical quantities.

## Section Review

**1** Write the name of each binary compound.
(a) $MgCl_2$
(b) $FeCl_3$
(c) $Cu_2O$
(d) $ZnS$
(e) $AlBr_3$
(f) $CF_4$

**2** Write the name of each of the following compounds.
(a) $NH_4NO_3$
(b) $K_2SO_4$
(c) $CaCO_3$
(d) $NaHCO_3$

**3** Write the formula of each compound.
(a) sodium iodide
(b) sodium hydrogen sulfate
(c) lithium hydroxide
(d) ammonium bromide
(e) calcium phosphate
(f) phosphorus pentachloride
(g) silicon tetrafluoride
(h) dinitrogen pentoxide

**4** Write a balanced equation to represent each of the following chemical reactions. Then classify each reaction.
(a) Sulfur dioxide gas reacts with oxygen gas to produce gaseous sulfur trioxide.
(b) Metallic sodium reacts with liquid water to produce hydrogen gas and aqueous sodium hydroxide.
(c) Aqueous barium chloride reacts with aqueous sodium sulfate to produce aqueous sodium chloride and solid barium sulfate.

**5** Balance and classify the following equations.
(a) $Al_{(s)} + O_{2(g)} \rightarrow Al_2O_{3(s)}$
(b) $Al_{(s)} + Fe_2O_{3(s)} \rightarrow Al_2O_3 + Fe_{(s)}$
(c) $Li_{(s)} + H_2O_{(\ell)} \rightarrow LiOH_{(aq)} + H_{2(g)}$
(d) $Cu(NO_3)_{2(s)} \rightarrow CuO_{(s)} + NO_{2(g)} + O_{2(g)}$
(e) $C_2H_{6(g)} + O_{2(g)} \rightarrow CO_{2(g)} + H_2O_{(g)}$
(f) $(NH_4)_3PO_{4(aq)} + Ba(OH)_{2(aq)} \rightarrow Ba_3(PO_4)_{2(s)} + NH_4OH_{(aq)}$

# CHAPTER 1 Review

## Reflecting on Chapter 1

Summarize this chapter in the format of your choice. Here are a few ideas to use as guidelines:
- Give examples of physical and chemical properties.
- Give examples of physical and chemical changes.
- Distinguish between qualitative and quantitative physical properties.
- Describe how matter can be categorized.
- Identify the subatomic particles that make up atoms, as well as the theory that chemists use to explain the composition and behaviour of atoms.
- Use the periodic law to discuss the arrangement of electrons in atoms.
- Describe how to use the periodic table to determine the number of valence electrons that an atom has.
- Summarize the naming rules for molecular and ionic compounds.
- Summarize the guidelines for balancing chemical equations.
- List the different types of chemical reactions and provide examples of each type.

## Reviewing Key Terms

atom
atomic mass unit
atomic number
atomic symbol
binary compounds
binary ionic compound
binary molecular compounds
chemical change
chemical property
chemistry
compound
element
hydrocarbons
ion
mass number
matter
periodic law
periodic table
periodic trends
physical change
physical property
properties
pure substance
systematic name
valence electrons
zero sum rule

## Knowledge/Understanding

1. Identify each property as either physical or chemical.
   (a) Hydrogen gas is extremely flammable.
   (b) The boiling point of ethanol is 78.5°C.
   (c) Chlorine gas is pale green in colour.
   (d) Aluminum reacts quickly with oxygen in air.

2. Name the property that each change depends on. Then classify the property as either chemical or physical.
   (a) You separate a mixture of gravel and road salt by adding water to it.
   (b) You add baking soda to vinegar, and the mixture bubbles and froths.
   (c) You use a strong magnet to locate iron nails that were dropped in a barn filled knee-deep with straw.
   (d) Carbon dioxide gas freezes at a temperature of −78°C.
   (e) You recover salt from a solution of saltwater by heating the solution until all the water has evaporated.
   (f) The temperature of a compost pile rises as the activity of the bacteria inside the pile increases.

3. Identify each change as either physical or chemical.
   (a) Over time, an iron swing set becomes covered with rust.
   (b) Sugar crystals seem to "disappear" when they are stirred into a glass of water.
   (c) Litmus paper turns pink when exposed to acid.
   (d) Butter melts when you spread it on hot toast.

4. Explain the difference between an atom and an element.

5. Compare protons, neutrons, and electrons in terms of their charge, their mass, and their size.

6. A cobalt atom has an atomic mass of 59 and an atomic number of 27. How many neutrons does it have? How many electrons does it have?

7. In your notebook, copy the table below and fill in the missing information.

| Symbol | Protons | Neutrons | Electrons | Charge |
|---|---|---|---|---|
| $^{14}_{7}N^{3-}$ | (a) | (b) | (c) | (d) |
| (e) | 34 | 45 | 36 | (f) |
| $^{52}_{24}(g)^{3+}$ | (h) | (i) | (j) | (k) |
| $^{(l)}_{(m)}(F)$ | (n) | 10 | (o) | 0 |

8. Write the chemical formula for each of the following compounds.
   (a) tin(II) fluoride
   (b) barium sulfate
   (c) magnesium hydroxide

(d) cesium bromide
(e) ammonium hydrogen carbonate
(f) sodium hydrogen sulfate
(g) potassium phosphate
(h) iron(III) sulfide

9. Write the systematic chemical name for each compound.
   (a) $NaHCO_3$
   (b) $FeO$
   (c) $CuCl_2$
   (d) $PbO_2$
   (e) $SnCl_4$
   (f) $P_2O_5$
   (g) $Al_2(SO_4)_3$

10. Explain why atoms are conserved in a chemical reaction. How does a balanced chemical reaction reflect the conservation of atoms?

## Inquiry

11. (a) Design an investigation to discover some of the physical and chemical properties of hydrogen peroxide, $H_2O_2$.
    (b) List the materials you need to carry out your investigation.
    (c) What specific physical and/or chemical properties does your investigation test for?
    (d) What variables are held constant during your investigation? What variables are changed? What variables are measured?
    (e) If you have time, obtain some hydrogen peroxide from a drugstore. With the permission of your teacher, perform your investigation, and record your observations.

12. Balance each of the following chemical equations.
    (a) $PdCl_{2(aq)} + HNO_{3(aq)} \rightarrow Pd(NO_3)_{2(aq)} + HCl_{(aq)}$
    (b) $Cr_{(s)} + HCl_{(aq)} \rightarrow CrCl_{2(aq)} + H_{2(g)}$
    (c) $K_{(s)} + H_2O_{(\ell)} \rightarrow KOH_{(aq)} + H_{2(g)}$

13. Classify each of the following reactions as synthesis, decomposition, single displacement, double displacement, or hydrocarbon combustion. Then balance each equation.
    (a) $H_{2(g)} + CuO_{(s)} \rightarrow Cu_{(s)} + H_2O_{(g)}$
    (b) $Ag_{(s)} + S_{8(s)} \rightarrow Ag_2S_{(s)}$
    (c) $C_4H_{8(g)} + O_{2(g)} \rightarrow CO_{2(g)} + H_2O_{(g)}$
    (d) $NH_{3(aq)} + HCl_{(aq)} \rightarrow NH_4Cl_{(aq)}$
    (e) $Mg_{(s)} + O_{2(g)} \rightarrow MgO_{(s)}$

14. Write a balanced chemical equation corresponding to each word equation.
    (a) The reaction of aqueous sodium hydroxide and iron(III) nitrate produces aqueous sodium nitrate and solid iron(III) hydroxide.
    (b) Powdered antimony reacts with chlorine gas to produce antimony trichloride.
    (c) Mercury(II) oxide may be prepared from its elements.
    (d) Ammonium nitrate decomposes to form nitrogen gas and water.
    (e) Aluminum metal reacts with a solution of zinc nitrate to form dissolved aluminum nitrate and metallic zinc.

15. Consider the unbalanced chemical equation corresponding to the formation of solid lead(II) chromate, $PbCrO_4$.
    $$Pb_{(s)} + Cr_{(s)} + O_{2(g)} \rightarrow PbCrO_{4(s)}$$
    (a) What type of chemical reaction is this?
    (b) Balance the equation.

16. Calcium chloride is often used to melt ice on roads and sidewalks, or to prevent ice from forming. Calcium chloride can be made by reacting hydrochloric acid with calcium carbonate. Write the balanced chemical equation corresponding to this reaction, and classify the reaction.

## Communication

17. In your notebook, draw a flowchart or concept web that illustrates the connections between the following words:
    - pure substance
    - homogeneous
    - heterogeneous
    - aluminum
    - apple juice
    - mixture
    - solution
    - matter
    - water
    - cereal

18. Is salad dressing a homogeneous mixture or a heterogeneous mixture? Use diagrams to explain.

19. Element A, with three electrons in its outer energy level, is in Period 4 of the periodic table. How does the number of its valence electrons compare with that of Element B, which is in Group 13 and Period 6? Use Bohr-Rutherford diagrams to help you express your answer.

20. Suggest reasons why it is important for chemists worldwide to decide on a system for naming and writing formulas for compounds.

21. Explain how you would design a database to display information about the atomic numbers, atomic masses, the number of subatomic particles, and the number of electrons in the valence energy levels of the main-group elements. If you have access to spreadsheet software, construct this table.

## Making Connections

22. At the beginning of this chapter, you saw how water, a very safe chemical compound, can be misrepresented to appear very dangerous. Issues about toxic and polluting chemicals are sometimes reported in newspapers or on television. List some questions you might ask to help you determine whether or not an issue was being misrepresented.

23. Have you ever heard someone refer to aluminum foil as "tin foil"? At one time, the foil was, in fact, made from elemental tin. Find out why manufacturers phased out tin in favour of aluminum. Compare their chemical and physical properties. Identify and classify the products made from or with aluminum. What are the technological costs and benefits of using aluminum? Write a brief report to assess the economic, social, and environmental impact of our use of aluminum.

**Answers to Practice Problems and Short Answers to Section Review Questions:**

**Practice Problems: 1.(a)** physical **(b)** physical **(c)** chemical **(d)** physical **(e)** chemical **(f)** physical **2.(a)** qualitative **(b)** qualitative **(c)** quantitative **(d)** quantitative **(e)** qualitative **3.(a)** quantitative **(b)** qualitative **4.(a)** chemical **(b)** chemical **(c)** physical **(d)** physical **(e)** physical **(f)** chemical **(g)** physical **(h)** physical **5.(a)** boron **(b)** 5 **(c)** 6 **(d)** lead **(e)** 82 **(f)** 126 **(g)** $^{184}_{74}$W **(h)** 74 **(i)** $^4_2$He **(j)** 2 **(k)** plutonium **(l)** 94 **(m)** 145 **(n)** Fe **(o)** iron **(p)** 30 **(q)** $^{126}_{83}$Bi **(r)** 83 **(s)** $^{107}_{47}$Ag **(t)** silver **(u)** Ne **(v)** neon **(w)** 10 **(x)** 10 **6.(a)** carbon, C **(b)** copper, Cu **(c)** radon, Rn **(d)** hydrogen, H **(e)** cadmium, Cd **(f)** calcium, Ca **(g)** iodine, I **(h)** aluminum, Al **7.** Diagrams should show same number of occupied electron energy levels as their period number. Diagrams should show number of valence electrons increasing from 1 to 2 across period 1, and from 1 to 8 across periods 2 and 3. **8.(a)** 7 **(b)** 2 **(c)** 3 **(d)** 2 **(e)** 1 **(f)** 4 **(g)** 5 **(h)** 6 **(i)** 5 **(j)** 8 **9.(a)** carbon tetrachloride **(b)** phosphorus pentachloride **(c)** sulfur hexafluoride **(d)** diphosphorus pentoxide **10.(a)** SO$_2$ **(b)** CO **(c)** As$_2$S$_3$ **(d)** SiCl$_4$ **11. (a)** sulfur monoxide **(b)** sulfur dioxide **(c)** sulfur trioxide **(d)** disulfur trioxide **12.** silicon monocarbide **13.(a)** potassium chloride **(b)** magnesium bromide **(c)** lithium oxide **(d)** sodium fluoride **14.(a)** titanium(IV) oxide **(b)** cobalt(II) chloride **(c)** nickel(II) bromide **(d)** mercury(II) oxide **(e)** tin(II) chloride **15.(a)** K$_2$S **(b)** Mg$_3$N$_2$ **(c)** AgCl **(d)** AlF$_3$ **16.(a)** Hg$_3$N$_2$ **(b)** HgS **(c)** SnCl$_4$ **(d)** CuBr **17.(a)** calcium carbonate **(b)** barium nitrate **(c)** sodium sulfate **(d)** ammonium hydroxide **18.(a)** copper(I) hydroxide **(b)** iron(III) sulfate **(c)** lead(II) nitrate **(d)** cobalt(II) hydrogen sulfate **19.(a)** NaOH **(b)** Na$_2$CO$_3$ **(c)** NH$_4$HCO$_3$ **(d)** Mg$_3$(PO$_4$)$_2$ **20.(a)** Cu(OH)$_2$ **(b)** Mn(HCO$_3$)$_2$ **(c)** Co(NO$_3$)$_3$ **(d)** Pb(HSO$_4$)$_2$ **21.(a)** 2(NH$_4$)$_3$PO$_{4(aq)}$ + 3Ba(OH)$_{2(aq)}$ → Ba$_3$(PO$_4$)$_{2(s)}$ + 6NH$_4$OH$_{(aq)}$ **(b)** 2Li$_{(s)}$ + 2H$_2$O$_{(\ell)}$ → 2LiOH$_{(aq)}$ + H$_{2(g)}$ **(c)** 4NH$_{3(g)}$ + 3O$_{2(g)}$ → 2N$_{2(g)}$ + 6H$_2$O$_{(\ell)}$ **(d)** 2Al(NO$_3$)$_{3(aq)}$ + 3H$_2$SO$_{4(aq)}$ → Al$_2$(SO$_4$)$_{3(aq)}$ + 6HNO$_{3(aq)}$ **(e)** 2Cu(NO$_3$)$_{2(s)}$ → 2CuO$_{(s)}$ + 4NO$_{2(g)}$ + O$_{2(g)}$ **(f)** 2C$_6$H$_{6(\ell)}$ + 15O$_{2(g)}$ → 12CO$_{2(g)}$ + 6H$_2$O$_{(g)}$ **22.** Cu(NO$_3$)$_{2(aq)}$ + 2KOH$_{(aq)}$ → 2KNO$_{3(aq)}$ + Cu(OH)$_2$ **23.(a)** 2C$_8$H$_{18(\ell)}$ + 25O$_{2(g)}$ → 16CO$_{2(g)}$ + 18H$_2$O$_{(g)}$, complete combustion **(b)** Pb$_{(s)}$ + H$_2$SO$_{4(aq)}$ → PbSO$_{4(s)}$ + H$_{2(g)}$, single displacement **(c)** Al$_2$(SO$_4$)$_{3(aq)}$ + 3K$_2$CrO$_{4(aq)}$ → 3K$_2$SO$_{4(aq)}$ + Al$_2$(CrO$_4$)$_{3(s)}$, double displacement **(d)** 3Mg$_{(s)}$ + N$_{2(g)}$ → Mg$_3$N$_{2(s)}$, synthesis **(e)** N$_2$O$_{4(g)}$ → 2NO$_{2(g)}$, decomposition **(f)** (NH$_4$)$_2$S$_{(aq)}$ + Pb(NO$_3$)$_{2(aq)}$ → PbS$_{(s)}$ + 2NH$_4$NO$_{3(aq)}$, double displacement **(g)** Sn$_{(s)}$ + 2AgNO$_{3(aq)}$ → Sn(NO$_3$)$_{2(aq)}$ + 2Ag$_{(s)}$, single displacement **(h)** Cu$_{(s)}$ + 2AgNO$_{3(aq)}$ → Cu(NO$_3$)$_{2(aq)}$ + 2Ag$_{(s)}$, single displacement **(i)** 2CH$_{4(g)}$ + 3O$_{2(g)}$ → 2CO$_{(g)}$ + 4H$_2$O$_{(g)}$, incomplete combustion **24.(a)** 4Al + 3O$_2$ → 2Al$_2$O$_3$, synthesis **(b)** C$_3$H$_8$ + 5O$_2$ → 3CO$_2$ + 4H$_2$O, complete combustion **(c)** 3Mg + 2Fe(NO$_3$)$_3$ → 3Mg(NO$_3$)$_2$ + 2Fe, single displacement **(d)** HCl + KOH → KCl + H$_2$O, double displacement

**Section Review: 1.2: 1.(a)** silver **(b)** 47 **(c)** 47 **(d)** 61 **(e)** arsenic **(f)** 33 **(g)** 75 **(h)** 33 **(i)** bromine **(j)** 80 **(k)** 35 **(l)** 35 **(m)** gold **(n)** 79 **(o)** 79 **(p)** 100 **(q)** tin **(r)** 50 **(s)** 119 **(t)** 50 **2.(a)** 3$p^+$; 2$e^-$ **(b)** 20$p^+$; 18$e^-$ **(c)** 35$p^+$; 36$e^-$ **(d)** 16$p^+$; 18$e^-$ **3.(a)** Group 1; sodium (Na), lithium (Li) **(b)** Group 18; argon (Ar), xenon (Xe) **(c)** Group 17; fluorine (F), bromine (Br) **(d)** Group 2; magnesium (Mg), calcium (Ca) **4.(a)** Si **(b)** Sb **(c)** Bi **(d)** Br **6.(a)** 8 **(b)** 7 **(c)** 6 **(d)** 2 **(e)** 1 **(f)** 7 **(g)** 2 **(h)** 4 **(i)** 2 **(j)** 4 **1.3: 1.(a)** magnesium chloride **(b)** iron(III) chloride **(c)** copper(I) oxide **(d)** zinc sulfide **(e)** aluminum bromide **(f)** carbon tetrafluoride **2.(a)** ammonium nitrate **(b)** potassium sulfate **(c)** calcium carbonate **(d)** sodium hydrogen carbonate **3.(a)** NaI **(b)** NaHSO$_4$ **(c)** LiOH **(d)** NH$_4$Br **(e)** Ca$_3$(PO$_4$)$_2$ **(f)** PCl$_5$ **(g)** SiF$_4$ **(h)** N$_2$O$_5$ **4.(a)** 2SO$_{2(g)}$ + O$_{2(g)}$ → 2SO$_{3(g)}$, synthesis **(b)** Na$_{(s)}$ + H$_2$O$_{(l)}$ → H$_{2(g)}$ + NaOH$_{(aq)}$, single displacement **(c)** BaCl$_{2(aq)}$ + Na$_2$SO$_{4(aq)}$ → 2NaCl$_{(aq)}$ + BaSO$_{4(s)}$, double displacement **5.(a)** 4Al$_{(s)}$ + 3O$_{2(g)}$ → 2Al$_2$O$_{3(s)}$ **(b)** 2Al$_{(s)}$ + Fe$_2$O$_{3(s)}$ → Al$_2$O$_3$ + 2Fe$_{(s)}$ **(c)** 2Li$_{(s)}$ + 2H$_2$O$_{(\ell)}$ → 2LiOH$_{(aq)}$ + H$_{2(g)}$ **(d)** 2Cu(NO$_3$)$_{2(s)}$ → 2CuO$_{(s)}$ + 4NO$_{2(g)}$ + O$_{2(g)}$ **(e)** 2C$_2$H$_{6(g)}$ + 7O$_{2(g)}$ → 4CO$_{2(g)}$ + 6H$_2$O$_{(g)}$ **(f)** 2(NH$_4$)$_3$PO$_{4(aq)}$ + 3Ba(OH)$_{2(aq)}$ → Ba$_3$(PO$_4$)$_{2(s)}$ + 6NH$_4$OH$_{(aq)}$

# UNIT 1

# Stoichiometry

**UNIT 1 CONTENTS**

**CHAPTER 2**
The Mole

**CHAPTER 3**
Chemical Proportions in Compounds

**CHAPTER 4**
Quantities in Chemical Reactions

**UNIT 1 OVERALL OUTCOMES**

- What is the mole? Why is it important for analyzing chemical systems?
- How are the quantitative relationships in balanced chemical reactions used for experiments and calculations?
- Why are quantitative chemical relationships important in the home and in industries?

In the 1939 film *The Wizard of Oz*, Dorothy and her companions collapsed in sleep in a field of poppies. This scene is not realistic, however. Simply walking in a field of poppies will not put you into a drugged sleep.

Poppy seeds do, however, contain a substance called opium. Opium contains the drugs morphine, codeine, and heroin, collectively known as opiates.

While you are unlikely to experience any physiological effects from eating the poppy seeds on a bagel, they could cost you your job! For some safety sensitive-jobs, such as nursing and truck-driving, you may be required to take a drug test as part of the interview process.

Each gram of poppy seeds may contain 2 mg to 18 mg of morphine and 0.6 mg to 2.4 mg of codeine. Eating foods with large amounts of poppy seeds can cause chemists to detect opiates in urine. The opiates may be at levels above employers' specified limits.

Knowing about quantities in chemical reactions is crucial to interpreting the results of drug tests. Policy-makers and chemists need to understand how the proportions of codeine to morphine caused by eating poppy seeds differ from the proportions caused by taking opiates.

In this unit, you will carry out experiments and calculations based on the quantitative relationships in chemical formulas and reactions.

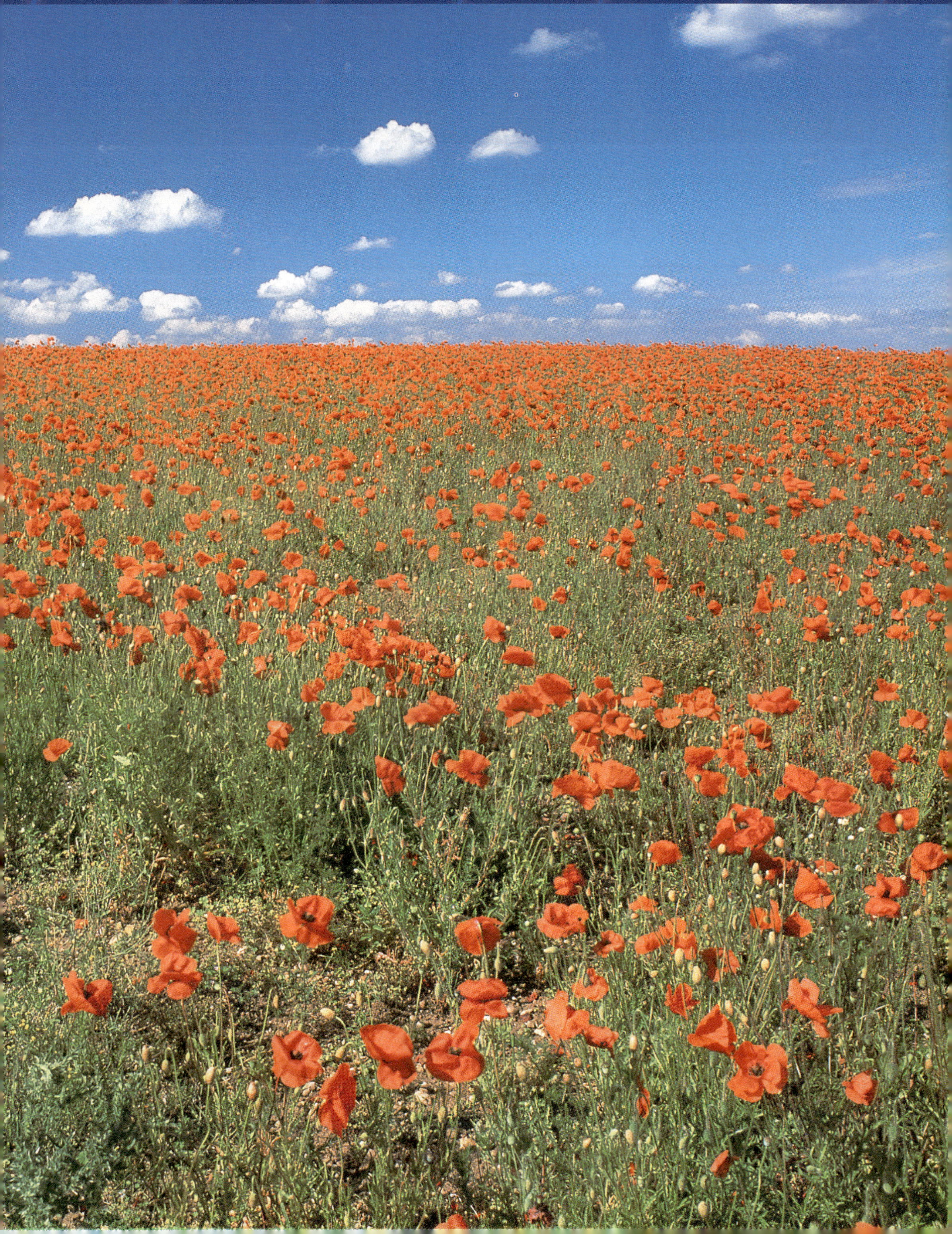

# CHAPTER 2

# The Mole

**Chapter Preview**

- 2.1 Isotopes and Average Atomic Mass
- 2.2 The Avogadro Constant and the Mole
- 2.3 Molar Mass
- 2.4 Molar Volume

**Prerequisite Concepts and Skills**

Before you begin this chapter, review the following concepts and skills:

- defining and describing the relationships among atomic number, mass number, and atomic mass (Chapter 1)
- writing chemical formulas and equations (Chapter 1)
- balancing chemical equations by inspection (Chapter 1)

A recipe for chocolate chip muffins tells you exactly what ingredients you will need. One recipe might call for flour, butter, eggs, milk, baking soda, sugar, and chocolate chips. It also tells you how much of each ingredient you will need, using convenient units of measurement. Which of the ingredients do you measure by counting? Which do you measure by volume or by mass? The recipe does not tell you exactly how many chocolate chips or grains of sugar you will need. It would take far too long to count individual chocolate chips or grains of sugar. Instead, the amounts are given in millilitres or grams—the units of volume or mass.

In some ways, chemistry is similar to baking. To carry out a reaction successfully—in chemistry or in baking—you need to know how much of each reactant you will need. When you mix vinegar and baking soda, for example, the baking soda reacts with acetic acid in the vinegar to produce carbon dioxide gas. When you bake with baking soda, similar reactions help the batter rise. The chemical equation for the reaction between baking soda and acetic acid is:

$$NaHCO_{3(s)} + CH_3COOH_{(aq)} \rightarrow NaCH_3COO_{(aq)} + H_2O_{(\ell)} + CO_{2(g)}$$

According to the balanced equation, one formula unit of baking soda reacts with one molecule of acetic acid to form a salt, water, and carbon dioxide. If you wanted to carry out the reaction, how would you know the amount of baking soda and vinegar to use? The particles are much too small and numerous to be counted like eggs.

In this chapter, you will learn how chemists organize large numbers of atoms into convenient, measurable groups. You will also learn how to relate the number of atoms in a substance to its mass.

You can count the number of eggs you need for muffin batter. How do you know how many atoms are in a sample of an element or compound?

# Isotopes and Average Atomic Mass

## 2.1

How does the mass of a substance relate to the number of atoms in the substance? To answer this question, you first need to understand how the relative masses of individual atoms relate to the masses of substances that you can measure on a balance.

### Isotopes

You may remember from previous science courses that the **mass number** (symbol *A*) of an atom tells you the total number of protons and neutrons in its nucleus. For example, an oxygen atom that has 8 protons and 8 neutrons in its nucleus has a mass number of 16.

All neutral atoms of the same element contain the same number of protons and the same number of electrons. The number of neutrons can vary, however. Therefore, atoms of the same element do not necessarily have the same mass number.

For example, most oxygen atoms have a mass number of 16. As you can see in Figure 2.1, however, there are two other naturally occurring forms of oxygen. One of these has nine neutrons, so $A = 17$. The other has ten neutrons, so $A = 18$. These three forms of oxygen are called isotopes. Isotopes are atoms of an element that have the same number of protons but different numbers of neutrons. The three isotopes of oxygen are called oxygen-16, oxygen-17, and oxygen-18. Oxygen-16 has the same meaning as $^{16}_{8}O$. Similarly, oxygen-17 has the same meaning as $^{17}_{8}O$ and oxygen-18 has the same meaning as $^{18}_{8}O$.

The isotopes of an element have chemical properties that are essentially identical. They differ in mass, however, because they have different numbers of neutrons.

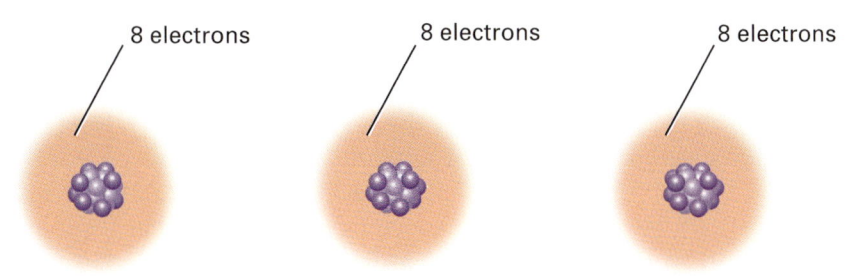

### Section Preview/Outcomes

In this section, you will

- **define** an isotope and use isotopic notation
- **explain** the relative nature of atomic mass
- **communicate** your understanding of the following terms: *mass number, isotope, atomic mass units, isotopic abundance, average atomic mass*

atom of oxygen-16
(8 protons + 8 neutrons)

atom of oxygen-17
(8 protons + 9 neutrons)

atom of oxygen-18
(8 protons + 10 neutrons)

**Figure 2.1** Oxygen has three naturally occurring isotopes.

### The Relative Mass of an Atom

The mass number tells you the total number of neutrons and protons in an atom. You can use the mass number as a rough way to compare the masses of atoms. How do you express and compare the mass of single atoms more precisely?

The mass of an atom is expressed in **atomic mass units (u)**. Atomic mass units are a relative measure, defined by the mass of carbon-12. By definition, one atom of carbon-12 is assigned a mass of 12 u. Stated another way, 1 u = $\frac{1}{12}$ of the mass of one atom of carbon-12.

Chapter 2 The Mole • MHR    43

**CHEM FACT**

The only elements with only one naturally occurring isotope are beryllium, sodium, aluminum, and phosphorus.

The masses of all other atoms are defined by their relationship to carbon-12. For example, oxygen-16 has a mass that is 133% of the mass of carbon-12. Hence the mass of an atom of oxygen-16 is $\frac{133}{100} \times 12.000 \text{ u} = 16.0 \text{ u}$.

## Isotopic Abundance

Because all the atoms in a given element do not have the same number of neutrons, they do not all have the same mass. For example, magnesium has three naturally occurring isotopes. It is made up of 79% magnesium-24, 10% magnesium-25, and 11% magnesium-26. Whether the magnesium is found in a supplement tablet (like the ones on the right) or in seawater as $Mg(OH)_2$, it is always made up of these three isotopes in the same proportion. The relative amount in which each isotope is present in an element is called the **isotopic abundance**. It can be expressed as a percent or as a decimal fraction. When chemists consider the mass of a sample containing billions of atoms, they must take the isotopic abundance into account.

## Average Atomic Mass and the Periodic Table

The **average atomic mass** of an element is the average of the masses of all the element's isotopes. It takes into account the abundance of each isotope of the element. The average atomic mass is the mass that is given for each element in the periodic table.

Examine Figure 2.2. Since the atomic mass unit is based on carbon-12, why does the periodic table show a value of 12.01 u, instead of exactly 12 u? Carbon is made up of several isotopes, not just carbon-12. Naturally occurring carbon contains carbon-12, carbon-13, and carbon-14. If all these isotopes were present in equal amounts, you could simply find the average of the masses of the isotopes. This average mass would be about 13 u, since the masses of carbon-13 and carbon-14 are about 13 u and 14 u respectively.

The isotopes, though, are *not* present in equal amounts. Carbon-12 comprises 98.9% of all carbon, while carbon-13 accounts for 1.1%. Carbon-14 is present in a very small amount—about $1 \times 10^{-10}$%. It makes sense that the average mass of all the isotopes of carbon is 12.01 u—very close to 12—since carbon-12 is by far the predominant isotope. Note that no one atom of carbon has a mass of 12.01 u.

**CHEM FACT**

Magnesium plays a variety of roles in the body. It is involved in energy production, nerve function, and muscle relaxation, to name just a few. The magnesium in these tablets, like all naturally occurring magnesium, is made up of three isotopes.

Thus chemists need to know an element's isotopic abundance and the mass of each isotope to calculate the average atomic mass. How do chemists determine the isotopic abundance associated with each element? How do they find the mass of each isotope? They use a **mass spectrometer**, a powerful instrument that generates a magnetic field to obtain data about the mass and abundance of atoms and molecules. Work through the following Sample Problem to learn how to work with isotopic abundance.

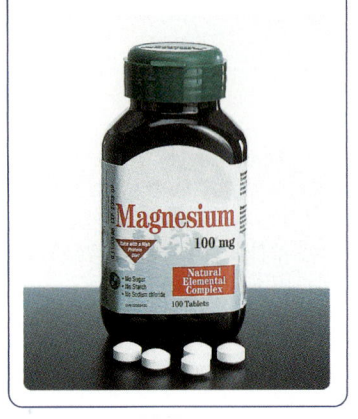

**Figure 2.2** The atomic mass that is given in the periodic table represents the average mass of all the naturally occurring isotopes of the element. It takes into account their isotopic abundances.

6
C
12.01
average atomic mass (u)

## Sample Problem
### Average Atomic Mass

**Problem**

Naturally occurring silver exists as two isotopes. From the mass of each isotope and the isotopic abundance listed below, calculate the average atomic mass of silver.

| Isotope | Atomic mass (u) | Relative abundance (%) |
|---|---|---|
| $^{107}_{47}Ag$ | 106.9 | 51.8 |
| $^{109}_{47}Ag$ | 108.9 | 48.2 |

**What Is Required?**

You need to find the average atomic mass of silver.

**What Is Given?**

You are given the relative abundance and the atomic mass of each isotope.

**Plan Your Strategy**

Multiply the atomic mass of each isotope by its relative abundance, expressed as a decimal. That is, 51.8% expressed as a decimal is 0.518 and 48.2% is 0.482.

**Act on Your Strategy**

Average atomic mass of Ag = 106.9 u (0.518) + 108.9 u (0.482)
= 107.9 u

**Check Your Solution**

In this case, the abundance of each isotope is close to 50%. An average atomic mass of about 108 u seems right, because it is between 106.9 u and 108.9 u. Checking the periodic table reveals that the average atomic mass of silver is indeed 107.9 u.

### CHEM FACT

Why are the atomic masses of individual isotopes not exact whole numbers? After all, $^{12}C$ has a mass of exactly 12 u. Since carbon has 6 neutrons and 6 protons, you might assume that protons and neutrons have masses of exactly 1 u each. In fact, protons and neutrons have masses that are close to, but slightly different from, 1 u. As well, the mass of electrons, while much smaller than the masses of protons and neutrons, must still be taken into account.

### CONCEPT CHECK

Why is carbon-12 the only isotope with an atomic mass that is a whole number?

## Practice Problems

1. The two stable isotopes of boron exist in the following proportions: 19.78% $^{10}_{5}B$ (10.01 u) and 80.22% $^{11}_{5}B$ (11.01 u). Calculate the average atomic mass of boron.

2. In nature, silicon is composed of three isotopes. These isotopes (with their isotopic abundances and atomic masses) are $^{28}_{14}Si$ (92.23%, 27.98 u), $^{29}_{14}Si$ (4.67%, 28.97 u), and $^{30}_{14}Si$ (3.10%, 29.97 u). Calculate the average atomic mass of silicon.

3. Copper exists as two naturally occurring isotopes: copper-63 (62.93 u) and copper-65 (64.93 u). These isotopes have isotopic abundances of 69.1% and 30.9% respectively. Calculate the average atomic mass of copper.

4. Rubidium has two isotopes: rubidium-85 (84.91 u) and rubidium-87 (86.91 u). If the average atomic mass of rubidium is 85.47 u, determine the percentage abundance of each isotope.

## Section Summary

In this section, you learned about the relationship between average atomic mass and isotopic abundance. You thought about elements on the atomic scale. In section 2.2, you will begin to think macroscopically. You will learn how chemists group atoms into amounts that are convenient to work with.

## Section Review

**1** Define the term "isotope" and give examples to illustrate your definition.

**2** Copy and complete the following chart.

| Isotope | Isotope symbol | Mass number | Number of neutrons | Number of protons |
|---------|----------------|-------------|--------------------|--------------------|
| oxygen-17 | | | | |
| hydrogen-2 | | | | |
| | $^{138}_{57}La$ | | | |
| | | 23 | | 11 |

**3** The average atomic mass of potassium is 39.1 u. Explain why no single atom of potassium has a mass of 39.1 u.

**4** Naturally occurring magnesium exists as a mixture of three isotopes. These isotopes (with their isotopic abundances and atomic masses) are Mg-24 (78.70%, 23.985 u), Mg-25 (10.13%, 24.985 u), and Mg-26 (11.17%, 25.983 u). Calculate the average atomic mass of magnesium.

**5** Hydrogen found in nature has two isotopes: hydrogen-1 and hydrogen-2. Hydrogen-2 is also called deuterium. Water containing deuterium instead of hydrogen-1 is used in nuclear reactors.

(a) Water that contains deuterium is called "heavy" water. Explain why this name is appropriate.

(b) Use print or internet resources to find out the role of "heavy" water in CANDU reactors.

**6** Suppose that a carbon-12 atom were assigned a mass of exactly 20 u instead of 12 u. What is the average atomic mass of each of the following elements, based on this standard?

(a) hydrogen, H

(b) argon, Ar

(c) potassium, K

(d) mercury, Hg

# The Avogadro Constant and the Mole

## 2.2

In section 2.1, you learned how to use isotopic abundances and isotopic masses to find the average atomic mass of an element. You can use the average atomic mass, found in the periodic table, to describe the average mass of an atom in a large sample.

Why is relating average atomic mass to the mass of large samples important? In a laboratory, as in everyday life, we deal with macroscopic samples. These samples contain incredibly large numbers of atoms or molecules. Can you imagine a cookie recipe calling for six septillion molecules of baking soda? What if copper wire in a hardware store were priced by the atom instead of by the metre, as in Figure 2.3? What if we paid our water bill according to the number of water molecules that we used? The numbers involved would be ridiculously inconvenient. In this section, you will learn how chemists group large numbers of atoms into amounts that are easily measurable.

### Section Preview/Outcomes

In this section, you will

- **identify** the Avogadro constant as the factor for converting between the mole and number of atoms, ions, formula units, or molecules
- **perform** calculations converting between the number of particles and number of moles
- **communicate** your understanding of the following terms: *mole, Avogadro constant* ($N_A$)

**Figure 2.3** Copper wire is often priced by the metre because the metre is a convenient unit. What unit do chemists use to work with large numbers of atoms?

## Grouping for Convenience

In a chemistry lab, as well as in other contexts, it is important to be able to measure amounts accurately and conveniently. When you purchase headache tablets from a drugstore, you are confident that each tablet contains the correct amount of the active ingredient. Years of testing and development have determined the optimum amount of the active ingredient that you should ingest. If there is too little of the active ingredient, the tablet may not be effective. If there is too much, the tablet may be harmful. When the tablets are manufactured, the active ingredient needs to be weighed in bulk. When the tablets are tested, however, to ensure that they contain the right amount of the active ingredient, chemists need to know how many molecules of the substance are present. How do chemists group particles so that they know how many are present in a given mass of substance?

On its own, the mass of a chemical is not very useful to a chemist. The chemical reactions that take place depend on the number of atoms present, not on their masses. Since atoms are far too small and numerous to count, you need a way to relate the numbers of atoms to masses that can be measured.

When many items in a large set need to be counted, it is often useful to work with groups of items rather than individual items. When you hear the word "dozen," you think of the number 12. It does not matter what the items are. A dozen refers to the quantity 12 whether the items are eggs or pencils or baseballs. Table 2.1 lists some common quantities that we use to deal with everyday items.

**Table 2.1** Some Common Quantities

| Item | Quantity | Amount |
|---|---|---|
| gloves | pair | 2 |
| soft drinks | six-pack | 6 |
| eggs | dozen | 12 |
| pens | gross (12 dozen) | 144 |
| paper | ream | 500 |

You do not buy eggs one at a time. You purchase them in units of a dozen. Similarly, your school does not order photocopy paper by the sheet. The paper is purchased in bundles of 500 sheets, called a ream. It would be impractical to sell sheets of paper individually.

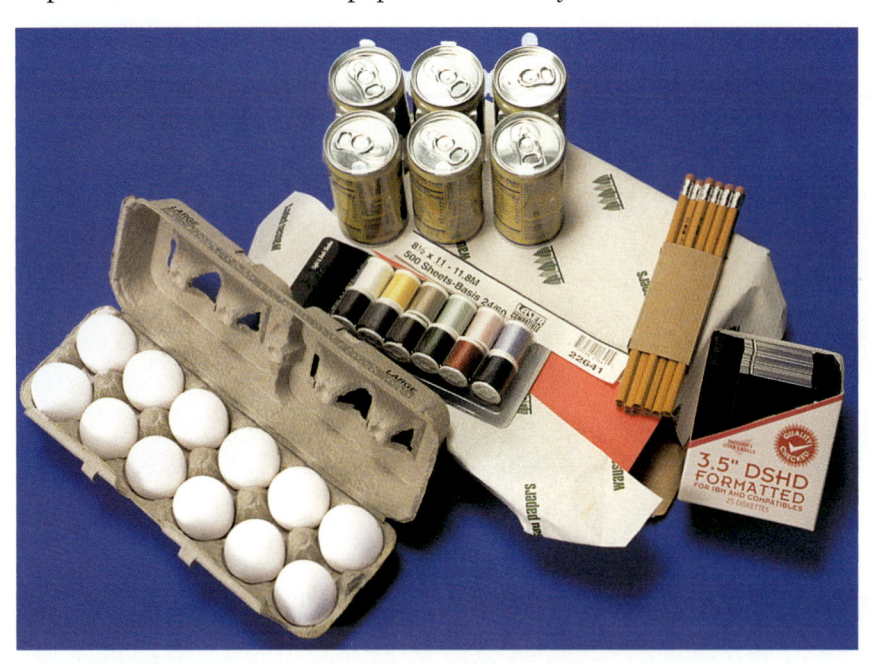

**Figure 2.4** Certain items, because of their size, are often handled in bulk. Would you rather count reams of paper or individual sheets?

### The Definition of the Mole

Convenient, or easily measurable, amounts of elements contain huge numbers of atoms. Therefore chemists use a quantity that is much larger than a dozen or a ream to group atoms and molecules together. This quantity is the mole (symbol **mol**).

- The mole is defined as the amount of substance that contains as many particles—atoms, molecules, or formula units—as exactly 12 g of carbon-12.
- One mole (1 mol) of a substance contains $6.022\,141\,99 \times 10^{23}$ particles of the substance. The constant $6.022\,141\,99 \times 10^{23}$ mol$^{-1}$ is called the **Avogadro constant**. Its symbol is $N_A$.

Note that the number $6.022\,141\,99 \times 10^{23}$ without the units mol$^{-1}$ is called Avogadro's number, not the Avogadro constant.

48 MHR • Unit 1 Stoichiometry

*The Avogadro constant is determined by experiment.* Chemists continually devise more accurate methods to determine how many atoms are in exactly 12 g of carbon-12. This means that the accepted value has changed slightly over the years since it was first defined. You will rarely need the precision of nine significant digits when working with Avogadro's number. In most cases, three significant digits will suffice.

For example, one mole of carbon contains $6.02 \times 10^{23}$ carbon atoms. One mole of sodium chloride contains $6.02 \times 10^{23}$ formula units of NaCl. One mole of hydrofluoric acid contains $6.02 \times 10^{23}$ molecules of HF.

### The Chemist's Dozen

In some ways, the mole is the chemist's dozen. While egg farmers and grocers use the dozen (a unit of 12) to count eggs, chemists use the mole (a much larger number) to represent $6.02 \times 10^{23}$ atoms, molecules, or formula units. When farmers think of two dozen eggs, they are also thinking of 24 eggs.

$$(2 \text{ dozen}) \times \left(\frac{12 \text{ eggs}}{\text{dozen}}\right) = 24 \text{ eggs}$$

You can convert between moles and particles in a similar way. For example, 1 mol of aluminum has $6.02 \times 10^{23}$ atoms of Al. Thus 2 mol of aluminum atoms contain $12.0 \times 10^{23}$ atoms of Al.

$$2 \text{ mol} \times \left(6.02 \times 10^{23} \frac{\text{atoms}}{\text{mol}}\right) = 1.20 \times 10^{24} \text{ atoms}$$

### How Big Is Avogadro's Number?

Avogadro's number is enormous. Its magnitude becomes easier to visualize if you imagine it in terms of ordinary items. For example, suppose that you created a stack of $6.02 \times 10^{23}$ loonies, as in Figure 2.5. To determine the height of the stack, you could determine the height of one loonie and multiply by $6.02 \times 10^{23}$. Avogadro's number needs to be this huge to group single atoms into convenient amounts. What does 1 mol of a substance look like? Figure 2.6 shows some samples of elements. Each sample contains $6.02 \times 10^{23}$ atoms. Notice that each sample has a different mass. You will learn why in section 2.3. In the following ThoughtLab, practise working with Avogadro's number.

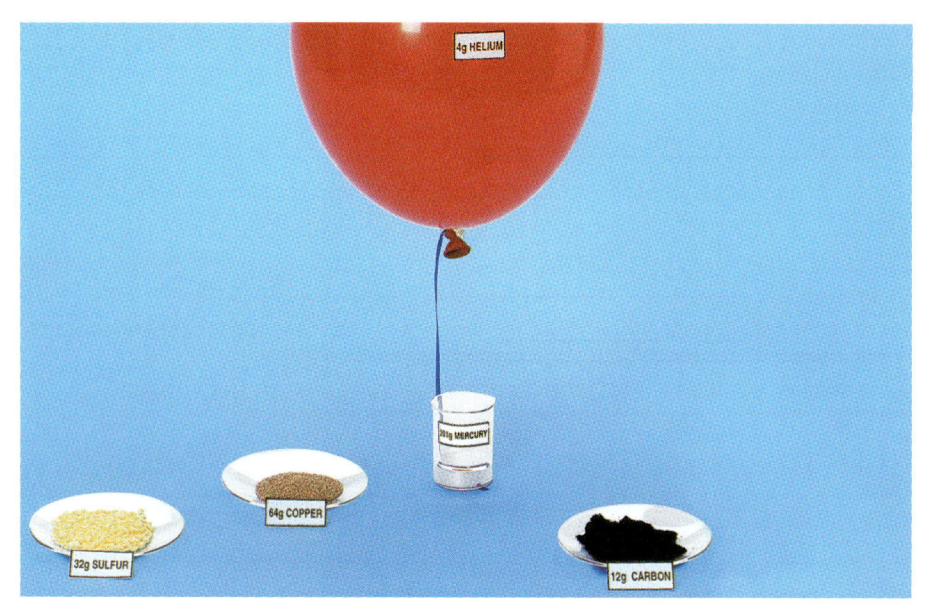

---

**Web LINK**

www.mcgrawhill.ca/links/atlchemistry

Chemists have devised various ways to determine the Avogadro constant. To learn more about how this constant has been found in the past and how it is found today, go to the web site above and click on **Web Links**. What are some methods that chemists have used to determine the number of particles in a mole? How has the accepted value of the Avogadro constant changed over the years?

**Figure 2.5** Measure the height of a pile of five loonies. How tall, in kilometres, would a stack of $6.02 \times 10^{23}$ loonies be?

**Figure 2.6** Each sample contains 1.00 mol, or $6.02 \times 10^{23}$ atoms. Why do you think the mass of each sample is different?

# ThoughtLab: The Magnitude of $6.02 \times 10^{23}$

In this activity, you will solve problems involving the number $6.02 \times 10^{23}$ and everyday items such as golf balls or apples. These problems are examples of *Fermi problems*, which involve large numbers (like Avogadro's number) and give approximate answers. The Italian physicist, Enrico Fermi, liked to pose and solve these types of questions.

## Procedure

1. Working in small groups, brainstorm to determine what information you will need to solve each of the five problems.
2. Use print and Internet resources to find the required information. You may also obtain your information empirically (e.g., by determining the average length of a five-dollar bill).
3. Use the information to solve each problem.
4. Write a detailed report that includes a list of the data you used to solve each problem, references for your data, and fully worked solutions to each problem.

## Analysis

1. If you covered Canada's land mass with $6.02 \times 10^{23}$ golf balls, how deep would the layer of golf balls be?
2. Suppose that you put $6.02 \times 10^{23}$ five-dollar bills end to end. How many round trips from Earth to the Moon would they make?
3. If you could somehow remove $6.02 \times 10^{23}$ teaspoons of water from the world's oceans, would you completely drain the oceans? Explain.
4. What is the mass of $6.02 \times 10^{23}$ apples? How does this compare with the mass of Earth?
5. How many planets would we need for $6.02 \times 10^{23}$ people, if each planet's population were limited to the current population of Earth?

### Technology LINK

Scientific calculators are made to accommodate scientific notation easily. To enter Avogadro's number, for example, type 6.02, followed by the key labelled "EE" or "EXP." (The label on the key depends on the make of calculator you have.) Then enter 23. The 23 will appear at the far right of the display, *without* the exponential base of 10. Your calculator *understands* the number to be in scientific notation.

## Converting Moles to Number of Particles

In the Thought Lab, you practised working with the number $6.02 \times 10^{23}$ by relating it to familiar items. The mole, however, is used to group atoms, molecules, and formula units. For example, chemists know that 1 mol of barium contains $6.02 \times 10^{23}$ atoms of barium. Similarly, 2 mol of barium sulfate contain $2 \times (6.02 \times 10^{23}) = 12.0 \times 10^{23}$ formula units of $BaSO_4$.

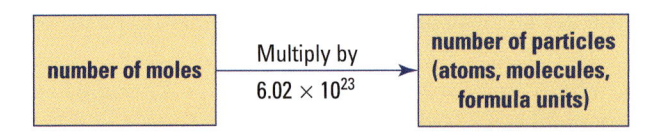

number of moles → Multiply by $6.02 \times 10^{23}$ → number of particles (atoms, molecules, formula units)

The mole is used to help us "count" atoms and molecules. The relationship between moles, number of particles, and the Avogadro constant is as follows:

$$N = n \times N_A$$

where
$N$ = number of particles
$n$ = amount (mol)
$N_A$ = Avogadro constant (mol$^{-1}$)

Work through the next Sample Problem to learn how the amount of a substance in moles relates to the number of particles in the substance. Following the Sample Problem, there are some Practice Problems for you to try.

## Sample Problem

### Moles to Atoms

**Problem**

A sample contains 1.25 mol of nitrogen dioxide, $NO_2$.

**(a)** How many molecules are in the sample?

**(b)** How many atoms are in the sample?

**What Is Required?**

You need to find the number of atoms and molecules in the sample.

**What Is Given?**

The sample consists of 1.25 mol of nitrogen dioxide molecules. Each nitrogen dioxide molecule is made up of three atoms: one nitrogen atom and two oxygen atoms.

$$N_A = 6.02 \times 10^{23} \text{ molecules/mol}$$

**Plan Your Strategy**

**(a)** Find the number of $NO_2$ *molecules* in 1.25 mol of nitrogen dioxide.

**(b)** A molecule of $NO_2$ contains three atoms. Multiply the number of molecules by 3 to arrive at the total number of *atoms* in the sample.

**Act on Your Strategy**

**(a)** Number of molecules of $NO_2$

$$= (1.25 \text{ mol}) \times \left(6.02 \times 10^{23} \frac{\text{molecules}}{\text{mol}}\right)$$

$$= 7.52 \times 10^{23} \text{ molecules}$$

Therefore there are $7.52 \times 10^{23}$ molecules in 1.25 mol of $NO_2$.

**(b)** $(7.52 \times 10^{23} \text{ molecules}) \times \left(3 \frac{\text{atoms}}{\text{molecule}}\right) = 2.26 \times 10^{24}$ atoms

Therefore there are $2.26 \times 10^{24}$ atoms in 1.25 mol of $NO_2$.

**Check Your Solution**

Work backwards. One mol contains $6.02 \times 10^{23}$ atoms. How many moles represent $2.2 \times 10^{24}$ atoms?

$$2.2 \times 10^{24} \text{ atoms} \times \frac{1 \text{ mol}}{6.02 \times 10^{23} \text{ atoms}} = 3.7 \text{ mol}$$

There are 3 atoms in each molecule of $NO_2$

$$3.7 \text{ mol atoms} \times \frac{1 \text{ mol molecule}}{3 \text{ mol atoms}} = 1.2 \text{ molecules}$$

This is close to the value of 1.25 mol of molecules, given in the question.

### Practice Problems

5. A small pin contains 0.0178 mol of iron, Fe. How many atoms of iron are in the pin?

6. A sample contains $4.70 \times 10^{-4}$ mol of gold, Au. How many atoms of gold are in the sample?

*Continued ...*

---

**Language LINK**

The term *order of magnitude* refers to the size of a number—specifically to its exponent when in scientific notation. For example, a scientist will say that 50 000 ($5 \times 10^4$) is two orders of magnitude ($10^2$ times) larger than 500 ($5 \times 10^2$). How many orders of magnitude greater is the Avogadro constant than one billion?

**Web LINK**

www.mcgrawhill.ca/links/atlchemistry

Go to the web site above for a video clip that describes the principles behind the Avogadro constant and the mole. Click on **Electronic Learning Partner**.

> Continued...
>
> 7. How many formula units are in 0.21 mol of magnesium nitrate, Mg(NO₃)₂?
>
> 8. A litre of water contains 55.6 mol of water. How many molecules of water are in this sample?
>
> 9. Ethyl acetate, C₄H₈O₂, is frequently used in nail polish remover. A typical bottle of nail polish remover contains about 2.5 mol of ethyl acetate.
>    (a) How many molecules are in the bottle of nail polish remover?
>    (b) How many atoms are in the bottle?
>    (c) How many carbon atoms are in the bottle?
>
> 10. Consider a 0.829 mol sample of sodium sulfate, Na₂SO₄.
>    (a) How many formula units are in the sample?
>    (b) How many sodium ions, Na⁺, are in the sample?
>    (c) How many sulfate ions, SO₄²⁻, are in the sample?
>    (d) How many oxygen atoms are in the sample?
>
> 11. A certain coil of pure copper wire has a mass of 5.00 g. This mass of copper corresponds to 0.0787 mol of copper.
>    (a) How many atoms of copper are in the coil?
>    (b) How many atoms of copper would there be in a coil of pure copper wire with a mass of 10.00 g?
>    (c) How many atoms of copper would there be in a coil of pure copper wire with a mass of 1.00 g?

## Converting Number of Particles to Moles

Chemists very rarely express the amount of a substance in number of particles. As you have seen, there are far too many particles to work with conveniently. For example, you would never say that you had dissolved $3.21 \times 10^{23}$ molecules of sodium chloride in water. You might say, however, that you had dissolved 0.533 mol of sodium chloride in water. When chemists communicate with each other about amounts of substances, they usually use units of moles (see Figure 2.7). To convert the number of particles in a substance to the number of moles, rearrange the equation you learned previously.

$$N = N_A \times n$$
$$n = \frac{N}{N_A}$$

To learn how many moles are in a substance when you know how many particles are present, find out how many times the Avogadro constant goes into the number of particles.

Examine the next Sample Problem to learn how to convert the number of atoms, formula units, or molecules in a substance to the number of moles.

**Figure 2.7** Chemists rarely use the number of particles to communicate how much of a substance they have. Instead, they use moles.

## Sample Problem

### Molecules to Moles

**Problem**

How many moles are present in a sample of carbon dioxide, $CO_2$, made up of $5.83 \times 10^{24}$ molecules?

**What Is Required?**

You need to find the number of moles in $5.83 \times 10^{24}$ molecules of carbon dioxide.

**What Is Given?**

You are given the number of molecules in the sample.

$$N_A = 6.02 \times 10^{23} \text{ molecules } CO_2/\text{mol } CO_2$$

**Plan Your Strategy**

$$n = \frac{N}{N_A}$$

**Act on Your Strategy**

$$n = \frac{5.83 \times 10^{24} \text{ molecules } CO_2}{(6.02 \times 10^{23} \text{ molecules } CO_2/\text{mol } CO_2)}$$
$$= 9.68 \text{ mol } CO_2$$

There are 9.68 mol of $CO_2$ in the sample.

**Check Your Solution**

$5.83 \times 10^{24}$ molecules is approximately equal to $6 \times 10^{24}$ molecules. Since the number of molecules is about ten times larger than the Avogadro constant, it makes sense that there are about 10 mol in the sample.

### History LINK

The Avogadro constant is named to honour the Italian chemist Amedeo Avogadro. In 1811, Avogadro postulated what is now known as Avogadro's hypothesis: Equal volumes of gases, at equal temperatures and pressures, contain the same number of molecules. You will learn more about Avogadro's law in the next section.

## Practice Problems

12. A sample of bauxite ore contains $7.71 \times 10^{24}$ molecules of aluminum oxide, $Al_2O_3$. How many moles of aluminum oxide are in the sample?

13. A vat of cleaning solution contains $8.03 \times 10^{26}$ molecules of ammonia, $NH_3$. How many moles of ammonia are in the vat?

14. A sample of hydrogen cyanide, HCN, contains $3.33 \times 10^{22}$ atoms. How many moles of hydrogen cyanide are in the sample? **Hint:** Find the number of molecules of HCN first.

15. A sample of pure acetic acid, $CH_3COOH$, contains $1.40 \times 10^{23}$ carbon atoms. How many moles of acetic acid are in the sample?

### Web LINK

www.mcgrawhill.ca/links/atlchemistry

The International System of Units (SI) is based on seven base units for seven quantities. (One of these quantities is the mole.) To learn about the seven base units, go to the web site above and click on **Web Links**. All other SI units are derived from the seven base units.

## Section Summary

In section 2.1, you learned about the average atomic mass of an element. Then, in section 2.2, you learned how chemists group particles using the mole. In section 2.3, you will learn how to use the average atomic masses of the elements to determine the mass of a mole of any substance. You will learn about a relationship that will allow you to relate the mass of a sample to the number of particles it contains.

## Section Review

**1** In your own words, define the mole. Use three examples.

**2** Imagine that $6.02 \times 10^{23}$ were evenly distributed among $6 \times 10^9$ people. How much money would each person receive?

**3** A typical adult human heart beats an average of 60 times per minute. If you were allotted $6.02 \times 10^{23}$ heartbeats, how long, in years, could you expect to live? You may assume each year has 365 days to simplify the calculation.

**4** Calculate the number of atoms in 3.45 mol of iron, Fe.

**5** A sample of carbon dioxide, $CO_2$, contains $2.56 \times 10^{24}$ molecules.
  (a) How many moles of carbon dioxide are present?
  (b) How many moles of atoms are present?

**6** A balloon is filled with 0.50 mol of helium. How many atoms of helium are in the balloon?

**7** A sample of benzene, $C_6H_6$, contains 5.69 mol.
  (a) How many molecules are in the sample?
  (b) How many hydrogen atoms are in the sample?

**8** Aluminum oxide, $Al_2O_3$, forms a thin coating on aluminum when aluminum is exposed to the oxygen in the air. Consider a sample made up of 1.17 mol of aluminum oxide.
  (a) How many molecules are in the sample?
  (b) How many atoms are in the sample?
  (c) How many oxygen atoms are in the sample?

**9** Why do you think chemists chose to define the mole the way they did?

**10** A sample of zinc oxide, ZnO, contains $3.28 \times 10^{24}$ formula units of zinc oxide. A sample of zinc metal contains 2.78 mol of zinc atoms. Which sample contains more zinc: the compound or the element? Show your work.

# Molar Mass

## 2.3

In section 2.2, you explored the relationship between the number of atoms or particles and the number of moles in a sample. Now you are ready to relate the number of moles to the mass, in grams. Then you will be able to determine the number of atoms, molecules, or formula units in a sample by finding the mass of the sample.

### Section Preview/Outcomes

In this section, you will
- **define** molar mass
- **calculate** the molar mass of compounds
- **perform** calculations converting between the number of particles, moles, and mass of various substances
- **communicate** your understanding of the following term: *molar mass*

## Mass and the Mole

You would never express the mass of a piece of gold, like the one in Figure 2.8, in atomic mass units. You would express its mass in grams. How does the mole relate the number of atoms to measurable quantities of a substance? The definition of the mole pertains to relative atomic mass, as you learned in section 2.1. One atom of carbon-12 has a mass of exactly 12 u. Also, by definition, one mole of carbon-12 atoms ($6.02 \times 10^{23}$ carbon-12 atoms) has a mass of exactly 12 g.

The Avogadro constant is the factor that converts the relative mass of individual atoms or molecules, expressed in atomic mass units, to amounts, expressed in moles.

How can you use this relationship to relate mass and moles? The periodic table tells us the average mass of a single atom in atomic mass units (u). For example, zinc has an average atomic mass of 65.39 u. One mole of zinc atoms has a mass of 65.39 g.

*One mole of an element has a mass expressed in grams that is numerically equivalent to the element's average atomic mass expressed in atomic mass units.*

## What is Molar Mass?

The mass of one mole of a substance is called its **molar mass** (symbol *M*). Molar mass is expressed in g/mol. For example, the average atomic mass of gold, as given in the periodic table, is 196.97 u. Thus the molar mass of gold is 196.97 g/mol. Table 2.2 gives some additional examples of molar masses.

**Table 2.2** Average Atomic Mass and Molar Mass of Four Elements

| Element | Average atomic mass (u) | Molar mass (g) |
|---|---|---|
| beryllium, Be | 9.01 | 9.01 |
| boron, B | 10.81 | 10.81 |
| sodium, Na | 22.99 | 22.99 |
| argon, Ar | 39.95 | 39.95 |

Express relative mass in atomic mass units.

Express mass in grams.

**Figure 2.8** The Avogadro constant is a factor that converts from atomic mass to molar mass.

> **CHEM FACT**
> The National Institute of Standards and Technology (NIST) and most other standardization bodies use *M* to represent molar mass. You may see other symbols, such as *MM* or *mm*, used to represent molar mass.

## Finding the Molar Mass of Compounds

While you can find the molar mass of an element just by looking at the periodic table, you need to do some calculations to find the molar mass of a compound. For example, 1 mol of beryllium oxide, BeO, contains 1 mol of beryllium and 1 mol of oxygen. To find the molar mass of BeO, add the molar mass of each element that it contains.

$$M_{BeO} = 9.01 \text{ g/mol} + 16.00 \text{ g/mol}$$
$$= 25.01 \text{ g/mol}$$

Examine the following Sample Problem to learn how to determine the molar mass of a compound. Then try the following Practice Problems.

### Sample Problem
### Molar Mass of a Compound

**Problem**
What is the molar mass of calcium phosphate, $Ca_3(PO_4)_2$?

**What Is Required?**
You need to find the molar mass of calcium phosphate.

**What Is Given?**
You know the formula of calcium phosphate. You also know, from the periodic table, the average atomic mass of each atom that makes up calcium phosphate.

**Plan Your Strategy**
Find the total mass of each element present to determine the molar mass of calcium phosphate. Find the mass of 3 mol of calcium, the mass of 2 mol of phosphorus, and the mass of 8 mol of oxygen. Then add these results together.

**Act on Your Strategy**
Multiply the molar mass of calcium by three.
$$M_{Ca} \times 3 = (40.08 \text{ g/mol}) \times 3$$
$$= 120.24 \text{ g/mol}$$
Multiply the molar mass of phosphorus by two.
$$M_P \times 2 = (30.97 \text{ g/mol}) \times 2$$
$$= 61.94 \text{ g/mol}$$
Multiply the molar mass of oxygen by eight.
$$M_O \times 8 = (16.00 \text{ g/mol}) \times 8$$
$$= 128.00 \text{ g/mol}$$
$$M_{Ca_3(PO_4)_2} = 120.24 \text{ g/mol} + 61.94 \text{ g/mol} + 128.00 \text{ g/mol}$$
$$= 310.18 \text{ g/mol}$$
Therefore the molar mass of calcium phosphate is 310.18 g/mol.

**Check Your Solution**
The units are correct. Using round numbers for a quick check, you get:
$$(40 \times 3) + (30 \times 2) + (15 \times 8) = 300$$
This estimate is close to the answer of 310.18 g/mol. Your answer is reasonable.

> **PROBLEM TIP**
> Once you are used to calculating molar masses, you may want to do the four calculations at right all at once. Try solving the Sample Problem using only one line of calculations.

### Practice Problems

**16.** State the molar mass of each element.
 (a) xenon, Xe
 (b) osmium, Os
 (c) barium, Ba
 (d) tellurium, Te

**17.** Determine the molar mass of each compound.
 (a) ammonia, $NH_3$
 (b) glucose, $C_6H_{12}O_6$
 (c) potassium dichromate, $K_2Cr_2O_7$
 (d) iron(III) sulfate, $Fe_2(SO_4)_3$

**18.** Strontium may be found in nature as celestite, $SrSO_4$. Find the molar mass of celestite.

**19.** What is the molar mass of the ion $[Cu(NH_3)_4]^{2+}$?

## From Number of Particles to Mass

Using the mole concept and the periodic table, you can determine the mass of one mole of a compound. You know, however, that one mole represents $6.02 \times 10^{23}$ particles. Therefore you can use a balance to determine the number of atoms, molecules, or formula units in a sample.

For example, consider carbon dioxide, $CO_2$. One mole of carbon dioxide has a mass of 44.0 g and contains $6.02 \times 10^{23}$ molecules. You can set up the following relationship:

$6.02 \times 10^{23}$ molecules of $CO_2$ → 1 mol of $CO_2$ → 44.0 g of $CO_2$

Figure 2.9 shows the relationships among mass, amount, and number of atoms, molecules, or formula units of a substance.

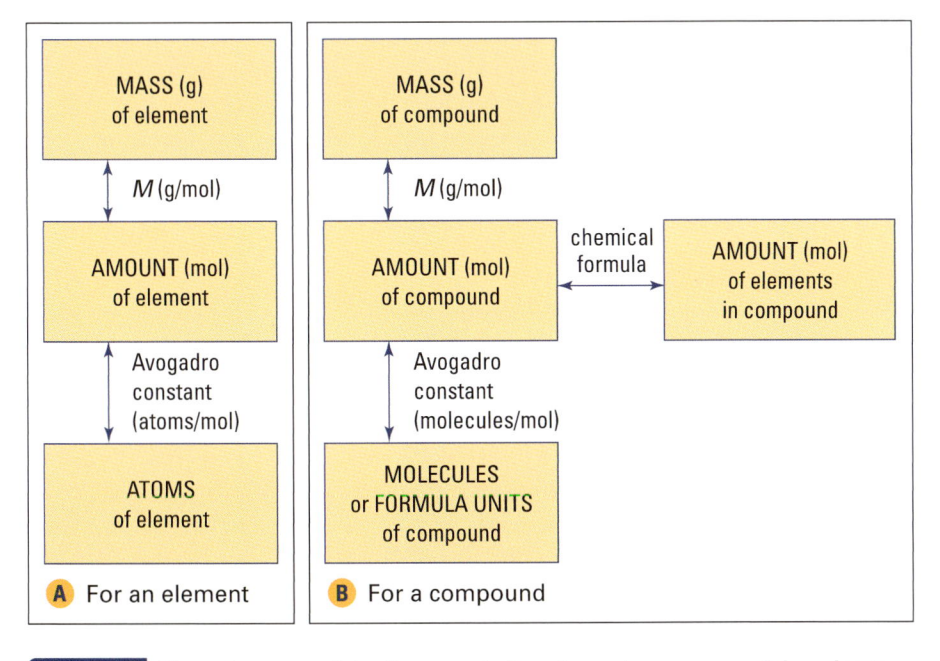

**Figure 2.9** The molar mass relates the amount of an element or a compound, in moles, to its mass. Similarly, the Avogadro constant relates the number of particles to the molar amount.

## Converting from Moles to Mass

**CONCEPT CHECK**

Suppose you have a solution of calcium chloride and a solution of ammonium sulfate. You mix the solutions. What reaction will occur? Write a balanced chemical equation to show the reaction. Note that ammonium chloride is soluble, but calcium sulfate is not.

Suppose that you want to carry out a reaction involving ammonium sulfate and calcium chloride. You have been told that the first step is to obtain one mole of each chemical. How do you decide how much of each chemical to use? You convert the molar amount to mass. Then you use a balance to determine the mass of the proper amount of each chemical.

The following equation can be used to solve problems involving mass, molar mass, and number of moles:

$$\text{Mass} = \text{Number of moles} \times \text{Molar mass}$$
$$m = n \times M$$

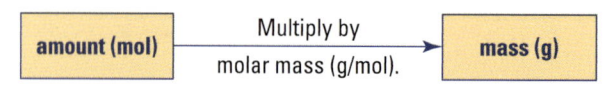

amount (mol) — Multiply by molar mass (g/mol). → mass (g)

The following Sample Problem shows how to convert an amount in moles to a mass in grams. After working through the Sample Problem, try the Practice Problems that follow.

### Sample Problem
#### Moles to Mass

**Problem**

A flask contains 0.750 mol of carbon dioxide gas, $CO_2$. What mass of carbon dioxide gas is in this sample?

**What Is Required?**

You need to find the mass of carbon dioxide.

**What Is Given?**

The sample contains 0.750 mol. You can determine the molar mass of carbon dioxide from the periodic table.

**Plan Your Strategy**

In order to convert moles to grams, you need to determine the molar mass of carbon dioxide from the periodic table.

Multiply the molar mass of carbon dioxide by the number of moles of carbon dioxide to determine the mass.

$$m = n \times M$$

**Act on Your Strategy**

Determine the molar mass of carbon dioxide.
$$M_{CO_2} = 2 \times (16.00 \text{ g/mol}) + 12.01 \text{ g/mol}$$
$$= 44.01 \text{ g/mol}$$

Determine the mass of 0.750 mol of carbon dioxide.
$$m = n \times M$$
$$= (0.750 \text{ mol}) \times (44.01 \text{ g/mol})$$
$$= 33.0 \text{ g}$$

The mass of 0.75 mol of carbon dioxide is 33.0 g.

**Check Your Solution**

The units are correct. 1 mol of carbon dioxide has a mass of 44 g. You need to determine the mass of 0.75 mol, or 75% of a mole. 33 g is equal to 75% of 44 g. Your answer is reasonable.

### Practice Problems

20. Calculate the mass of each molar amount.
    (a) 3.90 mol of carbon, C
    (b) 2.50 mol of ozone, $O_3$
    (c) $1.75 \times 10^7$ mol of propanol, $C_3H_8O$
    (d) $1.45 \times 10^{-5}$ mol of ammonium dichromate, $(NH_4)_2Cr_2O_7$

21. For each group, which sample has the largest mass?
    (a) 5.00 mol of C
    1.50 mol of $Cl_2$
    0.50 mol of $C_6H_{12}O_6$
    (b) 7.31 mol of $O_2$
    5.64 mol of $CH_3OH$
    12.1 mol of $H_2O$

22. A litre, 1000 mL, of water contains 55.6 mol. What is the mass of a litre of water?

23. To carry out a particular reaction, a chemical engineer needs 255 mol of styrene, $C_8H_8$. How many kilograms of styrene does the engineer need?

## Converting from Mass to Moles

In the previous Sample Problem, you saw how to convert moles to mass. Often, however, chemists know the mass of a substance but are more interested in knowing the number of moles. Suppose that a reaction produces 223 g of iron and 204 g of aluminum oxide. The masses of the substances do not tell you very much about the reaction. You know, however, that 223 g of iron is 4 mol of iron. You also know that 204 g of aluminum oxide is 2 mol of aluminum oxide. You may conclude that the reaction produces twice as many moles of iron as it does moles of aluminum oxide. You can perform the reaction many times to test your conclusion. If your conclusion is correct, the mole relationship between the products will hold. To calculate the number of moles in a sample, find out how many times the molar mass goes into the mass of the sample.

amount (mol) ← Divide by molar mass (g/mol). ← mass (g)

$$\text{Amount} = \frac{\text{Mass}}{\text{Molar mass}}$$

$$n = \frac{m}{M}$$

You can use unit analysis to ensure you are performing the correct operation to convert the mass of a sample to the amount of a sample.

$$\frac{\text{units for mass}}{\text{units for molar mass}} = \frac{g}{g/mol}$$

$$= g \times \frac{mol}{g}$$

$$= mol \text{ (units for amount)}$$

The following Sample Problem explains how to convert from the mass of a sample to the number of moles it contains.

## Sample Problem

### Mass to Moles

**Problem**

How many moles of acetic acid, $CH_3COOH$, are in a 23.6 g sample?

**What Is Required?**

You need to find the number of moles in 23.6 g of acetic acid.

**What Is Given?**

You are given the mass of the sample.

**Plan Your Strategy**

To obtain the number of moles of acetic acid, divide the mass of acetic acid by its molar mass.

**Act on Your Strategy**

The molar mass of $CH_3COOH$ is
$(12.01 \times 2) + (16.00 \times 2) + (1.01 \times 4) = 60.06 \text{ g}$.

$$n = \frac{m}{M_{CH_3COOH}}$$
$$= \frac{23.6 \text{ g}}{60.06 \text{ g/mol}}$$
$$= 0.393 \text{ mol}$$

Therefore there are 0.393 mol of acetic acid in 23.6 g of acetic acid.

**Check Your Solution**

Work backwards. There are 60.06 g in each mol of acetic acid. So in 0.393 mol of acetic acid, you have 0.393 mol × 60.06 g/mol = 23.6 g of acetic acid. This value matches the question.

## Practice Problems

24. Calculate the number of moles in each sample.
    (a) 103 g of Mo
    (b) $1.32 \times 10^4$ g of Pd
    (c) 0.736 kg of Cr
    (d) 56.3 mg of Ge

25. Express each of the following as a molar amount.
    (a) 39.2 g of silicon dioxide, $SiO_2$
    (b) 7.34 g of nitrous acid, $HNO_2$
    (c) $1.55 \times 10^5$ kg of carbon tetrafluoride, $CF_4$
    (d) $8.11 \times 10^{-3}$ mg of 1-iodo-2,3-dimethylbenzene, $C_8H_9I$

26. Sodium chloride, NaCl, can be used to melt snow. How many moles of sodium chloride are in a 10 kg bag?

27. Octane, $C_8H_{18}$, is a principal ingredient of gasoline. Calculate the number of moles in a 20.0 kg sample of octane.

# Chemistry Bulletin

## Chemical Amounts in Vitamin Supplements

Vitamins and minerals (micronutrients) help to regulate your metabolism. They are the building blocks of blood and bone, and they maintain muscles and nerves. In Canada, a standard called *Recommended Nutrient Intake* (*RNI*) outlines the amounts of micronutrients that people should ingest each day. Eating a balanced diet is the best way to achieve your RNI. Sometimes, however, you may need to take multivitamin supplements when you are unable to attain your RNI through diet alone.

The label on a bottle of supplements lists all the vitamins and minerals the supplements contain. It also lists the form and source of each vitamin and mineral, and the amount of each. The form of a mineral is especially important to know because it affects the quantity your body can use. For example, a supplement may claim to contain 650 mg of calcium carbonate, $CaCO_3$, per tablet. This does not mean that there is 650 mg of calcium in the tablet. The amount of actual calcium, or *elemental calcium*, in 650 mg of in calcium carbonate is only 260 mg. Calcium carbonate has more elemental calcium than the same amount of calcium gluconate, which only has 58 mg for every 650 mg of the compound. Calcium gluconate may be easier for your body to absorb, however.

### Quality Control

Multivitamin manufacturers employ chemists, or analysts, to ensure that the products they make have the right balance of micronutrients. Manufacturers have departments devoted to *quality control* (*QC*). QC chemists analyze all the raw materials in the supplements, using standardized tests. Most manufacturers use tests approved by a "standardization body," such as the US Pharmacopoeia. Such standardization bodies have developed testing guidelines to help manufacturers ensure that their products contain what the labels claim, within strict limits.

To test for quality, QC chemists prepare samples of the raw materials from which they will make the supplements. They label the samples according to the "lot" of materials from which the samples were taken. They powder and weigh the samples. Then they extract the vitamins. At the same time, they prepare standard solutions containing a known amount of each vitamin.

Next the chemists compare the samples to the standards by subjecting both to the same tests. One test that is used is *high-performance liquid chromatography* (*HPLC*). HPLC produces a spectrum, or "fingerprint," that identifies each compound. HPLC analysis can also determine the quantity of a substance in a sample.

Analysts test tablets and capsules for dissolution and disintegration properties. The analyst may use solutions that simulate the contents of the human stomach or intestines for these tests. Only when the analysts are sure that the tablets pass all the necessary requirements are the tablets shipped to retail stores.

### Making Connections

1. Why might consuming more of the daily RNI of a vitamin or mineral be harmful?

2. The daily RNI of calcium for adolescent females is 700 to 1100 mg. A supplement tablet contains 950 mg of calcium citrate. Each gram of calcium citrate contains $5.26 \times 10^{-3}$ mol calcium. How many tablets would a 16-year-old female have to take to meet her daily RNI?

**An analyst tests whether tablets will dissolve sufficiently within a given time limit.**

## Converting Between Moles, Mass, and Number of Particles

You can use what you now know about the mole to carry out calculations involving molar mass and the Avogadro constant. One mole of any compound or element contains $6.02 \times 10^{23}$ particles. The compound or element has a mass, in grams, that is determined from the periodic table.

Now that you have learned how the number of particles, number of moles, and mass of a substance are related, you can convert from one value to another. Usually chemists convert from moles to mass and from mass to moles. Mass is a property that can be measured easily. The following graphic shows the factors used to convert between particles, moles, and mass.

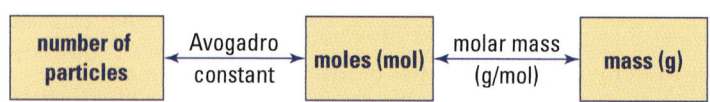

number of particles ←—Avogadro constant—→ moles (mol) ←—molar mass (g/mol)—→ mass (g)

To be certain you understand the relationship among particles, moles, and mass, examine the following Sample Problem.

### Math LINK
Average atomic mass values in some periodic tables can have five or more significant digits. How do you know how many significant digits to use? When you enter values, such as average atomic mass, into your calculator, be sure that you use at least one more significant digit than is required in your answer.

### Sample Problem
#### Particles to Mass

**Problem**

What is the mass of $5.67 \times 10^{24}$ molecules of cobalt(II) chloride, $CoCl_2$?

**What Is Required?**

You need to find the mass of $5.67 \times 10^{24}$ molecules of cobalt(II) chloride.

**What Is Given?**

You are given the number of molecules.

**Plan Your Strategy**

Convert the number of molecules into moles by dividing by the Avogadro constant. Then convert the number of moles into grams by multiplying by the molar mass of cobalt(II) chloride.

**Act on Your Strategy**

First determine the amount by dividing the number of molecules by the Avogadro constant.

$$n = \frac{5.67 \times 10^{24} \text{ molecules}}{6.02 \times 10^{23} \text{ molecules/mol}}$$

$$= 9.42 \text{ mol}$$

Next, determine the mass by multiplying the amount by the molar mass of $CoCl_2$. The molar mass of $CoCl_2$ is
58.93 g/mol + 2(35.45 g/mol) = 129.84 g/mol.

$$m = 9.42 \text{ mol} \times 129.84 \text{ g/mol}$$

$$= 1.22 \times 10^3 \text{ g}$$

**Check Your Solution**

$5.67 \times 10^{24}$ molecules is roughly 10 times the Avogadro constant. This means that you have about 10 mol of cobalt(II) chloride. The molar mass of cobalt(II) chloride is about 130 g, and 10 times 130 g is 1300 g.

## Practice Problems

**28.** Determine the mass of each sample.
   (a) $6.02 \times 10^{24}$ formula units of $ZnCl_2$
   (b) $7.38 \times 10^{21}$ formula units of $Pb_3(PO_4)_2$
   (c) $9.11 \times 10^{23}$ molecules of $C_{15}H_{21}N_3O_{15}$
   (d) $1.20 \times 10^{29}$ molecules of $N_2O_5$

**29.** What is the mass of lithium in 254 formula units of lithium chloride, LiCl?

**30.** A sample of potassium oxide, $K_2O$, contains $3.99 \times 10^{24}$ $K^+$ ions. What is the mass of the sample?

**31.** A sample of ammonium chloride, $NH_4Cl$, contains $9.03 \times 10^{23}$ hydrogen atoms. What is the mass of the sample?

**32.** Express the average mass of a single atom of titanium, Ti, in grams.

**33.** Vitamin $B_2$, $C_{17}H_{20}N_4O_6$, is also called riboflavin. What is the average mass, in grams, of a single molecule of riboflavin?

What if you wanted to compare amounts of substances, and you only knew their masses? You would probably convert their masses in grams to their amounts in moles. The Avogadro constant relates the molar amount to the number of particles. Examine the next Sample Problem to learn how to convert mass to number of particles.

## Sample Problem

### Mass to Particles

**Problem**

Chlorine gas, $Cl_2$, can react with iodine, $I_2$, to form iodine chloride, ICl. How many molecules of iodine chloride are contained in a $2.74 \times 10^{-1}$ g sample?

**What Is Required?**

You need to find the number of molecules in $2.74 \times 10^{-1}$ g of iodine.

**What Is Given?**

You are given the mass of the sample.

**Plan Your Strategy**

First convert the mass to moles, using the molar mass of iodine. Multiplying the number of moles by the Avogadro constant will yield the number of molecules.

**Act on Your Strategy**

The molar mass of ICl is 162.36 g.
Dividing the given mass of ICl by the molar mass gives

$$n = \frac{2.74 \times 10^{-1} \text{ g}}{162.36 \text{ g/mol}}$$
$$= 1.69 \times 10^{-3} \text{ mol}$$

*Continued...*

> **Continued...**
>
> Now multiply the number of moles by the Avogadro constant. This gives the number of molecules in the sample.
>
> $$(1.69 \times 10^{-3} \text{ mol}) \times \frac{(6.02 \times 10^{23} \text{ molecules})}{1 \text{ mol}} = 1.01 \times 10^{21} \text{ molecules}$$
>
> Therefore there are $1.01 \times 10^{21}$ molecules in $2.74 \times 10^{-1}$ g of iodine chloride.
>
> ### Check Your Solution
>
> Work backwards. Each mole of iodine chloride has a mass of 162.36 g/mol. Therefore $1.01 \times 10^{21}$ molecules of iodine chloride have a mass of:
>
> $$1.01 \times 10^{21} \text{ molecules} \times \frac{1 \text{ mol}}{6.02 \times 10^{23} \text{ molecules}} \times \frac{162.36 \text{ g}}{1 \text{ mol}}$$
> $$= 2.72 \times 10^{-1} \text{ g}$$
>
> The answer is close to the value given in the question. Your answer is reasonable.

### Practice Problems

**34.** Determine the number of molecules or formula units in each sample.
   - **(a)** 10.0 g of water, $H_2O$
   - **(b)** 52.4 g of methanol, $CH_3OH$
   - **(c)** 23.5 g of disulfur dichloride, $S_2Cl_2$
   - **(d)** 0.337 g of lead(II) phosphate, $Pb_3(PO_4)_2$

**35.** How many atoms of hydrogen are in $5.3 \times 10^4$ molecules of sodium glutamate, $NaC_5H_8NO_4$?

**36.** How many molecules are in a 64.3 mg sample of tetraphosphorus decoxide, $P_4O_{10}$?

**37. (a)** How many formula units are in a $4.35 \times 10^{-2}$ g sample of potassium chlorate, $KClO_3$?
   **(b)** How many ions (chlorate and potassium) are in this sample?

### Concept Organizer: Relationships Among Particles, Moles, and Mass

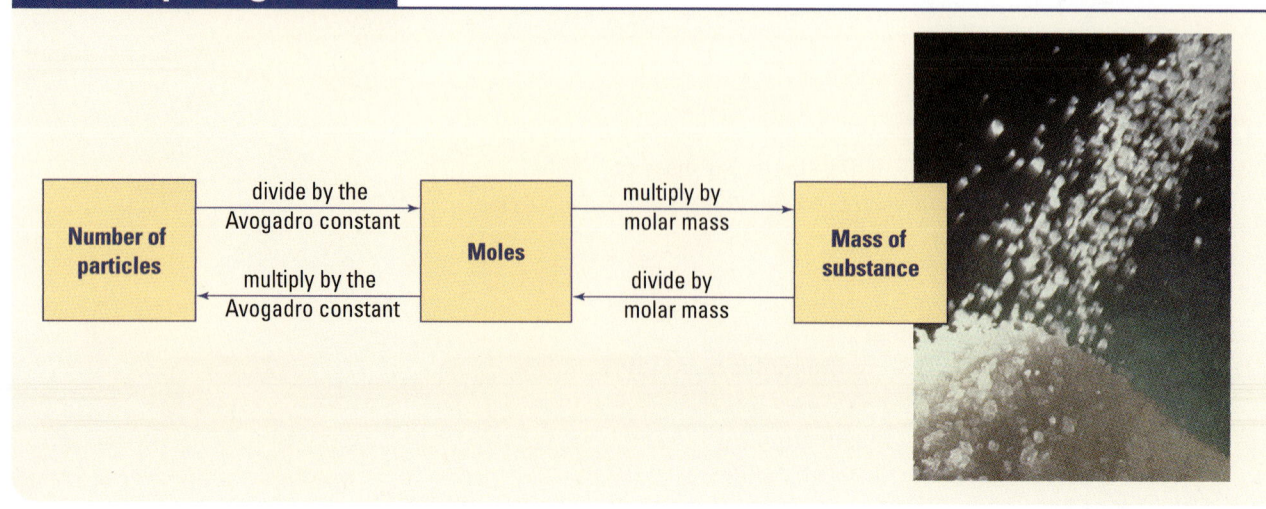

Number of particles — divide by the Avogadro constant → Moles — multiply by molar mass → Mass of substance

Number of particles ← multiply by the Avogadro constant — Moles ← divide by molar mass — Mass of substance

## Section Summary

You have now learned about the relationships among the number of particles in a substance, the amount of a substance in moles, and the mass of a substance in grams. The relationship between amount and mass is very useful for working with solids or liquids. When dealing with gases, however, it is often more convenient to determine the volume than the mass. In section 2.4, you will learn about the relationship between the volume a gas occupies and the amount of gas that is present.

## Section Review

**1** Draw a diagram that shows the relationship between the atomic mass and molar mass of an element and the Avogadro constant.

**2** Consider a 78.6 g sample of ammonia, $NH_3$.
  (a) How many moles of ammonia are in the sample?
  (b) How many molecules of ammonia are in the sample?

**3** Use your understanding of the mole to answer the following questions.
  (a) What is the average mass, in grams, of a single atom of silicon, Si?
  (b) What is the mass, in atomic mass units, of 1 mol of silicon atoms?

**4** Consider a 0.789 mol sample of sodium chloride, NaCl.
  (a) What is the mass of the sample?
  (b) How many formula units of sodium chloride are in the sample?
  (c) How many ions are in the sample?

**5** A 5.00 carat diamond has a mass of 1.00 g. How many carbon atoms are in a 5.00 carat diamond?

**6** A bottle of mineral supplement tablets contains 100 tablets and a total of 200 mg of copper. The copper is found in the form of copper(II) oxide. What mass of copper(II) oxide is contained in each tablet?

# 2.4 Molar Volume

### Section Preview/Outcomes

In this section, you will

- **state** Avogadro's hypothesis, and explain how it contributes to our understanding of the reactions of gases
- **define** STP and the molar volume of a gas at STP
- **perform** calculations converting between the number of particles, moles, mass, and volume of a gas at STP
- **communicate** your understanding of the following terms: *pressure, pascal* (Pa), *kilopascals* (kPa), *standard temperature and pressure* (STP), *law of combining volumes, Avogadro's hypothesis, molar volume, ideal gas*

An automobile air bag, shown in Figure 2.10, fills up with about 65 L of nitrogen gas in approximately 27 ms. The air bag can prevent a driver or passenger from being seriously injured. What amount of nitrogen gas corresponds to 65 L? As you will learn in this section, there is a way to relate the volume of a gas to the amount of gas in moles.

The volume of a gas, however, is also related to both its temperature and its pressure. Therefore, before examining the relationship between the amount of a gas and its volume, you must consider both its temperature and pressure.

## Gas Pressure

The earliest use of the word "pressure" in English referred to a burden or worry troubling a person's mind. Scientists found this concept a useful mental model to picture what happens when force is applied to a specific area. They adopted the word pressure to describe any application of force over an area.

More specifically, **pressure** is defined as the force exerted on an object per unit of surface area ($P = F/A$). The SI unit for pressure is the **pascal (Pa)**, equal to 1 N/m$^2$. More often, pressure is reported in **kilopascals (kPa)**, equal to 1000 Pa.

The pressure of a gas is determined by the motion of its particles. Picture hundreds of billions of gas molecules in random motion, each striking the inner surface of their container. Each collision exerts a force on the container's inner surface.

Figure 2.11 shows an inflated basketball. The air particles, in constant motion, collide with the basketball's inner surface. Each collision exerts a force. The number of collisions and the force of each collision form the overall gas pressure. Since the particles move at random in all directions, the net pressure exerted is equal throughout.

**Figure 2.10** A chemical reaction is triggered when sensors connected to an air bag detect sudden deceleration. The reaction produces nitrogen gas, which rapidly inflates the bag.

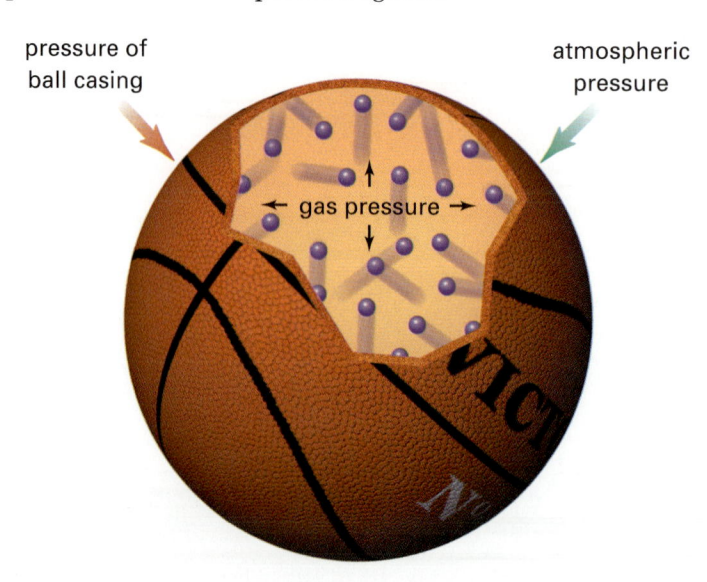

**Figure 2.11** Gas particles exert force as they collide with the inner surface of a basketball.

66 MHR • Unit 1 Stoichiometry

## Temperature, Pressure, and Volume of Gases

Pressure and temperature both affect the volume of gases. For example, if you put an inflated balloon in the freezer, the balloon will shrink. In other words, when the temperature of a gas decreases, the volume of the gas also decreases.

Deep-sea divers, shown in Figure 2.12, carry gas tanks to allow them to breathe underwater. The oxygen and nitrogen in the tank is under high pressure. At atmospheric pressure (101.3 kPa), the gas would occupy many litres. At high pressures, a large amount of gas fits into the small volume of the tank. In other words, the gas pressure is increased and the volume decreases.

## Standard Temperature and Pressure

Since changes in temperature or pressure affect the volume of a gas, you must specify the temperature and pressure of a gas when stating its volume. Scientists have designated standard conditions, described below, which allow them to compare different gas volumes meaningfully.

- The average pressure of the atmosphere at sea level is taken as standard pressure (101.3 kPa).
- The freezing point of water (0°C or 273 K) is defined as standard temperature.

Together, these conditions are referred to as **standard temperature and pressure** (**STP**).

**Figure 2.12** For about every 10 m of depth, water pressure on a diver's body adds the equivalent of 101.3 kPa. The air the diver breathes must equal the pressure of the surrounding water to inflate the diver's lungs.

## The Molar Volume of Gases

Near the beginning of the nineteenth century, French scientist Joseph Louis Gay-Lussac (1778–1853) measured the volumes of gases before and after a reaction. His research led him to devise the **law of combining volumes**: When gases react, the volumes of the reactants and the products, measured at equal temperatures and pressures, are always in whole number ratios. For example, Guy-Lussac made the following observations:

- 2 volumes of hydrogen gas react with 1 volume of oxygen gas to produce 2 volumes of water vapour.
- 1 volume of hydrogen gas reacts with 1 volume of chlorine gas to form 2 volumes of hydrogen chloride gas.
- 3 volumes of hydrogen gas react with 1 volume of nitrogen gas to form 2 volumes of ammonia gas.

Around the same time, John Dalton examined the masses of substances before and after a reaction. Dalton found that the masses of elements that combine to form compounds can be expressed as small whole number ratios.

Based on these findings, Amedeo Avogadro (1776–1856) realized that he could relate the *volume* of a gas to the *amount* that is present (calculated from the mass). He realized that the number of molecules of gas present corresponded to the volume ratios that Gay-Lussac had obtained.

For example, 2 L of hydrogen gas react with 1 L of oxygen gas to produce 2 L of water vapour. Avogadro hypothesized that there must be the same number of molecules in each litre of gas. Thus, **Avogadro's hypothesis** was formulated: *Equal volumes of all ideal gases at the same temperature and pressure contain the same number of molecules.*

Chapter 2 The Mole • MHR **67**

Avogadro's hypothesis allowed other scientists to develop the concept of the mole. The mole concept connects the *amount* of a substance to the *mass* of a substance. This connection allows chemists to write balanced chemical equations. Chemists use chemical equations to make accurate predictions about the quantities of products and reactants involved in chemical reactions. In Chapter 4, you will learn how to make predictions using chemical equations.

Figures 2.13, 2.14, and 2.15 show the three reactions that produced Gay-Lussac's observations, along with their chemical equations. You can see that the ratios of moles of products and reactants involved in the reactions are the same as the volume ratios. Today our knowledge of atoms and molecules helps us explain Gay-Lussac's results. We know that many gases are made of molecules that contain more than one atom.

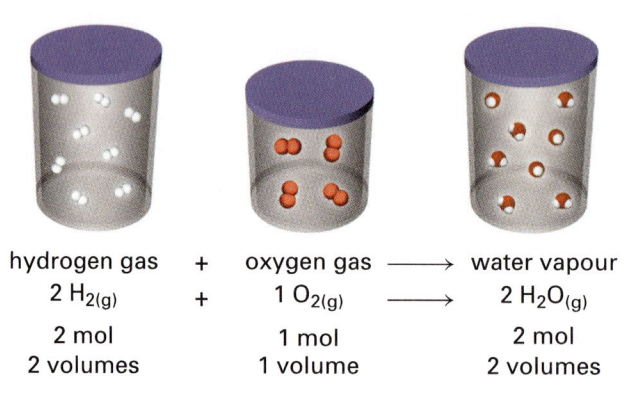

hydrogen gas + oxygen gas ⟶ water vapour
2 $H_{2(g)}$ + 1 $O_{2(g)}$ ⟶ 2 $H_2O_{(g)}$
2 mol      1 mol      2 mol
2 volumes    1 volume    2 volumes

**Figure 2.13** Hydrogen and oxygen gases combine to form water vapour.

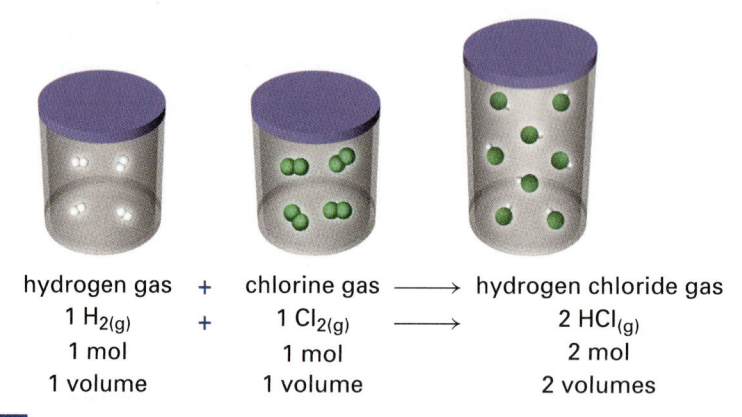

hydrogen gas + chlorine gas ⟶ hydrogen chloride gas
1 $H_{2(g)}$ + 1 $Cl_{2(g)}$ ⟶ 2 $HCl_{(g)}$
1 mol      1 mol      2 mol
1 volume    1 volume    2 volumes

**Figure 2.14** Hydrogen and chlorine gases combine to form hydrogen chloride gas.

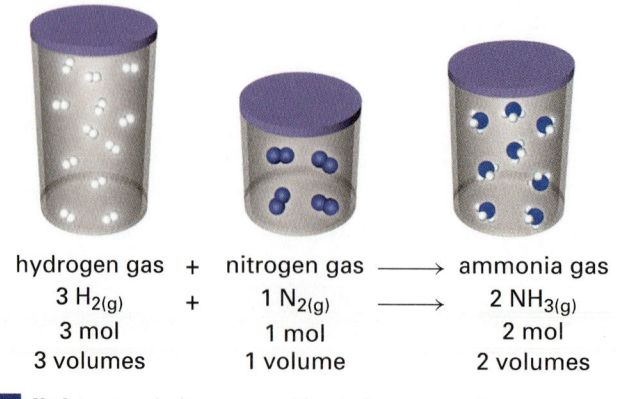

hydrogen gas + nitrogen gas ⟶ ammonia gas
3 $H_{2(g)}$ + 1 $N_{2(g)}$ ⟶ 2 $NH_{3(g)}$
3 mol      1 mol      2 mol
3 volumes    1 volume    2 volumes

**Figure 2.15** Hydrogen and nitrogen combine to form ammonia gas.

Avogadro's hypothesis can be written as a mathematical law, shown here.

> **Avogadro's Law**
>
> Avogadro's law gives us a mathematical relationship between the volume of a gas (V) and the number of moles of gas present (n).
>
> $$n \propto V \quad \text{or} \quad n = kV \quad \text{or} \quad \frac{n_1}{V_1} = \frac{n_2}{V_2}$$
>
> where $n$ = number of moles
> $V$ = volume
> $k$ = a constant

Based on Avogadro's law, one mole of a gas occupies the same volume as one mole of another gas at the same temperature and pressure. The molar volume of a gas is the space that is occupied by one mole of the gas. **Molar volume** sometimes represented by the symbol ($MV$) is measured in units of L/mol. You can find the molar volume of a gas by dividing its volume by the number of moles that are present ($\frac{V}{n}$). Complete the following Thought Lab to find the molar volumes of carbon dioxide gas, oxygen gas, and methane gas at STP.

## ThoughtLab — Molar Volume of Gases

Two students decided to calculate the molar volumes of carbon dioxide, oxygen, and methane gas. First they measured the mass of an empty 150 mL syringe under vacuum conditions. This ensured that the syringe did not contain any air. Next they filled the syringe with 150 mL of carbon dioxide gas. They measured and recorded the mass of the syringe plus the gas. The students repeated their procedure for oxygen gas and for methane gas.

The students were able to carry out their experiments in a room maintained at STP (273 K and 101.3 kPa). Their results are provided in the table.

### Procedure

1. Copy the table into your notebook.
2. Calculate the molar volume of carbon dioxide gas at STP. Record your calculations and answers in the table.
3. Do the same calculations for oxygen and methane gas. Record your calculations and answers in the table.

### Analysis

1. Compare the three molar volumes at STP. What do you observe?
2. The accepted molar volume of a gas at STP is 22.4 L/mol. Based on this value, calculate the percent error in your experimental data for each gas.

**Three Gases at STP**

| Gas | carbon dioxide $CO_{2(g)}$ | oxygen $O_{2(g)}$ | methane $CH_{4(g)}$ |
|---|---|---|---|
| Volume of gas (*V*) | 150 mL | 150 mL | 150 mL |
| Mass of empty syringe | 25.081 g | 25.085 g | 25.082 g |
| Mass of gas + syringe | 25.383 g | 25.304 g | 25.197 g |
| Mass of gas (*m*) | | | |
| Molar mass of gas (*M*) | | | |
| Number of moles of gas (*n* = *m*/*M*) | | | |
| Molar volume of gas at STP (*MV* = *V*/*n*) | | | |

## Standard Molar Volume

In the Thought Lab, you found that the molar volumes of carbon dioxide, oxygen, and methane were close to 22.4 L/mol. In fact, this finding extends to all gases.

> The molar volume of a gas at STP is 22.4 L/mol.

Figure 2.16 shows a balloon with a volume of 22.4 L compared to some common objects. This is a fairly large volume of gas. Compare the size of the balloon to the size of a basketball, which has a volume of 7.5 L.

**Figure 2.16** One mole of any gas at STP occupies 22.4 L (22.4 dm³). The yellow balloon has a volume of 22.4 L. The other objects are shown for comparison.

In fact, the molar volume of 22.4 L/mol at STP is considered true for a hypothetical gas called an **ideal gas**. The particles of an ideal gas are considered to have zero volume. They are also considered not to attract one another.

The particles of real gases, however, *do* take up space and *do* attract one another to varying degrees. The molar volumes of real gases at STP, therefore, vary slightly, as shown in Table 2.3. For the purposes of this textbook, however, you can treat all gases as though they have the properties of ideal gases and a molar volume of 22.4 L at STP.

If you assume that real gases have the properties of ideal gases, you can use Avogadro's law to solve gas problems involving moles and volume at STP.

**Table 2.3** Molar Volume of Several Real Gases at STP

| Gas | Molar volume (L/mol) |
|---|---|
| helium, He | 22.398 |
| neon, Ne | 22.401 |
| argon, Ar | 22.410 |
| hydrogen, $H_2$ | 22.430 |
| nitrogen, $N_2$ | 22.413 |
| oxygen, $O_2$ | 22.414 |
| carbon dioxide, $CO_2$ | 22.414 |
| ammonia, $NH_3$ | 22.350 |

### Sample Problem
#### Volume of Gases

**Problem**

What is the volume of 3.0 mol of nitrous oxide, $NO_{2(g)}$, at STP?

**What Is Required?**

You need to find the volume of nitrous oxide at STP ($V_f$).

**What Is Given?**

You know that 1.00 mol of a gas occupies 22.4 L at STP.
∴ $n_i$ = 1.0 mol
$V_i$ = 22.4 L
$n_f$ = 3.0 mol of $NO_2$

> **PROBLEM TIP**
>
> In the Sample Problems in this section, you will see two different methods of solving the problem: the algebraic method and the ratio method. Choose the method you prefer to solve this type of problem.

70   MHR • Unit 1 Stoichiometry

**Plan Your Strategy**

**Algebraic method**

Use Avogadro's law: $\dfrac{n_i}{V_i} = \dfrac{n_f}{V_f}$

Cross multiply to solve for $V_f$, the unknown volume of $NO_2$.

**Ratio method**

There are 3 mol of nitrous oxide. Thus, the volume of nitrous oxide at STP must be larger than the volume of 1 mol of gas at STP. Multiply by a mole-to-mole ratio that is greater than 1.

**Act on Your Strategy**

**Algebraic method**

$$V_f = \dfrac{n_f V_i}{n_i}$$
$$= \dfrac{3.0 \text{ mol} \times 22.4 \text{ L}}{1.00 \text{ mol}}$$
$$= 67 \text{ L}$$

**Ratio method**

$$V_f = 22.4 \text{ L} \times \dfrac{3.0 \text{ mol}}{1.0 \text{ mol}}$$
$$= 67 \text{ L}$$

Therefore, there are 67 L of nitrous oxide.

**Check Your Solution**

The significant digits and the units are all correct.
The volume of nitrous oxide is three times the volume of 1 mol of gas at STP. This makes sense, since there are 3 mol of nitrous oxide.

## Sample Problem

### Moles of Gas

**Problem**

Suppose that you have 44.8 L of methane gas at STP.
**(a)** How many moles are present?
**(b)** What is the mass (in g) of the gas?
**(c)** How many molecules of gas are present?

**What Is Required?**

**(a)** You need to calculate the number of moles.
**(b)** You need to calculate the mass of the gas.
**(c)** You need to calculate the number of molecules.

*Continued ...*

**Continued ...**

### What Is Given?
The gas is at STP. Thus 1.00 mol of gas has a volume of 22.4 L. You know that one mole contains $6.02 \times 10^{23}$ molecules. There are 44.8 L of gas.

### Plan Your Strategy

#### Algebraic method
(a) Use Avogadro's law. Solve for the number of moles by cross multiplying.
(b) Multiply the number of moles ($n$) by the molar mass ($M$) to find the mass of the gas ($m$).
$$m = n \times M$$
(c) Multiply the number of moles ($n$) by the Avogadro constant ($6.02 \times 10^{23}$) to find the number of molecules.
$$\text{number of molecules} = n \times 6.02 \times 10^{23} \text{ molecules/mol}$$

#### Ratio method
(a) The volume of the unknown gas is 44.8 L. Since the volume is greater than 22.4 L, there is more than 1 mol of gas. To find the unknown number of moles ($n$), multiply by a volume ratio that is greater than 1.
(b) Multiply the number of moles ($n$) by the molar mass ($M$) to find the mass of the gas ($m$).
$$m = n \times M$$
(c) Multiply the number of moles ($n$) by the Avogadro constant ($6.02 \times 10^{23}$) to find the number of molecules.
$$\text{number of molecules} = n \times 6.02 \times 10^{23} \text{ molecules/mol}$$

### Act on Your Strategy

#### Algebraic method
(a) $\dfrac{n_i}{V_i} = \dfrac{n_f}{V_f}$

$n_f = \dfrac{n_i V_f}{V_i}$

$= \dfrac{1.00 \text{ mol} \times 44.8 \text{ L}}{22.4 \text{ L}}$

$= 2.00 \text{ mol}$

(b) Find the molar mass of methane, $CH_4$.

$\quad 1C = 1 \times 12.01 \text{ g/mol}$
$+ \; \underline{4H = 4 \times 1.01 \text{ g/mol}}$
$\quad M_{CH_4} = 16.05 \text{ g/mol}$

$m = n \times M$
$= 2.00 \text{ mol} \times 16.05 \text{ g/mol}$
$= 32.1 \text{ g}$

(c) number of molecules $= 2.00 \text{ mol} \times 6.02 \times 10^{23} \; \dfrac{\text{molecules}}{\text{mol}}$

$= 1.20 \times 10^{24} \text{ molecules}$

**Ratio method**

(a) $n = 1.00 \text{ mol} \times \dfrac{44.8 \cancel{L}}{22.4 \cancel{L}}$

   $= 2.00 \text{ mol}$

(b) Find the molar mass of methane, $CH_4$.

$$\begin{aligned} 1C &= 1 \times 12.01 \text{ g/mol} \\ + \; 4H &= 4 \times 4.04 \text{ g/mol} \\ \hline M_{CH_4} &= 16.05 \text{ g/mol} \end{aligned}$$

$m = n \times M$

   $= 2.00 \text{ mol} \times 16.05 \text{ g/mol}$

   $= 32.1 \text{ g}$

(c) number of molecules $= 2.00 \cancel{\text{mol}} \times 6.02 \times 10^{23} \dfrac{\text{molecules}}{\cancel{\text{mol}}}$

   $= 1.20 \times 10^{24}$ molecules

Therefore, 2.00 mol of methane are present. The mass of the gas is 32.1 g. $1.20 \times 10^{24}$ molecules are present.

**Check Your Solution**

The significant digits are correct.
The volume of methane is double the volume of 1 mol of gas. It makes sense that 2 mol of methane are present. It also makes sense that the number of molecules present is double the Avogadro constant.

## Practice Problems

38. A balloon contains 2.0 L of helium gas at STP. How many moles of helium are present?

39. How many moles of gas are present in 11.2 L at STP? How many molecules?

40. What is the volume, at STP, of 3.45 mol of argon gas?

41. At STP, a container holds 14.01 g of nitrogen gas, 16.00 g of oxygen gas, 66.00 g of carbon dioxide gas, and 17.04 g of ammonia gas. What is the volume of the container?

42. (a) What volume do 2.50 mol of oxygen occupy at STP?
    (b) How many molecules are present in this volume of oxygen?
    (c) How many oxygen atoms are present in this volume of oxygen?

43. What volume do $2.00 \times 10^{24}$ atoms of neon occupy at STP?

> **PROBLEM TIP**
>
> To solve many of these problems, try setting up a proportion and solving by cross multiplication.

## Section Summary

In this section, you learned that ideal gases have a molar volume of 22.4 L at STP. Now, given the mass or the volume of a gas at STP, you can determine the amount of the substance and the number of particles it contains, by assuming it has the properties of an ideal gas. In Chapter 3, you will explore the mole concept further. You will learn how the mass proportions of elements in compounds relate to their formulas.

## Section Review

**1** How many grams of sulfur dioxide are in 36.2 L at STP?

**2** Hydrogen sulfide, $H_2S_{(g)}$, is present in some sources of natural gas. Natural gas that contains hydrogen sulfide is called sour gas because of its powerful stench. What is the mass of 20 L of hydrogen sulfide at STP?

**3** Helium-filled balloons are buoyant in air. Answer the following questions about helium.

(a) What mass of helium, $He_{(g)}$, occupies 1.00 L at STP?

(b) What is the density of helium at STP, in g/L?

(c) Do you predict that the density of air is greater than or less than the density of helium? Explain your answer.

(d) Estimate the density of air by determining the density of nitrogen, $N_{2(g)}$, at STP. Compare your result to your prediction.

**4** An analyst knows that a sample of gas is either carbon monoxide, $CO_{(g)}$, or carbon dioxide, $CO_{2(g)}$. The analyst determines that 250 mL of the gas has a mass of 0.31 g. Is the gas carbon monoxide or is it carbon dioxide?

**5** What was Avogadro's hypothesis? Explain how Avogadro contributed to our understanding of gases and the relationships among the properties of gases.

**6** Your teacher shows your class a cylinder of argon, $Ar_{(g)}$, similar to the one shown in Figure 2.17. The volume of the tank is 50.0 L, and the gas within the tank is at a pressure of $5 \times 10^3$ kPa. Your classmate says that there are 2.23 mol of argon in the tank. Do you agree? Explain why or why not.

**Figure 2.17** Pressure relief valves prevent a compressed gas cylinder from exploding if the temperature, and thus the pressure, increases.

# CHAPTER 2 Review

## Reflecting on Chapter 2
Summarize this chapter in the format of your choice. Here are a few ideas to use as guidelines:
- Describe the relationships among isotopic abundance, isotopic masses, and average atomic mass.
- Explain how to determine average atomic mass using the mass and isotopic abundance of isotopes.
- Describe how and why chemists group atoms and molecules into molar amounts.
- Explain how chemists define the mole and why this definition is useful.
- Use the Avogadro constant to convert between moles and particles.
- Explain the relationship between average atomic mass and the mole.
- Find a compound's molar mass using the periodic table.
- State Avogadro's hypothesis and explain its importance.
- Define molar volume and give examples.
- Explain how to convert among particles, moles, volume, and mass.

## Reviewing Key Terms
For each of the following terms, write a sentence that shows your understanding of its meaning.

average atomic mass
isotopic abundance
mole
kilopascals (kPa)
law of combining volumes
ideal gas
Avogadro constant
molar mass
pressure
pascal (Pa)
standard temperature and pressure (STP)
Avogadro's hypothesis
molar volume

## Knowledge/Understanding
1. Distinguish between atomic mass and average atomic mass. Give examples to illustrate each term.
2. What are isotopes? Give several examples in your definition.
3. The periodic table lists the average atomic mass of chlorine as 35.45 u. Are there any chlorine atoms that have a relative mass of 35.45 u? Explain your answer.
4. Explain how the Avogadro constant, average atomic mass, and molar mass are related.
5. (a) Describe the relationship between the mole, the Avogadro constant, and carbon-12.
   (b) Why do chemists use the concept of the mole to deal with atoms and molecules?
6. How is the molar mass of an element related to average atomic mass?
7. Explain what the term molar mass means for each of the following, using examples.
   (a) a metallic element
   (b) a diatomic element
   (c) a compound
8. "The volume of 1 mol of an ideal gas is always 22.4 L."
   (a) This statement is incorrect. Explain why.
   (b) Rewrite the statement so that it is correct.

## Inquiry
9. A particular isotope of barium has 81 neutrons. What is the mass number of this isotope?
10. How many neutrons are in the nucleus of an atom of xenon-131?
11. Lithium exists as two isotopes: lithium-7 (7.015 u) and lithium-6 (6.015 u). Lithium-6 accounts for only 7.4% of all lithium atoms. Calculate the average atomic mass of lithium. Show your calculations.
12. Potassium exists as two naturally occurring isotopes: K-39 and K-41. These isotopes have atomic masses of 39.0 u and 41.0 u respectively. If the average atomic mass of potassium is 39.10 u, calculate the relative abundance of each isotope.
13. How many moles of the given substance are present in each sample below?
    (a) 0.453 g of $Fe_2O_3$
    (b) 50.7 g of $H_2SO_4$
    (c) $1.24 \times 10^{-2}$ g of $Cr_2O_3$
    (d) $8.2 \times 10^2$ g of $C_2Cl_3F_3$
    (e) 12.3 g of $NH_4Br$
14. Convert each quantity to an amount in moles.
    (a) $4.27 \times 10^{21}$ atoms of He
    (b) $7.39 \times 10^{23}$ molecules of ICl
    (c) $5.38 \times 10^{22}$ molecules of $NO_2$
    (d) $2.91 \times 10^{23}$ formula units of $Ba(OH)_2$
    (e) $1.62 \times 10^{24}$ formula units of KI
    (f) $5.58 \times 10^{20}$ molecules of $C_3H_8$

Answers to questions highlighted in red type are provided in Appendix A.

15. Copy the following table into your notebook and complete it.

| Sample | Molar mass (g/mol) | Mass of sample (g) | Amount of substance (mol) | Number of molecules or formula units | Number of atoms |
|---|---|---|---|---|---|
| NaCl | 58.4 | 58.4 | 1.00 | $6.02 \times 10^{23}$ | $1.20 \times 10^{24}$ |
| $NH_3$ | (a) | 24.8 | (b) | (c) | (d) |
| $H_2O$ | (e) | (f) | (g) | $5.28 \times 10^{22}$ | (h) |
| $Mn_2O_3$ | (i) | (j) | (k) | (l) | $2.00 \times 10^{23}$ |
| $K_2CrO_4$ | (m) | $9.67 \times 10^{-1}$ | (n) | (o) | (p) |
| $C_8H_8O_3$ | (q) | (r) | (s) | $7.90 \times 10^{24}$ | (t) |
| $Al(OH)_3$ | (u) | (v) | $8.54 \times 10^2$ | (w) | (x) |

16. Calculate the molar mass of each compound.
    (a) $PtBr_2$
    (b) $C_3H_5O_2H$
    (c) $Na_2SO_4$
    (d) $(NH_4)_2Cr_2O_7$
    (e) $Ca_3(PO_4)_2$
    (f) $Cl_2O_7$

17. Express each amount as a mass in grams.
    (a) 3.70 mol of $H_2O$
    (b) $8.43 \times 10^{23}$ molecules of $PbO_2$
    (c) 14.8 mol of $BaCrO_4$
    (d) $1.23 \times 10^{22}$ molecules of $Cl_2$
    (e) $9.48 \times 10^{23}$ molecules of HCl
    (f) $7.74 \times 10^{19}$ molecules of $Fe_2O_3$

18. How many atoms of C are contained in 45.6 g of $C_6H_6$?

19. How many atoms of F are contained in 0.72 mol of $BF_3$?

20. Calculate the following.
    (a) the mass (u) of one atom of xenon
    (b) the mass (g) of one mole of xenon atoms
    (c) the mass (g) of one atom of xenon
    (d) the mass (u) of one mole of xenon atoms
    (e) the number of atomic mass units in one gram

21. How many atoms of C are in a mixture containing 0.237 mol of $CO_2$ and 2.38 mol of $CaC_2$?

22. How many atoms of hydrogen are in a mixture of $3.49 \times 10^{23}$ molecules of $H_2O$ and 78.1 g of $CH_3OH$?

23. How many nitrate ions are in a solution that contains $3.76 \times 10^{-1}$ mol of calcium nitrate, $Ca(NO_3)_2$?

24. Ethanol, $C_2H_5OH$, is frequently used as the fuel in wick-type alcohol lamps. One molecule of $C_2H_5OH$ requires three molecules of $O_2$ for complete combustion. What mass of $O_2$ is required to react completely with 92.0 g of $C_2H_5OH$?

25. Examine the following double displacement reaction.

    $NaCl_{(aq)} + AgNO_{3(aq)} \rightarrow AgCl_{(s)} + NaNO_{3(aq)}$

    In this reaction, one formula unit of NaCl reacts with one formula unit of $AgNO_3$. How many moles of NaCl would react with one mole of $AgNO_3$? Explain your answer.

26. Determine the mass of each of the following samples of gases at STP.
    (a) 22.4 L xenon, $Xe_{(g)}$
    (b) 500 L ammonia, $NH_{3(g)}$
    (c) 3.3 L oxygen, $O_{2(g)}$
    (d) 0.230 mL helium, $He_{(g)}$
    (e) 2.5 mL butane, $C_4H_{10(g)}$
    (f) 40.0 dm³ propane, $C_3H_{8(g)}$

27. A chemist isolates 2.99 g of a gas. The sample occupies 800 mL at STP. Determine the molar mass of the gas. Is the gas most likely to be oxygen, $O_{2(g)}$, krypton, $Kr_{(g)}$, neon, $Ne_{(g)}$, or fluorine, $F_{2(g)}$?

## Communication

28. Use the definition of the Avogadro constant to explain why its value must be determined by experiment.

29. Why is carbon-12 the only isotope with an atomic mass that is a whole number?

30. Draw a concept map for the conversion of mass (g) of a sample to amount (mol) of a sample to number of molecules in a sample to number of atoms in a sample. Be sure to include proper units.

76 MHR • Unit 1 Stoichiometry

31. Answer the following questions about the volume of a gas.
    (a) Explain how a change in temperature affects the volume of a gas. Use an example.
    (b) Explain how a change in pressure affects the volume of a gas. Use an example.
    (c) What does STP stand for? What are the conditions of STP?

## Making Connections

32. The RNI (Recommended Nutrient Intake) of iron for women is listed as 14.8 mg per day. Ferrous gluconate, $Fe(C_6H_{11}O_7)_2$ is often used as an iron supplement for those who do not get enough iron in their diet because it is relatively easy for the body to absorb. Some iron-fortified breakfast cereals contain elemental iron metal as their source of iron.
    (a) Calculate the number of moles of elemental iron, Fe, required by a woman, according to the RNI.
    (b) What mass, in milligrams, of ferrous gluconate, would satisfy the RNI for iron?
    (c) The term *bioavailability* refers to the extent that the body can absorb a certain vitamin or mineral supplement. There is evidence to suggest that the elemental iron in these iron-fortified cereals is absorbed only to a small extent. If this is the case, should cereal manufacturers be allowed to add elemental iron at all? How could cereal manufacturers assure that the consumer absorbs an appropriate amount of iron? Would adding more elemental iron be a good solution? List the pros and cons of adding more elemental iron, then propose an alternative solution.

33. Vitamin $B_3$, also known as niacin, helps maintain the normal function of the skin, nerves, and digestive system. The disease pellagra results from a severe niacin deficiency. People with pellagra experience mouth sores, skin irritation, and mental deterioration. Niacin has the following formula: $C_6H_5NO_2$. Often vitamin tablets contain vitamin $B_3$ in the form of niacinamide, $C_6H_6N_2O$, which is easier for the body to absorb.

    (a) A vitamin supplement tablet contains 100 mg of niacinamide. What mass of niacin contains an equivalent number of moles as 100 mg of niacinamide?
    (b) Do some research to find out how much niacin an average adult should ingest each day.
    (c) Do some research to find out what kinds of food contain niacin.
    (d) What are the consequences of ingesting too much niacin?
    (e) Choose another vitamin to research. Find out its chemical formula, its associated recommended nutrient intake, and where it is found in our diet. Prepare a poster to communicate your findings.

**Answers to Practice Problems and Short Answers to Section Review Questions**

**Practice Problems:** 1. 10.81 u  2. 28.09 u  3. 63.55 u
4. 72%, 28%  5. $1.07 \times 10^{22}$  6. $2.83 \times 10^{20}$  7. $1.3 \times 10^{23}$
8. $3.35 \times 10^{25}$  9.(a) $1.5 \times 10^{24}$ (b) $2.1 \times 10^{25}$ (c) $6.0 \times 10^{24}$
10.(a) $4.99 \times 10^{23}$ (b) $9.98 \times 10^{23}$ (c) $4.99 \times 10^{23}$ (d) $2.00 \times 10^{24}$
11.(a) $4.74 \times 10^{22}$ (b) $9.48 \times 10^{22}$ (c) $9.48 \times 10^{21}$  12. 12.8 mol
13. $1.33 \times 10^3$ mol  14. $1.84 \times 10^{-2}$ mol  15. $1.16 \times 10^{-1}$ mol
16.(a) 131.29 g/mol (b) 190.23 g/mol (c) 137.33 g/mol
(d) 127.60 g/mol  17.(a) 17.04 g/mol (b) 180.2 g/mol
(c) 294.2 g/mol (d) 399.9 g/mol  18. 183.69 g/mol
19. 131.71 g/mol  20.(a) 46.8 g (b) $1.20 \times 10^2$ g (c) $1.05 \times 10^9$ g
(d) $3.66 \times 10^{-3}$ g  21.(a) 1.5 mol $Cl_2$ (b) 7.31 mol $O_2$
22. $1.00 \times 10^3$ g  23. 26.6 kg  24.(a) 1.07 mol (b) $1.24 \times 10^2$ mol
(c) 14.2 mol (d) $7.75 \times 10^{-4}$ mol  25.(a) 0.652 mol (b) 0.156 mol
(c) $1.76 \times 10^6$ mol (d) $3.49 \times 10^{-8}$  26. $1.7 \times 10^2$ mol
27. $1.75 \times 10^2$ mol  28.(a) $1.36 \times 10^3$ g (b) 9.98 g
(c) $7.30 \times 10^2$ g (d) $2.15 \times 10^7$ g  29. $2.93 \times 10^{-21}$ g
30. 312 g  31. 20.1 g  32. $7.95 \times 10^{-23}$ g  33. $6.25 \times 10^{-22}$ g
34.(a) $3.34 \times 10^{23}$ molecules (b) $9.81 \times 10^{23}$ molecules
(c) $1.05 \times 10^{23}$ molecules (d) $2.50 \times 10^{20}$ formula units
35. $4.2 \times 10^5$ atoms  36. $1.36 \times 10^{20}$  37.(a) $2.14 \times 10^{20}$
(b) $4.28 \times 10^{20}$  38. 0.089 mol  39. 0.500 mol;
$3.01 \times 10^{23}$ molecules  40. 77.3 L  41. 78.40 L  42.(a) 56.0 L
(b) $1.51 \times 10^{24}$ molecules (c) $3.01 \times 10^{24}$ atoms  43. 74.4 L
**Section Review: 2.1:** 4. 24.31 u  6.(a) 1.68 u (b) 66.58 u
(c) 65.17 u (d) 334.32 u  **2.2:** 2. $1.00 \times 10^{14}$  3. $1.9 \times 10^{16}$ years
4. $2.08 \times 10^{24}$ atoms  5.(a) 4.25 mol (b) 12.8 mol
6. $3.0 \times 10^{23}$ atoms  7.(a) $3.43 \times 10^{24}$ molecules
(b) $2.06 \times 10^{25}$ atoms  8.(a) $7.04 \times 10^{23}$ molecules
(b) $3.52 \times 10^{24}$ atoms (c) $2.11 \times 10^{24}$ atoms  10. compound
**2.3:** 2.(a) 4.61 mol (b) $2.78 \times 10^{24}$  3.(a) $4.67 \times 10^{-23}$ g
(b) $1.69 \times 10^{25}$ u  4.(a) 46.1 g (b) $4.75 \times 10^{23}$ formula units
(c) $9.50 \times 10^{23}$  5. $5.01 \times 10^{22}$ atoms  6. $2.50 \times 10^{-3}$ g
**2.4:** 1. 104 g  2. 30 g  3.(a) 0.178 g (b) 0.178 g/L (c) greater; balloon floats (d) 1.25 g/L  4. carbon monoxide

# CHAPTER 3
# Chemical Proportions in Compounds

## Chapter Preview

- 3.1 Percentage Composition
- 3.2 The Empirical Formula of a Compound
- 3.3 The Molecular Formula of a Compound
- 3.4 Finding Empirical and Chemical Formulas by Experiment

## Prerequisite Concepts and Skills

Before you begin this chapter, review the following concepts and skills:

- naming chemical compounds (Chapter 1)
- understanding the mole (Chapter 2)
- explaining the relationship between the mole and molar mass (Chapter 2)
- solving problems involving number of particles, amount in moles, mass, and volume (Chapter 2)

How do chemists use what they know about molar masses? In Chapter 2, you learned how to use the periodic table and the mole to relate the mass of a compound to the number of particles in the compound. Chemists can use the concept of molar mass to find out important information about compounds.

Sometimes chemists analyze a compound that is found in nature to learn how to produce it more cheaply in a laboratory. For example, consider the flavour used in vanilla ice cream, which may come from natural or artificial vanilla extract. Natural vanilla extract is made from vanilla seed pods. The seed pods must be harvested and processed before being sold as vanilla extract. The scent and flavour of synthetic vanilla come from a compound called vanillin, which can be produced chemically in bulk. Therefore its production is much cheaper. Similarly, many medicinal chemicals that are similar or identical to those found in nature can be produced relatively cheaply and efficiently in a laboratory.

Suppose that you want to synthesize a compound such as vanillin in a laboratory. You must first determine the elements in the compound. Then you need to know the proportion of each element that is present. This information, along with your understanding of molar mass, will help you determine the chemical formula of the compound. Once you know the chemical formula, you are on your way to finding out how to produce the compound.

In this chapter, you will learn about the relationships between chemical formulas, molar masses, and the masses of elements in compounds.

The chemical formula of vanillin is $C_8H_8O_3$. How did chemists use information about the masses of carbon, hydrogen, and oxygen in the compound to determine this formula?

# Percentage Composition

## 3.1

When you calculate and use the molar mass of a compound, such as water, you are making an important assumption. You are assuming that every sample of water contains hydrogen and oxygen in the ratio of two hydrogen atoms to one oxygen atom. Thus you are also assuming that the masses of hydrogen and oxygen in pure water always exist in a ratio of 2 g:16 g. This may seem obvious to you, because you know that the molecular formula of water is always $H_2O$ regardless of whether it comes from any of the sources shown in Figure 3.1. When scientists first discovered that compounds contained elements in fixed mass proportions, they did not have the periodic table. In fact, the discovery of fixed mass proportions was an important step toward the development of atomic theory.

### The Law of Definite Proportions

In the late eighteenth century, Joseph Louis Proust, a French chemist, analyzed many samples of copper(II) carbonate, $CuCO_3$. He found that the samples contained the same proportion of copper, carbon, and oxygen, regardless of the source of the copper(II) carbonate. This discovery led Proust to propose the **law of definite proportions**: *the elements in a given chemical compound are always present in the same proportions by mass.*

**Section Preview/Outcomes**

In this section, you will

- **calculate** the percentage composition from a compound's formula
- **communicate** your understanding of the following terms: *law of definite proportions, mass percent, percentage composition*

**mind STRETCH**

If a bicycle factory has 1000 wheels and 400 frames, how many bicycles can be made? How many wheels does each bicycle have? Is the number of wheels per bicycle affected by any extra wheels that the factory may have in stock? Relate these questions to the law of definite proportions.

**Figure 3.1** Suppose that you distil pure water from each of these sources to purify it. Are the distilled water samples the same or different? What is the molar mass of the distilled water from each source?

Chapter 3 Chemical Proportions in Compounds • MHR

### CHEM FACT

Chemical formulas such as CO and CO₂ reflect an important law called the *law of multiple proportions*. This law applies when two elements (such as carbon and oxygen) combine to form two or more different compounds. In these cases, the masses of the element (such as O₂ in CO and CO₂) that combine with a fixed amount of the second element are in ratios of small whole numbers. For example, two moles of carbon can combine with one mole of oxygen to form carbon monoxide, or with two moles of oxygen to form carbon dioxide. The ratio of the two different amounts of oxygen that combine with the fixed amount of carbon is 1:2.

The mass of an element in a compound, expressed as a percent of the total mass of the compound, is the element's **mass percent**. The mass percent of hydrogen in water from any of the sources shown in Figure 6.1 is 11.2%. Similarly, the mass percent of oxygen in water is always 88.8%. Whether the water sample is distilled from a lake, an ice floe, or a drinking fountain, the hydrogen and oxygen in pure water are always present in these proportions.

## Different Compounds from the Same Elements

The law of definite proportions does not imply that elements in compounds are always present in the same relative amounts. It is possible to have different compounds made up of different amounts of the same elements. For example, water, $H_2O$, and hydrogen peroxide, $H_2O_2$, are both made up of hydrogen and oxygen. Yet, as you can see in Figure 3.2, each compound has unique properties. Each compound has a different mass percent of oxygen and hydrogen. You may recognize hydrogen peroxide as a household chemical. It is an oxidizing agent that is used to bleach hair and treat minor cuts. It is also sold as an alternative to chlorine bleach.

Figure 3.3 shows a molecule of benzene, $C_6H_6$. Benzene contains 7.76% hydrogen and 92.2% carbon by mass. Octane, $C_8H_{18}$, is a major component of the fuel used for automobiles. It contains 84.1% carbon and 15.9% hydrogen.

**Figure 3.2** Water contains hydrogen and oxygen, but it does not decompose in the presence of manganese(IV) oxide. Hydrogen peroxide is also composed of hydrogen and oxygen. It decomposes vigorously in the presence of a catalyst such as manganese(IV) oxide.

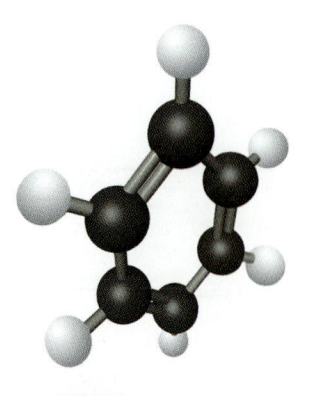

**Figure 3.3** Benzene, $C_6H_6$, is made up of six carbon atoms and six hydrogen atoms. Why does benzene not contain 50% of each element by mass?

Similarly, carbon monoxide, CO, and carbon dioxide, $CO_2$, are both made up of carbon and oxygen. Yet each compound is unique, with its own physical and chemical properties. Carbon dioxide is a product of cellular respiration and the complete combustion of fossil fuels. Carbon monoxide is a deadly gas formed when insufficient oxygen is present during the combustion of carbon-containing compounds. Carbon monoxide always contains 42.88% carbon by mass. Carbon dioxide always contains 27.29% carbon by mass.

## Percentage Composition

Carbon dioxide and carbon monoxide contain the same elements but have different proportions of these elements. In other words, they are composed differently. Chemists express the composition of compounds in various ways. One way is to describe how many moles of each element make up a mole of a compound. For example, one mole of carbon dioxide contains one mole of carbon and two moles of oxygen. Another way is to describe the mass percent of each element in a compound.

The **percentage composition** of a compound refers to the relative mass of each element in the compound. In other words, percentage composition is a statement of the values for mass percent of every element in the compound. For example, the compound vanillin, $C_8H_8O_3$, has a percentage composition of 63.1% carbon, 5.3% hydrogen, and 31.6% oxygen, as shown in Figure 3.4.

A compound's percentage composition is an important piece of information. For example, percentage composition can be determined experimentally, and then used to help identify the compound.

Examine the following Sample Problem to learn how to calculate the percentage composition of a compound from the mass of the compound and the mass of the elements that make up the compound. Then do the Practice Problems to try expressing the composition of substances as mass percents.

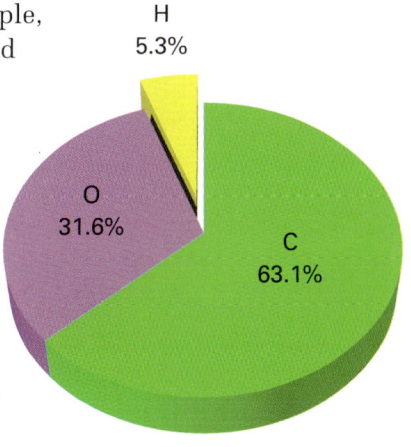

**Figure 3.4** This pie graph shows the percentage composition of vanillin.

### Sample Problem

#### Percentage Composition from Mass Data

**Problem**
A sample of a compound has a mass of 48.72 g. The sample is found to contain 32.69 g of zinc and 16.03 g of sulfur. What is the percentage composition of the compound?

**What Is Required?**
You need to find the mass percents of zinc and sulfur in the compound.

**What Is Given?**
You know the mass of the compound. You also know the mass of each element in the compound.
Mass of compound = 48.72 g
Mass of Zn = 32.69 g
Mass of S = 16.03 g

*Continued ...*

**mind STRETCH**

Does the unknown compound in the Sample Problem contain any elements other than zinc and sulfur? How do you know? Use the periodic table to predict the formula of the compound. Does the percentage composition support your prediction?

## mind STRETCH

Iron is commonly found as two oxides, with the general formula $Fe_xO_y$. One oxide is 77.7% iron. The other oxide is 69.9% iron. Use the periodic table to predict the formula of each oxide. Match the given values for the mass percent of iron to each compound. How can you use the molar mass of iron and the molar masses of the two iron oxides to check the given values for mass percent?

### Plan Your Strategy

To find the percentage composition of the compound, find the mass percent of each element. To do this, divide the mass of each element by the mass of the compound and multiply by 100%.

### Act on Your Strategy

$$\text{Mass percent of Zn} = \frac{\text{Mass of Zn}}{\text{Mass of compound}} \times 100\%$$

$$= \frac{32.69 \text{ g}}{48.72 \text{ g}} \times 100\%$$

$$= 67.10\%$$

$$\text{Mass percent of S} = \frac{\text{Mass of S}}{\text{Mass of compound}} \times 100\%$$

$$= \frac{16.03 \text{ g}}{48.72 \text{ g}} \times 100\%$$

$$= 32.90\%$$

The percentage composition of the compound is 67.10% zinc and 32.90% sulfur.

### Check Your Solution

The mass of zinc is about 32 g per 50 g of the compound. This is roughly 65%, which is close to the calculated value.

## Practice Problems

1. A sample of a compound is analyzed and found to contain 0.90 g of calcium and 1.60 g of chlorine. The sample has a mass of 2.50 g. Find the percentage composition of the compound.

2. Find the percentage composition of a pure substance that contains 7.22 g nickel, 2.53 g phosphorus, and 5.25 g oxygen only.

3. A sample of a compound is analyzed and found to contain carbon, hydrogen, and oxygen. The mass of the sample is 650 mg, and the sample contains 257 mg of carbon and 50.4 mg of hydrogen. What is the percentage composition of the compound?

4. A scientist analyzes a 50.0 g sample and finds that it contains 13.3 g of potassium, 17.7 g of chromium, and a gaseous element. Later the scientist learns that the sample is potassium dichromate, $K_2Cr_2O_7$. Potassium dichromate is a bright orange compound that is used in the production of safety matches.
   (a) What is the percentage composition of potassium dichromate?
   (b) If the chemist was able to decompose the sample and collect all the oxygen that was in the compound, what volume would the oxygen occupy at STP?

### Web LINK

www.mcgrawhill.ca/links/atlchemistry

Vitamin C is the common name for ascorbic acid, $C_6H_8O_6$. To learn about this vitamin, go to the web site above and click on **Web Links**. Do you think there is a difference between natural and synthetic vitamin C? Both are ascorbic acid. Natural vitamin C comes from foods we eat, especially citrus fruits. Synthetic vitamin C is made in a laboratory. Why do the prices of natural and synthetic products often differ? Make a list to show the pros and cons of vitamins from natural and synthetic sources.

It is important to understand clearly the difference between percent by mass and percent by number. In the ThoughtLab that follows, you will investigate the distinction between these two ways of describing composition.

# ThoughtLab — Percent by Mass and Percent by Number

A company manufactures gift boxes that contain two pillows and one gold brick. The gold brick has a mass of 20.0 kg. Each pillow has a mass of 1.0 kg.

## Procedure

1. You are a quality control specialist at the gift box factory. You need to know the following information:
   (a) What is the percent of pillows, in terms of the number of items, in the gift box?
   (b) What is the percent of pillows, by mass, in the gift box?
   (c) What is the percent of gold, by mass, in the gift box?

2. You have a truckload of gift boxes to inspect. You now need to know this information:
   (a) What is the percent of pillows, in terms of the number of items, in the truckload of gift boxes?
   (b) What is the percent of pillows, by mass, in the truckload of gift boxes?
   (c) What is the percent of gold, by mass, in the truckload of gift boxes?

## Analysis

1. The truckload of gift boxes, each containing one heavy gold brick and two light pillows, is an analogy for a pure substance such as water. One molecule of water is made up of one relatively massive oxygen atom and two less massive hydrogen atoms.
   (a) What is the percent of hydrogen, in terms of the number of atoms, in 1 mol of water?
   (b) What is the mass percent of hydrogen in 1 mol of water?
   (c) What is the mass percent of oxygen in 1 mol of water?

2. Would the mass percent of hydrogen or oxygen in question 1 change if you had 25 mol of water? Explain.

3. Why do you think chemists use mass percent rather than percent by number of atoms?

## Calculating Percentage Composition from a Chemical Formula

In the previous Practice Problems, you used mass data to calculate percentage composition. This skill is useful for interpreting experimental data when the chemical formula is unknown. Often, however, the percentage composition is calculated from a known chemical formula. This is useful when you are interested in extracting a certain element from a compound. For example, many metals, such as iron and mercury, exist in mineral form. Mercury is most often found in nature as mercury(II) sulfide, HgS. Knowing the percentage composition of HgS helps a metallurgist predict the mass of mercury that can be extracted from a sample of HgS.

When determining the percentage composition by mass of a homogeneous sample, the size of the sample does not matter. According to the law of definite proportions, there is a fixed proportion of each element in the compound, no matter how much of the compound you have. This means that you can choose a convenient sample size when calculating percentage composition from a formula.

If you assume that you have one mole of a compound, you can use the molar mass and chemical formula of the compound to calculate its percentage composition. For example, suppose that you want to find the percentage composition of HgS. You can assume that you have one mole of HgS and find the mass percents of mercury and sulfur in that amount.

## History LINK

Before AD 1500, many alchemists thought that matter was composed of two "elements": mercury and sulfur. To impress their patrons, they performed an experiment with mercury sulfide, HgS, also called cinnabar. They heated the red cinnabar, which drove off the sulfur and left the shiny liquid mercury. On further heating, the mercury reacted to form a red compound again. Alchemists wrongly thought that the mercury had been converted back to cinnabar. What Hg(II) compound do you think was really formed when the mercury was heated in the air? What is the mass percent of mercury in this new compound? What is the mass percent of mercury in cinnabar?

$$\text{Mass percent of Hg in HgS} = \frac{\text{Mass of Hg in 1 mol of HgS}}{\text{Mass of 1 mol of HgS}} \times 100\%$$

$$= \frac{200.6 \text{ g}}{228.68 \text{ g}} \times 100\%$$

$$= 87.72\%$$

Mercury(II) sulfide is 87.72% mercury by mass. Since there are only two elements in HgS, you can subtract the mass percent of mercury from 100 percent to find the mass percent of sulfur.

$$\text{Mass percent of S in HgS} = 100\% - 87.72\% = 12.28\%$$

Therefore, the percentage composition of mercury(II) sulfide is 87.72% mercury and 12.28% sulfur.

Sometimes there are more than two elements in a compound, or more than one atom of each element. This makes determining percentage composition more complex than in the example above. Work through the Sample Problem below to learn how to calculate the percentage composition of a compound from its chemical formula.

### Sample Problem
#### Finding Percentage Composition from a Chemical Formula

**Problem**

Cinnamaldehyde, $C_9H_8O$, is responsible for the characteristic odour of cinnamon. Determine the percentage composition of cinnamaldehyde by calculating the mass percents of carbon, hydrogen, and oxygen.

**What Is Required?**

You need to find the mass percents of carbon, hydrogen, and oxygen in cinnamaldehyde.

**What Is Given?**

The molecular formula of cinnamaldehyde is $C_9H_8O$.

Molar mass of C = 12.01 g/mol
Molar mass of H = 1.01 g/mol
Molar mass of O = 16.00 g/mol

**Plan Your Strategy**

From the molar masses of carbon, hydrogen, and oxygen, calculate the molar mass of cinnamaldehyde.

Then find the mass percent of each element. To do this, divide the mass of each element in 1 mol of cinnamaldehyde by the molar mass of cinnamaldehyde, and multiply by 100%. Remember that there are 9 mol carbon, 8 mol hydrogen, and 1 mol oxygen in each mole of cinnamaldehyde.

**Act on Your Strategy**

$M_{C_9H_8O}$
$= (9 \times M_C) + (8 \times M_H) + (M_O)$
$= (9 \times 12.01 \text{ g}) + (8 \times 1.01 \text{ g}) + 16.00 \text{ g}$
$= 132.2 \text{ g}$

Mass percent of C = $\dfrac{9 \times M_C}{M_{C_9H_8O}} \times 100\%$

$= \dfrac{9 \times 12.01 \text{ g/mol}}{132.2 \text{ g/mol}} \times 100\%$

$= 81.76\%$

Mass percent of H = $\dfrac{8 \times M_H}{M_{C_9H_8O}} \times 100\%$

$= \dfrac{8 \times 1.01 \text{ g/mol}}{132.2 \text{ g/mol}} \times 100\%$

$= 6.11\%$

Mass percent of O = $\dfrac{1 \times M_O}{M_{C_9H_8O}} \times 100\%$

$= \dfrac{1 \times 16.00 \text{ g/mol}}{132.2 \text{ g/mol}} \times 100\%$

$= 12.10\%$

The percentage composition of cinnamaldehyde is 81.76% carbon, 6.11% hydrogen, and 12.10% oxygen.

### Check Your Solution
The mass percents add up to 100%.

## Practice Problems

5. Calculate the mass percent of nitrogen in each compound.
   (a) $N_2O$
   (b) $Sr(NO_3)_2$
   (c) $NH_4NO_3$
   (d) $HNO_3$

6. Ammonia, $NH_3$, has a strong, distinctive odour. Determine the percentage composition of ammonia.

7. Potassium nitrate, $KNO_3$, is used to make fireworks. What is the mass percent of oxygen in potassium nitrate?

8. A mining company wishes to extract manganese metal from pyrolusite ore, $MnO_2$.
   (a) What is the percentage composition of pyrolusite ore?
   (b) Use your answer from part (a) to calculate the mass of pure manganese that can be extracted from 250 kg of pyrolusite ore.

### CONCEPT CHECK
When it is heated, solid potassium nitrate reacts to form solid potassium oxide, gaseous nitrogen, and gaseous oxygen. Write a balanced chemical equation for this reaction. What type of reaction is it?

## Section Summary
In this section, you learned that you can calculate percentage composition using the strategy summarized below.

- Use the chemical formula to determine the molar mass.
- Assume you have a one mole sample of the substance.
- Use the chemical formula to determine the mass of each element in one mole of the substance.
- Divide the mass of each element by the molar mass and multiply by 100% to obtain the percentage composition.

In section 3.2, you will learn about the first step in using the percentage composition of a compound to determine its chemical formula.

### mind STRETCH
You know that both elements and compounds are pure substances. Write a statement, using the term "percentage composition," to distinguish between elements and compounds.

## Section Review

1. Acetylene, $C_2H_2$, is the fuel in a welder's torch. It contains an equal number of carbon and hydrogen atoms. Explain why acetylene is not 50% carbon by mass.

2. When determining percentage composition, why is it acceptable to work with either molar quantities, expressed in grams, or average molecular (or atomic or formula unit) quantities, expressed in atomic mass units?

3. Indigo, $C_{16}H_{10}N_2O_2$, is the common name of the dye that gives blue jeans their characteristic colour. Calculate the mass of oxygen in 25.0 g of indigo.

4. Potassium perchlorate, $KClO_4$, is used extensively in explosives. Calculate the mass of oxygen in a 24.5 g sample of potassium perchlorate.

5. The label on a box of baking soda (sodium hydrogen carbonate, $NaHCO_3$) claims that there are 137 mg of sodium per 0.500 g of baking soda. Comment on the validity of this claim.

6. A typical soap molecule consists of a polyatomic anion associated with a cation. The polyatomic anion contains hydrogen, carbon, and oxygen. One particular soap molecule has 18 carbon atoms. It contains 70.5% carbon, 11.5% hydrogen, and 10.4% oxygen by mass. It also contains one alkali metal cation. Identify the cation.

7. Examine the photographs below. When concentrated sulfuric acid is added to sucrose, $C_{12}H_{22}O_{11}$, a column of pure carbon is formed, as well as some water vapour. How would you find the mass percent of carbon in sucrose using this reaction? You may assume that all the carbon in the sucrose is converted to carbon. Design an experiment to determine the mass percent of carbon in sucrose, based on this reaction. What difficulties might you encounter? **CAUTION** Do not attempt to carry out your procedure. Concentrated sulfuric acid is highly corrosive, as are the fumes that result from the reaction.

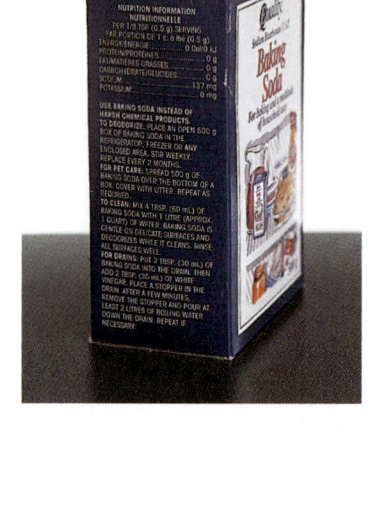

# 3.2 The Empirical Formula of a Compound

As part of his atomic theory, John Dalton stated that atoms combine with one another in simple whole number ratios to form compounds. For example, the molecular formula of benzene, $C_6H_6$, indicates that one molecule of benzene contains 6 carbon atoms and 6 hydrogen atoms. The **empirical formula** (also known as the simplest formula) of a compound shows the lowest whole number ratio of the elements in the compound. The **molecular formula** (also known as the actual formula) describes the number of atoms of each element that make up a molecule. Benzene, with a molecular formula of $C_6H_6$, has an empirical formula of CH. Table 3.1 shows the molecular formulas of several compounds, along with their empirical formulas.

**Table 3.1** Comparing Molecular Formulas and Empirical Formulas

| Name of compound | Molecular (actual) formula | Empirical (simplest) formula | Lowest ratio of elements |
|---|---|---|---|
| hydrogen peroxide | $H_2O_2$ | HO | 1:1 |
| glucose | $C_6H_{12}O_6$ | $CH_2O$ | 1:2:1 |
| benzene | $C_6H_6$ | CH | 1:1 |
| acetylene (ethyne) | $C_2H_2$ | CH | 1:1 |
| aniline | $C_6H_7N$ | $C_6H_7N$ | 6:7:1 |
| water | $H_2O$ | $H_2O$ | 2:1 |

It is possible for different compounds to have the same empirical formula, as you can see in Figure 3.5. For example, benzene and acetylene both have the empirical formula CH. Benzene, $C_6H_6$, is a clear liquid with a molar mass of 78 g/mol and a boiling point of 80°C. Acetylene, $C_2H_2$, has a molar mass of 26 g/mol. It is a highly flammable gas, commonly used in a welder's torch. There is, in fact, no existing compound with the molecular formula CH. The empirical formula of a compound shows the lowest whole number ratio of the atoms in the compound.

Many compounds have molecular formulas that are the same as their empirical formulas. One example is ammonia, $NH_3$. Try to think of three other examples.

Note that ionic compounds do not have molecular formulas, because they do not consist of molecules. Because ionic compounds consist of arrays of ions, their chemical formula reflects the ratio of ions in the compound. Therefore, the empirical formula of an ionic compound is usually the same as its chemical formula.

**Figure 3.5** The same empirical formula can represent more than one compound. These two compounds are different—at room temperature, one is a gas and one is a liquid. Yet they have the same empirical formula, CH.

### Section Preview/Outcomes

In this section, you will
- **determine** the empirical formula from percentage composition data
- **perform** an experiment to determine the percentage composition and the empirical formula of a compound
- **communicate** your understanding of the following terms: *empirical formula, molecular formula*

### Language LINK

The word "empirical" comes from the Greek word *empeirikos*, meaning, roughly, "by experiment." Why do you think the simplest formula of a compound is called its empirical formula?

| Math | LINK |
|---|---|

In mathematics, you frequently need to reduce an expression to lowest terms. For example, $\frac{4x^2}{x}$ is equivalent to $4x$. A ratio of 5:10 is equivalent to 1:2. In chemistry, however, the "lowest terms" version of a chemical formula is not equivalent to its "real" molecular formula. Why not?

The relationship between the molecular formula of a compound and its empirical formula can be expressed as

Molecular formula subscripts = $n$ × Empirical formula subscripts, where $n$ = 1, 2, 3...

This relationship shows that the molecular formula of a compound is the same as its empirical formula when $n = 1$. What information do you need in order to determine whether the molecular formula of a compound is the same as its empirical formula?

## Determining a Compound's Empirical Formula

In the previous section, you learned how to calculate the percentage composition of a compound from its chemical formula. Now you will do the reverse. You will use the percentage composition of a compound, along with the concept of the mole, to calculate the empirical formula of the compound. Since the percentage composition can often be determined by experiment, chemists use this calculation when they want to identify a compound.

The following Sample Problem illustrates how to use percentage composition to obtain the empirical formula of a compound.

### mind STRETCH

How do the molar masses of $C_6H_6$ and $C_2H_2$ compare with the molar mass of their empirical formula? How does the molar mass of water compare with the molar mass of its empirical formula? Describe the relationship between the molar mass of a compound and the molar mass of the empirical formula of the compound.

### Sample Problem

#### Finding a Compound's Empirical Formula from Percentage Composition: Part A

**Problem**

Calculate the empirical formula of a compound that is 85.6% carbon and 14.4% hydrogen.

**What Is Required?**

You need to find the empirical formula of the compound.

**What Is Given?**

You know the percentage composition of the compound. You have access to a periodic table.

**Plan Your Strategy**

Since you know the percentage composition, it is convenient to assume that you have 100 g of the compound. This means that you have 85.6 g of carbon and 14.4 g of hydrogen. Convert each mass to moles. The number of moles can then be converted into a lowest terms ratio of the elements to get the empirical formula.

**Act on Your Strategy**

Number of moles of C in 100 g sample = $\frac{85.6 \text{ g}}{12.01 \text{ g/mol}}$ = 7.13 mol

Number of moles of H in 100 g sample = $\frac{14.4 \text{ g}}{1.01 \text{ g/mol}}$ = 14.3 mol

Now determine the lowest whole number ratio. Divide both molar amounts by the lowest molar amount.

$C_{\frac{7.13}{7.13}} H_{\frac{14.3}{7.13}} \rightarrow C_{1.00}H_{2.01} \rightarrow CH_2$

Alternatively, you can set up your solution as a table.

| Element | Mass percent (%) | Grams per 100 g sample (g) | Molar mass (g/mol) | Number of moles (mol) | Molar amount ÷ lowest molar amount |
|---------|------------------|----------------------------|--------------------|-----------------------|-------------------------------------|
| C       | 85.6             | 85.6                       | 12.01              | 7.13                  | $\frac{7.13}{7.13} = 1$             |
| H       | 14.4             | 14.4                       | 1.01               | 14.3                  | $\frac{14.3}{7.13} = 2.01$          |

> **PROBLEM TIP**
>
> It is acceptable to round 2.01 to the 2 in $CH_2$. The percentage composition is often determined by experiment, so it is unlikely to be exact.

The empirical formula of the compound is $CH_2$.

### Check Your Solution

Work backward. Calculate the percentage composition of $CH_2$.

Mass percent of C = $\frac{12.01 \text{ g/mol}}{14.03 \text{ g/mol}} \times 100\%$

= 85.6%

Mass percent of H = $\frac{2 \times 1.01 \text{ g/mol}}{14.03 \text{ g/mol}} \times 100\%$

= 14.0%

The percentage composition calculated from the empirical formula closely matches the given data. The formula is reasonable.

### Practice Problems

9. A compound consists of 17.6% hydrogen and 82.4% nitrogen. Determine the empirical formula of the compound.

10. Find the empirical formula of a compound that is 46.3% lithium and 53.7% oxygen.

11. What is the empirical formula of a compound that is 15.9% boron and 84.1% fluorine?

12. Determine the empirical formula of a compound made up of 52.51% chlorine and 47.48% sulfur.

## Tips for Solving Empirical Formula Problems

In the Sample Problem above, the numbers were rounded at each step to simplify the calculation. To calculate an empirical formula successfully, however, you should not round the numbers until you have completed the calculation. Use the maximum number of significant digits that your calculator will allow, throughout the calculation. Rounding too soon when calculating an empirical formula may result in getting the wrong answer.

Often only one step is needed to determine the number of moles in an empirical formula. This is not always the case, however. Since you must divide by the lowest number of moles, initially one of your ratio terms will always be 1. If your other terms are quite close to whole numbers, as in the last Sample Problem, you can round them to the closest whole numbers. If your other terms are not close to whole numbers, you will need to do some additional steps. This is because empirical formulas do not always contain the subscript 1. For example, $Fe_2O_3$ contains only the subscripts 2 and 3.

**Table 3.2** Converting Subscripts in Empirical Formulas

| When you see this decimal... | Try multiplying all subscripts by... |
|---|---|
| x.80 ($\frac{4}{5}$) | 5 |
| x.75 ($\frac{3}{4}$) | 4 |
| x.67 ($\frac{2}{3}$) | 3 |
| x.60 ($\frac{3}{5}$) | 5 |
| x.40 ($\frac{2}{5}$) | 5 |
| x.50 ($\frac{1}{2}$) | 2 |
| x.33 ($\frac{1}{3}$) | 3 |
| x.25 ($\frac{1}{4}$) | 4 |
| x.20 ($\frac{1}{5}$) | 5 |
| x.17 ($\frac{1}{6}$) | 6 |

Numbers that end in decimals from .95 to .99 can be rounded up to the nearest whole number. Numbers that end in decimals from .01 to .05 can be rounded down to the nearest whole number. Other decimals require additional manipulation. What if you have the empirical formula $C_{1.5}H_3O_1$? To convert all subscripts to whole numbers, multiply each subscript by 2. This gives you the empirical formula $C_3H_6O_2$. Thus, a ratio that involves a decimal ending in .5 must be doubled. What if a decimal ends in .45 to .55? Round the decimal so that it ends in .5, and then double the ratio.

Table 3.2 gives you some strategies for converting subscripts to whole numbers. The variable $x$ stands for any whole number. Examine the following Sample Problem to learn how to convert the empirical formula subscripts to the lowest possible whole numbers.

### Sample Problem

#### Finding a Compound's Empirical Formula from Percentage Composition: Part B

**Problem**

The percentage composition of a fuel is 81.7% carbon and 18.3% hydrogen. Find the empirical formula of the fuel.

**What Is Required?**

You need to determine the empirical formula of the fuel.

**What Is Given?**

You know the percentage composition of the fuel. You have access to a periodic table.

**Plan Your Strategy**

Convert mass percent to mass, then to number of moles. Then find the lowest whole number ratio.

**Act on Your Strategy**

| Element | Mass percent (%) | Grams per 100 g sample (g) | Molar mass (g/mol) | Number of moles (mol) | Molar amount ÷ lowest molar amount |
|---|---|---|---|---|---|
| C | 81.7 | 81.7 | 12.0 | 6.81 | $\frac{6.81}{6.81} = 1$ |
| H | 18.3 | 18.3 | 1.01 | 18.1 | $\frac{18.1}{6.81} = 2.66$ |

You now have the empirical formula $C_1H_{2.66}$. Convert the subscript 2.66 ($\frac{8}{3}$) to a whole number. $C_{1\times3}H_{2.66\times3} = C_3H_8$.

**Check Your Solution**

Work backward. Calculate the percentage composition of $C_3H_8$.

$$\text{Mass percent of C} = \frac{3 \times 12.01 \text{ g/mol}}{44.09 \text{ g/mol}} \times 100\%$$
$$= 81.7\%$$

$$\text{Mass percent of H} = \frac{8 \times 1.008 \text{ g/mol}}{44.09 \text{ g/mol}} \times 100\%$$
$$= 18.3\%$$

The percentage composition calculated from the empirical formula matches the percentage composition given in the problem.

> **PROBLEM TIP**
> Notice that Table 3.2 suggests multiplying by 3 when you obtain a subscript ending in .67, which is very close to .66.

## Practice Problems

**13.** An oxide of chromium is made up of 68.4% chromium and 31.6% oxygen. What is the empirical formula of this oxide?

**14.** Phosphorus reacts with oxygen to give a compound that is 43.7% phosphorus and 56.4% oxygen. What is the empirical formula of the compound?

**15.** An inorganic salt is composed of 17.6% sodium, 39.7% chromium, and 42.8% oxygen. What is the empirical formula of this salt?

**16.** Compound X contains 69.9% carbon, 6.86% hydrogen, and 23.3% oxygen. Determine the empirical formula of compound X.

## Determining the Empirical Formula by Experiment

In practice, you can determine a compound's empirical formula by analyzing its percentage composition. There are numerous different ways to determine the percentage composition of a compound. One way is to use a synthesis reaction in which a sample of an element with a known mass reacts with another element to form a compound.

For example, if you heat copper strongly in pure oxygen, the copper will react with the oxygen to form a black substance. This substance is predominantly a certain oxide of copper. Suppose you have a 5.0 g sample of copper shavings. You heat the copper shavings in oxygen until the metal has reacted completely. If the resulting substance has a mass of 6.3 g, you know that the compound contains 5.0 g of copper and 1.3 g of oxygen. Is the compound CuO or $Cu_2O$? Try determining the percentage composition of the compound and finding its empirical formula. You should find that the compound has the empirical formula CuO.

Since copper(II) oxide is black, you might wonder why copper roofs turn green instead of black. The copper in the roof of the Parliament buildings, shown in Figure 3.6, reacts with oxygen in combination with other dissolved substances in rain. Depending on the substances in the rain, several different green compounds may be produced.

In Investigation 3-A, you will use a synthesis reaction to determine the empirical formula of magnesium oxide by experiment.

**Figure 3.6** One of the compounds that gives the roof of the Parliament buildings its green colour is $CuCO_3 \cdot Cu(OH)_2$.

Chapter 3 Chemical Proportions in Compounds • MHR  **91**

# Investigation 3-A

**SKILL FOCUS**
- Predicting
- Performing and recording
- Analyzing and interpreting
- Communicating results

## Determining the Empirical Formula of Magnesium Oxide

When magnesium metal is heated over a flame, it reacts with oxygen in the air to form magnesium oxide, $Mg_xO_y$:

$$Mg_{(s)} + O_{2(g)} \rightarrow Mg_xO_{y(s)}$$

In this investigation, you will react a strip of pure magnesium metal with oxygen, $O_2$, in the air to form magnesium oxide. Then you will measure the mass of the magnesium oxide produced to determine the percentage composition of magnesium oxide. You will use this percentage composition to calculate the empirical formula of magnesium oxide. **CAUTION** Do not perform this investigation unless welder's goggles are available.

### Question
What is the percentage composition and empirical formula of magnesium oxide?

### Predictions
Using what you know about bonding, predict the empirical formula and percentage composition of magnesium oxide.

### Materials
electronic balance
small square of sandpaper or emery paper
8 cm strip of magnesium ribbon
laboratory burner
sparker
retort stand
ring clamp
clay triangle
clean crucible with lid
crucible tongs
ceramic pad
distilled water
wash bottle
disposal beaker
welder's goggles

**Note:** Make sure that the mass of the magnesium ribbon is at least 0.10 g.

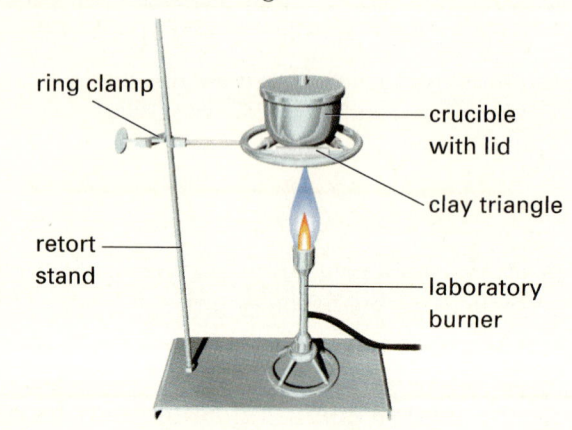

### Safety Precautions

- Do not look directly at the burning magnesium.
- Do not put a hot crucible on the bench or the balance.

### Procedure

1. Make a table like the one below.

**Observations**

| Mass of clean, empty crucible and lid | |
|---|---|
| Mass of crucible, lid, and magnesium | |
| Mass of crucible and magnesium oxide | |

2. Assemble the apparatus as shown in the diagram.

3. Obtain a strip of magnesium, about 8 cm long, from your teacher. Clean the magnesium strip with sandpaper or emery paper to remove any oxide coating.

4. Measure and record the mass of the empty crucible and lid. Add the strip of cleaned magnesium to the crucible. Record the mass of the crucible, lid, and magnesium.

92 MHR • Unit 1 Stoichiometry

5. With the lid off, place the crucible containing the magnesium on the clay triangle. Heat the crucible with a strong flame. Using the crucible tongs, hold the lid of the crucible nearby. **CAUTION** When the magnesium ignites, quickly cover the crucible with the lid. Continue heating for about 1 min.

6. Carefully remove the lid. **CAUTION** Heat the crucible until the magnesium ignites once more. Again, quickly cover the crucible. Repeat this heating and covering of the crucible until the magnesium no longer ignites. Heat for a further 4 to 5 min with the lid off.

7. Using the crucible tongs, put the crucible on the ceramic pad to cool.

8. When the crucible is cool enough to touch, put it on the bench. Carefully grind the product into small particles using the glass rod. Rinse any particles on the glass rod into the crucible with distilled water from the wash bottle.

9. Add enough distilled water to the crucible to thoroughly wet the contents. The white product is magnesium oxide. The yellowish-orange product is magnesium nitride.

10. Return the crucible to the clay triangle. Place the lid slightly ajar. Heat the crucible gently until the water begins to boil. Continue heating until all the water has evaporated, and the product is completely dry. Allow the crucible to cool on the ceramic pad.

11. Using the crucible tongs, carry the crucible and lid to the balance. Measure and record the mass of the crucible and lid.

12. Do not put the magnesium oxide in the garbage or in the sink. Put it in the disposal beaker designated by your teacher.

## Analysis

1. (a) What mass of magnesium did you use in the reaction?
   (b) What mass of magnesium oxide was produced?
   (c) Calculate the mass of oxygen that reacted with the magnesium.
   (d) Use your data to calculate the percentage composition of magnesium oxide.
   (e) Determine the empirical formula of magnesium oxide. Remember to round your empirical formula to the nearest whole number ratio, such as 1:1, 1:2, 2:1, or 3:3.

2. (a) Verify your empirical formula with your teacher. Use the empirical formula of magnesium oxide to determine the mass percent of magnesium in magnesium oxide.
   (b) Calculate your percent error (PE) by finding the difference between the experimental mass percent (EP) of magnesium and the actual mass percent (AP) of magnesium. Then divide the difference by the actual mass percent of magnesium and multiply by 100%.

$$PE = \frac{EP - AP}{AP} \times 100\%$$

3. Why did you need to round the empirical formula you obtained to a whole number ratio?

## Conclusion

4. Compare the empirical formula you obtained with the empirical formula you predicted.

## Applications

5. Write a balanced chemical equation for the reaction of magnesium with oxygen gas, $O_2$.

6. (a) Suppose that you had allowed some magnesium oxide smoke to escape during the investigation. How would the Mg:O ratio have been affected? Would the ratio have increased, decreased, or remained unchanged? Explain using sample calculations.
   (b) How would your calculated value for the empirical formula of magnesium oxide have been affected if all the magnesium in the crucible had not burned? Support your answer with sample calculations.
   (c) Could either of the situations mentioned in parts (a) and (b) have affected your results? Explain.

## Section Summary

In this section, you learned how to calculate the empirical formula of a compound based on percentage composition data obtained by experiment. In section 3.3, you will learn how chemists use the empirical formula of a molecular compound and its molar mass to determine its molecular formula.

## Section Review

**1 (a)** Why is the empirical formula of a compound also referred to as its simplest formula?

**(b)** Explain how the empirical formula of a compound is related to its molecular formula.

**2** Methyl salicylate, or oil of wintergreen, is produced by the wintergreen plant. It can also be prepared easily in a laboratory. Methyl salicylate is 63.1% carbon, 5.31% hydrogen, and 31.6% oxygen. Calculate the empirical formula of methyl salicylate.

**3** Determine the empirical formula of the compound that is formed by each of the following reactions.

**(a)** 0.315 mol chlorine atoms react completely with 1.1 mol oxygen atoms

**(b)** 4.90 g silicon react completely with 24.8 g chlorine

**4** Muscle soreness from physical activity is caused by a buildup of lactic acid in muscle tissue. Analysis of lactic acid reveals it to be 40.0% carbon, 6.71% hydrogen, and 53.3% oxygen by mass. Calculate the empirical formula of lactic acid.

**5** Imagine that you are a lawyer. You are representing a client charged with possession of a controlled substance. The prosecutor introduces, as forensic evidence, the empirical formula of the substance that was found in your client's possession. How would you deal with this evidence as a lawyer for the defence?

**6** Olive oil is used widely in cooking. Oleic acid, a component of olive oil, contains 76.54% carbon, 12.13% hydrogen and 11.33% oxygen by mass. What is the empirical formula of oleic acid?

**7** Phenyl valerate is a colourless liquid that is used as a flavour and odorant. It contains 74.13% carbon, 7.92% hydrogen and 17.95% oxygen by mass. Determine the empirical formula of phenyl valerate.

**8** Ferrocene is the common name given to a unique compound. Molecules of the compound consist of one iron atom sandwiched between two rings containing hydrogen and carbon. This orange, crystalline solid is added to fuel oil to improve combustion efficiency and eliminate smoke. As well, it is used as an industrial catalyst and a high-temperature lubricant.

**(a)** Elemental analysis reveals ferrocene to be 64.56% carbon, 5.42% hydrogen and 30.02% iron by mass. Determine the empirical formula of ferrocene.

**(b)** Read the description of ferrocene carefully. Does this description provide enough information for you to determine the molecular formula of ferrocene? Explain your answer.

# The Molecular Formula of a Compound

## 3.3

Determining the identity of an unknown compound is important in all kinds of research. It can even be used to solve crimes. **Forensic scientists** specialize in analyzing evidence for criminal and legal cases. To understand why forensic scientists might need to find out the molecular formula of a compound, consider the following example.

Suppose that a suspect in a theft investigation is a researcher in a biology laboratory. The suspect frequently works with formaldehyde, $CH_2O$. Police officers find traces of a substance at the crime scene, and send samples to the Centre for Forensic Science. The forensic analysts find that the substance contains a compound that has an empirical formula of $CH_2O$. Will this evidence help to convict the suspect? Not necessarily.

As you can see from Table 3.3, there are many compounds that have the empirical formula $CH_2O$. The substance might be formaldehyde, but it could also be lactic acid (found in milk) or acetic acid (found in vinegar). Neither lactic acid nor acetic acid connect the theft to the suspect. Further information is required to prove that the substance is formaldehyde. Analyzing the physical properties of the substance would help to discover whether it is formaldehyde. Another important piece of information is the molar mass of the substance. Continue reading to find out why.

### Section Preview/Outcomes

In this section, you will

- **determine** the molecular formula from percentage composition and molar mass data
- **communicate** your understanding of the following term: *forensic scientists*

**Table 3.3** Six Compounds with the Empirical Formula $CH_2O$

| Name | Molecular formula | Whole-number multiple | $M$ (g/mol) | Use or function |
|---|---|---|---|---|
| formaldehyde | $CH_2O$ | 1 | 30.03 | disinfectant; biological preservative |
| acetic acid | $C_2H_4O_2$ | 2 | 60.05 | acetate polymers; vinegar (5% solution) |
| lactic acid | $C_3H_6O_3$ | 3 | 90.08 | causes milk to sour; forms in muscles during exercise |
| erythrose | $C_4H_8O_4$ | 4 | 120.10 | forms during sugar metabolism |
| ribose | $C_5H_{10}O_5$ | 5 | 150.13 | component of many nucleic acids and vitamin $B_2$ |
| glucose | $C_6H_{12}O_6$ | 6 | 180.16 | major nutrient for energy in cells |

$CH_2O$  $C_2H_4O_2$  $C_3H_6O_3$  $C_4H_8O_4$  $C_5H_{10}O_5$  $C_6H_{12}O_6$

## Determining a Molecular Formula

Recall the equation

> Molecular formula subscripts = $n \times$ Empirical formula subscripts, where $n = 1, 2, 3...$

Additional information is required to obtain the molecular formula of a compound, given its empirical formula. We can use the molar mass and build on the above equation, as follows:

> Molar mass of compound = $n \times$ Molar mass of empirical formula, where $n = 1, 2, 3...$

**Table 3.4** Relating Molecular and Empirical Formulas

| Formula | Molar Mass (g) | Ratio |
|---|---|---|
| C$_6$H$_6$ molecular | 78 | $\frac{78}{13} = 6$ |
| CH empirical | 13 | |

Thus, the molar mass of a compound is a whole number multiple of the "molar mass" of the empirical formula.

Chemists can use a mass spectrometer to determine the molar mass of a compound. They can use the molar mass, along with the "molar mass" of a known empirical formula, to determine the compound's molecular formula. For example, the empirical formula CH has a "molar mass" of 13 g/mol. Acetylene, C$_2$H$_2$, and benzene, C$_6$H$_6$, both have the empirical formula CH. Suppose it is determined, through mass spectrometry, that a sample has a molar mass of 78 g/mol. The compound must be C$_6$H$_6$, since $6 \times 13$ g = 78 g, as shown in Table 3.4.

Examine the Sample Problem on the following page to learn how to determine the molecular formula of a compound.

# Careers in Chemistry

## Analytical Chemistry

Ben Johnson, Steve Vezina, Eric Lamaze—all of these athletes tested positive for performance-enhancing substances that are banned by the International Olympic Committee (IOC). Who conducts the tests for these substances? Meet Dr. Christiane Ayotte, head of Canada's Doping Control Laboratory since 1991.

### The Doping Control Lab

What happens to a urine sample after it arrives at the doping control lab? Technicians and scientists must be careful to ensure careful handling of the sample. Portions of the sample are taken for six different analytical procedures. More than 150 substances are banned by the IOC. These substances are grouped according to their physical and chemical properties. There are two main steps for analyzing a sample:

1. purification, which involves steps such as filtration and extraction using solvents, and
2. analysis by either gas chromatography, mass spectrometry, or high-performance liquid chromatography. Chromatography refers to certain methods by which chemists separate mixtures into pure substances.

For most substances, just their presence in a urine sample means a positive result. Other substances must be present in an amount higher than a certain threshold. According to Dr. Ayotte, a male athlete would have to consume "10 very strong French coffees within 15 min" to go over the 12 mg/L limit for caffeine. Ephedrines and pseudoephedrines, two decongestants that are found in cough remedies and that act as stimulants, have a cut-off level. This allows athletes to take them up to one or two days before a competition.

### Challenges

Dr. Ayotte and her team face many challenges. They look for reliable tests for natural substances, develop new analytical techniques, and determine the normal levels of banned substances for male and female athletes. Dr. Ayotte must defend her tests in hearings and with the press, especially when high-profile athletes get positive results. Her dreams include an independent international doping control agency and better drug-risk education for athletes.

For Dr. Ayotte, integrity and a logical mind are essential aspects of being a good scientist.

### Make Career Connections

1. For information about careers in analytical chemistry, contact university and college chemistry departments.
2. For information about doping control and the movement for drug-free sport, contact the Canadian Centre for Ethics in Sport (CCES), the World Anti-Doping Agency (WADA), and the Centre for Sport and Law.

## Sample Problem
### Determining a Molecular Formula

**Problem**

The empirical formula of ribose (a sugar) is $CH_2O$. In a separate experiment, using a mass spectrometer, the molar mass of ribose was determined to be 150 g/mol. What is the molecular formula of ribose?

**What Is Required?**

You need to find the molecular formula of ribose.

**What Is Given?**

You know the empirical formula and the molar mass of ribose.

**Plan Your Strategy**

Divide the molar mass of ribose by the "molar mass" of the empirical formula. The answer you get is the factor by which you multiply the empirical formula.

**Act on Your Strategy**

The "molar mass" of the empirical formula $CH_2O$, determined using the periodic table, is

$$12 \text{ g/mol} + 2(1) \text{ g/mol} + 16 \text{ g/mol} = 30 \text{ g/mol}$$

The molar mass of ribose is 150 g/mol.

$$\frac{150 \text{ g/mol}}{30 \text{ g/mol}} = 5$$

Molecular formula subscripts = 5 × Empirical formula subscripts
$$= C_{1 \times 5} H_{2 \times 5} O_{1 \times 5}$$
$$= C_5 H_{10} O_5$$

Therefore, the molecular formula of ribose is $C_5H_{10}O_5$.

**Check Your Solution**

Work backward by calculating the molar mass of $C_5H_{10}O_5$.
$(5 \times 12.01 \text{ g/mol}) + (10 \times 1.01 \text{ g/mol}) + (5 \times 16.00 \text{ g/mol}) = 150 \text{ g/mol}$
The calculated molar mass matches the molar mass that is given in the problem. The answer is reasonable.

---

> **CHEM FACT**
>
> Three classifications of food are proteins, fats, and carbohydrates. Many carbohydrates have the empirical formula $CH_2O$. This empirical formula looks like a hydrate of carbon, hence the name "carbohydrate." Glucose, fructose, galactose, mannose, and sorbose all have the empirical formula $CH_2O$ since they all have the same molecular formula, $C_6H_{12}O_6$. What makes these sugars different is the way in which their atoms are bonded to one another.

## Practice Problems

17. Oxalic acid has the empirical formula $CHO_2$. Its molar mass is 90 g/mol. What is the molecular formula of oxalic acid?

18. The empirical formula of codeine is $C_{18}H_{21}NO_3$. If the molar mass of codeine is 299 g/mol, what is its molecular formula?

19. A compound's molar mass is 240.28 g/mol. Its percentage composition is 75.0% carbon, 5.05% hydrogen, and 20.0% oxygen. What is the compound's molecular formula?

20. Lab technicians determine that a gas collected from a marshy area is made up of 80.0% carbon and 20.0% hydrogen. They also find that a 4.60 g sample occupies a volume of 3.44 L at STP. What is the molecular formula of the gas?

## Section Summary

In Investigation 3-A, you explored one technique for finding the percentage composition, and hence the empirical formula, of a compound containing magnesium and oxygen. In section 3.4, you will learn about another technique that chemists use to determine the empirical formula of compounds containing carbon and hydrogen. You will learn how chemists combine this technique with mass spectrometry to determine the compound's molecular formula. You will also perform an experiment to determine the chemical formula of a compound.

## Section Review

**1** Explain the role that a mass spectrometer plays in determining the molecular formula of an unknown compound.

**2** Tartaric acid, also known as cream of tartar, is used in baking. Its empirical formula is $C_2H_3O_3$. If 1.00 mol of tartaric acid contains $3.61 \times 10^{24}$ oxygen atoms, what is the molecular formula of tartaric acid?

**3** Why is the molecular formula of a compound much more useful to a forensic scientist than the empirical formula of the compound?

**4** Vinyl acetate, $C_4H_6O_2$, is an important industrial chemical. It is used to make some of the polymers in products such as adhesives, paints, computer discs, and plastic films.

(a) What is the empirical formula of vinyl acetate?

(b) How does the molar mass of vinyl acetate compare with the molar mass of its empirical formula?

**5** A compound has the formula $C_{6x}H_{5x}O_x$, where $x$ is a whole number. Its molar mass is 186 g/mol; what is its molecular formula?

**6** A gaseous compound contains 92.31% carbon and 7.69% hydrogen by mass. 4.35 g of the gas occupies 3.74 L at STP. Determine the molecular formula of the gas.

# Finding Empirical and Chemical Formulas by Experiment

## 3.4

You have learned how to calculate the percentage composition of a compound using its formula. Often, however, the formula of a compound is not known. Chemists must determine the percentage composition and molar mass of an unknown compound through experimentation. Then they use this information to determine the chemical formula of the compound. Determining the chemical formula is an important step in understanding the properties of the compound and developing a way to synthesize it in a laboratory.

In Investigation 3-A, you reacted a known mass of magnesium with oxygen and found the mass of the product. Then you determined the percentage composition and empirical formula of magnesium oxide. This is just one method for determining percentage composition. It is suitable for analyzing simple compounds that react in predictable ways. Chemists have developed other methods for analyzing different types of compounds, as you will learn in this section.

### The Carbon-Hydrogen Combustion Analyzer

A large number of important chemicals are composed of hydrogen, carbon, and oxygen. The **carbon-hydrogen combustion analyzer** is a useful instrument for analyzing these chemicals. It allows chemists to determine the percentage composition of compounds that are made up of carbon, hydrogen, and oxygen. The applications of this instrument include forensic science, food chemistry, pharmaceuticals and academic research—anywhere that an unknown compound needs to be analyzed.

The carbon-hydrogen combustion analyzer works because compounds containing carbon and hydrogen will burn in a stream of pure oxygen, $O_2$, to yield only carbon dioxide and water. If you can find the mass of the carbon dioxide and water separately, you can determine the mass percent of carbon and hydrogen in the compound.

Examine Figure 3.7 to see how a carbon-hydrogen combustion analyzer works. A sample is placed in a furnace. The sample is heated and simultaneously reacted with a stream of oxygen. If the sample contains carbon and hydrogen only, the sample is completely combusted to yield only water vapour and carbon dioxide.

**Section Preview/Outcomes**

In this section, you will

- **identify** real-life situations in which the analysis of unknown substances is important
- **determine** the chemical formula of a hydrate through experimentation
- **explain** how a carbon-hydrogen analyzer can be used to determine the empirical formula of a compound
- **communicate** your understanding of the following terms: carbon-hydrogen combustion analyzer, hydrate, anhydrous

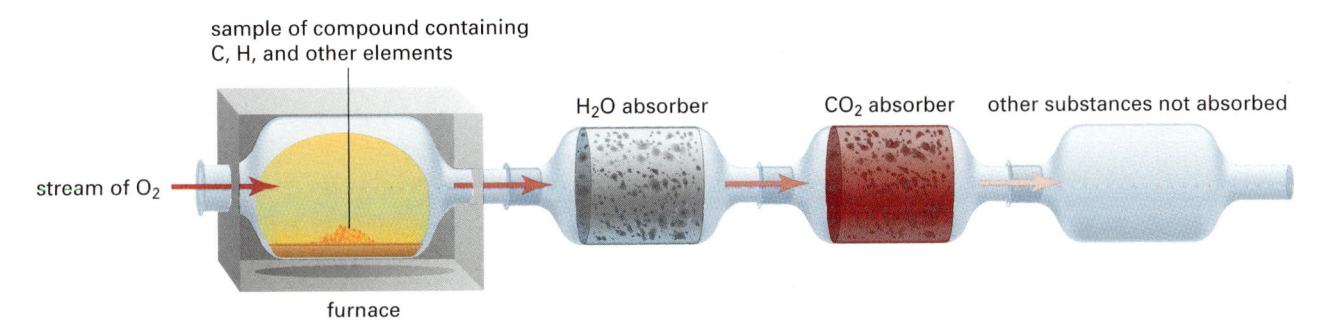

**Figure 3.7** A schematic diagram of a carbon-hydrogen combustion analyzer. After the combustion, all the carbon in the sample is contained in the carbon dioxide. All the hydrogen in the sample is contained in the water.

## CONCEPT CHECK

Carbon dioxide reacts with sodium hydroxide to form sodium carbonate and water. Write a balanced chemical equation for this reaction.

## PROBEWARE

If you have access to probeware, do the Determining Molecular Mass investigation, or a similar lab available from a probeware company.

The water vapour is collected by passing it through a tube that contains magnesium perchlorate, $Mg(ClO_4)_2$. The magnesium perchlorate absorbs all of the water. The mass of the tube is determined before and after the reaction. The difference is the mass of the water that is produced in the reaction. All the hydrogen in the sample is converted to water. Therefore, the percentage composition of hydrogen in water is used to determine the mass of the hydrogen in the sample.

The carbon dioxide is captured in a second tube, which contains sodium hydroxide, NaOH. The mass of this tube is also measured before and after the reaction. The increase in the mass of the tube corresponds to the mass of the carbon dioxide that is produced. All the carbon in the sample reacts to form carbon dioxide. Therefore, the percentage composition of carbon in carbon dioxide is used to determine the mass of the carbon in the sample.

The carbon-hydrogen combustion analyzer can also be used to find the empirical formula of a compound that contains carbon, hydrogen, and one other element, such as oxygen. The difference between the mass of the sample and the mass of the hydrogen and carbon it contains is the mass of the third element.

Examine the following Sample Problem to learn how to determine the empirical formula of a compound based on carbon-hydrogen combustion data.

### Sample Problem

#### Carbon-Hydrogen Combustion Analyzer Calculations

**Problem**

A 1.000 g sample of a pure compound, containing only carbon and hydrogen, was combusted in a carbon-hydrogen combustion analyzer. The combustion produced 0.6919 g of water and 3.338 g of carbon dioxide.

**(a)** Calculate the mass of carbon and hydrogen in the sample.

**(b)** Find the empirical formula of the compound.

**What Is Required?**

You need to find

**(a)** the mass of carbon and hydrogen in the sample

**(b)** the empirical formula of the compound

**What Is Given?**

You know the mass of the sample. You also know the masses of the water and the carbon dioxide produced in the combustion of the sample.

**Plan Your Strategy**

All the hydrogen in the sample was converted to water. Multiply the mass percent (as a decimal) of hydrogen in water by the mass of the water to get the mass of the hydrogen in the sample.

Similarly, all the carbon in the sample has been incorporated into the carbon dioxide. Multiply the mass percent (as a decimal) of carbon in carbon dioxide by the mass of the carbon dioxide to get the mass of carbon in the sample. Convert to moles and determine the empirical formula.

### Act on Your Strategy

(a) Mass of H in sample
$$= \frac{2.02 \text{ g H}_2}{18.02 \text{ g H}_2\text{O}} \times 0.6919 \text{ g H}_2\text{O} = 0.077\,56 \text{ g H}_2$$

Mass of C in sample $= \frac{12.01 \text{ g C}}{44.01 \text{ g CO}_2} \times 3.338 \text{ g CO}_2 = 0.9109 \text{ g C}$

The sample contained 0.077 56 g of hydrogen and 0.9109 g of carbon.

(b) Moles of H in sample $= \frac{0.077\,56 \text{ g}}{1.008 \text{ g/mol}} = 0.076\,94 \text{ mol}$

Moles of C in sample $= \frac{0.9109 \text{ g}}{12.01 \text{ g/mol}} = 0.075\,84 \text{ mol}$

Empirical formula $= \text{C}_{\frac{0.075\,84}{0.075\,84}} \text{H}_{\frac{0.075\,84}{0.076\,94}}$

$= \text{C}_{1.0}\text{H}_{1.0}$

$= \text{CH}$

### Check Your Solution
The sum of the masses of carbon and hydrogen is
0.077 56 g + 0.9109 g = 0.988 46 g. This is close to the mass of the sample. Therefore your answers are reasonable.

### Practice Problems

21. A 0.539 g sample of a compound that contained only carbon and hydrogen was subjected to combustion analysis. The combustion produced 1.64 g of carbon dioxide and 0.807 g of water. Calculate the percentage composition and the empirical formula of the sample.

22. An 874 mg sample of cortisol was subjected to carbon-hydrogen combustion analysis. 2.23 g of carbon dioxide and 0.652 g of water were produced. The molar mass of cortisol was found to be 362 g/mol. If cortisol contains carbon, hydrogen, and oxygen, determine its molecular formula.

**CHEM FACT**
Cortisol is an important steroid hormone. It helps your body synthesize protein. Cortisol can also reduce inflammation, and is used to treat allergies and rheumatoid arthritis.

## Hydrated Ionic Compounds

In some cases, chemists know most of the chemical formula of a compound, but one significant piece of information is missing.

For example, many ionic compounds crystallize from a water solution with water molecules incorporated into their crystal structure, forming a **hydrate**. Hydrates have a specific number of water molecules chemically bonded to each formula unit. A chemist may know the formula of the ionic part of the hydrate but not how many water molecules are present for each formula unit.

Epsom salts, for example, consist of crystals of magnesium sulfate heptahydrate, $\text{MgSO}_4 \cdot 7\text{H}_2\text{O}$. Every formula unit of magnesium sulfate has seven molecules of water weakly bonded to it. A raised dot in a chemical formula, in front of one or more water molecules, denotes a hydrated compound. Note that the dot does not stand for multiplication, but rather a weak bond between an ionic compound and one or more water molecules. Some other examples of hydrates are shown in Table 3.4 on the next page.

**Figure 3.8** Alabaster is a compact form of gypsum often used in sculpture. Gypsum is the common name for calcium sulfate dihydrate, $CaSO_4 \cdot 2H_2O$.

Ionic compounds that have no water molecules incorporated into them are called anhydrous to distinguish them from their hydrated forms. For example, a chemist might refer to $CaSO_4$ as anhydrous calcium sulfate. This is because it is often found in hydrated form as calcium sulfate dihydrate, shown in Figure 3.8.

**Table 3.4** Selected Hydrates

| Formula | Chemical name |
|---|---|
| $CaSO_4 \cdot 2H_2O$ | calcium sulfate dihydrate (gypsum) |
| $CaCl_2 \cdot 2H_2O$ | calcium chloride dihydrate |
| $MgSO_4 \cdot 7H_2O$ | magnesium sulfate heptahydrate (Epsom salts) |
| $Ba(OH)_2 \cdot 8H_2O$ | barium hydroxide octahydrate |
| $KAl(SO_4)_2 \cdot 12H_2O$ | potassium aluminum sulfate dodecahydrate (alum) |

The molar mass of a hydrated compound must include the mass of any water molecules that are in the compound. For example, the molar mass of magnesium sulfate heptahydrate includes the mass of 7 mol of water. It is very important to know whether a compound exists as a hydrate. If a chemical reaction calls for 0.25 mol of copper(II) chloride, you need to know whether you have anhydrous copper(II) chloride or copper(II) chloride dihydrate, shown in Figure 3.9. The mass of 0.25 mol of $CuCl_2$ is 33.61 g. The mass of 0.25 mol of $CuCl_2 \cdot 2H_2O$ is 38.11 g.

Calculations involving hydrates involve using the same techniques you have already practised for determining percent by mass and empirical formulas.

The following Sample Problem shows how to determine the chemical formula of a hydrate. In Investigation 3-B, you will heat a hydrate to determine its chemical formula.

**Figure 3.9** If you need 5 mol of $CuCl_2$, how much of the compound above would you use? Examine the label closely.

### Sample Problem
### Determining the Formula of a Hydrate

**Problem**

A hydrate of barium hydroxide, $Ba(OH)_2 \cdot xH_2O$, is used to make barium salts and to prepare certain organic compounds. Since it reacts with $CO_2$ from the air to yield barium carbonate, $BaCO_3$, it must be stored in tightly stoppered bottles.

(a) A 50.0 g sample of the hydrate contains 27.2 g of $Ba(OH)_2$. Calculate the percent, by mass, of water in $Ba(OH)_2 \cdot xH_2O$.

(b) Find the value of $x$ in $Ba(OH)_2 \cdot xH_2O$.

**What Is Required?**

(a) You need to calculate the percent, by mass, of water in the hydrate of barium hydroxide.

(b) You need to find how many water molecules are bonded to each formula unit of $Ba(OH)_2$.

**What Is Given?**

The formula of the sample is $Ba(OH)_2 \cdot xH_2O$.
The mass of the sample is 50.0 g.
The sample contains 27.2 g of $Ba(OH)_2$.

102 MHR • Unit 1 Stoichiometry

### Plan Your Strategy

**(a)** To find the mass of water in the hydrate, find the difference between the mass of barium hydroxide and the total mass of the sample. Divide by the total mass of the sample and multiply by 100%.

**(b)** Find the number of moles of barium hydroxide in the sample. Then find the number of moles of water in the sample. To find out how many water molecules bond to each formula unit of barium hydroxide, divide each answer by the number of moles of barium hydroxide.

### Act on Your Strategy

**(a)** Mass percent of water in $Ba(OH)_2 \cdot xH_2O$

$$= \frac{(\text{Total mass of sample}) - (\text{Mass of } Ba(OH)_2 \text{ in sample})}{(\text{Total mass of sample})} \times 100\%$$

$$= \frac{50.0 \text{ g} - 27.2 \text{ g}}{50.0 \text{ g}} \times 100\%$$

$$= 45.6\%$$

**(b)** Moles of $Ba(OH)_2 = \frac{\text{Mass of } Ba(OH)_2}{\text{Molar mass of } Ba(OH)_2}$

$$= \frac{27.2 \text{ g}}{171.3 \text{ g/mol}} = 0.159 \text{ mol } Ba(OH)_2$$

Moles of $H_2O = \frac{\text{Mass of } H_2O}{\text{Molar mass of } H_2O}$

$$= \frac{50.0 \text{ g} - 27.2 \text{ g}}{18.02 \text{ g/mol}} = 1.27 \text{ mol } H_2O$$

$\frac{0.159}{0.159}$ mol $Ba(OH)_2$ : $\frac{1.27}{0.159}$ mol $H_2O$ = 1.0 mol $Ba(OH)_2$ : 8.0 mol $H_2O$

The value of $x$ in $Ba(OH)_2 \cdot xH_2O$ is 8.
Therefore, the chemical formula of the hydrate is $Ba(OH)_2 \cdot 8H_2O$.

> **PROBLEM TIP**
> This step is similar to finding an empirical formula based on percentage composition.

### Check Your Solution

Work backward. According to the formula, the percent by mass of water in $Ba(OH)_2 \cdot 8H_2O$ is:

$$\frac{144.16 \text{ g/mol}}{315.51 \text{ g/mol}} \times 100\% = 45.7\%$$

According to the question, the percent by mass of water in the hydrate of $Ba(OH)_2$ is:

$$\frac{(50.0 \text{ g} - 27.2 \text{ g})}{50.0 \text{ g}} \times 100\% = 45.6\%$$

Therefore, your answer is reasonable.

### Practice Problems

**23.** What is the percent by mass of water in magnesium sulfite hexahydrate, $MgSO_3 \cdot 6H_2O$?

**24.** A 3.34 g sample of a hydrate has the formula $SrS_2O_3 \cdot xH_2O$, and contains 2.30 g of $SrS_2O_3$. Find the value of $x$.

**25.** A hydrate of zinc chlorate, $Zn(ClO_3)_2 \cdot xH_2O$, contains 21.5% zinc by mass. Find the value of $x$.

> **CONCEPT CHECK**
> Write an equation that shows what happens when you heat magnesium sulfate hexahydrate enough to convert it to its anhydrous form.

# Investigation 3-B

**SKILL FOCUS**
- Predicting
- Performing and recording
- Analyzing and interpreting
- Communicating results

## Determining the Chemical Formula of a Hydrate

Many ionic compounds exist as hydrates. Often you can convert hydrates to anhydrous ionic compounds by heating them.

In this investigation, you will find the percent by mass of water in a hydrate of copper(II) sulfate hydrate, $CuSO_4 \cdot xH_2O$. The crystals of the hydrate are blue, while anhydrous copper(II) sulfate is white.

### Question
What is the chemical formula of the hydrate of copper(II) sulfate, $CuSO_4 \cdot xH_2O$?

### Prediction
Predict what reaction will occur when you heat the hydrate of copper(II) sulfate.

### Materials
400 mL beaker (if hot plate is used)
tongs
scoopula
electronic balance, precise to two decimal places
glass rod
hot pad
3 g to 5 g hydrated copper(II) sulfate

### Safety Precautions
Heat the hydrate at a low to medium temperature only.

### Procedure
**Note:** If you are using a hot plate as your heat source, use the 400 mL beaker. If you are using a laboratory burner, use the porcelain evaporating dish.

1. Make a table like the one below, for recording your observations.

**Observations**

| | |
|---|---|
| Mass of empty beaker or evaporating dish | |
| Mass of beaker or evaporating dish + hydrated copper(II) sulfate | |
| Mass of beaker or evaporating dish + anhydrous copper(II) sulfate | |

2. Measure the mass of the beaker and stirring rod. Record the mass in your table.

3. Add 3 g to 5 g hydrated copper(II) sulfate to the beaker.

**A hydrate of copper(II) sulfate (far left) is light blue. It loses its colour on heating.**

4. Measure the mass of the beaker with the hydrated copper(II) sulfate. Record the mass in your table.

5. If you are using a hot plate, heat the beaker with the hydrated copper(II) sulfate until the crystals lose their blue colour. You may need to stir occasionally with the glass rod. Be sure to keep the heat at a medium setting. Otherwise, the beaker may break.

6. When you see the colour change, stop heating the beaker. Turn off or unplug the hot plate. Remove the beaker with the beaker tongs. Allow the beaker and crystals to cool on a hot pad.

7. Find the mass of the beaker with the white crystals. Record the mass in your table.

8. Return the anhydrous copper(II) sulfate to your teacher when you are finished. Do not put it in the sink or in the garbage.

## Analysis

1. (a) Determine the percent by mass of water in your sample of hydrated copper(II) sulfate. Show your calculations clearly.

   (b) Do you expect the mass percent of water that you determined to be similar to the mass percents that other groups determined? Explain.

2. (a) On the chalkboard, write the mass of your sample of hydrated copper(II) sulfate, the mass of the anhydrous copper(II) sulfate, and the mass percent of water that you calculated.

   (b) How do your results compare with other groups' results?

## Conclusion

3. Based on your observations, determine the chemical formula of $CuSO_4 \cdot xH_2O$.

## Applications

4. Suppose that you heated a sample of a hydrated ionic compound in a test tube. What might you expect to see inside the test tube, near the mouth of the test tube? Explain.

5. You obtained the mass percent of water in the copper sulfate hydrate.

   (a) Using your observations, calculate the percentage composition of the copper(II) sulfate hydrate.

   (b) In the case of a hydrate, and assuming you know the formula of the associated anhydrous ionic compound, do you think it is more useful to have the mass percent of water in the hydrate or the percentage composition? Explain your answer.

6. Compare the formula that you obtained for the copper(II) sulfate hydrate with the formulas that other groups obtained. Are there any differences? Provide some possible explanations for any differences.

7. Suppose that you did not completely convert the hydrate to the anhydrous compound. Explain how this would affect

   (a) the calculated percent by mass of water in the compound

   (b) the chemical formula you determined

8. Suppose the hydrate was heated too quickly and some of it was lost as it spattered out of the container. Explain how this would affect

   (a) the calculated percent by mass of water in the compound

   (b) the chemical formula you determined

9. Suggest a source of error (not already mentioned) that would result in a value of $x$ that is

   (a) higher than the actual value

   (b) lower than the actual value

## Section Summary

In section 3.4, you learned several practical methods for determining empirical and chemical formulas of compounds. You may have noticed that these methods work because you can often predict how compounds will react. For example, you learned that a compound containing carbon and hydrogen reacts with oxygen to produce water and carbon dioxide. From the mass of the products, you can determine the amount of carbon and hydrogen in the reactant. You also learned that a hydrate decomposes when it is heated to form water and an anhydrous compound. Again, the mass of one of the products of this reaction helps you identify the reactant. In Chapter 4, you will learn how to use information from chemical equations in order to do quantitative calculations.

## Section Review

**1** Many compounds that contain carbon and hydrogen also contain nitrogen. Can you find the nitrogen content by carbon-hydrogen analysis, if the nitrogen does not interfere with the combustion reaction? If so, explain how. If not, explain why not.

**2** What would be the mass of a bag of anhydrous magnesium sulfate, $MgSO_4$, if it contained the same amount of magnesium as a 1.00 kg bag of Epsom salts, $MgSO_4 \cdot 7H_2O$? Give your answer in grams.

**3** A compound that contains carbon, hydrogen, chlorine, and oxygen is subjected to carbon-hydrogen analysis. Can the mass percent of oxygen in the compound be determined using this method? Explain your answer.

**4** Imagine that you are an analytical chemist. You are presented with an unknown compound, in the form of a white powder, for analysis. Your job is to determine the chemical formula of the compound. Create a flow chart that outlines the questions that you would ask and the analyses you would carry out. Briefly explain why each question or analysis is needed.

**5** A carbon-hydrogen analyzer uses a water absorber (which contains magnesium perchlorate, $Mg(ClO_4)_2$, and a carbon dioxide absorber (which contains sodium hydroxide, NaOH). The water absorber is always located in front of the carbon dioxide absorber. What does this suggest about the sodium hydroxide that is contained in the $CO_2$ absorber?

**6** A hydrate of zinc nitrate has the formula $Zn(NO_3)_2 \cdot xH_2O$. If the mass of 1 mol of anhydrous zinc nitrate is 63.67% of the mass of 1 mol of the hydrate, what is the value of $x$?

**7** A 2.524 g sample of a compound contains carbon, hydrogen, and oxygen. The sample is subjected to carbon-hydrogen analysis. 3.703 g of carbon dioxide and 1.514 g of water are collected.

(a) Determine the empirical formula of the compound.

(b) If one molecule of the compound contains 12 atoms of hydrogen, what is the molecular formula of the compound?

# CHAPTER 3 Review

## Reflecting on Chapter 3
Summarize this chapter in the format of your choice. Here are a few ideas to use as guidelines:
- Explain how to determine the mass percent of each element in a compound.
- Predict the empirical formula of a compound using the periodic table, and test your prediction through experimentation.
- Describe how to use experimental data to determine the empirical formula of a compound.
- Describe how to use the molar mass and empirical formula of a compound to determine the molecular (actual) formula of the compound.
- Explain how to determine experimentally the percent by mass of water in a hydrate and determine its chemical formula.
- Explain how a carbon-hydrogen combustion analyzer can be used to determine the mass percent of carbon, hydrogen, and oxygen in a compound.

## Reviewing Key Terms
For each of the following terms, write a sentence that shows your understanding of its meaning.

anhydrous
carbon-hydrogen combustion analyzer
empirical formula
forensic scientists
hydrate
law of definite proportions
mass percent
molecular formula
percentage composition

## Knowledge/Understanding
1. When determining the percentage composition of a compound from its formula, why do you base your calculations on a one mole sample?
2. The main engines of the space shuttle burn hydrogen and oxygen, with water as the product. Is this synthetic (human-made) water the same as water found in nature? Explain.
3. (a) What measurements need to be taken during a carbon-hydrogen combustion analysis?
   (b) Acetylene, $C_2H_2$, and benzene, $C_6H_6$, both have the same empirical formula. How would their results compare in a carbon-hydrogen combustion analysis? Explain your answer.
4. If you know the molar mass of a substance, and the elements that make up the substance, can you determine its molecular formula? Explain your answer.

## Inquiry
5. A 5.00 g sample of borax (sodium tetraborate decahydrate, $Na_2B_4O_7 \cdot 10H_2O$) was thoroughly heated to remove all the water of hydration. What mass of anhydrous sodium tetraborate remained?
6. Determine the percentage composition of each compound.
   (a) freon-12, $CCl_2F_2$
   (b) white lead, $Pb_3(OH)_2(CO_3)_2$
7. (a) What mass of water is present in 25.0 g of $MgCl_2 \cdot 2H_2O$?
   (b) What mass of manganese is present in 5.00 g of potassium permanganate, $KMnO_4$?
8. Silver nitrate, $AgNO_3$, can be used to test for the presence of halide ions in solution. It combines with the halide ions to form a silver halide precipitate. In medicine, it is used as an antiseptic and an antibacterial agent. Silver nitrate drops are placed in the eyes of newborn babies to protect them against an eye disease.
   (a) Calculate the mass percent of silver in silver nitrate.
   (b) What mass of silver is contained in $2.00 \times 10^2$ kg of silver nitrate?
9. Barium sulfate, $BaSO_4$, is opaque to X-rays. For this reason, it is sometimes given to patients before X-rays of their intestines are taken. What is the mass percent of barium in barium sulfate?
10. Bismuth nitrate, $Bi(NO_3)_2$, is used in the production of some luminous paints. What is the mass percent of bismuth in bismuth nitrate?
11. The molar mass of a compound is approximately 121 g. The empirical formula of the compound is $CH_2O$. What is the molecular formula of the compound?

Answers to questions highlighted in red type are provided in Appendix A.

12. A complex organic compound, with the name 2,3,7,8-tetrachlorodibenza-para-dioxin, belongs to a family of toxic compounds called *dioxins*. The empirical formula of a certain dioxin is $C_6H_2OCl_2$. If the molar mass of this dioxin is 322 g/mol, what is its molecular formula?

13. A student obtains an empirical formula of $C_1H_{2.67}$ for a gaseous compound.
    (a) Why is this not a valid empirical formula?
    (b) Use the student's empirical formula to determine the correct empirical formula.
    (c) At STP, 4.92 g of the gas occupies 2.50 L. What is the molecular formula of the gas?

14. Progesterone, a hormone, is made up of 80.2% carbon, 10.18% oxygen, and 9.62% hydrogen. Determine the empirical formula of progesterone.

15. An inorganic salt is composed of 17.6% sodium, 39.7% chromium, and 42.8% oxygen. What is the empirical formula of this salt?

16. What is the empirical formula of a compound that contains 67.6% mercury, 10.8% sulfur, and 21.6% oxygen?

17. (a) An inorganic salt is made up of 38.8% calcium, 20.0% phosphorus, and 41.2% oxygen. What is the empirical formula of this salt?
    (b) On further analysis, each formula unit of this salt is found to contain two phosphate ions. Predict the chemical formula of this salt.

18. Capsaicin is the compound that is responsible for the "hotness" of chili peppers. Chemical analysis reveals capsaicin to contain 71.0% carbon, 8.60% hydrogen, 15.8% oxygen, and 4.60% nitrogen. Each molecule of capsaicin contains one atom of nitrogen. What is the molecular formula of capsaicin?

19. A compound has the molecular formula $X_2O_5$, where X is an unknown element. The compound is 44.0% oxygen by mass. What is the identity of element X?

20. A 1.254 g sample of an organic compound that contains only carbon, hydrogen, and oxygen reacts with a stream of chlorine gas, $Cl_{2(g)}$. After the reaction, 4.730 g of HCl and 9.977 g of $CCl_4$ are obtained. Determine the empirical formula of the organic compound.

21. A 2.78 g sample of hydrated iron(II) sulfate, $FeSO_4 \cdot xH_2O$, was heated to remove all the water of hydration. The mass of the anhydrous iron(II) sulfate was 1.52 g. Calculate the number of water molecules associated with each formula unit of $FeSO_4$.

22. Citric acid is present in citrus fruits. It is composed of carbon, hydrogen, and oxygen. When a 0.5000 g sample of citric acid was subjected to carbon-hydrogen combustion analysis, 0.6871 g of carbon dioxide and 0.1874 g of water were produced. Using a mass spectrometer, the molar mass of citric acid was determined to be 192 g/mol.
    (a) What are the percentages of carbon, hydrogen, and oxygen in citric acid?
    (b) What is the empirical formula of citric acid?
    (c) What is the molecular formula of citric acid?

23. Methanol, $CH_3OH$ (also known as methyl alcohol), is a common laboratory reagent. It can be purchased at a hardware store under the name "methyl hydrate" or "wood alcohol." If 1.00 g of methanol is subjected to carbon-hydrogen combustion analysis, what masses of carbon dioxide and water are produced?

24. Copper can form two different oxides: copper(II) oxide, CuO, and copper(I) oxide, $Cu_2O$. Suppose that you find a bottle labelled "copper oxide" in the chemistry prep room. You call this mystery oxide $Cu_xO$. Design an experiment to determine the empirical formula of $Cu_xO$. Assume that you have a fully equipped chemistry lab at your disposal. Keep in mind the following information:
    - Both CuO and $Cu_2O$ react with carbon to produce solid copper and carbon dioxide gas:
    $$Cu_xO_{(s)} + C_{(s)} \rightarrow Cu_{(s)} + CO_{2(g)}$$
    This reaction proceeds with strong heating.
    - Carbon reacts with oxygen to produce carbon dioxide gas:
    $$C_{(s)} + O_{2(g)} \rightarrow CO_{2(g)}$$
    This reaction also proceeds with strong heating.
    - Carbon is available in the form of activated charcoal.
    (a) State at least one safety precaution that you would take.
    (b) State the materials required, and sketch your apparatus.

(c) Outline your procedure.
(d) What data do you need to collect?
(e) State any assumptions that you would make.

25. Magnesium sulfate, $MgSO_4$, is available as anhydrous crystals or as a heptahydrate. Assume that you are given a bottle of $MgSO_4$, but you are not sure whether or not it is the hydrate.
    (a) What method could you use, in a laboratory, to determine whether this is the hydrate?
    (b) If it is the hydrate, what results would you expect to see?
    (c) If it is the anhydrous crystals, what results would you expect to see?

## Communication

26. Draw a concept map to relate the following terms: molar mass of an element, molar mass of a compound, percentage composition, empirical formula, and molecular formula. Use an example for each term.

27. Draw a schematic diagram of a carbon-hydrogen combustion analyzer. Write a few sentences to describe each stage of the analysis as dimethyl ether, $C_2H_6O$, passes through the apparatus.

## Making Connections

28. For many years, tetraethyl lead, $Pb(C_2H_5)_4$, a colourless liquid, was added to gasoline to improve engine performance. Over the last 20 years it has been replaced with non-lead-containing additives due to health risks associated with exposure to lead. Tetraethyl lead was added to gasoline up to 2.0 mL per 3.8 L of gasoline.
    (a) Calculate the mass of tetraethyl lead in 1.0 L of gasoline. The density of $Pb(C_2H_5)_4$ is 1.653 g/mL.
    (b) Calculate the mass of elemental lead in 1.0 L of gasoline.

29. Natron is the name of the mixture of salts that was used by the ancient Egyptians to dehydrate corpses before mummification. Natron is composed of $Na_2CO_3$, $NaHCO_3$, $NaCl$, and $CaCl_2$. The $Na_2CO_3$ absorbs water from tissues to form $Na_2CO_3 \cdot 7H_2O$.
    (a) Name the compound $Na_2CO_3 \cdot 7H_2O$.
    (b) Calculate the mass percent of water in $Na_2CO_3 \cdot 7H_2O$.
    (c) What mass of anhydrous $Na_2CO_3$ is required to dessicate (remove all the water) from an 80 kg body that is 78% water by mass?

30. Imagine that you are an analytical chemist at a pharmaceutical company. One of your jobs is to determine the purity of the acetylsalicylic acid (ASA), $C_9H_8O_4$. ASA is prepared by reacting salicylic acid (SA), $C_7H_6O_3$, with acetic anhydride, $C_4H_6O_3$. Acetic acid, $C_2H_3O_2H$, is also produced

    $C_7H_6O_3 + C_4H_6O_3 \rightarrow C_9H_8O_4 + C_2H_3O_2H$
    SA       acetic        ASA     acetic acid
             anhydride

    ASA often contains unreacted SA. Since it is not acceptable to sell ASA contaminated with SA, one of your jobs is to analyze the ASA to check purity. Both ASA and SA are white powders.
    (a) You analyze a sample that you believe to be pure ASA, but which is actually contaminated with some SA. How will this affect the empirical formula that you determine for the sample?
    (b) Another sample contains ASA contaminated with 0.35 g SA. The mass of the sample is 5.73 g. What empirical formula will you obtain?

### Answers to Practice Problems and Short Answers to Section Review Questions:

**Practice Problems:** 1. 36% Ca; 64% Cl  2. 48.1% Ni; 16.9% P; 35.0% O  3. 39.5% C; 7.8% H; 52.7% O  4.(a) 26.6% K; 35.4% Cr; 38.0% O  (b) 13.3 L  5.(a) 63.65% N  (b) 13.24% N  (c) 35.00% N  (d) 22.23% N  6. 82.22% N; 17.78% H  7. 47.47% O  8.(a) 63.19% Mn; 36.81% O  (b) 158 kg  9. $NH_3$  10. $Li_2O$  11. $BF_3$  12. $SCl$  13. $Cr_2O_3$  14. $P_2O_5$  15. $Na_2Cr_2O_7$  16. $C_{12}H_{14}O_3$  17. $C_2H_2O_4$  18. $C_{18}H_{21}NO_3$  19. $C_{15}H_{12}O_3$  20. $C_2H_6$  21. $C_5H_{12}$; 83.3% C; 16.7% H.  22. $C_{21}H_{30}O_5$  23. 50.9%  24. 5  25. 4

**Section Review: 3.1:** 3. 3.05 g  4. 11.3 g  6. $Na^+$
**3.2:** 2. $C_8H_8O_3$  3.(a) $Cl_2O_7$  (b) $SiCl_4$  4. $CH_2O$  6. $C_9H_{17}O$  7. $C_{11}H_{14}O_2$  8.(a) $FeC_{10}H_{10}$  (b) Yes.  **3.3:** 2. $C_4H_6O_6$  4.(a) $C_2H_3O$  (b) Double  5. $C_{12}H_{10}O_2$  6. $C_2H_2$  **3.4:** 1. Yes  2. 488 g  3. No  6. 6  7.(a) $CH_2O$  (b) $C_6H_{12}O_6$

# CHAPTER 4

# Quantities in Chemical Reactions

**Chapter Preview**

- 4.1 Introducing Stoichiometry
- 4.2 The Limiting Reactant
- 4.3 Percentage Yield

**Prerequisite Concepts and Skills**

Before you begin this chapter, review the following concepts and skills:

- balancing chemical equations (Chapter 1)
- understanding the Avogadro constant and the mole (Chapter 2)
- explaining the relationship between the mole, molar mass, and molar volume (Chapter 2)
- solving problems involving number of particles, amount, mass, and volume (Chapter 2)

A spacecraft requires a huge amount of fuel to supply the thrust needed to launch it into orbit. Engineers work very hard to minimize the launch mass of a spacecraft because each kilogram requires additional fuel. As well, each kilogram costs thousands of dollars to launch.

In 1969, the *Apollo 11* space mission was the first to land astronauts on the Moon. The engineers on the project faced a challenge when deciding on a fuel for the lunar module. The lunar module took the astronauts from the Moon, back to the command module that was orbiting the Moon. The engineers chose a fuel consisting of hydrazine, $N_2H_4$, and dinitrogen tetroxide, $N_2O_4$. These compounds, when mixed, reacted instantaneously and produced the energy needed to launch the lunar module from the Moon.

How do engineers know how much of each reactant they need for a chemical reaction? In this chapter, you will use the concept of the mole to calculate the amounts of reactants that are needed to produce given amounts of products. You will learn how to predict the amounts of products that will be produced in a chemical reaction. You will also learn how to apply this knowledge to any chemical reaction for which you know the balanced chemical equation. Finally, you will learn how calculated amounts deviate from the amounts in real-life situations.

How can you use the balanced equation below to calculate the amount of fuel needed to propel a lunar module back to a command module?

$$2N_2H_{4(\ell)} + N_2O_{4(\ell)} \rightarrow 3N_{2(g)} + 4H_2O_{(g)}$$

# Introducing Stoichiometry

## 4.1

Balanced chemical equations are essential for doing calculations and making predictions related to chemical reactions. To understand why, consider the following analogy.

Suppose that you are making a clubhouse sandwich for a very picky friend. Your friend insists that each sandwich must include exactly three slices of toast, two slices of turkey, and four strips of bacon. Figure 4.1 shows how you can express this sandwich recipe as an equation.

**Section Preview/Outcomes**

In this section, you will

- **identify** the mole ratios of reactants and products in a chemical reaction
- **define** mole ratio, and use mole ratios to represent the relative amounts of reactants and products involved in a chemical reaction
- **explain** how the law of conservation of mass allows chemists to write chemical equations and make accurate predictions using chemical equations
- **perform** stoichiometric calculations to convert among moles, mass, and volume of a gas at STP
- **communicate** your understanding of the following terms: *mole ratios, gravimetric stoichiometry, stoichiometry*

3 slices of toast + 2 slices of turkey + 4 strips of bacon → 1 sandwich

6 slices of toast + 4 slices of turkey + 8 strips of bacon → 2 sandwiches

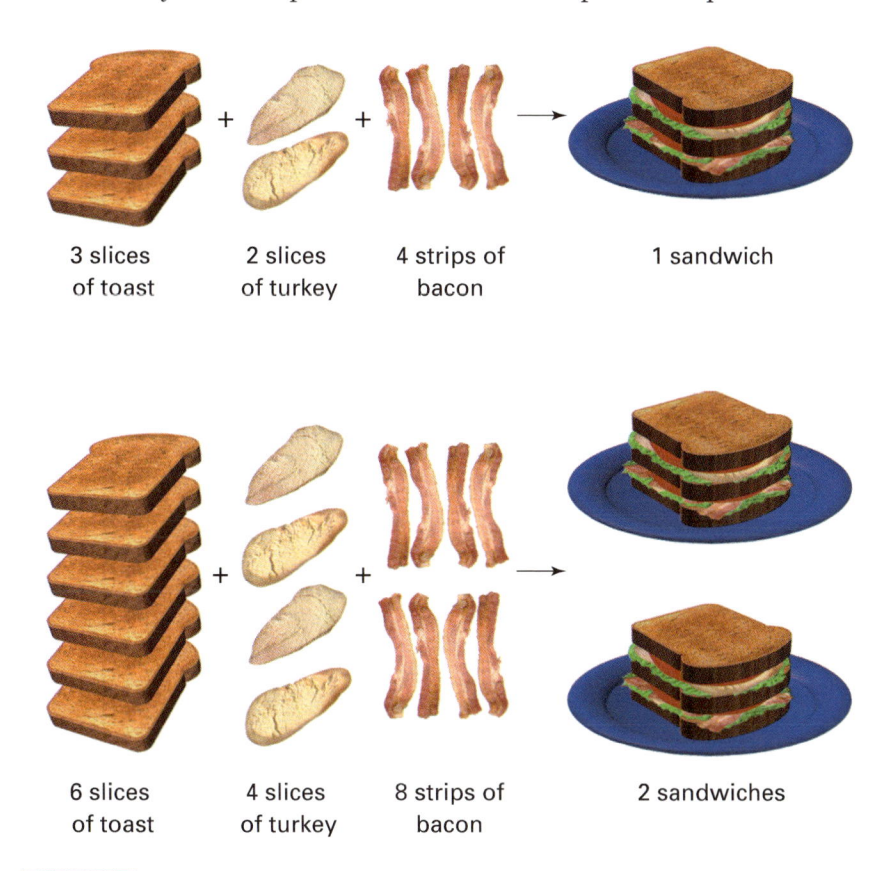

**Figure 4.1** A sandwich analogy showing how equations can be multiplied

Now imagine that you are making two sandwiches for your friend. How much of each ingredient do you need? You need twice the quantity that you used to make one sandwich, as shown in Figure 4.1.

How many sandwiches can you make if you have nine slices of bread, six slices of turkey, and twelve strips of bacon? According to the "sandwich equation," you can make three sandwiches.

You can get the same kind of information from a balanced chemical equation. In Chapters 2 and 3, you learned how chemists relate the number of particles in a substance to the amount of the substance in moles and grams. In this section, you will use your knowledge to interpret the information in a chemical equation, in terms of particles, moles, and mass. Try the following ExpressLab to explore the molar relationships between products and reactants.

## ExpressLab: Mole Relationships in a Chemical Reaction

The following balanced equation shows the reaction between sodium hydrogen carbonate, $NaHCO_3$, and hydrochloric acid, HCl.

$$NaHCO_{3(s)} + HCl_{(aq)} \rightarrow CO_{2(g)} + H_2O_{(\ell)} + NaCl_{(aq)}$$

In this Express Lab, you will determine the mole relationships between the products and reactants in the reaction. Then you will compare the mole relationships with the balanced chemical equation.

### Safety Precautions

Be careful when using concentrated hydrochloric acid. It burns skin and clothing. Do not inhale its vapour.

### Procedure

1. Obtain a sample of sodium hydrogen carbonate that is approximately 1.0 g.
2. Place a 24-well microplate on a balance. Measure and record its mass.
3. Place all the sodium hydrogen carbonate in well A4 of the microplate. Measure and record the mass of the microplate and sample.
4. Fill a thin-stem pipette with 8 mol/L hydrochloric acid solution.
5. Wipe the outside of the pipette. Stand it, stem up, in well A3.
6. Measure and record the total mass of the microplate and sample.
7. Add the hydrochloric acid from the pipette to the sodium hydrogen carbonate in well A4. Allow the gas to escape after each drop.
8. Continue to add the hydrochloric acid until all the sodium hydrogen carbonate has dissolved and the solution produces no more bubbles.
9. Return the pipette, stem up, to well A3. Again find the total mass of the microplate and samples.
10. Dispose of the reacted chemicals as directed by your teacher.

### Analysis

1. Calculate the number of moles of sodium hydrogen carbonate used.
2. Find the difference between the total mass of the microplate and samples before and after the reaction. This difference represents the mass of carbon dioxide gas produced.
3. Calculate the number of moles of carbon dioxide produced.
4. Express your answers to questions 1 and 3 as a mole ratio of mol $NaHCO_3$:mol $CO_2$.
5. According to the balanced equation, how many formula units of sodium hydrogen carbonate react to form one molecule of carbon dioxide?
   (a) Express your answer as a ratio.
   (b) Compare this ratio to your mole ratio in question 4.
6. How many moles of carbon dioxide do you think would be formed from 4.0 mol of sodium hydrogen carbonate?

---

You can use your understanding of the relationship between moles and number of particles to see how chemical equations communicate information about how many moles of products and reactants are involved in a reaction.

## Particle Relationships in a Balanced Chemical Equation

The coefficients in front of the formulas and symbols for compounds and elements in chemical equations tell you the ratios of particles involved in a reaction. A chemical equation can tell you much more, however. Consider, for example, the equation that describes the production of ammonia. Ammonia is an important industrial chemical. Several of its uses are shown in Figure 4.2 on the following page.

**Figure 4.2** Ammonia can be applied directly to the soil as a fertilizer. An aqueous (water) solution of ammonia can be used as a household cleaner.

Ammonia can be prepared industrially from its elements, using a process called the Haber Process. The Haber Process is based on the balanced chemical equation below.

$$N_{2(g)} + 3H_{2(g)} \rightarrow 2NH_{3(g)}$$

This equation tells you that one molecule of nitrogen gas reacts with three molecules of hydrogen gas to form two molecules of ammonia gas.

As you can see in Figure 4.3, there is the same number of each type of atom on both sides of the equation.

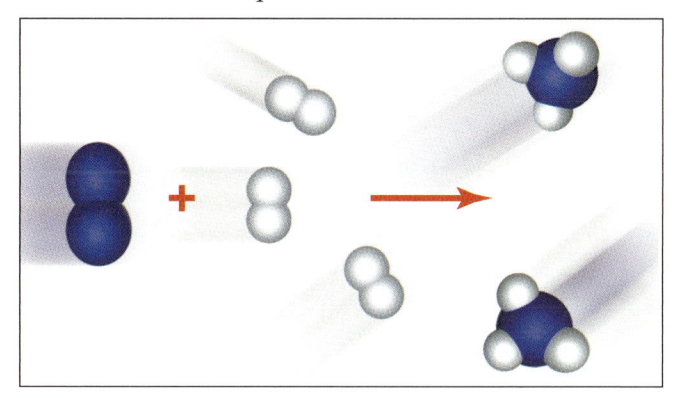

**Figure 4.3** The reaction of nitrogen gas with hydrogen gas.

You can use a ratio to express the numbers of molecules in the equation, as follows:

1 molecule $N_2$ : 3 molecules $H_2$ : 2 molecules $NH_3$

What happens if you multiply the ratio by 2? You get

2 molecules $N_2$ : 6 molecules $H_2$ : 4 molecules $NH_3$

This means that two molecules of nitrogen gas react with six molecules of hydrogen gas to produce four molecules of ammonia gas. Multiplying the original ratio by one dozen gives the following relationship:
1 dozen molecules $N_2$ : 3 dozen molecules $H_2$ : 2 dozen molecules $NH_3$

Suppose that you want to produce 20 molecules of ammonia. How many molecules of nitrogen do you need? You know that you need one molecule of nitrogen for every two molecules of ammonia produced. In other words, the number of molecules of nitrogen that you need is one half the number of molecules of ammonia that you want to produce.

$$20 \text{ molecules } NH_3 \times \frac{1 \text{ molecule } N_2}{2 \text{ molecules } NH_3} = 10 \text{ molecules } N_2$$

Try the following problems to practise working with ratios in balanced chemical equations.

### Practice Problems

1. Consider the following reaction.
$$2H_{2(g)} + O_{2(g)} \rightarrow 2H_2O_{(\ell)}$$
   (a) Write the ratio of $H_2$ molecules: $O_2$ molecules: $H_2O$ molecules.
   (b) How many molecules of $O_2$ are required to react with 100 molecules of $H_2$, according to your ratio in part (a)?
   (c) How many molecules of water are formed when 2478 molecules of $O_2$ react with $H_2$?
   (d) How many molecules of $H_2$ are required to react completely with $6.02 \times 10^{23}$ molecules of $O_2$?

2. Iron reacts with chlorine gas to form iron(III) chloride, $FeCl_3$.
$$2Fe_{(s)} + 3Cl_{2(g)} \rightarrow 2FeCl_{3(s)}$$
   (a) How many atoms of Fe are needed to react with three molecules of $Cl_2$?
   (b) How many molecules of $FeCl_3$ are formed when 150 atoms of Fe react with sufficient $Cl_2$?
   (c) How many $Cl_2$ molecules are needed to react with $1.204 \times 10^{24}$ atoms of Fe?
   (d) How many formula units of $FeCl_3$ are formed when $1.806 \times 10^{24}$ molecules of $Cl_2$ react with sufficient Fe?

3. Consider the following reaction.
$$Ca(OH)_{2(aq)} + 2HCl_{(aq)} \rightarrow CaCl_{2(aq)} + 2H_2O_{(\ell)}$$
   (a) How many formula units of calcium chloride, $CaCl_2$, would be produced by $6.7 \times 10^{25}$ molecules of hydrochloric acid, HCl?
   (b) How many molecules of water would be produced in the reaction in part (a)?

## Mole Relationships in Chemical Equations

Until now, you have assumed that the coefficients in a chemical equation represent particles. They can, however, also represent moles. Consider the following ratio to find out why.

1 molecule $N_2$ : 3 molecules $H_2$ : 2 molecules $NH_3$

You can multiply the above ratio by the Avogadro constant to obtain
$1 \times N_A$ molecules $N_2$ : $3 \times N_A$ molecules $H_2$ : $2 \times N_A$ molecules $NH_3$
This is the same as

1 mol $N_2$ : 3 mol $H_2$ : 2 mol $NH_3$

So the chemical equation $N_{2(g)} + 3H_{2(g)} \rightarrow 2NH_{3(g)}$ also means that 1 mol of nitrogen molecules reacts with 3 mol of hydrogen molecules to form 2 mol of ammonia molecules. The relationships between moles in a balanced chemical equation are called **mole ratios**. For example, the mole ratio of nitrogen to hydrogen in the equation above is 1 mol $N_2$ : 3 mol $H_2$. The mole ratio of hydrogen to ammonia is 3 mol $H_2$ : 2 mol $NH_3$.

You can manipulate mole ratios in the same way that you can manipulate ratios involving molecules. For example, suppose that you want to know how many moles of ammonia you can produce if you have 2.8 mol of hydrogen. You know that you can obtain 2 mol of ammonia for every 3 mol of hydrogen. Therefore, you multiply the number of moles of hydrogen by the mole ratio of ammonia to hydrogen. Another way to think about this is to equate the known mole ratio of hydrogen to ammonia to the unknown mole ratio of hydrogen to ammonia and solve for the unknown.

$$\underset{\text{unknown ratio}}{\frac{n \text{ mol NH}_3}{2.8 \text{ mol H}_2}} = \underset{\text{known ratio}}{\frac{2 \text{ mol NH}_3}{3 \text{ mol H}_2}}$$

$$(2.8 \text{ mol H}_2) \frac{n \text{ mol NH}_3}{2.8 \text{ mol H}_2} = (2.8 \text{ mol H}_2) \frac{2 \text{ mol NH}_3}{3 \text{ mol H}_2}$$

$$n \text{ mol NH}_3 = 1.9 \text{ mol NH}_3$$

Try the following problems to practise working with mole ratios.

## Practice Problems

4. Aluminum bromide can be prepared by reacting small pieces of aluminum foil with liquid bromine at room temperature. The reaction is accompanied by flashes of red light.

    $$2Al_{(s)} + 3Br_{2(\ell)} \rightarrow 2AlBr_{3(s)}$$

    How many moles of $Br_2$ are needed to produce 5 mol of $AlBr_3$, if sufficient Al is present?

5. Hydrogen cyanide gas, $HCN_{(g)}$, is used to prepare clear, hard plastics, such as Plexiglas™. Hydrogen cyanide is formed by reacting ammonia, $NH_3$, with oxygen and methane, $CH_4$.

    $$2NH_{3(g)} + 3O_{2(g)} + 2CH_{4(g)} \rightarrow 2HCN_{(g)} + 6H_2O_{(g)}$$

    (a) How many moles of $O_2$ are needed to react with 1.2 mol of $NH_3$?

    (b) How many moles of $H_2O$ can be expected from the reaction of 12.5 mol of $CH_4$? Assume that sufficient $NH_3$ and $O_2$ are present.

6. Ethane gas, $C_2H_6$, is present in small amounts in natural gas. It undergoes complete combustion to produce carbon dioxide and water.

    $$2C_2H_{6(g)} + 7O_{2(g)} \rightarrow 4CO_{2(g)} + 6H_2O_{(g)}$$

    (a) How many moles of $O_2$ are required to react with 13.9 mol of $C_2H_6$?

    (b) How many moles of $H_2O$ would be produced by 1.40 mol of $O_2$ and sufficient ethane?

7. Magnesium nitride reacts with water to produce magnesium hydroxide and ammonia gas, $NH_3$ according to the balanced chemical equation

    $$Mg_3N_{2(s)} + 6H_2O_{(\ell)} \rightarrow 3Mg(OH)_{2(s)} + 2NH_{3(g)}$$

    (a) How many molecules of water react with 2.3 mol $Mg_3N_2$?

    (b) How many formula units of $Mg(OH)_2$ will be expected in part (a)?

### CHEM FACT

Because the coefficients of a balanced chemical equation can represent moles, it is acceptable to use fractions in an equation. For example, you can write the equation

$$2H_{2(g)} + O_{2(g)} \rightarrow 2H_2O_{(\ell)}$$

as

$$H_{2(g)} + \frac{1}{2}O_{2(g)} \rightarrow H_2O_{(\ell)}$$

Half an oxygen molecule is an oxygen atom, which does not accurately reflect the reaction. Half a mole of oxygen molecules, however, makes sense.

### Technology LINK

In many areas, it is mandatory for every home to have a carbon monoxide detector, like the one shown below.

A carbon monoxide detector emits a sound when the amount of carbon monoxide in the air exceeds a certain limit. Find out how a carbon monoxide detector works, and where it should be placed. Present your findings as a public service announcement.

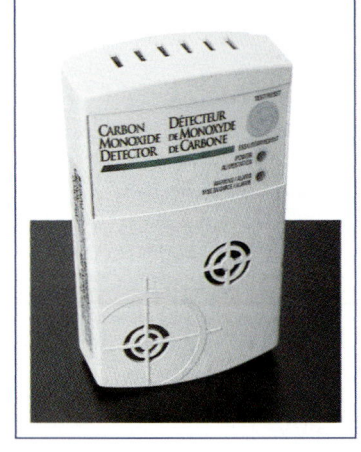

## Different Ratios of Reactants

The relative amounts of reactants are important. Different mole ratios of the same reactants can produce different products. For example, carbon can combine with oxygen in two different ratios, forming either carbon monoxide or carbon dioxide. In the following reaction, the mole ratio of carbon to oxygen is 2 mol C:1 mol $O_2$.

$$2C_{(s)} + O_{2(g)} \rightarrow 2CO_{(g)}$$

In the next reaction, the mole ratio of carbon to oxygen is 1 mol C:1 mol $O_2$.

$$C_{(s)} + O_{2(g)} \rightarrow CO_{2(g)}$$

Thus, carbon dioxide forms if carbon and oxygen are present in a mole ratio of about 1 mol C:1 mol $O_2$. Carbon dioxide is a product of cellular respiration in animals and humans, and it is a starting material for photosynthesis. It is also one of the products of the complete combustion of a hydrocarbon fuel.

If there is a relative shortage of oxygen, however, and the mole ratio of carbon to oxygen is closer to 2 mol C:1 mol O, carbon monoxide forms. Carbon monoxide is colourless, tasteless, and odourless. It is also highly poisonous. Carbon monoxide can escape from any fuel-burning appliance: a furnace, a water heater, a fireplace, a wood stove, or a space heater. If you have one of these appliances in your home, make sure that it has a good supply of oxygen to avoid the formation of carbon monoxide. The photograph on the left shows a carbon monoxide detector.

There are many reactions in which different mole ratios of the reactants result in different products. The following Sample Problem will help you understand how to work with these reactions.

### Sample Problem
### Mole Ratios of Reactants

#### Problem
Vanadium can form several different compounds with oxygen, including $V_2O_5$, $VO_2$, and $V_2O_3$. Determine the number of moles of oxygen that are needed to react with 0.56 mol of vanadium to form vanadium(V) oxide, $V_2O_5$.

#### What Is Required?
You need to find the number of moles of oxygen that are needed to react with 0.56 mol of vanadium to form vanadium(V) oxide.

#### What Is Given?
Reactant: vanadium, V → 0.56 mol
Reactant: oxygen, $O_2$
Product: vanadium(V) oxide, $V_2O_5$

#### Plan Your Strategy
Write a balanced chemical equation for the formation of vanadium(V) oxide. Use the known mole ratio of vanadium to oxygen to calculate the unknown amount of oxygen.

### Act on Your Strategy

The balanced equation is

$$4V_{(s)} + 5O_{2(g)} \rightarrow 2V_2O_{5(s)}$$

To determine the number of moles of oxygen required, equate the known ratio of oxygen to vanadium from the balanced equation to the unknown ratio from the question.

$$\underset{\text{unknown ratio}}{\frac{n \text{ mol } O_2}{0.56 \text{ mol V}}} = \underset{\text{known ratio}}{\frac{5 \text{ mol } O_2}{4 \text{ mol V}}}$$

Multiply both sides of the equation by 0.56 mol V.

$$(\cancel{0.56 \text{ mol V}}) \frac{n \text{ mol } O_2}{\cancel{0.56 \text{ mol V}}} = (0.56 \text{ mol V}) \frac{5 \text{ mol } O_2}{4 \text{ mol V}}$$

$$n \text{ mol } O_2 = (0.56 \,\cancel{\text{mol V}}) \frac{5 \text{ mol } O_2}{4 \,\cancel{\text{mol V}}}$$

$$= 0.70 \text{ mol } O_2$$

### Check Your Solution

The units are correct. The mole ratio of vanadium to oxygen is 4 mol V:5 mol $O_2$. Multiply 0.70 mol by 4/5, and you get 0.56 mol. The answer is therefore reasonable.

## Practice Problems

**8.** Refer to the Sample Problem above.
  (a) How many moles of V are needed to produce 7.47 mol of $VO_2$? Assume that sufficient $O_2$ is present. Assume that V and $O_2$ react to form $VO_2$ only.
  (b) How many moles of V are needed to react with 5.39 mol of $O_2$ to produce $V_2O_3$? Assume that V and $O_2$ react to form $V_2O_3$ only.

**9.** Nitrogen, $N_2$, can combine with oxygen, $O_2$, to form several different oxides of nitrogen. These oxides include $NO_2$, and $N_2O$.

$$2N_{2(g)} + O_{2(g)} \rightarrow 2N_2O_{(g)}$$
$$N_{2(g)} + 2O_{2(g)} \rightarrow 2NO_{2(g)}$$

  (a) How many moles of $O_2$ are required to react with $9.35 \times 10^{-2}$ mol of $N_2$ to form $N_2O$?
  (b) How many moles of $O_2$ are required to react with $9.35 \times 10^{-2}$ mol of $N_2$ to form $NO_2$?

**10.** When heated in a nickel vessel to 400°C, xenon can be made to react with fluorine to produce colourless crystals of xenon tetrafluoride according to the following equation.

$$Xe_{(g)} + 2F_{2(g)} \rightarrow XeF_{4(g)}$$

  (a) How many moles of fluorine gas, $F_2$, would be required to react with $3.54 \times 10^{-1}$ mol of xenon?
  (b) Under somewhat similar reaction conditions, xenon hexafluoride can also be obtained. How many moles of fluorine would be required to react with the amount of xenon given in part (a) to produce xenon hexafluoride?

## CONCEPT CHECK

Calculate the volume at STP of each of the following:

1 mol $N_2$

3 mol $H_2$

2 mol $NH_3$

## Mass Relationships in Chemical Equations

As you have learned, the coefficients in a balanced chemical equation represent moles as well as particles. Therefore, you can use the molar masses of reactants and products to determine the mass ratios for a reaction. For example, consider the equation for the formation of ammonia:

$$N_{2(g)} + 3H_{2(g)} \rightarrow 2NH_{3(g)}$$

You can find the mass of each substance using the equation $m = M \times n$ as follows:

$$1 \text{ mol } N_2 \times 28.0 \text{ g/mol } N_2 = 28.0 \text{ g } N_2$$
$$3 \text{ mol } H_2 \times 2.02 \text{ g/mol } H_2 = 6.1 \text{ g } H_2$$
$$2 \text{ mol } NH_3 \times 17.0 \text{ g/mol } NH_3 = 34.1 \text{ g } NH_3$$

In Table 4.1, you can see how particles, moles, and mass are related in a chemical equation.

**Table 4.1** What a Balanced Chemical Equation Tells You

| Balanced equation | $N_{2(g)} + 3H_{2(g)}$ | $\longrightarrow$ | $2NH_{3(g)}$ |
|---|---|---|---|
| Number of particles (molecules) | 1 molecule $N_2$ + 3 molecules $H_2$ | $\longrightarrow$ | 2 molecules $NH_3$ |
| Amount (mol) | 1 mol $N_2$ + 3 mol $H_2$ | $\longrightarrow$ | 2 mol $NH_3$ |
| Mass (g) | 28.0 g $N_2$ + 6.1 g $H_2$ | $\longrightarrow$ | 34.1 g $NH_3$ |
| Total mass (g) | 34.1 g reactants | $\longrightarrow$ | 34.1 g product |

## The Law of Conservation of Mass

In Table 4.1, you will notice that the mass of the products is the same as the mass of the reactants. This illustrates an important law, first stated during the late eighteenth century by French scientist Antoine Lavoisier (1743–1794), shown with his wife in Figure 4.4.

Lavoisier conducted numerous chemical experiments. One of his most successful and influential techniques as a scientist was his careful measurement of the mass of the reactants and products of a reaction. He emphasized the importance of measuring the mass of all the substances involved in a chemical change. By generalizing his observations, Lavoisier stated the **law of conservation of mass**:

*During a chemical reaction, the total mass of the reactants is always equal to the total mass of the products.*

**Figure 4.4** Marie-Anne Lavoisier read scientific articles in English and translated the articles she thought would interest her husband.

In other words, in a chemical reaction, the total mass of the substances involved does not change. If you determine the total mass of substances on each side of a balanced chemical reaction, you will find that the mass of the products is always equal to the mass of the reactants.

## Stoichiometric Calculations

A balanced chemical equation allows you to determine the number of particles and number of moles of products and reactants involved in a chemical reaction. A balanced chemical equation also allows you to determine the mass of products and reactants involved in a chemical reaction, in agreement with the law of conservation of mass. How do you use this information? If you know the number of moles of one substance, the balanced equation tells you the number of moles of all the other substances. In Chapters 2 and 3, you learned how to convert between particles, moles, mass, and volume. Therefore, *if you know the quantity of one substance in a chemical reaction (in particles, moles, grams, or litres), you can calculate the quantity of any other substance in the reaction (in particles, moles, grams or litres), using the information in the balanced chemical equation.*

You can see that a balanced chemical equation is therefore a powerful tool. In the next few pages you will explore its predictive power.

**Stoichiometry** is the study of the relative quantities of reactants and products in chemical reactions. Stoichiometric analysis involving mass is called **gravimetric stoichiometry**. Stoichimetry involving the volume of gases is called **gas stoichiometry**. Stoichiometric calculations are used for many purposes. One purpose is determining how much of a reactant is needed to carry out a reaction. This kind of knowledge is useful for any chemical reaction, and it can even be a matter of life or death.

In a spacecraft, for example, carbon dioxide is produced as the astronauts breathe (Figure 4.5). To maintain a low level of carbon dioxide, air in the cabin is passed continuously through canisters of lithium hydroxide granules. The carbon dioxide reacts with the lithium hydroxide in the following way:

$$CO_{2(g)} + 2LiOH_{(s)} \rightarrow Li_2CO_{3(s)} + H_2O_{(g)}$$

The canisters are changed periodically as the lithium hydroxide reacts. Engineers must calculate the amount of lithium hydroxide needed to ensure that the carbon dioxide level is safe. As you learned earlier, every kilogram counts in space travel. Therefore, a spacecraft cannot carry much more than the minimum amount.

To determine how much lithium hydroxide is needed, engineers need to ask and answer two important questions:

- How much carbon dioxide is produced per astronaut each day?
- How much lithium hydroxide is needed per kilogram of carbon dioxide?

Engineers can answer the first question by experimenting. To answer the second question, they can make a prediction using stoichiometric calculations. Examine the following Sample Problems to see how these calculations would be done.

**Figure 4.5** A spacecraft is a closed system. All chemical reactions must be taken into account when engineers design systems to keep the air breathable.

---

**Language LINK**

The word "stoichiometry" is derived from two Greek words: *stoikheion*, meaning "element," and *metron*, meaning "to measure." What other words might be derived from the Greek word *metron*?

**History LINK**

The concept of stoichiometry was first described in 1792 by the German scientist Jeremias Benjamin Richter (1762–1807). He stated that "stoichiometry is the science of measuring the quantitative proportions or mass ratios in which chemical elements stand to one another." Can you think of another reason why Richter was famous?

## Sample Problem
### Calculations for Reactants

**Problem**

Carbon dioxide that is produced by astronauts can be removed with lithium hydroxide. The reaction produces lithium carbonate and water. An astronaut produces an average of $1.00 \times 10^3$ g of carbon dioxide each day. What mass of lithium hydroxide should engineers put on board a spacecraft, per astronaut, for each day?

**What Is Required?**

You need to find the mass of lithium hydroxide that is needed to react with $1.00 \times 10^3$ g of carbon dioxide.

**What Is Given?**

Reactant: carbon dioxide, $CO_2 \rightarrow 1.00 \times 10^3$ g
Reactant: lithium hydroxide, LiOH
Product: lithium carbonate, $Li_2CO_3$
Product: water, $H_2O$

**Plan Your Strategy**

**Step 1** Write a balanced chemical equation.

**Step 2** Convert the given mass of carbon dioxide to the number of moles of carbon dioxide.

**Step 3** Calculate the number of moles of lithium hydroxide based on the mole ratio of lithium hydroxide to carbon dioxide.

**Step 4** Convert the amount of moles of lithium hydroxide to volume at STP.

**Act on Your Strategy**

The balanced chemical equation is

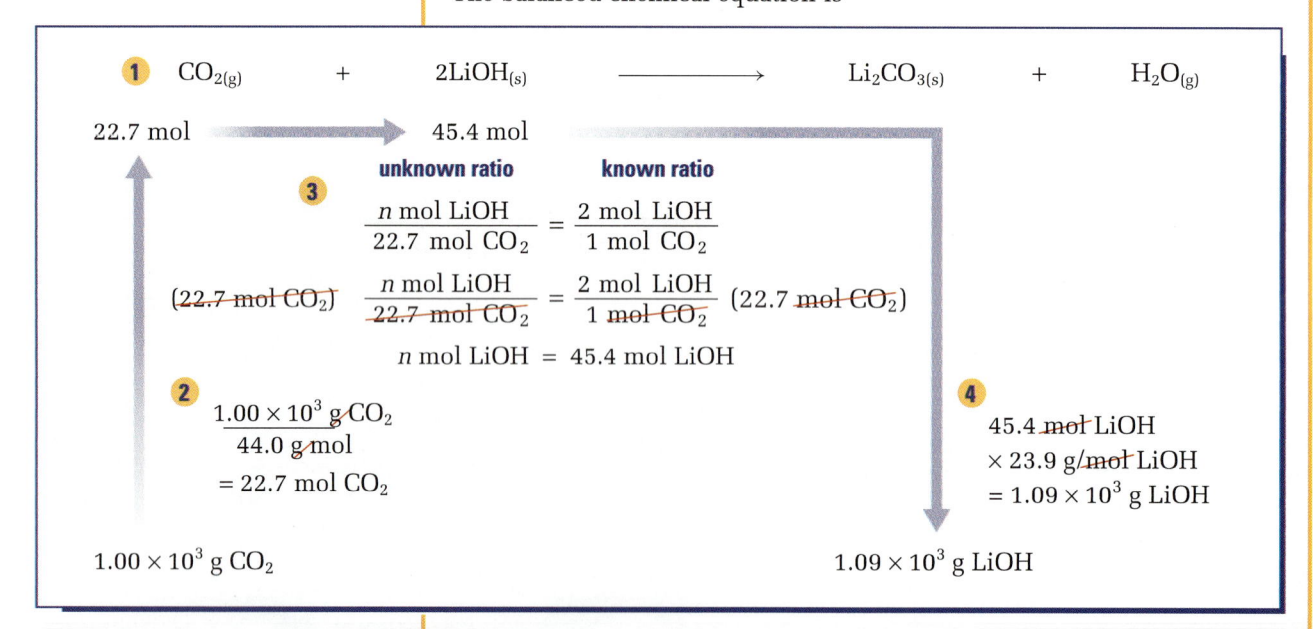

**①** $CO_{2(g)}$ + $2LiOH_{(s)}$ ⟶ $Li_2CO_{3(s)}$ + $H_2O_{(g)}$

22.7 mol → 45.4 mol

unknown ratio | known ratio

**③** $\dfrac{n \text{ mol LiOH}}{22.7 \text{ mol } CO_2} = \dfrac{2 \text{ mol LiOH}}{1 \text{ mol } CO_2}$

$(22.7 \text{ mol } CO_2) \dfrac{n \text{ mol LiOH}}{22.7 \text{ mol } CO_2} = \dfrac{2 \text{ mol LiOH}}{1 \text{ mol } CO_2} (22.7 \text{ mol } CO_2)$

$n$ mol LiOH = 45.4 mol LiOH

**②**
$\dfrac{1.00 \times 10^3 \text{ g } CO_2}{44.0 \text{ g/mol}}$
= 22.7 mol $CO_2$

**④**
45.4 mol LiOH
× 23.9 g/mol LiOH
= $1.09 \times 10^3$ g LiOH

$1.00 \times 10^3$ g $CO_2$ → $1.09 \times 10^3$ g LiOH

Therefore, $1.09 \times 10^3$ g LiOH are required.

### Check Your Solution

The units are correct. Lithium hydroxide has a molar mass that is about half of carbon dioxide's molar mass, but there are twice as many moles of lithium hydroxide. Therefore it makes sense that the mass of lithium hydroxide required is about the same as the mass of carbon dioxide produced.

## Practice Problems

**11.** Ammonium sulfate, $(NH_4)_2SO_4$, is used as a source of nitrogen in some fertilizers. It reacts with sodium hydroxide to produce sodium sulfate, water and ammonia.

$$(NH_4)_2SO_{4(s)} + 2NaOH_{(aq)} \rightarrow Na_2SO_{4(aq)} + 2NH_{3(g)} + 2H_2O_{(\ell)}$$

What mass of sodium hydroxide is required to react completely with 15.4 g of $(NH_4)_2SO_4$?

**12.** Iron(III) oxide, also known as rust, can be removed from iron by reacting it with hydrochloric acid to produce iron(III) chloride and water.

$$Fe_2O_{3(s)} + 6HCl_{(aq)} \rightarrow 2FeCl_{3(aq)} + 3H_2O_{(\ell)}$$

What mass of hydrogen chloride is required to react with $1.00 \times 10^2$ g of rust?

**13.** Iron reacts slowly with hydrochloric acid to produce iron(II) chloride and hydrogen gas.

$$Fe_{(s)} + 2HCl_{(aq)} \rightarrow FeCl_{2(aq)} + H_{2(g)}$$

What mass of HCl is required to react with 3.56 g of iron?

**14.** Dinitrogen pentoxide is a white solid. When heated it decomposes to produce nitrogen dioxide and oxygen.

$$2N_2O_{5(s)} \rightarrow 4NO_{2(g)} + O_{2(g)}$$

What volume of oxygen gas at STP will be produced in this reaction when 2.34 g of $NO_2$ are made?

> **PROBLEM TIP**
>
> In Practice Problem 14, you are asked to determine the volume of a substance, instead of the mass. Use the same steps shown in the Sample Problem, but in the final step, convert the amount of oxygen to volume instead of to mass.

## Sample Problem

### Calculations for Products and Reactants

### Problem

In the Chapter 4 opener, you learned that a fuel mixture consisting of hydrazine, $N_2H_4$, and dinitrogen tetroxide, $N_2O_4$, was used to launch a lunar module. These two compounds react to form nitrogen gas and water vapour. If 150.0 g of hydrazine reacts with sufficient dinitrogen tetroxide, what volume of nitrogen gas at STP is formed?

### What Is Required?

You need to find the volume of nitrogen gas at STP that is formed from 150.0 g of hydrazine.

> **Web LINK**
>
> www.mcgrawhill.ca/links/atlchemistry
>
> For a video clip showing a stoichiometry experiment, go to the web site above and click on **Electronic Learning Partner**.

*Continued ...*

### What Is Given?

Reactant: hydrazine, $N_2H_4 \rightarrow 150.0$ g
Reactant: dinitrogen tetroxide, $N_2O_4$
Product: nitrogen, $N_2$
Product: water, $H_2O$

### Plan Your Strategy

**Step 1** Write a balanced chemical equation.

**Step 2** Convert the mass of hydrazine to the number of moles of hydrazine.

**Step 3** Calculate the number of moles of nitrogen, using the mole ratio of hydrazine to nitrogen.

**Step 4** Convert the amount of nitrogen to volume at STP.

### Act on Your Strategy

The balanced chemical equation is

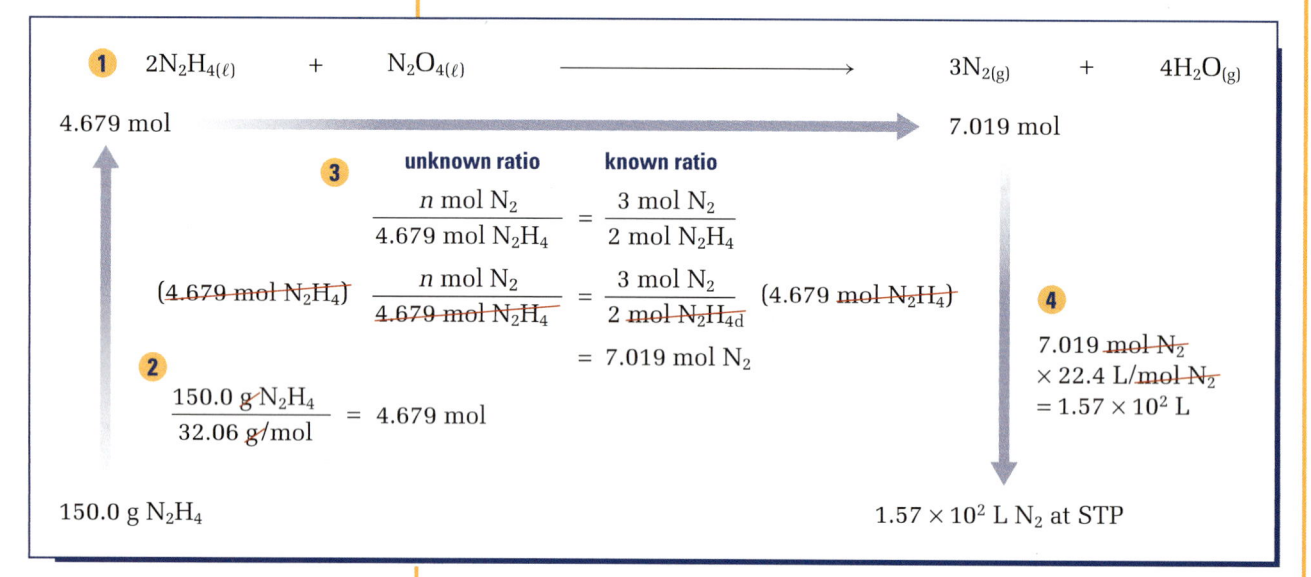

**1** $2N_2H_{4(\ell)} + N_2O_{4(\ell)} \longrightarrow 3N_{2(g)} + 4H_2O_{(g)}$

4.679 mol → 7.019 mol

**3** unknown ratio = known ratio

$$\frac{n \text{ mol } N_2}{4.679 \text{ mol } N_2H_4} = \frac{3 \text{ mol } N_2}{2 \text{ mol } N_2H_4}$$

$$(4.679 \text{ mol } N_2H_4) \frac{n \text{ mol } N_2}{4.679 \text{ mol } N_2H_4} = \frac{3 \text{ mol } N_2}{2 \text{ mol } N_2H_4} (4.679 \text{ mol } N_2H_4)$$

$$= 7.019 \text{ mol } N_2$$

**2** $\frac{150.0 \text{ g } N_2H_4}{32.06 \text{ g/mol}} = 4.679 \text{ mol}$

**4** $7.019 \text{ mol } N_2 \times 22.4 \text{ L/mol } N_2 = 1.57 \times 10^2 \text{ L}$

150.0 g $N_2H_4$ → $1.57 \times 10^2$ L $N_2$ at STP

Therefore, 196.6 g of nitrogen are formed.

### Check Your Solution

The units are correct. The mass of hydrazine is 150.0 g, and the molar mass is close to 30 g/mol. So there are about 5 mol of hydrazine. Multiply 5 mol by the mole ratio of nitrogen to hydrazine (3:2) to get 7.5 mol nitrogen. Multiply 7.5 mol by the molar volume at STP (22 L/mol) to get 165 L, which is close to the calculated answer, 157 L. The answer is reasonable.

## Practice Problems

**15.** Powdered zinc reacts rapidly with powdered sulfur in a highly exothermic reaction.

$$8Zn_{(s)} + S_{8(s)} \rightarrow 8ZnS_{(s)}$$

What mass of zinc sulfide is expected when 32.0 g of $S_8$ reacts with sufficient zinc?

16. The addition of concentrated hydrochloric acid to manganese(IV) oxide leads to the production of chlorine gas.

$$4HCl_{(aq)} + MnO_{2(s)} \rightarrow MnCl_{2(aq)} + Cl_{2(g)} + 2H_2O_{(\ell)}$$

What volume of chlorine at STP can be obtained when $4.76 \times 10^{-2}$ g of HCl react with sufficient $MnO_2$?

17. Aluminum carbide, $Al_4C_3$, is a yellow powder that reacts with water to produce aluminum hydroxide and methane.

$$Al_4C_{3(s)} + 12H_2O_{(\ell)} \rightarrow 4Al(OH)_{3(s)} + 3CH_{4(g)}$$

What volume of methane gas at STP is produced when water reacts with 25.0 g of aluminum carbide?

18. Magnesium oxide reacts with phosphoric acid, $H_3PO_4$, to produce magnesium phosphate and water.

$$3MgO_{(s)} + 2H_3PO_{4(aq)} \rightarrow Mg_3(PO_4)_{2(s)} + 3H_2O_{(\ell)}$$

How many grams of magnesium oxide are required to react completely with 33.5 g of phosphoric acid?

**PROBLEM TIP**

You will need to use your gas stoichiometry skills to answer questions 16 and 17.

## Canadians in Chemistry

**Dr. Stephen Beauchamp**

As a chemist with Environment Canada's Atmospheric Science Division in Dartmouth, Nova Scotia, Dr. Stephen Beauchamp studies toxic chemicals, such as mercury. Loons in Nova Scotia's Kejimkujik National Park are among the living creatures that he studies. Kejimkujik loons have higher blood mercury levels (5 mg Hg/1 g blood) than any other North American loons (2 mg Hg/1 g blood). Mercury is also found in high levels in the fish the loons eat. Mercury causes behavioural problems in the loons. As well, it may affect the loons' reproductive success and immune function.

Bacteria convert environmental mercury into methyl mercury, $CH_3Hg$. This is the form that is most easily absorbed into living organisms. Beauchamp examines forms and concentrations of mercury in the air, soil, and water.

Mercury emission sources include electrical power generation, manufacturing, and municipal waste incineration. Sources such as these, however, do not totally account for the high mercury levels found in Kejimkujik loons and other area wildlife. Beauchamp is working to discover what other factors are operating so that he will be able to recommend ways to improve the situation.

Dr. Stephen Beauchamp in Halifax Harbour. The flux chamber beside him helps him measure the changing concentrations of mercury in the air and water.

## A General Process for Solving Stoichiometric Problems

You have just solved several stoichiometric problems. In these problems, masses or volumes of products and reactants were given, and masses or volumes were also required for the answers. Chemists usually need to know what mass or volume of reactants they require and what mass or volume of products they can expect. Sometimes, however, a question requires you to work with the number of moles or particles. Use the same process for solving stoichiometric problems, whether you are working with mass, volume, amount, or number of particles:

> **Step 1** Write a balanced chemical equation.
>
> **Step 2** If you are given the mass, volume, or number of particles of a substance, convert it to the number of moles.
>
> **Step 3** Calculate the number of moles of the required substance based on the number of moles of the given substance, using the appropriate mole ratio.
>
> **Step 4** If required, convert the number of moles of the required substance to mass, volume, or number of particles.

Examine the following Sample Problem to see how to work with mass and particles.

### CHEM FACT
In this chapter, you are working with stoichiometry problems involving the mass of gases, liquids, and solids, and the volume of gases at STP. In Unit 3, you will encounter stoichiometry problems involving the concentration of solutions.

### Sample Problem
#### Mass and Particle Stoichiometry

**Problem**

Passing chlorine gas through molten sulfur produces liquid disulfur dichloride. How many molecules of chlorine react to produce 50.0 g of disulfur dichloride?

**What Is Required?**

You need to determine the number of molecules of chlorine gas that produce 50.0 g of disulfur dichloride.

**What Is Given?**

Reactant: chlorine, $Cl_2$
Reactant: sulfur, S
Product: disulfur dichloride, $S_2Cl_2$ → 50.0 g

**Plan Your Strategy**

**Step 1** Write a balanced chemical equation.

**Step 2** Convert the given mass of disulfur dichloride to the number of moles.

**Step 3** Calculate the number of moles of chlorine gas using the mole ratio of chlorine to disulfur dichloride.

**Step 4** Convert the number of moles of chlorine gas to the number of particles of chlorine gas.

## Act on Your Strategy

**1.** $Cl_{2(g)} + 2S_{(\ell)} \longrightarrow S_2Cl_{2(\ell)}$

0.370 mol ← ← ← ← ← ← ← ← ← ← ← ← ← ← ← ← ← ← ← ← ← 0.370 mol

**3.**
$$\frac{\text{unknown ratio}}{\text{amount } Cl_2}{0.370 \text{ mol } S_2Cl_2} = \frac{\text{known ratio}}{1 \text{ mol } Cl_2}{1 \text{ mol } S_2Cl_2}$$

$$(0.370 \text{ mol } S_2Cl_2) \frac{\text{amount } Cl_2}{0.370 \text{ mol } S_2Cl_2} = (0.370 \text{ mol } S_2Cl_2) \frac{1 \text{ mol } Cl_2}{1 \text{ mol } S_2Cl_2}$$

$$\text{amount } Cl_2 = 0.370 \text{ mol } Cl_2$$

**2.**
$$\frac{50.0 \text{ g } S_2Cl_2}{135 \text{ g/mol}}$$
$$= 0.370 \text{ mol } S_2Cl_2$$

**4.**
$$0.370 \text{ mol } Cl_2 \times 6.02 \times 10^{23} \frac{\text{molecules } Cl_2}{\text{mol } Cl_2}$$
$$= 2.22 \times 10^{23} \text{ molecules } Cl_2$$

$2.22 \times 10^{23}$ molecules $Cl_2$          50.0 g $S_2Cl_2$

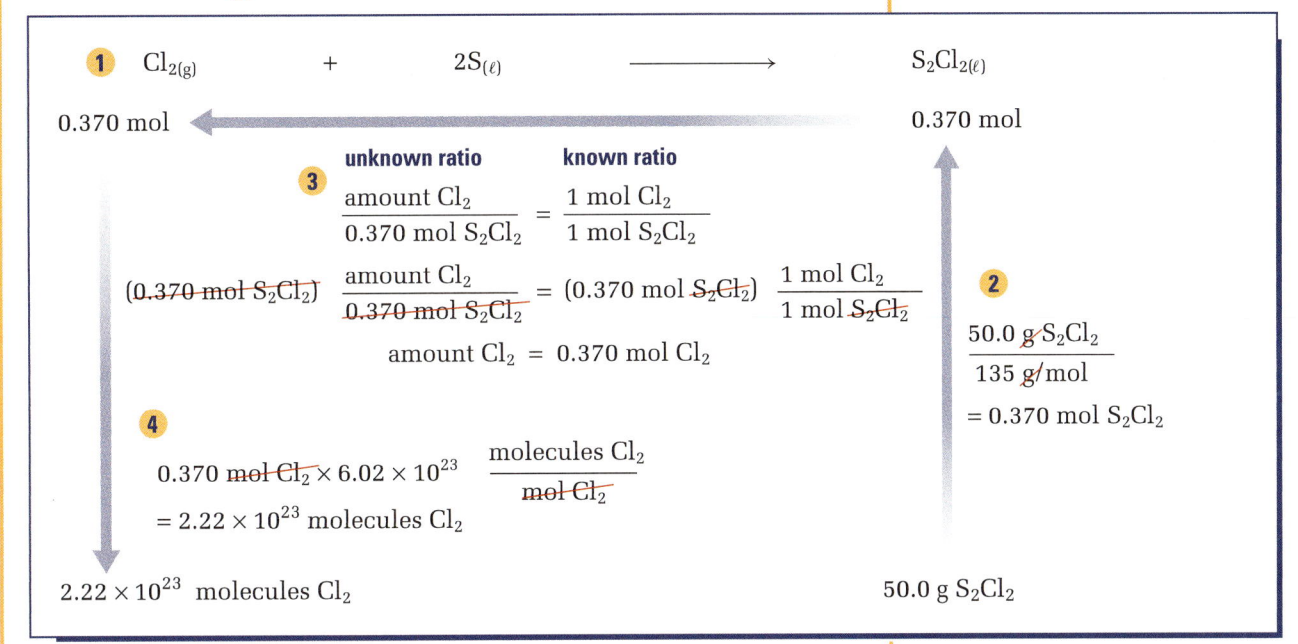

Therefore, $2.22 \times 10^{23}$ molecules of chlorine gas are required.

### Check Your Solution
The units are correct. $2.0 \times 10^{23}$ is about 1/3 of a mole, or 0.33 mol. One-third of a mole of disulfur dichloride has a mass of 45 g, which is close to 50 g. The answer is reasonable.

## Practice Problems

**19.** Nitrogen gas is produced in an automobile air bag. It is generated by the decomposition of sodium azide, $NaN_3$.

$$2NaN_{3(s)} \rightarrow 3N_{2(g)} + 2Na_{(s)}$$

(a) To inflate the air bag on the driver's side of a certain car, 80.0 g of $N_2$ are required. What mass of $NaN_3$ is needed to produce 80.0 g of $N_2$?

(b) How many atoms of Na are produced when 80.0 g of $N_2$ are generated in this reaction?

**20.** The reaction of iron(III) oxide with powdered aluminum is known as the thermite reaction (Figure 4.6).

$$2Al_{(s)} + Fe_2O_{3(s)} \rightarrow Al_2O_{3(s)} + 2Fe_{(\ell)}$$

(a) Calculate the mass of aluminum oxide, $Al_2O_3$, that is produced when $1.42 \times 10^{24}$ atoms of Al react with $Fe_2O_3$.

(b) How many formula units of $Fe_2O_3$ are needed to react with 0.134 g of Al?

**Figure 4.6** The thermite reaction generates enough heat to melt the elemental iron that is produced.

*Continued ...*

21. The thermal decomposition of ammonium dichromate is an impressive reaction. When heated with a Bunsen burner or propane torch, the orange crystals of ammonium dichromate slowly decompose to green chromium(III) oxide in a volcano-like display. Colourless nitrogen gas and water vapour are also given off.

$$(NH_4)_2Cr_2O_{7(s)} \rightarrow Cr_2O_{3(s)} + N_{2(g)} + 4H_2O_{(g)}$$

(a) Calculate the number of formula units of $Cr_2O_3$ that are produced from the decomposition of 10.0 g of $(NH_4)_2Cr_2O_7$.

(b) In a different reaction, 16.9 g of $N_2$ is produced when a sample of $(NH_4)_2Cr_2O_7$ is decomposed. How many water molecules are also produced in this reaction?

(c) How many formula units of $(NH_4)_2Cr_2O_7$ are needed to produce 1.45 g of $H_2O$?

22. Ammonia gas reacts with oxygen to produce water and nitrogen oxide. This reaction can be catalyzed, or sped up, by $Cr_2O_3$, produced in the reaction shown in problem 21.

$$4NH_{3(g)} + 5O_{2(g)} \rightarrow 4NO_{(g)} + 6H_2O_{(\ell)}$$

(a) How many molecules of oxygen are required to react with 34.0 g of ammonia?

(b) What volume of nitrogen monoxide at STP is expected from the reaction of $8.95 \times 10^{24}$ molecules of oxygen with sufficient ammonia?

## Section Summary

You have learned how to do stoichiometric calculations, using balanced chemical equations to find amounts of reactants and products. In these calculations, you assumed that the reactants and products occurred in the exact molar ratios shown by the chemical equation. In real life, however, reactants are often not present in these exact ratios. Similarly, the amount of product that is predicted by stoichiometry is not always produced. In the next two sections, you will learn how to deal with these challenges.

## Section Review

**1** Why is a balanced chemical equation needed to solve stoichiometric calculations?

**2** The balanced chemical equation for the formation of water from its elements is sometimes written as

$$H_{2(g)} + \frac{1}{2}O_{2(g)} \rightarrow H_2O_{(\ell)}$$

Explain why it is acceptable to use fractional coefficients in a balanced chemical equation.

**3** In the following reaction, does 1.0 g of sodium react completely with 0.50 g of chlorine? Explain your answer.

$$Na_{(s)} + \frac{1}{2}Cl_{2(g)} \rightarrow NaCl$$

**4** Sulfur and oxygen can combine to form sulfur dioxide, $SO_2$, and sulfur trioxide, $SO_3$.

(a) Write a balanced chemical equation for the formation of $SO_2$ from S and $O_2$.

(b) Write a balanced chemical equation for the formation of $SO_3$ from S and $O_2$.

(c) What amount of $O_2$ must react with 1 mol of S to form 1 mol of $SO_3$?

(d) What mass of $O_2$ is needed to react with 32.1 g of S to form $SO_3$?

**5** The balanced chemical equation for the combustion of propane is
$$C_3H_{8(g)} + 5O_{2(g)} \rightarrow 3CO_{2(g)} + 4H_2O_{(g)}$$

(a) Write the mole ratios for the reactants and products in the combustion of propane.

(b) How many moles of $O_2$ are needed to react with 0.500 mol of $C_3H_8$?

(c) How many molecules of $O_2$ are needed to react with 44.8 L of $C_3H_8$ at STP?

(d) If 3.00 mol of $C_3H_8$ burn completely in $O_2$, what volume of $CO_2$ at STP is produced?

**6** Phosphorus pentachloride, $PCl_5$, reacts with water to form phosphoric acid, $H_3PO_4$, and hydrochloric acid, HCl.
$$PCl_{5(s)} + 4H_2O_{(\ell)} \rightarrow H_3PO_{4(aq)} + 5HCl_{(aq)}$$

(a) What mass of $PCl_5$ is needed to react with an excess quantity of $H_2O$ to produce 23.5 g of $H_3PO_4$?

(b) How many molecules of $H_2O$ are needed to react with 3.87 g of $PCl_5$?

**7** A chemist has a beaker containing lead nitrate, $Pb(NO_3)_2$, dissolved in water. The chemist adds a solution containing sodium iodide, NaI, and a bright yellow precipitate is formed. The chemist continues to add NaI until no further yellow precipitate is formed. The chemist filters the precipitate, dries it in an oven, and finds it has a mass of 1.43 g.

(a) Write a balanced chemical equation to describe what happened in this experiment. Hint: compounds with sodium ions are always soluble.

(b) Use the balanced chemical equation to determine what mass of lead nitrate, $Pb(NO_3)_2$, was dissolved in the water in the beaker.

**8** Re-examine Figure 4.5 on page 119. This photo shows the Apollo-13 mission overcame an astonishing number of difficulties on its return to Earth. One problem the astronauts encountered was removing carbon dioxide from the air they were breathing. Do some research to answer the following questions.

(a) What happened to lead to an unexpected accumulation of carbon dioxide?

(b) What did the astronauts do to overcome this difficulty?

## 4.2 The Limiting Reactant

**Section Preview/Outcomes**

In this section, you will

- **perform** calculations involving limiting reactants in chemical reactions
- **investigate** to determine the limiting reactant in a chemical reaction
- **communicate** your understanding of the following terms: *stoichiometric coefficients, stoichiometric amounts, limiting reactant, excess reactant*

A balanced chemical equation shows the mole ratios of the reactants and products. To emphasize this, the coefficients of equations are sometimes called **stoichiometric coefficients**. Reactants are said to be present in **stoichiometric amounts** when they are present in a mole ratio that corresponds exactly to the mole ratio predicted by the balanced chemical equation. This means that when a reaction is complete, there are no reactants left. In practice, however, there often *are* reactants left.

In the previous section, you looked at an "equation" (shown below) for making a clubhouse sandwich for a picky friend. You looked at situations in which you had the right quantities of ingredients to make one or more sandwiches, with no leftover ingredients.

3 slices of toast + 2 slices of turkey + 4 strips of bacon → 1 sandwhich

What if you have six slices of toast, 12 slices of turkey, and 20 strips of bacon, as shown in Figure 4.7? How many sandwiches can you make for your friend? Because each sandwich requires three slices of toast, you can only make two sandwiches. Here the quantity of toast that you have limits the number of sandwiches that you can make. Some of the other two ingredients are left over.

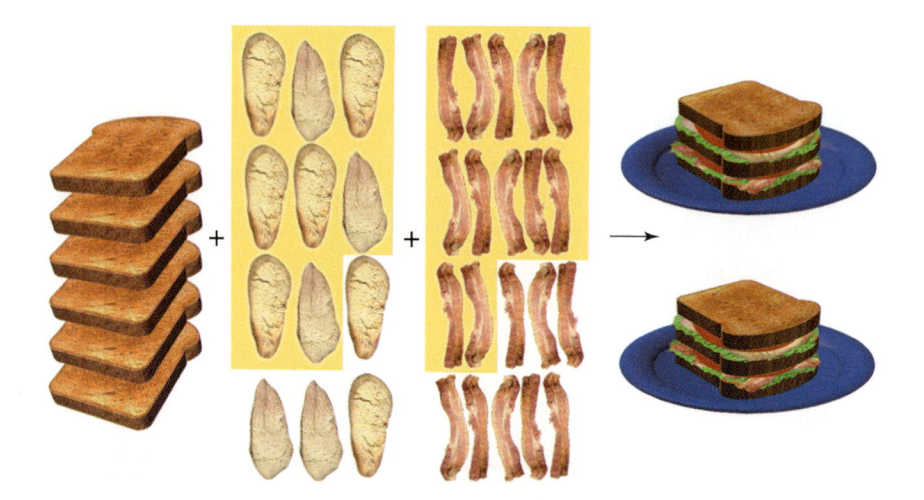

**Figure 4.7** Which ingredient limits how many sandwiches you can make?

Chemical reactions often work in the same way. For example, consider the first step in extracting zinc from zinc oxide:

$$ZnO_{(s)} + C_{(s)} \rightarrow Zn_{(s)} + CO_{(g)}$$

If you were carrying out this reaction in a laboratory, you could obtain samples of zinc oxide and carbon in a 1:1 mole ratio. In an industrial setting, however, it is impractical to spend time and money ensuring that zinc oxide and carbon are present in stoichiometric amounts. It is also unnecessary. In an industrial setting, engineers add more carbon, in the form of charcoal, than is necessary for the reaction. All the zinc oxide reacts, but there is carbon left over.

Having one or more reactants in excess is very common. Another example is seen in gasoline-powered vehicles. Their operation depends on the reaction between fuel and oxygen. Normally, the fuel-injection system regulates how much air enters the combustion chamber, and oxygen is the limiting reactant. When the fuel is very low, however, fuel becomes the limiting reactant and the reaction cannot proceed, as in Figure 4.8.

In nature, reactions almost never have reactants in stoichiometric amounts. Think about respiration, represented by the following chemical equation:

$$C_6H_{12}O_{6(s)} + 6O_{2(g)} \rightarrow 6CO_{2(g)} + 6H_2O_{(\ell)}$$

When an animal carries out respiration, there is an unlimited amount of oxygen in the air. The amount of glucose, however, depends on how much food the animal has eaten.

**Figure 4.8** All the gasoline in this car's tank has reacted. Thus, even though there is still oxygen available in the air, the combustion reaction cannot proceed.

## ThoughtLab: The Limiting Item

Imagine that you are in the business of producing cars. A simplified equation for making a car is
1 car body + 4 wheels + 2 wiper blades → 1 car

### Procedure

1. Assume that you have 35 car bodies, 120 wheels, and 150 wiper blades in your factory. How many complete cars can you make?

2. (a) Which item limits the number of complete cars that you can make? Stated another way, which item will "run out" first?

(b) Which items are present in excess amounts?

(c) How much of each excess item remains after the "reaction"?

### Analysis

1. Does the amount that an item is in excess affect the quantity of the product that is made? Explain.

2. There are fewer car bodies than wheels and wiper blades. Explain why car bodies are not the limiting item, in spite of being present in the smallest amount.

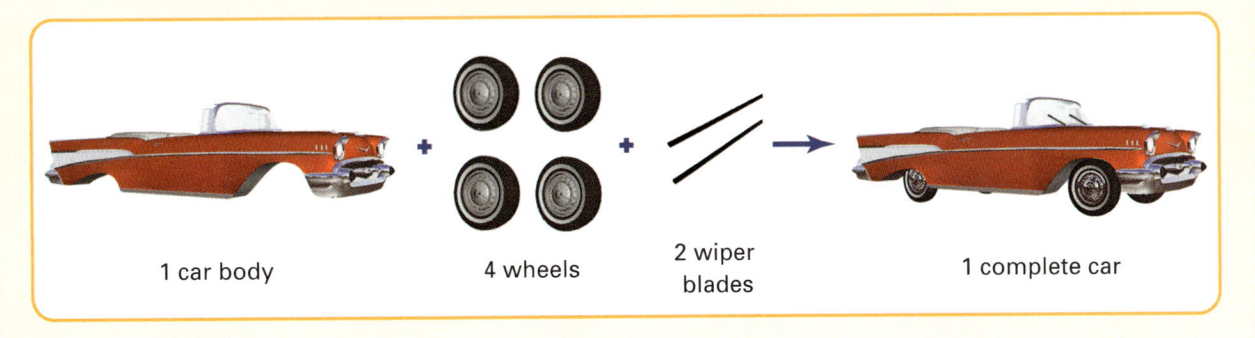

1 car body + 4 wheels + 2 wiper blades → 1 complete car

## Determining the Limiting Reactant

The reactant that is completely used up in a chemical reaction is called the **limiting reactant.** In other words, the limiting reactant determines how much product is produced. When the limiting reactant is used up, the reaction stops. In real-life situations, there is almost always a limiting reactant.

A reactant that remains after a reaction is over is called the **excess reactant.** Once the limiting reactant is used, no more product can be made, regardless of how much of the excess reactants may be present.

Chapter 4 Quantities in Chemical Reactions • MHR **129**

When you are given amounts of two or more reactants to solve a stoichiometric problem, you first need to identify the limiting reactant. One way to do this is to find out how much product would be produced by each reactant if the other reactant were present in excess. The reactant that produces the least amount of product is the limiting reactant. Examine the following Sample Problem to see how to use this approach to identify the limiting reactant.

## Sample Problem

### Identifying the Limited Reactant

**Problem**

Lithium nitride reacts with water to form ammonia and lithium hydroxide, according to the following balanced chemical equation:

$$Li_3N_{(s)} + 3H_2O_{(\ell)} \rightarrow NH_{3(g)} + 3LiOH_{(aq)}$$

If 4.87 g of lithium nitride reacts with 5.80 g of water, find the limiting reactant.

**What Is Required?**

You need to determine whether lithium nitride or water is the limiting reactant.

**What Is Given?**

Reactant: lithium nitride, $Li_3N \rightarrow 4.87$ g
Reactant: water, $H_2O \rightarrow 5.80$ g
Product: ammonia, $NH_3$
Product: lithium hydroxide, LiOH

**Plan Your Strategy**

Convert the given masses into moles. Use the mole ratios of reactants and products to determine how much ammonia is produced by each amount of reactant. The limiting reactant is the reactant that produces the smaller amount of product.

**Act on Your Strategy**

$$n \text{ mol } Li_3N = \frac{4.87 \text{ g } Li_3N}{34.8 \text{ g/mol}}$$
$$= 0.140 \text{ mol } Li_3N$$

$$n \text{ mol } H_2O = \frac{5.80 \text{ g } H_2O}{18.0 \text{ g/mol}}$$
$$= 0.322 \text{ mol } H_2O$$

Calculate the amount of $NH_3$ produced, based on the amount of $Li_3N$.

$$n \text{ mol of } NH_3 = \frac{1 \text{ mol } NH_3}{1 \text{ mol } Li_3N} (0.140 \text{ mol } Li_3N)$$
$$= 0.140 \text{ mol } NH_3$$

Calculate the amount of $NH_3$ produced, based on the amount of $H_2O$.

$$n \text{ mol } NH_3 = \frac{1 \text{ mol } NH_3}{3 \text{ mol } H_2O} \times (0.322 \text{ mol } H_2O)$$
$$= 0.107 \text{ mol } NH_3$$

> **PROBLEM TIP**
>
> To determine the limiting reactant, you can calculate how much of either ammonia or lithium hydroxide would be produced by the reactants. In this problem, ammonia was chosen because only one mole is produced, simplifying the calculation.

130 MHR • Unit 1 Stoichiometry

The water would produce less ammonia than the lithium nitride. Therefore, the limiting reactant is water and the excess reactant is lithium nitride. Notice that there is more water than lithium nitride, in terms of mass and moles. Water is the limiting reactant, however, because 3 mol of water are needed to react with 1 mol of lithium nitride.

### Check Your Solution

According to the balanced chemical equation, the ratio of lithium nitride to water is 1/3. The ratio of lithium nitride to water, based on the mole amounts calculated, is 0.14:0.32. Divide this ratio by 0.14 to get 1.0:2.3. For each mole of lithium nitride, there are only 2.3 mol water. However, 3 mol are required by stoichiometry. Therefore, water is the limiting reactant.

## Practice Problems

**23.** The following balanced chemical equation shows the reaction of aluminum with copper(II) chloride. If 0.25 g of aluminum reacts with 0.51 g of copper(II) chloride, determine the limiting reactant.

$$2Al_{(s)} + 3CuCl_{2(aq)} \rightarrow 3Cu_{(s)} + 2AlCl_{3(aq)}$$

**24.** Hydrogen fluoride, HF, is a highly toxic gas. It is produced by the double displacement reaction of calcium fluoride, $CaF_2$, with pure sulfuric acid, $H_2SO_4$.

$$CaF_{2(s)} + H_2SO_{4(\ell)} \rightarrow 2HF_{(g)} + CaSO_{4(s)}$$

Determine the limiting reactant when 10.0 g of $CaF_2$ reacts with 15.5 g of $H_2SO_4$.

**25.** Acrylic, a common synthetic fibre, is formed from acrylonitrile, $C_3H_3N$. Acrylonitrile can be prepared by the reaction of propylene, $C_3H_6$, with nitrogen monoxide, NO.

$$4C_3H_{6(g)} + 6NO_{(g)} \rightarrow 4C_3H_3N_{(g)} + 6H_2O_{(g)} + N_{2(g)}$$

What is the limiting reactant when 126 g of $C_3H_6$ reacts with 131 L of NO at STP?

**26.** 3.76 g of zinc reacts with $8.93 \times 10^{23}$ hydrogen ions, as shown in the following equation.

$$Zn_{(s)} + 2H^+_{(aq)} \rightarrow Zn^{2+}_{(aq)} + H_{2(g)}$$

Which reactant is present in excess?

You now know how to use a balanced chemical equation to find the limiting reactant. Can you find the limiting reactant by experimenting? You know that the limiting reactant is completely consumed in a reaction, while any reactants in excess remain after the reaction is finished. In Investigation 4-A, you will observe a reaction and identify the limiting reactant, based on your observations.

# Investigation 4-A

**SKILL FOCUS**
Predicting
Performing and recording
Analyzing and interpreting
Communicating results

## Limiting and Excess Reactants

In this investigation, you will predict and observe a limiting reactant. You will use the single replacement reaction of aluminum with aqueous copper(II) chloride:

$$2Al_{(s)} + 3CuCl_{2(aq)} \rightarrow 3Cu_{(s)} + 2AlCl_{3(aq)}$$

Note that copper(II) chloride, $CuCl_2$, is light blue in aqueous solution. This is due to the $Cu^{2+}_{(aq)}$ ion. Aluminum chloride, $AlCl_{3(aq)}$, is colourless in aqueous solution.

### Question

How can observations tell you which is the limiting reactant in the reaction of aluminum with aqueous copper(II) chloride?

### Prediction

Your teacher will give you a beaker that contains a 0.25 g piece of aluminum foil and 0.51 g of copper(II) chloride. Predict which one of these reactants will be the limiting reactant.

### Materials

100 mL beaker or 125 mL Erlenmeyer flask
stirring rod
0.51 g $CuCl_2$
0.25 g Al foil

### Safety Precautions

The reaction mixture may get hot. Do not hold the beaker as the reaction proceeds.

### Procedure

1. To begin the reaction, add about 50 mL of water to the beaker that contains the aluminum foil and copper(II) chloride.

2. Record the colour of the solution and any metal that is present at the beginning of the reaction.

3. Record any colour changes as the reaction proceeds. Stir occasionally with the stirring rod.

4. When the reaction is complete, return the beaker, with its contents, to your teacher for proper disposal. Do not pour anything down the drain.

### Analysis

1. According to your observations, which reactants were the limiting and excess reactants?

2. How does your prediction compare with your observations?

3. Do stoichiometric calculations to support your observations of the limiting reactant. Refer to the previous ThoughtLab if you need help.

4. If your prediction of the limiting reactant was incorrect, explain why.

### Conclusions

5. Summarize your findings. Did your observations support your prediction? Explain.

### Applications

6. Magnesium ($Mg_{(s)}$) and hydrochloric acid ($HCl_{(aq)}$) react according to the following unbalanced equation:

$$Mg_{(s)} + HCl_{(aq)} \rightarrow MgCl_{2(aq)} + H_{2(g)}$$

(a) Balance the equation.

(b) Examine the equation carefully. What evidence would you have that a reaction was taking place between the hydrochloric acid and the magnesium?

(c) You have a piece of magnesium of unknown mass, and a beaker of water containing an unknown amount of hydrochloric acid. Design an experiment to determine which reactant is the limiting reactant. If your teacher approves, carry out your procedure.

## The Limiting Reactant in Stoichiometric Problems

You are now ready to use what you know about finding the limiting reactant to predict the amount of product that is expected in a reaction. This type of prediction is a routine part of a chemist's job, both in academic research and industry. To produce a compound, for example, chemists need to know how much product they can expect from a given reaction. In analytical chemistry, chemists often analyze an impure substance by allowing it to react in a known reaction. They predict the expected mass of the product(s) and compare it with the actual mass of the product(s) obtained. Then they can determine the purity of the compound.

Since chemical reactions usually occur with one or more of the reactants in excess, you often need to determine the limiting reactant before you carry out stoichiometric calculations. You can incorporate this step into the process you have been using to solve stoichiometric problems, as shown in Figure 4.9.

---

Write a balanced chemical equation.

↓

Identify the limiting reactant. Express it as an amount in moles.

↓

Calculate the amount of the required substance based on the amount of the limiting reactant.

↓

Convert the amount of the required substance to mass, volume, or number of particles, as directed by the question.

**Figure 4.9** Be sure to determine the limiting reactant in any stoichiometric problem before you solve it.

**PROBEWARE**

If you have access to probeware, do the Stoichiometry investigation now.

---

### Sample Problem

#### The Limiting Reactant in a Stoichiometric Problem

**Problem**

White phosphorus consists of a molecule made up of four phosphorus atoms. It burns in pure oxygen to produce tetraphosphorus decaoxide.

$$P_{4(s)} + 5O_{2(g)} \rightarrow P_4O_{10(s)}$$

A 1.00 g piece of phosphorus is burned in a flask filled with $2.60 \times 10^{23}$ molecules of oxygen gas. What mass of tetraphosphorus decaoxide is produced?

**What Is Required?**

You need to find the mass of $P_4O_{10}$ that is produced.

**What Is Given?**

You know the balanced chemical equation. You also know the mass of phosphorus and the number of oxygen molecules that are initially present.

**Plan Your Strategy**

First convert each reactant to moles and find the limiting reactant. Using the mole to mole ratio of the limiting reactant to the product, determine the number of moles of tetraphosphorus decaoxide that is expected. Convert this number of moles to grams.

**Act on Your Strategy**

$$n \text{ mol } P_4 = \frac{1.00 \text{ g } P_4}{123.9 \text{ g/mol } P_4}$$
$$= 8.07 \times 10^{-3} \text{ mol } P_4$$

$$n \text{ mol } O_2 = \frac{2.60 \times 10^{23} \text{ molecules}}{6.02 \times 10^{23} \text{ molecules/mol}}$$
$$= 0.432 \text{ mol } O_2$$

Calculate the amount of $P_4O_{10}$ that would be produced by the $P_4$.

*Continued...*

**Continued ...**

$$\frac{n \text{ mol } P_4O_{10}}{8.07 \times 10^{-3} \text{ mol } P_4} = \frac{1 \text{ mol } P_4O_{10}}{1 \text{ mol } P_4}$$

$$(8.07 \times 10^{-3} \text{ mol } P_4)\frac{n \text{ mol } P_4O_{10}}{8.07 \times 10^{-3} \text{ mol } P_4} = \frac{1 \text{ mol } P_4O_{10}}{1 \text{ mol } P_4}(8.07 \times 10^{-3} \text{ mol } P_4)$$

$$= 8.07 \times 10^{-3} \text{ mol } P_4O_{10}$$

Calculate the amount of $P_4O_{10}$ that would be produced by the $O_2$.

$$\frac{n \text{ mol } P_4O_{10}}{0.432 \text{ mol } O_2} = \frac{1 \text{ mol } P_4O_{10}}{5 \text{ mol } O_2}$$

$$(0.432 \text{ mol } O_2)\frac{n \text{ mol } P_4O_{10}}{0.432 \text{ mol } O_2} = \frac{1 \text{ mol } P_4O_{10}}{5 \text{ mol } O_2}(0.432 \text{ mol } O_2)$$

$$= 8.64 \times 10^{-2} \text{ mol } P_4O_{10}$$

Since the $P_4$ would produce less $P_4O_{10}$ than the $O_2$ would, $P_4$ is the limiting reactant.

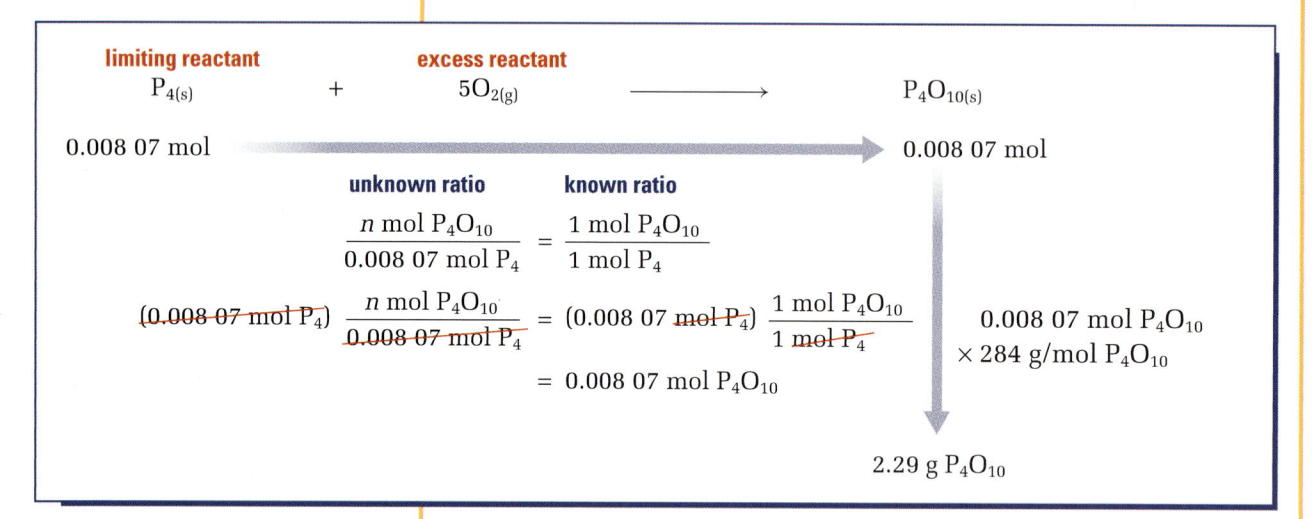

---

**limiting reactant**  **excess reactant**
$P_{4(s)}$  +  $5O_{2(g)}$  $\longrightarrow$  $P_4O_{10(s)}$

0.008 07 mol                              0.008 07 mol

**unknown ratio**          **known ratio**

$$\frac{n \text{ mol } P_4O_{10}}{0.008 \text{ 07 mol } P_4} = \frac{1 \text{ mol } P_4O_{10}}{1 \text{ mol } P_4}$$

$$(0.008 \text{ 07 mol } P_4)\frac{n \text{ mol } P_4O_{10}}{0.008 \text{ 07 mol } P_4} = (0.008 \text{ 07 mol } P_4)\frac{1 \text{ mol } P_4O_{10}}{1 \text{ mol } P_4}$$

$$= 0.008 \text{ 07 mol } P_4O_{10}$$

0.008 07 mol $P_4O_{10}$ × 284 g/mol $P_4O_{10}$

2.29 g $P_4O_{10}$

---

**Check Your Solution**

There were more than 5 times as many moles of $O_2$ as moles of $P_4$, so it makes sense that $P_4$ was the limiting reactant. An expected mass of 2.29 g of tetraphosphorus decaoxide is reasonable. It is formed in a 1:1 ratio from phosphorus. It has a molar mass that is just over twice the molar mass of phosphorus.

## Practice Problems

**27.** Chlorine dioxide, $ClO_2$, is a reactive oxidizing agent. It is used to purify water.

$$6ClO_{2(g)} + 3H_2O_{(\ell)} \rightarrow 5HClO_{3(aq)} + HCl_{(aq)}$$

**(a)** If 71.00 g of $ClO_2$ is mixed with 19.00 g of water, what is the limiting reactant?

**(b)** What mass of $HClO_3$ is expected in part (a)?

**(c)** How many molecules of HCl are expected in part (a)?

**28.** Hydrazine, $N_2H_4$, reacts exothermically with hydrogen peroxide, $H_2O_2$.

$$N_2H_{4(\ell)} + 7H_2O_{2(aq)} \rightarrow 2HNO_{3(g)} + 8H_2O_{(g)}$$

**(a)** 120 g of $N_2H_4$ reacts with an equal mass of $H_2O_2$. Which is the limiting reactant?

(b) What mass of HNO₃ is expected?

(c) What mass, in grams, of the excess reactant remains at the end of the reaction?

29. In the textile industry, chlorine is used to bleach fabrics. Any of the toxic chlorine that remains after the bleaching process is destroyed by reacting it with a sodium thiosulfate solution, Na₂S₂O₃(aq).

$$Na_2S_2O_{3(aq)} + 4Cl_{2(g)} + 5H_2O_{(\ell)} \rightarrow 2NaHSO_{4(aq)} + 8HCl_{(aq)}$$

135 kg of Na₂S₂O₃ reacts with 50.0 kg of Cl₂ and 238 kg of water. How many grams of NaHSO₄ are expected?

30. Manganese(III) fluoride can be formed by the reaction of manganese(II) iodide with fluorine.

$$2MnI_{2(s)} + 13F_{2(g)} \rightarrow 2MnF_{3(s)} + 4IF_{5(\ell)}$$

(a) 1.23 g of MnI₂ reacts with 14.7 L of F₂ at STP. What mass of MnF₃ is expected?

(b) How many molecules of IF₅ are produced in part (a)?

(c) What mass of the excess reactant remains at the end of the reaction?

### CHEM FACT

Carbon disulfide, CS₂, is an extremely volatile and flammable substance. It is so flammable that it can ignite when exposed to boiling water! Because carbon disulfide vapour is more than twice as dense as air, it can "blanket" the floor of a laboratory. There have been cases where the spark from an electrical motor has ignited carbon disulfide vapour in a laboratory, causing considerable damage. For this reason, specially insulated electrical motors are required in laboratory refrigerators and equipment.

## Section Summary

You now know how to identify a limiting reactant. This allows you to predict the amount of product that will be formed in a reaction. Often, however, your prediction will not accurately reflect reality. When a chemical reaction occurs—whether in a laboratory, in nature, or in industry—the amount of product that is formed is often different from the amount that was predicted by stoichiometric calculations. You will learn why this happens, and how chemists deal with it, in section 4.3.

## Section Review

**1** Why do you not need to consider reactants that are present in excess amounts when carrying out stoichiometric calculations? Use an everyday analogy to explain the idea of excess quantity.

**2** (a) Magnesium reacts with oxygen gas, O₂, from the air. Which reactant do you think will be present in excess?

(b) Gold is an extremely unreactive metal. Gold does react, however, with *aqua regia* (a mixture of concentrated nitric acid, HNO₃(aq), and hydrochloric acid, HCl(aq)). The complex ion AuCl₄⁻, as well as NO₂ and H₂O, are formed. This reaction is always carried out with *aqua regia* in excess. Why would a chemist not have the gold in excess?

(c) In general, what characteristics or properties of a chemical compound or atom make it suitable to be used as an excess reactant?

**3** Copper is a relatively inert metal. It is unreactive with most acids. It does, however, react with nitric acid.

$$3Cu_{(s)} + 8HNO_{3(aq)} \rightarrow 3Cu(NO_3)_{2(aq)} + 2NO_{(g)} + 4H_2O_{(\ell)}$$

What volume of NO at STP is produced when 57.4 g of Cu reacts with 165 g of HNO₃?

**Nitric acid reacts with copper metal to produce poisonous, brown nitrogen dioxide, NO₂, gas.**

**4** Iron can be produced when iron(III) oxide reacts with carbon monoxide gas.

$$Fe_2O_{3(s)} + 3CO_{(g)} \rightarrow 2Fe_{(s)} + 3CO_{2(g)}$$

$4.33 \times 10^{22}$ formula units of $Fe_2O_3$ react with 97.9 L of CO at STP. What mass of Fe is expected?

**5** The reaction of an aqueous solution of iron(III) sulfate with aqueous sodium hydroxide produces aqueous sodium sulfate and a solid precipitate, iron(III) hydroxide.

$$Fe_2(SO_4)_{3(aq)} + 6NaOH_{(aq)} \rightarrow 3Na_2SO_{4(aq)} + 2Fe(OH)_{3(s)}$$

What mass of $Fe(OH)_3$ is produced when 10.0 g of $Fe_2(SO_4)_3$ reacts with an equal mass of NaOH?

**6** Carbon disulfide is used as a solvent for water-insoluble compounds, such as fats, oils, and waxes. Calculate the mass of carbon disulfide that is produced when 17.5 g of carbon reacts with 78.7 L of sulfur dioxide at STP according to the following equation:

$$5C_{(s)} + 2SO_{2(g)} \rightarrow CS_{2(\ell)} + 4CO_{(g)}$$

**7** A chemist adds some zinc shavings to a beaker containing a blue solution of copper(II) chloride. The contents of the beaker are stirred. After several hours, the chemist observes that the blue colour has almost, but not completely, disappeared.

(a) Write a balanced chemical equation to describe this reaction.

(b) What other observations would you expect the chemist to make?

(c) According to the chemist's observations, which reactant was the limiting reactant?

(d) The beaker contained 3.12 g of copper(II) chloride dissolved in water. What does this tell you, quantitatively, about the amount of zinc that was added?

# Percentage Yield

## 4.3

When you write an examination, the highest grade that you can earn is usually 100%. Most people, however, do not regularly earn a grade of 100%. A percentage on an examination is calculated using the following equation:

$$\text{Percentage grade} = \frac{\text{Marks earned}}{\text{Maximum possible marks}} \times 100\%$$

Similarly, in baseball, a batter does not succeed at every swing. A batter's success rate is expressed as a decimal fraction. The decimal can be converted to a percent by multiplying by 100%, as shown in Figure 4.10. In this section, you will learn about a percentage that chemists use to predict and express the "success" of reactions.

**Figure 4.10** A baseball player's batting average is calculated as hits/attempts. For example, a player with 6 hits for 21 times at bat has a batting average of 6/21 = 0.286. This represents a success rate of 28.6%.

### Section Preview/Outcomes

In this section, you will

- **perform** calculations involving theoretical yield, actual yield, and percent difference

- **compare**, using experimental results, the theoretical yield with the actual yield, and suggest ways to improve the percentage yield

- **communicate** your understanding of the following terms: *theoretical yield, actual yield, competing reaction, percentage yield, percentage purity*

## Theoretical Yield and Actual Yield

Chemists use stoichiometry to predict the amount of product that can be expected from a chemical reaction. The amount of product that is predicted by stoichiometry is called the **theoretical yield**. This predicted yield, however, is not always the same as the amount of product that is actually obtained from a chemical reaction. The amount of product that is obtained in an experiment is called the **actual yield**.

### Why Actual Yield and Theoretical Yield Are Often Different

The actual yield of chemical reactions is usually less than the theoretical yield. This is caused by a variety of factors. For example, sometimes less than perfect collection techniques contribute to a lower than expected yield.

A reduced yield may also be caused by a **competing reaction**: a reaction that occurs at the same time as the principal reaction and involves one or more of its reactants or products. For example, phosphorus reacts with chlorine to form phosphorus trichloride. Some of the phosphorus trichloride, however, can then react with chlorine to form phosphorus pentachloride.

### CHEM FACT

Actual yield is a *measured* quantity. Theoretical yield is a *calculated* quantity.

Here are the chemical equations for these competing reactions:
$$2P_{(s)} + 3Cl_{2(g)} \rightarrow 2PCl_{3(\ell)}$$
$$PCl_{3(\ell)} + Cl_{2(g)} \rightarrow PCl_{5(s)}$$

Since some of the phosphorus trichloride reacts to form phosphorus pentachloride, the actual yield of phosphorus trichloride is less than the theoretical yield.

Experimental design and technique may affect the actual yield, as well. For example, suppose that you need to obtain a product by filtration. Some of the product may remain in solution and therefore not be caught on the filter paper.

Another common cause of reduced yield is impure reactants. Theoretical yield calculations are usually based on the assumption that reactants are pure. You will learn about the effects of impure reactants on page 144.

The *accuracy* with which you determine the yield of a chemical reaction can be decreased by:

- an unexpected competing reaction
- poor experimental technique or design
- impure reactants
- faulty measuring devices

The *precision* of the results refers to their reproducibility. In other words, can you get the same results again and again? Precision depends on:

- the precision of your instruments (e.g., a balance that displays four decimal places is more precise than a balance that displays one decimal place)
- your experimental technique

## Calculating Percentage Yield

The **percentage yield** of a chemical reaction compares the mass of product obtained by experiment (the actual yield) with the mass of product determined by stoichiometric calculations (the theoretical yield).

$$\text{Percentage yield} = \left(\frac{\text{Actual yield}}{\text{Theoretical yield}}\right) \times 100\%$$

In section 4.1, you looked at the reaction of hydrogen and nitrogen to produce ammonia. You assumed that all the nitrogen and hydrogen reacted. Under certain conditions of temperature and pressure, this is a reasonable assumption. When ammonia is produced industrially, however, temperature and pressure are manipulated to maximize the speed of production. Under these conditions, the actual yield is much less than the theoretical yield. Examine the next Sample Problem to learn how to calculate percentage yield.

### Sample Problem

### Calculating Percentage Yield

**Problem**

Ammonia can be prepared by reacting nitrogen gas, taken from the atmosphere, with hydrogen gas.
$$N_{2(g)} + 3H_{2(g)} \rightarrow 2NH_{3(g)}$$

When 7.5 g of nitrogen reacts with sufficient hydrogen, the theoretical yield of ammonia is 9.10 g. (You can verify this by doing the stoichiometric calculations.) If 1.72 g of ammonia is obtained by experiment, what is the percentage yield of the reaction?

### What Is Required?
You need to find the percentage yield of the reaction.

### What Is Given?
actual yield = 1.72 g
theoretical yield = 9.10 g

### Plan Your Strategy
Divide the actual yield by the theoretical yield, and multiply by 100%.

### Act on Your Strategy
$$\text{Percentage yield} = \frac{\text{Actual yield}}{\text{Theoretical yield}} \times 100\%$$
$$= \frac{1.72 \text{ g}}{9.10 \text{ g}} \times 100\%$$
$$= 18.9\%$$

The percentage yield of the reaction is 18.9%.

### Check Your Solution
By inspection, you can see that 1.72 g is roughly 20% of 9.10 g.

## Practice Problems

**31.** 20.0 g of bromic acid, $HBrO_3$, is reacted with excess HBr.
$$HBrO_{3(aq)} + 5HBr_{(aq)} \rightarrow 3H_2O_{(\ell)} + 3Br_{2(aq)}$$
(a) What is the theoretical yield of $Br_2$ for this reaction?
(b) If 47.3 g of $Br_2$ are produced, what is the percentage yield of $Br_2$?

**32.** Barium sulfate forms as a precipitate in the following reaction:
$$Ba(NO_3)_{2(aq)} + Na_2SO_{4(aq)} \rightarrow BaSO_{4(s)} + 2NaNO_{3(aq)}$$
When 35.0 g of $Ba(NO_3)_2$ are reacted with excess $Na_2SO_4$, 29.8 g of $BaSO_4$ are recovered by the chemist.
(a) Calculate the theoretical yield of $BaSO_4$.
(b) Calculate the percentage yield of $BaSO_4$.

**33.** Yeasts can act on a sugar, such as glucose, $C_6H_{12}O_6$, to produce ethyl alcohol, $C_2H_5OH$, and carbon dioxide.
$$C_6H_{12}O_{6(s)} \rightarrow 2C_2H_5OH_{(\ell)} + 2CO_{2(g)}$$
If 223 g of ethyl alcohol are recovered after 1.63 kg of glucose react, what is the percentage yield of the reaction?

Sometimes chemists know what percentage yield to expect from a chemical reaction. This is especially true of an industrial reaction, where a lot of experimental data are available. Examine the next Sample Problem to learn how to predict the actual yield of a reaction from a known percentage yield.

## Sample Problem
### Predicting Actual Yield Based on Percentage Yield

**Problem**

Calcium carbonate can be thermally decomposed to calcium oxide and carbon dioxide.

$$CaCO_{3(s)} \rightarrow CaO_{(s)} + CO_{2(g)}$$

Under certain conditions, this reaction proceeds with a 92.4% yield of calcium oxide. How many grams of calcium oxide can the chemist expect to obtain if 12.4 g of calcium carbonate is heated?

**What Is Required?**

You need to calculate the amount of calcium oxide, in grams, that will be formed in the reaction.

**What Is Given?**

Percentage yield CaO = 92.4%
$m$ CaCO$_3$ = 12.4 g

**Plan Your Strategy**

Calculate the theoretical yield of calcium oxide using stoichiometry. Then multiply the theoretical yield by the percentage yield to predict the actual yield.

**Act on Your Strategy**

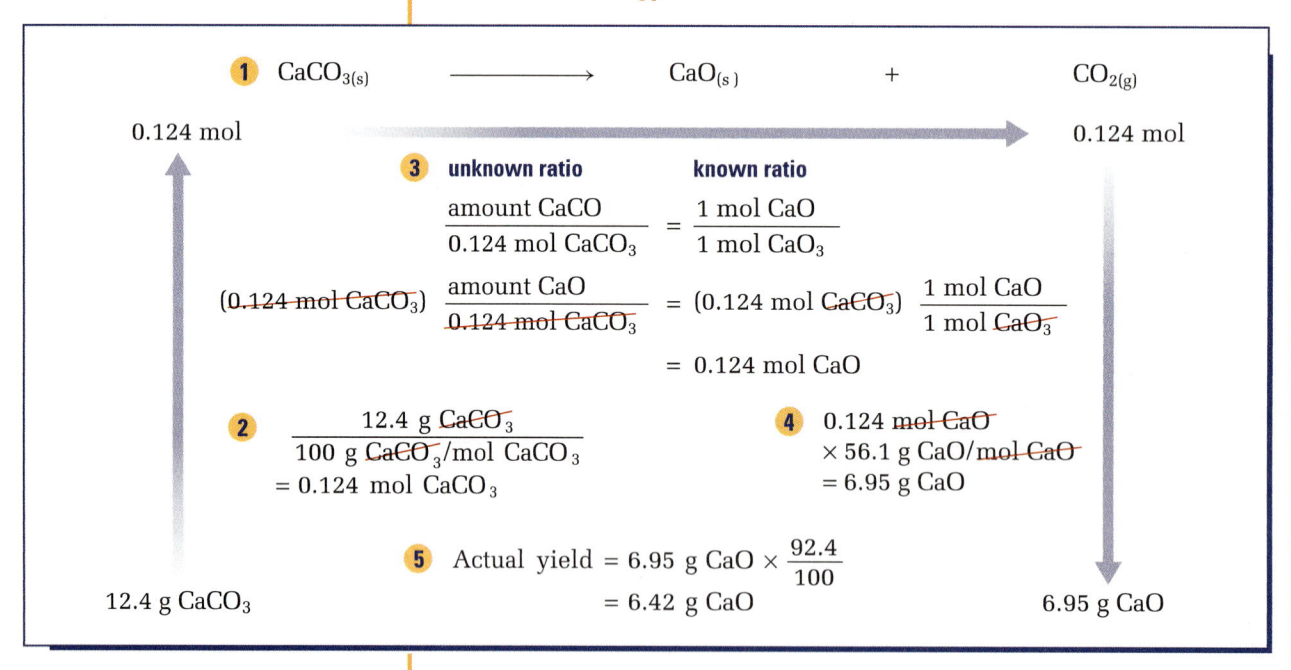

**1** $CaCO_{3(s)} \longrightarrow CaO_{(s)} + CO_{2(g)}$

0.124 mol → 0.124 mol

**3** unknown ratio = known ratio

$$\frac{\text{amount CaCO}}{0.124 \text{ mol CaCO}_3} = \frac{1 \text{ mol CaO}}{1 \text{ mol CaO}_3}$$

$$(0.124 \text{ mol CaCO}_3) \frac{\text{amount CaO}}{0.124 \text{ mol CaCO}_3} = (0.124 \text{ mol CaCO}_3) \frac{1 \text{ mol CaO}}{1 \text{ mol CaO}_3}$$

$$= 0.124 \text{ mol CaO}$$

**2** $\dfrac{12.4 \text{ g CaCO}_3}{100 \text{ g CaCO}_3/\text{mol CaCO}_3}$
= 0.124 mol CaCO$_3$

**4** 0.124 mol CaO
× 56.1 g CaO/mol CaO
= 6.95 g CaO

12.4 g CaCO$_3$          6.95 g CaO

**5** Actual yield = 6.95 g CaO × $\dfrac{92.4}{100}$
= 6.42 g CaO

**Check Your Solution**

92.5% of 6.95 g is about 6.4 g. The answer is reasonable.

## Practice Problems

**34.** The following reaction proceeds with a 70% yield.

$$C_6H_{6(\ell)} + HNO_{3(aq)} \rightarrow C_6H_5NO_{2(\ell)} + H_2O_{(\ell)}$$

Calculate the mass of $C_6H_5NO_2$ expected if 12.8 g of $C_6H_6$ reacts with excess $HNO_3$.

**35.** The reaction of toluene, $C_7H_8$, with potassium permanganate, $KMnO_4$, gives less than a 100% yield.

$$C_7H_{8(\ell)} + 2KMnO_{4(aq)} \rightarrow KC_7H_5O_{2(aq)} + 2MnO_{2(s)} + KOH_{(aq)} + H_2O_{(\ell)}$$

(a) 8.60 g of $C_7H_8$ is reacted with excess $KMnO_4$. What is the theoretical yield, in grams, of $KC_7H_5O_2$?

(b) If the percentage yield is 70.0%, what mass of $KC_7H_5O_2$ can be expected?

(c) What mass of $C_7H_8$ is needed to produce 13.4 g of $KC_7H_5O_2$, assuming a yield of 60%?

**36.** Marble is made primarily of calcium carbonate. When calcium carbonate reacts with hydrochloric acid, $HCl_{(aq)}$, it forms calcium chloride, carbon dioxide and water. If this reaction occurs with 81.5% yield, what volume of carbon dioxide at STP will be collected if 15.7 g of $CaCO_3$ is added to sufficient hydrochloric acid?

**37.** Mercury, in its elemental form or in a chemical compound is highly toxic. Water-soluble mercury compounds, such as mercury(II) nitrate, can be removed from industrial wastewater by adding sodium sulfide to the water, which forms a precipitate of mercury(II) sulfide, which can then be filtered out.

$$Hg(NO_3)_{2(aq)} + Na_2S_{(aq)} \rightarrow HgS_{(s)} + 2NaNO_{3(aq)}$$

If $3.45 \times 10^{23}$ formula units of $Hg(NO_3)_2$ react with excess $Na_2S$, what mass of HgS can be expected if this process occurs with 97.0% yield?

## Applications of Percentage Yield

The percentage yield of chemical reactions is extremely important in industrial chemistry and the pharmaceutical industry. For example, the synthesis of certain drugs involves many sequential chemical reactions. Often each reaction has a low percentage yield. This results in a tiny overall yield. Research chemists, who generally work with small quantities of reactants, may be satisfied with a poor yield. Chemical engineers, on the other hand, work with very large quantities. They may use hundreds or even thousands of kilograms of reactants! A difference of 1% in the yield of a reaction can translate into thousands of dollars.

The work of a chemist in a laboratory can be likened to making spaghetti for a family. The work of a chemical engineer, by contrast, is like making spaghetti for 10 000 people! You can learn more about chemical engineers in Careers in Chemistry on page 144. In Investigation 4–B you will determine the percentage yield of a reaction.

# Investigation 4-B

**SKILL FOCUS**
Predicting
Performing and recording
Analyzing and interpreting
Communicating results

## Determining the Percentage Yield of a Chemical Reaction

The percentage yield of a reaction is determined by numerous factors: The nature of the reaction itself, the conditions under which the reaction was carried out, and the nature of the reactants used.

In this investigation, you will determine the percentage yield of the following chemical reaction:

$$Fe_{(s)} + CuCl_{2(aq)} \rightarrow FeCl_{2(aq)} + Cu_{(s)}$$

You will use steel wool, since it is virtually pure iron.

### Question
What is the percentage yield of the reaction of iron and copper(II) chloride when steel wool and copper(II) chloride dihydrate are used as reactants?

### Predictions
Once you have determined the mass of the steel wool, calculate the mass of copper that will be produced, assuming the steel wool is 100% iron. Also assume the iron reacts *completely* with a solution containing excess $CuCl_2$. Then predict the percentage yield and actual yield, giving reasons for your prediction.

### Materials
1 beaker (250 mL)
Erlenmeyer flask
plastic funnel
retort stand
wash bottle
drying oven (if available)
stirring rod
ring clamp
filter paper
centigram electronic balance
distilled water

1.00 g–1.20 g rust-free, degreased steel wool
5.00 g copper chloride dihydrate, $CuCl_2 \cdot 2H_2O$
20 mL 1 mol/L hydrochloric acid, HCl

### Safety Precautions
If you get either $CuCl_{2(aq)}$ or $HCl_{(aq)}$ solution on your skin, flush with plenty of cold water.

### Procedure
1. Copy the table below into your notebook.

**Observations**

| | |
|---|---|
| Mass of filter paper | |
| Mass of steel wool | |
| Mass of filter paper and solid product | |

2. Place about 50 mL of distilled water in a 250 mL beaker. Add 5.00 g of copper(II) chloride dihydrate to the water. Stir to dissolve.

3. Determine the mass of your sample of steel wool. Record the mass in your table.

4. Add the steel wool to the copper(II) chloride solution in the beaker. Allow the mixture to sit until all the steel wool has reacted. The reaction could take up to 20 min.

5. While the reaction is proceeding, set up your filtration apparatus as shown on the following page. **Note:** Be sure to determine the mass of your piece of filter paper before folding and wetting it. Record the mass of the filter paper in your Observations table.

6. When you believe that the reaction is complete, carefully decant most the liquid in the beaker through the filter paper. Pouring the liquid down a stirring rod, as shown in the diagram, helps you to avoid losing any solid.

7. Pour the remaining liquid and solid through the filter paper.

8. Rinse the beaker and stirring rod several times with small quantities of water. Pour the rinse water through the filter paper. Ensure there is no solid product remaining in the beaker or on the stirring rod.

9. Rinse the filter paper and solid with about 10 mL of 1 mol/L $HCl_{(aq)}$. Then rinse the solid with water to remove the hydrochloric acid.

10. Repeat step 9 once.

11. After all the liquid has drained from the funnel, carefully remove your filter paper and place it on a labelled watch glass. Be careful not to lose any solid product.

12. Place the watch glass in a drying oven overnight. (If no drying oven is available, place the watch glass in a safe place for several days.) Return the Erlenmeyer flask and its contents to your teacher for proper disposal.

13. Determine the mass of the dried filter paper and product.

14. Bring the filter paper and product to your teacher for proper disposal.

## Analysis

1. (a) Using the mass of the iron (steel wool) you used, calculate the theoretical yield of copper, in grams.
   (b) How does the mass of the product you collected compare with the expected theoretical yield?
2. Based on the amount of iron that you used, prove that the 5.00 g of $CuCl_2 \cdot 2H_2O$ was the excess reactant.

## Conclusion

3. Calculate the percentage yield for this reaction.

## Applications

4. If your percentage yield was not 100%, suggest sources of error.
5. How would you attain an improved percentage yield if you performed this reaction again? Consider your technique and materials.
6. Consider the precision and accuracy of your results.
   (a) How precise was your determination of the maximum percentage yield of the reaction of iron with copper(II) chloride? Explain your answer.
   (b) Suggest how you could have improved the precision of your determination.
   (c) How accurate was your determination of the maximum percentage yield of the reaction of iron with copper(II) chloride? Explain your answer and list factors that affected the accuracy of your determination.
   (d) Suggest how you could have improved the accuracy of your determination.

---

**1** Fold a piece of fluted filter paper.

  (a) Fold the filter paper in half. 
  (b) Make creases in the half to divide it into eight sections of equal size.  
  (c) Flip the piece over. Make a fan shape by folding each section in the direction opposite to the previous direction. 
  (d) Open up the two halves. You have now "fluted" your filter paper.

**2** Place your fluted filter paper in the plastic funnel. Use your wash bottle to add a little distilled water to the centre of the filter paper so that it will stay in place.

**3** Set up the filtration apparatus as shown. The diagram also shows how to pour the liquid down a stirring rod to ensure no product is lost.

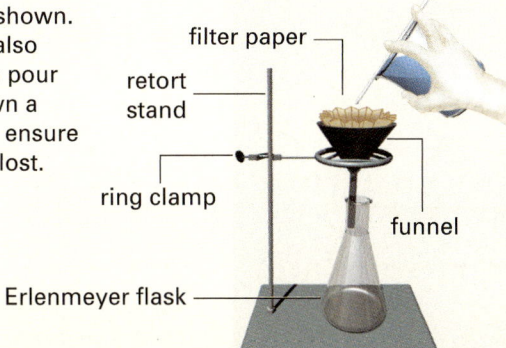

filter paper
retort stand
ring clamp
funnel
Erlenmeyer flask

# Careers in Chemistry

## Chemical Engineer

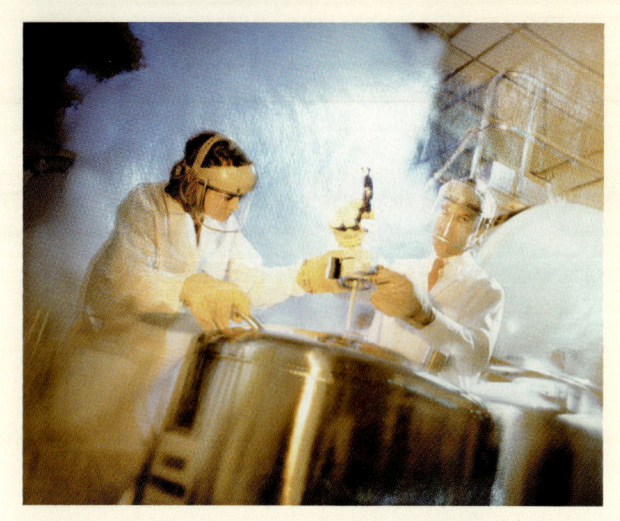

Chemical engineers are sometimes described as "universal engineers" because of their unique knowledge of math, physics, engineering, and chemistry. This broad knowledge allows them to work in a variety of areas, from designing paint factories to developing better tasting, more nutritious foods. Canadian chemical engineers are helping to lead the world in making cheap, long-lasting, and high-quality CDs and DVDs. In addition to designing and operating commercial plants, chemical engineers can be found in university labs, government agencies, and consulting firms.

### Producing More for Less

Once chemists have developed a product in a laboratory, it is up to chemical engineers to design a process to make the product in commercial quantities as efficiently as possible. "Scaling up" production is not just a matter of using larger beakers. Chemical engineers break down the chemical process into a series of smaller "unit operations" or processes and techniques. They use physics, chemistry, and complex mathematical models. For example, making liquid pharmaceutical products (such as syrups, solutions, and suspensions) on a large scale involves adding specific amounts of raw materials to large mixing tanks. Then the raw materials are heated to a set temperature and mixed at a set speed for a given amount of time. The final product is filtered and stored in holding tanks. Chemical engineers ensure that each process produces the maximum amount of product.

### Becoming a Chemical Engineer

To become a chemical engineer, you need a bachelor's degree in chemical engineering. Most provinces also require a Professional Engineer (P. Eng.) designation. Professional engineers must have at least four years of experience and must pass an examination. As well, they must commit to continuing their education to keep up with current developments. Chemical engineers must be able to work well with people and to communicate well.

### Make Career Connections

1. Discuss engineering studies and careers with working engineers, professors, and engineering students. Look for summer internship programs and job shadowing opportunities. Browse the Internet. Contact your provincial engineering association, engineering societies, and universities for more information.

2. Participate in National Engineering Week in Canada in March of each year. This is when postsecondary institutions, companies, science centres, and other organizations hold special events, including engineering contests and workshops.

## Impurities

Often impure reactants are the cause of a percentage yield of less than 100%. Impurities cause the mass data to be incorrect. For example, suppose that you have 1.00 g of sodium chloride and you want to carry out a reaction with it. You think that the sodium chloride may have absorbed some water, so you do not know exactly how much pure sodium chloride you have. If you calculate a theoretical yield for your reaction based on 1.00 g of sodium chloride, your actual yield will be less. There is not 1.00 g of sodium chloride in the sample.

# Chemistry Bulletin

**Science** | **Technology** | **Environment** | **Society**

## Nickel Mining at Voisey's Bay

In 1993, two Newfoundland prospectors chipped samples from an iron-stained rock outcrop at Voisey's Bay in eastern Labrador. When they saw brassy yellow veins of chalcopyrite ($CuFeS_2$) shooting through the dark-coloured host rock, they knew they were onto an important mineral discovery.

By July 1995, drilling samples had confirmed the presence of a major nickel-copper-cobalt sulphide ore body at the site. It has been estimated that the Voisey's Bay deposit may contain 150 million tonnes of ore grade material, making it one of the most economically significant geological discoveries in Canada in the last thirty years.

The chalcopyrite ($CuFeS_2$) or "fool's gold" that caught the prospectors' eyes makes up only about 8% of the massive bulk of the ore at Voisey's Bay. The remaining composition is approximately: 75% pyrrhotite (FeS); 12% pentlandite ($(Fe, Ni)_9S_8$); and 5% magnetite ($Fe_3O_4$). The nickel, copper, and cobalt for which the ore is prized occur in only trace amounts dispersed throughout the rock. Although these three metals are the main goal of mining operations, they are found in levels of: 2.83% nickel, 1.68% copper and 0.12% cobalt.

*The photo shows an open pit mine. Because the ore at Voisey's Bay is of high grade and can be extracted using open pit mines, the cost to extract the ore is relatively low.*

Nickel occurs in two different types of ore. *Laterite ores* (found in tropical regions) contain over 80% of the world's nickel resources, but sulphide ores (such as the one at Voisey's Bay) have provided more of the world's nickel. Part of the reason for this is the high amount of energy needed to process laterites and their relatively low yield of the desired metals. Maximum percentage yield of nickel from laterite is 80–85%, and typically much less. A low percentage yield can mean that large reserves of metal are left in the ground because it is not economical to mine and process them.

Improved technology can significantly influence a mining operation's profitability. Chemical engineers know that the theoretical yields are unlikely to be obtained in the real world. What is missing from the equation is the exact conditions needed to make a reaction occur. The only way to find out the importance of, say, maintaining a specific temperature or the influence of particle size, is by experimentation.

Experiments reveal how actual yield under given conditions compares with the theoretical yield. Experiments need to be carried out on a large, industrial scale rather than in the lab. Such experiments helped develop a new breed nickel refineries in the 1990s. The new refineries use a pressure acid leaching (PAL) process that has a much reduced cost over older methods. The process also increases the percentage yield of nickel produced from laterite ores.

### Making Connections

1. Describe an example of how a mining engineer might use knowledge of percentage yield in a report on a prospector's ore samples.

2. What do you think the notation (Fe, Ni) in $((Fe, Ni)_9S_8)$ signifies? Do some research to find out. Is it possible to determine the percentage composition of pentlandite based on its formula? Explain.

## Determining Percentage Purity

In the mining industry, metals are usually recovered in the form of an ore. An ore is a naturally occurring rock that contains a high concentration of one or more metals. Whether an ore can be profitably mined depends on several factors: the cost of mining and refining the ore (Figure 4.11), the price of the extracted metal, and the cost of any legal and environmental issues related to land use. The inaccurate chemical analysis of an ore sample can cost investors millions of dollars if the ore deposit does not yield what was expected.

The **percentage purity** of a sample describes what proportion, by mass, of the sample is composed of a specific compound or element. For example, suppose that a sample of gold has a percentage purity of 98%. This means that every 100 g of the sample contains 98 g of gold and 2 g of impurities.

You can apply your knowledge of stoichiometry and percentage yield to solve problems related to percentage purity.

**Figure 6.10** Copper is removed from mines like this one in the form of an ore. There must be sufficient copper in the ore to make the mine economically viable.

### Sample Problem
### Finding Percentage Purity

#### Problem
Iron pyrite, $FeS_2$, is known as "fool's gold" because it looks similar to gold. Suppose that you have a 13.9 g sample of *impure* iron pyrite. (The sample contains a non-reactive impurity.) You heat the sample in air to produce iron(III) oxide, $Fe_2O_3$, and sulfur dioxide, $SO_2$.

$$4FeS_{2(s)} + 11O_{2(g)} \rightarrow 2Fe_2O_{3(s)} + 8SO_{2(g)}$$

If you obtain 8.02 g of iron(III) oxide, what was the percentage of iron pyrite in the original sample? Assume that the reaction proceeds to completion. That is, all the available iron pyrite reacts completely.

#### What Is Required?
You need to determine the percentage purity of the iron pyrite sample.

#### What Is Given?
The mass of $Fe_2O_3$ is 8.02 g. The reaction proceeds to completion. You can assume that sufficient oxygen is present.

#### Plan Your Strategy

**Steps 1–4** Use your stoichiometry problem-solving skills to find the mass of $FeS_2$ expected to have produced 8.02 g $Fe_2O_3$.

**Step 5** Determine percentage purity of the $FeS_2$ using the following formula:
$$\frac{\text{theoretical mass (g)}}{\text{sample size (g)}} \times 100\%$$

### Act on Your Strategy

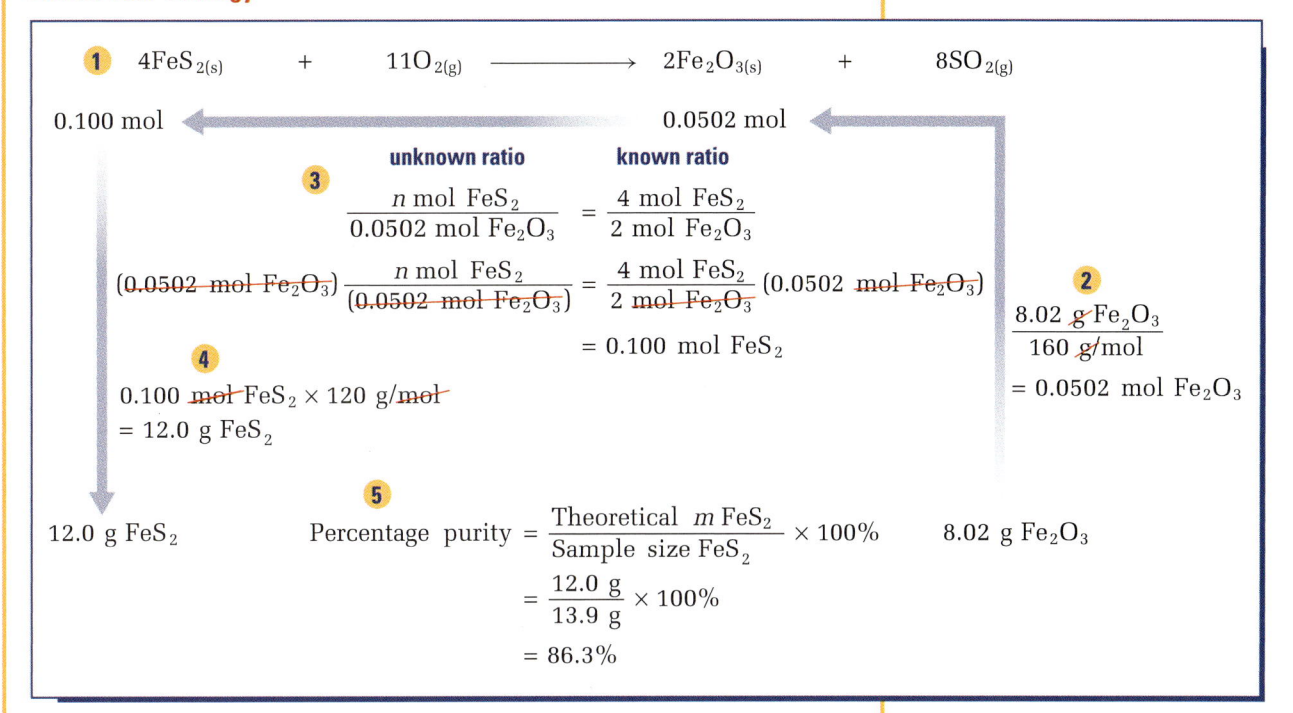

**1)** $4FeS_{2(s)} + 11O_{2(g)} \longrightarrow 2Fe_2O_{3(s)} + 8SO_{2(g)}$

0.100 mol ⟵ 0.0502 mol ⟵

**unknown ratio** | **known ratio**

**3)** $\dfrac{n \text{ mol } FeS_2}{0.0502 \text{ mol } Fe_2O_3} = \dfrac{4 \text{ mol } FeS_2}{2 \text{ mol } Fe_2O_3}$

$(0.0502 \text{ mol } Fe_2O_3)\dfrac{n \text{ mol } FeS_2}{(0.0502 \text{ mol } Fe_2O_3)} = \dfrac{4 \text{ mol } FeS_2}{2 \text{ mol } Fe_2O_3}(0.0502 \text{ mol } Fe_2O_3)$

$= 0.100 \text{ mol } FeS_2$

**2)** $\dfrac{8.02 \text{ g } Fe_2O_3}{160 \text{ g/mol}} = 0.0502 \text{ mol } Fe_2O_3$

**4)** $0.100 \text{ mol } FeS_2 \times 120 \text{ g/mol}$
$= 12.0 \text{ g } FeS_2$

12.0 g FeS₂    **5)** Percentage purity $= \dfrac{\text{Theoretical } m \text{ FeS}_2}{\text{Sample size FeS}_2} \times 100\%$    8.02 g Fe₂O₃

$= \dfrac{12.0 \text{ g}}{13.9 \text{ g}} \times 100\%$

$= 86.3\%$

Therefore, the percentage purity of the iron pyrite is 86.3%.

### Check Your Solution

The units are correct. The molar mass of iron pyrite is 3/4 the molar mass of iron(III) oxide. Mutiplying this ratio by the mole ratio of iron pyrite to iron(III) oxide (4/2) and 8 g gives 12 g. The answer is reasonable.

## Practice Problems

**38.** An impure sample of silver nitrate, $AgNO_3$, has a mass 0.340 g. It is dissolved in water and then treated with excess hydrochloric acid, $HCl_{(aq)}$. This results in the formation of a precipitate of silver chloride, AgCl.

$$AgNO_{3(aq)} + HCl_{(aq)} \rightarrow AgCl_{(s)} + HNO_{3(aq)}$$

The silver chloride is filtered, and any remaining hydrogen chloride is washed away. Then the silver chloride is dried. If the mass of the dry silver chloride is measured to be 0.213 g, what mass of silver nitrate was contained in the original (impure) sample?

**39.** Copper metal is mined as one of several copper-containing ores. One of these ores contains copper in the form of malachite. Malachite exists as a double salt, $Cu(OH)_2 \cdot CuCO_3$. It can be thermally decomposed at 200°C to yield copper(II) oxide, carbon dioxide gas, and water vapour.

$$Cu(OH)_2 \cdot CuCO_{3(s)} \rightarrow 2CuO_{(s)} + CO_{2(g)} + H_2O_{(g)}$$

**(a)** 5.000 kg of malachite ore, containing 5.20% malachite, $Cu(OH)_2 \cdot CuCO_3$, is thermally decomposed. Calculate the mass of copper(II) oxide that is formed.

*Continued...*

(b) Suppose that the reaction had a 78.0% yield, due to incomplete decomposition. How many grams of CuO would be produced?

40. Ethylene oxide, $C_2H_4O$, is a multi-purpose industrial chemical used, among other things, as a rocket propellant. It can be prepared by reacting ethylene bromohydrin, $C_2H_5OBr$, with sodium hydroxide.

$$C_2H_5OBr_{(aq)} + NaOH_{(aq)} \rightarrow C_2H_4O_{(aq)} + NaBr_{(aq)} + H_2O_{(\ell)}$$

If this reaction proceeds with an 89% yield, what mass of $C_2H_4O$ can be obtained when $3.61 \times 10^{23}$ molecules of $C_2H_5OBr$ react with excess sodium hydroxide?

## Section Summary

In this section, you have learned how the amount of products formed by experiment relates to the theoretical yield predicted by stoichiometry. You have learned about many factors that affect actual yield, including the nature of the reaction, experimental design and execution, and the purity of the reactants. Usually, when you are performing an experiment in a laboratory, you want to maximize your percentage yield. To do this, you need to be careful not to contaminate your reactants or lose any products. Either might affect your actual yield.

## Section Review

**1** When calculating the percentage yield of a reaction, what units should you use: grams, moles, or number of particles? Explain.

**2** Methyl salicylate, otherwise known as oil of wintergreen, is produced by the wintergreen plant. It can also be synthesized by heating salicylic acid, $C_7H_6O_3$, with methanol, $CH_3OH$.

$$C_7H_6O_{3(s)} + CH_3OH_{(\ell)} \rightarrow C_8H_8O_{3(\ell)} + H_2O_{(\ell)}$$

A chemist reacts 3.50 g of salicylic acid with excess methanol. She calculates the theoretical yield of methyl salicylate to be 3.86 g. If 2.84 g of methyl salicylate are recovered, what is the percentage yield?

**3** Unbeknownst to a chemist, the limiting reactant in a certain chemical reaction is impure. How will this affect the percentage yield of the reaction? Explain.

**4** You have a sample of copper that is impure, and you wish to determine its purity. You have some silver nitrate, $AgNO_3$, at your disposal. You also have some copper that you know is 100.0% pure.

(a) Design an experiment to determine the purity of the copper sample.

(b) Even with pure copper, the reaction may not proceed with 100% yield. How will you address this issue?

# CHAPTER 4 Review

## Reflecting on Chapter 4
Summarize this chapter in the format of your choice. Here are a few ideas to use as guidelines:
- Use the coefficients of a balanced chemical equation to determine the mole ratios between reactants and products.
- Predict quantities required or produced in a chemical reaction.
- Calculate the limiting reactant in cases where the amount of various reactants was given.
- Calculate the percentage yield of a chemical reaction based on the amount of product(s) obtained relative to what was predicted by stoichiometry.
- Use the percentage yield of a reaction to predict the amount of product(s) formed.
- Determine the percentage purity of a reactant based on the actual yield of a reaction.
- Distinguish between precision and accuracy in the context of carrying out a filtration to determine percentage yield.

## Reviewing Key Terms
For each of the following terms, write a sentence that shows your understanding of its meaning.

actual yield
excess reactant
mole ratios
percentage yield
stoichiometric coefficients
competing reaction
limiting reactant
percentage purity
stoichiometric amounts
stoichiometry
theoretical yield

## Knowledge/Understanding
1. Explain the different interpretations of the coefficients in a balanced chemical equation.
2. (a) State the law of conservation of mass.
   (b) Explain how the law of conservation of mass relates to balanced chemical equations.
   (c) Explain how the law of conservation of mass allows chemists to make accurate predictions using balanced chemical equations.
3. In what cases would it not be necessary to determine the limiting reactant before beginning any stoichiometric calculations?
4. (a) State the relationship between theoretical yield, actual yield, and percentage yield.
   (b) Use a sample calculation to demonstrate the relationship among the three terms.
   (c) Suggest three factors that could affect the percentage yield of a reaction.
5. A student is trying to determine the mass of aluminum oxide that is produced when aluminum reacts with excess oxygen.
$$4Al_{(s)} + 3O_{2(g)} \rightarrow 2Al_2O_{3(s)}$$
The student states that 4 g of aluminum reacts with 3 g of oxygen to produce 2 g of aluminum oxide. Is the student's reasoning correct? Explain your answer.

## Inquiry
6. A freshly exposed aluminum surface reacts with oxygen to form a tough coating of aluminum oxide. The aluminum oxide protects the metal from further corrosion.
$$4Al_{(s)} + 3O_{2(g)} \rightarrow 2Al_2O_{3(s)}$$
How many grams of oxygen are needed to react with 0.400 mol of aluminum?

7. Calcium metal reacts with chlorine gas to produce calcium chloride.
$$Ca_{(s)} + Cl_{2(g)} \rightarrow CaCl_{2(s)}$$
How many formula units of $CaCl_2$ are expected from 5.3 g of calcium and excess chlorine?

8. Propane is a gas at room temperature, but it exists as a liquid under pressure in a propane tank. It reacts with oxygen in the air to form carbon dioxide and water vapour.
$$C_3H_{8(\ell)} + 5O_{2(g)} \rightarrow 3CO_{2(g)} + 4H_2O_{(g)}$$
What mass of carbon dioxide gas is expected when 97.5 g of propane reacts with sufficient oxygen?

9. Powdered zinc and sulfur react in an extremely rapid, exothermic reaction. The zinc sulfide that is formed can be used in the phosphor coating on the inside of a television tube.
$$Zn_{(s)} + S_{(s)} \rightarrow ZnS_{(s)}$$
A 6.00 g sample of Zn is allowed to react with 3.35 g of S.
(a) Determine the limiting reactant.
(b) Calculate the mass of ZnS expected.
(c) How many grams of the excess reactant will remain after the reaction?

10. Titanium(IV) chloride reacts violently with water vapour to produce titanium(IV) oxide and hydrogen chloride gas. Titanium(IV) oxide, when finely powdered, is extensively used in paint as a white pigment.

Answers to questions highlighted in red type are provided in Appendix A.

$TiCl_{4(s)} + 2H_2O_{(\ell)} \rightarrow TiO_{2(s)} + 4HCl_{(g)}$

The reaction has been used to create smoke screens. In moist air, the TiCl₄ reacts to produce a thick smoke of suspended TiO₂ particles. What mass of TiO₂ can be expected when 85.6 g of TiCl₄ is reacted with excess water vapour?

11. Silver reacts with hydrogen sulfide gas, which is present in the air. (Hydrogen sulfide has the odour of rotten eggs.) The silver sulfide, Ag₂S, that is produced forms a black tarnish on the silver.

$4Ag_{(s)} + 2H_2S_{(g)} + O_{2(g)} \rightarrow 2Ag_2S_{(s)} + 2H_2O_{(g)}$

How many grams of silver sulfide are formed when 1.90 g of silver reacts with 0.280 g of hydrogen sulfide and 0.160 g of oxygen?

12. 20.8 g of calcium phosphate, Ca₃(PO₄)₂, 13.3 g of silicon dioxide, SiO₂, and 3.90 g of carbon react according to the following equation:

$2Ca_3(PO_4)_{2(s)} + 6SiO_{2(s)} + 10C_{(s)} \rightarrow$
$\qquad P_{4(s)} + 6CaSiO_{3(s)} + 10CO_{(g)}$

Determine the mass of calcium silicate, CaSiO₃, that is produced.

13. 1.56 g of As₂S₃, 0.140 g of H₂O, 1.23 g of HNO₃, and 3.50 g of NaNO₃ are reacted according to the equation below:

$3As_2S_{3(s)} + 4H_2O_{(\ell)} + 10HNO_{3(aq)} + 18NaNO_{3(aq)}$
$\qquad \rightarrow 9Na_2SO_{4(aq)} + 6H_3AsO_{4(aq)} + 28NO_{(g)}$

What mass of H₃AsO₄ is produced?

14. 2.85 × 10² g of pentane, C₅H₁₂, reacts with 3.00 g of oxygen gas according to the following equation: $C_5H_{12(\ell)} + 8O_{2(g)} \rightarrow 5CO_{2(g)} + 6H_2O_{(\ell)}$ What mass of carbon dioxide gas is produced?

15. Methanol has the potential to be used as an alternative fuel. It burns in the presence of oxygen to produce carbon dioxide and water.

$CH_3OH_{(\ell)} + O_{2(g)} \rightarrow CO_{2(g)} + H_2O_{(g)}$

 (a) Balance this equation.
 (b) 10 L of oxygen is completely consumed at STP. What volume of CO₂ at STP is produced?
 (c) What mass of methanol is consumed in this reaction?

16. A student wants to prepare carbon dioxide using sodium carbonate and dilute hydrochloric acid.

$Na_2CO_{3(s)} + 2HCl_{(aq)} \rightarrow 2NaCl_{(aq)} + CO_{2(g)} + H_2O_{(\ell)}$

The student produced 0.919 L of carbon dioxide at STP. What mass of sodium carbonate did the student use?

17. A scientist makes hydrogen gas in the laboratory by reacting calcium metal with an excess of hydrochloric acid.

$Ca_{(s)} + 2HCl_{(aq)} \rightarrow CaCl_{2(aq)} + H_{2(g)}$

A scientist reacts 5.00 g of calcium with excess hydrochloric acid. What volume of hydrogen at STP is produced?

18. Ammonia may be produced by the following reaction:

$CH_{4(g)} + H_2O_{(\ell)} + N_2O_{(g)} \rightarrow 2NH_{3(g)} + CO_{2(g)}$

500.0 g of methane reacts with excess H₂O and N₂O. What volume of ammonia gas is produced at STP?

19. Hydrochloric acid dissolves limestone, as shown in the following chemical equation:

$CaCO_{3(s)} + 2HCl_{(aq)} \rightarrow CaCl_{2(aq)} + CO_{2(g)} + H_2O_{(\ell)}$

12.0 g of CaCO₃ reacts with 0.138 mol HCl. At 101.3 kPa and 0.00°C, what volume of carbon dioxide at STP is produced?

20. Silica (also called silicon dioxide), along with other silicates, makes up about 95% of Earth's crust—the outermost layer of rocks and soil. Silicon dioxide is also used to manufacture transistors. Silica reacts with hydrofluoric acid to produce silicon tetrafluoride and water vapour.

$SiO_{2(s)} + 4HF_{(aq)} \rightarrow SiF_{4(g)} + 2H_2O_{(g)}$

 (a) 12.2 g of SiO₂ is reacted with an excess of HF. What is the theoretical yield, in grams, of H₂O?
 (b) If the actual yield of water is 2.50 g, what is the percentage yield of the reaction?
 (c) Assuming the yield obtained in part (b), what mass of SiF₄ is formed?

21. An impure sample of barium chloride, BaCl₂, with a mass of 4.36 g, is added to an aqueous solution of sodium sulfate, Na₂SO₄.

$BaCl_{2(s)} + Na_2SO_{4(aq)} \rightarrow BaSO_{4(s)} + 2NaCl_{(aq)}$

After the reaction is complete, the solid barium sulfate, BaSO₄, is filtered and dried. Its mass is found to be 2.62 g. What is the percentage purity of the original barium chloride?

22. Benzene reacts with bromine to form bromobenzene, C₆H₅Br.

$C_6H_{6(\ell)} + Br_{2(\ell)} \rightarrow C_6H_5Br_{(\ell)} + HBr_{(g)}$

 (a) What is the maximum amount of C₆H₅Br that can be formed from the reaction of 7.50 g of C₆H₆ with excess Br₂?
 (b) A competing reaction is the formation of dibromobenzene, C₆H₄Br₂.

$C_6H_{6(\ell)} + 2Br_{2(\ell)} \rightarrow C_6H_4Br_{2(\ell)} + 2HBr_{(g)}$
If 1.25 g of $C_6H_4Br_2$ was formed by the competing reaction, how much $C_6H_6$ was *not* converted to $C_6H_5Br$?
(c) Based on your answer to part (b), what was the actual yield of $C_6H_5Br$? Assume that all the $C_6H_5Br$ that formed was collected.
(d) Calculate the percentage yield of $C_6H_5Br$.

## Communication

23. Develop a new analogy for the concept of limiting and excess reactant.
24. Examine the balanced chemical "equation".
$$2A + B \rightarrow 3C + D$$
Using a concept map, explain how to calculate the number of grams of C that can be obtained when a given mass of A reacts with a certain number of molecules of B. Assume that you know the molar mass of A and C. Include proper units. Assume that A is limiting, but don't forget to show how to determine the limiting reactant.

## Making Connections

25. You must remove mercury ions present as mercury(II) nitrate in the waste water of an industrial facility. You have decided to use sodium sulfide in the reaction below. Write a short essay that addresses the following points. Include a well-organized set of calculations where appropriate.
$Hg(NO_3)_{2(aq)} + Na_2S_{(aq)} \rightarrow HgS_{(s)} + 2NaNO_{3(aq)}$
(a) Explain why the chemical reaction above can be used to remove mercury ions from the waste water. What laboratory technique must be used in order that this reaction is as effective as possible for removing mercury from the waste stream?
(b) Why is mercury(II) sulfide less of an environmental concern than mercury(II) nitrate?
(c) What assumptions are being made regarding the toxicity of sodium sulfide and sodium nitrate relative to either mercury(II) nitrate or mercury(II) sulfide?
(d) Every litre of waste water contains approximately 0.03 g of $Hg(NO_3)_2$. How many kg of $Na_2S$ will be required to remove the soluble mercury ions from 10 000 L of waste water?
(e) What factors would a company need to consider in adopting any method of cleaning its wastewater?

26. Complex carbohydrates are starches that your body can convert to glucose, a type of sugar. Simple carbohydrate foods contain glucose, ready for immediate use by the human body. Breathing and burning glucose, $C_6H_{12}O_6$, produces energy in a jogger's muscles, according to the following unbalanced equation:
$C_6H_{12}O_{6(aq)} + O_{2(g)} \rightarrow CO_{2(g)} + H_2O_{(g)}$
Just before going on a winter run, Myri eats two oranges. The oranges give her body 25 g of glucose to make energy. The temperature outside is 0.00°C, and the atmospheric pressure is 101.3 kPa. Although 21% (by volume) of the air Myri breathes in is oxygen, she breathes out about 16% of this oxygen. (In other words, she only uses about 5%.)
(a) How many litres of air does Myri breathe in while running to burn up the glucose she consumed?
(b) How many litres of carbon dioxide does she produce?

### Answers to Practice Problems and Short Answers to Section Review Questions

**Practice Problems:** 1.(a) 2:1:2 (b) 50 (c) 4956 (d) $1.20 \times 10^{24}$
2.(a) 2 (b) 150 (c) $1.806 \times 10^{24}$ (d) $1.204 \times 10^{24}$
3.(a) $3.4 \times 10^{25}$ (b) $6.7 \times 10^{25}$ 4. 7.5 mol 5.(a) 1.8 mol
(b) 37.5 mol 6.(a) 48.7 mol (b) 1.20 mol 7.(a) $8.3 \times 10^{24}$
(b) $4.2 \times 10^{24}$ 8.(a) 7.47 mol (b) 7.19 mol
9.(a) $4.68 \times 10^{-2}$ mol (b) 0.187 mol 10.(a) 0.708 mol
(b) 1.06 mol 11. 9.28 g 12. 137 g 13. 4.63 g 14. 0.284 L
15. 97.5 g 16. 7.35 mL 17. 11.7 L 18. 20.7 g 19.(a) 124 g
(b) $1.14 \times 10^{24}$ 20.(a) 120 g (b) $1.49 \times 10^{21}$ 21.(a) $2.39 \times 10^{22}$
(b) $1.45 \times 10^{24}$ (c) $1.21 \times 10^{22}$ 22.(a) $1.50 \times 10^{24}$ (b) 266 L
23. $CuCl_2$ 24. $CaF_2$ 25. $C_3H_6$ 26. HCl 27.(a) $ClO_2$ (b) 74.11 g
(c) $1.056 \times 10^{23}$ molecules 28.(a) $H_2O_2$ (b) 63.6 g (c) 104 g
29. $4.23 \times 10^4$ g 30.(a) 0.446 g (b) $4.79 \times 10^{21}$ molecules
(c) $F_2$, 24.0 g 31.(a) 74.4 g (b) 63.7% 32.(a) 31.3 g (b) 95.2%
33. 26.7% 34. 14.1 g 35.(a) 14.9 g (b) 10.5 g (c) 12.8 g
36. 2.87 L 37. 129 g 38. 0.253 g 39.(a) 188 g (b) 147 g
40. 23.5 g
**Section Review: 4.1:** 4.(a) $S + O_2 \rightarrow SO_2$
(b) $2S + 3O_2 \rightarrow 2SO_3$ (c) 1.5 mol (d) 48.0 g
5.(a) 1:5:3:4 (b) 2.50 mol (c) $6.02 \times 10^{24}$ (d) 202 L
6.(a) 50.0 g (b) $4.48 \times 10^{22}$
7.(a) $Pb(NO_3)_{2(aq)} + 2NaI_{(aq)} \rightarrow PbI_{2(s)} + 2NaNO_{3(aq)}$
(b) 1.03 g **4.2:** 2.(a) oxygen 3. 13.5 L 4. 8.04 g 5. 5.34 g
6. 22.2 g 7.(a) $Zn_{(s)} + CuCl_{2(aq)} \rightarrow ZnCl_{2(aq)} + Cu_{(s)}$
(b) zinc gone (c) zinc (d) less than 1.52 g Zn **4.3:** 2. 73.6%

# UNIT 1 Review

## Knowledge/Understanding

### True/False

In your notebook, indicate whether each statement is true or false. If a statement is false, rewrite it to make it true.

1. The molecular formula of a compound is the same as its empirical formula.
2. A 2.02 g sample of hydrogen, $H_2$, contains the same number of molecules as 32.0 g of oxygen, $O_2$.
3. The average atomic mass of an element is equal to the mass of its most abundant isotope.
4. The numerical value of the molar mass of a compound (expressed in atomic mass units) is the same as its molar mass (expressed in grams).
5. The fundamental unit for chemical quantity is the gram.
6. The mass of 1.00 mol of any chemical compound is always the same.
7. 1.00 mol of any chemical compound or element contains $6.02 \times 10^{23}$ particles.
8. The value of the Avogadro constant depends on temperature.
9. The empirical formula of an unknown compound must be determined by experiment.
10. The actual yield of most chemical reactions is less than 100%.
11. The theoretical yield of a chemical reaction must be determined by experiment.
12. Stoichiometric calculations are used to determine the products of a chemical reaction.

### Multiple Choice

In your notebook, write the letter for the best answer to each question.

13. The number of molecules in 2.0 mol of nitrogen gas, $N_{2(g)}$, is
    (a) $1.8 \times 10^{24}$
    (b) $2.4 \times 10^{23}$
    (c) $1.2 \times 10^{24}$
    (d) $1.2 \times 10^{23}$
    (e) $4.0 \times 10^{23}$

14. The molar mass of a compound with the empirical formula $CH_2O$ has a mass of approximately 121 g. What is the molecular formula of the compound?
    (a) $C_4H_8O_4$
    (b) $C_2H_6O_2$
    (c) $C_3H_3O_6$
    (d) $C_3H_6O_3$
    (e) $CH_6O$

15. Read the following statements about balancing chemical equations. Which of these statements is true?
    (a) To be balanced, an equation must have the same number of moles on the left side and the right side.
    (b) A chemical formula may be altered in order to balance a chemical equation.
    (c) To be balanced, a chemical equation must have the same number of each type of atom on both sides.
    (d) It is unacceptable to use fractional coefficients when balancing a chemical equation.
    (e) The coefficients represent the mass of reactants and products.

16. What is the molar mass of ammonium dichromate, $(NH_4)_2Cr_2O_7$?
    (a) 248 g/mol
    (b) 234 g/mol
    (c) 200 g/mol
    (d) 252 g/mol
    (e) 200 g/mol

17. A sample of benzene, $C_6H_6$, contains $3.0 \times 10^{23}$ molecules of benzene. How many atoms are in the sample?
    (a) $36 \times 10^{24}$
    (b) $1.8 \times 10^{23}$
    (c) $3.6 \times 10^{24}$
    (d) $2.5 \times 10^{22}$
    (e) $3.0 \times 10^{23}$

18. What is the molar mass of zinc sulfate heptahydrate, $ZnSO_4 \cdot 7H_2O$?
    (a) 161 g/mol
    (b) 288 g/mol
    (c) 182 g/mol
    (d) 240 g/mol
    (e) 312 g/mol

Answers to questions highlighted in red type are provided in Appendix A.

19. The molecular formula of citric acid monohydrate is $C_6H_8O_7 \cdot H_2O$. Its molar mass is as follows:
    (a) 192 g/mol
    (b) 210 g/mol
    (c) 188 g/mol
    (d) 206 g/mol
    (e) 120 g/mol

20. The relative mass of one isotope of sulfur is 31.9721 u. Its abundance is 95.02%. Naturally occurring elemental sulfur has a relative atomic mass of 32.066 u. The mass number of the one other isotope of sulfur is
    (a) 31
    (b) 32
    (c) 33
    (d) 34
    (e) 35

21. A sample of ethane, $C_2H_{6(g)}$, has a volume of 6.9 L at STP. It contains the same number of *atoms* as
    (a) 23.0 g of sodium, Na
    (b) 32.0 g of oxygen, $O_2$
    (c) 39.36 g of ozone, $O_3$
    (d) 30.0 g of formaldehyde, $CH_2O$
    (e) 14.0 g of nitrogen gas, $N_2$

22. A sample of ozone, $O_3$, has a mass of 48.0 g. It contains the same number of atoms as
    (a) 58.7 g of nickel
    (b) 27.0 g of aluminum
    (c) 38.0 g of fluorine
    (d) 3.02 g of hydrogen
    (e) 32.0 g of oxygen

23. Which substance contains $9.03 \times 10^{23}$ atoms?
    (a) 16.0 g of oxygen, $O_2$
    (b) 4.00 g of helium, He
    (c) 28.0 g of nitrogen, $N_2$
    (d) 22.0 g of carbon dioxide, $CO_2$
    (e) 8.0 g of methane, $CH_4$

24. Examine the following formulas. Which formula is an empirical formula?
    (a) $C_2H_4$
    (b) $C_6H_6$
    (c) $C_2H_2$
    (d) $H_2O_2$
    (e) $Na_2Cr_2O_7$

25. A sample of sulfur trioxide, $SO_{3(g)}$, has a volume of 5.6 L at STP. How many moles are in the sample?
    (a) 0.20
    (b) 0.25
    (c) 0.50
    (d) 0.75
    (e) 0.80

26. How many molecules are in 1.00 mg of glucose, $C_6H_{12}O_6$?
    (a) $2.18 \times 10^{18}$
    (b) $3.34 \times 10^{18}$
    (c) $2.18 \times 10^{21}$
    (d) $3.34 \times 10^{21}$
    (e) $3.34 \times 10^{20}$

27. A sample that contains carbon, hydrogen, and oxygen is analyzed in a carbon-hydrogen combustion analyzer. All the oxygen in the sample is
    (a) converted to the oxygen in carbon dioxide
    (b) converted to oxygen in water
    (c) mixed with the excess oxygen used to combust the sample
    (d) converted to oxygen in carbon dioxide and/or water
    (e) both (c) and (d)

28. A compound that contains carbon, hydrogen, and oxygen is going to be analyzed in a carbon-hydrogen combustion analyzer. Before beginning the analysis, which of the following steps must be carried out?
    I. Find the mass of the unknown sample.
    II. Add the precise amount of oxygen that is needed for combustion.
    III. Find the mass of the carbon dioxide and water absorbers.
    (a) I only
    (b) I and II only
    (c) I, II, and III
    (d) I and III only
    (e) none of the above

## Short Answer

29. Answer the following questions, which are related to the concept of the mole.
    (a) How many $N_2$ molecules are in a 1.00 mol sample of $N_2$? How many N atoms are in this sample?
    (b) How many $PO_4^{3-}$ ions are in 2.5 mol of $Ca_3(PO_4)_2$?
    (c) How many O atoms are in 0.47 mol of $Ca_3(PO_4)_2$?

30. Explain how a balanced chemical equation follows the law of conservation of mass. Use an example to illustrate your explanation.

31. List all the information that can be obtained from a balanced chemical equation.

32. Answer the following questions, which are related to the limiting reactant.
    (a) Explain the concept of the limiting reactant. Use a real-life analogy that is not used in this textbook.
    (b) What is the opposite of a limiting reactant?
    (c) Explain why, in many chemical reactions, the reactants are not present in stoichiometric amounts.

33. Consider a 7.35 g sample of propane, $C_3H_8$.
    (a) How many moles of propane are in this sample?
    (b) How many molecules of propane are in this sample?
    (c) How many atoms of carbon are in this sample?

34. How many atoms are in 10.0 g of white phosphorus, $P_4$?

35. A 2.00 g sample of the mineral troegerite, $(UO_2)_3(AsO_4)_2 \cdot 12H_2O$, has $1.38 \times 10^{21}$ uranium atoms. How many oxygen atoms are present in 2.00 g of troegerite?

36. Fuels that contain hydrogen can be classified according to their mass percent of hydrogen. Which of the following compounds has the greatest mass percent of hydrogen: ethanol, $C_2H_5OH$, or cetyl palmitate, $C_{32}H_{64}O_2$? Explain your answer.

37. Methyl tertiary butyl ether, or MTBE, is currently used as an octane booster in gasoline. It has replaced the environmentally unsound tetraethyl lead. MTBE has the formula $C_5H_{12}O$. What is the percentage composition of each element in MTBE?

38. Ammonia can be produced in the laboratory by heating ammonium chloride with calcium hydroxide.
    $2NH_4Cl_{(s)} + Ca(OH)_{2(s)} \rightarrow CaCl_{2(s)} + 2NH_{3(g)} + 2H_2O_{(g)}$
    8.93 g of ammonium chloride is heated with 7.48 g of calcium hydroxide. What volume of ammonia, $NH_3$, can be expected at STP? Assume that the reaction has 100% yield.

## Inquiry

39. The chemical equation below describes what happens when a match is struck against a rough surface to produce light and heat.
    $P_4S_{3(s)} + O_{2(g)} \rightarrow P_4O_{10(g)} + SO_{2(g)}$
    (a) Balance this chemical equation.
    (b) If 5.3 L of oxygen gas at STP were consumed, what volume of sulfur dioxide at STP would be produced?
    (c) What mass of $P_4S_{3(s)}$ would be consumed in the same reaction described in (b)?

40. An anesthetic used in hospitals after World War II was made up of 64.8% carbon, 13.67% hydrogen, and 21.59% oxygen. It was found that a 5.0 L sample of this anesthetic had a mass of 16.7 g at STP. What is the molecular formula of this gas?

41. Design an experiment to determine the value of x in a hydrate of sodium thiosulfate, $Na_2S_2O_3 \cdot xH_2O$. Include an outline of your procedure. Describe the data that you need to collect. What assumptions do you need to make?

42. Design an experiment to determine the mole-to-mole ratio of lead(II) nitrate, $Pb(NO_3)_2$, to potassium iodide, KI, in the reaction:
    $Pb(NO_3)_{2(aq)} + KI_{(aq)} \rightarrow PbI_{2(s)} + KNO_{3(aq)}$
    Assume that you have solutions of lead(II) nitrate and potassium iodide. Both of these solutions contain 0.0010 mol of solute per 10 mL of solution.

43. The following reaction can be used to obtain lead(II) chloride, $PbCl_2$. Lead(II) chloride is moderately soluble in warm water.
    $Pb(NO_3)_{2(aq)} + 2NaCl_{(aq)} \rightarrow PbCl_{2(s)} + 2NaNO_{3(aq)}$

Explain why carrying out this reaction in a warm aqueous solution is unlikely to produce a 100% yield of lead(II) chloride.

44. Imagine that you are given a sheet of aluminum foil that measures 10.0 cm × 10.0 cm. It has a mass of 0.40 g.
    (a) The density of aluminum is 2.70 g/cm³. Determine the thickness of the aluminum foil, in millimeters.
    (b) Using any of the above information, determine the radius of an aluminum atom, in nanometers. Assume that each aluminum atom is cube-shaped.
    (c) How will your answer to part (b) change if you assume that each aluminum atom is spherical?
    (d) What question(s) do your answers to parts (b) and (c) raise?

45. Consider the double displacement reaction below.
    $CaCl_{2(aq)} + Na_2SO_{4(aq)} \rightarrow CaCO_{3(s)} + 2NaCl_{(aq)}$
    (a) Design an experiment to determine the percentage yield of this reaction. Clearly indicate the measurements that need to be taken, along with suggested amounts.
    (b) How could the skills of a chemist influence the outcome of this experiment?

## Communication

46. Does there exist an atom of neon with a mass of exactly 20.18 u? Explain your answer.

47. The molecular mass of a compound is measured in atomic mass units but its molar mass is measured in grams. Explain why this is true.

48. Explain the relationship between an empirical formula and a molecular formula. Use the chemical and empirical formulas for sodium tartrate, $Na_2C_4H_4O_6$, and cyanocolabamin, $C_{63}H_{88}C \cdot N_{14}O_{14}P$ (vitamin $B_{12}$), to illustrate your answer.

49. Explain why an empirical formula can represent many different compounds.

50. Chemists need to know the percentage yield of a reaction. Why is this true, particularly for industrial reactions?

51. Examine the following reaction. List the steps needed to calculate the number of grams of C that can be expected when a given mass of A reacts with a given mass of B. Include proper units for each step. Express the answer in terms of A, B, C, and/or D as necessary.
    $2A + 3B \rightarrow 4C + D$

## Making Connections

52. Reread the Unit 1 opener.
    (a) Suppose that you ate a dessert containing poppy seeds. As a result, you tested positive for opiates when you applied for a summer job. What can you do?
    (b) Now suppose that you are a policy-writer for a manufacturing company that uses large, dangerous machines. What do you need to consider when you write a policy that deals with employee drug testing? What factors influence whether drug testing is warranted, how often it is warranted, and what substances should be tested for? How will you decide on levels that are acceptable? Do the federal and provincial Human Rights Commissions have anything to say about these issues?

53. The combustion of gasoline in an automobile engine can be represented by the equation
    $2C_8H_{18(g)} + 25O_{2(g)} \rightarrow 16CO_{2(g)} + 18H_2O_{(g)}$
    (a) In a properly tuned engine with a full tank of gas, what reactant do you think is limiting? Explain your reasoning.
    (b) A car that is set to inject the correct amount of fuel at sea level will run poorly at higher altitudes, where the air is less dense. Explain why.
    (c) The reaction of atmospheric oxygen with atmospheric nitrogen to form nitrogen monoxide, NO, occurs in a car's engine along with the combustion of fuel.
    $N_{2(g)} + O_{2(g)} \rightarrow 2NO_{(g)}$
    What adjustments need to be made to a vehicle's fuel injectors (which control the amount of fuel and air that are mixed) to compensate for this reaction? Explain your answer.

# UNIT 2

# Structure and Properties

## UNIT 2 CONTENTS

**CHAPTER 5**
Chemical Bonding

**CHAPTER 6**
Structure and Properties of Substances

## UNIT 2 OVERALL OUTCOMES

- What are ionic, covalent, and metallic bonds and how can you predict which type of bond will form between atoms of specific elements?

- What type of interactions cause molecules to attract each other and how do these attractive forces differ in liquids and solids?

- How can an understanding of bonds between atoms and interactions between molecules allow you to explain and predict the structure and behaviour of substances?

In 1958, an astonished television audience watched as Dr. Harry Coover used a tiny drop of a substance called cyanoacrylate to lift a game show host off the ground. Cyanoacrylate is an amazingly powerful adhesive. This stunt helped launch the commercial career of superglue, now used for everything from building circuit boards to sealing human tissue in surgery.

Superglue is just one example of an astounding variety of compounds with useful properties. Whether synthesized by design or discovered by trial and error, each new compound does more than provide a potentially useful product. It also contributes to scientists' understanding of the relationship between structure at the molecular level and properties at the macroscopic level.

In this unit, you will investigate the structure and properties of atoms, ions, and molecules, and the natural and synthetic products that result from their interactions.

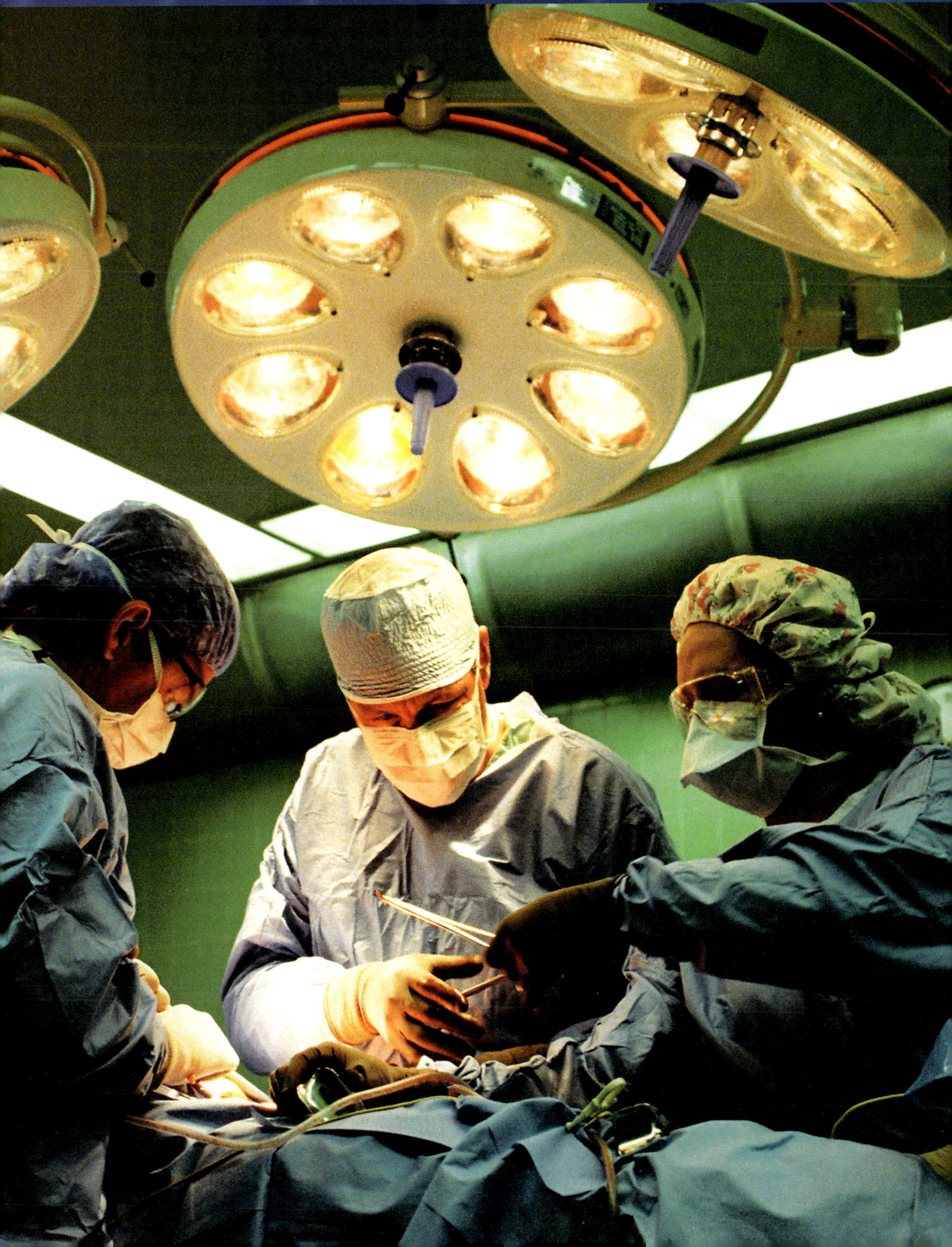

# CHAPTER 5
# Chemical Bonding

**Chapter Preview**
- 5.1 Elements and Compounds
- 5.2 Bond Formation
- 5.3 Bonds as a Continuum

The year was 1896. A chance discovery sent a message echoing from Yukon's far north to the southern reaches of the United States: "Gold!" People migrated in great numbers to the Yukon Territory, hoping to make their fortunes. Within two years, these migrants transformed a small fishing village into bustling Dawson City—one of Canada's largest cities at the time. They also launched the country's first metal-mining industry.

Gold, like all metals, is shiny, malleable, ductile, and a good conductor of electricity and heat. Unlike most metals and other elements, however, gold is found in nature in its pure form, as an element. Most elements are chemically combined in the form of compounds. Why is this so? Why do atoms of some elements join together as compounds, while others do not? In this chapter, you will use the periodic trends to help you answer these questions. You will also learn about the bonds that hold elements together in compounds.

**Prerequisite Concepts and Skills**

Before you begin this chapter, review the following concepts and skills:
- identifying elements by name and by symbol (Chapter 1)
- using the periodic table to get information about the elements

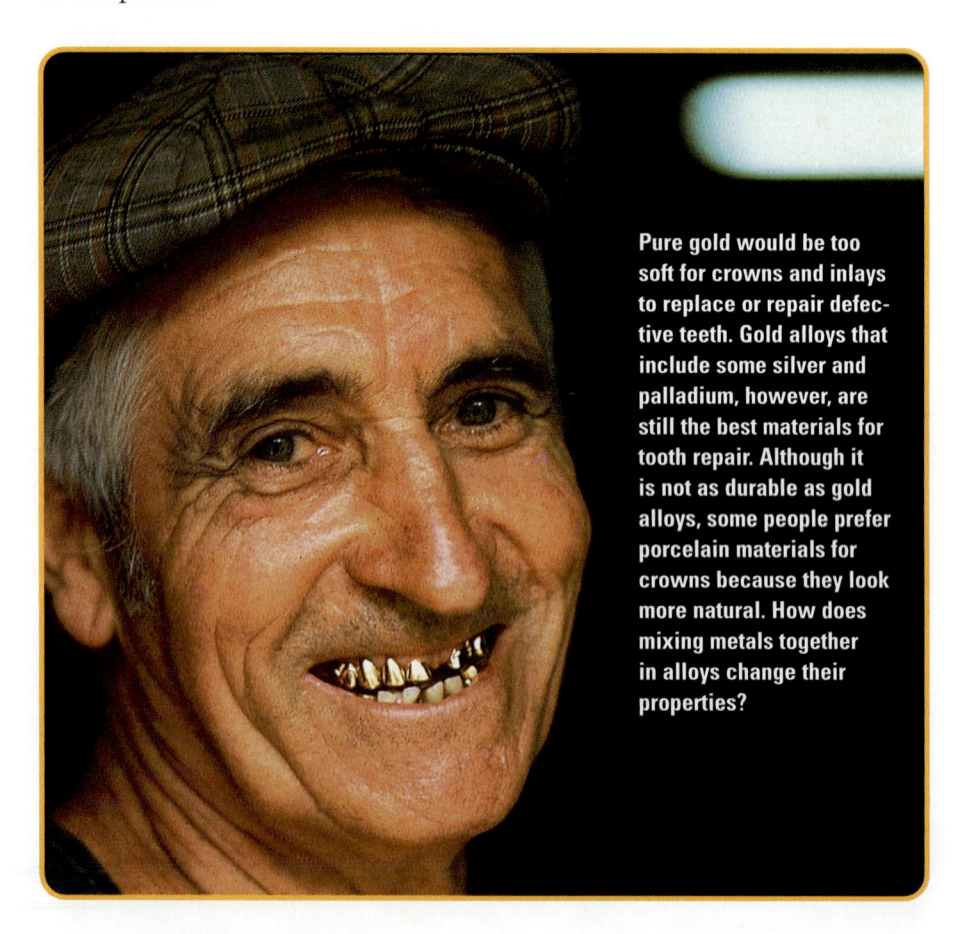

Pure gold would be too soft for crowns and inlays to replace or repair defective teeth. Gold alloys that include some silver and palladium, however, are still the best materials for tooth repair. Although it is not as durable as gold alloys, some people prefer porcelain materials for crowns because they look more natural. How does mixing metals together in alloys change their properties?

# Elements and Compounds

## 5.1

As you learned in the chapter opener, most elements do not exist in nature in their pure form, as elements. Gold, silver, and platinum are three metals that can be found in Earth's crust as elements. They are called "precious metals" because this occurrence is so rare. Most other metals, and most other elements, are found in nature only as compounds.

As the prospectors in the Yukon gold rush were searching for the element gold, they were surrounded by compounds. The streams they panned for gold ran with water, $H_2O$, a compound that is essential to the survival of nearly every organism on this planet. To sustain their energy, the prospectors ate food that contained, among other things, starch. Starch is a complex compound that consists of carbon, hydrogen, and oxygen. To flavour their food, they added sodium chloride, NaCl, which is commonly called table salt. Sometimes a compound called pyrite, also known as "fool's gold," tricked a prospector. Pyrite (iron disulfide, $FeS_2$) looks almost exactly like gold, as you can see in Figure 5.1. Pyrite, however, will corrode, and it is not composed of rare elements. Thus, it was not valuable to a prospector.

**Section Preview/Outcomes**

In this section, you will

- **identify** the properties of ionic, molecular, and metallic compounds
- **perform** a ThoughtLab to classify compounds as ionic or molecular according to their properties
- **draw** Lewis structures

**Figure 5.1** Prospectors used the physical properties of gold and pyrite to distinguish between them. Can you tell which of these photos shows gold and which shows pyrite?

## Physical Properties of Elements and Compounds

Although only a few elements are found pure in nature, those few elements have a wide variety of properties. For example, gold, silver, and platinum, mentioned above, are solid at standard temperatures. They all conduct electrical current in the solid and liquid states, and are ductile and malleable. Carbon is also found pure in nature in the form of diamonds. This form of carbon is solid but does not conduct electrical current and is the hardest naturally occurring material known. Yet, carbon in the form of graphite conducts electricity and is soft enough to use as a pencil. Some other elements that are found pure in nature are oxygen, nitrogen, hydrogen, helium, and argon. These substances are all gases at standard temperatures and pressures. In this chapter, you will learn why pure elements have such diverse physical properties. First, in the following ExpressLab, you will learn one way to determine whether an element has formed a compound with another element.

## ExpressLab: A Metal and a Compound

Humans have invented ways to extract iron from its compounds in order to take advantage of its properties. Does iron remain in its uncombined elemental form once it has been extracted? No, it doesn't. Instead, it forms rust, or iron(III) oxide, $Fe_2O_3$. How do we know that rust and iron are different substances? One way to check is to test a physical property, such as magnetism. In this activity, you will use magnetism to compare the properties of iron and rust.

### Safety Precautions

### Procedure

1. Obtain a new iron nail and a rusted iron nail from your teacher.
2. Obtain a thin, white piece of cardboard and a magnet. Wrap your magnet in plastic to keep it clean.
3. Test the iron nail with the magnet. Record your observations.
4. Gently rub the rusted nail with the other nail over the cardboard. Some rust powder will collect on the cardboard.
5. Hold up the cardboard horizontally. Move the magnet back and forth under the cardboard. Record your observations.

### Analysis

1. How did the magnet affect the new iron nail? Based on your observations, is iron magnetic?
2. What did you observe when you moved the magnet under the rust powder?
3. What evidence do you have to show that iron and rust are different substances?
4. Consider what you know about iron and rust from your everyday experiences. Is it more likely that rust will form from iron, or iron from rust?

## Properties of Ionic and Molecular Compounds

The 92 naturally occurring elements as well as the few artificial elements combine to form thousands of compounds. As you will see as you continue reading this chapter, these compounds, both natural and synthetic, have an astonishing variety of properties.

Based on their physical properties, compounds can be classified into two groups: ionic compounds and molecular compounds. Some of the properties of ionic and molecular compounds are summarized in Table 5.1.

**Table 5.1** Comparing Ionic and Molecular Compounds

| Property | Ionic compound | Molecular compound |
|---|---|---|
| state at room temperature | crystalline solid | liquid, gas, solid |
| melting point | high | low |
| electrical conductivity as a liquid | yes | no |
| solubility in water | most have high solubility | most have low solubility |
| conducts electricity when dissolved in water | yes | not usually |

In the following ThoughtLab, you will use the properties of various compounds to classify them as molecular or ionic.

# ThoughtLab: Ionic or Molecular

Imagine that you are a chemist. A colleague has just carried out a series of tests on the following compounds:

ethanol, C₂H₆O
carbon tetrachloride, CCl₄
glucose, C₆H₁₂O₆
table salt (sodium chloride), NaCl
water, H₂O
potassium permanganate, KMnO₄

You take the results home to organize and analyze them. Unfortunately your colleague labelled the tests by sample number and forgot to write down which compound corresponded to each sample number. You realize, however, that you can use the properties of the compounds to identify them. Then you can use the compounds' properties to decide whether they are ionic or molecular.

## Procedure

1. Copy the following table into your notebook.

| Sample | Compound name | Dissolves in water? | Conductivity as a liquid or when dissolved in water | Melting point | Appearance | Molecular or ionic? |
|---|---|---|---|---|---|---|
| 1 | | yes | high | 801°C | clear, white crystalline solid | |
| 2 | | yes | low | 0.0°C | clear, colourless liquid | |
| 3 | | yes | high | 240°C | purple, crystalline solid | |
| 4 | | yes | low | 146°C | white powder | |
| 5 | | no | low | −23°C | clear, colourless liquid | |
| 6 | | yes | low | −114°C | clear, colourless liquid | |

2. Based on what you know about the properties of compounds, decide which compound corresponds to each set of properties. Write your decisions in your table. Once you have identified the samples, share your results as a class and come to a consensus. **Hint:** Carbon tetrachloride is not soluble in water.

3. Examine the properties associated with each compound. Decide whether each compound is ionic or molecular. If you are unsure, leave the space blank. Discuss your results as a class, and come to a consensus.

## Analysis

1. Write down the reasoning you used to identify each compound, based on the properties given.
2. Write down the reasoning you used to decide whether each compound was ionic or molecular.
3. Were you unsure how to classify any of the compounds? Which ones, and why?
4. Think about the properties in the table you filled in, as well as your answers to questions 1 to 3. Which property is most useful for deciding whether a compound is ionic or molecular?
5. Suppose that you could further subdivide the molecular compounds into two groups, based on their properties. Which compounds would you group together? Explain your answer.

## Applications

6. Write the formulas for the compounds that you categorized as ionic compounds. Refer to a periodic chart and determine whether each of the elements in the ionic compounds is a metal or a nonmetal.
7. Carry out the procedure described in step 6 for the molecular compounds. Look for a pattern in the results.
8. Write a statement that summarizes the types of elements found in ionic and molecular compounds.

## Atoms and Elements

Did you conclude, in your ThoughtLab, that ionic compounds are composed of both metals and nonmetals while molecular compounds contain only nonmetals? From this finding, you probably concluded that the bonds between metal atoms and nonmetal atoms differ from the bonds between two or more nonmetal atoms. It is the difference between these types of bonds that give the compounds their diverse physical properties.

Chemical bonds are attractive electrostatic forces that hold atoms together. It is not difficult to understand how the positively charged nucleus and the surrounding negatively charged electrons attract each other in a single atom. A more detailed analysis is needed, however, to explain how the positive and negative charges of one atom interact with those of another. To study bond formation and bond properties, you need to have a clear picture of individual atoms and a way to visualize the electrons that participate in bonding.

In Chapter 1, you reviewed the concept that the electrons in atoms exist in specific energy levels. In addition, each energy level has a limit to the number of electrons that can occupy it. For example, only two electrons are allowed to occupy the lowest energy level while eight electrons can exist in the second energy level. Chemists often call these energy levels "shells." When the maximum number of electrons is in an energy level, it is said to be a filled energy level or "closed shell."

It is only the electrons in the outermost energy level, or shell, that participate in bonding. These outer electrons are called **valence electrons**. To assist in visualizing atoms, chemists have developed diagrams, called **Lewis structures**, that show only valence electrons. Each valence electron is represented by a dot and the dots are drawn around the symbol for the element as shown in Figure 5.2. Examine Figure 5.2 to see how Lewis structures are related to the energy level diagrams that you have used in previous science courses. You will find that Lewis structures give the most important information in a very simple way.

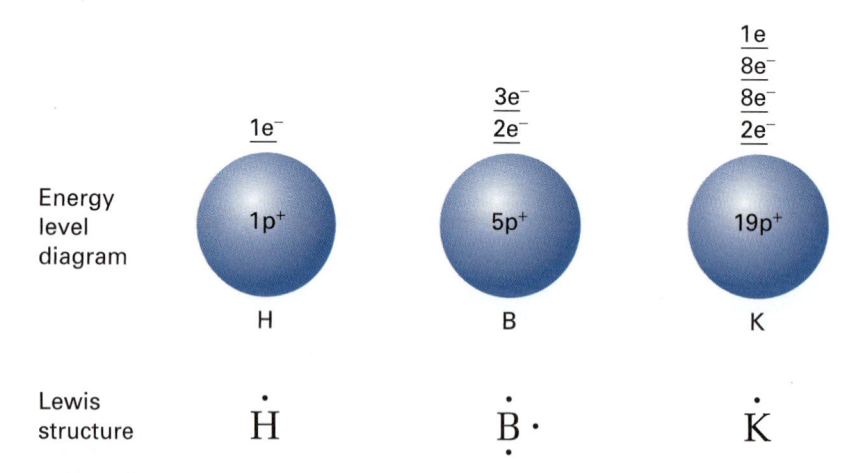

**Figure 5.2** In Lewis structures, the chemical symbol represents the nucleus and all of the filled energy levels. The dots represent the valence electrons.

For the first 20 elements, you can easily determine the number of valence electrons from the periodic table. Figure 5.3 shows a portion of the periodic table for the first 20 elements using the old system of numbering for the A groups. As one energy level becomes "filled," additional electrons move into the higher energy levels. Examination of Figure 5.3 will show you that, with the exception of helium, the number of valence electrons is the same as the number of the A group. For example, the number of valence electrons in Group VIIA is seven. All of the halogens have seven valence electrons.

| Group | IA | IIA | IIIA | IVA | VA | VIA | VIIA | VIIIA |
|---|---|---|---|---|---|---|---|---|
| Energy Level<br>1st element | 1<br>H | | | | | | | 2<br>He |
| 2nd<br>1st element | 1<br>2<br>Li | 2<br>2<br>Be | 3<br>2<br>B | 4<br>2<br>B | 5<br>2<br>N | 6<br>2<br>O | 7<br>2<br>F | 8<br>2<br>Ne |
| 3rd<br>2nd<br>1st element | 1<br>8<br>2<br>Na | 2<br>8<br>2<br>Mg | 3<br>8<br>2<br>Ai | 4<br>8<br>2<br>Si | 5<br>8<br>2<br>P | 6<br>8<br>2<br>S | 7<br>8<br>2<br>Cl | 8<br>8<br>2<br>Ar |
| 4th<br>3rd<br>2nd<br>1st element | 1<br>8<br>8<br>2<br>K | 2<br>8<br>8<br>2<br>Ca | | | | | | |

**Figure 5.3** The number of electrons in each energy level for the first 20 elements is shown in this portion of the periodic table.

Some patterns in the number of valence electrons can also be found beyond the first 20 elements. For example, all of the elements in Group IA have one valence electron. All elements in Group IIA have two valence electrons and all elements in Group VIIA have seven valence electrons.

For many elements with more than twenty electrons, the procedure for finding the number of valence electrons is more complex. If you encounter these in a problem, you will be given more information about the atoms involved in the problem.

When placing the dots on a Lewis structure, chemists often start with the first dot at the top as shown in Figure 5.4. Chemists then add dots clockwise or to the right, then bottom, then left. When there are more than four electrons in the outermost energy level, a second electron is added to the top to form a pair. Therefore, when there are five electrons in the outermost energy level—for example phosphorus—a second electron is added to the top to form an **electron pair**. The single electrons—for example phosphorous has three single electrons—are called **unpaired electrons**. Unpaired electrons always participate in bonding while electron pairs are often *not* involved in bonding.

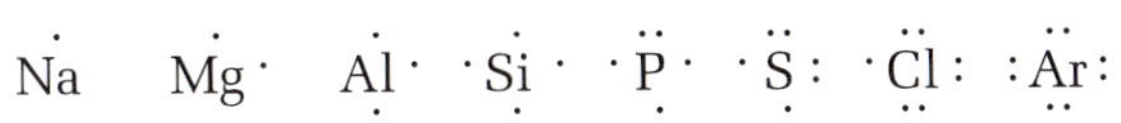

$$\text{Na} \quad \text{Mg}\cdot \quad \overset{\cdot}{\text{Al}}\cdot \quad \cdot\overset{\cdot}{\text{Si}}\cdot \quad \cdot\overset{\cdot\cdot}{\text{P}}\cdot \quad \cdot\overset{\cdot\cdot}{\text{S}}: \quad \cdot\overset{\cdot\cdot}{\text{Cl}}: \quad :\overset{\cdot\cdot}{\text{Ar}}:$$

**Figure 5.4** The Lewis structures for the elements in period three are shown here.

### Practice Problems

1. In your notebook, draw the first 20 squares of the periodic table similar to the outline below. In your periodic table, draw Lewis structures of each of the first twenty elements.

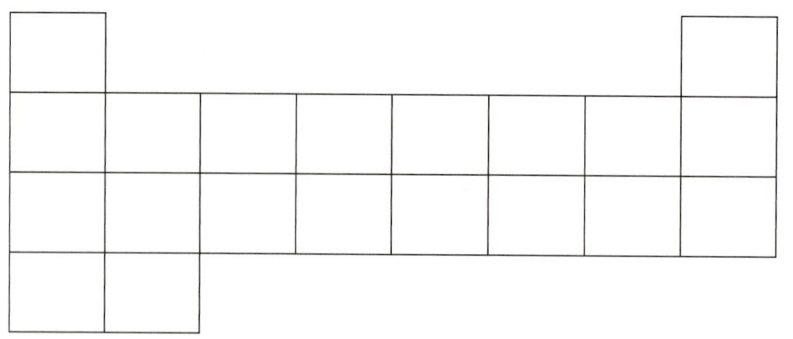

2. **(a)** Draw Lewis diagrams of all of the Group IA elements.
   **(b)** Strontium is called a "bone seeker" because it behaves much like calcium in the human body. Draw its Lewis structure.
   **(c)** All of the halogens have the same number of valence electrons. Draw the Lewis structure for bromine and iodine.

## Section Summary

Although there are only 92 naturally occurring elements, they combine to form the thousands of compounds with widely differing properties that are found in nature. Thousands more have been artificially synthesized in laboratories. Lewis structures provide an excellent way to analyze and understand bond formation. Lewis structures show only the valence electrons of the atoms.

### Section Review

**1** List some physical properties of substances found in nature that consist of a single element.

**2** Which electrons are more likely to participate in bonds between atoms?

**3** Explain how you would determine the number of valence electrons in atoms of sulfur.

**4** Draw Lewis structures for lithium, nitrogen, argon, cesium, barium, iodine, and radium.

# Bond Formation

## 5.2

Why do bonds form? Another way to ask that question would be, "Why are most substances found in nature as compounds and not pure elements?" In addition, the atoms of most of the elements that are found pure in nature are actually bonded to other atoms of the same element. In order for a substance to be solid, something must hold the atoms together. Therefore, in gold, silver, platinum, and carbon, the atoms are bonded to each other. Oxygen, nitrogen, and hydrogen are all gases in which the molecules are made of two atoms bonded together. It is only atoms of the group VIIIA or noble gases—helium, neon, argon, krypton, xenon, and radon—that are never found in nature bonded to any other atoms.

You can find a very important clue about bonding by examining the atoms that do *not* form bonds. What do atoms of the noble gases have in common that is unlike atoms of all other elements? The outer energy level of all of the noble gases is filled. For helium, the outer level is the first energy level and has two electrons. In all of the other noble gases, the outer energy level has eight electrons. This observation strongly indicates that a filled outer energy level makes atoms very stable and creates no tendency to form bonds with other atoms. In fact, atoms of all elements other than the noble gases form bonds in a way that will create a noble gas configuration—a filled outer energy level. This tendency is the basis of the **octet rule**, which states that when bonds form, atoms gain, lose, or share electrons in a way that creates an octet or filled outer energy level for the atoms involved in bonding. The term, octet, does not apply to hydrogen and helium for which the outer energy level is filled when two electrons are present.

### Ionic Bonding

Ions, as you know, are charged atoms or molecules. Thus an ionic bond is an attractive interaction between positive ions and negative ions. To form an ionic bond, the atoms must become ionized. Since the octet rule applies to the formation of ions, atoms gain or lose electrons until the outer energy level is filled with eight electrons. However, lithium ions and beryllium ions have only two electrons in the outer energy level because, when ionized, the outer energy level is the first level. Hydrogen is unique because it can gain or lose an electron. If hydrogen loses an electron, it has no electrons. Positively ionized hydrogen is simply a proton. If hydrogen gains an electron, it has two electrons in its filled outer energy level. The negatively charged hydrogen ion is called a hydride ion.

Recall that ionic compounds nearly always have a metal atom and a nonmetal atom. When you inspect a periodic table, you will notice that metals usually have three or fewer electrons in their outer energy levels. They can achieve an octet or noble gas configuration by losing these electrons. Conversely, nonmetals have four or more electrons in their outer energy level. Nonmetals can achieve an octet by gaining four or fewer electrons.

**Section Preview/Outcomes**

In this section, you will
- **explain** the octet rule
- **illustrate** and explain the formation of ionic bonds
- **illustrate** and explain the formation of covalent bonds
- **illustrate** and explain the formation of metallic bonds

### CHEM FACT

Helium was first discovered on the Sun. The name, helium, was taken from the Greek word *helios* which means sun. How did chemists discover helium on the Sun?

## CHEM FACT

The Group 18 elements in the periodic table are currently called the noble gases. In the past, however, they were referred to as the inert gases. They were believed to be totally unreactive. Scientists have found that this is not true. Some of them can be made to react with reactive elements, such as fluorine, under the proper conditions. In 1962, the synthesis of the first compound that contained a noble gas was reported. Since then, a number of noble gas compounds have been prepared, mostly from xenon. A few compounds of krypton, radon, and argon have also been prepared.

In the early 1960s, Neil Bartlett, of the University of British Columbia, synthesized the first compound that contained a noble gas.

Atoms of all Group 1 (IA) elements have one electron in the valence level. If they lose this electron, the remaining outer level is filled and the resulting ion is stable. Similarly, all atoms of Group 17 (VIIA) elements have seven electrons in the outer level. If they gain an electron, the outer level becomes filled and the atoms are stable. As shown in Figure 5.5, sodium can give an electron to chlorine and form positive sodium ions and negative chloride ions. The oppositely charged ions are then attracted to each other by electrostatic forces thus forming sodium chloride. This attraction is an ionic bond.

**Figure 5.5** The positive sodium ion and the negative chloride ion attract each other forming an ionic bond.

Group 2 (IIA) atoms can lose two electrons to form a closed shell and Group 16 (VIA) can gain two electrons to form a closed shell. The formation of the ionic compound magnesium oxide is shown in Figure 5.6.

**Figure 5.6** Magnesium loses two electrons and oxygen gains two electrons becoming doubly charged ions. An ionic bond then forms between them.

How can atoms of fluorine, that can gain only one electron to form an octet, form ionic bonds with atoms such as calcium that must lose two electrons in order to have a closed outer energy level? As shown in Figure 5.7, two atoms of fluorine accept one electron each from an atom of calcium. The resulting compound is represented by the symbol, $CaF_2$ to show that there are two fluoride ions for every one calcium ion.

**Figure 5.7** Calcium donates an electron to two different fluorine atoms that then become fluoride ions. Both fluoride ions are attracted to the calcium ion.

In previous science courses, you have written empirical formulas for ionic compounds such as $CaF_2$ by combining the charges on the ions and ensuring that they add to zero net charge. Now, because you have learned about valence electrons, the stability of filled energy levels, and Lewis structures, you can understand why specific elements form ions with the charges they have and why these positive and negative ions combine to form ionic compounds with the formulas you have been predicting using these charges.

Try the following practice problems to assess your understanding of the formation of ionic bonds.

**166** MHR • Unit 2 Structure and Properties

## Practice Problems

3. Using Lewis diagrams predict the correct formulas for the following ionic compounds.
   (a) potassium chloride
   (b) barium oxide
   (c) sodium fluoride
   (d) calcium oxide
   (e) magnesium chloride
   (f) sodium oxide
   (g) lithium oxide
   (h) potassium sulfide
   (i) aluminum oxide (Al has three valence electrons.)

> **PROBLEM TIP**
>
> When drawing Lewis structures to show the formation of a bond, you can use different colours or symbols to represent the electrons from different atoms. For example, you could use x's and o's, or open and closed circles as shown in Figure 5.7.

## What is an Ionic Bond?

Have you noticed that the terms "molecule" and "molecular" are never used in reference to ionic compounds? Chemists have a very good reason to avoid the use of these terms. To understand why, think about a crystal of solid sodium chloride with a mass of about 1 mg. That crystal contains about $1 \times 10^{21}$ sodium ions and an exactly equal number of chloride ions. Due to their opposite charge, an attractive force exists between every sodium ion and every chloride ion which surrounds it, and vice versa. In a crystal, the ions pack together is a way that allows the positive and negative charges to be positioned as close together as possible. This three dimensional array of alternating positive and negative ions is called a **crystal lattice**. Examine Figure 5.8 to visualize the arrangement of the ions in a sodium chloride crystal. The diagram on the left is a ball and stick model that allows you to see the positions of the ions relative to each other. You can see that each sodium ion is equally attracted to six different chloride ions and each chloride ion is equally attracted to six different sodium ions. There are no distinct pairs of sodium and chloride ions that you could identify as a "molecule." When you write the chemical formula, NaCl, you are simply stating that the ratio of sodium to chloride ions is one to one.

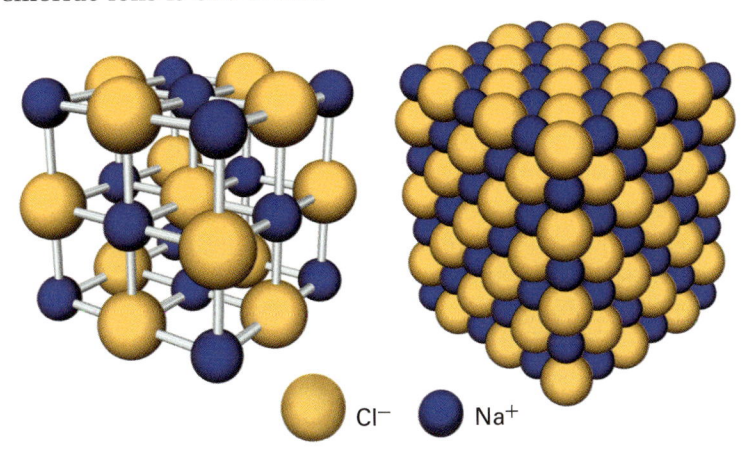

**Figure 5.8** Both of the diagrams represent a sodium chloride crystal. The ball and stick model on the left is easier to visualize. The diagram on the right is a space filling model that is a more accurate representation of the compound but more difficult to use to visualize the interactions among the ions.

Many ionic compounds that have a one to one ratio of ions have a crystalline structure similar to sodium chloride. However, compounds such as calcium fluoride, $CaF_2$, have a two to one ratio of ions. As you can see in the diagram of a calcium fluoride crystal in Figure 5.9, the structure of these crystalline solids is more complex. In any case, the ions pack together in a way that allows the positive and negative ions to be as close together as possible.

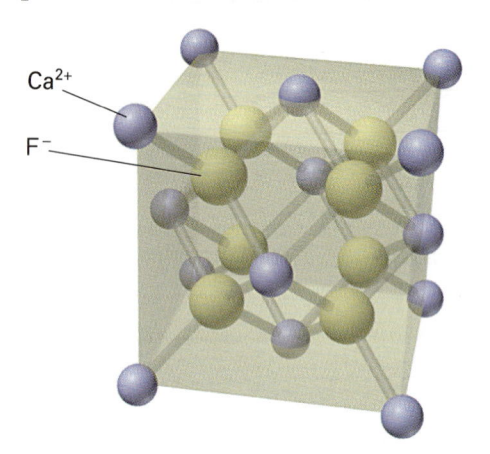

**Figure 5.9** The structure diagrammed here is the smallest repeating unit of a crystal of calcium fluoride. Such a structure is often called a unit cell.

**CONCEPT CHECK**

What does the formula of calcium bromide, $CaBr_2$, represent?

In ionic compounds, the unit cells do not function as independent units. Thus, there is no specific group of ions that you could call a molecule. Ionic bonds are attractive forces among many positive and negative ions. The chemical formulas that you write for ionic compounds are called empirical formulas. These formulas represent the whole number ratio between the positive and negative ions in the smallest neutral unit of an ionic crystal lattice—the formula unit.

## Covalent Bonding

In contrast to ionic compounds, individual units in molecular compounds are correctly called molecules. The atoms in molecules are connected by covalent bonds. To understand how covalent and ionic bonds differ, consider the difference between the elements they contain. Recall that molecular compounds contain atoms of nonmetal elements only. Nonmetals tend to accept electrons to fill their outermost energy levels. However, if there are no metals to lose electrons, as there are in ionic compounds, how can the nonmetals accept electrons? The answer is that the atoms in molecular compounds neither gain nor lose electrons but instead share them. Molecular compounds also follow the octet rule. They share electrons in a way that allows both atoms in a bond to have an octet of valence electrons. The only exception is hydrogen atoms which need only two electrons to be have a filled outer energy level. You can think of the shared electrons as orbiting both of the nuclei.

As shown in Figure 5.10, you can indicate diagrammatically that electrons are shared by drawing Lewis structures of the atoms beside each other with the unpaired valence electrons forming a shared electron pair. Each chlorine atom (Figure 5.10B) has achieved a noble gas configuration—an octet—by sharing their unpaired electrons. Since hydrogen atoms have only one electron, two hydrogen atoms share both of their unpaired electrons forming a filled first energy level of two electrons. (Figure 5.10A).

**Figure 5.10** The shared electron pair between the chlorine atoms and between the hydrogen atoms constitute a covalent bond.

As you can see in Figure 5.10, some of the electron pairs are shared by the two atoms and some are not shared. Names given to these pairs of electrons, as shown in Figure 5.11, are **bonding electron pairs** for the shared pairs and **lone pairs** for unshared electrons that fill a valence level for only one atom. A **covalent bond** is the electrostatic attraction between the nuclei of two adjacent atoms and a pair of shared bonding electrons.

$$:\!\ddot{F}\!:\!\ddot{F}\!:$$

bonding pair — lone pairs

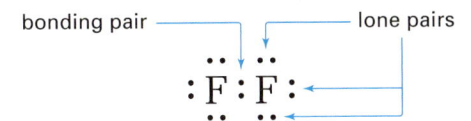

**Figure 5.11** The bonding electron pair is involved in the covalent bond while the lone pairs complete the octet for each fluorine atom.

Frequently, an atom will share electrons with more than one other atom in order to achieve a filled valence level. Figure 5.12 shows carbon sharing an electron with each of four hydrogen atoms. The result provides carbon with eight electrons in its valence level and the hydrogen atoms each have two electrons in their valence levels.

## Multiple Bonds

When nonmetal atoms have less than seven electrons in their valence level, they have more than one unpaired electron available to form covalent bonds. The result is the sharing of more than one pair of electrons with another atom in order to achieve an octet. For example, oxygen atoms have six valence electrons, two electron pairs and two unpaired electrons, and form oxygen molecules by sharing two electrons with another oxygen atom. Nitrogen atoms have only five valence electrons and form nitrogen molecules by sharing three electrons with a second nitrogen atom. Figure 5.13 shows how two oxygen atoms share four electrons (two bonding pairs) and two nitrogen atoms share six electrons (three bonding pairs) to achieve octets.

**Figure 5.13** All of the electrons drawn between two atoms in a Lewis structure are shared by both atoms.

One pair of shared electrons is considered to be one bond and is therefore called a **single covalent bond**. When atoms share four electrons, they are sharing two pairs of electrons and this combination is called a **double covalent bond**. Similarly, six shared electrons or three pair of shared electrons form a **triple covalent bond**.

Atoms of unlike elements can also form double and triple bonds with each other. As shown in Figure 5.14, in carbon dioxide one carbon atom shares two pairs of electrons with each of two oxygen atoms. The central carbon atom is bonded to two oxygen atoms by double bonds.
If you have trouble determining whether each atom in a Lewis diagram has an octet, remember to count all of the shared electrons involved in a bond with both of the elements that are connected by that bond.

> **CONCEPT CHECK**
>
> Individual atoms of hydrogen and fluorine are highly reactive and readily bond together to form molecules of hydrogen fluoride. Draw a Lewis structure for hydrogen fluoride. Label the bonding and lone pairs and explain why this molecule is stable.

**Figure 5.12** You might recognize this molecule as methane—one carbon atom bonded to four hydrogen atoms.

**Figure 5.14** In this carbon dioxide molecule, the carbon atom shares four electrons with each of two oxygen atoms.

**Web LINK**

www.mcgrawhill.ca/links/atlchemistry

To see several animations that show ionic and covalent bonding, go to the web site above and click on **Electronic Learning Partner**.

## Sample Problem

### Drawing Lewis Structures

**Problem**

One carbon atom is bonded to one oxygen atom and two hydrogen atoms. Draw a Lewis structure to represent the bonds in this molecule.

**Plan Your Strategy**

A carbon atom can form a total of four bonds. These bonds can all be single covalent bonds or combinations of single, double, or triple covalent bonds. Oxygen can form two single covalent bonds or one double covalent bond. Hydrogen can form only one single covalent bond. These bonds must be arranged in a way that will create two electrons in the valence level for hydrogen and eight valence electrons for carbon and oxygen. Draw each of the atoms alone and find a way to fit them together to satisfy all of the criteria.

**Act on Your Strategy**

The hydrogen atoms each share one electron with carbon. The carbon and oxygen atoms share four electrons in a double bond.

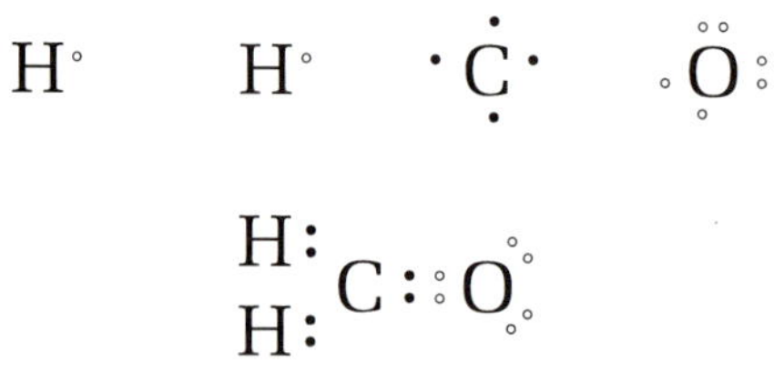

H·    H·    ·C·    ·O·

H:
   C::O:
H:

**Check Your Solution**

The outer energy level in each hydrogen atom is filled with two electrons. The outer energy levels of the carbon and oxygen atoms are both filled with eight electrons.

## Practice Problems

Draw Lewis structures to represent each of the following bonds.

4. One carbon atom is bonded to two sulfur atoms.

5. A central carbon atom is bonded to a hydrogen atom and to a nitrogen atom.

6. Two carbon atoms and two hydrogen atoms are bonded together forming a molecule. Each atom has a filled valence level. Use a Lewis structure to show how this can be accomplished.

Lewis structures are critical for analyzing the way that atoms bond together. However, they can become cumbersome and time consuming when drawing structures with which you are already quite familiar. To draw more simplified structures, use one single line to represent a single bond (—), two lines for a double bond (=), and three lines for a triple bond (≡). Lone pairs are usually omitted from the structures. However, they are still important to remember when you consider the overall shape of the molecules. Figure 5.15 shows several of the molecules that appear in the figures on the previous pages, written in the simplified form. Such structures are referred to as **structural formulas**.

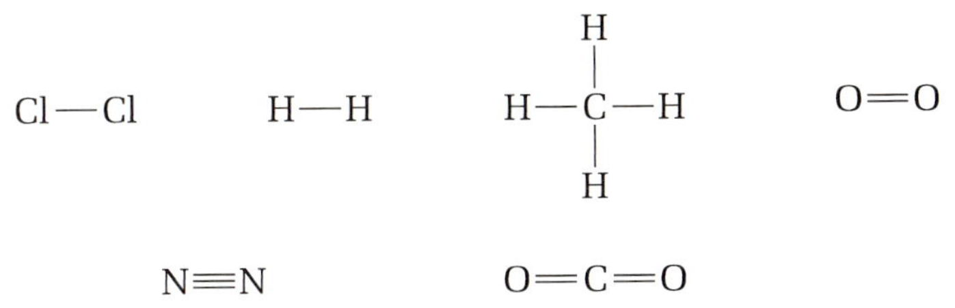

$$Cl-Cl \qquad H-H \qquad H-\underset{\underset{H}{|}}{\overset{\overset{H}{|}}{C}}-H \qquad O=O$$

$$N\equiv N \qquad O=C=O$$

**Figure 5.15** Find the Lewis structure of each of the compounds shown here and compare the diagrams. As you compare the structures, remember that one line represents a pair of shared electrons.

Molecular substances contain molecules in which all of the atoms are bonded together by covalent bonds. Sometimes, these molecules can form crystals by stacking on each other in an orderly array. They differ from ionic substances, however, because you can always identify individual molecules by the covalent bonds. Molecular substances vary tremendously in size and function. As you saw above, some molecules have only two or three atoms. Conversely, some biological molecules such as proteins can have thousands of atoms in one molecule.

## Metallic Bonding

While discussing ionic and molecular compounds, you never encountered substances that contain only metal atoms. You have, however, read about solid metals such as gold, silver, and platinum. These substances have properties that distinguish them from both ionic and molecular compounds. Since they are solids, there must be some type of bond holding the atoms together. These bonds cannot be ionic bonds because metals atoms usually tend to lose electrons and form positive ions. However, if there are no atoms to accept or gain these electrons to form negative ions, there cannot be any ionic bonds. Neither can the bonds between metal atoms be covalent because if two metal atoms shared all of their electrons, there would still be less than eight and they would not form an octet.

It should not be surprising that bonding in metals cannot be explained by the models for ionic or covalent bonding because the properties of pure metals differ from those of ionic and molecular compounds. You probably recall, from previous science courses, that metals conduct electric current in solid or liquid form. Metals also conduct heat and are malleable, ductile, and lustrous.

Because metals have a unique set of properties compared to ionic compounds and molecular compounds a different model of bonding was devised to explain them. The valence electrons in metals are somewhat loosely held and free to move from one atom to the next. The model, shown in Figure 5.16, is called the **free-electron model**. Since these freely moving valence electrons spend most of the time between one metal atom and the next you can visualize metals as positive ions embedded in a "sea" of valence electrons. After loosing their valence electrons, the ions share all of the electrons. The electrostatic attractive force between the positively charged metal ions and the "sea" of negative electrons constitutes a **metallic bond**.

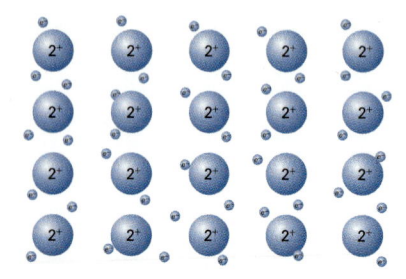

**Figure 5.16** The diagram represents magnesium atoms that have released their electrons and are embedded in a sea of electrons

## Section Summary

Noble gases are very stable atoms and in nature are never found bonded to other atoms. The noble gas configuration is so stable that all atoms form bonds in a way that achieves a noble gas configuration of eight electrons in their outer energy levels. This tendency is called the octet rule. The only exception is hydrogen, which is stable with two electrons.

Metal atoms usually form ions by loosing the electrons in their outer energy level. Nonmetals usually form ions by gaining electrons to fill their outer energy level. The electrostatic attraction between positive and negative ions constitutes ionic bonds.

When two atoms of nonmetal elements form bonds, they usually share electrons to achieve filled outer energy levels of both atoms. The electrostatic attraction between the positive nuclei of adjacent atoms and a pair of shared electrons constitutes a covalent bond. Atoms can share two, four, or six electrons thus forming single, double, or triple covalent bonds.

The model that explains the bonding between metal atoms is called the free-electron model. Metal atoms lose their valance electrons and move in a "sea" of electrons. The electrostatic attractive forces between the free electrons and the positive metal ions constitute metallic bonds.

## Section Review

**1** Read the following statement and then decide whether you agree with it. Explain why or why not. "In general, the farther away two elements are from each other on the periodic table, the more likely they are to participate in ionic bonding."

**2** Use Lewis structures to predict the formula for the ionic compounds formed between the following pairs of elements.

(a) magnesium and fluorine
(b) potassium and bromine
(c) rubidium and chlorine
(d) calcium and oxygen

**3** Covalent bonding and metallic bonding both involve electron sharing. Explain how covalent bonding is different from metallic bonding.

**4** Use Lewis structures to show how the following elements form covalent bonds.

(a) one silicon atom and two oxygen atoms
(b) one carbon atom, three hydrogen atoms, and one chlorine atom
(c) two iodine atoms
(d) two carbon atoms bonded together, three hydrogen atoms are bonded to one of the carbon atoms, and one hydrogen atom and one oxygen atom are bonded to the other carbon atom

# Careers in Chemistry

## Metallurgist

**Alison Dickson**

Alison Dickson is a metallurgist at Polaris, the world's northernmost mine. Polaris is located on Little Cornwallis Island in Nunavut. It is a lead and zinc mine, operated by Cominco Ltd., the world's largest producer of zinc concentrate.

After ore is mined at Polaris, metallurgists must separate the valuable lead- and zinc-bearing compounds from the waste or "slag." First the ore is crushed and ground with water to produce flour-like particles. Next a process called *flotation* is used to separate the minerals from the slag. In flotation, chemicals are added to the metal-containing compounds. The chemicals react with the lead and zinc to make them very insoluble in water, or *hydrophobic*. Air is then bubbled through the mineral and water mixture. The hydrophobic particles attach to the bubbles and float to the surface. They form a stable froth, or concentrate, which is collected. The concentrate is filtered and dried, and then stored for shipment.

Dickson says that she decided on metallurgy as a career because she wanted to do something that was "hands-on." After completing her secondary education in Malaysia, where she grew up, Dickson moved to Canada. She studied mining and mineral process engineering at the University of British Columbia.

Dickson says that she also wanted to do something adventurous. She wanted to travel and live in different cultures. As a summer student, Dickson worked at a Chilean copper mine. Her current job with Cominco involves frequent travel to various mines. "Every day provides a new challenge," Dickson says. When she is at Polaris, Dickson enjoys polar bear sightings on the tundra.

### Making Career Connections

Are you interested in a career in mining and metallurgy? Here are two ways that you can get information:

1. Explore the web site of The Canadian Institute of Mining, Metallurgy and Petroleum. Go to www.mcgrawhill.ca/links/atlchemistry to find out where to go next. It has a special section for students who are interested in mining and metallurgy careers. This section lists education in the field, scholarships and bursaries, and student societies for mining and metallurgy.

2. To discover the variety of jobs that are available for metallurgists, search for careers at Infomine. Go to www.mcgrawhill.ca/links/atlchemistry to find out where to go next. Many of the postings are for jobs overseas.

# 5.3 Bonds as a Continuum

**Section Preview/Outcomes**

In this section, you will

- **describe** bonding as a continuum
- **compare** the strengths of ionic and covalent bonds
- **explain** in simple terms the energy changes of bond breaking and bond formation

As you have examined ionic and covalent bonds in this chapter, you might have formed the impression that the two types of bonds are quite distinct from each other. As you continue to analyze covalent bonds, however, you will discover that they have a wide variety of properties such as bond lengths and bond energies. You will also discover that there is a significant separation of positive and negative charge in some covalent bonds. In this section, you will investigate covalent and ionic bonds in more detail and discover the relationships between them.

## Electronegativity of Elements

Fundamentally, bonds form because the positively charged nucleus of each atom attracts the electrons of another atom. Deeper inspection of bonding shows, however, that when involved in a bond, atoms of some elements attract the shared electrons to a much greater extent than do atoms of other elements. Chemists have named this property **electronegativity** (*EN*) and have experimentally determined the values for each element. These values are given in Figure 5.17, in the form of a periodic table.

Electronegativities

| 1 H 2.20 | | | | | | | | | | | | | | | | | 2 He - |
|---|---|---|---|---|---|---|---|---|---|---|---|---|---|---|---|---|---|
| 3 Li 0.98 | 4 Be 1.57 | | | | | | | | | | | 5 B 2.04 | 6 C 2.55 | 7 N 3.04 | 8 O 3.44 | 9 F 3.98 | 10 Ne - |
| 11 Na 0.93 | 12 Mg 1.31 | | | | | | | | | | | 13 Al 1.61 | 14 Si 1.90 | 15 P 2.19 | 16 S 2.58 | 17 Cl 3.16 | 18 Ar - |
| 19 K 0.82 | 20 Ca 1.00 | 21 Sc 1.36 | 22 Ti 1.54 | 23 V 1.63 | 24 Cr 1.66 | 25 Mn 1.55 | 26 Fe 1.83 | 27 Co 1.88 | 28 Ni 1.91 | 29 Cu 1.90 | 30 Zn 1.65 | 31 Ga 1.81 | 32 Ge 2.01 | 33 As 2.18 | 34 Se 2.55 | 35 Br 2.96 | 36 Kr - |
| 37 Rb 0.82 | 38 Sr 0.95 | 39 Y 1.22 | 40 Zr 1.33 | 41 Nb 1.6 | 42 Mo 2.16 | 43 Tc 2.10 | 44 Ru 2.2 | 45 Rh 2.28 | 46 Pd 2.20 | 47 Ag 1.93 | 48 Cd 1.69 | 49 In 1.78 | 50 Sn 1.96 | 51 Sb 2.05 | 52 Te 2.1 | 53 I 2.66 | 54 Xe - |
| 55 Cs 0.79 | 56 Ba 0.89 | 72 Hf 1.3 | 73 Ta 1.5 | 74 W 1.7 | 75 Re 1.9 | 76 Os 2.2 | 77 Ir 2.2 | 78 Pt 2.2 | 79 Au 2.4 | 80 Hg 1.9 | 81 Tl 1.8 | 82 Pb 1.8 | 83 Bi 1.9 | 84 Po 2.0 | 85 At 2.2 | 86 Rn - |
| 87 Fr 0.7 | 88 Ra 0.9 | 104 Rf - | 105 Db - | 106 Sg - | 107 Bh - | 108 Hs - | 109 Mt - | 110 Uun - | 111 Uuu - | 112 Uub - | 113 - | 114 Uuq - | 115 - | 116 Uuh - | | | |

| 57 La 1.10 | 58 Ce 1.12 | 59 Pr 1.13 | 60 Nd 1.14 | 61 Pm - | 62 Sm 1.17 | 63 Eu - | 64 Gd 1.20 | 65 Tb - | 66 Dy 1.22 | 67 Ho 1.23 | 68 Er 1.24 | 69 Tm 1.25 | 70 Yb - | 71 Lu 1.0 |
|---|---|---|---|---|---|---|---|---|---|---|---|---|---|---|
| 89 Ac 1.1 | 90 Th 1.3 | 91 Pa 1.5 | 92 U 1.7 | 93 Np 1.3 | 94 Pu 1.3 | 95 Am - | 96 Cm - | 97 Bk - | 98 Cf - | 99 Es - | 100 Fm - | 101 Md - | 102 No - | 103 Lr - |

**Figure 5.17** In the main group elements, electronegativities increase across the periods from left to right and increase as you go down the groups.

Inspection of the table shows that electronegativity is a periodic property. As you look across any period, at the main group elements, you will see that the electronegativity increases until you reach the noble gases. Since noble gases do not normally form bonds, values of electronegativity are not listed. As you look down any given group, the electronegativity decreases. Fluorine, at the top right, has the highest electronegativity of all elements. Chemists have determined experimentally that it is most reactive of all the nonmetals. Francium, at the bottom left, has the lowest electronegativity of any element and is also extremely reactive. Remembering these two extremes allows you to easily determine the periodic trends in electronegativity in periods and groups. These trends in electronegativity are opposite to the trends in the size of atoms as shown in Figure 5.18.

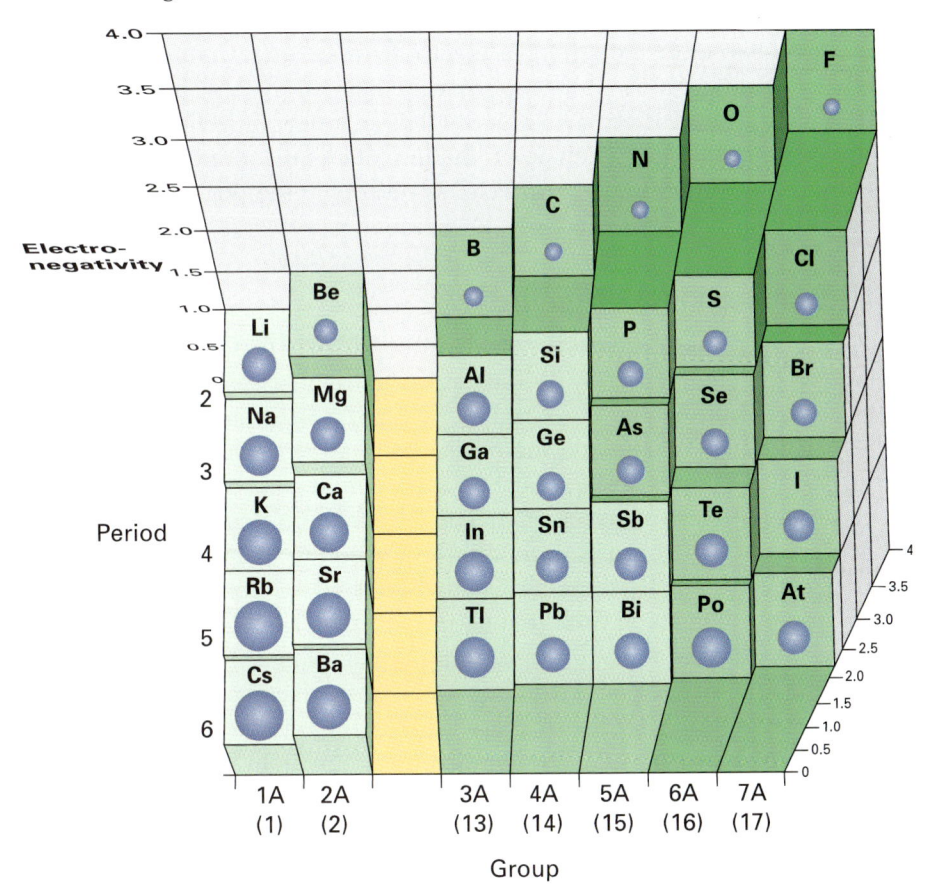

**Figure 5.18** The height of the bars represents the electronegativity and the size of the spheres represents the relative sizes of the neutral atoms.

How could size influence electronegativity? Consider properties of the elements from lithium to fluorine—left to right across period two. At first, it might seem strange that as the numbers of protons, neutrons, and electrons increase, the atoms become smaller. Consider, however, the fact that most of an atom is "empty space" and the energy level of the electrons accounts for the effective size of an atom. All of the valence electrons in period two are in the same energy level. As the number of protons in the nucleus increases, the attractive force on the electrons increases, pulling them closer to the nucleus. Therefore, when the nucleus of the atoms of the nonmetal elements in period two—carbon, nitrogen, oxygen, and fluorine—attract electrons of another atom with which it is bonding, those electrons can get close to the nucleus and are thus attracted strongly. Conversely, the valence electrons on the metals—lithium and beryllium—are further from the nucleus and the nucleus has a smaller charge than the nonmetals in the period. Therefore, the electrons are attracted much less strongly. Similar arguments could be made for all of the periods.

Now analyze the nature of the atoms as you look down a group, for example, Group 16 (VIA). As you go down, the electrons in each period are in a higher energy level than those in the previous period, thus they are farther from the nucleus. For example, compare oxygen and polonium. Electrons in lower energy levels in polonium form a negatively charged "cloud" around the nucleus that shields the outer electrons from the positive charge of the nucleus. Consequently, electrons from another atom that is bonded to polonium would also be further from the nucleus and thus would be not be attracted strongly. Electrons in an atom that is bonded to oxygen could get closer to the positively charged nucleus and would be attracted quite strongly. The size of an atom determines how close electrons of another atom can get to the nucleus of the atom with which it is bonded and therefore size affects how strongly the positively charged nucleus can attract those electrons in the bond.

> **CONCEPT CHECK**
>
> Which element is the most electronegative? Excluding the noble gases, which element is the least electronegative? How does the location of these two elements allow you to recognize the trends in electronegativity within periods and families of elements in the periodic table?

### Bond Type and Electronegativity

What can electronegativity tell you about bonds between atoms? Begin to analyze the connection by looking at the extreme cases. First consider metals versus nonmetals. The Group 1 (IA) and 2 (IIA) metals have the lowest electronegativities and therefore do not attract electrons very strongly. The Group 16 (VIA) and 17 (VIIA) nonmetals have the highest electronegativities and attract valance electrons more strongly than any other elements. This observation might cause you to expect that electron transfer would occur from metals to nonmetals when atoms of these elements interact. This is in excellent agreement with the observations that you have already made about ionic compounds. Metals lose electrons and become positively charged while nonmetals gain electrons and become negatively charged. After the ions are formed, there is an attractive electrostatic force between them so the nucleus of one atom is still attracted to the electrons of the other to some extent.

Now consider elements with the same electronegativities. You know that atoms of several elements exist as diatomic molecules such as hydrogen, oxygen, and nitrogen. Atoms that have the same electronegativity attract the valence electrons they share to exactly the same extent. Such covalent bonds resulting from equal sharing of bonding electron pairs are called **nonpolar covalent bonds**. Neither atom will release its electrons to the other. This situation is, once again, in agreement with your observations about molecular compounds. Atoms of nonmetals share electrons forming covalent bonds.

Thus far, electronegativity does not appear to provide any more information about bonds than you originally discovered by just applying the octet rule. However, you can learn more about bonds when you apply the concept of electronegativity quantitatively. Chemists have developed a scheme by which you can use the *difference* in the electronegativities to predict the type of bond that will form between atoms of any two elements. First you select the two elements—call them A and B. Then calculate the difference in their electronegativities or $\Delta EN = EN_A - EN_B$. If the difference in the electronegativities of two elements is greater than 1.7, the bond between atoms of the elements will be mostly ionic. The term "mostly" is included because there is always some attraction between the nucleus of one atom and the electrons of the other. If the electronegativity difference between two atoms is less than 0.5, the bond between the atoms will be only very slightly polar and are often referred to as covalent bonds. Of course, if the difference in the electronegativities is zero the bonds are **nonpolar covalent bonds**. The electronegativity difference between 0.5 and 1.7 is the region that creates a need to define a new category of bonding. Bonds that fit into this region are classified as **polar covalent bonds** as shown in Figure 5.19. Examples of bonds will help you understand the nature of the bonds in these three categories.

$\Delta EN$

| 3.3 | 1.7 | 0.5 | 0 |

Mostly Ionic | Polar covalent | Slightly polar covalent | non-polar covalent

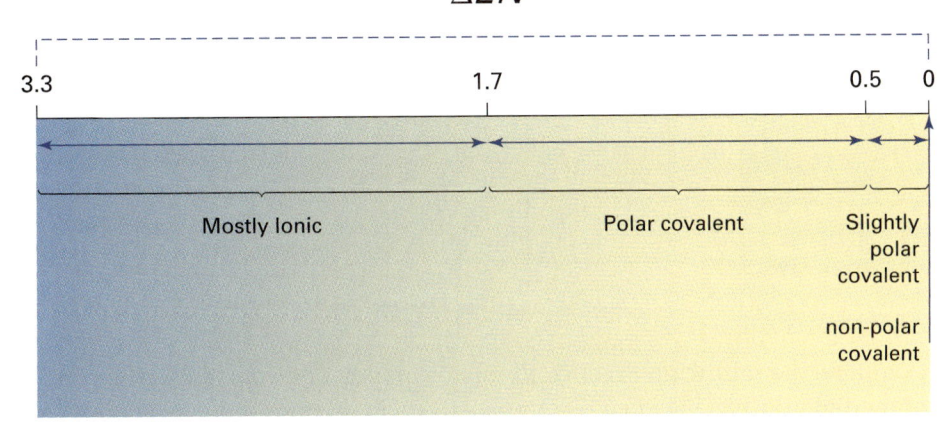

**Figure 5.19** The nature of chemical bonds change in a continuous way creating a broad range of characteristics.

First, analyze the bond between potassium and fluorine. The electronegativity of fluorine is 3.98 and of potassium is 0.82. The electronegativity difference is calculated as shown on the right.

$\Delta EN = EN_F - EN_K$
$\Delta EN = 3.98 - 0.82$
$\Delta EN = 3.16$

Since 3.16 is much greater than 1.7, it is clear that the bond in potassium fluoride is ionic. Now consider the case in which the two oxygen atoms bond together. The electronegativity of oxygen is 3.44. The electronegativity difference, as shown on the right, is zero. Therefore, the bond between oxygen atoms is certainly a nonpolar covalent bond.

$\Delta EN = EN_O - EN_O$
$\Delta EN = 3.44 - 3.44$
$\Delta EN = 0$

Finally consider the bonds between carbon and chlorine in a compound such as chloroform ($CHCl_3$). The electronegativity of carbon is 2.55 and of chlorine is 3.16. The electronegativity difference, as shown on the right, is 0.61 which is smaller than 1.7 but higher than 0.5. Therefore, the bond fits into the category of polar covalent bonds. To develop an understanding of the nature of polar covalent bonds, analyze the effects that the attractive forces of the nuclei will have on the shared electrons. Since the chlorine atom has a stronger attraction for the electrons than the carbon atom, the electrons will spend more time nearer to the chlorine than the carbon. The result is a slight separation of positive and negative charge. The region around the chlorine atom will be slightly negative and around the carbon it will be slightly positive. Chemists call these "partial charges" and denote them with the lowercase Greek letter, delta ($\delta$). The symbol, $\delta^+$ represent a partial charge less than +1 and $\delta^-$ represents a partial negative charge less than −1. Since the charge in the bond is polarized into a positive area and a negative area the bond is

$\Delta EN = EN_{Cl} - EN_C$
$\Delta EN = 3.16 - 2.55$
$\Delta EN = 0.61$

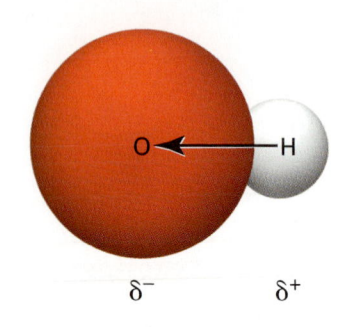

**Figure 5.20** The O—H bond is very polar. This polarity affects the properties of all of the molecules that have this bond. Water has two O—H bonds.

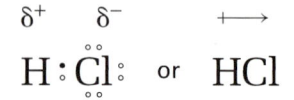

**Figure 5.21** Hydrogen chloride is an especially interesting compound. The electronegativity difference between hydrogen and chlorine is 0.96 which indicates that it is a polar covalent molecule. However, when dissolved in water, it conducts electricity like an ionic compound. Classification of bonds is not always simple.

### PROBEWARE

www.mcgrawhill.ca/links/atlchemistry

If you have access to probeware, go the website above and click on Probeware to find an investigation on the properties of bonds.

sometimes called a **bond dipole**. Two more examples of polar covalent bonds are the bond between oxygen and hydrogen atoms (O—H) and between hydrogen and chlorine atoms (H—Cl). The arrow that points toward the more electronegative atom as shown in Figure 5.20 for the O—H bond points in the direction of the more negative end of the bond. You can also represent a polar bond by drawing the Lewis structure and adding the symbols for the partial charges as shown for the H—Cl bond on the left side of Figure 5.21. Chemists often use an arrow above the chemical symbol to indicate the direction of the negative end of the polar covalent bond as shown on the right side of Figure 5.21.

Since a polar covalent bond has some similarities with an ionic bond, chemists sometimes describe them stating that the covalent bond has a certain "percent ionic character." Conversely, you can also describe a bond as having a "percent covalent character." Table 5.1 lists the electronegativities differences that represent the percent of ionic and covalent character.

**Table 5.1** Character of Bonds

| Electronegativity difference | 0.00 | 0.65 | 0.94 | 1.19 | 1.43 | 1.67 | 1.91 | 2.19 | 2.54 | 3.03 |
|---|---|---|---|---|---|---|---|---|---|---|
| Percent ionic character | 0% | 10% | 20% | 30% | 40% | 50% | 60% | 70% | 80% | 90% |
| Percent covalent character | 100% | 90% | 80% | 70% | 60% | 50% | 40% | 30% | 20% | 10% |

Complete the following Practice Problems to enhance your understanding of the relationship between electronegativity and bond type.

### Practice Problems

7. For each of the following pairs of atoms, predict whether a bond between them will be nonpolar covalent, polar covalent, or ionic.
   (a) carbon and fluorine
   (b) oxygen and nitrogen
   (c) chlorine and chlorine
   (d) copper and oxygen
   (e) silicon and hydrogen
   (f) sodium and fluorine
   (g) iron and oxygen
   (h) manganese and oxygen

8. For each of the polar bonds in problem 7, indicate the locations of the partial positive charges. Use an arrow over the element symbols to indicate the bond dipole.

9. Arrange the bonds in each set in order of increasing polarity. (For this problem, assume that an ionic bond is a completely polarized bond.)
   (a) hydrogen bonded to chlorine, oxygen bonded to oxygen, nitrogen bonded to nitrogen, sodium bonded to chlorine
   (b) carbon bonded to chlorine, magnesium bonded to chlorine, phosphorous bonded to oxygen, nitrogen bonded to nitrogen

## Bond Length and Strength

How long is an average chemical bond? Obviously, they are extremely short. Nevertheless, there is a significant amount of variation depending on the number and distribution of charges in the atoms. Figure 5.22 illustrates the interactions that determine the length of a bond. The nuclei repel each other and the electrons repel each other. Conversely, the nuclei and the electrons attract each other. The bond length is the point at which the difference between the attractive forces and the repulsive forces is greatest.

Average bond lengths for some commonly occurring bonds are listed in Table 5.2. The values are given in picometres (1 pm = $1 \times 10^{-12}$ m). The values in the table are averages of the bond type from several different molecules. The other bonds and atoms in the molecule will have a small effect on bond length of any given bond. As you would probably predict, double bonds are shorter than single bonds between the same two atoms and triple bonds are shorter than double bonds. What other characteristics would you predict would affect bond length?

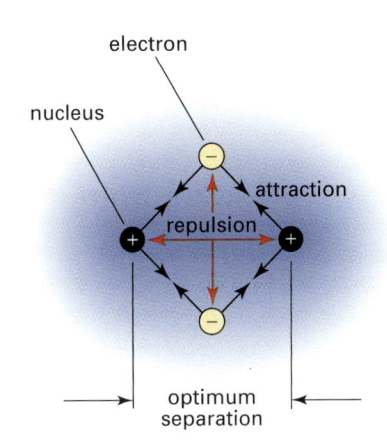

**Figure 5.22** The magnitude of the charges in the nuclei, the energy level of the electrons, and the number of shared electrons all influence the length of a chemical bond.

**Table 5.2** Average Bond Lengths

| Bond | Average bond length (pm) |
|---|---|
| C—C | 154 |
| C=C | 134 |
| C≡C | 121 |
| Si—Si | 234 |
| C—H | 109 |
| Si—H | 148 |
| H—H | 74 |
| C—O | 143 |
| C=O | 123 |
| O—H | 96 |
| O=O | 121 |

**Table 5.3** Average Bond Energies

| Bond | Average bond energy (kJ/mol) |
|---|---|
| C—C | 346 |
| C=C | 610 |
| C≡C | 835 |
| Si—Si | 226 |
| C—H | 413 |
| Si—H | 318 |
| H—H | 432 |
| C—O | 358 |
| C=O | 745 |
| O—H | 467 |
| O=O | 498 |

As you read at the beginning of the chapter, most atoms found in nature are found in compounds, bonded to other atoms. This observation is a strong indication that compounds are more stable than free atoms. Chemists have verified this prediction and determined that energy is released when bonds are formed. Clearly, then, energy would be required to break the bonds. Strong bonds would require more energy to break than weaker bonds. The amount of energy that would be required to break a specific bond in one mole of molecules ($6.02 \times 10^{23}$ molecules) is called the **bond energy**. Once again, you would probably predict that triple bonds have a greater bond energy than double bonds between the same atoms. As well, you might predict that double bonds have a larger bond energy than single bonds. Verify these predictions by comparing values in Table 5.3, Average Bond Energies. Values for bond energies and bond lengths for more bonds can be found in Appendix E.

Knowing bond energies allows you to predict whether a chemical reaction will be endothermic—require energy in order to proceed—or exothermic—release energy. You can compare the bond energies of the reactants and the products in a reaction. If the products have total bond energies larger than the reactants, energy would be released—the reaction would be exothermic. If the reactants have total bond energies larger than the products, more energy would be required to break the bonds of the reactants than would be released when the products were formed. The reaction would be endothermic.

### Section Summary

Bond categories are not distinct properties. When you compare the properties of a large number of bonds, you discover that they form a continuum—from almost completely ionic, to very polar but covalent, to completely non-polar covalent. The difference of the electronegativities of the atoms between which the bond forms provides a way to predict the nature of the bond.

Bond lengths vary from just under 100 pm to just under 300 pm. Bond lengths are roughly related to bond energies. Bond energy is the amount of energy required to break the bond. Shorter bonds are often stronger than longer bonds.

### Section Review

**1** Define electronegativity.

**2** In your own words, describe and explain the periodic trends for electronegativity.

**3** Arrange the elements in each set in order of increasing attraction for electrons in a bond.

(a) Li, Br, Zn, La, Si

(b) P, Ga, Cl, Y, Cs

**4** Determine $\Delta EN$ for the following pairs of elements. State the type of bond that forms between the atoms of the pairs of elements.

(a) N and O

(b) Mn and O

(c) H and Cl

(d) Ca and Cl

**5** Define bond energy.

# CHAPTER 5 Review

## Reflecting on Chapter 5

Summarize this chapter in the form of your choice. Here are a few ideas to use as guidelines:
- Elements combine to form a wide variety of compounds.
- Lewis structures can represent the formation of ionic and molecular compounds according to the octet rule.
- Compare the bonding between a metal and a nonmetal with the bonding between nonmetals.
- Individual "molecules" do not exist within ionic compounds.
- Metal atoms bond by losing their valance electrons and then the metal ions are embedded in a "sea" of electrons in the free-electron model.
- The electronegativity difference between elements can be used to predict the kinds of bonds that will form between atoms of the elements.
- Bond types form a continuum.

## Reviewing Key Terms

For each of the following terms, write a sentence that shows your understanding of its meaning.

valance electrons
electron pair
octet rule
crystal lattice
bonding electrons
single covalent bond
triple covalent bond
free-electron model
electronegativity
polar covalent bond
Lewis structure
unpaired electron
ionic bond
covalent bond
lone pair
double covalent bond
structural formula
metallic bond
non-polar covalent bond
bond energy

## Knowledge/Understanding

1. Give two examples of ionic compounds and two examples of molecular compounds.
2. Since there are only 92 naturally occurring elements, how can the thousands of compounds have such a wide variety of properties? How does a property of noble gases lead to the octet rule?
3. Explain the meaning of the letters and the dots in a Lewis structure.
4. Use Lewis structures to predict the formulas of the following ionic compounds:
   (a) potassium bromide
   (b) calcium fluoride
   (c) magnesium oxide
   (d) lithium oxide
5. When you look at the Lewis structure of an element, how can you predict the number of electrons that will most likely be involved in bonding and the number of electrons that will most likely *not* be involved in bonding?
6. Draw Lewis structures to represent each covalent compound.
   (a) $SiO_2$
   (b) $OBr_2$
   (c) $ClF$
   (d) $NF_3$
7. Describe, in detail, what happens when an ionic bond forms between calcium and chlorine. Use Lewis structures to illustrate the formation of the bond.
8. Describe the structure of an ionic compound such as sodium chloride and explain the concept of an ionic bond as it related to that structure.
9. What does the formula $K_2S$ tell you about a crystal of potassium sulfide?
10. Explain the meaning of "lone pair" and "bonding pair" as these terms apply to covalently bonded atoms.
11. Since every oxygen atom needs three additional electrons to form a stable octet, how can two oxygen atoms bond together to make an oxygen molecule in which both oxygen atoms have an octet of electrons?
12. Explain the relationship between structural formulas for a molecule and a Lewis structure for the same molecule.
13. Explain why bonding in pure metals cannot be explained by either ionic bonds or covalent bonds.
14. Describe the current model for bonding in metals.
15. Define electronegativity.
16. Discuss the periodic trends in electronegativity. In other words, as you go across a period on the periodic table, how are electronegativities of the elements changing? As you go down a group on the periodic table, how are the electronegativities changing?

Answers to questions highlighted in red type are provided in Appendix A.

17. How does the motion of bonding electrons differ between a non-polar covalent bond and a polar covalent bond?
18. Calculate $\Delta EN$ for bonds between each of the following pairs of elements.
    (a) zinc and oxygen
    (b) magnesium and iodine
    (c) cobalt and chlorine
    (d) nitrogen and oxygen
19. Indicate whether each bond in question 18 is ionic or polar covalent.
20. Distinguish between a nonpolar covalent bond and a polar covalent bond.
21. Distinguish between an ionic bond and a polar covalent bond.
22. Explain why units of an ionic compound should not be called molecules.
23. Explain why metal substances are not included in the continuum from ionic to covalent bonds.
24. Without calculating $\Delta EN$, arrange each set of bonds from most polar to least polar. Then calculate $\Delta EN$ for each bond to check your arrangements.
    (a) Mn and O, Mn and N, Mn and F
    (b) Be and F, Be and Cl, Be and Br
    (c) Ti and Cl, Fe and Cl, Cu and Cl, Ag and Cl, Hg and Cl
25. Explain the meaning of the title of Section 5.3, "Bonds as a Continuum."
26. Describe the interactions that determine the length of a covalent bond.
27. The total bond energies of the reactants in a reaction is 1432 kJ/mol and the total bond energies of the produces is 1874 kJ/mol. Is the reaction endothermic or exothermic? Explain your reasoning.

## Inquiry

28. The name "Lewis structure" was named for Gilbert Newton Lewis. Do library or internet research to learn about Lewis and his contributions to chemistry.
29. Do research to find out how chemists were able to determine bond lengths.

## Communication

30. Explain how you would predict the number of valance electrons in the elements of the second period (Li, Be, B, C, N, O, F, and Ne) if you did not have access to the periodic table. Use Lewis structures to illustrate your explanation.
31. Create a concept map to summarize what you learned in this chapter about the nature of chemical bonds.
32. Compare and contrast ionic bonding and metallic bonding. Include the following ideas.
    - Metals do not bond to other metals in definite ratios. Metals do bond to nonmetals in definite ratios.
    - Ionic compounds do not conduct electricity but metallic compounds do.
    
    Make a poster that explains and gives examples of the concept that bonding is a continuum.

## Making Connections

33. Locate three cleaning products in your home and read the labels. For each product, predict whether it is an ionic compound or a molecular compound.
34. The bonds in water are very polar. Table salt (NaCl) dissolves in water. Vegetable oil has only nonpolar bonds and does not dissolve in water. Make a prediction about the way that polar bonds influence solubility of a compound in water.

**Answers to Practice Problems and Short Answers to Section Review Questions**

**Practice Problems**

1.

| H | | | | | | | He |
|---|---|---|---|---|---|---|---|
| Li | Be· | B· | ·C· | ·N· | ·O: | ·F: | :Ne: |
| Na | Mg· | Al· | ·Si· | ·P· | ·S: | ·Cl: | :Ar: |
| K | Ca· | | | | | | |

2. (a) H, Li, Na, K, Rb, Cs, Fr   (b) Sr·   (c) ·Br:, ·I:

3. (a) K⌐·Cl:   [K]⁺[:Cl:]⁻   KCl
   (b) Ba⌐·Ö:   [Ba]²⁺[:Ö:]²⁻   BaO
   (c) Na⌐·F:   [Na]⁺[:F:]⁻   NaF

(d)  Ca⋯Ö:  [Ca]²⁺ [:Ö:]²⁻  CaO

(e) 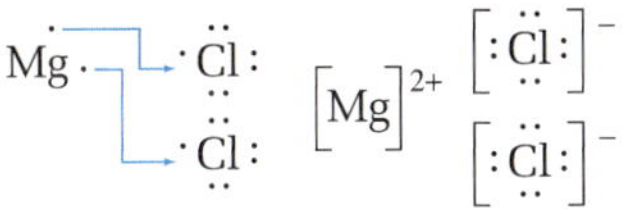 Mg⋯·Cl:  [Mg]²⁺ [:Cl:]⁻  MgCl₂
        ·Cl:         [:Cl:]⁻

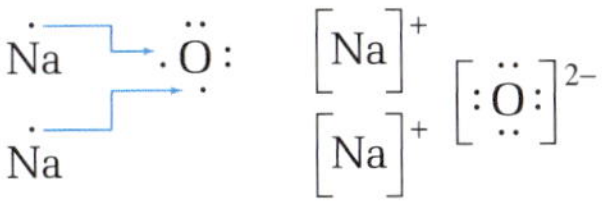

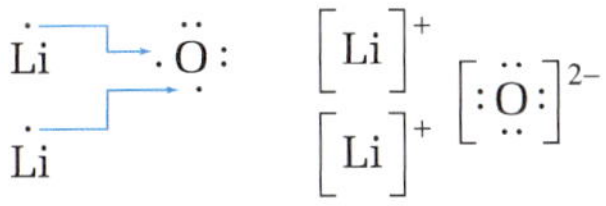

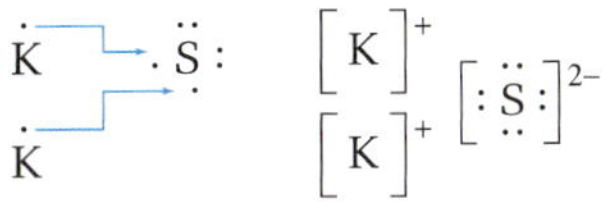

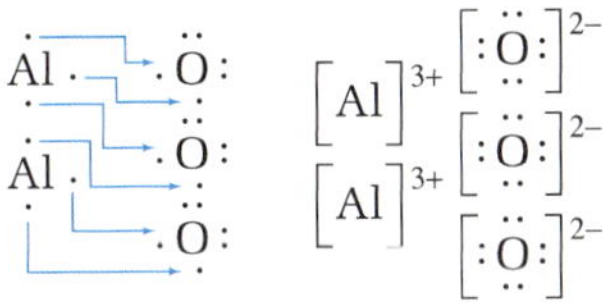

(f) Na⋯·Ö:  [Na]⁺ [:Ö:]²⁻  Na₂O
    Na      [Na]⁺

(g) Li⋯·Ö:  [Li]⁺ [:Ö:]²⁻  Li₂O
    Li      [Li]⁺

(h) K⋯·S:  [K]⁺ [:S:]²⁻  K₂S
    K      [K]⁺

(i) Al⋯·Ö:  [Al]³⁺ [:Ö:]²⁻
    Al⋯·Ö:  [Al]³⁺ [:Ö:]²⁻  Al₂O₃
       ·Ö:         [:Ö:]²⁻

**4.** :S::C::S:  **5.** H:C:::N:  **6.** H:C:::C:H

**7.(a)** polar covalent **(b)** slightly polar covalent **(c)** non-polar covalent **(d)** polar covalent **(e)** slightly polar covalent **(f)** ionic **(g)** polar covalent **(h)** ionic  **8.(a)** C⟶F **(d)** Cu⟶O **(g)** Fe⟶O
**9.(a)** O—O = N—N, H—Cl, Na—Cl **(b)** N—N, C—Cl, P—O, Mg—Cl

**Section Review 5.1** **1.** gold/silver-solid are malleable, ductile, good conductors of electric current and heat; nitrogen, oxygen, hydrogen, noble gases are gases, odorless, colourless **2.** unpaired electrons **3.** check periodic table, it is a group VIA element therefore it has six valence electrons.

**4.** ·Li  ·N·  :Ar:  Cs  Ba·  ·I:  Ra·

**5.2: 1.** The statement, "In general, the farther away two elements are from each other on the periodic table, the more likely they are to participate in ionic bonding." is correct. In general, the lowest electronegativities are on the left side of the periodic table and the highest are on the right, neglecting Group VIIIA.

**2.(a)**
Mg⋯·F:  [Mg]²⁺ [:F:]⁻  MgF₂
   ·F:          [:F:]⁻

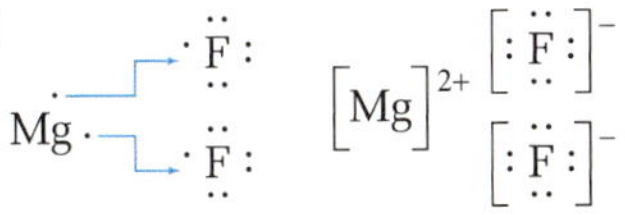

**(b)**  K⋯·Br:  [K]⁺ [:Br:]⁻  KBr

**(c)**  Rb⋯·Cl:  [Rb]⁺ [:Cl:]⁻  RbCl

**(d)**  Ca⋯·Ö:  [Ca]²⁺ [:Ö:]²⁻  CaO

**3.** In covalent bonding, two specific electrons are shared between two specific nuclei. In metallic bonding, all of the electrons are shared by all of the nuclei.

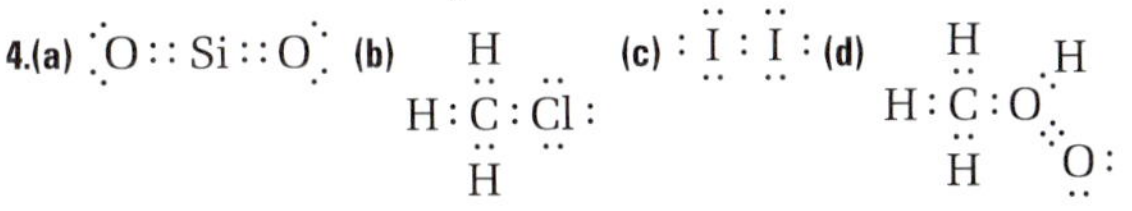

**4.(a)** :O::Si::O: **(b)** 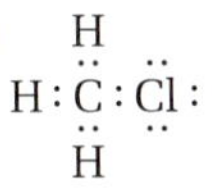 H:C:Cl: with H above and below **(c)** :I:I: **(d)** 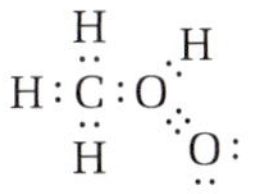 H:C:O: with H's and O arrangement

**5.3: 1.** Electronegativity is the degree to which a nucleus attracts the electrons in a chemical bond. **2.** In general, neglecting Group VIIIA, electronegativity increases from left to right across the periodic table and decreases from top to bottom. **3.(a)** Li, La, Zn, Si, Br **(b)** Cs, Y, Ga, P, Cl
**4.(a)** 0.40, slightly polar covalent **(b)** 1.89, mostly ionic
**(c)** 0.96, polar covalent **(d)** 2.16, mostly ionic

Chapter 5 Chemical Bonding • MHR  **183**

# CHAPTER 6
# Structure and Properties of Substances

## Chapter Preview

- 6.1 Covalent Bonds and Structures
- 6.2 Intermolecular Forces
- 6.3 Structure Determines Properties

## Prerequisite Concepts and Skills

Before you begin this chapter, review the following concepts and skills:

- predicting the number of valence electrons of main group metals and of nonmetals using periodic table (Chapter 5)
- drawing Lewis structures for atoms and simple molecules (Chapter 5)
- determining bond polarity (Chapter 5)

A water molecule has a bent shape, while a carbon dioxide molecule is linear. An ammonia molecule looks like a pyramid, and sulfur hexafluoride (one of the most dense and most stable gaseous compounds) is shaped like an octahedron. In fact, all molecules in nature have a specific shape, which is important to their chemistry.

Why does the shape of a molecule matter? To answer this question, consider that, right now, each nerve cell in your brain is communicating with adjacent nerve cells by releasing molecules called neurotransmitters that diffuse from one cell to the next. Enzymes are assisting in the chemical breakdown of food in your digestive system. The aroma of cologne, desk wood, or cleansers that you may be smelling is a result of odorous molecules migrating from their sources to specific sites in your nasal passages.

Each of these situations depends on the ability of one molecule with a specific shape to "fit" into a precise location with a corresponding shape. The properties of substances are determined by the ways in which particles bond together, the forces that act within and among the compounds they form, and the shapes that result from these interactions.

In this chapter, you will discover how and why each molecule has a characteristic shape, and how molecular shape is linked to the properties of substances. You will also consider the importance of molecular shape to the development of materials with specific applications such as the Kevlar™ in the photograph below.

What atomic and molecular properties explain the macroscopic properties of the fabric used to make these protective gloves?

# Covalent Bonds and Structures

## 6.1

In Chapter 5, you learned about the different types of bonds and their individual properties. In this section, you will look at entire molecules and how the properties of the bonds influence the shape of the molecule as a whole. You have already had a glance into the structure of substances stabilized by ionic bonds. The ions form a crystal lattice in a way that allows the positive and negative charges to be as close to each other as possible. When you look at a macroscopic object, you can frequently determine from its appearance, whether it is a crystal. Large ionic crystals, such as those in Figure 6.1, usually have some flat faces and distinct patterns created by the orderliness of the atoms. These crystals exist in a variety of shapes from nearly cubical to needle-like. However, molecular compounds exist as single molecules, which have a large variety of shapes. In this section, you will focus on the way that covalent bonds determine the structure of molecules.

### Section Preview/Outcomes

In this section, you will

- **draw** Lewis dot diagrams and structural formulas for simple molecules
- **determine** the shapes about central atoms by applying VSEPR theory to electron dot diagrams
- **build** models depicting the shape of simple covalent molecules
- **illustrate** and explain the formation of network covalent solids

**Figure 6.1** The macroscopic structure of a crystal often reveals the shape of the tiny repeating units within the crystal.

## Covalent Bonds and Molecules

Molecular compounds have a much greater variety of shapes than ionic compounds. Molecules are distinct from each other because the atoms in one molecule are connected by covalent bonds. As you know, the properties of covalent bonds vary over a wide range from nearly ionic to non-polar covalent. These varying properties have a great influence on the shapes of whole molecules. While studying this section, keep in mind the fact that in addition to molecules many polyatomic ions contain nonmetal atoms held together by covalent bonds. In Chapter 1, you reviewed the composition of several of these common polyatomic ions, eg. $NH_4^+$, $CO_3^{2-}$. You will learn that they have the same shapes as neutral molecules but the charges influence their properties.

As you begin to analyze and predict structures and properties of molecules, Lewis structures are extremely helpful. In Chapter 5, you practiced drawing Lewis structures for some very simple molecules. You might have already discovered that the "guess and check" method can become quite perplexing, even for simple molecules. The method and Sample Problems below will give you a consistent method for drawing Lewis structures for many relatively small molecules.

### Drawing Lewis Structures for Simple Molecules and Ions with a Central Atom

**Step 1** Determine the total number of valence electrons in all of the atoms in the molecule.

**Step 2** Draw a skeleton structure for the molecule by choosing the atom with the largest number of unpaired electrons as the central atom. Join the atoms with one pair of bonding electrons.

**Step 3** Place lone pairs of electrons around all of the atoms *except* the central atom to form an octet of electrons. Hydrogen, of course, has only two valence electrons.

**Step 4** (a) If all of the valence electrons determined in Step 1 have not been used, add a one or more lone pairs around the central atom to complete an octet of electrons.

(b) If all of the valence electrons have been used up but the central atom still does not have an octet of electrons, move one or more of the lone pairs to form double or triple bonds between the central atom and an adjacent atom.

### Sample Problem

#### Drawing the Lewis Structure of a Molecule

**Problem**
Draw the Lewis structure for methanal (formaldehyde), $CH_2O$.

**Solution**
The molecular formula, $CH_2O$, tells you the number of each kind of atom in the molecule. Following steps 1 to 4 from the procedure outlined above:

**Step 1** Determine the total number of valence electrons in all of the atoms in the molecule.

$$\left(1 \text{ C atom} \times \frac{4e^-}{\text{C atom}}\right) + \left(1 \text{ O atom} \times \frac{6e^-}{\text{O atom}}\right) + \left(2 \text{ H atoms} \times \frac{1e^-}{\text{H atom}}\right)$$
$$= 4e^- + 6e^- + 2e^-$$
$$= 12e^-$$

**Step 2** Draw a skeleton structure of the atom with one pair of bonding electrons between each of the atoms. To decide which atom is the central atom, note that carbon has four unpaired electrons, oxygen has two unpaired electrons, and hydrogen has one unpaired electron. Therefore, choose carbon as the central atom.

O
H:C:H

**Step 3** Place lone pairs of electrons around the oxygen atom to complete an octet of electrons.

:Ö:
H:C:H

**Step 4** Compare the number of electrons in the structure so far with the number determined in Step 1. There are 12 electrons in the structure, which is the same as the total number of valence electrons determined in Step 1. However, carbon does not have an octet of electrons. Therefore, move one of the lone pairs around the oxygen atom to the position between the oxygen and carbon atoms to form a double bond.

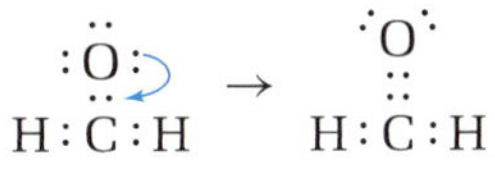

$$\begin{array}{c} :\ddot{O}: \\ H:C:H \end{array} \rightarrow \begin{array}{c} \cdot\ddot{O}\cdot \\ :: \\ H:C:H \end{array}$$

### Check Your Solution

Each atom has achieved a noble gas electron configuration. Thus, you can be confident that this is a reasonable Lewis structure.

## Co-ordinate Covalent Bonds

You have learned that a covalent bond is the sharing of a pair of electrons between two atoms. Normally, each atom contributes one electron to the shared pair. In some cases, such as the ammonium ion, $NH_4^+$, however, one atom contributes both of the electrons to the shared pair. The bond in these cases is called a **co-ordinate covalent bond**. Once a co-ordinate covalent bond has formed, it behaves in the same way as any other single covalent bond. The next Sample Problem involves a polyatomic ion with a co-ordinate covalent bond.

### Sample Problem

#### Lewis Structures with a Co-ordinate Covalent Bond

**Problem**

Draw the Lewis structure for the ammonium ion, $NH_4^+$.

**Solution**

The formula, $NH_4^+$, tells you the number of atoms of each element in the ion. Follow steps 1 to 4 in the procedure outlined above.

**Step 1** Determine the total number of valence electrons in one nitrogen atom and four hydrogen atoms. Subtract one electron to give the ion its charge.

$$\left(1 \text{ N atom} \times \frac{5e^-}{\text{N atom}}\right) + \left(4 \text{ H atoms} \times \frac{1e^-}{\text{H atom}}\right) - 1e^-$$
$$= 8e^-$$

**Step 2** Draw a skeleton structure of the atom with one pair of bonding electrons between each of the atoms. To decide which atom is the central atom, note that nitrogen has three unpaired electrons and hydrogen has one unpaired electron. Therefore, choose nitrogen as the central atom.

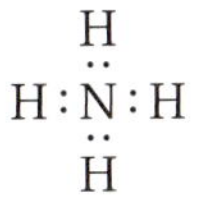

$$\begin{array}{c} H \\ H:\ddot{N}:H \\ H \end{array}$$

**Step 3** There are no lone pairs in hydrogen so no more electrons should be added at this step.

*Continued ...*

> **PROBLEM TIP**
> Because you are working with a polyatomic cation, you must account for the positive charge. Remember to subtract one electron from the total number of valence electrons.

**Step 4** Compare the number of electrons in the structure so far with the number determined in Step 1. There are eight electrons in the structure, which is the same as the total number of valence electrons determined in Step 1. The nitrogen atom has an octet of electrons and each hydrogen atom has its full complement of two electrons. However, to correctly write the structure, you must indicate that it is an ion with a charge of +1 as shown below.

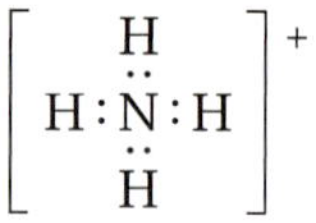

$$\left[\begin{array}{c} H \\ H:\overset{..}{N}:H \\ H \end{array}\right]^+$$

**Check Your Solution**
Each atom has achieved a noble gas electron configuration. The positive charge on the ion is included. This is a reasonable Lewis structure for $NH_4^+$.

Note that the Lewis structure for $NH_4^+$ does *not* indicate which atom provides each shared pair of electrons around the central nitrogen atom. However, if you think of the ammonium ion as the addition of a hydrogen ion ($H^+$) to an ammonia molecule ($NH_3$), you can see (Figure 6.2) that the two electrons that are shared by the ammonia molecule and the added hydrogen ion, are provided by the nitrogen atom. Once the hydrogen ion has combined with the ammonia molecule, it is not possible to distinguish any of the hydrogen atoms from each other. The 1+ charge becomes a property of the entire ammonium ion.

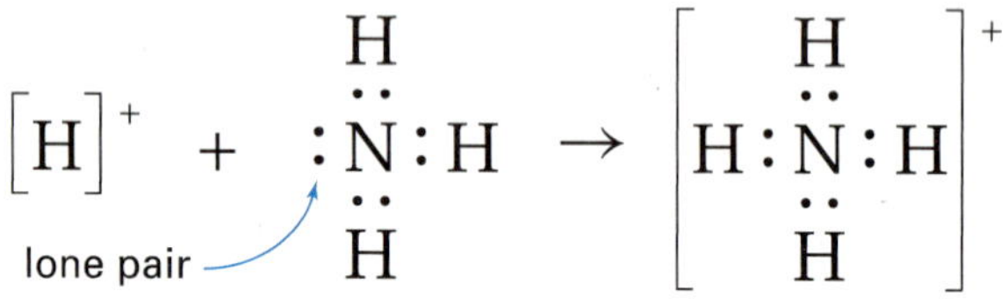

$$[H]^+ \ + \ :\overset{..}{N}:H \ \rightarrow \ \left[\begin{array}{c} H \\ H:\overset{..}{N}:H \\ H \end{array}\right]^+$$

lone pair

**Figure 6.2** The hydrogen ion is attracted to the lone pair on the ammonia molecule. A reaction occurs in which the lone pair becomes a bonding pair. The ammonium ion is now symmetrical and the charge is a property of the entire ion.

### Resonance Structures: More Than One Possible Lewis Structure

Imagine that you are asked to draw the Lewis structure for sulfur dioxide, $SO_2$. A typical answer would look like this:

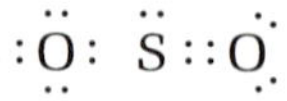

$$:\overset{..}{O}: \ \overset{..}{S}::\overset{..}{O}:$$

This Lewis structure suggests that $SO_2$ contains a single bond and a double bond. However, experimental measurements of bond lengths indicate that the bonds between the S and each O are identical. The two bonds have properties that are somewhere between a single bond and a double bond. In effect, the $SO_2$ molecule contains two "one-and-a-half" bonds.

To communicate the bonding in $SO_2$ more accurately, chemists draw two different resonance Lewis structures side by side separated by a double-headed arrow. **Resonance structures** are models that give the same relative position of atoms as in Lewis structures, but show different placing of their bonding pairs and lone pairs.

$$:\overset{..}{O}:\overset{..}{S}::\overset{..}{O}: \ \longleftrightarrow \ :\overset{..}{O}::\overset{..}{S}:\overset{..}{O}:$$

Many molecules and ions—especially organic ones—require resonance structures to represent their bonding. It is essential to bear in mind, however, that resonance structures do *not* exist in reality. For example, SO₂ does not shift back and forth from one structure to the other. An actual SO₂ molecule is a combination—a hybrid—of its two resonance structures. It is more properly called a *resonance hybrid*. When drawing bonds as lines, chemists sometimes draw the "half bond" as a dotted line as shown here. You can imagine one pair of electrons as resonating across the entire molecule from one oxygen atom to the other.

$$O\cdots S\cdots O$$

## Practice Problems

1. Draw Lewis structures for each of the following molecules.
   - (a) NH₃
   - (b) CH₄
   - (c) CF₄
   - (d) AsH₃
   - (e) BrO⁻
   - (f) H₂S
   - (g) H₂O₂
   - (h) ClNO
   - (i) C₂H₄
   - (j) CO
   - (k) HOCl
   - (l) C₂H₂

2. Draw Lewis structures for each of the following ions. (**Note:** Consider resonance structures.)
   - (a) CO₃²⁻
   - (b) NO⁺
   - (c) ClO₃⁻
   - (d) SO₃²⁻

3. Dichlorofluoroethane, CH₃CFCl₂, has been proposed as a replacement for chlorofluorocarbons (CFCs). The presence of hydrogen in CH₃CFCl₂ markedly reduces the ozone-depleting ability of the compound. Draw a Lewis structure for this molecule.

4. Draw Lewis structures for the following molecules. (**Note:** Neither of these molecules has a single central atom.)
   - (a) N₂H₄
   - (b) N₂F₂

5. Although Group 18 (VIIIA) elements are never found in compounds in nature, chemists are able to synthesize compounds of several noble gases, including Xe. Draw a Lewis structure for the XeO₄ molecule. Indicate if co-ordinate covalent bonding is likely a part of the bonding in this molecule.

Lewis structures provide important information about molecules. You have learned how to predict single, double, and triple bonds, co-ordinate covalent bonds, and resonance structures. However, Lewis structures do not give you any information about the shape of the molecule. To analyze and predict the shape, you need a technique by which you can represent molecules in three dimensions. The following topic presents one such method.

### Valence-Shell Electron-Pair Repulsion (VSEPR) Theory

In 1957, an English chemist, Ronald Gillespie, joined Hamilton, Ontario's McMaster University for a long, distinguished career as a professor of chemistry. Earlier that year, Gillespie, in collaboration with a colleague, Ronald Nyholm, had developed a model for predicting the shape of molecules. Like chemistry students around the world since that time, you are about to learn the central features of this model, called the Valence-Shell Electron-Pair Repulsion theory. This is usually abbreviated to VSEPR (pronounced "vesper") theory. The fundamental principle of the **Valence-Shell Electron-Pair Repulsion theory** is that the bonding pairs and lone pairs of electrons in the valence level of an atom repel one another due to their negative charges. The pairs of electrons take up positions as far from each other as possible about the spherical central atom while remaining in the molecule. A lone pair (LP) will spread out more than a bonding pair. Therefore, the repulsion is greatest between lone pairs (LP—LP). Bonding pairs (BP) are more localized between the atomic nuclei, so they spread out less than lone pairs. Therefore, the BP—BP repulsions are smaller than the LP—LP repulsions. The repulsion between a bonding pair and a lone-pair (BP—LP) has a magnitude intermediate between the other two. The order of decreasing repulsion can be expresses as follows.

$$LP-LP > LP-BP > BP-BP$$

Many molecules contain double bonds, two bonding pairs, or triple bonds, three bonding pairs, which act as a single electron group when repelling other electrons about the central atom. By analyzing the types of electron groupings—single bonds, double bonds, triple bonds, or lone pairs—around one central atom, you can predict the shape of the molecule around that atom.

For central atoms that have an octet of electrons in the valence level, you can classify most shapes into five categories. You will start with central atoms for which all of the eight valance level electrons are in bonding pairs. Consequently, all of the electron groups repel each other to the same extent. The central atom can have two, three, or four electron groupings.

A combination of two double bonds or a single bond and a triple bond results in two electron groupings. As shown in Figure 6.3A, two electron groupings results in a **linear** shape. Examples of molecules with the two possible sets of electron groupings are also shown in the figure. Carbon dioxide ($CO_2$) has two double bonds and hydrogen cyanide (HCN) has one double bond and one triple bond. The **bond angle**, the angle formed by two bonds, in linear molecules is 180°.

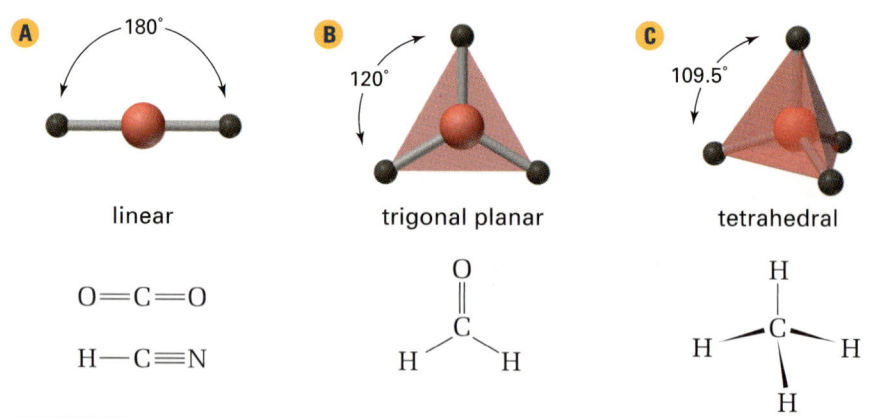

**Figure 6.3** When all eight valance electrons in a central atom are part of bonding pairs, the three shapes that molecules can assume are linear, trigonal planar, and tetrahedral. (A) Carbon dioxide and hydrogen cyanide are linear molecules. (B) Methanal is a trigonal planar molecule and (C) methane is a tetrahedral molecule.

Three electron groupings around a central atom can be achieved by a combination of two single bonds and one double bond resulting in a **trigonal planar** shape, as shown in Figure 6.3 B. Methanal ($CH_2O$) is an example of a trigonal planar molecule. Notice in the figure that all three bond angles are 120°.

The third shape with all bonding pairs has four single bonds. The resulting shape is called **tetrahedral**. Methane (CH₄) or natural gas is a tetrahedral molecule. The molecule is completely symmetrical having four bond angles that are each 109.5°. As you can see in Figure 6.3, linear and trigonal planar shapes are two-dimensional and can easily be drawn using chemical symbols and lines for bonds. Tetrahedral shapes are three-dimensional and require other techniques for illustrating the shape of the molecule. One method that chemists sometimes use is shown in Figure 6.3 C. Wedge shaped lines, indicating that the bond is coming from the central atom out of the plane of the page, start with a point at the central atom and become wider as they go toward the bonded atom. To indicate that a bond goes into the plane of the page, the wide end of the wedge shaped line is larger at the central atom and narrows to a point at the bonded atom.

As you have seen, shapes involving only bonding pairs are very symmetrical because the repulsive forces between all electron groupings are the same. When lone pairs are introduced into a shape, the forces between electron groupings are no longer the same resulting is a distortion of the symmetry. The two shapes that include lone pairs are shown in Figure 6.4. Both shapes are derived from a tetrahedral shape because they have four electron groupings extending from the central atom. One lone pair and three single bonds result in a **pyramidal** shape as shown in Figure 6.4 A. The lone pair exerts a stronger force on the bonded pairs than they do on each other, reducing the bond angle to less than the 109.5° found in the tetrahedral shape. However, it is not possible to predict exactly how much the angle will be changed because the result depends on the properties of the elements and the bonds between the atoms. In the example given in the figure, ammonia, the bond angle is 107.3°.

> **Biology LINK**
>
> Summer in Canada brings mosquitoes—and mosquito repellents. The key chemical used in DEET (the active ingredient in most insect repellents) is N,N-diethyl-3-methyl-benzamide (N,N-dimethyl-m-toluamide). This molecule does not, in fact, repel biting insects but blocks their chemoreceptors for carbon dioxide and lactic acid—two substances that attract mosquitoes to people and other warm-blooded hosts. Can you find other examples in which molecular shape determines biological responses?

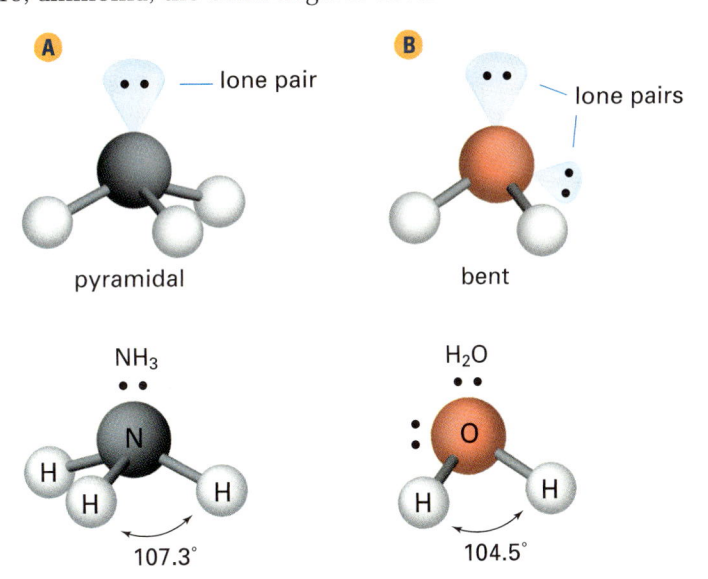

**Figure 6.4** Pyramidal and bent shapes are related to tetrahedral shapes but bond angles are smaller than the 109.5° found in tetrahedral molecules that have all bonding pairs.

When the central atom has two lone pairs and two single bonds, the resulting shape is **bent**. The two lone pairs exert a greater force on each other than on the bonding pairs and reduce the angle between the two bonds even more than in the pyramidal shape. In water, as shown in Figure 6.4 B, you see that the bond angle is 104.5°.

## Predicting Molecular Shape

You have seen and simulated the shapes of molecules. Now you can use the steps below to help you predict the shape of a molecule (or polyatomic ion) that has one central atom. Refer to these steps as you work through the Sample Problem and the Practice Problems that follow. You can also refer to Table 6.1, which contains a summary of the properties of the five molecular shapes.

**Table 6.1** Summary of Molecular Shapes

| Number of electron groups | Name of molecular shape | Type of electron pairs | Shape | Example |
|---|---|---|---|---|
| 2 | linear | all BP | | $CO_2$ |
| 3 | trigonal planar | all BP | | $CH_2O$ |
| 4 | tetrahedral | all BP | | $CH_4$ |
| 4 | pyramidal | 3 BP, 1LP | | $NH_3$ |
| 4 | bent | 2 BP, 2LP | | $H_2O$ |

1. Draw a preliminary Lewis structure of the molecule based on the formula given.
2. Determine the total number of electron groups around the central atom. Remember that a double bond or a triple bond is counted as one electron group.
3. Determine the types of the electron groupings (bonding pairs or lone pairs).
4. Determine which one of the five shapes will accommodate this combination of electron groups.

> **Web LINK**
>
> www.mcgrawhill.ca/links/atlchemistry
>
> To find help visualizing the geometry of molecular shapes, go to the web site above and click on **Electronic Learning Partner**.

### Sample Problem

#### Predicting Molecular Shape for a Simple Compound

**Problem**

Determine the molecular shape of the hydronium ion, $H_3O^+$. (**Note:** Hydrogen ions, $H^+$, do not exist free in water solutions. They bond to water molecules producing hydronium ions.)

**Plan Your Strategy**

Follow the four-step procedure that helps you to predict molecular shape. Use Table 6.1 for a summary of the molecular shapes.

**Act on Your Strategy**

**Step 1** A possible Lewis structure for $H_3O^+$ is:

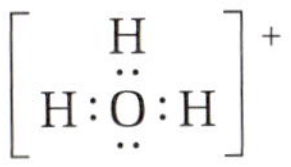

$$\left[ \begin{array}{c} H \\ H : \overset{..}{\underset{..}{O}} : H \end{array} \right]^+$$

**Step 2** The Lewis structure shows 4 electron groupings around the central oxygen atom.

**Step 3** The Lewis structure shows 3 bonding groups and 1 lone pair (LP) around the central oxygen atom.

**Step 4** For 3 bonding groups and 1 lone pair (LP), the molecular shape is pyramidal.

**Check Your Solution**

The answer is in agreement with the shapes in Table 6.1.

### Practice Problems

6. Use VSEPR theory to predict the molecular shape for each of the following:
   (a) HCN   (c) $SO_3$   (e) $AsCl_3$   (g) $CH_3F$
   (b) $SO_2$   (d) $SO_4{}^{2-}$   (f) $SI_2$

7. Use VSEPR theory to predict the molecular shape for each of the following:
   (a) $CH_2F_2$   (b) $NH_4{}^+$   (c) $NO_2{}^+$   (d) $NO_2{}^-$

# Tools & Techniques

## AIM Theory and Electron Density Maps

The microscopic world of atoms is difficult to imagine, let alone visualize in detail. Chemists and chemical engineers employ different molecular modelling tools to study the structure, properties, and reactivity of atoms, and the way they bond to one another. Richard Bader, a chemistry professor at McMaster University, has invented an interpretative theory that is gaining acceptance as an accurate method to describe molecular behaviour and predict molecular properties. According to Dr. Bader, shown below, small molecules are best represented using topological maps, where contour lines (which are commonly used to represent elevation on maps) represent the electron density of molecules.

**The diagrams above show two different ways to model the electron density in a plane of a molecule of ethylene.**

In the early 1970s, Dr. Bader invented the theory of "Atoms in Molecules," otherwise known as AIM theory. This theory links the mathematics of quantum mechanics to the atoms and bonds in a molecule. AIM theory adopts electron density, which is related to the Schrödinger description of the atom, as a starting point to mapping molecules.

Chemists often focus on the energetic, geometric, and spectroscopic properties of molecules. However, since the electron density exists in ordinary three-dimensional space, electron density maps of molecules can be used as tools to unearth a wealth of information about the molecule. This information includes, but is not limited to, a molecule's magnetic properties, per-atom electron population, and bond types.

Using diagrams and computer graphics, chemists use AIM theory to construct three-dimensional electron density maps of molecules. Chemists can then use the maps as tools to perform experiments on the computer as if they were performing the same experiments in the laboratory, but with higher efficiency. Many chemists are excited by the prospect of using computer graphics, both inside and outside of the classroom, as teaching tools for students and the public at large.

In addition to being able to simulate very simple isolated molecules and their reactions, AIM theory provides a physical basis for the theory of Lewis electron pairs and the VSEPR model of molecular geometry. Equipped with computers and computer-generated, three-dimensional electron density maps, scientists are able to view molecules and predict molecular phenomena without even having to get off their chairs!

## Polar Bonds and Molecular Shapes

You have learned about polar bonds but are all molecules that have polar bonds polar molecules? An analysis of two very common compounds—water and carbon dioxide—will help you answer this question.

Water is a liquid at standard temperature and pressure. Droplets of water assume a nearly spherical shape when falling as you can see in Figure 6.5. This shape has the smallest surface for the given volume. These facts indicate that water molecules are strongly attracted to each other.

**Figure 6.5** Surface tension, caused by the attractive forces between water molecules accounts for the spherical droplets and the sheeting effect observed in moving water.

Conversely, carbon dioxide is a gas at standard temperature and pressure. However, molecules of water and carbon dioxide each have three atoms. In addition, a molecule of carbon dioxide has over twice the mass of a water molecule. Nevertheless, carbon dioxide molecules appear to have little attraction for each other. To determine the difference in the properties of these two compounds, it would seem logical to compare their bonds. The bonds between carbon and oxygen are polar and the bonds between hydrogen and oxygen are polar. This information does not provide a solution to the problem. Finally, consider the shapes of the two molecules. As you can see in Figure 6.6, the direction of the arrows indicating the bond dipoles of the C—O bonds in carbon dioxide are in exactly opposite directions because carbon dioxide is a linear molecule. One bond cancels the polar effect of the other making molecules of $CO_2$ **non-polar**. Water is a bent molecule so the arrows representing bond polarity are at an angle with each other. This bent shape results in is a net polarity in the water molecule giving it a partially positive end near the H atoms and a partially negative end near the highly electronegative O atom. As a result, water molecules have **molecular dipoles** because they have two partially charged ends—a positive end and a negative end. Water molecules are **polar** molecules.

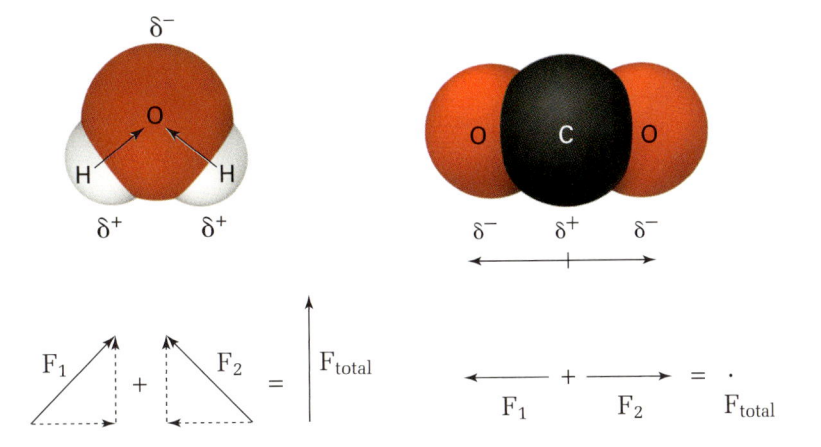

**Figure 6.6** To find out if polar bonds cause a molecule to be polar, analyze the arrows that indicate the direction of the polarity of the individual bonds.

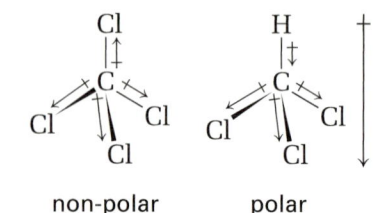

**Figure 6.7** The shapes of the molecules of carbon tetrachloride and trichloromethane are nearly identical yet one is polar and the other is not.

You can predict the polarity of molecules by drawing arrows to represent the bond dipoles in each polar bond and then adding the arrows by vector addition as shown in Figure 6.6 below the molecular models. Notice in the case of the water molecule, that the parts, or components, of the arrows that point toward each other cancel each other while the components that both point upward add together. Another interesting case, shown in Figure 6.7, is a comparison between carbon tetrachloride (CCl$_4$) and trichloromethane (CHCl$_3$) also called chloroform. Both molecules are tetrahedral and all of the bonds are polar. However, the C—Cl bonds are *more* polar than the C—H bond. The result is that CHCl$_3$ is polar while CCl$_4$ is not.

Table 6.2 summarizes most of the possible molecular shapes with polar bonds and shows which result in polar molecules and which produce non-polar molecules. In the table, the letter A represents the central atom while X and Y represent atoms that are bonded to the central atom. Refer to the table while completing the practice problems that follow.

**Table 6.2** Effect of Molecular Shapes on Molecular Polarity

| Molecular shape | Bond polarity | Molecular polarity |
|---|---|---|
| linear | X—A—X | non-polar |
| linear | X—A—Y | polar |
| bent | A with two X | polar |
| trigonal planar | A with three X | non-polar |
| trigonal planar | A with two X and one Y | polar |
| tetrahedral | A with four X | non-polar |
| tetrahedral | A with three X and one Y | polar |

### Practice Problems

8. Use VSEPR theory to predict the shape of each of the following molecules. From the molecular shape and the polarity of the bonds, determine whether or not the molecule is polar.
   (a) CH$_3$F      (b) CH$_2$O      (c) AsI$_3$      (d) H$_2$O$_2$

9. Freon-12, CCl$_2$F$_2$, was used as a coolant in refrigerators until it was suspected to be a cause of ozone depletion. Determine the molecular shape of CCl$_2$F$_2$ and discuss the possibility that the molecule will be polar.

10. Which is more polar NF$_3$ or NCl$_3$? Justify your answer.

# Investigation 6-A

**SKILL FOCUS**
Performing and recording
Analyzing and interpreing

## Modelling Molecules

You cannot see molecules with your eyes or with a light microscope but you can predict their shapes, based on what you know about their electron configurations. In this investigation, you will work with soap bubbles and a kit to build models of molecules.

### Question

How can models of molecules help you predict their shape and polarity?

### Material

**Part 1 Soap Bubbles**

soap solution (mixture of 80 mL distilled water, 15 mL detergent, and 5 mL glycerin)
100 mL beaker
straw
protractor
hard, flat surface
paper towels

**Part 2 Molecular Model Kit**

molecular model kit
pen
paper

### Safety Precautions

- Ensure that each person uses a clean straw.
- Clean up all spills immediately.

### Procedure

**Part 1 Soap Bubbles**

1. Obtain approximately 25 mL of the prepared soap solution in a 100 mL beaker. **CAUTION** Clean up any spills immediately.

2. Wet a hard, flat surface, about 10 cm × 10 cm in area, with the soap solution.

3. Dip a straw into the soap solution in the beaker and blow a bubble on the wetted surface. **CAUTION** Each person must use a clean straw. Blow a second bubble of the same size onto the same surface to touch the first bubble. Record the shape of this simulated molecule and measure the bond angle between the centres of the two bubbles where the nuclei of the atoms would be located.

4. Repeat step 3 with three bubbles of the same size (to simulate three bonding pairs). Record the shape and the bond angle between the centres of the bubbles.

5. Repeat step 3, but this time make one bubble slightly larger than the other two. This will simulate a lone pair of electrons. Record what happens to the bond angles.

6. On top of the group of three equal-sized bubbles, blow a fourth bubble of the same size that touches the centre. Estimate the angle that is formed where bubbles meet.

7. On top of the group of bubbles you made in step 5, make a fourth bubble that is larger than the other three (to simulate a lone pair). Record what happens to the bond angles.

*continued...*

Chapter 6 Structure and Properties of Substances • MHR 197

**continued**

8. Clean your work area using dry paper towels first. Then wipe the area with wet paper towels.

### Part 2 Molecular Model Kit

9. Obtain a model kit from your teacher.

10. Draw a Lewis structure for each molecule below.
    (a) hydrogen bonded to a hydrogen: $H_2$
    (b) chlorine bonded to a chlorine: $Cl_2$
    (c) oxygen bonded to two hydrogens: $H_2O$
    (d) carbon bonded to two oxygens: $CO_2$
    (e) nitrogen bonded to three hydrogens: $NH_3$
    (f) carbon bonded to four chlorines: $CCl_4$
    (g) boron bonded to three fluorines: $BF_3$

11. Build a three-dimensional model of each molecule using your model kit.

12. Sketch the molecular models you have built.

13. In your notebook, make a table like the one below. Give it a title, fill in your data, and exchange your table with a classmate.

| Compound | Lewis structure for compound | Sketch of predicted shape of molecule |
|---|---|---|
|  |  |  |
|  |  |  |

## Analysis

### Part 1 Soap Bubbles

1. What shapes and bond angles were associated with two, three, and four same-sized bubbles? Give an example of a molecule that matches each of these shapes.

2. Give an example of a molecule that matches each of the shapes in steps 5 and 7, where you made one bubble larger than the others.

3. If they are available, use inflated balloons to construct each of the arrangements that you created with soap bubbles. How does this model compare with your bubble models?

### Part 2 Molecular Model Kit

4. Compare your models with the models that your classmates built. Discuss any differences.

5. How did your Lewis structures help you predict the shape of each molecule?

## Conclusion

6. What property of soap bubbles causes them to assume shapes similar to atoms in a molecule?

7. Summarize the strengths and limitations of creating molecular models using molecular model kits.

## Applications

8. Calculate the electronegativity difference for each bond in the molecules you built. Show partial charges. Based on the electronegativity difference and the predicted shape of each molecule, decide whether the molecule is polar or non-polar.

9. Look back through Chapter 6, and locate some different simple molecules. Build models of these molecules. Predict whether they are polar or non-polar.

198 MHR • Unit 2 Structure and Properties

## Network Solids

Do you recall reading in Chapter 5 that pure carbon can be found in two distinct forms—diamond and graphite? Did you wonder how a pure element assumes different forms that have such different properties? The answer lies in yet another example of substances stabilized by covalent bonds. The atoms in these substances, called **network solids**, bond covalently in continuous three-dimensional arrays. Carbon can form several different network solids because the bonding patterns differ. Figure 6.8 shows four allotropic forms of carbon. **Allotropes** are different crystalline or molecular forms of the same element that differ in physical and chemical properties.

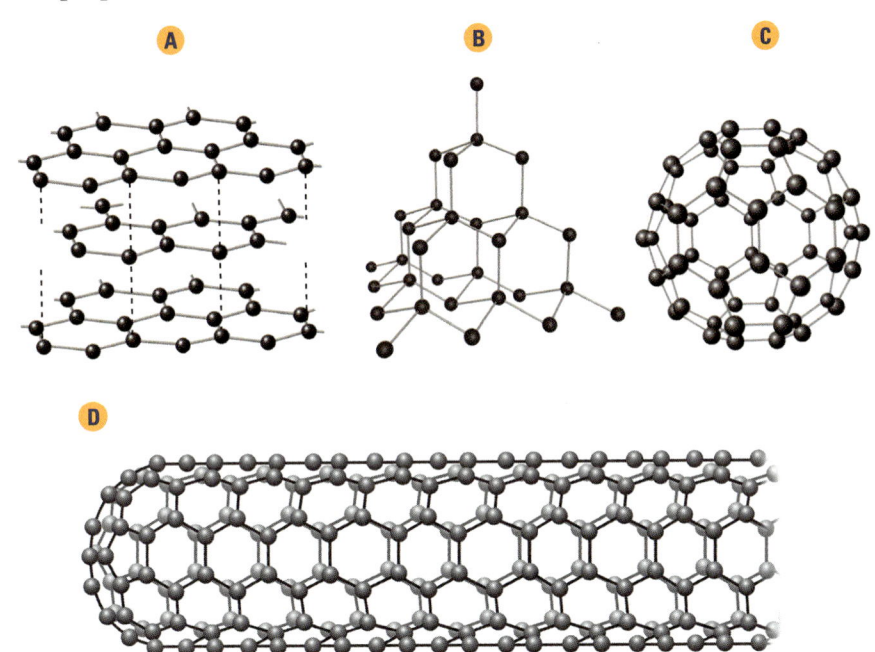

**Figure 6.8** The four different network solids formed by carbon—(A) graphite, (B) diamond, (C) buckminsterfullerene, and (D) nanotube—have dramatically different properties.

The graphite shown in Figure 6.8A is a soft, two-dimensional network solid that is a good conductor of electricity. It has a high melting point, which suggests strong bonds, but its softness indicates the presence of weak bonds as well. Each carbon forms three strong covalent bonds with three of its nearest neighbours in a trigonal planar pattern.

This structure gives stability to the layers. The fourth bond, between the layers, is longer than the other three and is so long and weak that it is easily broken, allowing the layers to slide by one another. Graphite feels slippery as a result of this characteristic. In fact, many industrial processes use graphite as a high temperature lubricant. Also, the "lead" in pencils is actually graphite. The layers of graphite slide past each other so easily that, as you push the pencil across paper, the layers slide off of the pencil onto the paper.

In diamond, each carbon atom forms four strong covalent bonds with four other carbon atoms. This three-dimensional array makes diamond the hardest naturally occurring substance. The valence electrons are in highly localized bonds between carbon atoms. Therefore, diamond does not conduct electricity. The planes of atoms within the diamond reflect light, which gives diamond its brilliance and sparkle. The regular arrangement of the carbon atoms in a crystalline structure, coupled with the strength of the covalent C—C bonds, gives diamond its extreme hardness, and makes it inert to corrosive chemicals.

> **CHEM FACT**
>
> At temperatures around 600°C, and extremely low pressures, methane will decompose and deposit a thin film of carbon, in the form of diamond, on a surface. The process, known as chemical vapour deposition, or CVD, can be used to coat a wide variety of surfaces. The hardness of this diamond film can be used for applications that require non-scratch surfaces such as cookware, eyeglasses, and razor blades.

*Fullerenes* make up a group of spherical allotropes of carbon. Figure 6.8C shows $C_{60}$, which is one type of fullerene discovered in 1985. It was given the name buckminsterfullerene because it resembles the geodesic-domed structure designed by architect R. Buckminster Fuller. Also known as "buckyballs," $C_{60}$ is just one of several fullerenes that have been discovered. Others have been shown to have the formula $C_{70}$, $C_{74}$, and $C_{82}$. Due to their spherical shape, researchers have speculated that fullerenes might make good lubricants. Recently, microscopic-level research has developed very small carbon networks called nanotubes. As you can see in Figure 6.8D, nanotubes are like a fullerene network that has been stretched into a cylinder shape. Nanotubes of $C_{400}$ and higher may have applications in the manufacture of high-strength fibres. In the year 2000, researchers built a nanotube with a diameter of $4 \times 10^{-8}$ m. At that time, the nanotube was the smallest structure assembled.

Many network solids are compounds, not elements. Examples include silica, $SiO_2$, and various metal silicates. If you were asked to draw a Lewis structure representing $SiO_2$, you might draw a molecule that looks similar to $CO_2$. However, unlike the bonding that occurs in $CO_2$, the silicon atom in silica does not form a double bond with an oxygen atom. Instead, each silicon atom bonds to four oxygen atoms to give a tetrahedral network. Each oxygen atom is then bonded to another silicon atom, as shown in Figure 6.9 The network is represented by the formula $(SiO_2)_n$, indicating that $SiO_2$ is the simplest repeating unit. Quartz is a familiar example of a material composed of silica.

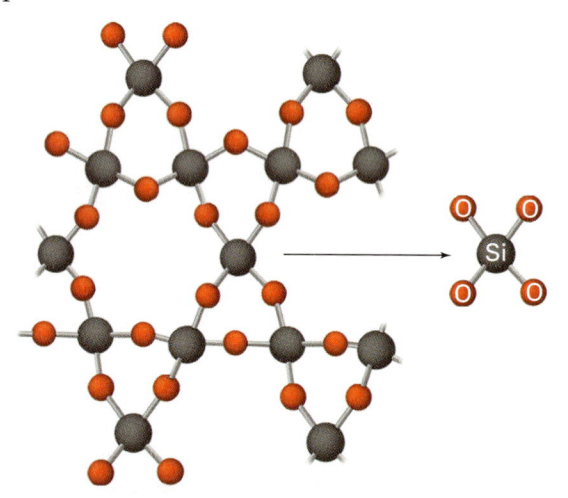

**Figure 6.9** Quartz $(SiO_2)_n$ is a network solid with repeating units of one sulfur and two oxygen atoms.

## Section Summary

In this section, you extended your ability to draw Lewis structures to small molecules. You then applied VSEPR theory to the Lewis structures in order to determine the shape of small molecules. Finally, you learned how to combine information about the polarity of bonds and the shape of molecules to determine whether a molecule was a dipole. In the next two sections, you will use the knowledge you have gained about the structure of some compounds to explain their macroscopic properties.

## Section Review

**1** For each of the following molecules, draw a Lewis structure, and determine if the molecule is polar or non-polar.

(a) $AsCl_3$  (b) $CH_3CN$  (c) $Cl_2O$  (d) $SiCl_4$

**2** Discuss the validity of the statement: "All polar molecules must have polar bonds and all non-polar molecules must have non-polar bonds."

**3** What similarities and what differences would you expect in the molecular shape and the polarity of $CH_4$ and $CH_3OH$? Explain your answer.

**4** The molecules $BF_3$ and $NH_3$ are known to undergo a combination reaction in which the boron and nitrogen atoms of the respective molecules join together. Sketch a Lewis structure of the molecule that you would expect to be formed from this reaction. Give a reason for your answer.

**5** Use VSEPR theory to predict the molecular shape and the polarity of $SF_2$. Explain your answer.

**6** Draw a Lewis structure for $PCl_4^+$. Use VSEPR theory to predict the molecular shape of this ion.

**7** Identify and explain the factors that determine the structure and polarity of molecules.

## 6.2 Intermolecular Forces

**Section Preview/Outcomes**

In this section, you will

- **illustrate** hydrogen bonds and van der Waals forces
- **define** London dispersion forces, dipole-dipole forces, and hydrogen bonding and explain how they form
- **compare** and contrast intramolecular forces with intermolecular forces in terms of strength and species involved
- **describe** how the type of attractions account for the properties of molecular compounds

Water and hydrogen sulfide have the same molecular shape: bent. Both are polar molecules. However, H$_2$O, with a molar mass of 18 g, is a liquid at room temperature, while H$_2$S, with a molar mass of 34 g, is a gas. Water's boiling point is 100°C, while hydrogen sulfide's boiling point is −61°C. These property differences are difficult to explain using the bonding models you have learned so far. As you know, pure covalent compounds are not held together by ionic bonds in lattice structures. They do form liquids and solids at low temperatures, however. Something must hold the molecules together when a covalent compound is in its liquid or solid state. The forces that bond the *atoms* to each other within a molecule are called **intramolecular forces**. Covalent bonds are intramolecular forces. In comparison, the forces that bond *molecules* to each other are called **intermolecular forces**.

You can see the difference between intermolecular forces and intramolecular forces in Figure 6.10. Because pure covalent compounds have low melting and boiling points, you know that the intermolecular forces must be very weak compared with the intramolecular forces. It does not take very much energy to break the bonds that hold the molecules to each other.

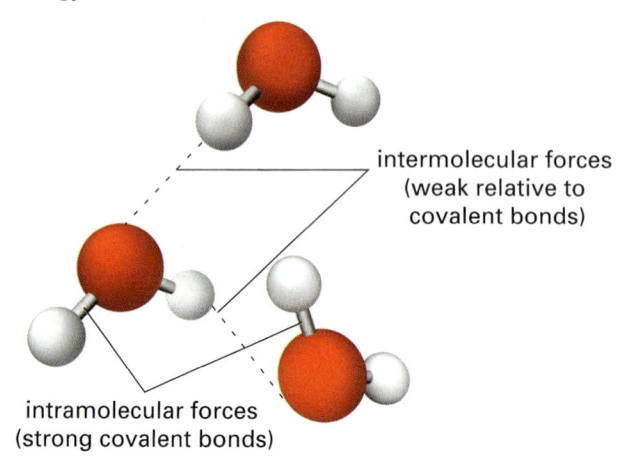

**Figure 6.10** Strong intramolecular forces (covalent bonds) hold the atoms in molecules together. Relatively weak intermolecular forces act between molecules.

Intermolecular forces were extensively studied by the Dutch physicist Johannes van der Waals (1837–1923). To mark his contributions to the understanding of intermolecular attractions, these forces are often called van der Waals forces. In the pages that follow, you will read about these intermolecular forces. You will then apply your understanding of intermolecular forces to explain some physical properties of substances.

### Dipole-Dipole Forces

In the liquid state, polar molecules (dipoles) orient themselves so that oppositely charged ends of the molecules are near to one another. The electrostatic attractions between these oppositely charged ends of the polar molecules are called **dipole-dipole forces**. Figure 6.11 shows the orientation of polar molecules due to these forces in a liquid.

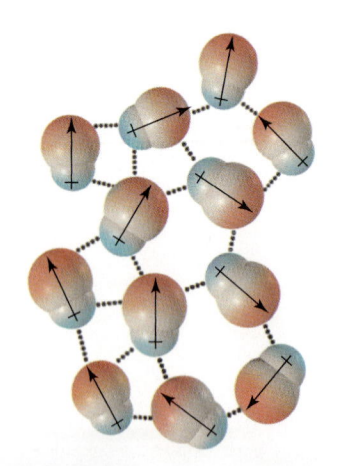

**Figure 6.11** Dipole-dipole forces among polar molecules in the liquid state

As a result of these dipole-dipole forces of attraction, polar molecules will tend to attract one another more at room temperature than similarly sized non-polar molecules would. The energy required to separate polar molecules from one another is therefore greater than that needed to separate non-polar molecules of similar molar mass. The energy that causes molecules that are attracted to each other, to be separated is their own kinetic energy. As you probably recall, the kinetic energy of atoms, ions, and molecules is directly related to temperature. Therefore, when molecules are attracted to each other, the strength of the attractive forces and the amount of energy needed to separate them is indicated by the temperatures at which the substance changes phase—its melting point and its boiling point. The melting points and boiling points of substances made of polar molecules are higher than for substances made of non-polar molecules.

### CONCEPT CHECK

Acetaldehyde (ethanal), $CH_3CHO$, is a polar molecule that has a boiling point of 20°C. Propane, on the other hand, is a non-polar molecule of similar size, number of electrons, and molar mass. The boiling point of propane, $CH_3CH_2CH_3$, is −42°C. Use the concept of dipole-dipole forces to explain these property differences.

## Ion-Dipole Forces

An **ion-dipole force** is the force of attraction between an ion and a polar molecule (a dipole). For example, NaCl dissolves in water because the attractions between the $Na^+$ and $Cl^-$ ions and the partial charges on the water molecules are strong enough to overcome the forces that bind the ions together. Figure 6.12 shows how ion-dipole forces dissolve any type of soluble ionic compound.

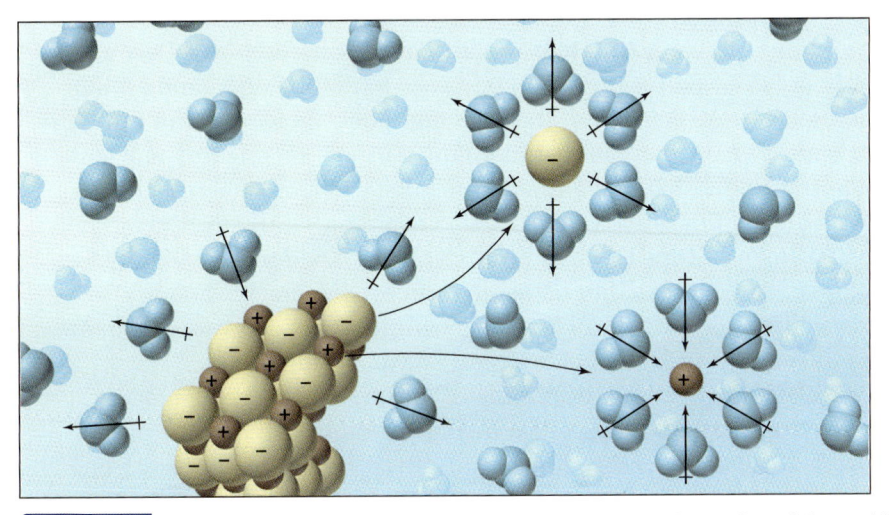

**Figure 6.12** In aqueous solution, ionic solids dissolve as the negative ends and the positive ends of the water molecules become oriented with the corresponding oppositely charged ions that make up the ionic compound, pulling them away from the solid into solution.

## Induced Intermolecular Forces

Induction of electric charge occurs when a charge on one object causes a change in the distribution of charge on a nearby object. A familiar example can be found in the children's trick of rubbing a balloon on some cloth or their hair and sticking it to a wall. When you rub the balloon, the friction causes it to loose electrons and become charged. When you hold it near a wall, the positive charge on the balloon cause the charge distribution in the molecules in the wall to change so that negative charges are closer to the balloon. Then the ions on the balloon are attracted to the negative charges on the wall. This process is illustrated in Figure 6.13.

There are two types of charge-induced dipole forces. An **ion-induced dipole force** results when an ion in close proximity to a non-polar molecule distorts the electron density of the non-polar molecule. The molecule then becomes momentarily polarized, and the two species are attracted to each other. This force is active during every moment of your life, in the bonding between non-polar $O_2$ molecules and the $Fe^{2+}$ ion in hemoglobin. Ion-induced dipole forces, therefore, are part of the process that transports vital oxygen throughout your body.

A **dipole-induced dipole force** is similar to that of an ion-induced dipole force. In this case, however, the charge on a polar molecule is responsible for inducing the charge on the non-polar molecule. Non-polar gases such as oxygen and nitrogen dissolve, sparingly, in water because of dipole-induced dipole forces.

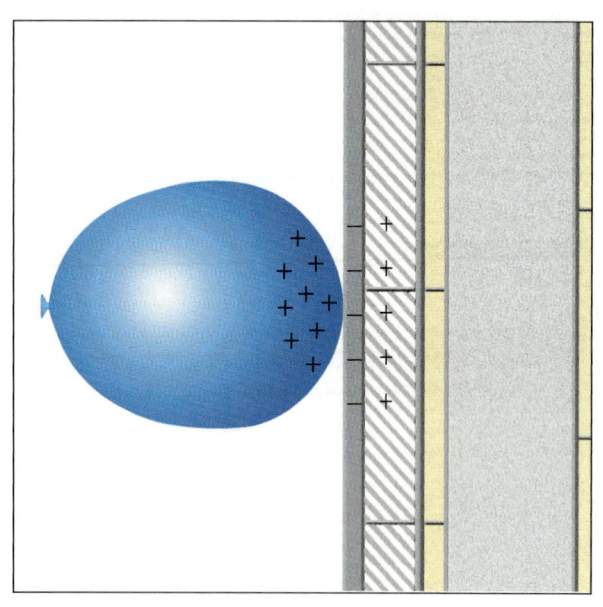

**Figure 6.13** The charges on the balloon induce the molecules in the wall to become dipoles. Then the opposite charges attract each other and the balloon sticks to the wall.

## Dispersion (London) Forces

The shared pairs of electrons in covalent bonds are constantly vibrating. While this applies to all molecules, it is of particular interest for molecules that are non-polar. The bond vibrations, which are part of the normal condition of a non-polar molecule, cause momentary, uneven distributions of charge. In other words, a non-polar molecule becomes slightly polar for an instant, and continues to do this on a random but on-going basis. At the instant that one non-polar molecule is in a slightly polar condition, it is capable of inducing a dipole in a nearby molecule. An intermolecular force of attraction results. This force of attraction is called a **dispersion force**. Chemists also commonly refer to it as a London force, in honour of the German physicist, Fritz London, who studied this force. Unlike other intermolecular forces, dispersion (London) forces act between any particles, polar or otherwise. They are the main intermolecular forces that act between non-polar molecules.

Two factors affect the magnitude of dispersion forces. One is the number of electrons in the molecule and the other is the shape of the molecule. Vibrations within larger molecules that have more electrons than smaller molecules can easily cause an uneven distribution of charge. The dispersion forces between these larger molecules are thus stronger, which has the effect of raising the boiling point for larger molecules. A molecule with a spherical shape has a smaller surface area than a straight chain molecule that has the same number of electrons. The smaller surface area allows less opportunity for the molecule to induce a charge on a nearby molecule. Therefore, for two substances with molecules that have a similar number of electrons, the substance with molecules that have a more spherical shape will have weaker dispersion forces and a lower boiling point.

London dispersion forces are responsible for the formation and stabilization of the biological membranes that surround every living cell. The molecules that make up membranes are in the class called lipids. Membrane lipids have two long non-polar chains attached to a polar head as illustrated in Figure 6.14. The non-polar chains attract each other with London dispersion forces and form a bilayer of fatty material with charged heads on both ends that interact with water.

### CONCEPT CHECK

The boiling point of $F_2$ is −188.14°C and the boiling point of $Cl_2$ is −34.6°C. Explain the difference in the boiling points of these halogen gases based on the factors that affect the magnitude of dispersion forces. The boiling point of butane, $C_4H_{10}$, is −0.5°C. Explain the difference in the boiling points of butane and chlorine, based on the factors that affect the magnitude of dispersion forces.

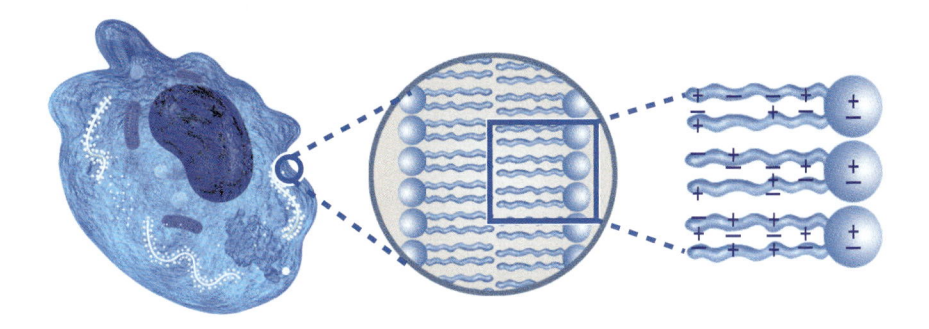

**Figure 6.14** Although London dispersion forces are very weak, there are so many interactions holding membranes together that the structure is very stable. Proteins are also dispersed throughout the membranes.

## Hydrogen Bonding

One specific dipole-dipole interaction is so important in both biological systems and in water that it is given a category of its own—hydrogen bonding. To form a hydrogen bond, a hydrogen atom must be bonded to a highly electronegative atom such as oxygen, nitrogen, or fluorine. Such bonds are very polar, and since hydrogen has no other electrons, the positive proton which makes up its nucleus is exposed and can become strongly attracted to the negative end of another dipole nearby. This partial negative charge is often found on an oxygen, nitrogen, or fluorine atom that is bonded to carbon or another atom with low electronegativity. A **hydrogen bond** is an electrostatic attraction between the nucleus of a hydrogen atom, bonded to fluorine, oxygen, or nitrogen, and the negative end of a dipole nearby. In biological systems, these polar bonds are often parts of much larger molecules. For example, an N—H bond and C=O bond are found in many biological molecules. A hydrogen bond between these two groups is shown in Figure 6.15.

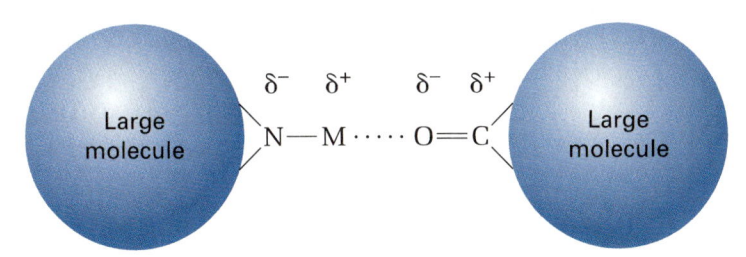

**Figure 6.15** The N—H bond is very polar with the hydrogen carrying a partial positive charge. A C=O bond is also polar with the oxygen having a partial negative charge. The attraction between these partial charges is a hydrogen bond.

Every water molecule has two O—H bonds, which are polar. Hydrogen bonds between the hydrogen atoms in one water molecule and the oxygen atom in another account for many unique properties of water. In liquid water, each water molecule is hydrogen bonded to at least four other water molecules. The molecules can move past each other because the molecules have enough energy to cause the hydrogen bonds to break and reform with another water molecule. Although individual hydrogen bonds are only about 5% as strong as covalent bonds, the large number of bonds between water molecules makes the net attractive force quite strong.

These forces are responsible for the relatively high boiling point of water. Hydrogen bonding in liquid water is illustrated in Figure 6.16.

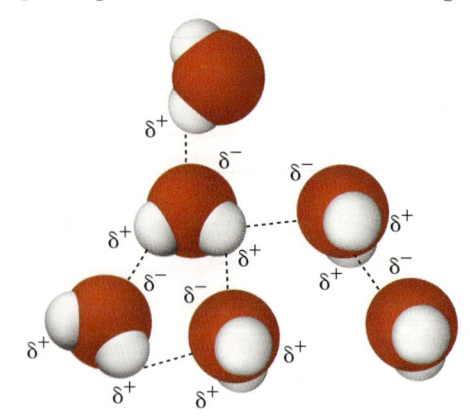

**Figure 6.16** A few of the hydrogen bonds between water molecules are shown here.

### CONCEPT CHECK

The boiling point of water, $H_2O$, is 100°C under standard conditions and the boiling point of hydrogen sulfide, $H_2S$, is −61°C. Use a combination of London dispersion forces and hydrogen bonding to explain why the boiling point of water is so much higher than the boiling point of hydrogen sulfide.

Hydrogen bonding can also explain why solid water, ice, floats on liquid water. To float, an object must be less dense than the liquid. In liquid water the molecules are able to get closer together because their kinetic energy is high enough to overcome some of the attractive force of the hydrogen bonds. When water cools below 4°C the kinetic energy can no longer overcome the hydrogen bonds and the molecules are forced into the special hexagonal, crystalline structure of ice, shown in Figure 6.17A. When the hydrogen bonds are oriented along a line from the centre of one oxygen atom, through the covalently bonded hydrogen atom, and to the centre of the oxygen atom of the next water molecule, the hydrogen bonds have their maximum strength. This orientation is shown in Figure 6.17B. The water molecules are farther apart than they are in liquid water making ice less dense than liquid water.

If water molecules behaved as do most molecules, then lake water would freeze from the bottom up, rather than from the top down because ice formed at the cold surface would immediately sink to the bottom. The consequences for aquatic life, including all the ground fish species found in the coastal waters of Atlantic Canada, and on organisms (like us) that depend on aquatic ecosystems, would be profound. A lake, or ocean area, that freezes from the bottom up would solidify completely, killing most of the life forms it contains. Because ice floats on water, it insulates the water beneath it, preventing it from freezing. Our national pastimes of hockey and skating could not have developed without this property of ice caused by hydrogen bonding because our winters would have been over before most lakes and ponds froze all the way to the surface to provide a solid surface.

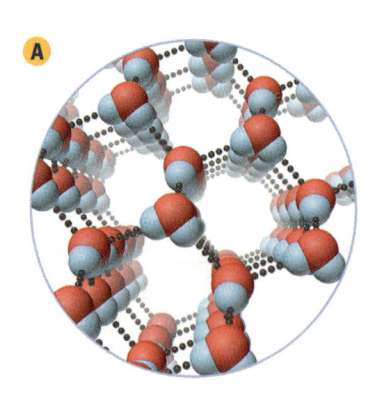

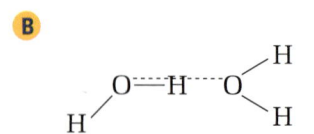

**Figure 6.17** Every hydrogen bond in ice (A) is oriented as shown in B. This orientation gives the hydrogen bonds their maximum strength.

# Investigation 6-B

**SKILL FOCUS**
Modelling concepts
Analyzing and interpreing
Communicating results

## Investigating the Properties of Water

In this investigation, you will infer a property of water from an observation. You will then use that property to explain other observations.

### Question
How can you explain the properties of water based on its shape and charge properties?

### Safety Precautions
- Use care in handling sharp objects.

### Materials
acetate strip
cotton cloth
vinyl strip
wool cloth
vegetable oil
500 mL beaker
2 shallow dishes
water
ethanol
sewing needles
2 small glasses
150 ml beaker
pepper
liquid dish detergent

### Procedure
1. Turn on a water tap so that a very small steady stream of water is flowing.
2. Vigorously rub the end of the acetate strip with the cotton cloth.
3. Slowly move the acetate strip near to the stream of water without touching the water. Record your observations
4. Repeat Steps 2 and 3 with the vinyl strip and wool cloth.
5. Have one lab partner pour vegetable oil into the 500 mL beaker, forming a small steady stream of vegetable oil similar to the water in step 1.
6. Repeat Steps 2 through 4 with the vegetable oil. Record any differences between the behaviour of the oil and the water.
7. Place water in one shallow dish. Very carefully place a pin or needle onto the surface of the water in a horizontal position. (Your teacher might choose to do this step.) Record the appearance of the needle.
8. Gently touch the needle. Record your observations.
9. Place ethanol in the second shallow dish. Try to repeat Steps 5 and 6 with the ethanol instead of water. Record your observations.
10. Place one of the glasses in an empty shallow dish. Pour water into the glass from a beaker. When the glass is nearly full, pour very slowly and carefully. Observe and record the shape of the top of the water just before it starts to spill over into the dish.
11. Repeat Step 8 with ethanol. Record your observations.

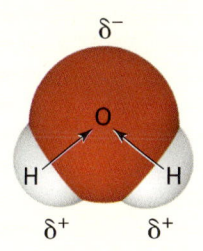

*continued...*

**continued**

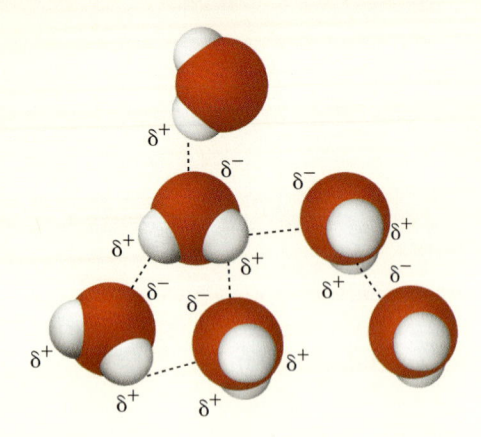

12. Fill the 150 mL beaker about two thirds full of water.
13. Sprinkle pepper onto the water until a thin film of pepper forms on the water.
14. Add a few drops of liquid dish detergent to the water. Observe the response of the film of pepper. Record your observations.

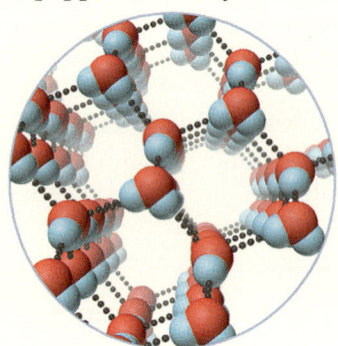

### Analysis

1. The cotton-rubbed acetate strip was positively charged and the wool-rubbed vinyl strip was negatively charged. Was the stream of water attracted or repelled by the positive charge? Was the stream of water attracted or repelled by the negative charge? Was the stream of vegetable oil attracted or repelled by the positive charge? Was the stream of vegetable oil attracted or repelled by the negative charge? What do these observations tell you about the electrical properties of water and vegetable oil?

2. What happened when you touched the needles that were floating on the water? Is the needle more dense than water? Use the property of water that you described in your answer to Question 1 to explain the observations of the needle.

3. What happened when you attempted the same procedure with ethanol? Explain your observations.

4. Describe the shape of the water in the glass just before it started to run over? How does the property of water that you described in Question 1 account for this shape?

5. How did filling the glass with ethanol differ from filling it with water? Explain the difference in the two liquids.

6. How did the dish detergent affect the film of pepper on the water? The molecules in dish detergent have a long non-polar end and a charged head. Use this information to explain the effect on the pepper film that you described?

### Conclusion

7. Describe the differences in the physical properties of ethanol and water. What is the significance of these properties of water to living systems considering the fact that living cells contain a large percentage of water?

**Table 6.3** Comparing Types of Bonding

| Bond Type | Model | Nature of Attraction | Energy (kJ/mol) | Example |
|---|---|---|---|---|
| ionic | | cation-anion | 400–4000 | NaCl |
| covalent | | nuclei-shared electron pair | 150–1100 | H—H |
| metallic | | cations-delocalized electrons | 75–1000 | Fe |
| ion-dipole | | ion charge-dipole charge | 40–600 | $Na^+$----O(H)(H) |
| hydrogen bond | —A—H·······:B— | polar bond to hydrogen-dipole charge (lone pair, high EN of N, O, F) | 10–40 | :Ö—H----:Ö—H with H below each |
| dipole-dipole | | dipole charges | 5–25 | I—Cl----I—Cl |
| ion-induced dipole | | ion charge-polarizable electrons | 3–15 | $Fe^{2+}$----$O_2$ |
| dipole-induced dipole | | dipole charge-polarizable electrons | 2–10 | H—Cl----Cl—Cl |
| dispersion (London) | | polarizable electrons | 0.05–40 | F—F----F—F |

You can also see the effect of hydrogen bonding in the boiling point data of the binary hydrides of Groups 14 to 17 (IVA to VIIA), shown in Figure 6.18. In Group 14, the trend in boiling point is as expected. The increase in boiling point from methane, $CH_4$, to tin(IV) hydride, $SnH_4$, is due to an increase in the number of electrons. Therefore, there are larger dispersion forces, since more electrons are able to temporarily shift from one part of the molecule to another. In $SnH_4$, this results in greater intermolecular forces of attraction. There is no hydrogen bonding in these substances, because there are no lone pairs on the molecules.

However, the hydrides of Groups 15, 16, and 17 (VA to VIIA) only show the expected trend based on dispersion forces for the larger mass hydrides. The smallest mass hydrides, $NH_3$, $H_2O$, and HF, have relatively high boiling points, indicating strong intermolecular forces caused by hydrogen bonding.

Why is there no hydrogen bonding in $H_2S$ and $H_2Se$, which have unshared electron pairs on their central atoms? Neither S nor Se is small enough or electronegative enough to support hydrogen bonding. Thus, H₂O (with its small and very electronegative O atom) has a much higher boiling point than H₂S and H₂Se, as you can see in Figure 6.18.

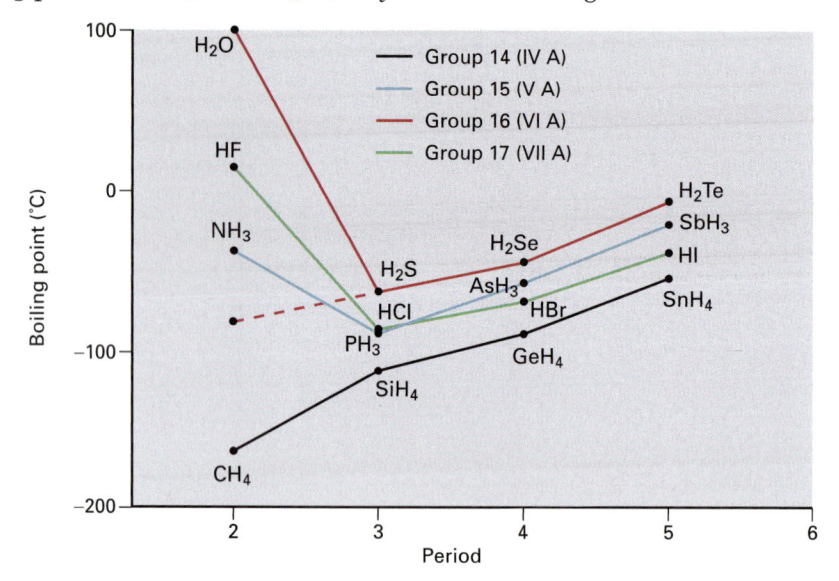

**Figure 6.18** The boiling points of four groups of hydrides. The break in the trends for NH₃, H₂O, and HF is due to hydrogen bonding. The dashed line shows the likely boiling point of water, if no hydrogen bonding were present.

Table 6.3 summarizes many of the properties and contains information about the ranges of bond strengths of all the types of bonding about which you have learned. Compare the relative strengths of the various types of bonds and keep them in mind as you carry out the ThoughtLab that follows.

## ThoughtLab — Properties of Liquids

As you know, intermolecular forces act between both like and unlike molecules. *Adhesive forces* are intermolecular forces between two different kinds of molecules. *Cohesive forces* are intermolecular forces between two molecules of the same kind. Adhesive and cohesive forces affect the physical properties of many common phenomena in the world around you.

### Procedure

1. Consider each of the following observations. For each, infer the intermolecular (and, where appropriate, intramolecular) forces that are active, and explain how they account for the observations.
   - Why are bubbles in a soft drink spherical?
   - How can water striders walk on water?
   - Why does water form beaded droplets on surfaces such as leaves, waxed car hoods, and waxed paper?
   - How does a ballpoint pen work?
   - How does a pencil work?
   - How does a towel work?
   - Why does water form a concave meniscus in a tube, while mercury forms a convex meniscus?
   - Why does liquid honey flow so slowly from a spoon?
   - Why do some motorists change to different motor oils over the course of a year?

2. Based on your inferences and explanations, develop definitions for the following terms: surface tension, capillarity, and viscosity.

3. Verify your inferences and definitions by consulting print and electronic resources. Then identify the property or properties that best explain each of the observation statements.

### Analysis

1. Compare your inferences and explanations with those that you found in reference sources. Identify those cases in which your ideas matched or closely matched what you found. Explain where your reasoning may have led you astray, for those cases in which your ideas did not match closely with reference-source information.

2. Re-design the Concept Organizer on page xxx so that it includes practical applications and explanations such as those you have investigated in this activity.

210 MHR • Unit 2 Structure and Properties

## Bonding in Biological Molecules

The structure and consequently the function of two of the most important biological molecules—DNA and proteins—depend on several types of bonds acting together in the same molecule. Although van der Waals forces are considered intermolecular forces, they can exist between two different parts of the same large molecules. Take a closer look at the structures of DNA and proteins.

The structure of DNA, the molecules that carry genetic information, is shown in Figure 6.19. DNA consists of long strands of units called nucleotides. Each nucleotide contains a sugar, a phosphate group, and a nitrogenous base all covalently bonded together. The sugar and phosphate groups of the nucleotides are also covalently bonded in a very long, linear chain with the nitrogenous bases protruding out from the sugar. The genetic information is contained within the sequence of the four bases—represented by A, C, G, and T—along the chain. Due to the precise structures of the bases, hydrogen bonds form between C and G and between A and T as shown in Figure 6.19. Since individual DNA strands contain thousands of bases, the large number of hydrogen bonds holds the strands together securely. As you can see in the figure, the two strands coil around each other forming the famous "double helix." The structure is further stabilized by London dispersion forces. Although it is not clear in the figure, the flat planes of the C, G, A, and T bases lie perpendicular to the direction of the backbone, sugar-phosphate strands. This orientation places the bases above and below each other. London dispersion forces between the ring shaped structures draw the bases together, stabilizing the helical structure.

DNA carries the information necessary for the cell to synthesize proteins and then proteins do the work of the cell. Proteins are linear chains of units called amino acids. When in a protein molecule, each amino acid has an N—H group and a C=O group as shown in Figure 6.20. The 20 amino acids found in proteins are distinguished only by differences in chemical groups called R groups.

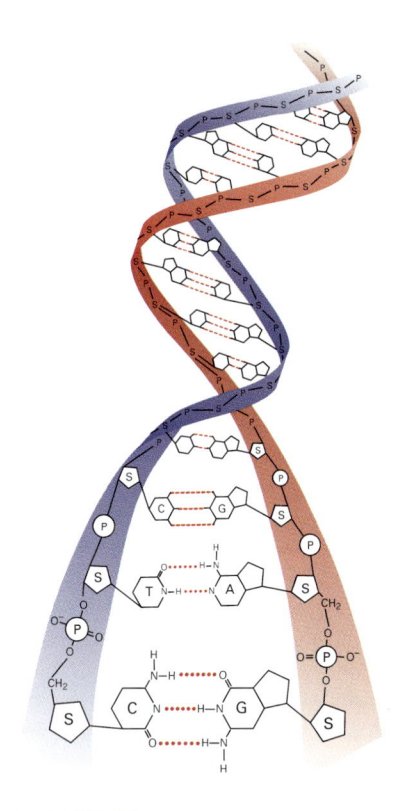

**Figure 6.19** DNA (deoxyribonucleic acid) is stabilized by hydrogen bonds and London dispersion forces. If you could stretch all of the DNA in the nucleus of one human cell into a straight line, it would be over 1.5 m long.

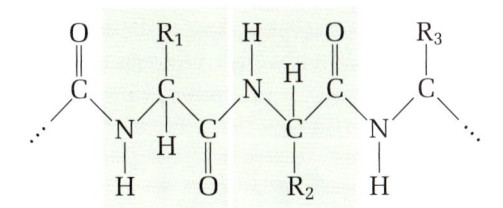

**Figure 6.20** Boxes are drawn around individual amino acids in this part of a protein chain. The subscripts on the Rs indicate that they are different chemical groups.

Proteins vary tremendously in size. Some are as short as about 50 amino acids long and others have over 1000 amino acids. The sequence of the amino acids that are covalently bonded together is called the *primary structure*. Proteins can carry out their function only if the amino acid chain is folded in its precisely correct three-dimensional structure. Hydrogen bonds stabilize two common structures within many proteins—the alpha helix and the beta structure. In an alpha helix, the chain coils back on itself and the N—H group of one amino acid hydrogen bonds to the C=O group of another amino acid three groups along the chain.

A section of alpha helix is illustrated in Figure 6.21A. To form a beta structure, different parts of the chain of a protein fold back and lie beside each other and hydrogen bonds form between the chains as shown in Figure 6.21B. Silk contains large amounts of beta structure. These structures that are stabilized by hydrogen bonds are called the *secondary structure* of a protein.

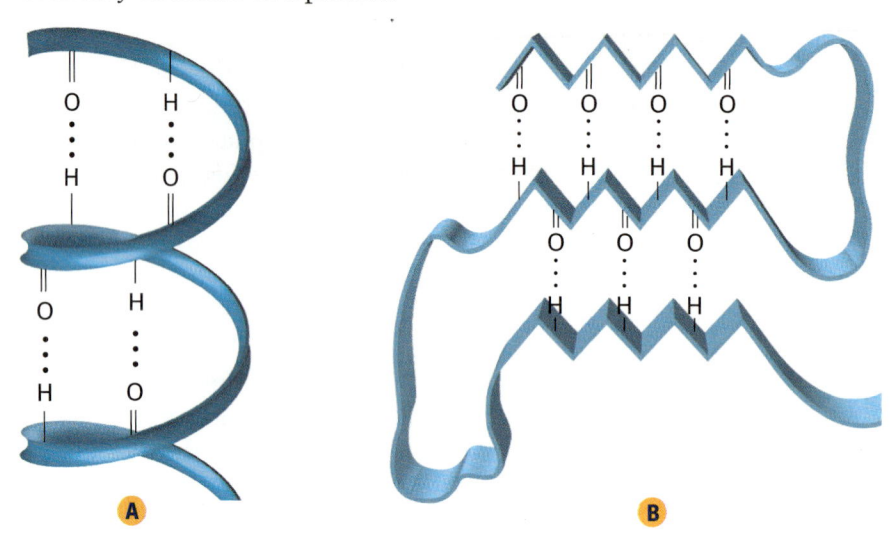

**Figure 6.21** (A) The carbon and nitrogen atoms in the chain are not shown so that the hydrogen bonds can be easily observed in this alpha helix. (B) When a beta structure forms, the protein chains bend up and down in a regular pattern. Thus, the beta structure is often called a "pleated sheet."

To complete the three-dimensional structure of a protein, the segments that have formed an alpha helix or a beta structure fold back on each other. The R groups of the amino acids that are non-polar are drawn together in the centre of the protein and are stabilized by London dispersion forces. The level of protein structure that is stabilized by London dispersion forces is called the *tertiary structure*. An example of a protein that is correctly folded is shown in Figure 6.22. This protein is myoglobin, a protein that binds oxygen and stores it in muscle cells. About 70% of all of the amino acids in myoglobin are part of a segment of alpha helix. The segments of alpha helix are held in place by London dispersion forces. Myoglobin contains no beta structure.

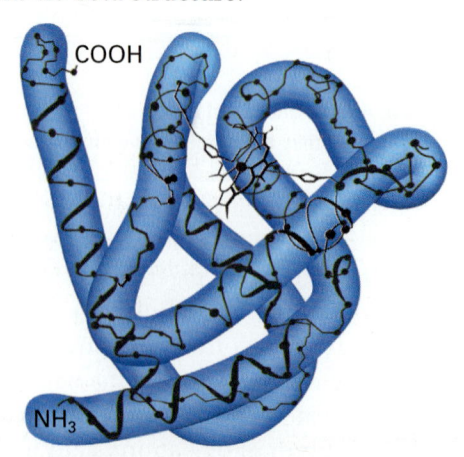

**Figure 6.22** The dots in the chain indicate the points where individual amino acids are joined together. As you can see, myoglobin contains about six segments of alpha helix.

# Canadians in Chemistry

## Dr. R.J. Le Roy

You usually study science as discrete fields of inquiry: chemistry, physics, and biology, for example. Each of these fields has numerous sub-fields, such as geophysics and biochemistry. One of the scientists at the forefront of chemical physics is Canada's Dr. R.J. Le Roy.

During a fourth-year undergraduate research project at the University of Toronto, Dr. R.J. Le Roy pursued an answer to the following question: If you use photodissociation to separate iodine molecules into atoms, what fraction of the light is absorbed, and how does that fraction depend on the colour (frequency) of the light and on the temperature of the molecules? (Photodissociation is a process that involves the use of light to decompose molecules.)

The light causes a bonded molecule in its ground state to undergo a transition into an excited state. In this new state, the atoms repel one another, causing the molecule to break apart. The fraction of light absorbed, as measured spectroscopically, depends on the population of molecules in the initial state.

Chemists can use the fraction of light absorbed by a sample to monitor the concentration of specific molecules in a closed, controlled environment. To do this, chemists also need to know the value for a key proportionality constant that is unique to each substance, known as the *molar absorption coefficient*. These coefficients can be measured at room temperature, but not at high temperatures.

"In principle, we have known the quantum mechanics to describe these processes," Dr. Le Roy explains, "But to use that information, we have to understand the forces atoms experience as the atoms come together or fall apart."

For his master's thesis at the University of Toronto, Dr. Le Roy developed computer programs to perform the exact quantum mechanical calculation of the absorption coefficients governing photodissociation processes. Later, when doing his doctoral thesis at the University of Wisconsin, he and his thesis supervisor, Dr. Richard Bernstein, developed the Le Roy-Bernstein theory. This theory describes the patterns of energies and other properties for vibrational levels of molecules lying very close to dissociation.

To determine when to trust the equations of the Le Roy-Bernstein theory, Dr. Le Roy developed what he calls "a little rule of thumb," more generally known as the Le Roy radius. A calculation for the Le Roy radius, $R_{LR}$, is:

$$R > R_{LR} = 2[<r_A^2>^{1/2} + <r_B^2>^{1/2}]$$

where $R$ is the internuclear distance between two atoms A and B.

The Le Roy radius is the minimum distance at which the equation describing the vibrations of molecules close to dissociation is thought to be valid. The $<r_A^2>$ and $<r_B^2>$ values are the squares of the radii of the atoms of the two elements (or of the two atoms of the same element) in their ground states.

Is there a practical outcome to mathematical chemistry? Dr. Le Roy strongly encourages "curiosity-oriented science," as opposed to a focus on short-term, practical results, partly because practical results do come when processes are fully understood.

For example, the study of global warming is hindered, in his view, because the intermolecular interactions are not well known. "To properly model the processes, we have to understand the exchange of energy during molecular collisions. If we know them, we can predict an immense amount."

### Making Connections

1. In what ways are chemistry and physics related to each other? Design a graphic organizer to outline their similarities and differences.

2. Research and development funding often encourages chemists and chemical engineers to focus on practical applications. How does this differ from Dr. Le Roy's perspective? With which perspective do you agree more? Write an editorial outlining your opinions.

## A Material Made to Order: Intermolecular and Intramolecular Forces in Action

The glove in the photograph at the start of this chapter is anything but ordinary. The thin, lightweight fabric can withstand startling impacts and piercing forces, including small projectiles such as bullets. In fact, this same fabric is used to make bulletproof vests. The long strands of fibre used to make such protective gear are interlaced to form a dense "net" that is capable of absorbing large amounts of energy. You have probably heard the trade name for this astounding fibre: KEVLAR®. It was invented by an American research chemist, Stephanie Kwolek, in 1965.

The secret to stopping a bullet with a fairly soft fabric is that it behaves in much the same way as the netting on the goal frame used in hockey. When a hockey puck strikes a net at 100 km/h, its kinetic energy is transferred to the lengths of twine of the goal net. This twine is an interlaced mesh that has strands which stretch horizontally and vertically to disperse the energy of impact over a wide area.

The same principle applies to KEVLAR®. It belongs to a class of fibres called aramids, which are polymers containing aromatic and amide groups. The polymer forms in the condensation reaction between terephthalic acid and 1,6-hexamethylenediamine, as shown in Figure 6.23A. The strength of the polymer comes from the intramolecular forces in the aromatic group that limit bond rotation in the straight chains, and the amide linkage that leads to intermolecular hydrogen bonding between the straight chains, as shown in Figure 6.23B.

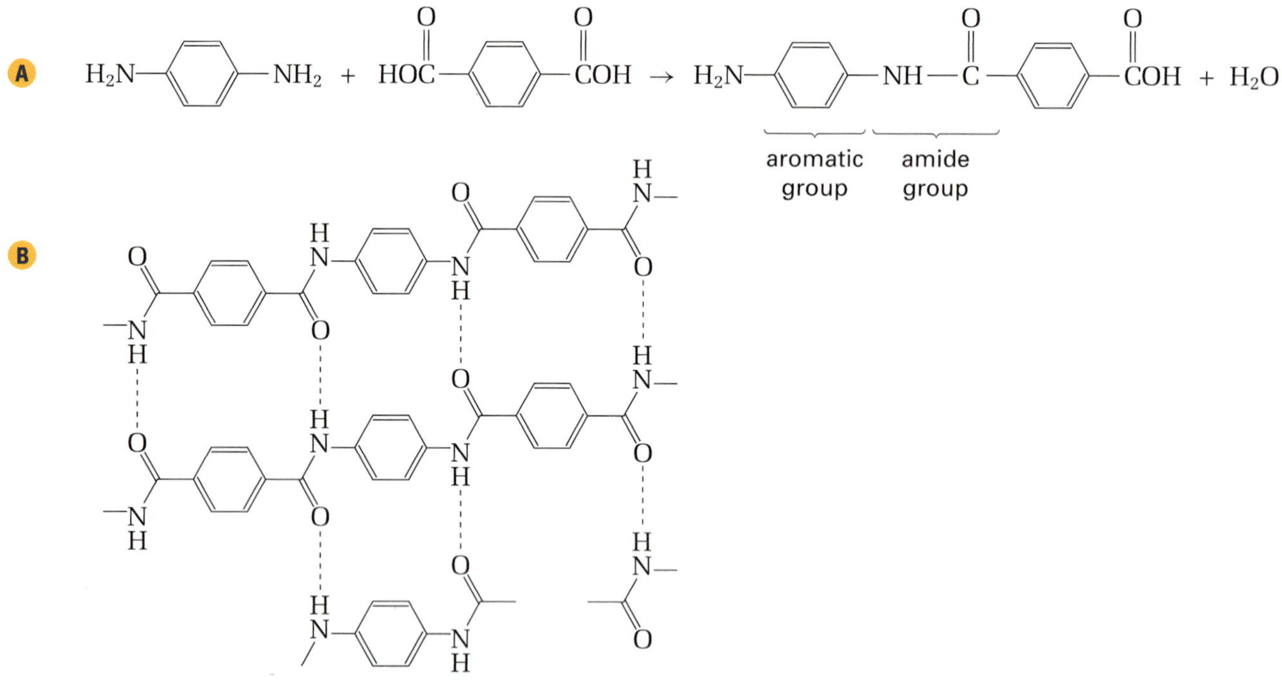

**Figure 6.23** (A) A condensation reaction between terephthalic acid and 1,6-hexamethylenediamine to form the monomer used to make KEVLAR®; (B) Structure of KEVLAR® showing hydrogen bonds between straight chains

In addition to stopping a bullet from penetrating the armour, a bulletproof vest must also protect the body from the blunt force of impact. A bulletproof vest cannot have as much "give" as a hockey net (that is, it should not stretch back as much as hockey nets do upon impact).

Otherwise, its wearer would still be injured, even though the bullet might not pierce the fabric. To accomplish this, the properties of KEVLAR® are enhanced by twisting the strands to increase their density and thickness. The strands are then woven very tightly. A resin coating is applied and sandwiched between two layers of plastic film. Each layer acts as a net and slows down the bullet a little at a time until it has been stopped.

## Section Summary

In this section, you focussed on the interactions between molecules. Without these forces, molecular compounds would exist only as gases because there would be nothing to hold the molecules together. All intermolecular forces are caused by electrostatic attractive forces between opposite charges. You compared dipole-dipole forces, ion-induced dipole forces, and dipole-induced dipole forces. If no permanent charge of dipole exists in a class of molecules, they are attracted to each other by London dispersion forces. These forces are caused by temporary changes in the distribution of charges in a molecule. A "temporary dipole" in one molecule induces another "temporary dipole" in the neighbouring molecule. Hydrogen bonds are a form of dipole-dipole interactions but are so common and usually stronger than other dipole-dipole forces that they are considered as a separate class of intermolecular force. All forms of intermolecular forces are critical to the structure and function of most biological molecules.

## Section Review

1. Describe the intermolecular forces that cause an ionic compound to dissolve in water.

2. Explain how the size (number of electrons) and the shape of a non-polar molecule affect the strength of London dispersion forces between the molecules.

3. How can an ion "induce" another molecule to become a dipole?

4. What characteristics of molecules are necessary in order for hydrogen bonds to form between them?

5. Compare the strengths of intermolecular bonds with ionic bonds, covalent bonds, and metallic bonds bonds.

6. Explain how life on Earth would be different if water was a non-polar molecule.

# 6.3 Structure Determines Properties

**Section Preview/Outcomes**

In this section, you will

- **compare** the melting points and boiling points of simple substances
- **explain** the general properties of ionic compounds
- **explain** why metals are malleable, ductile, and good conductors of electric current

Water is the only pure substance that exists in nature in all three phases at the same time. Scenes such as the one in Figure 6.23, showing ice, liquid water, and clouds indicating the presence of water vapour, are so common that it might seem that the statement above could not be true. Nevertheless, it is correct. The boiling point and melting point of water are unusual for molecules of its size. What, exactly, is happening in a substance that determines melting points and boiling points?

## Melting Points and Boiling Points

In a solid, the particles—atoms, ions, or molecules—remain beside each other. In crystalline solids, the packing of particles is very uniform as you saw in Chapter 5. Solid substances begin to melt when the average kinetic energy of the particles is great enough to overcome the force of attraction with neighbouring particles and allow them more freedom to move about. Particles in liquids are always interacting with their "nearest neighbour" but they can easily "change partners." In order for a liquid to boil or evaporate, the particles must have enough kinetic energy to break all bonds with all neighbouring particles. When no attractive forces exist with other particles the substance is in the gas phase. These conditions are summarized in Figure 6.24.

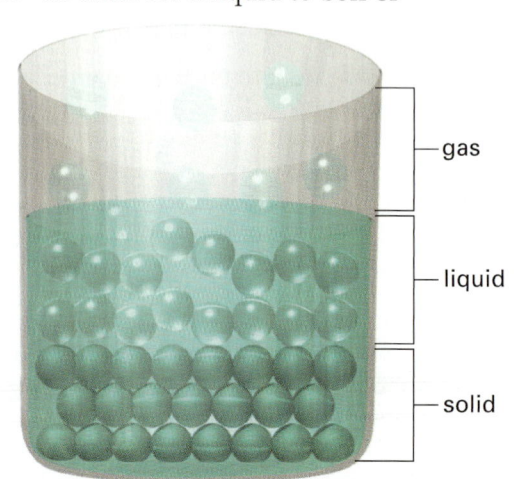

**CHEM FACT**

Differences in melting points and boiling points are due in part to the type of bonding in a compound. They are also due to different masses of the individual molecules.

**Figure 6.24** The state of a substance depends on the ability of the particles to break free from each other.

The melting points and boiling points of substances clearly depends on the bonds holding the particles together. For metals or ionic compounds to melt or boil, they must break the ionic bonds or metallic bonds. When molecular substances melt or boil, the intramolecular covalent bonds between the atoms do not break but, instead, intermolecular bonds between the particles must break. Table 6.4 shows the melting and boiling points of some common metals, ionic compounds, and molecular compounds.

**Table 6.4** Melting Points and Boiling Points of Some Common Substances

| Metals ||| Ionic Compounds ||| Molecular Substances |||
|---|---|---|---|---|---|---|---|---|
| Substance | Melting Point (°C) | Boiling Point (°C) | Substance | Melting Point (°C) | Boiling Point (°C) | Substance | Melting Point (°C) | Boiling Point (°C) |
| Li | 180 | 1347 | CsBr | 636 | 1300 | $H_2$ | −259 | −253 |
| Sn | 232 | 2623 | NaI | 661 | 1304 | $Cl_2$ | −101 | −34 |
| Al | 660 | 2467 | $MgCl_2$ | 714 | 1412 | $H_2O$ | 0 | 100 |
| Ag | 961 | 2155 | NaCl | 801 | 1413 | $C_6H_6$ | 6 | 80 |
| Cu | 1083 | 2570 | MgO | 2852 | 3600 | $C_6H_{12}O_6$ | 142 | decompose |

Look at the ranges of melting points and boiling points in the three categories of substances. Then go back to Table 6.3, Comparing Types of Bonding, on page 209 and review the bond energies for ionic bonds, metallic bonds, and the intermolecular forces. The bond energies, as you know, are the amounts of energy required to break the bonds. To increase the kinetic energy of an ion or molecule, you must increase the temperature of the substance. With these concepts in mind, consider the different types of intermolecular forces that account for the attractions between particles of the substances in Table 6.4. Is there a close relationship between the bond energies and the melting and boiling points?

Ionic compounds have extremely high melting points because the electrostatic attractive force between whole charges is a very strong force. As you read earlier, the ions in crystals align in a pattern that brings the oppositely charged ions as close together as possible thus the attractive forces are as great as possible. In addition, the electrostatic force is a long range force. Although the force decreases with distance between the charges, ions attract not only the nearest neighbour, they also attract opposite charges farther away in the crystal. All of these features act together making the attraction among the ions extremely large.

The fact that the strength of the ionic bonds vary with the size of the ions and the size of the positive and negative charges on these ions explains why some ionic compounds have higher melting points than others have. For example, compare the melting points of sodium iodide (661°C) and sodium chloride (801°C). Chloride ions are much smaller than iodide ions therefore the sodium and chloride ions are able to get closer together than the sodium and iodide ions. Consequently, the attractive force between the sodium and chloride ions is stronger than the force between the sodium and iodide ions and NaCl has a higher melting point. If you compare the melting points of sodium iodide (661°C) and magnesium oxide (2852°C) you can see that the melting point of MgO is extremely high compared to the melting point of NaCl. There is a size difference in the ions which causes a small difference in melting points, however the major reason for the huge difference is that the magnesium ions have 2+ charges and the oxide ions have 2⁻ charges compared to the 1⁺ charges on the sodium ions and the 1⁻ charges on the chloride ions. Thus the melting point of MgO is much higher than the melting point of NaCl.

A similar case can be made for the variation in melting points for metals as for the case of ionic compounds. Within a group on the periodic table, the melting point decreases as the size of atoms increase. The freely moving valence electrons can get closer to the smaller ions than larger ions, therefore the attractive forces between ions and electrons is greater. For example, lithium with small ions has a melting point of 180°C whereas cesium, with much larger ions, has a melting point of 28.4°C. However, Group 2 (IIA) metals have much higher melting points than Group 1 (IA) metals. Not only are Group 2 (IIA) metal atoms smaller, their ions also have twice the charge that the ions of Group 1 (IA) metals have and have twice as many valence electrons moving about. The larger charge attracts the greater number of freely moving valence electrons with a greater force.

Melting points and boiling points are just one set of properties that are determined by bonding between particles in a substance. Before looking at some other physical properties, review Table 6.5 that summarizes many properties of solids.

**Table 6.5** Summary of Properties of Different Types of Solids

| Type of Crystalline Solid | Particles Involved | Primary Forces of Attraction Between Particles | Boiling Point | Electrical Conductivity in Liquid State | Other Physical Properties of Crystals | Conditions Necessary for Formation | Examples |
|---|---|---|---|---|---|---|---|
| Nonmetal | atoms | dispersion | low | very low | very soft | formed between atoms with no electronegativity difference | all Group 18 (VIIIA) elements |
| Molecular | non-polar molecules or polar molecules | dispersion, dipole-dipole, hydrogen bonds | generally low (non-polar); intermediate polar | very low | non-polar: very soft; soluble in non-polar solvents | non-polar: formed from symmetrical molecules containing covalent bonds between atoms with small electronegativity differences | non-polar: $Br_2$, $CH_4$, $CO_2$, $N_2$, $P_4$, $S_8$ |
|  |  |  |  |  | polar: somewhat hard, but brittle; many are soluble in water | polar: formed from asymmetrical molecules containing polar covalent bonds. Electronegativity difference between atoms < 1.7 | polar: $H_2O$, $NH_3$, $CHCl_3$, $CH_3COOH$, $CO$, $SO_2$, many organic compounds |
| Covalent Network | atoms | covalent bonds | very high | low | hard crystals that are insoluble in most liquids | formed usually from elements belonging to Group 14 (IV A) | graphite, diamond, $SiO_2$ |
| Ionic | cations and anions | electrostatic attraction between oppositely charged ions | high | high | hard and brittle; many dissolve in water | formed between atoms with electronegativity difference > 1.7 | NaCl, $CaF_2$, $Cs_2S$, MgO, $MgF_2$ |
| Metallic | atoms | metallic bonds | most high | very high | all have a lustre, are malleable and ductile, and are good electrical and thermal conductors; they dissolve in other metals to form alloys | formed by metals with low electronegativity | Hg, Cu, Fe, Ca, Zn, Pb |

## Mechanical Properties of Solids

You have probably read, in many science books, that metals are malleable and ductile while ionic compounds are hard and brittle. Now you can explain these properties on the basis of bonding. Examine Figure 6.25 and envision the free-electron model of bonding in metals. The malleability of metals can be explained by viewing metallic bonds as being non-directional. The positive metal cations are often layered as fixed arrays (like soldiers lined up for inspection). When stress is applied to a metal, such as bending or hammering, one layer of positive ions can slide over another layer. The layers move without breaking the array because the delocalized, freely moving valence electrons continue to exert a uniform attraction on the positive ions. For this reason, metals do not shatter immediately along a clearly defined point of stress but bend to a new shape or spread out when hammered. This is why many metals such as copper, can be bent and hammered to create ornate designs as shown in Figure 6.26 and to make all the metallic products we use.

In contrast to metals, ionic compounds are brittle. The structure of the compounds and the nature of ionic bonds can easily explain this property as shown in Figure 6.27. When bent or hammered, like charges can become aligned and repel each other. The crystal breaks along the line where the charges are repelling. The conditions in polar molecular compounds are similar. For example, consider crystals of table sugar. Each sugar molecule has several O—H groups making them polar so they attract each other with dipole-dipole forces and hydrogen bonds. However, they are often not as hard as ionic compounds because the charges on polar molecules are partial charges and therefore these intermolecular attractive forces between the molecules are not as strong as they are in ionic compounds.

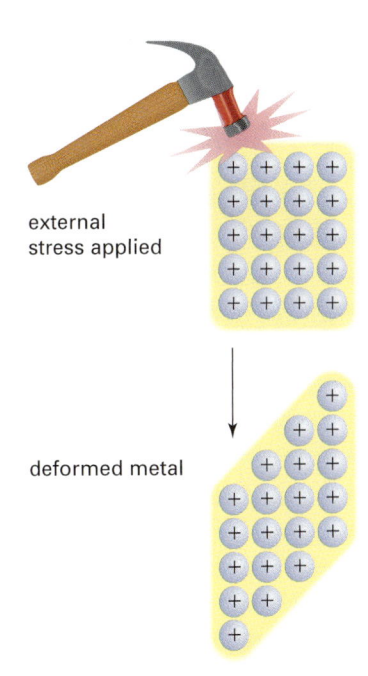

**Figure 6.25** Metals are easily deformed because one layer can slide over another while the free electrons (yellow cloud) continue to bind the metal ions together.

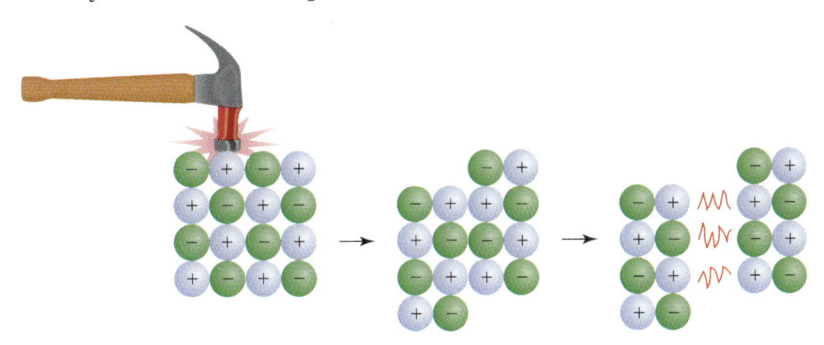

**Figure 6.27** One reason that crystals have some flat surfaces is that, when stressed, they break along the plane at which like charges are aligned and repelling each other.

Non-polar molecular compounds with large molecules can be solid at room temperature but they are very soft. Paraffin wax, $C_{20}H_{42}$, is an excellent example. These solids have only weak London dispersion forces acting between their molecules. These forces can act between induced dipoles in any part of one molecule and the opposite ends of induced dipoles in any part of another molecule because they are random and constantly changing. Consequently, the molecules can easily slide along beside each other, making paraffin wax soft and easily broken.

**Figure 6.26** The malleability of copper makes it a material of choice for some artists who hammer designs into the metal sheet.

# Canadians in Chemistry

## Dr. Geoffrey Ozin

His work has flown on a Space Shuttle, and it has been hailed as art. It may well be part of the next computing revolution.

What does Dr. Geoffrey Ozin do? As little as possible, for he believes in letting the atoms do most of the work. This approach has made him one of the more celebrated chemists in Canada. Time and again, he has brought together organic and inorganic molecules, polymers, and metals in order to create materials with just the right structure for a specific purpose.

Self-assembly is the key. Atoms and molecules are driven into pre-designed shapes by intermolecular forces and geometrical constraints. At the University of Toronto, Dr. Ozin teaches his students the new science of intentional design, instead of the old trial-and-error methods.

Born in London, England, in 1943, Geoffrey Ozin earned a doctorate in chemistry at Oxford University. He joined the University of Toronto in 1969. Ozin's father was a tailor. In a way, Ozin is continuing the family tradition. Ozin, however, uses ionic and covalent bonds, atoms and molecules, acids, gases, and solutions to fashion his creations.

In 1996, Dr. Ozin demonstrated the self-assembly of crystals with a porous structure in space, under the conditions (such as microgravity) found aboard a Space Shuttle. Since then, he has shown how the self-assembly of many materials can be controlled to produce their structure.

Dr. Ozin's latest achievement involves structure. Ozin was part of an international research team that created regular microscopic cavities inside a piece of silicon. This material can transmit light photons in precisely regulated ways. In the future, this material might be used to build incredibly fast computers that function by means of photons instead of electrons!

## Electrical Conductivity

The ability to conduct an electrical current requires that negative and/or positive charges can move freely and independently of each other. The only substances that are good electrical conductors in the solid state are metals. Ionic compounds are good conductors in the liquid state or when dissolved in water. Molecular substances are nonconductors as solids, liquids, or when dissolved in water. Once again, these properties can be explained on the basis of bonding.

Electric current is a flow of charge carried by moving electrons or ions. Metals are good conductors of electricity because the valence electrons can move freely throughout the metallic structure carrying electric charge from one place to another. This freedom of movement is not possible in solid ionic compounds, because the valence electrons are held within the individual ions in the lattice and these charged ions are not free to move about. In molten sodium chloride, for example, the rigid crystalline lattice structure is disrupted. Although the ions still attract each other, they can move past each other in opposite directions as shown in Figure 6.28. When ionic compounds are dissolved in water, the ions are surrounded by water molecules but they still carry a net charge. The ions can easily move through water toward oppositely charged electrodes.

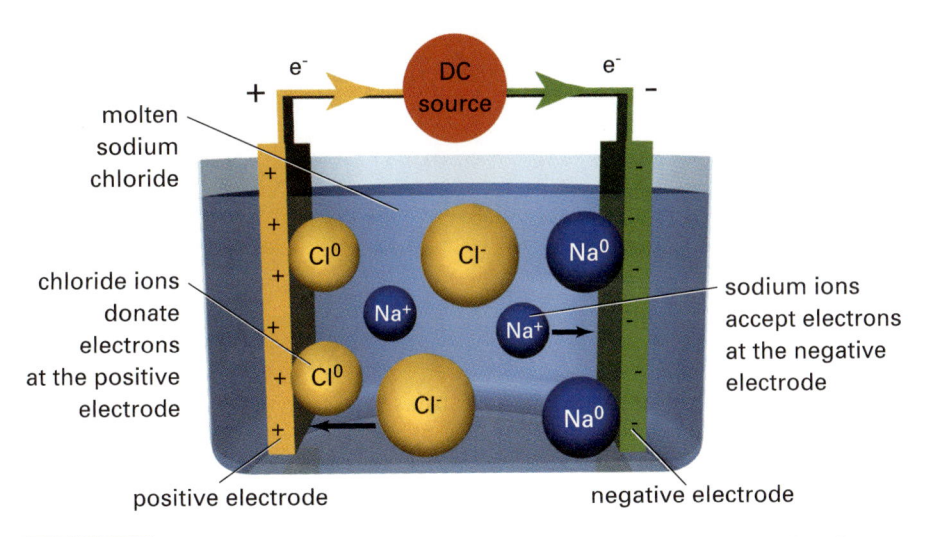

**Figure 6.28** When molten sodium chloride is placed between charged electrodes, the positive sodium ions are attracted to the negative electrode. When they reach the electrode, they pick up an electron and become neutrally charged. Likewise, negative chloride ions are attracted to the positive electrode where they release the extra electron and become neutrally charged.

When you read about network solids, you learned that diamonds cannot conduct electrical current but graphite can. In diamonds, all electrons are localised in strong covalent bonds and cannot move. In graphite however, three of the four bonds of each carbon atom are strong and the shared electrons cannot move away from the atoms. However, the bonds between the layers of graphite are long and weak. Since the bonds are weak, when graphite is placed between charged electrodes, the electrons in the weak bonds can be induced to "jump" from one bond to the next and move toward the positive electrode.

In general, molecular compounds cannot conduct electrical current at all because the electrons are all localised within the molecules. Although polar molecules have partially charged ends, the charges cannot leave the molecule and the net charge on the molecule is zero.

## Thermal Conductivity

The same properties that account for the ability of metals to conduct electric current also account for their high thermal conductivity. Since the valence electrons are free to move, they can easily collide with particles of adjacent hot objects and receive kinetic energy in the collisions. The electrons then move freely throughout the metal and pass the kinetic energy on to other particles by colliding with them.

Now you can use what you have learned about the properties of substances in the following Investigation.

# Investigation 6-C

**SKILL FOCUS**
Performing and recording
Analyzing and interpreing

## Properties of Substances

In this investigation, you will study the properties of five different types of solids: non-polar covalent, polar covalent, ionic, network, and metallic. You will be asked to identify each substance as one of the five types. In some cases, this will involve making inferences and drawing on past knowledge and experience. In others, this may involve process-of-elimination. The emphasis is on the skills and understandings you use to make your decisions. Later, you will be able to assess the validity of your decisions.

### Question

What are some of the properties of the different types of solids?

### Safety Precautions

- Tie back long hair and any loose clothing.
- Before you light the candle or Bunsen burner, check that there are no flammable solvents nearby.
- Use only a pinch of $SiO_2$ in steps 6 and 7.

### Materials

samples of paraffin wax, $C_{20}H_{42}$ (shavings and a block)
samples of sucrose, $C_{12}H_{22}O_{11}$ (granular and a large piece, such as a candy)
samples of sodium chloride, NaCl (granular and rock salt)
samples of silicon dioxide, $SiO_2$ (sand and quartz crystals)
samples of tin, Sn (granular and foil)
distilled water
100 mL beaker and stirrer
metal plate (iron or aluminum)
conductivity tester
candle
Bunsen burner
retort stand with ring clamp
timer

### Procedure

1. Read the entire Procedure before you begin. Design a table to record your observations.

2. Rub one solid onto the surface of another. Rank the relative hardness of each solid on a scale of 1 to 5. A solid that receives no scratch marks when rubbed by the other four is the hardest (5). A solid that can be scratched by the other four is the softest (1).

3. One at a time, place a small quantity (about 0.05 g) of each solid into 25 mL of distilled water in a 100 mL beaker. Observe and record the solubility of each solid.

4. For solids that dissolved in step 3, test the solution for electrical conductivity.

5. Test each of the five large solid samples for electrical conductivity.

6. One at a time, put a small sample of each solid on the metal plate. Place the plate on the ring stand and heat it with the burning candle. The plate should be just above the flame. Observe how soon the solid melts.

7. Repeat step 6 with a Bunsen flame. Observe how soon the solid melts.

### Analysis

1. Based on what you know about bonding, classify each solid as non-polar covalent, polar covalent, ionic, network, or metallic. Give reasons to support your decision.

2. Determine the electronegativity difference between the elements in each of the substances. How does this difference relate to the properties you observed?

### Conclusion

3. Based on the properties you observed, write a working definition of each type of solid.

222 MHR • Unit 2 Structure and Properties

# Chemistry Bulletin

**Science**    **Technology**    **Science**    **Environment**

## Ionic Liquids: A Solution to the Problem of Solutions

Although organic solvents are commonplace in industry, they are highly volatile and dangerous in the large quantities that manufacturing processes require. Disposing of used solvents is a major environmental concern.

Until recently, the idea of a recyclable, non-volatile solvent seemed like the stuff of science fiction. But in the 1980s, researchers in the United States were trying to create a new electrolyte for batteries. Instead, they created a colourless, odorless liquid that was composed of nothing but ions—an *ionic liquid*.

An ionic liquid is a salt in liquid form. Ionic liquids do not usually find practical applications, since ionic compounds have such high boiling points. For example, sodium chloride does not begin to melt until it reaches a temperature of about 800°C.

In an ionic liquid, the cations are much larger than in typical ionic substances. As a result, the anions and cations cannot be packed together in an orderly way that balances both the sizes and distances of the ions with the charges between them. As a result, they remain in a loosely-packed, liquid form. Because the charges on the cations have the capacity to bind more anions, an ionic liquid has a net positive charge. Notice the large cation of the ionic liquid molecule shown below. This is what makes ionic liquids suitable solvents. The net charge can attract molecules of other compounds, dissolving them.

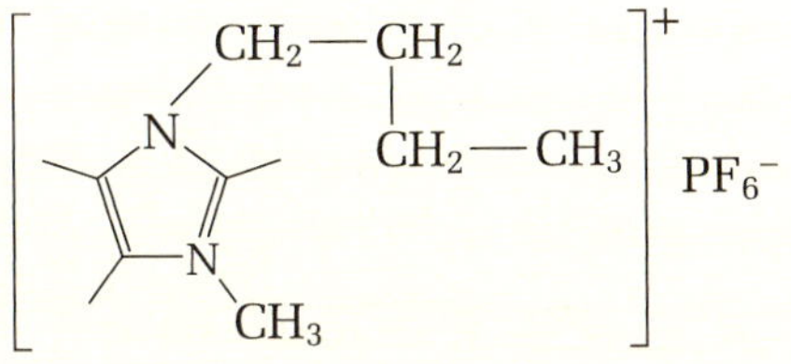

Many properties of ionic liquids make them more desirable solvents than organic solvents. For example, ionic liquids require a lot of energy to change their state—they remain as liquids even at temperatures of 200°C.

In addition, ionic liquids evaporate sparingly. Organic solvents, on the other hand, easily release fumes that are harmful, often toxic, and that can eventually oxidize and create carbon dioxide, a notorious greenhouse gas.

As solvents, ionic liquids can be uniquely "tuned" to a particular purpose by adjusting the anion/cation ratio. To decaffeinate coffee, for example, you could create an ionic liquid that would just dissolve caffeine and nothing else. Current research suggests that ionic liquids can be recovered from solution and reused.

Yet, there are still many hurdles that must be overcome before these substances can be put to widespread use. Because ionic liquids boil at high temperatures, conventional methods of separating reaction products from solution cannot be used. Pharmaceutical manufacturing processes, for instance, often involve the process of distillation. The higher temperatures required to distill using ionic liquids (as opposed to organic liquids) would decompose many of the desired pharmaceutical compounds.

The discovery of ionic liquids is proving a fertile field for researchers. Ionic solvents are revealing new mechanisms of reaction and enhancing our understanding of the molecular world. As well, their potential applications in both industry and the home have encouraged several companies, world-wide, to further explore and develop these unusual substances.

### Making Connections

1. Investigate some of the industrial processes where solvents are currently used. Could ionic liquids be used in any of these industries instead? Why or why not? Look into some of the ideas that ionic liquid researchers have proposed for dealing with the unique challenges posed by different industries.

## Section Summary

In this section, you learned how to predict many physical properties of substances based of the type of bonding in the substances. Since the average kinetic energy of the molecules in a substance is directly related to its temperature, higher temperatures are required to loosen or break the forces of attraction between particles and cause melting or boiling of substances with higher bond energies. Solids are brittle and break easily if stress causes opposite charges to become aligned. To conduct electrical current, electrons or ions must be able to move independently of oppositely charged ions.

Study the concept organizer below and make other similar organizers for the various properties that you studied in this section.

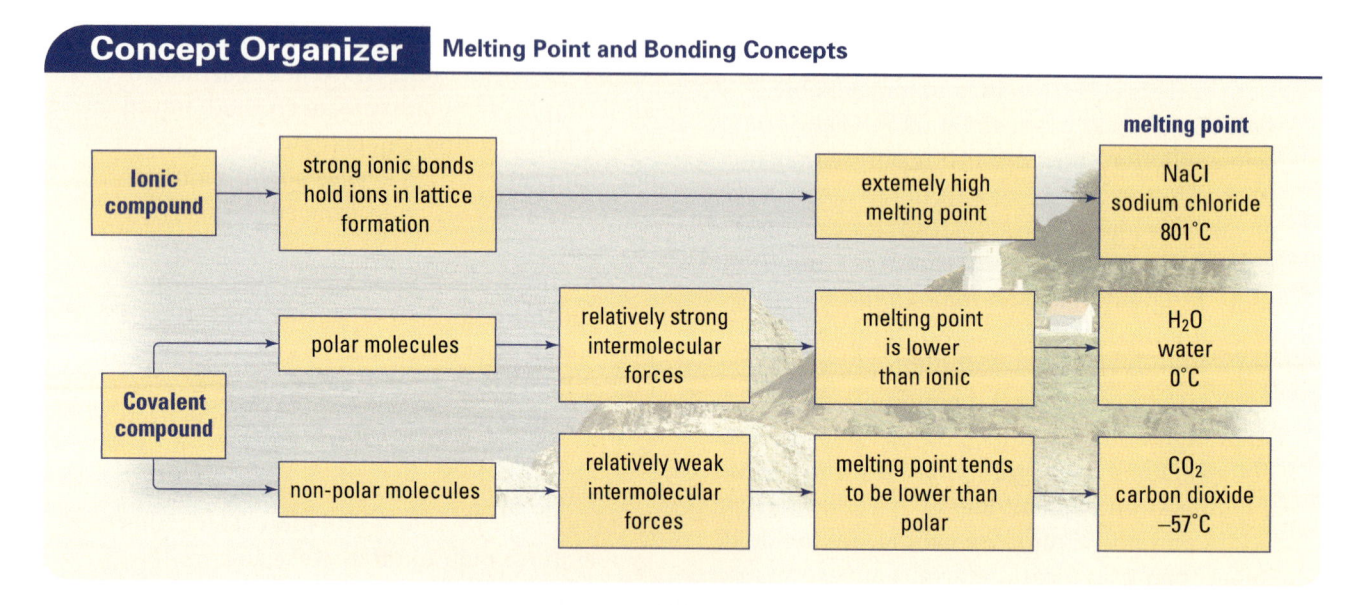

### Concept Organizer: Melting Point and Bonding Concepts

- Ionic compound → strong ionic bonds hold ions in lattice formation → extremely high melting point → NaCl sodium chloride 801°C
- Covalent compound → polar molecules → relatively strong intermolecular forces → melting point is lower than ionic → H₂O water 0°C
- Covalent compound → non-polar molecules → relatively weak intermolecular forces → melting point tends to be lower than polar → CO₂ carbon dioxide −57°C

## Section Review

**1** A molecule of chloroform, $CHCl_3$, has the same shape as a molecule of methane, $CH_4$. However, methane's boiling point is −164°C and chloroform's boiling point is 62°C. Explain the difference between the two boiling points.

**2** Describe the relationship between bond energy of a compound and its melting and boiling points.

**3** A solid substance is found to be soluble in water and has a melting point of 140°C. In order to classify this solid as ionic, covalent, metallic, or network, what additional test(s) should be carried out?

**4** Ionic compounds are extremely hard. They hold their shape extremely well. Explain these properties. Give two reasons why it is not practical to make tools out of ionic compounds.

**5** A chemist analyzes a white, solid compound and finds that it does not dissolve in water. When the compound is melted, it does not conduct electricity.

(a) What would you predict about this compound's melting point?

(b) Are the atoms that make up this compound joined by covalent or ionic bonds?

# CHAPTER 6 Review

## Reflecting on Chapter 6

Summarize this chapter in the form of your choice. Here are a few ideas to use as guidelines:
- VSEPR theory allows you to predict the three-dimensional shape of molecules based on Lewis diagrams.
- By predicting the shapes of molecules, you can predict their polarity.
- Forces that are usually classed as intermolecular forces can act between different parts of the same large biological molecules.
- The type of bonding that holds atoms and molecules together can explain many of the hardness, malleability, and electrical conductivity of substances.
- Intermolecular forces determine state of substances and many of their physical properties.
- The polarity of molecules can be used to explain the range of boiling points and melting points among compounds that contain molecules of similar masses.
- Illustrate how metallic bonding determines uses of metals.

## Reviewing Key Terms

For each of the following terms, write a sentence that shows your understanding of its meaning.

co-ordinate covalent bonding
resonance structures
valance-shell electron-pair repulsion theory
linear
trigonal planar
tetrahedral
pyramidal
bent
network solid
intramolecular force
intermolecular force
dipole-dipole force
ion-induced dipole force
dipole-induced dipole force
dispersion force
hydrogen bond

## Knowledge/Understanding

1. Use a diagram to illustrate each of the following ideas.
   (a) Water is a molecule that has a bent shape.
   (b) Each oxygen atom in carbon dioxide had two bonding pairs and two non-bonding pairs.
   (c) Silicon tetrachloride is a tetrahedral molecule.

2. Diatomic molecules such as oxygen, nitrogen, and chlorine, tend to exist as gases at room temperature. Explain this observation using bonding theory.

3. A solid molecular compound has both intermolecular and intramolecular forces. Do solid ionic compounds contain these types of forces? Explain your answer.

4. Describe the intermolecular forces between the molecules of hydrogen halides (HF, HCl, HBr, HI) and explain the differences in their boiling points.

5. What is the difference between a permanent molecular dipole in a polar molecule and an induced dipole in a non-polar molecule?

6. Cooking oil, a non-polar liquid, has a boiling point in excess of 200°C. Water boils at 100°C. How can you explain these facts, given the strength of water's hydrogen bonds?

7. What types of intermolecular forces must be broken to melt solid samples of the following?
   (a) $NH_3$
   (b) NaI
   (c) Fe
   (d) $CH_4$

8. In which compound, $H_2O$ or $NH_3$, will the hydrogen bonding be stronger? Explain.

9. List the following substances in order of increasing boiling points: $C_2H_5OH$, $SiO_2$, $C_3H_8$, K. Give a reason for your answer.

10. What are the molecular shapes of $CCl_4$, $CH_3Cl$, and $CHCl_3$? In liquid samples of these compounds, which would have dipole-dipole forces between molecules? Predict the order of their boiling points from lowest to highest.

11. Distinguish between dipole-dipole attractive forces and an ionic bond.

Answers to questions highlighted in red type are provided in Appendix A.

12. How can a molecule with polar covalent bonds be non-polar?

13. Predict which substance will have the higher boiling point in each of the following pairs. Explain your prediction using bonding theories.
    (a) $NH_2Cl$ or $PH_2F$
    (b) SiC or $AsH_3$
    (c) Ne or Xe
    (d) $CH_3OH$ or $C_2H_5NH_2$
    (e) $CH_3F$ or $F_2$
    (f) $AlCl_3$ or $AsCl_3$
    (g) $C_4H_{10}$ or $Cl_2$
    (h) $NH_3$ or $PH_3$
    (i) $C_5H_{12}$ or $C_4H_9F$
    (j) ammonia or methane
    (k) carbon dioxide or silicon dioxide
    (l) neon or krypton
    (m) KCl or ICl
    (n) KBr or ClBr
    (o) $NH_2Cl$ or $PH_2Cl$
    (p) $NH_4Cl$ or $CH_3Br$
    (q) $C_2H_5F$ or $CH_3Cl$
    (r) $CH_3F$ or $CH_3NH_2$
    (s) $C_2F_2$ or $C_2HCl$
    (t) $H_2O$ or $H_2S$
    (u) $SO_2$ or $SiO_2$
    (v) NaCl or CO

14. Explain each of the following using bonding theory.
    (a) Glycerol, $C_3H_5(OH)_3$, flows much more slowly than water, $H_2O$.
    (b) $CaCl_2$ has no molecules.
    (c) Silver can be bent to make chains.
    (d) Graphite is used as a high temperature lubricant.
    (e) Diamonds are nonconductors of electricity.
    (f) Molten potassium chloride conducts electricity.
    (g) Metals have variable m.p.'s.
    (h) Diamond is the hardest known substance.
    (i) A water filled bottle breaks when placed in a freezer overnight.
    (j) Graphite is used to make tennis rackets and golf clubs.

## Inquiry

15. Suppose that you have two colourless compounds. You know that one is an ionic compound and the other is a molecular compound. Design an experiment to determine which compound is which. Describe the tests you would perform and the results you would expect.

16. You have two liquids, A and B. You know that one liquid contains polar molecules and the other liquid contains non-polar molecules. You do not know which is which. You pour each liquid so that it falls in a steady, narrow stream. As you pour, you hold a negatively charged ebonite rod to the stream. The stream of liquid A is deflected toward the rod. The rod does not affect the stream of liquid B. Which liquid is good polar? Explain your answer.

17. Water reaches its maximum density at 4°C. At temperatures above or below 4°C, the density is lower. Discuss what is happening to the water molecules at temperatures around 4°C.

18. In which liquid, $HF_{(\ell)}$ or $H_2O_{(\ell)}$, would the hydrogen bonding be stronger? Based upon your prediction, which of these two liquids would have a higher boiling point? Refer to a reference book to find the boiling points of these two liquids to check your predictions. Account for any difference between your predictions and the actual boiling points.

## Communication

19. Compare the molecules $SF_4$ and $SiF_4$ with respect to molecular shape and molecular polarity.

20. Compare the forces of attraction that must be overcome to melt samples of NaI and HI.

21. (a) Contrast the physical properties of diamond, a covalent network solid with the molecular compound 2,2-dimethylpropane shown here.

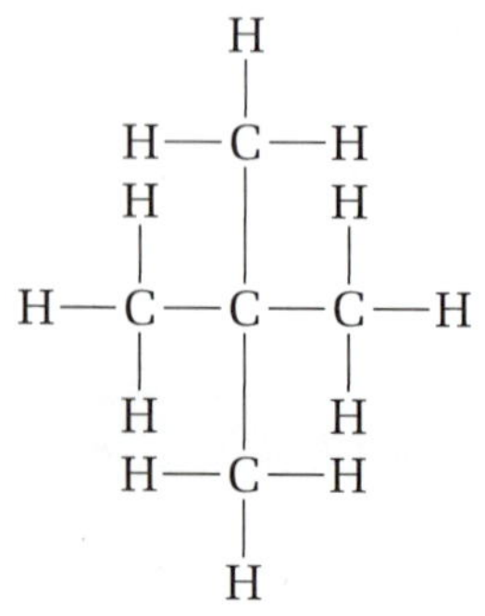

(b) In diamond, covalent bonds join the carbons atoms. In a molecule of 2,2-dimethylpropane, a central carbon atom is covalently bonded to four –CH₃ groups. How are the covalent bonds in the two substances different from one another?

(c) How are the physical properties of these two substances related to the covalent bonds in them?

22. At room temperature, carbon dioxide, CO₂, is a gas, while silica, SiO₂, is a hard solid. Compare the bonding in these two compounds to account for this difference in physical states.

23. Compare the bonding and molecular polarity of SeO₃ and SeO₂.

24. Discuss the intermolecular and intramolecular forces in N₂H₄ and C₂H₄. Based upon the bonding between molecules, which of these two compounds would have a lower boiling point?

## Making Connections

25. Many advanced materials, such as KEVLAR®, have applications that depend on their chemical inertness. What hazards do such materials pose for the environment? In your opinion, do the benefits of using these materials outweigh the long-term risks associated with their use? Give reasons to justify your answer.

26. Considering the changes of states that occur with water in the environment, suggest how intermolecular forces of attraction influence the weather.

**Answers to Practice Problems and Short Answers to Section Review Questions**

**Practice Problems** 1.(a) H:N:H with H below (b) H:C:H with H above and below (c) :F:C:F: with :F: above and below

(d) H:As:H with H below (e) [:Br:O:]⁻ (f) H:S:H (g) H:O:O:H

(h) :Cl:N::O: (i) H,H C::O H,H (j) :C:::O: (k) H:O:Cl:

(l) H:C:::C:H

2.(a) [:O:C:O: with :O: below]²⁻ ↔ [:O:C:O: with :O: below]²⁻ ↔ [:O:C:O: with :O: below]²⁻

(b) [:N:::O:]⁺ (c) [:O:Cl:O: with :O: below]⁻ (d) [:O:S:O: with :O: below]²⁻

3. H:Cl: C:C: C:F: H:Cl: (both)

4.(a) H,H N::N H,H (b) :F:N:N:F:

5. :O: :O:Xe:O: :O:

6.(a) H:C:::N two electron groupings, all BP, linear

(b) :O:S::O: ↔ :O::S:O: Both resonance structures have three electron groupings. two BP, one LP, bent

(c) :O::S:O: with :O: below ↔ :O:S:O: with :O: below ↔ :O:S::O: with :O: below
All three resonance structures have three electron groupings. All BP, trigonal planar

(d) [:O:S:O: with :O: above and below]²⁻ four electron groupings, all BP, tetrahedral

(e) :Cl:As:Cl: with :Cl: below four electron groupings, three BP, one LP, pyramidal

(f) :I:S:I: four electron groupings, two BP, two LP, bent

(g) H:C:F: with H above and below four electron groupings, all BP, tetrahedral

7.(a) :F:C:F: with H above and below four electron groupings, all BP, tetrahedral

(b) [H:N:H with H above and below]⁺ four electron groupings, all BP, tetrahedral

(c) :O::N::O: two electron groupings, all BP, linear

(d) [:O::N:O:]⁻ ↔ [:O:N::O:]⁻ Both resonance structures have three electron groupings, two BP and one LP, bent

8.(a) H:S:F: with H above and below four electron groupings, all BP, C—F bond is much more polar than C—H bonds, all bonds in same general direction, polar

(b) H,H C::O three electron groupings, all BP, trigonal planar, C—O much more polar than C—H bonds, all in same general direction, polar

Chapter 6 Structure and Properties of Substances • MHR  227

(c) four electron groupings, three BP and one LP, pyramidal, polar bonds add from top to bottom of pyramid, polar

(d) H:O:O:H four electron groupings around each O, two lone pairs and two BP, bent around each oxygen, polar

9. four electron groupings, all BP, tetrahedral, C—F bonds more polar than C—Cl bonds, polar 10. Both molecules have the same pyramidal shape with the nitrogen atom at the top of the pyramid. The arrow representing the net polarity points downwards and is the sum of the downward portion of the polarity of each of the three bonds. The $\Delta EN$ for the N—F bonds is 0.94 and $\Delta EN$ for the N—Cl bonds is 0.12. Since each individual bond in $NF_3$ is more polar than in $NCl_3$, the $NF_3$ is the more polar molecule.

**Section Review 6.1:**

1.(a) four electron groupings, three BP, one LP, polar

(b) four electron groupings groupings around C in $CH_3$, all BP, pyramidal around the C in $CH_3$; two electron groupings around the C in CN, all BP, linear: C—N bond is much more polar than C—H bonds, polar

(c) four electron groupings, two BP, two LP, bent, polar

(d) four electron groupings, all BP, tetrahedral, all polar bonds equal and cancel each other, non-polar 2. The first part of the statement is true. In order for a molecule to be polar, one or more of the bonds must be polar. However, the second part of the statement is not true. A molecule can have polar bonds without being polar. The polarity of the bonds of a symmetrical molecule can exactly cancel each other making the molecule as a whole non-polar.

3. Both molecules have a tetrahedral shape around the central carbon atom. However, the $CH_3OH$ (methanol) molecule is bent around the oxygen atom. The small polarity of the C—H bonds in $CH_4$ (methane) cancel each other so the molecule is non-polar. The O—H bond in $CH_3OH$ is very polar and since it is much more polar than the other bonds, the molecule is polar.

4. The octet rule is obeyed. The BN bond is co-ordinate covalent.

5. four electron groupings, two BP, two LP, bent, polar

6. four electron groupings, all BP, tetrahedral, symmetrical, non-polar

7. The number of electron groupings and whether the groupings consist of bonding pairs or lone pairs determines the shape of the molecule. Lone pairs spread out more than bonding pairs and therefore repel each other and bonding pairs more than bonding pairs repel each other. The presence and orientation of polar bonds along with the shape determine the polarity of the molecule as a whole. If the polarities of the bonds cancel each other, the molecular is non-polar. If there is a non-zero vector sum of bond polarities, the molecule is polar. **6.2: 1.** The negative ends of the polar water molecules are attracted to and surround the positive ions and pull them away from the solid crystal. The positive ends of the polar water molecules are attracted to and surround the negative ions and pull them away from the solid crystal. **2.** As the number of electrons in a molecule increases, the chance that the vibration of electrons will create a transient, local dipole increases. As the shape of a molecule becomes less spherical, the surface area increases. As the surface area of a molecule increases, the chance that a transient dipole will occur somewhere on the surface, increases. **3.** The charge on a positive ion attracts, and on a negative ion repels, the electrons in an adjacent non-polar molecule. Thus the charge on the ion causes the electron cloud around that molecule to become distorted, "inducing" a temporary dipole. **4.** A hydrogen atom in one molecule must be bonded to a very electronegative atom such as O, N, or F. The positive nucleus of the hydrogen atom is exposed, giving the atom a partial positive charge. The second molecule must have a strongly electronegative atom such as O, N, or F, that will be attracted to the partially positive hydrogen atom of the first molecule.
**5.** Intermolecular (covalent) bonds are, in general, stronger than intermolecular (van der Walls) bonds. **6.** If water was non-polar, it would be a gas under nearly all conditions that occur on Earth. Since all forms of life contain a large percentage of liquid water in their cells, life, as we know it, could not exist. It is possible that another substance similar to water might have made it possible for some form of life to exist. **6.3: 1.** Methane ($CH_4$) is perfectly symmetrical and therefore non-polar. The C—Cl bonds in chloroform ($CHCl_3$) are very polar and add vectorally to make the molecule polar. Dipole-dipole forces cause the oppositely charged ends of the chloroform molecules to be attracted to each other. Much more molecular kinetic energy is required to separate the chloroform molecules than the methane molecules. This required energy is reflected in higher melting and boiling points. **2.** The ionic and metallic bonds in these compounds must be broken in order for the compounds to melt or vapourize. The energy that breaks the bonds is the kinetic energy of the particles. Temperature is a measure of the average kinetic energy of atoms, ions, or molecules. Therefore, high melting points and boiling points indicates, that the bond energies are strong. In molecular substances,

228 MHR • Unit 2 Structure and Properties

covalent bonds do not break when a substance melts or vapourizes. It is the van der Walls attractions that must be broken to cause melting or evapouration. **3.** The solid and the water solution of the compound should be tested for the ability to conduct electric current. The hardness, brittleness, malleability, and ductility should also be tested. **4.** The ionic bonds between ions in an ionic solid are very strong making the compound very hard. However, the brittleness of ionic compounds makes them impractical for making tools. Ionic compounds are neither malleable nor ductile. Shaping them into tools would be too difficult. **5.(a)** The melting point is probably relatively low. **(b)** The atoms are joined by covalent bonds.

# UNIT 2 Review

## Knowledge/Understanding

### Multiple Choice

In your notebook, write the letter for the best answer to each question.

1. Which of the following is *not* a physical property?
   (a) melting point
   (b) electronegativity
   (c) electrical conductivity
   (d) malleability
   (e) boiling point

2. The electrons in the outer shell of an atom are called
   (a) unpaired electrons
   (b) valance electrons
   (c) bonding electrons
   (d) lone pairs
   (e) octet of electrons

3. Ionic compounds consist of
   (a) metal atoms only
   (b) nonmetal atoms only
   (c) Group IA and Group VII A only
   (d) Group VIII atoms
   (e) metal and nonmetal atoms

4. Which of the following is an example of a network solid?
   (a) sodium, $Na_{(s)}$
   (b) sucrose, $C_{12}H_{22}O_{11(s)}$
   (c) graphite, $C_{(s)}$
   (d) silica, $SiO_{2(s)}$
   (e) magnesium fluoride, $MgF_{2(s)}$

5. The two electrons in a bonding pair are
   (a) always donated by the same atom
   (b) never donated by the same atom
   (c) usually donated by two different atoms
   (d) always donated by two different atoms
   (e) never donated by two different atoms

6. A crystalline substance that is hard, unmalleable, has a high melting point, and is a non-conductor of electricity could be
       I an ionic crystal
       II a polar covalent solid
       III a metal
       IV a network solid
   (a) I or II
   (b) I or III
   (c) III or IV
   (d) II or IV
   (e) I or IV

7. Which of the following molecules does not have a linear shape?
   (a) $Cl_2O$
   (b) $CO_2$
   (c) $XeF_2$
   (d) OCS
   (e) $BeF_2$

8. The bond between atoms having an electronegativity difference of 1.0 is
   (a) ionic
   (b) mostly ionic
   (c) polar coavlent
   (d) slightly polar covalent
   (e) non-polar covalent

9. A substance that has a melting point of 1850°C and a boiling point of 2700°C is insoluble in water and is a good insulator of electricity. The substance is most likely
   (a) an ionic solid
   (b) a polar covalent solid
   (c) a metal
   (d) a network solid
   (e) a molecular solid

10. The main factor that leads to the formation of a chemical bond between atoms is
    (a) the lowering of melting point and boiling point when a compound is formed
    (b) a tendency in nature to achieve a minimum energy for a system
    (c) the formation of a shape consistent with VSEPR theory
    (d) a tendency to make the attractions equal to the repulsions between atoms
    (e) a tendency to reduce the number of existing particles

### Short Answers

In your notebook, write a sentence or a short paragraph to answer each question.

11. List three properties of ionic and molecular compounds that differ. Explain how they differ.

12. Draw Lewis structures for the following neutral atoms.
    (a) neon
    (b) beryllium
    (c) phosphorous
    (d) iodine
    (e) lithium

Answers to questions highlighted in red type are provided in Appendix A.

13. Indicate whether the following statement is correct. If not, restate the sentences correctly. "In Lewis structures of atoms, some electrons are drawn in pairs when the atom has more than four valence electrons. This method of drawing electrons is simply for convenience. There is no significance in drawing two electrons as a pair."

14. Explain how can you predict whether a compound is ionic or molecular by looking at a formula such as MgO or $NH_3$.

15. Use Lewis structures to illustrate the formation of potassium iodide and magnesium bromide.

16. How do the lone pairs of electrons around the central atom of a molecule affect the bond angle between two bonding pairs of electrons?

17. Why is it incorrect to refer to an ionic compound as a "molecule?"

18. Chlorine molecules ($Cl_2$) have a single covalent bond while oxygen molecules ($O_2$) have one double bond and nitrogen molecules ($N_2$) have a triple bond. Explain why these diatomic molecules have different types of bonds.

19. In both metallic and molecular compounds, atoms share electrons. Why was it necessary to develop a different model of bonding for the two types of substances?

20. Explain why, at room temperature, $CO_2$ is a gas but $CS_2$ is a liquid.

21. Explain the process by which you can use the electronegativities of two atoms to determine the type of bond that will form between the atoms.

22. If you read that a certain bond had 70% ionic character, what properties would you expect that bond to have? What would be the percent covalent character of the bond?

23. What might you predict about the length of a bond that had a very large bond energy?

24. Draw Lewis structures for each of the following molecules.
    (a) $OCF_2$
    (b) $H_2NNH_2$
    (c) HCN
    (d) $Cl_2O$
    (e) $HCO_3^-$
    (f) $NO_3^-$

25. The molecule $NO_3$ requires three resonance structures to show the Lewis structure. Using double ended arrows, draw the Lewis structure of $NO_3$.

26. Use VSEPR theory to predict the shape of the following.
    (a) $NO_2$
    (b) $COCl_2$
    (c) NOF

27. Distinguish between dipole-dipole attraction forces and dispersion (London) forces. Is one type of these intermolecular forces stronger than the other? Explain.

28. Describe the properties of graphite that make it a good lubricant at high temperatures.

29. Describe some similarities and differences between metallic compounds and network solids.

30. Explain why the boiling point of a pure molecular compound is an indication of the strength of intermolecular attractive forces.

31. Explain the difference between a dipole-dipole force and a dipole-induced dipole force.

32. London dispersion forces are extremely weak compared to all other intermolecular forces. How can they create stability of important structures such as cell membranes?

33. State the two factors that affect the strength of dispersion forces between two molecules.

34. What are the requirements for the formation of hydrogen bonds?

35. Under what conditions are hydrogen bonds at their maximum strength?

36. Hydrogen bonds are usually considered intermolecular forces. Describe one case in which hydrogen bonds occur within the same molecule.

37. Why does water have a much higher melting point and boiling point than other molecules of similar size such as ammonia and methane?

38. Explain the reason for the malleability of metals.

39. Explain why ionic solids are brittle.

40. Why do ionic compounds conduct electric current when in the molten state but not when they are solid?

Unit 2 Review • MHR **231**

## Inquiry

41. Which ionic solid would you expect to have a higher melting point: LiI or LiBr? Justify your answer.

42. Copy the following bond pairs into your notebook. Use electronegativity values to indicate the polarity of the bond in each case.
    (a) N—H
    (b) F—N
    (c) I—Cl

43. Arrange the following sets of bond pairs in order of increasing polarity. Identify the direction of bond polarity for the bond pairs in each set.
    (a) Cl—F, Br—Cl, Cl—Cl
    (b) Si—Cl, P—Cl, S—Cl, Si—Si

44. Formulate a procedure by which you could experimentally determine whether an unknown solid was a network solid, an ionic solid, a molecular solid, or a metallic solid.

## Communication

45. Draw a Lewis structure for the ionic compound $Ba(OH)_2$.

46. The ammonium ion, $NH_4^+$, is charged but not polar. The ammonia molecule, $NH_3$, is polar but not charged. Explain why these statements are true.

47. Use VSEPR theory to predict the molecular shape of $CH_2Cl_2$. Draw a sketch to indicate the polarity of the bonds around the central atom to verify that this is a polar molecule.

48. Do you agree or disagree with the following statement. Explain why. "If there were no intermolecular forces, all molecular compounds would be gases."

49. The group trend for boiling point is the same as the trend for atomic radius. For the compounds formed between hydrogen and the first three elements of group 16 (VIA), $H_2S$ has a lower boiling point than both $H_2O$ and $H_2Se$. Explain this diversion from the trend you would expect.

50. Define the following terms, and identify the term that best describes the H—O bond in a water molecule: non-polar covalent, polar covalent, ionic.

51. For a solid metallic element:
    (a) Identify four physical properties.
    (b) Identify two chemical properties that would enable you to classify the element as metallic.

52. You can use electronegativity differences to think of chemical bonds as having a percent ionic or a percent covalent character. The graph below plots percent ionic character versus $\Delta EN$ for a number of gaseous binary molecules. Use this graph to answer the following questions.

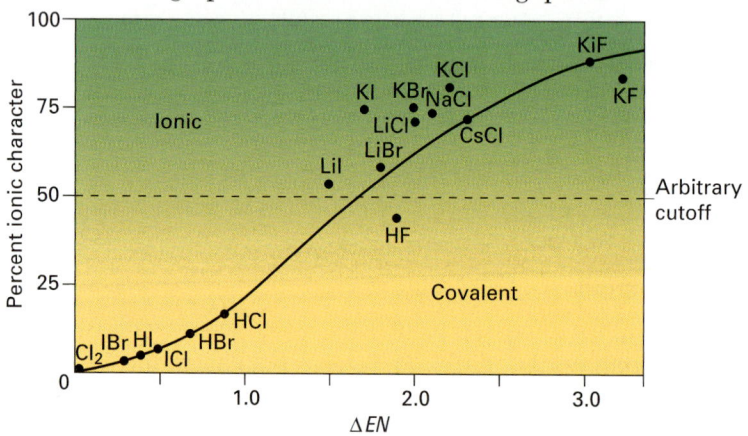

    (a) Describe the ionic character of the molecules as a function of $\Delta EN$.
    (b) Which molecule has a zero percent ionic character? What can you infer about the interactions among the electrons of this molecule?
    (c) Do any molecules have a 100 percent ionic character? What can you infer about the interactions among the electrons of ionic compounds?
    (d) Chemists often assign the value of 50 percent as an arbitrary cutoff for separating ionic compounds from covalent compounds. Based on your answer in part (c), what does the use of the term "arbitrary" suggest about the nature of chemical bonds?

53. In 1906, the Nobel Prize in chemistry was awarded to a French chemist, Ferdinand Moissan, for isolating fluorine in its pure elemental form. Why would this achievement be deserving of such a prestigious honour? Use your understanding of atomic properties as well as chemical bonds to explain your answer.

54. Use a graphic organizer such as a Venn diagram to compare ionic bonding with metallic bonding.

## Making Connections

55. Some of the most important discoveries of the last few decades have been made serendipitously, that is, by accident or by chance. For example, when the principle behind lasers was discovered and lasers were first built, no one hand any idea that there would be a use for them. Such is the case with fullerenes or "buckyballs." Do research in print or in the Internet to find out how fullerenes were discovered. Then find out about the types of research that are ongoing to learn about possible applications for fullerenes. Focus on a type of research and application that you would enjoy pursuing. What application would you like to develop? What reason do you have to believe that this application is feasable?

56. Many useful products, such as superglue and fire-retardant materials, are domestic "spin-offs" from research that is related to military applications. Military expense budgets are large, in part, because of the money required to support this research. In your opinion, should military budgets be exempt from cost-conservation measures that governments typically must consider? Debate this question with your classmates, and issue a statement that summarizes your decisions.

# UNIT 3

# Solutions and Solubility

**UNIT 3 CONTENTS**

**CHAPTER 7**
Solutions and Their Concentrations

**CHAPTER 8**
Solution Stoichiometry

**UNIT 3 OVERALL OUTCOMES**

- What are the properties of solutions, and what methods are used to describe their concentrations?
- Why do some substances dissolve and others not?
- What skills are involved in working experimentally with solutions, and in solving quantitative solution problems?

Keeping fish as pets seems like a straightforward task. You start off with a suitable fish tank, some water, and some fish. Once you have added a few water-plants for decoration, all you have to do is feed the fish daily. It sounds simple.

In fact, providing a safe, life-sustaining environment for fish involves a variety of complex, interdependent factors. Many of these are directly related to chemistry. For example, the presence of ammonia, $NH_3$, is one of the most common causes of death in poorly maintained fish tanks. Ammonia is produced when uneaten fish food decays. This compound dissolves readily in water. It is toxic to fish in even minute concentrations.

Chlorine is another chemical that can harm fish when it is dissolved in the tank water. Unfortunately, almost all tap water is treated with chlorine. Another factor that affects the quality of water for fish is temperature. For example, warm water contains too little dissolved oxygen. This will cause the fish to suffocate.

Why do ammonia and chlorine dissolve so easily in water? How can we tell what the concentration of any dissolved substance in a solution is? And why would we want to know? In this unit, you will review solutions. You will also learn how to calculate their concentrations and apply your stoichiometry skills to chemical reactions that involve solutions.

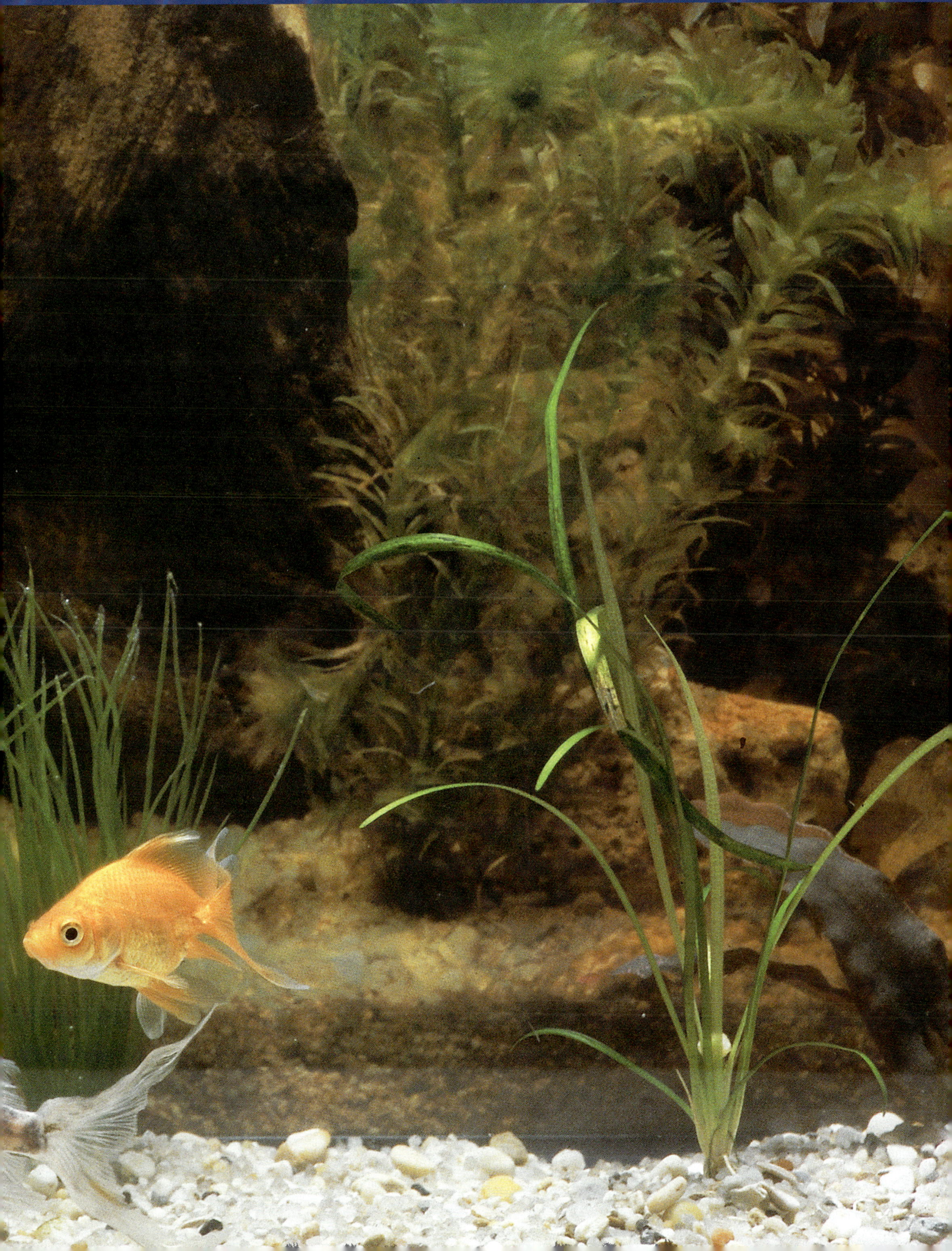

# CHAPTER 7

# Solutions and Their Concentrations

## Chapter Preview

- **7.1** Types of Solutions
- **7.2** Factors That Affect Rate of Dissolving and Solubility
- **7.3** The Concentration of Solutions
- **7.4** Preparing Solutions

## Prerequisite Concepts and Skills

Before you begin this chapter, review the following concepts and skills:

- predicting molecular polarity (Chapter 6)
- calculating molar mass (Chapter 2)
- calculating molar amounts (Chapter 2)

Your environment is made up of many important solutions. The air you breathe and the liquids you drink are solutions. So are many of the metallic objects that you use every day. The quality of a solution, such as tap water, depends on the substances that are dissolved in it. "Clean" water may contain small amounts of dissolved substances, such as iron and chlorine. "Dirty" water may have dangerous chemicals dissolved in it.

The difference between clean water and undrinkable water often depends on concentration: the amount of a dissolved substance in a particular quantity of a solution. For example, tap water contains a low concentration of fluoride to help keep your teeth healthy. Water with a high concentration of fluoride, however, is harmful to your health.

Water dissolves many substances. You may have noticed, however, that grease-stained clothing cannot be cleaned by water alone. Grease is one substance that does not dissolve in water. Why doesn't it dissolve? In this chapter, you will find out. You will learn how solutions form. You will explore factors that affect a substance's ability to dissolve. You will find out more about the concentration of solutions, and you will have a chance to prepare your own solutions as well.

As water runs through soil and rocks, it dissolves minerals such as iron, calcium, and magnesium. What makes water such a good solvent?

# Types of Solutions

## 7.1

A **solution** is a homogeneous mixture. It is uniform throughout. If you analyze any two samples of a solution, you will find that they contain the same substances in the same relative amounts. The simplest solutions contain two substances. Most common solutions contain many substances.

A **solvent** is any substance that has other substances dissolved in it. In a solution, the substance that is present in the largest amount (whether by volume, mass, or number of moles) is usually referred to as the solvent. The other substances that are present in the solution are called the **solutes**.

Pure substances (such as pure water, $H_2O$) have fixed composition. You cannot change the ratio of hydrogen, H, to oxygen, O, in water without producing an entirely different substance. Solutions, on the other hand, have variable composition. This means that different ratios of solvent to solute are possible. For example, you can make a dilute or a concentrated solution of sugar and water, depending on how much sugar you add. Figure 7.1 shows a concentrated solution of tea and water on the left, and a dilute solution of tea and water on the right. The ratio of solvent to solute in a **concentrated solution** is different from the ratio of solvent to solute in a **diluted solution**.

### Section Preview/Outcomes

In this section, you will

- **identify** examples of solutions and name their solvents and solutes
- **communicate** your understanding of the following terms: *solution, solvent, solutes, variable composition, concentrated solution, diluted solution, aqueous solution, miscible, immiscible, alloys, solubility, saturated solution, unsaturated solution*

**Figure 7.1** How can a solution have variable composition yet be uniform throughout?

When a solute dissolves in a solvent, no chemical reaction occurs. Therefore, the solute and solvent can be separated using physical properties, such as boiling point or melting point. For example, water and ethanol have different boiling points. Using this property, a solution of water and ethanol can be separated by the process of distillation. What physical properties, besides boiling point, can be used to separate the components of solutions and other mixtures?

**Figure 7.2** Can you identify the components of some of these solutions?

A solution can be a gas, a liquid, or a solid. Figure 7.2 shows some examples of solutions. Various combinations of solute and solvent states are possible. For example, a gas can be dissolved in a liquid, or a solid can be dissolved in another solid. Solid, liquid, and gaseous solutions are all around you. Steel is a solid solution of carbon in iron. Juice is a liquid solution of sugar and flavouring dissolved in water. Air is an example of a gaseous solution. The four main components of dry air are nitrogen (78%), oxygen (21%), argon (0.9%), and carbon dioxide (0.03%). Table 7.1 lists some other common solutions.

**Table 7.1** Types of Solutions

| Original state of solute | Solvent | Examples |
|---|---|---|
| gas | gas | air; natural gas; oxygen-acetylene mixture used in welding |
| gas | liquid | carbonated drinks; water in rivers and lakes containing oxygen |
| gas | solid | hydrogen in platinum |
| liquid | gas | water vapour in air; gasoline-air mixture |
| liquid | liquid | alcohol in water; antifreeze in water |
| liquid | solid | amalgams, such as mercury in silver |
| solid | gas | mothballs in air |
| solid | liquid | sugar in water; table salt in water; amalgams |
| solid | solid | alloys, such as the copper-nickel alloy used to make coins |

You are probably most familiar with liquid solutions, especially aqueous solutions. An **aqueous solution** is a solution in which water is the solvent. Because aqueous solutions are so important, you will focus on them in the next two sections and again in Chapter 8.

Some liquids, such as water and ethanol, dissolve readily in each other in any proportion. That is, any amount of water dissolves in any amount of ethanol. Similarly, any amount of ethanol dissolves in any amount of water. Liquids such as these are said to be **miscible** with each other. Miscible liquids can be combined in any proportions. Thus, either ethanol or water can be considered to be the solvent. Liquids that do not readily dissolve in each other, such as oil and water, are said to be **immiscible**.

As you may know, solid solutions of metals are called **alloys**. Adding even small quantities of another element to a metal changes the properties of the metal. Technological advances throughout history have been linked closely to the discovery of new alloys. For example, bronze is an alloy of copper and tin. Bronze contains only about 10% tin, but it is much stronger than copper and more resistant to corrosion. Also, bronze can be melted in an ordinary fire so that castings can be made, as shown in Figure 7.3.

## Solubility and Saturation

The ability of a solvent to dissolve a solute depends on the forces of attraction between the particles. There is always some attraction between solvent and solute particles, so some solute always dissolves. The **solubility** of a solute is the mass of solute that dissolves in a given quantity of solvent, at a certain temperature. For example, the solubility of sodium chloride in water at 20°C is 36 g per 100 mL of water. (Sometimes, solubility is expressed in terms of moles instead. In such cases, the *molar solubility* of a solution refers to the amount, in moles, of dissolved solute in 1 L of solution.

A **saturated solution** is formed when no more solute will dissolve in a solution, and excess solute is present. For example, 100 mL of a saturated solution of table salt (sodium chloride, NaCl) in water at 20°C contains 36 g of sodium chloride. The solution is saturated with respect to sodium chloride. If more sodium chloride is added to the solution, it will not dissolve. The solution may still be able to dissolve other solutes, however.

An **unsaturated solution** is a solution that is not yet saturated. Therefore, it can dissolve more solute. For example, a solution that contains 20 g of sodium chloride dissolved in 100 mL of water at 20°C is unsaturated. This solution has the potential to dissolve another 16 g of salt, as Figure 7.4 on the next page demonstrates.

**Web LINK**

www.mcgrawhill.ca/links/atlchemistry

To learn more about the properties of water, go to the web site above and click on **Electronic Learning Partner**.

**CHEM FACT**

An alloy that is made of a metal dissolved in mercury is called an *amalgam*. A traditional dental amalgam, used to fill cavities in teeth, contains 50% mercury. Due to concern over the use of mercury, which is toxic, dentists now use other materials, such as ceramic materials, to fill dental cavities.

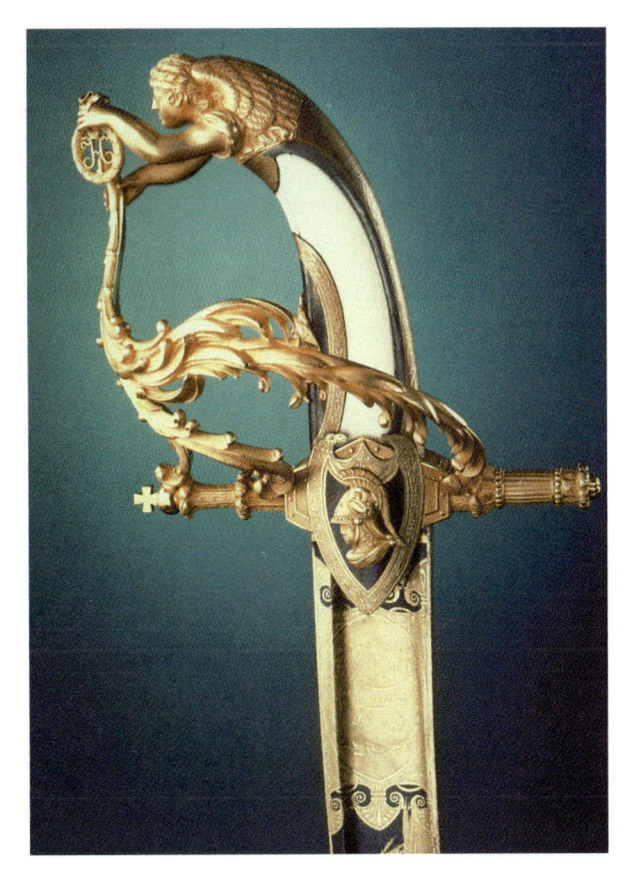

**Figure 7.3** The introduction of the alloy bronze around 3000 BCE led to the production of better-quality tools and weapons.

**Figure 7.4** At 20°C, the solubility of table salt in water is 36 g/100 mL.

**A** 20 g of NaCl dissolve to form an unsaturated solution.
**B** 36 g of NaCl dissolve to form a saturated solution.
**C** 40 g of NaCl are added to 100 mL of water. 36 g dissolve to form a saturated solution. 4 g of undissolved solute are left.

### CHEM FACT

When ionic solids such as NaCl dissolve in water, they dissociate—they separate into their component ions. Thus, in Figure 7.4C, 36 g of NaCl have dissociated completely, and 4 g of solute remain, undissolved in crystalline form. To the eye, this situation appears static. At a molecular level, however, there is a lot of activity taking place in the beaker. As free ions of sodium and chlorine collide with the crystalline mass of undissolved solute, some ions are attracted back into the crystal lattice and re-crystallize. At the same time, dissociation continues, releasing ions back into solution. At a certain point, the rate at which ions from the solid are dissociating is equal to the rate at which dissociated ions are re-crystallizing. This ongoing condition, in which there is no net change in the concentration of the components, is called a *dynamic equilibrium*. A solution that is in dynamic equilibrium with undissolved solute is saturated.

Suppose that a solute is described as *soluble* in a particular solvent. This generally means that its solubility is greater than 1 g per 100 mL of solvent. If a solute is described as *insoluble*, its solubility is less than 0.1 g per 100 mL of solvent. Substances with solubility between these limits are called *sparingly soluble*, or *slightly soluble*. Solubility is a relative term, however. Even substances such as oil and water dissolve in each other to some extent, although in very tiny amounts.

The general terms that are used to describe solubility for solids and liquids do not apply to gases in the same way. For example, oxygen is described as soluble in water. Oxygen from the air dissolves in the water of lakes and rivers. The solubility of oxygen in fresh water at 20°C is only 9 mg/L, or 0.0009 g/100 mL. This small amount of oxygen is enough to ensure the survival of aquatic plants and animals. A solid solute with the same solubility, however, would be described as insoluble in water.

### Identifying Suitable Solvents

Water is a good solvent for many compounds, but it is a poor solvent for others. If you have grease on your hands after adjusting a bicycle chain, you cannot use water to dissolve the grease and clean your hands. You need to use a detergent, such as soap, to help dissolve the grease in the water. You can also use another solvent to dissolve the grease. How can you find a suitable solvent? How can you predict whether a solvent will dissolve a particular solute? Try the ThoughtLab on the next page to find out for yourself.

# ThoughtLab — Matching Solutes and Solvents

Although there is a solvent for every solute, not all mixtures produce a solution. Table salt dissolves in water but not in kerosene. Oil dissolves in kerosene but not in water. What properties must a solvent and a solute share in order to produce a solution?

In an investigation, the bottom of a Petri dish was covered with water, as shown in photo A. An equal amount of kerosene was added to a second Petri dish. When a crystal of iodine was added to the water, it did not dissolve. When a second crystal of iodine was added to the kerosene, however, it did dissolve.

## Procedure

Classify each compound as ionic (containing ions), polar (containing polar molecules), or non-polar (containing non-polar molecules).

(a) iodine, $I_2$

(b) cobalt(II) chloride, $CoCl_2$

(c) sucrose, $C_{12}H_{22}O_{11}$ (**Hint:** Sucrose contains 8 O–H bonds.)

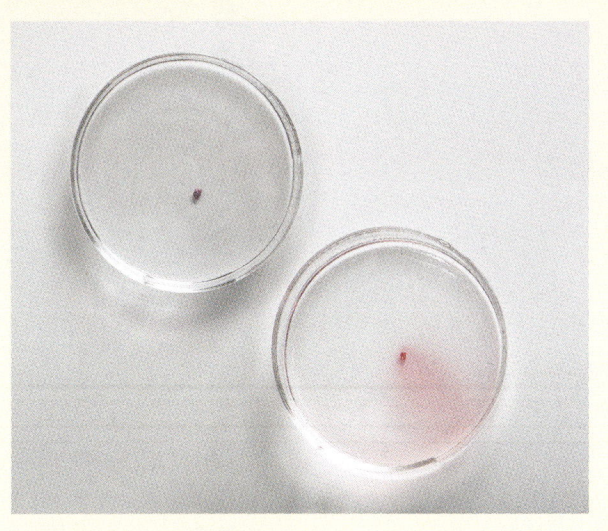

In photo B, the same experiment was repeated with crystals of cobalt(II) chloride. This time, the crystal dissolved in the water but not in the kerosene.

## Analysis

1. Water is a polar molecule. Therefore, it acts as a polar solvent.
   (a) Think about the compounds you classified in the Procedure. Which compounds are soluble in water?
   (b) Assume that the interaction between solutes and solvents that you examine here applies to a wide variety of substances. Make a general statement about the type of solute that dissolves in polar solvents.

2. Kerosene is non-polar. It acts as a non-polar solvent.
   (a) Which of the compounds you classified in the Procedure is soluble in kerosene?
   (b) Make a general statement about the type of solute that dissolves in non-polar solvents.

## Section Summary

In this section, you reviewed the meanings of several important terms, such as *solvent*, *solute*, *saturated solution*, *unsaturated solution*, *aqueous solution*, and *solubility*. You need to know these terms in order to understand the material in the rest of the chapter. In section 7.2, you will examine the factors that affect the rate at which a solute dissolves in a solvent. You will also learn about factors that affect solubility.

## Section Review

**1** Name the two basic components of a solution.

**2** Give examples of each type of solution.
   (a) solid solution
   (b) liquid solution
   (c) gaseous solution (at room temperature)

**3** Explain the term "homogeneous mixture."

**4** You add 37 g of sodium chloride to a glass of water. Describe what the solution looks like when the salt has finished dissolving.

**5** Imagine you are given a filtered solution of sodium chloride. How can you decide whether the solution is saturated or unsaturated?

**6** Distinguish between the following terms: soluble, miscible, and immiscible.

**7** What type of solute dissolves in a polar solvent, such as water? Give an example.

**8** Potassium bromate, $KBrO_3$, is sometimes added to bread dough to make it easier to work with. Suppose that you are given an aqueous solution of potassium bromate. How can you determine if the solution is saturated or unsaturated?

**9** Two different clear, colourless liquids were gently heated in an evaporating dish. Liquid A left no residue, while liquid B left a white residue. Which liquid was a solution, and which was a pure substance? Explain your answer.

**10** You are given three liquids. One is a pure substance, and the second is a solution of two miscible liquids. The third is a solution composed of a solid solute dissolved in a liquid solvent. Describe the procedure you would follow to distinguish among the three solutions.

**11** In 1989, the oil tanker Exxon Valdez struck a reef in Prince William Sound, Alaska. The accident released 40 million litres of crude oil. The oil eventually covered 26 000 km² of water.
   (a) Explain why very little of the spilt oil dissolved in the water. Why, then, was this spill such an environmental disaster?
   (b) How do you think most of the oil from a tanker accident is dispersed over time? Why would this have been a slow process in Prince William Sound?

# Factors That Affect Rate of Dissolving and Solubility

## 7.2

As you learned in section 7.1, the solubility of a solute is the amount of solute that dissolves in a given volume of solvent at a certain temperature. Solubility is determined by the intermolecular attractions between solvent and solute particles. You will learn more about solubility and the factors that affect it later in this section. First, however, you will look at an important property of a solution: the **rate of dissolving**, or how quickly a solute dissolves in a solvent.

The rate of dissolving depends on several factors, including temperature, agitation, and particle size. You have probably used these factors yourself when making solutions like the fruit juice shown in Figure 7.5.

### Section Preview/Outcomes

In this section, you will

- **explain** solution formation in terms of intermolecular forces between polar, ionic, and non-polar substances
- **communicate** your understanding of the following terms: *rate of dissolving, dipole, dipole-dipole attraction, hydrogen bonding, ion-dipole attractions, hydrated*

**Figure 7.5** Fruit juice is soluble in water. The concentrated juice in this photograph, however, will take a long time to dissolve. Why?

## Factors That Affect the Rate of Dissolving

You may have observed that a solute, such as sugar, dissolves faster in hot water than in cold water. In fact, *for most solid solutes, the rate of dissolving is greater at higher temperatures.* At higher temperatures, the solvent molecules have greater kinetic energy. Thus, they collide with the undissolved solid molecules more frequently. This increases their rate of dissolving.

Suppose that you are dissolving a spoonful of sugar in a cup of hot coffee. How can you make the sugar dissolve even faster? You can stir the coffee. *Agitating a mixture by stirring or by shaking the container increases the rate of dissolving.* Agitation brings fresh solvent into contact with undissolved solid.

Finally, you may have noticed that a large lump of solid sugar dissolves more slowly than an equal mass of powdered sugar. *Decreasing the size of the particles increases the rate of dissolving.* When you break up a large mass into many smaller masses, you increase the surface area that is in contact with the solvent. This allows the solid to dissolve faster. Figure 7.6 shows one way to increase the rate of dissolving.

**Figure 7.6** Chemists often grind solids into powders using a mortar and pestle. This increases the rate of dissolving.

Chapter 7 Solutions and Their Concentrations • MHR  **243**

> **Math LINK**
>
> Calculate the surface area of a cube with the dimensions 5.0 cm × 5.0 cm × 5.0 cm. Now imagine cutting this cube to form smaller cubes with the dimensions 1.0 cm × 1.0 cm × 1.0 cm. How many smaller cubes could you make? Calculate their total surface area.

## Solubility and Particle Attractions

By now, you are probably very familiar with the process of dissolving. You already know what it looks like when a solid dissolves in a liquid. Why, however, does something dissolve? What is happening at the molecular level?

The reasons why a solute may or may not dissolve in a solvent are related to the forces of attraction between the solute and solvent particles. These forces include the attractions between two solute particles, the attractions between two solvent particles, and the attractions between a solute particle and a solvent particle. When the forces of attraction between *different* particles in a mixture are stronger than the forces of attraction between *like* particles in the mixture, a solution forms. The strength of each attraction influences the solubility, or the amount of a solute that dissolves in a solvent.

To make this easier to understand, consider the following three steps in the process of dissolving a solid in a liquid.

> **The Process of Dissolving at the Molecular Level**
>
> **Step 1** The forces between the particles in the solid must be broken. This step always requires energy. In an ionic solid, the forces that are holding the ions together must be broken. In a molecular solid, the forces between the molecules must be broken.
>
> **Step 2** Some of the intermolecular forces between the particles in the liquid must be broken. This step also requires energy.
>
> **Step 3** There is an attraction between the particles of the solid and the particles of the liquid. This step always gives off energy.

The solid is more likely to dissolve in the liquid if the energy change in step 3 is greater than the sum of the energy changes in steps 1 and 2.

## Polar and Non-Polar Substances

In the ThoughtLab in section 7.1, you observed that solid iodine is insoluble in water. Only a weak attraction exists between the non-polar iodine molecules and the polar water molecules. On the other hand, the intermolecular forces between the water molecules are very strong. As a result, the water molecules remain attracted to each other rather than attracting the iodine molecules.

You also observed that iodine is soluble in kerosene. Both iodine and kerosene are non-polar substances. The attraction that iodine and kerosene molecules have for each other is greater than the attraction between the iodine molecules in the solid and the attraction between the kerosene molecules in the liquid.

The Concept Organizer shown on the next page summarizes the behaviour of polar and non-polar substances in solutions. You will learn more about polar and non-polar substances later in this section.

> **CONCEPT CHECK**
>
> Remember that the rate at which a solute dissolves is different from the *solubility* of the solute. In your notebook, explain briefly and clearly the difference between rate of dissolving and solubility.

## Concept Organizer — Polar and Non-Polar Compounds

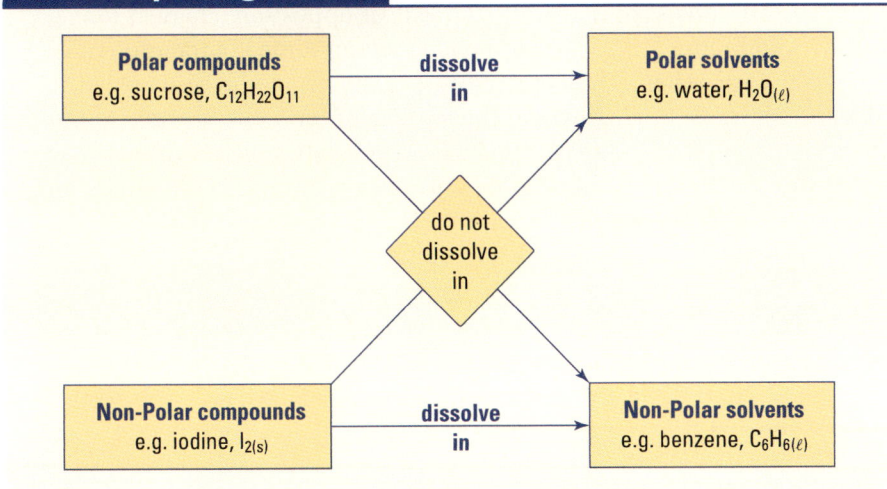

**Polar compounds** e.g. sucrose, $C_{12}H_{22}O_{11}$ — dissolve in → **Polar solvents** e.g. water, $H_2O_{(\ell)}$

**Non-Polar compounds** e.g. iodine, $I_{2(s)}$ — dissolve in → **Non-Polar solvents** e.g. benzene, $C_6H_{6(\ell)}$

Polar and non-polar do not dissolve in each other.

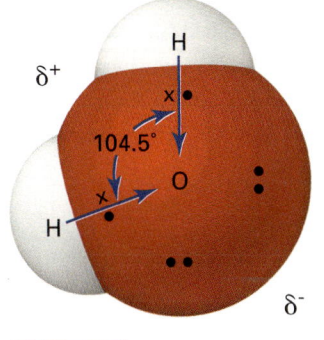

**Figure 7.7** The bent shape and polar bonds of a water molecule give it a permanent dipole.

## Solubility and Intermolecular Forces

You have learned that solubility depends on the forces between particles. Thus, polar substances dissolve in polar solvents, and non-polar substances dissolve in non-polar solvents. What are these forces that act between particles?

You know that a water molecule is polar. It has a relatively large negative charge on the oxygen atom, and positive charges on both hydrogen atoms. Molecules such as water, which have charges separated into positive and negative regions, are said to have a permanent dipole. A **dipole** consists of two opposite charges that are separated by a short distance. Figure 7.7 shows the dipole of a water molecule.

### Dipole-Dipole Attractions

The attraction between the opposite charges on two different polar molecules is called a **dipole-dipole attraction**. Dipole-dipole attractions are intermolecular—they act *between* molecules. Usually they are only about 1% as strong as an ionic or covalent bond. In water, there is a special dipole-dipole attraction called **hydrogen bonding**. It occurs between the oxygen atom on one molecule and the hydrogen atoms on a nearby molecule. Hydrogen bonding is much stronger than an ordinary dipole-dipole attraction. It is much weaker, however, than the covalent bond between the oxygen and hydrogen atoms in a water molecule. Figure 7.8 illustrates hydrogen bonding between water molecules.

**Web LINK**

www.mcgrawhill.ca/links/atlchemistry

Hydrogen bonding leads to water's amazing surface tension. Find out how by going to the web site above and clicking on **Electronic Learning Partner**.

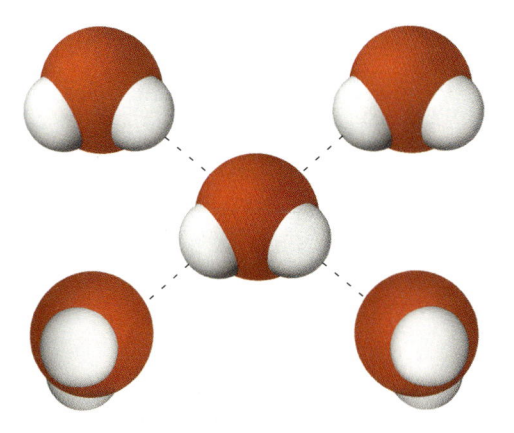

**Figure 7.8** Hydrogen bonding between water molecules is shown as dotted lines. The H atoms on each molecule are attracted to O atoms on other water molecules.

Chapter 7 Solutions and Their Concentrations • MHR  245

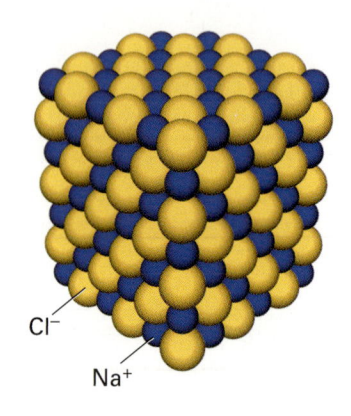

**Figure 7.9** Ionic crystals have very ordered structures.

## Ion-Dipole Attractions

Ionic crystals consist of repeating patterns of oppositely charged ions, as shown in Figure 7.9. What happens when an ionic compound comes in contact with water? The negative end of the dipole on some water molecules attracts the cations on the surface of the ionic crystal. At the same time, the positive end of the water dipole attracts the anions. These attractions are known as **ion-dipole attractions**: attractive forces between an ion and a polar molecule. If ion-dipole attractions can replace the ionic bonds between the cations and anions in an ionic compound, the compound will dissolve. *Generally an ionic compound will dissolve in a polar solvent.* For example, table salt (sodium chloride, NaCl) is an ionic compound. It dissolves well in water, which is a polar solvent.

When ions are present in an aqueous solution, each ion is **hydrated**. This means that it is surrounded by water molecules. Hydrated ions can move through a solution and conduct electricity. As you know, a solute that forms an aqueous solution with the ability to conduct electricity is called an **electrolyte**. Figure 7.10 shows hydrated sodium chloride ions, which are electrolytes.

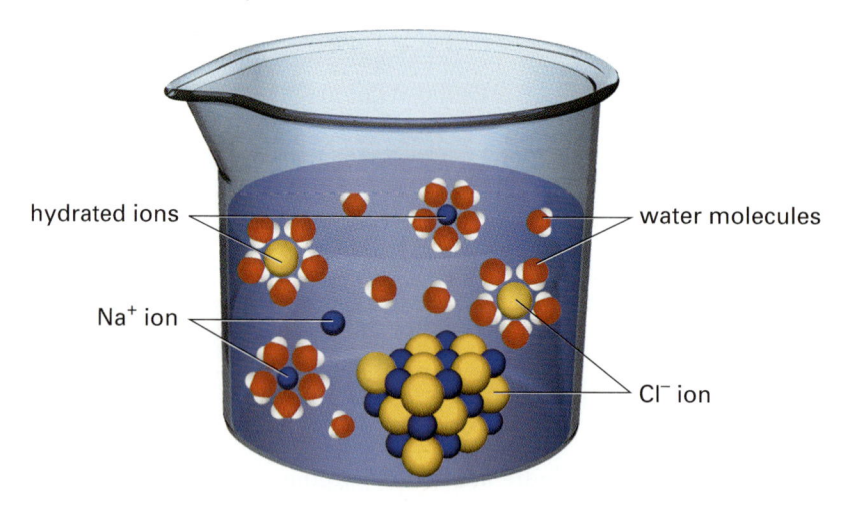

**Figure 7.10** Ion-dipole attractions help to explain why sodium chloride dissolves in water. Look closely at the hydrated ions. In what way are they different? Why are they different?

### An Exception: Insoluble Ionic Compounds

Although most ionic compounds are soluble in water, some are not very soluble at all. The attraction between ions is difficult to break. As a result, compounds with very strong ionic bonds, such as silver chloride, tend to be less soluble in water than compounds with weak ionic bonds, such as sodium chloride.

## Using Electronegativity Differences to Predict Solubility

You can predict the solubility of a binary compound, such as mercury(II) sulfide, HgS, by comparing the electronegativity of each element in the compound. If there is a large difference in the two electronegativities, the bond between the elements is polar or even ionic. This type of compound probably dissolves in water. If there is only a small difference in the two electronegativities, the bond is not polar or ionic. This type of compound probably does not dissolve in water. For example, the electronegativity of mercury is 1.9. The electronegativity of sulfur is 2.5. The difference in these two electronegativities is small, only 0.6. Therefore, you can predict that mercury(II) sulfide is insoluble in water. In Chapter 8, you will learn another way to predict the solubility of ionic compounds in water.

## The Solubility of Covalent Compounds

Many covalent compounds do not have negative and positive charges to attract water molecules. Thus they are not soluble in water. There are some exceptions, however. Methanol (a component of windshield washer fluid), ethanol (the "alcohol" in alcoholic beverages), and sugars (such as sucrose) are examples of covalent compounds that are extremely soluble in water. These compounds dissolve because their molecules contain polar bonds, which are able to form hydrogen bonds with water.

For example, sucrose molecules have a number of sites that can form a hydrogen bond with water to replace the attraction between the sucrose molecules. (See Figure 7.11.) The sucrose molecules separate and become hydrated, just like dissolved ions. The molecules remain neutral, however. As a result, sucrose and other soluble covalent compounds do not conduct electricity when dissolved in water. They are **non-electrolytes**.

**CONCEPT CHECK**

Look back at the Concept Organizer on page 245. Where do ionic compounds belong in this diagram?

**Web LINK**

www.mcgrawhill.ca/links/atlchemistry

If you are having difficulty visualizing particle attractions, go to the web site above and click on **Electronic Learning Partner** to see a video clip showing how water dissovles ionic and some covalent compounds.

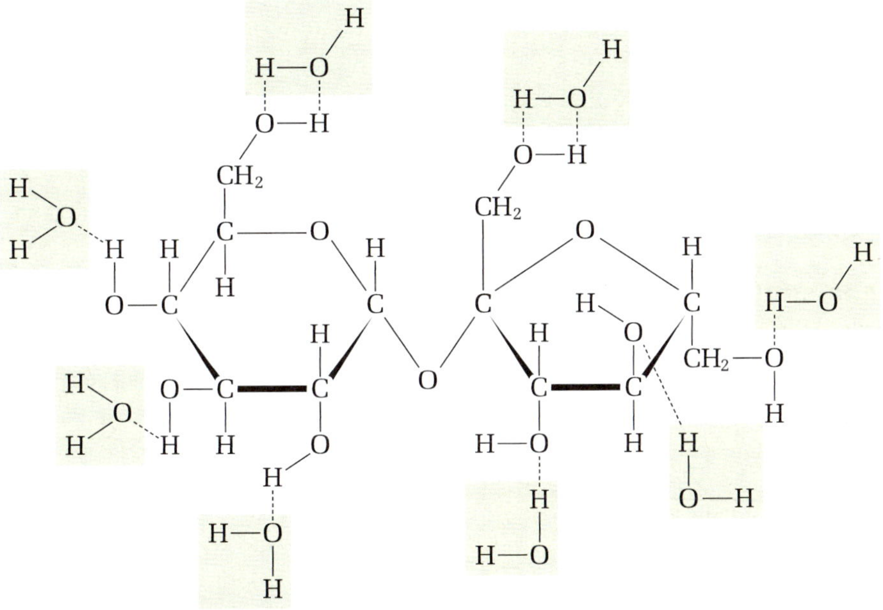

**Figure 7.11** A sucrose molecule contains several O—H atom connections. The O—H bond is highly polar, with the H atom having the positive charge. The negative charges on water molecules form hydrogen bonds with a sucrose molecule, as shown by the dotted lines.

### Insoluble Covalent Compounds

The covalent compounds that are found in oil and grease are insoluble in water. They have no ions or highly polar bonds, so they cannot form hydrogen bonds with water molecules. Non-polar compounds tend to be soluble in non-polar solvents, such as benzene or kerosene. The forces between the solute molecules are replaced by the forces between the solute and solvent molecules.

In general, *ionic solutes and polar covalent solutes both dissolve in polar solvents. Non-polar solutes dissolve in non-polar solvents.* The phrase *like dissolves like* summarizes these observations. It means that solutes and solvents that have similar properties form solutions.

If a compound has both polar and non-polar components, it may dissolve in both polar and non-polar solvents. For example, acetic acid, $CH_3COOH$, is a liquid that forms hydrogen bonds with water. It is fully miscible with water. Acetic acid also dissolves in non-polar solvents, such as benzene and carbon tetrachloride, because the $CH_3$ component is non-polar.

## Factors That Affect Solubility

You have taken a close look at the attractive forces between solute and solvent particles. Now that you understand why solutes dissolve, you can examine the three factors that affect solubility: molecule size, temperature, and pressure. Notice that these three factors are similar to the factors that affect the rate of dissolving. Be careful not to confuse them.

### Molecule Size and Solubility

Small molecules are often more soluble than larger molecules. Methanol, $CH_3OH$, and ethanol, $CH_3CH_2OH$, are both completely miscible with water. These compounds have OH groups that form hydrogen bonds with water. Larger molecules with the same OH group but more carbon atoms, such as pentanol, $CH_3CH_2CH_2CH_2CH_2OH$, are far less soluble. All three compounds form hydrogen bonds with water, but the larger pentanol is less polar overall, making it less soluble. Table 7.2 compares five molecules by size and solubility.

**Table 7.2** Solubility and Molecule Size

| Name of compound | methanol | ethanol | propanol | butanol | pentanol |
|---|---|---|---|---|---|
| Chemical formula | $CH_3OH$ | $CH_3CH_2OH$ | $CH_3CH_2CH_2OH$ | $CH_3(CH_2)_3OH$ | $CH_3(CH_2)_4OH$ |
| Solubility | infinitely soluble | infinitely soluble | very soluble | 9 g/100 mL (at 25°C) | 3 g/100 mL (at 25°C) |

### Temperature and Solubility

At the beginning of this section, you learned that temperature affects the rate of dissolving. Temperature also affects solubility. You may have noticed that solubility data always include temperature. The solubility of a solute in water, for example, is usually given as the number of grams of solute that dissolve in 100 mL of water at a specific temperature. (See Table 7.2 for two examples.) Specifying temperature is essential, since the solubility of a substance is very different at different temperatures.

When a solid dissolves in a liquid, energy is needed to break the strong bonds between particles in the solid. At higher temperatures, more energy is present. Thus, *the solubility of most solids increases with temperature.* For example, caffeine's solubility in water is only 2.2 g/100 mL at 25°C. At 100°C, however, caffeine's solubility increases to 40 g/100 mL.

The bonds between particles in a liquid are not as strong as the bonds between particles in a solid. When a liquid dissolves in a liquid, additional energy is not needed. Thus, *the solubility of most liquids is not greatly affected by temperature.*

Gas particles move quickly and have a great deal of kinetic energy. When a gas dissolves in a liquid, it loses some of this energy. At higher temperatures, the dissolved gas gains energy again. As a result, the gas comes out of solution and is less soluble. Thus, *the solubility of gases decreases with higher temperatures.*

In the next investigation, you will observe and graph the effect of temperature on the solubility of a solid dissolved in a liquid solvent, water. As you have learned, most solid solutes become more soluble at higher temperatures. By determining the solubility of a solute at various temperatures, you can make a graph of solubility against temperature. The curve of best fit, drawn through the points, is called the *solubility curve*. You can use a solubility curve to determine the solubility of a solute at any temperature in the range shown on the graph.

---

**CHEM FACT**

The link between cigarette smoking and lung cancer is well known. Other cancers are also related to smoking. It is possible that a smoker who consumes alcohol may be at greater risk of developing stomach cancer. When a person smokes, a thin film of tar forms inside the mouth and throat. The tar from cigarette smoke contains many carcinogenic (cancer-causing) compounds. These compounds are non-polar and do not dissolve in saliva. They are more soluble in alcohol, however. As a result, if a smoker drinks alcohol, carcinogenic compounds can be washed into the stomach.

# Investigation 7-A

**SKILL FOCUS**
Predicting
Performing and recording
Analyzing and interpreting

## Plotting Solubility Curves

In this investigation, you will determine the temperature at which a certain amount of potassium nitrate is soluble in water. You will then dilute the solution and determine the solubility again. By combining your data with other students' data, you will be able to plot a solubility curve.

### Question
What is the solubility curve of $KNO_3$?

### Prediction
Draw a sketch to show the shape of the curve you expect for the solubility of a typical solid dissolving in water at different temperatures. Plot solubility on the y-axis and temperature on the x-axis.

### Safety Precautions

- Before lighting the Bunsen burner, check that there are no flammable solvents nearby. If you are using a Bunsen burner, tie back long hair and loose clothing. Be careful of the open flame.
- After turning it on, be careful not to touch the hot plate.

### Materials
large test tube
balance
stirring wire
two-hole stopper to fit the test tube, with a thermometer inserted in one hole
400 mL beaker
graduated cylinder or pipette or burette
hot plate or Bunsen burner with ring clamps and wire gauze
retort stand and thermometer clamp
potassium nitrate, $KNO_3$
distilled water

### Procedure
1. Read through the steps in this Procedure. Prepare a data table to record the mass of the solute, the initial volume of water, the total volume of water after step 9, and the temperatures at which the solutions begin to crystallize.

2. Put the test tube inside a beaker for support. Place the beaker on a balance pan. Set the reading on the balance to zero. Then measure 14.0 g of potassium nitrate into the test tube.

3. Add one of the following volumes of distilled water to the test tube, as assigned by your teacher: 10.0 mL, 15.0 mL, 20.0 mL, 25.0 mL, 30.0 mL. (If you use a graduated cylinder, remember to read the volume from the bottom of the water meniscus.)

4. Pour about 300 mL of tap water into the beaker. Set up a hot-water bath using a hot plate, retort stand, and thermometer clamp. Alternatively, use a Bunsen burner, retort stand, ring clamp, thermometer clamp, and wire gauze.

5. Put the stirring wire through the second hole of the stopper. Insert the stopper, thermometer, and wire into the test tube. Make sure that the thermometer bulb is below the surface of the solution. (Check the diagram on the next page to make sure that you have set up the apparatus properly thus far.)

6. Place the test tube in the beaker. Secure the test tube and thermometer to the retort stand, using clamps. Begin heating the water bath gently.

7. Using the stirring wire, stir the mixture until the solute completely dissolves. Turn the heat source off, and allow the solution to cool.

*continued...*

## continued

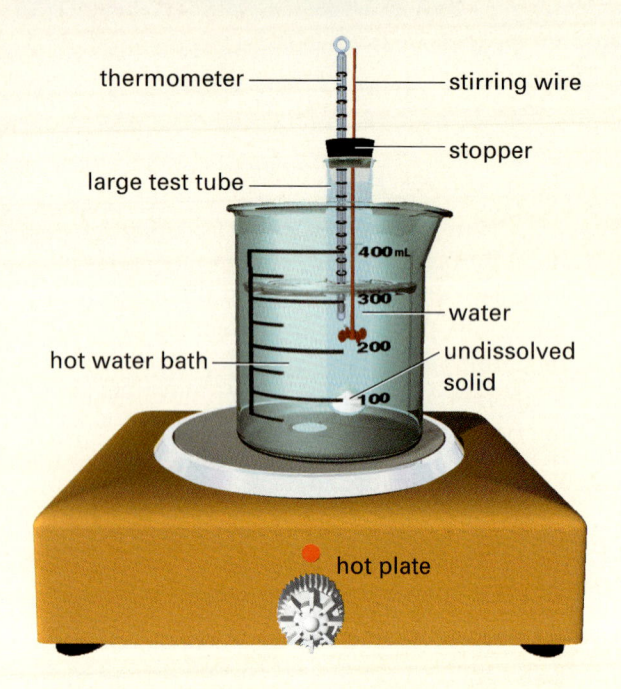

*Diagram labels: thermometer, stirring wire, stopper, large test tube, hot water bath, water, undissolved solid, hot plate*

8. Continue stirring. Record the temperature at which crystals begin to appear in the solution.

9. Remove the stopper from the test tube. Carefully add 5.0 mL of distilled water. The solution is now more dilute and therefore more soluble. Crystals will appear at a lower temperature.

10. Put the stopper, with the thermometer and stirring wire, back in the test tube. If crystals have already started to appear in the solution, begin warming the water bath again. Repeat steps 7 and 8.

11. If no crystals are present, stir the solution while the water bath cools. Record the temperature at which crystals first begin to appear.

12. Dispose of the aqueous solutions of potassium nitrate into the labelled waste container.

### Analysis

1. Use the volume of water assigned by your teacher to calculate how much solute dissolved in 100 mL of water. Use the following equation to help you:

$$\frac{x \text{ g}}{100 \text{ mL}} = \frac{14.0 \text{ g}}{\text{your volume}}$$

This equation represents the solubility of $KNO_3$ at the temperature at which you recorded the first appearance of crystals. Repeat your calculation to determine the solubility after the solution was diluted. Your teacher will collect and display all the class data for this investigation.

2. Some of your classmates were assigned the same volume of water that you were assigned. Compare the temperatures they recorded for their solutions with the temperatures you recorded. Comment on the precision of the data. Should any data be removed before averaging?

3. Average the temperatures at which crystal formation occurs for solutions that contain the same volume of water. Plot these data on graph paper. Set up your graph sideways on the graph paper (landscape orientation). Plot solubility on the vertical axis. (The units are grams of solute per 100 mL of water.) Plot temperature on the horizontal axis.

4. Draw the best smooth curve through the points. (Do not simply join the points.) Label each axis. Give the graph a suitable title.

### Conclusions

5. Go back to the sketch you drew to predict the solubility of a typical solid dissolving in water at different temperatures. Compare the shape of your sketch with the shape of your graph.

6. Use your graph to *interpolate* the solubility of potassium nitrate at
   (a) 60°C  (b) 40°C

7. Use your graph to *extrapolate* the solubility of potassium nitrate at
   (a) 80°C  (b) 20°C

### Application

8. At what temperature can 40 mL of water dissolve the following quantities of potassium nitrate?
   (a) 35.0 g  (b) 20.0 g

## Heat Pollution: A Solubility Problem

For most solids, and almost all ionic substances, solubility increases as the temperature of the solution increases. Gases, on the other hand, always become *less* soluble as the temperature increases. This is why a refrigerated soft drink tastes fizzier than the same drink at room temperature. The warmer drink contains less dissolved carbon dioxide than the cooler drink.

This property of gases makes heat pollution a serious problem. Many industries and power plants use water to cool down overheated machinery. The resulting hot water is then returned to local rivers or lakes. Figure 7.12 shows steam rising from a "heat-polluted" river. Adding warm water into a river or lake does not seem like actual pollution. The heat from the water, however, increases the temperature of the body of water. As the temperature increases, the dissolved oxygen in the water decreases. Fish and other aquatic wildlife and plants may not have enough oxygen to breathe.

The natural heating of water in rivers and lakes can pose problems, too. Fish in warmer lakes and rivers are particularly vulnerable in the summer. When the water warms up even further, the amount of dissolved oxygen decreases.

**Figure 7.12** This image shows the result of heat pollution. Warmer water contains less dissolved oxygen.

## ExpressLab: The Effect of Temperature on Soda Water

In this Express Lab, you will have a chance to see how a change in temperature affects the dissolved gas in a solution. You will be looking at the pH of soda water. A low pH (1–6) indicates that the solution is acidic. You will learn more about pH in Unit 6.

### Safety Precautions

- If using a hot plate, avoid touching it when it is hot.
- If using a Bunsen burner, check that there are no flammable solvents nearby.

### Procedure

1. Open a can of cool soda water. (Listen for the sound of excess carbon dioxide escaping.) Pour about 50 mL into each of two 100 mL beakers. Note the rate at which bubbles form. Record your observations.
2. Add a few drops of universal indicator to both beakers. Record the colour of the solutions. Then estimate the pH.
3. Measure and record the mass of each beaker. Measure and record the temperature of the soda water.
4. Place one beaker on a heat source. Heat it to about 50°C. Compare the rate of formation of the bubbles with the rate of formation in the beaker of cool soda water. Record any change in colour in the heated solution. Estimate its pH.
5. Allow the heated solution to cool. Again record any change in colour in the solution. Estimate its pH.
6. Measure and record the masses of both beakers. Determine any change in mass by comparing the final and initial masses.

### Analysis

1. Which sample of soda water lost the most mass? Explain your observation.
2. Did the heated soda water become more or less acidic when it was heated? Explain why you think this change happened.

## Pressure and Solubility

The final factor that affects solubility is pressure. Changes in pressure have hardly any effect on solid and liquid solutions. Such changes do affect the solubility of a gas in a liquid solvent, however. The solubility of the gas is directly proportional to the pressure of the gas above the liquid. For example, the solubility of oxygen in lake water depends on the air pressure above the lake.

When you open a carbonated drink, you can observe the effect of pressure on solubility. Figure 7.13 shows this effect. Inside a soft drink bottle, the pressure of the carbon dioxide gas is very high: about 400 kPa. When you open the bottle, you hear the sound of escaping gas as the pressure is reduced. Carbon dioxide gas escapes quickly from the bottle, since the pressure of the carbon dioxide in the atmosphere is much lower: only about 0.03 kPa. The solubility of the carbon dioxide in the liquid soft drink decreases greatly. Bubbles begin to rise in the liquid as gas comes out of solution and escapes. It takes a while for all the gas to leave the solution, so you have time to enjoy the taste of the soft drink before it goes "flat."

Figure 7.14 illustrates another example of dissolved gases and pressure. As a scuba diver goes deeper underwater, the water pressure increases. The solubility of nitrogen gas, which is present in the lungs, also increases. Nitrogen gas dissolves in the diver's blood. As the diver returns to the surface, the pressure acting on the diver decreases. The nitrogen gas in the blood comes out of solution. If the diver surfaces too quickly, the effect is similar to opening a soft drink bottle. Bubbles of nitrogen gas form in the blood. This leads to a painful and sometimes fatal condition known as "the bends."

**Figure 7.13** What happens when the pressure of the carbon dioxide gas in a soft drink bottle is released? The solubility of the gas in the soft drink solution decreases.

### CHEM FACT

Do you crack your knuckles? The sound you hear is another example of the effect of pressure on solubility. Joints contain fluid. When a joint is suddenly pulled or stretched, the cavity that holds the fluid gets larger. This causes the pressure to decrease. A bubble of gas forms, making the sound you hear. You cannot repeatedly crack your knuckles because it takes some time for the gas to re-dissolve.

**Figure 7.14** Scuba divers must heed the effects of decreasing water pressure on dissolved nitrogen gas in their blood. They must surface slowly to avoid "the bends."

# Chemistry Bulletin

### Science  Technology  Society  Environment

## Solvents and Coffee: What's the Connection?

The story of coffee starts with the coffee berry. First the pulp of the berry is removed. This leaves two beans, each containing 1% to 2% caffeine. The beans are soaked in water and natural enzymes to remove the outer parchment husk and to start a slight fermentation process. Once the beans have been fermented, they are dried and roasted. Then the coffee is ready for grinding. Grinding increases the surface area of the coffee. Thus, finer grinds make it easier to dissolve the coffee in hot water.

Decaffeinated coffee satisfies people who like the smell and taste of coffee but cannot tolerate the caffeine. How is caffeine removed from coffee?

All the methods of extracting caffeine take place before the beans are roasted. Caffeine and the other organic compounds that give coffee its taste are mainly non-polar. (Caffeine does contain some polar bonds, however, which allows it to dissolve in hot water.) Non-polar solvents, such as benzene and trichloroethene, were once used to dissolve and remove caffeine from the beans. These chemicals are now considered to be too hazardous. Today most coffee manufacturers use water or carbon dioxide as solvents.

In the common Swiss Water Process, coffee beans are soaked in hot water. This dissolves the caffeine and the flavouring compounds from the beans. The liquid is passed through activated carbon filters. The filters retain the caffeine, but let the flavouring compounds pass through. The filtered liquid, now caffeine-free, is sprayed back onto the beans. The beans reabsorb the flavouring compounds. Now they are ready for roasting.

Carbon dioxide gas is a normal component of air. In the carbon dioxide decaffeination process, the gas is raised to a temperature of at least 32°C. Then it is compressed to a pressure of about 7400 kPa. At this pressure, it resembles a liquid but can flow like a gas. The carbon dioxide penetrates the coffee beans and dissolves the caffeine. When the pressure returns to normal, the carbon dioxide reverts to a gaseous state. The caffeine is left behind.

What happens to the caffeine that is removed by decaffeination? Caffeine is so valuable that it is worth more than the cost of taking it out of the beans. It is extensively used in the pharmaceutical industry, and for colas and other soft drinks.

### Making Connections

1. As you have read, water is a polar liquid and the soluble fractions of the coffee grounds are non-polar. Explain, in chemical terms, how caffeine and the coffee flavour and aroma are transferred to hot coffee.
2. Why does hot water work better in the brewing process than cold water?
3. In chemical terms, explain why fine grinds of coffee make better coffee.

How can caffeine form hydrogen bonds with water?

## Section Summary

In this section, you examined the factors that affect the rate of dissolving: temperature, agitation, and particle size. Next you looked at the forces between solute and solvent particles. Finally, you considered three main factors that affect solubility: molecule size, temperature, and pressure. In section 7.3, you will learn about the effects of differing amounts of solute dissolved in a certain amount of solvent.

## Section Review

**1** Describe the particle attractions that occur as sodium chloride dissolves in water.

**2** When water vaporizes, which type of attraction, intramolecular or intermolecular, is broken? Explain.

**3** Describe the effect of increasing temperature on the solubility of
 (a) a typical solid in water
 (b) a gas in water

**4** Sugar is more soluble in water than salt. Why does a salt solution (brine) conduct electricity, while a sugar solution does not?

**5** Dissolving a certain solute in water releases heat. Dissolving a different solute in water absorbs heat. Explain why.

**6** The graph shows the solubility of various substances plotted against the temperature of the solution.

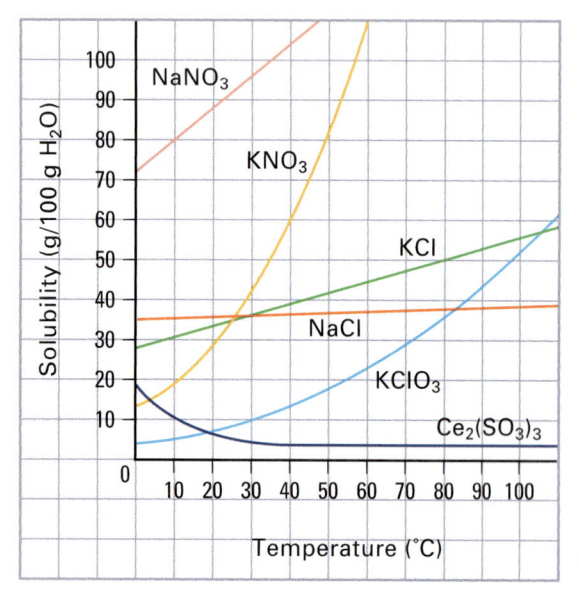

 (a) Which substance decreases in solubility as the temperature increases?
 (b) Which substance is least soluble at room temperature? Which substance is most soluble at room temperature?
 (c) The solubility of which substance is least affected by a change in temperature?
 (d) At what temperature is the solubility of potassium chlorate equal to 40 g/100 g of water?
 (e) 20 mL of a saturated solution of potassium nitrate at 50°C is cooled to 20°C. Approximately what mass of solid will precipitate from the solution? Why is it not possible to use the graph to interpolate an accurate value?

**7** A saturated solution of potassium nitrate was prepared at 70°C and then cooled to 55°C. Use your graph from Investigation 8-A to predict the fraction of the dissolved solute that crystallized out of the solution.

**8** Would you expect to find more mineral deposits near a thermal spring or near a cool mountain spring? Explain.

# The Concentration of Solutions

## 7.3

**Material Safety Data Sheet**

| Component Name | CAS Number |
|---|---|
| PHENOL, 100% Pure | 108952 |

**SECTION III: Hazards Identification**
- Very hazardous in case of ingestion, inhalation, skin contact, or eye contact.
- Product is corrosive to internal membranes when ingested.
- Inhalation of vapours may damage central nervous system. Symptoms: nausea, headache, dizziness.
- Skin contact may cause itching and blistering.
- Eye-contact may lead to corneal damage or blindness.
- Severe over-exposure may lead to lung-damage, choking, or coma.

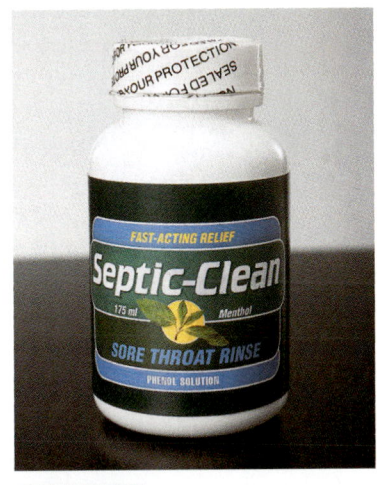

**Figure 7.15** Should phenol be banned from drugstores?

**Section Preview/Outcomes**

In this section, you will

- **solve problems** involving the concentration of solutions
- **calculate** concentration as grams per 100 mL, mass and volume percents, parts per million and billion, and moles per litre
- **communicate** your understanding of the following terms: *concentration, mass/volume percent, mass/mass percent, volume/volume percent, parts per million, parts per billion, molar concentration*

Figure 7.15 shows a portion of an MSDS for pure, undiluted phenol. It is a hazardous liquid, especially when it is at room temperature. Inhaling phenol adversely affects the central nervous system, and can lead to a coma. Coma and death have been known to occur within 10 min after phenol has contacted the skin. Also, as little as 1 g of phenol can be fatal if swallowed.

Would you expect to find such a hazardous chemical in over-the-counter medications? Check your medicine cabinet at home. You may find phenol listed as an ingredient in throat sprays and in lotions to relieve itching. You may also find it used as an antiseptic or disinfectant. Is phenol a hazard or a beneficial ingredient in many medicines? This depends entirely on **concentration**: the amount of solute per quantity of solvent. At high concentrations, phenol can kill. At low concentrations, it is a safe component of certain medicines.

Modern analytical tests allow chemists to detect and measure almost any chemical at extremely low concentrations. In this section, you will learn ways that chemists express the concentration of a solution. As well, you will find the concentration of a solution by experiment.

## Concentration as a Mass/Volume Percent

Recall that the solubility of a compound at a certain temperature is often expressed as the mass of the solute per 100 mL of solvent. For example, you know that the solubility of sodium chloride is 36 g/100 mL of water at room temperature. The final volume of the sodium chloride solution may or may not be 100 mL. It is the volume of the solvent that is important.

Chemists often express the concentration of an *unsaturated* solution as the mass of solute dissolved per volume of the *solution*. This is different from solubility. It is usually expressed as a percent relationship. A **mass/volume percent** gives the mass of solute dissolved in a volume of solution, expressed as a percent. The mass/volume percent is also referred to as the *percent (m/v)*.

> **PROBEWARE**
>
> www.mcgrawhill.ca/links/atlchemistry
>
> Go to the above link and click on **Probeware** for a lab in which you can investigate the concentration of solutions.

$$\text{Mass/volume percent} = \frac{\text{Mass of solute (in g)}}{\text{Volume of solution (in mL)}} \times 100\%$$

Suppose that a hospital patient requires an intravenous drip to replace lost body fluids. The intravenous fluid may be a saline solution that contains 0.9 g of sodium chloride dissolved in 100 mL of solution, or 0.9% (m/v). *Notice that the number of grams of solute per 100 mL of solution is numerically equal to the mass/volume percent.* Explore this idea further in the following problems.

## Sample Problem

### Solving for a Mass/Volume Percent

#### Problem

A pharmacist adds 2.00 mL of distilled water to 4.00 g of a powdered drug. The final volume of the solution is 3.00 mL. What is the concentration of the drug in g/100 mL of solution? What is the percent (m/v) of the solution?

#### What Is Required?

You need to calculate the concentration of the solution, in grams of solute dissolved in 100 mL of solution. Then you need to express this concentration as a mass/volume percent.

#### What Is Given?

The mass of the dissolved solute is 4.00 g. The volume of the solution is 3.00 mL.

#### Plan Your Strategy

There are two possible methods for solving this problem.

**Method 1**

Use the formula

$$\text{Mass/volume percent} = \frac{\text{Mass of solute (in g)}}{\text{Volume of solution (in mL)}} \times 100\%$$

**Method 2**

Let $x$ represent the mass of solute dissolved in 100 mL of solution. The ratio of the dissolved solute, $x$, in 100 mL of solution must be the same as the ratio of 4.00 g of solute dissolved in 3.00 mL of solution. The concentration, expressed in g/100 mL, is numerically equal to the percent (m/v) of the solution.

#### Act on Your Strategy

**Method 1**

$$\text{Percent (m/v)} = \frac{4.00 \text{ g}}{3.00 \text{ mL}} \times 100\%$$
$$= 133\%$$

256  MHR • Unit 3 Solutions

### Method 2

$$\frac{x}{100 \text{ mL}} = \frac{4.00 \text{ g}}{3.00 \text{ mL}}$$

$$\frac{x}{100 \text{ mL}} = 1.33 \text{ g/mL}$$

$$x = 100 \text{ mL} \times 1.33 \text{ g/mL}$$

$$= 133 \text{ g}$$

The concentration of the drug is 133 g/100 mL of solution, or 133% (m/v).

### Check Your Solution

The units are correct. The numerical answer is large, but this is reasonable for an extremely soluble solute.

## Sample Problem

### Finding Mass for an (m/v) Concentration

### Problem

Many people use a solution of trisodium phosphate, $Na_3PO_4$ (commonly called TSP), to clean walls before putting up wallpaper. The recommended concentration is 1.7% (m/v). What mass of TSP is needed to make 2.0 L of solution?

### What Is Required?

You need to find the mass of TSP needed to make 2.0 L of solution.

### What Is Given?

The concentration of the solution should be 1.7% (m/v). The volume of solution that is needed is 2.0 L.

### Plan Your Strategy

There are two different methods you can use.

**Method 1**

Use the formula for (m/v) percent. Rearrange the formula to solve for mass. Then substitute in the known values.

**Method 2**

The percent (m/v) of the solution is numerically equal to the concentration in g/100 mL. Let $x$ represent the mass of TSP dissolved in 2.0 L of solution. The ratio of dissolved solute in 100 mL of solution must be the same as the ratio of the mass of solute, $x$, dissolved in 2.0 L (2000 mL) of solution.

### Act on Your Strategy

**Method 1**

$$(m/v) \text{ percent} = \frac{\text{Mass of solute (in g)}}{\text{Volume of solution (in mL)}} \times 100\%$$

$$\therefore \text{Mass of solute} = \frac{(m/v) \text{ percent} \times \text{Volume of solution}}{100\%}$$

$$= \frac{1.7\% \times 2000 \text{ mL}}{100\%}$$

$$= 34 \text{ g}$$

*Continued ...*

### Method 2

A TSP solution that is 1.7% (m/v) contains 1.7 g of solute dissolved in 100 mL of solution.

$$\frac{1.7 \text{ g}}{100 \text{ mL}} = \frac{x}{2000 \text{ mL}}$$

$$0.017 \text{ g/mL} = \frac{x}{2000 \text{ mL}}$$

$$x = 0.017 \text{ g/mL} \times 2000 \text{ mL}$$

$$= 34 \text{ g}$$

Therefore, 34 g of TSP are needed to make 2.0 L of cleaning solution.

### Check Your Solution

The units are appropriate for the problem. The answer appears to be reasonable.

### Practice Problems

1. What is the concentration in percent (m/v) of each solution?
   (a) 14.2 g of potassium chloride, KCl (used as a salt substitute), dissolved in 450 mL of solution
   (b) 31.5 g of calcium nitrate, $Ca(NO_3)_2$ (used to make explosives), dissolved in 1.80 L of solution
   (c) 1.72 g of potassium permanganate, $KMnO_4$ (used to bleach stone-washed blue jeans), dissolved in 60 mL of solution

2. A solution of hydrochloric acid was formed by dissolving 1.52 g of hydrogen chloride gas in enough water to make 24.1 mL of solution. What is the concentration in percent (m/v) of the solution?

3. At 25°C, a saturated solution of carbon dioxide gas in water has a concentration of 0.145% (m/v). What mass of carbon dioxide is present in 250 mL of the solution?

4. Ringer's solution contains three dissolved salts in the same proportions as they are found in blood. The salts and their concentrations (m/v) are as follows: 0.86% NaCl, 0.03% KCl, and 0.033% $CaCl_2$. Suppose that a patient needs to receive 350 mL of Ringer's solution by an intravenous drip. What mass of each salt does the pharmacist need to make the solution?

## Concentration as a Mass/Mass Percent

The concentration of a solution that contains a solid solute dissolved in a liquid solvent can also be expressed as a mass of solute dissolved in a mass of solution. This is usually expressed as a percent relationship. A **mass/mass percent** gives the mass of a solute divided by the mass of solution, expressed as a percent. The mass/mass percent is also referred to as the *percent (m/m)*, or the *mass percent*. It is often inaccurately referred to as a weight (w/w) percent, as well. Look at your tube of toothpaste, at home.

The percent of sodium fluoride in the toothpaste is usually given as a w/w percent. This can be confusing, since weight ($w$) is not the same as mass ($m$). In fact, this concentration should be expressed as a mass/mass percent.

$$\text{Mass/mass percent} = \frac{\text{Mass of solute (in g)}}{\text{Mass of solution (in g)}} \times 100\%$$

For example, 100 g of seawater contains 0.129 g of magnesium ion (along with many other substances). The concentration of $Mg^{2+}$ in seawater is 0.129 (m/m). *Notice that the number of grams of solute per 100 g of solution is numerically equal to the mass/mass percent.*

The concentration of a solid solution, such as an alloy, is usually expressed as a mass/mass percent. Often the concentration of a particular alloy may vary. Table 7.3 gives typical compositions of some common alloys.

**Table 7.3** The Composition of Some Common Alloys

| Alloy | Uses | Typical percent (m/m) composition |
|---|---|---|
| brass | ornaments, musical instruments | Cu (85%) Zn (15%) |
| bronze | statues, castings | Cu (80%) Zn (10%) Sn (10%) |
| cupronickel | "silver" coins | Cu (75%) Ni (25%) |
| dental amalgam | dental fillings | Hg (50%) Ag (35%) Sn (15%) |
| duralumin | aircraft parts | Al (93%) Cu (5%) other (2%) |
| pewter | ornaments | Sn (85%) Cu (7%) Bi (6%) Sb (2%) |
| stainless steel | cutlery, knives | Fe (78%) Cr (15%) Ni (7%) |
| sterling silver | jewellery | Ag (92.5%) Cu (7.5%) |

Figure 7.16, on the following page, shows two objects made from brass that have distinctly different colours. The difference in colours reflects the varying concentrations of the copper and zinc that make up the objects.

**Figure 7.16** Brass can be made using any percent from 50% to 85% copper, and from 15% to 50% zinc. As a result, two objects made of brass can look very different.

## Sample Problem

### Solving for a Mass/Mass Percent

**Problem**

Calcium chloride, $CaCl_2$, can be used instead of road salt to melt the ice on roads during the winter. To determine how much calcium chloride had been used on a nearby road, a student took a sample of slush to analyze. The sample had a mass of 23.47 g. When the solution was evaporated, the residue had a mass of 4.58 g. (Assume that no other solutes were present.) What was the mass/mass percent of calcium chloride in the slush? How many grams of calcium chloride were present in 100 g of solution?

**What Is Required?**

You need to calculate the mass/mass percent of calcium chloride in the solution (slush). Then you need to use your answer to find the mass of calcium chloride in 100 g of solution.

**What Is Given?**

The mass of the solution is 23.47 g. The mass of calcium chloride that was dissolved in the solution is 4.58 g.

**Plan Your Strategy**

There are two methods that you can use to solve this problem.

**Method 1**

Use the formula for mass/mass percent.

$$\text{Mass/mass percent} = \frac{\text{Mass of solute (in g)}}{\text{Mass of solution (in g)}} \times 100\%$$

The mass of calcium chloride in 100 g of solution will be numerically equal to the mass/mass percent.

**Method 2**

Use ratios, as in the previous Sample Problems.

260 MHR • Unit 3 Solutions

### Act on Your Strategy

**Method 1**

Mass/mass percent = $\dfrac{4.58 \text{ g}}{23.47 \text{ g}} \times 100\%$

$= 19.5\%$

**Method 2**

$\dfrac{x \text{ g}}{100 \text{ g}} = \dfrac{4.58 \text{ g}}{23.47 \text{ g}}$

$\dfrac{x \text{ g}}{100 \text{ g}} = 0.195$

$x = 19.5\%$

The mass/mass percent was 19.5% (m/m). 19.5 g of calcium chloride was dissolved in 100 g of solution.

### Check Your Solution

The mass units divide out properly. The final answer has the correct number of significant digits. It appears to be reasonable.

### Practice Problems

5. Calculate the mass/mass percent of solute for each solution.
   (a) 17 g of sulfuric acid in 65 g of solution
   (b) 18.37 g of sodium chloride dissolved in 92.2 g of water
       **Hint:** Remember that a solution consists of both solute and solvent.
   (c) 12.9 g of carbon tetrachloride dissolved in 72.5 g of benzene

6. If 55 g of potassium hydroxide is dissolved in 100 g of water, what is the concentration of the solution expressed as a mass/mass percent?

7. Steel is an alloy of iron and about 1.7% carbon. It also contains small amounts of other materials, such as manganese and phosphorus. What mass of carbon is needed to make a 5.0 kg sample of steel?

8. Stainless steel is a variety of steel that resists corrosion. Your cutlery at home may be made of this material. Stainless steel must contain at least 10.5% chromium. What mass of chromium is needed to make a stainless steel fork with a mass of 60.5 g?

9. 18-carat white gold is an alloy. It contains 75% gold, 12.5% silver, and 12.5% copper. A piece of jewellery, made of 18-carat white gold, has a mass of 20 g. How much pure gold does it contain?

## Concentration as a Volume/Volume Percent

When mixing two liquids to form a solution, it is easier to measure their volumes than their masses. A **volume/volume percent** gives the volume of solute divided by the volume of solution, expressed as a percent. The volume/volume percent is also referred to as the *volume percent concentration*, *volume percent*, *percent* (v/v), or the percent by volume. You can see this type of concentration on a bottle of rubbing alcohol from a drugstore. (See Figure 7.17.)

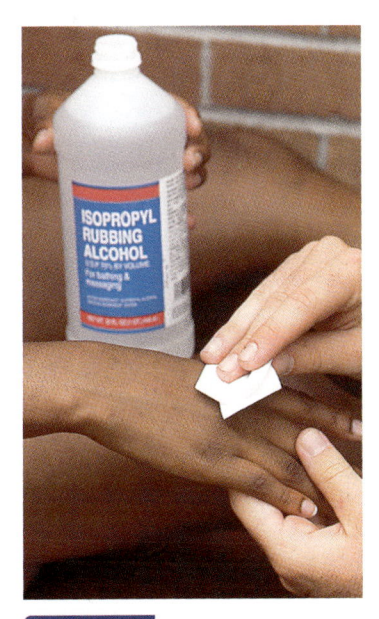

**Figure 7.17** The concentration of this solution of isopropyl alcohol in water is expressed as a volume/volume percent.

$$\text{Volume/volume percent} = \frac{\text{Volume of solute (in mL)}}{\text{Volume of solution (in mL)}} \times 100\%$$

Read through the Sample Problem below, and complete the Practice Problems that follow. You will then have a better understanding of how to calculate the volume/volume percent of a solution.

### Sample Problem
### Solving for a Volume/Volume Percent

**Problem**

Rubbing alcohol is commonly used as an antiseptic for small cuts. It is sold as a 70% (v/v) solution of isopropyl alcohol in water. What volume of isopropyl alcohol is used to make 500 mL of rubbing alcohol?

**What Is Required?**

You need to calculate the volume of isopropyl alcohol (the solute) used to make 500 mL of solution.

**What Is Given?**

The volume/volume percent is 70% (v/v). The final volume of the solution is 500 mL.

**Plan Your Strategy**

**Method 1**

Rearrange the following formula to solve for the volume of the solute. Then substitute the values that you know into the rearranged formula.

$$\text{Volume/volume percent} = \frac{\text{Volume of solute}}{\text{Volume of solution}} \times 100\%$$

**Method 2**

Use ratios to solve for the unknown volume.

**Act on Your Strategy**

**Method 1**

$$\text{Volume/volume percent} = \frac{\text{Volume of solute}}{\text{Volume of solution}} \times 100\%$$

$$\text{Volume of solute} = \frac{\text{Volume/volume percent} \times \text{Volume of solution}}{100\%}$$

$$= \frac{70\% \times 500 \text{ mL}}{100\%}$$

$$= 350 \text{ mL}$$

**Method 2**

$$\frac{x \text{ mL}}{500 \text{ mL}} = \frac{70 \text{ mL}}{100 \text{ mL}}$$

$$x = 0.7 \times 500 \text{ mL}$$

$$= 350 \text{ mL}$$

Therefore, 350 mL of isopropyl alcohol is used to make 500 mL of 70% (v/v) rubbing alcohol.

**Check Your Solution**

The answer seems reasonable. It is expressed in appropriate units.

---

### CHEM FACT

Archaeologists can learn a lot about ancient civilizations by chemically analyzing the concentration of ions in the soil where the people lived. When crops are grown, the crops remove elements such as nitrogen, magnesium, calcium, and phosphorus from the soil. Thus, soil with a lower-than-average concentration of these elements may have held ancient crops. Chlorophyll, which is present in all plants, contains magnesium ions. In areas where ancient crops were processed, the soil has a higher concentration of $Mg^{2+}$.

## Practice Problems

**10.** 60 mL of ethanol is diluted with water to a final volume of 400 mL. What is the percent by volume of ethanol in the solution?

**11.** Milk fat is present in milk. Whole milk usually contains about 5.0% milk fat by volume. If you drink a glass of milk with a volume of 250 mL, what volume of milk fat have you consumed?

**12.** Both antifreeze (shown in Figure 7.18) and engine coolant contain ethylene glycol. A manufacturer sells a concentrated solution that contains 75% (v/v) ethylene glycol in water. According to the label, a 1:1 mixture of the concentrate with water will provide protection against freezing down to a temperature of −37°C. A motorist adds 1 L of diluted solution to a car radiator. What is the percent (v/v) of ethylene glycol in the diluted solution?

**13.** The average adult human body contains about 5 L of blood. Of this volume, only about 0.72% consists of leukocytes (white blood cells). These essential blood cells fight infection in the body. What volume of pure leukocyte cells is present in the body of a small child, with only 2.5 L of blood?

**14.** Vinegar is sold as a 5% (v/v) solution of acetic acid in water. How much water should be added to 15 mL of pure acetic acid (a liquid at room temperature) to make a 5% (v/v) solution of acetic acid? **Note:** Assume that when water and acetic acid are mixed, the total volume of the solution is the sum of the volumes of each.

**Figure 7.18** Antifreeze is a solution of ethylene glycol and water.

## Concentration in Parts per Million and Parts per Billion

The concentration of a very small quantity of a substance in the human body, or in the environment, can be expressed in **parts per million (ppm)** and **parts per billion (ppb)**. Both parts per million and parts per billion are usually mass/mass relationships. They describe the amount of solute that is present in a solution. Notice that parts per million does not refer to the number of particles, but to the *mass* of the solute compared with the *mass* of the solution.

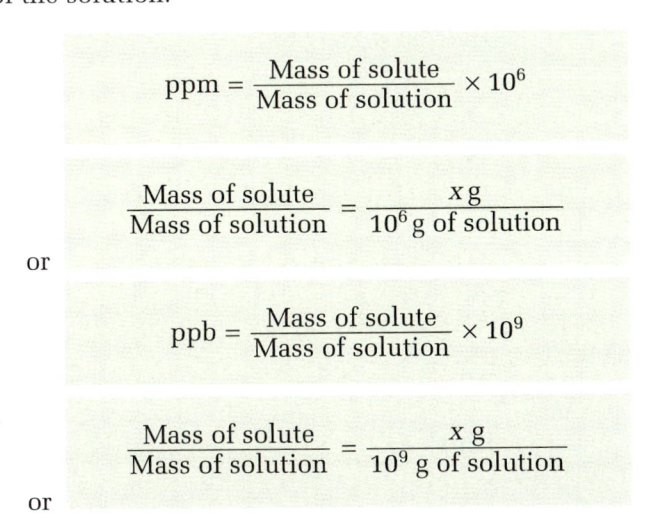

$$\text{ppm} = \frac{\text{Mass of solute}}{\text{Mass of solution}} \times 10^6$$

or

$$\frac{\text{Mass of solute}}{\text{Mass of solution}} = \frac{x \text{ g}}{10^6 \text{ g of solution}}$$

$$\text{ppb} = \frac{\text{Mass of solute}}{\text{Mass of solution}} \times 10^9$$

or

$$\frac{\text{Mass of solute}}{\text{Mass of solution}} = \frac{x \text{ g}}{10^9 \text{ g of solution}}$$

### Math LINK

One part per million is equal to 1¢ in $10 000. One part per billion is equal to 1 s in almost 32 years.

What distance (in km) would you travel if 1 cm represented 1 ppm of your journey?

A swimming pool has the dimensions 10 m × 5 m × 2 m. If the pool is full of water, what volume of water (in cm³) would represent 1 ppb of the water in the pool?

## Sample Problem

### Parts per Billion in Peanut Butter

**Problem**

A fungus that grows on peanuts produces a deadly toxin. When ingested in large amounts, this toxin destroys the liver and can cause cancer. Any shipment of peanuts that contains more than 25 ppb of this dangerous fungus is rejected. A company receives 20 t of peanuts to make peanut butter. What is the maximum mass (in g) of fungus that is allowed?

**What Is Required?**

You need to find the allowed mass (in g) of fungus in 20 t of peanuts.

**What Is Given?**

The allowable concentration of the fungus is 25 ppb. The mass of the peanut shipment is 20 t.

**Plan Your Strategy**

**Method 1**

Convert 20 t to grams. Rearrange the formula below to solve for the allowable mass of the fungus.

$$\text{ppb} = \frac{\text{Mass of fungus}}{\text{Mass of peanuts}} \times 10^9$$

**Method 2**

Use ratios to solve for the unknown mass.

**Act on Your Strategy**

**Method 1**

First convert the mass in tonnes into grams.

$$20 \text{ t} \times 1000 \text{ kg/t} \times 1000 \text{ g/kg} = 20 \times 10^6 \text{ g}$$

Next rearrange the formula and find the mass of the fungus.

$$\text{ppb} = \frac{\text{Mass of fungus}}{\text{Mass of peanuts}} \times 10^9$$

$$\therefore \text{Mass of fungus} = \frac{\text{ppb} \times \text{Mass of peanuts}}{10^9}$$

$$= \frac{25 \text{ ppb} \times (20 \times 10^6 \text{ g})}{10^9}$$

$$= 0.5 \text{ g}$$

**Method 2**

$$\frac{x \text{ g solute}}{20 \times 10^6 \text{ g solution}} = \frac{25 \text{ g solute}}{1 \times 10^9 \text{ g solution}}$$

$$x \text{ g} = (20 \times 10^6 \text{ g solution}) \times \frac{25 \text{ g solute}}{1 \times 10^9 \text{ g solution}}$$

$$= 0.5 \text{ g}$$

The maximum mass of fungus that is allowed is 0.5 g.

**Check Your Solution**

The answer appears to be reasonable. The units divided correctly to give grams. **Note:** Parts per million and parts per billion have no units. The original units, g/g, cancel out.

## Practice Problems

**15.** Symptoms of mercury poisoning become apparent after a person has accumulated more than 20 mg of mercury in the body.
   (a) Express this amount as parts per million for a 60 kg person.
   (b) Express this amount as parts per billion.
   (c) Express this amount as a (m/m) percent.

**16.** The use of the pesticide DDT has been banned in Canada since 1969 because of its damaging effect on wildlife. In 1967, the concentration of DDT in an average lake trout, taken from Lake Simcoe in Ontario, was 16 ppm. Today it is less than 1 ppm. What mass of DDT would have been present in a 2.5 kg trout with DDT present at 16 ppm?

**17.** The concentration of chlorine in a swimming pool is generally kept in the range of 1.4 to 4.0 mg/L. The water in a certain pool has 3.0 mg/L of chlorine. Express this value as parts per million. (**Hint:** 1 L of water has a mass of 1000 g.)

**18.** Water supplies with dissolved calcium carbonate greater than 500 mg/L are considered unacceptable for most domestic purposes. Express this concentration in parts per million.

## Careers in Chemistry

### Product Development Chemist

A solvent keeps paint liquefied so that it can be applied to a surface easily. After the paint has been exposed to the air, the solvent evaporates and the paint dries. Product development chemists develop and improve products such as paints. To work in product development, they require at least one university chemistry degree.

Chemists who work with paints must examine the properties of many different solvents. They must choose solvents that dissolve paint pigments well, but evaporate quickly and pose a low safety hazard.

Product development chemists must consider human health and environmental impact when choosing between solvents. Many solvents that have been used in the past, such as benzene and carbon tetrachloride, are now known to be harmful to health and/or the environment. A powerful new solvent called *d-limonene* has been developed from the peel of oranges and lemons. This solvent is less harmful than many older solvents. It has been used successfully in domestic and industrial cleaning products and as a pesticide. Chemists are now studying new applications for d-limonene.

### Make Career Connections

1. Use reference books or the Internet to find the chemical structure of d-limonene. What else can you discover about d-limonene?

2. To learn more about careers involving work with solvents, contact the Canadian Chemical Producers Association (CCPA).

## Molar Concentration

The most useful unit of concentration in chemistry is molar concentration. **Molar concentration** is the number of moles of solute in 1 L of solution. Notice that the volume of the *solution* in *litres* is used, rather than the volume of the *solvent* in *millilitres*. Molar concentration is also known as *molarity*.

$$\text{Molar concentration (in mol/L)} = \frac{\text{Amount of solute (in mol)}}{\text{Volume of solution (in L)}}$$

This formula can be shortened to give

$$C = \frac{n}{V}$$

Molar concentration is particularly useful to chemists because it is related to the number of particles in a solution. None of the other measures of concentration are related to the number of particles. If you are given the molar concentration and the volume of a solution, you can calculate the amount of dissolved solute in moles. This allows you to solve problems involving quantities in chemical reactions, such as the ones on the following pages.

### Sample Problem

### Calculating Molar Concentration

**Problem**

A saline solution contains 0.90 g of sodium chloride, NaCl, dissolved in 100 mL of solution. What is the molar concentration of the solution?

**What Is Required?**

You need to find the molar concentration of the solution in mol/L.

**What Is Given?**

You know that 0.90 g of sodium chloride is dissolved in 100 mL of solution.

**Plan Your Strategy**

**Step 1** To find the amount (in mol) of sodium chloride, first determine its molar mass. Then divide the amount of sodium chloride (in g) by its molar mass (in g/mol).

**Step 2** Convert the volume of solution from mL to L using this formula:

$$\text{Volume (in L)} = \text{Volume (in mL)} \times \frac{1.000 \text{ L}}{1000 \text{ mL}}$$

**Step 3** Use the following formula to calculate the molar concentration:

$$\text{Molar concentration (in mol/L)} = \frac{\text{Amount of solute (in mol)}}{\text{Volume of solution (in L)}}$$

**Act on Your Strategy**

**Step 1** Molar mass of NaCl = 22.99 + 35.45
= 58.44 g/mol

$$\text{Amount of NaCl} = \frac{0.90 \text{ g}}{58.44 \text{ g/mol}}$$
$$= 1.54 \times 10^{-2} \text{ mol}$$

---

**PROBLEM TIP**

When you convert from millilitres to litres, remember that you are dividing by 1000, or $10^3$. In practice, this means you are moving the decimal point three places to the left:

1.000.

**Step 2** Convert the volume from mL to L.

$$\text{Volume} = 100 \text{ mL} \times \frac{1.000 \text{ L}}{1000 \text{ mL}}$$
$$= 0.100 \text{ L}$$

**Step 3** Calculate the molar concentration.

$$\text{Molar concentration} = \frac{1.54 \times 10^{-2} \text{ mol}}{0.100 \text{ L}}$$
$$= 1.54 \times 10^{-1} \text{ mol/L}$$

The molar concentration of the saline solution is 0.15 mol/L.

### Check Your Solution

The answer has the correct units for molar concentration.

## Sample Problem

### Using Molar Concentration to Find Mass

### Problem

At 20°C, a saturated solution of calcium sulfate, $CaSO_4$, has a concentration of 0.0153 mol/L. A student takes 65 mL of this solution and evaporates it. What mass (in g) is left in the evaporating dish?

### What Is Required?

You need to find the mass (in g) of the solute, calcium sulfate.

### What Is Given?

The molar concentration is 0.0153 mol/L. The volume of the solution is 65 mL.

### Plan Your Strategy

**Step 1** Convert the volume from mL to L using the formula

$$\text{Volume (in L)} = \text{Volume (in mL)} \times \frac{1.000 \text{ L}}{1000 \text{ mL}}$$

**Step 2** Rearrange the following formula to solve for the amount of solute (in mol).

$$\text{Molar concentration (in mol/L)} = \frac{\text{Amount of solute (in mol)}}{\text{Volume of solution (in L)}}$$

**Step 3** Determine the molar mass of calcium sulfate. Use the molar mass to find the mass in grams, using the formula below:

Mass (in g) of $CaSO_4$
= Amount (in mol) × Molar mass of $CaSO_4$ (in g/mol)

### Act on Your Strategy

**Step 1** Convert the volume from mL to L.

$$\text{Volume} = 65 \text{ mL} \times \frac{1.000 \text{ L}}{1000 \text{ mL}}$$
$$= 0.065 \text{ L}$$

**Step 2** Rearrange the formula to solve for the amount of solute.

$$\text{Molar concentration} = \frac{\text{Amount of solute}}{\text{Volume of solution}}$$

> **PROBLEM TIP**
>
> Your teacher may show you another way to solve molar concentration problems, using an equation that looks like this:
>
> $$C_i V_i = C_f V_f$$
>
> Make sure you understand how to use the strategy shown in this book if you use this other method.

*Continued...*

∴ Amount of solute = Molar concentration × Volume of solution
= 0.0153 mol/L × 0.065 L
= 9.94 × 10⁻⁴ mol

**Step 3** Determine the molar mass. Then find the mass in grams.

Molar mass of $CaSO_4$ = 40.08 + 32.07 + (4 × 16.00)
= 136.15 g/mol

Mass (in g) of $CaSO_4$ = 9.94 × 10⁻⁴ mol × 136 g/mol
= 0.135 g

Therefore, 0.14 g of calcium sulfate are left in the evaporating dish.

### Check Your Solution

The answer has the correct units and the correct number of significant figures.

## Practice Problems

19. What is the molar concentration of each solution?
    (a) 0.50 mol of NaCl dissolved in 0.30 L of solution
    (b) 0.289 mol of iron(III) chloride, $FeCl_3$, dissolved in 120.0 mL of solution
    (c) 0.0877 mol of copper(II) sulfate, $CuSO_4$, dissolved in 0.07 L of solution
    (d) 4.63 g of sugar, $C_{12}H_{22}O_{11}$, dissolved in 16.8 mL of solution
    (e) 1.2 g of $NaNO_3$ dissolved in 80 mL of solution

20. What mass of solute is present in each aqueous solution?
    (a) 1.00 L of 0.045 mol/L calcium hydroxide, $Ca(OH)_2$, solution
    (b) 500.0 mL of 0.100 mol/L silver nitrate, $AgNO_3$, solution
    (c) 2.5 L of 1.00 mol/L potassium chromate, $K_2CrO_4$, solution
    (d) 0.040 L of 6.0 mol/L sulfuric acid, $H_2SO_4$, solution
    (e) 4.24 L of 0.775 mol/L ammonium nitrate, $NH_4NO_3$, solution

21. A student dissolves 30.46 g of silver nitrate, $AgNO_3$, in water to make 500.0 mL of solution. What is the molar concentration of the solution?

22. What volume of 0.25 mol/L solution can be made using 14 g of sodium hydroxide, NaOH?

23. A 100.0 mL bottle of skin lotion contains a number of solutes. One of these solutes is zinc oxide, ZnO. The concentration of zinc oxide in the skin lotion is 0.915 mol/L. What mass of zinc oxide is present in the bottle?

24. Formalin is an aqueous solution of formaldehyde, HCHO, used to preserve biological specimens. What mass of formaldehyde is needed to prepare 1.5 L of formalin with a concentration of 10.0 mol/L?

You have done many calculations for the concentration of various solutions. Now you are in a position to do some hands-on work with solution concentration. In the following investigation, you will use what you have learned to design your own experiment to determine the concentration of a solution.

# Investigation 7-B

**SKILL FOCUS**
Initiating and planning
Performing and recording
Communicating results
Predicting

## Determining the Concentration of a Solution

Your teacher will give you a sample of a solution. Design and perform an experiment to determine the concentration of the solution. Express the concentration as

(a) mass of solute dissolved in 100 mL of *solution*

(b) mass of solute dissolved in 100 g of *solvent*

(c) amount of solute (in mol) dissolved in 1 L of *solution*

### Safety Precautions

When you have designed your investigation, think about the safety precautions you will need to take.

### Materials

any apparatus in the laboratory

solution containing a solid dissolved in water

**Note:** Your teacher will tell you the name of the solute.

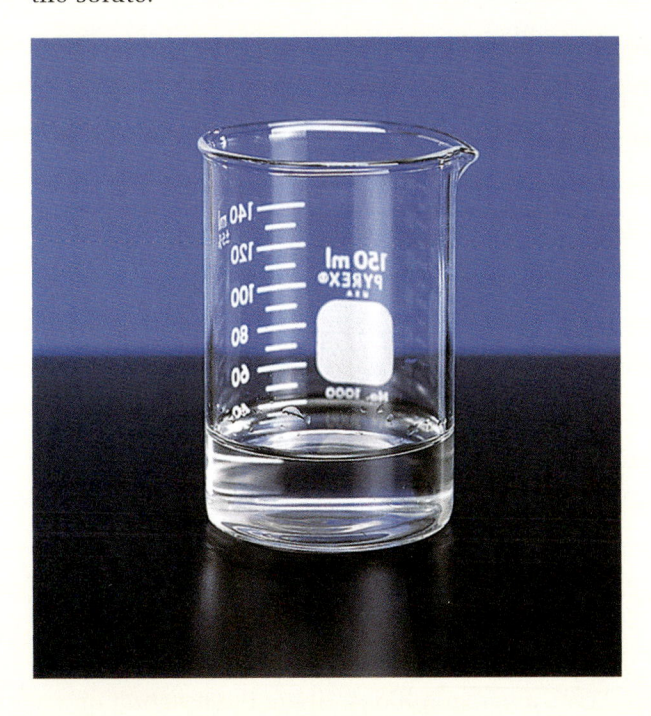

### Procedure

1. Think about what you need to know in order to determine the concentration of a solution. Then design your experiment so that you can measure each quantity you need. Assume that the density of pure water is 1.00 g/mL.

2. Write the steps that will allow you to measure the quantities you need. Design a data table for your results. Include a space for the name of the solute in your solution.

3. When your teacher approves your procedure, complete your experiment.

4. Dispose of your solution as directed by your teacher.

### Analysis

1. Express the concentration of the solution you analyzed as

    (a) mass of solute dissolved in 100 mL of *solution*

    (b) mass of solute dissolved in 100 g of *solvent*

    (c) molar concentration

    Show your calculations.

### Conclusions

2. List at least two important sources of error in your measurements.

3. List at least two important ways that you could improve your procedure.

4. Did the solute partially decompose on heating, producing a gas and another solid? If so, how do you think this affected the results of your experiment?

## Concept Organizer — Summary of Sections 7.1 to 7.3

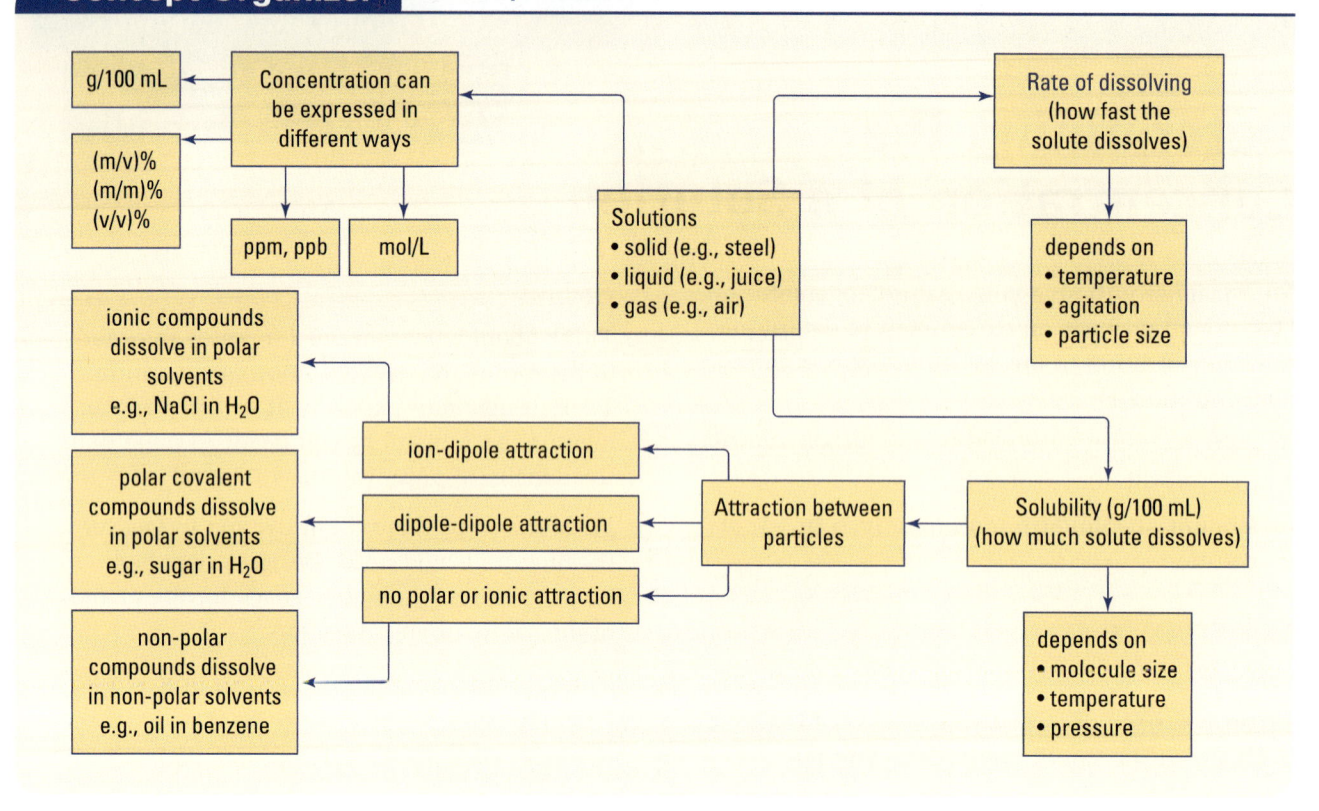

**Concentration can be expressed in different ways**
- g/100 mL
- (m/v)%
- (m/m)%
- (v/v)%
- ppm, ppb
- mol/L

**Solutions**
- solid (e.g., steel)
- liquid (e.g., juice)
- gas (e.g., air)

**Rate of dissolving (how fast the solute dissolves)** depends on
- temperature
- agitation
- particle size

**Solubility (g/100 mL) (how much solute dissolves)** depends on
- molecule size
- temperature
- pressure

**Attraction between particles:**
- ion-dipole attraction → ionic compounds dissolve in polar solvents e.g., NaCl in H₂O
- dipole-dipole attraction → polar covalent compounds dissolve in polar solvents e.g., sugar in H₂O
- no polar or ionic attraction → non-polar compounds dissolve in non-polar solvents e.g., oil in benzene

### Section Summary

You have learned about several different ways in which chemists express concentration: mass/volume, mass/mass, and volume/volume percent; parts per million and parts per billion; and molar concentration. The Concept Organizer above summarizes what you have learned in this chapter so far.

In section 7.4, you will learn how standard solutions of known concentration are prepared. You will also learn how to dilute a standard solution.

### Section Review

**1** Ammonium chloride, $NH_4Cl$, is a very soluble salt. 300 g of ammonium chloride are dissolved in 600 mL of water. What is the percent (m/m) of the solution?

**2** A researcher measures 85.1 mL of a solution of liquid hydrocarbons. The researcher then distills the sample to separate the pure liquids. If 20.3 mL of the hydrocarbon hexane are recovered, what is its percent (v/v) in the sample?

**3** A stock solution of phosphoric acid is 85.0% (m/v) $H_3PO_4$ in water. What is its molar concentration?

**4** Cytosol is an intracellular solution containing many important solutes. Research this solution. Write a paragraph describing the function of cytosol, and the solutes it contains.

# Preparing Solutions

## 7.4

What do the effectiveness of a medicine, the safety of a chemical reaction, and the cost of an industrial process have in common? They all depend on solutions that are made carefully with known concentrations. A solution with a known concentration is called a **standard solution**. There are two ways to prepare an aqueous solution with a known concentration. You can make a solution by dissolving a measured mass of pure solute in a certain volume of solution. Alternatively, you can dilute a solution of known concentration.

**Section Preview/Outcomes**

In this section, you will

- **prepare** solutions and calculate their concentrations
- **communicate** your understanding of the following terms: *standard solution, volumetric flask*

### Using a Volumetric Flask

A **volumetric flask** is a pear-shaped glass container with a flat bottom and a long neck. Volumetric flasks like the ones shown in Figure 7.19 are used to make up standard solutions. They are available in a variety of sizes. Each size can measure a fixed volume of solution to ±0.1 mL at a particular temperature, usually 20°C. When using a volumetric flask, you must first measure the mass of the pure solute. Then you transfer the solute to the flask using a funnel, as shown in Figure 7.20. At this point, you add the solvent (usually water) to dissolve the solute, as in Figure 7.21. You continue adding the solvent until the bottom of the meniscus appears to touch the line that is etched around the neck of the flask. See Figure 7.22. This is the volume of the solution, within ±0.1 mL. If you were performing an experiment in which significant digits and errors were important, you would record the volume of a solution in a 500 mL volumetric flask as 500.0 mL ±0.1 mL. Before using a volumetric flask, you need to rinse it several times with a small quantity of distilled water and discard the washings. *Standard solutions are never stored in volumetric flasks*. Instead, they are transferred to another bottle that has a secure stopper or cap.

**Figure 7.19** These volumetric flasks, from left to right, contain solutions of chromium(III) salts, iron(III) salts, and cobalt(II) salts.

**Figure 7.20** Transfer a known mass of solid solute into the volumetric flask. Alternatively, dissolve the solid in a small volume of solvent. Then add the liquid to the flask.

**Figure 7.21** Add distilled water until the flask is about half full. Swirl the mixture around in order to dissolve the solute completely. Rinse the beaker that contained the solute with solvent. Add the rinsing to the flask.

**Figure 7.22** Add the rest of the water slowly. When the flask is almost full, add the water drop by drop until the bottom of the meniscus rests at the etched line.

Chapter 7 Solutions and Their Concentrations • MHR  **271**

## Diluting a Solution

You can make a less concentrated solution of a known solution by adding a measured amount of additional solvent to the standard solution. The number of molecules, or moles, of solute that is present remains the same before and after the dilution. (See Figure 7.23.)

To reinforce these ideas, read through the Sample Problem below. Then try the Practice Problems that follow.

**Figure 7.23** When a solution is diluted, the volume increases. However, the amount of solute remains the same.

### Sample Problem

#### Diluting a Standard Solution

**Problem**

For a class experiment, your teacher must make 2.0 L of 0.10 mol/L sulfuric acid. This acid is usually sold as an 18 mol/L concentrated solution. How much of the concentrated solution should be used to make a new solution with the correct concentration?

**What Is Required?**

You need to find the volume of concentrated solution to be diluted.

**What Is Given?**

Initial concentration = 18 mol/L
Concentration of diluted solution = 0.10 mol/L
Volume of diluted solution = 2.0 L

**Plan Your Strategy**

**Note:** Amount of solute (mol) after dilution = Amount of solute (mol) before dilution

**Step 1** Calculate the amount of solute (in mol) that is needed for the final dilute solution.

**Step 2** Calculate the volume of the concentrated solution that will provide the necessary amount of solute.

**Act on Your Strategy**

**Step 1** Calculate the amount of solute that is needed for the final dilute solution.

$$\text{Molar concentration (in mol/L)} = \frac{\text{Amount of solute (in mol)}}{\text{Volume of solution (in L)}}$$

∴ Amount of solute = Molar concentration × Volume of solution

For the final dilute solution,

Amount of solute = 0.10 mol/L × 2.0 L
= 0.20 mol

**Step 2** Calculate the volume of the original concentrated solution that is needed.

Rearrange and use the molar concentration equation. Substitute in the amount of solute you calculated in step 1.

$$\text{Volume of solution (in L)} = \frac{\text{Amount of solute (in mol)}}{\text{Molar concentration (in mol/L)}}$$

$$= \frac{0.20 \text{ mol}}{18 \text{ mol/L}}$$

= 0.011 L

Therefore, 0.011 L, or 11 mL, of the concentrated 18 mol/L solution should be used to make 2.0 L of 0.10 mol/L sulfuric acid.

**Check Your Solution**

The units are correct. The final solution must be much less concentrated. Thus, it is reasonable that only a small volume of concentrated solution is needed.

### Practice Problems

**25.** Suppose that you are given a solution of 1.25 mol/L sodium chloride in water, $NaCl_{(aq)}$. What volume must you dilute to prepare the following solutions?

(a) 50.0 mL of 1.00 mol/L $NaCl_{(aq)}$

(b) 200.0 mL of 0.800 mol/L $NaCl_{(aq)}$

(c) 250.0 mL of 0.300 mol/L $NaCl_{(aq)}$

**26.** What concentration of solution is obtained by diluting 50.0 mL of 0.720 mol/L aqueous sodium nitrate, $NaNO_{3(aq)}$, to each volume?

(a) 120.0 mL  (b) 400.0 mL  (c) 5.00 L

**27.** A solution is prepared by adding 600.0 mL of distilled water to 100.0 mL of 0.15 mol/L ammonium nitrate. Calculate the molar concentration of the solution. Assume that the volume quantities can be added together.

Now that you understand how to calculate standard solutions and dilution, it is time for you to try it out for yourself. In the following investigation, you will prepare and dilute standard solutions.

# Investigation 7-C

**SKILL FOCUS**
Initiating and planning
Performing and recording
Analyzing and interpreting

## Estimating Concentration of an Unknown Solution

Copper(II) sulfate, $CuSO_4$, is a soluble salt. It is sometimes added to pools and ponds to control the growth of fungi. Solutions of this salt are blue in colour. The intensity of the colour increases with increased concentration. In this investigation, you will prepare copper(II) sulfate solutions with known concentrations. Then you will estimate the concentration of an unknown solution by comparing its colour intensity with the colour intensities of the known solutions.

Copper(II) sulfate pentahydrate is a *hydrate*. Hydrates are ionic compounds that have a specific amount of water molecules associated with each ion pair.

### Question

How can you estimate the concentration of an unknown solution?

### Part 1 Making Solutions with Known Concentrations

### Safety Precautions

- Copper(II) sulfate is poisonous. Wash your hands at the end of this investigation.
- If you spill any solution on your skin, wash it off immediately with copious amounts of cool water.

### Materials

graduated cylinder
6 beakers
chemical balance
stirring rod
copper(II) sulfate pentahydrate, $CuSO_4 \cdot 5H_2O$
distilled water
labels or grease marker

### Procedure

1. With your partner, develop a method to prepare 100 mL of 0.500 mol/L aqueous $CuSO_4 \cdot 5H_2O$ solution. Include the water molecules that are hydrated to the crystals, as given in the molecular formula, in your calculation of the molar mass. Show all your calculations. Prepare the solution.

2. Save some of the solution you prepared in step 1, to be tested in Part 2. Use the rest of the solution to make the dilutions in steps 3 to 5. Remember to label the solutions.

3. Develop a method to dilute part of the 0.500 mol/L $CuSO_4$ solution, to make 100.0 mL of 0.200 mol/L solution. Show your calculations. Prepare the solution.

4. Show your calculations to prepare 100.0 mL of 0.100 mol/L solution, using the solution you prepared in step 3. (You do not need to describe the method because it will be similar to the method you developed in step 3. Only the volume diluted will be different.) Prepare the solution.

5. Repeat step 4 to make 100.0 mL of 0.050 mol/L $CuSO_4$, by diluting part of the 0.100 mol/L solution you made. Then make 50 mL of 0.025 mol/L solution by diluting part of the 0.050 mol/L solution.

### Part 2 Estimating the Concentration of an Unknown Solution

### Materials

paper towels
6 clean, dry, identical test tubes
medicine droppers
5 prepared solutions from Part 1
10 mL of copper(II) sulfate, $CuSO_4$, solution with an unknown concentration

274 MHR • Unit 3 Solutions

### Procedure

1. You should have five labelled beakers containing $CuSO_4$ solutions with the following concentrations: 0.50 mol/L, 0.20 mol/L, 0.10 mol/L, 0.05 mol/L, and 0.025 mol/L. Your teacher will give you a sixth solution of unknown concentration. Record the letter or number that identifies this solution.

2. Label each test tube, one for each solution. Pour a sample of each solution into a test tube. The height of the solution in the test tubes should be the same. Use a medicine dropper to add or take away solution as needed. (Be careful not to add water, or a solution of different concentration, to a test tube.)

3. The best way to compare colour intensity is by looking down through the test tube. Wrap each test tube with a paper towel to stop light from entering the side. Arrange the solutions of known concentration in order.

4. Place the solutions over a diffuse light source such as a lightbox. Compare the colour of the unknown solution with the colours of the other solutions.

5. Use your observations to estimate the concentration of the unknown solution.

6. Pour the solutions of $CuSO_4$ into a beaker supplied by your teacher. Wash your hands.

### Analysis

1. Describe any possible sources of error for Part 1 of this investigation.

2. What is your estimate of the concentration of the unknown solution?

### Conclusion

3. Obtain the concentration of the unknown solution from your teacher. Calculate the percentage error in your estimate.

### Applications

4. Use your estimated concentration of the unknown solution to calculate the mass of $CuSO_4 \cdot 5H_2O$ that your teacher would need to prepare 500 mL of this solution.

5. If your school has a spectrometer or colorimeter, you can measure the absorption of light passing through the solutions. By measuring the absorption of solutions of copper(II) sulfate with different concentrations, you can draw a graph of absorption against concentration.

## Diluting Concentrated Acids

The acids that you use in your investigations are bought as concentrated standard solutions. Sulfuric acid is usually bought as an 18 mol/L solution. Hydrochloric acid is usually bought as a 12 mol/L solution. These acids are far too dangerous for you to use at these concentrations. Your teacher dilutes concentrated acids, following a procedure that minimizes the hazards involved.

Concentrated acids should be diluted in a fume hood because breathing in the fumes causes acid to form in air passages and lungs. As shown in Figure 7.24 rubber gloves must be used to protect the hands. A lab coat is needed to protect clothing. Even small splashes of a concentrated acid will form holes in fabric. Safety goggles, or even a full-face shield, are essential.

Mixing a strong, concentrated acid with water is a very exothermic process. A concentrated acid is denser than water. Therefore, when it is poured into water, it sinks into the solution and mixes with the solution. The heat that is generated is spread throughout the solution. This is the only safe way to mix an acid and water. If you added water to a concentrated acid, the water would float on top of the solution. The heat generated at the acid-water layer could easily boil the solution and splatter highly corrosive liquid. The sudden heat generated at the acid-water boundary could crack the glassware and lead to a very dangerous spill. Figure 7.24 illustrates safety precautions needed to dilute a strong acid.

**Figure 7.24** When diluting acid, always add the acid to the water—never the reverse. Rubber gloves, a lab coat, and safety goggles or a face shield protect against acid splashes.

## Section Summary

In this section, you learned how to prepare solutions by dissolving a solid solute and then diluting a concentrated solution. In the next chapter, you will see how water is used as a solvent in chemistry laboratories. Many important reactions take place in water. You will learn how to use your stoichiometry skills to calculate quantitative information about these reactions.

## Section Review

**1** What mass of potassium chloride, KCl, is used to make 25.0 mL of a solution with a concentration of 2.00 mol/L?

**2** A solution is prepared by dissolving 42.5 g of silver nitrate, $AgNO_3$ in a 1 L volumetric flask. What is the molar concentration of the solution?

**3** The solution of aqueous ammonia that is supplied to schools has a concentration of 14 mol/L. Your class needs 3.0 L of a solution with a concentration of 0.10 mol/L.

   **(a)** What procedure should your teacher follow to make up this solution?

   **(b)** Prepare an instruction sheet or a help file for your teacher to carry out this dilution.

**4** 47.9 g of potassium chlorate, $KClO_3$, is used to make a solution with a concentration of 0.650 mol/L. What is the volume of the solution?

**5** Water and 8.00 mol/L potassium nitrate solution are mixed to produce 700.0 mL of a solution with a concentration of 6.00 mol/L. What volumes of water and potassium nitrate solution are used?

# CHAPTER 7 Review

## Reflecting on Chapter 7
Summarize this chapter in the format of your choice. Here are a few ideas to use as guidelines:
- Describe the difference between a saturated and an unsaturated solution.
- What factors affect the rate of dissolving?
- What factors affect solubility?
- How does temperature affect the solubility of a solid, a liquid, and a gas?
- Describe how particle attractions affect solubility.
- Explain how to plot a solubility curve.
- Write the formulas for (m/v) percent, (m/m) percent, (v/v) percent, ppm, ppb, and molar concentration.
- Explain how you would prepare a standard solution using a volumetric flask.

## Reviewing Key Terms
For each of the following terms, write a sentence that shows your understanding of its meaning.

solution
solvent
solutes
variable composition
dilute
aqueous solution
miscible
immiscible
alloys
solubility
saturated solution
unsaturated solution
rate of dissolving
dipole
dipole-dipole attraction
hydrogen bonding
ion-dipole attractions
hydrated
electrolyte
non-electrolytes
concentration
mass/volume percent
mass/mass percent
volume/volume percent
parts per million
parts per billion
molar concentration
standard solution
volumetric flask

## Knowledge/Understanding

1. Identify at least two solutions in your home that are
   (a) beverages
   (b) found in the bathroom or medicine cabinet
   (c) solids

2. How is a solution different from a pure compound? Give specific examples.

3. Mixing 2 mL of linseed oil and 4 mL of turpentine makes a binder for oil paint. What term is used to describe liquids that dissolve in each other? Which liquid is the solvent?

4. How does the bonding in water molecules account for the fact that water is an exellent solvent?

5. Why does an aqueous solution of an electrolyte conduct electricity, but an aqueous solution of a non-electrolyte does not?

6. Use the concept of forces between particles to explain why oil and water are immiscible.

7. Explain the expression "like dissolves like" in terms of intermolecular forces.

8. What factors affect the rate of dissolving of a solid in a liquid?

9. Which of the following substances would you predict to be soluble in water? Briefly explain each answer.
   (a) potassium chloride, KCl
   (b) carbon tetrachloride, $CCl_4$
   (c) sodium sulfate, $Na_2SO_4$
   (d) butane, $C_4H_{10}$

10. Benzene, $C_6H_6$, is a liquid at room temperature. It is sometimes used as a solvent. Which of the following compounds is more soluble in benzene: naphthalene, $C_{10}H_8$, or sodium fluoride, NaF? Would you expect ethanol, $CH_3CH_2OH$, to be soluble in benzene? Explain your answers.

## Inquiry

11. Boric acid solution is used as an eyewash. What mass of boric acid is present in 250.0 g of solution that is 2.25% (m/m) acid in water?

12. 10% (m/m) sodium hydroxide solution, $NaOH_{(aq)}$, is used to break down wood fibre to make paper.
    (a) What mass of solute is needed to make 250.0 mL of 10% (m/m) solution?
    (b) What mass of solvent is needed?
    (c) What is the molar concentration of the solution?

13. What volume of pure ethanol is needed to make 800.0 mL of a solution of ethanol in water that is 12% (v/v)?

14. Some municipalities add sodium fluoride, NaF, to drinking water to help protect the teeth of children. The concentration of sodium fluoride is maintained at $2.9 \times 10^{-5}$ mol/L. What mass (in mg) of sodium fluoride is dissolved in 1.00 L of water? Express this concentration in ppm.

Answers to questions highlighted in red type are provided in Appendix A.

15. A saturated solution of sodium acetate, NaCH₃COO, can be prepared by dissolving 4.65 g in 10.0 mL of water at 20°C. What is the molar concentration of the solution?

16. What is the molar concentration of each of the following solutions?
    (a) 7.25 g of silver nitrate, AgNO₃, dissolved in 100.0 mL of solution
    (b) 80 g of glucose, C₆H₁₂O₆, dissolved in 70.0 mL of solution

17. Calculate the mass of solute that is needed to prepare each solution below.
    (a) 250 mL of 0.250 mol/L calcium acetate, Ca(CH₃COO)₂
    (b) 1.8 L of 0.35 mol/L ammonium sulfate, (NH₄)₂SO₄

18. Calculate the molar concentration of each solution formed after dilution.
    (a) 20 mL of 6.0 mol/L hydrochloric acid, HCl$_{(aq)}$, diluted to 70 mL
    (b) 300.0 mL of 12.0 mol/L ammonia, NH$_{3(aq)}$, diluted to 2.50 L

19. Calculate the molar concentration of each solution. Assume that the volumes can be added.
    (a) 85.0 mL of 1.50 mol/L ammonium chloride, NH₄Cl$_{(aq)}$, added to 250 mL of water
    (b) a 1:3 dilution of 1.0 mol/L calcium phosphate (that is, one part stock solution mixed with three parts water)

20. A standard solution of 0.250 mol/L calcium ion is prepared by dissolving solid calcium carbonate in an acid. What mass of calcium carbonate is needed to prepare 1.00 L of the solution?

21. Suppose that your teacher gives you three test tubes. Each test tube contains a clear, colourless liquid. One liquid is an aqueous solution of an electrolyte. Another liquid is an aqueous solution of a non-electrolyte. The third liquid is distilled water. Outline the procedure for an experiment to identify which liquid is which.

22. Fertilizers for home gardeners may be sold as aqueous solutions. Suppose that you want to begin a company that sells an aqueous solution of potassium nitrate, KNO₃, fertilizer. You need a solubility curve (a graph of solubility versus temperature) to help you decide what concentration to use for your solution. Describe an experiment that you might perform to develop a solubility curve for potassium nitrate. State which variables are controlled, which are varied, and which must be measured.

23. Potassium alum, KAl(SO₄)₂·12H₂O, is used to stop bleeding from small cuts. The solubility of potassium alum, at various temperatures, is given in the following table.

Solubility of Potassium Alum

| Solubility (g/100 g water) | Temperature (°C) |
|---|---|
| 4 | 0 |
| 10 | 10 |
| 15 | 20 |
| 23 | 30 |
| 31 | 40 |
| 49 | 50 |
| 67 | 60 |
| 101 | 70 |
| 135 | 80 |

(a) Plot a graph of solubility against temperature.
(b) From your graph, interpolate the solubility of potassium alum at 67°C.
(c) By extrapolation, estimate the solubility of potassium alum at 82°C.
(d) Look at your graph. At what temperature will 120 g of potassium alum form a saturated solution in 100 g of water?

24. Use the graph on the next page to answer questions 24 and 25. At 80°C, what mass of sodium chloride dissolves in 1.0 L of water?

25. What minimum temperature is required to dissolve 24 g of potassium nitrate in 40 g of water?

26. A teacher wants to dilute 200.0 mL of 12 mol/L hydrochloric acid to make a 1 mol/L solution. What safety precautions should the teacher take?

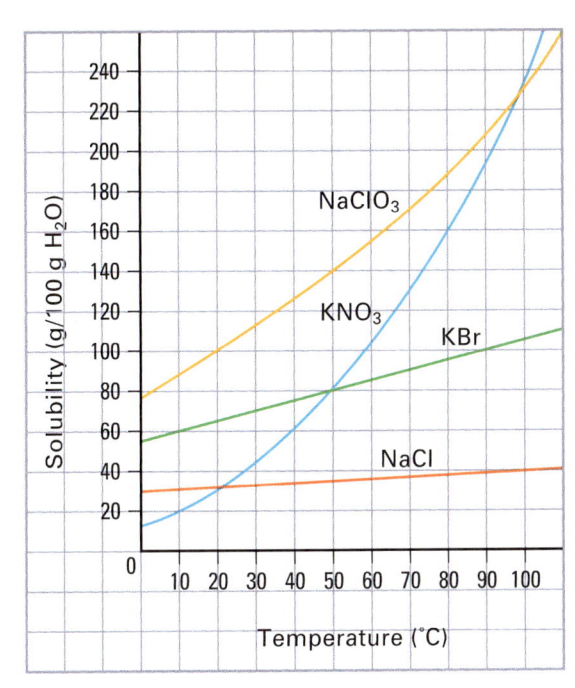

This graph shows the solubility of four salts at various temperatures. Use it to answer questions 24 and 25.

## Communication

27. Suppose that you make a pot of hot tea. Later, you put a glass of the tea in the refrigerator to save it for a cool drink. When you take it out of the refrigerator some hours later, you notice that it is cloudy. How could you explain this to a younger brother or sister?

28. Define each concentration term.
    (a) percent (m/v)
    (b) percent (m/m)
    (c) percent (v/v)
    (d) parts per million, ppm
    (e) parts per billion, ppb

29. The concentration of iron in the water that is supplied to a town is 0.25 mg/L. Express this in ppm and ppb.

30. Ammonia is a gas at room temperature and pressure, but it can be liquefied easily. Liquid ammonia is probably present on some planets. Scientists speculate that it might be a good solvent. Explain why, based on the structure of the ammonia molecule shown above.

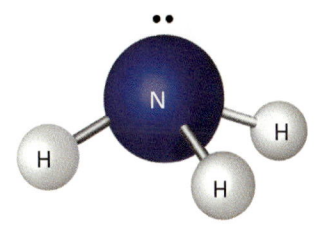

31. At 20°C, the solubility of oxygen in water is more than twice that of nitrogen. A student analyzed the concentration of dissolved gases in an unpolluted pond. She found that the concentration of nitrogen gas was greater than the concentration of oxygen. Prepare an explanation for the student to give to her class.

32. What is the concentration of pure water?

## Making Connections

33. A bright red mineral called cinnabar has the chemical formula HgS. It can be used to make an artist's pigment, but it is a very insoluble compound. A saturated solution at 25°C has a concentration of $2 \times 10^{-27}$ mol/L. In the past, why was heavy metal poisoning common in painters? Why did painters invariably waste more cinnabar than they used?

34. Vitamin A is a compound that is soluble in fats but not in water. It is found in certain foods, including yellow fruit and green vegetables. In parts of central Africa, children frequently show signs of vitamin A deficiency, although their diet contains a good supply of the necessary fruits and vegetables. Why?

**Answers to Practice Problems and Short Answers to Section Review Questions:**
**Practice Problems: 1.**(a) 3.16% (b) 1.75% (c) 2.9% **2.** 6.31%
**3.** 0.362 g **4.** 3.0 g, 0.1 g, 0.12 g **5.**(a) 26% (b) 16.61%
(c) 15.1% **6.** 35% **7.** 85 g **8.** 6.35 g **9.** 15 g **10.** 15%
**11.** 12 mL **12.** 38% **13.** 18 mL **14.** 285 mL **15.**(a) 0.33 ppm
(b) $3.3 \times 10^2$ ppb (c) 0.000033% **16.** 0.040 g **17.** 3.0 ppm
**18.** 500 ppm **19.**(a) 1.7 mol/L (b) 2.41 mol/L (c) 1.2 mol/L
(d) 0.804 mol/L (e) 0.18 mol/L **20.**(a) 3.3 g (b) 8.49 g
(c) $4.9 \times 10^2$ g (d) 24 g (e) 263 g **21.** 0.3589 mol/L **22.** 1.4 L
**23.** 7.45 g **24.** $4.5 \times 10^2$ g **25.**(a) 40 mL (b) 128 mL
(c) 60.0 mL **26.**(a) 0.300 mol/L (b) $9.00 \times 10^{-2}$ mol/L
(c) $7.20 \times 10^{-3}$ mol/L **27.** $2.1 \times 10^{-2}$ mol/L
**Section Review: 7.1: 1.** solute, solvent **7.** polar, ionic
**7.2: 3.**(a) increases (b) decreases **6.**(a) $Ce_2(SO_4)_3$ (b) $Ce_2(SO_4)_3$, $NaNO_3$ (c) NaCl (d) 84°C (e) 10 g **7.3: 1.** 33.3 **2.** 23.8% **3.** 8.67 mol/L **7.4: 1.** 3.73 g **2.** 0.25 mol/L **4.** 601 mL
**5.** 175 mL, 525 mL

# CHAPTER 8

# Solution Stoichiometry

### Chapter Preview

- **8.1** Making Predictions About Solubility
- **8.2** Reactions in Aqueous Solutions
- **8.3** Stoichiometry in Solution Chemistry

### Prerequisite Concepts and Skills

Before you begin this chapter, review the following concepts and skills:

- identifying factors that affect solubility (Chapter 7)
- performing stoichiometry calculations (Chapter 4)

At this moment, in your body, in your home, in the environment, and in industries of all kinds, there are millions of chemical reactions taking place. The majority of these reactions take place in aqueous solution. This should not be surprising. Water is the most abundant compound on Earth. Water covers about 70% of the Earth's surface, and occupies a mind-numbing global volume in excess of 130 billion (130 000 000 000) cubic kilometres!

You need to replenish the supply of water in your body constantly, because water is essential to life. It dissolves ions and molecules such as sodium and glucose, enabling them to pass into and out of cells. It transports ions and molecules throughout the body. Virtually all chemical reactions that occur in the body require water either as a reactant or as a solvent. For example, glucose is the key reactant in the energy-releasing process of cellular respiration, but glucose must be dissolved in aqueous solution in order to begin this life-sustaining chemical reaction.

Aqueous solutions are also common in the non-living world. Bakers mix baking powder with water to obtain the gas-forming reaction they require. Plaster and cement need water to form a new chemical compound when they solidify. Sulfuric acid must be mixed with water to be used for car batteries. What other examples can you think of?

Dissolving substances in aqueous solution often results in a chemical change. However, this is not the only possible outcome. The substances may simply remain in solution after mixing. In this chapter, you will learn how to predict which substances are soluble in water and what, if any, reaction they will undergo. You will also apply your stoichiometry skills to examine aqueous reactions quantitatively.

In what ways are aqueous solutions a part of your daily life?

# Making Predictions About Solubility

## 8.1

In Chapter 7, you examined factors that affect the solubility of a compound. As well, you learned that the terms "soluble" and "insoluble" are relative, because no substance is completely insoluble in water. "Soluble" generally means that more than about 1 g of solute will dissolve in 100 mL of water at room temperature. "Insoluble" means that the solubility is less than 0.1 g per 100 mL.

While many ionic compounds are soluble in water, many others are not. Cooks, chemists, farmers, pharmacists, and gardeners need to know which compounds are soluble and which are insoluble. (See Figure 8.1.)

**Section Preview/Outcomes**

In this section, you will

- **describe** and **identify** combinations of aqueous solutions that result in the formation of precipitates
- **communicate** your understanding of the following terms: *precipitate, general solubility guidelines*

**Figure 8.1** Plant food may come in the form of a liquid or a powder. Gardeners dissolve the plant food in water, and then either spray or water the plant with the resulting solution. What could happen if the plant food is too dilute or too concentrated?

### CHEM FACT

Some textbooks give different definitions and units of concentration to describe the terms "soluble" and "insoluble." Here are a few examples:

**Soluble:**
- more than 1 g in 100 mL; *or*
- greater than 0.1 mol/L

**Insoluble:**
- less than 0.1 g in 100 mL; *or*
- less than 0.01 mol/L

**Partly or slightly soluble:**
- between 1 g and 0.1 g in 100 mL; *or*
- between 0.1 mol/L and 0.01 mol/L

## Factors That Affect the Solubility of Ionic Substances

Nearly all alkali metal compounds are soluble in water. Sulfide and phosphate compounds are usually insoluble. How, then, do you account for the fact that sodium sulfide and potassium phosphate are soluble, while iron sulfide and calcium phosphate are insoluble? Why do some ions form soluble compounds, while other ions form insoluble compounds?

### The Effect of Ion Charge on Solubility

Compounds of ions with small charges tend to be soluble. Compounds of ions with large charges tend to be insoluble. Why? Increasing the charge increases the force that holds the ions together. For example, phosphates (compounds of $PO_4^{3-}$) tend to be insoluble. On the other hand, the salts of alkali metals are soluble. Alkali metal cations have a single positive charge, so the force that holds the ions together is less.

Chapter 8 Solution Stoichiometry • MHR **281**

## CONCEPT CHECK

Sketch an outline of the periodic table. Add labels and arrows to indicate what you think are the trends for ionic size (radius) across a period and down a group. Then suggest reasons for these trends. Specifically, identify the factors that you think are responsible for the differences in the sizes of ions and their "parent" atoms.

### The Effect of Ion Size on Solubility

When an atom gives up or gains an electron, the size of the ion that results is different from the size of the original atom. In Figure 8.2A, for example, you can see that the sodium ion is smaller than the sodium atom. In general, the ions of metals tend to be smaller than their corresponding neutral atoms. For the ions of non-metals, the reverse is true. The ions of non-metals tend to be larger than their corresponding neutral atoms.

Small ions bond more closely together than large ions. Thus, the bond between small ions is stronger than the bond between large ions with the same charge. As a result, compounds with small ions tend to be less soluble than compounds with large ions. Consider the ions of elements from Group 17 (VIIA), for example. As you can see in Figure 8.2B, the size of the ions increases as you go down a family in the periodic table. Therefore, you would expect that fluoride compounds are less soluble than chloride, bromide, and iodide compounds. This tends to be the case.

**A**

**B**

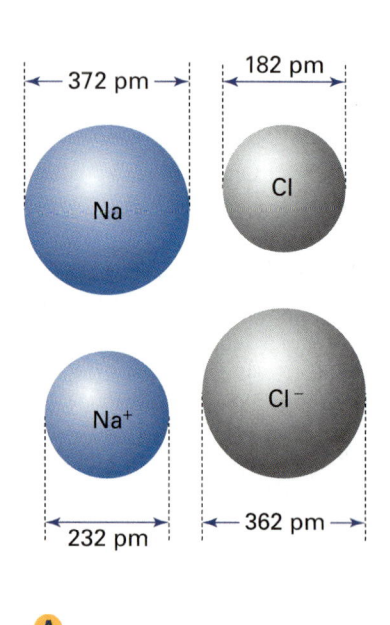

**Figure 8.2A** The size of the cation of a metal, such as sodium, tends to be smaller than the size of the atom from which it was formed. The size of the anion of a non-metal, such as phosphorus, tends to be larger than the size of its atom.

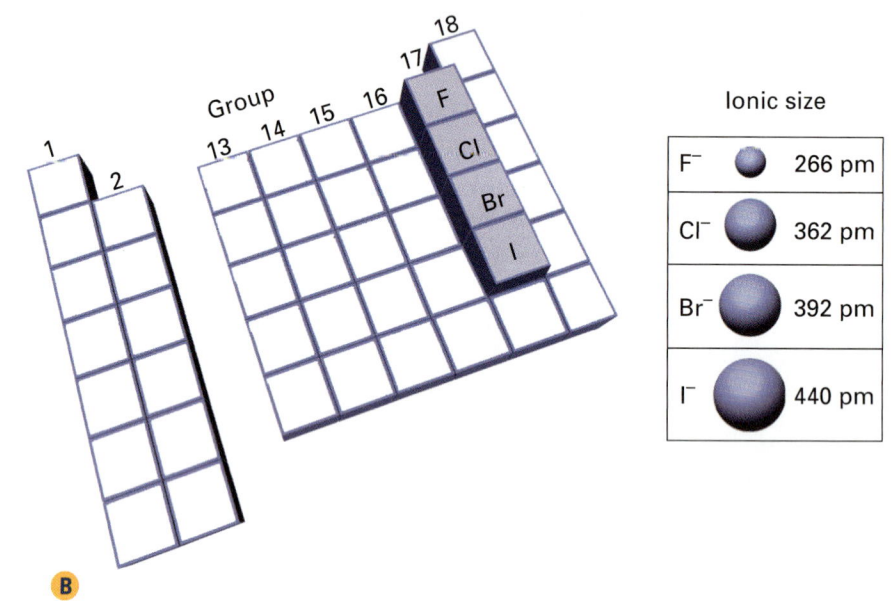

**Figure 8.2B** In Group 17 (VIIA), fluoride ions are smaller than chloride, bromide, and iodide ions. How does this periodic trend affect the solubility of compounds that are formed from these elements?

### Making Predictions About Solubility

Sulfides (compounds of $S^{2-}$) and oxides (compounds of $O^{2-}$) are influenced by both ion size and ion charge. These compounds tend to be insoluble because their ions have a double charge *and* are relatively small. Even so, a few sulfides and oxides are soluble, as you will discover in Investigation 8-A.

Many interrelated factors affect the solubility of substances in water. This makes it challenging to predict which ionic substances will dissolve in water. By performing experiments, chemists have developed guidelines to help them make predictions about solubility. In Investigation 8-A, you will perform your own experiments to develop guidelines about the solubility of ionic compounds in water.

# Investigation 8-A

**SKILL FOCUS**
Predicting
Performing and recording
Analyzing and interpreting
Communicating results

## The Solubility of Ionic Compounds

In this investigation, you will work with a set of solutions. You will chemically combine small quantities, two at a time. This will help you determine which combinations react to produce a **precipitate**. A precipitate is an insoluble solid that may result when two aqueous solutions chemically react. *The appearance of a precipitate indicates that an insoluble compound is present.* Then you will compile your data with the data from other groups to develop some guidelines about the solubility of several ionic compounds.

### Problem

How can you develop guidelines to help you predict the solubility of ionic compounds in water?

### Prediction

Read the entire Procedure. Predict which combination of anions and cations will likely be soluble and which combination will likely be insoluble. Justify your prediction by briefly explaining your reasoning.

### Materials

12-well or 24-well plate, or spot plate
toothpicks
cotton swabs or coarse paper towelling
wash bottle with distilled water
piece of black paper
piece of white paper
labelled dropper bottles of aqueous solutions that contain the following cations:
 $Al^{3+}$, $NH_4^+$, $Ba^{2+}$, $Ca^{2+}$, $Cu^{2+}$, $Fe^{2+}$, $Mg^{2+}$, $Ag^+$, $Na^+$, $Zn^{2+}$
labelled dropper bottles of aqueous solutions that contain the following anions:
 $CH_3COO^-$, $Br^-$, $CO_3^{2-}$, $Cl^-$, $OH^-$, $PO_4^{3-}$, $SO_4^{2-}$, $S^{2-}$

### Safety Precautions

- Do not contaminate the dropper bottles. The tip of a dropper should not make contact with either the plate or another solution. Put the cap back on the bottle immediately after use.
- Dispose of solutions as directed by your teacher.
- Make sure that you are working in a well-ventilated area.
- If you accidentally spill any of the solutions on your skin, wash the area immediately with plenty of cool water.

### Procedure

1. Your teacher will give you a set of nine solutions to test. Each solution includes one of five cations or four anions. Design a table to record the results of all the possible combinations of cations with anions in your set of solutions.

2. Decide how to use the well plate or spot plate to test systematically all the combinations of cations with anions in your set. If your plate does not have enough wells, you will need to clean the plate before you can test all the possible combinations. To clean the plate, first discard solutions into the container provided by your teacher. Then rinse the plate with distilled water, and clean the wells using a cotton swab.

3. To test each combination of anion and cation, add one or two drops into your well plate or spot plate. Then stir the mixture using a toothpick. Rinse the toothpick with running water before each stirring. Make sure that you keep track of the combinations of ions in each well or spot.

Chapter 8 Solution Stoichiometry • MHR **283**

continued

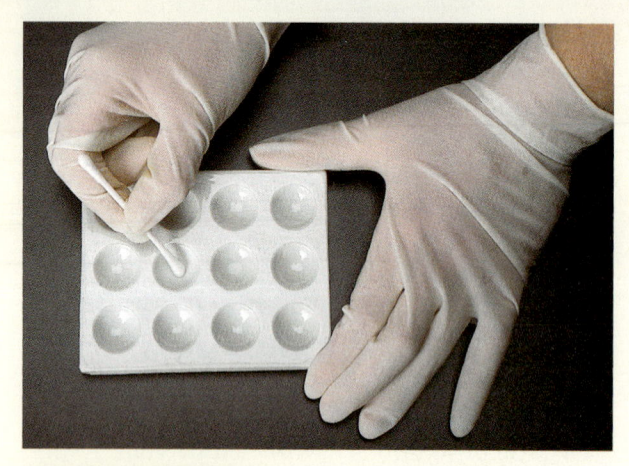

**Why is it necessary to clean the well or spot plate as described in step 2?**

4. Examine each mixture for evidence of a precipitate. Place the plate on a sheet of white or black paper. (Use whichever colour of paper helps you see a precipitate best.) Any cloudy appearance in the mixture is evidence of a precipitate. Many precipitates are white.
   - If you can see that a precipitate has formed, enter "I" in your table. This indicates that the combination of ions produces an insoluble substance.
   - If you cannot see a precipitate, enter "S" to indicate that the ion you are testing is soluble.
5. Repeat steps 2 to 4 for each cation solution.
6. Discard the solutions and precipitates into the container provided by your teacher. Rinse the plate with water, and clean the wells using a cotton swab.
7. If time permits, your teacher may give you a second set of solutions to test.
8. Add your observations to the class data table. Use your completed copy of the class data table to answer the questions below.

### Analysis

1. Identify any *cations* that
   (a) always appear to form soluble compounds
   (b) always appear to form insoluble compounds
2. Identify any *anions* that
   (a) always appear to form soluble compounds
   (b) always appear to form insoluble compounds
3. Based on your observations, which sulfates are insoluble?
4. Based on your observations, which phosphates are soluble?
5. Explain why each reagent solution you tested must contain both cations and anions.
6. Your teacher prepared the cation solutions using compounds that contain the nitrate ion. For example, the solution marked $Ca^{2+}$ was prepared by dissolving $Ca(NO_3)_2$ in water. Why were nitrates used to make these solutions?

### Conclusions

7. Which group in the periodic table most likely forms cations with salts that are usually soluble?
8. Which group in the periodic table most likely forms anions with salts that are usually soluble?
9. Your answers to questions 7 and 8 represent a preliminary set of guidelines for predicting the solubility of the compounds you tested. Many reference books refer to guidelines like these as "solubility rules." Why might "solubility guidelines" be a better term to use for describing solubility patterns?

### Application

10. Predict another combination of an anion and a cation (not used in this investigation) that you would expect to be soluble. Predict another combination that you would expect to be insoluble. Share your predictions, and your reasons, with the class. Account for any agreement or disagreement. **Do not test your predictions without your teacher's approval.**

## Soluble or Insoluble: General Solubility Guidelines

As you have seen, nearly all salts that contain the ammonium ion or an alkali metal are soluble. This observed pattern does not tell you how soluble these salts are, however. As well, it does not tell you whether ammonium chloride is more or less soluble than sodium chloride. Chemists rely on published data for this information. (See Figure 8.3.)

**Figure 8.3** Many web sites on the Internet provide chemical and physical data for tens of thousands of compounds. Print resources, such as *The CRC Handbook of Chemistry and Physics*, provide these data as well.

Many factors affect solubility. Thus, predicting solubility is neither straightforward nor simple. Nevertheless, the **general solubility guidelines** in Table 8.1 are a useful summary of ionic-compound interactions with water. To use Table 8.1, look for the anion (negative ion) of the compound in the columns of the table. Then look to see if the cation (positive ion) of the compound corresponds to high solubility or low solubility in water. For example, use the table to predict the solubility of KCl in water. The chloride anion, $Cl^-$, appears in the fourth column of the table. The potassium cation, $K^+$, is not listed in the slightly soluble row of this column. Therefore, you can predict that KCl is soluble in water.

You will be referring to the general solubility guidelines often in this chapter. They will help you identify salts that are soluble and insoluble in aqueous solutions. Always keep in mind, however, that water is a powerful solvent. Even an "insoluble" salt may dissolve enough to present a serious hazard if it is highly poisonous.

**Table 8.1** General Solubility Guidelines for Some Common Ionic Compounds in Water at 25°C

| Ion | Group 1 (IA) $NH_4^+$, $H^+$ ($H_3O^+$) | $ClO_3^-$ $NO_3^-$ $ClO_4^-$ | $Cl^-$ $Br^-$ $I^-$ | $CH_3COO^-$ | $SO_4^{2-}$ | $S^{2-}$ | $OH^-$ | $PO_4^{3-}$ $SO_3^{2-}$ $CO_3^{2-}$ |
|---|---|---|---|---|---|---|---|---|
| Very soluble (solubility greater than or equal to 0.1 mol/L) | all | all | most | most | most | Group 1 (IA) Group (2 IIA) $NH_4^+$ | Group 1 (IA) $NH_4^+$ $Sr^{2+}$ $Ba^{2+}$ $Tl^+$ | Group 1 (IA) $NH_4^+$ |
| Slightly soluble (solubility less than 0.1 mol/L) | none | none | $Ag^+$ $Hg^+$ $Tl^+$ | $Ag^+$ $Pb^{2+}$ $Hg^+$ $Cu^+$ | $Ca^{2+}$ $Sr^{2+}$ $Ba^{2+}$ $Ra^{2+}$ $Pb^{2+}$ $Ag^+$ | most | most | most |

## Web Link

www.mcgrawhill.ca/links/atlchemistry

Different references present solubility guidelines in different ways. During a 1 h "surf session" on the Internet, a student collected ten different versions. See how many versions you can find. Compare their similarities and differences. Which version(s) do you prefer, and why? To start your search, go to the web site above. Click on **Web Links** to find out where to go next.

### Practice Problems

1. Decide whether each of the following salts is soluble or insoluble in distilled water. Give reasons for your answer.
   (a) lead(II) chloride, $PbCl_2$ (a white crystalline powder used in paints)
   (b) zinc oxide, $ZnO$ (a white pigment used in paints, cosmetics, and calamine lotion)
   (c) silver acetate, $AgCH_3COO$ (a whitish powder that is used to help people quit smoking because of the bitter taste it produces)

2. Which of the following compounds are soluble in water? Explain your reasoning for each compound.
   (a) potassium nitrate, $KNO_3$ (used to manufacture gunpowder)
   (b) lithium carbonate, $Li_2CO_3$ (used to treat people who suffer from depression)
   (c) lead(II) oxide, $PbO_2$ (used to make crystal glass)

3. Which of the following compounds are insoluble in water?
   (a) calcium carbonate, $CaCO_3$ (present in marble and limestone)
   (b) magnesium sulfate, $MgSO_4$ (found in the hydrated salt, $MgSO_4 \cdot 7H_2O$, also known as Epsom salts; used for the relief of aching muscles and as a laxative)
   (c) aluminum phosphate, $AlPO_4$ (found in dental cements)

## Section Summary

In Chapter 7, you focussed mainly on physical changes that involve solutions. In the first section of this chapter, you observed that mixing aqueous solutions of ionic compounds may result in either a physical change (dissolving) or a chemical change (a reaction that forms a precipitate). Chemical changes that involve aqueous solutions, especially ionic reactions, are common. They occur in the environment, in your body, and in the bodies of other organisms. In the next section, you will look more closely at reactions that involve aqueous solutions. As well, you will learn how to represent these reactions using a special kind of chemical equation, called an ionic equation.

# Section Review

**1 (a)** Name the two factors that affect the solubility of an ionic compound in water.

**(b)** Briefly explain how each factor affects solubility.

**2** Which would you expect to be less soluble: sodium fluoride, NaF (used in toothpaste), or sodium iodide, NaI (added to table salt to prevent iodine deficiency in the diet)? Explain your answer.

**3** Which of the following compounds are soluble in water?

**(a)** calcium sulfide, CaS (used in skin products)

**(b)** iron(II) sulfate, $FeSO_4$ (used as a dietary supplement)

**(c)** magnesium chloride, $MgCl_2$ (used as a disinfectant and a food tenderizer)

**4** Which of the following compounds are insoluble in water? For each compound, relate its solubility to the use described.

**(a)** barium sulfate, $BaSO_4$ (can be used to obtain images of the stomach and intestines because it is opaque to X-rays)

**(b)** aluminum hydroxide, $Al(OH)_3$ (found in some antacid tablets)

**(c)** zinc carbonate, $ZnCO_3$ (used in suntan lotions)

**5** Calcium nitrate is used in fireworks. Silver nitrate turns dark when exposed to sunlight. When freshly made, both solutions are clear and colourless. Imagine that someone has prepared both solutions but has not labelled them. You do not want to wait for the silver nitrate solution to turn dark in order to identify the solutions. Name a chemical that can be used to precipitate a silver compound with the silver nitrate solution, but will produce no precipitate with the calcium nitrate solution. State the reason for your choice.

**6** Suppose that you discover four dropper bottles containing clear, colourless liquids in your school laboratory. The following four labels lie nearby:

- barium, $Ba^{2+}$
- chloride, $Cl^-$
- silver, $Ag^+$
- sulfate, $SO_4^{2-}$.

Unfortunately the labels have not been attached to the bottles. You decide to number the bottles 1, 2, 3, and 4. Then you mix the solutions in pairs. Three combinations give white precipitates: bottles 1 and 2, 1 and 4, and 2 and 3. Which ion does each bottle contain?

---

**Language LINK**

This poem is by an unknown author. It is not great literature. For decades, however, it has helped many students remember the solubility guidelines. Maybe it will do the same for you. If not, you can try writing your own poem!

Potassium, sodium, and ammonium salts
Whatever they may be,
Can always be relied upon
For solubility.

Every single sulfate
Is soluble it's said,
Except barium and calcium
And strontium and lead.

Most every chloride's soluble,
That's what we've always read,
Save silver, mercurous mercury,
And (slightly) chloride of lead.

When asked about the nitrates,
The answer's always clear;
They each and all are soluble,
That's all we want to hear.

Metallic bases won't dissolve,
That is, all but three;
Potassium, sodium, and ammonium
Dissolve quite readily.

But then you must remember,
You must surely not forget,
Calcium and barium
Dissolve a little bit.

Carbonates are insoluble,
It's lucky that it's so,
Or all our marble buildings
Would melt away like snow.

# 8.2 Reactions in Aqueous Solutions

**Section Preview/Outcomes**

In this section, you will

- **describe** combinations of aqueous solutions that result in the formation of precipitates
- **write** net ionic equations for aqueous ionic reactions
- **communicate** your understanding of the following terms: *spectator ions, total ionic equation, net ionic equation, qualitative analysis*

When you mix two aqueous ionic compounds together, there are two possible outcomes. Either the compounds will remain in solution without reacting, or one aqueous ionic compound will chemically react with the other. How can you predict which outcome will occur? Figure 8.4 shows what happens when an aqueous solution of lead(II) nitrate is added to an aqueous solution of potassium iodide. As you can see, a yellow solid—a precipitate—is forming. This is a double displacement reaction. Recall that a double displacement reaction is a chemical reaction that involves the exchange of ions to form two new compounds.

$$WX + YZ \rightarrow WZ + YX$$

In the reaction shown in Figure 8.4, for example, the lead cation is exchanged with the iodide anion.

You can usually recognize a double displacement reaction by observing one of these possible results:

- the formation of a precipitate (so that ions are removed from solution as an insoluble solid)
- the formation of a gas (so that ions are removed from solution in the form of a gaseous product)
- the formation of water (so that $H^+$ and $OH^-$ ions are removed from solution as water)

In this section, you will examine each of these results. At the same time, you will learn how to represent a double displacement reaction using a special kind of chemical equation: a net ionic equation.

**Figure 8.4** Lead(II) nitrate and potassium iodide are clear, colourless aqueous solutions. Mixing them causes a double displacement reaction. An insoluble yellow precipitate (lead(II) iodide) and a soluble salt (potassium nitrate) are produced.

## Double Displacement Reactions That Produce a Precipitate

A double displacement reaction that results in the formation of an insoluble substance is often called a *precipitation reaction*. Figure 8.4 is a clear example of a precipitation reaction. What if you did not have this photograph, however, and you were unable to do an experiment? Could you have predicted that mixing Pb(NO$_3$)$_{(aq)}$ and 2KI$_{(aq)}$ would result in an insoluble compound? Yes. When you are given (on paper) a pair of solutions to be mixed together, start by thinking about the exchange of ions that may occur. Then use the general solubility guidelines (Table 8.1) to predict which compounds, if any, are insoluble.

For example, consider lead(II) nitrate, Pb(NO$_3$)$_2$, and potassium iodide, KI. Lead(II) nitrate contains Pb$^{2+}$ cations and NO$_3^-$ anions. Potassium iodide contains K$^+$ cations and I$^-$ anions. Exchanging positive ions results in lead(II) iodide, PbI$_2$, and potassium nitrate, KNO$_3$. From the solubility guidelines, you know that all potassium salts and nitrates are soluble. Thus, potassium nitrate is soluble. The I$^-$ ion is listed in Table 8.1 as a soluble anion. The Pb$^{2+}$ ion is listed as a slightly soluble cation in the column that includes the I$^-$ ion. Thus, you can predict that lead(II) iodide is insoluble. It will form a precipitate when the solutions are mixed. The balanced chemical equation for this reaction is

$$Pb(NO_3)_{2(aq)} + 2KI_{(aq)} \rightarrow 2KNO_{3(aq)} + PbI_{2(s)}$$

### Sample Problem
### Predicting the Formation of a Precipitate

**Problem**

Which of the following pairs of aqueous solutions produce a precipitate when mixed together? Write the balanced chemical equation if you predict a precipitate. Write "NR" if you predict that no reaction takes place.

(a) potassium carbonate and copper(II) sulfate

(b) ammonium chloride and zinc sulfate

**What Is Required?**

You need to predict whether or not each pair of aqueous solutions forms an insoluble product (a precipitate). If it does, you need to write a balanced chemical equation.

**What Is Given?**

You know the names of the compounds in each solution.

**Plan Your Strategy**

Start by identifying the ions in each pair of compounds. Then exchange the positive ions in the two compounds. Compare the resulting compounds against the solubility guidelines, and make your prediction.

*Continued ...*

*Continued ...*

### Act on Your Strategy

**(a)** Potassium carbonate contains K⁺ and CO₃²⁻ ions. Copper(II) sulfate contains Cu²⁺ and SO₄²⁻ ions. Exchanging positive ions results in potassium sulfate, K₂SO₄, and copper(II) carbonate, CuCO₃.

All potassium salts are soluble, so these ions remain dissolved in solution.

The carbonate anion is listed in the last column of Table 8.1. The copper(II) cation is not in Group 1 (IA), so copper(II) carbonate should be insoluble. Thus, you can predict that a precipitate forms. The balanced chemical equation for this reaction is

K₂CO₃₍aq₎ + CuSO₄₍aq₎ → K₂SO₄₍aq₎ + CuCO₃₍s₎

**(b)** Ammonium chloride contains NH₄⁺ and Cl⁻ ions. Zinc sulfate consists of Zn²⁺ and SO₄²⁻ ions. Exchanging positive ions results in ammonium sulfate, (NH₄)₂SO₄, and zinc chloride, ZnCl₂.

Since all ammonium salts are soluble, the ammonium sulfate stays dissolved in solution.

Zinc chloride consists of a soluble anion. The cation does not appear as an exception, so zinc chloride should be soluble. Thus, you can predict that no precipitate forms.

NH₄Cl₍aq₎ + ZnSO₄₍aq₎ → NR

### Check Your Solution

An experiment is always the best way to check a prediction. If possible, obtain samples of these solutions from your teacher, and mix them together.

## Practice Problems

**4.** Predict the result of mixing each pair of aqueous solutions. Write a balanced chemical equation if you predict that a precipitate forms. Write "NR" if you predict that no reaction takes place.

(a) sodium sulfide and iron(II) sulfate
(b) sodium hydroxide and barium nitrate
(c) cesium phosphate and calcium bromide
(d) sodium carbonate and sulfuric acid
(e) sodium nitrate and copper(II) sulfate
(f) ammonium iodide and silver nitrate
(g) potassium carbonate and iron(II) nitrate
(h) aluminum nitrate and sodium phosphate
(i) potassium chloride and iron(II) nitrate
(j) ammonium sulfate and barium chloride
(k) sodium sulfide and nickel(II) sulfate
(l) lead(II) nitrate and potassium bromide

## Double Displacement Reactions That Produce a Gas

Double displacement reactions are responsible for producing a number of gases. (See Figure 8.5.) These gases include
- hydrogen
- hydrogen sulfide (a poisonous gas that smells like rotten eggs)
- sulfur dioxide (a reactant in forming acid rain)
- carbon dioxide
- ammonia

### A Reaction that Produces Hydrogen Gas

The alkali metals form bonds with hydrogen to produce compounds called hydrides. Hydrides react readily with water to produce hydrogen gas. Examine the following equation for the reaction of lithium hydride, LiH, with water. If you have difficulty visualizing the ion exchange that takes place, rewrite the equation for yourself using HOH instead of $H_2O$.

$$LiH_{(s)} + H_2O_{(\ell)} \rightarrow LiOH_{(aq)} + H_{2(g)}$$

**Figure 8.5** You can easily identify limestone and marble by their reaction with hydrochloric acid. The gas that is produced by this double displacement reaction is carbon dioxide.

### A Reaction that Produces Hydrogen Sulfide Gas

Sulfides react with certain acids, such as hydrochloric acid, to produce hydrogen sulfide gas.

$$K_2S_{(aq)} + 2HCl_{(aq)} \rightarrow 2KCl_{(aq)} + H_2S_{(g)}$$

### A Reaction that Produces Sulfur Dioxide Gas

Some reactions produce a compound that, afterward, decomposes into a gas and water. Sodium sulfite is used in photography as a preservative. It reacts with hydrochloric acid to form sulfurous acid. The sulfurous acid then breaks down into sulfur dioxide gas and water. The net reaction is the sum of both changes. If the same compound appears on both sides of an equation (as sulfurous acid, $H_2SO_3$, does here), it can be eliminated. This is just like eliminating terms from an equation in mathematics.

$$Na_2SO_{3(aq)} + 2HCl_{(aq)} \rightarrow 2NaCl_{(aq)} + \cancel{H_2SO_{3(aq)}}$$
$$\cancel{H_2SO_{3(aq)}} \rightarrow SO_{2(g)} + H_2O_{(\ell)}$$

Therefore, the net reaction is

$$Na_2SO_{3(aq)} + 2HCl_{(aq)} \rightarrow 2NaCl_{(aq)} + SO_{2(g)} + H_2O_{(\ell)}$$

### A Reaction that Produces Carbon Dioxide Gas

The reaction of a carbonate with an acid produces carbonic acid. Carbonic acid decomposes rapidly into carbon dioxide and water.

$$Na_2CO_{3(aq)} + 2HCl_{(aq)} \rightarrow 2NaCl_{(aq)} + \cancel{H_2CO_{3(aq)}}$$
$$\cancel{H_2CO_{3(aq)}} \rightarrow CO_{2(g)} + H_2O_{(\ell)}$$

The net reaction is

$$Na_2CO_{3(aq)} + 2HCl_{(aq)} \rightarrow 2NaCl_{(aq)} + CO_{2(g)} + H_2O_{(\ell)}$$

### A Reaction that Produces Ammonia Gas

Ammonia gas is very soluble in water. You can detect it easily, however, by its sharp, pungent smell. Ammonia gas can be prepared by the reaction of an ammonium salt with a base.

$$NH_4Cl_{(aq)} + NaOH_{(aq)} \rightarrow NaCl_{(aq)} + NH_{3(aq)} + H_2O_{(\ell)}$$

## Double Displacement Reactions That Produce Water

The neutralization reaction between an acid and a base is a very important double displacement reaction. In a neutralization reaction, water results when an H⁺ ion from the acid bonds with an OH⁻ ion from the base.

$$H_2SO_{4(aq)} + 2NaOH_{(aq)} \rightarrow Na_2SO_{4(aq)} + 2H_2O_{(\ell)}$$

Most metal oxides are bases. Therefore, a metal oxide will react with an acid in a neutralization reaction to form a salt and water.

$$2HNO_{3(aq)} + MgO_{(s)} \rightarrow Mg(NO_3)_{2(aq)} + H_2O_{(\ell)}$$

Non-metal oxides are acidic. Therefore, a non-metal oxide will react with a base. This type of reaction is used in the space shuttle. Cabin air is circulated through canisters of lithium hydroxide (a base) to remove the carbon dioxide before it can reach dangerous levels.

$$2LiOH_{(s)} + CO_{2(g)} \rightarrow Li_2CO_{3(aq)} + H_2O_{(\ell)}$$

## Representing Aqueous Ionic Reactions with Net Ionic Equations

Mixing a solution that contains silver ions with a solution that contains chloride ions produces a white precipitate of silver chloride. There must have been other ions present in each solution, as well. You know this because it is impossible to have a solution of just a cation or just an anion. Perhaps the solution that contained silver ions was prepared using silver nitrate or silver acetate. Similarly, the solution that contained chloride ions might have been prepared by dissolving NaCl in water, or perhaps NH₄Cl or another soluble chloride. Any solution that contains Ag⁺$_{(aq)}$ will react with any other solution that contains Cl⁻$_{(aq)}$ to form a precipitate of AgCl$_{(s)}$. The other ions in the solutions are not important to the net result. Like spectators at a sports event, with no net impact on the result of the game, these ions are like passive onlookers. They are called **spectator ions**.

The reaction between silver nitrate and sodium chloride can be represented by the following chemical equation:

$$AgNO_{3(aq)} + NaCl_{(aq)} \rightarrow NaNO_{3(aq)} + AgCl_{(s)}$$

This equation does not show the change that occurs, however. It shows the reactants and products as intact compounds. In reality, soluble ionic compounds *dissociate* into their respective ions in solution. So chemists often use a **total ionic equation** to show the dissociated ions of the soluble ionic compounds.

$$Ag^+_{(aq)} + NO_3^-_{(aq)} + Na^+_{(aq)} + Cl^-_{(aq)} \rightarrow Na^+_{(aq)} + NO_3^-_{(aq)} + AgCl_{(s)}$$

Notice that the precipitate, AgCl, is still written as an ionic formula. This makes sense because precipitates are insoluble, so they do not dissociate into ions. Also notice that the spectator ions appear on both sides of the equation. Here is the total ionic equation again, with slashes through the spectator ions.

$$Ag^+_{(aq)} + \cancel{NO_3^-_{(aq)}} + \cancel{Na^+_{(aq)}} + Cl^-_{(aq)} \rightarrow \cancel{Na^+_{(aq)}} + \cancel{NO_3^-_{(aq)}} + AgCl_{(s)}$$

If you eliminate the spectator ions, the equation becomes

$$Ag^+_{(aq)} + Cl^-_{(aq)} \rightarrow AgCl_{(s)}$$

An ionic equation that is written this way, without the spectator ions, is called a **net ionic equation**. Before you try writing your own net ionic equations, examine the guidelines in Table 8.2 below.

**Table 8.2** Guidelines for Writing a Net Ionic Equation

1. Include only ions and compounds that have reacted. Do not include spectator ions.

2. Write the soluble ionic compounds as ions. For example, write $NH_4^+{}_{(aq)}$ and $Cl^-{}_{(aq)}$, instead of $NH_4Cl_{(aq)}$.

3. Write insoluble ionic compounds as formulas, not ions. For example, zinc sulfide is insoluble, so you write it as $ZnS_{(s)}$, not $Zn^{2+}$ and $S^{2-}$.

4. Since covalent compounds do not dissociate in aqueous solution, write their molecular formulas. Water is a common example, because it dissociates only very slightly into ions. When a reaction involves a gas, always include the gas in the net ionic equation.

5. Write strong acids (discussed in Chapter 14) in their ionic form. There are six strong acids:
    - hydrochloric acid (write as $H^+{}_{(aq)}$ and $Cl^-{}_{(aq)}$, not $HCl_{(aq)}$)
    - hydrobromic acid (write as $H^+{}_{(aq)}$ and $Br^-{}_{(aq)}$)
    - hydroiodic acid (write as $H^+{}_{(aq)}$ and $I^-{}_{(aq)}$)
    - sulfuric acid (write as $H^+{}_{(aq)}$ and $SO_4^{2-}{}_{(aq)}$)
    - nitric acid (write as $H^+{}_{(aq)}$ and $NO_3^-{}_{(aq)}$)
    - perchloric acid (write as $H^+{}_{(aq)}$ and $ClO_4^-{}_{(aq)}$)

    All other acids are weak and form few ions. Therefore, write them in their molecular form.

6. Finally, check that the net ionic equation is balanced for charges as well as for atoms.

### Sample Problem

#### Writing Net Ionic Equations

**Problem**

A chemical reaction occurs when the following aqueous solutions are mixed: sodium sulfide and iron(II) sulfate. Identify the spectator ions. Then write the balanced net ionic equation.

**What Is Required?**

You need to identify the spectator ions and write a balanced net ionic equation for the reaction between sodium sulfide and iron(II) sulfate.

**What Is Given?**

You know the chemical names of the compounds.

*Continued ...*

> **Continued ...**
>
> ### Plan Your Strategy
>
> **Step 1** Start by writing the chemical formulas of the given compounds.
>
> **Step 2** Then write the complete chemical equation for the reaction, using your experience in predicting the formation of a precipitate.
>
> **Step 3** Once you have the chemical equation, you can replace the chemical formulas of the soluble ionic compounds with their dissociated ions.
>
> **Step 4** This will give you the total ionic equation. Next you can identify the spectator ions (the ions that appear on both sides of the equation).
>
> **Step 5** Finally, by rewriting the total ionic equation *without* the spectator ions, you will have the net ionic equation.
>
> ### Act on Your Strategy
>
> **Steps 1 and 2** The chemical equation for the reaction is
> $$Na_2S_{(aq)} + FeSO_{4(aq)} \rightarrow Na_2SO_{4(aq)} + FeS_{(s)}$$
>
> **Step 3** The total ionic equation is
> $$\cancel{2Na^+_{(aq)}} + S^{2-}_{(aq)} + Fe^{2+}_{(aq)} + \cancel{SO_4^{2-}_{(aq)}} \rightarrow \cancel{2Na^+_{(aq)}} + \cancel{SO_4^{2-}_{(aq)}} + FeS_{(s)}$$
>
> **Step 4** Therefore, the spectator ions are $Na^+_{(aq)}$ and $SO_4^{2-}_{(aq)}$.
>
> **Step 5** The net ionic equation is
> $$Fe^{2+}_{(aq)} + S^{2-}_{(aq)} \rightarrow FeS_{(s)}$$
>
> ### Check Your Solution
>
> Take a final look at your net ionic equation to make sure that no ions are on both sides of the equation.

### Practice Problems

> 5. Mixing each pair of aqueous solutions results in a chemical reaction. Identify the spectator ions. Then write the balanced net ionic equation.
>
>    (a) sodium carbonate and hydrochloric acid
>
>    (b) sulfuric acid and sodium hydroxide
>
> 6. Identify the spectator ions for the reaction that takes place when each pair of aqueous solutions is mixed. Then write the balanced net ionic equation.
>
>    (a) ammonium phosphate and zinc sulfate
>
>    (b) lithium carbonate and nitric acid
>
>    (c) sulfuric acid and barium hydroxide

## Identifying Ions in Aqueous Solution

Suppose that you have a sample of water. You want to know what, if any, ions are dissolved in it. Today technological devices, such as the mass spectrometer, make this investigative work fairly simple. Before such devices, however, chemists relied on *wet chemical techniques*: experimental tests, such as submitting a sample to a series of double displacement reactions. Chemists still use wet chemical techniques. With each reaction, insoluble compounds precipitate out of the solution. (See Figure 8.6.) This enables the chemist to determine, eventually, the identity of one or several ions in the solution. This ion-identification process is an example of **qualitative analysis**.

Chemists use a range of techniques for qualitative analysis. For example, the colour of an aqueous solution can help to identify one of the ions that it contains. Examine Table 8.3. However, the intensity of ion colour varies with its concentration in the solution. Also keep in mind that many ions are colourless in aqueous solution. For example, the cations of elements from Groups 1 (IA) and 2 (IIA), as well as aluminum, zinc, and most anions, are colourless. So there are limits to the inferences you can make if you rely on solution colour alone.

Another qualitative analysis technique is a flame test. A dissolved ionic compound is placed in a flame. Table 8.4 lists the flame colours associated with several ions. Notice that all the ions are metallic. The flame test is only useful for identifying metallic ions in aqueous solution.

Qualitative analysis challenges a chemist's creative imagination and chemical understanding. Discover this for yourself in Investigation 8-B.

**Table 8.3** The Colour of Some Common Ions in Aqueous Solution

| | Ions | Symbol | Colour |
|---|---|---|---|
| Cations | chromium (II) <br> copper(II) | $Cr^{2+}$ <br> $Cu^{2+}$ | blue |
| | chromium(III) <br> copper(I) <br> iron(II) <br> nickel(II) | $Cr^{3+}$ <br> $Cu^+$ <br> $Fe^{2+}$ <br> $Ni^{2+}$ | green |
| | iron(III) | $Fe^{3+}$ | pale yellow |
| | cobalt(II) <br> manganese(II) | $Co^{2+}$ <br> $Mn^{2+}$ | pink |
| Anions | chromate | $CrO_4^{2-}$ | yellow |
| | dichromate | $Cr_2O_7^{2-}$ | orange |
| | permanganate | $MnO_4^-$ | purple |

**Table 8.4** The Flame Colour of Selected Metallic Ions

| Ion | Symbol | Colour |
|---|---|---|
| lithium | $Li^+$ | red |
| sodium | $Na^+$ | yellow |
| potassium | $K^+$ | violet |
| cesium | $Cs^+$ | violet |
| calcium | $Ca^{2+}$ | red |
| strontium | $Sr^{2+}$ | red |
| barium | $Ba^{2+}$ | yellowish-green |
| copper | $Cu^{2+}$ | bluish-green |
| boron | $B^{2+}$ | green |
| lead | $Pb^{2+}$ | bluish-white |

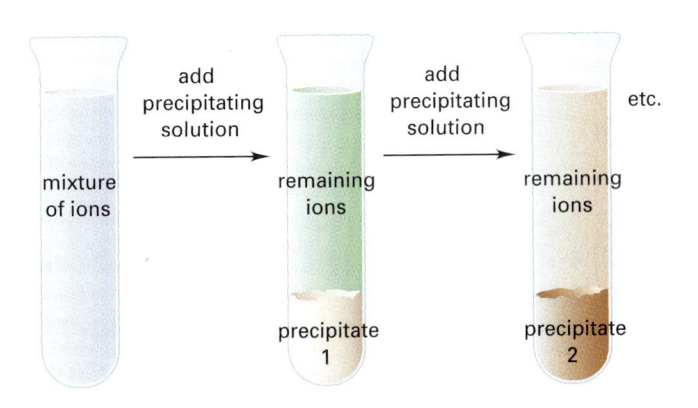

**Figure 8.6** This illustration shows the basic idea behind a qualitative analysis for identifying ions in an aqueous solution. At each stage, the resulting precipitate is removed.

# Investigation 8-B

**SKILL FOCUS**
- Predicting
- Performing and recording
- Analyzing and interpreting

## Qualitative Analysis

In this investigation, you will apply your knowledge of chemical reactions and the general solubility guidelines to identify unknown ions.

### Question
How can you identify ions in solution?

### Predictions
Read the entire Procedure. Can you predict the results of any steps? Write your predictions in your notebook. Justify each prediction.

### Materials
**Part 1**
12-well or 24-well plate, or spot plate
toothpicks
cotton swabs or coarse paper towelling
unknowns: 4 dropper bottles (labelled A, B, C, and D) of solutions that include $Na^+_{(aq)}$, $Ag^+_{(aq)}$, $Ca^{2+}_{(aq)}$, and $Cu^{2+}_{(aq)}$
reactants: 2 labelled dropper bottles, containing dilute $HCl_{(aq)}$ and dilute $H_2SO_{4(aq)}$

**Part 2**
cotton swabs
Bunsen burner
heat-resistant pad
unknowns: 4 dropper bottles, containing the same unknowns that were used in Part 1
reactants: 4 labelled dropper bottles, containing $Na^+_{(aq)}$, $Ag^+_{(aq)}$, $Ca^{2+}_{(aq)}$, and $Cu^{2+}_{(aq)}$

**Part 3**
12-well or 24-well plate, or spot plate
toothpicks
cotton swabs
unknowns: 3 dropper bottles (labelled X, Y, and Z), containing solutions of $SO_4^{2-}_{(aq)}$, $CO_3^{2-}_{(aq)}$, and $I^-_{(aq)}$
reactants: 3 labelled dropper bottles, containing $Ba^{2+}_{(aq)}$, $Ag^+_{(aq)}$, and $HCl_{(aq)}$

### Safety Precautions

- Be careful not to contaminate the dropper bottles. The tip of a dropper should not make contact with either the plate or another solution. Put the cap back on the bottle immediately after use.
- Hydrochloric acid and sulfuric acid are corrosive. Wash any spills on your skin with plenty of cool water. Inform your teacher immediately.
- Part 2 of this investigation requires an open flame. Tie back long hair, and confine any loose clothing.

### Procedure

**Part 1 Using Acids to Identify Cations**

1. Read steps 2 and 3 below. Design a suitable table for recording your observations.

2. Place one or two drops of each unknown solution into four different wells or spots. Add one or two drops of hydrochloric acid to each unknown. Record your observations.

3. Repeat step 2. This time, test each unknown solution with one or two drops of sulfuric acid. Record your observations.

4. Answer Analysis questions 1 to 5.

**Part 2 Using Flame Tests to Identify Cations**

Note: Your teacher may demonstrate this part or provide you with an alternative version.

1. Design tables to record your observations.

2. Observe the appearance of each known solution. Record your observations. Repeat for each unknown solution. Some cations have a characteristic colour. (Refer to Table 8.4.) If you think that you can identify one of the unknowns, record your identification.

296  MHR • Unit 3 Solutions

3. Flame tests can identify some cations. Set up the Bunsen burner and heat-resistant pad. Light the burner. Adjust the air supply to produce a hot flame with a blue cone.

4. Place a few drops of solution containing Na$^+_{(aq)}$ on one end of a cotton swab.

   **CAUTION** Carefully hold the saturated tip so it is just in the Bunsen burner flame, near the blue cone. You may need to hold it in this position for as long as 30 s to allow the solution to vaporize and mix with the flame. Record the colour of the flame.

5. Not all cations give colour to a flame. The sodium ion *does* give a distinctive colour to a flame, however. It is often present in solutions as a contaminant. For a control, repeat step 4 with water and record your observations. You can use the other end of the swab for a second test. Dispose of used swabs in the container your teacher provides.

6. Repeat the flame test for each of the other known solutions. Then test each of the unknown solutions.

7. Answer Analysis question 6.

### Part 3 Identifying Anions

1. Place one or two drops of each unknown solution into three different wells or spots. Add one or two drops of Ba$^{2+}_{(aq)}$ to each unknown solution. Stir with a toothpick. Record your observations.

2. Add a drop of hydrochloric acid to any well or spot where you observed a precipitate in step 1. Stir and record your observations.

3. Repeat step 1, adding one or two drops of Ag$^+_{(aq)}$ to each unknown solution. Record the colour of any precipitate that forms.

4. Answer Analysis questions 7 to 9.

## Analysis

1. (a) Which of the cations you tested should form a precipitate with hydrochloric acid? Write the net ionic equation.

   (b) Did your results support your predictions? Explain.

2. (a) Which cation(s) should form a precipitate when tested with sulfuric acid? Write the net ionic equation.

   (b) Did your results support your predictions? Explain.

3. Which cation(s) should form a soluble chloride and a soluble sulfate?

4. Which cation has a solution that is not colourless?

5. Based on your analysis so far, tentatively identify each unknown solution.

6. Use your observations of the flame tests to confirm or refute the identifications you made in question 5. If you are not sure, check your observations and analysis with other students. If necessary, repeat some of your tests.

7. Which anion(s) should form a precipitate with Ba$^{2+}$? Write the net ionic equation.

8. Which precipitate should react when hydrochloric acid is added? Give reasons for your prediction.

9. Tentatively identify each anion. Check your observations against the results you obtained when you added hydrochloric acid. Were they what you expected? If not, check your observations and analysis with other students. If necessary, repeat some of your tests.

## Conclusion

10. Identify the unknown cations and anions in this investigation. Explain why you do, or do not, have confidence in your decisions. What could you do to be more confident?

### Section Summary

Qualitative analysis helps you identify ions that may be present in a solution. It does not, however, tell you how much of these ions are present. In other words, it does not provide any quantitative information about the quantity or concentration of ions in solution. In the next section, you will find out how to calculate this quantitative information, using techniques you learned in Unit 1.

## Section Review

**1** Briefly compare the relationships among a chemical formula, a total ionic equation, and a net ionic equation. Use sentences or a graphic organizer.

**2** Write a net ionic equation for each double displacement reaction in aqueous solution.
   (a) tin(II) chloride with potassium phosphate
   (b) nickel(II) chloride with sodium carbonate
   (c) chromium(III) sulfate with ammonium sulfide

**3** For each reaction in question 2, identify the spectator ions.

**4** Would you expect a qualitative analysis of a solution to give you the amount of each ion present? Explain why or why not.

**5** A solution of limewater, $Ca(OH)_{2(aq)}$, is basic. It is used to test for the presence of carbon dioxide. Carbon dioxide is weakly acidic and turns limewater milky. Use a chemical equation to explain what happens during the test. What type of reaction occurs?

**6** State the name and formula of the precipitate that forms when aqueous solutions of copper(II) sulfate and sodium carbonate are mixed. Write the net ionic equation for the reaction. Identify the spectator ions.

**7** All the solutions in this photograph have the same concentration: 0.1 mol/L. Use Table 8.3 to infer which ion causes the colour in each solution. How much confidence do you have in your inferences? What could you do to increase your confidence?

298 MHR • Unit 3 Solutions

# Stoichiometry in Solution Chemistry

## 8.3

**Section Preview/Outcomes**

In this section, you will

- **write** net ionic equations for aqueous ionic reactions
- **solve** stoichiometry problems that involve aqueous solutions
- **communicate** your understanding of the following terms: *solution stoichiometry, dissociation equation*

Recall that stoichiometry involves calculating the amounts of reactants and products in chemical reactions. **Solution stoichiometry** involves calculating quantities of substances in chemical reactions that take place in solutions.

You will use solution stoichiometry in the remainder of this chapter, as well as in Units 5 and 6. Therefore, it is important that you understand how to perform mole-ratio calculations that involve solutions, and can determine whether or not your answers are reasonable. To begin, you first need to make sure that you know how to write and interpret dissociation equations for dissolved substances.

### Writing and Interpreting Dissociation Equations

As you know, when an ionic compound dissolves in water, it dissociates, separating into individual hydrated ions. For example, sodium chloride, NaCl, dissociates in water to form $Na^+_{(aq)}$ and $Cl^-_{(aq)}$. You can represent this *physical* process of dissolving with a dissociation equation, shown below. A **dissociation equation** is a balanced chemical equation that shows all the ions that are produced when an ionic compound dissolves. Figure 8.7 shows a molecular view of the same process.

$$NaCl_{(s)} \rightarrow Na^+_{(aq)} + Cl^-_{(aq)}$$

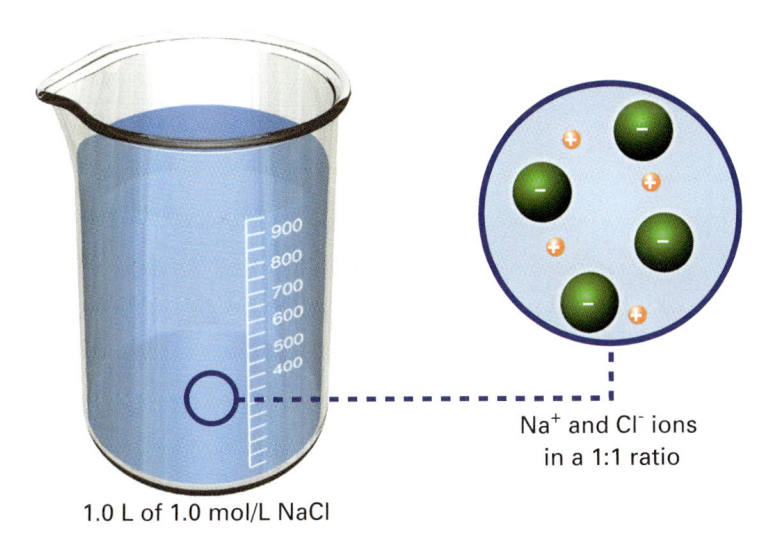

**Figure 8.7** Sodium chloride is an ionic compound in which sodium cations bond with chloride anions in a 1:1 ratio. In 1.0 L of a 1.0 mol/L solution of NaCl, 1.0 mol $NaCl_{(s)}$ dissociates to produce 1.0 mol/L $Na^+_{(aq)}$ ions and 1.0/L mol $Cl^-_{(aq)}$ ions.

Dissociation equations are helpful because you can use them to determine the concentration of the individual ions in solution. How is this different from the concentration of the solution? Recall that the concentration of a solution is the number of moles of *solute* in one litre of solution. For example, to prepare a 1.0 mol/L solution of sodium chloride, you dissolve 1.0 mol of solid sodium chloride in a sufficient amount water to make 1.0 L of solution. The solution contains 1.0 mol/L $Na^+$ ions and 1.0 mol/L $Cl^-$ ions. It does not contain 1.0 mol/L NaCl formula units.

Figure 8.8 shows the dissociation of $MgCl_2$. Examine the illustration, along with the dissociation equation below, to make sure you understand why a solution of 1.0 mol/L $MgCl_2$ contains 1.0 mol/L $Mg^{2+}$ ions and 2.0 mol/L $Cl^-$ ions. Then use the Practice Problems that follow to reinforce your understanding.

$$MgCl_{2(s)} \rightarrow Mg^{2+}_{(aq)} + 2Cl^-_{(aq)}$$

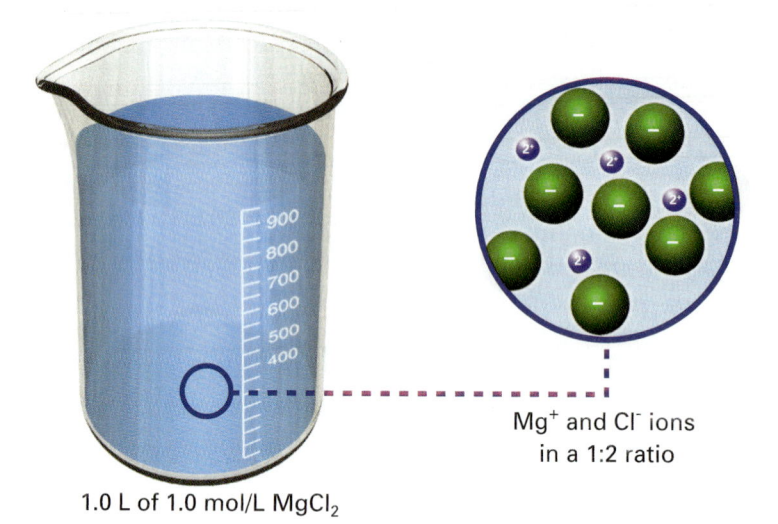

**Figure 8.8** Magnesium chloride is an ionic compound in which magnesium cations bond with chloride anions in a 1:2 ratio. In 1.0 L of a 1.0 mol/L solution of $MgCl_2$, 1.0 mol $MgCl_{2(s)}$ dissociates to produce 1.0 mol/L $Mg^{2+}_{(aq)}$ ions and 2.0 mol/L $Cl^-_{(aq)}$ ions.

1.0 L of 1.0 mol/L $MgCl_2$

$Mg^+$ and $Cl^-$ ions in a 1:2 ratio

### Practice Problems

7. Write dissociation equations for the following ionic compounds in water, and state the concentration of the dissociated ions in each aqueous solution. Assume that each compound dissolves completely.
   (a) 1 mol/L lithium fluoride
   (b) 0.5 mol/L sodium bromide
   (c) 1.5 mol/L sodium sulfide
   (d) 1 mol/L calcium nitrate

8. Write dissociation equations for the following compounds, and state the concentration of the dissociated ions in each aqueous solution. Assume that each compound dissolves completely.
   (a) 1.75 mol/L ammonium chloride
   (b) 0.25 mol/L ammonium sulfide
   (c) 6 mol/L ammonium sulfate
   (d) 2 mol/L silver nitrate

9. 18.2 g of $CaCl_2$ is dissolved in 300.0 mL of water to make a clear, colourless solution. Give the dissociation equation for this physical change, and state the concentration of the dissociated ions.

10. Cobalt(II) chloride is a pale blue compound that dissolves easily in water. It is often used to detect excess humidity in air, because the compound dissolves and turns pink.
    (a) 14.2 g of cobalt(II) chloride is dissolved in 1.50 L of water. Determine the concentration of $CoCl_{2(aq)}$.
    (b) Write the dissociation equation for this change.
    (c) Determine the concentration of all ions in the final solution.

In the preceding Practice Problems, you considered concentrations involving only a single solution. How do the concentrations of an ion change if two solutions containing that ion are mixed together? Work through the next Sample Problem, and then try the Practice Problems that follow it.

## Sample Problem
### The Concentration of Ions

**Problem**

Calculate the concentration (in mol/L) of chloride ions in each solution.

(a) 19.8 g of potassium chloride dissolved in 100 mL of solution

(b) 26.5 g of calcium chloride dissolved in 150 mL of solution

(c) a mixture of the two solutions in parts (a) and (b), assuming that the volumes are additive

**What Is Required?**

(a) and (b) You need to find the concentration (in mol/L) of chloride ions in two different solutions.

(c) You need to find the concentration of chloride ions when the two solutions are mixed.

**What Is Given?**

You know that 19.8 g of potassium chloride is dissolved in 100 mL of solution. You also know that 26.5 g of calcium chloride is dissolved in 150 mL of solution.

**Plan Your Strategy**

(a) and (b) For each solution, determine the molar mass. Find the amount (in mol) using the mass and the molar mass. Write equations for the dissociation of the substance. (That is, write the total ionic equation.) Use the coefficients in the dissociation equation to determine the amount (in mol) of chloride ions present. Calculate the concentration (in mol/L) of chloride ions from the amount and volume of the solution.

(c) Add the amounts of chloride ions in the two solutions to find the total. Add the volumes of the solutions to find the total volume. Calculate the concentration of chloride ions (in mol/L) using the total amount (in mol) divided by the total volume (in L).

*Continued...*

**Continued ...**

**Act on Your Strategy**

(a) and (b)

| Solution | KCl | CaCl$_2$ |
|---|---|---|
| **Molar mass** | 39.10 + 35.45 = 74.55 g | 40.08 + (2 × 35.45) = 110.98 g |
| **Amount (mol)** | $19.8 \text{ g} \times \dfrac{1 \text{ mol}}{74.55 \text{ g}} = 0.266$ mol | $26.5 \text{ g} \times \dfrac{1 \text{ mol}}{110.98 \text{ g}} = 0.239$ mol |
| **Dissociation equation** | KCl$_s$ → K$^+_{(aq)}$ + Cl$^-_{(aq)}$ | CaCl$_{2(s)}$ → Ca$^{2+}_{(aq)}$ + 2Cl$^-_{(aq)}$ |
| **Amount of Cl$^-$** | $0.266 \text{ mol KCl} \times \dfrac{1 \text{ mol Cl}^-}{1 \text{ mol KCl}} = 0.266$ mol | $0.239 \text{ mol CaCl}_2 \times \dfrac{2 \text{ mol Cl}^-}{1 \text{ mol CaCl}_2} = 0.478$ mol |
| **Concentration of Cl$^-$** | $\dfrac{0.266 \text{ mol}}{0.100 \text{ L}} = 2.66$ mol/L | $\dfrac{0.478 \text{ mol}}{0.150 \text{ L}} = 3.19$ mol/L |

The concentration of chloride ions when 19.8 g of potassium chloride is dissolved in 100 mL of solution is 2.66 mol/L. The concentration of chloride ions when 26.5 g of calcium chloride is dissolved in 150 mL of solution is 3.19 mol/L.

(c) Total amount of Cl$^-_{(aq)}$ = 0.266 + 0.478 mol
  = 0.744 mol

Total volume of solution = 0.100 + 0.150 L
  = 0.250 L

Total concentration of Cl$^-_{(aq)}$ = $\dfrac{0.744 \text{ mol}}{0.250 \text{ L}}$
  = 2.98 mol/L

The concentration of chloride ions when the solutions are mixed is 2.98 mol/L.

**Check Your Solution**

The units for amount and concentration are correct. The answers appear to be reasonable. When the solutions are mixed, the concentration of the chloride ions is not a simple average of the concentrations of the two solutions. Why? The volumes of the two solutions were different.

### Practice Problems

11. Equal volumes of 0.150 mol/L lithium nitrate and 0.170 mol/L sodium nitrate are mixed together. Determine the concentration of nitrate ions in the mixture.

12. Equal volumes of 1.25 mol/L magnesium sulfate and 0.130 mol/L ammonium sulfate are mixed together. What is the concentration of sulfate ions in the mixture?

13. A solution is prepared by adding 25.0 mL of 0.025 mol/L sodium chloride with 35.0 mL of 0.050 mol/L barium chloride. What is the concentration of chloride ions in the resulting solution, assuming that the volumes are additive?

14. What is the concentration of nitrate ions in a solution that results by mixing 350 mL of 0.275 mol/L potassium nitrate with 475 mL of 0.345 mol/L magnesium nitrate? Assume the volumes are additive.

So far, you have been working with solution stoichiometry involving physical changes. Chemists also use solution stoichiometry extensively for chemical changes that occur in solutions, just as gravimetric stoichiometry is used for known masses in chemical reactions. The key difference between solution stoichiometry and gravimetric stoichiometry is your use of the equation $n = CV$ to calculate the amount (in mol) of solute that reacts.

## Sample Problem

### Finding the Minimum Volume to Precipitate

**Problem**

Aqueous solutions that contain silver ions are usually treated with chloride ions to recover silver chloride. What is the minimum volume of 0.25 mol/L magnesium chloride, $MgCl_{2(aq)}$, needed to precipitate all the silver ions in 60 mL of 0.30 mol/L silver nitrate, $AgNO_{3(aq)}$? Assume that silver chloride is completely insoluble in water.

**What Is Required?**

You need to find the minimum volume of magnesium chloride that will precipitate all the silver ions.

**What Is Given?**

You know the volumes and concentrations of the silver nitrate (volume = 60 mL; concentration = 0.30 mol/L). The concentration of the magnesium chloride solution is 0.25 mol/L.

**Plan Your Strategy**

Find the amount (in mol) of silver nitrate from the volume and concentration of solution. Write a balanced chemical equation for the reaction. Use mole ratios from the coefficients in the equation to determine the amount (in mol) of magnesium chloride that is needed. Use the amount (in mol) of magnesium chloride and the concentration of solution to find the volume that is needed.

**Act on Your Strategy**

Amount $AgNO_3$ = 0.060 L × 0.30 mol/L
= 0.018 mol

$MgCl_{2(aq)} + 2AgNO_{3(aq)} \rightarrow 2AgCl_{(s)} + Mg(NO_3)_{2(aq)}$

The mole ratio of $MgCl_2$ to $AgNO_3$ is 1:2.

$n$ mol $MgCl_2$ = 0.018 mol $AgNO_3$ × $\dfrac{1 \text{ mol } MgCl_2}{2 \text{ mol } AgNO_3}$

= 0.0090 mol

Volume of 0.25 mol/L $MgCl_2$ needed = $\dfrac{0.0090 \text{ mol}}{0.25 \text{ mol/L}}$

= 0.036 L

The minimum volume of 0.25 mol/L magnesium chloride that is needed is 36 mL.

**Check Your Solution**

The answer is in millilitres, an appropriate unit of volume. The amount appears to be reasonable.

## Practice Problems

**15.** Food manufacturers sometimes add calcium acetate to puddings and sweet sauces as a thickening agent. What volume of 0.500 mol/L calcium acetate, $Ca(CH_3COO)_{2(aq)}$, contains 0.300 mol of acetate ions?

**16.** In a neutralization reaction, hydrochloric acid reacts with sodium hydroxide to produce sodium chloride and water:

$$HCl_{(aq)} + NaOH_{(aq)} \rightarrow NaCl_{(aq)} + H_2O$$

What is the minimum volume of 0.676 mol/L HCl that is needed to combine exactly with 22.7 mL of 0.385 mol/L aqueous sodium hydroxide?

**17.** Sodium sulfate and water result when sulfuric acid, $H_2SO_4$, reacts with sodium hydroxide. What volume of 0.320 mol/L sulfuric acid is needed to react with 47.3 mL of 0.224 mol/L aqueous sodium hydroxide?

**18.** Your stomach secretes hydrochloric acid to help you digest the food you have eaten. If too much HCl is secreted, however, you may need to take an antacid to neutralize the excess. One antacid product contains the compound magnesium hydroxide, $Mg(OH)_2$.

  **(a)** Predict the reaction that takes place when magnesium hydroxide reacts with hydrochloric acid. (**Hint:** This is a double-displacement reaction.)

  **(b)** Imagine that you are a chemical analyst testing the effectiveness of antacids. If 0.10 mol/L $HCl_{(aq)}$ serves as your model for stomach acid, how many litres will react with an antacid that contains 0.10 g of magnesium hydroxide?

## Limiting Reactant Problems in Aqueous Solutions

In Chapter 4, you learned how to solve limiting reactant problems. You can always recognize a limiting reactant problem because you are always given the amounts of both reactants. A key step in a limiting reactant problem is determining which one of the two reactants is limiting. In aqueous solutions, this usually means finding the amount (in mol) of a reactant, given the volume and concentration of the solution.

### Sample Problem

#### Finding the Mass of a Precipitated Compound

**Problem**

Mercury salts have a number of important uses in industry and in chemical analysis. Because mercury compounds are poisonous, however, the mercury ions must be removed from the waste water. Suppose that 25.00 mL of 0.085 mol/L aqueous sodium sulfide is added to 56.5 mL of 0.10 mol/L mercury(II) nitrate. What mass of mercury(II) sulfide, $HgS_{(s)}$, precipitates?

**What Is Required?**

You need to find the mass of mercury(II) sulfide that precipitates.

### What Is Given?
You know the volumes and concentrations of the sodium sulfide and mercury(II) nitrate solutions.

### Plan Your Strategy
Write a balanced chemical equation for the reaction. Find the amount (in mol) of each reactant, using its volume and concentration. Identify the limiting reactant. Determine the amount (in mol) of mercury(II) sulfide that forms. Calculate the mass of mercury(II) sulfide that precipitates.

### Act on Your Strategy
The chemical equation is

$Hg(NO_3)_{2(aq)} + Na_2S_{(aq)} \rightarrow 2NaNO_{3(aq)} + HgS_{(s)}$

Calculate the amount (in mol) of each reactant.

Amount of $Hg(NO_3)_2$ = 0.056 L × 0.10 mol/L
            = 0.0056 mol

Amount of $Na_2S$ = 0.0250 L × 0.085 mol/L
            = 0.0021 mol

The reactants are in a 1:1 ratio. Because $Na_2S$ is present in the smallest amount, it is the limiting reactant.

The equation indicates that each mol of $Na_2S$ reacts to produce the same amount of $HgS_{(s)}$ precipitate. This amount is 0.002 12 mol.

Molar mass of HgS = 200.6 + 32.1
            = 232.7

Mass of $HgS_{(s)}$ = Amount × Molar mass

$= 0.0021 \text{ mol} \times \dfrac{232.7 \text{ g}}{1 \text{ mol}}$

= 0.49 g

The mass of mercury(II) sulfide that precipitates is 0.49 g

### Check Your Solution
The answer has appropriate units of mass. This answer appears to be reasonable, given the values in the problem.

> **PROBLEM TIP**
>
> Chemists solve limiting reactant problems in different ways. The method used here is different from the one you used in Chapter 4. Note that this strategy depends on a 1:1 mole ratio between the reactants, and can be used only in *this* situation.

## Sample Problem

### Finding the Mass of Another Precipitated Compound

#### Problem
Silver chromate, $Ag_2CrO_4$, is insoluble. It forms a brick-red precipitate. Calculate the mass of silver chromate that forms when 50.0 mL of 0.100 mol/L silver nitrate reacts with 25.0 mL of 0.150 mol/L sodium chromate.

#### What Is Required?
You need to find the mass of silver chromate that precipitates.

#### What Is Given?
You know the volumes and concentrations of the silver nitrate and sodium chromate solutions.

*Continued...*

**Continued ...**

### Plan Your Strategy

Write a balanced chemical equation for the reaction. Find the amount (in mol) of each reactant, using its volume and concentration. Identify the limiting reactant. Determine the amount (in mol) of silver chromate that forms. Calculate the mass of silver chromate that precipitates.

### Act on Your Strategy

The chemical equation is

$2AgNO_{3(aq)} + Na_2CrO_{4(aq)} \rightarrow Ag_2CrO_{4(s)} + 2NaNO_{3(aq)}$

Calculate the amount (in mol) of each reactant.

Amount of $AgNO_3$ = 0.0500 L × 0.100 mol/L
= $5.00 \times 10^{-3}$ mol

Amount of $Na_2CrO_4$ = 0.0250 L × 0.150 mol/L
= $3.75 \times 10^{-3}$ mol

To identify the limiting reactant, divide by the coefficient in the equation and find the smallest result.

$AgNO_3 : 5.00 \times \dfrac{10^{-3} \text{ mol}}{2} = 2.50 \times 10^{-3}$ mol

$Na_2CrO_4 : 3.75 \times \dfrac{10^{-3} \text{ mol}}{1} = 3.75 \times 10^{-3}$ mol

Since the smallest result is given for $AgNO_3$, this reactant is the limiting reactant.

Using the coefficients in the balanced equation, 2 mol of $AgNO_3$ react for each mole of $Ag_2CrO_4$ formed.

Amount of $Ag_2CrO_4$ = $5.00 \times 10^{-3}$ mol $AgNO_3 \times \dfrac{1 \text{ mol } Ag_2CrO_4}{2 \text{ mol } AgNO_3}$

= $2.50 \times 10^{-3}$ mol $Ag_2CrO_4$

The molar mass of $Ag_2CrO_4$ is 331.7 g/mol.

Mass of precipitate = $2.50 \times 10^{-3}$ mol $\times \dfrac{331.7 \text{ g}}{1 \text{ mol}}$

= 0.829 g

The mass of silver chromate that precipitates is 0.829 g.

### Check Your Solution

The answer has appropriate units of mass. The answer appears to be reasonable, given the values in the problem.

## Practice Problems

**19.** 8.76 g of sodium sulfide is added to 350.0 mL of 0.250 mol/L lead(II) nitrate solution. Calculate the maximum mass of precipitate that can form.

**20.** 25.0 mL of 0.400 mol/L $Pb(NO_3)_{2(aq)}$ is mixed with 300.0 mL of 0.220 mol/L $KI_{(aq)}$. What is the maximum mass of precipitate that can form?

**21.** Zoe mixes 15.0 mL of 0.250 mol/L aqueous sodium hydroxide with 20.0 mL of 0.400 mol/L aqueous aluminum nitrate.

  **(a)** Write the chemical equation for the reaction.

(b) Calculate the maximum mass of precipitate that Zoe can expect to obtain from this experimental design.

(c) How can Zoe make her experiment more cost-effective without decreasing her theoretical yield?

22. In an experiment, Kendra mixed 40.0 mL of 0.552 mol/L lead(II) nitrate with 50.0 mL of 1.22 mol/L hydrochloric acid. A white precipitate formed, which Kendra filtered and dried. She determined the mass to be 5.012 g.

(a) What is the formula of the precipitate that formed?

(b) Write the chemical equation for the reaction.

(c) Which is the limiting reactant?

(d) What is the theoretical yield?

(e) Calculate the percentage yield.

## Section Summary

In the last two sections, you have used qualitative and quantitative techniques to investigate ions in aqueous solution. You will use these skills, as well as your understanding of solutions chemistry, in Unit 5 and 6 of this book.

## Section Review

**1** Even though lead is toxic, many lead compounds are still used as paint pigments (colourings). What volume of 1.50 mol/L lead(II) acetate contains 0.400 mol $Pb^{2+}$ ions.

**2** Ammonium phosphate can be used as a fertilizer. 6.0 g of ammonium phosphate is dissolved in sufficient water to produce 300 mL of solution. What are the concentrations (in mol/L) of the ammonium ions and phosphate ions present?

**3** Equal volumes of 0.120 mol/L potassium nitrate and 0.160 mol/L iron(III) nitrate are mixed together. What is the concentration of nitrate ions in the mixture?

**4** Suppose that you want to remove the barium ions from 120 mL of 0.0500 mol/L aqueous barium nitrate solution. What is the minimum mass of sodium carbonate that you should add?

**5** An excess of aluminum foil is added to a certain volume of 0.675 mol/L aqueous copper(II) sulfate solution. The mass of solid copper that precipitates is measured and found to be 4.88 g. What was the volume of the copper(II) sulfate solution?

**6** To generate hydrogen gas, a student adds 5.77 g of mossy zinc to 80.1 mL of 4.00 mol/L hydrochloric acid in an Erlenmeyer flask. When the reaction is over, what is the concentration of aqueous zinc chloride in the flask?

**7** Tap water containing calcium hydrogen carbonate is referred to as hard water. Hard water can be softened by adding calcium hydroxide to precipitate calcium carbonate, which is easily removed by filtration. The balanced equation for the process is:

$$Ca(HCO_3)_{2(aq)} + Ca(OH)_{2(aq)} \rightarrow 2CaCO_{3(s)} + 2H_2O_{(\ell)}$$

What volume of 0.00500 mol/L calcium hydroxide solution would be required to soften 4.00 L of hard water with a calcium hydrogen carbonate concentration of $1.05 \times 10^{-4}$ mol/L?

**8** When a clear, yellow solution of potassium chromate is added to a clear, colourless solution of silver nitrate, a brick-red precipitate forms:

$$2AgNO_{3(aq)} + K_2CrO_{4(aq)} \rightarrow Ag_2CrO_{4(s)} + 2KNO_{3(aq)}$$

A laboratory technician added 30.0 mL of 0.150 mol/L potassium chromate to 45.0 mL of 0.145 mol/L silver nitrate.

(a) Perform calculations to show which of the two reagents is limiting.

(b) Calculate the theoretical yield of the precipitate in grams.

(c) Which reagent is in excess? By how many millilitres is the reagent in excess?

(d) Only 0.22 g of dry precipitate is collected. What is the percent yield of the reaction?

**9** Copper can be recovered from scrap metal by adding sulfuric acid. Soluble copper sulfate is formed. The copper sulfate then reacts with metallic iron in a single displacement reaction. To simulate this reaction, a student places 1.942 g of iron wool in a beaker that contains 136.3 mL of 0.0750 mol/L aqueous copper(II) sulfate. What mass of copper is formed?

# Chemistry Bulletin

### Science  Technology  Society  Environment

## Water Quality

In the town of Walkerton, Ontario, in 2000, seven people died and several thousand became ill as a result of exposure to *E. coli* bacteria that were present in the town's drinking water supply. The maximum acceptable concentration for *E. coli* bacteria is zero organisms detectable in 100 mL of drinking water. To achieve this strict target, water is commonly chlorinated to kill *E. coli* and other organisms in drinking water. Chlorination involves adding sodium hypochlorite, $NaOCl_{(s)}$, as a source of chloride ions.

Chlorination was first used to disinfect water in Britain in 1904, after a typhoid epidemic. (Typhoid is a water-borne, contagious illness that is caused by a species of Salmonella bacteria.) Strict limits are necessary because chlorination is ineffective when chloride concentration is less than 0.1 mg/L. It gives water an unpleasant taste at concentrations above 1.0 mg/L.

Despite receiving two warnings from the Ontario Ministry of the Environment, the town's water manager chose not to increase the concentration of chlorine in the water supply. As a result of this incident, the Ontario government drafted legislation to enforce strict adherence to water quality guidelines. Provincial governments across the country were swift to examine their own enforcement of water quality guidelines to avoid similar occurrences.

Although widely used in water treatment facilities around the world, chlorination is not a "magic bullet" for ensuring water quality. It does, in fact, have a disadvantage. Chloride ions can react with other chemicals in the water to form poisonous compounds, such as chloroform, $CHCl_3$, and a family of compounds called trihalomethanes (THMs). These chemicals may remain in solution even after the water has been treated by other means.

Since the 1990s, many water treatment plants have explored alternatives to chlorination. These include ozone, $O_3$, chloramine, $NH_2Cl$, and a technique called sedimentation.

Ultimately, the most effective approach to large-scale water treatment depends on the substances that the facility must remove. Other factors such as cost, available technology, and public response play important roles as well. The responsibility for assessing health risks of specific substances in drinking water rests with the federal government. Guidelines for acceptable concentrations of each substance are then established in partnership with provincial and territorial governments. Corrective action, as in the case of Walkerton, may be taken when guidelines are violated.

## Making Connections

1. In light of problems with chlorination and THMs, many communities are in the process of modifying and upgrading their water treatment systems. Research how your community treats the local water supply, making note of the alternative method of water treatment if chlorination is not used.

# CHAPTER 8 Review

## Reflecting on Chapter 8
Summarize this chapter in the format of your choice. Here are a few ideas to use as guidelines:
- Predict combinations of aqueous solutions that result in the formation of precipitates.
- Represent a double-displacement reaction using its net ionic equation.
- Write balanced chemical equations and net ionic equations for double-displacement reactions.
- Write dissociation equations for soluble ionic compounds, and determine the resulting ion concentrations.
- Apply your understanding of stoichiometry to solve quantitative problems involving solutions.

## Reviewing Key Terms
For each of the following terms, write a sentence that shows your understanding of its meaning.

| | |
|---|---|
| precipitate | net ionic equation |
| general solubility guidelines | qualitative analysis |
| spectator ions | solution stoichiometry |
| total ionic equation | dissociation equation |

## Knowledge/Understanding

1. In your own words, define the terms "dissociation equation," "spectator ion," and "net ionic equation."

2. Identify the spectator ions in the following skeleton equation. Then write the balanced ionic equation for the reaction.
$Al(NO_3)_{3(aq)} + NH_4OH_{(aq)} \rightarrow Al(OH)_{3(s)} + NH_4NO_{3(aq)}$

3. Hydrogen sulfide gas can be prepared by the reaction of sulfuric acid with sodium sulfide.
$H_2SO_{4(aq)} + Na_2S_{(aq)} \rightarrow H_2S_{(g)} + Na_2SO_{4(aq)}$
Write the net ionic equation for this reaction.

4. Each of the following combinations of reagents results in a double displacement reaction. In your notebook, complete the chemical equation. Then identify the spectator ions, and write the net ionic equation.
   (a) copper(II) chloride$_{(aq)}$ + ammonium phosphate$_{(aq)}$ →
   (b) aluminum nitrate$_{(aq)}$ + barium hydroxide$_{(aq)}$ →
   (c) sodium hydroxide$_{(aq)}$ + magnesium chloride$_{(aq)}$ →

5. Use the general solubility guidelines to name three reagents that will combine with each ion below to form a precipitate. Assume that the reactions take place in aqueous solution. For each reaction, write the net ionic equation.
   (a) bromide ion
   (b) carbonate ion
   (c) lead(II) ion
   (d) iron(III) ion

6. The transition metals form insoluble sulfides, often with a characteristic colour. Write the net ionic equation for the precipitation of each ion by the addition of an aqueous solution of sodium sulfide.
   (a) $Cr^{3+}_{(aq)}$ (Note: $Cr_2S_{3(s)}$ is brown-black.)
   (b) $Ni^{2+}_{(aq)}$ (Note: $NiS_{(s)}$ is black.)
   (c) $Mn^{4+}_{(aq)}$ (Note: $MnS_{2(s)}$ is green or red, depending on the arrangement of ions in the solid.)

7. (a) Many liquid antacids contain magnesium hydroxide, $Mg(OH)_2$. Why must the bottle be shaken before a dose is poured?
   (b) Stomach acid contains hydrochloric acid. Excess acid that backs up into the esophagus is the cause of "heartburn." Write the chemical equation and the net ionic equation for the reaction that takes place when someone with heartburn swallows a dose of liquid antacid.

8. Aqueous solutions of iron(III) chloride and ammonium sulfide react in a double displacement reaction.
   (a) Write the name and formula of the substance that precipitates.
   (b) Write the chemical equation for the reaction.
   (c) Write the net ionic equation.

## Inquiry

9. A reference book states that the solubility of silver sulfate is 0.57 g in 100 mL of cold water. You decide to check this by measuring the mass of a silver salt precipitated from a known volume of saturated silver sulfate solution.

Answers to questions highlighted in red type are provided in Appendix A.

Solubility data show that silver chloride is much less soluble than silver nitrate. Explain why you should not use barium chloride to precipitate the silver ions. Suggest a different reagent, and write the net ionic equation for the reaction.

10. The presence of copper(II) ions in solution can be tested by adding an aqueous solution of sodium sulfide. The appearance of a black precipitate indicates that the test is positive. A solution of copper(II) bromide is tested this way. What precipitate is formed? Write the net ionic equation for the reaction.

11. An old home-gardening "recipe" for fertilizer suggests adding 15 g of Epsom salts (magnesium sulfate heptahydrate, $MgSO_4 \cdot 7H_2O$) to 4.0 L of water. What will be the concentration of magnesium ions?

12. Calculate the concentration (in mol/L) of each aqueous solution.
    (a) 7.37 g of table sugar, $C_{12}H_{22}O_{11}$, dissolved in 125 mL of solution
    (b) 15.5 g of ammonium phosphate, $(NH_4)_3PO_4$, dissolved in 180.0 mL of solution
    (c) 76.7 g of glycerol, $C_3H_8O_3$, dissolved in 1.20 L of solution

13. 50.0 mL of 0.200 mol/L $Ca(NO_3)_{2(aq)}$ is mixed with 200.0 mL of 0.180 mol/L $K_2SO_{4(aq)}$. What is the concentration of sulfate ions in the final solution?

14. Suppose that 1.00 L of 0.200 mol/L $KNO_{3(aq)}$ is mixed with 2.00 L of 0.100 mol/L $Ca(NO_3)_{2(aq)}$. Determine the concentrations of the major ions in the solution.

15. Equal masses of each of the following salts are dissolved in equal volumes of water: sodium chloride, calcium chloride, and iron(III) chloride. Which salt produces the largest concentration of chloride ions?

16. Imagine that you are the chemist at a cement factory. You are responsible for analyzing the factory's waste water. If a 50.0 mL sample of waste water contains 0.090 g of $Ca^{2+}_{(aq)}$ and 0.029 g of $Mg^{2+}_{(aq)}$, calculate
    (a) the concentration of each ion in mol/L
    (b) the concentration of each ion in ppm

17. The concentration of calcium ions, $Ca^{2+}$, in blood plasma is about $2.5 \times 10^{-3}$ mol/L. Calcium ions are important in muscle contraction and in regulating heartbeat. If the concentration of calcium ions falls too low, death is inevitable. In a television drama, a patient is brought to hospital after being accidentally splashed with hydrofluoric acid. The acid readily penetrates the skin, and the fluoride ions combine with the calcium ions in the blood. If the patient's volume of blood plasma is 2.8 L, what amount (in mol) of fluoride ions would completely combine with all the calcium ions in the patient's blood?

18. A double displacement reaction occurs in aqueous solution when magnesium phosphate reacts with lead(II) nitrate. If 20.0 mL of 0.750 mol/L magnesium phosphate reacts, what is the maximum mass of precipitate that can be formed?

## Communication

19. Phosphate ions act as a fertilizer. They promote the growth of algae in rivers and lakes. They can enter rivers and lakes from fields that are improperly fertilized or from untreated waste water that contains phosphate detergents. How can the water be treated to remove the phosphate ions?

20. A chemist analyzes the sulfate salt of an unknown alkaline earth metal. The chemist adds 1.273 g of the salt to excess barium chloride solution. After filtering and drying, the mass of precipitate is found to be 2.468 g.
    (a) Use the formula $MSO_4$ to represent the unknown salt. Write the molecular and net ionic equations for the reaction.
    (b) Calculate the amount (in mol) of $MSO_4$ used in the reaction.
    (c) Determine the molar mass of the unknown salt.
    (d) What is the likely identity of the unknown metal cation? What test might the chemist perform to help confirm this conclusion?

21. The same volume of solution is made using the same masses of two salts: rubidium carbonate and calcium carbonate. Which salt gives the larger concentration of aqueous carbonate ions?

22. A prospector asks you to analyze a bag of silver ore. You measure the mass of the ore and add excess nitric acid to it. Then you add excess sodium chloride solution. You filter and dry the precipitate. The mass of the ore is 856.1 g, and 1.092 g of silver chloride is collected.
   (a) Why did you first treat the ore with excess nitric acid?
   (b) Calculate the mass percent of silver in the ore. The ore that is extracted at a silver mine typically contains about 0.085% silver by mass. Should the prospector keep looking or begin celebrating?

## Making Connections

23. The hydrogen peroxide sold in pharmacies is not a concentrated solution. Over time, concentrated aqueous hydrogen peroxide decomposes according to this equation:
   $2H_2O_{2(aq)} \rightarrow 2H_2O_{(l)} + O_{2(g)}$
   If 125 mL of a solution of 0.90 mol/L hydrogen peroxide purchased in a pharmacy completely decomposes under standard pressure and temperature:
   (a) What volume of oxygen gas would be produced?
   (b) What mass of water would be produced?
   (c) What volume of water would be produced? (**Hint:** Think carefully here.)

24. Examine the following table, which shows the acceptable concentrations of various substances in drinking water, as determined by Health Canada.
   (a) Which chemical in the table has the lowest acceptable concentration?
   (b) Which has the highest acceptable concentration?
   (c) Re-design the table so that the substances are organized by concentration, from lowest to highest, rather than alphabetically by name.
   (d) What kinds of information would be required to arrive at the quantities listed in this table?

**Answers to Practice Problems and Short Answers to Section Review Questions:**
**Practice Problems: 1.(a)** insoluble **(b)** insoluble **(c)** insoluble **2.(a)** soluble **(b)** soluble **(c)** insoluble **3.(a)** insoluble **(b)** soluble **(c)** insoluble **4.(a)** $Na_2S_{(aq)} + FeSO_{4(aq)} \rightarrow Na_2SO_{4(aq)} + FeS_{(s)}$ **(b)** $NaOH_{(aq)} + Ba(NO_3)_{2(aq)} \rightarrow NR$

### Acceptable Concentrations of Selected Ions and Compounds in Drinking Water

| Ion or compound | Maximum Acceptable Concentration (MAC) (mg/L) | Interim Maximum Acceptable Concentration (mg/L) | Aesthetic Objectives (AO) (mg/L) |
|---|---|---|---|
| aldrin and dieldrin (organic insecticides*) | 0.0007 | 0.1 | |
| aluminum | | | |
| arsenic | | 0.025 | |
| benzene (organic component of gasoline) | 0.005 | | |
| cadmium (component of batteries) | 0.005 | | |
| chloride | | | ≤ 250 |
| fluoride | 1.5 | | |
| iron | | | ≤ 0.3 |
| lead | 0.010 | | |
| malathion (organic insecticide) | 0.19 | | |
| mercury | 0.001 | | |
| selenium | 0.01 | | |
| sulfide (as $H_2S$) | | | ≤ 0.05 |
| toluene (organic solvent) | | | ≤ 0.024 |
| uranium | 0.1 | | |

*As you will learn in Unit 4, the term "organic" refers to most compounds structured around the element carbon. Toluene belongs to a large class of petroleum-related compounds called hydrocarbons.
** Health-based guidelines have not yet been established. The concentrations that are listed depend on the method of treatment. They are noted as a precautionary measure.

(c) $2Cs_3PO_{4(aq)} + 3CaBr_{2(aq)} \rightarrow 6CsBr_{(aq)} + Ca_3(PO_4)_{2(s)}$
(d) $Na_2CO_{3(aq)} + H_2SO_{4(aq)} \rightarrow Na_2SO_{4aq} + H_{2(g)} + H_2O_{(l)}$
(e) $NaNO_{3(aq)} + CuSO_{4(aq)} \rightarrow NR$
(f) $NH_4I_{(aq)} + AgNO_{3(aq)} \rightarrow AgI_{(s)} + NH_4NO_{3(aq)}$
(g) $K_2CO_{3(aq)} + Fe(NO_3)_{2(aq)} \rightarrow FeCO_{3(s)} + 2KNO_{3(aq)}$
(h) $Al(NO_3)_{3(aq)} + Na_3PO_{4(aq)} \rightarrow AlPO_{4(s)} + 3NaNO_{3(aq)}$
(i) $KCl_{(aq)} + Fe(NO_3)_{2(aq)} \rightarrow NR$
(j) $(NH_4)_2SO_{4(aq)} + BaCl_{2(aq)} \rightarrow BaSO_{4(s)} + 2NH_4Cl_{(aq)}$
(k) $Na_2S_{(aq)} + NiSO_{4(aq)} \rightarrow NiS_{(s)} + Na_2SO_{4(aq)}$
(l) $Pb(NO_3)_2 + 2KBr_{(aq)} \rightarrow PbBr_{2(s)} + 2KNO_{3(aq)}$

**5.(a)** spectator ions: $Na^+_{(aq)}$ and $Cl^-_{(aq)}$; net ionic equation: $CO_3^{2-}{}_{(aq)} + 2H^+{}_{(aq)} \rightarrow CO_{2(g)} + H_2O_{(l)}$ **(b)** spectator ions: $Na^+_{(aq)}$ and $SO_4^{2-}{}_{(aq)}$; net ionic equation: $H^+{}_{(aq)} + OH^-{}_{(aq)} \rightarrow H_2O_{(l)}$

**6.(a)** spectator ions: $NH_4^+{}_{(aq)}$ and $SO_4^{2-}{}_{(aq)}$; net ionic equation: $3Zn^{2+}{}_{(aq)} + 2PO_4^{3-}{}_{(aq)} \rightarrow Zn_3(PO_4)_{2(s)}$ **(b)** spectator ions: $Li^+{}_{(aq)}$ and $NO^{3-}{}_{(aq)}$; net ionic equation: $CO_3^{2-}{}_{(aq)} + 2H^+{}_{(aq)} \rightarrow CO_{2(g)} + H_2O_{(l)}$
**(c)** spectator ions: none; net ionic equation: $2H^+{}_{(aq)} + SO_4^{2-}{}_{(aq)} + Ba^{2+}{}_{(aq)} + 2OH^-{}_{(aq)} \rightarrow BaSO_{4(s)} + H_2O_{(l)}$

**7.(a)**
$LiF_{(s)} \rightarrow Li^+{}_{(aq)} + F^-{}_{(aq)}$
1 mol/L     1 mol/L     1 mol/L
**(b)**
$NaBr_{(s)} \rightarrow Na^+{}_{(aq)} + Br^-{}_{(aq)}$
0.5 mol/L    0.5 mol/L    0.5 mol/L
**(c)**
$Na_2S_{(s)} \rightarrow 2Na^+{}_{(aq)} + S^{2-}{}_{(aq)}$
1.5 mol/L    3.0 mol/L    1.5 mol/L
**(d)**
$Ca(NO_3)_{2(s)} \rightarrow Ca^{2+}{}_{(aq)} + 2NO_3^-{}_{(aq)}$
1.0 mol/L    1.0 mol/L    2.0 mol/L

**8.(a)**
$NH_4Cl_{(s)} \rightarrow NH_4^+{}_{(aq)} + Cl^-{}_{(aq)}$
1.75 mol/L   1.75 mol/L   1.75 mol/L
**(b)**
$(NH_4)_2S_{(s)} \rightarrow 2NH_4^+{}_{(aq)} + S^{2-}{}_{(aq)}$
0.25 mol/L   0.50 mol/L   0.25 mol/L
**(c)**
$(NH_4)_2SO_{4(s)} \rightarrow 2NH_4^+{}_{(aq)} + SO_4^{2-}{}_{(aq)}$
6.0 mol/L    12.0 mol/L   6.0 mol/L
**(d)**
$AgNO_{3(s)} \rightarrow Ag^+{}_{(aq)} + NO_3^-{}_{(aq)}$
2.0 mol/L    2.0 mol/L    2.0 mol/L

**9.**
$CaCl_{2(s)} \rightarrow Ca^{2+}{}_{(aq)} + 2Cl^-{}_{(aq)}$
0.547 mol/L  0.547 mol/L  1.09 mol/L

**10.(a)** 0.109 mol $CoCl_2$
**(b)** $CoCl_{2(s)} \rightarrow Co^{2+}{}_{(aq)} + 2Cl^-{}_{(aq)}$
**(c)** 0.0727 mol/L; 0.0727 mol/L; 0.145 mol/L
**11.** 0.160 mol/L
**12.** 0.690 mol/L
**13.** 0.069 mol/L
**14.** 0.514 mol/L

**15.** 300 mL
**16.** 0.0129 L
**17.** 0.0166 L
**18.(a)** $2HCl_{(aq)} + Mg(OH)_{2(aq)} \rightarrow MgCl_{2(aq)} + 2H_2O_{(l)}$
**(b)** 0.034 L
**19.** 20.9 g of PbS
**20.** 4.61 g $PbI_2$
**21.(a)** $3NaOH_{(aq)} + Al(NO_3)_{3(aq)} \rightarrow Al(OH)_{3(s)} + 3NaNO_{3(aq)}$
**(b)** 0.0975 g
**22.(a)** $PbCl_2$
**(b)** $2HCl_{(aq)} + Pb(NO_3)_{2(aq)} \rightarrow 2HNO_{3(aq)} + PbCl_{2(s)}$
**(c)** $Pb(NO_3)_2$ is limiting reagent
**(d)** 6.15 g $PbCl_2$
**(e)** 81.5%

**Section Review: 8.1: 2.** NaF less soluble, because $F^-$ is smaller than $I^-$. **3.(a)** insoluble **(b)** soluble **(c)** soluble.
**4.** all insoluble. **5.** Any reagent containing $Cl^-$, $Br^-$, or $I^-$ will precipitate silver ion but leave calcium ion in solution.
**6.** 1 = $Ag^+$, 2 = $SO_4^{2-}$, 3 = $Ba^{2+}$, 4 = $Cl^-$.
**8.2: 2.(a)** $3Sn^{2+}{}_{(aq)} + 2PO_4^{3-}{}_{(aq)} \rightarrow Sn_3(PO_4)_{2(s)}$
**(b)** $Ni^{2+}{}_{(aq)} + CO_3^{2-(aq)} \rightarrow NiCO_{3(s)}$
**(c)** $2Cr^{3+}{}_{(aq)} + 3S^{2-}{}_{(aq)} \rightarrow Cr_2S_{3(s)}$ **3.(a)** $Cl^-{}_{(aq)}$ and $K^+{}_{(aq)}$
**(b)** $Cl^-{}_{(aq)}$ and $Na^+{}_{(aq)}$ **(c)** $NH_4^+{}_{(aq)}$ and $SO_4^{2-}{}_{(aq)}$ **6.** copper(II) carbonate, $CuCO_3$; $Cu^{2+}{}_{(aq)} + CO_3^{2-}{}_{(aq)} \rightarrow CuCO_{3(s)}$; spectator ions: $SO_4^{2-}{}_{(aq)}$ and $Na^+{}_{(aq)}$.
**8.3: 1.** 0.267 L **2.** concentration of $NH_4^+{}_{(aq)}$ is 0.40 mol/L; concentration of $PO_4^{3-}{}_{(aq)}$ is 0.13 mol/L **3.** 0.300 mol/L nitrate ion **4.** 0.636 g $Na_2CO_{3(s)}$ **5.** 114 mL **6.** 1.09 mol/L $ZnCl_2$ **7.** 0.0840 L **8.(a)** $AgNO_3$ is the limiting reagent
**(b)** 1.08 g $Ag_2CrO_4$ **(c)** $K_2CrO_4$ is in excess by 8.27 mL
**(d)** 20% **9.** 0.648 g Cu

# UNIT 3 Review

## Knowledge/Understanding

### Multiple Choice

In your notebook, write the letter for the best answer to each question.

1. Select the answer that best describes what happens when an ionic compound dissolves in water.
   (a) The ions dissociate.
   (b) The solution conducts an electric current.
   (c) The solution may or may not be saturated.
   (d) (b) and (c)
   (e) (a), (b), and (c)

2. If 1.00 g of solid sodium chloride is dissolved in enough water to make 350 mL of solution, what is the molar concentration of the solution?
   (a) 5.98 mol/L
   (b) $1.67 \times 10^{-1}$ mol/L
   (c) $4.89 \times 10^{-2}$ mol/L
   (d) $5.98 \times 10^{-3}$ mol/L
   (e) $4.88 \times 10^{-5}$ mol/L

3. What volume of $5.00 \times 10^{-2}$ mol/L $Ca(NO_3)_2$ solution will contain $2.50 \times 10^{-2}$ mol of nitrate ions?
   (a) 200 mL
   (b) 250 mL
   (c) 500 mL
   (d) 750 mL
   (e) 1.00 L

4. If 40.0 mL of 6.00 mol/L sulfuric acid is diluted to 120 mL by the addition of water, what is the molar concentration of the sulfuric acid after dilution?
   (a) $5.00 \times 10^{-2}$ mol/L
   (b) $7.50 \times 10^{-2}$ mol/L
   (c) 1.00 mol/L
   (d) 2.0 mol/L
   (e) 4.0 mol/L

5. When solutions of sodium chloride, NaCl, and silver nitrate, $AgNO_3$, are mixed, what is the net ionic equation for the reaction that results?
   (a) $Na^+_{(aq)} + NO_3^-_{(aq)} \rightarrow NaNO_{3(aq)}$
   (b) $Ag^+_{(aq)} + Cl^-_{(aq)} \rightarrow AgCl_{(s)}$
   (c) $Na^+_{(aq)} + NO_3^-_{(aq)} + Ag^+_{(aq)} + Cl^-_{(aq)} \rightarrow AgCl_{(s)} + NaNO_{3(aq)}$
   (d) $Na^+_{(aq)} + NO_3^-_{(aq)} + Ag^+_{(aq)} + Cl^-_{(aq)} \rightarrow AgCl_{(s)} + Na^+_{(aq)} + NO_3^-_{(aq)}$
   (e) $Na^+_{(aq)} + NO_3^-_{(aq)} + Ag^+_{(aq)} + Cl^-_{(aq)} \rightarrow NaNO_{3(aq)} + Ag^+_{(aq)} + Na^+_{(aq)}$

6. A beaker contains an aqueous saturated solution of sodium chloride, as well as a small amount of solid sodium chloride crystals. Which of the following statements are true?
   (a) The sodium chloride crystals are inert.
   (b) Sodium chloride is coming out of solution.
   (c) Sodium chloride is dissolving.
   (d) answers (b) and (c)
   (e) answers (a) and (c)

7. Bromine dissolves in carbon tetrachloride and in benzene, but not in water. This can probably be explained by saying that
   (a) like dissolves like.
   (b) bromine, carbon tetrachloride, and benzene are non-polar substances.
   (c) water is a polar substance.
   (d) answers (b) and (c), but not (a)
   (e) answers (a), (b), and (c)

8. Dilution is
   (a) the process of making a solution stronger by adding more solute.
   (b) the process of making a solution less concentrated by adding more solute.
   (c) the process of making a solution less concentrated by adding more solvent.
   (d) the process of making a solvent more concentrated by adding more solute.
   (e) the process of making a solvent less concentrated by removing water.

9. What is the molar concentration of the solution formed when 0.90 g of sodium chloride dissolves in 100 mL of water?
   (a) 0.90 mol/L
   (b) 0.15 mol
   (c) 0.15 L
   (d) 0.15 mol/L
   (e) 0.90 L

10. Identify the spectator ions in a solution formed by the reaction of aqueous sodium sulfide with aqueous iron(II) sulfate.
    (a) $FeS_{(s)}$, $S^{2-}_{(aq)}$, and $Na^+_{(aq)}$
    (b) $Na^-_{(aq)}$ and $SO_4^{2+}_{(aq)}$
    (c) $Fe^{2+}_{(aq)}$ and $S^{2-}_{(aq)}$
    (d) $Na^+_{(aq)}$ and $S^{2-}_{(aq)}$
    (e) $SO_4^{2-}_{(aq)}$ and $Na^+_{(aq)}$

Answers to questions highlighted in red type are provided in Appendix A.

## Short Answers

In your notebook, write a sentence or a short paragraph to answer each question.

11. Is a saturated solution always a concentrated solution? Give an example to explain your answer.
12. How can a homogeneous mixture be distinguished from a heterogeneous mixture? Give one example of each.
13. List three different ways in which the concentration of a solution could be described.
14. What would you observe if a saturated solution of sodium carbonate (commonly called washing soda) at room temperature was cooled to 5°C?
15. Explain why calcium hydroxide (solubility 0.165 g per 100 g water at 20°C) is much more soluble than magnesium hydroxide (solubility 0.0009 g per 100 g water at 20°C).
16. Iron concentrations of 0.2 to 0.3 parts per million in water can cause fabric staining when washing clothes. A typical wash uses 12 L of water. What is the maximum mass of iron that can be present so that the clothes will not be stained?
17. Is a 1% solution of table salt, $NaCl_{(aq)}$, more concentrated, less concentrated, or at the same concentration as a 1% solution of sugar, $C_{12}H_{22}O_{11(aq)}$? Explain.
18. Which of the following would you expect to be soluble in water? Explain why.
    (a) sodium hydroxide
    (b) potassium chloride
    (c) ethanol
    (d) benzene
19. Your teacher is going to demonstrate dilution by diluting a 10 mol/L solution to a 1.00 mol/L solution. Your teacher will do this by adding 900.0 mL of water to 100.0 mL of 10.0 mol/L solution. Should you correct your teacher, or sit back and watch the demonstration? Explain.
20. Box I (top of next column) represents a unit volume of a solution. Which of boxes II and III represents the same unit volume of solution for each of the following situations:
    (a) more solvent added
    (b) more solute added
    (c) lower molar concentration
    (d) higher molar concentration

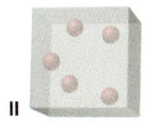

21. Chloroform and diethyl ether were among the first substances used as anaesthetics. Both are non-polar substances.
    (a) Would you expect either or both of these substances to be soluble in water? Explain.
    (b) Write a sentence or two to describe how you think these substances are able to get from the lungs to the brain.

## Inquiry

22. A chemist has a large beaker containing ice-cold water, and another containing boiling water. The laboratory is well-equipped with other apparatus.
    (a) Explain how the chemist could maximize the solubility of the following solutes in water (following appropriate safety precautions):
        (i) magnesium chloride, $MgCl_2$, used to fire-proof wood
        (ii) benzene, a non-polar liquid used by the industry and found in gasoline
        (iii) carbon monoxide, CO, a poisonous gas formed by incomplete combustion of hydrocarbons
    (b) Explain how you could minimize the solubility of the same solutes in water.
23. Water that contains high concentrations of calcium and magnesium ions, $Ca^{2+}$ and $Mg^{2+}$, is called "hard" water. If these ions are present in lower concentrations, the water is considered to be "soft." Distilled water has few ions of any kind, and no $Ca^{2+}$ or $Mg^{2+}$. Sodium oxalate, $Na_2C_2O_4$, is a compound that causes $Ca^{2+}$ to precipitate as calcium oxalate, $CaC_2O_4$, and $Mg^{2+}$ to precipitate as magnesium oxalate, $MgC_2O_4$.
    (a) A student adds 1 mL of hard water to one test tube, 1 mL of soft water to another, and 1 mL of distilled water to a third. The student then put two drops of 0.1 mol/L sodium oxalate solution into each test tube, and mixed the contents. Infer what the student observed in each test tube. Write the net

ionic equation if you predict that a precipitate formed, and NR if you think that no reaction occurred.

(b) Imagine that you have the three water samples the student used. You do not have any sodium oxalate, because it is a poisonous substance. Suggest another way you could test the validity of your predictions, and explain why you think it will work.

24. Given the net ionic equation below, write at least four total ionic equations that would correspond to it.
$HC_2H_3O_{2(aq)} + OH^-_{(aq)} \rightarrow C_2H_3O_2^-_{(aq)} + H_2O_{(\ell)}$

25. For each of the following, predict whether a reaction will occur. If so, write the balanced total and net ionic equations for it. Write NR if no reaction will occur.
   (a) an aqueous solution of iron(III) chloride + an aqueous solution of cesium phosphate
   (b) an aqueous solution of sodium hydroxide + an aqueous solution of cadmium nitrate
   (c) an aqueous solution of ammonium perchlorate + an aqueous solution of sodium bromide

26. An aqueous solution of calcium hydroxide is mixed with an aqueous solution of nitric acid. Write the balanced molecular equation for this reaction, as well as the total and net ionic equations.

27. 35.0 mL of lead(II) nitrate solution reacts completely with excess sodium iodide solution. A precipitate with a mass of 0.628 g forms. What is the molar concentration of lead(II) ion in the original solution?

28. How many moles of hydrochloric acid are there in 155 mL of $1.25 \times 10^{-2}$ mol/L $HCl_{(aq)}$?

29. Vitamin C, $C_6H_8O_6$, reacts completely with aqueous bromine according to the following equation.
$C_6H_8O_{6(aq)} + Br_{2(aq)} \rightarrow 2HBr_{(aq)} + C_6H_6O_{6(aq)}$
Chemists use this reaction to analyze the Vitamin C content in diet supplements and drinks that have been fortified with Vitamin C. What is the concentration of Vitamin C in 125 mL of apple juice that has Vitamin C added to it if 22.65 mL of 0.134 mol/L bromine solution is required for complete reaction of all the Vitamin C in the juice?

30. Phosphoric acid and magnesium sulfate can be produced from magnesium phosphate and sulfuric acid. The phosphoric acid is then used to make fertilizers.
   (a) Write a complete balanced equation for the process.
   (b) If 50.0 mL of 0.100 mol/L sulfuric acid is used, what is the final concentration of phosphoric acid if the volume of the phosphoric acid is 45.0 mL when the reaction is complete?

31. When aqueous solutions of ammonium chloride and lead(II) nitrate are mixed, the result is a white precipitate, lead(II) chloride, and a clear solution of ammonium nitrate. At one time, before lead compounds were banned from most paint products, the white precipitate was used in colouring window blinds.
   (a) Write a balanced chemical reaction for the precipitation reaction.
   (b) If 21.0 mL of 0.125 mol/L ammonium chloride reacts, what mass of precipitate forms?

32. At 0°C, 119 g of sodium acetate will dissolve in 100 g of water. At 100°C, 170 g will dissolve.
   (a) You place 170 g of sodium acetate in 100 g of water. The temperature of the water is 0°C. How much solute dissolves?
   (b) What would you observe if the resulting solution were in front of you? Is it unsaturated, saturated, or supersaturated?
   (c) You heat the system to 100°C. What would you observe now? Is this a mixture or a solution? Is it unsaturated, saturated, or supersaturated?
   (d) How would your answer to part (c) change if you were using 145 g of sodium acetate?
   (e) How would you answer to part (c) change if you used 180 g of sodium acetate?

33. Determine the number of moles of chloride ions in the following:
   (a) 25 mL of 1.25 mol/L NaCl
   (b) 1.70 L of $1.0 \times 10^{-3}$ mol/L $ZnCl_2$

## Communication

34. A chemist prepares a solution by dissolving 516.5 mg of oxalic acid, $C_2H_2O_4$, making 100.0 mL of solution. The chemist then dilutes a 10.00 mL portion to 250.0 mL. Calculate the molar concentration of this diluted solution.

35. Aqueous solutions of certain substances are electrolytes, meaning that they can conduct an electric current. Will either of the following aqueous solutions conduct a current? Why or why not?
    (a) 0.01 mol/L sodium iodide
    (b) 0.01 mol/L hydrogen bromide

36. Write balanced chemical equations, total ionic equations, and net ionic equations for the following reactions.
    (a) sodium hydroxide$_{(aq)}$ + zinc nitrate$_{(aq)}$
    (b) magnesium bromide$_{(aq)}$ + potassium acetate$_{(aq)}$
    (c) silver sulfate$_{(aq)}$ + barium chloride$_{(aq)}$
    (d) sodium sulfate$_{(aq)}$ + strontium nitrate$_{(aq)}$

37. Explain how you would prepare each of the following aqueous solutions.
    (a) 1.00 L of 3.00 mol/L NiCl$_2$
    (b) $2.50 \times 10^2$ mL of 4.00 mol/L CoCl$_2$
    (c) $5.00 \times 10^2$ mL of 0.133 mol/L MnSeO$_4$ (manganese(II) selenate)

38. State the molar concentration of each of the following solutions.
    (a) 5.23 g of Fe(NO$_3$)$_2$ in 100.0 mL of solution
    (b) 44.3 g of Pb(ClO$_4$)$_2$ in 250.0 mL of solution
    (c) 9.94 g of CoSO$_4$ in $2.50 \times 10^2$ mL of solution

39. Concentrated hydrochloric acid is sold industrially as muriatic acid. Its molar concentration is 11.7 mol/L. A local rock carver requires a diluted solution of the acid. Explain how the carver should dilute the high-concentration product to make 4.7 L of 3.5 mol/L muriatic acid.

40. You have a 1.5 L container, a 2.0 L container, and a 2.5 L container. Which of these containers would you need to hold 2.11 mol of 0.988 mol/L solute? Why would the other two containers be poor choices?

41. What number of moles of NaCl is produced by reacting 2.68 L of 2.11 mol/L HCl with 3.17 L of 2.28 mol/L NaOH?

42. (a) 1.68 mol of CaCl$_2$ is dissolved in enough water to make 4.00 L of solution. What is the molar concentration of CaCl$_2$?
    (b) What is the concentration of each type of ion in the solution?

43. A solution is made by mixing 3.00 L of 2.22 mol/L HCl with 4.00 L of 0.128 mol/L Ba(OH)$_2$. The solution is then diluted to 8.00 L. Determine the concentration of each ion.

44. A chemist pours 150 mL of 12 mol/L HCl$_{(aq)}$ into a beaker.
    (a) How many moles of HCl are in the beaker?
    (b) If the contents of the beaker are added to a 2.00 L volumetric flask holding 1.5 L of water, how many moles of HCl are in the larger container?
    (c) Determine the concentration of the HCl if it is added to a 2 L flask and enough water is added to fill the flask to 2.00 L.

45. You want to prepare 250 mL of a 0.850 mol/L solution of sodium chloride.
    (a) How many moles of NaCl do you need?
    (b) What mass of NaCl does this value represent?
    (c) Your teacher needs this solution for the next class. Outline the steps you would take to prepare this solution to help your teacher.

## Making Connections

46. When the space shuttle *Challenger* exploded in 1986, divers recovered the data recorder from the ocean several weeks after the accident. They were not able to analyze the tape at first because magnesium from the recorder case had reacted with the water. The tapes were coated with insoluble magnesium hydroxide, Mg(OH)$_2$, which had to be removed.
    (a) Write a balanced equation for the reaction.
    (b) Scientists removed the magnesium hydroxide by reacting it carefully with nitric acid, HNO$_{3(aq)}$. Write the balanced chemical equation for this reaction, and explain why magnesium hydroxide converted to magnesium nitrate.

47. In 1963, a treaty was signed by the United States, the United Kingdom, and the former Soviet Union to ban the atmospheric testing of atomic weapons. Previous testing of such weapons had added radioactive isotopes of strontium (Sr-90) and cesium (Cs-137) to the atmosphere. Eventually, these pollutants fell to the ground and may have entered the food chain.
    (a) Which would you expect to form compounds that are more soluble in water: strontium or cesium? Explain your answer.
    (b) State two factors that might help you determine the health risks of these isotopes.

# UNIT 4

# Organic Chemistry

**UNIT 4 CONTENTS**

**CHAPTER 9**
Hydrocarbons

**CHAPTER 10**
Hydrocarbon Derivatives

**CHAPTER 11**
Organic Compounds on a Larger Scale

**UNIT 4 OVERALL OUTCOMES**

- What are hydrocarbons, and how can you represent their structures? What physical and chemical properties are common to hydrocarbons?

- How can you represent the structures of other types of organic compounds? What chemical reactions are typical of these organic compounds?

- How do organic compounds affect your life? How do they affect the environment?

At this moment, you are walking, sitting, or standing in an "organic" body. Your skin, hair, muscles, heart, and lungs are all made from *organic compounds*, chemical compounds that are based on the carbon atom. In fact, the only parts of your body that are *not* mostly organic are your teeth and bones! When you study organic chemistry, you are studying the substances that make up your body and much of the world around you. Medicines, clothing, carpets, curtains, and plastic are all manufactured from organic chemicals. If you look out a window, the grass, trees, squirrels, and insects you may see are also composed of organic compounds.

Are you having a sandwich for lunch? Bread, butter, meat, and lettuce are made of organic compounds. Will you have dessert? Sugar, flour, vanilla, and chocolate are also organic. What about a drink? Milk and juice are solutions of water in which organic compounds are dissolved.

In this unit, you will study a variety of organic compounds. You will learn how to name them and how to draw their structures. You will also learn how these compounds react, and you will use your knowledge to predict the products of organic reactions. In addition, you will discover the amazing variety of organic compounds in your body and in your life.

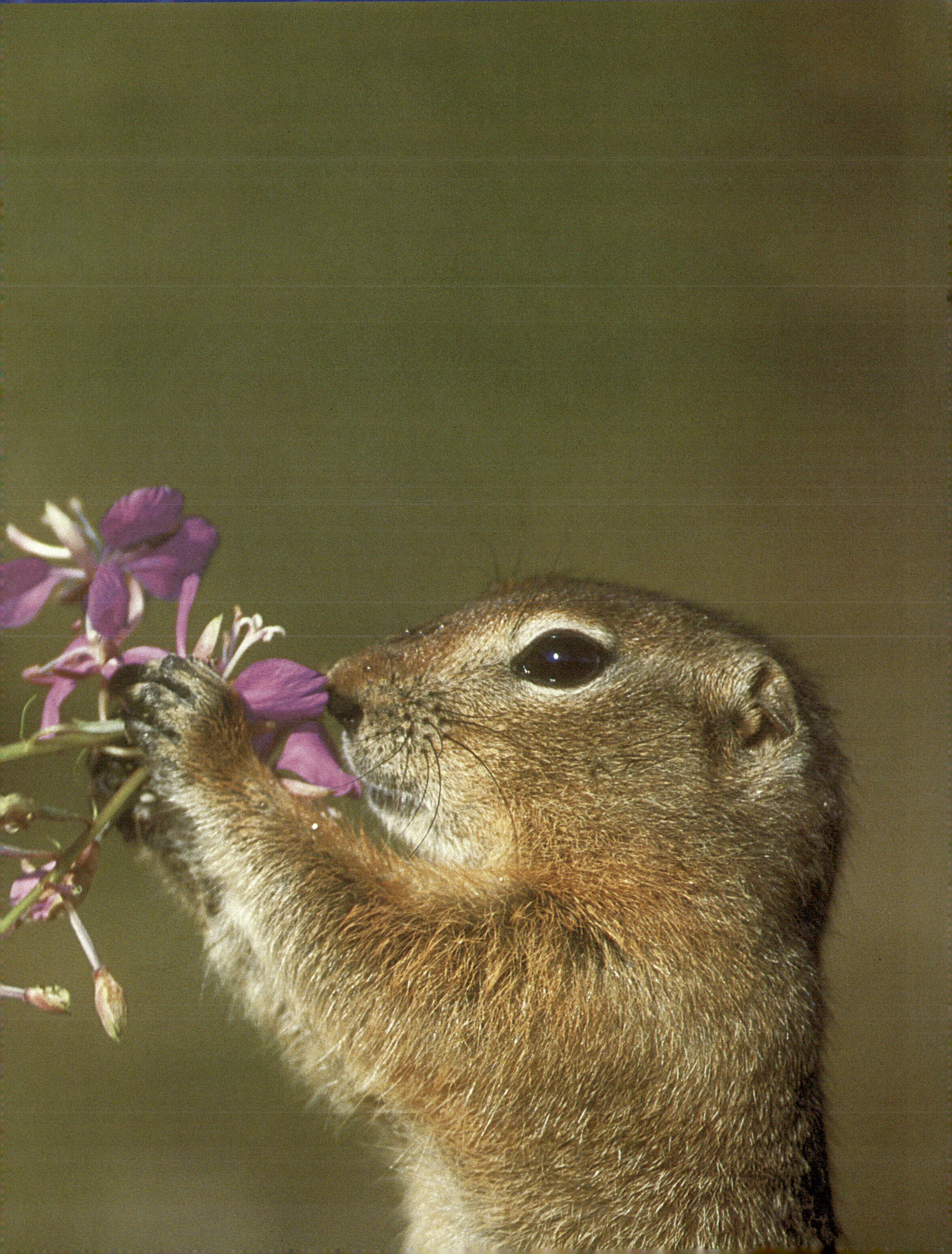

# CHAPTER 9
# Hydrocarbons

**Chapter Preview**

9.1 Introducing Organic Compounds

9.2 Representing Hydrocarbons

9.3 Classifying Hydrocarbons

9.4 Refining and Using Hydrocarbons

**Prerequisite Concepts and Skills**

Before you begin this chapter, review the following concepts and skills:

- understanding and writing molecular formulas (Chapter 3)
- identifying covalent and polar covalent bonds (Chapter 5)
- relating the structure of a molecule to its physical and chemical properties (Chapter 6)

**D**id you know that you have bark from a willow tree in your medicine cabinet at home? The model at the bottom right of this page shows a compound that is found naturally in willow bark. This chemical is called salicin. It is a source of pain relief for moose, deer, and other animals that chew the bark. For thousands of years, Aboriginal people in Canada and around the world have relied on salicin's properties for the same pain-relieving purpose.

The model at the bottom left of this page shows a close relative of salicin. Scientists made, or *synthesized*, this chemical near the end of the nineteenth century. It is called acetylsalicylic acid (ASA). You probably know it better by its brand name, Aspirin™.

Chemists refer to salicin, ASA, and more than ten million other chemicals like them as organic compounds. As described in the unit opener, an **organic compound** is a molecular compound of carbon. Despite the tremendous diversity of organic compounds, nearly all of them share something in common. They are structured from a "backbone" that consists of just two kinds of atoms: carbon and hydrogen.

Compounds that are formed from carbon and hydrogen are called **hydrocarbons**. In this chapter, you will explore the sources, structures, properties, and uses of hydrocarbons—an enormous class of compounds. As well, you will learn how scientists and engineers use the properties of hydrocarbons to produce a seemingly infinite variety of chemicals and products.

# Introducing Organic Compounds 9.1

As stated in the chapter opener, an organic compound is a molecular compound of carbon. There are a few exceptions to this definition, however. For example, scientists classify oxides of carbon, such as carbon dioxide and carbon monoxide, as *in*organic. However, the vast majority of carbon-containing compounds are organic.

## Organic Compounds: Natural and Synthetic

Organic compounds abound in the natural world. In fact, you probably ate sugar or starch at your last meal. Sugars, starches, and other carbohydrates are natural organic compounds. So are fats, proteins, and the enzymes that help you digest your food. Do you wear clothing made from wool, silk, or cotton? These are natural organic compounds, too. So are the molecules of DNA in the nuclei of your cells.

Until 1828, the only organic compounds on Earth were those that occur naturally. In that year, a German chemist named Friedrich Wohler synthesized urea—an organic compound found in mammal urine—from an inorganic compound, ammonium cyanate. (See Figure 9.1.) This was a startling achievement. Until then, chemists had assumed that only living or once-living organisms could be the source of organic compounds. They believed that living matter had an invisible "vital force." According to these early chemists, this vital (life) force made organic matter fundamentally different from inorganic (non-living) matter.

**Section Preview/Outcomes**

In this section, you will

- **identify** the origins and major sources of hydrocarbons and other organic compounds
- **communicate** your understanding of the following terms: *organic compound, hydrocarbons, petroleum*

**CHEM FACT**

A few carbon compounds are considered to be inorganic. These include carbon dioxide, $CO_2$, and carbon compounds containing complex negative ions (for example, $CO_3^{2-}$, $HCO_3^-$, and $OCN^-$).

**Language LINK**

Acetylsalicylic acid (ASA) was first produced commercially under the brand name Aspirin® by Frederick Bayer and Company in 1897. The word "aspirin" comes from "a," for acetyl, and "spir," for spirea. *Spirea* is a genus of plants that is another natural source of salicylic acid.

Urea

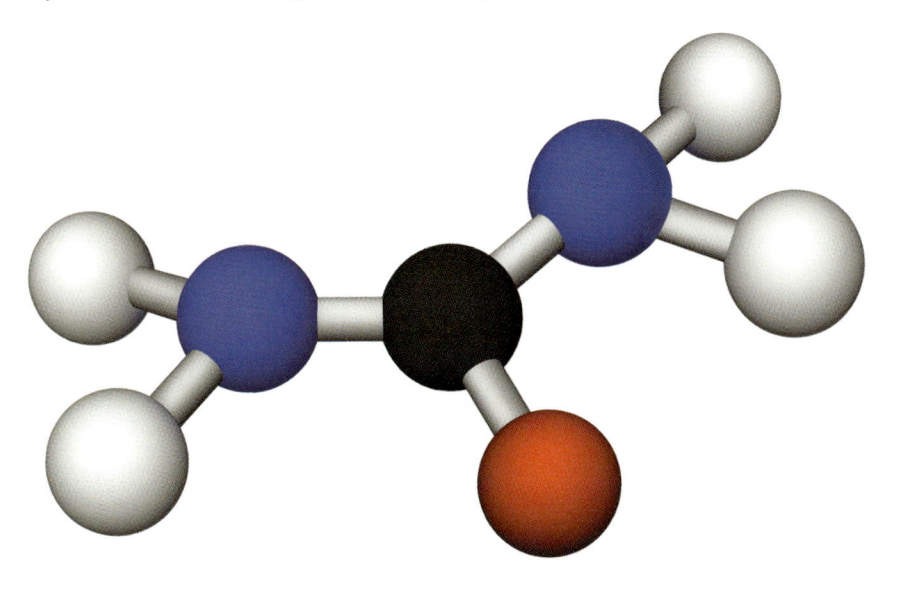

**Figure 9.1** When Friedrich Wohler (1800–1882) synthesized urea, he wrote a letter to his teacher. In his letter, he said, "I must tell you that I can make urea without the use of kidneys…Ammonium cyanate is urea." About 20 years earlier, Wohler's teacher, Jons Jakob Berzelius (1779–1848), had invented the system that distinguishes organic substances from inorganic substances.

During the mid-1850s, chemists synthesized other organic compounds, such as methane and ethanol, from inorganic chemicals. Eventually chemists abandoned their vital-force ideas. We still use the terms "organic" and "inorganic," however, to distinguish carbon-based compounds from other compounds. For example, sugar is an organic compound since it is carbon based. Salt is inorganic since it contains no carbon.

During the last century, the number of synthetic (human-made) organic compounds has skyrocketed. Chemists invent more than 250 000 new synthetic organic chemicals *each year*. With almost endless variations in properties, chemists can synthesize organic compounds to make products as diverse as life-giving drugs, and plastic toys. Nearly all medicines, such as painkillers, cough syrups, and antidepressants, are based on organic compounds. Perfumes, food flavourings, materials such as rubber and plastic, and fabrics such as nylon, rayon, and polyester are all organic compounds as well.

## The Origins of Hydrocarbons

Most hydrocarbons have their origins deep below Earth's present surface. In the past, as now, photosynthetic organisms used energy from the Sun to convert carbon dioxide and water into oxygen and carbohydrates, such as sugars, starches, and cellulose. When these organisms died, they settled to the bottom of lakes, rivers, and ocean beds, along with other organic matter. Bacterial activity removed most of the oxygen and nitrogen from the organic matter, leaving behind mainly hydrogen and carbon.

Over time, the organic matter was covered with layers of mud and sediments. As layer upon layer built up, heat and tremendous pressure transformed the sediments into shale and the organic matter into solid, liquid, and gaseous materials. These materials are the *fossil fuels*—coal, oil, and natural gas—that society depends on today. (See Figure 9.2.)

**Origin of Fossil Fuels**

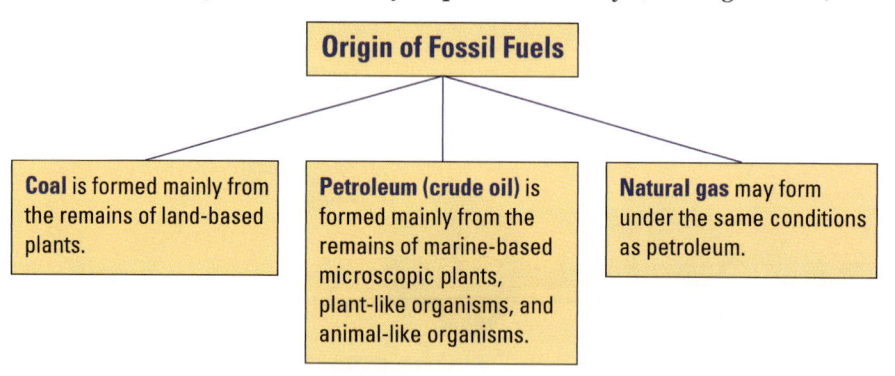

- **Coal** is formed mainly from the remains of land-based plants.
- **Petroleum (crude oil)** is formed mainly from the remains of marine-based microscopic plants, plant-like organisms, and animal-like organisms.
- **Natural gas** may form under the same conditions as petroleum.

**Figure 9.2** Ancient eras that had higher carbon dioxide concentrations, as well as warmer climates, gave rise to abundant plant and animal life on land and under water. Over time, as these organisms died, the organic substances that made up their bodies were chemically transformed into the materials known today as fossil fuels.

## Sources of Hydrocarbons

Sources of hydrocarbons include wood, the products that result from the fermentation of plants, and fossil fuels. However, one fossil fuel—petroleum—is the main source of the hydrocarbons that are used for fuels and many other products, such as plastics and synthetic fabrics.

**Petroleum**, sometimes referred to as crude oil, is a complex mixture of solid, liquid, and gaseous hydrocarbons. Petroleum also includes small amounts of other compounds that contain elements such as sulfur, nitrogen, and oxygen. To understand the importance of petroleum in our society, you need to become better acquainted with hydrocarbons. Your introduction begins, in the next section, with carbon—one of the most versatile elements on Earth.

## Section Review

**1** (a) Name three compounds that you know are organic.

(b) Name three compounds that you know are inorganic.

(c) Name three compounds that may be organic, but you do not know for sure.

**2** What are the origins of hydrocarbons and other organic compounds?

**3** Identify at least two sources of hydrocarbons and other organic compounds.

**4** Design a concept map (or another kind of graphic organizer) to show the meanings of the following terms and the relationships among them: organic compound, inorganic compound, hydrocarbon, fossil fuels, petroleum, natural gas.

**5** Copy the following compounds into your notebook. Identify each carbon as organic or inorganic, and give reasons for your answer. If you are not sure whether a compound is organic or inorganic, put a question mark beside it.

(a) $CH_4$

(b) $CH_3OH$

(c) $CO_2$

(d) $HCN$

(e) $C_6H_6$

(f) $NH_4SCN$

(g) $CH_3COOH$

(h) $CaCO_3$

> **Geology LINK**
>
> The origin of fossil fuels, depicted in Figure 9.2, is based on a theory called the *biogenic theory*. Most geologists accept this theory. A small minority of geologists have proposed an alternative theory, called the *abiogenic theory*. Use print and electronic resources to investigate the following:
>
> - the main points of each theory, and the evidence used to support these points
> - the reasons why one theory is favoured over the other
>
> Record your findings in the form of a brief report. Include your own assessment of the two theories.

## 9.2 Representing Hydrocarbons

**Section Preview/Outcomes**

In this section, you will

- **demonstrate** an understanding of the bonding characteristics of the carbon atom
- **draw** structural representations of aliphatic hydrocarbons
- **demonstrate** the arrangement of atoms in isomers of hydrocarbons using molecular models
- **communicate** your understanding of the following terms: *expanded molecular formula, isomers, structural model, structural diagram*

Examine the three substances in Figure 9.3. It is hard to believe that they have much in common. Yet each substance is composed entirely of carbon atoms. Why does the carbon atom lead to such diversity in structure? Why do carbon compounds outnumber all other compounds so dramatically? The answers lie in carbon's atomic structure and behaviour.

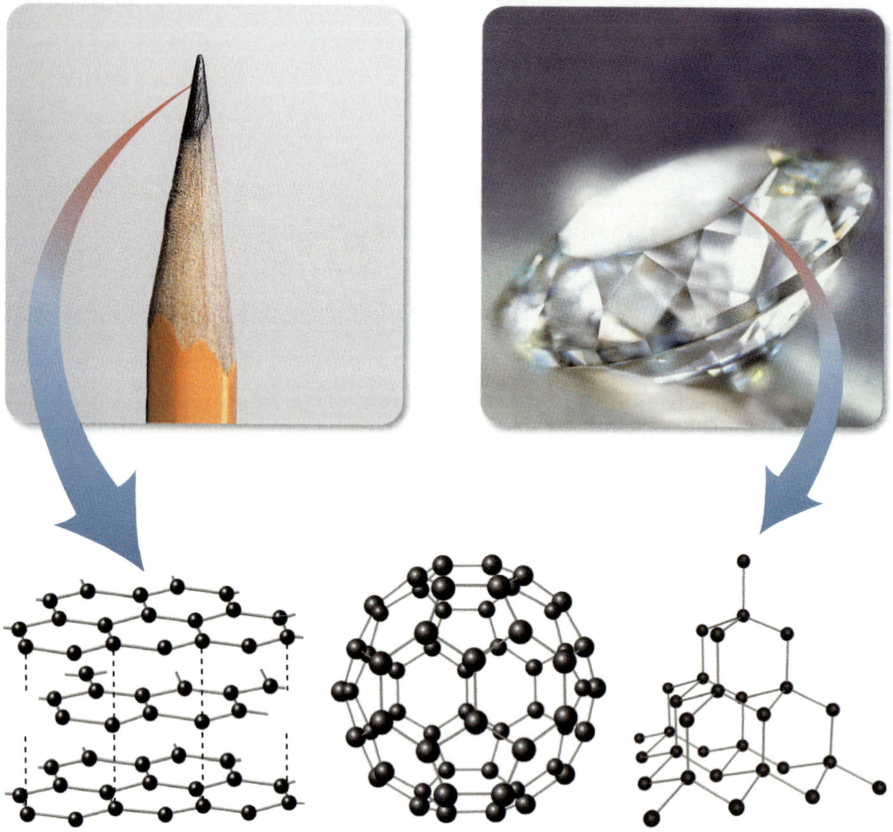

**Figure 9.3** Each of these substances is pure carbon. What makes carbon a "chemical chameleon?"

### CONCEPT CHECK

What are the differences between organic compounds and inorganic compounds?

- organic compounds have structures based on the carbon atom, while inorganic compounds do not
- many more organic compounds exist than inorganic compounds
- organic compounds usually have larger molecular size and greater molecular mass than inorganic compounds

Use what you learn in this section to explain why each of these three points is true.

### The Carbon Atom

There are several million organic compounds, but only about a quarter of a million inorganic compounds (compounds that are not based on carbon). Why are there so many more organic compounds than inorganic compounds? The answer lies in the bonding properties of the carbon atom.

As shown in Figure 9.4, *each carbon atom usually forms a total of four covalent bonds.* Thus, a carbon atom can connect to as many as four other atoms. Carbon can bond to many other types of atoms, including hydrogen, oxygen, and nitrogen.

$$\cdot \overset{\cdot}{\underset{\cdot}{C}} \cdot \; + \; 4H \cdot \; \rightarrow \; H : \overset{\overset{H}{\cdot \cdot}}{\underset{\underset{H}{\cdot \cdot}}{C}} : H$$

**Figure 9.4** This Lewis structure shows methane, the simplest organic compound. The carbon atom has four valence electrons, and it obtains four more electrons by forming four covalent bonds with the four hydrogen atoms.

## Properties of Carbon

**Carbon has four bonding electrons.** This electron structure enables carbon to form four strong covalent bonds. As a result, carbon may bond to itself, as well as to many different elements (mainly hydrogen, oxygen, and nitrogen, but also phosphorus, sulfur, and halogens such as chlorine).

**Carbon can form strong single, double, and triple bonds with itself.** This allows carbon to form long chains of atoms—something that very few other atoms can do. In addition, the resulting compounds are fairly stable under standard conditions of temperature and pressure.

**Carbon atoms can bond together to form a variety of geometrical structures.** These structures include straight chains, branched chains, rings, sheets, tubes, and spheres. No other atom can do this.

**Figure 9.5** Three key properties of carbon

In addition, *carbon atoms can form strong single, double, or triple bonds with other carbon atoms.* Carbon's unique bonding properties allow the formation of a variety of structures, including chains and rings of many shapes and sizes. Figure 9.5 summarizes these important properties of carbon.

### Web LINK

www.mcgrawhill.ca/links/atlchemistry

To see an animation comparing structural diagrams with molecular formulas, go to the web site above and click on **Electronic Learning Partner**.

## Representing Structures and Bonding

You have written chemical formulas for inorganic compounds such as ammonia, $NH_3$, and calcium carbonate, $CaCO_3$. As well, you have represented these compounds using Lewis diagrams, and perhaps other models. Such compounds are fairly small, so they are easy to represent using these methods. Many organic compounds are quite large and structurally complex. Therefore chemists have devised other methods to represent them, as explained below.

### Using Structural Diagrams to Represent Hydrocarbons

A **structural diagram** is a two-dimensional representation of the structure of a compound. (In some chemistry textbooks, structural diagrams are called structural formulas.) There are three kinds of structural diagrams: *complete structural diagrams, condensed structural diagrams,* and *line structural diagrams.*

A **complete structural diagram** as shown in Figure 9.6 A, shows all the atoms in a structure and the way they are bonded to one another. Straight lines represent the bonds between the atoms.

A **condensed structural diagram** simplifies the presentation of the structure. As you can see in Figure 9.6 B it shows the bonds between the carbon atoms but not the bonds between the carbon and hydrogen atoms. Chemists assume that these bonds are present. Notice how much cleaner and clearer this diagram is, compared with the complete structural diagram.

A **line structural diagram** as shown in Figure 9.6 C, is even simpler than a condensed structural diagram. The end of each line, and the points at which the lines meet, represent carbon atoms. Hydrogen atoms are not included in the diagram but are assumed to be present to account for the remainder of the four bonds on each carbon atom. This kind of diagram gives you a sense of the three-dimensional nature of a hydrocarbon. **Note:** Line structural diagrams are used only for hydrocarbons, and hydrocarbon portions of other organic compounds.

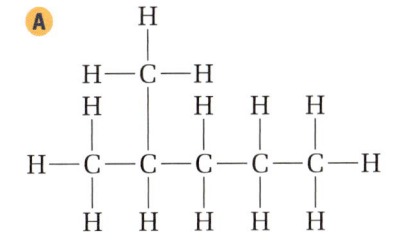

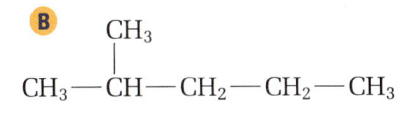

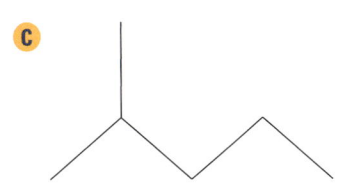

**Figure 9.6** Comparing (A) complete, (B) condensed, and (C) line structural diagrams

### Math LINK

Graph the data in the table below. For example, you could graph carbon bond energies versus equivalent silicon bond energies.

**Some Average Bond Energies**

| Bond | Bond energy (kJ/mol) |
|---|---|
| C—C | 346 |
| C=C | 610 |
| C≡C | 835 |
| Si—Si | 226 |
| Si=Si | 318 (estimate) |
| C—H | 413 |
| Si—H | 318 |

Infer a relationship between the stability of a compound and bond energy. Noting that both carbon atoms and silicon atoms can form four bonds, suggest a reason why there are many more carbon-based compounds than silicon-based compounds.

## Using Expanded Molecular Formulas to Represent Hydrocarbons

Structural diagrams give a clear picture of the way in which carbon atoms are bonded to each other but they take space and cannot be written on one line. When you understand the meaning of the structural diagrams clearly, you can write them in an even more condensed form that gives the same amount of information. An **expanded molecular formula** shows groupings of atoms without drawing bonds for lines. Brackets are used to indicate the locations of branched chains. For example, to write the expanded molecular formula for the compound in Figure 9.6, write all of the groupings in the straight chain directly after each other without lines between them. Put the side chain in brackets after the second carbon in the chain to show that it is branching from that carbon. The molecule is written $CH_3CH(CH_3)CH_2CH_2CH_3$. If the chain has no branches, no brackets are needed. For example, the expanded molecular formula for propane is $CH_3CH_2CH_3$.

Writing expanded molecular formulas becomes more helpful when you are writing several formulas. For example, $C_6H_{14}$ is a component of gasoline. It is also used as a solvent for extracting oils from soybeans and other edible oil seeds. Depending on how the carbon and hydrogen atoms are bonded together, $C_6H_{14}$ can have any of the five structural arrangements shown in Figure 9.7. Each arrangement has a different name.

$CH_3CH_2CH_2CH_2CH_2CH_3$

$CH_3CH(CH_3)CH_2CH_2CH_3$

$CH_3C(CH_3)_2CH_2CH_3$

The Five Isomers of $C_6H_{14}$

$CH_3CH_2CH(CH_3)CH_2CH_3$

$CH_3CH(CH_3)CH(CH_3)CH_3$

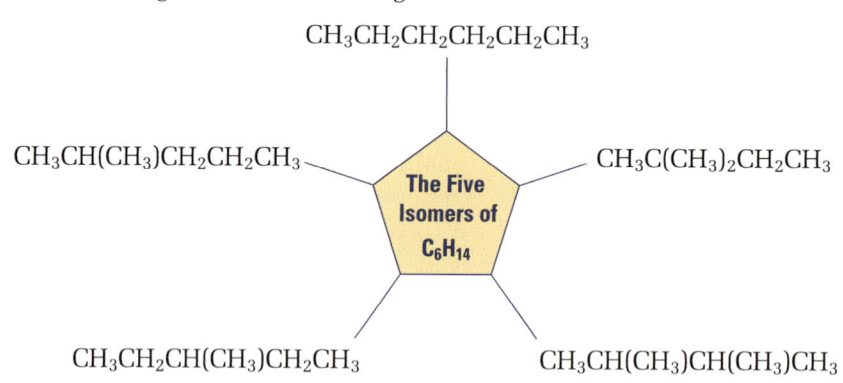

**Figure 9.7** Expanded molecular formulas for five structural arrangements of $C_6H_{14}$

Keep in mind that all five of these arrangements have the same chemical formula: $C_6H_{14}$. Compounds that have the same formula, but different structural arrangements, are called **isomers**. Hexane, for example, is one isomer of $C_6H_{14}$. You will learn more about isomers as you study this chapter.

## Using Structural Models to Represent Hydrocarbons

A structural model is a three-dimensional representation of the structure of a compound. There are two kinds of structural models: *ball-and-stick models* and *space-filling models*. Figure 9.8 shows ball-and-stick models for two of the five isomers of $C_6H_{14}$. Notice that they show how the carbon and hydrogen atoms are bonded within the structures.

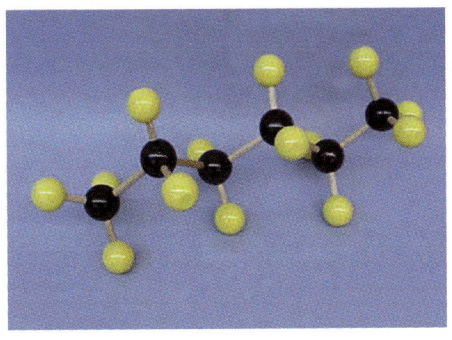

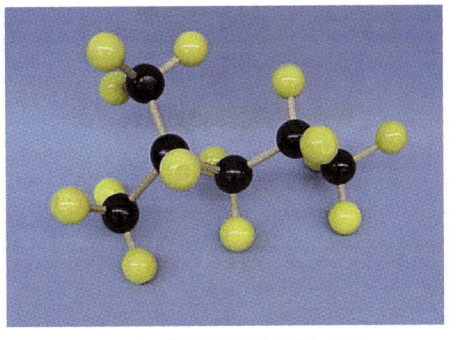

CH_3CH_2CH_2CH_2CH_2CH_3        CH_3CH(CH_3)CH_2CH_2CH_3

**Figure 9.8** These are two of the five possible structural arrangements for $C_6H_{14}$. Try using a molecular model kit to build the other three structural arrangements.

A space-filling model, such as the one in Figure 9.9, also shows the arrangement of the atoms in a compound. As well, it represents the molecular shape and the amount of space that each atom occupies within the structure.

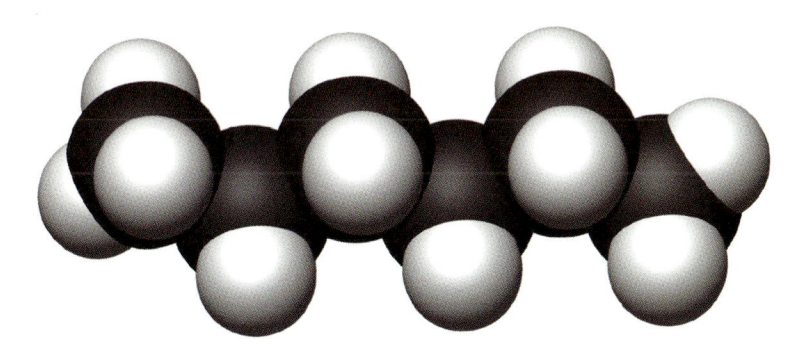

**Figure 9.9** A space-filling model for hexane, one of the isomers of $C_6H_{14}$

The following Investigation will give you an opportunity to build molecular models and visualize the structures of some hydrocarbons.

### CHEM FACT

The number of isomers of an organic compound increases greatly as the number of carbon atoms increases. For example, $C_5H_{12}$ has three isomers, $C_6H_{14}$ has five isomers, and $C_8H_{18}$ has 18 isomers. $C_{30}H_{62}$, a large hydrocarbon, has over four billion isomers!

# Investigation 9-A

**SKILL FOCUS**
Predicting
Analyzing and interpreting
Communicating results
Modelling concepts

## Modelling Organic Compounds

Figure 9.7 showed you that an organic compound can be arranged in different structural shapes, called isomers. All the isomers of a compound have the same molecular formula. In this investigation, you will make two-dimensional and three-dimensional models of isomers. Your models will help you explore the arrangements of the atoms in organic compounds.

### Question

How do models help you visualize the isomers of organic compounds?

### Predictions

Predict the complete, condensed, and line structural diagrams for the three isomers of $C_5H_{12}$. Then predict the complete and condensed structural diagrams for at least five isomers of $C_7H_{16}$.

### Materials

paper and pencil
molecular modelling kit

These representations of the hydrocarbon ethane, $C_2H_6$, were made using different modelling kits. Your school may have one or more of these kits available.

### Procedure

1. Construct three-dimensional models of the three isomers of $C_5H_{12}$. Use your predictions to help you. As you complete each model, draw a careful diagram of the structure. Your diagram might be similar to the one below.

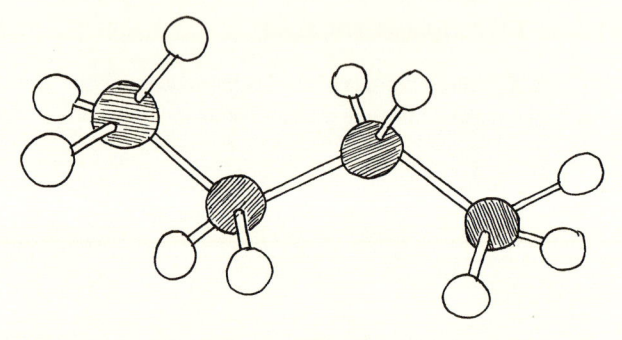

2. Repeat step 1 for at least five isomers of $C_7H_{16}$.

### Analysis

1. In what ways were your completed models similar to your predictions? In what ways were they different?

2. How do the models of each compound help you understand the concept of isomers?

### Conclusion

3. How accurately do you think your models represent the real-life structural arrangements of $C_5H_{12}$ and $C_7H_{16}$?

### Applications

4. In earlier units, you considered how the structure and polarity of molecules can affect the boiling point of a compound. For each compound you studied in this investigation, predict which isomer has the higher boiling point. Explain your prediction.

5. You made five or more isomers for $C_7H_{16}$. How many isomers are possible?

Carbon forms double and triple bonds in addition to single bonds. Carbon compounds in which carbon forms only single bonds have a different shape than compounds in which carbon forms double or triple bonds. In the following ExpressLab, you will see how each type of bond affects the shape of a molecule. Table 9.1 reviews some shapes that will be helpful in your study of organic chemistry.

## ExpressLab: Molecular Shapes

The type of bonding affects the shape and movement of a molecule. In this ExpressLab, you will build several molecules to examine the shape and character of their bonds.

### Procedure

1. Build a model for each of the following compounds. Use a molecular model kit or a chemical modelling computer program.

   $CH_3 — CH_2 — CH_2 — CH_3$    $H_2C = CH — CH_2 — CH_3$
   butane                          1–butene

   $H_3C — C \equiv C — CH_3$
   2–butyne

2. Identify the different types of bonds.

3. Try to rotate each of the bonds in the molecules. Which bonds allow rotation, and which prevent it?

4. Examine the shape of the molecule around each carbon atom. Draw diagrams to show your observations.

### Analysis

1. Which bond or bonds allow rotation to occur? Which bond or bonds are fixed in space?

2. (a) Describe the shape of the molecule around a carbon atom with only single bonds.

   (b) Describe the shape of the molecule around a carbon atom with one double bond and two single bonds.

   (c) Describe the shape of the molecule around a carbon atom with a triple bond and a single bond.

**Table 9.1** Common Molecular Shapes in Organic Molecules

| Central atom | Shape | Diagram |
|---|---|---|
| carbon with four single bonds | The shape around this carbon atom is **tetrahedral**. That is, the carbon atom is at the centre of an invisible tetrahedron, with the other four atoms at the vertices of the tetrahedron. This shape results because the electrons in the four bonds repel each other. In the tetrahedral position, the four bonded atoms and the bonding electrons are as far apart from each other as possible. | 109.5° tetrahedron with H atoms at vertices and C at centre |
| carbon with one double bond and two single bonds | The shape around this carbon atom is **trigonal planar**. The molecule lies flat in one plane around the central carbon atom, with the three bonded atoms spread out, as if to touch the corners of a triangle. | 120° angles around C=C and around C=O |
| carbon with two double bonds or one triple bond and one single bond | The shape around this carbon atom is **linear**. The two atoms bonded to the carbon atom are stretched out to either side to form a straight line. | 180°  $H — C \equiv C — CH_3$ |

Chapter 9 Hydrocarbons • MHR  **329**

## Three-Dimensional Structural Diagrams

Two-dimensional structural diagrams of organic compounds, such as condensed structural diagrams and line structural diagrams, work well for flat molecules. As shown in the table on the previous page, however, molecules containing single-bonded carbon atoms are not flat.

You can use a three-dimensional structural diagram to draw the tetrahedral shape around a single-bonded carbon atom. In a three-dimensional diagram, wedges are used to give the impression that an atom or group is coming forward, out of the page. Dashed or dotted lines are used to show that an atom or group is receding, or being pushed back into the page. In Figure 9.10, the Cl atom is coming forward, and the Br atom is behind. The two H atoms are flat against the surface of the page.

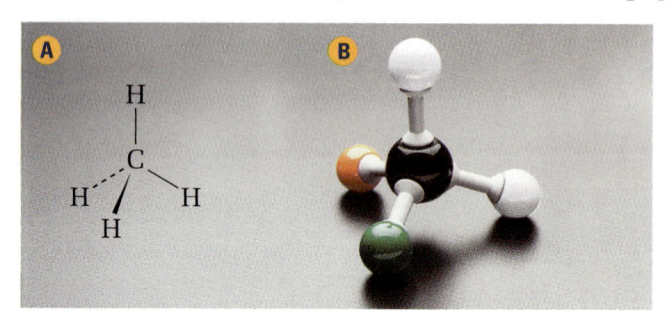

**Figure 9.10** (A) Three-dimensional structural diagram of the methane molecule (B) Ball-and-stick model

## Section Summary

In this section, you were introduced to several ways of representing hydrocarbons: expanded molecular formulas, structural models, and structural diagrams. In the next section, you will discover how to calssify and name hydrocarbon compounds.

### Section Review

**1** Identify the three properties of carbon that allow it to form such a great variety of compounds.

**2** Choose one of the hydrocarbon molecules that was shown in this section. Sketch the molecule in as many different ways as you can. Label each sketch to identify the type of model you were using.

**3** You have seen the expanded molecular formulas for the five isomers of $C_6H_{14}$. Draw the condensed, complete, and line structural diagrams for each of these isomers.

**4** Many organic compounds contain elements such as oxygen, nitrogen, and sulfur, as well as carbon and hydrogen. For example, think about ethanol, $CH_3CH_2OH$. Draw complete and condensed structural diagrams for ethanol. Can you draw a line structural diagram for ethanol? Explain your answer.

**5** Compare the shape around a single-bonded carbon atom with the shape around a double-bonded carbon atom. Use diagrams to illustrate your answer.

# Classifying Hydrocarbons

## 9.3

Chemists group hydrocarbons and other organic compounds into the categories shown in Figure 9.11. The International Union of Pure and Applied Chemistry (IUPAC) has developed a comprehensive set of rules for naming the compounds within each category. Using these rules, you will be able to classify and name all the hydrocarbon compounds that you will encounter in this unit.

The names that are based on the IUPAC rules are called *systematic names*. During your study of organic chemistry, you will also run across many common names for organic compounds. For example, the systematic name for the organic acid $CH_3CO_2H$ is ethanoic acid. You are probably more familiar with its common name: vinegar.

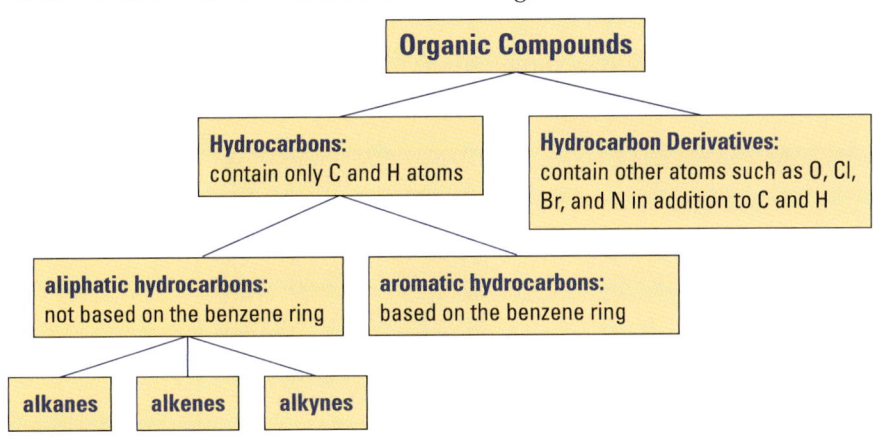

**Figure 9.11** This concept map illustrates a system for classifying organic compounds.

### Section Preview/Outcomes

In this section, you will

- **demonstrate** an understanding of the carbon atom by classifying hydrocarbons and by analyzing the bonds that carbon forms in aliphatic hydrocarbons
- **name** alkanes, alkenes, and alkynes, and **draw** structural representations for them
- **describe** some of the physical properties of hydrocarbons
- **predict** the products of reactions of hydrocarbons
- **determine** through experimentation some of the characteristic properties of saturated and unsaturated hydrocarbons
- **communicate** your understanding of the following terms: *alkanes, aliphatic hydrocarbons, saturated hydrocarbons, alkyl group, homologous series, complete combustion, incomplete combustion, alkenes, unsaturated hydrocarbons, cis-trans isomer, addition reaction, Markovnikov's Rule, alkynes, cyclic hydrocarbons, aromatic hydrocarbon, substitution reaction*

## Alkanes

**Alkanes** are hydrocarbon molecules that contain only *single* covalent bonds. They are the simplest hydrocarbons. Methane, $CH_4$, is the simplest alkane. It is the main component of natural gas. Alkanes are **aliphatic hydrocarbons**: organic compounds in which carbon atoms form chains and non-aromatic rings.

Figure 9.12 on the next page compares the structural formulas of methane and the next three members of the alkane family. Notice three facts about these alkanes:

1. Each carbon atom is bonded to the maximum possible number of atoms (either carbon or hydrogen atoms). As a result, chemists refer to alkanes as **saturated hydrocarbons**.

2. Each molecule differs from the next molecule by the structural unit $-CH_2-$. A series of molecules like this, in which each member increases by the same structural unit, is called a **homologous series**.

3. A mathematical pattern underlies the number of carbon and hydrogen atoms in each alkane. All alkanes have the general formula $C_nH_{2n+2}$, where $n$ is the number of carbon atoms. For example, propane has 3 carbon atoms. Using the general formula, we find that

$$2n + 2 = 2(3) + 2 = 8$$

Thus propane should have the formula $C_3H_8$, which it does.

### Language LINK

The name "aliphatic" comes from the Greek word *aleiphatos*, meaning "fat." Early chemists found these compounds to be less dense than water and insoluble in water, like fats. "Aliphatic" now refers to the classes of hydrocarbons called alkanes, alkenes, and alkynes.

Chapter 9 Hydrocarbons • MHR **331**

**CONCEPT CHECK**

Why is methane the simplest of all the millions of hydrocarbons? **Hint:** Recall what you know about chemical bonding and the common valences of elements.

methane

ethane

propane

butane

**Figure 9.12** Carefully examine these four molecules. They are the first four alkanes. In what ways are they similar? In what ways are they different?

Figure 9.13 illustrates these three important facts about alkanes. Study the two alkanes, then complete the Practice Problems that follow.

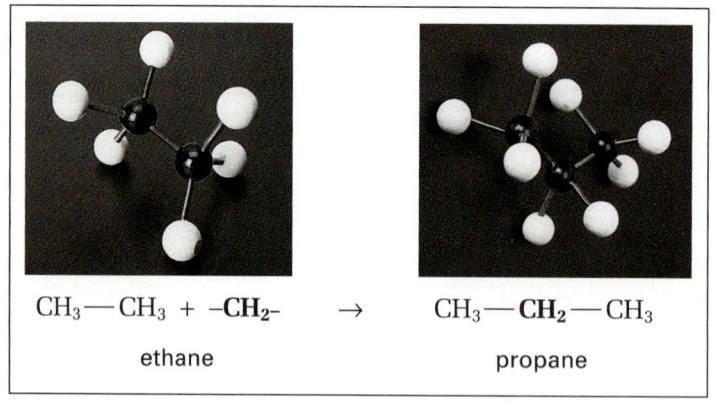

$CH_3$—$CH_3$ + –$CH_2$–   →   $CH_3$—$CH_2$—$CH_3$
ethane                                        propane

**Figure 9.13** How are these two alkanes similar? How are they different? Use the ideas and terms you have just learned to help you answer these questions.

### Practice Problems

1. Heptane has 7 carbon atoms. What is the chemical formula of heptane?
2. Nonane has 9 carbon atoms. What is its chemical formula?
3. An alkane has 4 carbon atoms. How many hydrogen atoms does it have?
4. Candle wax contains an alkane with 52 hydrogen atoms. How many carbon atoms does this alkane have?

### Naming Alkanes

The IUPAC system for naming organic compounds is very logical and thorough. The rules for naming alkanes are the basis for naming the other organic compounds that you will study. Therefore it is important that you understand how to name alkanes.

## Straight-Chain Alkanes

Recall that carbon can bond to form long, continuous, chain-like structures. Alkanes that bond in this way are called *straight-chain alkanes*. (They are also called *unbranched alkanes*.) Straight-chain alkanes are the simplest alkanes. Table 9.2 lists the names of the first ten straight-chain alkanes.

**Table 9.2** The First Ten Straight-Chain Alkanes

| Name | Number of carbon atoms | Expanded molecular formula |
| --- | --- | --- |
| methane | 1 | $CH_4$ |
| ethane | 2 | $CH_3CH_3$ |
| propane | 3 | $CH_3CH_2CH_3$ |
| butane | 4 | $CH_3(CH_2)_2CH_3$ |
| pentane | 5 | $CH_3(CH_2)_3CH_3$ |
| hexane | 6 | $CH_3(CH_2)_4CH_3$ |
| heptane | 7 | $CH_3(CH_2)_5CH_3$ |
| octane | 8 | $CH_3(CH_2)_6CH_3$ |
| nonane | 9 | $CH_3(CH_2)_7CH_3$ |
| decane | 10 | $CH_3(CH_2)_8CH_3$ |

The *root* of each name (highlighted in colour) serves an important function. It tells you the number of carbon atoms in the chain. The *suffix* -ane tells you that these compounds are alkane hydrocarbons. Thus the root and the suffix of one of these simple names provide the complete structural story of the compound.

## Branched-Chain Alkanes

The names for straight-chain alkanes can, with a few additions, help you recognize and name other organic compounds. You now know that the name of a straight-chain alkane is composed of a root (such as meth-) plus a suffix (-ane). Earlier in the chapter, you saw the isomers of $C_6H_{14}$. Figure 9.14 shows one of them, called 2-methylpentane.

$$CH_3-\overset{\overset{\displaystyle CH_3}{|}}{CH}-CH_2-CH_2-CH_3$$

**Figure 9.14** 2-methylpentane

Notice four important facts about 2-methylpentane:

1. Its structure is different from the structure of a straight-chain alkane. Like many hydrocarbons, this isomer of $C_6H_{14}$ has a branch-like structure. Alkanes such as 2-methylpentane are called *branched-chain alkanes*. (The branch is sometimes called a *side-chain*.)

2. The name of this alkane has a *prefix* (2-methyl-) as well as a root and a suffix. Many of the hydrocarbons you will name from now on have a prefix.

3. This alkane has a single $CH_3$ unit that branches off from the main (parent) chain of the compound.

### CHEM FACT

When is an isomer not an isomer? When it is *exactly* the same compound! The hydrocarbon $C_6H_{14}$ has only five isomers. Examine these two representations for one of them, hexane.

$$CH_3-CH_2-CH_2-CH_2-CH_2-CH_3$$

$$\begin{array}{c} CH_2-CH_2-CH_3 \\ | \\ CH_2 \\ | \\ CH_3-CH_2 \end{array}$$

You might think that these are different isomers, but they are actually the same. The structure on the bottom is just a bent version of the structure on the top. Think about a length of chain. You can lay it out straight, or you can bend it. The chain is still a chain in either case. It just looks different. Naming a compound helps you recognize the difference between a true isomer and a structure that just looks like an isomer.

### CONCEPT CHECK

The root and the suffix of an alkane name do not tell you directly about the number of hydrogen atoms in the compound. If you did not know the molecular formula of heptane, for example, how would you still know that heptane contains 16 hydrogen atoms?

## Naming Alkanes

The names of branched-chain alkanes (and most other organic compounds) have the same general format, as shown below. This format will become clearer as you learn and practise the rules for naming hydrocarbons.

## prefix + root + suffix

### The Root: How Long is the Main Chain?

The root of a compound's name indicates the number of carbon atoms in the main (parent) chain or ring. Table 9.2 on the previous page lists the roots for hydrocarbon chains that are up to ten carbons long.

### The Suffix: What Family Does the Compound Belong To?

The suffix indicates the type of organic compound. As you progress through this unit, you will learn the suffixes for different chemical families. You have already learned that the suffix -ane indicates an alkane.

### The Prefix: What is Attached to the Main Chain?

The prefix indicates the name and location of each branch or other type of group on the main carbon chain. Many organic compounds have hydrocarbon branches, called alkyl groups, attached to the main chain. An **alkyl group** is obtained by removing one hydrogen atom from an alkane. To name an alkyl group, start with the name of the corresponding alkane, and then change the -ane suffix to -yl. For example, the methyl group, —$CH_3$, is the alkyl group that is derived from methane, $CH_4$. Table 9.3 gives the names of the most common alkyl groups.

**Table 9.3** Common Alkyl Groups

| methyl | ethyl | propyl | isopropyl |
|---|---|---|---|
| —$CH_3$ | —$CH_2CH_3$ | —$CH_2CH_2CH_3$ | 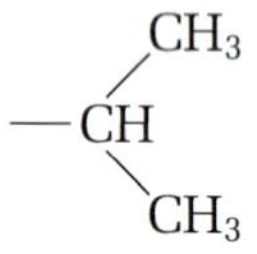 —CH($CH_3$)$CH_3$ |
| **butyl** | **sec-butyl** | **iso-butyl** | **tert-butyl** |
| —$CH_2CH_2CH_2CH_3$ | 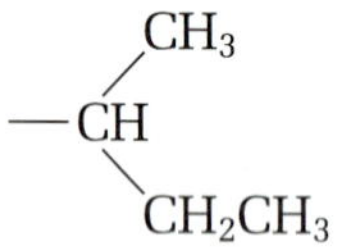 —CH($CH_3$)$CH_2CH_3$ | 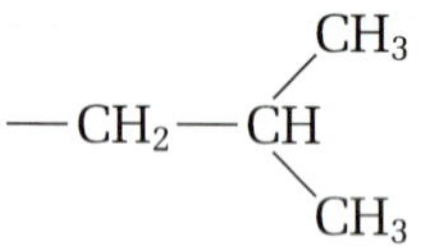 —$CH_2$—CH($CH_3$)$CH_3$ | 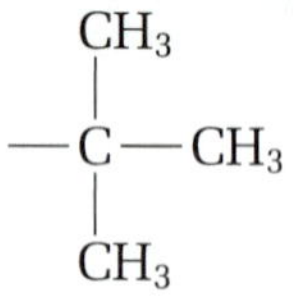 —C($CH_3$)($CH_3$)$CH_3$ |

The steps on the next page illustrate the process of naming alkanes. The same steps, with a few variations, also apply for many other organic compounds. Figure 9.15 illustrates these rules.

## Rules for Naming Alkanes

**Step 1 Find the root:** Identify the longest continuous chain or ring in the hydrocarbon. Count the number of carbon atoms in the main chain to obtain the root.
- Remember that the main chain can be bent; it does not have to look like a straight line.
- If more than one chain could be the main chain (because they are the same length), choose the chain that has the most branches attached.

**Step 2 Find the suffix:** If the compound is an alkane, use the suffix *-ane*. Later in this unit, you will learn the suffixes for other chemical families.

**Step 3 Give position numbers:** Identify any branches that are present. Then give a consecutive number to every carbon atom in the main chain. Number the main chain starting from the end that gives the lowest number to the first location at which branching occurs. See Figure 9.15, part (b).

**Step 4 Find the prefix:** Name each branch as an alkyl group. Give each branch a position number. The number indicates which carbon in the main chain the branch is bonded to. See Figure 9.15, part (c).
- If more than one branch is present, write the names of the branches in alphabetical order. Determine the alphabetical order by using the first letter of the prefix (e.g., methyl or ethyl), not the multiplying prefix (e.g., di-, tri-, tetra-, etc.).
- If there are two or more of the same type of branch, use multiplying prefixes such as di- (meaning 2), tri- (meaning 3), and tetra- (meaning 4) to indicate how many of each type of branch are present.
- Use hyphens to separate words from numbers. Use commas to separate numbers.
- When possible, put numbers in ascending order. For example, you would write 2,2,5-trimethylpentane instead of 5,2,2-trimethylpentane.

**Step 5 Put the name together:** prefix + root + suffix

The following Sample Problem gives two examples of naming alkanes. Complete the Practice Problems that follow to apply what you have learned.

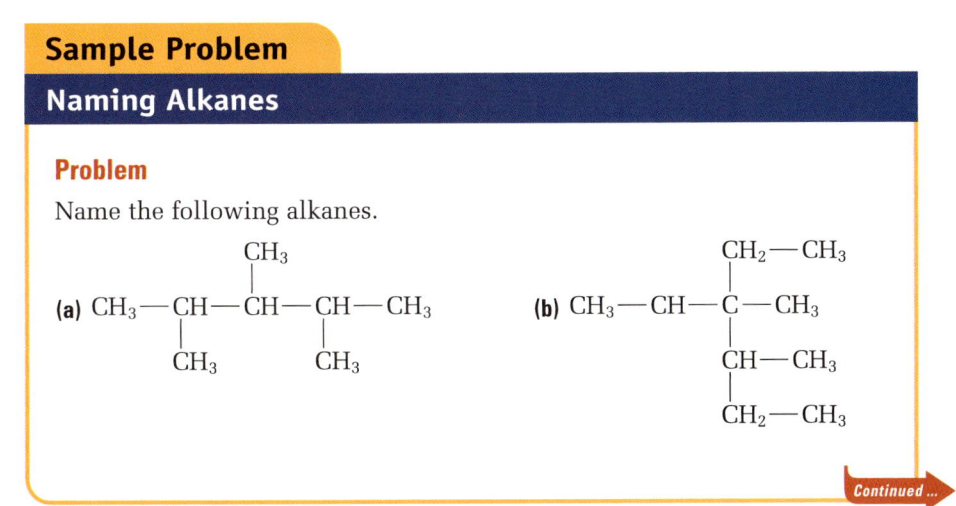

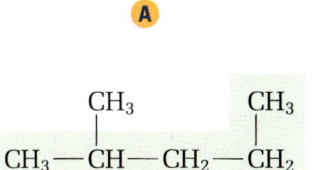

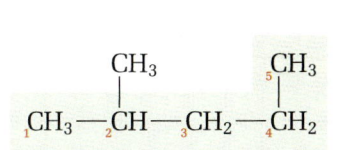

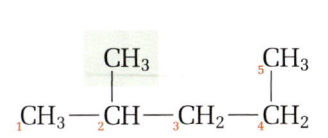

**Figure 9.15** (A) There are five carbons in the longest chain, as highlighted. The root of the name is *-pent-*. (B) There is only one branch. Since the branch is closer to the left end of the molecule, number the carbon at the left end "1." Then number the other main chain carbons consecutively. (C) The branch is a methyl group. The branch is bonded to the number 2 carbon in the main chain, so it has the position number 2. The prefix is *2-methyl*. The full name is 2-methylpentane.

### Sample Problem

#### Naming Alkanes

**Problem**

Name the following alkanes.

(a) $CH_3-CH(CH_3)-CH(CH_3)-CH(CH_3)-CH_3$

(b) $CH_3-CH(CH_2-CH_3)-C(CH_3)(CH_2-CH_3)-CH_3$ with CH—CH₃ and CH₂—CH₃ branches

*Continued...*

**Solution**

(a) **Step 1 Find the root:** The longest chain has 5 carbon atoms, so the root is -pent-.

**Step 2 Find the suffix:** The suffix is -ane.

**Step 3 Give position numbers:** It doesn't matter which end you start numbering at, the result will be the same.

**Step 4 Find the prefix:** Three methyl groups are attached to the main chain at positions 2, 3, and 4. The prefix is 2,3,4-trimethyl-.

**Step 5** The full name is 2,3,4-trimethylpentane.

(b) **Steps 1 and 2 Find the root and suffix:** There are 6 carbon atoms in the main chain. (Note that the main chain is bent at two places.) The root is -hex-. The suffix is -ane.

**Step 3 Give position numbers:** Number from the left so there are two branches at position 3 and one branch at position 4.

**Step 4 Find the prefix:** There is an ethyl group and a methyl group at position 3, and another methyl group at position 4. The prefix is 3-ethyl-3,4-dimethyl-.

**Step 5** The full name is 3-ethyl-3,4-dimethylhexane.

## Practice Problems

5. Name each compound.

(a) 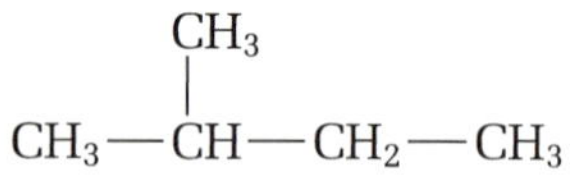 $CH_3-CH(CH_3)-CH_2-CH_3$

(b) 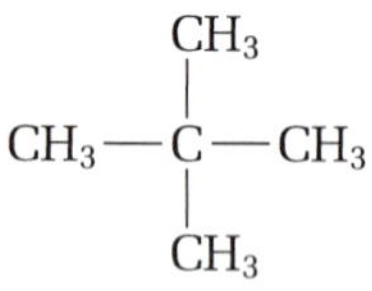 $CH_3-C(CH_3)(CH_3)-CH_3$

(c) $CH_3-CH(CH_2CH_3)-CH_2-CH(CH_2CH_3)-CH(CH_3)-CH_3$
   with a $CH_2-CH_3$ branch

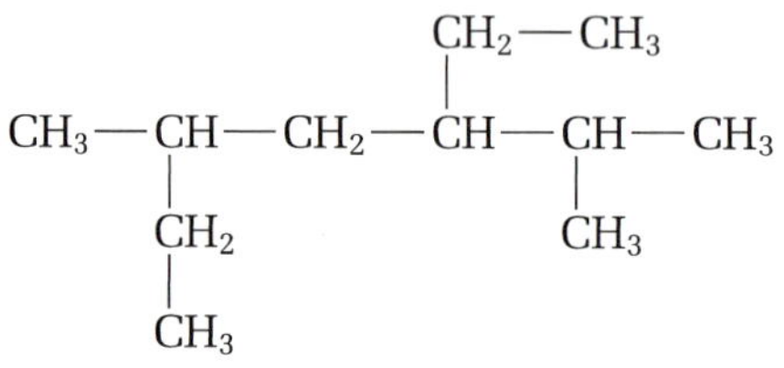

(d) 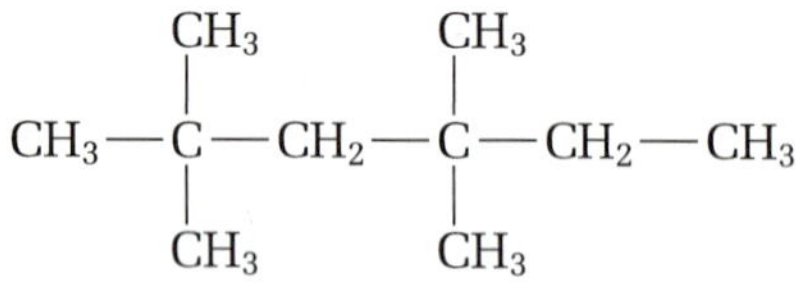 $CH_3-C(CH_3)(CH_3)-CH_2-C(CH_3)(CH_3)-CH_2-CH_3$

(e) 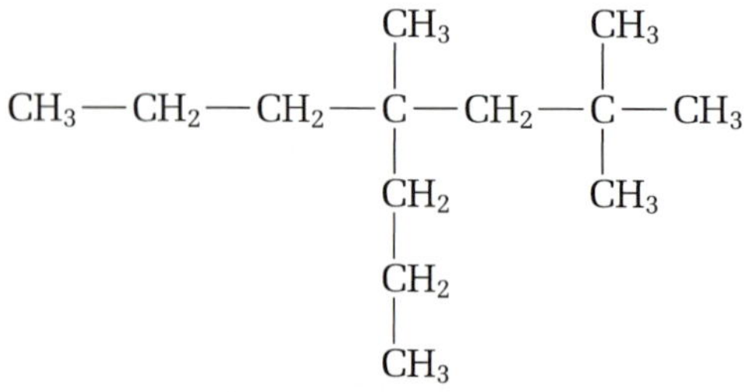 $CH_3-CH_2-CH_2-C(CH_3)(CH_2CH_2CH_3)-CH_2-C(CH_3)(CH_3)-CH_3$

6. Identify any errors in the name of each hydrocarbon.
   (a) 2,2,3-dimethylbutane
   (b) 2,4-diethyloctane
   (c) 3-methyl-4,5-diethyl-nonane

7. Name each compound.
   (a)
   (b)
   (c)

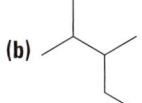

## Drawing Alkanes

As you learned earlier in this chapter, three kinds of diagrams can be used to represent the structure of a hydrocarbon. The easiest kind is probably the condensed structural diagram. When you are asked to draw a condensed structural diagram for an alkane, such as 2,3-dimethylhexane, you can follow several simple rules. These rules are listed below. After you have studied the rules, use the Practice Problems to practise your alkane-drawing skills.

### Rules for Drawing Condensed Structural Diagrams

**Step 1** Identify the root and the suffix of the name. In 2,3-dimethylhexane, for example, the root and suffix are -hexane. The *-hex-* tells you that there are six carbons in the main chain. The *-ane* tells you that the compound is an alkane. Therefore this compound has single carbon-carbon bonds only.

**Step 2** Draw the main chain first. Draw it straight, to avoid mistakes caused by a fancy shape. Do not include any hydrogen atoms. You will need to add branches before you finalize the number of hydrogen atoms on each carbon. Leave space beside each carbon on the main chain to write the number of hydrogen atoms later.

C —C —C —C —C —C

**Step 3** Choose one end of your carbon chain to be carbon number 1. Then locate the carbon atoms to which the branches must be added. Add the appropriate number and size of branches, according to the prefix in the name of the compound. In this example, *2,3-dimethyl* tells you that there is one methyl (single carbon-containing) branch on the second carbon of the main chain, and another methyl branch on the third carbon of the main chain. It does not matter whether you place both branches above the main chain, both below, or one above and one below. The compound will still be the same.

```
              CH₃
              |
   C — C — C — C — C — C
              |
              CH₃
```

**Step 4** Finish drawing your diagram by adding the appropriate number of hydrogen atoms beside each carbon. Remember that each carbon has a valence of four. So if a carbon atom has one other carbon atom bonded to it, you need to add three hydrogen atoms. If a carbon atom has two other carbon atoms bonded to it, you need to add two hydrogen atoms, and so on.

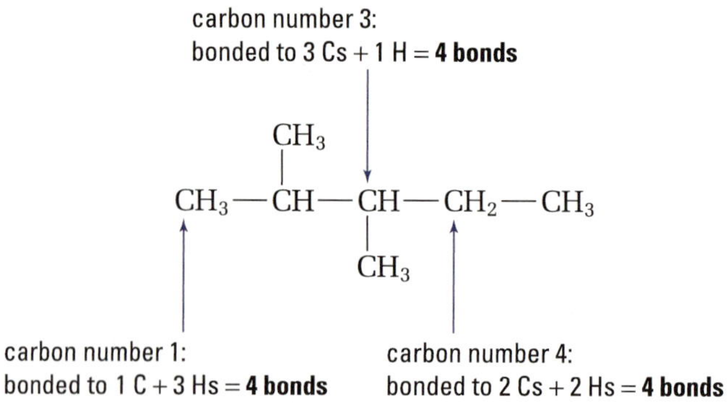

carbon number 3:
bonded to 3 Cs + 1 H = **4 bonds**

```
        CH₃
         |
CH₃—CH—CH—CH₂—CH₃
         |
        CH₃
```

carbon number 1:
bonded to 1 C + 3 Hs = **4 bonds**

carbon number 4:
bonded to 2 Cs + 2 Hs = **4 bonds**

## Sample Problem

### Drawing an Alkane

**Problem**

Draw a condensed structural diagram for 3-ethyl-2-methylheptane.

**Solution**

**Step 1** The root and suffix are -heptane. The -hept tells you that there are 7 carbons in the main chain. The -ane tells you that this compound is an alkane, with only single carbon-carbon bonds.

**Step 2** Draw 7 carbon atoms in a row, leaving spaces for hydrogen atoms.

**Step 3** The ethyl group is attached to carbon 3, and the methyl group to carbon 2.

**Step 4** Add hydrogen atoms so each carbon atom forms 4 bonds.

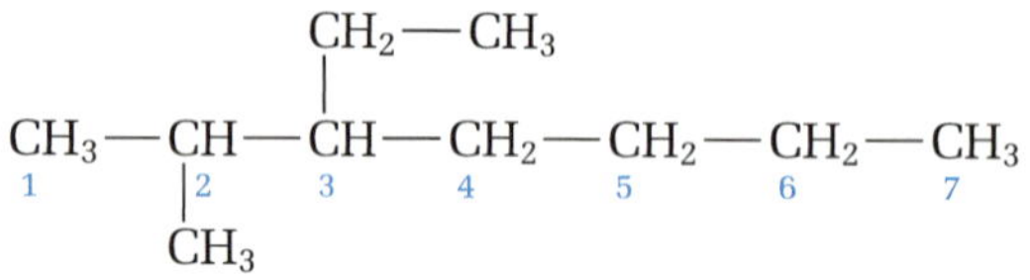

```
              CH₂—CH₃
               |
CH₃—CH—CH—CH₂—CH₂—CH₂—CH₃
 1   |2   3   4    5    6    7
    CH₃
```

## Practice Problems

8. Draw a condensed structural diagram for each hydrocarbon.

    (a) propane
    (b) 4-ethyl-3-methylheptane
    (c) 3-methyloctane

9. Use each *incorrect* name to draw the corresponding hydrocarbon. Examine your drawing, and rename the hydrocarbon correctly.

    (a) 3-propyl-butane
    (b) 1,3-dimethyl-hexane
    (c) 4-methylpentane

10. Draw a line structural diagram for each alkane.

    (a) 3-ethyl-3,4-dimethyloctane
    (b) 2,3,4-trimethylhexane
    (c) 5-ethyl-3,3-dimethylheptane
    (d) 6-isobutyl-4-ethyl-5-methyldecane

11. One way to assess how well you have learned a new skill is to identify mistakes. Examine the following compounds and their names. Identify any mistakes, and correct the names.

(a) 4-ethyl-2-methylpentane

$$CH_3-CH(CH_3)-CH_2-CH(CH_2CH_3)-CH_3$$

(b) 4,5-methylhexane

$$CH_3-CH_2-CH_2-CH(CH_3)-CH(CH_3)-CH_3$$

(c) 3-methyl-3-ethylpentane

$$CH_3-CH_2-C(CH_3)(CH_2CH_2CH_3)-CH_2-CH_3$$

## CHEM FACT

You have learned how the non-polar nature of alkanes affects their boiling point. This non-polarity also affects another physical property: the solubility of alkanes in water. For example, the solubility of pentane in water is only $5.0 \times 10^{-3}$ mol/L at 25°C. Hydrocarbon compounds, such as those found in crude oil, do not dissolve in water. Instead they float on the surface. This physical property helps clean-up crews minimize the devastating effects of an oil spill.

## Physical Properties of Alkanes

Alkanes (and all other aliphatic compounds) have an important physical property. They are non-polar. Therefore, alkanes are soluble in benzene and other non-polar solvents. They are not soluble in water and other polar solvents. The intermolecular forces between non-polar molecules are fairly weak. As a result, hydrocarbons such as alkanes have relatively low boiling points. As the number of atoms in the hydrocarbon molecule increases, the boiling point increases. Because of this, alkanes exist in a range of states under standard conditions.

Table 9.4 compares the sizes (number of carbon atoms per molecule) and boiling points of alkanes. Notice how the state changes as the size increases.

## CONCEPT CHECK

Many industries rely on alkane hydrocarbons. The states of these hydrocarbons can affect how they are stored at industrial sites. For example, methane is a gas under standard conditions. In what state would you expect a large quantity of methane to be stored? What safety precautions would be necessary?

**Table 9.4** Comparing the Sizes and Boiling Points of Alkanes

| Size (number of carbon atoms per molecule) | Boiling point range (°C) | Examples of products |
|---|---|---|
| 1 to 5 | below 30 | gases: used for fuels to cook and heat homes |
| 5 to 16 | 30 to 275 | liquids: used for automotive, diesel, and jet engine fuels; also used as raw materials for the petrochemical industry |
| 16 to 22 | over 250 | heavy liquids: used for oil furnaces and lubricating oils; also used as raw materials to break down more complex hydrocarbons into smaller molecules |
| over 18 | over 400 | semi-solids: used for lubricating greases and paraffin waxes to make candles, waxed paper, and cosmetics |
| over 26 | over 500 | solid residues: used for asphalts and tars in the paving and roofing industries |

**Figure 9.16** Propane burning in a propane torch: A blue flame indicates that complete combustion is occurring.

## Reactions of Alkanes

One of the most common uses of alkanes and other hydrocarbons is as fuel. The combustion of hydrocarbon fossil fuels gives us the energy we need to travel and to keep warm in cold climates. Fossil fuel combustion is also an important source of energy in the construction and manufacturing industries. As well, many power plants burn natural gas when generating electricity. At home, some people burn an alkane, the methane in natural gas, to cook food.

How do we get energy from these compounds? As you will discover, complete and incomplete combustion can be expressed as chemical equations. Combustion in the presence of oxygen is a chemical property of alkanes and all other hydrocarbons.

### Complete and Incomplete Combustion

During a typical combustion reaction, an element or a compound reacts with oxygen to produce oxides of the element (or elements) found in the compound.

Alkanes and other hydrocarbon compounds will burn in the presence of air to produce oxides. This is a chemical property of all hydrocarbons. **Complete combustion** occurs if enough oxygen is present. A hydrocarbon that undergoes complete combustion produces carbon dioxide and water vapour. The following equation shows the complete combustion of propane. (See also Figure 9.16.)

$$C_3H_{8(g)} + 5O_{2(g)} \rightarrow 3CO_{2(g)} + 4H_2O_{(g)}$$

If you burn a fuel, such as propane, in a barbecue, you want complete combustion to occur. Complete combustion ensures that you are getting maximum efficiency from the barbecue. More importantly, toxic gases can result from **incomplete combustion**: combustion that occurs when not enough oxygen is present. During incomplete combustion, other products (besides carbon dioxide and water) can form. The equation below shows the incomplete combustion of propane. Note that unburned carbon, $C_{(s)}$, and carbon monoxide, $CO_{(g)}$, are produced as well as carbon dioxide and water.

$$2C_3H_{8(g)} + 7O_{2(g)} \rightarrow 2C_{(s)} + 2CO_{(g)} + 2CO_{2(g)} + 8H_2O_{(g)}$$

Figure 9.17 shows another example of incomplete combustion. Go back to the equation for the complete combustion of propane. Notice that the mole ratio of oxygen to propane for the complete combustion (5 mol oxygen to 1 mol propane) is higher than the mole ratio for the incomplete combustion (7 mol oxygen to 2 mol propane, or 3.5 mol oxygen to 1 mol propane). These ratios show that the complete combustion of propane used up more oxygen than the incomplete combustion. In fact, the incomplete combustion probably occurred because not enough oxygen was present. You just learned that incomplete combustion produces poisonous carbon monoxide. This is why you should never operate a gas barbecue or gas heater indoors, where there is less oxygen available. This is also why you should make sure that any natural gas or oil-burning furnaces and appliances in your home are working at peak efficiency, to reduce the risk of incomplete combustion. Carbon monoxide detectors are a good safeguard. They warn you if there is dangerous carbon monoxide in your home, due to incomplete combustion.

**Figure 9.17** The yellow flame of this candle indicates that incomplete combustion is occurring. Carbon, $C_{(s)}$, emits light energy in the yellow wavelength region of the visible spectrum.

### Balancing Combustion Equations

Butane is a flammable gas that is used as a fuel in cigarette lighters. How do you write the balanced equation for the complete combustion of an alkane such as butane? Complete hydrocarbon combustion reactions follow a general format:

### hydrocarbon + oxygen → carbon dioxide + water vapour

You can use this general format for the complete combustion of an alkane or any other hydrocarbon, no matter how large or small. For example, both propane and butane burn completely to give carbon dioxide and water vapour. Each hydrocarbon, however, produces different amounts, or mole ratios, of carbon dioxide and water.

You have seen, written, and balanced several types of reaction equations so far in this textbook. In the following Sample Problem, you will learn an easy way to write and balance hydrocarbon combustion equations.

## Sample Problem
### Complete Combustion of Butane

#### Problem
Write the balanced equation for the complete combustion of butane.

#### What Is Required?
You need to write the equation, then balance the atoms of the reactants and products.

#### What Is Given?
You know that butane, $C_4H_{10}$, and oxygen, $O_2$, are the reactants. Since the reaction is a complete combustion, carbon dioxide, $CO_2$, and water, $H_2O$, are the products.

#### Plan Your Strategy
**Step 1** Write the equation.
**Step 2** Balance the carbon atoms first.
**Step 3** Balance the hydrogen atoms next.
**Step 4** Balance the oxygen atoms last.

#### Act on Your Strategy
**Step 1** Write the chemical formulas and states for the reactants and products.
$$C_4H_{10(g)} + O_{2(g)} \rightarrow CO_{2(g)} + H_2O_{(g)}$$
**Step 2** Balance the carbon atoms first.
$$C_4H_{10(g)} + O_{2(g)} \rightarrow 4CO_{2(g)} + H_2O_{(g)}$$
(four carbons)    (four carbons)
**Step 3** Balance the hydrogen atoms next.
$$C_4H_{10(g)} + O_{2(g)} \rightarrow 4CO_{2(g)} + 5H_2O_{(g)}$$
(ten hydrogens)    (ten hydrogens)

*Continued...*

**Step 4** Balance the oxygen atoms last. The product coefficients are now set. Therefore, count the total number of oxygen atoms on the product side. Then place an appropriate coefficient in front of the reactant oxygen.

$$C_4H_{10(g)} + \frac{13}{2}O_{2(g)} \rightarrow 4CO_{2(g)} + 5H_2O_{(g)}$$
**(thirteen oxygens)         (thirteen oxygens)**

You end up with thirteen oxygen atoms on the product side of the equation. This is an odd number. When this happens, use a fractional coefficient so that the reactant oxygen balances, as shown in the equation above. Alternatively, you may prefer to balance the equation with whole numbers. If so, multiply everything by a factor that is equivalent to the denominator of the fraction. Since the fractional coefficient of $O_2$ has a 2 in the denominator, multiply *all* the coefficients by 2 to get whole number coefficients.

$$2C_4H_{10(g)} + 13O_{2(g)} \rightarrow 8CO_{2(g)} + 10H_2O_{(g)}$$

### Check Your Solution

The same number of carbon atoms, hydrogen atoms, and oxygen atoms appear on both sides of the equation.

## Sample Problem

### Incomplete Combustion of 2,2,4-Trimethylpentane

### Problem

2,2,4-trimethylpentane is a major component of gasoline. Write one possible equation for the incomplete combustion of 2,2,4-trimethylpentane.

### What Is Required?

You need to write the equation for the incomplete combustion of 2,2,4-trimethylpentane. Then you need to balance the atoms of the reactants and the products. For an incomplete combustion reaction, more than one balanced equation is possible.

### What Is Given?

You know that 2,2,4-trimethylpentane and oxygen are the reactants. Since the reaction is an incomplete combustion reaction, the products are unburned carbon, carbon monoxide, carbon dioxide, and water vapour.

### Plan Your Strategy

Draw the structural diagram for 2,2,4-trimethylpentane to find out how many hydrogen and oxygen atoms it has. Then write the equation and balance the atoms. There are many carbon-containing products but only one hydrogen-containing product, water. Therefore, you need to balance the hydrogen atoms first. Next balance the carbon atoms, and finally the oxygen atoms.

**Act on Your Strategy**

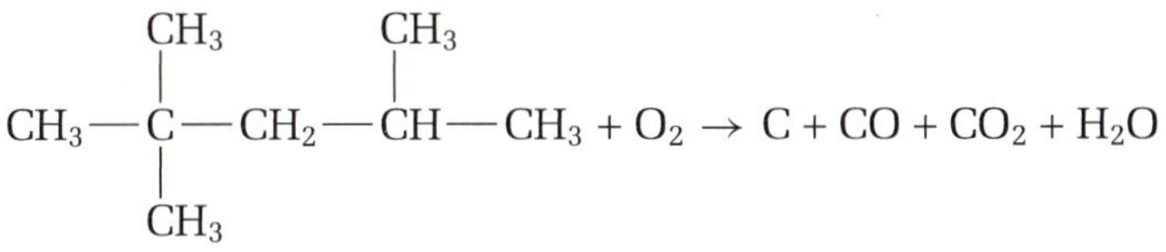

$$CH_3-\underset{\underset{CH_3}{|}}{\overset{\overset{CH_3}{|}}{C}}-CH_2-\underset{}{\overset{\overset{CH_3}{|}}{CH}}-CH_3 + O_2 \rightarrow C + CO + CO_2 + H_2O$$

Count the carbon and hydrogen atoms, and balance the equation. Different coefficients are possible for the carbon-containing product molecules. Follow the steps you learned in the previous sample problem to obtain the balanced equation shown below.

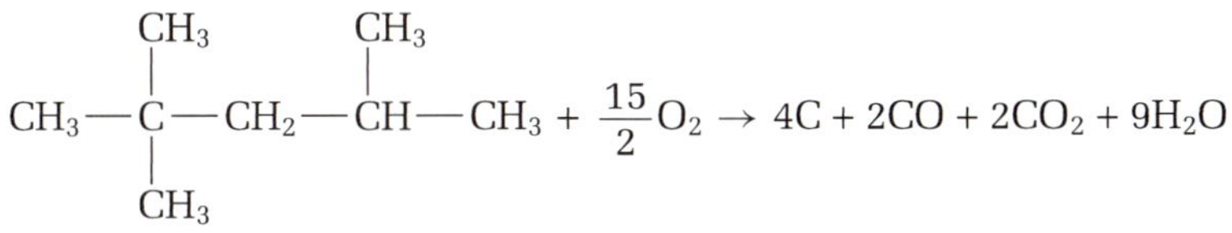

$$CH_3-\underset{\underset{CH_3}{|}}{\overset{\overset{CH_3}{|}}{C}}-CH_2-\overset{\overset{CH_3}{|}}{CH}-CH_3 + \frac{15}{2}O_2 \rightarrow 4C + 2CO + 2CO_2 + 9H_2O$$

*or*

$$2\left(CH_3-\underset{\underset{CH_3}{|}}{\overset{\overset{CH_3}{|}}{C}}-CH_2-\overset{\overset{CH_3}{|}}{CH}-CH_3\right) + 15O_2 \rightarrow 8C + 4CO + 4CO_2 + 18H_2O$$

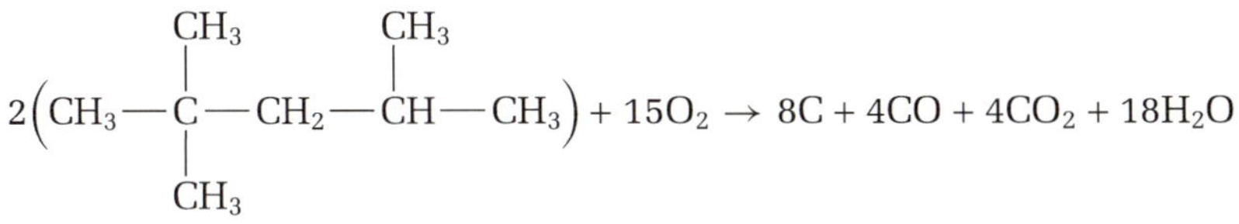

**Check Your Solution**

The same number of carbon atoms, hydrogen atoms, and oxygen atoms appear on both sides of the equation.

## Practice Problems

**12.** The following equation shows the combustion of 3-ethyl-2,5-dimethylheptane:

$C_{11}H_{24} + 17O_2 \rightarrow 11CO_2 + 12H_2O$

(a) Does this equation show *complete* or *incomplete* combustion?

(b) Draw the structural formula for 3-ethyl-2,5-dimethylheptane.

**13.** (a) Write a balanced equation for the complete combustion of pentane, $C_5H_{12}$.

(b) Write a balanced equation for the complete combustion of octane, $C_8H_{18}$.

(c) Write two possible balanced equations for the incomplete combustion of ethane, $C_2H_6$.

**14.** The flame of a butane lighter is usually yellow, indicating *incomplete* combustion of the gas. Write a balanced chemical equation to represent the incomplete combustion of butane in a butane lighter. Use the condensed structural formula for butane.

**15.** The paraffin wax in a candle burns with a yellow flame. If it had sufficient oxygen to burn with a blue flame, it would burn rapidly and release a lot of energy. It might even be dangerous! Write the balanced chemical equation for the complete combustion of candle wax, $C_{25}H_{52(s)}$.

**16.** 4-propyldecane burns to give solid carbon, water vapour, carbon monoxide, and carbon dioxide.

*Continued...*

> **Continued...**
>
> (a) Draw the structural formula for 4-propyldecane.
> (b) Write two different balanced equations for the reaction described in this problem.
> (c) Name the type of combustion. Explain.

## Substitution Reactions of Alkanes

Alkanes react with very few compounds other than oxygen. They are often considered to be almost inert. However, in the presence of ultraviolet light, they will react with halogens in substitution reactions. The reactions are hard to control and result in a mixture of products. For example, both of the following reactions occur in a mixture of butane and chlorine when they are exposed to ultraviolet light.

$$CH_3CH_2CH_2CH_3 + Cl_2 \xrightarrow{UV} CH_3CH_2CH_2CH_2Cl + HCl$$
$$CH_3CH_2CH_2CH_3 + Cl_2 \xrightarrow{UV} CH_3CH_2CHClCH_3 + HCl$$

Analysis of the products shows that 30% of the time, the chlorine atom bonds to a carbon on the end of the butane molecule and 70% of the time, it bonds to a carbon within the chain. You will also find small amounts of products in which chlorine atoms have substituted for more than one hydrogen atom. If enough chlorine is present in a mixture of chlorine and methane, all four possible products will be present as show below.

$$CH_4 + Cl_2 \xrightarrow{UV} CH_3Cl + HCl$$
$$+ Cl_2 \xrightarrow{UV} CH_2Cl_2 + HCl$$
$$+ Cl_2 \xrightarrow{UV} CHCl_3 + HCl$$
$$+ Cl_2 \xrightarrow{UV} CCl_4 + HCl$$

These reactions are not very useful because chemists usually want a specific product. If there are other reactions that will produce the product that they require, chemists will chose the more specific reaction.

So far in this section, you have learned how to name and draw alkanes, the most basic hydrocarbon family. You also examined some physical and chemical properties of alkanes. The next family of hydrocarbons that you will encouter is very similar to the alkane family. Much of what you have already learned can be applied to this next family as well.

## Alkenes

Did you know that bananas are green when they are picked? How do they become yellow and sweet by the time they reach your grocer's produce shelf or your kitchen? Food retailers rely on a hydrocarbon to make the transformation from bitter green fruit to delicious ripe fruit. The hydrocarbon is ethene. It is the simplest member of the second group of aliphatic compounds: the alkenes.

**Alkenes** are hydrocarbons that contain one or more double bonds. Like alkanes, alkenes can form continuous chain and branched-chain structures. They also form a homologous series. As well, they are non-polar, which gives them physical properties similar to those of alkanes.

Alkenes are different from alkanes, however, in a number of ways. First, their bonds are different, as indicated by their suffixes. As you will recall, the -ane ending tells you that alkane compounds are joined by single bonds. *The -ene ending for alkenes tells you that these compounds have one or more double bonds.* A double bond involves four bonding electrons between two carbon atoms, instead of the two bonding electrons in all alkane bonds. Examine Figure 9.18 to see how the presence of a double bond affects the number of hydrogen atoms in an alkene.

**Figure 9.18** This diagram shows how an ethane molecule can become an ethene molecule. The two hydrogen atoms that are removed from ethane often form hydrogen gas, $H_{2(g)}$.

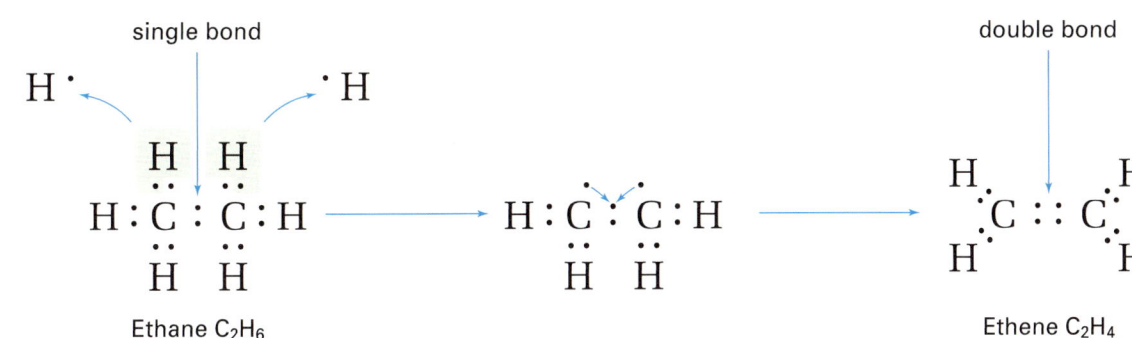

Ethane $C_2H_6$

2 hydrogen atoms must be removed in order to free up 2 bonding electrons

the 2 bonding electrons are then shared between the 2 carbon atoms

Ethene $C_2H_4$

the molecule then contains a double bond with a new ratio of carbon atoms to hydrogen atoms

The general formula for an alkene is $C_nH_{2n}$. You can check this against the next two members of the alkene series. Propene has three carbon atoms, so you would expect it to have six hydrogens. Butene has four carbon atoms, so it should have eight hydrogens. The formulas for propene and butene are $C_3H_6$ and $C_4H_8$, so the general formula is accurate. Note, however, that the general formula applies only if there is one double bond per molecule. You will learn about alkenes with multiple double bonds in future chemistry courses.

**Biology** LINK

According to many nutrition scientists, fats that contain double or triple bonds (unsaturated compounds) are healthier for us than fats that contain single bonds (saturated compounds). Research the implications of including unsaturated and saturated fats and oils in your diet. Use a library, the Internet, or any other sources of information. Decide on a suitable format in which to present your findings.

## Naming Alkenes

The names of alkenes follow the same format as the names of alkanes: **prefix + root + suffix**. The prefixes and the steps for locating and identifying branches are the same, too. The greatest difference involves the double bond. The suffix -ene immediately tells you that a compound has at least one double bond. The number of carbon atoms in the main chain is communicated in the root but the location of the double bond is included in the prefix. Follow the steps below to find out how to name alkenes.

### Rules for Naming Alkenes

**Step 1 Find the root:** Identify the longest continuous chain that contains the double bond. (This step represents the main difference from naming alkanes.)

**Step 2 Find the suffix:** If the compound is an alkene, use the suffix *-ene*.

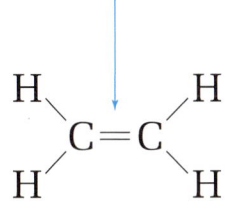

**Step 3 Give position numbers:** Number the main chain from the end that is closest to the double bond. (The double bond should have the lowest possible position number, even if this means the branches have higher position numbers.) Identify the lowest position number that locates the double bond. If the double bond is between carbons 3 and 4, for example, the position number of the double bond is 3.

**Step 4 Find the prefix:** Identify and locate the branches. As you did when naming alkanes, put the branches in alphabetical order. Use multiplying prefixes such as di- and tri- as required.

**Step 5 Put the name together:** prefix + root + suffix

The following Sample Problem gives an example of naming an alkene. Complete the Practice Problems that follow to apply what you have just learned.

### Sample Problem
### Naming an Alkene

**Problem**

Name the following alkene.

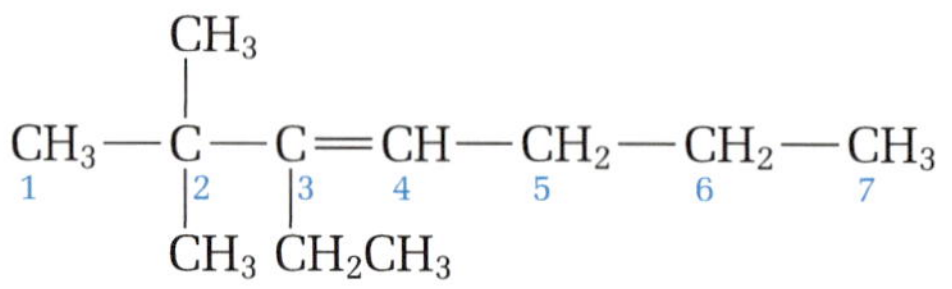

$$\begin{array}{c} \quad\quad\quad CH_3 \\ \quad\quad\quad | \\ CH_3-C-C=CH-CH_2-CH_2-CH_3 \\ 1 \quad |2 \quad |3 \quad 4 \quad\quad 5 \quad\quad 6 \quad\quad 7 \\ \quad\quad CH_3 \ CH_2CH_3 \end{array}$$

**Solution**

**Step 1** Find the root: The longest chain in the molecule has seven carbon atoms. The root is -hept-.

**Step 2** Find the suffix: The suffix is -ene. The root and suffix together are -heptene.

**Step 3** Numbering the chain from the left, in this case, gives the smallest position number for the double bond.

**Step 4** Find the prefix: Two methyl groups are attached to carbon number 2. One ethyl group is attached to carbon number 3. There is a double bond at position 3. The prefix is 3-ethyl-2,2-dimethyl-3-.

**Step 5** The full name is 3-ethyl-2,2-dimethyl-3-heptene.

In the preceding Sample Problem, you learned how to name 3-ethyl-2,2-dimethyl-3-heptene. In this structure, the double bond and the branches are all at the same end of the main chain. What happens if the double bond is close to one end of the main chain, but the branches are close to the other end? For example, how would you name the compound in Figure 9.19?

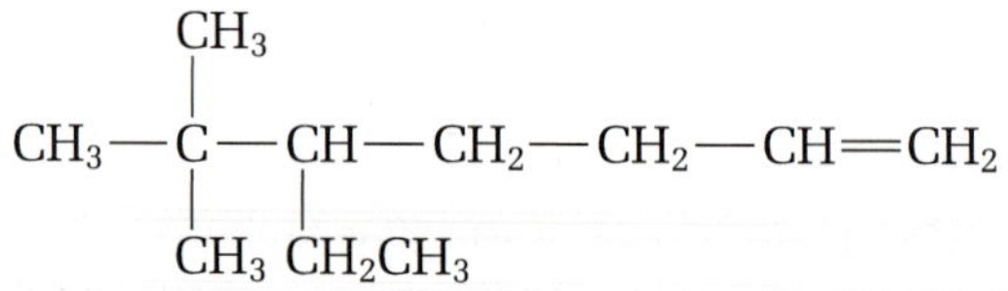

$$\begin{array}{c} \quad\quad\quad CH_3 \\ \quad\quad\quad | \\ CH_3-C-CH-CH_2-CH_2-CH=CH_2 \\ \quad\quad | \quad | \\ \quad\quad CH_3 \ CH_2CH_3 \end{array}$$

**Figure 9.19** How does this structure differ from the one in the sample problem?

The branches are close to the left end of the main chain. The double bond is close to the right end. Therefore, you need to follow the rule in Step 3: *The double bond should have the lowest possible position number.* Now you can name the compound as follows.

- Number the 7 carbon main chain from the right. Give the double bond position number 1. The root and suffix are heptene and the prefix includes a 1.
- There are two methyl groups at position 6 and an ethyl group at position 5. The prefix is 5-ethyl-6, 6-dimethyl-1-.
- The full name is 5-ethyl-6, 6-dimethyl-1-heptene. Although the compound you have just named has a different structural formula than the compound in the Sample Problem above, they both have the same molecular formula: $C_{11}H_{22}$. These compounds are both isomers of $C_{11}H_{22}$. You worked with structural isomers earlier in this chapter. You have seen that isomers can be made by rearranging carbon and hydrogen atoms and creating new branches. If you carefully analyze the structures in the sample problem and Figure 9.19, you will see that these isomers were made by rearranging the double bonds.

## Drawing Alkenes

You draw alkenes using the same method you learned for drawing alkanes. There is only one difference: you have to place the double bond in the main chain. Remember the valence of carbon, and be careful to count to four for each carbon atom on the structure. Figure 9.20 illustrates this point. Be especially careful with the carbon atoms on each side of the double bond. The double bond is worth two bonds for each carbon. Now complete the Practice Problems to reinforce what you have learned about naming and drawing alkenes.

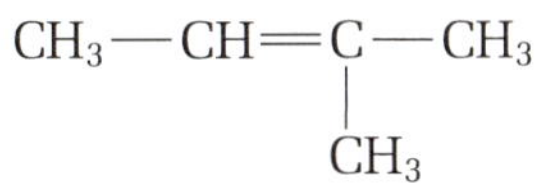

$CH_3-CH=C-CH_3$
              $|$
             $CH_3$

**Figure 9.20** Each carbon atom is bonded four times, once for each valence electron.

### Practice Problems

17. Name each hydrocarbon.
    (a) $CH_3-CH_2-CH=CH-CH_2-CH_3$
    (b) $CH_3-CH_2-CH_2-CH_2-C=CH-CH_3$
                                   $|$
                                  $CH_2$
                                   $|$
                                  $CH_2$
                                   $|$
                                  $CH_3$

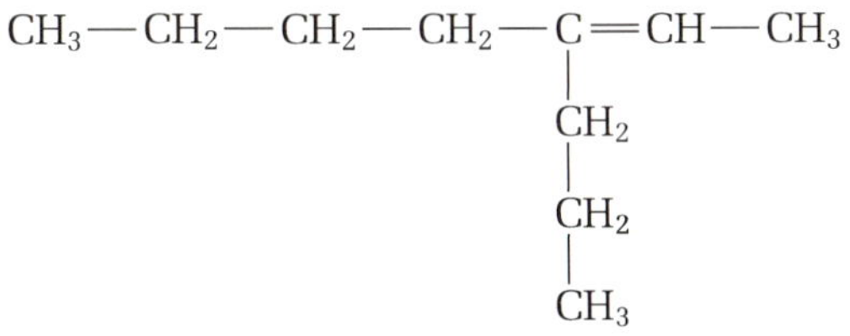

    (c) $CH_3-CH-CH-C-CH_2-CH_3$
            $|$   $|$ $||$
           $CH_3$ $CH_3$ $CH-CH_2-CH_2-CH_3$
           (with $CH_3$ above the third carbon)

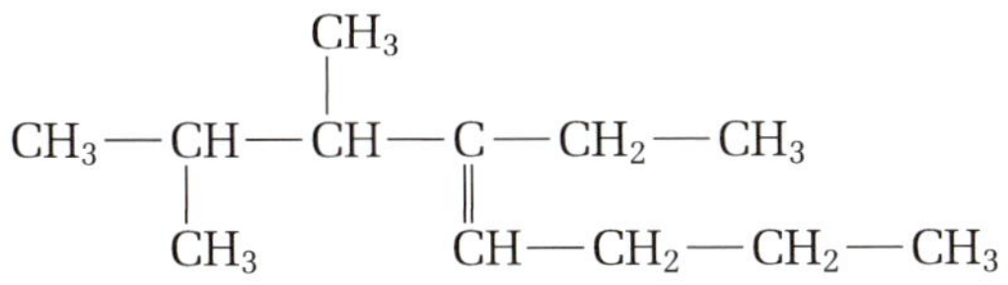

18. Draw a condensed structural diagram for each compound.
    (a) 2-methyl-1-butene
    (b) 5-ethyl-3,4,6-trimethyl-2-octene

19. You have seen that alkenes, such as $C_{11}H_{22}$, can have isomers. Draw condensed structural formulas for the isomers of $C_4H_8$. Then name the isomers.

**PROBLEM TIP**

The easiest way to tell whether or not isomers are true isomers or different sketches of the same compound is to name them. Two structures that look different may turn out to have the same name. If this happens, they are not true isomers.

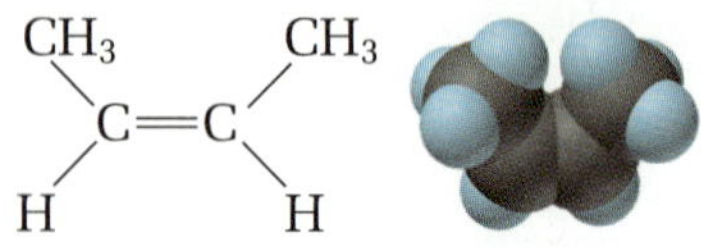

cis-2-Butene

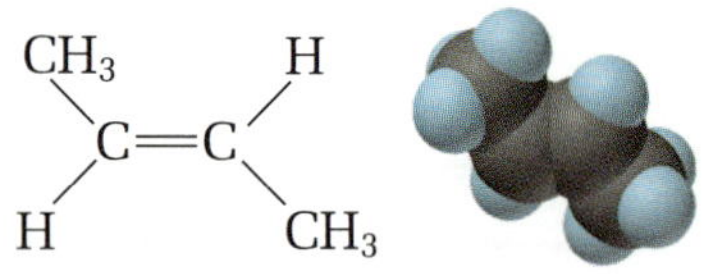

trans-2-Butene

**Figure 9.21** These diagrams show the cis and trans isomers of 2-butene. Notice that the larger methyl groups are on the same side of the double bond (both above) in the cis isomer. They are on opposite sides (one above and one below) in the trans isomer.

**Web LINK**

www.mcgrawhill.ca/links/atlchemistry

Click on the web site given above and go to the **Electronic Learning Partner** to find out more about cis-trans isomers.

## Cis-Trans (Geometric) Isomers

You have seen that isomers result from rearranging carbon atoms and double bonds in alkenes. Another type of isomer results from the presence of a double bond. It is called a **cis-trans isomer** (or **geometric isomer**). Cis-trans isomers occur when different groups of atoms are arranged around the double bond. Unlike the single carbon-carbon bond, which can rotate, the double carbon-carbon bond remains fixed. As a result, groups bonded to the carbon atoms that are joined by a double bond, remain on the same side of the double bond. Figure 9.21 shows the geometric isomers of 2-butene.

Remember these general rules:

- To have a cis-trans (geometric) isomer, each carbon in the C=C double bond must be attached to two different groups.
- In a *cis isomer*, the two larger groups that are attached to the double bonded carbon atoms are on the same side of the double bond.
- In a *trans isomer*, the two larger groups that are attached to the double bonded carbon atoms are on opposite sides of the double bond.

Like all isomers, cis-trans isomers have different physical and chemical properties. For example, the cis-2-butene isomer has a boiling point of 3.7°C, while the trans-2-butene isomer has a boiling point of 0.9°C.

### Practice Problems

20. Draw and name the cis-trans isomers for $C_5H_{10}$.

21. Why can 1-butene not have cis-trans isomers? Use a structural diagram to explain.

22. Like other isomers, two cis-trans isomers have the same atomic weight. They also yield the same elements when decomposed. How might you distinguish between two such isomers in the lab?

23. $C_6H_{12}$ has four possible pairs of cis-trans isomers. Draw and name all four pairs.

## Physical Properties of Alkenes

The general formula for alkenes implies that at least two carbon atoms in any alkene compound have fewer than four bonded atoms. As a result, chemists refer to alkenes as unsaturated compounds. Unlike saturated compounds, **unsaturated hydrocarbons** contain carbon atoms that can potentially bond to additional atoms.

Unsaturated compounds have physical properties that differ from those of saturated compounds. For example, the boiling points of alkenes are usually slightly lower than the boiling points of similar-sized alkanes (alkanes with the same number of carbon atoms). This difference reflects the fact that the forces between molecules are slightly smaller for alkenes than for alkanes. For example, the boiling point of ethane is −89°C, whereas the boiling point of ethene is −104°C. On the other hand, both alkenes and alkanes have a low solubility in water and other non-polar solvents. Alkenes, like all aliphatic compounds, are non-polar. They dissolve in non-polar solvents.

Earlier in this section, you learned that ethene is used to ripen fruit. How does this work? When a fruit ripens, enzymatic reactions in the fruit begin to produce ethene gas which acts as a plant hormone. The ethene is responsible for the colour change, as well as the softening and sweetening that occur as the fruit ripens. Food chemists have learned that they can suppress ethene production (and delay ripening) by keeping fruits at a low temperature while they are transported. Once the fruits reach their final destination, ethene can be pumped into the fruit containers to hasten the ripening process.

> **PROBEWARE**
>
> www.mcgrawhill.ca/links/atlchemistry
>
> If you have access to probeware, go to the above web site and do the Properties of Hydrocarbons lab, or do a similar lab available from a probeware company.

## Reactions of Alkenes

Like alkanes, alkenes react with oxygen in complete and incomplete combustion reactions. You can use the same steps you learned earlier for alkanes to write and balance equations for the combustion of alkenes.

As you will see, however, alkenes are much more reactive than alkanes. The reactivity of an alkene takes place around the double bond. Alkenes commonly undergo addition reactions. In an **addition reaction**, atoms are added to a multiple bond. One bond of the multiple bond breaks so that two new bonds can form. To recognize an addition reaction, remember that *two* compounds usually react to form *one* major product or two products that are isomers of each other. The product has more atoms bonded to carbon atoms than the organic reactant did. A general example of addition to an alkene is given below.

$$-CH=CH- \; + \; XY \; \rightarrow \; -\underset{\underset{\text{}}{|}}{\overset{\overset{X}{|}}{C}H}-\underset{\underset{\text{}}{|}}{\overset{\overset{Y}{|}}{C}H}-$$

The letters XY represent a small molecule that adds to the double bond in the alkene. During the reaction, this small molecule is split up into two parts. Each part of the molecule bonds to one of the carbon atoms that originally formed a double bond. Some common atoms and groups of atoms that can be added to a double bond include:

- H and OH (the reacting molecule is $H_2O$)
- H and Cl (the reacting molecule is HCl. HBr and HI can react in the same way)
- Cl and Cl– (the reacting molecule is $Cl_2$. $Br_2$ and $I_2$ can react in the same way)
- H and H (the reacting molecule is $H_2$)

In the following investigation, you will compare the reactivity of alkanes and alkenes.

# Investigation 9-B

**SKILL FOCUS**
Predicting
Performing and recording
Analyzing and interpreting
Conducting research

## Comparing the Reactivity of Alkanes and Alkenes

Because of differences in reactivity, you can use aqueous potassium permanganate, $KMnO_{4(aq)}$, to distinguish alkanes from alkenes. When the permanganate ion comes in contact with unsaturated compounds, such as alkenes, a reaction occurs. The permanganate ion changes to become manganese dioxide. This is shown by a colour change, from purple to brown. When the permanganate ion comes in contact with saturated compounds, such as alkanes, that have only single bonds, no reaction occurs. The colour of the permanganate does not change.

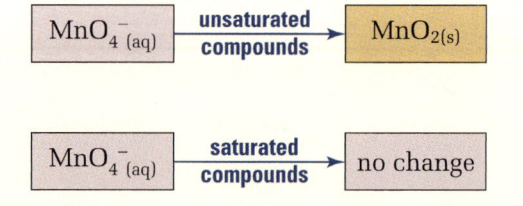

$MnO_4^-{}_{(aq)}$ →(unsaturated compounds)→ $MnO_{2(s)}$

$MnO_4^-{}_{(aq)}$ →(saturated compounds)→ no change

Aqueous potassium permanganate was added to the two test tubes on the left. One test tube contains an alkane compound. The other test tube contains an alkene compound. Which is which?

### Question
How can you use aqueous potassium permanganate in a test to identify unsaturated compounds in fats and oils?

### Predictions
Predict whether each substance to be tested will react with aqueous potassium permanganate. (In part, you can base your predictions on what you know about saturated and unsaturated fats from the media, as well as from biology and health classes.)

### Safety Precautions
Do not spill any $KMnO_{4(aq)}$ on your clothing or skin, because it will stain. If you do accidentally spill $KMnO_{4(aq)}$ on your skin, remove the stain using a solution of sodium bisulfite.

### Materials
test tubes (13 × 100 mm)
test tube rack
stoppers
medicine droppers
hot plate
5.0 mmol/L $KMnO_{4(aq)}$
water bath
samples of vegetable oils, such as varieties of margarine, corn oil, and coconut oil
samples of animal fats, such as butter and lard

## Procedure

1. Read steps 2 and 3. Design a table to record your predictions, observations, and interpretations. Give your table a title.

2. Place about two full droppers of each test substance in the test tubes. Use a different dropper for each substance, or clean the dropper each time. Melt solids, such as butter, in a warm water bath (40°C to 50°C). Then test them as liquids.

3. Use a clean dropper to add one dropper full of potassium permanganate solution to each substance. Record your observations.

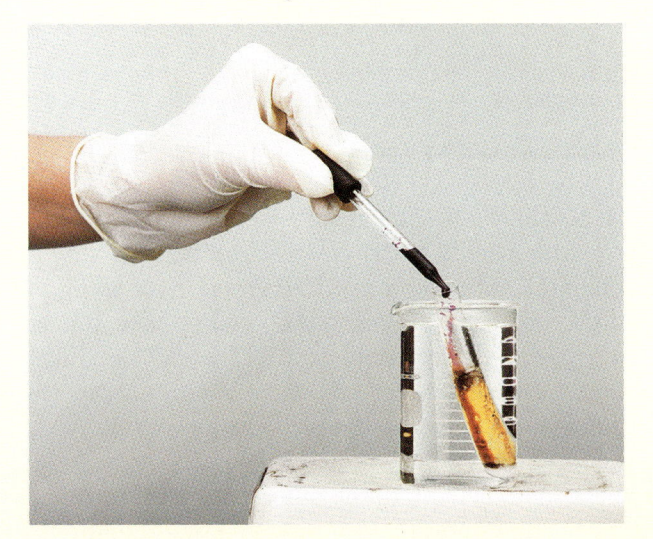

4. Dispose of the reactants as directed by your teacher.

## Analysis

1. On what basis did you make your predictions? How accurate were they?

2. Did any of your results surprise you? Explain your answer.

## Conclusion

3. Is there a connection between your results and the unsaturated or saturated compounds in the substances you tested? Explain your answer.

## Application

4. Investigate the possible structures of some of the compounds (fatty acids) that are contained in various fats and oils. Use both print and electronic resources. See if you can verify a link between the tests you performed and the structures of these fats and oils. Check out compounds such as trans-fats (fats containing trans-fatty acids) and cis-fats (fats containing cis-fatty acids). Record your findings, and compare them with your classmates' findings.

## Symmetrical and Asymmetrical Addition to Alkenes

The product of an addition reaction depends on the symmetry of the reactants. A *symmetrical alkene* has identical groups on either side of the double bond. Ethene, $CH_2=CH_2$, is an example of a symmetrical alkene. An alkene that has different groups on either side of the double bond is called an *asymmetrical alkene*. Propene, $CH_3CH=CH_2$, is an example of an asymmetrical alkene.

The molecules that are added to a multiple bond can also be classified as symmetrical or asymmetrical. For example, chlorine, $Cl_2$, breaks into two identical parts when it adds to a multiple bond. Therefore, it is a symmetrical reactant. Water, HOH, breaks into two different groups (H and OH) when it adds to a multiple bond, so it is an asymmetrical reactant.

In Figures 9.22 and 9.23, at least one of the reactants is symmetrical. When one or more reactants in an addition reaction are symmetrical, only one product is possible. (You will learn how to name the products of these reactions in the next chapter.)

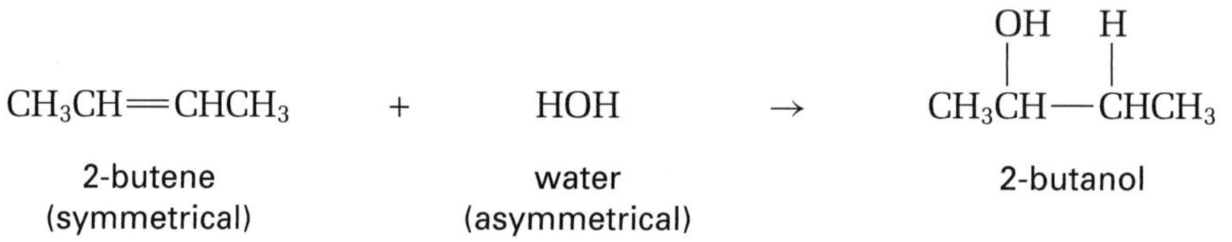

$$CH_3CH=CHCH_3 \ + \ HOH \ \rightarrow \ CH_3\underset{|}{\overset{OH}{C}}H-\underset{|}{\overset{H}{C}}HCH_3$$

2-butene (symmetrical)     water (asymmetrical)     2-butanol

**Figure 9.22** The addition of water to 2-butene

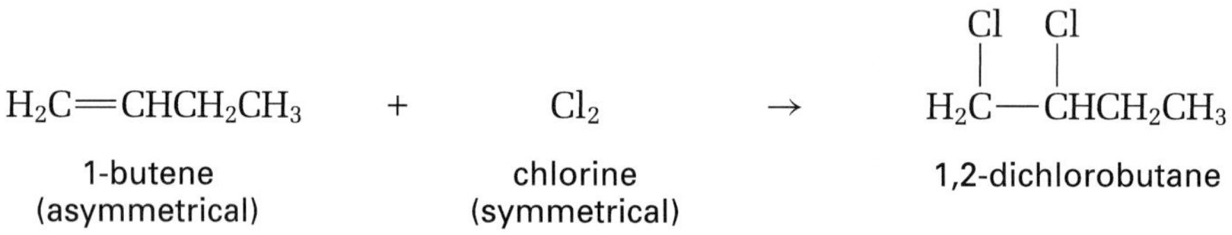

$$H_2C=CHCH_2CH_3 \ + \ Cl_2 \ \rightarrow \ H_2\underset{|}{\overset{Cl}{C}}-\underset{|}{\overset{Cl}{C}}HCH_2CH_3$$

1-butene (asymmetrical)     chlorine (symmetrical)     1,2-dichlorobutane

**Figure 9.23** The addition of chlorine to 1-butene

When both reactants are asymmetrical, however, more than one product is possible. This is shown in Figure 9.24. The two possible products are isomers of each other.

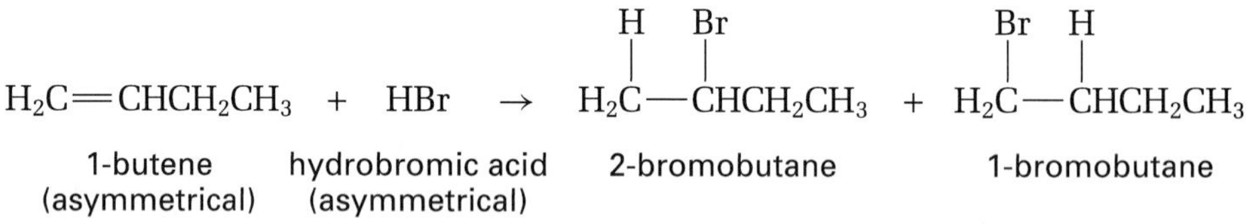

$$H_2C=CHCH_2CH_3 \ + \ HBr \ \rightarrow \ H_2\underset{|}{\overset{H}{C}}-\underset{|}{\overset{Br}{C}}HCH_2CH_3 \ + \ H_2\underset{|}{\overset{Br}{C}}-\underset{|}{\overset{H}{C}}HCH_2CH_3$$

1-butene (asymmetrical)    hydrobromic acid (asymmetrical)    2-bromobutane    1-bromobutane

**Figure 9.24** The addition of hydrobromic acid to 1-butene

Although both products are possible, 2-bromobutane is the observed product. This observation is explained by Markovnikov's rule. **Markovnikov's rule** states that *the halogen atom or OH group in an addition reaction is usually added to the more substituted carbon atom—the carbon atom that is bonded to the largest number of other carbon atoms*. Think of the phrase "the rich get richer." The carbon with the most hydrogen atoms receives even more hydrogen atoms in an addition reaction. According to Markovnikov's rule, the addition of two asymmetrical reactants forms primarily one product. Only a small amount of the other isomer is formed. The following Sample Problem shows how to use Markovnikov's rule to predict the products of addition reactions.

## Sample Problem

### Using Markovnikov's Rule

**Problem**

Draw the reactants and products of the following incomplete reaction.

2-methyl-2-pentene + hydrochloric acid →?

Use Markovnikov's rule to predict which of the two isomeric products will form in a greater amount.

**Solution**

According to Markovnikov's rule, *the hydrogen atom will go to the double-bonded carbon with the larger number of hydrogen atoms.* Thus, the chlorine atom will go to the other carbon, which has the larger number of C—C bonds. The product with the chlorine atom on the number 2 carbon (left) will form in the greater amount. Small amounts with the chlorine on the number 3 carbon (right) will also form. (You will learn to name these products in Chapter 10.)

$$CH_3-\underset{\underset{CH_3}{|}}{C}=CH-CH_2-CH_3 \quad + \quad HCl \quad \rightarrow$$

2-methyl-2-pentene      hydrochloric acid

$$CH_3-\underset{\underset{Cl}{|}}{\overset{\overset{CH_3}{|}}{C}}-CH_2-CH_2-CH_3 \quad + \quad CH_3-\underset{}{\overset{\overset{CH_3}{|}}{CH}}-\underset{\underset{Cl}{|}}{CH}-CH_2-CH_3$$

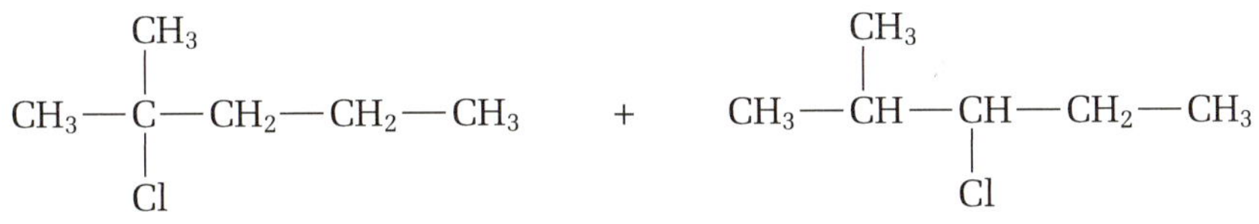

(major product)      (minor product)

## Practice Problems

**24.** Draw the reactants and products of the following reaction.

3-ethyl-2-heptene + HOH →

Use Markovnikov's rule to predict which of the two products will form in the greater amount.

**25.** Use Markovnikov's rule to predict which of the two products will form in the greater amount.

**(a)** $CH_2=CHCH_2CH_2CH_2CH_3 + HBr \rightarrow$

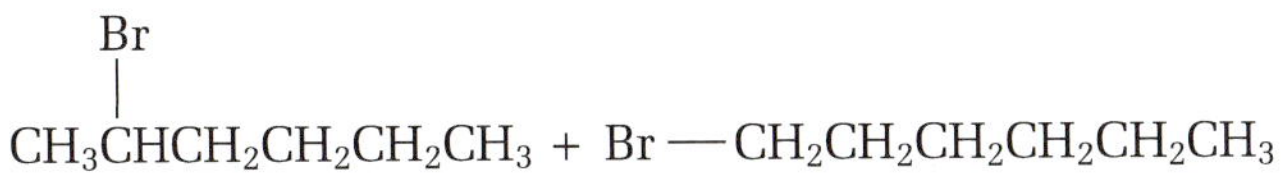

$$\underset{}{\overset{\overset{Br}{|}}{CH_3CH}}CH_2CH_2CH_2CH_3 \quad + \quad Br-CH_2CH_2CH_2CH_2CH_2CH_3$$

**(b)** 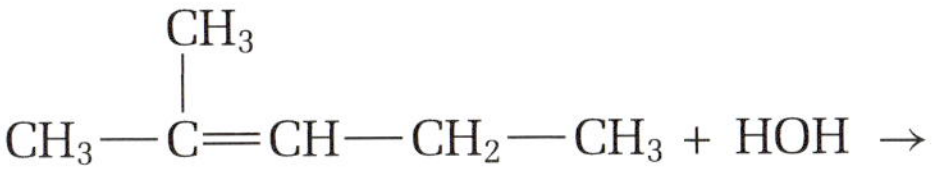 $CH_3-\underset{\underset{}{}}{\overset{\overset{CH_3}{|}}{C}}=CH-CH_2-CH_3 + HOH \rightarrow$

$$CH_3-\underset{\underset{OH}{|}}{\overset{\overset{CH_3}{|}}{C}}-CH_2-CH_2-CH_3 \quad + \quad CH_3-\overset{\overset{CH_3}{|}}{CH}-\underset{\underset{OH}{|}}{CH}-CH_2-CH_3$$

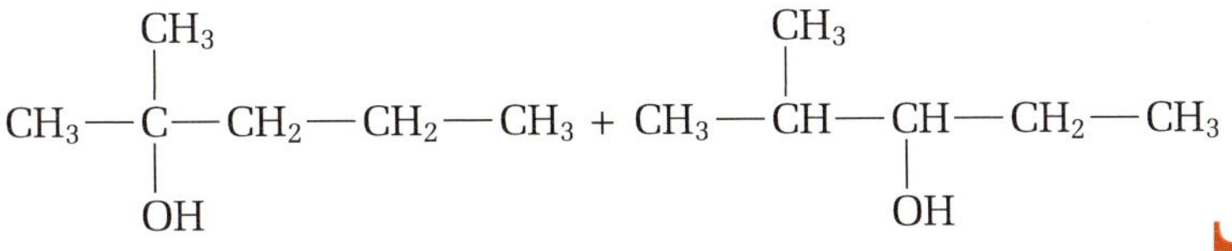

*Continued...*

> **Continued...**
>
> **26.** Draw the major product of each reaction.
>   - **(a)** $CH_3CH=CH_2 + Br_2 \rightarrow$
>   - **(b)** $CH_2=CH_2 + HOH \rightarrow$
>   - **(c)** $CH_2=CHCH_2CH_3 + HBr \rightarrow$
>   - **(d)** $(CH_3)_2C=CHCH_2CH_2CH_3 + HCl \rightarrow$
>
> **27.** For each reaction, name and draw the reactants that are needed to produce the given product.
>   - **(a)** $? + ? \rightarrow CH_3CH(Cl)CH_3$
>   - **(b)** $? + ? \rightarrow Br-CH_2CH_2-Br$
>   - **(c)** $? + HOH \rightarrow CH_3CH_2\underset{\underset{CH_3}{|}}{\overset{\overset{OH}{|}}{C}}CH_2CH_3$
>   - **(d)** $CH_2=CHCH_3 + ? \rightarrow CH_3CH_2CH_3$

## Alkynes

Carbon and hydrogen atoms can be arranged in many ways to produce a great variety of compounds. Yet another way involves triple bonds in the structure of compounds. This bond structure creates a class of aliphatic compounds called alkynes. **Alkynes** are aliphatic compounds that contain one or more triple bonds.

## Naming and Drawing Alkynes

Both double and triple bonds are multiple bonds. Therefore alkynes are unsaturated hydrocarbons, just as alkenes are. To name alkynes and draw their structures, follow the same rules that you used for alkenes. The only difference is the suffix *-yne*, which you use when naming alkyne compounds. Also, remember to count the number of bonds for each carbon. An alkyne bond counts as three bonds.

As you might expect, the shapes of alkynes are different from the shapes of alkanes and alkenes. A structure with a triple bond must be linear around the bond. (See Figure 9.25.)

Alkynes are similar to both alkanes and alkenes because they form a homologous series. Alkynes have the general formula of $C_nH_{2n-2}$. So, for example, the first member of the alkyne series, ethyne, has the formula $C_2H_2$. (You may know this compound by its common name: acetylene.) The next member, propyne, has the formula $C_3H_4$. The Practice Problems on the next page give you an opportunity to name and draw some alkynes.

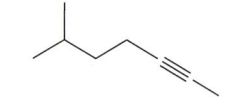

**Figure 9.25** Alkynes are linear around the triple bond, as these examples show.

> ### Practice Problems
>
> **28.** Name each alkyne.
>   - **(a)** $CH_3-\underset{\underset{CH_3}{|}}{\overset{\overset{CH_3}{|}}{C}}-C{\equiv}C-CH_3$

(b) CH₃—CH—CH₂—C—C≡CH
     |         |
    CH₃      CH₂—CH₃ (with CH₃ above C, and CH₂—CH₂—CH₃ below C)

**29.** Draw a condensed structural diagram for each compound.
   (a) 2-pentyne
   (b) 4,5-dimethyl-2-heptyne
   (c) 3-ethyl-4-methyl-1-hexyne
   (d) 2,5,7-trimethyl-3-octyne

## Physical Properties of Alkynes

Alkynes are non-polar compounds, like other hydrocarbons. They are not very soluble in water. Alkynes with low molecular masses exist as gases.

The boiling points of alkynes are in the same range as those of alkenes and alkanes. Interestingly, however, the boiling points of alkynes are slightly higher than either alkenes or alkanes that have the same number of carbon atoms. The boiling points of a few straight chain alkanes, alkenes, and alkynes are given in Table 9.5 for comparison.

| Alkanes | Boiling Point (°C) | Alkenes | Boiling Point (°C) | Alkynes | Boiling Point (°C) |
|---|---|---|---|---|---|
| Ethane | −89 | ethene | −104 | ethyne | −84 |
| Propane | −42 | propene | −47 | propyne | −23 |
| Butane | −0.5 | 1-butene | −6.3 | 1-butyne | 8.1 |
| Pentane | 36 | 1-pentene | 30 | 1-pentyne | 39 |
| Hexane | 69 | 1-hexene | 63 | 1-hexyne | 71 |

### Reactions of Alkynes

The presence of the triple bond makes alkynes even more reactive than alkenes. In fact, alkynes are so reactive that few of these compounds occur naturally. Like other hydrocarbons, alkynes are flammable compounds that react with oxygen in combustion reactions. Alkynes also undergo addition reactions similar to those of alkenes.

### Addition Reactions of Alkynes

Since alkynes have triple bonds, two addition reactions can take place in a row. If one mole of a reactant, such as HCl, Br₂, or H₂O, is added to one mole of an alkyne, the result is a substituted alkene.

H—C≡C—CH₃   +   Br₂   →   H—C=C—CH₃ (with Br above and Br below)

propyne (1 mol)   bromine (1 mol)   1,2-dibromopropene

**Math LINK**

www.mcgrawhill.ca/links/atlchemistry

Go to the web site above and click on **Electronic Learning Partner** for more information about the bonding in ethyne.

If two moles of the reactant are added to one mole of an alkyne, the reaction continues one step further. A second addition reaction takes place, producing an alkane.

$$H-C\equiv C-CH_3 + 2Br_2 \rightarrow H-\underset{\underset{Br}{|}}{\overset{\overset{Br}{|}}{C}}-\underset{\underset{Br}{|}}{\overset{\overset{Br}{|}}{C}}-CH_3$$

propyne (1 mol)　　bromine (2 mol)　　1,1,2,2-tetrabromopropane

Like alkenes, asymmetrical alkynes follow Markovnikov's rule when an asymmetrical molecule, such as $H_2O$ or HBr, is added to the triple bond. An example is given below.

$$H-C\equiv C-CH_2CH_3 + HBr \rightarrow H_2C=\overset{\overset{Br}{|}}{C}-CH_2CH_3 + HC=\overset{\overset{Br}{|}}{CH}-CH_2CH_3$$

1-butyne (1 mol)　　hydrobromic acid (1 mol)　　2-bromo-1-butene (major product)　　1-bromo-1-butene (minor product)

## Cyclic Hydrocarbons

You have probably heard the term "steroid" used in the context of athletics. (See Figure 9.26.) Our bodies contain steroids, such as testosterone (a male sex hormone) and estrone (a female sex hormone). Steroids also have important medicinal uses. For example, budesonide is a steroid that is used to treat asthma. One of the most common steroids is cholesterol. This compound is essential to your normal body functions, but it has been linked to blocked artery walls and heart disease, as well.

Steroids have also been associated with misuse, especially at the Olympics and other sporting events. Some athletes have tried to gain an advantage by using steroids to increase their muscle mass.

**Figure 9.26** Steroids are organic compounds. Our bodies make steroids naturally. Steroids may also be synthesized in chemical laboratories. What do the structures of these steroids have in common?

356　MHR • Unit 4 Organic Chemistry

What do steroids have to do with hydrocarbons? Steroids are unsaturated compounds. Although they are complex organic molecules, their basic structure centres on four rings of carbon atoms. In other words, steroids are built around ring structures of alkanes and alkenes.

Hydrocarbon ring structures are called **cyclic hydrocarbons**. They occur when the two ends of a hydrocarbon chain join together. In order to do this, a hydrogen atom from each end carbon must be removed, just as in the formation of a multiple bond. (See Figure 9.27 on the next page.)

## Naming and Drawing Cyclic Hydrocarbons

To draw the structure of a cyclic hydrocarbon, use a line diagram in a ring-like shape, such as the one shown in Figure 9.27. Each carbon-carbon bond is shown as a straight line. Each corner of the ring represents a carbon atom. Hydrogen atoms are not shown, but they are assumed to be present in the correct numbers. Cyclohexane is a member of the alkane family. Notice, however, that cycloalkanes, such as cyclohexane, have two fewer hydrogen atoms compared with other alkanes. Thus they have the general formula $C_nH_{2n}$. (This is the same as the general formula for alkenes.)

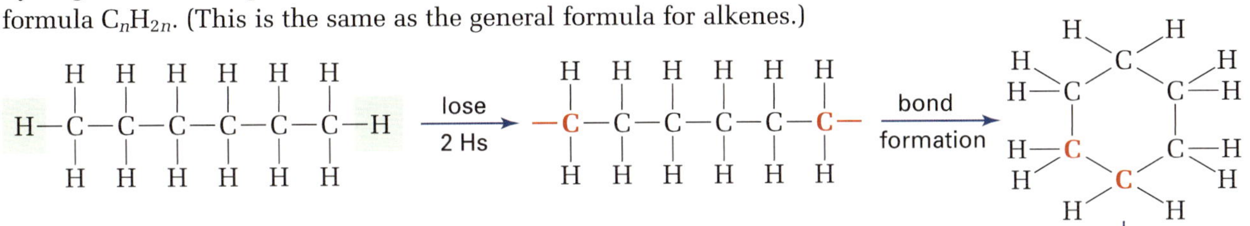

**Figure 9.27** How hexane, $C_6H_{14}$, can become cyclohexane, $C_6H_{12}$

Because of the ring structure, the naming rules for cyclic hydrocarbons, including cycloalkanes and cycloalkenes, are slightly different from those for alkanes and alkenes. Below are four examples.

To draw cyclic hydrocarbons, start with the rules you learned for drawing other types of compounds. To place multiple bonds and branches, you have the option of counting in either direction around the ring.

### Naming Cyclic Hydrocarbons

**Example 1:** Use the general formula: prefix + root + suffix. In Figure 9.28, there are only single carbon-carbon bonds. There are also five corners (carbon atoms) in the ring, which is the main chain. Since there are no branches, the name of this compound is cyclopentane. Notice the addition of *cyclo-* to indicate the ring structure.

**Example 2:** When naming cycloalkanes, all carbon atoms in the ring are treated as equal. This means that any carbon can be carbon number 1. In Figure 9.29, only one branch is attached to the ring. Therefore the carbon that the branch is attached to is carbon number 1. Because this branch automatically gets the lowest possible position number, no position number is required in the name. Thus the name of this compound is methylcyclohexane.

**Example 3:** When two or more branches are on a ring structure, each must have the lowest possible position number. Which way do you count the carbons around the ring? You can count in either direction around the ring. In Figure 9.30, a good choice is to make the ethyl branch carbon number 1 and then count counterclockwise. This allows you to sequence the branches in alphabetical order and to add the position numbers in ascending order. So the name of this structure is 1-ethyl-3-methylcyclohexane.

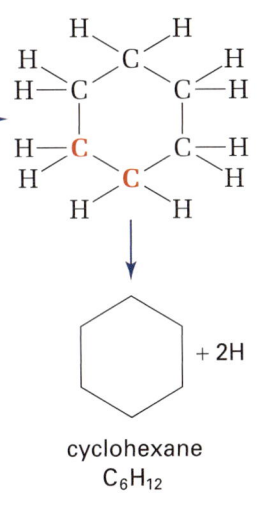

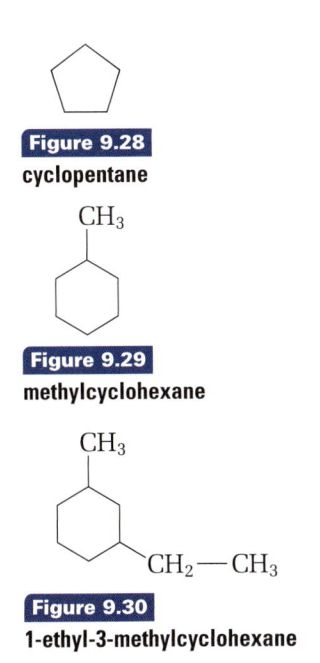

**Figure 9.28** cyclopentane

**Figure 9.29** methylcyclohexane

**Figure 9.30** 1-ethyl-3-methylcyclohexane

Chapter 9 Hydrocarbons • MHR **357**

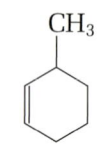

**Figure 9.31**
3-methyl-1-cyclohexene

**Example 4:** In Figure 9.31, there is a double bond, represented by the extra vertical line inside the ring structure. You must follow the same rules as for alkenes. That is, the double bond gets priority for the lowest number. This means that one of the carbon atoms, on either end of the double bond, must be carbon number 1. The carbon atom at the other end must be carbon number 2. Next you have to decide in which direction to count so that the branch gets the lowest possible position number. In this compound, the carbon atom on the bottom end of the double bond is carbon number 1. Then you can count clockwise so that the methyl group on the top carbon of the ring has position number 3. (Counting in the other direction would give a higher locating number for the branch.) The name of this structure is 3-methyl-1-cyclohexene.

### Practice Problems

**30.** Name each compound.

**31.** Draw a condensed structural diagram for each compound.
- (a) 1,2,4-trimethylcycloheptane
- (b) 2-ethyl-3-propyl-1-cyclobutene
- (c) 3-methyl-2-cyclopentene
- (d) cyclopentene
- (e) 1,3-ethyl-2-methylcyclopentane
- (f) 4-butyl-3-methyl-1-cyclohexene
- (g) 1,1-dimethylcyclopentane
- (h) 1,2,3,4,5,6-hexamethyl cyclohexane

### Physical Properties and Reactions of Cyclic Hydrocarbons

Cyclic hydrocarbons are non-polar, and have similar physical properties to alkanes, alkenes, and alkynes. They are flammable, and react with oxygen in combustion reactions. Cycloalkenes and cycloalkynes react in addition reactions following the same patterns as alkenes and alkynes. In the next investigation, you will develop a more thorough understanding of aliphatic compounds by examining their structural and physical properties.

# Investigation 9-C

**SKILL FOCUS**
Predicting
Modelling concepts
Analyzing and interpreting

## Structures and Properties of Aliphatic Compounds

To compare the properties of alkanes, alkenes, and alkynes, you will be working with compounds that have the same number of carbon atoms. First, you will construct and compare butane, trans-2-butene, 2-butyne, and cyclobutane. You will use a graph to compare the boiling points of each compound. Next, you will use what you have just observed to predict the relative boiling points of pentane, trans-2-pentene, 2-pentyne, and cyclopentane. You will construct and compare these structures and graph their boiling points.

### Question
How can constructing models of butane, trans-2-butene, 2-butyne, and cyclobutane help you understand and compare their physical properties?

### Predictions
Predict the structural formula for each compound. After completing steps 1 to 4, predict what the graph of the boiling points of pentane, trans-2-pentene, 2-pentyne, and cyclopentane will look like.

### Materials
molecular model kits
reference books

### Procedure
1. Construct models of butane, trans-2-butene, 2-butyne, and cyclobutane.
2. Examine the structure of each model. In your notebook, write molecular formulas and draw condensed structural diagrams of each model.
3. If possible, rotate the molecule around each carbon-carbon bond to see if this changes the appearance of the structure.
4. Look up the boiling points of these four compounds in the table below. Draw a bar graph to compare the boiling points.
5. Repeat steps 1 to 3 for pentane, trans-2-pentene, 2-pentyne, and cyclopentane.
6. Predict the relative boiling points for these four compounds. Use a reference book to find and graph the actual boiling points.

**Comparing the Boiling Points of Four-Carbon Compounds**

| Compound | Boiling point (°C) |
|---|---|
| butane | −0.5 |
| trans-2-butene | 0.9 |
| 2-butyne | 27 |
| cyclobutane | 12 |

### Analysis
1. What are the differences between the multiple-bond compounds and the alkanes?
2. You have been comparing straight chain alkanes, cycloalkanes, trans-2-alkenes and 2-alkynes having the same number of carbon atoms. For these compounds, which type has the highest boiling point? Which type has the lowest boiling point?

### Conclusion
3. Identify possible reasons for the differences in boiling points between compounds.

### Application
4. Compare the boiling points of cyclopentane and cyclobutane. Use this information to put the following compounds in order from highest to lowest boiling point: cyclohexane, cyclobutane, cyclopropane, cyclopentane. Use a reference book to check your order.

Chapter 9 Hydrocarbons • MHR  359

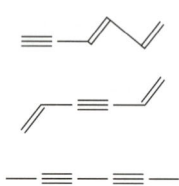

**Figure 9.32** All these structures have the molecular formula C₆H₆. The properties of benzene show, however, that none of these structures can be correct.

**Figure 9.33** Kekule's representation of the benzene molecule shows three single bonds and three double bonds.

### History LINK

One story describes Kekule's discovery like this: After an exhausting effort to work out benzene's structure, Kekule fell asleep. He dreamt about a snake that took its tail in its mouth to become a ring. Kekule woke up with his new idea for benzene: a ring structure!

**Figure 9.34** This structure shows a more accurate way to represent the delocalized bonding electrons in benzene.

## Aromatic Hydrocarbons

At the beginning of this section, you learned that hydrocarbons can be classified into two groups: aliphatic hydrocarbons, and aromatic hydrocarbons. An **aromatic hydrocarbon** is a hydrocarbon based on the aromatic benzene group. The molecular formula of benzene is $C_6H_6$. For many years, scientists could not determine the structure of benzene. Because its molecular formula follows the pattern $C_nH_n$, they reasoned that the compound should contain two double bonds and one triple bond, or even two triple bonds. Figure 9.32 gives three different possible structures for benzene.

In 1865, a German chemist named August Kekulé proposed a new structure for benzene, shown in Figure 9.33. This structure was ring-shaped and symmetrical. It matched the physical and chemical properties of benzene more closely than any of the other structures that had been suggested. Two properties of benzene, however, still did not match the proposed structure.

- As you have just learned, compounds with double and triple bonds undergo addition reactions. However, *benzene does not undergo addition reactions under ordinary conditions.*

- The length of a single carbon-carbon bond is 148 pm. The length of a double carbon-carbon bond is 134 pm. However, *the length of each carbon-carbon bond in benzene is 140 pm.*

Based on these two facts, scientists reasoned that benzene cannot contain ordinary double or triple bonds.

We know today that benzene is a flat, cyclic compound with the equivalent of three double bonds and three single bonds, as Kekule's structure shows. However, the electrons that form the double bonds in benzene are spread out and shared over the whole molecule. Electrons that behave in this way are called *delocalized electrons*. Thus, benzene actually has six identical bonds, each one half-way between a single and a double bond. These bonds are much more stable than ordinary double bonds and do not react in the same way. Figure 9.34 shows a more accurate way to represent the bonding in benzene. Molecules with this type of special electron sharing are called *aromatic compounds*. Benzene is the simplest aromatic compound. Almost all aromatic compounds contain one or more benzene rings.

Figure 9.35 illustrates some common aromatic compounds. To name an aromatic compound, follow the steps below. Figure 9.36 gives an example.

methylbenzene (toluene)

phenylethene (styrene)

2,4,6-trinitromethylbenzene (trinitrotoluene, TNT)

**Figure 9.35** The common name for methylbenzene is toluene. Toluene is used to produce explosives, such as trinitrotoluene (TNT). Phenylethene, with the common name styrene, is an important ingredient in the production of plastics and rubber.

### Naming an Aromatic Hydrocarbon

**Step 1** Number the carbons in the benzene ring. If more than one type of branch is attached to the ring, start numbering at the carbon with the highest priority (or most complex) group. (See the Problem Tip.)

**Step 2** Name any branches that are attached to the benzene ring. Give these branches position numbers. If only one branch is attached to a benzene ring, you do not need to include a position number.

**Step 3** Place the branch numbers and names as a prefix before the root, benzene.

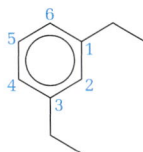

**Figure 9.36** Two ethyl groups are present. They have the position numbers 1 and 3. The name of this compound is 1,3-diethylbenzene.

Chemists do not always use position numbers to describe the branches that are attached to a benzene ring. When a benzene ring has only two branches, the prefixes ortho-, meta-, and para- are sometimes used instead of numbers.

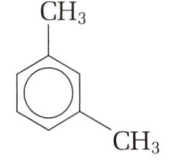

1,2-dimethylbenzene
ortho-dimethylbenzene
(common name: ortho-xylene)

1,3-dimethylbenzene
meta-dimethylbenzene
(common name: meta-xylene)

1,4-dimethylbenzene
para-dimethylbenzene
(common name: para-xylene)

### PROBLEM TIP

Often an organic compound has more than one type of branch. When possible, number the main chain or ring of the compound to give the most important branches the lowest possible position numbers. The table below ranks some branches (and other groups) you will encounter in this chapter, from the highest priority to the lowest priority.

The Priority of Branches

| Highest priority | —OH |
| --- | --- |
| | —NH$_2$ |
| | —F, —Cl, —Br, —I |
| | —CH$_2$CH$_2$CH$_3$ |
| | —CH$_2$CH$_3$ |
| Lowest priority | —CH$_3$ |

### CHEM FACT

Kathleen Yardley Lonsdale (1903–1971) used X-ray crystallography to prove that benzene is a flat molecule. All six carbon atoms lie in one plane, forming a regular hexagonal shape. The bonds are all exactly the same length. The bond angles are all 120°.

### Practice Problems

32. Name the following aromatic compound.

33. Draw a structural diagram for each aromatic compound.
    (a) 1-ethyl-3-methylbenzene
    (b) 2-ethyl-1,4-dimethylbenzene
    (c) para-dichlorobenzene (**Hint:** *Chloro* refers to the chlorine atom, Cl.)

34. Give another name for the compound in question 33(a).

35. Draw and name three aromatic isomers with the molecular formula C$_{10}$H$_{14}$.

## Reactions of Aromatic Compounds

Aromatic compounds do not react in the same way that compounds with double or triple bonds do. Benzene's stable ring does not usually accept the addition of other atoms. Instead, aromatic compounds undergo substitution reactions. In a **substitution reaction**, a hydrogen atom or a group of atoms is replaced by a different atom or group of atoms. To help you recognize this type of reaction, remember that *two* compounds usually react to form *two* different products. The organic reactant(s) and the organic product(s) have the same number of atoms bonded to carbon. Figure 9.37 shows two reactions for benzene. Notice that iron(III) bromide, $FeBr_3$, is used as a catalyst in the substitution reaction.

$$-\overset{|}{\underset{|}{C}}-X \ + \ AY \ \rightarrow \ -\overset{|}{\underset{|}{C}}-Y \ + \ AX$$

An addition reaction does *not* occur because the product of this reaction would be less stable than benzene.

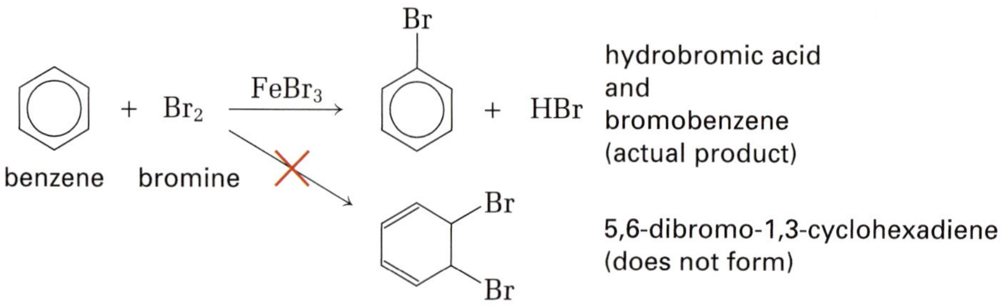

**Figure 9.37** The bromine does not add to benzene in an addition reaction. Instead, one of the H atoms on the benzene ring is replaced with a Br atom in a substitution reaction.

Another example of a substitution reaction for benzene is shown below, in Figure 9.38.

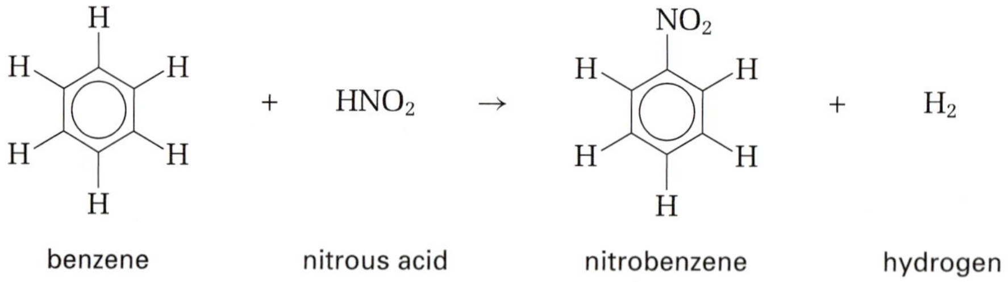

**Figure 9.38** A substitution reaction of benzene. Hydrogen atoms are included in the diagram to make it clear that the -NO₂ group is not added to the benzene ring but is displacing a hydrogen atom.

### Section Summary

In this section, you learned to name and draw aliphatic and aromatic hydrocarbons. You also learned to predict the products of reactions of these compounds. In the next section, you will look at the role played by hydrocarbons in the petrochemical industry.

---

**Web LINK**

www.mcgrawhill.ca/links/atl-chemistry

Go to the web site above and click on **Electronic Learning Partner** for more information about aspects of material covered in this section of the chapter.

## Section Review

**1 (a)** What are the names of the three types of aliphatic compounds that you studied in this section?

**(b)** Which of these are saturated compounds, and which are unsaturated compounds? How does this difference affect their properties?

**(c)** Use Lewis dot diagrams to show how ethane can become ethene, and how ethene can become ethyne. **Hint:** See Figure 9.18.

**2** List the roots used to name the first ten members of the alkane homologous series. Indicate the number of carbon atoms that each represents.

**3** If water and octane are mixed, does the octane dissolve in the water? Explain.

**4 (a)** Name each compound.

(a) CH$_3$—CH(CH$_3$)—CH(CH$_2$CH$_3$)—CH(CH$_3$)—CH$_2$—CH$_3$

(b) CH$_3$—C(CH$_3$)(CH$_3$)—CH$_2$—C(CH$_3$)=CH$_2$

(c) CH≡C—CH(CH$_2$CH$_2$CH$_2$CH$_3$)—CH(CH$_3$)—CH$_2$—CH$_3$ (with CH$_2$—CH$_3$ branch)

(d) cycloheptene with CH$_3$, CH$_3$, CH$_3$, and CH$_3$—CH$_2$ substituents

(e) ethylbenzene (CH$_2$CH$_3$ on benzene ring)

**5** Draw a condensed structural diagram for each compound.

(a) 2,4-dimethyl-3-hexene
(b) 5-ethyl-4-propyl-2-heptyne
(c) 3,5-diethyl-2,4,7,8-tetramethyl-5-propyldecane
(d) trans-4-methyl-3-heptene
(e) 1,4-dimethylbenzene
(f) 2-ethyl-1-dimethylbenzene

**6** Draw and name all the isomers that are represented by each molecular formula. Include any cis-trans isomers and ring structures.

(a) C$_3$H$_4$   (b) C$_5$H$_{12}$   (c) C$_5$H$_{10}$

**7** Identify any mistakes in the name and/or structure of each compound.

(a) 3-methyl-2-butene
CH$_3$—C≡C—CH$_3$ (with CH$_3$ branch)

(b) 2-ethyl-4,5-methyl-1-hexene
cyclohexene with CH$_3$, CH$_3$, and CH$_2$—CH$_3$ substituents

**8** Draw a condensed structural diagram for each of the following compounds.
 (a) propene
 (b) 2-methyl-1-butene
 (c) pentylcyclohexane

**9** Draw two different structures for the benzene molecule. Explain why one structure is more accurate than the other.

**10** What is the difference between incomplete and complete hydrocarbon combustion reactions?

**11** (a) Write the balanced equation for the complete combustion of heptane, $C_7H_{16}$.
 (b) Write a balanced equation for the incomplete combustion of 1-pentene, $C_5H_{10}$.

**12** Identify each reaction as an addition reaction or a substitution reaction. Explain your answer.

(a) 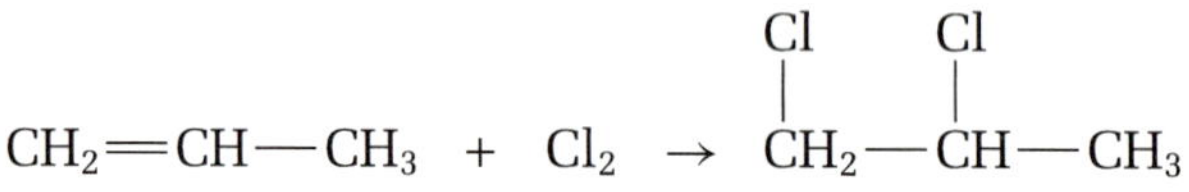 $CH_2{=}CH{-}CH_3 + Cl_2 \rightarrow CH_2Cl{-}CHCl{-}CH_3$

(b) 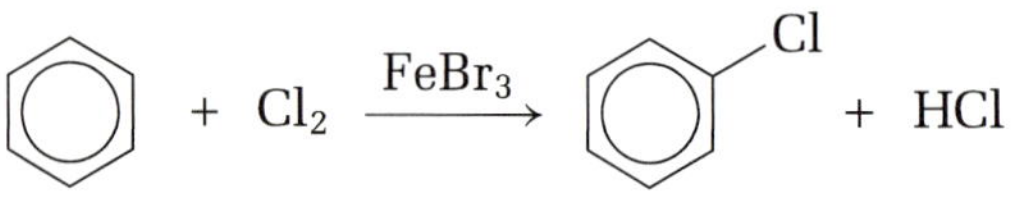 $C_6H_6 + Cl_2 \xrightarrow{FeBr_3} C_6H_5Cl + HCl$

(c) $CH_3CH_2CH(CH_3)CH_3 + Br_2 \xrightarrow{UV} CH_3CH_2CBr(CH_3)CH_3 + HBr$

(d) $HC{\equiv}CCH_2CH_2CH_3 + HCl \rightarrow H_2C{=}CClCH_2CH_2CH$

**13** Draw the product(s) of each reaction.
 (a) $CH_3CH{=}CHCH_3 + Br_2 \rightarrow$
 (b) $CH_2{=}CHCH_2CH_3 + HCl \rightarrow$
 (c) benzene + chlorine $\xrightarrow{FeBr_3}$

**14** (a) State Markovnikov's rule. Under what circumstances is this rule important?
 (b) Which product of the following reaction is formed in the greater amount?

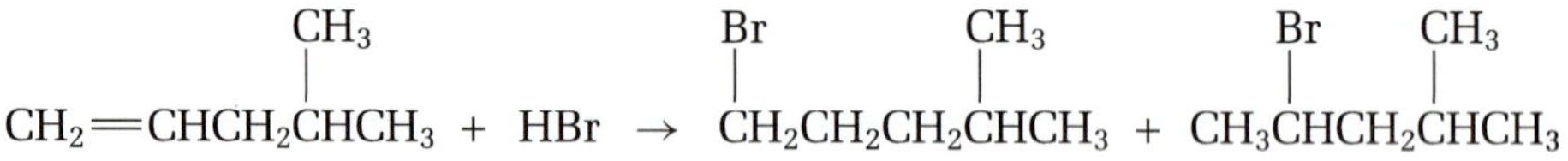

$CH_2{=}CHCH_2CH(CH_3)CH_3 + HBr \rightarrow CH_2BrCH_2CH_2CH(CH_3)CH_3 + CH_3CHBrCH_2CH(CH_3)CH_3$

**15** Draw all of the possible products of the following reaction. Assume that an excess of chlorine is present. That is, the value of "n" can be anything from 1 to 6.

$CH_3CH_3 + nCl_2 \xrightarrow{UV}$

**16** Draw two possible products of the following reaction and indicate which one will be the major product.

$CH_3CH(CH_3)C{\equiv}CH + HBr \rightarrow$

# Refining and Using Hydrocarbons

## 9.4

Is it possible to take hydrocarbons straight out of the ground and use them as they are? Since the earliest recorded history, people have done just that. In the past, people used crude oil that seeped up to Earth's surface to waterproof boats and buildings. They also used it to grease wheels and even dress the wounds of animals and humans. As well, people burned natural gas, mainly to supply lighting for temples and palaces.

Today hydrocarbons are extracted from the ground at well sites, then processed further at refineries. (See Figure 9.39.) The first commercial oil well in North America began production in 1858. At that time, kerosene (which was used to fuel lamps) was the principal focus of the young petroleum industry. Paraffin (for making candles) and lubricating oils were also produced, but there was little demand for other hydrocarbon materials, such as gasoline.

**Section Preview/Outcomes**

In this section, you will

- **describe** the steps involved in refining petroleum to obtain gasoline and other useful fractions
- **explain** the importance of hydrocarbons in the petrochemical industry
- **communicate** your understanding of the following terms: *petrochemicals, fractional distillation, cracking, reforming*

**Figure 9.39** Hydrocarbons are processed at oil refineries like this one.

Reliance on hydrocarbons has increased substantially since the nineteenth century. Our society now requires these compounds for fuel, as well as for the raw materials that are used to synthesize petrochemicals. **Petrochemicals** are basic hydrocarbons, such as ethene and propene, that are converted into plastics and other synthetic materials. Petroleum is the chief source of the petrochemicals that drive our cars and our economy. Petroleum is not a pure substance, however. Rather, petroleum is a complex mixture of hydrocarbons—mainly alkanes and alkenes—of varying molecular sizes and states. Because petroleum is a mixture, its composition varies widely from region to region in the world. An efficient process is essential for separating and collecting the individual, pure hydrocarbon components. Read on to learn about this process in greater detail.

## Using Properties to Separate Petroleum Components

**Fractional distillation** is a process for separating petroleum into its hydrocarbon components. This process relies on a physical property—boiling point. Each of the hydrocarbon components, called *fractions*, has its own range of boiling points. At an oil refinery, the separating (refining) of petroleum begins in a large furnace. The furnace vaporizes the liquid components. The fluid mixture then enters a large fractionation tower. Figure 9.40 outlines how the various hydrocarbon fractions are separated in the tower. (**Note:** The temperatures shown are approximate boiling points for the hydrocarbon fractions.)

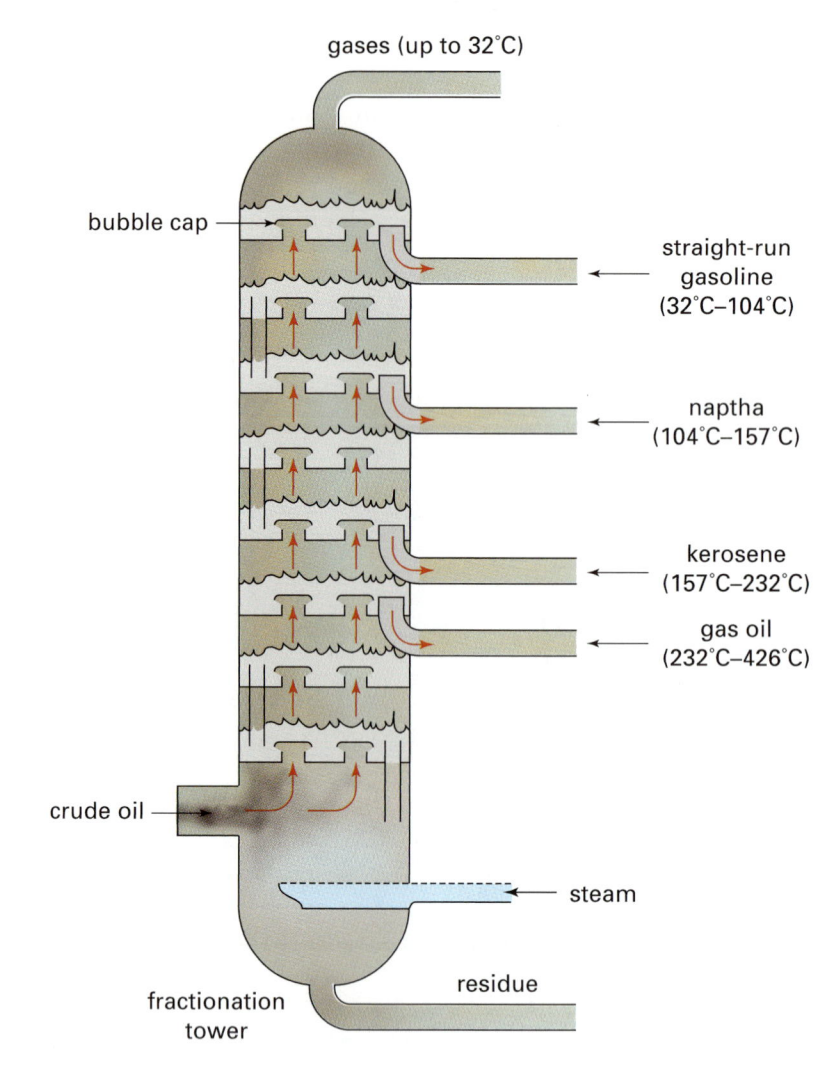

### Web LINK

www.mcgrawhill.ca/links/atlchemistry

Canada's land mass has experienced dynamic changes over tens of millions of years. Climatic conditions, along with the deposition of countless remains of organisms, created the areas in which petroleum is found today. These areas are called sedimentary basins. Where are the sedimentary basins? Are they all active sites for oil and gas extraction? How much oil and gas do scientists estimate there is? How do Canada's reserves compare with those of other petroleum-producing nations? Go to the web site above to "tap" into Canada's and the world's petroleum resources. Decide on a suitable format in which to record your findings.

**Figure 9.40** This diagram shows how petroleum is separated into its hydrocarbon fractions. Each fraction has a different range of boiling points. The tower separates the fractions by a repeated process of heating, evaporating, cooling, and condensing.

Labels on diagram: gases (up to 32°C); bubble cap; straight-run gasoline (32°C–104°C); naptha (104°C–157°C); kerosene (157°C–232°C); gas oil (232°C–426°C); crude oil; steam; residue; fractionation tower

Perforated plates, which are fitted with bubble caps, are placed at various levels in the tower. As each fraction reaches a plate where the temperature is just below its boiling point, it condenses and liquefies. The liquid fractions are taken from the tower by pipes. Other fractions that are still vapours continue to pass up through the plates to higher levels.

Several plates are needed to collect each fraction. Heavier hydrocarbons (larger molecules) have higher boiling points. They condense first and are removed in the lower sections of the tower. The lighter hydrocarbons, with lower boiling points, reach the higher levels of the tower before they are separated.

## Cracking and Reforming

Once the fractions are removed from the distillation tower, they may be chemically processed or purified further to make them marketable. (See Figure 9.41.) There has been a tremendous increase in the demand for a variety of petroleum products in the early twentieth century. This demand has forced the oil industry to develop new techniques to increase the yield from each barrel of oil. These techniques are called cracking and reforming.

**Petroleum Fractionation Products**

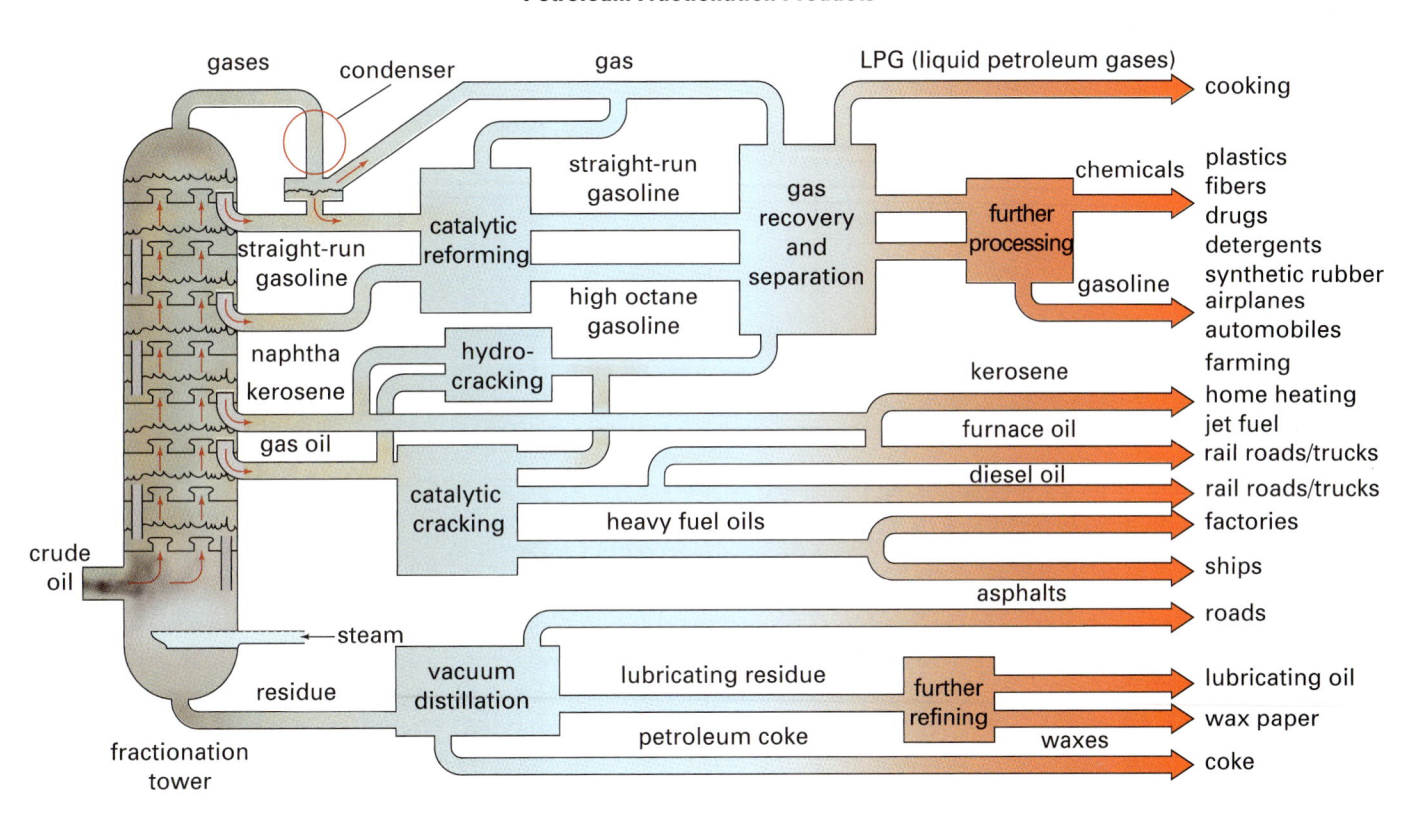

**Figure 9.41** Selected uses of petroleum fractions

**Cracking** was first introduced in Sarnia, Ontario. This process uses heat to break larger hydrocarbon molecules into smaller gasoline molecules. Cracking is done in the absence of air and can produce different types of hydrocarbons. For example, the cracking of propane can produce methane and ethene, as well as propene and hydrogen. (See Figure 9.42.)

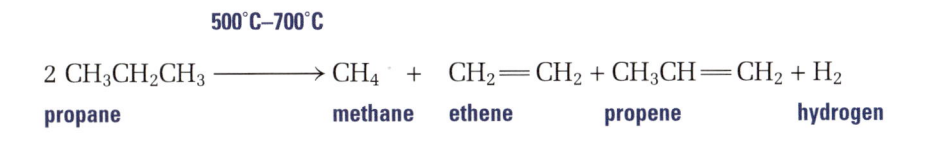

$$2\ CH_3CH_2CH_3 \xrightarrow{500°C-700°C} CH_4 + CH_2{=}CH_2 + CH_3CH{=}CH_2 + H_2$$

propane → methane + ethene + propene + hydrogen

**Figure 9.42** This chemical equation summarizes the cracking of propane.

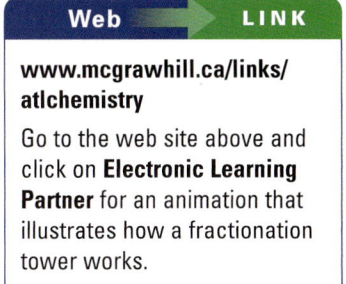

**Web LINK**

www.mcgrawhill.ca/links/atlchemistry

Go to the web site above and click on **Electronic Learning Partner** for an animation that illustrates how a fractionation tower works.

> **Language → LINK**
> What is a catalyst? Use reference books to find out.

**Reforming** is another technique that uses heat, pressure, and catalysts to convert straight chain alkanes to branched chains and also convert cyclic alkanes into aromatic compounds. Branched chain alkanes are desirable because they burn more efficiently in internal combustion engines than straight chains. For example, octane causes knocking in an engine while 2,2,4-trimethyl pentane does not. Also, aromatic compounds are rarely found in the fractions of petroleum but are needed for many applications in the petrochemical industry. Reforming can convert methylcyclohexane into toluene and hydrogen. Some straight chain alkanes can even be converted into aromatic compounds in the reforming process. Figure 9.43 summarizes the processes of cracking and reforming.

### Cracking

Larger Hydrocarbon — Heat, Catalyst → Smaller Hydrocarbon + Smaller Hydrocarbon

### Reforming

Straight chain and cyclic alkanes — Heat, Pressure, Catalyst → Branched chain and aromatic compounds

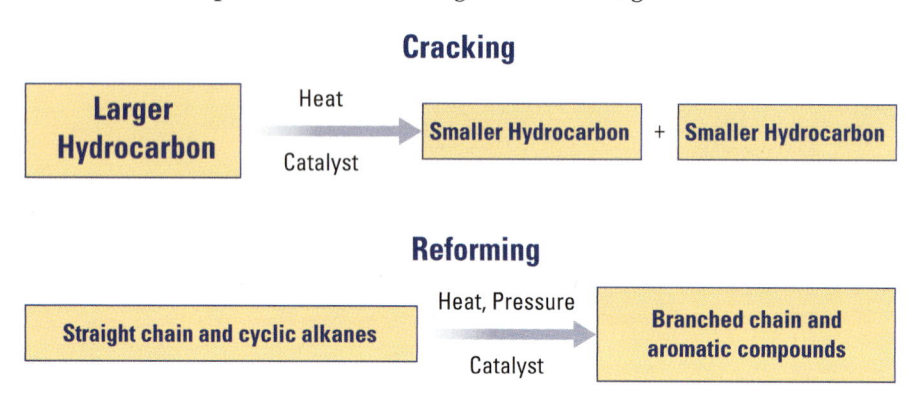

**Figure 9.43** Comparing cracking and reforming

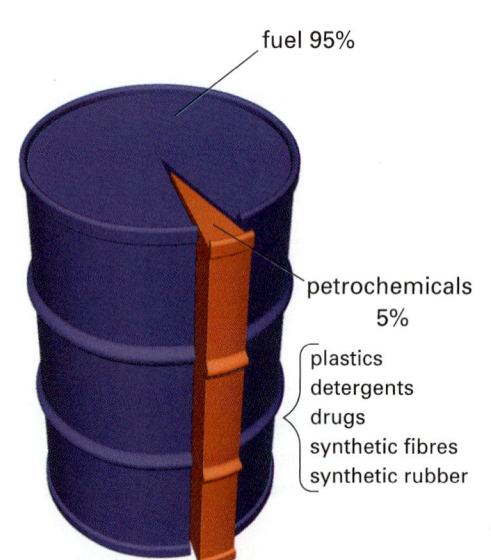

Some of the products of the refining process are transported to petrochemical plants. These plants convert complex hydrocarbons, such as naphtha, into simple chemical compounds, or a small number of compounds, for further processing by other industries. Canadian petrochemical plants produce chemicals such as methanol, ethylene, propylene, styrene, butadiene, butylene, toluene, and xylene. These chemicals are used as building blocks in the production of other finished products. Nearly every room in your school, your home, and your favourite shopping centre contains, or is made from, at least one petrochemical product. Yet, as you can see in Figure 9.44, petrochemicals represent only a small fraction of society's uses of petroleum.

In the next chapter, you will learn about many compounds that are derived from hydrocarbon. You will begin to see why there are millions of types of organic compounds.

Petroleum
- fuel 95%
- petrochemicals 5%
  - plastics
  - detergents
  - drugs
  - synthetic fibres
  - synthetic rubber

**Figure 9.44** A tremendous number of petrochemical products enrich your life. Yet they still make up only about 5% of what is produced from a barrel of crude oil. A staggering 95% of all petroleum is used as fuel.

## Section Summary

In this section, you learned that petrochemicals, an essential part of our society and technology, are obtained from petroleum. You discovered how petroleum is separated into its components by fractional distillation, cracking, and reforming.

In this chapter, you learned a great deal about hydrocarbons. You learned how hydrocarbons are formed in section 9.1. In sections 9.2 and 9.3, you learned to draw, name, and predict the properties of different hydrocarbons. Finally, you learned about the practical side of hydrocarbons—how they are produced and used in everyday life.

# Tools & Techniques

## Oil Refining in Newfoundland and Labrador

The oil refinery at Come By Chance, Newfoundland and Labrador, can process over 100,000 barrels of crude oil per day. The refining process converts the thick, black mixture of hydrocarbons into more usable materials such as petroleum, jet fuel, diesel, and heating oil. Oil refining involves similar chemical techniques to those you may have seen in your school laboratory, such as fractional distillation.

Fractional distillation in an oil refinery is somewhat different from the same process carried out in a laboratory. In a laboratory, the mixture is slowly heated. As each liquid component in the mixture reaches its boiling point, it evaporates and is drawn off. In an oil refinery, however, the entire mixture is quickly heated to a vapour at the base of the tower, using extremely high temperatures. The oil vapour rises up the tower. As each component of the vapour cools and condenses at a different temperature, fractions are drawn off at different heights.

Engineers who design oil refineries work to improve the safety and efficiency of fractional distillation and other processes. To avoid unwanted high pressures and the risk of explosion due to the presence of hot, flammable, toxic gases, equipment is designed to maintain a steady flow of materials. For example, vent stacks purge excess gases while the plant is operating. Gas emissions from these stacks pollute the environment, however. Engineers use large-scale experiments and mathematical modelling to determine the best balance of purging and flow rates needed to maintain safety while minimizing polluting emissions.

Natural oil includes various amounts and types of minerals and gases as well as hydrocarbons. These must be removed from oil and its products to meet market purity standards. For example, sulfur is a common component of diesel and gas. It produces sulfur oxides when these fuels are burned.

Improvements in technology have brought many changes to oil refineries in the past decade. Emissions of sulfur dioxide have been reduced by as much as 80 percent by the use of furnaces which provide more complete combustion. In addition, regular air and water monitoring at refineries reduces the risk of undetected pollution.

## Section Review

**1 (a)** What physical property allows various fractions of crude oil to be separated in a fractionation tower?

**(b)** What is this process of separation called? Briefly describe it. Include a diagram in your answer if you wish.

**2** Our society's demand for petroleum products has increased dramatically over the last century. Describe two techniques that can be used to transform the petroleum fractions from a fractionation tower, into other products to meet this demand.

**3** List five types of petrochemicals that can be produced from refined hydrocarbons. Describe briefly how each type has affected your life.

**4** Draw two possible cracking reactions in which the reactant is $CH_3(CH_2)_{14}CH_3$.

**5** Draw possible products of reforming reactions starting with:

**(a)** $CH_3(CH_2)_6CH_3$     **(b)** 1,3-dimethyl cyclohexane

# Canadians in Chemistry

## Dusanka Filipovic

When Dusanka Filipovic was a teenager, she liked hiking, playing volleyball, and working as a camp counsellor. Now she is a successful chemical engineer who develops and markets technology that helps the environment. What do her early pursuits have to do with her engineering career? "Both require creativity, persistence, and an ability to work under pressure," says Filipovic.

Filipovic was born and grew up in Belgrade, in the former Yugoslavia. She attended the University of Belgrade, and graduated with a degree in chemical engineering. Attracted by Canada's technical advances, she came here after winning a scholarship to study at McMaster University in Hamilton, Ontario. Canada has become her home.

Filipovic has worked hard to break down barriers for women in the fields of engineering and business. In 1974, she became the first female professional engineer employed by a major chemical producing company. The National Museum of Science and Technology in Ottawa, Ontario, has featured her work as part of an exhibit on women inventors.

Since the 1980s, Filipovic has concentrated on developing environmentally friendly technologies. She is the co-inventor of a patented process known as Blue Bottle™ technology. This process is used to recover and recycle halogenated hydrocarbons, such as chlorofluorocarbons (CFCs) and other ozone-depleting substances (ODS), from damaged or unused residential refrigerators and automotive air conditioners.

Using adsorption, the binding of molecules or particles to a surface, Blue Bottle™ technology acts as a selective molecular sieve to capture refrigerant gases so that they can be safely stored and re-used. A non-pressurized Blue Bottle™ cylinder, packed with an adsorbent, synthetic zeolite called Halozite™, is connected to the back of a refrigeration or air-conditioning unit. Zeolites are porous aluminosilicate minerals that commonly contain sodium and calcium as major cations (positively charged ions) and are capable of ion exchange. (Zeolites contain water molecules that allow reversible dehydration, and they are often used as water softeners.) The Halozite™ adsorbs the refrigerants that are released from the unit at ambient temperatures, under atmospheric pressure.

Once the adsorbent is saturated, the Blue Bottle™ cylinder is sent to a central reclamation facility, where the refrigerants are reclaimed and stored. The Blue Bottle™ cylinders can be re-used after the refrigerants have been collected. Reclaimed halogenated hydrocarbons can be used as refrigerants, solvents, cleaners, fumigants, and fire retardants.

In 1991, Filipovic formed her own company to commercialize Blue Bottle™ technology. In 1999, she founded a new company, which uses a process similar to Blue Bottle™ technology to capture and convert greenhouse gas emissions from hospital operating rooms. These emissions had previously been discharged into the atmosphere.

Filipovic has this advice for aspiring engineers and scientists: "Once you recognize what it is you want to pursue, make sure you take advantage of every training opportunity offered to you. I became an engineer because I wanted to be able to create something new, and monitor and enjoy the results of my work . . . I am working to make a significant contribution that will live on after me."

# CHAPTER 9 Review

## Reflecting on Chapter 9
Summarize this chapter in the format of your choice. Here are a few ideas to use as guidelines:
- Identify the origins and sources of hydrocarbons and other organic compounds.
- Describe the characteristics that enable carbon to form so many, varied compounds.
- Distinguish among complete, condensed, and line structural diagrams.
- Identify, draw, and name at least two examples of each kind of hydrocarbon you studied: alkanes, alkenes, alkynes, cyclic hydrocarbons, and aromatics
- Demonstrate, using suitable examples, true isomers of a hydrocarbon.
- Compare physical properties, such as boiling point, of aliphatic compounds.
- Describe the processes and techniques that the petrochemical industry depends on. Identify at least ten products of this industry.

## Reviewing Key Terms
For each of the following terms, write a sentence that shows your understanding of its meaning.

organic compound
hydrocarbons
petroleum
expanded molecular formula
isomers
structural model
structural diagram
alkanes
aliphatic hydrocarbons
alkyl group
homologous series
complete combustion
unsaturated hydrocarbons
alkenes
cis-trans isomer
addition reaction
Markovnikov's Rule
alkynes
cyclic hydrocarbons
aromatic hydrocarbon
substitution reaction
petrochemicals
fractional distillation
cracking
reforming
incomplete combustion

## Knowledge/Understanding
1. (a) What are the origins of most hydrocarbons and other organic compounds?
   (b) List three sources of hydrocarbons.
   (c) What is the main source of hydrocarbons used for fuels?
2. List three factors that are required to change biological matter into petroleum.
3. What are the key properties of the carbon atom that allow it to form such diverse compounds?
4. What are isomers? Give one example of a set of isomers for a molecular formula.
5. Describe how the boiling point changes as the chain length of an aliphatic compound increases. Explain why this happens.
6. Briefly compare alkanes, alkenes, and alkynes. (Give both similarities and differences.)
7. Describe the difference between structural isomers and cis-trans isomers.
8. Compare organic compounds and inorganic compounds, based on the following criteria:
   - the presence of carbon
   - the variety of compounds
   - relative size and mass of molecules
9. Describe the origins of the term "organic."
10. (a) What is an aromatic hydrocarbon?
    (b) How does it differ from an aliphatic hydrocarbon?
11. Write the general formula for:
    (a) an alkane
    (b) an alkene
    (c) an alkyne
    (d) a cycloalkane
    (e) a cycloalkene
12. Based on the general formula, identify each compound as a straight-chain alkane, alkene, or alkyne.
    (a) $C_5H_{10}$
    (b) $C_2H_6$
    (c) $C_{15}H_{28}$
    (d) $C_{13}H_{28}$
13. Classify each hydrocarbon as aliphatic or aromatic.
    (a) $CH_3-CH_3-CH_3$
    (b) methylbenzene (benzene ring with $CH_3$)
    (c) cyclopentane with $CH_3$ and $CH_2CH_3$ substituents
    (d) $CH_2=CH-CH_3$ (line structure)
    (e) styrene (benzene ring with $CH=CH_2$)

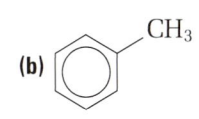

Answers to questions highlighted in red type are provided in Appendix A.

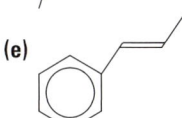

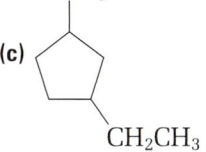

14. Classify each hydrocarbon as an alkane, alkene, or alkyne.

   (a) HC≡CH

   (b) CH₃—CH=CH₂

   (c) CH₃—C≡CH

   (d) CH₃—CH—CH—CH₃
              |      |
              CH₃  CH₃

   (e) [cyclohexene ring]

15. (a) What is the difference between a saturated hydrocarbon and an unsaturated hydrocarbon?

   (b) Classify each hydrocarbon in question 14 as saturated or unsaturated.

16. Name each compound.

   (a) CH₃—CH₂—CH₂—CH₃

   (b) CH₃—CH₃

   (c) CH₃—CH—CH—CH—CH₃
              |     |     |
              CH₃ CH₃ CH₃ (middle branch is CH₃ below center)

   (d) CH₃—CH₂—CH—CH—CH₃
                      |     |
                      CH₂  CH₂
                      |     |
                      CH₃  CH₃

17. For each structural diagram, write the IUPAC name and identify the type of compound.

   (a) CH₂=C—CH₃
              |
              CH₃

   (b) [cyclohexane ring with CH₂CH₃ and CH₃ substituents]

   (c) CH≡C—CH—CH₂—CH₃
                |
                CH₃

   (d) CH₃     CH₃
          \   /
           C=C
          /   \
         H    CH₂—CH₃

   (e) CH₃—CH₂—C—CH₃
                      |
                      CH₂
                      |
                      CH₃
              and
         CH₃—CH₂—CH—CH₃ (lower branch from central C)

   (f) [benzene ring with CH₂CH₃ substituent]

18. Name the following isomers of 2-heptene.

   H            CH₂—CH₂—CH₂—CH₃
    \          /
     C=C
    /          \
   CH₃         H

   CH₃          CH₂—CH₂—CH₂—CH₃
      \         /
       C=C
      /         \
     H           H

19. Match the following names and structural diagrams. Note that only four of the six names match.

   (a) cis-3-methyl-3-hexene
   (b) trans-3-hexene
   (c) trans-3-methyl-2-hexene
   (d) trans-3-methyl-3-hexene
   (e) cis-3-methyl-2-hexene
   (f) cis-3-hexene

   (A)  CH₃          CH₂—CH₃
            \       /
             C=C
            /       \
   CH₃—CH₂           H

   (B)  H             H
          \          /
           C=C
          /          \
   CH₃—CH₂         CH₂—CH₃

   (C)  CH₃          CH₃
            \       /
             C=C
            /       \
           H         CH₂—CH₂—CH₃

   (D)  CH₃          H
            \       /
             C=C
            /       \
   CH₃—CH₂         CH₂—CH₃

20. Is 1-methyl-2-cyclobutene the correct name for a compound? Draw the structural diagram for the compound, and rename it if necessary.

21. Explain why the rule "like dissolves like" is very useful when cleaning up an oil spill on a body of water.

22. Why is it impossible for the correct name of a linear alkane to begin with 1-methyl?

23. Which combinations of the following structural diagrams represent true isomers of C₆H₁₄? Which structural diagrams represent the same isomer?

372 MHR • Unit 4 Organic Chemistry

(a) CH$_3$—CH(CH$_3$)—CH(CH$_3$)—CH$_3$ with CH$_3$ on second carbon

(b) CH$_3$—CH(CH$_3$)—CH$_2$—CH$_2$—CH$_3$

(c) CH$_3$—CH(CH$_3$)—CH(CH$_3$)—CH$_3$

(d) CH$_3$—CH(CH$_3$)—CH$_2$—CH$_2$—CH$_3$

(e) CH$_3$—C(CH$_3$)(CH$_3$)—CH$_2$—CH$_3$

**24.** Give IUPAC names for the following compounds.

(a) benzene with CH$_2$CH$_3$
(b) benzene with CH$_2$CH$_2$CH$_3$ and CH$_2$CH$_3$ (meta)
(c) benzene with CH$_3$ and CH$_2$CH$_3$ (ortho)
(d) benzene with CH$_3$ and CH$_3$ (para)

## Inquiry

**25.** Someone has left two colourless liquids, each in an unlabelled beaker, on the lab bench. You know that one liquid is an alkane, and one is an alkene. Describe a simple test that you can use to determine which liquid is which.

**26.** Suppose that you were given a liquid in a beaker labelled "C$_6$H$_{12}$." Discuss how you would determine whether the substance was an alkane, a cycloalkane, an alkene, or an alkyne.

**27.** In Investigation 9-C, you constructed cyclobutane. In order to construct this isomer, you had to use springs for bonds.
   (a) Why did you have to use springs?
   (b) What do you think the bulging springs tell you about the stability of cyclobutane in real life?

**28.** Draw the missing product for each reaction.
   (a) CH$_3$CH=CHCH$_3$ + Br$_2$ →
   (b) CH$_3$CH=CHCH$_2$CH$_3$ + H$_2$ →
   (c) CH$_3$CH$_2$C≡CH + Cl$_2$ → (i) + Cl$_2$ → (ii)
   (d) H$_2$C=CHCH$_2$CH(CH$_3$)$_2$ + HOH →
        (i) (major product) + (ii) (minor product)
   (e) benzene + Cl$_2$ $\xrightarrow{FeBr_3}$

**29.** Draw and name the missing reactant for each reaction.

   (a) ? + Cl$_2$ → H—CCl$_2$—CCl$_2$—H (with H's) — i.e., CHCl$_2$—CHCl$_2$

   (b) HC≡C—CH$_3$ + ? → CBr(H)=CBr(CH$_3$)

   (c) ? + Br$_2$ $\xrightarrow{FeBr_3}$ bromobenzene + HBr

   (d) C$_4$H$_{10}$ + ? → 4CO$_2$ + 5H$_2$O

   (e) CH$_3$—CH$_2$—CH(CH$_3$)—CH$_3$ + ? $\xrightarrow{UV}$ CH$_3$—CHCl—CCl(CH$_3$)—CH$_3$ + 2HCl

   (f) ? + HBr → CH$_3$—CBr=CH—CH$_2$—CH$_3$

**30.** Balance the following equations. Identify each reaction as a complete combustion or an incomplete combustion.
   (a) C$_5$H$_{12(\ell)}$ + O$_{2(g)}$ → H$_2$O$_{(\ell)}$ + CO$_{2(g)}$
   (b) C$_8$H$_{16(\ell)}$ + O$_{2(g)}$ → H$_2$O$_{(\ell)}$ + CO$_{2(g)}$ + CO$_{(g)}$ + C$_{(s)}$
   (c) C$_{25}$H$_{50(x\ell)}$ + O$_{2(g)}$ → H$_2$O$_{(\ell)}$ + CO$_{2(g)}$ + CO$_{(g)}$ + C$_{(s)}$

Chapter 9 Hydrocarbons • MHR   **373**

31. Draw possible products for the following two reactions. State the type of reaction that you have completed.
    (a) $CH_3-CH_2-CH_2-CH_3 + 2Br_2 \xrightarrow{UV}$
    (b) $HC\equiv C-\underset{\underset{CH_3}{|}}{CH}-CH_2-CH_3 + 2HCl \rightarrow$

32. (a) Draw structural diagrams of all organic reactants and products for the following reaction:
    2 propane $\xrightarrow{500°C - 700°C}$ methane + ethane + propene + hydrogen
    (b) What process does this equation represent?

33. The following chemical reaction shows a reforming technique used by oil refineries to produce aromatic hydrocarbons. Draw structural diagrams of all organic reactants and products for this reaction.
    heptane $\xrightarrow[\text{catalyst}]{\text{heat, pressure}}$ toluene + hydrogen

34. The following reaction can produce more than one isomer. Draw structures for all the expected products of this reaction.

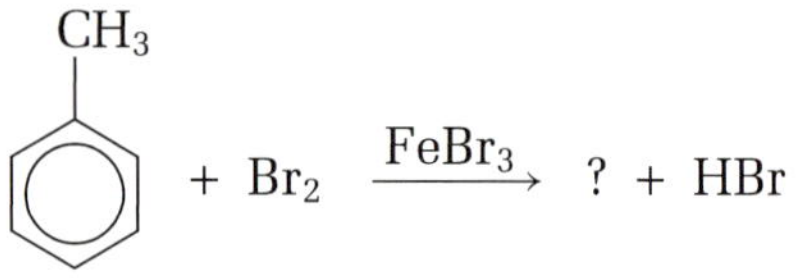

    (methylbenzene) + $Br_2 \xrightarrow{FeBr_3}$ ? + HBr

## Communication

35. Draw a complete structural diagram for each of these compounds.
    (a) 3-ethylhexane
    (b) 1-butene
    (c) 2,3-dimethylpentane

36. Draw a condensed structural diagram for each of these compounds.
    (a) methylpropane
    (b) 1-ethyl-3-propylcyclopentane
    (c) cis-3-methyl-3-heptene
    (d) 3-butyl-4-methyl-1-octyne
    (e) 4-ethyl-1-cyclooctene

37. (a) Draw all the structural isomers of $C_5H_{12}$.
    (b) Draw all the structural isomers of $C_5H_{10}$.
    (c) Draw four cis-trans isomers of $C_8H_{16}$.
    (d) Draw all the structural isomers of $C_4H_8$.

38. In a fractionation tower, the gaseous fractions are removed from the top and the solid residues are removed from the bottom.
    (a) Explain why the fractions separate like this.
    (b) A sample of crude oil was tested and found to contain mostly smaller hydrocarbon molecules, less than 15 carbon atoms long. From what part of the tower would most of the fractions of this sample be removed? Why?

39. Use a diagram to describe the steps involved in refining petroleum to obtain gasoline.

40. The following compounds are found in fractions from fractional distillation of petroleum. For each compound, draw a possible reaction that would yield a product that is more useful than the reactant. State the type of each reaction that that you have drawn.
    (a) 1,3,5-trimethylcyclohexane
    (b) long-chain alkane
    (c) branched alkane

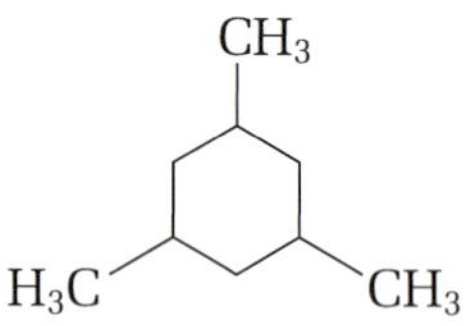

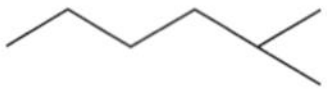

41. Draw a concept map to illustrate how hydrocarbon molecules can start as crude oil and end up as synthetic rubber.

42. Imagine that you are the owner of an oil refinery. Your main supply of crude oil contains a high percent of longer-chain hydrocarbons (greater than 15 carbon atoms per molecule). A new customer is looking for a large supply of gasoline, which contains 7- and 8-carbon molecules. How will you meet your customer's needs?

43. Place the following alkanes in order from lowest to highest boiling point. Explain your reasoning.
    (a) $CH_3-CH_2-CH_2-CH_2-CH_3$
    (b) $CH_3-CH_2-CH_3$
    (c) $CH_3-\underset{\underset{CH_3}{|}}{CH}-CH_3$
    (d) $CH_3-CH_2-CH_2-CH_3$

44. Draw the complete, condensed, and line structural diagrams for 2,3,4-trimethylpentane. Discuss the main advantage of each type of structural diagram.

45. For each compound, write or draw all of the following:
    - a molecular formula
    - a structural formula
    - a complete structural diagram
    - a condensed structural diagram
    - a line diagram
    
    (a) 3-ethyl-2,2-dimethylhexane
    (b) 2-methyl-2-pentene
    (c) 5-ethyl-1-heptyne

46. (a) Describe the bonding in benzene.
    (b) What evidence shows that benzene does not have alternating single and double bonds?
    (c) Is 1,3,5-cyclohexatriene an acceptable IUPAC name for benzene? Why or why not?

## Making Connections

47. List two ways in which ethene is important in everyday life.

48. Research the similarities and differences in drilling for oil offshore versus onshore. Write a report of your findings.

49. The National Energy Board has estimated that Canada's original petroleum resources included $4.3 \times 10^{11}$ m³ (430 billion cubic metres) of oil and bitumen (a tar-like substance) and $1.7 \times 10^{14}$ m³ (170 trillion cubic metres) of natural gas. Canada's petroleum-producing companies have used only a small fraction of these resources. Why, then, do you think many people are so concerned about exhausting them?

**Answers to Practice Problems and Short Answers to Section Review Questions:**

**Practice Problems:** 1. $C_7H_{16}$ 2. $C_9H_{20}$ 3. 10 4. 25
5.(a) 2-methylbutane (b) 2,2-dimethylpropane
(c) 3-ethyl-2,5-dimethylheptane (d) 2,2,4,4-tetramethylhexane
(e) 2,2,4-trimethyl-4-propylheptane 6.(a) 2,2,3-trimethylbutane
(b) 5-ethyl-3-methylnonane (c) 4,5-diethyl-3-methylnonane
7.(a) heptane (b) 2,3-dimethylpentane
(c) 4-ethyl-2,3-dimethylhexane 9.(a) 3-methylhexane
(b) 4-methylheptane (c) 2-methylpentane
11.(a) 2,4-dimethylhexane (b) 2,3-dimethylhexane
(c) 3-ethyl-3-methylhexane 12.(a) complete

13.(a) $C_5H_{12} + 8O_2 \rightarrow 5CO_2 + 6H_2O$
(b) $2C_8H_{18} + 25O_2 \rightarrow 16CO_2 + 18H_2O$
(c) $6C_2H_6 + 15O_2 \rightarrow 4CO_2 + 4CO + 4C + 18H_2O$,
$3C_2H_6 + 6O_2 \rightarrow CO_2 + CO + 4C + 9H_2O$
14. $C_4H_{10} + 4O_2 \rightarrow CO_2 + CO + 2C + 5H_2O$
15. $C_{25}H_{52} + 38O_2 \rightarrow 25CO_2 + 26H_2O$
16.(b) $C_{13}H_{28} + 9O_2 \rightarrow CO_2 + 2CO + 10C + 14H_2O$,
$C_{13}H_{28} + 10O_2 \rightarrow 2CO_2 + 2CO + 9C + 14H_2O$ (c) incomplete
17.(a) 3-hexene (b) 3-propyl-2-heptene (c) 4-ethyl-2,3-dimethyl-4-octene 19. 2-butene, 1-butene, 2-methyl-1-propene 20. cis-2-pentene, trans-2-pentene 23. cis-2-hexene, trans-2-hexene; cis-3-hexene, trans-3-hexene; 3-methyl-cis-2-pentene, 3-methyl-trans-2-pentene; 4-methyl-cis-2-pentene, 4-methyl-trans-2-pentene 24. 3-ethyl-3-heptanol
25.(a) 1-bromohexane (minor), 2-bromohexane (major)
(b) 2-methyl-2-pentanol (major), 2-methyl-3-pentanol (minor)
27.(a) propene + hydrochloric acid (b) ethene + bromine
(c) 3-methyl-2-pentene (d) hydrogen gas
28.(a) 4,4-dimethyl-2-pentyne (b) 3-ethyl-5-methyl-3-propyl-1-hexyne 30.(a) 1-ethyl-3-methylcyclopentane
(b) 1,2,3,4-tetramethylcyclohexane (c) methylcyclobutane
(d) 3-methyl-5-propyl-1-cyclopentene
(e) 4-methyl-1-cyclooctene (f) 5-ethyl-3,4-dimethyl-1-cyclononene (g) 3,5-diethyl-1-cyclohexene
(h) 1-methyl-2-pentylcyclopentane 32. 1,3,5-trimethylbenzene
34. meta-ethylmethylbenzene 35. some isomers include ortho-, meta-, and para-diethyl-benzene; 1,2,3,4,-tetramethylbenzene; and 1-methyl-2-propylbenzene

**Section Review 9.1:** 5.(a) organic (b) organic (c) inorganic (d) inorganic (e) organic (f) inorganic (g) organic (h) inorganic
**9.2:** 4. Only for part of the molecule. 5. tetrahedral v.s. trigonal planar **9.3:** 1.(a) alkane, alkene, alkyne (b) saturated—alkanes; unsaturated—alkenes, alkynes 2. meth (1), eth (2), prop (3), but (4), pent (5), hex (6), hept (7), oct (8), non (9), dec (10) 3. No. 4.(a) 3-ethyl-2,4-dimethylhexane
(b) 2,4,4-trimethyl-1-pentene (c) 4-ethyl-3-propyl-1-hexyne
(d) 2-ethyl-3,4,5,6-tetramethyl-1-cycloheptene (e) ethylbenzene
6.(a) propyne, cyclopropene (b) pentane, 2-methylbutane, 2,2-dimethylpropane (c) 1-pentene, cis-2-pentene, trans-2-pentene, 2-methyl-2-butene, 2-methyl-1-butene, cyclopentane 7.(a) 2-methyl-2-butyne (b) 6-ethyl-3,4-dimethyl-1-cyclohexene, or 3-ethyl-5,6-dimethyl-1-cyclohexene
11.(a) $C_7H_{16} + 11O_2 \rightarrow 7CO_2 + 8H_2O$
(b) $C_5H_{10} + 6O_2 \rightarrow 3CO_2 + CO + C + 5H_2O$ 12.(a) addition
(b) substitution (c) substitution (d) addition 14.(b) the second product **9.4:** 1.(a) boiling point (b) fractional distillation

# CHAPTER 10
# Hydrocarbon Derivatives

**Chapter Preview**

10.1 Functional Groups

10.2 Single-Bonded Functional Groups

10.3 Functional Groups with the C=O Bond

**Prerequisite Concepts and Skills**

Before you begin this chapter, review the following concepts and skills:

- naming and drawing organic compounds (Chapter 9)

Cancer is one of the leading causes of death in Canada. It is a disease in which cells mutate and grow at uncontrolled rates, disrupting the body's normal functions. TAXOL™, shown at the bottom of the page, is an organic compound that is found in the bark of the Pacific yew tree. After many tests and reviews, TAXOL™ was approved for use in treating ovarian cancer.

The bark of a large yew tree yields only enough TAXOL™ for a single treatment. However, cancer patients require repeated treatments over a long period of time. Working in a laboratory, chemists found the first solution to this problem. They developed several different methods to make, or *synthesize*, TAXOL™, from simple, widely available chemicals. Unfortunately, these methods are expensive and time-consuming.

When studying the Pacific yew's close relative, the European yew, chemists made an exciting discovery that led to the second solution. The needles and twigs of the European yew contain 10-deacetylbaccatin III, shown at the bottom of the page. A few reaction steps can transform 10-deacetylbaccatin III into TAXOL™. Instead of destroying an entire Pacific yew tree for a single treatment of TAXOL™, scientists can now use the needles from a European yew. The parent tree is not harmed.

As you are about to learn, the study of organic chemistry is really the study of functional groups. Organic compounds are classified according to their functional groups. Each functional group undergoes specific, predictable reactions. Chemists used their knowledge of functional groups to design the reactions that convert 10-deacetylbaccatin III into TAXOL™. In this chapter, you will examine the physical and chemical properties of several different classes of organic compounds.

TAXOL™

10-deacetylbaccatin III

376 MHR • Unit 4 Organic Chemistry

# Functional Groups

## 10.1

### Section Preview/Outcomes

In this section, you will

- **discover** the importance of functional groups in organic chemistry
- **predict** the molecular polarity of organic molecules
- **describe** the effects of intermolecular forces on the physical properties of organic compounds
- **identify** some common types of organic reactions
- **communicate** your understanding of the following terms: hydroxyl group, functional group, general formula, addition reaction, substitution reaction, elimination reaction, condensation reaction, hydrolysis reaction

When you cut yourself, it is often a good idea to swab the cut with rubbing alcohol to disinfect it. Most rubbing alcohols that are sold in drugstores are based on 2-propanol (common name: isopropanol), $C_3H_8O$. You can also swab a cut with a rubbing alcohol based on ethanol, $C_2H_6O$. It can be hard to tell the difference between these two compounds. Both have a sharp smell, and both evaporate quickly. Both are effective at killing bacteria and disinfecting wounds. What is the connection between these compounds? Why is their behaviour so similar?

## Functional Groups

Both 2-propanol and ethanol contain the same reactive group of atoms: an oxygen atom bonded to a hydrogen atom. This group of atoms is called the **hydroxyl group**, written as —OH. The hydroxyl group is the *functional group* for both 2-propanol and ethanol, shown in Figure 10.10. Because ethanol and 2-propanol have the same functional group, their behaviour is similar.

$$CH_3-CH-CH_3 \qquad CH_3-CH_2-OH$$
$$\qquad \; | $$
$$\qquad OH$$

2-propanol ethanol

**Figure 10.1** Ethanol and 2-propanol both belong to the alcohol family.

A **functional group** is a reactive group of bonded atoms that appears in all the members of a chemical family. The functional group is the reactive part of an organic molecule. Each functional group behaves and reacts in a characteristic way. Thus, functional groups help to determine the physical and chemical properties of compounds. For example, the reactive double bond is the functional group for an alkene. In this chapter, you will encounter many different functional groups. Although it is possible for an organic compound to have more than one functional group, you will focus on compounds with only one functional group in this course.

The **general formula** for a family of simple organic compounds is $R + functional\ group$. The letter $R$ stands for any alkyl group. (If more than one alkyl group is present, $R'$ and $R''$ are also used.) For example, the general formula R—OH refers to any of the following compounds:

$CH_3OH$, $CH_3CH_2OH$, $CH_3CH_2CH_2OH$, $CH_3CH_2CH_2CH_2OH$, etc.

Organic compounds are named according to their functional group. Generally, the suffix of a compound's name indicates the most important functional group in the molecule. For example, the suffix -ene indicates the presence of a double bond, and the suffix -ol indicates the presence of a hydroxyl group.

### CONCEPT CHECK

Each organic family follows a set pattern. You have just seen that you can represent the hydrocarbon part of a functional family by the letter $R$. All the structures below belong to the primary amine family. What is the functional group for this family? Write the general formula for an amine.

$$CH_3-NH_2$$

$$CH_3-CH_2-NH_2$$

$$CH_3-CH_2-CH_2-NH_2$$

Chapter 10 Hydrocarbon Derivatives • MHR  377

Functional groups are a useful way to classify organic compounds, for two reasons:

1. *Compounds with the same functional group often have similar physical properties.* In this chapter, you will learn to recognize various functional groups. You will use functional groups to help you predict the physical properties of compounds.

2. *Compounds with the same functional group react chemically in very similar ways.* In this chapter, you will learn how compounds with each functional group react.

Table 10.1 lists some of the most common functional groups.

**Table 10.1** Common Functional Groups

| Type of compound | Suffix | Functional group | Example |
|---|---|---|---|
| alkane | -ane | none | propane |
| alkene | -ene | C=C | propene |
| alkyne | -yne | —C≡C— | propyne |
| alcohol | -ol | —C—OH | propanol |
| amine | -amine | —C—N | propanamine |
| aldehyde | -al | —C(=O)—H | propanal |
| ketone | -one | —C(=O)— | propanone |
| carboxylic acid | -oic acid | —C(=O)—OH | propanoic acid |
| ester | -oate | —C(=O)—O— | methyl propanoate |
| amide | -amide | —C(=O)—N | propanamide |

How does a functional group help determine the physical and chemical properties of an organic compound? One answer is that functional groups usually contain polar covalent bonds. You learned about polar bonds in Chapter 5. Examples of these bonds that you will find in functional groups on organic compounds include O—H, C—O, C=O, C—N, and N—H. The hydrocarbon parts of substituted hydrocarbons, of course, contain many C—H bonds that have a $\Delta$EN of 0.35. If you refer back to Figure 5.19 on page 177, you will find that this bond is classified as "slightly polar covalent" thus it is a weak dipole. Also, in many hydrocarbon derivatives, there are several C—H bonds that partially cancel each other. Therefore, C—H bonds contribute very little to the polarity of hydrocarbon derivatives and are often neglected when considering the polarity of the molecule as a whole.

To refresh your memory about polar molecules, complete the following practice problems. If you have any difficulty solving the problems, review the information on molecular shape and polar bonds in Chapter 6.

### Practice Problems

1. Predict and sketch the three-dimensional shape around each single-bonded atom.
   (a) C and O in $CH_3OH$
   (b) C in $CH_4$

2. Predict and sketch the three-dimensional shape of each multiple-bonded molecule.
   (a) $HC\equiv CH$
   (b) $H_2C=O$

3. Identify any polar bonds that are present in each molecule in questions 1 and 2.

4. For each molecule in questions 1 and 2, predict whether the molecule as a whole is polar or non-polar.

You have just discovered the importance of bonding in hydrocarbon derivatives. Now you can apply this knowledge to help you understand the basis of some physical properties of organic compounds.

## Physical Properties of Substituted Hydrocarbons

In Chapter 6, you learned about intermolecular forces such as hydrogen bonds, dipole-dipole interactions, and dispersion forces. Substituted hydrocarbons experience one or more of these interactions with each other and with solvent molecules. As you progress through this chapter, you will analyze several types of substituted hydrocarbons in detail and find out how these intermolecular forces contribute to the physical properties of each type. Below you will find some general principles to consider while you are making your detailed analyses.

### Hydrogen Bonds

If the molecules have O—H or N—H bonds, they can form hydrogen bonds with themselves and with water. *Such molecules will have higher boiling points than similar molecules that cannot form hydrogen bonds.* For example, alcohols can form hydrogen bonds but alkanes cannot. Therefore, alcohols have higher boiling points than similar sized alkanes.

If the molecules have an O, N, or F atom bonded to atoms other than hydrogen, they cannot form hydrogen bonds with themselves but they can form hydrogen bonds with water. *Molecules that can form hydrogen bonds with water are often soluble in water.* However, if an alcohol has a long hydrocarbon chain, it will likely not be soluble in water. For example, octanol, $CH_3CH_2CH_2CH_2CH_2CH_2CH_2CH_2OH$, is not soluble in water whereas ethanol, $CH_3CH_2OH$, is very soluble in water.

### Dipole Interactions

*Polar molecules usually have a higher boiling point than non-polar molecules of a similar size but not as high as molecules that can form hydrogen bonds.* For example, ethanol, $CH_3CH_2OH$, is polar and can form hydrogen bonds. Methoxymethane, $CH_3OCH_3$, is an isomer of ethanol and is also polar but cannot form hydrogen bonds. Ethanol has a higher boiling point than methoxymethane. Both compounds have a higher boiling point than the non-polar molecule, ethane, $CH_3CH_3$.

### Dispersion Forces

As you learned in Chapter 6, dispersion forces are very weak. However, many different dispersion interactions can occur at the same time in long hydrocarbon molecules. Therefore, *a molecule with a greater number of carbon atoms usually has a higher boiling point than the same type of molecule with fewer carbon atoms.* For example, hexane, $CH_3CH_2CH_2CH_2CH_2CH_3$, has a higher boiling point than ethane, $CH_3CH_3$.

The melting points of organic compounds follow approximately the same trend as do their boiling points. There are some anomalies, however, due to more complex forces of binding in solids. In the following ThoughtLab, you will use the information above to predict and compare the physical properties of some common organic compounds.

## ThoughtLab — Comparing Intermolecular Forces

Intermolecular forces affect the physical properties of compounds. In this ThoughtLab, you will compare the intermolecular forces of different organic compounds.

### Procedure

1. Draw three molecules of each compound below.
   (a) propane, $CH_3CH_2CH_3$
   (b) heptane, $CH_3CH_2CH_2CH_2CH_2CH_2CH_3$
   (c) 1-propanol, $CH_3CH_2CH_2OH$
   (d) 1-heptanol, $CH_3CH_2CH_2CH_2CH_2CH_2CH_2OH$

2. For each compound, consider whether or not hydrogen bonding can occur between its molecules. Use a dashed line to show any hydrogen bonding.

3. For each compound, consider whether or not any polar bonds are present.
   (a) Use a different-coloured pen to identify any polar bonds.
   (b) Which compounds are polar? Which compounds are non-polar? Explain your reasoning.

4. Compare your drawings of propane and heptane.
   (a) Which compound has stronger dispersion forces? Explain your answer.
   (b) Which compound is expected to have a higher boiling point? Explain your answer.

5. Compare your drawings of 1-propanol and 1-heptanol.
   (a) Which compound is more polar? Explain your answer.
   (b) Which compound is more soluble in water? Explain your answer.

### Analysis

1. Which compound has a higher solubility in water?
   (a) a polar compound or a non-polar compound
   (b) a compound that forms hydrogen bonds with water, or a compound that does not form hydrogen bonds with water
   (c) $CH_3CH_2CH_2OH$ or $CH_3CH_2CH_2CH_2CH_2OH$

2. Which compound has stronger attractions between molecules?
   (a) a polar compound or a non-polar compound
   (b) a compound without O—H or N—H bonds, or a compound with O—H or N—H bonds

3. Which compound is likely to have a higher boiling point?
   (a) a polar compound without O—H or N—H bonds, or a polar compound with O—H or N—H bonds
   (b) $CH_3CH_2CH_2OH$ or $CH_3CH_2CH_2CH_2CH_2OH$

4. Compare boiling points and solubilities in water for each pair of compounds. Explain your reasoning.
   (a) ammonia, $NH_3$, and methane, $CH_4$
   (b) pentanol, $C_5H_{11}OH$, and pentane, $C_5H_{12}$

## The Main Types of Organic Reactions

Addition reactions, substitution reactions, and elimination reactions are three common types of organic reactions. Most organic reactions can be classified as one of these three types.

- Recall that, in an *addition reaction*, atoms are added to a double or triple bond. One bond of the multiple bond breaks so that two new bonds are formed. As you saw in Chapter 9, addition reactions are common for alkenes and alkynes. Addition reactions can also occur at a C=O bond. Review page 349 of Chapter 9 for more information on addition reactions.

- Recall that, in a *substitution reaction*, a hydrogen atom or a functional group is replaced by a different functional group. Aromatic compounds commonly undergo substitution reactions. (See page 362 of Chapter 9 to review substitution reactions of aromatic compounds.) As you will discover in the next section, other organic compounds also undergo substitution reactions. Two examples are given below.

**Example 1**

$$CH_3-CH_2-OH + HI \rightarrow CH_3-CH_2-I + HOH$$

ethanol     hydroiodic acid     iodoethane     water

**Example 2**

$$CH_3-\underset{\underset{Br}{|}}{CH}-CH_2-CH_3 + NH_3 \rightarrow CH_3-\underset{\underset{NH_2}{|}}{CH}-CH_2-CH_3 + HBr$$

2-bromobutane     ammonia     2-butanamine     hydrobromic acid

- In an **elimination reaction**, atoms are removed from a molecule to form a double bond. This type of reaction is the reverse of an addition reaction. *One* reactant usually breaks up to give *two* products. The organic product typically has fewer atoms bonded to carbon atoms than the organic reactant did.

$$-\underset{|}{\overset{\overset{X}{|}}{C}}-\underset{|}{\overset{\overset{Y}{|}}{C}}- \rightarrow \phantom{x} C=C \phantom{x} + \phantom{x} XY$$

> **CONCEPT CHECK**
>
> You can express an addition reaction algebraically, using the equation $a + b \rightarrow ab$. Come up with similar equations for substitution and elimination reactions.

- Alcohols often undergo elimination reactions when they are heated in the presence of strong acids, such as sulfuric acid, $H_2SO_4$, which acts as a catalyst. (See the first example below.) Another class of compounds, called alkyl halides, undergoes elimination reactions as shown in Example 2. You will study alkyl halides in more detail in Section 10.2

**Example 1**

$$H-\underset{\underset{H}{|}}{\overset{\overset{H}{|}}{C}}-\underset{\underset{H}{|}}{\overset{\overset{OH}{|}}{C}}-\underset{\underset{H}{|}}{\overset{\overset{H}{|}}{C}}-H \xrightarrow{H_2SO_4} \phantom{x} C=C \phantom{x} + HOH$$

2-propanol     propene     water

> **CONCEPT CHECK**
>
> In the first example of an elimination reaction, the strong acid, $H_2SO_4$, is not a reactant. It is not directly involved in the reaction. It is a *catalyst*: a compound that speeds up a reaction but is not consumed by it.

Chapter 10 Hydrocarbon Derivatives • MHR    381

**Example 2**

$$\underset{\text{bromoethane}}{\text{H}_3\text{C}-\text{CHBr}-\text{H (with H's)}} + \underset{\text{hydroxide ion}}{\text{OH}^-} \rightarrow \underset{\text{ethene}}{\text{H}_2\text{C}=\text{CH}_2} + \underset{\text{water}}{\text{HOH}} + \underset{\text{bromide ion}}{\text{Br}^-}$$

## Other Important Organic Reactions

In Chapter 9, you wrote equations for the combustion of hydrocarbons. Most organic compounds undergo combustion reactions. In this chapter, you will also encounter the following classes of organic reactions: condensation reactions and hydrolysis reactions. Condensation and hydrolysis reactions are both types of substitution reactions.

**CHEM FACT**

When you see two arrows that point in opposite directions in a chemical equation, the reaction can proceed in both directions. This type of reaction is called an *equilibrium reaction*. You will learn more about equilibrium reactions in Unit 5.

- Recall that, in a *combustion reaction*, hydrocarbons react with oxygen to produce a mixture of carbon dioxide and water. Incomplete combustion produces solid carbon and carbon monoxide as additional products. Note that other products can also result from the combustion of inorganic compounds such as ammonia ($NH_3$) and hydrazine ($N_2H_4$) as well as organic compounds that contain atoms such as N, S, P, or halogen atoms.

- In a **condensation reaction**, two organic molecules combine to form a single organic molecule. A small molecule, usually water, is produced during the reaction. An example is given below.

$$\underset{\text{ethanoic acid}}{\text{H}_3\text{C}-\text{C(=O)}-\text{OH}} + \underset{\text{methanol}}{\text{HO}-\text{CH}_3} \underset{}{\overset{\text{H}_2\text{SO}_4}{\rightleftharpoons}} \underset{\text{methyl ethanoate}}{\text{H}_3\text{C}-\text{C(=O)}-\text{O}-\text{CH}_3} + \underset{\text{water}}{\text{H}_2\text{O}}$$

- In a **hydrolysis reaction**, water adds to a bond, splitting it in two. This reaction is the reverse of a condensation reaction.

$$\underset{\text{methyl ethanoate}}{\text{H}_3\text{C}-\text{C(=O)}-\text{O}-\text{CH}_3} + \underset{\text{water}}{\text{H}_2\text{O}} \underset{}{\overset{\text{H}_2\text{SO}_4}{\rightleftharpoons}} \underset{\text{ethanoic acid}}{\text{H}_3\text{C}-\text{C(=O)}-\text{OH}} + \underset{\text{methanol}}{\text{HO}-\text{CH}_3}$$

The following Sample Problem shows how to identify different types of organic reactions.

### Sample Problem

#### Identifying Organic Reactions

**Problem**

Identify each type of organic reaction.

(a) $HO-CH_2CH_2CH_3 \rightarrow CH_2=CHCH_3 + H_2O$

(b) $H_2C=CHCH_2CH_3 + H_2 \rightarrow CH_3CH_2CH_2CH_3$

(c) $CH_3CH(CH_3)CH_2CH_2Br + NaOH \rightarrow CH_3CH(CH_3)CH_2CH_2OH + NaBr$

(d) $CH_3CH_2CH_3 + 5O_2 \rightarrow 3CO_2 + 4H_2O$

(e) $CH_3CH_2OH + CH_3OH \rightarrow CH_3CH_2OCH_3 + H_2O$

(f) $CH_3CHClCH_3 + OH^- \rightarrow H_2C=CHCH_3 + HOH + Cl^-$

**Solution**

(a) A double bond is formed. One reactant becomes two products. The organic product has fewer atoms bonded to carbon. Thus, this reaction is an elimination reaction.

(b) A double bond becomes a single bond. Two reactants become one product. The organic product has more atoms bonded to carbon. Thus, this reaction is an addition reaction.

(c) No double bond is broken or formed. Two reactants form two products. An atom in the organic reactant (Br) is replaced by a different group of atoms (OH). Thus, this is a substitution reaction.

(d) An organic compound reacts with oxygen. Carbon dioxide and water are produced. Thus, this reaction is a combustion.

(e) Two organic molecules combine to form a larger organic molecule. A small molecule, water, is also produced. Thus, this reaction is a condensation reaction.

(f) The organic compound contains a halogen (Cl) and it is in the presence of a base. Water and a halogen ion are formed. Thus the reaction is an elimination reaction.

## Practice Problems

5. Identify each reaction as an addition, substitution, or elimination reaction.

(a) 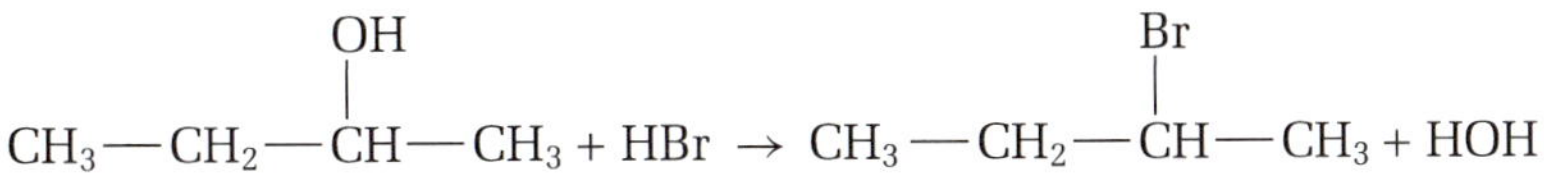 $CH_3-CH_2-\underset{\underset{OH}{|}}{CH}-CH_3 + HBr \rightarrow CH_3-CH_2-\underset{\underset{Br}{|}}{CH}-CH_3 + HOH$

(b) 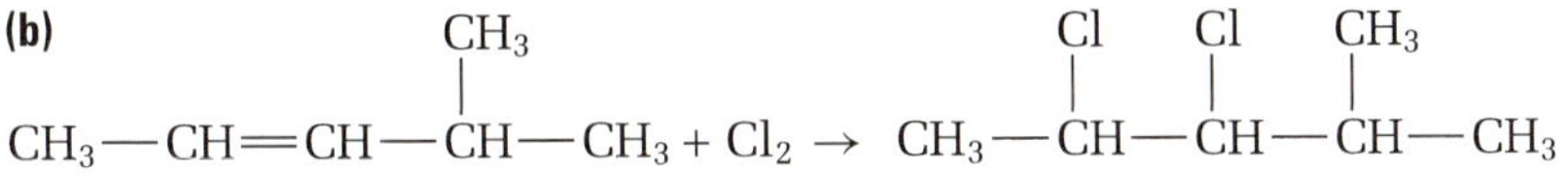 $CH_3-CH=CH-\underset{\underset{CH_3}{|}}{CH}-CH_3 + Cl_2 \rightarrow CH_3-\underset{\underset{Cl}{|}}{CH}-\underset{\underset{Cl}{|}}{CH}-\underset{\underset{CH_3}{|}}{CH}-CH_3$

(c) 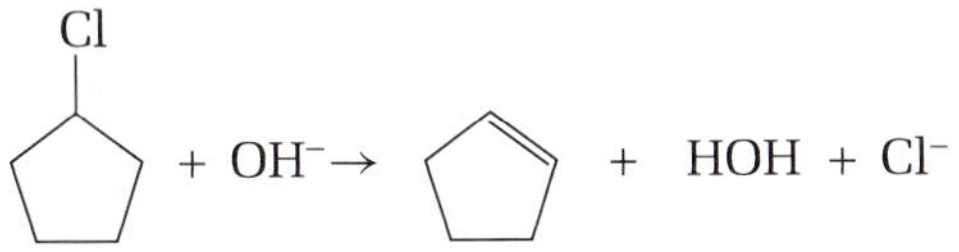 [cyclopentane with Cl] $+ OH^- \rightarrow$ [cyclopentene] $+ HOH + Cl^-$

(d) 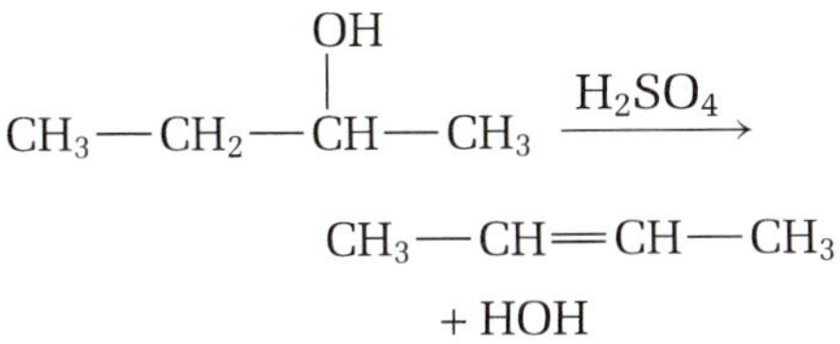 $CH_3-CH_2-\underset{\underset{OH}{|}}{CH}-CH_3 \xrightarrow{H_2SO_4}$
$CH_3-CH=CH-CH_3$
$+ HOH$

6. Identify each reaction as a condensation or a hydrolysis.

(a) 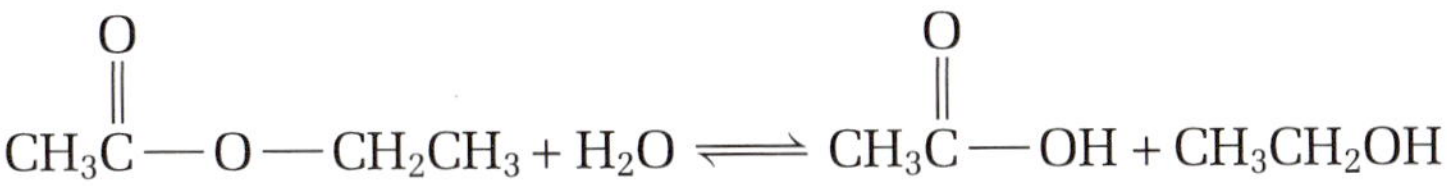 $CH_3\overset{\overset{O}{\|}}{C}-O-CH_2CH_3 + H_2O \rightleftharpoons CH_3\overset{\overset{O}{\|}}{C}-OH + CH_3CH_2OH$

*Continued...*

(b) 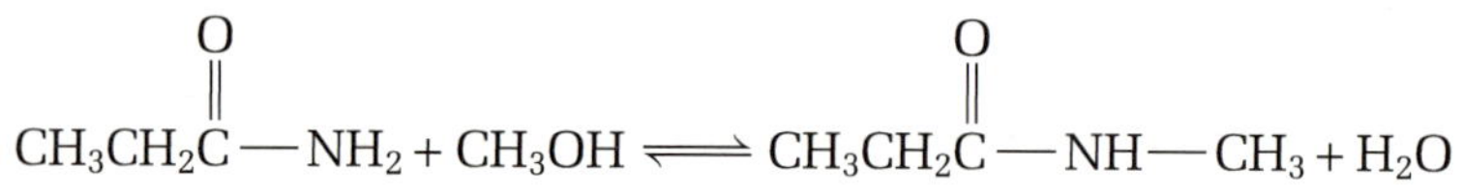 $CH_3CH_2\overset{\overset{O}{\|}}{C}-NH_2 + CH_3OH \rightleftharpoons CH_3CH_2\overset{\overset{O}{\|}}{C}-NH-CH_3 + H_2O$

(c) $CH_3CH_2CH_2OH + CH_3CH_2\underset{\underset{OH}{|}}{C}HCH_3 \rightarrow CH_3CH_2\underset{\underset{CH_3}{|}}{C}H-O-CH_2CH_2CH_3 + H_2O$

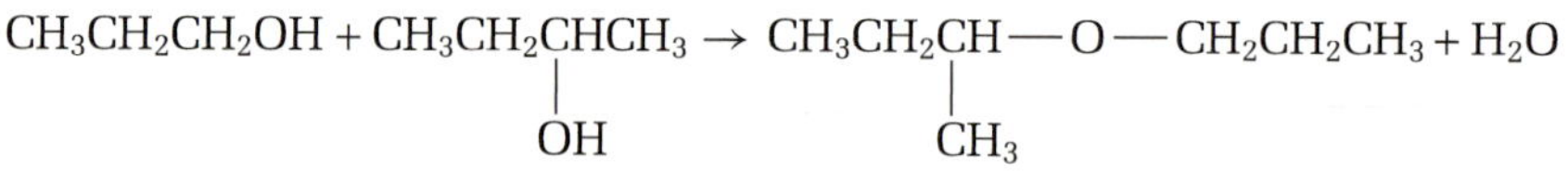

**7.** Identify the type of reaction.

(a) 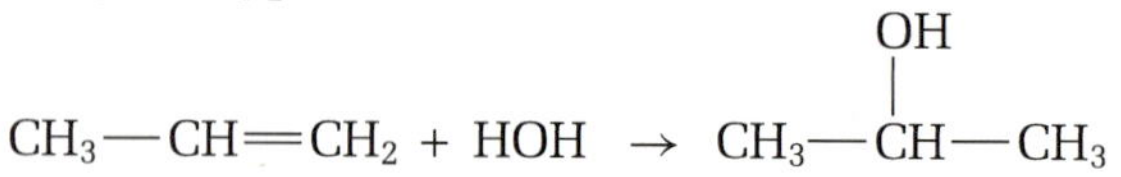 $CH_3-CH=CH_2 + HOH \rightarrow CH_3-\underset{\underset{}{}}{\overset{\overset{OH}{|}}{C}}H-CH_3$

(b) $CH_3-O-CH_2CH_3 + H_2O \rightleftharpoons CH_3OH + CH_3CH_2OH$

(c) $CH_3CH_2OH + 3O_2 \rightarrow 2CO_2 + 3H_2O$

(d) $CH_3CH_2CH_2OH \rightarrow CH_3CH=CH_2 + H_2O$

## Section Summary

In this section, you learned that organic compounds are classified according to functional groups. You discovered how to predict the molecular polarity and other physical properties of organic compounds. You also examined and reviewed some of the main types of organic reactions: addition, substitution, and elimination reactions; combustion reactions; and condensation and hydrolysis reactions. In the next section, you will be introduced to several different classes of organic compounds. You will apply what you have learned in this section to help you examine the physical and chemical properties of these compounds.

## Section Review

**1** Classify each bond as polar or non-polar.

(a) C—O   (b) C—C   (c) C—N   (d) C=C   (e) C=O

**2** Describe the shape of the molecule around the carbon atom that is highlighted. **Hint:** See Table 9.1 in Chapter 9.

(a) 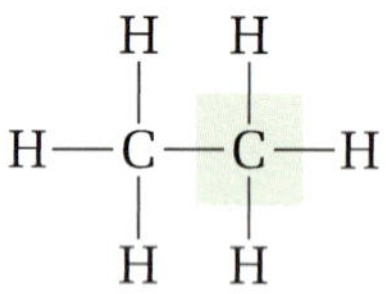 ethane structure with highlighted C—C bond

(b) 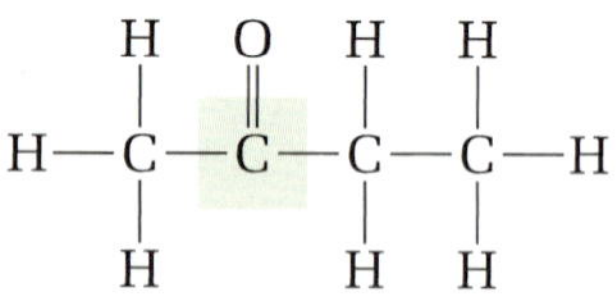 structure H—C—C(=O)—C—C—H with highlighted second carbon

**3** Identify each molecule in question 2 as either polar or non-polar. Explain your reasoning.

**4** Which compounds can form hydrogen bonds?

(a) $CH_3CH_3$

(b) $CH_3OH$

(c) $CH_4$

(d) $CH_3NH_2$

**5** For each part, predict which compound has a higher boiling point. Explain your reasoning.

(a) $CH_4$ or $CH_3CH_2CH_3$

(b) $CH_3CH_2NH_2$ or $CH_3CH_2CH_3$

(c) $CH_3CH_2CH_2CH_2OH$ or $CH_3CH_2OCH_2CH_3$

**6** For each part, predict which compound is more soluble in water. Explain your reasoning.

(a) $CH_3NH_2$ or $CH_3CH_3$

(b) $CH_3CH_2CH_2CH_2OH$ or $CH_3OH$

**7** Rank the following compounds from highest to lowest melting point. Explain your reasoning.

$CH_3CH_2OCH_3$     $CH_3CH_2CH_2CH_3$     $CH_3CH_2CH_2OH$

**8** Identify each reaction as an addition, substitution, or elimination reaction.

(a) $H_2C{=}CH_2 + Br_2 \rightarrow H_2\overset{|}{C}{-}\overset{|}{C}H_2$ (with Br, Br on the carbons)

(b) $CH_3CH_2\overset{OH}{\underset{|}{C}}HCH_3 \xrightarrow{H_2SO_4} CH_3CH{=}CHCH_3 + H_2O$

(c) 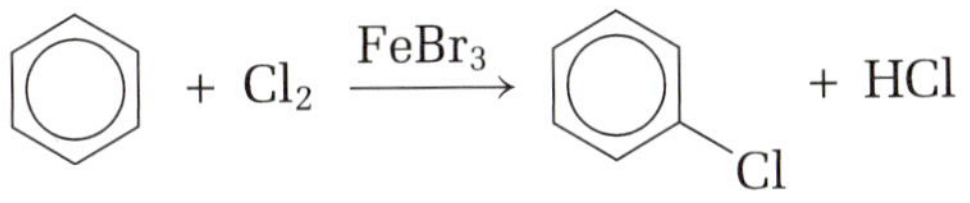 benzene $+ Cl_2 \xrightarrow{FeBr_3}$ chlorobenzene $+ HCl$

(d) $CH_3CH_2CH_2Br + H_2NCH_2CH_3 \rightarrow CH_3CH_2CH_2NHCH_2CH_3 + HBr$

(e) 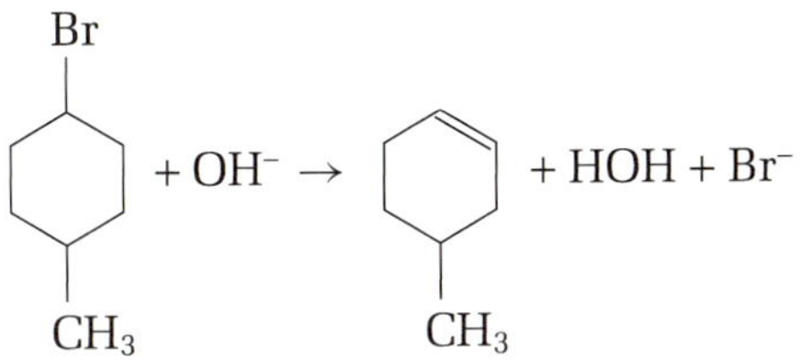 1-bromo-4-methylcyclohexane $+ OH^- \rightarrow$ 4-methylcyclohexene $+ HOH + Br^-$

**9** In your own words, describe each type of organic reaction. Include an example for each type.

(a) condensation

(b) hydrolysis

# 10.2 Single-Bonded Functional Groups

**Section Preview/Outcomes**

In this section, you will

- **name** and **draw** members of the alcohol, alkyl halide, ether, and amine families of organic compounds
- **describe** some physical and chemical properties of these families of compounds
- **communicate** your understanding of the following terms: *alcohol, parent alkane, alkyl halide, haloalkane, ether, amine*

You may have read murder mystery novels in which the detective discovers an unconscious victim lying beside a wad of cloth on the floor. The detective sniffs the cloth, notices a distinctive sickly-sweet smell, and announces: "Aha! Chloroform was used to knock out the victim." Chloroform, $CHCl_3$, is an organic compound that was used for many years as an anaesthetic during surgery. Diethyl ether, $CH_3CH_2OCH_2CH_3$, also known as "ether," is another organic compound that was once a common anaesthetic. Because of their toxic and irritating properties, chloroform and ether have been replaced in the operating room by other, less harmful, organic compounds such as methyl propyl ether, $CH_3OCH_2CH_2CH_3$. All three of these compounds have functional groups with single bonds.

In the previous section, you learned how functional groups affect the physical and chemical properties of organic compounds. In this section, you will be introduced to several classes of organic compounds that have single-bonded functional groups.

## Compounds With Single-Bonded Functional Groups

Alcohols, alkyl halides, ethers, and amines all have functional groups with single bonds. These families of compounds have many interesting uses in daily life. As you learn how to identify and name these types of compounds, think about how intermolecular forces affect their properties and uses.

## Alcohols

An **alcohol** is an organic compound that contains the —OH, or hydroxyl, functional group. Depending on the position of the hydroxyl group, an alcohol can be *primary*, *secondary*, or *tertiary*. Figure 10.3 gives some examples of alcohols.

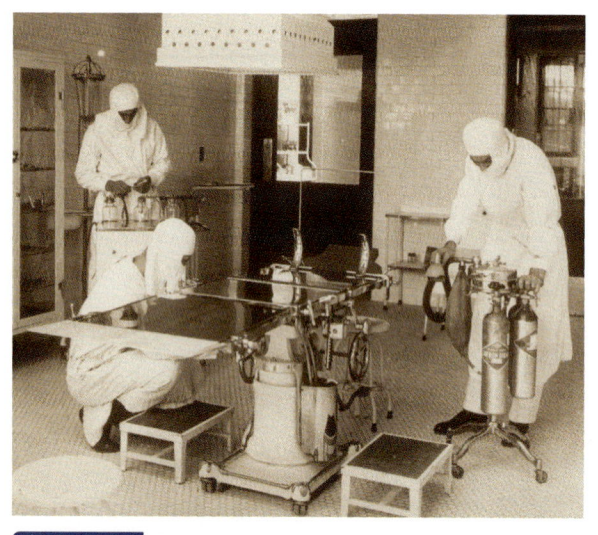

**Figure 10.2** Organic compounds with single-bonded functional groups have long been used as anaesthetics.

---

**primary alcohol**

HO—CH$_2$—CH$_2$—CH$_2$—CH$_3$

The hydroxyl group is bonded to a carbon that is bonded to only one other carbon atom.

**secondary alcohol**

$$CH_3-\underset{\underset{\displaystyle}{|}}{\overset{\overset{\displaystyle OH}{|}}{CH}}-CH_2-CH_3$$

The hydroxyl group is bonded to a carbon that is bonded to two other carbon atoms.

**tertiary alcohol**

$$CH_3-\underset{\underset{\displaystyle OH}{|}}{\overset{\overset{\displaystyle CH_3}{|}}{C}}-CH_3$$

The hydroxyl group is bonded to a carbon that is bonded to three other carbon atoms.

**Figure 10.3**

Table 10.2 lists some common alcohols and their uses. Alcohols are very widely used, and can be found in drug stores, hardware stores, liquor stores, and as a component in many manufactured products.

**Table 10.2** Common Alcohols and Their Uses

| Name | Common name(s) | Structure | Boiling point | Use(s) |
|---|---|---|---|---|
| methanol | wood alcohol, methyl alcohol | $CH_3-OH$ | 64.6°C | • solvent in many chemical processes<br>• component of automobile antifreeze |
| ethanol | grain alcohol, ethyl alcohol | $CH_3-CH_2-OH$ | 78.2°C | • solvent in many chemical processes<br>• component of alcoholic beverages<br>• antiseptic liquid |
| 2-propanol | isopropanol, isopropyl alcohol, rubbing alcohol | 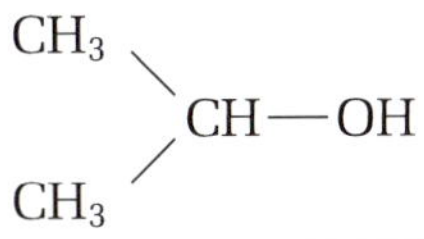 $(CH_3)_2CH-OH$ | 82.4°C | • antiseptic liquid |
| 1,2-ethanediol | ethylene glycol | $HO-CH_2-CH_2-OH$ | 197.6°C | • main component of automobile antifreeze |

Alcohols are named from the **parent alkane**: the alkane with the same basic carbon structure. Follow the steps below to name an alcohol. The Sample Problem that follows gives an example.

### How to Name an Alcohol

**Step 1** Locate the longest chain that contains an —OH group attached to one of the carbon atoms. Name the parent alkane.

**Step 2** Replace the -e at the end of the name of the parent alkane with -ol.

**Step 3** Add a position number before the root of the name to indicate the location of the —OH group. (Remember to number the main chain of the hydrocarbon so that the hydroxyl group has the lowest possible position number.) If there is more than one —OH group, leave the -e in the name of the parent alkane, and put the appropriate prefix (di-, tri-, or tetra-) before the suffix -ol.

**Step 4** Name and number any other branches on the main chain. Add the name of these branches to the prefix.

**Step 5** Put the name together: prefix + root + suffix.

## Sample Problem

### Naming an Alcohol

**Problem**

Name the following alcohol. Identify it as primary, secondary, or tertiary.

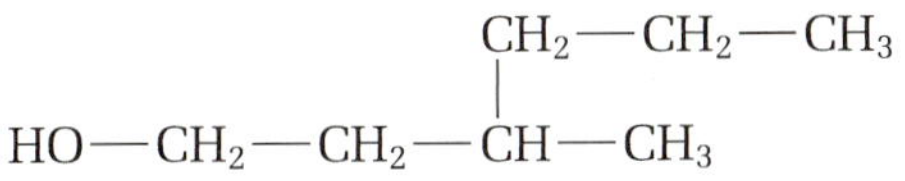

$HO-CH_2-CH_2-CH(CH_2-CH_2-CH_3)-CH_3$

*Continued...*

> **PROBLEM TIP**
> If an organic compound is complex, with many side branches, the main chain may not be obvious. Sketch the compound in your notebook or on scrap paper. Circle or highlight the main chain

## Math LINK

www.mcgrawhill.ca/links/atlchemistry

Methanol and ethanol are produced industrially from natural, renewable resources. Go to the web site above, and click on **Web Links** to find out where to go next. Research the processes that produce these important chemicals. From where do they obtain their raw materials?

---

*Continued ...*

**Solution**

**Step 1** The main chain has six carbon atoms. The name of the parent alkane is hexane.

**Step 2** Replacing -e with -ol gives hexanol.

**Step 3** Add a position number for the —OH group, to obtain 1-hexanol.

**Step 4** A methyl group is present at the third carbon. The prefix is 3-methyl.

**Step 5** The full name is 3-methyl-1-hexanol. This is a primary alcohol.

### Practice Problems

8. Name each alcohol. Identify it as primary, secondary, or tertiary.

   (a) CH₃—CH₂—CH₂—OH

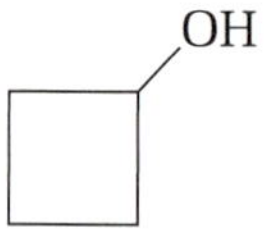

   (d) CH₃—CH—CH—CH₂—CH₃ with OH on second carbon and OH on third carbon

   (b) [structure with OH on second carbon of a 4-carbon chain]

   (e) 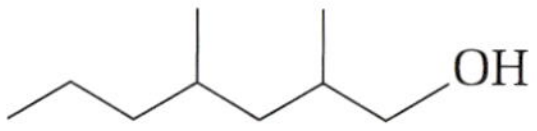 [branched chain with OH, methyl substituents]

   (c) [cyclobutane with OH]

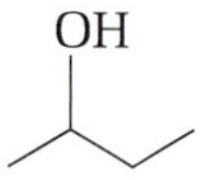

9. Draw each alcohol.
   (a) methanol
   (b) 2-propanol
   (c) 2,2-butanediol
   (d) 3-ethyl-4-methyl-1-octanol
   (e) 2,4-dimethyl-1-cyclopentanol

10. Identify any errors in each name. Give the correct name for the alcohol.

    (a) 1,3-heptanol
    
    HO—CH₂—CH₂—CH—CH₂—CH₃ with OH on the CH
    
    (b) 3-ethyl-4-ethyl-1-decanol 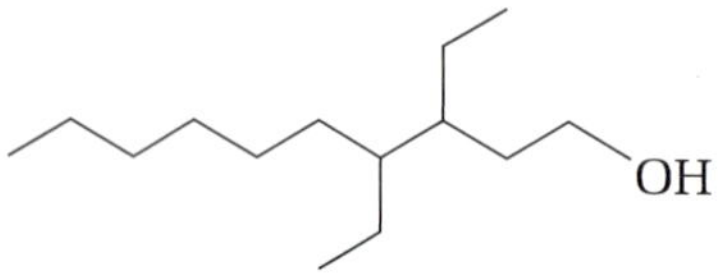
    
    [structure of branched alcohol with two ethyl groups]
    
    (c) 1,2-dimethyl-3-butanol
    
    CH₂—CH—CH—CH₃ with CH₃ on first carbon, CH₃ on second carbon, OH on third carbon

11. Sketch a three-dimensional diagram of methanol. **Hint:** Recall that the shape around an oxygen atom is *bent*.

Table 10.3 lists some common physical properties of alcohols. As you learned earlier in this chapter, alcohols are polar molecules that experience hydrogen bonding. The physical properties of alcohols depend on these characteristics.

**Table 10.3** Physical Properties of Alcohols

| Polarity of functional group | The O—H bond is very polar. As the number of carbon atoms in an alcohol becomes larger, the alkyl group's non-polar nature becomes more important than the polar O—H bond. Therefore small alcohols are more polar than alcohols with large hydrocarbon portions. |
|---|---|
| Hydrogen bonding | Alcohols experience hydrogen bonding with other alcohol molecules and with water. |
| Solubility in water | The capacity of alcohols for hydrogen bonding makes them extremely soluble in water. Methanol and ethanol are *miscible* (infinitely soluble) with water. The solubility of an alcohol decreases as the number of carbon atoms increases. |
| Melting and boiling points | Due to the strength of the hydrogen bonding, most alcohols have higher melting and boiling points than alkanes with the same number of carbon atoms. Most alcohols are liquids at room temperature. |

### Additional Characteristics of Alcohols

- Most alcohols are poisonous. Methanol can cause blindness or death when consumed. Ethanol is consumed widely in moderate quantities, but it causes impairment and/or death when consumed in excess.

## Reactions of Alcohols

Alcohols can react in several ways, depending on the reactants and on the conditions of the reaction. For example, alcohols can undergo combustion, substitution with halogen acids, and elimination to form alkenes.

### Combustion of Alcohols

Alcohols are extremely flammable, and should be handled with caution. Like hydrocarbons, alcohols react with oxygen in combustion reactions to produce carbon dioxide and water.

$$CH_3—CH_2—OH + 3O_2 \rightarrow 2CO_2 + 3H_2O$$

ethanol        oxygen    carbon dioxide   water

**Figure 10.4** The complete combustion of ethanol produces carbon dioxide and water.

### Substitution Reactions of Alcohols

When a halogen acid, such as HCl, HBr, or HI, reacts with an alcohol, the halogen atom is substituted for the OH group of the alcohol. This is shown in Figure 10.5.

$$CH_3—CH_2—OH + HCl \rightarrow CH_3—CH_2—Cl + H_2O$$

**Figure 10.5** Ethanol reacts with hydrochloric acid to produce chloroethane.

### Elimination Reactions of Alcohols

When an alcohol is heated in the presence of the strong acid and dehydrating agent, $H_2SO_4$, an elimination reaction takes place. This type of reaction is shown in Figure 10.6, below. The OH group and one H atom leave the molecule, and water is produced. As a result, the molecule forms a double bond. Because water is produced, this type of reaction is also called a *dehydration* (meaning "loss of water") reaction.

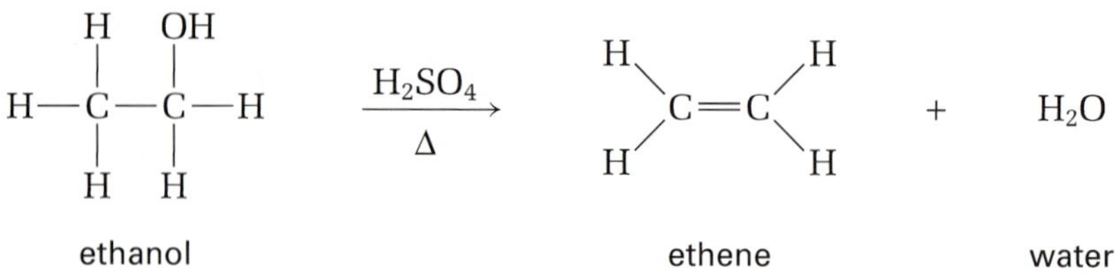

$$\underset{\text{ethanol}}{\text{H}_3\text{C}-\text{CH}_2\text{OH}} \xrightarrow{\underset{\Delta}{\text{H}_2\text{SO}_4}} \underset{\text{ethene}}{\text{H}_2\text{C}=\text{CH}_2} + \underset{\text{water}}{\text{H}_2\text{O}}$$

**Figure 10.6** The $\Delta$ symbol is used in chemistry to represent heat added to a reaction.

## Alkyl Halides

On the previous page, you saw that alcohols can undergo substitution reactions with halogen acids to produce a different kind of organic compound. An **alkyl halide** (also known as a **haloalkane**) is an alkane in which one or more hydrogen atoms have been replaced with halogen atoms, such as F, Cl, Br, or I. The general formula for alkyl halides is R—X, where X represents a halogen atom. Alkyl halides are similar in structure, polarity, and reactivity to alcohols. To name an alkyl halide, first name the parent hydrocarbon. Then use the prefix fluoro-, chloro-, bromo-, or iodo-, with a position number, to indicate the presence of a fluorine atom, chlorine atom, bromine atom, or iodine atom. The following Sample Problem shows how to name an alkyl halide.

---

**Sample Problem**

### Naming an Alkyl Halide

**Problem**

Name the following compound.

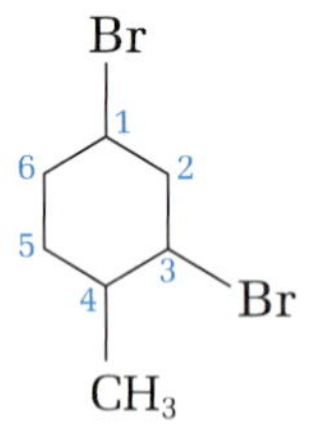

[Structure: cyclohexane ring with Br at position 1, Br at position 3, and CH₃ at position 4]

**Solution**

The parent hydrocarbon of this compound is cyclohexane. There are two bromine atoms attached at position numbers 1 and 3. Therefore, part of the prefix is 1,3-dibromo-. There is also a methyl group at position number 4. Because the groups are put in alphabetical order, the full prefix is 1,3-dibromo-4-methyl-. (The ring is numbered so that the two bromine atoms have the lowest possible position numbers. See the Problem Tip on page 361.) The full name of the compound is 1,3-dibromo-4-methylcyclohexane.

### Practice Problems

**12.** Draw a condensed structural diagram for each alkyl halide.
   (a) bromoethane
   (b) 2,3,4-triiodo-3-methylheptane

**13.** Name the alkyl halide at the right. Then draw a condensed structural diagram to represent it.

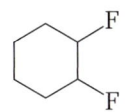

**14.** Draw and name an alkyl halide that has three carbon atoms and one iodine atom.

**15.** Draw and name a second, different alkyl halide that matches the description in the previous question.

## Reactions of Alkyl Halides

In Chapter 9, you learned that alkenes and alkynes can react with halogen compounds, such as $Br_2$ and $Cl_2$, or with halogen acids, such as HBr and HCl, to produce alkyl halides. In this section, you have just learned that alkyl halides can be made from alcohols by substitution reactions in an acidic solution. By changing the conditions of the reaction such as adding a strong base, you can cause the reaction to go in the opposite direction, as shown in Figure 10.7A. This reaction is a substitution reaction in which a hydroxyl group is substituted for the halogen atom. Another adjustment of reaction conditions can cause an entirely different reaction to occur. For example, you can raise the temperature and reduce the polarity of the solvent by carrying out the reaction in ethanol instead of water. As shown in Figure 10.7B, careful choice of reaction conditions can result in an elimination reaction. When an alkyl halide undergoes an elimination reaction in the presence of a base, an alkene, water, and a halogen ion are produced. Alkyl halides can also undergo substitution reactions to form different alkyl halides as shown in Figure 10.7C.

**A**    $CH_3-CH_2-Cl + OH^- \rightarrow CH_3-CH_2-OH + Cl^-$

**B**    $CH_3-CH_2-Cl + OH^- \rightarrow H_2C=CH_2 + HOH + Cl^-$

**C**    $CH_3-CH_2-Br + KI \rightarrow CH_3-CH_2-I + KBr$

**Figure 10.7** (A) In this substitution reaction, chloroethane reacts with the hydroxide ion to produce an alcohol and a chloride ion. This reaction is the inverse of the reaction shown in Figure 10.7. (B) By changing the reaction conditions, you can change the reaction to an elimination reaction. (C) This substitution reaction converts one alkyl halide, bromoethane, into another, iodoethane.

> **PROBLEM TIP**
> $Na^+$ does not participate in the substitution reaction in Figure 10.10. It is a *spectator ion*, an ion that is present, but does not participate in the reaction. Spectator ions are sometimes included in the written reaction, as in part (B), and sometimes omitted, as in part (A).

You have learned to recognize substitution and elimination reactions of alcohols and alkyl halides. To avoid confusion, for the remainder of this Unit, you can assume that the conditions of reactions involving an alkyl halide and a base have been adjusted to favour elimination reactions. Note that alcohols undergo elimination reactions in the presence of a strong acid while alkyl halides undergo elimination reactions in the presence of a strong base. Now examine the Sample Problems then complete the Practice Problems to enhance your understanding of these reactions.

> **CHEM FACT**
> In most reactions of organic compounds, more than one product is formed. Organic chemists modify reaction conditions to maximize the amount of the desired product.

## Sample Problem

### Predicting the Reaction of Alcohols and Alkyl Halides

**Problem**

Name each type of reaction. Then predict and name the products.

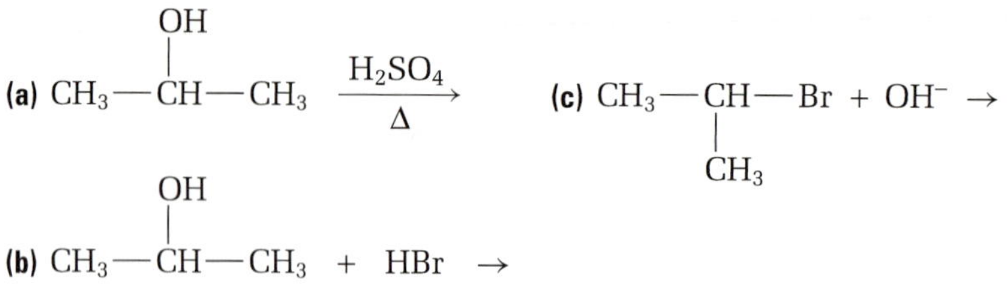

(a) $CH_3-\underset{\underset{OH}{|}}{CH}-CH_3 \xrightarrow[\Delta]{H_2SO_4}$

(b) $CH_3-\underset{\underset{OH}{|}}{CH}-CH_3 + HBr \rightarrow$

(c) $CH_3-\underset{\underset{CH_3}{|}}{CH}-Br + OH^- \rightarrow$

**Solution**

(a) This reaction takes place in the presence of heat and sulfuric acid, $H_2SO_4$. It is an elimination reaction. The product is an alkene, with the same number of carbon atoms as the reacting alcohol. Since this reaction is an elimination reaction, a small molecule (in this case, water) must be eliminated as the second product. The organic product is propene.

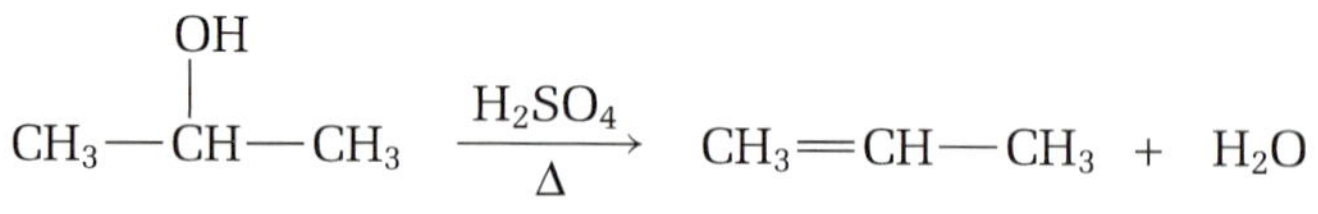

$CH_3-\underset{\underset{OH}{|}}{CH}-CH_3 \xrightarrow[\Delta]{H_2SO_4} CH_3=CH-CH_3 + H_2O$

2-propanol             propene     water

(b) In this reaction, an alcohol reacts with hydrobromic acid, HBr. This is a substitution reaction. The product is an alkyl halide, with the same carbon-carbon bonds, and the same number of carbon atoms, as the reacting alcohol. The alkyl halide is 2-bromopropane. The second product is water.

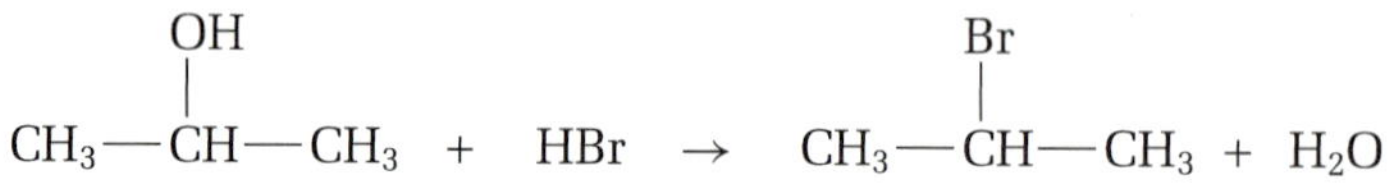

$CH_3-\underset{\underset{OH}{|}}{CH}-CH_3 + HBr \rightarrow CH_3-\underset{\underset{Br}{|}}{CH}-CH_3 + H_2O$

2-propanol    hydrobromic    2-bromopropane    water
               acid

(c) This reaction involves an alkyl halide reacting with the hydroxide ion from a base. It is an elimination reaction. The product is an alkene with the same number of carbon atoms, as the alkyl halide reactant. The alkene formed by this reaction is propene. The bromide ion is released by this reaction.

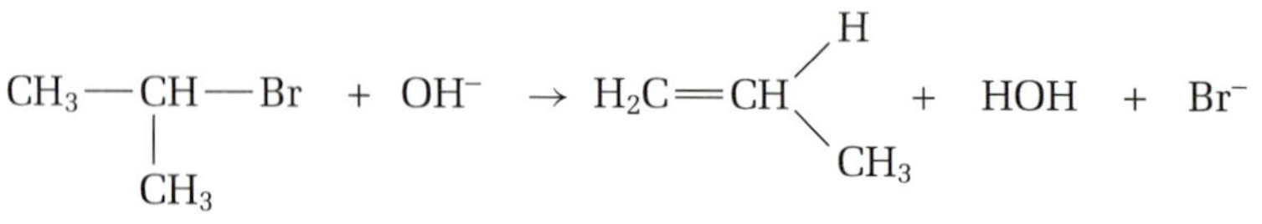

$CH_3-\underset{\underset{CH_3}{|}}{CH}-Br + OH^- \rightarrow H_2C=CH\underset{\diagdown CH_3}{\diagup H} + HOH + Br^-$

2-bromopropane   hydroxide          propene     water    bromide
                  ion                                                     ion

## Practice Problems

**16.** Name each type of reaction.
 (a) 1-propanol + HCl → 1-chloropropane + H$_2$O
 (b) 1-chlorobutane + KBr → 1-bromobutane + KCl
 (c) CH$_3$CH$_2$CH$_2$Cl + OH$^-$ → CH$_3$CH═CH$_2$ + HOH + Cl$^-$
 (d) CH$_3$—CH(OH)—CH$_2$—CH$_2$—CH$_3$ $\xrightarrow[\Delta]{H_2SO_4}$ CH$_3$—CH═CH—CH$_2$—CH$_3$ + H$_2$O

**17.** Draw the structures of the organic reactants and products in parts (a) and (b) of question 16.

**18.** Predict what type of reaction will occur.
 (a) CH$_3$—CH$_2$—CH$_2$—CH$_2$—OH $\xrightarrow[\Delta]{H_2SO_4}$
 (b) CH$_3$—CH(OH)—CH(CH$_3$)—CH$_3$ + HCl →
 (c) CH$_3$—CH(CH$_3$)—CH$_2$—I + OH$^-$ →
 (d) (cyclohexanol) + HBr →
 (e) (1,3-dimethylcyclopentan-... with OH) $\xrightarrow[\Delta]{H_2SO_4}$
 (f) CH$_3$—CH(CH$_3$)—CH(CH$_2$CH$_3$)—Cl + OH$^-$ →

**19.** Draw and name the products of each reaction in question 18.

**Figure 10.8** Many alkyl halides are used for pesticides.

## Ethers

Suppose that you removed the H atom from the —OH group of an alcohol. This would leave space for another alkyl group to attach to the oxygen atom.

$$CH_3CH_2—OH$$
$$\downarrow$$
$$CH_3CH_2—O—$$
$$\downarrow$$
$$CH_3CH_2—O—CH_3$$

The compound you have just made is called an ether. An **ether** is an organic compound that has two alkyl groups joined by an oxygen atom. The general formula of an ether is R—O—R'. You can think of alcohols and ethers as derivatives of the water molecule, as shown in Figure 10.9. Figure 10.10 gives two examples of ethers.

water    alcohol    ether

**Figure 10.9** An alcohol is equivalent to a water molecule with one hydrogen atom replaced by an alkyl group. Similarly, an ether is equivalent to a water molecule with both hydrogen atoms replaced by alkyl groups.

ethoxyethane
(common name:
diethyl ether)

1-methoxypropane
(common name:
methyl propyl ether)

**Figure 10.10** Until fairly recently, ethoxyethane was widely used as an anaesthetic. It had side effects, such as nausea, however. Compounds such as 1-methoxypropane are now used instead.

To name an ether, follow the steps below. The Sample Problem then shows how to use these steps to give an ether its IUPAC name and its common name.

**How to Name an Ether**

**IUPAC Name**

**Step 1** Choose the longest alkyl group as the parent alkane. Give it an alkane name.

**Step 2** Treat the second alkyl group, along with the oxygen atom, as an *alkoxy group* attached to the parent alkane. Name the alkoxy group by replacing the -yl ending of the corresponding alkyl group's name with -oxy. Give it a position number.

**Step 3** Put the prefix and suffix together: alkoxy group + parent alkane.

**Common Name**

**Step 1** List the alkyl groups that are attached to the oxygen atom, in alphabetical order.

**Step 2** Place the suffix -*ether* at the end of the name.

## Sample Problem

### Naming an Ether

**Problem**

Give the IUPAC name and the common name of the following ether.

$CH_3-CH_2-O-CH_2-CH_2-CH_3$

**Solution**

**IUPAC Name**

**Step 1** The longest alkyl group is based on propane.

$CH_3-CH_2-O-CH_2-CH_2-CH_3$

**Step 2** The alkoxy group is based on ethane (the ethyl group). It is located at the first carbon atom of the propane part. Therefore, the prefix is 1-ethoxy-.

**Step 3** The full name is 1-ethoxypropane.

**Common Name**

**Step 1** The two alkyl groups are ethyl and propyl.

**Step 2** The full name is ethyl propyl ether.

## Practice Problems

20. Use the IUPAC system to name each ether.

    (a) $H_3C-O-CH_3$

    (b) 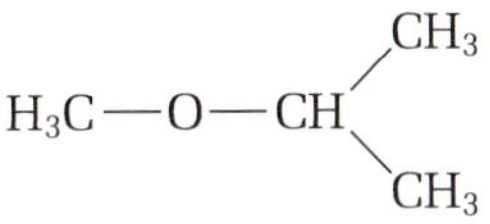 $H_3C-O-CH(CH_3)_2$ (with CH branching to two CH$_3$ groups)

    (c) $CH_3-CH_2-CH_2-CH_2-O-CH_3$

*Continued...*

> **21.** Give the common name for each ether.
>
> (a) H₃C—O—CH₂CH₃     (b) H₃C—O—CH₃
>
> **22.** Draw each ether.
>
> (a) 1-methoxypropane     (c) tert-butyl methyl ether
>
> (b) 3-ethoxy-4-methylheptane
>
> **23.** Sketch diagrams of an ether and an alcohol with the same number of carbon atoms. Generally speaking, would you expect an ether or an alcohol to be more soluble in water? Explain your reasoning.

Table 10.4 describes some physical properties of ethers. Like alcohols, ethers are polar molecules. Ethers, however, cannot form hydrogen bonds with themselves. The physical properties of ethers depend on these characteristics.

**Table 10.4** Physical Properties of Ethers

| Polarity of functional group | The bent shape around the oxygen atom in an ether means that the two C—O dipoles do not cancel each other. Because a C—O bond is less polar than an O—H bond, an ether is less polar than an alcohol. |
|---|---|
| Hydrogen bonding | Because there is no O—H bond in an ether, hydrogen bonding does not occur between ether molecules. However, the oxygen atom in ethers can form a hydrogen bond with a hydrogen atom in water. |
| Solubility in water | Ethers are usually soluble in water. The solubility of an ether decreases as the size of the alkyl groups increases. |
| Melting and boiling points | The boiling points of ethers are much lower than the boiling points of alcohols with the same number of carbon atoms. |

**Additional Characteristics of Ethers**

- Like alcohols, ethers are extremely flammable and should be used with caution. You will study the reactions of ethers in a later chemistry course.

## Amines

An organic compound with the functional group —NH₂, —NHR, or —NR₂ is called an **amine**. The letter N refers to the nitrogen atom. The letter R refers to an alkyl group attached to the nitrogen. The general formula of an amine is R—NR′₂. Amines can be thought of as derivatives of the ammonia molecule, NH₃. They are classified as *primary*, *secondary*, or *tertiary*, depending on how many alkyl groups are attached to the nitrogen atom. Note that the meanings of "primary," "seconday," and "tertiary" are slightly different from their meanings for alcohols. Figure 10.11 gives some examples of amines.

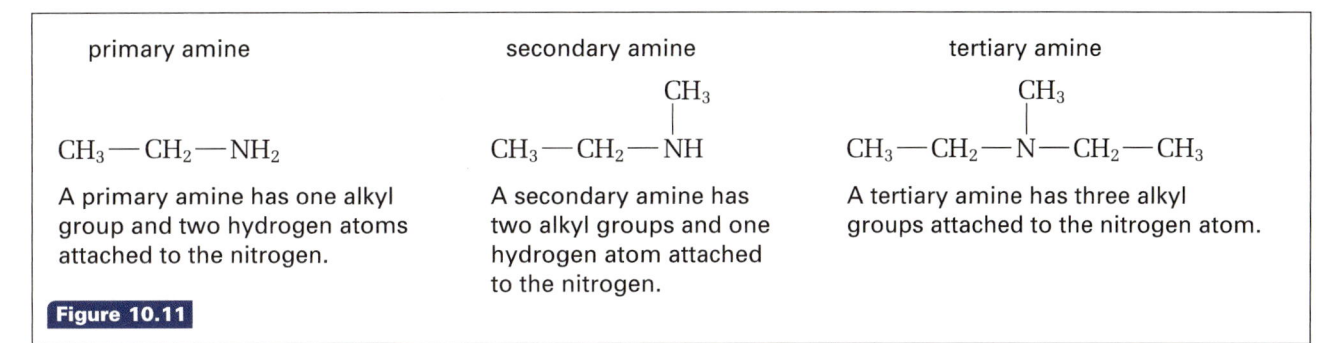

| primary amine | secondary amine | tertiary amine |
|---|---|---|
| CH₃—CH₂—NH₂ | CH₃—CH₂—NH—CH₃ | CH₃—CH₂—N(CH₃)—CH₂—CH₃ |
| A primary amine has one alkyl group and two hydrogen atoms attached to the nitrogen. | A secondary amine has two alkyl groups and one hydrogen atom attached to the nitrogen. | A tertiary amine has three alkyl groups attached to the nitrogen atom. |

**Figure 10.11**

To name an amine, follow the steps below. The Sample Problems illustrate how to use these steps to name primary and secondary amines.

### How to Name an Amine

**Step 1** Identify the largest hydrocarbon group attached to the nitrogen atom as the parent alkane.

**Step 2** Replace the -e at the end of the name of the parent alkane with the new ending -amine. Include a position number, if necessary, to show the location of the functional group on the hydrocarbon chain.

**Step 3** Name the other alkyl group(s) attached to the nitrogen atom. Instead of position numbers, use the letter N- to locate the group(s). (If two identical alkyl groups are attached to the nitrogen atom, use N,N-.) This is the prefix.

**Step 4** Put the name together: prefix + root + suffix.

### Sample Problem

#### Naming a Primary Amine

**Problem**
Name the following primary amine.

CH₃—CH₂—CH₂—CH(NH₂)—CH₂—CH₃

**Solution**

**Step 1** The only alkyl group is a hexanyl group. Therefore, the parent alkane is hexane.

**Step 2** Replacing the -e with -amine results in hexanamine. Numbering is started on the end that will make the substitution on the smallest number. In this case, it is 3.

⁶CH₃—⁵CH₂—⁴CH₂—³CH(NH₂)—²CH₂—¹CH₃

**Step 3** There are no other alkyl groups on the nitrogen so this is a primary amine. No prefix is needed.

**Step 4** The full name is 3-hexanamine.

## CHEM FACT

It is also common and correct to name amines with each alkyl branch listed as an attachment before the suffix -amine. In this system of nomenclature, the molecules in Figure 10.11 are ethylamine, methyl ethyl amine, and methyl diethyl amine. Several other methods of naming amines exist, but they will not be covered in this text.

## Sample Problem

### Naming a Secondary Amine

**Problem**

Name the following secondary amine.

$$H_3C-NH-\underset{CH_3}{\overset{CH_3}{\underset{|}{\overset{|}{CH}}}}$$

**Solution**

**Step 1** The propyl group is the largest of the two hydrocarbon groups attached to the nitrogen atom. Therefore, the parent alkane is propane.

**Step 2** Replacing the -e with -amine gives propanamine. The position number of the functional group in the propane chain is 2.

$$H_3C-NH-\underset{^3CH_3}{\overset{^1CH_3}{\underset{|}{\overset{|}{^2CH}}}}$$

**Step 3** A methyl group is also attached to the nitrogen atom. The corresponding prefix is N-methyl-.

**Step 4** The full name is N-methyl-2-propanamine.

## Practice Problems

24. Name each amine.

    (a) $CH_3-NH_2$

    (b) $CH_3-CH_2-\underset{CH_3}{\overset{|}{CH}}-NH_2$

    (c) $C(CH_3)_3CH_2-\underset{H}{\overset{|}{N}}-CH_2CH_3$

    (d) $CH_3-CH_2-\underset{NH_2}{\overset{CH_2-CH_3}{\underset{|}{\overset{|}{CH}}}}$

    (e) $CH_3-CH_2-\underset{NH_2}{\overset{|}{CH}}-CH_3$

    (f) cyclopentyl-$\underset{|}{\overset{CH_3}{N}}$-CH_3

25. Draw a condensed structural diagram for each amine.

    (a) 2-pentanamine
    (b) cyclohexanamine
    (c) N-methyl-1-butanamine
    (d) 3-heptanamine
    (e) N,N-diethyl-3-heptanamine
    (f) ethanamine

26. Classify each amine in the previous question as primary, secondary, or tertiary.

27. Draw and name all the isomers with the molecular formula $C_4H_{11}N$.

**Figure 10.12** Aniline is an aromatic amine that is useful for preparing dyes.

398 MHR • Unit 4 Organic Chemistry

Amines are polar compounds. Primary and secondary amines can form hydrogen bonds, but tertiary amines cannot. Table 10.5 lists some common physical properties of amines.

**Table 10.5** Physical Properties of Amines

| Polarity of functional group | C—N and N—H bonds are polar. Thus, amines are usually polar. |
|---|---|
| Hydrogen bonding | The presence of one or more N—H bonds allows hydrogen bonding to take place. |
| Solubility in water | Because of hydrogen bonding, amines with low molecular masses (four or less carbon atoms) are completely miscible with water. The solubility of an amine decreases as the number of carbon atoms increases. |
| Melting and boiling points | The boiling points of primary and secondary amines (which contain N—H bonds) are higher than the boiling points of tertiary amines (which do not contain an N—H bond). The higher boiling points are due to hydrogen bonding between amine molecules. |

### Additional Characteristics of Amines

- Amines are found widely in nature. They are often toxic. Many amines that are produced by plants have medicinal properties. (See Figure 10.13.)
- Amines with low molecular masses have a distinctive fishy smell. Also, many offensive odours of decay and decomposition are caused by amines. For example, cadavarine, $H_2NCH_2CH_2CH_2CH_2CH_2NH_2$, contributes to the odour of decaying flesh. This compound gets its common name from the word "cadaver," meaning "dead body."

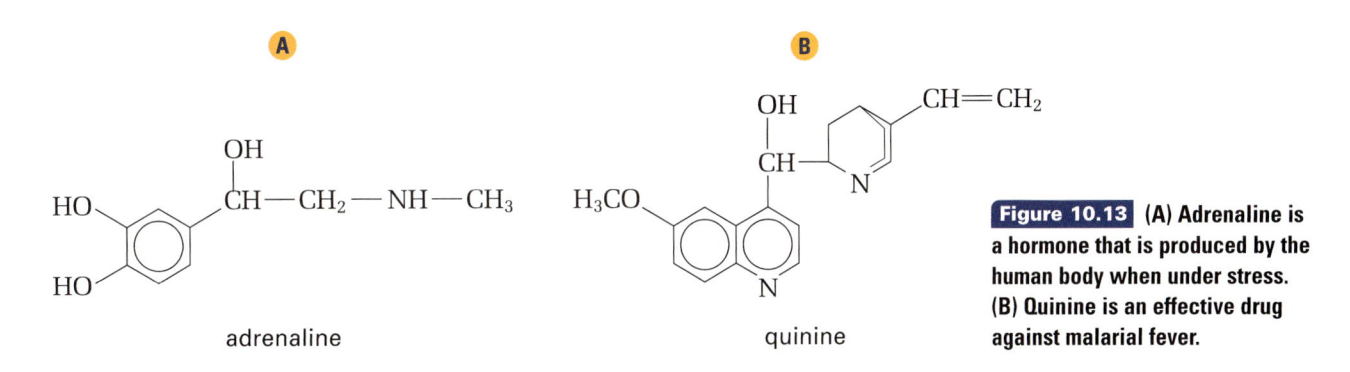

**Figure 10.13** (A) Adrenaline is a hormone that is produced by the human body when under stress. (B) Quinine is an effective drug against malarial fever.

## Reactions of Amines

Like ammonia, amines act as weak bases. Since amines are bases, adding an acid to an amine produces a salt, as shown in Figure 10.14. This explains why vinegar and lemon juice (both acids) can be used to neutralize the fishy smell of seafood, which is caused by amines.

**A**  $CH_3NH_2 + H_2O \rightarrow CH_3NH_3^+ + OH^-$

**B**  $CH_3NH_2 + HCl \rightarrow CH_3NH_3^+Cl^-$

**Figure 10.14** (A) Like other bases, amines react with water to produce hydroxide ions. (B) Amines react with acids to form salts.

## Section Summary

In this section, you learned how to recognize, name, and draw members of the alcohol, alkyl halide, ether, and amine families. You learned how to recognize some of the physical properties of these compounds and how to predict the products of reactions involving these compounds. In the next section, you will learn about families of organic compounds with functional groups that contain the C=O bond.

### Section Review

**1** Name each compound.

(a)  CH₃CH₂CH₂CH₂NH₂

(b) CH₃CH₂—O—CH(CH₃)CH₂CH₃

(c) cyclopentane with —OH, and two CH₃ substituents (H₃C and CH₃)

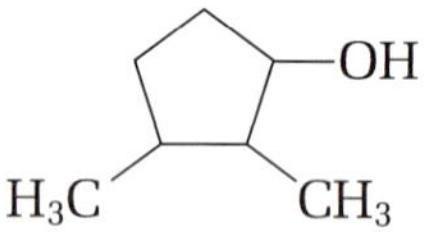

(d) CH₃CH₂CH₂CHBrCH₃

**2** Write the IUPAC name for each compound.

(a) $CH_3CH_2CH_2CH_2CH_2CH_2CH_2OH$
(b) $CH_3CH(OH)CH_2CH_3$
(c) $CH_3CH_2NH_2$
(d) $(CH_3)_2NH$
(e) $CH_3CH_2OCH_3$
(f) $CH_3CH(Cl)CH_3$

**3** Draw a condensed structural diagram for each compound.

(a) 3-heptanol
(b) N-ethyl-2-hexanamine
(c) 3-methoxypentane
(d) 2-iodobutane
(e) 1-propanamine
(e) bromoethane

**4** Draw and name three isomers that have the molecular formula $C_5H_{12}O$.

**5** Name the following compounds. Then rank them, from highest to lowest boiling point. Explain your reasoning.

(a) H₃C—CH₃   H₃C—O—CH₃   CH₃—CH₂—OH

(b) CH₃—CH₂—CH₃   CH₃—CH₂—CH₂—NH₂   CH₃—N(CH₃)—CH₃

**6** Draw cyclohexanol and cyclohexane. Which compound do you expect to be more soluble in water? Explain your reasoning.

**7** Name the following compounds. Which compound do you expect to be more soluble in benzene? Explain your reasoning.

(propanol and 2-propanol structures shown)

**8** Name these compounds. Then rank them, from highest to lowest molecular polarity. Explain your reasoning.

CH₃—CH₂—CH₂—CH₃

HO—CH₂—CH₂—CH₂—CH₃

CH₃CH₂—O—CH₂CH₃

**9** Identify each type of reaction.

(a) HO—CH$_2$—CH$_2$—CH(CH$_3$)—CH$_3$ $\xrightarrow{H_2SO_4, \Delta}$ CH$_2$=CH—CH(CH$_3$)—CH$_3$ + H$_2$O

(b) CH$_3$—CH(Br)—CH(CH$_3$)—CH$_2$—CH$_3$ + OH$^-$ → CH$_3$—CH=C(CH$_3$)—CH$_2$—CH$_3$ + Br$^-$ + HOH

(c) 2CH$_3$—CH(CH$_3$)—OH + 9O$_2$ → 6CO$_2$ + 8H$_2$O

(d) CH$_3$—CH(CH$_3$)—CH(CH$_3$)—CH(CH$_3$)—CH$_3$—Cl + OH$^-$ → CH$_3$—CH$_2$—CH(CH$_3$)—C(CH$_3$)=CH$_2$ + Cl$^-$

**10** Draw the products of each reaction.

(a) CH$_3$CH$_2$CH$_2$CH$_2$CH$_2$CH$_2$OH $\xrightarrow{H_2SO_4, \Delta}$

(b) CH$_3$CH$_2$CH$_2$CH$_2$CH$_2$OH + HBr →

(c) CH$_3$CH$_2$NH$_2$ + H$_2$O →

(d) (CH$_3$)$_2$CHCH(Br)CH$_3$ + OH$^-$ →

(e) (CH$_3$)$_2$CHCH(Cl)CH$_2$CH$_3$ (with OH branch) + OH$^-$ →

(f) cyclopentanol $\xrightarrow{H_2SO_4, \Delta}$

**11** Name and draw the missing organic reactants for each incomplete reaction.

(a) ? + OH$^-$ → CH$_3$—C(CH$_3$)=CH$_2$ + HOH + Br$^-$

(b) ? + HCℓ → CH$_3$—C(CH$_3$)(CH$_2$CH$_3$)—CH$_2$—NH$_3$$^+$Cl$^-$

(c) ? $\xrightarrow{H_2SO_4, \Delta}$ ☐ + H$_2$O

(d) ? + HI → (structure with I substituent) + H$_2$O

(e) ? + OH$^-$ → (cyclohexyl-substituted alkene) + Cl$^-$ + HOH

## 10.3 Functional Groups With the C=O Bond

**Section Preview/Outcomes**

In this section, you will

- **name** and **draw** members of the aldehyde, ketone, carboxylic acid, ester, and amide families of organic compounds
- **describe** some physical and chemical properties of these families of compounds
- **communicate** your understanding of the following terms: *carbonyl group, aldehyde, ketone, carboxylic acid, ester, amide*

Some of the most interesting and useful organic compounds belong to families you are about to encounter. For example, the sweet taste of vanilla and the spicy scent of cinnamon have something in common: a carbonyl group. A **carbonyl group** is composed of a carbon atom double-bonded to an oxygen atom. In this section, you will study the structures and properties of organic compounds that have the C=O group. You will also study the chemical reactions some of these compounds undergo.

### Aldehydes and Ketones

Aldehydes and ketones both have the carbonyl functional group. An **aldehyde** is an organic compound that has a double-bonded oxygen on the last carbon of a carbon chain. The functional group for an aldehyde is

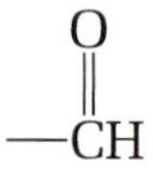

$$-\overset{\overset{\displaystyle O}{\|}}{C}H$$

The general formula for an aldehyde is R—CHO, where R is any alkyl group. Figure 10.14 shows the two simplest aldehydes.

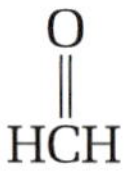

$$H\overset{\overset{\displaystyle O}{\|}}{C}H \qquad H_3C-\overset{\overset{\displaystyle O}{\|}}{C}H$$

methanal (common name: formaldehyde)    ethanal (common name: acetaldehyde)

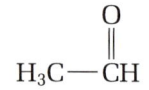

**Figure 10.14** Methanal is made from methanol. It is used to preserve animal tissues and to manufacture plastics. Ethanal is also used as a preservative, and in the manufacture of resins and dyes.

When the carbonyl group occurs within a hydrocarbon chain, the compound is a ketone. A **ketone** is an organic compound that has a double-bonded oxygen on any carbon within the carbon chain. The functional group of a ketone is

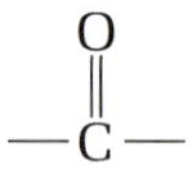

$$-\overset{\overset{\displaystyle O}{\|}}{C}-$$

The general formula for a ketone is RCOR′, where R and R′ are alkyl groups. Figure 10.15 shows the simplest ketone, propanone.

Like the other organic compounds you have encountered, the names of aldehydes and ketones are based on the names of the parent alkanes. To name an aldehyde, follow the steps below.

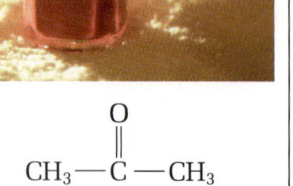

### How to Name an Aldehyde

**Step 1** Name the parent alkane. Always give the carbon atom of the carbonyl group the position number 1.

**Step 2** Replace the -e at the end of the name of the parent alkane with -al. The carbonyl group is always given position number 1. Therefore, you do not need to include a position number for it.

To name a ketone, follow the steps on the next page. The Sample Problem that follows gives examples for naming both aldehydes and ketones.

$$CH_3-\overset{\overset{\displaystyle O}{\|}}{C}-CH_3$$

propanone (common name: acetone)

**Figure 10.15** Propanone is the main component of most nail polish removers. It is used as a solvent, and in the manufacture of plastics.

402 MHR • Unit 4 Organic Chemistry

### How to Name a Ketone

**Step 1** Name the parent alkane. Remember that the main chain must contain the C=O group.

**Step 2** If there is one ketone group, replace the -e at the end of the name of the parent alkane with -one. If there is more than one ketone group, keep the -e suffix and add a suffix such as -dione or -trione.

**Step 3** For carbon chains that have more than four carbons, a position number is needed for the carbonyl group. Number the carbon chain so that the carbonyl group has the lowest possible number.

#### Sample Problem
#### Drawing and Naming Aldehydes and Ketones

**Problem**
Draw and name seven isomers with the molecular formula $C_5H_{10}O$.

**Solution**

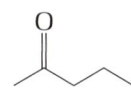

2-pentanone 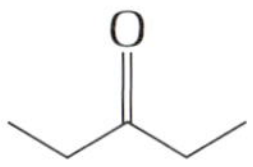     3-pentanone 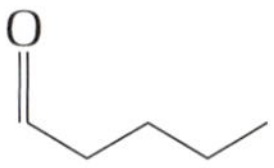     pentanal 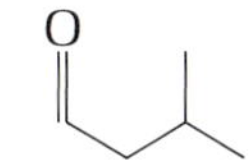     3-methylbutanal

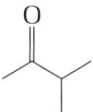

3-methyl-2-butanone 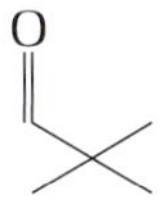     2,2-dimethylpropanal 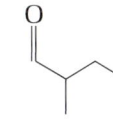     2-methylbutanal

#### Practice Problems

28. Name each aldehyde or ketone.

    (a) 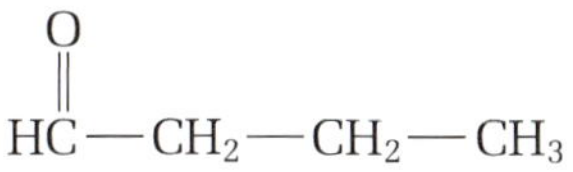 $HC{-}CH_2{-}CH_2{-}CH_3$ with =O on HC

    (b) 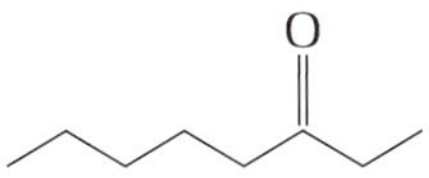 [structure]

    (c) 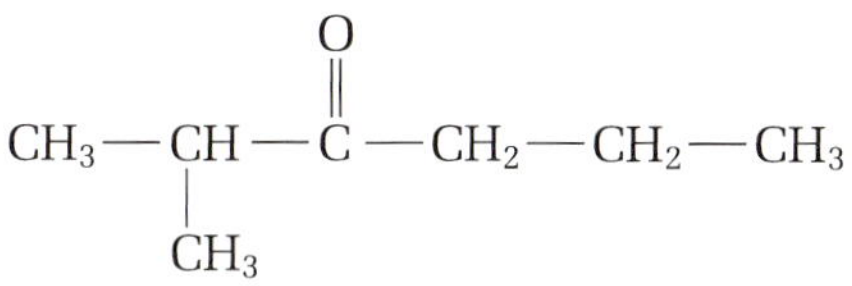 $CH_3{-}CH{-}C({=}O){-}CH_2{-}CH_2{-}CH_3$ with $CH_3$ branch

    (d) 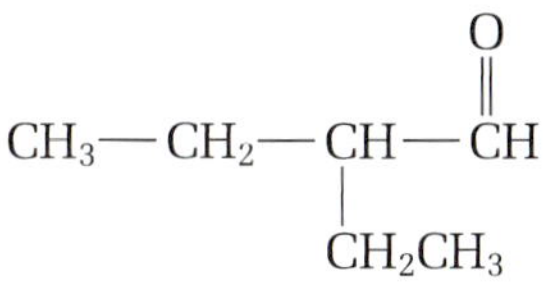 $CH_3{-}CH_2{-}CH{-}CH$ with =O on right CH and $CH_2CH_3$ branch

    (e) 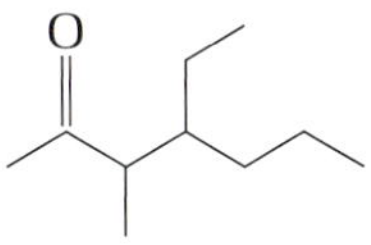 [structure]

29. Draw a condensed structural diagram for each aldehyde or ketone.

    (a) 2-propylpentanal
    (b) cyclohexanone
    (c) 4-ethyl-3,5-dimethyloctanal

30. Is a compound with a C=O bond and the molecular formula $C_2H_4O$ an aldehyde or a ketone? Explain.

31. Draw and name five ketones and aldehydes with the molecular formula $C_6H_{12}O$.

**Table 10.5** Physical Properties of Aldehydes and Ketones

| | |
|---|---|
| **Polarity of functional group** | The C=O bond is polar, so aldehydes and ketones are usually polar. |
| **Hydrogen bonding** | Hydrogen bonding cannot occur between molecules of these compounds, since there is no O—H bond. The oxygen atom, however, can accept hydrogen bonds from water, as shown here. |
| **Solubility in water** | Aldehydes and ketones with low molecular masses are very soluble in water. Aldehydes and ketones with a large non-polar hydrocarbon part are less soluble in water. |
| **Melting and boiling points** | The boiling points of aldehydes and ketones are lower than the boiling points of the corresponding alcohols. They are higher than the boiling points of the corresponding alkanes. |

### Additional Characteristics of Aldehydes and Ketones

- In general, aldehydes have a strong pungent smell, while ketones smell sweet. Aldehydes with higher molecular masses have a pleasant smell. For example, cinnamaldehyde gives cinnamon its spicy smell. (See Figure 10.16.) Aldehydes and ketones are often used to make perfumes. The rose ketones (shown in Figure 10.17) provide up to 90% of the characteristic rose odour. Perfumers mix organic compounds, such as the rose ketones, to obtain distinctive and attractive scents.
- Since aldehydes and ketones are polar, they can act as polar solvents. Because of the non-polar hydrocarbon part of their molecules, aldehydes and ketones can also act as solvents for non-polar compounds. For example, 2-propanone (common name: acetone) is an important organic solvent in the chemical industry.
- Table 10.6 compares the boiling points of an alkane, an alcohol, and an aldehyde with the same number of carbon atoms. You can see that the boiling point of an alcohol is much greater than the boiling point of an alkane or an aldehyde.
- The reactions of aldehydes and ketones will be left for a later course.

**Table 10.6** Comparing Boiling Points

| Compound | Structure | Boiling point |
|---|---|---|
| propane | CH$_3$CH$_2$CH$_3$ | 42.1°C |
| 1-propanol | CH$_3$CH$_2$CH$_2$OH | 97.4°C |
| propanal | CH$_3$CH$_2$CHO | 48.8°C |

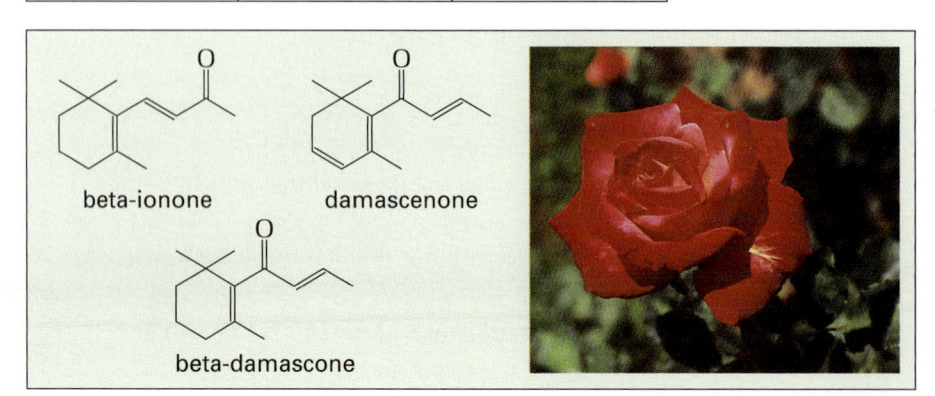

**Figure 10.16** Cinnemaldehyde gives cinnamon its spicy smell.

**Figure 10.17** The rose ketones provide the sweet smell of a rose.

## Carboxylic Acids

You are already familiar with one carboxylic acid. In fact, you may sprinkle it over your French fries or your salad, as shown in Figure 10.18. Vinegar is a 5% solution of acetic acid in water. The IUPAC name for acetic acid is ethanoic acid, $CH_3COOH$.

A **carboxylic acid** is an organic compound with the following functional group:

$$-\overset{\overset{O}{\|}}{C}-OH$$

This —COOH group is called the **carboxyl group**. The general formula for a carboxylic acid is R—COOH. Figure 10.19 shows some common carboxylic acids.

methanoic acid
(common name: formic acid)

ethanoic acid
(common name: acetic acid)

benzoic acid

citric acid

**Figure 10.18** Many salad dressings contain vinegar.

**Figure 10.19** Common carboxylic acids

To name a simple carboxylic acid, follow the steps below. Figure 10.20 gives some examples of carboxylic acid names.

### How to Name a Carboxylic Acid

**Step 1** Name the parent alkane.

**Step 2** Replace the -e at the end of the name of the parent alkane with -oic acid.

**Step 3** The carbon atom of the carboxyl group is always given position number 1. Name and number the branches that are attached to the compound.

butanoic acid

3-methylpentanoic acid

**Figure 10.20** Examples of carboxylic acid names

## CONCEPT CHECK

Organic compounds that have one or two carbon atoms are usually known by their common names. These common names are based on the Latin words *formica* (ant) and *acetum* (vinegar). Give the IUPAC names for formaldehyde, formic acid, acetaldehyde, and acetic acid.

### Practice Problems

32. Name each carboxylic acid.

    (a) 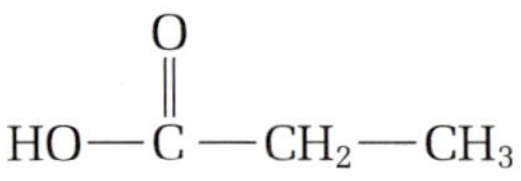 HO—C(=O)—CH$_2$—CH$_3$

    (b) CH$_3$—CH$_2$—C(CH$_3$)(CH$_3$)—C(=O)—OH

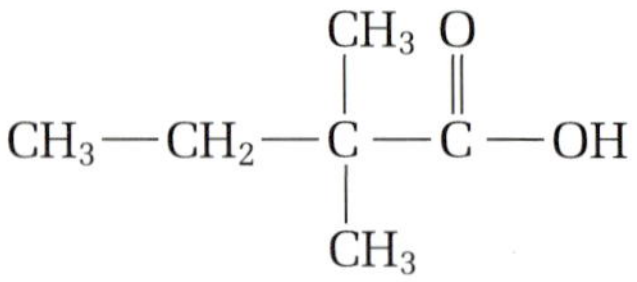

    (c) [structure showing isobutyl group attached to CH with ethyl branch and COOH]

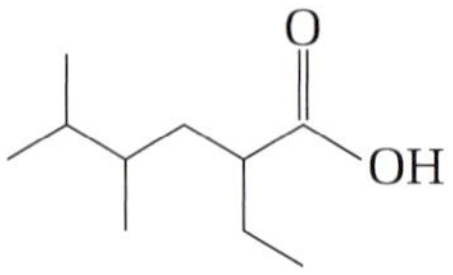

33. Draw a condensed structural diagram for each carboxylic acid.

    (a) hexanoic acid

    (b) 3-propyloctanoic acid

    (c) 3,4-diethyl-2,3,5-trimethylheptanoic acid

34. Draw a line structural diagram for each compound in question 33.

35. Draw and name two carboxylic acids with the molecular formula C$_4$H$_8$O$_2$.

Table 10.7 lists some of the physical properties of carboxylic acids. Notice that carboxylic acids have even stronger hydrogen bonding than alcohols.

**Table 10.7** Physical Properties of Carboxylic Acids

| | |
|---|---|
| **Polarity of functional group** | Due to the presence of the polar O—H and C=O bonds, carboxylic acids are polar compounds. |
| **Hydrogen bonding** | The hydrogen bonding between carboxylic acid molecules is strong, as shown here: 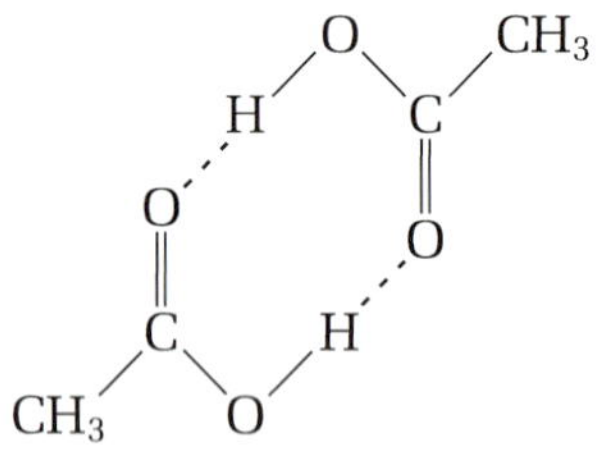 Hydrogen bonding also occurs between carboxylic acid molecules and water molecules. |
| **Solubility in water** | Carboxylic acids with low molecular masses are very soluble in water. The first four simple carboxylic acids (methanoic acid, ethanoic acid, propanoic acid, and butanoic acid) are miscible with water. Like alcohols, ketones, and aldehydes, the solubility of carboxylic acids in water decreases as the number of carbon atoms increases. |
| **Melting and boiling points** | Because of the strong hydrogen bonds between molecules, the melting and boiling points of carboxylic acids are very high. |

### Additional Characteristics of Carboxylic Acids

- Carboxylic acids often have unpleasant odours. For example, butanoic acid has the odour of stale sweat.
- The —OH group in a carboxylic acid does *not* behave like the basic hydroxide ion, OH⁻. Oxygen has a high electronegativity (attraction to electrons) and there are two oxygen atoms in the carboxylic acid functional group. These electronegative oxygen atoms help to carry the extra negative charge that is caused when a positive hydrogen atom dissociates. This is why the hydrogen atom in a carboxylic acid is able to dissociate, and the carboxylic acid behaves like an acid.

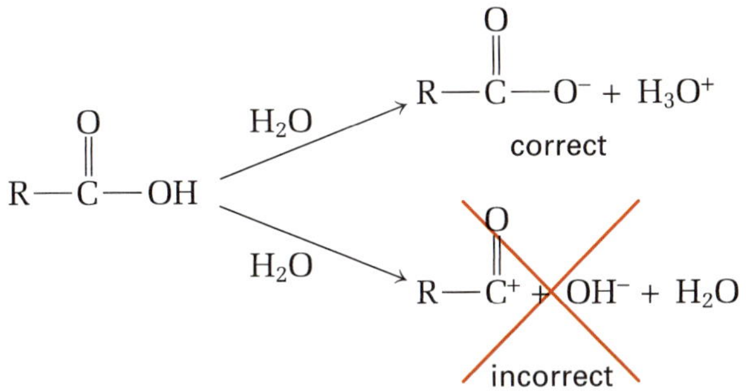

$$R-\overset{O}{\underset{\|}{C}}-OH \xrightarrow{H_2O} R-\overset{O}{\underset{\|}{C}}-O^- + H_3O^+ \text{ (correct)}$$

$$R-\overset{O}{\underset{\|}{C}}-OH \xrightarrow{H_2O} R-\overset{O}{\underset{\|}{C^+}} + OH^- + H_2O \text{ (incorrect)}$$

- Figure 10.21 compares the melting and boiling points of a carboxylic acid with the melting and boiling points of other organic compounds. As you can see, the melting and boiling points of the carboxylic acid are much higher than the melting and boiling points of the other compounds. This is due to the exceptionally strong hydrogen bonding between carboxylic acid molecules.

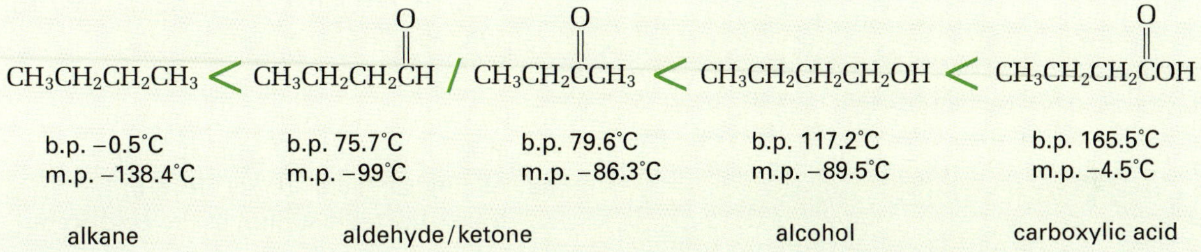

$$CH_3CH_2CH_2CH_3 < CH_3CH_2CH_2\overset{O}{\underset{\|}{C}}H \,/\, CH_3CH_2\overset{O}{\underset{\|}{C}}CH_3 < CH_3CH_2CH_2CH_2OH < CH_3CH_2CH_2\overset{O}{\underset{\|}{C}}OH$$

| b.p. −0.5°C | b.p. 75.7°C | b.p. 79.6°C | b.p. 117.2°C | b.p. 165.5°C |
| m.p. −138.4°C | m.p. −99°C | m.p. −86.3°C | m.p. −89.5°C | m.p. −4.5°C |

alkane       aldehyde/ketone       alcohol       carboxylic acid

**Figure 10.21**

### Reactions of Carboxylic Acids

Like other acids, a carboxylic acid reacts with a base to produce a salt and water.

$$CH_3-CH_2-CH_2-CH_2-\overset{O}{\underset{\|}{C}}-OH + NaOH \rightarrow CH_3-CH_2-CH_2-CH_2-\overset{O}{\underset{\|}{C}}-O^-Na^+ + H_2O$$

pentanoic acid (acid)     sodium hydroxide (base)     sodium pentanoate (salt)     water

### Derivatives of Carboxylic Acids

Exchanging the hydroxyl group of a carboxylic acid with a different group produces a *carboxylic acid derivative*. In the following investigation, you will react two carboxylic acids with primary alcohols to produce different organic compounds.

# Investigation 10-A

**SKILL FOCUS**
- Predicting
- Performing and recording
- Analyzing and interpreting

## Preparing a Carboxylic Acid Derivative

The reaction between a carboxylic acid and an alcohol produces an organic compound that you have not yet seen. In this investigation, you will prepare this new type of compound and observe one of its properties.

### Questions

What kind of compound will form in the reaction of a carboxylic acid with an alcohol? What observable properties will the new compounds have?

### Safety Precautions

- The products of these reactions are very flammable. Before starting, make sure that there is *no* open flame in the laboratory. Use a hot plate, *not* a Bunsen burner.
- Carry out all procedures in a well-ventilated area. Use a fume hood for steps involving butanoic acid.
- Concentrated sulfuric acid, ethanoic acid, and butanoic acid are all extremely corrosive. Wear goggles, gloves, and an apron while performing this investigation. Treat the acids with extreme care. If you spill any acid on your skin, wash it with plenty of cold water and notify your teacher.

### Materials

retort stand with two clamps
thermometer
hot plate
2 small beakers (50 or 100 mL)
100 mL beaker
2 large beakers (250 mL)
4 plastic micropipettes
2 graduated cylinders (10 mL)
medicine dropper
stopper or paper towel
distilled water
6 mol/L sulfuric acid
ethanoic acid (glacial acetic acid)
butanoic acid
ethanol
1-propanol
ice

### Procedure

1. Before starting the investigation, read through the Procedure. Then prepare your equipment as follows.

   (a) Make sure that all the glassware is very clean and dry.

   (b) Prepare a hot-water bath. Heat about 125 mL of tap water in a 250 mL beaker on the hot plate to 60°C. Adjust the hot plate so the temperature remains between 50°C and 60°C.

   (c) Prepare a cold-water bath. Place about 125 mL of a mixture of water and ice chips in the second 250 mL beaker. As long as there are ice chips in the water, the temperature of the cold-water bath will remain around 0°C.

   (d) Place about 5 mL of distilled water in a small beaker (or graduated cylinder). You will use this in step 8.

   (e) Set up the retort stand beside the hot-water bath. Use one clamp to hold the thermometer in the hot-water bath. Make sure that the thermometer is not touching the walls or the bottom of the beaker. Use a stopper or wrap a piece of paper towel around the thermometer to hold the thermometer in place carefully but firmly. You will use the other clamp to steady the micropipettes after you fill them with reaction mixture and place them in the hot-water bath.

**408** MHR • Unit 4 Organic Chemistry

(f) Label three pipettes according to each trial:
- ethanoic acid + ethanol
- ethanoic acid + 1-propanol
- butanoic acid + ethanol

(g) Cut off the bulb of one of the micropippettes, halfway along the wide part of the bulb (see the diagram below.) You will use the bulb as a cap to prevent vapours from escaping during the reactions.

2. Use the graduated cylinder to measure 1.0 mL of ethanol into the 100 mL beaker. As you do so, you will get a whiff of the odour of the alcohol. (**CAUTION** Do not inhale the alcohol vapour directly.) Record your observations.

3. Use the graduated cylinder to add 1.0 mL of ethanoic acid to the ethanol. As you do so, you will get a whiff of the odour of the acid. (**CAUTION** Do not inhale the acid directly.) Record your observations.

4. Your teacher will *carefully* add 4 drops of sulfuric acid to the alcohol/acid mixture.

5. Suction the mixture up into the appropriately labelled micropipette. Invert the micropipette. Place it, bulb down, in the hot-water bath. (See the diagram below.) Place the cap over the tip of the pipette. Use a clamp to hold the pipette in place.

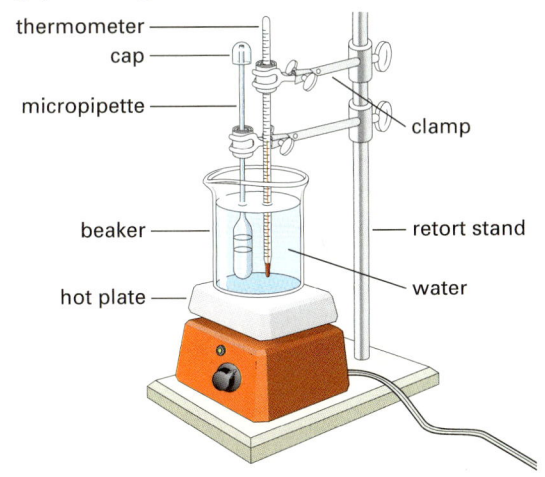

6. Leave the pipette in the hot water for about 10 min to 15 min. Use the thermometer to monitor the temperature of the hot water. The temperature should stay between 50°C and 60°C.

7. After 10 to 15 min in the hot-water bath, place the pipette in the cold-water bath. Allow it to cool for about 5 min.

8. Carefully squeeze a few drops of the product onto a watch glass. Mix it with a few drops of distilled water. To smell the odour of the compound, take a deep breath. Use your hand to wave the aroma toward your nose as you breathe out. Record your observations of the aroma.

9. Repeat the procedure for the other two trials. (**CAUTION** Butanoic acid has a strong odour. Do not attempt to smell it.)
   **Note:** While the first reaction mixture is being heated, you may wish to prepare the other mixtures and put them in the hot-water bath. Working carefully, you should be able to stabilize more than one micropipette with the clamp. If you choose to do this, make sure that your materials are clearly labelled. Also remember to keep a record of the time at which each micropipette was introduced to the hot-water bath.

10. Dispose of all materials as your teacher directs. Clean all glassware thoroughly with soap and water.

### Analysis

1. Make up a table to organize your data. What physical property did you observe?

2. How do you know that a new product was formed in each reaction? Explain.

### Conclusion

3. Describe the odour of each product that was formed. Compare the odours to familiar odours (such as the odours of plants, flowers, fruits, and animals) to help you describe them.

### Application

4. Research the organic compounds that are responsible for the smell and taste of oranges, pineapples, pears, oil of wintergreen, and apples. Find and record the chemical structure of each compound.

## Carboxylic Acid Derivatives

The strong-smelling compounds you prepared in Investigation 10-A do not fit into any of the organic families you have studied so far. According to their molecular formulas, however, they are isomers of carboxylic acids. They are *esters*. Because an ester is obtained by replacing the —OH group of a carboxylic acid with a different group, it is called a **derivative** of a carboxylic acid. Carboxylic acids have several important derivatives. Later in this section, you will study two of these derivatives: esters and amides.

As you saw in the investigation, a carboxylic acid reacts with an alcohol to produce an ester. Water is the second product of this reaction. A strong acid, such as $H_2SO_4$, is used to catalyze (speed up) the reaction. The reverse reaction can also occur, as you will see later in this section.

$$R-\underset{\text{carboxylic acid}}{\overset{\overset{O}{\|}}{C}-OH} + \underset{\text{alcohol}}{HO-R'} \underset{}{\overset{H_2SO_4}{\rightleftharpoons}} R-\underset{\text{ester}}{\overset{\overset{O}{\|}}{C}-O-R'} + \underset{\text{water}}{H_2O}$$

The reaction of a carboxylic acid with an alcohol to form an ester is called an **esterification reaction**. An esterification reaction is one type of condensation reaction.

## Esters

An **ester** is an organic compound that has the following functional group:

$$-\overset{\overset{O}{\|}}{C}-O-$$

The general formula for an ester is RCOOR′, where R is a hydrogen atom or a hydrocarbon, and R′ is a hydrocarbon. You can think of an ester as the product of a reaction between a carboxylic acid and an alcohol, as shown in Figure 10.22.

$$\underset{\text{carboxylic acid}}{CH_3CH_2CH_2\overset{\overset{O}{\|}}{C}-OH} + \underset{\text{alcohol}}{HO-CH_2CH_3} \rightarrow \underset{\text{ester}}{CH_3CH_2CH_2\overset{\overset{O}{\|}}{C}-O-CH_2CH_3} + \underset{\text{water}}{H_2O}$$

**Figure 10.22** When heated, butanoic acid reacts with ethanol to produce an ester called ethyl butanoate.

To name an ester, you must recognize that an ester can be thought of as having two distinct parts. The main part of the ester contains the —COO group. When numbering the main chain of a carboxylic acid, the carbon atom in the carboxyl group is always given position number 1. The second part of an ester is the alkyl group.

To name an ester, follow the steps below.

### How to Name an Ester

**Step 1** Identify the main part of the ester, which contains the C=O group. This part comes from the parent acid. Begin by naming the parent acid.

**Step 2** Replace the *-oic acid* ending of the name of the parent acid with *-oate*.

> **PROBEWARE**
> 
> www.mcgrawhill.ca/links/atlchemistry
> 
> If you have access to probeware, go to the website above for a probeware activity covering some of the material in this section.

**Step 3** The second part of an ester is the alkyl group that is attached to the oxygen atom. Name this as you would name any other alkyl group.

**Step 4** Put the two names together. Note that esters are named as two words. (See Figure 10.23.)

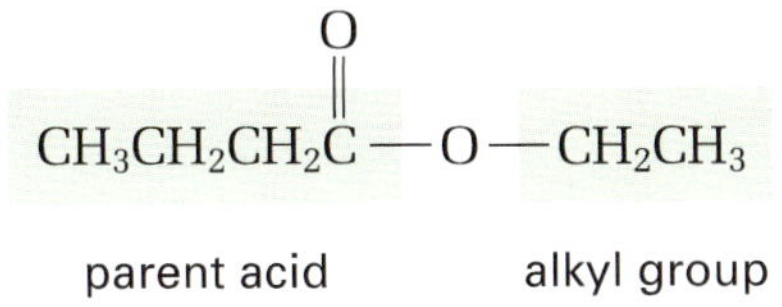

$$\underbrace{CH_3CH_2CH_2\overset{\overset{O}{\|}}{C}}_{\text{parent acid}} - O - \underbrace{CH_2CH_3}_{\text{alkyl group}}$$

**Figure 10.23** The main part of this molecule is based on butanoic acid. The other part of the ester is an ethyl group. The full name is ethyl- + -butanoate → ethyl butanoate

## Sample Problem

### Naming and Drawing Esters

**Problem**
Name and draw three esters that have the molecular formula $C_4H_8O_2$.

**Solution**

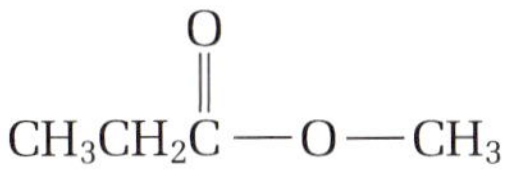

$$CH_3CH_2\overset{\overset{O}{\|}}{C} - O - CH_3$$
methyl propanoate

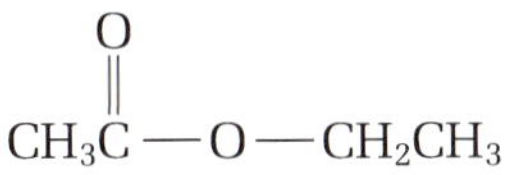

$$CH_3\overset{\overset{O}{\|}}{C} - O - CH_2CH_3$$
ethyl ethanoate

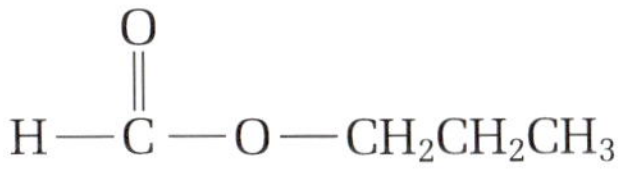

$$H - \overset{\overset{O}{\|}}{C} - O - CH_2CH_2CH_3$$
propyl methanoate

### Practice Problems

36. Name each ester.

    (a) 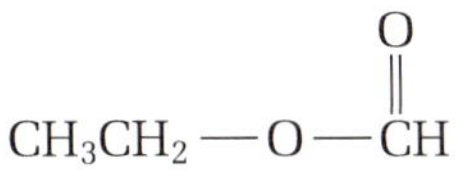 $CH_3CH_2 - O - \overset{\overset{O}{\|}}{C}H$

    (b) $CH_3CH_2CH_2\overset{\overset{O}{\|}}{C} - O - CH_3$

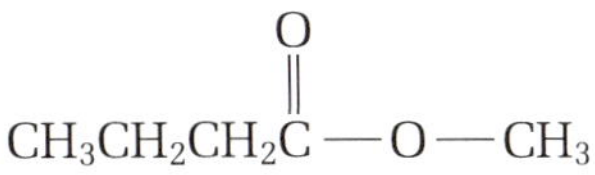

    (c) $CH_3CH_2CH_2CH_2\overset{\overset{O}{\|}}{C} - O - CH_2CH_2CH_2CH_2CH_3$

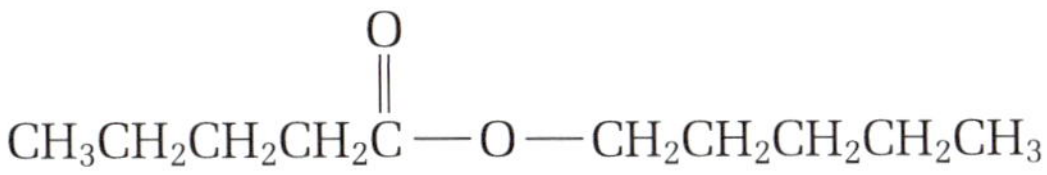

37. For each ester in the previous question, name the carboxylic acid and the alcohol that are needed to synthesize it.

38. Draw each ester.
    (a) methyl pentanoate
    (b) heptyl methanoate
    (c) butyl ethanoate
    (d) propyl octanoate
    (e) ethyl 3,3-dimethylbutanoate

*Continued...*

> *Continued ...*
>
> 39. Write the molecular formula of each ester in the previous question. Which esters are isomers of each other?
>
> 40. Draw and name five ester isomers that have the molecular formula $C_5H_{10}O_2$.

Table 10.8, below, describes some of the physical properties of esters. As you will see, esters have different physical properties than carboxylic acids, even though esters and carboxylic acids are isomers of each other.

**Table 10.8** Physical Properties of Esters

| Polarity of functional group | Like carboxylic acids, esters are usually polar molecules. |
|---|---|
| Hydrogen bonding | Esters do not have an O—H bond. Therefore, they cannot form hydrogen bonds with other ester molecules. |
| Solubility in water | Esters can accept hydrogen bonds from water. Therefore, esters with very low molecular masses are soluble in water. Esters with carbon chains that are longer than three or four carbons are not soluble in water. |
| Melting and boiling points | Because esters cannot form hydrogen bonds, they have low boiling points. They are usually volatile liquids at room temperature. |

**Figure 10.24** Octyl ethanoate is found in oranges.

### Additional Characteristics of Esters

- Esters often have pleasant odours and tastes, so they are used to produce perfumes and artificial flavours. In fact, the characteristic tastes and smells of many fruits come from esters. (See Figure 10.24.)

### Reactions of Esters

Esters undergo hydrolysis reactions. In a hydrolysis reaction, the ester bond is *cleaved*, or split in two, to form two products. Acid hydrolysis of an ester is the reverse of an esterification reaction: it produces a carboxylic acid and an alcohol. Hydrolysis usually requires heat. There are two methods of hydrolysis: acidic hydrolysis and basic hydrolysis. Both methods are shown in Figure 10.25. Soap is made by the basic hydrolysis of ester bonds in vegetable oils or animal fats.

**A** $\quad CH_3CH_2\overset{O}{\overset{\|}{C}}-O-CH_3 \; + \; H_2O \; \underset{\Delta}{\overset{H_2SO_4}{\rightleftharpoons}} \; CH_3CH_2\overset{O}{\overset{\|}{C}}-OH \; + \; HO-CH_3$

methyl propanoate (ester)    water    propanoic acid (carboxylic acid)    methanol (alcohol)

**B** $\quad CH_3CH_2CH_2\overset{O}{\overset{\|}{C}}-O-CH_2CH_3 \; + \; NaOH \; \overset{\Delta}{\rightarrow} \; CH_3CH_2CH_2\overset{O}{\overset{\|}{C}}-O^-Na^+ \; + \; CH_3CH_2OH$

ethyl butanoate (ester)    sodium hydroxide (base)    sodium butanoate (salt of a carboxylic acid)    ethanol (alcohol)

**Figure 10.25** (A) The acidic hydrolysis of an ester produces a carboxylic acid and an alcohol. (B) The basic hydrolysis of an ester produces the salt of a carboxylic acid and an alcohol. To name the salt, place the name of the cation first (e.g., sodium) then follow it with the root and ending of the acid as if you were naming an ester.

## Amides

An ester can be thought of as the combination of a carboxylic acid and an alcohol. Similarly, you can think of an amide as the combination of a carboxylic acid and ammonia or an amine. An **amide** is an organic compound that has a carbon atom double-bonded to an oxygen atom and single-bonded to a nitrogen atom.

Amides have the functional group below:

$$-\underset{\underset{}{}}{\overset{\overset{O}{\|}}{C}}-\underset{|}{N}-$$

The general formula for an amide is R—CO—NR$_2$. R can stand for a hydrogen atom or an alkyl group. Figure 10.26 gives some examples of amides.

HC(=O)—NH$_2$
methanamide

CH$_3$C(=O)—NH$_2$
ethanamide

CH$_3$CH$_2$C(=O)—NH$_2$
propanamide

CH$_3$CH$_2$C(=O)—NH—CH$_3$
N-methylpropanamide

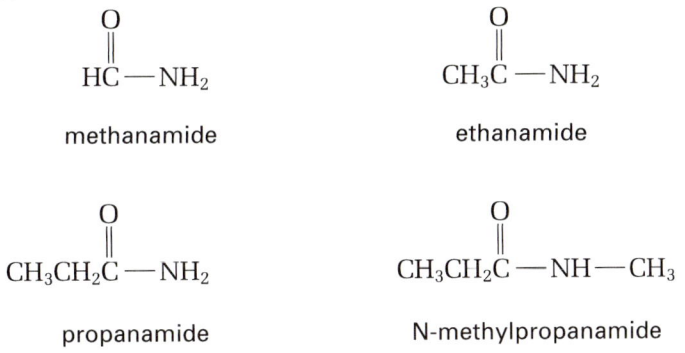

**Figure 10.26** Examples of amides

To name an amide, follow the steps below. The Sample Problem that follows illustrates how to use these steps. Later, Table 10.9 describes some physical properties of amides.

### How to Name an Amide

**Step 1** Locate the part of the amide that contains the C=O group. Name the parent carboxylic acid from which this part is derived. **Note:** The carbon in the C=O group is always given position number 1.

**Step 2** Replace the -oic acid ending of the name of the parent acid with the suffix -amide.

**Step 3** Decide whether the compound is a primary, secondary, or tertiary amide:
- If there are two hydrogen atoms (and no alkyl groups) attached to the nitrogen atom, the compound is a *primary amide* and needs no other prefixes.
- If there is one alkyl group attached to the nitrogen atom, the compound is a *secondary amide*. Name the alkyl group, and give it location letter N- to indicate that it is bonded to the nitrogen atom.
- If there are two alkyl groups, the compound is a *tertiary amide*. Place the alkyl groups in alphabetical order. Use location letter N- before each group to indicate that it is bonded to the nitrogen atom. If the two groups are identical, use N,N-.

**Step 4** Put the name together: prefix + root + suffix.

## Sample Problem

### Naming a Secondary Amide

**Problem**

Name the following compound.

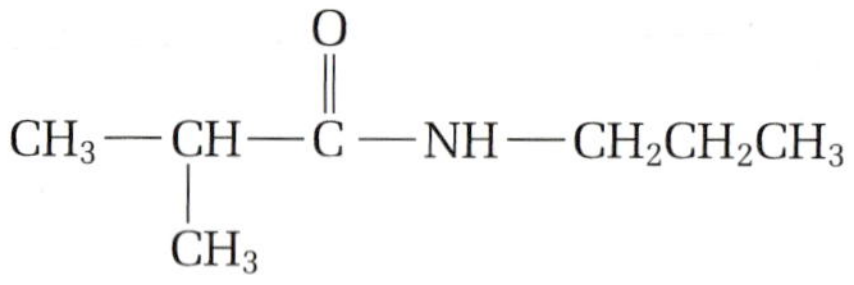

$$\text{CH}_3-\underset{\underset{\text{CH}_3}{|}}{\text{CH}}-\overset{\overset{\text{O}}{\|}}{\text{C}}-\text{NH}-\text{CH}_2\text{CH}_2\text{CH}_3$$

**Solution**

**Step 1** The parent acid is 2-methylpropanoic acid.

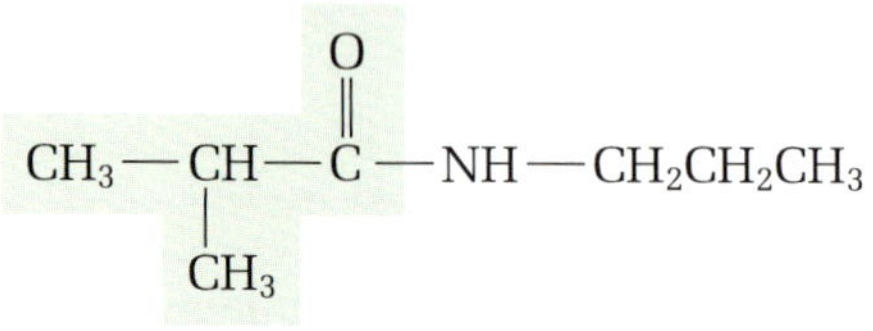

$$\text{CH}_3-\underset{\underset{\text{CH}_3}{|}}{\text{CH}}-\overset{\overset{\text{O}}{\|}}{\text{C}}-\text{NH}-\text{CH}_2\text{CH}_2\text{CH}_3$$

**Step 2** The base name for this compound is 2-methylpropanamide. (The root is -propan-, and the suffix is -amide.)

**Step 3** A propyl group is attached to the nitrogen atom. Thus, the compound's prefix is N-propyl-.

**Step 4** The full name of the compound is N-propyl-2-methylpropanamide.

## Practice Problems

**41.** Name each amide.

(a) 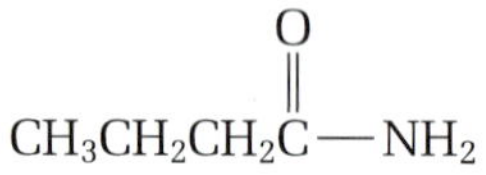 $\text{CH}_3\text{CH}_2\text{CH}_2\overset{\overset{\text{O}}{\|}}{\text{C}}-\text{NH}_2$

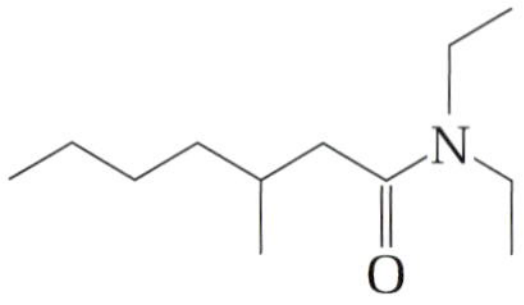

(b)  $\text{H}_3\text{C}-\text{NH}-\overset{\overset{\text{O}}{\|}}{\text{C}}\text{CH}_2\text{CH}_2\text{CH}_2\text{CH}_2\text{CH}_3$

(c) [line structural diagram]

**42.** Draw each amide.

(a) nonanamide
(b) N-methyloctanamide
(c) N-ethyl-N-propylpropanamide
(d) N-ethyl-2,4,6-trimethyldecanamide

**43.** Name each amide.

(a) $\text{CH}_3\text{CONH}_2$
(b) $\text{CH}_3\text{CH}_2\text{CH}_2\text{CH}_2\text{CH}_2\text{CH}_2\text{CONHCH}_3$
(c) $(\text{CH}_3)_2\text{CHCON}(\text{CH}_3)_2$

**44.** Draw a line structural diagram for each amide in the previous question.

## Table 10.9 Physical Properties of Amides

| | |
|---|---|
| **Polarity of functional group** | Because the nitrogen atom attracts electrons more strongly than carbon or hydrogen atoms, the C–N and N–H bonds are polar. As a result, the physical properties of amides are similar to the physical properties of carboxylic acids. |
| **Hydrogen bonding** | Because primary amides have two N–H bonds, they have even stronger hydrogen bonds than carboxylic acids. Secondary amides also experience hydrogen bonding. |
| **Solubility in water** | Amides are soluble in water. Their solubility decreases as the non-polar hydrocarbon part of the molecule increases in size. |
| **Melting and boiling points** | Primary amides have much higher melting and boiling points than carboxylic acids. Many simple amides are solid at room temperature. |

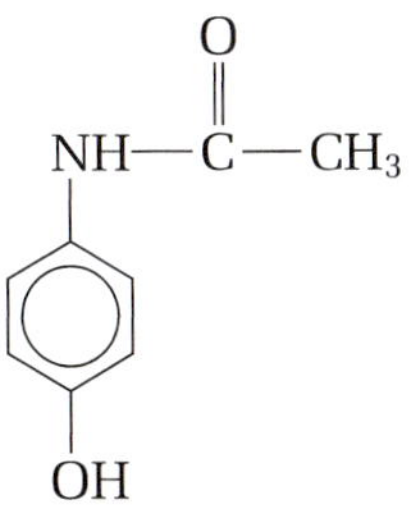

### Additional Characteristics of Amides

- An amide called acetaminophen is a main component of many painkillers.
- Urea, another common example of an amide, is made from the reaction between carbon dioxide gas, $CO_2$, and ammonia, $NH_3$. Urea was the first organic compound to be synthesized in a laboratory. It is found in the urine of many mammals, including humans, and it is used as a fertilizer.

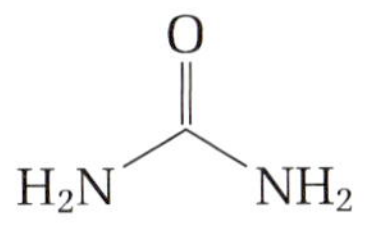

acetaminophen

urea

## Reactions of Amides

Like esters, amides undergo hydrolysis reactions. In acidic hydrolysis, the amide reacts with water in the presence of an acid, such as $H_2SO_4$. In basic hydrolysis, the amide reacts with the $OH^-$ ion, from water or a base, in the presence of a base, such as NaOH. Figure 10.27 illustrates these reactions.

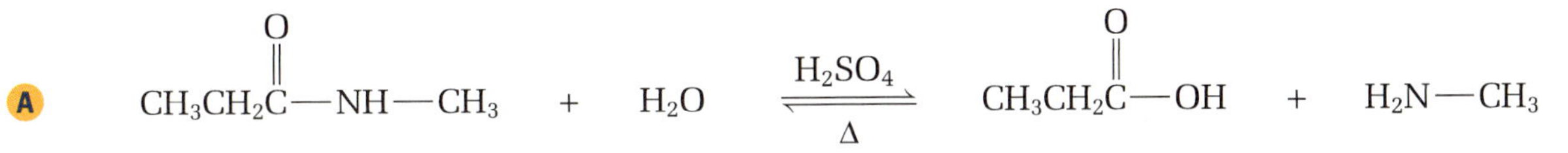

**A**  $CH_3CH_2\overset{O}{\underset{\|}{C}}-NH-CH_3$ + $H_2O$ $\underset{\Delta}{\overset{H_2SO_4}{\rightleftharpoons}}$ $CH_3CH_2\overset{O}{\underset{\|}{C}}-OH$ + $H_2N-CH_3$

methyl propanoate (ester)　　water　　　　　　　propanoic acid (carboxylic acid)　　methanamine (amine)

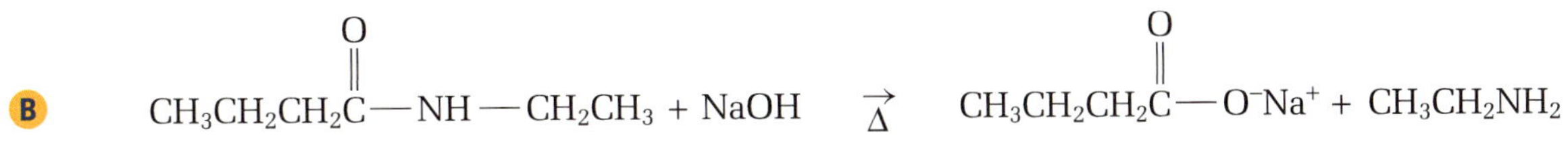

**B**  $CH_3CH_2CH_2\overset{O}{\underset{\|}{C}}-NH-CH_2CH_3$ + NaOH $\underset{\Delta}{\rightarrow}$ $CH_3CH_2CH_2\overset{O}{\underset{\|}{C}}-O^-Na^+$ + $CH_3CH_2NH_2$

N-ethylbutanamide (amide)　　sodium hydroxide (base)　　sodium butanoate (salt of a carboxylic acid)　　ethanamine (amine)

**Figure 10.27** (A) The acidic hydrolysis of an amide produces a carboxylic acid and an amine. (B) The basic hydrolysis of an amide produces the salt of a carboxylic acid and an amine.

## Sample Problem

### Predicting the Products of More Organic Reactions

**Problem**

Identify each type of reaction. Then predict and name the product(s).

(a) $\text{CH}_3\overset{\displaystyle O}{\overset{\|}{\text{C}}}-\text{NH}-\text{CH}_3 + \text{H}_2\text{O} \underset{\Delta}{\overset{\text{H}_2\text{SO}_4}{\rightleftharpoons}}$

(b) $\text{CH}_3(\text{CH}_2)_5\overset{\displaystyle O}{\overset{\|}{\text{C}}}-\text{O}-\text{CH}_2(\text{CH}_2)_3\text{CH}_3 + \text{NaOH} \xrightarrow{\Delta}$

(c) $\text{CH}_3\overset{\displaystyle O}{\overset{\|}{\text{C}}}\text{OH} + \text{CH}_3\text{CH}_2\text{OH} \rightleftharpoons$

**Solution**

(a) This reaction involves an amide and water being heated in the presence of an acid. Thus, it is a hydrolysis reaction. (Because it takes place in acid, it is an acidic hydrolysis.) The products of this reaction are a carboxylic acid and an amine.

$\text{CH}_3\overset{\displaystyle O}{\overset{\|}{\text{C}}}-\text{NH}-\text{CH}_3 + \text{H}_2\text{O} \underset{\Delta}{\overset{\text{H}_2\text{SO}_4}{\rightleftharpoons}} \text{CH}_3\overset{\displaystyle O}{\overset{\|}{\text{C}}}-\text{OH} + \text{H}_2\text{N}-\text{CH}_3$

N-methylethanamide  water         ethanoic acid   methanamine

(b) This reaction involves an ester and a base. It is a basic hydrolysis reaction. The products of this reaction are a salt of a carboxylic acid and an alcohol.

$\text{CH}_3(\text{CH}_2)_5\overset{\displaystyle O}{\overset{\|}{\text{C}}}-\text{O}-\text{CH}_2(\text{CH}_2)_3\text{CH}_3 + \text{NaOH} \xrightarrow{\Delta}$
1-pentanyl heptanoate                Sodium
                                    hydroxide

$\text{CH}_3(\text{CH}_2)_5\overset{\displaystyle O}{\overset{\|}{\text{C}}}-\text{O}^-\text{Na}^+ + \text{CH}_3(\text{CH}_2)\text{OH}$
Sodium              1-pentanol
heptanoate

(c) This reaction occurs between a carboxylic acid and an alcohol. It is an esterification, or condensation, reaction. The product of this reaction is an ester called ethyl ethanoate.

$\text{CH}_3\overset{\displaystyle O}{\overset{\|}{\text{C}}}\text{OH} + \text{CH}_3\text{CH}_2\text{OH} \rightleftharpoons \text{CH}_3\overset{\displaystyle O}{\overset{\|}{\text{C}}}\text{OCH}_2\text{CH}_3 + \text{H}_2\text{O}$
ethanoic acid    ethanol            ethyl ethanoate    water

## Practice Problems

**45.** Identify each type of reaction.

(a) $HC(=O)-OH + CH_3(CH_2)_{14}CH_2OH \rightleftharpoons HC(=O)-O-(CH_2)_{15}CH_3 + H_2O$

(b) 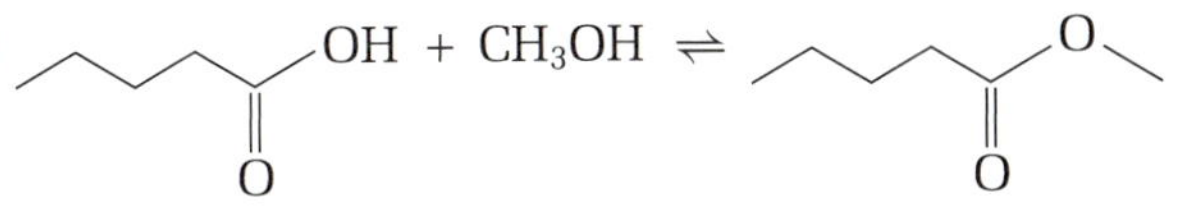 [butanoic acid structure]–OH + $CH_3OH \rightleftharpoons$ [methyl butanoate structure]

(c) N-propylmethanamide + NaOH → sodium methanoate + 1-propanamine

(d) hexanoic acid + ethanol ⇌ ethyl hexanoate + water

(e) butyl methanoate + water ⇌ methanoic acid + 1-butanol

**46. (a)** Name the reactants and products of the first two reactions in question 45.

**(b)** Draw the reactants and products of the last three reactions in question 45.

**47.** Name and draw the product(s) of each reaction.

(a) butanoic acid + octanol $\xrightleftharpoons{H_2SO_4}$

(b) methyl hexanoate + NaOH $\xrightarrow{\Delta}$

(c) propanoic acid + methanol ⇌

(d) propyl ethanoate + water $\xrightleftharpoons[\Delta]{H_2SO_4}$

(e) ethanoic acid + pentanol $\xrightleftharpoons{H_2SO_4}$

(f) 2-propanol + 3-methylpentanoic acid ⇌

**48.** Name and draw the reactant(s) in the following reaction.

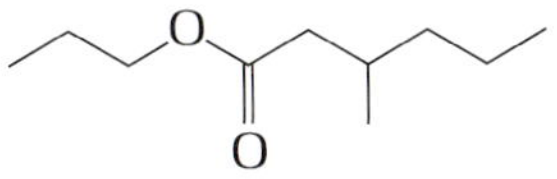

? + ? → [ester structure]

## Concept Organizer: Organic Reactions

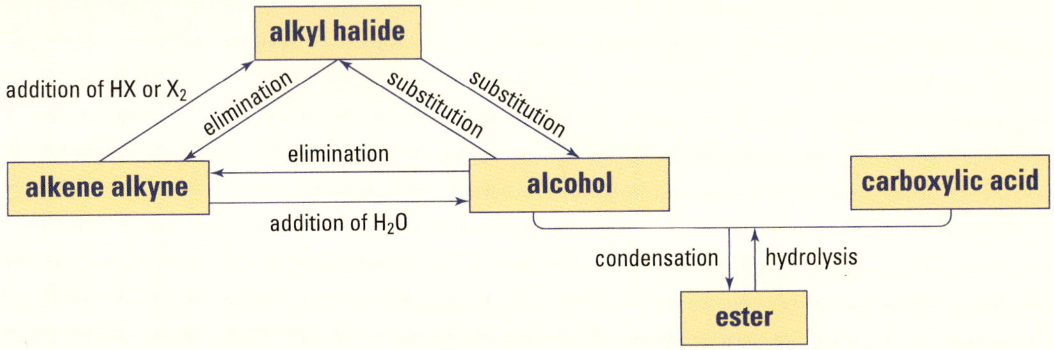

alkyl halide — addition of HX or X₂ / elimination ↔ alkene alkyne
alkyl halide — substitution ↔ alcohol
alkene alkyne — elimination / addition of H₂O ↔ alcohol
alcohol — condensation / hydrolysis ↔ ester ↔ carboxylic acid

## Comparing Physical Properties

In this chapter, you have learned how to recognize many different types of organic compounds. In the first section, you learned how to use polar bonds and the shape of a molecule to determine its molecular polarity. The following investigation allows you to apply what you have learned to predict and compare the physical properties of various organic compounds.

# Investigation 10-B

**SKILL FOCUS**
Predicting
Performing and recording

# Comparing Physical Properties

In this investigation, you will examine the differences between molecules that contain different functional groups. As you have learned, the polarity and hydrogen bonding abilities of each functional group affect how these molecules interact among themselves and with other molecules. You will examine the shape of each molecule and the effects of intermolecular forces in detail to make predictions about properties.

## Prediction
How can you predict the physical properties of an organic compound by examining its structure?

## Materials
molecular model kit

## Procedure
1. Choose a parent alkane that has four to ten carbon atoms (butane to decane).

2. By adding functional groups to your parent alkane, build a model of a molecule from each class of organic compounds:
   - aromatic hydrocarbons (section 1.2)
   - alcohols, ethers, amines (section 1.3)
   - aldehydes, ketones, carboxylic acids, esters, amides (section 1.4)

3. Draw a condensed structural diagram and a line structural diagram of each compound. Include the IUPAC name of the compound.

## Analysis
1. Look at your aromatic hydrocarbon, your alcohol, your ether, and your carboxylic acid. Rank these compounds from
   - least polar to most polar
   - lowest to highest boiling point

   Record your rankings, and explain them.

2. Compare the alcohol you made with an alcohol that has a different number of carbon atoms, made by a classmate. Which alcohol is more soluble in water, yours or your classmate's? Which alcohol is more polar? Explain your reasoning.

3. Choose two of your compounds at random. Which compound do you think has the higher melting point? Explain your reasoning.

4. Consider all the compounds you made. Which compound do you think is the most soluble in water? Explain your reasoning.

5. Which compound do you think is the most soluble in benzene? Explain your reasoning.

6. Where possible, predict the odour of each organic compound you made.

## Conclusions
7. Look up the boiling point, melting point, and solubility of each compound in a reference book, such as *The CRC Handbook of Chemistry and Physics*. Compare your findings with the predictions you made in the Analysis.

8. Challenge a classmate to name each compound you made and identify its functional group.

## Application
9. Use reference books and the Internet to discover any useful applications of the compounds you made. Prepare a short report for your class, explaining how you think each compound's physical properties may affect its usefulness in real life.

## Section Summary

In this chapter, you learned how to recognize, name, and predict the physical properties and chemical reactions of organic compounds that belong to the alcohol, ether, amine, aldehyde, ketone, carboxylic acid, ester, and amide families. You discovered many important uses for organic compounds. You know that 2-propanol (isopropyl alcohol) is used as an antiseptic. Acetone (a ketone) is the main component of nail polish remover. Esters give many fruits and processed foods their distinctive odours and tastes. In the next chapter, you will take a detailed look at some common organic compounds in your life, and learn about some of the benefits and risks of using organic compounds.

**Web LINK**

www.mcgrawhill.ca/links/altchemistry

If you would like more practice in naming substituted hydrocarbons, go to the web site above and click on **Electronic Learning Partner**.

## Section Review

**1** Write the IUPAC name for each compound.

(a)  $CH_3CH_2CH_2\overset{\overset{O}{\|}}{C}H$

(b) $CH_3\overset{\overset{O}{\|}}{C}CH_2CH_2CH_2CH_3$

(c) $CH_3\overset{\overset{O}{\|}}{C}-O-CH_2CH_2CH_2CH_3$

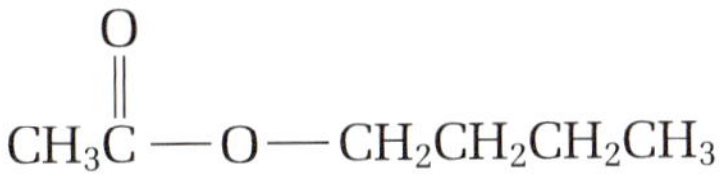

(d) 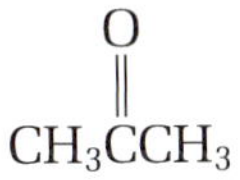 $CH_3\overset{\overset{O}{\|}}{C}CH_3$

(e) 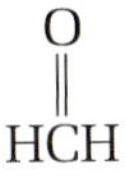 $H\overset{\overset{O}{\|}}{C}H$

(f) 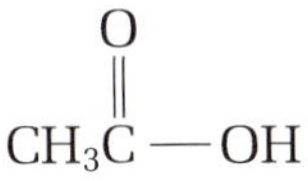 $CH_3\overset{\overset{O}{\|}}{C}-OH$

(g) 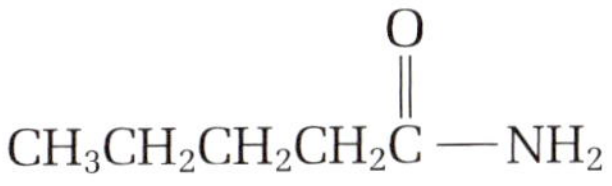 $CH_3CH_2CH_2CH_2\overset{\overset{O}{\|}}{C}-NH_2$

**2** Write the common name for d, e, and f in question 1.

**3** Name each compound.

(a) 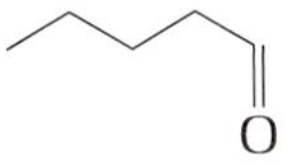 [structure: CH3CH2CH2CH2CHO drawn as zigzag with =O]

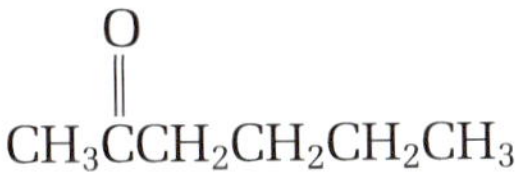

(b) 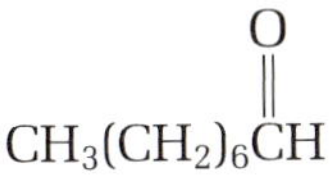 $CH_3(CH_2)_6\overset{\overset{O}{\|}}{C}H$

(c) $HO-\overset{\overset{O}{\|}}{C}-\underset{\underset{CH_2CH_3}{|}}{\overset{\overset{CH_2CH_3}{|}}{C}H}-CH-CH_2-CH_2-CH_3$

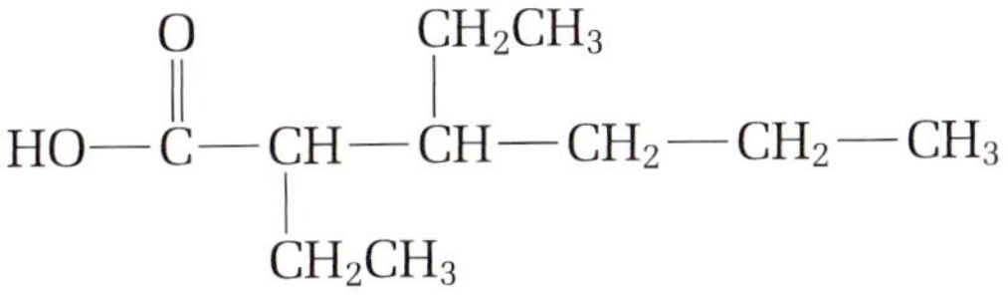

(d) 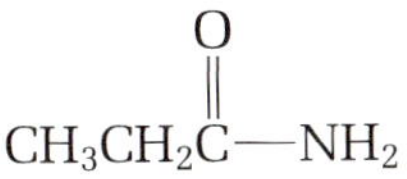 $CH_3CH_2\overset{\overset{O}{\|}}{C}-NH_2$

(e) 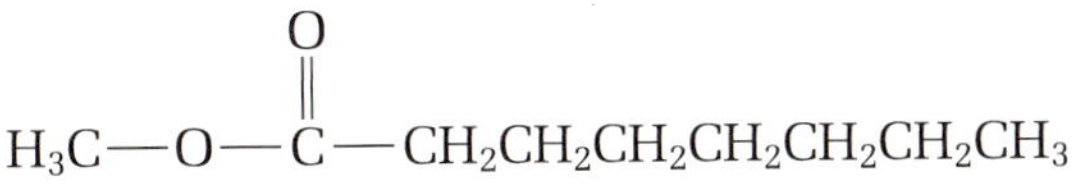 $H_3C-O-\overset{\overset{O}{\|}}{C}-CH_2CH_2CH_2CH_2CH_2CH_2CH_3$

**4** Identify the family that each organic compound belongs to.

(a) $CH_3CH_2CONH_2$

(b) $CH_3CH_2CH(CH_3)CH_2CH_2COOH$

(c) $CH_3CH_2C(CH_3)_2CH_2CHO$

(d) $CH_3COOCH_3$

**5** Name each organic compound in question 4.

**6** Draw and name one carboxylic acid and one ester with the molecular formula C₆H₁₂O₂.

**7** Draw and name one primary amide, one secondary amide, and one tertiary amide with the molecular formula C₆H₁₃ON.

**8** Identify the functional groups in each molecule.

(a) vanillin, a flavouring agent

(b) DEET, an insect repellant

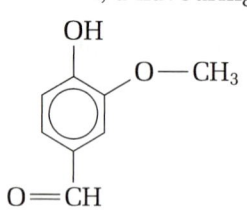

(c) penicillin V, an antibiotic (This compound also contains a functional group that is unfamiliar to you. Identify the atom(s) in the unfamiliar functional group as part of your answer.)

**9** Consider the compounds CH₃CH₂COH, CH₃CH₂COOH, and CH₃COOCH₃. Which compound has the highest boiling point? Use diagrams to explain your reasoning.

**10** How could you use your sense of smell to distinguish an amine from a ketone?

**11** Name and draw the reactant(s) in each reaction.

(a) ? + ? → HC(=O)—O—CH₂CH₂CH₂CH₂CH₃ + HOH

(b) ? + ? → 2-butyl methanoate + water

(c) ? + ? → methyl pentanoate + water

# CHAPTER 10 Review

## Reflecting on Chapter 10
Summarize this chapter in the format of your choice. Here are a few ideas to use as guidelines:
- Describe the shape and structure of a benzene molecule.
- Draw and name an example of an alcohol, an alkyl halide, an ether, and an amine.
- Draw and name an example of an aldehyde, a ketone, a carboxylic acid, an ester, and an amide.
- Describe the different types of organic reactions you learned, and give an example of each type.

## Reviewing Key Terms
For each of the following terms, write a sentence that shows your understanding of its meaning.

hydroxyl group
functional group
general formula
elimination reaction
condensation reaction
hydrolysis reaction
alcohol
parent alkane
alkyl halide
haloalkane
ether
amine
carbonyl group
aldehyde
ketone
carboxylic acid
ester
amide

## Knowledge/Understanding

1. Examine the following molecules. Then answer the questions below.

   $CH_4$    $CH_3-\overset{\overset{O}{\|}}{C}-CH_3$    $CH_3-CH=CH_2$

   (a) Predict the shape around the central carbon atom in each molecule.
   (b) Identify any polar bonds that are present in each molecule.
   (c) Describe each molecule as polar or non-polar.

2. Classify each molecule as polar or non-polar.
   (a) $CH_3OH$    (c) $CH_3NH_2$    (e) $H_2C=O$
   (b) $CH_3CH_3$    (d) $H_2C=CH_2$

3. Describe, in words, the functional group for each type of organic compound.
   (a) alkene    (c) amine    (e) carboxylic acid
   (b) alcohol    (d) ketone

4. Give the general formula for each type of organic compound.
   (a) alcohol    (c) ester
   (b) amine    (d) amide

5. What is the difference between each two organic compounds?
   (a) alcohol and ether
   (b) amine and amide
   (c) carboxylic acid and ester

6. Use the suffix to name the functional family of each type of organic compound.
   (a) ethanamine
   (b) 2,3-dimethyloctane
   (c) cyclohexanol
   (d) methyl propanoate
   (e) N-methylpentanamide
   (f) 3-ethyl heptanoic acid

7. Identify the functional group(s) in each molecule.

   (a) $CH_3CH_2\overset{\overset{OH}{|}}{C}HCH_2CH_3$    (b) (cyclohexene with $NH_2$ substituent)

   (c) $CH_3-CH=CH-\underset{\underset{\underset{\underset{O-CH_3}{|}}{\underset{O=C}{|}}}{\underset{CH_2}{|}}}{CH}-CH_3$

   (d) $HC\equiv C-CH_2-\overset{\overset{O}{\|}}{C}-CH_2-CH_2-O-CH_3$

   (e) (benzene ring with Cl substituent and $-CH_2CH_2C(=O)NH-$ ethyl chain)

   (f) $CH_3CH_2\overset{\overset{COOH}{|}}{C}HCH_2CH_2CHO$

Answers to questions highlighted in red type are provided in Appendix A

8. Classify each type of organic compound.
   (a) CH₃—NH₂
   (b) CH₃—CH₂—CH=O
   (c) CH₃—O—C(=O)—CH₂—CH(CH₃)—CH₃
   (d) HO—C(=O)—CH(CH₂—CH₃)—CH₂—CH₃
   (e) CH₃CH(CH₃)CH₂CCH₃ (with C=O)
   (h) CH₃—CH₂—CH₂—CH=O
   (i) CH₃—CHBr—CH₃
   (j) cyclopentanol (cyclopentane with OH)
   (k) HO—C(=O)—CH₂—CH₂—CH₂—CH₃
   (l) CH₃—CH₂—C(=O)—O—CH₂—CH₃
   (m) CH₃—CH(CH₃)—C(CH₃)(CH₂CH₃)—CH₂—C(=O)—CH₂—CH₃
   (n) H(I)C=C(Br)H

9. (a) Suppose that you have an alcohol and a ketone with the same molecular mass. Which compound has a higher boiling point?
   (b) Use your understanding of hydrogen bonding to explain your answer to part (a).

10. (a) Benzene is a non-polar solvent. Is an alcohol or an alkane more soluble in benzene?
    (b) Use molecular polarity to explain your answer to part (a).

11. Give the IUPAC name for each compound.
    (a) CH₃—CH₂—CH₂—OH
    (b) CH₃—CH₂—CH(OH)—CH₂—CH₂—CH₂—CH₂—CH₃
    (c) CH₃—CH₂—CH₂—CH₂—CH(NH₂)—CH₃
    (d) CH₃—CH₂—CH₂—CH₂—O—CH₃
    (e) CH₃—C(=O)—CH₃
    (f) CH₃—O—C(=O)—CH₂—CH₂—CH₂—CH₃
    (g) CH₃—CH₂—CH₂—C(OH)(CH₂—CH₃)—CH₃

12. Name each compound. Then identify the family of organic compounds that it belongs to.
    (a) ethyl ether structure
    (b) heptanal chain with C=O
    (c) amide with NH₂
    (d) alcohol with branched chain, OH
    (e) carboxylic acid, OH
    (f) amine, NH₂
    (g) amide, NH with propyl

13. Give the common name for each compound.
    (a) CH₃—CH₂—OH
    (b) CH₃—OH
    (c) H—C(=O)—H
    (d) CH₃—C(=O)—OH
    (e) CH₃—C(=O)—CH₃
    (f) CH₃—CH(OH)—CH₃

422 MHR • Unit 4 Organic Chemistry

14. Describe each type of organic reaction, and give an example.
    (a) addition
    (b) substitution
    (c) elimination
    (d) condensation
    (e) hydrolysis

15. Identify each reaction as an addition reaction, a substitution reaction, or an elimination reaction.
    (a) $CH_3CH_2CH_2OH \rightarrow CH_3CH=CH_2 + H_2O$
    (b) $CH_3CH_2OH + HCl \rightarrow CH_3CH_2Cl + H_2O$
    (c) $CH_3CH(CH_3)CH=CH_2 + HBr$
        $\rightarrow CH_3CH(CH_3)CH(Br)CH_3$
    (d) $CH_3CH_2Br + NH_3 \rightarrow CH_3CH_2NH_2 + HBr$

## Inquiry

16. Suppose that you are working with five unknown compounds in a chemistry laboratory. Your teacher tells you that these compounds are ethane, ethanol, methoxymethane, ethanamine, and ethanoic acid.
    (a) Use the following table of physical properties to identify each unknown compound.

| Compound | Solubility in water | Hydrogen bonding | Boiling point | Odour | Molecular polarity |
|---|---|---|---|---|---|
| A | infinitely soluble | strong | 17°C | fishy | polar |
| B | not soluble | none | −89°C | odourless | non-polar |
| C | soluble | accepts hydrogen bonds from water, but cannot form hydrogen bonds between its molecules | −25°C | sweet | polar |
| D | infinitely soluble | very strong | 78°C | sharp, antiseptic smell | very polar |
| E | infinitely soluble | extremely strong | 118°C | sharp, vinegar smell | very polar |

   (b) Draw a complete structural diagram for each compound.

17. Draw and name the product(s) of each incomplete reaction. (**Hint:** Do not forget to include any second products, such as $H_2O$ or HBr.)
    (a) $HO-CH_2CH_2CH_2CH_2CH_3 + HBr \rightarrow$
    (b) $HO-CH_2CH_2CH_3 \xrightarrow[\Delta]{H_2SO_4}$
    (c) $CH_3CH_2\overset{\overset{O}{\|}}{C}OH + HOCH_3 \xrightarrow{H_2SO_4}$
    (d) $CH_3-\overset{\overset{O}{\|}}{C}-O-CH_2CH_3 + H_2O \xrightarrow[\Delta]{H_2SO_4}$

18. Draw the product(s) of each incomplete reaction. **Hint:** Do not forget to include the second product, such as $H_2O$ or HBr, for a substitution reaction.
    (a) $CH_3CH_2-\overset{\overset{O}{\|}}{C}-OH + CH_3CH_2CH_2OH \rightarrow$
    (b) cyclopentanol $\xrightarrow[\Delta]{H_2SO_4}$
    (c) pentanoic acid ethyl ester $+ H_2O \xrightarrow[\Delta]{H_2SO_4}$
    (d) $CH_3CH_2CH(Br)CH_3 + OH^- \rightarrow$
    (e) $HO-CH_2CH_2CH_2CH_3 + H\overset{\overset{O}{\|}}{C}-OH \rightarrow$

19. Draw and name the reactant(s) in each reaction.
    (a) $? + ? \rightarrow CH_3CH_2\overset{\overset{O}{\|}}{C}OCH_2CH_2CH_3 + HOH$
    (b) $? \xrightarrow[\Delta]{H_2SO_4} \square + H_2O$

20. In Investigation 10-A, you carried out a condensation reaction to produce an ester called ethyl ethanoate. In this chapter, you also learned about the reverse reaction: hydrolysis.
    (a) Write a balanced chemical equation, showing the reactants and the products of the hydrolysis of ethyl ethanoate.
    (b) Write a step-by-step procedure to carry out the acid hydrolysis of ethyl ethanoate.
    (c) List the materials and equipment you would need to carry out this reaction.

21. One type of condensation reaction (which you did not study in this chapter) takes place when an acid, usually $H_2SO_4$, is added to an alcohol at temperatures that are lower than the temperatures for an elimination reaction. In this type of condensation reaction, two alcohol

Chapter 10 Hydrocarbon Derivatives • MHR **423**

molecules combine to form an ether. A water molecule is eliminated as a second product of the reaction.
(a) Write a chemical equation to show the condensation reaction of ethanol to produce ethoxyethane.
(b) Classify this condensation reaction as an addition reaction, a substitution reaction, or an elimination reaction. Explain your answer.
(c) A different reaction occurs if the acid HCl is used instead of $H_2SO_4$. Name this type of reaction, and write the equation for it.

## Communication

22. Draw a condensed structural diagram for each compound.
    (a) 1-propanamine
    (b) 3-ethylpentane
    (c) 4-heptanol
    (d) propanoic acid
    (e) cyclobutanol
    (f) methoxyethane
    (g) 1,1-dibromobutane
    (h) 2-methyl-3-octanone
    (i) hexanal
    (j) N-ethylpropanamide
    (k) methyl butanoate

23. Draw a structural diagram for each compound.
    (a) 3,4-dimethylheptanoic acid
    (b) 3-ethyl-3-methyl-1-pentyne
    (c) N-ethyl-2,2-dimethyl-3-octanamine
    (d) N-ethyl-N-methylhexanamide
    (e) 1,3-dibromo-5-chlorobenzene

24. Draw a structural diagram for each compound. Then identify the family of organic compounds that it belongs to.
    (a) 4-ethylnonane
    (b) 4-propylheptanal
    (c) 3,3-dimethyl-2-hexanamine
    (d) 2-methoxypentane
    (e) *trans*-1,2-diiodo ethene
    (f) *cis*-1-chloro-2-fluoroethene

25. Identify the error in each name. Then correct the name.
    (a) 2-pentanal
    (b) N,N-diethyl-N-methylpentanamide
    (c) 1-methylpropanoic acid

26. Draw and name all the isomers that have the molecular formula $C_4H_{10}O$.

27. Draw and name six isomers that have the molecular formula $C_4H_8O_2$.

## Making Connections

28. List at least three common organic compounds. Describe how these compounds are used in everyday life.

29. The term "organic" has a slightly different meaning in science than it does in everyday life. Compare the use of the term "organic" in the phrases "organic vegetables" and "organic compound." Where might each phrase be used?

### Answers to Practice Problems and Short Answers to Section Review Questions:

**Practice Problems: 3.** C—O, O—H, C=O  **4.** polar, non-polar, non-polar, polar  **5.(a)** substitution **(b)** addition **(c)** elimination **(d)** elimination  **6.(a)** hydrolysis **(b)** condensation **(c)** condensation  **7.(a)** addition **(b)** hydrolysis **(c)** combustion **(d)** elimination  **8.(a)** 1-propanal; primary **(b)** 2-butanol; secondary **(c)** cyclobutanol; secondary **(d)** 2,3 pentanediol; secondary **(e)** 2,4-dimethyl-1-heptanol; primary  **10.(a)** 1,3-pentanediol **(b)** 3,4-diethyl-1-decanol **(c)** 3-methyl-2-pentanol  **13.** 1,2-difluorocyclohexane  **14.** 1-iodopropane  **15.** 2-iodopropane  **16.(a)** substitution **(b)** substitution **(c)** elimination **(d)** elimination  **18.(a)** elimination **(b)** substitution **(c)** substitution **(d)** substitution **(e)** elimination **(f)** elimination  **19.(a)** 1-butene + water **(b)** 2-chloro-3-methylbutane + water **(c)** 2-methyl-1-propanol + I⁻ **(d)** bromocyclohexane + water **(e)** 3,4-dimethyl-1-cyclopentene **(f)** 3-chloro-4-methyl-2-pentene  **20.(a)** methoxymethane **(b)** 2-methoxypropane **(c)** 1-methoxybutane  **21.(a)** ethyl methyl ether **(b)** dimethyl ether  **23.** alcohol  **24.(a)** methanamine **(b)** 2-butanamine **(c)** N-ethyl-2,2-dimethyl-1-propanamine **(d)** 3-pentanamine **(e)** 2-butanamine **(f)** N,N-dimethylcyclopentanamine  **26.(a)** primary **(b)** primary **(c)** secondary **(d)** primary **(e)** tertiary **(f)** primary  **27.** 1-butanamine, 2-butanamine, 2-methyl-1-propanamine, N-methyl-1-propanamine, N-methyl-2-propanamine, N ethylethanamine, N,N-dimethylethanamine, 2-methyl-2-propanamine  **28.(a)** butanal **(b)** 3-octanone **(c)** 2-methyl-3-hexanone **(d)** 2-ethylbutanal **(e)** 4-ethyl-3-methyl-2-heptanone  **30.** aldehyde  **31.** hexanal, 2-hexanone, 3-hexanone, 3-methyl-2-pentanone, 3,3-dimethylbutanal; many other isomers are possible  **32.(a)** propanoic acid **(b)** 2,2-dimethylbutanoic acid **(c)** 2-ethyl-4,5-dimethylhexanoic acid  **35.** butanoic acid, 2-methylpropanoic acid  **36.(a)** ethyl methanoate **(b)** methyl butanoate **(c)** pentyl pentanoate  **39.(a)** and **(c)**; **(b)** and **(e)**  **40.** methyl butanoate, ethyl propanoate, propyl ethanoate, butyl methanoate, methyl 2-methylpropanoate; other isomers exist  **41.(a)** butanamide **(b)** N-methylhexanamide **(c)** N,N-diethyl-3-methylheptanamide

**43.(a)** ethanamide **(b)** N-methylheptanamide
**(c)** N,N-dimethyl-2-methylpropanamide **45.(a)** esterification or condensation **(b)** esterification **(c)** basic hydrolysis
**(d)** esterification **(e)** hydrolysis **46.(a)** methanoic acid + 1-hexadecanol → hexadecanyl methanoate + water
**(b)** pentanoic acid methanol, methyl pentanoate
**47.(a)** octyl pentanoate **(b)** methanol + sodium hexanoate
**(c)** methyl propanoate + water
**(d)** 1-propanol + ethanoic acid **(e)** pentyl ethanoate
**(f)** 2-propyl (or isopropyl) 3-methylpentanoate
**48.** 1-propanol, 3-methylhexanoic acid
**Section Review 10.1: 1.(a)** polar **(b)** non-polar **(c)** polar
**(d)** non-polar **(e)** polar **2.(a)** tetrahedral **(b)** trigonal planar
**3.(a)** non-polar **(b)** polar **4. (b)** and **(d)** **5.(a)** $CH_3CH_2CH_3$
**(b)** $CH_3CH_2CH_2NH_2$ **(c)** $CH_3CH_2CH_2CH_2OH$ **6.(a)** $CH_3NH_2$
**(b)** $CH_3OH$ **7.** $CH_3CH_2CH_2OH$, $CH_3CH_2OCH_3$, $CH_3CH_2CH_3$
**8.(a)** addition **(b)** elimination **(c)** substitution **(d)** substitution
**(e)** elimination **10.2: 1.(a)** butanamine **(b)** 2-ethoxybutane
**(c)** 2,3-dimethylcyclopentanol **(d)** 2-bromopentane
**2.(a)** 1-heptanol **(b)** 2-butanol **(c)** ethanamine
**(d)** N-methylmethanamine **(e)** methoxyethane
**(f)** 2-chloropropane **4.** ethoxypropane, 1-pentanol, methoxybutane **5.(a)** ethanol > methoxymethane > ethane
**(b)** 1-propanamine > N,N-dimethylmethanamine > propane
**6.** cyclohexanol **7.** propane, 2-propanol; propane
**8.** 1-butanol > ethoxyethane > butane **9.(a)** elimination
**(b)** elimination **(c)** combustion **(d)** elimination
**11.(a)** 2-methyl-2-bromopropane **(b)** 2,2-dimethyl-1-butanamine
**(c)** cyclobutanol **(d)** 3-methyl-4-heptanol
**(e)** 3-chloro-5-ethyl-4-methyloctane **10.3: 1.(a)** butanal
**(b)** 2-hexanone **(c)** butyl ethanoate **(d)** propanone
**(e)** methanal **(f)** ethanoic acid **(g)** pentanamide
**2.(d)** acetone **(e)** formaldehyde **(f)** acetic acid **3.(a)** pentanal
**(b)** octanal **(c)** 2,3-diethylhexanoic acid **(d)** propanamide
**(e)** methyl octanoate **4.(a)** amide **(b)** carboxylic acid
**(c)** aldehyde **(d)** ester **5.(a)** propanamide
**(b)** 4-methylhexanoic acid **(c)** 3,3-dimethylpentanal
**(d)** methyl ethanoate **6.** hexanoic acid, propyl propanoate
**7.** hexanamide, N-methylpentanamide, N,N-dimethylbutanamide **8.(a)** benzene ring, methoxy group, aldehyde group, hydroxyl group **(b)** benzene ring, amide group **(c)** benzene ring, ether linkage, two amide groups, acid group, sulfur atom **9.** $CH_3CH_2COOH$
**10.** An amine has a distinctive "fishy" smell. A ketone smells sweet. **11.(a)** methanoic acid and 1-pentanol
**(b)** 2-butanol + methanoic acid **(c)** pentanoic acid and methanol

# CHAPTER 11
# Organic Compounds on a Larger Scale

**Chapter Preview**

- 11.1 Polymer Chemistry
- 11.2 Natural Polymers and Other Biomolecules
- 11.3 Organic Compounds and Everyday Life

**Prerequisite Concepts and Skills**

Before you begin this chapter, review the following concepts and skills:

- addition reactions of alkenes (Chapter 9)
- condensation reactions (Chapter 10)

In St. John's, Newfoundland and Labrador, a large factory building stands in the industrial area of the city. Every morning, employees of this company, named Enviro-Shred, pick up bales of recycled newsprint from local newspaper publishers. In the factory, the newsprint is finely ground and processed to yield a fuzzy fibre with low density. This fibre is manufactured into environmentally-friendly insulation and agricultural mulch. In this way, waste newspaper is recycled into useful and marketable products.

What is this fuzzy fibre? Why does its presence make shredded newsprint and other wood products so useful? Where do the strength and endurance, the ability to absorb and store moisture, and the insulating capabilities of wood products come from? The answer lies deep in the structure of trees and other plants. Plant cell walls are built of a fibrous compound named *cellulose*. Cellulose molecules are long and thin, composed of many identical units linked together to form a chain. Large molecules, such as cellulose, that are composed of many smaller units linked together are called *polymers*. Every day, you wear, eat, and use substances and materials made from polymers.

In this chapter, you will explore some of the many synthetic polymers that have been designed and built in imitation of naturally occurring polymers. You will also look at the natural polymers that are used and manufactured by your own body. Finally, you will examine the risks and benefits of using organic compounds to meet our needs.

What natural organic compounds exist within trees and other living things? How do we use our knowledge of these compounds to improve our quality of life?

# Polymer Chemistry

## 11.1

Try to imagine a day without plastics. From simple plastic bags to furniture and computer equipment, plastics are an intergral part of our homes, schools, and work places. So far in this unit, you have encountered fairly small organic molecules. Many of the organic molecules that are used industrially, such as plastics, are much larger in size.

A **polymer** is a very long molecule that is made by linking together many smaller molecules called **monomers**. To picture a polymer, imagine taking a handful of paper clips and joining them into a long chain. Each paper clip represents a monomer. The long chain of repeating paper clips represents a polymer. Many polymers are made of just one type of monomer. Other polymers are made from a combination of two or more different monomers. Figure 11.1 shows an example of joined monomers in a polymer structure.

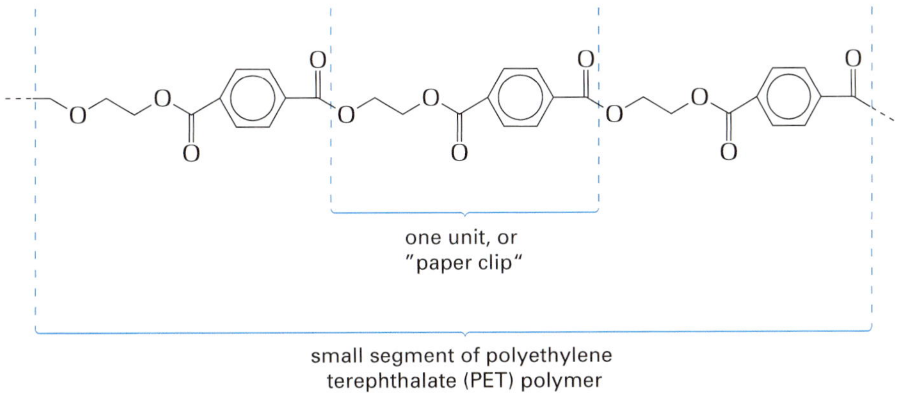

**Figure 11.1** Polyethylene terephthalate (PET) is a plastic that is used to make soft drink bottles.

Polymers that can be heated and moulded into specific shapes and forms are commonly known as **plastics**. All plastics are synthetic (artificially made) polymers. For example, polyethene is a common synthetic polymer that is used to make plastic bags. Ethene, $CH_2 = CH_2$, is the monomer for polyethene. Adhesives, rubber, chewing gum, and Styrofoam™ are other important materials that are made from synthetic polymers.

Natural polymers are found in living things. For example, glucose, $C_6H_{12}O_6$, is the monomer for the natural polymer starch. You will learn more about natural polymers in the next section.

Some polymers, both synthetic and natural, can be spun into long, thin fibres. These fibres are woven into natural fabrics (such as cotton, linen, and wool) or synthetic fabrics (such as rayon, nylon, and polyester). Figure 11.2, on the next page, shows some polymer products.

### Section Preview/Outcomes

In this section, you will

- **describe** some common polymers
- **communicate** your understanding of addition and condensation polymerization reactions
- **draw** structures and chemical equations for monomers, polymers, and polymerization reactions
- **synthesize** a polymer
- **analyze** health and environmental issues connected with the use of polymers
- **form, present,** and **defend** your own opinion, and **listen** to the opinions of others regarding the risks and benefits of polymer use and recycling
- **communicate** your understanding of the following terms: *polymer, monomers, plastics, addition polymerization, condensation polymerization, nylons, polyamides, polyesters*

### CONCEPT CHECK

As directed in the text, use paper clips to make a model of a polymer. What can this model tell you about the flexibility and strength of polymers? How is this model an accurate depiction of monomers and polymers, and how is it inaccurate? Come up with your own model to represent a polymer.

**Web LINK**

www.mcgrawhill.ca/links/
atlchemistry

There are two main types of plastics: *thermoplastics* and *thermosets*. What is the difference between them? What is each type of plastic used for?

To answer these questions, go to the web site above and click on **Web Links**.

**Figure 11.2** Both synthetic and natural polymers are used to make clothing.

## Making Synthetic Polymers: Addition and Condensation Polymerization

Synthetic polymers are extremely useful and valuable. Many polymers and their manufacturing processes have been patented as corporate technology. Polymers form by two of the reactions you have already learned: addition reactions and condensation reactions.

The name of a polymer is usually written with the prefix *poly-* (meaning "many") before the name of the monomer. Often the common name of the monomer is used, rather than the IUPAC name. For example, the common name of ethene is ethylene. Polyethene, the polymer that is made from ethene, is often called polyethylene. Similarly, the polymer that is made from chloroethene (common name: vinyl chloride) is named polyvinylchloride (PVC). The polymer that is made from propene monomers (common name: propylene) is commonly called polypropylene, instead of polypropene.

**Addition polymerization** is a reaction in which monomers with double bonds are joined together through multiple addition reactions to form a polymer. Figure 11.3 illustrates the addition polymerization reaction of ethene to form polyethene. Table 11.1, on the next page, gives the names, structures, and uses of some common polymers that are formed by addition polymerization.

**CONCEPT CHECK**

Why does "polyethene" end with the suffix -ene, if it does not contain any double bonds?

$$H_2C\!=\!CH_2 + H_2C\!=\!CH_2 \rightarrow -\underset{H}{\overset{H}{C}}-\underset{H}{\overset{H}{C}}-\underset{H}{\overset{H}{C}}-\underset{H}{\overset{H}{C}}- \xrightarrow{H_2C=CH_2}$$

$$-\underset{H}{\overset{H}{C}}-\underset{H}{\overset{H}{C}}-\underset{H}{\overset{H}{C}}-\underset{H}{\overset{H}{C}}-\underset{H}{\overset{H}{C}}-\underset{H}{\overset{H}{C}}- \xrightarrow{H_2C=CH_2} \text{etc.}$$

**Figure 11.3** The formation of polyethene from ethene

**Table 11.1** Examples of Addition Polymers

| Name | Structure of monomer | Structure of polymer | Uses |
|---|---|---|---|
| polyethene | H₂C=CH₂<br>ethene | polyethene (–CH₂–CH₂–CH₂–CH₂–CH₂–CH₂–) | • plastic bags<br>• plastic milk, juice, and water bottles<br>• toys |
| polystyrene | H₂C=CH–(C₆H₅)<br>styrene | –CH₂–CH(C₆H₅)–CH₂–CH(C₆H₅)–<br>polystyrene | • styrene and Styrofoam™ cups<br>• insulation<br>• packaging |
| polyvinylchloride (PVC, vinyl) | H₂C=CHCl<br>vinyl chloride | –CH₂–CHCl–CH₂–CHCl–<br>polyvinylchloride (PVC) | • building and construction materials<br>• sewage pipes<br>• medical equipment |
| polyacrylonitrile | H₂C=CH–CN<br>acrylonitrile | –CH₂–CH(CN)–CH₂–CH(CN)–<br>polyacrylonitrile | • paints<br>• yarns, knit fabrics, carpets, and wigs |

In **condensation polymerization**, monomers are combined to form a single organic polymer, with a second smaller product, usually water, produced by this reaction. For condensation polymerization to occur, each monomer must have two functional groups (usually one at each end of the molecule). To recognize that condensation polymerization has occurred, look for the formation of an ester or amide bond (see Table 11.2).

Nylon-66 is made by the condensation polymerization of the dicarboxylic acid adipic acid, and 1,6-diaminohexane, an amine. (The number 66 comes from the fact that each of the two reactants contains six carbon atoms.) This reaction results in the formation of amide bonds between monomers, as shown in Figure 11.4. Condensation polymers that contain amide bonds are called **nylons** or **polyamides**. Condensation polymers that contain ester bonds are called **polyesters**. Polyesters result from the esterification of diacids and dialcohols.

> **Web LINK**
>
> www.mcgrawhill.ca/links/atlchemistry
>
> Go to the web site above and click on **Electronic Learning Partner** for more information about Nylon-66.

H–N(H)–(CH₂)₆–N(H)–H   HO–C(=O)–(CH₂)₄–C(=O)–OH   H–N(H)–(CH₂)₆–N(H)–H

1,6-diaminohexane          adipic acid            1,6-diaminohexane

↓

···–NH–(CH₂)₆–NH–C(=O)–(CH₂)₄–C(=O)–NH–(CH₂)₆–NH–··· + H₂O

                   amide bond         amide bond                              water

**Figure 11.4** Nylon-66 is made from adipic acid and 1,6-diaminohexane.

Table 11.2 shows two common polymers that are formed by condensation. Notice that Dacron™, a polyester, contains ester linkages between monomers. Nylon-6, a polyamide, contains amide linkages between monomers.

**Table 11.2** Examples of Condensation Polymers

| Name | Structure | Uses |
|---|---|---|
| Dacron™ (a polyester) | 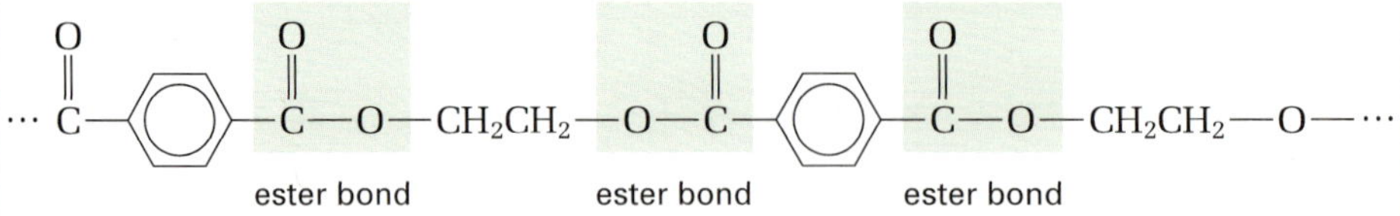 ···C(=O)—⟨benzene⟩—C(=O)—O—CH₂CH₂—O—C(=O)—⟨benzene⟩—C(=O)—O—CH₂CH₂—O—··· (ester bonds) | • synthetic fibres used to make fabric for clothing and surgery |
| Nylon-6 (a polyamide) | 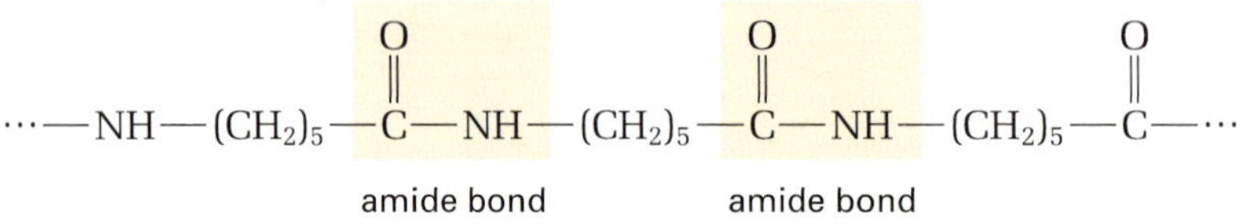 ···—NH—(CH₂)₅—C(=O)—NH—(CH₂)₅—C(=O)—NH—(CH₂)₅—C(=O)—··· (amide bonds) | • tires<br>• synthetic fibres used to make rope and articles of clothing, such as stockings |

The following Sample Problem shows how to classify a polymerization reaction.

## Sample Problem

### Classifying a Polymerization Reaction

**Problem**

Tetrafluoroethene polymerizes to form the slippery polymer that is commonly known as Teflon™. Teflon™ is used as a non-stick coating in frying pans, among other uses. Classify the following polymerization reaction, and name the product. (The letter $n$ indicates that many monomers are involved in the reaction.)

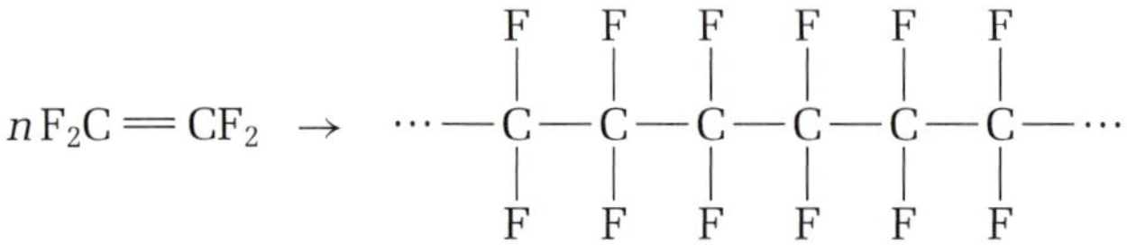

$$n\,F_2C=CF_2 \rightarrow \cdots-\underset{F}{\overset{F}{C}}-\underset{F}{\overset{F}{C}}-\underset{F}{\overset{F}{C}}-\underset{F}{\overset{F}{C}}-\underset{F}{\overset{F}{C}}-\underset{F}{\overset{F}{C}}-\cdots$$

**Solution**

The monomer reactant of this polymerization reaction contains a double bond. The product polymer has no double bond, so an addition reaction must have occurred. Thus, this reaction is an addition polymerization reaction. Since the monomer's name is tetrafluoroethene, the product's name is polytetrafluoroethene.

### Practice Problems

1. A monomer called methylmethacrylate polymerizes to form an addition polymer that is used to make bowling balls. What is the name of this polymer?

2. Classify each polymerization reaction as an addition or condensation polymerization reaction.

(a) $n\text{HO}-\overset{\overset{O}{\|}}{C}-\phantom{x}\bigcirc\phantom{x}-\overset{\overset{O}{\|}}{C}-\text{OH} + n\text{HO}-\text{CH}_2-\bigcirc-\text{CH}_2-\text{OH} \rightarrow$

$\cdots-\overset{\overset{O}{\|}}{C}-\bigcirc-\overset{\overset{O}{\|}}{C}-\text{O}-\text{CH}_2-\bigcirc-\text{CH}_2-\text{O}-\cdots$

(b) $n\text{CH}_2=\overset{\overset{\text{CN}}{|}}{\text{CH}} \rightarrow \cdots-\text{CH}_2-\overset{\overset{\text{CN}}{|}}{\text{CH}}-\text{CH}_2-\overset{\overset{\text{CN}}{|}}{\text{CH}}-\cdots$

(c) $n\text{HO}-\text{CH}_2-\overset{\overset{O}{\|}}{C}-\text{OH} \rightarrow \cdots-\text{O}-\text{CH}_2-\overset{\overset{O}{\|}}{C}-\text{O}-\text{CH}_2-\overset{\overset{O}{\|}}{C}-\cdots$

3. Draw the product of each polymerization reaction. Include at least two linkages for each product.

(a) $n\text{HO}-\text{CH}_2\text{CH}_2\text{CH}_2-\text{OH} + n\text{HO}-\overset{\overset{O}{\|}}{C}-\text{CH}_2-\overset{\overset{O}{\|}}{C}-\text{OH} \rightarrow$

(b) $n\text{H}_2\text{C}=\overset{\overset{\text{CH}_3}{|}}{\text{CH}} \rightarrow$

(c) $n\text{H}_2\text{NCH}_2-\bigcirc-\text{CH}_2\text{NH}_2 + n\text{HO}-\overset{\overset{O}{\|}}{C}(\text{CH}_2)_6\overset{\overset{O}{\|}}{C}-\text{OH} \rightarrow$

4. Classify each polymer as an *addition polymer* (formed by addition polymerization) or a *condensation polymer* (formed by condensation polymerization). Then classify each condensation polymer as either a polyester or a nylon (polyamide).

(a) $\cdots-\text{CH}_2-\overset{\overset{\text{Br}}{|}}{\text{CH}}-\text{CH}_2-\overset{\overset{\text{Br}}{|}}{\text{CH}}-\cdots$

(b) $\cdots-\text{NH}-\text{CH}_2-\text{NH}-\overset{\overset{O}{\|}}{C}-\text{CH}_2\text{CH}_2-\overset{\overset{O}{\|}}{C}-\text{NH}-\text{CH}_2-\text{NH}-\cdots$

(c) $\cdots-\text{O}-\text{CH}_2\text{CH}_2-\overset{\overset{O}{\|}}{C}-\text{O}-\text{CH}_2\text{CH}_2-\overset{\overset{O}{\|}}{C}-\text{O}-\text{CH}_2\text{CH}_2-\overset{\overset{O}{\|}}{C}-\cdots$

(d) $\cdots-\text{O}-\bigcirc-\text{O}-\overset{\overset{O}{\|}}{C}\text{CH}_2\overset{\overset{O}{\|}}{C}-\text{O}-\bigcirc-\text{O}-\overset{\overset{O}{\|}}{C}\text{CH}_2\overset{\overset{O}{\|}}{C}-\cdots$

5. Draw the structure of the repeating unit for each polymer in question 4. Then draw the structure of the monomer(s) used to prepare each polymer.

> **Web LINK**
>
> www.mcgrawhill.ca/links/atlchemistry
>
> Go to web site above and click on **Electronic Learning Partner** for more information about aspects of material covered in this section of the chapter.

In the following investigation, you will examine some of the properties of a water-absorbent polymer. You will also prepare a cross-linked polymer.

# Investigation 11-A

**SKILL FOCUS**
Performing and recording
Analyzing and interpreting

## Examining Polymers

In this investigation, you will examine three different polymers. First, you will examine the addition polymer sodium polyacrylate. This polymer contains sodium ions trapped inside the three-dimensional structure of the polymer. When placed in distilled water, the concentration of sodium ions inside the polymer is much greater than the concentration of sodium ions outside the polymer. The concentration imbalance causes water molecules to move by diffusion into the polymer. As a result, the polymer absorbs many times its own mass in distilled water.

$$\cdots-CH_2-CH-CH_2-CH-\cdots$$
$$\qquad\qquad\quad | \qquad\qquad |$$
$$\qquad\qquad\; C=O \qquad\; C=O$$
$$\qquad\qquad\quad | \qquad\qquad |$$
$$\qquad\qquad\; O^-Na^+ \qquad O^-Na^+$$

sodium polyacrylate

You will also examine polyvinyl alcohol, another addition polymer. Polyvinyl alcohol is made by the polymerization of hydroxyethene, also known as vinyl alcohol.

$$\cdots-CH_2-CH-CH_2-CH-\cdots$$
$$\qquad\qquad\quad | \qquad\qquad |$$
$$\qquad\qquad\; OH \qquad\quad OH$$

polyvinyl alcohol

Using an aqueous solution of polyvinyl alcohol, you will prepare a cross-linked polymer, commonly known as "Slime."

### Questions

How much water can sodium polyacrylate absorb? What properties do polymers have in common? How do polymers differ?

### Prediction

How can you tell if polymerization has occurred? List one or more changes you would expect to see in a liquid solution as polymerization takes place.

### Materials

**Part 1**

about 0.5 g sodium polyacrylate powder
distilled water
500 mL beaker
100 mL graduated cylinder
sodium chloride
balance

**Part 2**

10 mL graduated cylinder
50 mL beaker
20 cm × 20 cm piece of a polyvinyl alcohol bag (Alternative: Your teacher will prepare a solution of 40 g polyvinyl alcohol powder in 1 L of water. You will be given 20 mL of this solution.)
25 mL hot tap water
stirring rod
food colouring
5 mL of 4% borax solution (sodium tetraborate decahydrate)

### Safety Precautions

- Wear an apron, safety glasses, and gloves while completing this investigation.
- Wash your hands thoroughly after this investigation.
- Sodium polyacrylate is irritating to eyes and nasal membranes. Use only a small amount, and avoid inhaling the powder.

## Procedure

### Part 1 Examination of a Water-Absorbent Polymer

1. Examine the sodium polyacrylate powder. Record your observations.

2. Record the mass of a dry, empty 500 mL beaker. Place about 0.5 g of the powder in the bottom of the beaker. Record the mass of the beaker again, to find the exact mass of the powder.

3. Use a graduated cylinder to measure 100 mL of distilled water. Very slowly, add the water to the beaker. Stop frequently to observe what happens. Keep track of how much water you add, and stop when no more water can be absorbed. (You may need to add less or more than 100 mL.) Record the volume of water at which the powder becomes saturated and will no longer absorb water.

4. Add a small amount of sodium chloride to the beaker. Record your observations. Continue adding sodium chloride until no more changes occur. Record your observations.

5. Dispose of excess powder and the gelled material as directed by your teacher. Do not put the powder or gel down the sink!

### Part 2 Synthesis of a Cross-Linked Addition Polymer

1. Before starting, examine the polyvinyl alcohol bag. Record your observations.

2. Place 25 mL of hot tap water into a small beaker.

3. Add the piece of polyvinyl alcohol bag to the hot water. Stir the mixture until the polymer has dissolved.

4. Add a few drops of food colouring to the mixture, and stir again.

5. Add 5 mL of the borax solution, and stir.

6. Examine the "Slime" you have produced. Record your observations.

## Analysis

1. Calculate the mass of water that was absorbed by the sodium polyacrylate polymer. (**Note:** 1 mL of water = 1 g)

2. Use the mass of the sodium polyacrylate polymer and the mass of the water it absorbed to calculate the mass/mass ratio of the polymer to water.

3. What practical applications might the sodium polyacrylate polymer be used for?

4. What happened when you added the salt? Come up with an explanation for what you observed. Search for a discussion of sodium polyacrylate on the Internet to check your explanation.

## Conclusions

5. How do you think a plastic bag made of polyvinyl alcohol might be useful?

6. What change(s) alerted you to the fact that polymerization was occurring when you added borax to the dissolved solution of polyvinyl alcohol? Explain your observations.

7. Compare the properties of the polymers you observed in this investigation. Were there any similarities? How were they different?

## Applications

8. Sodium polyacrylate absorbs less tap water than distilled water. Design and carry out an investigation to compare the mass of tap water absorbed to the mass of distilled water absorbed by the same mass of powder.

9. Research the use of polyvinyl alcohol (PVA) bags in hospitals.

## Risks of the Polymer Industry

In the 1970s, workers at an American plastics manufacturing plant began to experience serious illnesses. Several workers died of liver cancer before the problem was traced to its source: prolonged exposure to vinyl chloride. Vinyl chloride (or chloroethene, $C_2H_3Cl$) is used to make polyvinylchloride (PVC). Unfortunately, vinyl chloride is a powerful carcinogen. Government regulations have now restricted workers' exposure to vinyl chloride. Trace amounts of this dangerous chemical are still present, however, as pollution in the environment.

The manufacture and disposal of PVC creates another serious problem. Dioxin, shown in Figure 11.5, is a highly toxic chemical. It is produced during the manufacture and burning of PVC plastics.

**Figure 11.5** Dioxin has been shown to be extremely carcinogenic to some animals. As well, it is suspected of disrupting hormone activity in the body, leading to reduced fertility and birth defects.

Many synthetic polymers, including most plastics, do not degrade in the environment. What can be done with plastic and other polymer waste? One solution may be the development and use of *degradable plastics*. These are polymers that break down over time when exposed to environmental conditions, such as light and bacteria. The Chemistry Bulletin on the next page gives more information about degradable plastics.

Another solution to the problem of polymer waste may be well-known to you already: recycling. Plastics make up approximately seven percent by mass of the garbage we generate. Because plastics are strong and resilient, they can be collected, processed, and recycled into a variety of useful products, including clothing, bags, bottles, synthetic lumber, and furniture. For example, polyethylene terephthalate (PET) plastic from pop bottles can be recycled to produce fleece fabric. This fabric is used to make jackets and sweatshirts for many major clothing companies.

Although it has many benefits, recycling does carry significant costs for transporting, handling, sorting, cleaning, processing, and storing. Along with recycling, it is essential that society learns to *reduce* its use of polymers, and to develop ways of *reusing* polymer products. Together with recycling, these directives are known as the "three R's:" reduce, reuse, and recycle.

---

**Web LINK**

www.mcgrawhill.ca/links/atlchemistry

Polymer recycling is becoming more and more widespread across Canada. Use the Internet and other sources of information to find answers to the following questions:

- Which specific polymers and polymer products can be recycled?
- What products are produced from recycled polymers?
- How does recycling polymers benefit society?
- What can you and others in your community do to promote the recycling of plastics and other polymers?

To start your search, go to the web site above and click on **Web Links**. After you have collected information on polymer recycling, create a poster or web page to share what you have learned.

**Figure 11.6** To produce this fleece fabric, plastic pop bottles are cleaned and chopped into small flakes. These flakes are melted down and extruded as fibre, which is then knitted and woven into fabric.

# Chemistry Bulletin

Science    Technology    Science    Environment

## Degradable Plastics: Garbage That Takes Itself Out

Did you know that the first plastics were considered too valuable to be thrown out? Today, many plastics are considered disposable. Plastics now take up nearly one third of all landfill space, and society's use of plastics is on the rise. Recycling initiatives are helping to reduce plastic waste. Another solution to this problem may involve the technology of degradable plastics.

All plastics are polymers, long molecules made of repeating units. Because polymers are extremely stable, they resist reaction. Most conventional plastics require hundreds or even thousands of years to break down.

*Degradable* plastics are an important step forward in technology. These new plastics can break down in a relatively short time—within six months to a year. There are two main kinds of degradable plastics: biodegradable plastics and photodegradable plastics. *Biodegradable plastics* are susceptible to decay processes that are similar to the decay processes of natural objects, such as plants. *Photodegradable plastics* disintegrate when they are exposed to certain wavelengths of light.

The ability of a degradable plastic to decay depends on the structure of its polymer chain. Biodegradable plastics are often manufactured from natural polymers, such as cornstarch and wheat gluten. Micro-organisms in the soil can break down these natural polymers. Ideally, a biodegradable plastic would break down completely into carbon dioxide, water, and biomass within six months, just like a natural material.

Photodegradable plastic is a Canadian contribution to the environmental health of the planet. It was developed in Canada more than 25 years ago by Dr. James Guillet, of the University of Toronto. This plastic incorporates light-absorbing, or *chromomorphic*, molecular groups as part of its backbone chain. The chromomorphic groups change their structure when they are exposed to particular frequencies of light (usually ultraviolet). The structural changes in these groups cause the backbone to break apart. Although the photodegradable process is not very effective at the bottom of a landfill, it is ideally suited to the marine environment. Plastics that are discarded in oceans and lakes usually float at the surface, and thus are exposed to light. The rings that are used to package canned drinks can strangle marine wildlife. These rings are now made with photodegradable technology in many countries.

Several nations are conducting further research into other marine uses for this technology.

### Making Connections

1. When a degradable plastic decays, the starch or cellulose backbone may break down completely into carbon dioxide, water, and organic matter. What do you think happens to F, Cl, or other heteroatom groups on the polymer chain as the backbone decays? Why might this be a problem?

2. Several notable scientists have cautioned against the use of biodegradable and photodegradable plastics in landfills. The bottom of a landfill has no light, or oxygen for the metabolism of bacteria. Why is this a problem? What changes must be made to landfill design, if biodegradable and photodegradable plastics are going to be included in landfills?

3. Do degradable plastics disappear as soon as they are discarded? What are the dangers of public confidence in the degradability of plastics?

## Section Review

**1** What is the difference between a synthetic polymer and a natural polymer? Give one example of each.

**2** Classify each polymerization reaction as an addition reaction or a condensation reaction.

(a) $n\text{H}_2\text{C}=\overset{\overset{\text{F}}{|}}{\text{CH}} \rightarrow \cdots-\text{CH}_2-\overset{\overset{\text{F}}{|}}{\text{CH}}-\text{CH}_2-\overset{\overset{\text{F}}{|}}{\text{CH}}-\cdots$

(b) $n\text{HO}-\text{C}_6\text{H}_4-\text{OH} + n\text{HO}-\overset{\overset{\text{O}}{\|}}{\text{C}}-(\text{CH}_2)_4-\overset{\overset{\text{O}}{\|}}{\text{C}}-\text{OH} \rightarrow$

$\cdots-\text{O}-\text{C}_6\text{H}_4-\text{O}-\overset{\overset{\text{O}}{\|}}{\text{C}}-(\text{CH}_2)_4-\overset{\overset{\text{O}}{\|}}{\text{C}}-\text{O}-\text{C}_6\text{H}_4-\text{O}-\cdots$

**3** Draw the product of each polymerization reaction, and classify the reaction. Then circle and identify any amide or ester bonds in the product.

(a) $n\text{H}_2\text{C}=\text{CH}_2 \rightarrow$

(b) $n\text{HO}-\text{CH}_2-\text{OH} + n\text{HO}-\overset{\overset{\text{O}}{\|}}{\text{C}}-\text{C}_6\text{H}_4-\overset{\overset{\text{O}}{\|}}{\text{C}}-\text{OH} \rightarrow$

(c) $n\text{HO}-\overset{\overset{\text{O}}{\|}}{\text{C}}-(\text{CH}_2)_3-\overset{\overset{\text{O}}{\|}}{\text{C}}-\text{OH} + n\text{H}_2\text{N}-\underset{\underset{\text{CH}_3}{|}}{\text{CH}}-\text{NH}_2 \rightarrow$

**4** What problems are caused by society's use of synthetic polymers? What benefits do we obtain from polymers? In your opinion, do the benefits of polymer use outweigh the risks? Write a short paragraph to answer these questions.

**5** Find out what plastic products are recycled in your community. In your opinion, does your community do a good job of recycling polymers? Why or why not?

**6** Find out what plastics are most commonly discarded in homes, schools, businesses, and industries in your community. Provide a list of suggestions on ways your community can improve its recycling of plastics. Your suggestions could include tips for individuals, schools, businesses, and industries, and ideas for new legislation involving recycling. Present your information to the class as a poster, brochure, or pamphlet. You may also wish to work with your class to prepare a single brochure to be handed out in your school or neighbourhood.

**7** Gore-tex™ is a popular fabric that provides protection from rain while allowing perspiration to escape. What polymer coating is used to make Gore-tex™ fabric? What special properties does this polymer have? What else is this polymer used for? Research the answers to these questions.

# Natural Polymers and Other Biomolecules

## 11.2

Much of our technology has been developed by observing and imitating the natural world. Synthetic polymers, such as those you encountered in the last section, were developed by imitating natural polymers. For example, the natural polymer *cellulose* provides most of the structure of plants, as you learned in the chapter opener. Wood, paper, cotton, and flax are all composed of cellulose fibres. Figure 11.7 shows part of a cellulose polymer. Figure 11.8 gives a close-up look at cellulose fibres.

### Section Preview/Outcomes

In this section, you will

- **describe** some natural polymers and monomers
- **identify** polymers and polymerization in natural systems
- **draw** structures for natural monomers and polymers

cellulose

**Figure 11.7** Like other polymers, cellulose is made of repeating units. It is the main structural fibre in plants, and it makes up the fibre in your diet.

**Biochemistry** is the study of the organic compounds and reactions that occur in living things. Proteins, starches, and nucleic acids are important biological polymers. Some biological monomers, like amino acids and sugars, are biologically important all on their own. You will also encounter lipids, a class of large biological molecules that are not polymers.

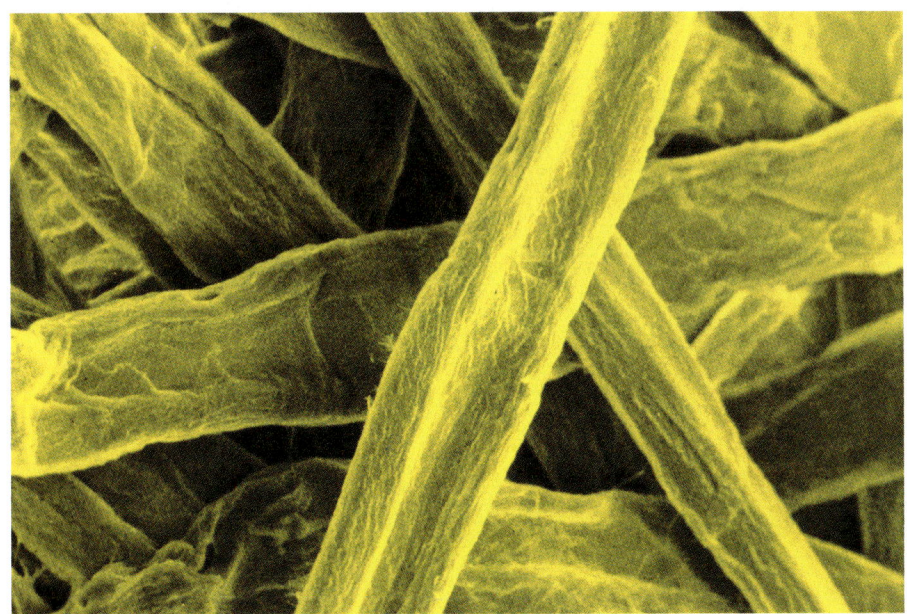

**Figure 11.8** This photograph shows cellulose fibres through a scanning electron microscope (SEM).

### History LINK

Charles Fenerty, born in Sackville, Nova Scotia, revolutionized both the publishing and forestry industries in 1844 when he revealed a new process for producing paper cheaply from spruce wood pulp. Before that time, the main ingredient in paper was cotton and flax fibres. Today, the pulp and paper industry in Atlantic Canada produces 12% of Canada's output, and researchers continue to make improvements in the way paper is made. To conserve forests, a growing source of fibre is materials once considered waste, such as wood chips, recycled paper and paperboard. Use your library and the Internet to learn more about Charles Fenerty, the evolving process of papermaking, and the University of New Brunswick's Industrial Research Centre in Pulping Technology.

## Amino Acids and Proteins

A **protein** is a natural polymer that is composed of monomers called **amino acids**. Proteins are found in meat, milk, eggs, and legumes. Wool, leather, and silk are natural materials that are made of proteins. Your fingernails, hair, and skin are composed of different proteins. Proteins carry out many important functions in your body, such as speeding up chemical reactions (enzymes), transporting oxygen in your blood (hemoglobin), and regulating your body responses (hormones).

Although many amino acids are found in living systems, only 20 common amino acids are incorporated into proteins. Amino acids contain a carboxylic acid group and an amino group. Amino acids can be linked by amide bonds to form proteins. Each amino acid has a different side chain, which is attached to the central carbon atom. Figure 11.9, below, shows the structure of an amino acid. The letter R represents the side chain of the amino acid. Examples of side chains include —CH$_3$ (for the amino acid alanine), —CH$_2$CH$_2$CONH$_2$ (for glutamine), and —CH$_2$OH (for serine).

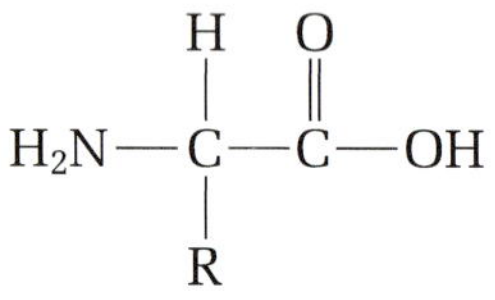

**Figure 11.9** The structure of an amino acid

### Biology LINK

Studies have shown that honey has the same quick absorption and long-lasting energy-burst effect that many sports drinks do. This is because honey is high in fructose, a simple fruit sugar that has been found to be the best sugar for athletes. Unpasteurized honey also contains hydrogen peroxide and enzymes, which aid in digestion.

The shape and biological function of a protein depends on its sequence of amino acids. The smallest protein, insulin, contains 51 amino acids. Some proteins contain over 1000 amino acids. Depending on the sequence of these amino acids, an infinite number of different proteins are possible. The DNA in your body contains the blueprints for making specific proteins for your body's structure and function.

## Carbohydrates

A **carbohydrate** (also called a **saccharide**) is a biological molecule that contains either an aldehyde or ketone group, along with two or more hydroxyl groups. The C=O group at one end of a simple saccharide reacts with an OH group at the other end to produce the most common formation, a ring. Figure 11.10 shows the ring forms of some of the simplest carbohydrates.

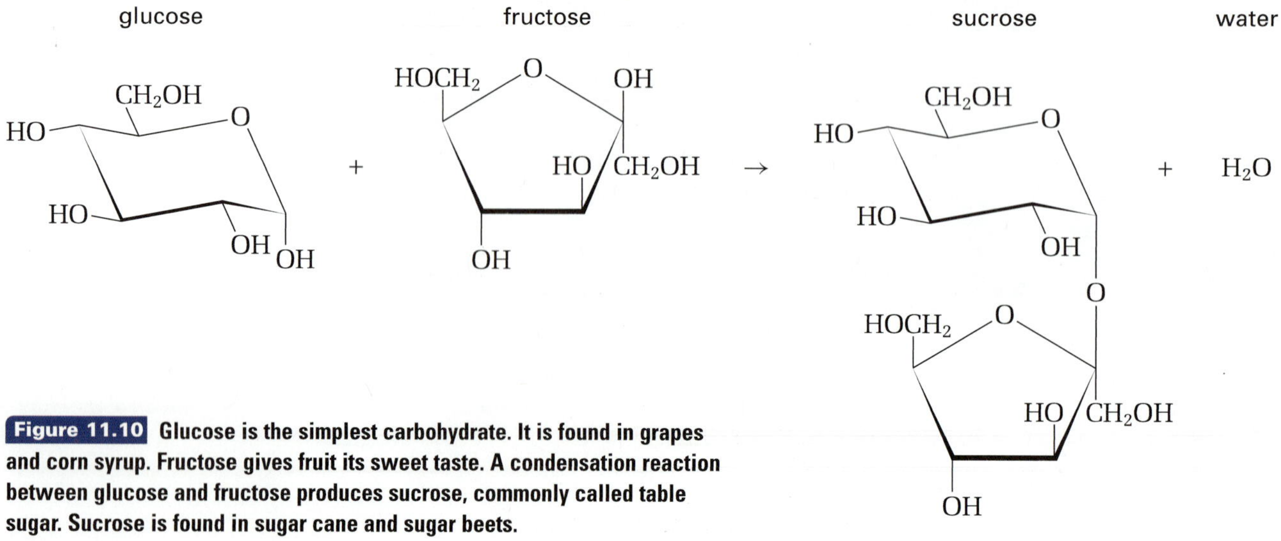

**Figure 11.10** Glucose is the simplest carbohydrate. It is found in grapes and corn syrup. Fructose gives fruit its sweet taste. A condensation reaction between glucose and fructose produces sucrose, commonly called table sugar. Sucrose is found in sugar cane and sugar beets.

Carbohydrates are found in foods such as bread, pasta, potatoes, and fruits. They are the primary source of energy for the body. In a process called *cellular respiration*, carbohydrates combine with inhaled oxygen and are oxidized to produce carbon dioxide and water, plus energy. As shown in Figure 11.11, carbon dioxide and water vapour from cellular respiration are expelled in your breath.

**Figure 11.11** On a cold day, the water vapour in your breath condenses on contact with the air, forming a visible cloud.

A **monosaccharide** is composed of one saccharide unit. Thus, it is the smallest molecule possible for a carbohydrate. A monosaccharide is often called a simple carbohydrate, or simple sugar. Glucose, seen in Figure 11.10, on the previous page, is one example of a monosaccharide. Other monosaccharides include fructose, galactose, and mannose.

Sucrose (shown in Figure 11.10) is a **disaccharide**, containing two saccharide monomers. To produce a disaccharide, an —OH group on one monomer reacts with an —OH group on another monomer to form a bond similar to an ether —O— linkage called a glycosidic linkage (also shown in Figure 11.10). This reaction is a condensation, because two molecules combine and produce a molecule of water. During digestion, carbohydrates undergo hydrolysis reactions, which break the bonds between the monomers.

**Polysaccharides** are polymers that contain many saccharide units. The most abundant polysaccharides are those in which glucose is the monomer. Plants use the glucose polysaccharide *cellulose* as a structural material. (See Figure 11.7 for the structure of cellulose.) They store energy in the starches *amylose* and *amylopectin*. A **starch** is a glucose polysaccharide that is used by plants to store energy. Humans can digest starches. Animals store energy in the glucose polysaccharide *glycogen*, illustrated in Figure 11.12

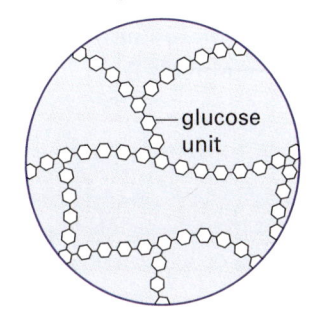

**Figure 11.12** Glycogen is the major storage polysaccharide in animals. This tiny portion of a tangled glycogen granule shows the highly branched polymer, made of glucose units.

Chapter 11 Organic Compounds on a Larger Scale • MHR **439**

## Nucleotides and Nucleic Acids

Humans and cats both begin life as fertilized eggs, which look fairly similar. Why does a fertilized cat egg develop into a kitten, while a fertilized human egg develops into a child (Figure 11.13)? A very special type of polymer, located in the cell nucleus, directs the formation of the proteins from which an organism develops.

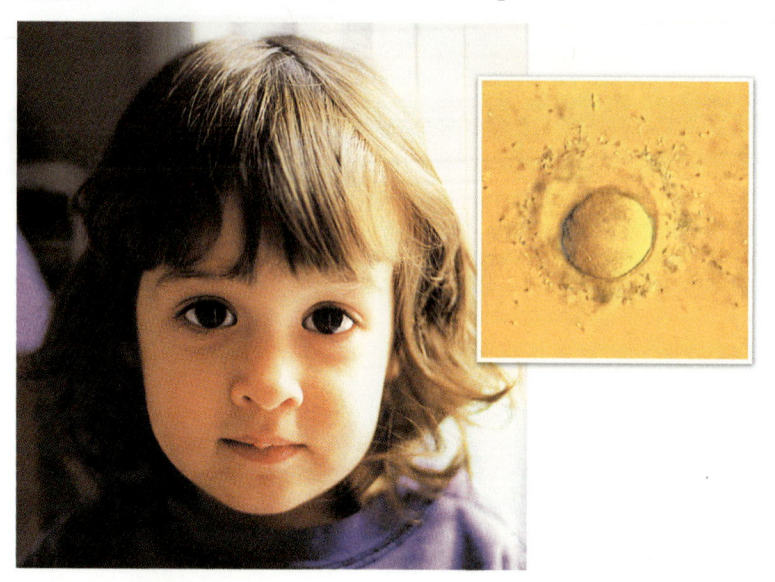

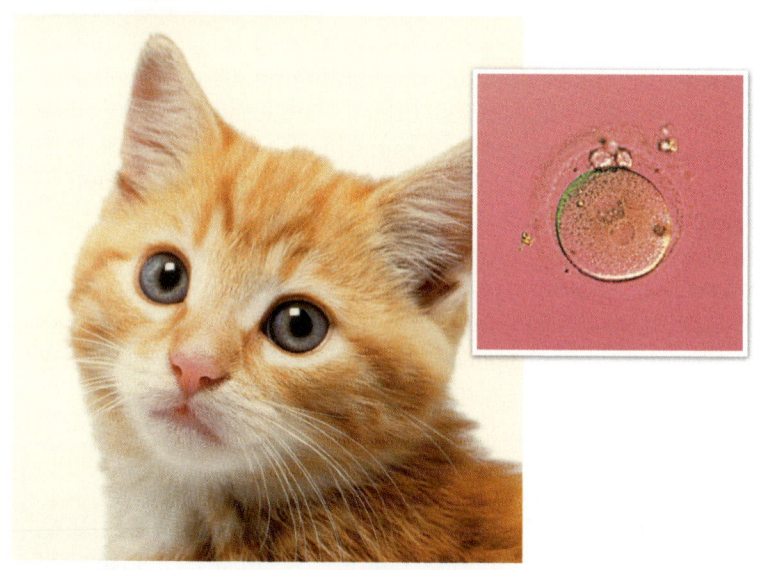

**Figure 11.13** DNA in the cells of your body direct the production of proteins. Since your bones, muscles, and organs are built from proteins that are specific to humans, your body is different from the body of a cat.

**DNA** (short for **2-deoxyribonucleic acid**) is the biological molecule that determines the shape and structure of all organisms. It is found mostly in the nuclei of cells. Each strand of DNA is a polymer that is composed of repeating units, called **nucleotides**. A single strand of DNA may have more than one million nucleotides. Each nucleotide consists of three parts: a sugar, a phosphate group, and a base.

DNA is one type of nucleic acid. The other type of nucleic acid is called **RNA** (short for **ribonucleic acid**). RNA is present throughout a cell. It works closely with DNA to produce the proteins in the body. Table 11.3, on the next page, shows the structures of RNA and DNA.

**Table 11.3** Structures of DNA and RNA

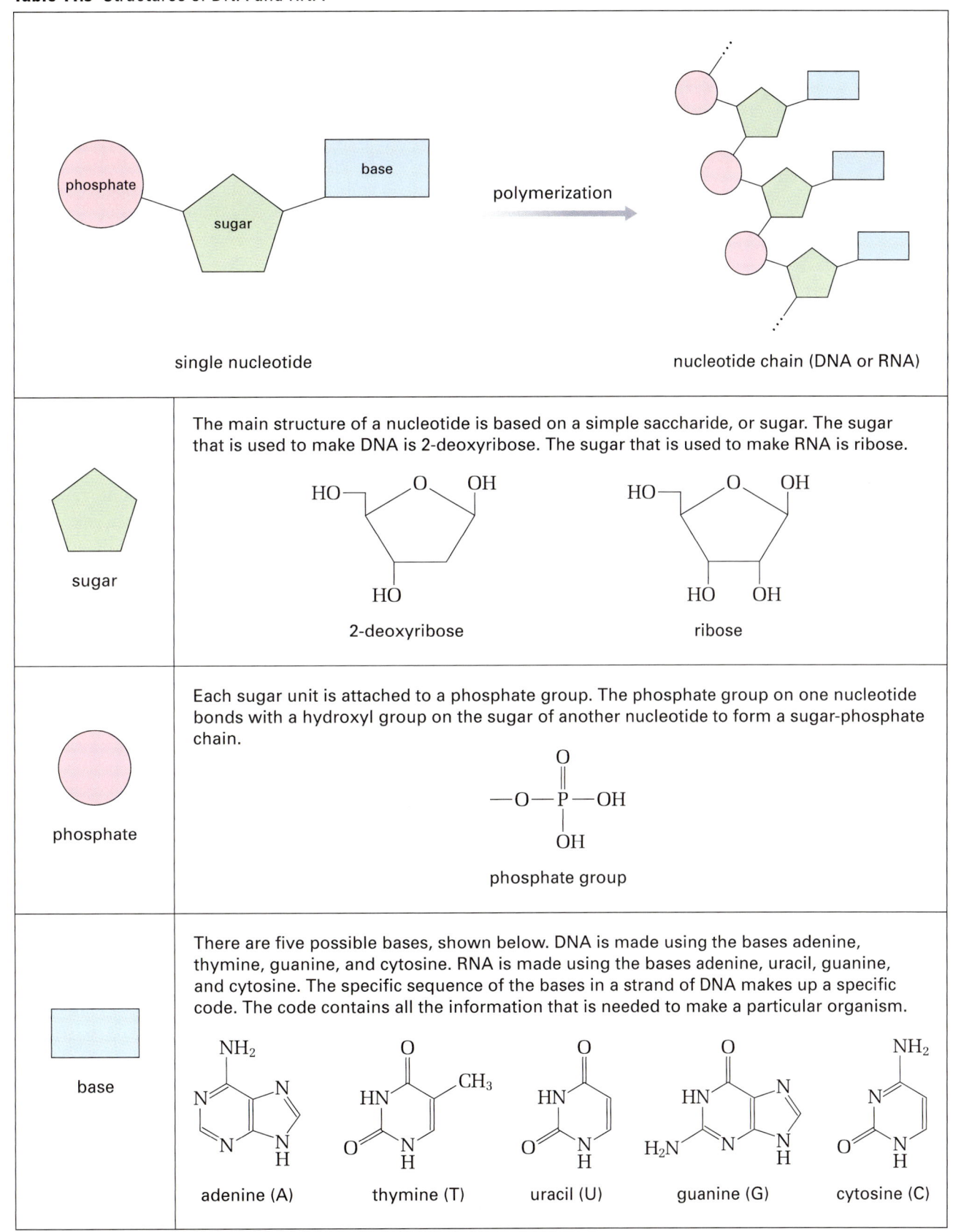

| | |
|---|---|
| | single nucleotide → polymerization → nucleotide chain (DNA or RNA) |
| sugar | The main structure of a nucleotide is based on a simple saccharide, or sugar. The sugar that is used to make DNA is 2-deoxyribose. The sugar that is used to make RNA is ribose.<br><br>2-deoxyribose      ribose |
| phosphate | Each sugar unit is attached to a phosphate group. The phosphate group on one nucleotide bonds with a hydroxyl group on the sugar of another nucleotide to form a sugar-phosphate chain.<br><br>phosphate group |
| base | There are five possible bases, shown below. DNA is made using the bases adenine, thymine, guanine, and cytosine. RNA is made using the bases adenine, uracil, guanine, and cytosine. The specific sequence of the bases in a strand of DNA makes up a specific code. The code contains all the information that is needed to make a particular organism.<br><br>adenine (A)    thymine (T)    uracil (U)    guanine (G)    cytosine (C) |

A complete molecule of DNA consists of two complementary strands. These strands are held together by hydrogen bonds between the bases on the nucleotide units, as shown in Figure 11.14.

DNA has two main functions in the body:
- It replicates itself exactly whenever a cell divides. Thus, every cell in an organism contains the same information.
- It contains codes that allow each type of protein that is needed by the body to be made by the body's cells. RNA assists in this process.

## Lipids

The last biological molecules that you will examine in this text are not polymers. They are, however, very large in size. **Lipids** are defined as biological molecules that are not soluble in water, but are soluble in non-polar solvents, such as benzene and hexanes.

From your previous studies of solubility, you might hypothesize that lipid molecules must have large hydrocarbon parts, since they are soluble in hydrocarbon solvents. Your hypothesis would be correct. Lipids contain large hydrocarbon chains, or other large hydrocarbon structures.

Several types of lipids may be familiar to you. Fats, oils, and some waxes are examples of lipids. **Fats** are lipids that contain a glycerol molecule, bonded by ester linkages to three long-chain carboxylic acids called fatty acids. **Oils** have the same structure as fats, but are classified as being liquid at room temperature, while fats are solids at room temperature. **Waxes** are esters of long-chain alcohols and long-chain carboxylic acids.

Other lipids include steroids, such as cholesterol, and terpenes, which are plant oils such as oil of turpentine, or oil of cedar. Figure 11.15 shows the lipid *glyceryl trioleate*, which is present in olive oil.

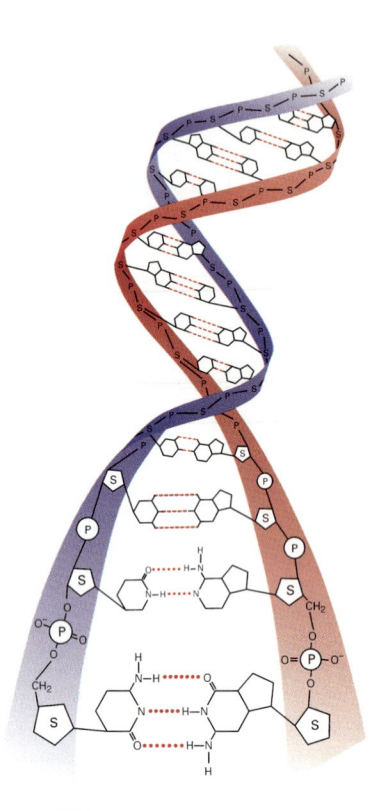

**Figure 11.14** Two strands of DNA are held together by hydrogen bonds to form a double-helix molecule.

**Figure 11.15** A fat or an oil is produced when three long-chain carboxylic acids, called *fatty acids*, bond with a glycerol unit to form three ester bonds. (Note that the three fatty acids are not always identical.) A hydrolysis reaction can split these three ester bonds to produce glycerol and a combination of the salts of fatty acids, better known as soap.

glycerol — 3 fatty acids

glyceryl trioleate

Lipids carry out many important functions in the body. One well known function is the long-term storage of energy. One gram of fat contains two-and-a-quarter times more energy than a gram of carbohydrate or protein. When an animal consumes more carbohydrates than are needed at the time, the body converts the excess carbohydrates to fat. Later, when insufficient carbohydrates are available for energy, the body may break down some of these fats to use for energy. Carbohydrates are still needed in your diet, however, for the proper functioning of your cells.

In addition to storing energy, lipids have other vital functions:
- Cell membranes are made of a double layer of *phospholipids*. These are lipids that are polar at one end and non-polar at the other end.
- Lipids such as testosterone and cortisol act as hormones to regulate body functions. (Not all hormones are lipids. Some hormones are proteins. Other hormones are neither lipids nor proteins.)
- Several vitamins (such as vitamins A, D, and E) are lipids. These vitamins are known as *fat-soluble*.
- The fatty tissues in your body are made from lipids. They act like protective "packaging" around fragile organs, such as the heart. They insulate the body from excessive heat or cold, and they act as padding to reduce impact from collisions.

## Canadians in Chemistry

### Dr. Raymond Lemieux

Raymond Lemieux was born into a carpenter's family in 1920 at Lac La Biche, Alberta. He obtained a B.Sc. degree at the University of Alberta and a Ph.D. at McGill University in Montréal. Lemieux then worked briefly at Ohio State University, where he met Dr. Virginia McConaghie. They were married and soon moved to Saskatoon, Saskatchewan. There Lemieux became a senior researcher at a National Research Council (NRC) laboratory.

Raymond Lemieux (1920–2000)

In 1953, while at the NRC lab, Lemieux conquered what some have called "the Mount Everest of organic chemistry." He became the first person to completely synthesize sucrose, or table sugar. Sucrose is a carbohydrate with the chemical formula $C_{12}H_{22}O_{11}$. It is the main sugar in the sap of plants such as sugar beets and sugar cane. Sucrose is related to glucose, $C_6H_{12}O_6$, and other sugars.

Lemieux continued his research at the University of Ottawa, and then at the University of Alberta in Edmonton. He was especially interested in how molecules "recognize" each other and interact in the human body. For example, different blood groups, such as group A and group B, are determined by carbohydrate molecules that differ by only a single sugar. The body is able to recognize these specific sugars and adapt its response to foreign substances, such as bacteria and transplanted organs.

Since it was hard to obtain natural samples of the body sugars that Lemieux wanted to study, he found ways to synthesize them. This was groundbreaking research. Seeing the practical applications of his research, Lemieux was instrumental in starting several chemical companies. Today these companies make products such as antibiotics, blood-group determinants, and immunizing agents that are specific to various human blood groups. They also make complex carbohydrates that absorb antibodies from the blood in order to prevent organ transplant rejection.

Lemieux and his wife had six children and a number of grandchildren. With all that he had accomplished in life, Lemieux said, "My proudest achievement is my family." His autobiography, titled *Explorations with Sugar: How Sweet It Was*, was published in 1990.

## Section Summary

In this section, you examined the structures and functions of several important biological molecules, such as proteins, amino acids, carbohydrates, DNA, and lipids. In the next section, you will examine the risks and benefits of manufacturing and using organic compounds.

## Section Review

**1** In this section, you learned about biological polymers.
   (a) Name three types of biological polymers.
   (b) Give an example of each type of biological polymer.
   (c) Name the monomer unit for each type of biological polymer.

**2** What function does each type of biological molecule perform in living organisms?
   (a) carbohydrate
   (b) protein
   (c) DNA
   (d) lipid

**3** Identify each type of biological molecule. Copy the structure into your notebook, and circle the functional group(s).

444  MHR • Unit 4 Organic Chemistry

# Organic Compounds and Everyday Life

## 11.3

You are already aware that organic compounds are an inescapable part of your daily life. As you know, much of your body is composed of organic compounds. The food you eat is made from organic compounds, as are the clothes you wear. The fuel that heats your home and powers buses and cars is also organic. When you are sick or injured, you may use organic compounds to fight infection or to reduce pain and swelling. Modern processed foods include many flavourings and colourings—another use of organic compounds. (See Figure 11.16.)

**Section Preview/Outcomes**

In this section, you will

- **describe** how the science and technology of organic chemistry is an essential part of your life, and of society
- **gather information** and **analyze** the risks and benefits of using organic compounds
- **identify** and **use** criteria to evaluate information you gather
- **examine** the benefits of funding scientific studies into uses of certain types of organic compounds over others
- **describe** and **evaluate** technological solutions involving organic chemistry
- **form and defend your own opinion** on society's use of organic compounds based on your findings

**Figure 11.16** In Chapter 10, you learned about esters that are used as flavourings. Not all flavourings belong to the ester family. Menthol is an organic compound that is found in the mint plant. It is used as a flavouring in many substances, including toothpaste, chewing gum, and cough syrups. What functional group is present in menthol? How might you guess this from the name?

In this section, you will examine the uses of some common organic compounds. You will also carry out an analysis of some of the risks and benefits that are involved in making and using organic compounds.

## Risk-Benefit Analysis

How can you make the most informed decision possible when analyzing the risks or benefits of a product or issue? The following steps are one way to carry out a risk-benefit analysis for the manufacture and use of an organic compound.

**Step 1** Identify possible risks and benefits of the organic product. Ask questions such as these:

- What are the direct uses of this product? (For example, is it used as a pharmaceutical drug for a specific disease or condition? Is it an insecticide for a specific insect "pest"?)
- What are the indirect uses of this product? (For example, is it used in the manufacture of a different organic or inorganic product?)
- What are the direct risks of making or using this product? (For example, is the product an environmental pollutant, or does it contain compounds that may be harmful to the environment? Can the workers who manufacture this product be harmed by exposure to it?)

## CONCEPT CHECK

When gathering and using information on any topic, how can you make sure the information you find is free from bias? Work with your class to come up with a list of criteria that will help you to recognize whether source material is accurate and unprejudiced.

- What are the indirect risks of making or using this product? (For example, are harmful by-products produced in the manufacture of this product? How much energy does the manufacture of this product require, and where does the energy come from? What, if any, side effects does this product cause?)

**Step 2** Research the answers to your questions from step 1. Ensure that your information comes from reliable sources. If there is controversy on the risks and benefits of the product, find information that covers both sides of the issue.

**Step 3** Decide on your own point of view. In your opinion, are the benefits greater than the risks? Do the risks outweigh the benefits?

**Step 4** Consider possible alternatives that may lessen the risks. Ask questions such as the following:

- Does a similar, less harmful, product exist?
- Is there a better way to manufacture or dispose of this product?
- What safety precautions can be taken to reduce the risks?

### Risks and Benefits of Organic Compounds

The next four tables outline common uses of various organic compounds, as well as possible complications resulting from each use. The ThoughtLab that follows provides an opportunity for you to work with and extend the information in the tables. Of course, the compounds listed are only a few of the many useful organic compounds that are available to society.

**Table 11.4** Uses and Possible Complications of Selected Pharmaceutical Drugs

| Organic compound | Source | Common uses | Side effects and/or concerns |
|---|---|---|---|
| salbutalmol (Ventolin®) | • developed from ephedrine, a compound that occurs naturally in the ma huang (ephedra) plant | • used as asthma medication | • possible side effects include tremors, nausea, heart palpitations, headaches, and nervousness |
| lidocaine | • developed from cocaine, a compound that occurs naturally in the coca leaf | • used as a local anaesthetic, in throat sprays and sunburn sprays | • possible side effects include skin rash or swelling |
| morphine | • occurs naturally in the opium poppy plant, which is still the simplest and cheapest source of morphine | • one of the strongest pain-killers known | • highly addictive<br>• possible side effects include dizziness, drowsiness, and breathing difficulty |
| acetylsalicylic acid (A.S.A., Aspirin®) | • developed from salicin, found in the bark of the willow tree | • reduces fever, inflammation, and pain<br>• used to prevent heart attacks by thinning blood | • possible side effects include stomach upset<br>• more rare side effects include stomach bleeding and Reye's syndrome |
| acetaminophen (Tylenol™) | • developed from acetyl salicylic acid | • reduces fever and pain | • possible side effects include stomach upset<br>• can cause liver damage in high doses |

**Table 11.5** Uses and Possible Complications of Selected Pesticides

| Pesticide | Common uses | Side effects and/or concerns |
|---|---|---|
| parathion | • insecticide<br>• kills insects by disrupting their nerve impulses | • one of the most toxic organophosphate pesticides<br>• highly toxic to humans, birds, aquatic invertebrates, and honeybees |
| carbaryl | • insecticide<br>• similar to parathion, but safer for humans | • more expensive than parathion<br>• generally low toxicity to humans and most other mammals<br>• highly toxic to fish, aquatic invertebrates, and honey bees |
| endrin | • insecticide<br>• also controls rodents, such as mice | • persists in soil up to 12 years<br>• highly toxic to fish<br>• banned in some countries and restricted use in many others |
| pyrethrin | • natural pesticide<br>• breaks down quickly in the environment | • irritating to skin, eyes, respiratory system<br>• highly toxic<br>• can be dangerous to humans and pets within range of the pesticide application<br>• very toxic to aquatic life |
| resmethrin | • synthetic pyrethroid pesticide<br>• 75 times less toxic to humans and pets than pyrethrin | • low toxicity to humans and other mammals (causing skin, eye, and respiratory system irritation)<br>• irritating to skin, eyes, and respiratory system<br>• highly toxic to fish and honey bees |

**Table 11.6** Uses and Possible Complications of Selected Food Additives

| Organic Compound | Common uses | Side effects and/or concerns |
|---|---|---|
| aspartame | • artificial sweetener | • significant controversy over possible health risks |
| menthol | • flavouring in many substances, including toothpaste, chewing gum, and cough syrups | • may cause allergic reaction |
| MSG (monosodium-glutamate) | • flavour enhancer for foods | • may cause allergic reaction known as MSG symptom complex, with symptoms such as nausea and headache<br>• may worsen already severe asthma |
| Red #40 | • food colouring | • may cause allergic reaction |

## CONCEPT CHECK

Re-read Question 6 from the ThoughtLab on this page. Should the government of Canada fund research into organic compounds? If your answer is yes, which types of organic compounds should receive funding for further study, and which types, if any, should not? Carry out a debate with your class on these topics.

**Table 11.7** Uses and Possible Complications of Selected Common Organic Compounds

| Organic compound | Common uses | Side effects and/or concerns |
|---|---|---|
| tetrachloroethene (Perc™) | • solvent used to dry-clean clothing | • toxic<br>• may damage central nervous system, kidneys, and liver |
| ethylene glycol | • automobile antifreeze | • toxic, especially to small pets and wildlife<br>• may damage central nervous system, kidneys, and heart |
| acetone | • nail polish remover<br>• industrial solvent | • highly flammable<br>• irritating to throat, nose, and eyes |

## ThoughtLab — Risk-Benefit Analyses of Organic Products

As you know, organic products provide many of the necessities and comforts of your life. They are also responsible for a great deal of damage, including environmental pollution and harm to human health. In this Thought Lab, you will perform risk-benefit analyses on three organic compounds.

### Procedure

1. Choose three of the organic compounds listed in Tables 11.4 through 11.7. Research and record the chemical structure of each compound.
2. Follow the four steps for a risk-benefit analysis, given at the beginning of this section. You will need to carry out additional research on the compounds of your choice.

### Analysis

For each compound, answer the following questions.

1. What are the practical purposes of this organic compound?
2. (a) What alternative compounds (if any) can be used for the same purpose?
   (b) What are the benefits of using your particular compound, instead of using the alternatives in part (a)?
3. (a) What are the direct risks of using this compound?
   (b) What are the indirect risks of using this compound?
4. In your opinion, should this compound continue to be used? Give reasons for your answer.
5. What safety precautions could help to reduce possible harm—to people as well as to the environment—resulting from this compound?
6. In your opinion, should the government fund scientific research into organic compounds that are used as food additives? Why or why not?

### Problem Solving With Organic Compounds

Many chemicals and products have been developed as solutions to health, safety, and environmental problems. Sometimes, however, a solution to one problem can introduce a different problem. Read the three articles on the next few pages. Then complete the ThoughtLab that follows.

# Replacing CFCs—At What Cost?

At the beginning of the twentieth century, refrigeration was a relatively new technology. Early refrigerators depended on the use of toxic gases, such as ammonia and methyl chloride. Unfortunately, these gases sometimes leaked from refrigerators, leading to fatal accidents. In 1928, a new, "miracle" compound was developed to replace these toxic gases. Dichlorodifluoromethane, commonly known as Freon®, was a safe, non-toxic alternative. Freon® and other chlorofluorocarbon compounds, commonly referred to as CFCs, were also used for numerous other products and applications. They were largely responsible for the development of many conveniences, such as air-conditioning, that we now take for granted.

Today we know that CFCs break up when they reach the ozone layer, releasing chlorine atoms. The chlorine atoms destroy ozone molecules faster than the ozone can regenerate from oxygen gas. Studies in the past ten years have shown dramatic drops in ozone concentration at specific locations. Since ozone protects Earth from the Sun's ultraviolet radiation, this decrease in ozone has led to increases in skin cancer, as well as damage to plants and animals. In addition, CFCs are potent greenhouse gases and contribute to global warming. Through the Montréal Protocol, and later "Earth Summit" gatherings, many countries—including Canada—have banned CFC production.

Substitutes for CFCs are available, but none provide a completely satisfactory alternative. Hydrofluorocarbons (HFCs) are organic compounds that behave like CFCs, but do not harm the ozone layer. For example, 1,1,1,2-tetrafluoroethane and 1,1-difluoroethane are HFCs that can be used to replace CFCs in refrigerators and air conditioners. Unfortunately, HFCs are also greenhouse gases.

Simple hydrocarbons can also be used as CFC substitutes. Hydrocarbons such as propane, 2-methylpropane (common name: isobutane), and butane are efficient aerosol propellants. These hydrocarbons are stable and inexpensive, but they are extremely flammable.

$H_3C-CH-CH_3$
     |
     $CH_3$

isobutane

$CH_3CH_2CH_3$

propane

$CH_3CH_2CH_2CH_3$

butane

$$F-\underset{\underset{H}{|}}{\overset{\overset{H}{|}}{C}}-\underset{\underset{F}{|}}{\overset{\overset{F}{|}}{C}}-F$$

1,1,1,2-tetrafluoroethane

$$F-\underset{\underset{H}{|}}{\overset{\overset{F}{|}}{C}}-\underset{\underset{H}{|}}{\overset{\overset{H}{|}}{C}}-H$$

1,1-difluoroethane

# Revisiting DDT: Why Did it Happen?

Dichlorodiphenyltrichloroethane, better known as DDT, is a well-known pesticide that has caused significant environmental damage. It is easy to point fingers, blaming scientists and manufacturers for having "unleashed" this organic compound on an unsuspecting world. It is more difficult, however, to understand why this environmental disaster took place.

Insects consume more than one third of the world's crops each year. In addition, insects such as mosquitoes spread life-threatening diseases, including malaria and encephalitis. Weeds reduce crop yields by taking over space, using up nutrients, and blocking sunlight. Some weeds even poison the animals that the crops are intended to feed. Crop damage and low crop yield are significant problems for countries undergoing food shortages and famines.

DDT was one of the first pesticides that was developed. For many years, it was used successfully to protect crops and fight disease epidemics. Paul Mueller, the scientist who discovered DDT's use as a pesticide, was awarded the Nobel Prize in 1948.

Later, however, tests revealed that DDT does not readily decompose in the environment. Instead, DDT remains present in the soil for decades after use. The hydrocarbon part of this molecule makes it soluble in the fatty tissues of animals. As it passes through the food chain, DDT accumulates. Animals that are higher in the food chain, such as large birds and fish, contain dangerous concentrations of this chemical in their tissues. A high DDT concentration in birds causes them to lay eggs with very thin shells, which are easily destroyed. Today DDT is no longer produced or used in Canada. It is still used, however, in many developing countries.

# Knocking on the Car Door

Automobile fuels are graded using *octane numbers*, which measure the combustibility of a fuel. A high octane number means that a fuel requires a higher temperature and/or higher pressure to ignite. Racing cars with high-compression engines usually run on pure methanol, which has an octane number of 120.

Gasoline with too low an octane number can cause "knocking" in the engine of a car, when the fuel ignites too easily and burns in an uncontrolled manner. Knocking lowers fuel efficiency, and it can damage the engine.

As early as 1925, two of the first automobile engineers became aware of the need to improve the octane number of fuels. Charles Kettering advocated the use of a newly developed compound called tetra-ethyl lead, $Pb(C_2H_5)_4$. This compound acts as a catalyst to increase the efficiency of the hydrocarbon combustion reaction. Henry Ford believed that ethanol, another catalyst, should be used instead of tetra-ethyl lead. Ethanol could be produced easily from locally grown crops. As we now know, ethanol is also much better for the environment.

Tetra-ethyl lead became the chosen fuel additive. Over many decades, lead emissions from car exhausts accumulated in urban ponds and water systems. Many waterfowl that live in urban areas experience lead poisoning. Lead is also dangerous to human health.

Leaded fuels are now banned across Canada. In unleaded gasoline, simple organic compounds are added instead of lead compounds. These octane-enhancing compounds include methyl-t-butyl ether, t-butyl alcohol, methanol, and ethanol. Like lead catalysts, these compounds help to reduce engine knocking. In addition, burning ethanol and methanol produces fewer pollutants than burning hydrocarbon fuels, which contain contaminants. Since they can be made from crops, these alcohols are a renewable resource.

2-methoxy-2-methyl propane (methyl tert-butyl ether)

1,1-dimethyl ethanol (tert-butyl alcohol)

# ThoughtLab: Problem Solving with Organic Compounds

In this ThoughtLab, you will consider several situations in which organic compounds were used to help solve a health, safety, or environmental problem.

## Procedure

1. Choose two of the situations that are discussed in the three newspaper articles. Come up with a third situation, involving organic compounds, that you have heard or read about recently. For the three situations you have chosen, answer the following questions.
   (a) What was the original problem?
   (b) How was this problem resolved? What part did organic compounds play in the solution?
   (c) What further problems (if any) were introduced by the solution in part (b)?
   (d) Have these additional problems been resolved? If so, how have they been resolved?
   (e) Do you foresee any new problems arising from the solutions in part (d)? Explain why or why not.

## Analysis

1. Use your knowledge of organic chemistry to describe how organic chemistry has helped to provide a solution to
   (a) a human health problem
   (b) a safety problem
   (c) an environmental problem

2. In your opinion, is it worth coming up with solutions to problems, if the solutions carry the possibility of more problems? Explain your answer.

# Careers in Chemistry

## Polymer Chemist

Bulletproof vests used to be heavy, bulky, and uncomfortable. Stephanie Kwolek changed that. Now police officers, police dogs, and soldiers, can wear light, strong bulletproof vests made of a synthetic fibre called Kevlar™. Kwolek (shown above) developed Kevlar™ while working for the DuPont chemical company. Kevlar™ first came into general use in 1971. It is five times stronger than steel, gram for gram, but almost as light as nylon. It is flame resistant, resists wear and tear, and does not conduct electricity. This versatile material is used not only in bulletproof vests, but also in other manufactured items, including hockey helmets, firefighters' suits, spacecraft shells, and surgeons' gloves.

Kwolek's branch of organic chemistry—polymer chemistry—specializes in creating synthetic materials that are cheaper, faster, and stronger than natural materials. Polymer chemists often work by stringing together thousands of atoms to form long molecules called polymers. Then they manipulate these polymers in various ways. Polymer chemists have invented an amazing array of materials. These include polyvinyl chloride (used to make garden hose and duct tape), polyurethane (used to stuff teddy bears and make spandex bicycle pants), and acrylonitrile-butadiene-styrene, or ABS (used to make brakes and other auto body parts).

### Make Career Connections

What kind of education do polymer chemists need for their jobs?

- Find out about polymer chemistry programs that are offered by a university near you.
- Research different companies that employ polymer chemists. If possible, interview a polymer chemist who is employed by one of these companies.

## Section Summary

In this section, you encountered some of the ways that organic compounds make our lives easier. As well, you learned about some of the risks involved in using organic compounds. You carried out risk-benefit analyses on several organic compounds. Finally, you examined some situations in which organic chemistry has helped to solve problems related to human health, safety, and the environment.

## Section Review

**1** Think about the risks and benefits of organic compounds.

   **(a)** Describe five specific benefits you obtain from organic compounds.

   **(b)** Describe two risks from organic compounds, and explain how these risks may affect you.

**2** Prepare a brochure, booklet, or web page explaining the practical uses of organic compounds to a younger age group. Your booklet should include the following information:

- a simple explanation of what an organic compound is
- a description of some benefits of organic compounds
- specific examples of at least three different organic compounds that you studied in this chapter, plus their uses
- specific examples of at least three different organic compounds that you did *not* study in this chapter, plus their uses

**3** Research persistent organic pollutants (POPs) on the Internet. Use your research to prepare a poster about POPs for your community. Include the following information:

- examples of three POPs, with descriptions of their negative effects on the environment and/or human health
- a description of what the Canadian government has done to address the problem of POPs
- suggestions for ways that your community can avoid or reduce harm from organic pollutants

**4** In section 11.1, you learned about some of the risks and benefits of polymers. Choose a synthetic polymer from section 11.1. Carry out a risk-benefit analysis of this polymer.

**5** Scientists and horticulturalists who work with and sell the synthetic pesticide *pyrethroid* are concerned about public perception. Many people buy the natural pesticide *pyrethrin*, even though it is more toxic than pyrethroid. Why do you think this happens? Do you think this happens for other products that have natural and synthetic alternatives? Write a brief editorial outlining your opinions and advice to consumers, to help them make informed product choices.

# CHAPTER 11 Review

## Reflecting on Chapter 11
Summarize this chapter in the format of your choice. Here are a few ideas to use as guidelines:
- Compare addition polymerization reactions with condensation polymerization reactions.
- List the biomolecules you learned about in this chapter, and give an example of each.
- List some benefits of the use of organic compounds.
- List some risks from the use of organic compounds.

## Reviewing Key Terms
For each of the following terms, write a sentence that shows your understanding of its meaning.

polymer  
plastics  
condensation polymerization  
polyesters  
protein  
carbohydrate  
monosaccharide  
polysaccharide  
DNA (2-deoxyribonucleic acid)  
RNA (ribonucleic acid)  
waxes  
monomers  
addition polymerization  
nylons  
polyamides  
biochemistry  
amino acids  
saccharide  
disaccharide  
starch  
nucleotides  
lipids  
oils  
fats  

## Knowledge/Understanding

1. Use your own words to define the following terms.
   (a) polymer
   (b) monomer
   (c) plastic

2. What is the difference between a polymer and any other large molecule?

3. Why are some natural molecules, such as proteins and starches, classified as polymers?

4. What type of polymer contains monomers that are linked by ester bonds? Give an example, including both the structure and the name of the compound.

5. Describe each type of polymerization, and give an example.
   (a) addition polymerization
   (b) condensation polymerization

6. A short section of the polymer *polypropene* is shown below.
   $\cdots - CH_2 - CH(CH_3) - CH_2 - CH(CH_3) - \cdots$
   (a) Draw and name the monomer that is used to make this polymer.
   (b) What type of polymer is polypropene? Explain your reasoning.
   (c) What is the common name of this polymer?

7. What is the difference between a nylon and a polyester? How are they similar?

8. Is a protein an example of a polyester or a polyamide? How do you know?

9. How are DNA and RNA different? How are they similar?

10. Fats and oils have the same basic chemical structure. Why are they classified into two separate groups?

11. (a) What is the difference between a protein and an amino acid?
    (b) How are proteins important to living organisms?

12. (a) Define the term "lipid."
    (b) Give three different examples of lipids.
    (c) List four foods you eat that contain lipids.
    (d) How are lipids important to your body?

13. (a) Distinguish between monosaccharides, disaccharides, and polysaccharides.
    (b) Draw and name an example of each.
    (c) How are carbohydrates important to living organisms?

14. (a) What is the generic monomer of a polysaccharide polymer?
    (b) What is the monomer of cellulose?

## Inquiry

15. Draw the product(s) of each incomplete polymerization reaction. **Hint:** Don't forget to include water as a second product when necessary.

    (a) $n CH_2 = CH_2 \xrightarrow{\text{polymerization}}$

    (b) $n H_2N - (CH_2)_4 - NH_2 +$
    $n HO - \overset{\overset{O}{\|}}{C} - (CH_2)_4 - \overset{\overset{O}{\|}}{C} - OH \xrightarrow{\text{polymerization}}$

Answers to questions highlighted in red type are provided in Appendix A.

(c) $n\text{HO}-\overset{\overset{\text{O}}{\|}}{\text{C}}-(\text{CH}_2)_7-\text{NH}_2 \xrightarrow{\text{polymerization}}$

(d) $n\text{CH}_2=\overset{\overset{\displaystyle\text{CH}}{|}}{\underset{\underset{\displaystyle\text{H}_3\text{C}\quad\text{CH}_3}{}}{\text{CH}}} \xrightarrow{\text{polymerization}}$

(e) $n\text{HO}-\overset{\overset{\text{O}}{\|}}{\text{C}}-\overset{|}{\underset{|}{\text{CH}}}-\text{NH}_2 \xrightarrow{\text{polymerization}}$ (with phenyl group attached to CH)

(f) $n\text{CH}_2=\overset{|}{\underset{\underset{\displaystyle\text{OH}}{}}{\text{CH}}} \xrightarrow{\text{polymerization}}$

(g) $n\text{HO}(\text{CH}_2)_7\text{OH} + n\text{HO}\overset{\overset{\text{O}}{\|}}{\text{C}}\text{CH}_2\overset{\overset{\text{O}}{\|}}{\text{C}}\text{OH} \xrightarrow{\text{polymerization}}$

**16.** Draw the monomer(s) in each reaction.

(a) ? $\xrightarrow{\text{polymerization}}$ $\cdots-\text{CH}_2-\text{CH}_2-\text{CH}_2-\text{CH}_2-\cdots$

(b) ? $\xrightarrow{\text{polymerization}}$ $\cdots-\overset{\overset{\text{O}}{\|}}{\text{C}}-\text{CH}_2-\overset{|}{\underset{\underset{\displaystyle\text{CH}_3}{}}{\text{CH}}}-\text{O}-\overset{\overset{\text{O}}{\|}}{\text{C}}-\text{CH}_2-\overset{|}{\underset{\underset{\displaystyle\text{CH}_3}{}}{\text{CH}}}-\text{O}-\cdots$

(c) ? $\xrightarrow{\text{polymerization}}$ $\cdots-\text{CH}_2-\overset{|}{\underset{\underset{\displaystyle\text{CH}_2\text{CH}_3}{}}{\text{CH}}}-\text{CH}_2-\overset{|}{\underset{\underset{\displaystyle\text{CH}_2\text{CH}_3}{}}{\text{CH}}}-\cdots$

(d) ? + ? $\xrightarrow{\text{polymerization}}$ $\cdots-\text{NH}-\text{CH}_2-\overset{|}{\underset{\underset{\displaystyle\text{CH}_3}{}}{\text{CH}}}-\text{NH}-\overset{\overset{\text{O}}{\|}}{\text{C}}-\text{CH}_2-\overset{\overset{\text{O}}{\|}}{\text{C}}-\text{NH}-\text{CH}_2-\overset{|}{\underset{\underset{\displaystyle\text{CH}_3}{}}{\text{CH}}}-\text{NH}-\cdots$

(e) ? + ? $\xrightarrow{\text{polymerization}}$ $\cdots-\text{O}-\text{CH}_2-\langle\text{C}_6\text{H}_4\rangle-\text{CH}_2-\text{O}-\overset{\overset{\text{O}}{\|}}{\text{C}}-\text{CH}_2-\overset{\overset{\text{O}}{\|}}{\text{C}}-\cdots$

## Communication

**17.** Use a molecular model kit to build an example of a DNA nucleotide monomer. Join your nucleotide to your classmates' nucleotides to build a short strand of DNA.

**18.** Make a list of five materials or substances you use every day that are made from organic compounds. Write a paragraph that describes how your life would be different without these five materials or substances.

**19.** How has organic chemistry helped to solve the problems caused by the use of the following substances?
(a) leaded gasoline      (b) CFCs

**20.** In this chapter, you carried out research using the Internet and other sources to find out about organic compounds. Not all sources contain accurate information, and many sources have a bias towards one side or another of an issue. How did you attempt to distinguish between reliable sources and unreliable sources? How could you improve your assessment of whether or not a source provides accurate and unbiased information?

**21.** What are some of the properties of plastics that make them so useful? Use your own observation of plastic products to answer this question.

**22.** Reducing our consumption of polymer products, and reusing polymer products also help to minimize polymer waste. Come up with suggestions for your school or community on ways to *reduce* and *reuse* polymer products.

## Making Connections

23. In the Web Link on page 434 of this chapter, you were asked to find out more about recycling polymers. In your opinion, does recycling solve the problem of accumulating polymer product waste? Why or why not?

24. The Chemistry Bulletin on page 435 gave a description of degradable plastics.
    (a) In your opinion, should all plastics that are made in the future be degradable? Why or why not?
    (b) What are some risks and benefits of using biodegradable and photodegradable plastics?

25. Biofuel for cars and other engines can be made from biological materials such as plants and animal fats. Biodiesel and ethanol are the two most common biological fuels. What factors do you think might be holding back the sale of biofuels on the Canadian market? What benefits do you think society could obtain from using biofuel to help replace hydrocarbon-based fuel? What negative results might impact the Canadian economy if biofuels become more common? What is your own opinion on the development and use of biofuels? Write a short paragraph answering these questions. You may need to do some preliminary research on the Internet to help you form an opinion.

26. Suppose that the Canadian government proposed a $3 million grant to scientists who are working on the development of synthetic pesticides. Would you support this grant? Write a brief paragraph that explains your point of view.

27. Styrofoam™ "clamshells" were used for packaging fast food until a few years ago. Today, large fast food corporations use paper packaging instead. Why has Styrofoam™ use been reduced? What problems does this product cause? Research Styrofoam™ (made from a polymer called polystyrene) to answer these questions.

28. Do you think Canada should cut back on the manufacture and use of synthetic polymers? Write a brief paragraph that explains your point of view. Include at least three good reasons to back up your point of view. If necessary, research the information you need.

29. Steroids are one type of lipid. Research answers to the following questions.
    (a) What is the structure of a steroid?
    (b) What are some common examples of steroids? How are steroids used in medicine?
    (c) How are steroids misused in sports?

30. Muscle-building athletes sometimes drink beverages that contain amino acids.
    (a) Why might this type of drink help to build muscles?
    (b) Research the benefits and risks of this type of drink.

### Answers to Practice Problems and Short Answers to Section Review Questions

**Practice Problems: 1.** polymethylmethacrylate (PMMA) **2.(a)** condensation **(b)** addition **(c)** condensation **4.(a)** addition **(b)** condensation; nylon **(c)** condensation; polyester **(d)** condensation; polyester
**Section Review: 11.1: 1.** artificially made versus found in nature; polyethylene and cellulose **2.(a)** addition **(b)** condensation **3.(a)** addition polymerization **(b)** condensation polymerization **(c)** condensation polymerization
**11.2 1.(a)** protein, starch, nucleic acid **(b)** insulin, amylose, DNA **(c)** amino acid, monosaccharide (glucose), nucleotide **3. (a)** amino acid **(b)** monosaccharide **(c)** amino acid **(d)** lipid **(e)** disaccharide

# UNIT 4 Review

## Knowledge/Understanding

### Multiple Choice

In your notebook, write the letter for the best answer to each question.

1. Carbon atoms can bond with each other to form
   (a) single bonds only
   (b) double bonds only
   (c) single and double bonds
   (d) single, double, and triple bonds
   (e) single, double, triple, and quadruple bonds

2. The three-dimensional shape around a single-bonded carbon atom is
   (a) linear
   (b) trigonal planar
   (c) tetrahedral
   (d) pyramidal
   (e) bent

3. An alkane contains 5 carbon atoms. How many hydrogen atoms does it contain?
   (a) 5 hydrogen atoms
   (b) 8 hydrogen atoms
   (c) 10 hydrogen atoms
   (d) 12 hydrogen atoms
   (e) 14 hydrogen atoms

4. An alkyne contains 12 hydrogen atoms. What is its chemical formula?
   (a) $C_5H_{12}$
   (b) $C_6H_{12}$
   (c) $C_7H_{12}$
   (d) $C_8H_{12}$
   (e) $C_9H_{12}$

5. Which set of properties best describes a small alkane, such as ethane?
   (a) non-polar, high boiling point, insoluble in water, extremely reactive
   (b) non-polar, low boiling point, insoluble in water, not very reactive
   (c) polar, low boiling point, soluble in water, not very reactive
   (d) polar, high boiling point, insoluble in water, not very reactive
   (e) non-polar, low boiling point, soluble in water, extremely reactive

6. Which equation shows the complete combustion of propane?
   (a) $2C_2H_6 + 7O_2 \rightarrow 4CO_2 + 6H_2O$
   (b) $2C_2H_6 + 5O_2 \rightarrow C + 2CO + CO_2 + 6H_2O$
   (c) $2C_3H_6 + 9O_2 \rightarrow 6CO_2 + 6H_2O$
   (d) $C_3H_8 + 5O_2 \rightarrow 3CO_2 + 4H_2O$
   (e) $2C_3H_8 + 7O_2 \rightarrow 2C + 2CO + 2CO_2 + 8H_2O$

7. The length of a carbon-carbon bond in a benzene molecule is
   (a) halfway between a single bond and a double bond
   (b) halfway between a double bond and a triple bond
   (c) the same as a single bond
   (d) the same as a double bond
   (e) the same as a triple bond

8. The functional group of an alcohol is
   (a) −OH
   (b) −C=C−
   (c) −NR$_2$
   (d) −C(O)NR$_2$
   (e) −OR

9. The functional group of an amide is
   (a) −OH
   (b) −C=C−
   (c) −NR$_2$
   (d) −C(O)NR$_2$
   (e) −OR

10. The compound $NH_2CH_2CH_2CH_3$ has the IUPAC name
    (a) methanamine
    (b) ethanamine
    (c) propanamine
    (d) butanamine
    (e) butanamide

11. The compound $CH_3CH(CH_3)CH_2CH_2OH$ has the IUPAC name
    (a) 1-pentanol
    (b) 1-butanol
    (c) 2-methyl-4-butanol
    (d) 3-methyl-1-butanol
    (e) 3-methyl-1-pentanol

12. The compound $CH_3CH_2CH_2C(O)CH_2CH_3$ has the IUPAC name
    (a) ethoxypentane
    (b) ethoxypropane
    (c) ethyl butanoate
    (d) 3-pentanone
    (e) 3-hexanone

Answers to questions highlighted in red type are provided in Appendix A.

13. CH₃CH₂C(CH₃)₂CH₂CH₂CH₂COOH has the IUPAC name
    (a) 5,5-dimethylheptanoic acid
    (b) 3,3-dimethylheptanoic acid
    (c) 4,4-dimethylheptanoic acid
    (d) 2,2-dimethylheptanoic acid
    (e) 3,3-dimethyl heptanoate

14. The esterification of methanol and butanoic acid produces
    (a) 1-methyl butanoic acid
    (b) 4-methyl butanoic acid
    (c) butyl methanoate
    (d) methyl butanoate
    (e) butanol and methanoic acid

15. The polymerization of propene, CH₃CH=CH₂, can be classified as
    (a) a hydrolysis reaction
    (b) an elimination reaction
    (c) a substitution reaction
    (d) a condensation reaction
    (e) an addition reaction

16. Choose the most correct definition.
    (a) Organic compounds are based on carbon, and they usually contain carbon-nitrogen and carbon-silicon bonds.
    (b) Organic compounds are based on nitrogen, and they usually contain carbon-nitrogen and carbon-hydrogen bonds.
    (c) Organic compounds are based on carbon, and they usually contain carbon-hydrogen and carbon-carbon bonds.
    (d) Organic compounds are based on hydrogen, and they usually contain carbon-hydrogen and carbon-oxygen bonds.
    (e) Organic compounds are based on oxygen, and they usually contain carbon-oxygen and carbon-carbon bonds.

17. Which statement is *not* correct?
    (a) A C–O bond is polar.
    (b) An N–H bond is polar.
    (c) An O–H bond is polar.
    (d) A C–C bond is polar.
    (e) A C–Cl bond is polar.

18. Which description of an addition reaction is accurate?
    (a) Two molecules are combined, and a small molecule, such as water, is produced as a second product.
    (b) A hydrogen atom or functional group is replaced with a different functional group.
    (c) Atoms are added to a double or triple carbon-carbon bond.
    (d) A double carbon-carbon bond is formed when atoms are removed from a molecule.
    (e) Carbon atoms form more bonds to oxygen or less bonds to hydrogen.

## Short Answer

In your notebook, write a sentence or a short paragraph to answer each question.

19. What is the difference between an amide and an amine? Use examples to explain your answer.
20. Do hydrocarbons have functional groups? Use examples to explain your answer.
21. Are the boiling points of carboxylic acids higher or lower than the boiling points of alcohols? Explain your answer.
22. A primary amine can form intermolecular hydrogen bonds, but a tertiary amine cannot. Explain this statement.
23. What is the common name for ethanol?
24. What is the IUPAC name for acetic acid?
25. The aldehyde CH₃CH₂CHO is named propanal, rather than 1-propanal. Why is the position number unnecessary?

## Inquiry

26. Write the IUPAC name for each compound.
    (a) CH₃CH₂CH₂COOH
    (b) CH₃OH
    (c) CH₃CH₂CH₂CH₂COOCH₃
    (d) CH₃CH₂C(O)CH₃
    (e) CH₃CH₂NHCH₃
    (f) CH₃CH₂CH(CH₃)CH(CH₃)CH₂CH₂CHO

27. Give a common name for each compound.
    (a) methanol
    (b) 2-propanol
    (c) ethoxyethane

28. Write the IUPAC name for each compound.
    
    (a) CH₃—CH₂—CH(CH₃)—CH(CH₂CH₃)—CH=CH₂

(b) CH₃—O—CH₂—CH₂—CH₃

(c) 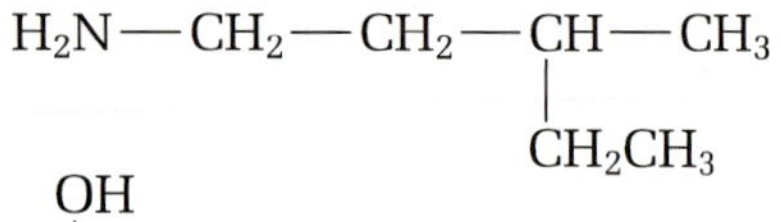 H₂N—CH₂—CH₂—CH(CH₂CH₃)—CH₃

(d) 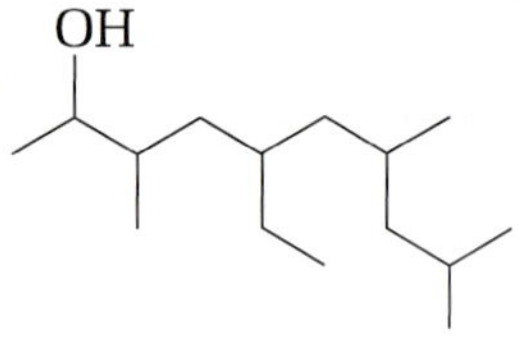 [structure with OH]

(e) 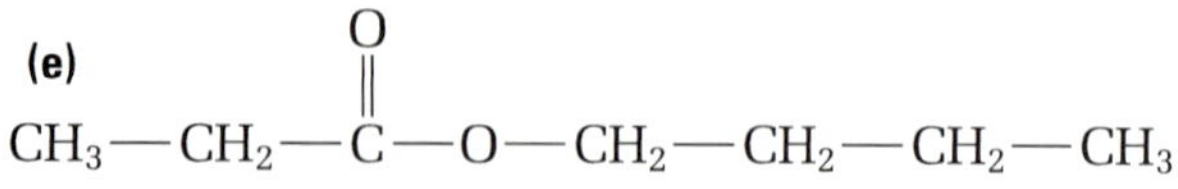 CH₃—CH₂—C(=O)—O—CH₂—CH₂—CH₂—CH₃

(i) CH₃—N(CH₂CH₃)—C(=O)—CH₃

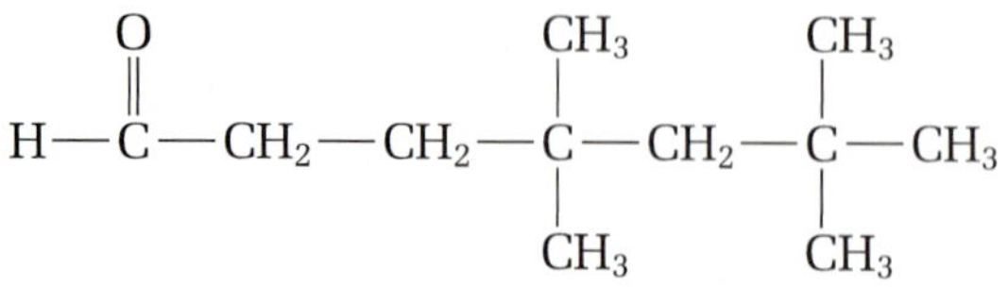

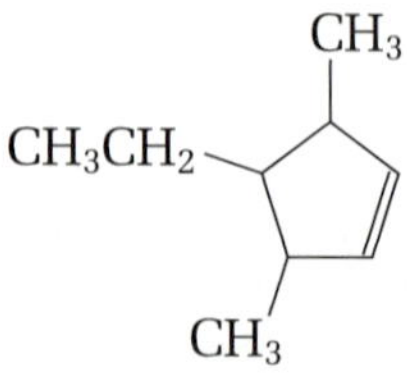

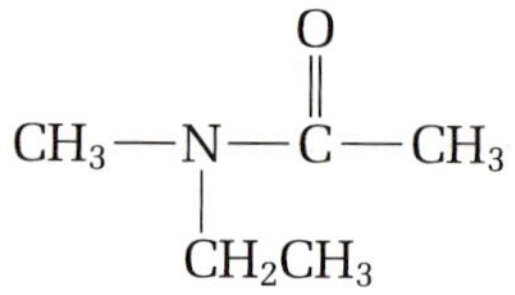

(j) [cyclic ketone structure with three CH₃ groups]

**29.** Name each compound.

(a) 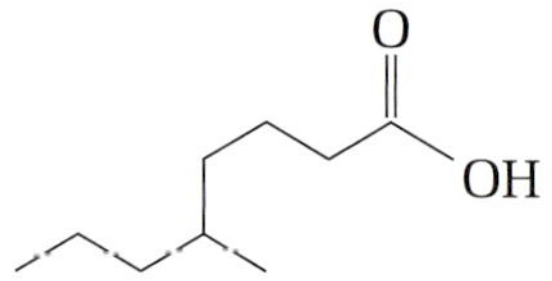 [carboxylic acid structure]

(b) CH₃—CH₂—NH—CH₂—CH₂—CH₂—CH₃

(c) 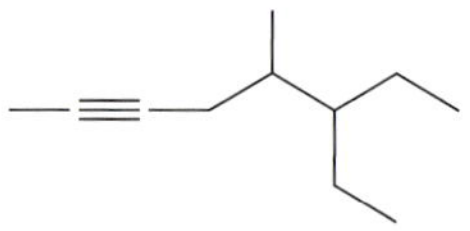 [alkyne structure]

(d) 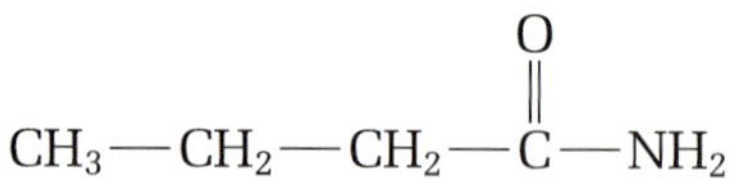 CH₃—CH₂—CH₂—C(=O)—NH₂

(e) 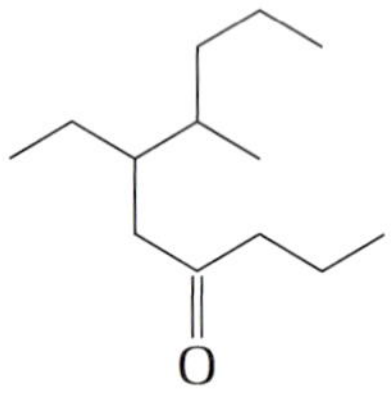 [ketone structure]

(f) 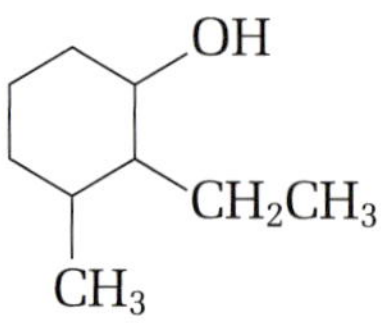 [cyclohexane with OH, CH₂CH₃, CH₃]

(g) H—C(=O)—CH₂—CH₂—C(CH₃)₂—CH₂—C(CH₃)₂—CH₃

(h) [cyclopentene with CH₃CH₂, CH₃, CH₃ substituents]

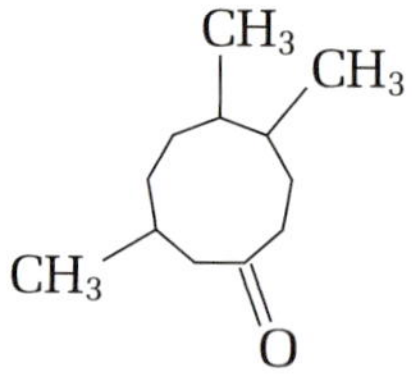

**30.** What is wrong with the following names? Use each incorrect name to draw a structure. Then correctly rename the structure you have drawn.
(a) 2-propyl-4-pentanone
(b) 3,3-trimethylhexane
(c) 2,5-octanol

**31.** Imagine that you dip one finger into a beaker of rubbing alcohol and another finger into a beaker of water. When you take your fingers out of the beakers, you observe that one finger feels colder than the other. Which finger feels colder, and why?

**32.** You are given two beakers. One contains 1-pentanol, and the other contains pentanoic acid. Describe the materials you would need to distinguish between the two liquids, in a laboratory investigation.

**33.** You are given three test tubes that contain colourless liquids. One test tube contains benzene, another contains ethanol, and the third contains 2,4-hexadiene. Design a procedure that will tell you the contents of each test tube. Describe your expected observations. **CAUTION** Do not try your procedure in a lab. Benzene is carcinogenic.

**34.** Predict the product(s) of each reaction.
(a) CH₃CH₂CH=CHCH₃ + I₂ →
(b) CH₃CH₂CH₂OH + HCl →
(c) 1-propanol + methanoic acid →
(d) 1-heptanol $\xrightarrow[\Delta]{H_2SO_4}$

**35.** Predict the product(s) of each reaction.
(a) CH₂=CHCH₃ + H₂O →
(b) propyl decanoate + water →
(c) CH₃CH(CH₃)CH₂CH₂Br + OH⁻ →
(d) CH₂=CHCH₂CH₂OH + 2HCl →

458 MHR • Unit 4 Organic Chemistry

36. Draw and name the missing reactant(s) in each reaction.
    (a) ? + HBr → $CH_2$=CHBr
    (b) ?+ → $CH_3CH_2CH_2C(CH_3)$=$CH_2$ + HOH + $Cl^-$
    (c) ? + ? → butyl heptanoate + $H_2O$

## Communication

37. Use a concept map to trace the path of a six-carbon hydrocarbon molecule as it goes from its source in the ground, through processing, into a can of paint. Describe each step in its processing. Explain any changes that the molecule undergoes.

38. Why do alkanes and alkynes, unlike alkenes, have no geometric isomers?

39. Explain why carbon is able to form so many more compounds than any other element.

40. Draw a structural diagram for each compound.
    (a) methanamine
    (b) 2-octanol
    (c) 1,2-dichlorobutane
    (d) pentanamide
    (e) methoxynonane

41. Draw a structural diagram for each compound.
    (a) propyl propanoate
    (b) 2,2,4,5-tetramethylheptanal
    (c) 2-methylpropanoic acid
    (d) N-ethyl-N-methyl butanamide
    (e) 4-ethyl-3,3-dimethyl-2-heptanone

42. Prepare a poster that describes condensation polymerization to students who do not take science. Use the polymerization of Nylon-6 as an example. Include information on the practical uses of Nylon-6. What natural product was nylon designed to replace? When nylon was first invented in the 1930s, what consumer products were made from it? What part did nylon play in World War II?

43. Draw a concept web that summarizes what you have learned about organic chemistry. Include the following topics:
    - functional groups
    - physical properties of organic compounds
    - reactivity of organic compounds
    - polymers, both natural and synthetic
    - biological molecules and their functions
    - practical applications of organic chemistry
    - risks and benefits of organic chemistry

44. In your opinion, are the risks of manufacturing and using pesticides greater or less than the benefits? Write a short (half-page) essay that explains your point of view. Include specific examples to back up your argument. (You may want to carry out further research on pesticides, and the effects of pesticides, before writing your essay.)

45. How are organic compounds important to your health and lifestyle? Write a short paragraph that describes any benefits you obtain from organic compounds.

46. Choose one of the biological molecules described in Chapter 11. Research this molecule until you discover three interesting facts about it, which you did not know before. Share your findings with your class.

## Making Connections

47. When was plastic cling wrap first invented? What problem was it created to solve? Use the Internet or other resources to find the answers to these questions.

48. Organic pollutants can build up in living things by a process called bioaccumulation.
    (a) Use the Internet to research bioaccumulation. Where do most organic pollutants accumulate in the body of an animal?
    (b) Explain your answer to the previous question, using your knowledge of the solubility of organic compounds.

49. Most dyes are organic compounds. Before artificial dyes were invented, people prepared natural dyes from roots, berries, leaves, and insects.
    (a) Research some dyes used in the past and in the present. Find the chemical structure of each dye. Identify the functional group(s) in each dye.
    (b) How do artificial dyes differ from natural dyes? How have artificial dyes affected the fashion industry?

# UNIT 5

# From Kinetics to Equilibrium

**UNIT 5 CONTENTS**

**CHAPTER 12**
Rates of Chemical Reactions

**CHAPTER 13**
Reversible Reactions and Chemical Equilibrium

**UNIT 5 OVERALL OUTCOMES**

- What factors affect the rate and the extent of a chemical reaction?
- How can you measure the position of equilibrium experimentally and predict the concentrations of chemicals in a system at equilibrium?
- How do chemical technologies depend on the rate and the extent of chemical reactions?

Photochromic lenses darken in sunlight and then gradually clear in shade and dim light. The change is permanent in constant light conditions, yet reversible if the light intensity changes. How?

The molten glass contains dissolved silver chloride. As the solution cools, silver chloride precipitates, forming tiny silver chloride crystals. Light striking these crystals produces silver and chlorine atoms. The silver atoms tend to clump together to form particles that are big enough to block light and darken the glass.

Adding copper(I) chloride to the molten glass makes the process reversible. When light intensity diminishes, copper ions remove electrons from silver atoms, converting the silver atoms into silver ions. The silver ions then migrate back to the silver chloride crystals. The glass becomes transparent again.

You are used to reading chemical equations from left to right, using stoichiometry to calculate the amounts of products formed. However, many chemical reactions are reversible—they are also able to proceed from right to left. In this unit, you will examine the factors that affect the rate at which a reaction proceeds, in either direction, and the concepts that describe reversible directions.

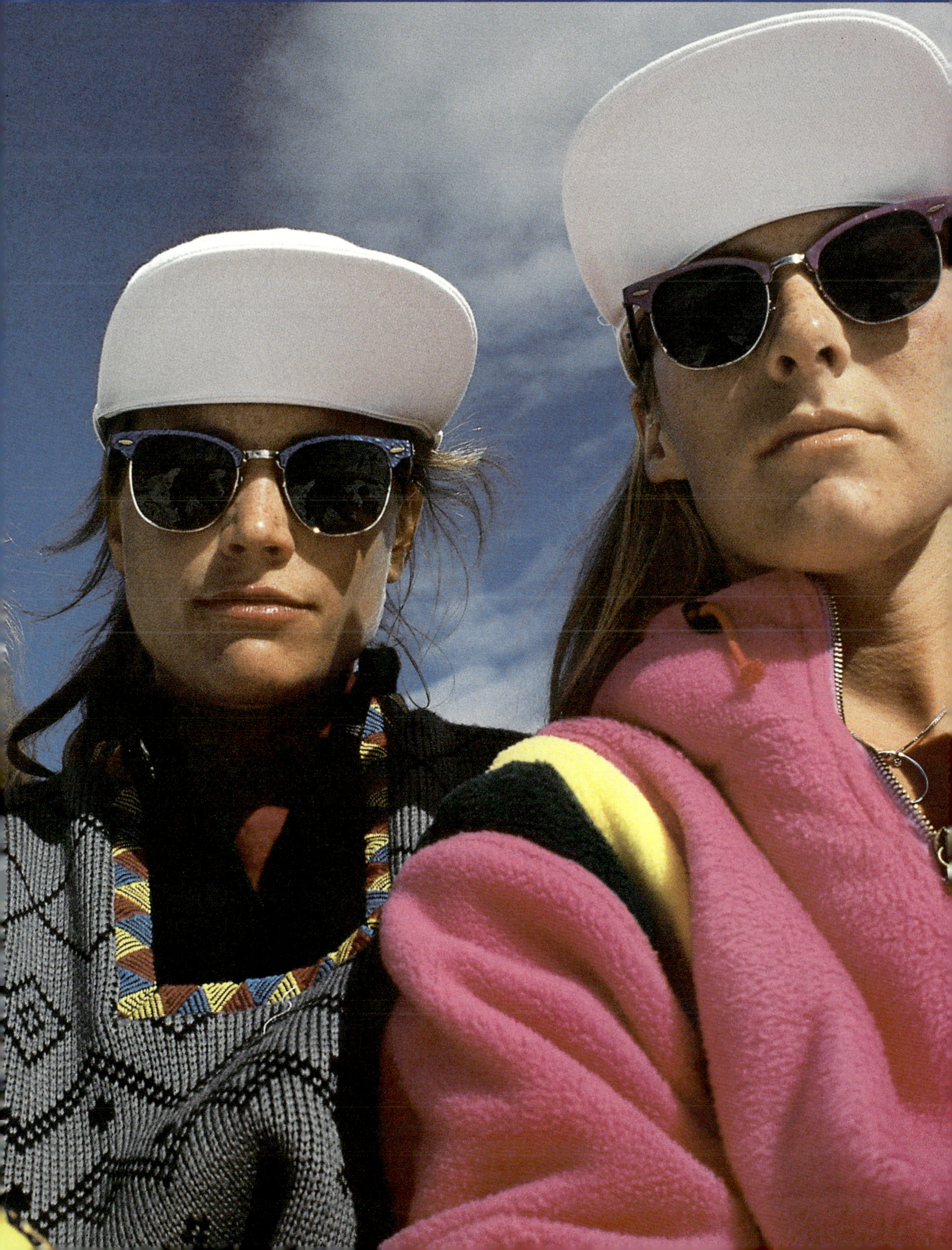

# CHAPTER 12
# Rates of Chemical Reactions

**Chapter Preview**

12.1 Factors that Affect Reaction Rates

12.2 Collision Theory and Reaction Mechanisms

**Prerequisite Concepts and Skills**

Before you begin this chapter, review the following concepts and skills:

- balancing chemical equations (Chapter 1)
- expressing concentration in units of mol/L (Chapter 7)

Racing cars can reach speeds that are well above 200 km/h. In contrast, the maximum speed of many farm tractors is only about 25 km/h. Just as some vehicles travel more quickly than others, some chemical reactions occur more quickly than others. For example, compare the two reactions that occur in vehicles: the decomposition of sodium azide in an air bag and the rusting of iron in steel.

When an automobile collision activates an air bag, sodium azide, $NaN_{3(g)}$, decomposes to form sodium, $Na_{(s)}$, and nitrogen gas, $N_{2(g)}$. (The gas inflates the bag.) This chemical reaction occurs almost instantaneously. It inflates the air bag quickly enough to cushion a driver's impact.

On the other hand, the reaction of iron with oxygen to form rust proceeds quite slowly. Most Canadians know that the combination of road salt and wet snow somewhat increases the rate of the reaction. Even so, it takes several years for a significant amount of rust to form on the body of a car. This is a good thing for car owners—if rusting occurred as fast as the reaction in an inflating air bag, cars would flake to pieces in seconds.

Why do some reactions occur slowly while others seem to take place instantaneously? How do chemists describe and compare the rates at which chemical reactions occur? What factors affect reaction rates? Can chemists predict and control the rate of a chemical reaction? These questions will be answered in Chapter 12.

The chemical reactions that produced the oil and natural gas reserves of the Hibernia oil field began well over 100 million years ago. Why do chemical reactions occur at different rates?

# Factors That Affect Reaction Rates     12.1

Nitroglycerin is an explosive that was used to clear the way for railroads across North America. It decomposes instantly. The reactions that cause fruit to ripen, then rot, take place over a period of days. The reactions that lead to human ageing take place over a lifetime.

How quickly a chemical reaction occurs is a crucial factor in how the reaction affects its surroundings. Therefore, knowing the rate of a chemical reaction is integral to understanding the reaction.

## Expressing Reaction Rates

The change in the amount of reactants or products over time is called the **reaction rate** or **rate of reaction**. How do chemists express reaction rates? Consider how the rates of other processes are expressed. For example, the Olympic sprinter in Figure 12.1 can run 100 m in about 10 s, resulting in an average running rate of 100 m/10 s or about 10 m/s.

The running rate of a sprinter is calculated by dividing the distance travelled by the interval of time the sprinter takes to travel this distance. In other words, running rate (speed) is expressed as a change in distance divided by a change in time. In general, a change in a quantity with respect to time can be expressed as follows.

$$\text{Rate} = \frac{\text{Change in quantity}}{\text{Change in time}}$$
$$= \frac{\text{Quantity}_{\text{final}} - \text{Quantity}_{\text{initial}}}{t_{\text{final}} - t_{\text{initial}}}$$
$$= \frac{\Delta \text{Quantity}}{\Delta t}$$

Chemists express reaction rates in several ways. For example, a reaction rate can be expressed as a change in the amount of reactant consumed or product made per unit of time, as shown below. (The letter A represents a compound.)

$$\text{Rate of reaction} = \frac{\text{Amount of A}_{\text{final}} - \text{Amount of A}_{\text{initial}} \text{ (in mol)}}{t_{\text{final}} - t_{\text{initial}} \text{ (in s)}}$$
$$= \frac{\Delta \text{ Amount of A}}{\Delta t} \text{ (in mol/s)}$$

When a reaction occurs between gaseous species or in solution, chemists usually express the reaction rate as a change in the concentration of the reactant or product per unit time. The concentration of a compound (in mol/L) is symbolized by placing square brackets, [ ], around the chemical formula.

$$\text{Rate of reaction} = \frac{\text{Concentration of A}_{\text{final}} - \text{Concentration of A}_{\text{initial}} \text{ (in mol/L)}}{t_{\text{final}} - t_{\text{initial}} \text{ (in s)}}$$
$$= \frac{\Delta[A]}{\Delta t} \text{ (in mol/(L} \cdot \text{s))}$$

You know that chemical reactions occur at widely differing rates. In Investigation 12-A, you will explore factors that affect the rate at which they occur.

**Section Preview/Outcomes**

In this section, you will
- **identify** factors that affect the rate of a reaction
- **identify** methods for measuring reaction rates
- **communicate** your understanding of the following term: *reaction rate (rate of reaction)*

**Figure 12.1** The running rate (speed) of a sprinter is expressed as a change in distance over time.

# Investigation 12-A

**SKILL FOCUS**
Predicting
Performing and recording
Analyzing and interpreting

## Factors Affecting the Rate of a Reaction

You have probably already encountered the reaction of vinegar with baking soda. The carbon dioxide that is produced can be used to simulate a volcano, for example, or to propel a toy car or rocket.

$CH_3COOH_{(aq)} + NaHCO_{3(s)} \rightarrow$
$\qquad NaCH_3COO_{(aq)} + CO_{2(g)} + H_2O_{(\ell)}$

Other carbonate-containing compounds, such as calcium carbonate, can also react with vinegar to produce $CO_2$.

In this investigation, you will determine reaction rates by recording the time taken to produce a fixed volume of $CO_2$. You will collect the $CO_2$ by downward displacement of water.

### Question

How do factors such as concentration, temperature, a different reactant, and surface area affect the rate of this reaction?

### Prediction

Read the Procedure. *Quantitatively* predict the effects of changes to the concentration and temperature. *Qualitatively* predict the effects of changes to the reactant and surface area.

### Materials

electronic balance
pneumatic trough
stopwatch
250 mL Erlenmeyer flask
retort stand and clamp
one-holed rubber stopper, fitted with a piece of glass tubing (must be airtight)
1 m rubber hose to fit glass tubing (must be airtight)
25 or 100 mL graduated cylinder
large test tube
weighing paper, weighing boat, or small beaker
100 mL vinegar (at room temperature)
10 g baking soda, $NaHCO_3$
2.0 g powdered $CaCO_3$
2.0 g solid $CaCO_3$ (marble chips)
scoopula
thermometer
wash bottle with distilled water (at room temperature)
warm-water bath (prepared using a large beaker and a hot plate or electric kettle)
ice bath (ice cubes and water)
paper towel

### Safety Precautions

- Beware of shock hazard if an electric kettle or hot plate is used.
- Wear safety glasses at all times.

### Procedure

#### Part 1 The Effect of Concentration

1. The distilled water and vinegar that you are going to use should be at room temperature. Measure and record the temperature of either the vinegar or the distilled water.

2. Assemble the apparatus for the collection of $CO_2$, by downward displacement of water, as shown below.

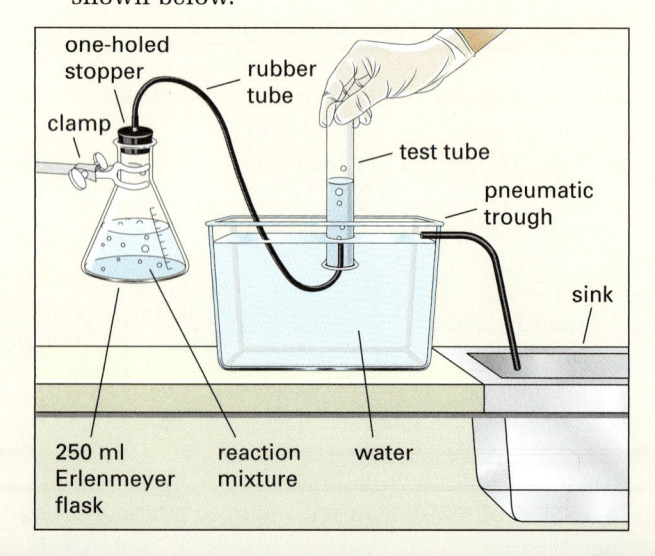

**Note:** To invert a test tube filled with water, place a piece of paper over the mouth of the filled test tube before inverting it.

3. Copy the table below into your notebook, to record your data.

| Trial | Mass of NaHCO₃ (g) | Volume of vinegar (mL) | Volume of distilled water (mL) | Time to fill test tube with CO₂ (s) | Average reaction rate (mL/s) |
|---|---|---|---|---|---|
| 1 | 1.00 | 20.0 | 0.0 | | |
| 2 | 1.00 | 15.0 | 5.0 | | |
| 3 | 1.00 | 10.0 | 10.0 | | |
| 4 | 1.00 | 5.0 | 15.0 | | |

4. For trial 1, add 20.0 mL of vinegar to the flask. Have the stopwatch ready. The end of the rubber tubing should be in place under the water-filled test tube in the pneumatic trough. Quickly add 1.00 g of NaHCO₃ to the flask, and put in the stopper. Record the time taken to fill the tube completely with $CO_2$.

5. Complete trials 2 to 4 with the indicated quantities.

6. Determine the volume of $CO_2$ you collected by filling the gas collection test tube to the top with water and then pouring the water into a graduated cylinder.

### Part 2 The Effect of Temperature

1. Repeat trial 3 in Part 1 using a mixture of 10 mL of water and 10 mL of vinegar that has been cooled to about 10°C below room temperature in an ice bath. Measure and record the temperature of the mixture immediately before the reaction. Record the time taken to fill the test tube with $CO_2$. Determine the average rate of production of $CO_2$ (in mL/s).

2. Use a hot-water bath to warm a mixture of 10 mL of distilled water and 10 mL of vinegar to about 10°C above room temperature. Repeat trial 3 in Part 1 using this heated mixture. Measure and record the temperature of the vinegar-water mixture immediately before the reaction. Record the time taken to fill the test tube with $CO_2$. Determine the average rate of production of $CO_2$ (in mL/s).

### Part 3 The Effect of Reactants and Surface Area

1. Repeat trial 3 in Part 1, using 1.00 g of powdered calcium carbonate, $CaCO_3$, instead of $NaHCO_3$. All the reactants should be at room temperature. Record the time taken to fill the tube with $CO_2$. Determine the average rate of production of $CO_2$ (in mL/s).

2. Repeat step 1, using 1.00 g of solid $CaCO_3$ (marble chips) instead of powdered $CaCO_3$.

### Analysis

1. Draw a graph to show your results for Part 1. Plot average reaction rate (in mL $CO_2$/s) on the y-axis. Plot [$CH_3COOH$] (in mol/L) on the x-axis. Vinegar is 5.0% (m/v) $CH_3COOH$.

2. As quantitatively as possible, state a relationship between [$CH_3COOH$] and the average rate of the reaction.

3. Compare the average reaction rate for corresponding concentrations of vinegar at the three different temperatures tested.

4. What effect did a 10°C temperature change have on the reaction rate? Be as quantitative as possible.

5. What effect did using $CaCO_3$ instead of $NaHCO_3$ have on the reaction rate?

6. What effect did the surface area of the $CaCO_3$ have on the reaction rate?

### Conclusion

7. State the effect of each factor on the reaction rate. Compare your results with your prediction.

# Tools & Techniques

## Methods for Measuring Reaction Rates

How do chemists collect the data they need to determine a reaction rate? To determine empirically the rate of a chemical reaction, chemists must monitor the concentration or amount of at least one reactant or product. There are a variety of techniques available. The choice of technique depends on the reaction under study and the equipment available.

### Monitoring Volume, Mass, pH, and Conductivity

Consider the reaction of magnesium with hydrochloric acid.

$$Mg_{(s)} + 2HCl_{(aq)} \rightarrow MgCl_{2(aq)} + H_{2(g)}$$

Hydrogen gas is released in the reaction. You can monitor and collect the volume of gas produced. Alternatively, you can track the decrease in mass, due to the escaping hydrogen, by carrying out the reaction in an open vessel on an electric balance. The decrease in mass can be plotted against time. Some electronic balances can be connected to a computer, with the appropriate software, to record mass and time data automatically as the reaction proceeds.

Another technique for monitoring the reaction above involves pH. Since HCl is consumed in the reaction, you can record changes in pH with respect to time.

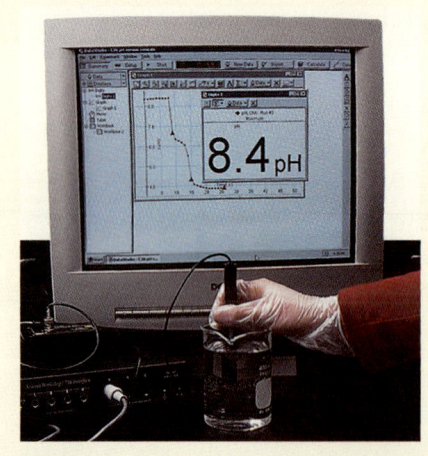

*If the concentration of $H_3O^+$ or $OH^-$ ions changes over the course of a reaction, a chemist can use a pH meter to monitor the reaction.*

A third technique involves electrical conductivity. Dissolved ions in aqueous solution conduct electricity. The electrical conductivity of the solution is proportional to the concentration of ions. Therefore, reactions that occur in aqueous solution, and involve a change in the quantity of dissolved ions, undergo a change in electrical conductivity. In the reaction above, hydrochloric acid is a mix of equal molar amounts of two ions: hydronium, $H_3O^+$, and chloride, $Cl^-$. The $MgCl_2$ that is produced exists as three separate ions in solution: one $Mg^{2+}$ ion and two $Cl^-$ ions. Since there is an increase in the concentration of ions as the reaction proceeds, the conductivity of the solution also increases with time.

### Monitoring Pressure

When a reaction involves gases, the pressure of the system often changes as the reaction progresses. Chemists can monitor this pressure change. For example, consider the decomposition of dinitrogen pentoxide, shown in the following chemical reaction.

$$2N_2O_{5(g)} \rightarrow 4NO_{2(g)} + O_{2(g)}$$

When 2 mol of $N_2O_5$ gas decompose, they form 5 mol of gaseous products. Therefore, the pressure of the system increases as the reaction proceeds, provided that the reaction is carried out in a closed container. Chemists use a pressure sensor to monitor pressure changes.

### Monitoring Colour

Colour change can also be used to monitor the progress of a reaction. The absorption of light by a chemical compound is directly related to the concentration of the compound. For example, suppose you add several drops of blue food colouring to a litre of water. If you add a few millilitres of bleach to the solution, the intensity of the colour of the food dye diminishes as it reacts. You can then monitor the colour change. (Do not try this experiment without your teacher's supervision.) For accurate measurements of the colour intensity of a solution, chemists use a device called a *spectrophotometer*.

*This photograph shows a simple spectrophotometer, which measures the amount of visible light that is absorbed by a coloured solution. More sophisticated devices can measure the absorption of ultraviolet and infrared radiation.*

## Factors That Affect Reaction Rate

Chemists have made the following observations about factors that affect reaction rate. You are already familiar with first three of these observations from Investigation 12-A. You will examine the fourth, the use of a catalyst, in section 12.2.

> **Summary of Some Factors That Affect Reaction Rate**
>
> 1. The rate of a reaction can be increased by increasing the temperature.
> 2. Increasing the concentrations of the reactants usually increases the rate of the reaction.
> 3. Increasing the available surface area of a reactant increases the rate of a reaction.
> 4. A *catalyst* is a substance that increases the rate of a reaction. The catalyst is regenerated at the end of the reaction and can be re-used.
> 5. The rate of a chemical reaction depends on what the reactants are. In other words, the reactivity of the reactants has a major impact on reactant rate.

Chemists and engineers use their understanding of these factors to manipulate the rate of a particular reaction to suit their needs. For example, consider the synthesis of ammonia, $NH_3$, from nitrogen and hydrogen.

$$N_{2(g)} + 3H_{2(g)} \rightarrow 2NH_{3(g)}$$

This reaction must be carried out with high concentrations of reactants, at a temperature of 400°C to 500°C, in the presence of a catalyst. Otherwise, the rate of production of ammonia is too slow to be economically feasible. (You will look more closely at this industrially important chemical reaction in Chapter 13.)

**CONCEPT CHECK**

Elemental sodium reacts immediately with oxygen and moisture in air to form a thin, dull-coloured coating. (Pure sodium is, otherwise, a shiny metal.) Elemental iron undergoes a similar reaction in air, but much more slowly. Explain how the nature of the reactants in these two reactions accounts for this difference.

## Section Summary

In this section, you learned the meaning of the rate of a reaction, and investigated factors that affect reaction rate. In the next section, you will explore two related theories that chemists developed to explain the effects of these factors.

## Section Review

**1** Explain why the rate of a chemical reaction is fastest at the beginning of the reaction.

**2** In your own words, describe the effects of the following factors on the rate of a chemical reaction, and use an example to illustrate your description.

(a) concentration

(b) temperature

(c) surface area

(d) nature of the reactants

**3** For the general reaction A$_{(g)}$ → B$_{(g)}$, sketch a graph that shows the consumption of reactant over time. On the same graph, show the formation of product over time.

**4** The table below gives the mass of magnesium left over after each second of reaction time, over a period of 5 s, for the reaction between solid magnesium and aqueous hydrochloric acid. Explain why the data in the table does or does not make sense. If it does not make sense to you, propose a way to make the data more sensible.

| Time (s) | Mass of magnesium (g) |
|---|---|
| 0 | 1.00 |
| 1 | 0.50 |
| 2 | 0.75 |
| 3 | 0.15 |
| 4 | 0.05 |

**5** In terms of a chemical reaction, explain what is happening in Figure 12.2.

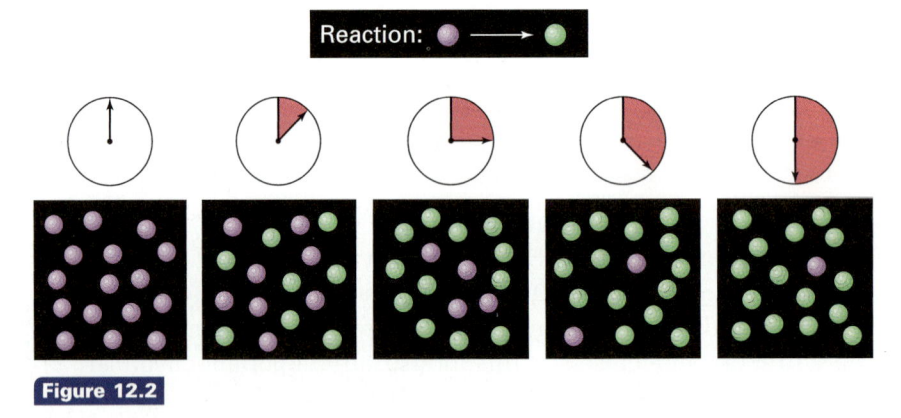

Figure 12.2

**6** In this section, you read about several techniques that chemists and chemical engineers may use to monitor the progress of a reaction. State the conditions that must change as a reaction proceeds, in order to allow each technique to be used.

(a) using a pH meter

(b) using a spectrophotometer

(c) using a conductivity meter

(d) monitoring pressure

# Theories of Reaction Rates

## 12.2

Why do factors such as temperature and concentration affect the rate of a reaction? To answer this question, chemists must first answer another question: What causes a reaction to occur? One way to answer this question involves a theory that you may recall from earlier studies—the **kinetic molecular theory**. According to this theory:

- All matter is made up of microscopic-sized particles (atoms, molecules, ions).
- These particles are in constant motion, because they possess kinetic (movement) energy.
- There are spaces between the particles of matter. The speed and spacing determine the physical state of matter. (See Figure 12.3.)
- Adding heat increases the speed of the moving particles, thus increasing their kinetic energy, as well as the space they occupy.

The kinetic molecular theory (KMT) explains many observations and events in chemistry. For example, gas particles move randomly in all directions, following a straight line path. Picture inflating a basketball. As you add more and more air to it, more air particles randomly collide with the inside wall of the basketball. Each collision exerts a force on the ball's inner surface area, as shown in Figure 12.4. The pressure inside the ball is the result of the collective number of collisions, as well as the strength of the force the particles exert.

### Section Preview/Outcomes

In this section, you will

- **explain**, using the kinetic molecular theory and collision theory, the factors that affect reaction rates
- **demonstrate** an understanding that most reactions occur as a series of elementary reactions in a reaction mechanism
- **draw** and **analyze** simple potential energy diagrams of chemical reactions
- **communicate** your understanding of the following terms: *kinetic molecular theory, collision theory, activation energy ($E_a$), transition state theory, potential energy diagram, transition state, activated complex, reaction mechanism, elementary reaction, reaction intermediates, molecularity, rate-determining steps, catalyst, homogeneous catalyst, heterogeneous catalyst*

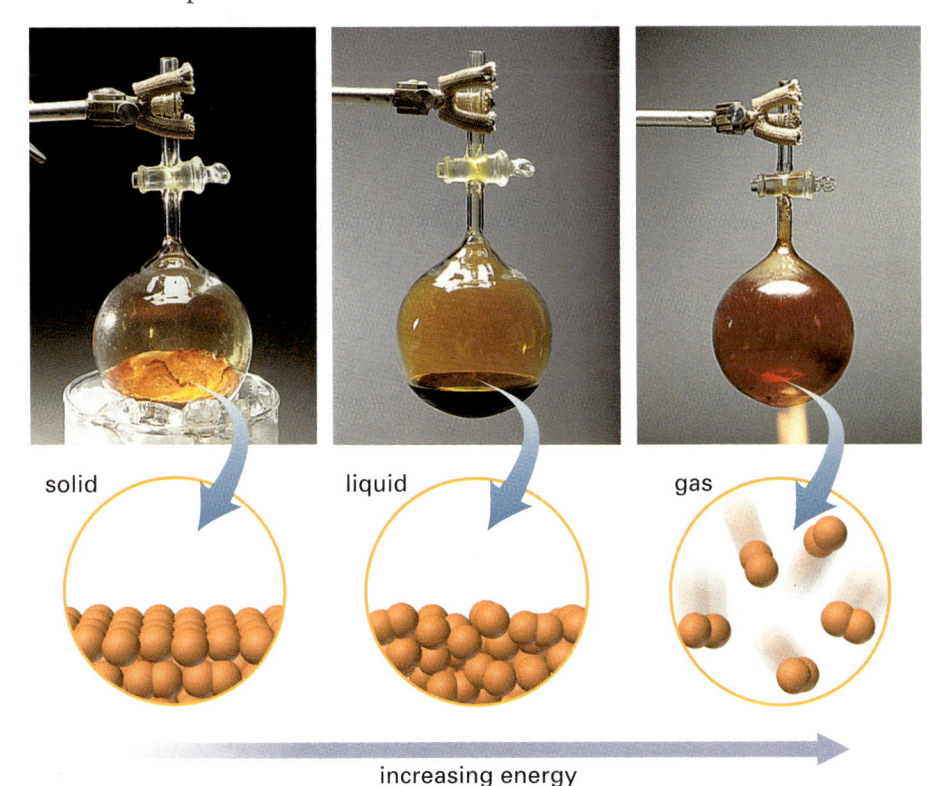

solid        liquid        gas

→ increasing energy

**Figure 12.3** Particles in a solid are spaced very closely together, with room only to vibrate in place. Particles in a liquid are spaced slightly farther apart than those of a solid, and can slide and flow over and around one another. The high kinetic energy of particles in a gas is greater than their attraction for one another, so they spread far apart. When gas particles collide with one another or with other forms of matter (such as a container), there is no loss of their kinetic energy.

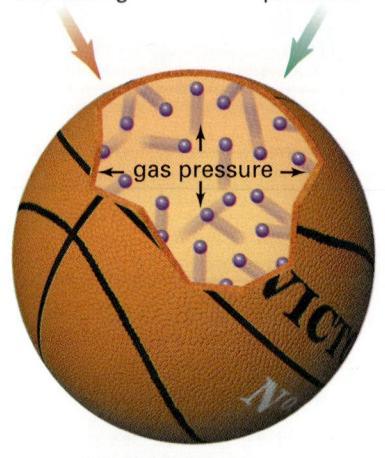

**Figure 12.4** Pressure is the force exerted on an object per unit of surface area. As they collide with the inner surface of a basketball, gas particles exert pressure.

**CHEM FACT**

At room temperature, oxygen molecules in the atmosphere travel at an average rate of 443 m/s. This is approximately 1600 km/h!

**CONCEPT CHECK**

In general, a chunk of solid matter will react more slowly than an equal mass of the solid in powdered form. Is this always the case? Magnesium burns in air with a very bright, white flame. Which do you think reacts faster: a strip of magnesium ribbon or an equal mass of powdered magnesium? Explain your answer.

The KMT also explains why you can smell a bottle of cologne when it is opened a distance away from you. Inside the bottle, the particles that make up the cologne are concentrated in a small space, moving in every direction. When the bottle is opened, the fast-moving particles are no longer contained within a small space. They spread out into the much larger volume of the air, and their motion eventually extends far enough to reach your nose. This spreading-out process, called diffusion, occurs in liquids as well.

## Collision Theory

Chemists have proposed an extension of the KMT to explain why chemical reactions occur. Chemists hypothesize that in order for a reaction to occur, reacting particles (atoms, molecules, or ions) must collide successfully with one another. This idea is called **collision theory**. You can use collision theory to begin to understand the factors that affect reaction rates.

### Collision Theory and Concentration

If a collision is necessary for a reaction to occur, then it makes sense that the rate of the reaction will increase if there are more collisions per unit time. More reactant particles in a given volume (that is, greater concentration) increases the number of collisions between the particles per second. (See Figure 12.5.) As a result, the reaction rate increases. Conversely, decreasing the concentration of reactants decreases the reaction rate. If the particles are those of a gaseous reactant, *increasing the pressure has the same effect as increasing the concentration*, because the same mass of particles is confined to a smaller volume. This effectively increases the concentration of the particles and, thus, the reaction rate. Similarly, decreasing the pressure decreases the reaction rate.

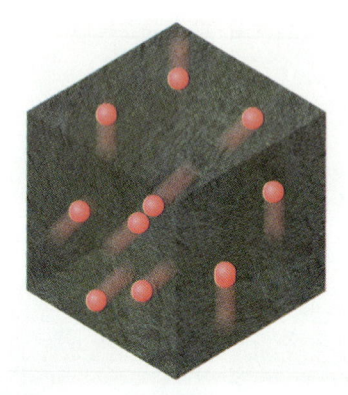

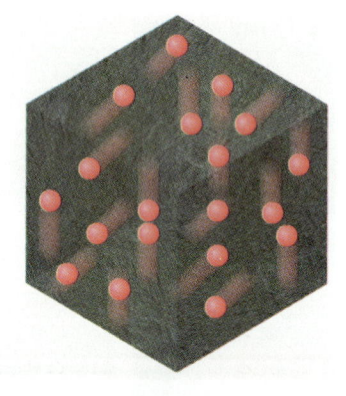

**Figure 12.5** At increased reactant concentrations, there is an increased number of collisions per second.

### Collision Theory and Surface Area

When a solid undergoes a chemical reaction, collisions can occur only at the solid's surface. You know from experience that increasing the surface area of a solid speeds up a reaction. For example, if you want to make a campfire, you will likely start with paper and small twigs, rather than with logs. The paper and twigs provide a greater surface area, which enables more collisions to occur.

## Collision Theory and the Nature of Reactants

The physical and chemical properties of reactants affect the degree to which they take part in chemical reactions. Their state, molecular structure, the type of bonding, and the strength and number of chemical bonds or electrostatic attractions affect the collisions that can take place between their particles. Many factors influence the rate at which any given substance reacts. However, some generalizations are possible.

- Reactions that involve ionic compounds and simple ions are generally faster than reactions involving molecular compounds. For example, compare the following two reactions. The first reaction involves hydrated ions that move freely in solution. There are no bonds to be broken, so ions approach and collide with one another easily to form new attractions. In the second reaction, covalent bonds must be broken in order for new bonds to form. This requires a greater number of effective collisions among particles. (You will learn more about this shortly.) As a result, the rate of the reaction is slower.

  $5Fe^{2+}_{(aq)} + MnO_4^-_{(aq)} + 8H^+_{(aq)} \rightarrow 5Fe^{3+}_{(aq)} + Mn^{2+}_{(aq)} + 4H_2O_{(\ell)}$    very fast at 25°C

  $2CO_{(g)} + O_{2(g)} \rightarrow 2CO_{2(g)}$    very slow at 25°C

- Reactions that involve breaking weaker bonds are generally faster than reactions that involve breaking stronger bonds. Fewer effective collisions are needed to break weaker bonds. For example, both ethane, $C_2H_6$, and ethene, $C_2H_4$, are highly combustible. However, the double carbon bond in ethane is stronger than the single carbon bond in ethene. As a result, the rate at which ethane reacts is faster than the rate at which ethene reacts, as you can see below.

  $2C_2H_{6(g)} + 7O_{2(g)} \rightarrow 4CO_{2(g)} + 6H_2$    very very fast at 25°C

  $C_2H_{4(g)} + 4O_{2(g)} \rightarrow 2CO_{2(g)} + 4H_2O_{(g)}$    very fast at 25°C

- Reactions that involve breaking fewer bonds are generally faster than reactions that involve breaking a greater number of bonds. Again, the number of effective collisions affects the difference in rates for reactions such as this:

  $2C_5H_{12(g)} + 11O_{2(g)} \rightarrow 5CO_{2(g)} + 12H_2O_{(g)}$    fast at 25°C

  $2C_{10}H_{22(g)} + 21O_2 \rightarrow 20CO_{2(g)} + 22H_2O_{(g)}$    slow at 25°C

## Collision Theory and Temperature

In general, increasing the temperature of a reaction results in an increase in the reaction rate. This seems logical—at higher temperatures, particles move more quickly, so they have greater kinetic energy and collide more frequently. However, the frequency of the collisions is not the only factor that affects rate. Otherwise, all chemical reactions involving gases, with their already-energetic particles, would occur in an instant.

Increasing the temperature also increases the *intensity* of the particle collisions. In order for a collision to result in a chemical reaction, the particles must collide with enough energy to break bonds in the reactants and start to form bonds in the products. In most reactions, only a small fraction of collisions has sufficient energy to activate this process. Thus, the **activation energy**, $E_a$, of a reaction is the minimum collision energy that is required for a successful reaction.

---

**Web LINK**

www.mcgrawhill.ca/links/atlchemistry

In 1992, an explosion in a coal mine in Nova Scotia killed 26 miners, and sparked a wave of criminal, judicial, and social responses. Do research on the Westray Mine explosion to find its causes and the effects it had on the miners' families, the people of Nova Scotia, and the people of Canada.

---

**CONCEPT CHECK**

Complete the following chemical equations, and predict which of these reactions has a faster rate. Use collision theory to explain your reasoning.

$AgNO_{3(s)} + KBr_{(s)} \rightarrow ?$

$AgNO_{3(aq)} + KBr_{(aq)} \rightarrow ?$

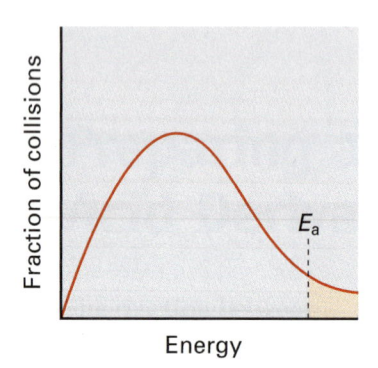

**Figure 12.6** The area under the curve represents the distribution of the kinetic energy of collisions at a constant temperature. At a given temperature, only a certain fraction of the particles in a substance has enough kinetic energy to activate a reaction.

## Reactions and Activation Energy

Activation energy (minimum collision energy) depends on the kinetic energy of the colliding particles. Recall that temperature is a measure of the *average* kinetic energy of the particles in a substance. In other words, at any given temperature, the particles in a substance have a range of kinetic energy. This means that their collisions also have a range of energy. If you plot the number of collisions in a substance at a given temperature against the kinetic energy of each collision, you get a curve like the one in Figure 12.6. The dotted line indicates the activation energy. The larger, unshaded area under the curve represents the fraction of collisions that *do not* have sufficient energy to react. The smaller shaded area indicates the collisions with energy that is equal to or greater than the activation energy. Only this small fraction of particles has sufficient energy to react.

How does the distribution of kinetic energy change as the temperature of a substance increases? Figure 12.7 shows the distribution of kinetic energy in a sample of reacting gases at two different temperatures, $T_1$ and $T_2$, where $T_2 > T_1$. Again, a dashed line indicates the activation energy. Notice that at both temperatures, only a small fraction of collisions has sufficient energy (activation energy) to result in a reaction. Notice also that at the higher temperature, the fraction of collisions with sufficient energy to cause a reaction increases significantly. In fact, for many reactions, the rate roughly doubles for every 10°C rise in temperature.

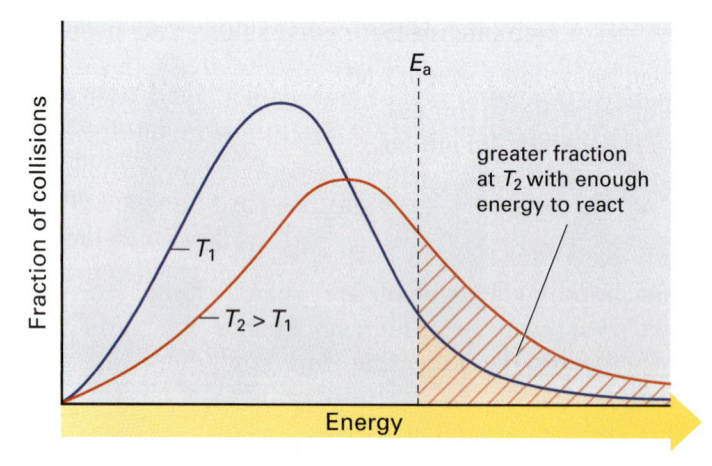

**Figure 12.7** At increased temperatures, more particles collide with enough energy to react.

## Reactions and Orientation of Reactants

In addition to activation energy, there is another factor that determines whether a collision will result in a reaction: the molecular structure of the reactants. Reacting particles must collide with the proper orientation in relation to one another. This is also known as having the correct collision geometry. Consider the following reaction, for example:

$$NO_{(g)} + NO_{3(g)} \rightarrow NO_{2(g)} + NO_{2(g)}$$

Figure 12.8 shows five of the many possible ways that NO and $NO_3$ can collide. Chemists have determined that only *one* orientation has the correct collision geometry for a reaction to occur. The probability of a successfully oriented collision for this reaction is about six collisions in every thousand.

**Figure 12.8** Only one of these five possible orientations of NO and $NO_3$ will lead to the formation of a product.

In summary, for a collision between reactants to result in a reaction, the collision must satisfy both of these two criteria:

- correct orientation of reactant particles
- sufficient collision energy

When both of these criteria are met, a collision is said to be *effective*. Only effective collisions result in the formation of products.

## Transition State Theory

Collision theory outlines what is necessary for particles to collide effectively. **Transition state theory** is used to explain what happens once the colliding particles react. It examines the transition, or change, from reactants to products. The kinetic energy of the reactants is transferred to potential energy as the reactants collide, due to the law of conservation of energy. This is analogous to a bouncing basketball. The kinetic energy of the ball is converted to potential energy, which is stored in the deformed ball as it hits the floor. The potential energy is converted to kinetic energy as the ball bounces away.

You can represent the increase in potential energy during a chemical reaction using a **potential energy diagram**: a diagram that charts the potential energy of a reaction against the progress of the reaction. Examples of potential energy diagrams are shown in Figures 12.9 and 12.10 on the next page. The *y*-axis represents potential energy. The *x*-axis, labelled "Reaction progress," represents the progress of the reaction through time.

The "hill" in each diagram illustrates the activation energy barrier of the reaction. A slow reaction has a high activation energy barrier. This indicates that relatively few reactants have sufficient kinetic energy for a successful reaction. A fast reaction, by contrast, has a low activation energy barrier.

Figure 12.9 shows a potential energy diagram for an exothermic reaction—a reaction that releases energy into its surroundings. The reactants at the start of the reaction have a greater total energy than the products. The overall difference in potential energy is the amount of energy that is released by the reaction, $\Delta E$. (In Unit 7, you will learn a more precise way to describe and represent changes in energy when matter undergoes chemical, as well as physical, changes.)

Figure 12.10 shows a potential energy diagram for an endothermic reaction—a reaction that absorbs energy from its surroundings. The reactants at the start of the reaction have a lower total energy than the products. The overall difference in potential energy is the amount of energy that is absorbed by the reaction, $\Delta E$.

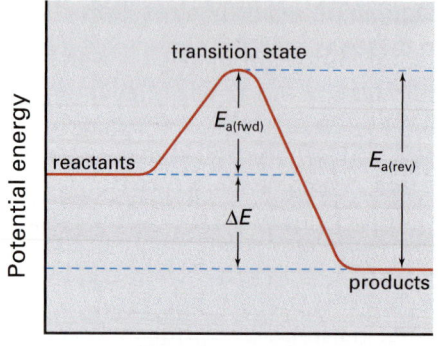

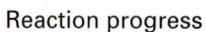

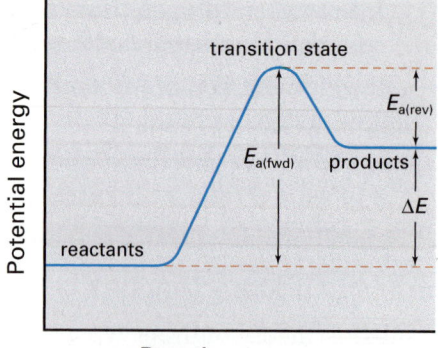

**Figure 12.9** A potential energy diagram for an exothermic reaction. Products are more energetically stable than reactants in exothermic reactions.

**Figure 12.10** A potential energy diagram for an endothermic reaction. Products are less energetically stable than reactants in endothermic reactions.

You may already know that many reactions can proceed in two directions. For example, hydrogen and oxygen react to form water. Water, however, can also undergo electrolysis, forming hydrogen and oxygen. This is the reverse of the first reaction.

$$H_{2(g)} + \frac{1}{2}O_{2(g)} \rightarrow H_2O_{(\ell)} \quad \Delta E = -285.8 \text{ kJ}$$

$$H_2O_{(\ell)} \rightarrow H_{2(g)} + \frac{1}{2}O_{2(\ell)} \quad \Delta E = 285.8 \text{ kJ}$$

The difference in the total energy of the first reaction is the same as the difference in the total energy of the second reaction, with the opposite sign. How are the activation energies of forward and reverse reactions related? For an exothermic reaction, the activation energy of the reverse reaction, $E_{a(rev)}$ equals $E_{a(fwd)} + \Delta E$. For an endothermic reaction, $E_{a(rev)}$ equals $E_{a(fwd)} - \Delta E$. Figures 12.9 and 12.10 show the activation energies of forward and reverse reactions.

The top of the activation energy barrier on a potential energy diagram represents the **transition state**, or change-over point, of the reaction. The chemical species that exists at the transition state is referred to as the **activated complex**. The activated complex is a transitional species that exists for a fraction of a moment and is neither product nor reactant. It has partial bonds and is highly unstable. There is a subtle difference between the transition state and the activated complex. The transition state refers to the top of the "hill" on a potential energy diagram. The *chemical species* that exists at this transition point is called the activated complex.

### Tracing a Reaction With a Potential Energy Diagram

Consider the substitution reaction between a hydroxide ion, $OH^-$, and methyl bromide, $BrCH_3$. Methanol, $CH_3OH$, and a bromide ion, $Br^-$, are formed.

$$BrCH_3 + OH^- \rightarrow CH_3OH + Br^-$$

Figure 12.11 is a potential energy diagram for this reaction. It includes several "snapshots" as the reaction proceeds.

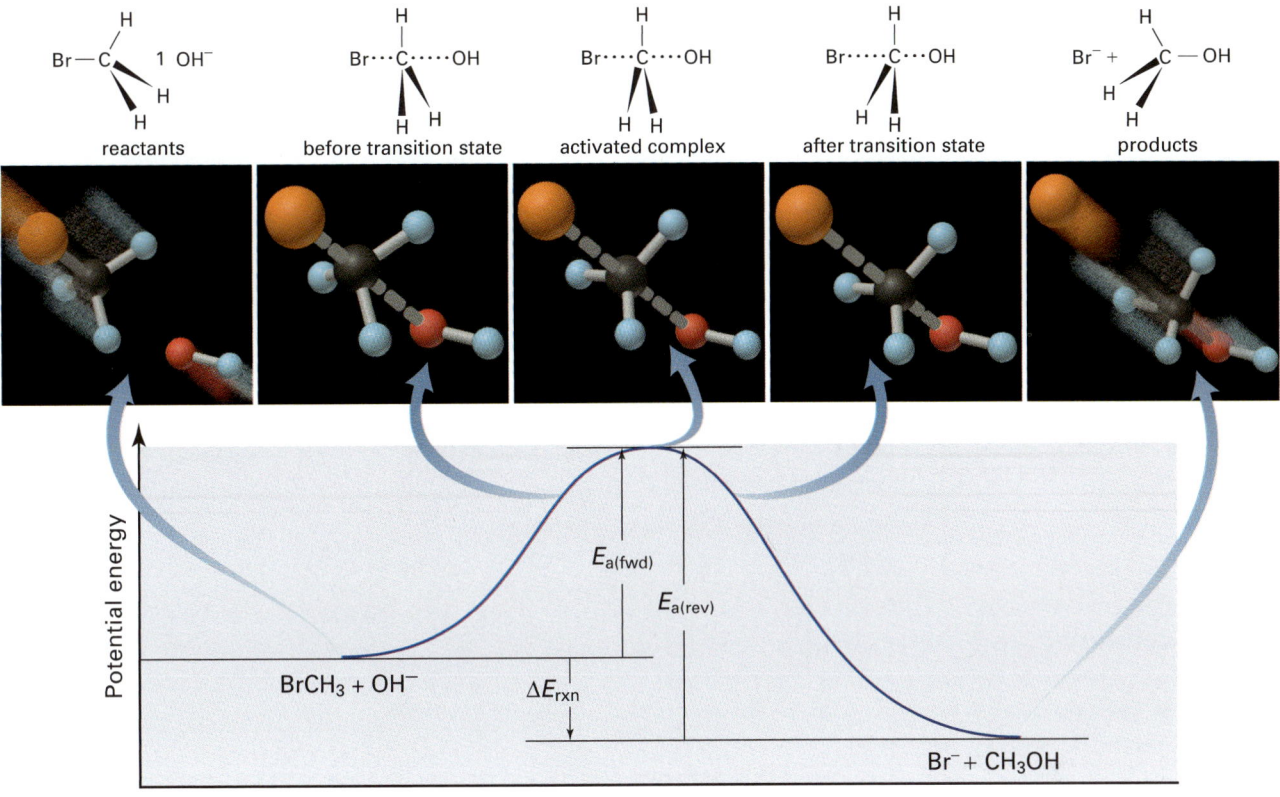

**Figure 12.11** As the reactants collide, chemical bonds break and form.

For a successful reaction to take place, BrCH₃ and OH⁻ must collide in a favourable orientation. The OH⁻ ion must approach BrCH₃ from the side that is *opposite* to the Br atom. When this occurs, a partial bond is formed between the O of the OH⁻ ion and the C atom. Simultaneously, the C—Br bond is weakened.

Because the activated complex contains partial bonds, it is highly unstable. It can either break down to form products or it can decompose to re-form the reactants. The activated complex is like a rock teetering on top of a mountain. It could fall either way.

## Sample Problem
### Drawing a Potential Energy Diagram

#### Problem
Carbon monoxide, CO, reacts with nitrogen dioxide, $NO_2$. Carbon dioxide, $CO_2$, and nitric oxide, NO, are formed. Draw a potential energy diagram to illustrate the progress of the reaction. (You do not need to draw your diagram to scale). Label the axes, the transition state, and the activated complex. Indicate the activation energy of the forward reaction, $E_{a(fwd)} = 134$ kJ, as well as $\Delta E = -226$ kJ. Calculate the activation energy of the reverse reaction, $E_{a(rev)}$, and show it on the graph.

*Continued ...*

### Solution

Since $E_{a(rev)} = \Delta E + E_{a(fwd)}$
$$E_{a(rev)} = 226 \text{ kJ} + 134 \text{ kJ}$$
$$= 360 \text{ kJ}$$

The activation energy of the reverse reaction is 360 kJ.

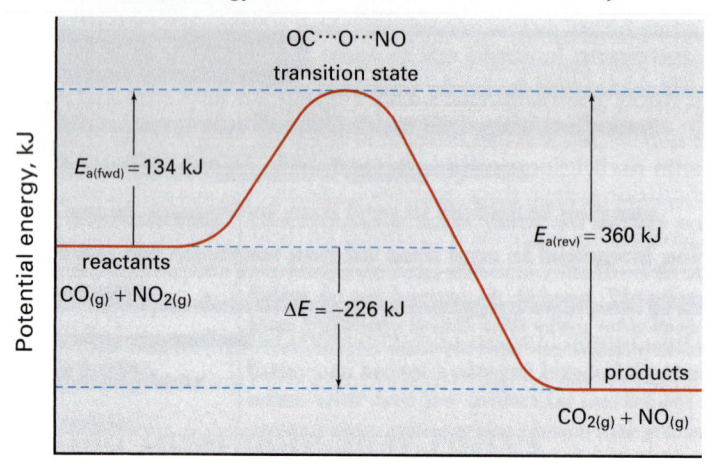

[Potential energy diagram showing reactants CO(g) + NO₂(g) at higher level, transition state OC···O···NO at top, products CO₂(g) + NO(g) at lower level. $E_{a(fwd)} = 134$ kJ, $E_{a(rev)} = 360$ kJ, $\Delta E = -226$ kJ. X-axis: reaction progress; Y-axis: Potential energy, kJ.]

### Check Your Solution

Look carefully at the potential energy diagram. Check that you have labelled it completely. Since the forward reaction is exothermic, the reactants should be at a higher energy level than the products, and they are. The value of $E_{a(rev)}$ is reasonable.

## Practice Problems

1. The following reaction is exothermic.

   $2ClO_{(g)} \rightarrow Cl_{2(g)} + O_{2(g)}$

   Draw and label a potential energy diagram for the reaction. Propose a reasonable activated complex.

2. Consider the following reaction.

   $AB + C \rightarrow AC + B$  $\Delta E = +65$ kJ, $E_{a(rev)} = 34$ kJ

   Draw and label a potential energy diagram for this reaction. Calculate and label $E_{a(fwd)}$. Include a possible structure for the activated complex.

3. Consider the reaction below.

   $C + D \rightarrow CD$  $\Delta E = -132$ kJ, $E_{a(fwd)} = 61$ kJ

   Draw and label a potential energy diagram for this reaction. Calculate and label $E_{a(rev)}$. Include a possible structure for the activated complex.

4. In the upper atmosphere, oxygen exists in forms other than $O_{2(g)}$. For example, it exists as ozone, $O_{3(g)}$, and as single oxygen atoms, $O_{(g)}$. Ozone and atomic oxygen react to form two molecules of oxygen. For this reaction, the energy change is −392 kJ and the activation energy is 19 kJ. Draw and label a potential energy diagram. Include a value for $E_{a(rev)}$. Propose a structure for the activated complex.

476 MHR • Unit 5 Kinetics and Equilibrium

## Reaction Mechanisms and Catalysts

Chemical reactions are like factories, where goods (products) are created from raw materials (reactants). An assembly line, like the one shown in Figure 12.12, involves many steps. An automobile is not formed from its components in just one step. Similarly, most chemical reactions do not proceed immediately from products to reactants. They take place via a number of steps. While you can go inside a factory to see all the steps that are involved in making an automobile or a piece of clothing, you cannot observe a chemical equation on a molecular scale as it proceeds. Chemists can experimentally determine the reactants and products of a reaction, but they must use indirect evidence to suggest the steps in-between.

## Elementary Reactions

A **reaction mechanism** is a series of steps that make up an overall reaction. Each step, called an **elementary reaction**, involves a single molecular event, such as a simple collision between atoms, molecules, or ions. An elementary step can involve the formation of different molecules or ions, or it may involve a change in the energy or geometry of the starting molecules. It cannot be broken down into further, simpler steps.

For example, consider the following reaction.

$$2NO_{(g)} + O_{2(g)} \rightarrow 2NO_{2(g)}$$

Chemists have proposed the following two-step mechanism for this reaction. Each step is an elementary reaction.

**Step 1** $NO_{(g)} + O_{2(g)} \rightarrow NO_{3(g)}$

**Step 2** $NO_{3(g)} + NO_{(g)} \rightarrow 2NO_{2(g)}$

When you add the two elementary reactions, you get the overall reaction. Notice that $NO_3$ is present in both elementary reactions, but it is not present in the overall reaction. It is produced in the first step and consumed in the second step.

Molecules (or atoms or ions) that are formed in an elementary reaction and consumed in a subsequent elementary reaction are called **reaction intermediates**. Even though they are not products or reactants in the overall reaction, reaction intermediates are essential for the reaction to take place. They are also useful for chemists who are trying to find evidence to support a proposed reaction mechanism.

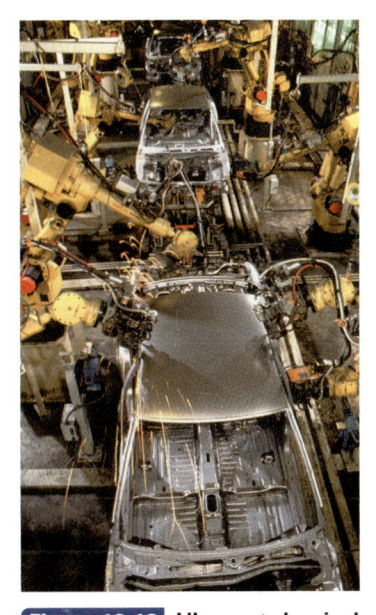

**Figure 12.12** Like most chemical reactions, building a car on an assembly line takes more than one step.

### Sample Problem
### Determining Reaction Intermediates

**Problem**

Chemists have proposed the following two-step mechanism for a certain reaction.

**Step 1** $NO_{2(g)} + NO_{2(g)} \rightarrow NO_{3(g)} + NO_{(g)}$

**Step 2** $NO_{3(g)} + CO_{(g)} \rightarrow NO_{2(g)} + CO_{2(g)}$

Write the overall balanced equation for this reaction and identify the reaction intermediate.

*Continued...*

### Solution

You know that each step is an elementary reaction. Add the two steps together, and cancel out chemical species that occur on both sides, to get the overall reaction. The reaction intermediate is the molecule that is produced in the first step and used in the second, so it does not appear in the overall balanced equation.

**Step 1** $NO_{2(g)} + NO_{2(g)} \rightarrow NO_{3(g)} + NO_{(g)}$

**Step 2** $NO_{3(g)} + CO_{(g)} \rightarrow NO_{2(g)} + CO_{2(g)}$

$\overline{NO_{2(g)} + \cancel{NO_{2(g)}} + \cancel{NO_{3(g)}} + CO_{(g)} \rightarrow \cancel{NO_{3(g)}} + NO_{(g)} + \cancel{NO_{2(g)}} + CO_{2(g)}}$

Therefore, the balanced equation is

$NO_{2(g)} + CO_{(g)} \rightarrow NO_{(g)} + CO_{2(g)}$

The reaction intermediate is $NO_{3(g)}$.

## Practice Problems

5. A chemist proposes the following mechanism. Write the overall balanced equation. Identify any reaction intermediate.

   **Step 1** $NO_{2(g)} + F_{2(g)} \rightarrow NO_2F_{(g)} + F_{(g)}$

   **Step 2** $NO_{2(g)} + F_{(g)} \rightarrow NO_2F_{(g)}$

6. A chemist proposes the following mechanism for a certain reaction. Write the equation for the chemical reaction that this mechanism describes, and identify any reaction intermediate.

   **Step 1** $A + B \rightarrow C$

   **Step 2** $C + A \rightarrow E + F$

7. A chemist proposes the following mechanism for a certain reaction. Write the equation for the chemical reaction that this mechanism describes, and identify any reaction intermediate.

   **Step 1** $X + Y \rightarrow XY$

   **Step 2** $XY + 2Z \rightarrow XYZ_2$

   **Step 3** $XYZ_2 \rightarrow XZ + YZ$

8. A chemist proposes the following mechanism for the reaction between 2-bromo-2-methylpropane, $(CH_3)_3CBr_{(aq)}$, and water, $H_2O_{(\ell)}$. Write the overall balanced equation. Identify any reaction intermediate.

   **Step 1** $(CH_3)_3CBr_{(aq)} \rightarrow (CH_3)_3C^+_{(aq)} + Br^-_{(aq)}$

   **Step 2** $(CH_3)_3C^+_{(aq)} + H_2O_{(\ell)} \rightarrow (CH_3)_3COH_2^+_{(aq)}$

   **Step 3** $(CH_3)_3COH_2^+_{(aq)} \rightarrow H^+_{(aq)} + (CH_3)_3COH_{(aq)}$

### The Molecularity of Elementary Reactions

The term **molecularity** refers to the number of reactant particles (molecules, atoms, or ions) that are involved in an elementary reaction.

In the reaction mechanism you just saw, each elementary reaction consisted of two molecules colliding and reacting. When two particles collide and react, the elementary reaction is said to be *bimolecular*.

A *unimolecular* elementary reaction occurs when one molecule or ion reacts. For example, when one molecule of chlorine absorbs ultraviolet light, the Cl—Cl bond breaks. The product is two chlorine atoms.

$$Cl_{2(g)} \xrightarrow{\text{UV light}} 2Cl_{(g)}$$

In rare cases, an elementary reaction may also involve three particles colliding in a *termolecular* reaction. Termolecular elementary steps are rare, because it is extremely unlikely that three particles will collide all at once. Think of it this way. You have probably bumped into someone accidentally, many times, on the street or in a crowded hallway. How many times, however, have you and two other people collided at exactly the same time? Figure 12.13 models unimolecular, bimolecular, and termolecular reactions.

## Proposing and Evaluating Mechanisms

You can take apart a wind-up clock to take a look at its mechanism. You can physically examine and take apart its gears and springs to determine how it works.

When chemists investigate the mechanism of a reaction, they are not so lucky. Determining the mechanism of a chemical reaction is a bit like figuring out how a clock works just by looking at its face and hands. For this reason, reaction mechanisms are *proposed* rather than definitively stated. Much of the experimental evidence that is obtained to support a mechanism is indirect. Researchers need a lot of creativity as they propose and test mechanisms.

One of the ways that researchers provide evidence for proposed mechanisms is by proving the existence of a reaction intermediate. Although a reaction intermediate usually appears for a very short time and cannot be isolated, there are other ways to show its presence. For example, if a reaction intermediate is coloured while other reactants are colourless, a spectrophotometer will show when the reaction intermediate is formed. You may even be able to see a fleeting colour change without a spectrophotometer.

When chemists propose a mechanism, the equations for the elementary steps must combine to give the equation for the overall reaction. As well, the proposed elementary steps must be reasonable. (In later chemistry courses, you will learn that the mechanism must also support an experimentally determined expression, the rate law, which relates the rate of the reaction to the concentrations of the reactants.)

### The Rate-Determining Step

Elementary reactions in mechanisms all have different rates. Usually one elementary reaction, called the **rate-determining step**, is much slower. Hence, it determines the overall rate.

To understand the rate-determining step, consider a two-step process. Suppose that you and a friend are making buttered toast for a large group of people. The first step is toasting the bread. The second step is buttering the toast. Suppose that toasting two slices of bread takes about two minutes, but buttering the toast takes only a few seconds.

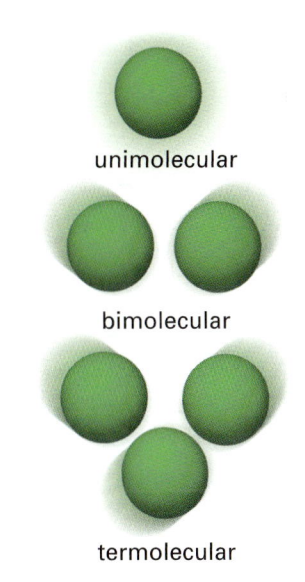

**Figure 12.13** This figure shows the molecularity of elementary reactions. Termolecular reactions are rare.

The rate at which the toast is buttered does not have any effect on the overall rate of making buttered toast, because it is much faster than the rate at which the bread is toasted. The overall rate of making buttered toast, therefore, depends only on the rate of toasting the bread. In other words, two pieces of buttered toast will be ready every two minutes. This is the same as the rate for the first step in the "buttered toast mechanism." Thus, toasting the bread is the rate-determining step in the mechanism.

**Step 1** Bread → Toast (slow, rate-determining)

**Step 2** Toast + Butter → Buttered toast (fast)

Now suppose that the butter is frozen nearly solid. It takes you about five minutes to scrape off enough butter for one piece of toast. Pieces of toast pile up, waiting to be buttered. In this case, the rate of making buttered toast depends on the rate of spreading the butter. The rate of toasting is fast relative to the rate of buttering.

**Step 1** Bread → Toast (fast)

**Step 2** Toast + Frozen butter → Buttered toast (slow, rate-determining)

Most chemical reactions involve two or more elementary steps. One of those steps will determine the rate at which the reaction proceeds. For example, consider the reaction below.

$$2NO_{(g)} + 2H_{2(g)} \rightarrow N_{2(g)} + 2H_2O_{(g)}$$

A chemist proposes the following two-step mechanism for this reaction.

**Step 1** $2NO_{(g)} + H_{2(g)} \rightarrow N_2O_{(g)} + H_2O_{(g)}$ (slow, rate-determining)

**Step 2** $N_2O_{(g)} + H_{2(g)} \rightarrow N_{2(g)} + H_2O_{(g)}$ (fast)

Because the first step is slow, it is the rate-determining step. The rate of the overall reaction is determined by the rate at which the reaction intermediate, $N_2O_{(g)}$ is formed during this step. A potential energy diagram for this reaction mechanism must take this reaction intermediate into account.

As you can see in Figure 12.14, the potential energy diagram for this mechanism presents two activation energy barriers: one for each proposed elementary step. As a result, there are two transition states and one reaction intermediate. Because the first step is slower than the second, the activation energy barrier for the first step must be greater than that of the second.

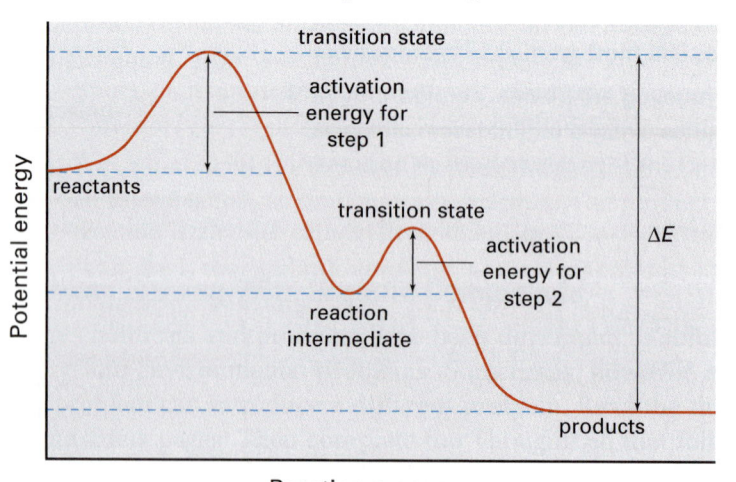

**Figure 12.14** A potential energy diagram for a two-step mechanism

# Catalysts

A **catalyst** is a substance that increases the rate of a chemical reaction without being consumed by the reaction. Catalysts are of tremendous importance in all facets of chemistry, from the laboratory to industry. Many industrial reactions, for example, would not be economically viable without catalysts. Well over three million tonnes of catalysts are produced annually in North America.

## How a Catalyst Works

A catalyst works by lowering the activation energy of a reaction so that a larger fraction of the reactants have sufficient energy to react. It lowers the activation energy by providing an alternative mechanism for the reaction. The potential energy diagram in Figure 12.15 shows the activation energy for an uncatalyzed reaction and the activation energy for the same reaction with the addition of a catalyst.

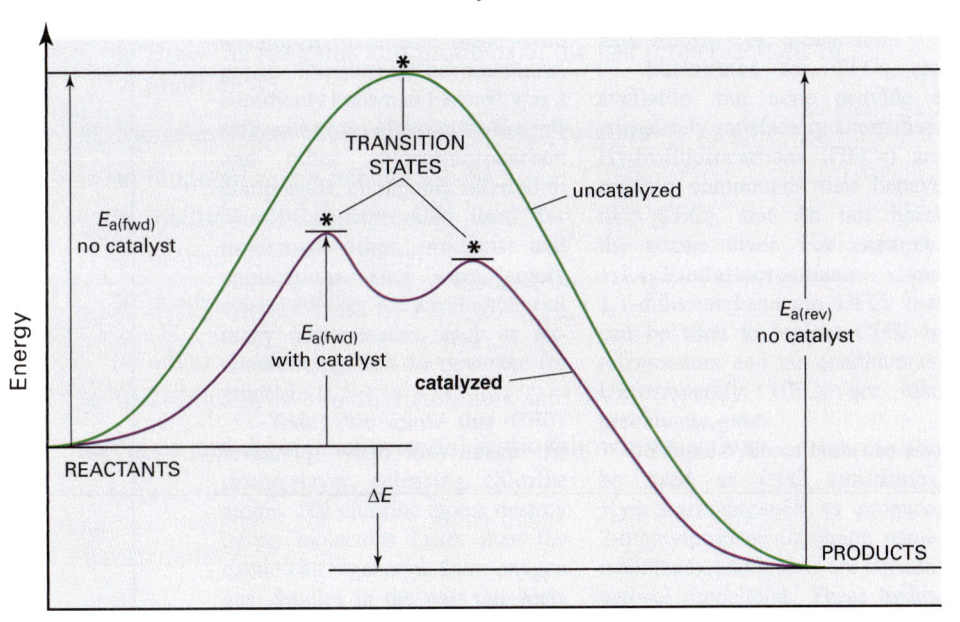

**Figure 12.15** A catalyst lowers the activation energy of a reaction by providing an alternative mechanism. A catalyst also increases the rate of the reverse reaction. What effect does a catalyst have on $\Delta E$ of a reaction?

In Figure 12.15, the catalyzed reaction consists of a two-step mechanism. The uncatalyzed reaction consists of a one-step mechanism. To see how a catalyst works, in general, consider a simple, one-step, bimolecular reaction:

$$A + B \rightarrow AB$$

A catalyst can increase the rate of this reaction by providing an alternative mechanism with lower activation energy. A possible mechanism for the catalyzed reaction is shown below.

| | |
|---|---|
| **Step 1** | A + catalyst → A—catalyst |
| **Step 2** | A—catalyst + B → AB + catalyst |
| **Overall reaction** | A + B → AB |

Both steps are faster than the original, uncatalyzed reaction. Therefore, although the overall reaction of the catalyzed mechanism has the same reactants and products as the uncatalyzed reaction, the catalyzed mechanism is faster. The chemical species A—catalyst is a reaction intermediate. It is produced in step 1 but consumed in step 2. By contrast, the catalyst is *regenerated* in the reaction. It appears as a reactant in step 1 and as a product in step 2. Although the catalyst changes *during* the overall reaction, it is regenerated unchanged at the *end* of the overall reaction.

Figure 12.16 shows the reaction of sodium potassium tartrate with $H_2O_2$, catalyzed by cobaltous chloride, $CoCl_2$. The $CoCl_2$ is pink in solution. Notice that the contents of the beaker briefly change colour to dark green, suggesting the formation of a reaction intermediate. Also notice that you can see the regeneration of the catalyst when the reaction is over.

## Homogeneous Catalysts and Heterogeneous Catalysts

Catalysts are divided into two categories, depending on whether or not they are in the same phase as the reactants. A **homogeneous catalyst** exists in the same phase as the reactants. Homogeneous catalysts most often catalyze gaseous and aqueous reactions. For example, aqueous zinc chloride, $ZnCl_2$, is used to catalyze the following reaction.

$$(CH_3)_2CHOH_{(aq)} + HCl_{(aq)} \xrightarrow{ZnCl_{2(aq)}} (CH_3)_2CHCl_{(aq)} + H_2O_{(\ell)}$$

The reaction takes place in aqueous solution, and the catalyst is soluble in water. Therefore, $ZnCl_2$ is a homogeneous catalyst when it is used with this reaction.

A **heterogeneous catalyst** exists in a phase that is different from the phase of the reaction it catalyzes. An important industrial use of heterogeneous catalysts is the addition of hydrogen to an organic compound that contains C=C double bonds. This process is called *hydrogenation*.

Consider the hydrogenation of ethylene, shown below.

$$H_2C=CH_{2(g)} + H_{2(g)} \xrightarrow{Pd_{(s)}} H_3C-CH_{3(g)}$$

Without a catalyst, the reaction is very slow. When the reaction is catalyzed by a metal such as palladium or platinum, however, the rate increases dramatically. The ethylene and hydrogen molecules form bonds with the metal surface. This weakens the bonds of the hydrogen and ethylene. The H—H bonds of the hydrogen molecules break, and the hydrogen atoms are somewhat stabilized because of their attraction to the metal. The hydrogen atoms react with the ethylene, forming ethane. Figure 12.17 shows the hydrogenation of ethylene to ethane.

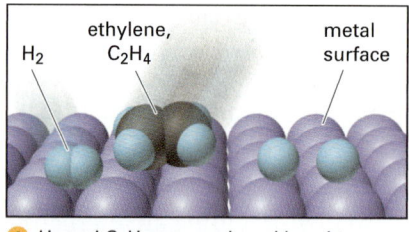

**Figure 12.16** The reaction of Rochelle's salt, sodium potassium tartrate, with hydrogen peroxide is catalyzed with $CoCl_2$ at 70°C.

Photograph A shows the reactants before mixing.

Photograph B was taken immediately after $CoCl_2$ was added. Notice the pink colour.

Photograph C was taken after about 20 s.

Photograph D was taken after about 2 min.

① $H_2$ and $C_2H_4$ approach and bond to metal surface.

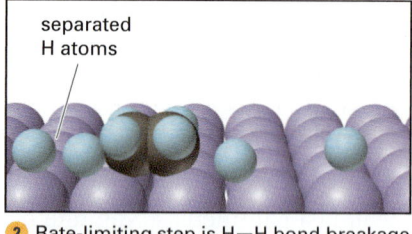

② Rate-limiting step is H—H bond breakage.

③ One H atom bonds to $C_2H_4$.

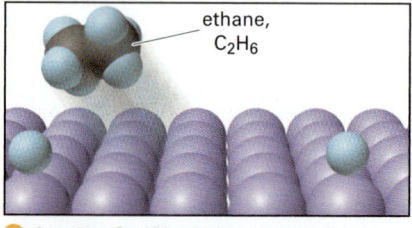

④ Another C—H bond forms, and $C_2H_6$ leaves the surface.

**Figure 12.17** Platinum metal is used as a heterogeneous catalyst for the hydrogenation of gaseous ethylene to ethane.

# Canadians in Chemistry

## Dr. Maud L. Menten

Dr. Maud L. Menten, a leading researcher in the field of biochemistry, was born in Port Lambton, Ontario, in 1879. She grew up in rural British Columbia, where she had to take a canoe across the Fraser River every day to get to school. Later, Menten returned to Ontario to attend the University of Toronto. Upon receiving her medical degree in 1911, she became one of the first female doctors in Canada.

Menten soon received international recognition for her study of enzymes (biological catalysts). From 1912 to 1913, she worked at Leonor Michaelis' lab at the University of Berlin. While conducting experiments on the breakdown of sucrose by the enzyme called invertase, Menten and Michaelis were able to refine the work of Victor Henri to explain how enzymes function. A few years earlier, Henri had proposed that enzymes bind directly to their reactant molecules. Michaelis and Menten obtained the precise measurements that were needed to support Henri's hypothesis. Using the recently developed concept of pH, they were able to buffer their chemical reactions and thereby control the conditions of their experiments more successfully than Henri.

Their findings, published in 1913, provided the first useful model of enzyme function.

In the ground-breaking scientific paper that presented their work, Menten and Michaelis also derived an important mathematical formula. This formula describes the rate at which enzymes break down their reactant molecules. It correlates the speed of the enzyme reaction with the concentrations of the enzyme and the molecule. Called the Michaelis-Menten equation, it remains fundamental to our understanding of how enzymes catalyze reactions.

Menten remained an avid researcher all her life. She worked long days, dividing her time between teaching at the University of Pittsburgh's School of Medicine and working as a pathologist at the Children's Hospital of Pittsburgh. She also worked with her peers and students on more than 70 papers that dealt with a wide range of medical topics. In addition to her work with Michaelis, Menten's contributions to science include pioneering techniques for studying the chemical composition of tissues and for investigating the nature of hemoglobin in human red blood cells.

It was not until 1949, only a year before retiring from the University of Pittsburgh, that Menten obtained a full professorship. Shortly after retiring, Menten joined the Medical Research Institute of British Columbia to do cancer research. Sadly, ill health forced her to cease her scientific work only a few years later. She spent her remaining days in Leamington, Ontario.

Although she was a brilliant medical researcher, Menten's enthusiasm overflowed beyond the boundaries of science. She was a world traveller, a mountain climber, a musician, and a student of languages. She was even a painter, whose work made its way into art exhibitions.

In honour of her lifetime of achievements, Maud L. Menten was inducted into the Canadian Medical Hall of Fame in 1998.

## Section Summary

In this section, you used the kinetic molecular theory, collision theory, and transition state theory to explain how reaction rates are affected by various factors. You also learned that chemical reactions usually proceed as a series of steps called elementary reactions. Finally, you learned how a catalyst controls the rate of a chemical reaction by providing a lower-energy reaction mechanism. In this chapter, you compared activation energies of forward and reverse reactions. In the next chapter, you will study, in detail, reactions that proceed in both directions.

## Section Review

**Math LINK**

www.mcgrawhill.ca/links/atlchemistry

You probably know that compounds called chlorofluorocarbons (CFCs) are responsible for depleting the ozone layer in Earth's stratosphere. Did you know, however, that CFCs do their destructive work by acting as homogeneous catalysts? Use the Internet to find out how CFCs catalyze the decomposition of ozone in the stratosphere. To start your research, go to the web site above and click on **Web Links**. Communicate your findings as a two-page press release.

**1** In your own words, describe how collision theory explains the factors that affect reaction rate.

**2** Suppose that you have a solid sample of an organic substance. You are going to burn the substance in an analytical furnace and determine whether or not any unburned matter remains. The substance is somewhat hard, and you receive it in a solid chunk.
   (a) Suggest two ways to increase the rate of burning so that your analysis will take less time.
   (b) Explain why your suggestions would work, using the theories you learned in this section.

**3** Consider the following reaction.
   $A_2 + B_2 \rightarrow 2AB$    $E_{a(fwd)} = 143$ kJ, $E_{a(rev)} = 75$ kJ
   (a) Is the reaction endothermic or exothermic in the forward direction?
   (b) Draw and label a potential energy diagram. Include a value for $\Delta E$.
   (c) Suggest a possible activated complex.

**4** Consider two exothermic reactions. Reaction (1) has a much smaller activation energy than reaction (2).
   (a) Sketch a potential energy diagram for each reaction, showing how the difference in activation energy affects the shape of the graph.
   (b) How do you think the rates of reactions (1) and (2) compare? Explain your answer.

**5** In your own words, describe an elementary reaction.

**6** Distinguish between an overall reaction and an elementary reaction.

**7** Why do chemists say that a reaction mechanism is *proposed*?

**8** Explain the difference between a reaction intermediate and an activated complex in a reaction mechanism.

**9** Consider the reaction below.
   $2A + B_2 \xrightarrow{C} D + E$

   A chemist proposes the following reaction mechanism.
   **Step 1**  $A + B_2 \rightarrow AB_2$
   **Step 2**  $AB_2 + C \rightarrow AB_2C$
   **Step 3**  $AB_2C + A \rightarrow A_2B_2 + C$
   **Step 4**  $A_2B_2 \rightarrow D + E$

   (a) What is the role of $AB_2C$ and $AB_2$?
   (b) What is the role of C?
   (c) Given a proposed reaction mechanism, how can you differentiate, in general, between a reaction intermediate and a catalyst?

**10** Chlorine gas reacts with aqueous hydrogen sulfide (also known as hydrosulfuric acid) to form elemental sulfur and hydrochloric acid.

$Cl_{2(g)} + H_2S_{(aq)} \rightarrow S_{(s)} + 2HCl_{(aq)}$

A proposed mechanism for this reaction is

**Step 1** $Cl_2 + H_2S \rightarrow Cl^+ + HCl + HS^-$ (slow)
**Step 2** $Cl^+ + HS^- \rightarrow HCl + S$ (fast)

Sketch a potential energy diagram for this reaction, with labels to identify activation energy, transition state, rate-determining step, and any reaction intermediate.

**11** Consider the following general endothermic reaction.

$A + B \rightarrow C + D$

(a) Explain why a catalyst has no effect on the change in total energy from reactants to products. Illustrate your answer with a potential energy diagram.

(b) Explain the effect of a catalyst on a reaction that has similar products and reactants, but is an exothermic reaction. Illustrate your answer with a potential energy diagram.

**12** The decomposition of hydrogen peroxide, $H_2O_2$, may be catalyzed with the iodide ion. Chemists propose this reaction occurs according to the following mechanism.

**Step 1** $H_2O_2 + I^- \rightarrow H_2O + OI^-$ (slow, rate-determining)
**Step 2** $H_2O_2 + OI^- \rightarrow H_2O + O_2 + I^-$ (fast)

(a) Show that these two steps are consistent with the overall stoichiometry of the reaction.

(b) How does this mechanism account for the fact that the iodide ion is a catalyst?

(c) Draw and label a potential energy diagram for this mechanism, including the rate-determining step, the catalyst, and any reaction intermediate.

(d) What effect would decreasing the concentration of hydrogen peroxide have on the overall rate?

# CHAPTER 12 Review

## Reflecting on Chapter 12

Summarize this chapter in the format of your choice. Here are a few ideas to use as guidelines:
- Use a graph to describe the rate of reaction as a function of the change of concentration of a reactant or product with respect to time.
- Use collision theory and transition state theory to explain how concentration, temperature, surface area, and the nature of reactants control the rate of a chemical reaction.
- Use a potential energy diagram to demonstrate your understanding of the relationships between activation energy, reactants, products, energy change, and the activated complex.
- Explain why most reactions occur as a series of elementary reactions.
- Explain how a catalyst increases the rate of a chemical reaction.

## Reviewing Key Terms

For each of the following terms, write a sentence that shows your understanding of its meaning.

reaction rate (rate of reaction)
activation energy ($E_a$)
transition state
activated complex
elementary reaction
molecularity
catalyst
heterogeneous catalyst
kinetic molecular theory
collision theory
transition state theory
potential energy diagram
reaction mechanism
reaction intermediates
rate-determining step
homogeneous catalyst

## Knowledge/Understanding

1. Explain how increasing the pressure in a reaction involving gases can affect the rate of the reaction.
2. You are conducting a reaction in aqueous solution. You add more water to the container in which you are carrying out the reaction. How will this affect the reaction rate?
3. Explain, using collision theory, the effect of raising the temperature on reaction rate.
4. Consider the following reaction.
   $CaCO_{3(s)} + 2HCl_{(aq)} \rightarrow CaCl_{2(aq)} + CO_{2(g)} + H_2O_{(\ell)}$
   Over the course of the reaction, explain what happens to:
   (a) the concentration of hydrogen chloride
   (b) the concentration of calcium chloride
   (c) the volume of carbon dioxide gas
5. In your own words, describe how collision theory explains why increased surface area increases the rate of a reaction.
6. Depending on the nature of the substance, agitation (stirring) often increases the rate of a reaction. Use collision theory to explain why.
7. State two requirements for an effective collision between particles of reactants.
8. For each of the following, state whether the change will increase, decrease, or have no effect on the rate of the reaction.
   (a) Increasing the concentration of a reactant that forms during an elementary step that occurs before a rate-determining step.
   (b) Increasing the concentration of a reactant that forms during an elementary step that occurs after a rate-determining step.
9. How can you use the idea of activation energy to explain why a catalyst increases the rate of a chemical reaction?

## Inquiry

10. Examine the potential energy diagram below.

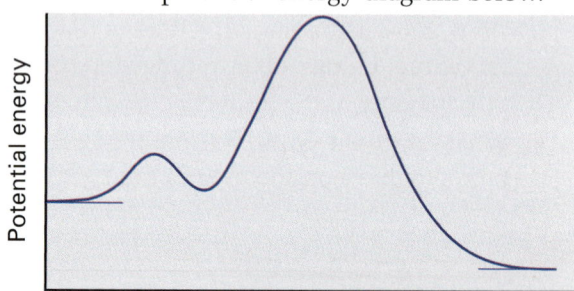

    (a) How many elementary steps does the reaction mechanism have?
    (b) Which step is the rate-determining step?
    (c) Is this reaction exothermic or endothermic? Explain how you know.

11. A chemist proposes the following mechanism for the decomposition of formic acid, $HCO_2H$.
    Step 1  $HCO_2H + H^+ \rightarrow HCO_2H_2^+$   (fast)
    Step 2  $HCO_2H_2^+ \rightarrow HCO^+ + H_2O$   (slow)
    Step 3  $HCO^+ \rightarrow CO + H^+$   (fast)
    (a) Write the overall balanced equation for this reaction.

Answers to questions highlighted in red type are provided in Appendix A.

(b) Sketch a potential energy diagram for this reaction, and identify any activation energy barriers, as well as the rate-determining step.

(c) Use a different colour to add any other labels to your sketch that you think are important or meaningful.

12. For each of the following pairs of reactions, identify the reaction that has a faster rate, and explain your reasoning.
    (a) $Pb^{2+}_{(aq)} + 2Cl^-_{(aq)} \rightarrow PbCl_{2(aq)}$;
    $2Na^+_{(aq)} + 2ClO^-_{(aq)} \rightarrow 2Na^+_{(aq)} + 2Cl^-_{(aq)} + O_{2(g)}$
    (b) $2Al_{(s)} + 3I_{2(s)} \rightarrow 2AlI_{3(s)}$;
    $Pb^{2+}_{(aq)} + 2ClO^-_{(aq)} \rightarrow Pb^{2+}_{(aq)} + 2Cl^-_{(aq)} + O_{2(g)}$
    (c) $2H_{2(g)} + O_{2(g)} \rightarrow 2H_2O_{(\ell)}$;
    $2Ag^+_{(aq)} + CO_3^{2-}_{(aq)} \rightarrow Ag_2CO_{3(s)}$

13. Suggest three ways to increase the rate of the following reaction: $C_{(s)} + O_{2(g)} \rightarrow CO_{2(g)}$

## Communication

14. People who have been submerged in very cold water, and presumed drowned, have sometimes been revived. By contrast, people who have been submerged for a similar period of time in warmer water have not survived. Suggest reasons for this difference.

15. Create a graphic organizer to summarize how the theories you have studied in this chapter explain the factors that affect reaction rates.

16. Use potential energy diagrams to compare an exothermic reaction with an endothermic reaction.

17. Consider the reaction below.
    $2A + B_2 \rightarrow D + E$
    A chemist proposes the following mechanism.
    **Step 1** $A + B_2 \rightarrow AB_2$
    **Step 2** $AB_2 + C \rightarrow AB_2C$
    **Step 3** $AB_2C + A \rightarrow A_2B_2 + C$
    **Step 4** $A_2B_2 \rightarrow D + E$
    Sketch potential energy diagrams for these situations:
    (a) Step 1 is the rate-determining step.
    (b) Step 2 is the rate-determining step.
    (c) Step 3 is the rate-determining step.

## Making Connections

18. Suppose you are using an instant camera outdoors during the winter. You want your photos to develop faster. Suggest a way to accomplish this.

19. Why are catalytic converters not very effective immediately after starting a vehicle in the winter? Suggest a way to correct this problem.

20. In a coal-fired electric generating plant, the coal is pulverized before being mixed with air in the incinerator. Use collision theory to explain why the coal is pulverized.

**Answers to Practice Problems and Short Answers to Section Review Questions**

**Practice Problems: 1.** Products are at lower energy than reactants. **2.** $E_{a(fwd)} = 99$ kJ **3.** $E_{a(rev)} = 193$ kJ **4.** $E_{a(rev)} = 411$ kJ
**5.** $2NO_{2(g)} + F_{2(g)} \rightarrow 2NO_2F_{(g)}$; $F_{(g)}$ **6.** $2A + B \rightarrow E + F$; C
**7.** $X + Y + 2Z \rightarrow XZ + YZ$; XY and $XYZ_2$
**8.** $(CH_3)_3CBr_{(aq)} + H_2O_{(\ell)} \rightarrow (CH_3)_3COH_{(aq)} + H^+_{(aq)} + Br^-_{(aq)}$;
$(CH_3)_3C^+_{(aq)}$ and $(CH_3)_3COH_2^+_{(aq)}$
**Section Review: 12.2: 3. (a)** endothermic **(b)** 68 kJ
**9.(a)** reaction intermediates **(b)** catalyst

# CHAPTER 13

# Reversible Reactions and Chemical Equilibrium

**Chapter Preview**

- 13.1 Recognizing Equilibrium
- 13.2 The Equilibrium Constant
- 13.3 Predicting the Direction of a Reaction

**Prerequisite Concepts and Skills**

Before you begin this chapter, review the following concepts and skills:

- using stoichiometry with chemical equations (Chapter 4)
- calculating molar concentrations (Chapter 7)

Your body contains over 1 kg of nitrogen, mainly in the proteins and nucleic acids that make up your cells. Your body, however, cannot absorb nitrogen directly from the atmosphere. You get the nitrogen-containing compounds you need by eating plants and other foods.

Plants cannot absorb nitrogen directly from the atmosphere, either. Their roots absorb nitrogen compounds—mainly nitrates—from the soil. Some of these nitrates result from atmospheric chemical reactions that involve nitrogen and oxygen. First, lightning supplies the energy that is needed to form nitric oxide from nitrogen and oxygen. The resulting nitric oxide combines with oxygen to form nitrogen dioxide, which reacts with rainwater to form nitric acid. Then, in the seas and the soil at Earth's surface, nitric acid ionizes to produce nitrate, $NO_3^-$. Plants and plant-like organisms can readily absorb nitrate.

Scientists call this chemical "circle of life" the *nitrogen cycle*. There are, as you know, similar, equally important cyclic processes in nature. Each process involves *reversible changes*: changes that may proceed in either direction, from reactants to products or from products to reactants.

In this chapter, you will study factors that affect reversible changes, notably those in chemical reactions. You will learn how to determine the amounts of reactants and products that are present when their proportions no longer change. You will also learn how to make qualitative predictions about the ways that chemists can change these proportions. Finally, you will see how industrial chemists apply their knowledge of reversible changes to increase the yield of chemicals that are important to society.

How is this dynamic situation similar to the activity of reactants and products in a chemical reaction?

# Recognizing Equilibrium

## 13.1

In Chapter 12, you learned that the rate of any reaction depends on the concentration of the reacting chemicals. As a reaction proceeds, the concentrations of the "product" chemicals increase, and the reverse reaction may re-form "reactants." Under certain conditions, the rate of the reverse reaction increases as the rate of the forward reaction decreases. Eventually, the rate of the forward reaction equals the rate of the reverse reaction. **Dynamic equilibrium** occurs when opposing changes, such as those just described, are occurring at the same time and at the same rate.

How can you recognize equilibrium? In the following paragraphs, you will read about physical changes that may or may not involve a system at equilibrium. As you consider these changes, note which systems are at equilibrium and what changes are taking place. Also note the conditions that would be needed for the changes to occur in the opposite direction at the same rate. The ExpressLab that follows these descriptions will add to your understanding of equilibrium. It will help you recognize systems that are at equilibrium and systems that are not.

- A puddle of water remains after a summer shower. The puddle evaporates because some of the water molecules near the surface have enough kinetic energy to escape from the liquid. At the same time, some water molecules in the air may condense back into the liquid. The chance of this happening is small, however. The puddle soon evaporates completely, as shown in Figure 13.1(A).

- Even if you put a lid on a jar of water, some of the water in the jar evaporates. Careful measurements show that the level of liquid water in the jar initially decreases because more water evaporates than condenses. As the number of water vapour molecules increases, however, some condense back into the liquid. Eventually, the number of evaporating molecules and the number of condensing molecules are equal. The level of water inside the jar remains constant. At this point, shown in Figure 13.1(B), equilibrium has been reached.

**Section Preview/Outcomes**

In this section, you will

- **identify** and **illustrate** equilibrium in various systems and the conditions that lead to dynamic equilibrium

- **communicate** your understanding of the following terms: *dynamic equilibrium, homogeneous equilibrium, heterogeneous equilibrium*

**Figure 13.1** Why is (B) an example of equilibrium, while (A) is not?

- Crystals of copper(II) sulfate pentahydrate, $CuSO_4 \cdot 5H_2O$, are blue. If you place a few small crystals in a beaker of water, water molecules break apart the ions and they enter into the solution. A few ions in the solution may re-attach to the crystals. Because more ions enter the solution than re-attach to the crystals, however, all the solid eventually dissolves. If you keep adding crystals of copper(II) sulfate pentahydrate, the solution eventually becomes saturated. Crystals remain at the bottom of the beaker, as shown in Figure 13.2.

**Figure 13.2** Which of these systems, (A) or (B), is at equilibrium? Will changing the temperature affect the concentration of the solution in the equilibrium system?

- Figure 13.3(A) shows a supersaturated solution of sodium acetate. It was prepared by adding sodium acetate to a saturated aqueous solution. The mixture was heated to dissolve the added crystals. Finally, the solution was left to return slowly to room temperature. In Figure 13.3(B), a single crystal of sodium acetate has been added to the supersaturated solution. As you can see, solute ions rapidly leave the solution, and solid forms.

**Figure 13.3** These photographs show a supersaturated solution of sodium acetate and the effect of adding a crystal to it. Is either of these systems at equilibrium?

# ExpressLab: Modelling Equilibrium

In this ExpressLab, you will model what happens when forward and reverse reactions occur. You will take measurements to gain quantitative insight into an equilibrium system. Then you will observe the effect of introducing a change to the equilibrium.

## Materials

2 graduated cylinders (25 mL)
2 glass tubes with different diameters (for example, 10 mm and 6 mm)
2 labels or a grease pencil
supply of water, coloured with food dye

## Procedure

1. Copy the following table into your notebook, to record your observations.

| Transfer number | Volume of water in reactant cylinder (mL) | Volume of water in product cylinder (mL) |
|---|---|---|
| 0 | 25.0 | 0.0 |
| 1 | | |
| 2 | | |

2. Label one graduated cylinder "reactant." Label the other "product."

3. Fill the reactant cylinder with coloured water, up to the 25.0 mL mark. Leave the product cylinder empty.

4. With your partner, transfer water simultaneously from one cylinder to the other as follows: Lower the larger-diameter glass tube into the reactant cylinder. Keep the top of the tube open. When the tube touches the bottom of the cylinder, cover the open end with a finger. Then transfer the liquid to the product cylinder. *At the same time* as you are transferring liquid into the product cylinder, your partner must use the smaller-diameter tube to transfer liquid from the product cylinder into the reactant cylinder.

5. Remove the glass tubes. Record the volume of water in each graduated cylinder, to the nearest 0.1 mL.

6. Repeat steps 4 and 5 until there is no further change in the volumes of water in the graduated cylinders.

7. Add approximately 5 mL of water to the reactant cylinder. Record the volume in each cylinder. Then repeat steps 4 and 5 until there is no further change in volume.

## Analysis

1. Plot a graph of the data you collected. Put transfer number on the *x*-axis and volume of water on the *y*-axis. Use different symbols or colours to distinguish between the reactant volume and the product volume. Draw the best smooth curve through the data.

2. In this activity, the volume of water is a model for concentration. How can you use your graph to compare the rate of the forward reaction with the rate of the reverse reaction? What happens to these rates as the reaction proceeds?

3. At the point where the two curves cross, is the rate of the forward reaction equal to the rate of the reverse reaction? Explain.

4. How can you recognize when the system is at equilibrium?

5. Were the volumes of water (that is, concentrations of reactants and products) in the two tubes equal at equilibrium? How do you know?

6. In a chemical reaction, what corresponds to the addition of more water to the reactant cylinder? How did the final volume of water in the product cylinder change as a result of adding more water to the reactant cylinder?

7. Determine the ratio $\dfrac{\text{Volume of product}}{\text{Volume of reactant}}$ at the end of the first equilibrium and at the end of the second equilibrium. Within experimental error, were these two ratios the same or different?

8. In this activity, what do you think determined the relative volumes of water in the graduated cylinders? In a real chemical reaction, what factors might affect the relative concentrations of reactants and products at equilibrium?

## Conditions That Apply to All Equilibrium Systems

The fundamental requirement for dynamic equilibrium is that opposing changes must occur at the same rate. There are many processes that reach equilibrium. Three physical processes that reach equilibrium are

- a solid in contact with a solution that contains this solid: for example, sugar crystals in a saturated aqueous sugar solution
- the vapour above a pure liquid: for example, a closed jar that contains liquid water
- the vapour above a pure solid: for example, mothballs in a closed drawer

Two chemical processes that reach equilibrium are

- a reaction with reactants and products in the same phase: for example, a reaction between two gases to produce a gaseous product. In this chapter, you will focus on reactions in which the reactants and products are in the same phase. The equilibrium they reach is called **homogeneous equilibrium**.
- a reaction in which reactants and products are in different phases: for example, an aqueous solution of ions, in which the ions combine to produce a slightly soluble solid that forms a precipitate. The equilibrium they reach is called **heterogeneous equilibrium**.

What is it that equilibrium systems like these have in common? What conditions are necessary for equilibrium to become established? As outlined in the box below, there are four conditions that apply to all equilibrium systems.

### The Four Conditions That Apply to All Equilibrium Systems

1. Equilibrium is achieved in a reversible process when the rates of opposing changes are equal. A double arrow, $\rightleftharpoons$, indicates reversible changes. For example:

$$H_2O_{(\ell)} \rightleftharpoons H_2O_{(g)}$$
$$H_{2(g)} + Cl_{2(g)} \rightleftharpoons 2HCl_{(g)}$$

2. The observable (macroscopic) properties of a system at equilibrium are constant. At equilibrium, there is no overall change in the properties that depend on the total quantity of matter in the system. Examples of these properties include colour, pressure, concentration, and pH.

You can summarize the first two equilibrium conditions by stating that *equilibrium involves dynamic change at the molecular level but no change at the macroscopic level.*

3. Equilibrium can only be reached in a closed system. A closed system is a system that does not allow the input or escape of any component of the equilibrium system, including energy. For this reason, a system can be at equilibrium only if it is at constant temperature. A common example of a closed system is carbon dioxide gas that is in equilibrium with dissolved carbon dioxide in a soda drink.

### CONCEPT CHECK

A decomposition reaction is taking place in a closed container, R → P. At equilibrium, does the concentration of reactant have to equal the concentration of product? Explain your answer.

The system remains at equilibrium as long as the container is not opened. Note that small changes to the components of a system are sometimes negligible. Thus, equilibrium principles can be applied if a system is not physically closed. For example, consider the equilibrium of a solid in a saturated aqueous solution, such as $CaO_{(s)} + H_2O_{(\ell)} \rightleftharpoons Ca^{2+}_{(aq)} + 2OH^-_{(aq)}$. You can neglect the small amount of water that vaporizes from the open beaker during an experiment.

4. Equilibrium can be approached from either direction. For example, the proportions of $H_{2(g)}$, $Cl_{2(g)}$, and $HCl_{(g)}$ (in a closed container at constant temperature) are the same at equilibrium, regardless of whether you started with $H_{2(g)}$ and $Cl_{2(g)}$ or whether you started with $HCl_{(g)}$.

## Section Summary

In this section, you learned how to recognize equilibrium. As well, you learned about the conditions that are needed for equilibrium to be reached. Later in this chapter, you will examine what happens when some equilibrium conditions in a system are changed. You will learn how chemists can control conditions to increase the yield of reactions. First, though, you will look more closely at the reactants and products of chemical systems and—in particular—learn how equilibrium is measured.

## Section Review

**1** Give two physical and two chemical processes that are examples of reversible changes that are not at equilibrium.

**2 (a)** Describe how the changes that take place in a reversible reaction approach equilibrium.

**(b)** Explain why the equilibrium that is established is dynamic.

**3** A sealed carbonated-drink bottle contains a liquid drink with a space above it. The space contains carbon dioxide at a pressure of about 405 kPa.

**(a)** What changes are taking place at the molecular level?

**(b)** Which macroscopic properties are constant?

**4** Agree or disagree with the following statement, and give reasons to support your answer: "In a sealed jar of water at equilibrium, the quantity of water molecules in the liquid state equals the quantity of water vapour molecules."

**5** Ice and slush are a feature of Canadian winters. Under what conditions do ice and water form an equilibrium mixture?

**6** A chemist places a mixture of NO and $Br_2$ in a closed flask. A product mixture of NO, $Br_2$, and NOBr is observed. Explain this observation.

### Web LINK

www.mcgrawhill.ca/links/atlchemistry

To observe a molecular view of dynamic equilibrium, go to the web site above and click on **Electronic Learning Partner**.

# 13.2 The Equilibrium Constant

**Section Preview/Outcomes**

In this section, you will

- **write** equilibrium expressions to express your understanding of the law of chemical equilibrium as it applies to concentrations of reactants and products at equilibrium
- **solve** equilibrium problems when $K_c$ and/or equilibrium concentrations are known
- **communicate** your understanding of the following terms: *law of chemical equilibrium, equilibrium constant ($K_c$), ICE table*

In this section, you will learn about the *extent* of a reaction: the relative concentrations of products to reactants at equilibrium.

## Opposing Rates and the Law of Chemical Equilibrium

In 1864, two Norwegian chemists, Cato Guldberg and Peter Waage, summarized their experiments on chemical equilibrium in the **law of chemical equilibrium**: *At equilibrium, there is a constant ratio between the concentrations of the products and reactants in any change.*

Figure 13.4 shows how the law of chemical equilibrium applies to one chemical system. Chemists have studied this system extensively. It involves the reversible reaction between two gases: dinitrogen tetroxide, which is colourless, and nitrogen dioxide, which is dark brown. The use of oppositely pointing arrows in the equation below means that the reaction is at equilibrium.

$$N_2O_{4(g)} \rightleftharpoons 2NO_{2(g)}$$
**colourless    brown**

By observing the intensity of the brown colour in the mixture, chemists can determine the concentration of nitrogen dioxide.

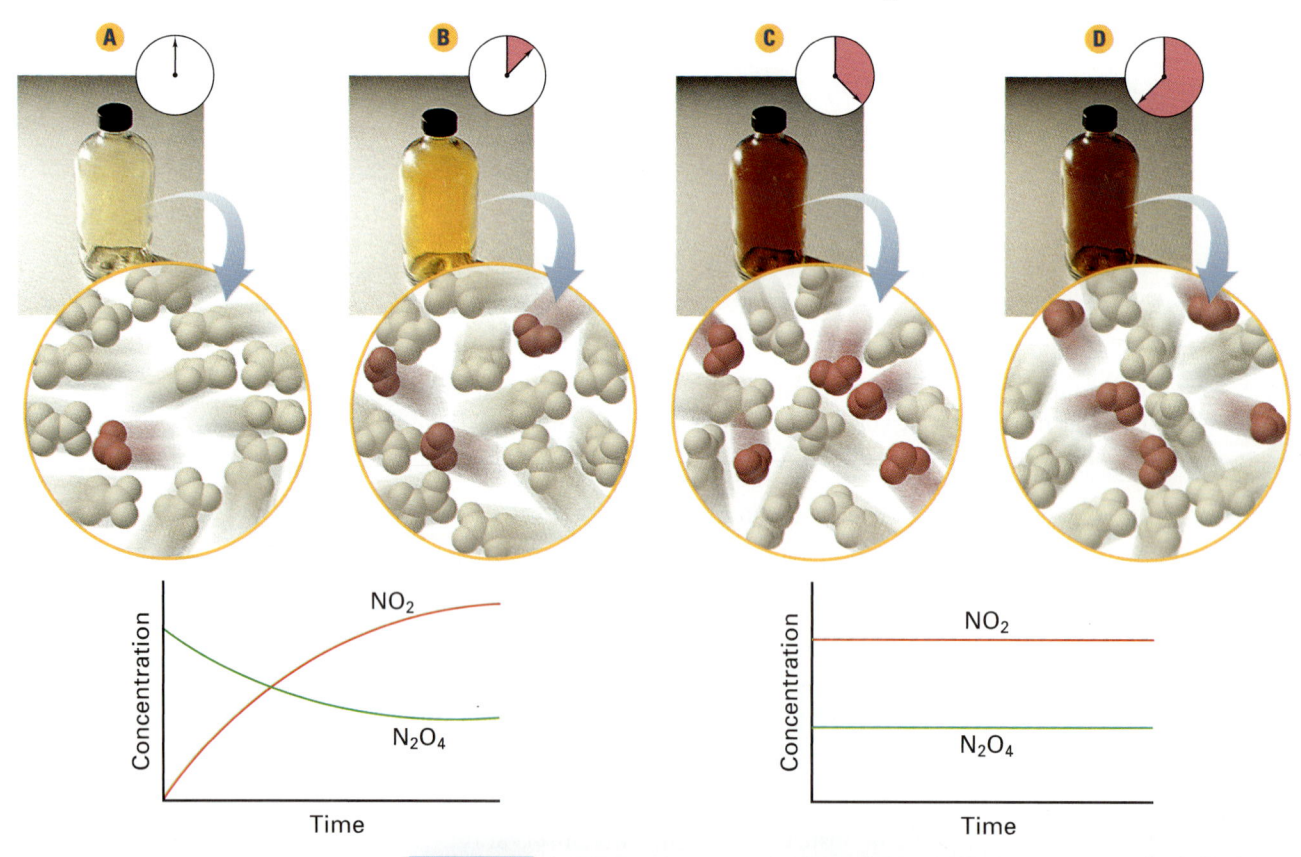

**Figure 13.4** As this system nears equilibrium, the rate of the forward reaction decreases and the rate of the reverse reaction increases. At equilibrium, the macroscopic properties of this system are constant. Changes at the molecular level take place at equal rates.

Dinitrogen tetroxide gas is produced by vaporizing dinitrogen tetroxide liquid. Dinitrogen tetroxide liquid boils at 21°C. If a small quantity is placed in a sealed flask at 100°C, it vaporizes, filling the flask with dinitrogen tetroxide gas.

Suppose that the initial concentration of $N_2O_{4(g)}$ is 0.0200 mol/L. The initial concentration of $NO_{2(g)}$ is zero. The initial rate of the forward reaction is relatively large, while the initial rate of the reverse reaction is zero. The initial conditions correspond to the first exchange in the ExpressLab in section 13.1, where the reactant cylinder was full and the product cylinder was empty. As the reaction proceeds, the rate of the forward reaction decreases because the concentration of $N_2O_{4(g)}$ decreases. At the same time, the rate of the reverse reaction increases because the concentration of $NO_{2(g)}$ increases. At equilibrium, the rate of the forward reaction equals the rate of the reverse reaction. There are no further changes in the relative amounts of $N_2O_4$ and $NO_2$.

> **CONCEPT CHECK**
>
> How did the ExpressLab model a reaction approaching equilibrium? How did it model the situation when equilibrium was reached?

## The Equilibrium Constant

To formulate the law of chemical equilibrium, Guldberg and Waage conducted their own experiments. They also reviewed the results of many experiments conducted by other chemists. Thus, they had a vast body of observations to examine. They discovered a particular relationship among the reactants and products of any chemical system. Consider, for example, the data in chart below. It compares several different initial and equilibrium concentration values for the $N_2O_4 - NO_2$ system. Notice the use of square brackets around products and reactants. This is a "short form" to indicate the molar concentration of a compound, in mol/L.

| Initial [$N_2O_4$] at 100°C | Initial [$NO_2$] at 100°C | Equilibrium [$N_2O_4$] at 100°C | Equilibrium [$NO_2$] at 100°C | Ratio: $\dfrac{[NO_2]^2}{[N_2O_4]}$ |
|---|---|---|---|---|
| 0.1000 | 0.0000 | 0.0491 | 0.1018 | 0.211 |
| 0.0000 | 0.1000 | 0.0185 | 0.0627 | 0.212 |
| 0.0500 | 0.0500 | 0.0332 | 0.0837 | 0.211 |
| 0.0750 | 0.0250 | 0.0411 | 0.0930 | 0.210 |

What do these data show? Start by focussing your attention only on the first four columns. You can see four different experiments involving different initial concentrations of reactants and products. As you might expect, at equilibrium, the concentrations of reactants and products are different for each experiment. There are no apparent trends in these data. However, look closely at the last column. If you allow for experimental error, this particular ratio of concentration terms, $\dfrac{[NO_2]^2}{[N_2O_4]}$, gives a value that is basically identical for all four experiments. You could repeat this experiment hundreds of times with different initial concentrations, and the value of this particular ratio would remain constant at 0.211. Therefore, for the reaction $N_2O_{4(g)} \rightleftharpoons NO_{2(g)}$, at 100°C, you can write the following expression:

$$\dfrac{[NO_2]^2}{[N_2O_4]} = 0.211$$

Observations from numerous experiments led Guldberg and Waage to show that for any chemical reaction at equilibrium, and at a given temperature, a specific ratio of concentration terms has a constant value. This constant value is achieved regardless of the initial concentrations of

> **CONCEPT CHECK**
>
> - Compare the graphs in Figure 13.4 with the values recorded in the chart on this page. How are they related?
> - Compare the expression, $\dfrac{[NO_2]^2}{[N_2O_4]}$, with the balanced chemical equation for the reaction between dinitrogen tetroxide and nitrogen dioxide. What do you notice?

the reactants and the products. It is called the **equilibrium constant**, $K_c$, for the reaction. The subscript "c" is used to show that the equilibrium constant is expressed in terms of molar concentration. (There are other equilibrium constants, each with its own subscript letter, as you will see in Unit 6.)

Thus, for any general equilibrium reaction,

$$a\text{P} + b\text{Q} \rightleftharpoons c\text{R} + d\text{S}$$

the value of the equilibrium constant is

$$K_c = \frac{[\text{R}]^c[\text{S}]^d}{[\text{P}]^a[\text{Q}]^b}$$

For this general equation, P, Q, R, and S represent chemical formulas and $a$, $b$, $c$, and $d$ represent their respective coefficients in the chemical equation. Again, square brackets indicate molar concentrations of the reactants and products.

This equilibrium expression depends only on the stoichiometry of the reaction. By convention, chemists always write the concentrations of the products in the numerator and the concentrations of the reactants in the denominator. Each concentration term is raised to the power of the coefficient in the chemical equation. The terms are multiplied, never added. As well, pure liquids and solids are *not* included in the equilibrium constant expression, because their concentrations are already constants.

> **CHEM FACT**
>
> The law of chemical equilibrium is sometimes known as the law of mass action. Before the term "concentration" was used, the concept of amount per unit volume was called "active mass."

### Sample Problem

#### Writing Equilibrium Expressions

**Problem**

One of the steps in the production of sulfuric acid involves the catalytic oxidation of sulfur dioxide.

$$2\text{SO}_{2(g)} + \text{O}_{2(g)} \rightarrow 2\text{SO}_{3(g)}$$

Write the equilibrium expression.

**What Is Required?**

You need to find an expression for $K_c$.

**What Is Given?**

You know the balanced chemical equation.

**Plan Your Strategy**

The expression for $K_c$ is a fraction. The concentration of the product is in the numerator, and the concentrations of the reactants are in the denominator. Each concentration term must be raised to the power of the coefficient in the balanced equation.

**Act on Your Strategy**

$$K_c = \frac{[\text{SO}_3]^2}{[\text{SO}_2]^2[\text{O}_2]}$$

**Check Your Solution**

The square brackets indicate concentrations. The product is in the numerator, and each term is raised to the power of the coefficient in the chemical equation. The coefficient or power of 1 is not written, thus following chemistry conventions.

## Practice Problems

Write the equilibrium expression for each homogeneous reaction.

1. The reaction between propane and oxygen to form carbon dioxide and water vapour:
$$C_3H_{8(g)} + 5O_{2(g)} \rightleftharpoons 3CO_{2(g)} + 4H_2O_{(g)}$$

2. The reaction between nitrogen gas and oxygen gas at high temperatures:
$$N_{2(g)} + O_{2(g)} \rightleftharpoons 2NO_{(g)}$$

3. The reaction between hydrogen gas and oxygen gas to form water vapour:
$$2H_{2(g)} + O_{2(g)} \rightleftharpoons 2H_2O_{(g)}$$

4. The reduction-oxidation equilibrium of iron and iodine ions in aqueous solution:
$$2Fe^{3+}_{(aq)} + 2I^-_{(aq)} \rightleftharpoons 2Fe^{2+}_{(aq)} + I_{2(aq)}$$

    **Note:** You will learn about reduction-oxidation reactions in Unit 8.

5. The oxidation of ammonia (one of the reactions in the production of nitric acid):
$$4NH_{3(g)} + 5O_{2(g)} \rightleftharpoons 4NO_{(g)} + 6H_2O_{(g)}$$

## The Equilibrium Constant and Temperature

Adding a chemical that is involved in a reaction at equilibrium increases the rate at which this chemical reacts. The rate, however, decreases as the concentration of the added chemical decreases. Eventually, equilibrium is re-established with the same equilibrium constant. *For a given system at equilibrium, the value of the equilibrium constant depends only on temperature.* Changing the temperature of a reacting mixture changes the rate of the forward and reverse reactions by different amounts, because the forward and reverse reactions have different activation energies. A reacting mixture at one temperature has an equilibrium constant whose value changes if the mixture is allowed to reach equilibrium at a different temperature.

The numerical value of the equilibrium constant does not depend on whether the starting point involves reactants or products. These are just labels that chemists use to identify particular chemicals in the reaction mixture. Also, remember that, at a given temperature, the value of $K_c$ does not depend on the starting concentrations. The reaction gives the same ratio of products and reactants according to the equilibrium law. Remember, however, that $K_c$ is calculated using *concentration* values when the system is at *equilibrium*.

## Sample Problem

### Calculating an Equilibrium Constant

**Problem**

A mixture of nitrogen and chlorine gases was kept at a certain temperature in a 5.0 L reaction flask.

$$N_{2(g)} + 3Cl_{2(g)} \rightleftharpoons 2NCl_{3(g)}$$

When the equilibrium mixture was analyzed, it was found to contain 0.0070 mol of $N_{2(g)}$, 0.0022 mol of $Cl_{2(g)}$, and 0.95 mol of $NCl_{3(g)}$. Calculate the equilibrium constant for this reaction.

**What Is Required?**

You need to calculate the value of $K_c$.

**What Is Given?**

You have the balanced chemical equation and the amount of each substance at equilibrium.

**Plan Your Strategy**

**Step 1** Calculate the molar concentration of each compound at equilibrium.

**Step 2** Write the equilibrium expression. Then substitute the equilibrium molar concentrations into the expression.

**Act on Your Strategy**

**Step 1** The reaction takes place in a 5.0 L flask. Calculate the molar concentrations at equilibrium.

$$[N_2] = \frac{0.0070 \text{ mol}}{5.0 \text{ L}} = 1.4 \times 10^{-3} \text{ mol/L}$$

$$[Cl_2] = \frac{0.0022 \text{ mol}}{5.0 \text{ L}} = 4.4 \times 10^{-4} \text{ mol/L}$$

$$[NCl_3] = \frac{0.95 \text{ mol}}{5.0 \text{ L}} = 1.9 \times 10^{-1} \text{ mol/L}$$

**Step 2** Write the equilibrium expression. Substitute the equilibrium molar concentrations into the expression.

$$K_c = \frac{[NCl_3]^2}{[N_2][Cl_2]^3}$$

$$= \frac{(1.9 \times 10^{-1})^2}{(1.4 \times 10^{-3}) \times (4.4 \times 10^{-4})^3}$$

$$= 3.0 \times 10^{11}$$

**Check Your Solution**

The equilibrium expression has the product terms in the numerator and the reactant terms in the denominator. The exponents in the equilibrium expression match the corresponding coefficients in the chemical equation. The molar concentrations at equilibrium were substituted into the expression.

> **PROBLEM TIP**
>
> Notice that units are not included when using or calculating the value of $K_c$. This is the usual practice. The units do not help you check your solution.

## Practice Problems

6. The following reaction took place in a sealed flask at 250°C.

   $PCl_{5(g)} \rightleftharpoons PCl_{3(g)} + Cl_{2(g)}$

   At equilibrium, the gases in the flask had the following concentrations: $[PCl_5] = 1.2 \times 10^{-2}$ mol/L, $[PCl_3] = 1.5 \times 10^{-2}$ mol/L, and $[Cl_2] = 1.5 \times 10^{-2}$ mol/L. Calculate the value of $K_c$ at 250°C.

7. Iodine and bromine react to form iodine monobromide, IBr.

   $I_{2(g)} + Br_{2(g)} \rightleftharpoons 2IBr_{(g)}$

   At 250°C, an equilibrium mixture in a 2.0 L flask contained 0.024 mol of $I_{2(g)}$, 0.050 mol of $Br_{2(g)}$, and 0.38 mol of $IBr_{(g)}$. What is the value of $K_c$ for the reaction at 250°C?

8. At high temperatures, carbon dioxide gas decomposes into carbon monoxide and oxygen gas. At equilibrium, the gases have the following concentrations: $[CO_{2(g)}] = 1.2$ mol/L, $[CO_{(g)}] = 0.35$ mol/L, and $[O_{2(g)}] = 0.15$ mol/L. Determine $K_c$ at the temperature of the reaction.

9. Hydrogen sulfide is a pungent, poisonous gas. At 1400 K, an equilibrium mixture was found to contain 0.013 mol/L hydrogen, 0.18 mol/L hydrogen sulfide, and an undetermined amount of sulfur in the form of $S_{2(g)}$. If the value of $K_c$ is $2.40 \times 10^{-4}$ at this temperature, what concentration of $S_{2(g)}$ is present at equilibrium?

10. Methane, ethyne, and hydrogen form the following equilibrium mixture.

    $2CH_{4(g)} \rightleftharpoons C_2H_{2(g)} + 3H_{2(g)}$

    While studying this reaction mixture, a chemist analyzed a 4.0 L sealed flask at 1700°C. The chemist found 0.46 mol of $CH_{4(g)}$, 0.64 mol of $C_2H_{2(g)}$, and 0.92 mol of $H_{2(g)}$. Calculate the value of $K_c$ at 1700°C.

## Measuring Equilibrium Concentrations

The equilibrium constant, $K_c$, is calculated by substituting equilibrium concentrations into the equilibrium expression. Experimentally, this means that a reaction mixture must come to equilibrium. Then one or more properties are measured. The properties that are measured depend on the reaction. Common examples for gaseous reactions include colour, pH, and pressure. From these measurements, the concentrations of the reacting substances can be determined. Thus, you do not need to measure all the concentrations in the mixture at equilibrium. You can determine some equilibrium concentrations if you know the initial concentrations of the reactants and the concentration of one product at equilibrium.

For example, a mixture of iron(III) nitrate and potassium thiocyanate, in aqueous solution, react to form the iron(III) thiocyanate ion, $Fe(SCN)^{2+}_{(aq)}$. The reactant solutions are nearly colourless. The product solution ranges in colour from orange to blood-red, depending on its concentration. The nitrate and potassium ions are spectators. Therefore, the net ionic equation is

$$Fe^{3+}_{(aq)} + SCN^-_{(aq)} \rightleftharpoons Fe(SCN)^{2+}_{(aq)}$$
**nearly colourless**      **red**

### CONCEPT CHECK

Think about a chemical system at equilibrium. Are the concentrations of the reactants and products in the same ratio as the coefficients in the chemical equation? Explain your answer.

Because the reaction involves a colour change, you can determine the concentration of $Fe(SCN)^{2+}_{(aq)}$ by measuring the intensity of the colour. You will find out how to do this in Investigation 13-A. For now, assume that it can be done. From the measurements of colour intensity, you can calculate the equilibrium concentration of $Fe(SCN)^{2+}_{(aq)}$. Then, knowing the initial concentration of each solution, you can calculate the equilibrium concentration of each ion using the chemical equation.

Suppose, for instance, that the initial concentration of $Fe^{3+}_{(aq)}$ is 0.0064 mol/L and the initial concentration of $SCN^-_{(aq)}$ is 0.0010 mol/L. When the solutions are mixed, the red complex ion forms. By measuring the intensity of its colour, you can determine that the concentration of $Fe(SCN)^{2+}_{(aq)}$ is $4.5 \times 10^{-4}$ mol/L. From the stoichiometry of the equation, each mole of $Fe(SCN)^{2+}_{(aq)}$ forms when equal amounts of $Fe^{3+}_{(aq)}$ and $SCN^-_{(aq)}$ react. So, if there is $4.5 \times 10^{-4}$ mol/L of $Fe(SCN)^{2+}_{(aq)}$ at equilibrium, then the same amounts of both $Fe^{3+}_{(aq)}$ and $SCN^-_{(aq)}$ must have reacted. This represents the change in their concentrations. The equilibrium concentration of a reacting species is the sum of its initial concentration and the change that results from the reaction. Therefore, the concentration of $Fe^{3+}_{(aq)}$ at equilibrium is (0.0064 − 0.000 45) mol/L = 0.005 95 mol/L, or 0.0060 mol/L. You can calculate the equilibrium concentration of $SCN^-_{(aq)}$ the same way, and complete the table below.

| Concentration (mol/L) | $Fe^{3+}_{(aq)}$ + | $SCN^-_{(aq)}$ ⇌ | $Fe(SCN)^{2+}_{(aq)}$ |
|---|---|---|---|
| Initial | 0.0064 | 0.0010 | 0 |
| Change | $-4.5 \times 10^{-4}$ | $-4.5 \times 10^{-4}$ | $4.5 \times 10^{-4}$ |
| Equilibrium | 0.0060 | $5.5 \times 10^{-4}$ | $4.5 \times 10^{-4}$ |

Finally, you can calculate $K_c$ by substituting the equilibrium concentrations into the equilibrium expression.

In the following investigation, you will collect experimental data to determine an equilibrium constant.

# Investigation 13-A

**SKILL FOCUS**
- Predicting
- Performing and recording
- Analyzing and interpreting
- Communicating results

## Measuring an Equilibrium Constant

The colour intensity of a solution is related to the concentration of the ions and the depth of the solution. By adjusting the depth of a solution with unknown concentration until it has the same intensity as a solution with known concentration, you can determine the concentration of the unknown solution. For example, if the concentration of a solution is lower than the standard, the depth of the solution has to be greater in order to have the same colour intensity. Thus, the ratio of the concentrations of two solutions with the same colour intensity is inversely proportional to the ratio of their depths.

In this investigation, you will examine the homogeneous equilibrium between iron(III) (ferric) ions, thiocyanate ions, and ferrithiocyanate ions, $Fe(SCN)^{2+}$.

$$Fe^{3+}_{(aq)} + SCN^{-}_{(aq)} \rightleftharpoons Fe(SCN)^{2+}_{(aq)}$$

You will prepare four equilibrium mixtures with different initial concentrations of $Fe^{3+}_{(aq)}$ ions and $SCN^{-}_{(aq)}$ ions. You will calculate the initial concentrations of these reacting ions from the volumes and concentrations of the stock solutions used, and the *total* volumes of the equilibrium mixtures. Then you will determine the concentration of $Fe(SCN)^{2+}$ ions in each mixture by comparing the colour intensity of the mixture with the colour intensity of a solution that has known concentration. After you find the concentration of $Fe(SCN)^{2+}$ ions, you will use it to calculate the concentrations of the other two ions at equilibrium. You will substitute the three concentrations for each mixture into the equilibrium expression to determine the equilibrium constant.

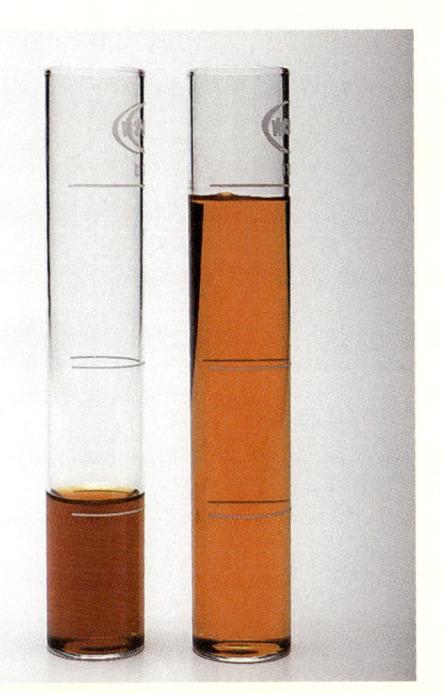

**Which solution is the least concentrated? Why is the colour intensity the same when you look vertically through the solutions?**

*continued...*

### Question

What is the value of the equilibrium constant at room temperature for the following reaction?

$$Fe^{3+}_{(aq)} + SCN^-_{(aq)} \rightleftharpoons Fe(SCN)^{2+}_{(aq)}$$

### Prediction

Write the equilibrium expression for this reaction.

### Safety Precautions

- The $Fe(NO_3)_3$ solution is acidified with nitric acid. It should be handled with care. Wash any spills on your skin or clothing with plenty of water, and inform your teacher.
- All glassware must be clean and dry before using it.

### Materials

3 beakers (100 mL)
5 test tubes (18 mm × 150 mm)
5 flat-bottom vials
test tube rack
labels or grease pencil
thermometer
stirring rod
strip of paper
diffuse light source, such as a light box
medicine dropper
3 Mohr pipettes (5.0 mL)
20.0 mL Mohr pipette
pipette bulb
30 mL 0.0020 mol/L $KSCN_{(aq)}$
30 mL 0.0020 mol/L $Fe(NO_3)_{3(aq)}$ (acidified)
25 mL 0.200 mol/L $Fe(NO_3)_{3(aq)}$ (acidified)
distilled water

### Procedure

1. Copy the following tables into your notebook, and give them titles. You will use the tables to record your measurements and calculations.

2. Label five test tubes and five vials with the numbers 1 through 5. Label three beakers with the names and concentrations of the stock solutions: $2.00 \times 10^{-3}$ mol/L KSCN, $2.00 \times 10^{-3}$ mol/L $Fe(NO_3)_3$, and 0.200 mol/L $Fe(NO_3)_3$. Pour about 30 mL of each stock solution into its labelled beaker. Be sure to distinguish between the different concentrations of the iron(III) nitrate solutions. Make sure that you choose the correct solution when needed in the investigation. Measure the volume of each solution as carefully as possible to ensure the accuracy of your results.

| Test tube | $Fe(NO_3)_3$ (mL) | $H_2O$ (mL) | KSCN (mL) | Initial $[SCN^-]$ (mol/L) |
|---|---|---|---|---|
| 2 | 5.0 | 3.0 | 2.0 | |
| 3 | 5.0 | 2.0 | 3.0 | |
| 4 | 5.0 | 1.0 | 4.0 | |
| 5 | 5.0 | 0 | 5.0 | |

| Vial | Depth of solution in vial (mm) | Depth of standard solution (mm) | $\dfrac{\text{Depth of standard solution}}{\text{Depth of solution in vial}}$ |
|---|---|---|---|
| 2 | | | |
| 3 | | | |
| 4 | | | |
| 5 | | | |

3. Prepare the standard solution of $FeSCN^{2+}$ in test tube 1. Use the 20 mL Mohr pipette to transfer 18.0 mL of 0.200 mol/L $Fe(NO_3)_{3(aq)}$ into the test tube. Then use a 5 mL pipette to add 2.0 mL of $2.00 \times 10^{-3}$ mol/L KSCN. The large excess of $Fe^{3+}_{(aq)}$ is to ensure that all the $SCN^-_{(aq)}$ will react to form $Fe(SCN)^{2+}_{(aq)}$.

4. Pipette 5.0 mL of $2.0 \times 10^{-3}$ mol/L $Fe(NO_3)_{3(aq)}$ into each of the other four test tubes (labelled 2 to 5).

5. Pipette 3.0 mL, 2.0 mL, 1.0 mL, and 0 mL of distilled water into test tubes 2 to 5.

6. Pipette 2.0 mL, 3.0 mL, 4.0 mL, and 5.0 mL of $2.0 \times 10^{-3}$ mol/L $KSCN_{(aq)}$ into test tubes 2 to 5. Each of these test tubes should now contain 10.0 mL of solution. Notice that the first table you prepared (in step 1) shows the volumes of the liquids you added to the test tubes. Use a stirring rod to mix each solution. (Remember to rinse the stirring rod with water and then dry it with a paper towel before you stir each solution. Measure and record the temperature of one of the solutions. Assume that all the solutions are at the same temperature.

7. Pour about 5 mL of the standard solution from test tube 1 into vial 1.

8. Pour some of the solution from test tube 2 into vial 2. Look down through vials 1 and 2. Add enough solution to vial 2 to make its colour intensity about the same as the colour intensity of the solution in vial 1. Use a sheet of white paper as background to make your rough colour intensity comparison.

9. Wrap a sheet of paper around vials 1 and 2 to prevent light from entering the sides of the solutions. Looking down through the vials over a diffuse light source, adjust the volume of the standard solution in vial 1 until the colour intensity in the vials is the same. Use a medicine dropper to remove or add standard solution. Be careful not to add standard solution to vial 2.

10. When the colour intensity is the same in both vials, measure and record the depth of solution in each vial as carefully as possible.

11. Repeat steps 9 and 10 using vials 3 to 5.

12. Discard the solutions into the container supplied by your teacher. Rinse the test tubes and vials with distilled water. Then return all the equipment. Wash your hands.

13. Copy the following table into your notebook, to summarize the results of your calculations.

    (a) Calculate the equilibrium concentration of $Fe(SCN)^{2+}$ in the standard solution you prepared in test tube 1. The $[Fe(SCN)^{2+}]_{standard}$ is essentially the same as the starting concentration of $SCN^-_{(aq)}$ in test tube 1. The large excess of $Fe^{3+}_{(aq)}$ ensured that the reaction of $SCN^-$ was almost complete. Remember, however, to include the volume of $Fe(NO_3)_{3(aq)}$ in the total volume of the solution for your calculation.

| Test tube | Initial concentrations | | Equilibrium concentrations | | | Equilibrium constant, $K_c$ |
|---|---|---|---|---|---|---|
| | $[Fe^{3+}]_i$ | $[SCN^-]_i$ | $[Fe^{3+}]_{eq}$ | $[SCN^-]_{eq}$ | $[Fe(SCN)^{2+}]_{eq}$ | |
| 2 | | | | | | |
| 3 | | | | | | |
| 4 | | | | | | |
| 5 | | | | | | |

*continued...*

(b) Calculate the initial concentration of $Fe^{3+}_{(aq)}$ in test tubes 2 to 5. $[Fe^{3+}]_i$ is the same in these four test tubes. They all contained the same volume of $Fe(NO_3)_{3(aq)}$, and the total final volume was the same. Remember to use the total volume of the solution in your calculation.

(c) Calculate the initial concentration of $SCN^-_{(aq)}$ in test tubes 2 to 5. $[SCN^-]_i$ is different in each test tube.

(d) Calculate the equilibrium concentration of $Fe(SCN)^{2+}$ in test tubes 2 to 5. Use the following equation:

$$[FeSCN^{2+}]_{eq} = \frac{\text{Depth of standard solution}}{\text{Depth of solution in vial}} \times [FeSCN^{2+}]_{standard}$$

(e) Based on the stoichiometry of the reaction, each mole of $Fe(SCN)^{2+}$ is formed by the reaction of one mole of $Fe^{3+}$ with one mole of $SCN^-$. Thus, you can find the equilibrium concentrations of these ions by using the equations below:

$[Fe^{3+}]_{eq} = [Fe^{3+}]_i - [Fe(SCN)^{2+}]_{eq}$

$[SCN^-]_{eq} = [SCN^-]_i - [Fe(SCN)^{2+}]_{eq}$

(f) Calculate four values for the equilibrium constant, $K_c$, by substituting the equilibrium concentrations into the equilibrium expression. Find the average of your four values for $K_c$.

## Analysis

1. How did the colour intensity of the solutions in test tubes 2 to 5 vary at equilibrium? Explain your observation.

2. How consistent are the four values you calculated for $K_c$? Suggest reasons that could account for any differences.

3. Use your four sets of data to evaluate the following expressions:

$[Fe(SCN)^{2+}][Fe^{3+}][SCN^-]$

$\dfrac{[Fe^{3+}] + [SCN^-]}{[Fe(SCN)^{2+}]}$

Do these expressions give a value that is more constant than your equilibrium constant? Make up another expression, and evaluate it using your four sets of data. How constant is its value?

4. Suppose that the following reaction was the equilibrium reaction.

$Fe^{3+}_{(aq)} + 2SCN^-_{(aq)} \rightleftharpoons Fe(SCN)_2^+_{(aq)}$

(a) Would the equilibrium concentration of the product be different from the concentration of the product in the actual reaction? Explain.

(b) Would the value of $K_c$ be different from the value you calculated earlier? Explain.

## Conclusion

5. Write a short conclusion, summarizing your results for this investigation.

## Equilibrium Calculations

The Sample Problems that follow all involve homogeneous equilibrium systems. Each problem illustrates a particular type of system and includes brief tips. Each problem also includes a table like the one on page 500 to organize the data. Because this table is used to record the **i**nitial, **c**hange, and **e**quilibrium values of the reacting species, it is often called an **ICE table**. In Unit 6, you will use ICE tables again to help you solve problems that involve heterogeneous equilibrium systems.

### Sample Problem
### Finding Equilibrium Amounts From Initial Amounts and $K_c$

#### Problem
The following reaction increases the proportion of hydrogen gas for use as a fuel.

$$CO_{(g)} + H_2O_{(g)} \rightleftharpoons H_{2(g)} + CO_{2(g)}$$

This reaction has been studied at different temperatures to find the optimum conditions. At 700 K, the equilibrium constant is 8.3. Suppose that you start with 1.0 mol of $CO_{(g)}$ and 1.0 mol of $H_2O_{(g)}$ in a 5.0 L container. What amount of each substance will be present in the container when the gases are at equilibrium, at 700 K?

#### What Is Required?
You need to find the amount (in moles) of $CO_{(g)}$, $H_2O_{(g)}$, $H_{2(g)}$, and $CO_{2(g)}$ at equilibrium.

#### What Is Given?
You have the balanced chemical equation. You know the initial amount of each gas, the volume of the container, and the equilibrium constant.

#### Plan Your Strategy
**Step 1** Calculate the initial concentrations.

**Step 2** Set up an ICE table. Record the initial concentrations you calculated in step 1 in your ICE table. Let the change in molar concentrations of the reactants be $x$. Use the stoichiometry of the chemical equation to write and record expressions for the equilibrium concentrations.

**Step 3** Write the equilibrium expression. Substitute the expressions for the equilibrium concentrations into the equilibrium expression. Solve the equilibrium expression for $x$.

**Step 4** Calculate the equilibrium concentration of each gas. Then use the volume of the container to find the amount of each gas.

#### Act on Your Strategy
**Step 1** The initial amount of CO is equal to the initial amount of $H_2O$.

$$\therefore [CO] = [H_2O] = \frac{1.0 \text{ mol}}{5.0 \text{ L}} = 0.20 \text{ mol/L}$$

> **PROBLEM TIP**
>
> Generally, it is best to let $x$ represent the substance with the smallest coefficient in the chemical equation. This helps to avoid fractional values of $x$ in the equilibrium expression. Fractional values make solving the expression more difficult.

*Continued ...*

**Step 2** Set up an ICE table.

| Concentration (mol/L) | CO$_{(g)}$ + | H$_2$O$_{(g)}$ ⇌ | H$_{2(g)}$ + | CO$_{2(g)}$ |
|---|---|---|---|---|
| Initial | 0.20 | 0.20 | 0 | 0 |
| Change | $-x$ | $-x$ | $+x$ | $+x$ |
| Equilibrium | $0.20 - x$ | $0.20 - x$ | $0 + x$ | $0 + x$ |

**Step 3**
$$K_c = \frac{[H_2][CO_2]}{[CO][H_2O]}$$

$$8.3 = \frac{(x)(x)}{(0.20 - x)(0.20 - x)} = \frac{(x)^2}{(0.20 - x)^2}$$

$$\pm 2.88 = \frac{(x)}{(0.20 - x)}$$

Solving for both values,
$x = 0.148$ and $x = 0.306$

**Step 4** The value $x = 0.306$ is physically impossible because it would result in a negative concentration of both CO and H$_2$O at equilibrium.

$\therefore [H_2] = [CO_2] = 0.15$ mol/L

$[CO] = [H_2O] = 0.20 - 0.15 = 0.05$ mol/L

To find the amount of each gas, multiply the concentration of each gas by the volume of the container (5.0 L).

Amount of H$_{2(g)}$ = CO$_{2(g)}$ = 0.75 mol

Amount of CO$_{(g)}$ = H$_2$O$_{(g)}$ = 0.25 mol

### Check Your Solution

The equilibrium expression has product terms in the numerator and reactant terms in the denominator. The amounts of chemicals at equilibrium are given in moles. Check $K_c$.

$$K_c = \frac{(0.75)^2}{(0.25)^2} = 9.0$$

This is close to the given value, $K_c = 8.3$. The difference is due to mathematical rounding.

> **PROBLEM TIP**
>
> In this problem, the right-hand side of the equilibrium expression is a perfect square. Noticing perfect squares, then taking the square root of both sides, makes solving the equation easier. It avoids solving a quadratic equation.

Many problems do not involve perfect squares. You may need to use a quadratic equation to solve the equilibrium expression.

### Sample Problem

### Solving an Equilibrium Expression Using a Quadratic Equation

#### Problem

The following reaction has an equilibrium constant of 25.0 at 1100 K.
$$H_{2(g)} + I_{2(g)} \rightleftharpoons 2HI_{(g)}$$
2.00 mol of H$_{2(g)}$ and 3.00 mol of I$_{2(g)}$ are placed in a 1.00 L reaction vessel at 1100 K. What is the equilibrium concentration of each gas?

### What Is Required?

You need to find [H$_2$], [I$_2$], and [HI].

### What Is Given?

You have the balanced chemical equation. You know the equilibrium constant for the reaction, $K_c$ = 25. You also know the concentrations of the reactants and product: [H$_2$]$_i$ = 2.00 mol/L, [I$_2$]$_i$ = 3.00 mol/L, and [HI]$_i$ = 0.

### Plan Your Strategy

**Step 1** Set up an ICE table. Let the change in molar concentrations of the reactants be $x$. Use the stoichiometry of the chemical equation to write expressions for the equilibrium concentrations. Record these expressions in your ICE table.

**Step 2** Write the equilibrium expression. Substitute the expressions for the equilibrium concentrations into the equilibrium expression. Rearrange the equilibrium expression into the form of a quadratic equation. Solve the quadratic equation for $x$.

**Step 3** Substitute $x$ into the equilibrium line of the ICE table to find the equilibrium concentrations.

### Act on Your Strategy

**Step 1** Set up an ICE table.

| Concentration (mol/L) | H$_{2(g)}$ + | I$_{2(g)}$ ⇌ | 2HI$_{(g)}$ |
|---|---|---|---|
| Initial | 2.00 | 3.00 | 0 |
| Change | $-x$ | $-x$ | $+2x$ |
| Equilibrium | 2.00 − $x$ | 3.00 − $x$ | $+2x$ |

**Step 2** Write and solve the equilibrium expression.

$$K_c = \frac{[HI]^2}{[H_2][I_2]}$$

$$25.0 = \frac{(2x)^2}{(2.00 - x)(3.00 - x)}$$

This equation does not involve a perfect square. It must be re-arranged into a quadratic equation.

$$0.840x^2 - 5.00x + 6.00 = 0$$

Recall that a quadratic equation of the form $ax^2 + bx + c = 0$ has the following solution:

$$x = \frac{-b \pm \sqrt{b^2 - 4ac}}{2a}$$

$$\therefore x = \frac{-(-5.00) \pm \sqrt{25.0 - 20.16}}{1.68}$$

$$= \frac{5.00 \pm 2.2}{1.68}$$

$x$ = 4.3 and $x$ = 1.7

> **PROBLEM TIP**
>
> In the equilibrium expression, the numerator is [HI]$^2$. The concentration is 2$x$, so the substitution is (2$x$)$^2$ = 4$x^2$. A common mistake is to write 2$x^2$. You can avoid this mistake by using brackets.

*Continued ...*

**Step 3** The value x = 4.3 is not physically possible. It would result in negative concentrations of H₂ and I₂ at equilibrium. The concentration of each substance at equilibrium is found by substituting x = 1.7 into the last line of the ICE table.

| Concentration (mol/L) | $H_{2(g)}$ | + | $I_{2(g)}$ | ⇌ | $2HI_{(g)}$ |
|---|---|---|---|---|---|
| Equilibrium | 2.00 − 1.7 | | 3.00 − 1.7 | | 3.4 |

Applying the rule for subtraction involving measured values,

[H₂] = 0.3 mol/L

[I₂] = 1.3 mol/L

[HI] = 3.4 mol/L

### Check Your Solution

The coefficients in the chemical equation match the exponents in the equilibrium expression. To check your concentrations, substitute them back into the equilibrium expression.

$$K_c = \frac{3.4^2}{0.3 \times 1.3} = 30$$

Solving the quadratic equation gives a value of x, correct to one decimal place. As a result, [H₂] can have only one significant figure. The calculated value of $K_c$ is equal to the given value, within the error introduced by rounding.

## Practice Problems

**11.** At a certain temperature, hydrogen fluoride gas dissociates.

$2HF_{(g)} \rightleftharpoons H_{2(g)} + F_{2(g)}$

At equilibrium in a 1.0 L reaction vessel, the mixture of gases contained 0.045 mol of $H_{2(g)}$, 0.045 mol of $F_{2(g)}$, and 0.022 mol of $HF_{(g)}$. What is the value of $K_c$?

**12.** At 25°C, the following reaction takes place.

$I_{2(g)} + Cl_{2(g)} \rightleftharpoons 2ICl_{(g)}$

A chemist determined that a 10 L container contained these amounts of gases at equilibrium: I₂ = 0.15 mol, $Cl_{2(g)}$ = 0.15 mol, and ICl = 1.4 mol. What is the value of $K_c$ for the reaction at 25°C?

**13.** A chemist was studying the following reaction.

$SO_{2(g)} + NO_{2(g)} \rightleftharpoons NO_{(g)} + SO_{3(g)}$

In a 1.0 L container, the chemist added $1.7 \times 10^{-1}$ mol of $SO_{2(g)}$ to $1.1 \times 10^{-1}$ mol of $NO_{2(g)}$. At equilibrium, the concentration of $SO_{3(g)}$ was found to be 0.089 mol/L. What is the value of $K_c$ for the reaction at this temperature?

**14.** Phosgene, $COCl_{2(g)}$, is an extremely toxic gas. It was used during World War I. Today it is used to manufacture pesticides, pharmaceuticals, dyes, and polymers. It is prepared by mixing carbon monoxide and chlorine gas.

$CO_{(g)} + Cl_{2(g)} \rightleftharpoons COCl_{2(g)}$

0.055 mol of CO$_{(g)}$ and 0.072 mol of Cl$_{2(g)}$ are placed in a 5.0 L container. At a certain temperature, the equilibrium constant is 20. What are the equilibrium concentrations of the mixture?

**15.** Hydrogen bromide decomposes at 700 K.

$2HBr_{(g)} \rightleftharpoons H_{2(g)} + Br_{2(g)} \quad K_c = 4.2 \times 10^{-9}$

0.090 mol of HBr is placed in a 2.0 L reaction vessel and heated to 700 K. What is the equilibrium concentration of each gas?

## Percent Reaction and $K_c$

Sometimes the extent to which a chemical reaction occurs is expressed as a *percent reaction*—the percentage of a reactant molecule that reacts to form products. Accordingly, if a chemist knows the percent reaction for one of the reactants in an equilibrium, this value may be multiplied by the initial reactant concentration to determine how the concentrations of reactants and products change as the system comes to equilibrium. The following Sample Problem shows how these problems are a variation of previous ICE table equilibrium calculations.

### Sample Problem

### Determining $K_c$ from Initial Reactant Concentrations and Percent Reaction

**Problem**

In a closed 3.00 L flask, 14.2 g of boron nitride, BN$_{(s)}$, is mixed with 2.70 mol of chlorine gas. The mixture is then heated to 373 K. Two gaseous equilibrium products result: boron trichloride and nitrogen. What is the value of the equilibrium constant if the percent reaction for chlorine is 16.5%?

**What is Required?**

You need to find the equilibrium constant, $K_c$.

**What is Given?**

You are given data that you can use to determine the balanced chemical equation and the initial concentration of chlorine gas. Notable among the given values is the percent reaction value, which indicates that you can use the specific problem solving strategy below.

**Plan your Strategy**

**Step 1** Write the balanced chemical equation.

**Step 2** Calculate the initial concentrations. Note that if any pure solids or liquids are present, it is unnecessary to calculate their amounts since these pure substances are not included in the equilibrium constant expression.

**Step 3** Use the percent reaction data given for the reactant to determine the change in concentration for that reactant.

**Step 4** Set up an ICE table. Record the initial concentrations you calculated in step 1 in your ICE table and the change in the reactant concentration

*Continued...*

Continued ...

calculated in step 3. Use the stoichiometry of the chemical equation to determine the change in concentration for all other substances contained in the equilibrium constant expression.

**Step 5** Calculate the equilibrium concentration of each substance in the equilibrium constant expression using the ICE table, and determine the $K_c$ value by substitution of these values into the $K_c$ expression.

## Act on Your Strategy

**Step 1** Balance the equation:

$$2BN_{(s)} + 3Cl_{2(g)} \rightleftharpoons 2BCl_{3(g)} + N_{2(g)}$$

Note that the [BN] initial is not to be calculated. It is a pure solid and is not included in the $K_c$ expression.

**Step 2** To calculate the change in concentration for $Cl_2$, start by converting to mol/L:

$[Cl_2]$ = 2.70 mol/3.00 L = 0.90 mol/L

**Step 3** Change in $[Cl_2]$ = 0.90 mol/L × 0.165 = 0.149 mol/L

Therefore, 0.149 mol/L $Cl_2$ has reacted.

**Step 4** Set up an ICE table.

| Concentration (mol/L) | 2BN | + | $3Cl_2$ | $\rightleftharpoons$ | $2BCl_3$ | + | $N_2$ |
|---|---|---|---|---|---|---|---|
| Initial | | | 0.90 | | 0 | | 0 |
| Change | | | −0.149 | | $+\frac{2}{3}(0.149)$ | | $+\frac{1}{3}(0.149)$ |
| Equilibrium | | | 0.90 − 0.149 | | $+\frac{2}{3}(0.149)$ | | $+\frac{1}{3}(0.149)$ |

$$\begin{aligned}[BCl_3]\text{ change} &= [Cl_2]\text{ change} \times \text{mole ratio}\\ &= 0.149\text{ mol/L } \cancel{Cl_2} \times \frac{2\ \cancel{\text{mol}}\ BCl_3}{3\ \cancel{\text{mol}}\ \cancel{Cl_2}}\\ &= 0.0993\text{ mol/L } BCl_3\end{aligned}$$

$$\begin{aligned}[N_2]\text{ change} &= [Cl_2]\text{ change} \times \text{mole ratio}\\ &= 0.149\text{ mol/L } \cancel{Cl_2} \times \frac{1\ \cancel{\text{mol}}\ N_2}{3\ \cancel{\text{mol}}\ \cancel{Cl_2}}\\ &= 0.0497\text{ mol/L } N_2\end{aligned}$$

**Step 5** At equilibrium:

$[Cl_2]$ = 0.751

$[BCl_3]$ = 0.0993

$[N_2]$ = 0.0497

Substitute into the equilibrium expression and solve.

$$\begin{aligned}K_c &= \frac{(0.0993)^2(0.0497)}{(0.751)^3}\\ &= 0.001157\end{aligned}$$

Therefore, $K_c = 1.16 \times 10^{-3}$

## Check Your Solution

The percent reaction for chlorine is 16.5%, which means that only a small amount of this reactant is converted into product. Therefore, a small value for $K_c$ is reasonable.

## Practice Problems

16. 2.50 mol of oxygen and 2.50 mol of fluorine gas are placed in a 2.00 L glass container at room temperature. The container is heated to 400 K and the following equilibrium is established:

    $O_{2(g)} + 2F_{2(g)} \rightleftharpoons 2OF_{2(g)}$

    If 37.2% of the fluorine reacts, what is the value of the equilibrium constant?

17. When 0.846 mol/L of dinitrogen tetrachloride gas is placed in a closed reaction vessel, 73.5% of it decomposes into dinitrogen dichloride gas and chlorine gas. Calculate the equilibrium constant.

18. When 3.00 mol/L of phosphorus heptabromide gas is initially placed in a sealed 3.00 L container at 295 K, it is the only gas present. Upon heating the container to 500 K and maintaining the temperature, an equilibrium is established and phosphorus tribromide gas and bromine vapour are observed products. Calculate the equilibrium constant, given the fact that 85.2% of the $PBr_{7(g)}$ reacts.

19. Nitrogen gas is relatively unreactive at room temperature, but combines with oxygen at higher temperatures:

    $N_{2(g)} + O_{2(g)} \rightleftharpoons 2NO_{(g)}$

    When equilibrium is attained at a temperature of 1800 K, 5.3% of nitrogen will react with oxygen. What is the value of $K_c$ at this temperature?

## Qualitatively Interpreting the Equilibrium Constant

The value for a percent reaction indicates the extent of the reaction. It is linked to the size of $K_c$. As you know, the equilibrium expression is always written with the product terms over the reactant terms. Therefore, a large $K_c$ means that the concentration of products is larger than the concentration of reactants at equilibrium. When referring to a reaction with a large $K_c$, chemists often say that the position of equilibrium lies to the right, or that it favours the products. Similarly, if $K_c$ is small, the concentration of reactants is larger than the concentration of products. Chemists say that the position of equilibrium lies to the left, or that it favours reactants. Thus, the following general statements are true:

- *When K > 1, products are favoured.* The equilibrium lies far to the right. Reactions where $K$ is greater than $10^{10}$ are usually regarded as going to completion.

- *When K ≈ 1, there are approximately equal concentrations of reactants and products at equilibrium.*

- *When K < 1, reactants are favoured.* The equilibrium lies far to the left. Reactions in which K is smaller than $10^{-10}$ are usually regarded as not taking place at all.

Notice that the subscript "c" has been left off $K$ in these general statements. This reflects the fact that there are other equilibrium constants that these statements apply to, not just equilibrium constants involving concentrations. You will learn about other types of equilibrium constants in the next two chapters. For the rest of this chapter, though, you will continue to see the subscript "c" used.

In the Sample Problem and Practice Problems below, you will consider how temperature affects the extent of a reaction. Keep in mind that the size of $K_c$ is not related to the time that a reaction takes to achieve equilibrium. Very large values of $K_c$ may be associated with reactions that take place extremely slowly. The time that a reaction takes to reach equilibrium depends on the rate of the reaction. This is determined by the size of the activation energy.

> **CONCEPT CHECK**
>
> Does a small value of $K_c$ indicate a slow reaction? Use an example to justify your answer.

## Sample Problem
### Temperature and the Extent of a Reaction

**Problem**

Consider the reaction of carbon monoxide and chlorine to form phosgene, $COCl_{2(g)}$.

$$CO_{(g)} + Cl_{2(g)} \rightleftharpoons COCl_{2(g)}$$

At 870 K, the value of $K_c$ is 0.20. At 370 K, the value of $K_c$ is $4.6 \times 10^7$. Based on only the values of $K_c$, is the production of $COCl_{2(g)}$ more favourable at the higher or lower temperature?

**What Is Required?**

You need to choose the temperature that favours the greater concentration of product.

**What Is Given?**

$K_c = 0.2$ at 870 K
$K_c = 4.6 \times 10^7$ at 370 K

**Plan Your Strategy**

Phosgene is a product in the chemical equation. A greater concentration of product corresponds to a larger value of $K_c$.

**Act on Your Strategy**

The value of $K_c$ is larger at 370 K. The position of equilibrium lies far to the right. It favours the formation of $COCl_2$ at the lower temperature.

**Check Your Solution**

The problem asked you to choose the temperature for a larger concentration of the product. $K_c$ is expressed as a fraction with product terms in the numerator. Therefore, the larger value of $K_c$ corresponds to the larger concentration of product.

## Practice Problems

20. For the reaction $H_{2(g)} + I_{2(g)} \rightleftharpoons 2HI_{(g)}$, the value of $K_c$ is 25.0 at 1100 K and $8.0 \times 10^2$ at room temperature, 300 K. Which temperature favours the dissociation of $HI_{(g)}$ into its component gases?

21. Three reactions, and their equilibrium constants, are given below.
    I. $N_{2(g)} + O_{2(g)} \rightleftharpoons 2NO_{(g)}$    $K_c = 4.7 \times 10^{-31}$
    II. $2NO_{(g)} + O_{2(g)} \rightleftharpoons 2NO_{2(g)}$    $K_c = 1.8 \times 10^{-6}$
    III. $N_2O_{4(g)} \rightleftharpoons 2NO_{2(g)}$    $K_c = 0.025$

Arrange these reactions in the order of their tendency to form products.

22. Identify each reaction as essentially going to completion or not taking place.
    (a) $N_{2(g)} + 3Cl_{2(g)} \rightleftharpoons 2NCl_{3(g)}$   $K_c = 3.0 \times 10^{11}$
    (b) $2CH_{4(g)} \rightleftharpoons C_2H_{6(g)} + H_{2(g)}$   $K_c = 9.5 \times 10^{-13}$
    (c) $2NO_{(g)} + 2CO_{(g)} \rightleftharpoons N_{2(g)} + 2CO_{2(g)}$   $K_c = 2.2 \times 10^{59}$

23. Most metal ions combine with other ions in solution. For example, in aqueous ammonia, silver(I) ions are at equilibrium with different complex ions.
    $[Ag(H_2O)_2]^+_{(aq)} + 2NH_{3(aq)} \rightleftharpoons [Ag(NH_3)_2]^+_{(aq)} + 2H_2O_{(\ell)}$
    At room temperature, $K_c$ for this reaction is $1 \times 10^7$. Which of the two silver complex ions is more stable? Explain your reasoning.

24. Consider the following reaction.
    $H_{2(g)} + Cl_{2(g)} \rightleftharpoons 2HCl_{(g)}$   $K_c = 2.4 \times 10^{33}$ at 25°C
    $HCl_{(g)}$ is placed in a reaction vessel. To what extent do you expect the equilibrium mixture to dissociate into $H_{2(g)}$ and $Cl_{2(g)}$?

## The Meaning of a Small Equilibrium Constant

Understanding the meaning of a small equilibrium constant can sometimes help to simplify a calculation that would otherwise involve a quadratic equation. When $K_c$ is small compared with the initial concentration, the value of the initial concentration minus $x$ is approximately equal to the initial concentration. Thus, you can ignore $x$. Of course, if the initial concentration of a substance is zero, any equilibrium concentration of the substance, no matter how small, is significant. In general, values of $K_c$ are not measured with accuracy better than 5%. Therefore, making the approximation is justified if the calculation error you introduce is less than 5%.

To help you decide whether or not the approximation is justified, divide the initial concentration by the value of $K_c$. If the answer is greater than 500, the approximation is justified. If the answer is between 100 and 500, it may be justified. If the answer is less than 100, it is not justified. The equilibrium expression must be solved in full.

### Sample Problem

#### Using the Approximation Method

**Problem**

The atmosphere contains large amounts of oxygen and nitrogen. The two gases do not react, however, at ordinary temperatures. They do react at high temperatures, such as the temperatures produced by a lightning flash or a running car engine. In fact, nitrogen oxides from exhaust gases are a serious pollution problem.

A chemist is studying the following equilibrium reaction.
$$N_{2(g)} + O_{2(g)} \rightleftharpoons 2NO_{(g)}$$

*Continued...*

> **Continued...**
>
> The chemist puts 0.085 mol of N$_{2(g)}$ and 0.038 mol of O$_{2(g)}$ in a 1.0 L rigid cylinder. At the temperature of the exhaust gases from a particular engine, the value of $K_c$ is $4.2 \times 10^{-8}$. What is the concentration of NO$_{(g)}$ in the mixture at equilibrium?
>
> ### What Is Required?
>
> You need to find the concentration of NO at equilibrium.
>
> ### What Is Given?
>
> You have the balanced chemical equation. You know the value of $K_c$ and the following initial concentrations: [N$_2$] = 0.085 mol/L and [O$_2$] = 0.038 mol/L.
>
> ### Plan Your Strategy
>
> **Step 1** Divide the smallest initial concentration by $K_c$ to determine whether you can ignore the change in concentration.
>
> **Step 2** Set up an ICE table. Let $x$ represent the change in [N$_2$] and [O$_2$].
>
> **Step 3** Write the equilibrium expression. Substitute the equilibrium concentrations into the equilibrium expression. Solve the equilibrium expression for $x$.
>
> **Step 4** Calculate [NO] at equilibrium.
>
> ### Act on Your Strategy
>
> **Step 1** $\dfrac{\text{Smallest initial concentration}}{K_c} = \dfrac{0.038}{4.2 \times 10^{-8}}$
>
> $\qquad\qquad\qquad\qquad\qquad\quad = 9.0 \times 10^5$
>
> Because this is well above 500, you can ignore the changes in [N$_2$] and [O$_2$].
>
> **Step 2**
>
> | Concentration (mol/L) | N$_{2(g)}$ + | O$_{2(g)}$ ⇌ | 2NO$_{(g)}$ |
> |---|---|---|---|
> | Initial | 0.085 | 0.038 | 0 |
> | Change | $-x$ | $-x$ | $+2x$ |
> | Equilibrium | $0.085 - x \approx 0.085$ | $0.038 - x \approx 0.038$ | $2x$ |
>
> **Step 3** $\qquad K_c = \dfrac{[\text{NO}]^2}{[\text{N}_2][\text{O}_2]}$
>
> $\qquad 4.2 \times 10^{-8} = \dfrac{(2x)^2}{0.085 \times 0.038}$
>
> $\qquad\qquad\qquad = \dfrac{4x^2}{0.003\,23}$
>
> $\qquad\qquad x = \sqrt{3.39 \times 10^{-11}}$
>
> $\qquad\qquad\quad = 5.82 \times 10^{-6}$
>
> **Step 4** [NO] = $2x$
>
> Therefore, the concentration of NO$_{(g)}$ at equilibrium is $1.2 \times 10^{-5}$ mol/L.

**Check Your Solution**

First, check your assumption that x is negligible compared with the initial concentrations. Your assumption is valid because, using the rules for subtracting measured quantities, $0.038 - (5.8 \times 10^{-6}) = 0.038$. Next, check the equilibrium values:

$$\frac{(1.2 \times 10^{-5})^2}{0.0085 \times 0.038} = 4.5 \times 10^{-8}$$

This is equal to the equilibrium constant, within rounding errors in the calculation.

## Practice Problems

**25.** The following equation represents the equilibrium reaction for the dissociation of phosgene gas.

$COCl_{2(g)} \rightleftharpoons CO_{(g)} + Cl_{2(g)}$

At 100°C, the value of $K_c$ for this reaction is $2.2 \times 10^{-8}$. The initial concentration of $COCl_{2(g)}$ in a closed container at 100°C is 1.5 mol/L. What are the equilibrium concentrations of $CO_{(g)}$ and $Cl_{2(g)}$?

**26.** Hydrogen sulfide is a poisonous gas with a characteristic, offensive odour. It dissociates at 1400°C, with $K_c$ equal to $2.4 \times 10^{-4}$.

$H_2S_{(g)} \rightleftharpoons 2H_{2(g)} + S_{2(g)}$

4.0 mol of $H_2S$ is placed in a 3.0 L container. What is the equilibrium concentration of $H_{2(g)}$ at 1400°C?

**27.** At a certain temperature, the value of $K_c$ for the following reaction is $3.3 \times 10^{-12}$.

$2NCl_{3(g)} \rightleftharpoons N_{2(g)} + 3Cl_{2(g)}$

A certain amount of nitrogen trichloride, $NCl_{3(g)}$, is put in a 1.0 L reaction vessel at this temperature. At equilibrium, $4.6 \times 10^{-4}$ mol of $N_{2(g)}$ is present. What amount of $NCl_{3(g)}$ was put in the reaction vessel?

**28.** At a certain temperature, the value of $K_c$ for the following reaction is $4.2 \times 10^{-8}$.

$N_{2(g)} + O_{2(g)} \rightleftharpoons 2NO_{(g)}$

0.45 mol of $N_{2(g)}$ and 0.26 mol of $O_{2(g)}$ are put in a 6.0 L reaction vessel. What is the equilibrium concentration of $NO_{(g)}$ at this temperature?

**29.** At a particular temperature, $K_c$ for the decomposition of carbon dioxide gas is $2.0 \times 10^{-6}$.

$2CO_{2(g)} \rightleftharpoons 2CO_{(g)} + O_{2(g)}$

3.0 mol of $CO_2$ is put in a 5.0 L container. Calculate the equilibrium concentration of each gas.

## Section Summary

In this section, you learned that the equilibrium constant, $K_c$, is a ratio of product concentrations to reactant concentrations. You used concentrations to find $K$, and you used $K$ to find concentrations. You also used an ICE table to track and summarize the initial, change, and equilibrium quantities in a reaction. You found that the value of $K_c$ is small for reactions that reach equilibrium with a high concentration of reactants, and the value of $K_c$ is large for reactions that reach equilibrium with a low concentration of reactants. In the next section, you will learn how to determine whether or not a reaction is at equilibrium, and, if it is not, in which direction it will go to achieve equilibrium.

## Section Review

**1** Write equilibrium expressions for each reaction.

(a) $SbCl_{5(g)} \rightleftharpoons SbCl_{3(g)} + Cl_{2(g)}$

(b) $2H_{2(g)} + 2NO_{(g)} \rightleftharpoons N_{2(g)} + 2H_2O_{(g)}$

(c) $2H_2S_{(g)} + CH_{4(g)} \rightleftharpoons 4H_{2(g)} + CS_{2(g)}$

(d) $P_{4(s)} + 3O_{2(g)} \rightleftharpoons 2P_2O_{3(g)}$

(e) $7N_{2(g)} + 2S_{8(s)} \rightleftharpoons 2N_3S_{4(s)} + 4N_2S_{2(g)}$

**2** When 1.0 mol of ammonia gas is injected into a 0.50 L flask, the following reaction proceeds to equilibrium.

$2NH_{3(g)} \rightleftharpoons N_{2(g)} + 3H_{2(g)}$

At equilibrium, 0.30 mol of hydrogen gas is present.

(a) Calculate the equilibrium concentrations of $N_{2(g)}$ and $NH_{3(g)}$.

(b) What is the value of $K_c$?

**3** At a certain temperature, $K_c$ for the following reaction between sulfur dioxide and nitrogen dioxide is 4.8.

$SO_{2(g)} + NO_{2(g)} \rightleftharpoons NO_{(g)} + SO_{3(g)}$

$SO_{2(g)}$ and $NO_{2(g)}$ have the same initial concentration: 0.36 mol/L. What amount of $SO_{3(g)}$ is present in a 5.0 L container at equilibrium?

**4** Phosphorus trichloride reacts with chlorine to form phosphorus pentachloride.

$PCl_{3(g)} + Cl_{2(g)} \rightleftharpoons PCl_{5(g)}$

0.75 mol of $PCl_3$ and 0.75 mol of $Cl_2$ are placed in a 8.0 L reaction vessel at 500 K. What is the equilibrium concentration of the mixture? The value of $K_c$ at 500 K is 49.

**5** Hydrogen gas has several advantages and disadvantages as a potential fuel. Hydrogen can be obtained by the thermal decomposition of water at high temperatures.

$2H_2O_{(g)} \rightleftharpoons 2H_{2(g)} + O_{2(g)} \quad K_c = 7.3 \times 10^{-18}$ at 1000°C

(a) The initial concentration of water in a reaction vessel is 0.055 mol/L. What is the equilibrium concentration of $H_{2(g)}$ at 1000°C?

(b) Comment on the practicality of the thermal decomposition of water to obtain $H_{2(g)}$.

# Predicting the Direction of a Reaction

## 13.3

So far in this chapter, you have worked with reactions that have reached equilibrium. What if a reaction has not yet reached equilibrium, however? How can you predict the direction in which the reaction must proceed to reach equilibrium? To do this, you substitute the concentrations of reactants and products into an expression that is identical to the equilibrium expression. Because these concentrations may not be the concentrations that the equilibrium system would have, the expression is given a different name: the reaction quotient. The **reaction quotient**, $Q_c$, is an expression that is identical to the equilibrium constant expression, *but* its value is calculated using concentrations that are not necessarily those at equilibrium.

### The Relationship Between the Equilibrium Constant and the Reaction Quotient

Recall the general reaction $aP + bQ \rightleftharpoons cR + dS$. The reaction quotient expression for this reaction is

$$Q_c = \frac{[R]^c[S]^d}{[P]^a[Q]^b}$$

You can calculate a value for $Q_c$ by substituting the concentration of each substance into this expression. If the value of $Q_c$ is equal to $K_c$, the system must be at equilibrium. If $Q_c$ is greater than $K_c$, the numerator must be very large. The concentrations of the chemicals on the right side of the equation must be more than their concentrations at equilibrium. In this situation, the system attains equilibrium by moving to the left. Conversely, if $Q_c$ is less than $K_c$, the system attains equilibrium by moving to the right. Figure 13.5 summarizes these relationships between $Q_c$ and $K_c$.

**Section Preview/Outcomes**

In this section, you will

- **explain** how to use $Q_c$ and Le Châtelier's principle to predict the direction in which a chemical system at equilibrium will shift when concentration changes
- **apply** Le Châtelier's principle to **make** and **test** predictions about how different factors affect a chemical system at equilibrium
- **communicate** your understanding of the following terms: *reaction quotient ($Q_c$), Le Châtelier's principle*

reactants ⟶ products          equilibrium          reactants ⟵ products

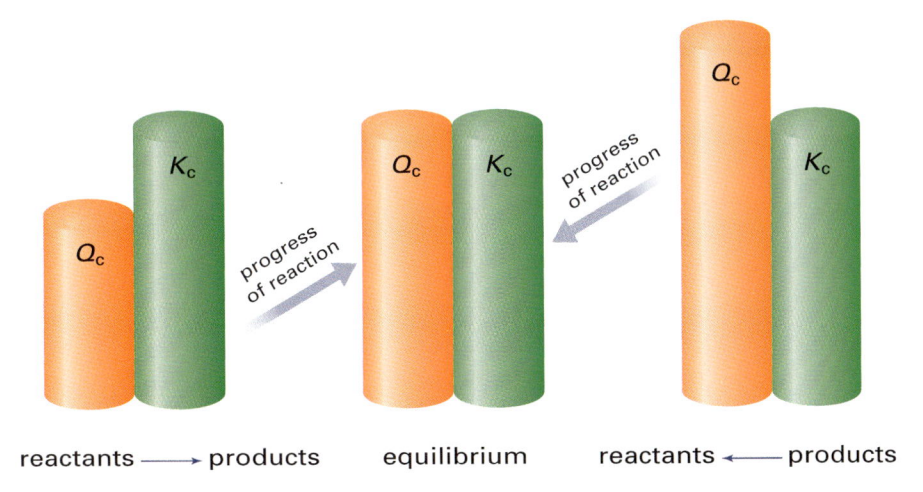

**Figure 13.5** This diagram shows how $Q_c$ and $K_c$ determine reaction direction. When $Q_c < K_c$, the system attains equilibrium by moving to the right, favouring products. When $Q_c = K_c$, the system is at equilibrium. When $Q_c > K_c$, the system attains equilibrium by moving to the left, favouring reactants.

The next Sample Problem shows you how to calculate $Q_c$ and interpret its value.

## Sample Problem

### Determining the Direction of Shift to Attain Equilibrium

**Problem**

Ammonia is one of the world's most important chemicals, in terms of the quantity manufactured. Some ammonia is processed into nitric acid and various polymers. Roughly 80% of ammonia is used to make fertilizers, such as ammonium nitrate. In the Haber process for manufacturing ammonia, nitrogen and hydrogen combine in the presence of a catalyst.

$$N_{2(g)} + 3H_{2(g)} \rightleftharpoons 2NH_{3(g)}$$

At 500°C, the value of $K_c$ for this reaction is 0.40. The following concentrations of gases are present in a container at 500°C: $[N_{2(g)}] = 0.10$ mol/L, $[H_{2(g)}] = 0.30$ mol/L, and $[NH_{3(g)}] = 0.20$ mol/L. Is this mixture of gases at equilibrium? If not, in which direction will the reaction go to reach equilibrium?

**What Is Required?**

You need to calculate $Q_c$ and interpret its value.

**What Is Given?**

You have the balanced chemical equation. You know that the value of $K_c$ is 0.40. You also know the concentrations of the gases: $[N_2] = 0.10$ mol/L, $[H_2] = 0.30$ mol/L, and $[NH_3] = 0.20$ mol/L.

**Plan Your Strategy**

Write the expression for $Q_c$, and then calculate its value. Compare the value of $Q_c$ with the value of $K_c$. Decide whether the system is at equilibrium and, if not, in which direction the reaction will go.

**Act on Your Strategy**

$$Q_c = \frac{[NH_3]^2}{[N_2][H_2]^3}$$

$$= \frac{(0.20)^2}{(0.10)(0.30)^3}$$

$$= 14.8$$

$\therefore Q_c > 0.40$

The system is not at equilibrium because the value of $Q_c$ is greater than the value of $K_c$. The reaction will proceed by moving to the left.

**Check Your Solution**

Check your calculation of $Q_c$. (Errors in evaluating fractions with exponents are common.) The value of $K_c$ is less than one, so you would expect the numerator to be less than the denominator. This is difficult to evaluate without a calculator, however, because of the powers.

## Practice Problems

**30.** The following reaction takes place inside a cylinder with a movable piston.

$2NO_{2(g)} \rightleftharpoons N_2O_{4(g)}$

At room temperature, the equilibrium concentrations inside the cylinder are $[NO_2] = 0.0206$ mol/L and $[N_2O_4] = 0.0724$ mol/L.

(a) Calculate the value of $K_c$.

(b) Calculate the concentration of each gas at the moment that the piston is used to halve the volume of the reacting mixture. Assume that the temperature remains constant.

(c) Determine the value of $Q_c$ when the volume is halved.

(d) Predict the direction in which the reaction will proceed to re-establish equilibrium.

**31.** Ethyl acetate is an ester that can be synthesized by reacting ethanoic acid (acetic acid) with ethanol. At room temperature, the equilibrium constant for this reaction is 2.2.

$CH_3COOH_{(\ell)} + CH_3CH_2OH_{(\ell)} \rightleftharpoons CH_3COOCH_2CH_{3(l)} + H_2O_{(\ell)}$

Various samples were analyzed. The concentrations are given in the table below. Decide whether each sample is at equilibrium. If it is not at equilibrium, predict the direction in which the reaction will proceed to establish equilibrium.

| Sample | [CH₃COOH] (mol/L) | [CH₃CH₂OH] (mol/L) | [CH₃COOCH₂CH₃] (mol/L) | [H₂O] (mol/L) |
|---|---|---|---|---|
| (a) | 0.10 | 0.10 | 0.10 | 0.10 |
| (b) | 0.084 | 0.13 | 0.16 | 0.28 |
| (c) | 0.14 | 0.21 | 0.33 | 0.20 |
| (d) | 0.063 | 0.11 | 0.15 | 0.17 |

**32.** In the past, methanol was obtained by heating wood without allowing the wood to burn. The products were collected, and methanol (sometimes called "wood alcohol") was separated by distillation. Today methanol is manufactured by reacting carbon monoxide with hydrogen gas.

$CO_{(g)} + 2H_{2(g)} \rightleftharpoons CH_3OH_{(g)}$

At 210°C, $K_c$ for this reaction is 14.5. A gaseous mixture at 210°C contains the following concentrations of gases: $[CO] = 0.25$ mol/L, $[H_2] = 0.15$ mol/L, and $[CH_3OH] = 0.36$ mol/L. What will be the direction of the reaction if the gaseous mixture reaches equilibrium?

## Le Châtelier's Principle

What happens to a system at equilibrium if the concentration of one of the reacting chemicals is changed? This question has practical importance, because many manufacturing processes are continual. Products are removed and more reactants are added without stopping the process. For example, consider the Haber process that was mentioned in the previous Sample Problem.

$$N_{2(g)} + 3H_{2(g)} \rightleftharpoons 2NH_{3(g)}$$

At 500°C, $K_c$ is 0.40. The gases have the following concentrations: [N$_2$] = 0.10 mol/L, [H$_2$] = 0.10 mol/L, and [NH$_3$] = 0.0063 mol/L. (You can check this by calculating $Q_c$ and making sure that it is equal to $K_c$.)

Ammonia can be removed from the equilibrium mixture by cooling because ammonia liquefies at a higher temperature than either nitrogen or hydrogen. What happens to an equilibrium mixture if some ammonia is removed? To find out, you need to calculate $Q_c$.

$$Q_c = \frac{[NH_3]^2}{[N_2][H_2]^3}$$

If ammonia is removed from the equilibrium mixture, the numerator will decrease, so $Q_c$ will be smaller than $K_c$. When $Q_c < K_c$, the system attains equilibrium by moving to the right. Why? The system must respond to the removal of ammonia by forming more ammonia. Similarly, if nitrogen, hydrogen, or both were added, the system would re-establish equilibrium by shifting to the right. As long as the temperature remains constant, the re-established equilibrium will have the same $K_c$ as the original equilibrium.

A French chemist, Henri Le Châtelier, experimented with various chemical equilibrium systems. (See Figure 13.6.) In 1888, Le Châtelier summarized his work on changes to equilibrium systems in a general statement called **Le Châtelier's principle**. It may be stated as follows: *A dynamic equilibrium tends to respond so as to relieve the effect of any change in the conditions that affect the equilibrium.*

Le Châtelier's principle predicts the way that an equilibrium system responds to change. For example, when the concentration of a substance in a reaction mixture is changed, Le Châtelier's principle qualitatively predicts what you can show quantitatively by evaluating the reaction quotient. If products are removed from an equilibrium system, more products must be formed to relieve the change. This is just as you would predict, because $Q_c$ will be less than $K_c$.

Le Châtelier's principle also predicts what will happen when other changes are made to an equilibrium system. For example, you can use Le Châtelier's principle to predict the effects of changing the volume of a cylinder that contains a mixture of gases, or the effects of changing the temperature of an equilibrium system. In the next investigation, you will use Le Châtelier's principle to predict the effects of making various changes to different equilibrium systems. The Chemistry Bulletin that follows the investigation considers applications of Le Châtelier's principle to areas outside of chemistry. Afterward, you will continue your studies of Le Châtelier's principle and equilibrium systems in chemistry.

**Figure 13.6** Henri Le Châtelier (1850–1936) specialized in mining engineering. He studied the thermodynamics involved in heating ores to obtain metals. His results led to important advances in the understanding of equilibrium systems.

**Web LINK**

www.mcgrawhill.ca/links/atlchemistry

If you would like to reinforce your understanding of Le Châtelier's principle, go to the web site above and click on **Electronic Learning Partner**.

# Investigation 13-B

**SKILL FOCUS**
Predicting
Performing and recording
Analyzing and interpreting

## Perturbing Equilibrium

In this investigation, you will use Le Châtelier's principle to predict the effect of changing one factor in each system at equilibrium. Then you will test your prediction using a colour change or the appearance (or disappearance) of a precipitate.

### Question
How can Le Châtelier's principle qualitatively predict the effect of a change in a chemical equilibrium?

### Predictions
Your teacher will give you a table that lists four equilibrium systems and the changes you will make to each system. In the appropriate column, record your predictions for each test. If you predict that the change will cause the system to re-attain equilibrium by shifting toward the reactants, record "left." If you predict that the system will re-establish equilibrium by shifting toward the products, record "right."

### Safety Precautions

- Potassium chromate and barium chloride are toxic.
- Hydrochloric acid and sodium hydroxide are corrosive. Wash any spills on your skin or clothing with plenty of cool water. Inform your teacher immediately.

### Materials

**Part 1**
3 test tubes
0.1 mol/L $K_2CrO_{4(aq)}$
1 mol/L $HCl_{(aq)}$
1 mol/L $NaOH_{(aq)}$
1 mol/L $Fe(NO_3)_{3(aq)}$
1 mol/L $BaCl_{2(aq)}$

**Part 2**
25 mL beaker
2 test tubes
test tube rack
scoopula
0.01 mol/L $NH_{3(aq)}$
6.0 mol/L $HCl_{(aq)}$
phenolphthalein solution
$NH_4Cl_{(s)}$

**Part 3**
4 small test tubes
$CoCl_{2(s)}$
ethanol
concentrated $HCl_{(aq)}$ in a dropper bottle
0.1 mol/L $AgNO_{3(aq)}$ in a dropper bottle
distilled water in a dropper bottle
25 mL or 50 mL beaker
test tube rack
test tube holder
hot-water bath (prepared by your teacher)
cold-water bath (prepared by your teacher)

**Part 4**
small piece of copper
concentrated nitric acid
test tube
test tube rack
one-hole stopper
glass delivery tube
short length of rubber tubing
syringe with a cap or rubber stopper to seal the tip
$NO_{2(g)}/N_2O_{4(g)}$ tubes
boiling water
ice water

*continued...*

continued

## Procedure

### Part 1 The Chromate/Dichromate Equilibrium

**Equilibrium System**

$$H^+_{(aq)} + 2CrO_4^{2-}_{(aq)} \rightleftharpoons Cr_2O_7^{2-}_{(aq)} + OH^-_{(aq)}$$

yellowish orange

1. Pour about 5 mL of 0.1 mol/L $K_2CrO_{4(aq)}$ into each of three test tubes. You will use test tube 1 as a colour reference.

2. Add 5 drops of $HCl_{(aq)}$ to test tube 2. Record any colour change.

3. Add 5 drops of $NaOH_{(aq)}$ to test tube 2. Record what happens.

4. Finally, add 5 drops of $Fe(NO_3)_{3(aq)}$ to test tube 2. A precipitate of $Fe(OH)_{3(s)}$ should form. Note the colour of the solution above the precipitate.

5. Add 5 drops of $BaCl_{2(aq)}$ to test tube 3. A precipitate of $BaCrO_{4(s)}$ should form. Record the colour of the solution above the precipitate.

6. Dispose of the chemicals as instructed by your teacher.

### Part 2 Changes to a Base Equilibrium System

**Safety Precautions**

Ammonia and hydrochloric acid are corrosive. Wash any spills on your skin or clothing with plenty of cool water. Inform your teacher immediately.

**Equilibrium System**

$$NH_{3(aq)} + H_2O_{(\ell)} \rightleftharpoons NH_4^+_{(aq)} + OH^-_{(aq)}$$

1. Pour about 10 mL of $NH_{3(aq)}$ into a small beaker. Place the beaker on a sheet of white paper. Add 2 drops of phenolphthalein indicator.

2. Divide the solution equally into two small test tubes. To one of the test tubes, add a few small crystals of $NH_4Cl_{(s)}$ on the end of a scoopula. Record your observations.

3. To the other test tube, add a few drops of $HCl_{(aq)}$ until you see a change. Again note the colour change. (The $H^+$ ions combine with the $OH^-$ ions, removing them from the equilibrium mixture.)

4. Dispose of the chemicals as instructed by your teacher.

### Part 3 Concentration and Temperature Changes

**Safety Precautions**

- Concentrated hydrochloric acid is hazardous to your eyes, skin, and clothing. Treat spills with baking powder and copious amounts of cool water. Inform your teacher immediately.
- Ethanol is flammable. Keep samples of ethanol and the supply bottle away from open flames.

**Equilibrium System**

$$Co(H_2O)_6^{2+}_{(aq)} + 4Cl^-_{(aq)} + 50 \text{ kJ/mol} \rightleftharpoons CoCl_4^{2-}_{(aq)} + 6H_2O_{(\ell)}$$

pink                                            blue or purple

1. Measure about 15 mL of ethanol into a small beaker.

2. Record the colour of the $CoCl_2$. Dissolve a small amount (about half the size of a pea) in the beaker of ethanol. The solution should be blue or purple. If it is pink, add drops of concentrated $HCl_{(aq)}$ until the solution is blue-purple.

3. Divide the cobalt solution equally among the four small test tubes. Put one of the test tubes aside as a control.

4. To each of the other three test tubes, add 3 drops of distilled water, one drop at a time. Stir after you add each drop. Record any change in colour that occurs with each drop.

5. To one of the test tubes from step 4, add 5 drops of concentrated HCl, one drop at a time with stirring. Record the results.

6. Silver and chloride ions combine to form a precipitate of AgCl. To the third test tube, add AgNO$_{3(aq)}$, one drop at a time, until no more precipitate appears. Record the colour of the solution as the chloride ions precipitate.

7. Record the colour of the liquid mixture in the fourth test tube. Use a test tube holder to immerse this test tube in the hot-water bath. Record any colour change.

8. Place the test tube from step 7 in the cold-water bath. Record any colour change.

9. Dispose of the chemicals as instructed by your teacher.

### Part 4 Investigating Gaseous Equilibria

**CAUTION** These are not student tests. Your teacher may demonstrate this equilibrium if a suitable fume hood is available for the first test, and if sealed tubes containing a mixture of nitrogen dioxide, NO$_{2(g)}$, and dinitrogen tetroxide, N$_2$O$_{4(g)}$, are available for the second test. If either or both tests are not demonstrated, refer to the photographs that show the changes.

### Safety Precautions

- Concentrated nitric acid is highly corrosive and a strong oxidizing agent.
- Nitrogen dioxide and dinitrogen tetroxide are poisonous gases.

### Equilibrium System

N$_2$O$_{4(g)}$ + 59 kJ/mol ⇌ 2NO$_{2(g)}$

**colourless      brown**

1. Your teacher will use sealed tubes that contain a mixture of N$_2$O$_{4(g)}$ and NO$_{2(g)}$. One tube will be placed in boiling water, and a second tube will be placed in ice water. A third tube (if available) will remain at room temperature as a control. Compare and record the colour of the gas mixture at each temperature.

2. NO$_{2(g)}$ can be prepared by reacting copper with concentrated nitric acid. The gas is poisonous. The reaction, if your teacher performs it, *must* take place in a fume hood.

3. By using a one-hole stopper, glass delivery tube, and a short length of rubber tubing, some NO$_{2(g)}$ can be collected in a syringe. The syringe is then sealed by attaching a cap or by pushing the needle into a rubber stopper.

4. Observe what happens when the syringe plunger is pressed down sharply, changing the volume of the equilibrium mixture. You will observe an immediate change in colour. Then, if the plunger is held in a fixed position, the colour will change over a few seconds as the system re-establishes equilibrium. Carefully record these colour changes.

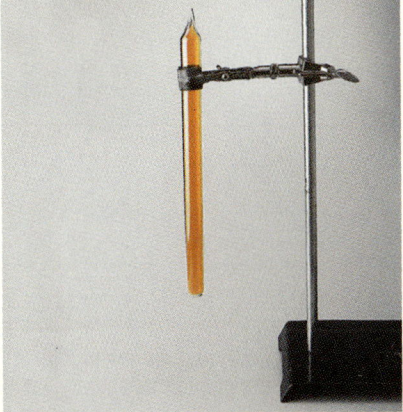

These three tubes contain a mixture of NO$_{2(g)}$ and N$_2$O$_{4(g)}$. The tube on the left is in an ice-water mixture. The centre tube is at room temperature. The tube on the right is in boiling water. Given that NO$_{2(g)}$ is brown, can you explain the shift in equilibrium? Think about Le Châtelier's principle and the energy change of the reaction between the two gases.

*continued...*

**continued**

### Analysis

1. Compare the predictions you made using Le Châtelier's principle with the observations you made in your tests. Account for any differences.

2. In which tests did you increase the concentration of a reactant or a product? Did your observations indicate a shift in equilibrium to form more or less of the reactant or product?

3. In which tests did you decrease the concentration of a reactant or product? Did your observations indicate a shift in equilibrium to form more or less of the reactant or product?

4. Look again at two of the systems you studied:

    $N_2O_{4(g)} + 59 \text{ kJ/mol} \rightleftharpoons 2NO_{2(g)}$

    $Co(H_2O)_6^{2+}{}_{(aq)} + 4Cl^-{}_{(aq)} + 50 \text{ kJ/mol} \rightleftharpoons CoCl_4^{2-}{}_{(aq)} + 6H_2O_{(\ell)}$

    (a) Are these systems endothermic or exothermic when read from left to right?

    (b) When heated, did these systems shift to the left or to the right? In terms of the energy change, was the observed shift in equilibrium toward the endothermic or exothermic side of the reaction?

    (c) Do you think the value of $K_c$ changed or remained the same when the equilibrium mixture was heated? Explain your answer.

5. Think about the $N_2O_{4(g)}/NO_{2(g)}$ equilibrium.

    (a) How was the total pressure of the mixture affected when the plunger was pushed down?

    (b) How was the pressure of the mixture affected by the total number of gas molecules in the syringe?

    (c) Explain the observed shift in equilibrium when the plunger was pushed down. In your explanation, refer to Le Châtelier's principle and the total amount of gas in the syringe.

    (d) What would be the effect, if any, on the following equilibrium system if the volume was reduced? Explain.

    $2IBr_{(g)} \rightleftharpoons I_{2(g)} + Br_{2(g)}$

### Conclusion

6. How did your results compare with your predictions? Discuss and resolve any discrepancies with your class.

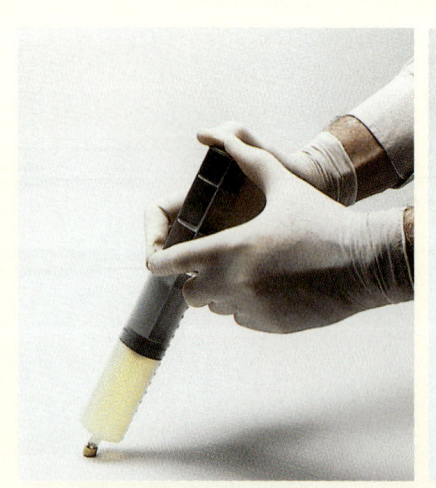

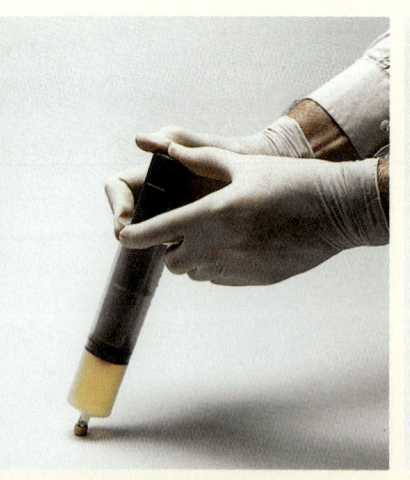

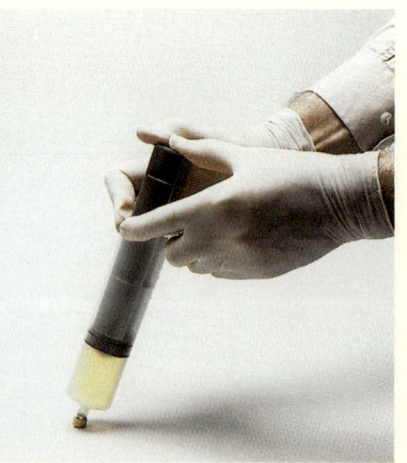

The sealed syringe contains a mixture of $NO_{2(g)}$ and $N_2O_{4(g)}$. The photograph on the left shows an equilibrium mixture at atmospheric pressure. The middle photograph shows that the plunger has been pushed down, increasing the pressure. Two changes cause the darker appearance of the gas mixture. First, the concentration of gases is greater. Second, decreasing the volume heats the gas. This causes a shift toward $NO_{2(g)}$. The photograph on the right shows the result a few seconds after the plunger was pushed down. The gas has cooled back to room temperature. The colour of the mixture is less brown, indicating a shift toward $N_2O_{4(g)}$.

# Chemistry Bulletin

**Science**    **Technology**    **Society**    **Environment**

## Le Châtelier's Principle: Beyond Chemistry

**How is Le Châtelier's principle related to predator-prey interactions in ecosystems?**

Le Châtelier's principle is a general statement about how any system in equilibrium—not just a chemical system—responds to change. Le Châtelier principle concerns the conservation of energy or matter. There are corresponding laws in several other areas of science.

Heinrich Lenz studied the direction of the current that is induced in a conductor as a result of changing the magnetic field near it. You can think of this as the change in a system in electromagnetic equilibrium. Lenz published his law in 1834. It states that when a conductor interacts with a magnetic field, there must be an induced current that opposes the interaction, because of the law of conservation of energy. Lenz's law is used to explain the direction of the induced current in generators, transformers, inductors, and many other systems.

In geology, John Pratt and George Airy introduced the idea of isostacy in 1855. Earth's crust is in gravitational equilibrium in almost all places. The crust responds to changes in the load on it, however, by slowly moving vertically. Examples of how the load on Earth's crust can change include the weathering of mountain ranges and the melting of ice caps. Geologists estimate that the ice over parts of Scandinavia and northern Europe was well over 2 km thick during the last ice age. As a result of the ice melting and the removal of the weight pushing down on Earth's crust, the central part of Scandinavia has risen over 500 m. This uplift is continuing. Here gravitational potential energy is being conserved.

In ecology, ecosystems are really examples of steady state systems. Ecosystems involve a flow of energy in only one direction, from the Sun. Nevertheless, populations of plants and animals develop stable numbers that react to external changes, such as disease and variations in the weather. For example, if the number of carnivores in an ecosystem increases, more herbivores are eaten and the herbivore population decreases. The carnivores then have difficulty finding food. *Their* populations decrease due to competition, emigration, and starvation until a new balance, however delicate, is attained.

Le Châtelier's principle also applies to biology. Homeostasis is the tendency of a body system to remain in a state of equilibrium. Examples of homeostasis include the maintenance of body temperature (homeothermy) and the pH balance of blood.

In economics, the law of supply and demand is similar to Le Châtelier's principle. When the price of a commodity, such as the price of a kilogram of apples, is constant, the market for the commodity is at equilibrium. If the supply of the commodity falls, the equilibrium is changed. The market adjusts by increasing the price, which tends to increase the supply.

### Making Connections

1. Identify and describe two other examples, in science or other fields, that illustrate Le Châtelier's principle.

2. Draw diagrams to show the above examples of Le Châtelier's principle in human physiology, ecology, and economics. Show how different conditions affect the equilibrium and how the systems react to establish a new equilibrium.

### The Effect of Ions on Aqueous Equilibrium Systems

Many important equilibrium systems involve ions in aqueous solution. The *common ion effect* applies Le Châtelier's principle to ions in aqueous solution. As its name suggests, the common ion effect involves adding an ion to a solution in which the ion is already present in solution. It is really a concentration effect. The equilibrium shifts away from the added ion, as predicted by Le Châtelier's principle.

An aqueous solution that contains certain ions can be added to another solution to form a precipitate, if the ions have low solubility. For example, adding silver nitrate solution to test for $Cl^-_{(aq)}$ is effective due to the very low solubility of silver chloride. You can use the precipitation of an insoluble salt to remove almost all of a particular ion from a solution and, as a result, cause a shift in the position of equilibrium of the original solution. The common ion effect is important in the solubility of salts. The precipitation of insoluble salts is used to identify the presence of unknown ions. You will learn more about the common ion effect in later chemistry courses.

### The Effect of Temperature Changes on the Position of Equilibrium

As you know, the value of the equilibrium constant changes with temperature, because the rates of the forward and reverse reactions are affected. Le Châtelier's principle still holds, however. It can be used to predict the effect of temperature changes on a system when the sign of the enthalpy change for the reaction is known. For example, the dissociation of sulfur trioxide is endothermic.

$$2SO_{3(g)} + 197 \text{ kJ} \rightleftharpoons 2SO_{2(g)} + O_{2(g)}$$

As the reaction proceeds from left to right, energy is absorbed by the chemical system and is converted to chemical potential energy. If an equilibrium mixture of these gases is heated, energy is added to the system. Le Châtelier's principle predicts a shift that will relieve the imposed change. Therefore, the shift will tend to remove the added energy. This will happen if the equilibrium shifts to the right, increasing the amount of $SO_{2(g)}$ and $O_{2(g)}$ formed. As a result, the value of $K_c$ will increase. The shift in equilibrium is consistent (as it must be) with the law of conservation of energy.

Suppose that the equilibrium of an endothermic reaction shifted to the left when the system was heated. From right to left, the reaction is exothermic. Thus, a shift to the left would release energy to the surroundings. This would make the mixture hotter. The equilibrium would shift still more to the left, releasing more and more energy. Clearly, this cannot happen without violating the law of conservation of energy.

The effect of temperature on the position of equilibrium can be summarized as follows:

- *Endothermic change* - An increase in temperature shifts the equilibrium to the right, forming more products. The value of $K_c$ increases. A decrease in temperature shifts the equilibrium to the left, forming more reactants. The value of $K_c$ decreases.

- *Exothermic change* - An increase in temperature shifts the equilibrium to the left, forming more reactants. The value of $K_c$ decreases. A decrease in temperature shifts the equilibrium to the right, forming more products. The value of $K_c$ increases.

---

**CHEM FACT**

Chickens have no sweat glands. When the temperature rises, they tend to breathe faster. This lowers the concentration of carbonate ions in their blood. Because eggshells are mostly calcium carbonate, faster-breathing chickens lay eggs with thinner shells. Rather than installing expensive air conditioning, chicken farmers can supply carbonated water for their chickens to drink when the temperature reaches "fowl" highs. How does this relate to Le Châtelier's principle?

## The Effects of Volume and Pressure Changes on Equilibrium

When the volume of a mixture of gases decreases, the pressure of the gases must increase. Le Châtelier's principle predicts a shift in equili-brium to relieve this change. Therefore, the shift must tend to reduce the pressure of the gases. Molecules striking the walls of a container cause gas pressure, so a reduction in gas pressure at constant temperature must mean fewer gas molecules. Consider the following reaction again.

$$2SO_{3(g)} \rightleftharpoons 2SO_{2(g)} + O_{2(g)}$$

There are two gas molecules on the left side of the equation. There are three gas molecules on the right side. Consequently, if the equilibrium shifts to the left, the pressure of the mixture will decrease. *Reducing the volume of an equilibrium mixture of gases, at constant temperature, causes a shift in equilibrium in the direction of fewer gas molecules.* What if there is the same number of gas molecules on both sides of the reaction equation? Changing the volume of the container, as shown in Figure 13.7, has no effect on the position of equilibrium. The value of $K_c$ will be unchanged, as long as there is no change in temperature.

It is possible to increase the total pressure of gases in a rigid container by injecting more gas. If the injected gas reacts with the other gases in the equilibrium mixture, the effect is the same as increasing the concentration of a reactant. The equilibrium shifts to reduce the amount of added gas. If the injected gas does not react with the other gases (if an "inert" gas is added), there is no effect on the equilibrium. This is because the added gas is not part of the equilibrium system. Its addition causes no change in the volume of the container. Nitrogen (because of its low reactivity) and the noble gases are often used as inert gases. If the container expands when an inert gas is added, the effect is the same as increasing the volume and therefore decreasing the pressure of the reacting gases.

## The Effect of a Catalyst on Equilibrium

A catalyst speeds up the *rate* of a reaction, either by allowing a different reaction mechanism or by providing additional mechanisms. The overall effect is to lower the activation energy, which increases the rate of reaction. The activation energy is lowered the same amount for the forward and reverse reactions, however. There is the same increase in reaction rates for both reactions. As a result, a catalyst does not affect the position of equilibrium. It only affects the time that is taken to achieve equilibrium.

Table 13.2 summarizes the effect of a catalyst, and other effects of changing conditions, on a system at equilibrium. The Sample Problem that follows provides an opportunity for you to use Le Châtelier's principle to predict the equilibrium shift in response to various conditions.

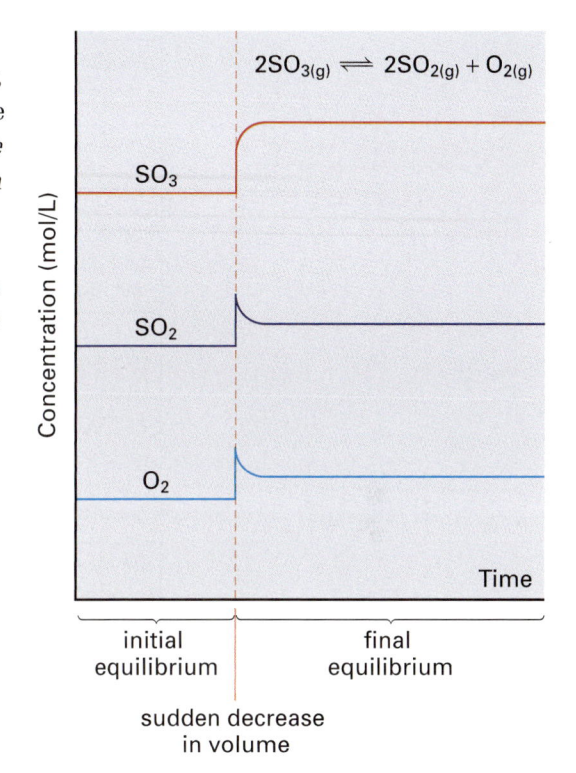

**Figure 13.7** Note the changes in the concentrations of the reacting gases. When the new equilibrium is established, the concentrations are different. Compare the concentrations of the gases in the first equilibrium and second equilibrium.

**Table 13.2** The Effects of Changing Conditions on a System at Equilibrium

| Type of reaction | Change to system | Effect on $K_c$ | Direction of change |
|---|---|---|---|
| all reactions | increasing any reactant concentration, or decreasing any product concentration | no effect | toward products |
| | decreasing any reactant concentration, or increasing any product concentration | no effect | toward reactants |
| | using a catalyst | no effect | no change |
| exothermic | increasing temperature | decreases | toward reactants |
| | decreasing temperature | increases | toward products |
| endothermic | increasing temperature | increases | toward products |
| | decreasing temperature | decreases | toward reactants |
| equal number of reactant and product gas molecules | changing the volume of the container, or adding a non-reacting gas | no effect | no change |
| more gaseous product molecules than reactant gaseous molecules | decreasing the volume of the container at constant temperature | no effect | toward reactants |
| | increasing the volume of the container at constant temperature, or adding a non-reacting gas at contstant pressure | no effect | toward products |
| fewer gaseous product molecules than reactant gaseous molecules | decreasing the volume of the container at constant temperature | no effect | toward products |
| | increasing the volume of the container at constant temperature | no effect | toward reactants |

## Sample Problem

### Using Le Châtelier's Principle

**Problem**

The following equilibrium takes place in a rigid container.

$$PCl_{5(g)} + 56 \text{ kJ} \rightleftharpoons PCl_{3(g)} + Cl_{2(g)}$$

In which direction does the equilibrium shift as a result of each change?

(a) adding phosphorus pentachloride gas
(b) removing chlorine gas
(c) decreasing the temperature
(d) increasing the pressure by adding helium gas
(e) using a catalyst

**What Is Required?**

You need to determine whether each change causes the equilibrium to shift to the left or the right, or whether it has no effect.

**What Is Given?**

You have the balanced chemical equation. You know that the reaction is endothermic.

**Plan Your Strategy**

Identify the change. Then use the chemical equation to determine the shift in equilibrium that will minimize the change.

**Act on Your Strategy**

(a) [PCl$_5$] increases. Therefore, the equilibrium must shift to minimize [PCl$_5$]. The reaction shifts to the right.

(b) [Cl$_2$] is reduced. Therefore, the equilibrium must shift to increase [Cl$_2$]. The reaction shifts to the right.

(c) The temperature decreases. Therefore, the equilibrium must shift in the direction in which the reaction is exothermic. From left to right, the reaction is endothermic. Thus, the reaction must be exothermic from right to left. The reaction shifts to the left if the temperature is decreased.

(d) Helium does not react with any of the gases in the mixture. The position of equilibrium does not change.

(e) A catalyst has no effect on the position of equilibrium.

**Check Your Solution**

Check the changes. Any change that affects the equilibrium reaction must result in a shift that minimizes it.

## Practice Problems

33. Consider the following reaction.

    $2HI_{(g)} \rightleftharpoons H_{2(g)} + I_{2(g)} + 52$ kJ

    In which direction does the equilibrium shift if there is an increase in temperature?

34. A decrease in the pressure of each equilibrium system below is caused by increasing the volume of the reaction container. In which direction does the equilibrium shift?

    (a) $CO_{2(g)} + H_{2(g)} \rightleftharpoons CO_{(g)} + H_2O_{(g)}$
    (b) $2NO_{2(g)} \rightleftharpoons N_2O_{4(g)}$
    (c) $2CO_{2(g)} \rightleftharpoons 2CO_{(g)} + O_{2(g)}$
    (d) $CH_{4(g)} + 2H_2S_{(g)} \rightleftharpoons CS_{2(g)} + 4H_{2(g)}$

35. The following reaction is exothermic.

    $2NO_{(g)} + 2H_{2(g)} \rightleftharpoons N_{2(g)} + 2H_2O_{(g)}$

    In which direction does the equilibrium shift as a result of each change?

    (a) removing the hydrogen gas
    (b) increasing the pressure of gases in the reaction vessel by decreasing the volume
    (c) increasing the pressure of gases in the reaction vessel by pumping in argon gas while keeping the volume of the vessel constant
    (d) increasing the temperature
    (e) using a catalyst

*Continued...*

**36.** In question 35, which changes affect the value of $K_c$? Which changes do not affect the value of $K_c$?

**37.** Toluene, $C_7H_8$, is an important organic solvent. It is made industrially from methyl cyclohexane.

$C_7H_{14(g)} \rightleftharpoons C_7H_{8(g)} + 3H_{2(g)}$

The forward reaction is endothermic. State three different changes to an equilibrium mixture of these reacting gases that would shift the equilibrium toward greater production of toluene.

## Applying Le Châtelier's Principle: Manufacturing Ammonia

Industrial processes must be run at optimal conditions to be economic. This means more than simply manipulating the reaction conditions to maximize the extent of the reaction. Business people, and the professionals they hire to advise them, must consider other factors as well. These factors include the rate of the reaction, safety, the location of the plant, the cost to build and operate the plant, the cost of acquiring reactant materials, the cost of transporting products, and the cost of hiring, educating, and maintaining plant workers.

The chapter introduction mentioned the importance of nitrogen to plants and animals. Despite nitrogen's abundance in the atmosphere, its low reactivity means that there is a limited supply in a form that organisms can use. In nature, bacteria and the energy of lightning supply the nitrates and other nitrogen compounds that plants need to survive. Humans use technology to produce these compounds for our extensive agricultural industry.

In this chapter, you learned about the Haber process for manufacturing ammonia. You used this process to help you understand various concepts related to equilibrium. As you can see in Figure 13.8, ammonia is a valuable industrial chemical. Its annual global production is well over 100 million tonnes. The vast majority of ammonia, roughly 80%, is used to make fertilizers. You will now examine how the equilibrium concepts you have been studying work together to provide society with a reliable, cost-effective supply of ammonia.

Before World War I, the main source of nitrates for human use was from large deposits of bird droppings in Peru and sodium nitrate from Chile. These sources were becoming scarce and expensive. Then Fritz Haber (1868–1934), a lecturer in a technical college in Germany, began to experiment with ways to manufacture ammonia. Haber knew that ammonia could be easily converted to nitrates and other useful nitrogen compounds. What he needed was a method for producing large quantities of ammonia at minimal cost.

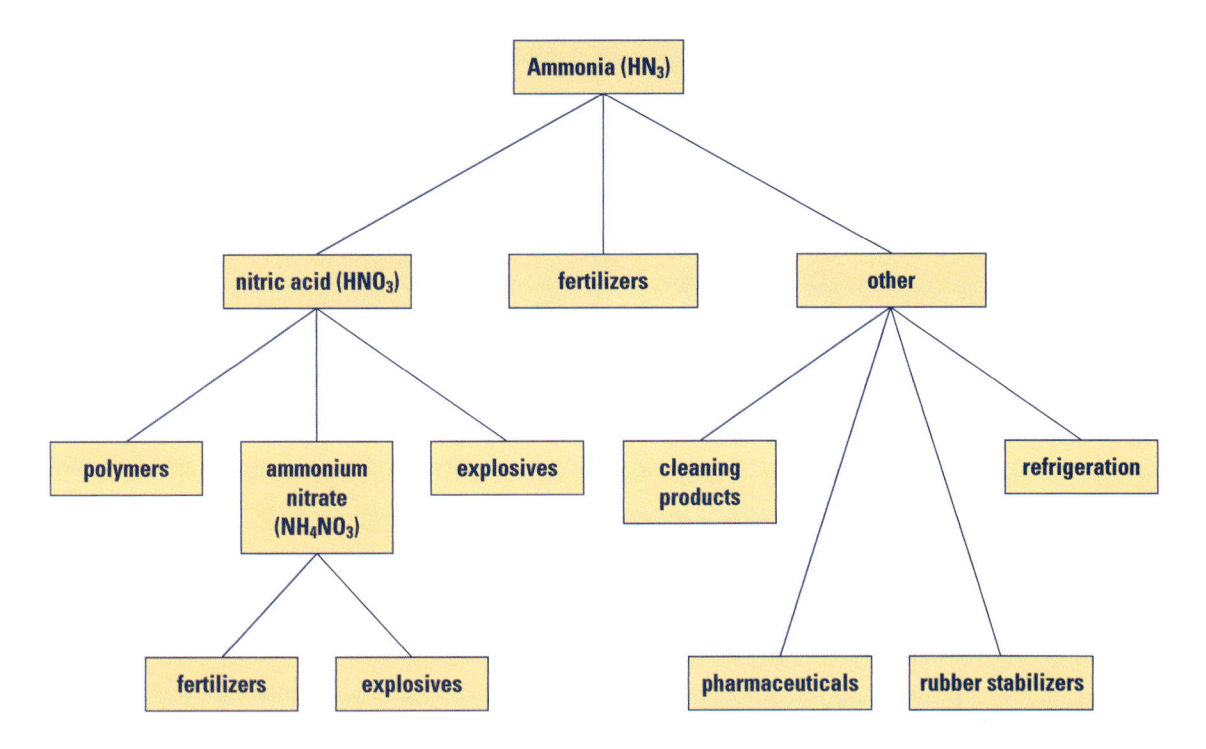

**Figure 13.8** Because ammonia is used to manufacture many essential products, it is one of the top five industrial chemicals in the world.

Haber experimented with the direct synthesis of ammonia.

$$N_{2(g)} + 3H_{2(g)} \rightleftharpoons 2NH_{3(g)} + 92 \text{ kJ/mol}$$

Because the reaction is exothermic, heat is released as the reaction proceeds. Le Châtelier's principle predicts that the yield of ammonia is greater at lower temperatures. Just as in the contact process for manufacturing sulfuric acid, however, high yield is not the only important factor. The rate of reaction for ammonia synthesis is too slow at low temperatures.

Le Châtelier's principle also predicts that the yield of ammonia is greater at higher pressures. High-pressure plants are expensive to build and maintain, however. In fact, the first industrial plant that manufactured ammonia had its reaction vessel blow up.

A German chemical engineer, Carl Bosch, solved this problem by designing a double-walled steel vessel that could operate at several hundred times atmospheric pressure. Modern plants operate at pressures in the range of 20 200 kPa to 30 400 kPa.

The gas mixture is kept pressurized in a condenser. (See Figure 13.9.) Ammonia is removed from the reaction vessel by cooling the gas mixture. Because of the hydrogen bonding between ammonia molecules, the gas condenses into a liquid while the nitrogen and hydrogen remain as gases. Nitrogen is removed to shift the equilibrium toward the production of more ammonia. Once the ammonia is removed, the gases are recycled back to the reaction vessel in a continuous operation.

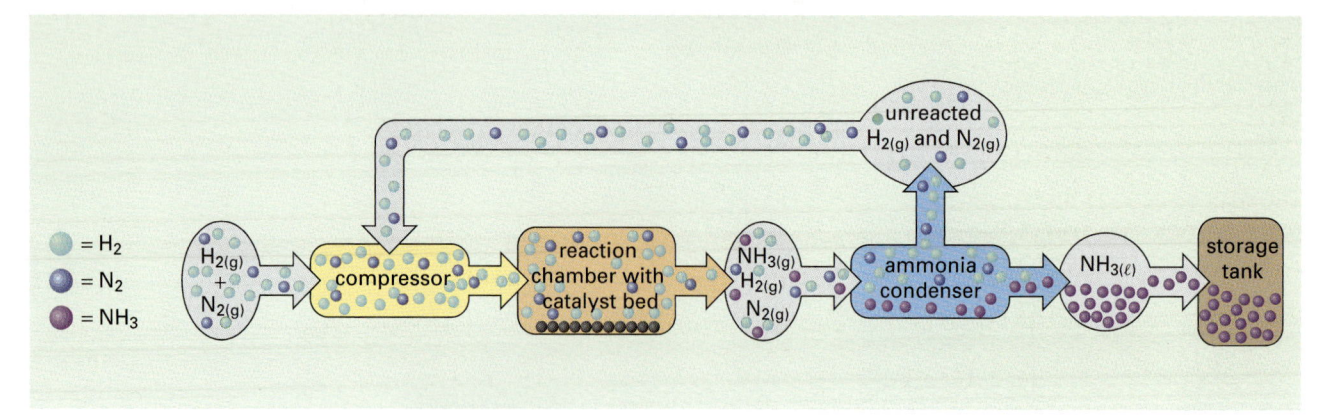

**Figure 13.9** This diagram shows the stages in the Haber-Bosch process for manufacturing ammonia. The catalyst is a mixture of MgO, Al$_2$O$_3$, and SiO$_2$, with embedded iron crystals.

By manipulating the pressure and removing ammonia from the reaction vessel, Haber successfully increased the yield of ammonia. To increase the rate of the reaction, Haber needed to find a catalyst. A catalyst would allow the reaction to proceed at higher temperatures—a compromise between rate and yield. Some historians claim that Haber performed more than 6500 experiments trying to find a suitable catalyst. He finally chose an iron catalyst. This catalyst works well at the relatively moderate temperature of 400°C that is used for the reaction. It lasts about five years before losing its effectiveness.

Haber received the Nobel prize for Chemistry in 1918. Bosch received a Nobel prize in 1931 for his work on high pressure reactions. Today the Haber-Bosch process is used to manufacture virtually all the ammonia that is produced in the world. Plants are usually located near a source of natural gas, which is used to obtain the hydrogen gas.

$$CH_{4(g)} + H_2O_{(g)} \rightarrow CO_{(g)} + 3H_{2(g)}$$

The other raw material is air, which provides an inexhaustible supply of nitrogen gas.

### CHEM FACT

During World War I, Haber helped to develop the technology for deploying phosgene, chlorine, and mustard gas as weapons of chemical warfare. His wife Clara, also a chemist, was disgusted by the use of science in war. When her husband refused to stop his support of the war effort, she committed suicide.

### Web LINK

www.mcgrawhill.ca/links/atlchemistry

Sulfuric acid, methanol, and polystyrene are other industrially important chemicals that depend on equilibrium reactions for their production. Choose one of these chemicals, or another industrial chemical. Research methods that are used to produce it, as well as the products that are derived from it. To start your research, go to the web site above and click on **Web Links**. Prepare a report that outlines what you learned.

## Section Summary

In this section, you learned that the expression for the reaction quotient is the same as the expression for the equilibrium constant. The concentrations that are used to solve these expressions may be different, however. When $Q_c$ is less than $K_c$, the reaction proceeds to form more products. When $Q_c$ is greater than $K_c$, the reaction proceeds to form more reactants. These changes continue until $Q_c$ is equal to $K_c$. Le Châtelier's principle describes this tendency of a chemical system to return to equilibrium after a change moves it from equilibrium. The industrial process for manufacturing ammonia illustrates how chemical engineers apply Le Châtelier's principle to provide the most economical yield of a valuable chemical product.

$K_c$ is only one of several equilibrium constants that chemists use to describe chemical systems. In the next chapter, you will learn how equilibrium applies to chemical systems that involve acids and bases.

## Section Review

**1** In which direction does the equilibrium shift as a result of the change to each homogeneous equilibrium system?

(a) Adding $Cl_{2(g)}$: $2Cl_{2(g)} + O_{2(g)} \rightleftharpoons 2Cl_2O_{(g)}$

(b) Removing $N_{2(g)}$: $2NO_{2(g)} \rightleftharpoons N_{2(g)} + 2O_{2(g)}$

(c) Using a catalyst: $CH_{4(g)} + H_2O_{(g)} \rightleftharpoons CO_{2(g)} + H_{2(g)}$

(d) Decreasing the total volume of the reaction container:
$2NO_{2(g)} \rightleftharpoons N_2O_{4(g)}$

(e) Increasing the temperature: $CO_{(g)} + 3H_{2(g)} \rightleftharpoons CH_{4(g)} + H_2O_{(g)} + 230$ kJ

**2** For each reversible reaction, determine whether the forward reaction is favoured by high temperatures or low temperatures.

(a) $N_2O_{4(g)} + 59$ kJ $\rightleftharpoons 2NO_{2(g)}$

(b) $2ICl_{(g)} \rightleftharpoons I_{2(g)} + Cl_{2(g)} + 35$ kJ

(c) $2CO_{2(g)} + 566$ kJ $\rightleftharpoons 2CO_{(g)} + O_{2(g)}$

(d) $2HF_{(g)} \rightleftharpoons H_{2(g)} + F_{2(g)} + 536$ kJ

**3** In each reaction, the volume of the reaction vessel is decreased. What is the effect (if any) on the position of equilibrium?

(a) $N_{2(g)} + O_{2(g)} \rightleftharpoons 2NO_{(g)}$

(b) $4HCl_{(g)} + O_{2(g)} \rightleftharpoons 2Cl_{2(g)} + 2H_2O_{(g)}$

(c) $2H_2S_{(g)} + CH_{4(g)} \rightleftharpoons 4H_{2(g)} + CS_{2(g)}$

(d) $2CH_3COOH_{(aq)} + Ba(OH)_{2(aq)} \rightleftharpoons Ba(CH_3COO)_{2(aq)} + 2H_2O_{(\ell)}$

**4** As you learned, the Haber process is used to produce ammonia.

(a) Write the chemical equation for the Haber process.

(b) Describe three changes that would increase the yield of ammonia, and explain why.

**5** A key step in the production of sulfuric acid is the conversion of sulfur dioxide to sulfur trioxide. This step is exothermic.

(a) Write the chemical equation for this step.

(b) Describe three changes that would increase the yield of sulfur trioxide.

(c) Explain why the reaction is carried out at a relatively high temperature.

(d) Why is a catalyst (vanadium pentoxide) used?

# Careers in Chemistry

## Anesthesiology: A Career in Pain Management

**Laila Rafik Karwa**

Dr. Laila Rafik Karwa is a specialist in pain. She prevents it, manages it, and alleviates it. She is an anesthesiologist.

Anesthesiologists are medical doctors who are responsible for inducing sleep before surgery, and for managing patients' pain before, during, and after surgery. Anesthesiologists are also involved in managing pain in patients who have severe illnesses, or acute or chronic pain. For example, Karwa's areas of interest include pediatric and obstetric anesthesia and acute pain management.

Anesthesiologists must have an intimate knowledge of the chemical and physical properties of gases. Many anesthetics are inhaled and are delivered to the bloodstream by diffusion. The speed at which diffusion occurs between the lungs, the blood, and other tissues of the body depends on a constant called the *partition coefficient*. This constant is a ratio that describes the equilibrium concentrations of a solute that is dissolved in two separate phases. The solute becomes separated (partitioned) between the two solvents in such a way that its concentration in one is directly proportional to its concentration in the other.

Gases diffuse from areas of high concentration to areas of low concentration. The speed at which diffusion occurs in the body depends on partition coefficients. The faster the concentration in the lung and brain tissues reaches the inhaled concentration of an anesthetic, the sooner a patient is induced.

Dr. Karwa enjoys the immediate gratification that comes with anesthesiology and the speed at which the anesthetics exact their effects. "What I like most about my job is the ability to let patients have their surgery being completely unaware of it, and to make them as pain-free as possible during and after surgery," she says. She chose medicine as a career because she was interested in biology and because she wanted to help other people with their health problems, after witnessing her father's struggles with rheumatic heart disease.

Dr. Karwa completed high school in Bombay, India, and later studied medicine at the University of Bombay. After medical school, she immigrated to Canada to complete her residency training in anesthesia, and now works as a staff anesthesiologist at a prominent healthcare centre.

In Canada, students who are interested in a career as an anesthesiologist must complete two to three years of university before applying to medical school. Many medical schools in Canada require applicants to have taken courses in specific subjects in the biological and physical sciences, and to have some background in the humanities or social sciences. This education also helps students prepare for the Medical College Admissions Test (MCAT), which is required for application to most medical schools. The MCAT tests an applicant's aptitude for science, verbal reasoning, and writing. Most medical school programs in Canada are four years long.

Medical school graduates who decide to do specialty training in anesthesiology must complete a postgraduate residency program, which usually takes at least five years.

### Making Career Connections

1. Although anesthesiologists tend to be associated with surgery, they also work outside the confines of the operating room. Research other locations where you might find an anesthesiologist.
2. Anesthetics are powerful medications. What risks to patients' health do anesthesiologists have to consider when they administer anesthetics?
3. Anesthetics are classified into three main groups: general, regional, and local. Research the criteria for this classification, and explain the distinctions.

# CHAPTER 13 Review

## Reflecting on Chapter 13

Summarize this chapter in the format of your choice. Here are a few ideas to use as guidelines:
- State the four conditions that apply to all equilibrium systems. Give examples to illustrate these conditions.
- Identify conditions that favour a reaction, and explain how they are related to equilibrium.
- Use experimental data to determine an equilibrium constant for concentration.
- Describe how to use an ICE table to solve problems that involve $K_c$.
- Compare $Q_c$ and $K_c$ to determine the direction of a chemical reaction.
- Explain the meaning of a small value of $K_c$.
- Use Le Châtelier's principle to predict the direction of a reaction.
- Outline the effects of changing conditions on a chemical system at equilibrium.
- Summarize the use of equilibrium and Le Châtelier's principle in industrial processes, such as the production of ammonia and sulfuric acid.

## Reviewing Key Terms

For each of the following terms, write a sentence that shows your understanding of its meaning.
equilibrium
homogeneous equilibrium
heterogeneous equilibrium
favourable change
law of chemical equilibrium
equilibrium constant ($K_c$)
ICE table
reaction quotient ($Q_c$)
Le Châtelier's principle

## Knowledge/Understanding

1. Explain the difference between the rate of a reaction and the extent of a reaction.
2. At equilibrium, there is no overall change in the concentrations of reactants and products. Why, then, is this state described as dynamic?
3. For a reaction that goes to completion, is $K_c$ very large or very small? Explain why.
4. Name the factors that can affect the equilibrium of a reaction.
5. The following reaction is at equilibrium. Which condition will produce a shift to the right: a decrease in volume or a decrease in temperature? Explain why.
$H_{2(g)} + Cl_{2(g)} \rightleftharpoons 2HCl_{(g)} + \text{heat}$
6. The following system is at equilibrium. Will an increase in pressure result in a shift to the left or to the right? How do you know?
$2CO_{2(g)} \rightleftharpoons 2CO_{(g)} + O_{2(g)}$
7. Consider the following reaction.
$CO_{(g)} + 3H_{2(g)} \rightleftharpoons CH_{4(g)} + H_2O_{(g)}$
   (a) The volume is changed to increase the pressure of the system. Explain how this affects the concentration of the reactants and products, and the direction in which the equilibrium shifts.
   (b) When equilibrium has been re-established, which substance(s) will show an increase in concentration?

## Inquiry

8. The following equation represents the dissociation of hydrogen iodide gas. At 430°C, the value of $K_c$ is 0.20.
$2HI_{(g)} \rightleftharpoons H_{2(g)} + I_{(g)}$
Some $HI_{(g)}$ is placed in a closed container at 430°C. Analysis at equilibrium shows that the concentration of $I_{2(g)}$ is $5.6 \times 10^{-4}$ mol/L. What are the equilibrium concentrations of $H_{2(g)}$ and $HI_{(g)}$?

9. The oxidation of sulfur dioxide to sulfur trioxide is an important reaction. At 1000 K, the value of $K_c$ is $3.6 \times 10^{-3}$.
$2SO_{2(g)} + O_{2(g)} \rightleftharpoons 2SO_{3(g)}$
A closed flask originally contains 1.7 mol/L $SO_{2(g)}$ and 1.7 mol/L $O_{2(g)}$. What is $[SO_3]$ at equilibrium when the reaction vessel is maintained at 1000 K?

10. Ethanol and propanoic acid react to form the ester ethyl propanoate, which has the odour of bananas.
$CH_3CH_2OH_{(aq)} + CH_3CH_2COOH_{(aq)} \rightleftharpoons CH_3CH_2COOCH_2CH_{3(aq)} + H_2O_{(\ell)}$
At 50°C, $K_c$ for this reaction is 7.5. If 30.0 g of ethanol is mixed with 40.0 g of propanoic acid, what mass of ethyl propanoate will be present in the equilibrium mixture at 50°C?

Answers to questions highlighted in red type are provided in Appendix A.

**Hint:** Calculate the initial amounts of the reactants. Then solve the equilibrium equation using amounts instead of concentrations. The volume of the mixture does not affect the calculation.

11. 0.50 mol of $CO_{(g)}$ and 0.50 mol of $H_2O_{(g)}$ are placed in a 10 L container at 700 K. The following reaction occurs.
    $CO_{(g)} + H_2O_{(g)} \rightleftharpoons H_{2(g)} + CO_{2(g)}$  $K_c = 8.3$
    What is the concentration of each gas that is present at equilibrium?

12. Sulfur atoms combine to form molecules that have different numbers of atoms depending on the temperature. At about 1050°C, the following dissociation occurs.
    $S_{8(g)} \rightleftharpoons 4S_{2(g)}$
    The initial concentration of $S_{8(g)}$ in a flask is $9.2 \times 10^{-3}$ mol/L, and the equilibrium concentration of the same gas is $2.3 \times 10^{-3}$ mol/L. What is the value of $K_c$?

13. Consider an equilibrium in which oxygen gas reacts with gaseous hydrogen chloride to form gaseous water and chlorine gas. At equilibrium, the gases have the following concentrations:
    $[O_2] = 8.6 \times 10^{-2}$ mol/L,
    $[HCl] = 2.7 \times 10^{-2}$ mol/L,
    $[H_2O] = 7.8 \times 10^{-3}$ mol/L,
    $[Cl_2] = 3.6 \times 10^{-3}$ mol/L.
    (a) Write a balanced chemical equation for this reaction.
    (b) Calculate the value of the equilibrium constant.

14. Sulfur trioxide gas reacts with gaseous hydrogen fluoride to produce gaseous sulfur hexafluoride and water vapour. The value of $K_c$ is $6.3 \times 10^{-3}$.
    (a) Write a balanced chemical equation for this reaction.
    (b) 2.9 mol of sulfur trioxide is mixed with 9.1 mol of hydrogen fluoride in a 4.7 L flask. Set up an equation to determine the equilibrium concentration of sodium hexafluoride.

15. The following results were collected for two experiments that involve the reaction, at 600°C, between gaseous sulfur dioxide and oxygen to form gaseous sulfur trioxide. Show that the value of $K_c$ was the same in both experiments.

| Experiment 1 | | Experiment 2 | |
|---|---|---|---|
| Initial concentration (mol/L) | Equilibrium concentration (mol/L) | Initial concentration (mol/L) | Equilibrium concentration (mol/L) |
| $[SO_2]$ = 2.00 | $[SO_2]$ = 1.50 | $[SO_2]$ = 0.500 | $[SO_2]$ = 0.590 |
| $[O_2]$ = 1.50 | $[O_2]$ = 1.25 | $[O_2]$ = 0 | $[O_2]$ = 0.0450 |
| $[SO_3]$ = 3.00 | $[SO_3]$ = 3.50 | $[SO_3]$ = 0.350 | $[SO_3]$ = 0.260 |

16. Write the chemical equation for the reversible reaction that has the following equilibrium expression.
    $$K_c = \frac{[NO]^4[H_2O]^6}{[NH_3]^4[O_2]^5}$$
    Assume that, at a certain temperature, [NO] and $[NH_3]$ are equal. Also assume that $[H_2O]$ = 2.0 mol/L and $[O_2]$ = 3.0 mol/L. What is the value of $K_c$ at this temperature?

## Communication

17. Equal amounts of hydrogen gas and iodine vapour are heated in a sealed flask.
    (a) Sketch a graph to show how $[H_{2(g)}]$ and $[HI_{(g)}]$ change over time.
    (b) Would you expect a graph of $[I_{2(g)}]$ and $[HI_{(g)}]$ to appear much different from your first graph? Explain why.
    (c) How does the value of $Q_c$ change over time for this reaction?

18. A younger student in your school wants to grow a crystal of copper(II) sulfate. Write a short explanation that outlines how the student can use a small crystal and a saturated solution of copper(II) sulfate to grow a larger crystal.

## Making Connections

19. At 25°C, the value of $K_c$ for the reaction between nitrogen and oxygen is $4.7 \times 10^{-31}$:
    $N_{2(g)} + O_{2(g)} \rightleftharpoons 2NO_{(g)}$  $K_c = 4.7 \times 10^{-31}$
    Assuming that air is composed of 80% nitrogen by volume and 20% oxygen by volume, estimate the concentration of nitric oxide to be expected in the atmosphere. Why is the actual [NO] usually greater?

20. One of the steps in the Ostwald process for the production of nitric acid involves the oxidation of ammonia.
    $4NH_{3(g)} + 5O_{2(g)} \rightleftharpoons 4NO_{(g)} + 6H_2O_{(g)} + 905$ kJ

(a) State the reaction conditions that favour the production of nitrogen monoxide.

(b) A rhodium/platinum alloy is used as a catalyst. What effect does the catalyst have on the rate of reaction? What effect does the catalyst have on the position of equilibrium?

(c) Explain why the reaction temperature is relatively high, typically about 900°C.

(d) A relatively low pressure of about 710 kPa is used. Suggest why.

(e) In the next step of the Ostwald process, nitrogen monoxide is mixed with air to form nitrogen dioxide.
$2NO_{(g)} + O_{2(g)} \rightleftharpoons 2NO_{2(g)} + 115$ kJ
Why are the gases cooled for this reaction? What do you think happens to the heat that is extracted?

(f) Finally, the nitrogen dioxide reacts with water to form nitric acid.
$3NO_{2(g)} + H_2O_{(\ell)} \rightleftharpoons 2HNO_{3(aq)} + NO_{(g)}$
What is done with the $NO_{(g)}$ that is formed? Name three uses of this product.

21. Polystyrene is one of our most useful polymers. Polystyrene resin is manufactured from styrene, which is made as follows:
$C_6H_5CH_2CH_{3(g)} + 123$ kJ $\rightleftharpoons C_6H_5CHCH_{2(g)} + H_{2(g)}$

(a) Predict the effects (if any) on this equilibrium if the following changes are made:
  • increasing the applied pressure
  • removing styrene
  • reducing the temperature
  • adding a catalyst
  • adding helium at constant volume

(b) Based on your predictions in part (a), select the conditions that would maximize the equilibrium yield of styrene.

(c) Why are the conditions you selected unlikely to be used by industry? What conditions must be used to maximize the yield of an industrial process?

(d) In practice, the pressure of the ethyl benzene is kept low. An inert substance—superheated steam—is added to keep the total pressure of the mixture at atmospheric pressure. What advantage does this method have over running the reaction without the steam at a pressure that is less than atmospheric pressure?

## Answers to Practice Problems and Short Answers to Section Review Questions

**Practice Problems:** 1. $K_c = \dfrac{[CO_2]^3[H_2O]^4}{[C_3H_8][O_2]^5}$
2. $K_c = \dfrac{[NO]^2}{[N_2][O_2]}$ 3. $K_c = \dfrac{[H_2O]^2}{[H_2]^2[O_2]}$ 4. $K_c = \dfrac{[Fe^{2+}]^2[I_2]}{[Fe^{3+}]^2[I^-]^2}$
5. $K_c = \dfrac{[NO]^4[H_2O]^6}{[NH_3]^4[O_2]^5}$ 6. $1.9 \times 10^{-2}$ 7. $1.2 \times 10^2$ 8. 0.013
9. 0.046 mol/L 10. 0.15 11. 4.2 mol/L 12. 87 13. 4.7
14. [CO] = 0.0088 mol/L; [Cl$_2$] = 0.012 mol/L; [COCl$_2$] = 0.0022 mol/L 15. [HBr] = 0.045 mol/L; [H$_2$] = [Br$_2$] = $2.9 \times 10^{-6}$ mol/L 16. $K_c$ = 0.344 17. $K_c$ = 1.58
18. $K_c$ = 153 19. $K_c$ = $1.3 \times 10^{-2}$ 20. 1100 K 21. III, II, I
22.(a) completion (b) no reaction (c) completion
23. Ag(NH$_3$)$_2^+$ $_{(aq)}$ 24. essentially no dissociation
25. [CO] = [Cl$_2$] = $1.8 \times 10^{-4}$ mol/L 26. [H$_2$] = $8.6 \times 10^{-2}$ mol/L
27. $6.1 \times 10^{-1}$ mol 28. [NO] = $1.2 \times 10^{-5}$ mol/L
29. [CO$_2$] = 0.59 mol/L; [CO] = $1.1 \times 10^{-2}$ mol/L; [O$_2$] = $5.6 \times 10^{-3}$ mol/L 30.(a) $K_c$ = 171
(b) [NO$_2$] = 0.0412 mol/L; [N$_2$O$_4$] = 0.145 mol/L (c) $Q_c$ = 85.3
(d) right 31.(a) right (b) left (c) at equilibrium (d) left 32. left
33. left 34.(a) no change (b) left (c) right (d) right 35.(a) left
(b) right (c) no change (d) left (e) no change 36. Only (d) changes the value of $K_c$. 37. adding methyl cyclohexane; removing toluene; decreasing the pressure; increasing the temperature

**Section Review: 13.2:** 1.(a) $\dfrac{[SbCl_3][Cl_2]}{[SbCl_5]}$ (b) $\dfrac{[N_2][H_2O]^2}{[H_2]^2[NO]^2}$
(c) $\dfrac{[H_2]^4[CS_2]}{[H_2S]^2[CH_4]}$ (d) $\dfrac{[P_2O_3]^2}{[O_2]^3}$ (e) $\dfrac{[N_2S_2]^4}{[N_2]^7}$

2.(a) [N$_2$] = 0.20 mol/L; [NH$_3$] = 1.6 mol/L (b) $K_c$ = 0.017
3. 1.2 mol 4. [PCl$_3$] = [Cl] = 0.035 mol/L; [PCl$_5$] = 0.059 mol/L 5.(a) [H$_2$] = $1.8 \times 10^{-7}$ mol/L (b) an impractical method 13.3: 1.(a) right (b) right (c) unchanged (d) right (e) left 2.(a) high temperatures (b) low temperatures (c) high temperatures (d) low temperatures 3.(a) unchanged (b) right (c) left (d) unchanged (because liquid volumes are largely unaffected by pressure changes)
4.(a) N$_2$ + 3H$_2$ $\rightleftharpoons$ 2NH$_3$ + 92 kJ (b) adding N$_2$ or H$_2$; removing NH$_3$; increasing the pressure; lowering the temperature (because the reaction is exothermic)
5.(a) 2SO$_{2(g)}$ + O$_{2(g)}$ $\rightleftharpoons$ 2SO$_{3(g)}$ + 197 kJ (b) increase concentration of SO$_{2(g)}$ and/or O$_{2(g)}$; increase pressure; lower temperature (c) high temperature increases rate of reaction and increases the pressure, favouring products (d) lowers activation energy, increases rate of reaction

# UNIT 5 Review

## Knowledge/Understanding

### Multiple Choice

In your notebook, write the letter for the best answer to each question.

1. Which statement about an activated complex is true?
   (a) It is a stable substance.
   (b) It has lower chemical potential energy than reactants or products.
   (c) It occurs only in endothermic reactions.
   (d) It occurs at the transition state of the reaction.
   (e) It always breaks down to form product molecules.

2. A catalyst changes the
   I. mechanism of a reaction
   II. change in total energy of a reaction
   III. activation energy of a reaction
   (a) I only
   (b) III only
   (c) I and III only
   (d) II and III only
   (e) I, II, and III

3. The overall rate of any chemical reaction is most related to
   (a) the number of steps in the reaction mechanism
   (b) the overall reaction
   (c) the fastest step in the reaction mechanism
   (d) the slowest step in the reaction mechanism
   (e) the average rate of all the steps in the reaction mechanism

4. Which of the following statements about the kinetic molecular theory is untrue?
   (a) All matter is made up of particles too small to be seen with the unaided eye.
   (b) Adding heat increases the speed of moving particles.
   (c) There are spaces between the particles of matter.
   (d) Atoms of different elements combine in specific proportions to form compounds.
   (e) Particles of matter are in constant motion.

5. Which reaction represents a heterogeneous equilibrium?
   (a) $N_2O_{4(g)} \rightleftharpoons 2NO_{2(g)}$
   (b) $2SO_{2(g)} + O_{2(g)} \rightleftharpoons 2SO_{3(g)}$
   (c) $3H_{2(g)} + N_{2(g)} \rightleftharpoons 2NH_{3(g)}$
   (d) $2NaHCO_{3(s)} \rightleftharpoons Na_2CO_{3(s)} + CO_{2(g)} + H_2O_{(g)}$
   (e) $HCO_3^-{}_{(aq)} + H^+{}_{(aq)} \rightleftharpoons H_2CO_{3(aq)}$

6. Consider the following equilibrium.
   $N_2O_{4(g)} \rightleftharpoons 2NO_{2(g)}$ $K_c = 4.8 \times 10^{-3}$
   Which set of concentrations represents equilibrium conditions?
   (a) $[N_2O_{4(g)}] = 4.8 \times 10^{-1}$ and $[NO_{2(g)}] = 1.0 \times 10^{-4}$
   (b) $[N_2O_{4(g)}] = 1.0 \times 10^{-1}$ and $[NO_{2(g)}] = 4.8 \times 10^{-4}$
   (c) $[N_2O_{4(g)}] = 1.0 \times 10^{-1}$ and $[NO_{2(g)}] = 2.2 \times 10^{-2}$
   (d) $[N_2O_{4(g)}] = 2.2 \times 10^{-2}$ and $[NO_{2(g)}] = 1.0 \times 10^{-1}$
   (e) $[N_2O_{4(g)}] = 5.0 \times 10^{-2}$ and $[NO_{2(g)}] = 1.1 \times 10^{-2}$

7. Choose the equilibrium in which products are favoured by a decrease in pressure but reactants are favoured by a decrease in temperature.
   (a) $H_{2(g)} + I_{2(g)} + 51.8$ kJ $\rightleftharpoons 2HI_{(g)}$
   (b) $2NO_{2(g)} \rightleftharpoons 2NO_{(g)} + O_{2(g)}$ $\Delta H = +54$ kJ
   (c) $N_{2(g)} + 3H_{2(g)} \rightleftharpoons 2NH_{3(g)} + 92.3$ kJ
   (d) $PCl_{3(g)} + Cl_{2(g)} \rightleftharpoons PCl_{5(g)}$ $\Delta H = -84.2$ kJ
   (e) $2SO_{2(g)} + O_{2(g)} \rightleftharpoons 2SO_{3(g)} +$ heat

8. The equilibrium constant is
   I. temperature-dependent
   II. related to the rate of a chemical reaction
   III. a ratio of product concentrations to reactant concentrations
   IV. a ratio of reactant concentrations to product concentrations
   (a) I only
   (b) II only
   (c) I and III
   (d) II and IV
   (e) I, II, and IV

9. When $Q_c$ is greater than $K_c$, a chemical system attains equilibrium by
   (a) moving to the left, favouring reactants
   (b) moving to the right, favouring reactants
   (c) remaining stationary
   (d) moving to the right, favouring products
   (e) moving to the left, favouring products

10. Which of the following statements about Le Châtelier's principle is true?
    (a) The rate of a chemical reaction depends on the value of $K_c$.

Answers to questions highlighted in red type are provided in Appendix A.

538 MHR • Unit 5 Kinetics and Equilibrium

(b) Any change that affects the equilibrium of a reaction must result in a shift that minimizes the effect of that change.
(c) A dynamic equilibrium only affects substances when they are in the same phase.
(d) The activation energy of a reaction is affected by temperature.
(e) A reaction mechanism applies only to chemical systems at equilibrium.

## Short Answer

In your notebook, write a sentence or a short paragraph to answer each question.

11. "The reactants have more potential energy than the products." What kind of reaction does this statement describe. Justify your answer.

12. List the factors that affect the rate of a chemical reaction.

13. Select two of the factors you listed in question 12, and use collision theory to explain them.

14. Draw and label a potential energy diagram for the following exothermic reaction. Include activation energy, the activated complex, and the total change in energy.
$A + B \rightarrow AB$

15. Consider the following equilibrium.
$2SO_{2(g)} + O_{2(g)} \rightleftharpoons 2SO_{3(g)} + 198.2$ kJ
Indicate if reactants or products are favoured, or if no change occurs, when
(a) the temperature is increased
(b) helium gas is added at constant volume
(c) helium gas is added at constant pressure

16. Explain the reasoning you used to answer question 15(b).

17. Describe how a decrease in volume will affect the following equilibrium.
$PCl_{3(g)} + Cl_{2(g)} \rightleftharpoons PCl_{5(g)}$

18. $K_c$ for the equilibrium $N_2O_4 \rightleftharpoons 2NO_2$ is $4.8 \times 10^{-3}$. What is $K_c$ for the equilibrium $\frac{1}{2}N_2O_4 \rightleftharpoons NO_2$?

19. For the following equilibrium, $K_c = 1.0 \times 10^{-15}$ at 25°C and $K_c = 0.05$ at 2200°C.
$N_{2(g)} + O_{2(g)} \rightleftharpoons 2NO_{(g)}$
Based on this information, is the reaction exothermic or endothermic? Briefly explain your answer.

20. $K_c$ for the equilibrium system
$2SO_{2(g)} + 2O_{2(g)} \rightleftharpoons SO_{3(g)}$ is 40. What is $K_c$ for $SO_{3(g)} \rightleftharpoons SO_{2(g)} + \frac{1}{2}O_{2(g)}$?

21. 1.0 mol of $NH_{3(g)}$ is placed initially in a 500 mL flask. The following equilibrium is established.
$2NH_{3(g)} \rightleftharpoons N_{2(g)} + 3H_{2(g)}$
Write an equilibrium expression that describes the system at equilibrium.

22. A mixture is prepared for the following equilibrium.
$I_{2(aq)} + I^-_{(aq)} \rightleftharpoons I_3^-_{(aq)}$
The initial concentrations of $I_{2(aq)}$ and $I^-_{(aq)}$ are both 0.002 mol/L. When equilibrium is established, $[I_{2(aq)}] = 5.0 \times 10^{-4}$ mol/L. What is $K_c$ for this equilibrium?

23. (a) Write the overall balanced equation from the following proposed three-step mechanism:
$Cl_2 \rightarrow 2Cl$
$Cl + CHCl_3 \rightarrow HCl + CCl_3$
$Cl + CCl_3 \rightarrow CCl_4$
(b) Identify all the reaction intermediates in the proposed mechanism.

## Inquiry

24. The reaction between hydrogen and carbon dioxide was being studied:
$H_{2(g)} + CO_{2(g)} \rightleftharpoons H_2O_{(g)} + CO_{(g)}$
When 3.60 mol of hydrogen gas and 5.20 mol of carbon dioxide were placed in a 2.00 L container, it was found that 75.0% of the hydrogen had reacted at equilibrium. What is the equilibrium constant?

25. When 2.38 mol of $IBr_{(g)}$ was placed in a 3.00 L container and heated to 150°C, the gas decomposed.
$2IBr_{(g)} \rightleftharpoons I_{2(g)} + Br_{2(g)}$
If 89.3% of IBr does not decompose at 150°C, what is the equilibrium constant at this temperature?

26. Nitrogen gas is relatively unreactive at room temperature, but combines with oxygen at higher temperatures.
$N_{2(g)} + O_{2(g)} \rightleftharpoons 2NO_{(g)}$
When equilibrium is attained at a temperature of 1800 K, 5.3% of nitrogen will react with oxygen. What is the value of $K_c$ at this temperature?

Unit 5 Review • MHR  539

27. $K_c$ for the following system is 1.05 at a certain temperature.
   PCl$_{5(g)}$ + heat ⇌ PCl$_{3(g)}$ + Cl$_{2(g)}$
   After 0.22 mol of PCl$_{3(g)}$ and 0.11 mol of Cl$_{2(g)}$ are added to a 500 mL flask, equilibrium is established.
   (a) Explain what happens to the rates of the forward and reverse reactions as equilibrium is established.
   (b) Calculate the equilibrium concentration of each component of this equilibrium.

28. (a) You are using a stopwatch to determine the rate of a reaction. However, the reaction rate is too quick for you to observe adequately. Assume the reactants involved are gases. How could you slow down the rate?
   (b) What, if anything, would you do differently to slow the rate of reaction if the reactants were in aqueous solution?

29. In the 1960s, chemists used spectroscopic analysis to propose a three-step mechanism for a reaction:
   I$_{2(g)}$ → 2I$_{(g)}$           (fast)
   H$_{2(g)}$ + I$_{(g)}$ → H$_2$I$_{(g)}$    (fast)
   H$_2$I$_{(g)}$ + I$_{(g)}$ → 2HI$_{(g)}$   (slow)
   (a) Before the 1960s, chemists believed that the overall balanced equation for this reaction was an elementary mechanism. Explain what this means, and write the overall equation.
   (b) Identify the reaction intermediate(s) in the three-step mechanism.

30. For the following equilibrium, $K_c$ = 0.81 at a certain temperature.
   $\frac{1}{2}$N$_{2(g)}$ + $\frac{3}{2}$H$_{2(g)}$ ⇌ NH$_{3(g)}$ + heat
   (a) Explain what will happen to the magnitude of $K_c$ if the temperature is increased.
   (b) The initial concentrations of the gases are [H$_{2(g)}$] = 0.76 mol/L, [N$_{2(g)}$] = 0.60 mol/L, and [NH$_{3(g)}$] = 0.48 mol/L. Determine if the concentration of [N$_{2(g)}$] will increase or decrease when equilibrium is established.

31. Consider the following equilibrium.
   2XO$_{(s)}$ + O$_{2(g)}$ ⇌ 2XO$_{2(g)}$ + heat
   (a) Predict whether reactants or products are favoured when the volume of the container is increased. Give a reason for your answer.
   (b) What do you predict will happen to the concentration of each substance when the volume of the container is increased?
   (c) At a certain temperature, $K_c$ = 1.00 × 10$^{-4}$ and the equilibrium concentrations are [O$_2$] = 1.99 mol/L and [XO$_2$] = 1.41 × 10$^{-2}$ mol/L. If the volume of the container is now doubled, calculate the concentrations of these gases when equilibrium is re-established. Assume that sufficient XO$_{(s)}$ is present to maintain equilibrium. Are the concentrations you calculate consistent with your prediction in part (b)?

32. Which of the following reactions likely will occur at a faster rate? Explain your reasoning.
   (a) C$_6$H$_{12(l)}$ + HBr$_{(aq)}$ → C$_6$H$_{13}$Br$_{9(aq)}$
   (b) 2AlCl$_{(aq)}$ + BaSO$_{4(aq)}$ → BaCl$_{2(aq)}$ + Al$_2$SO$_{4(s)}$

33. (a) A chemist adds 0.100 mol of CaCO$_3$ to a 1.00 L reaction vessel. What is the concentration of CO$_2$ when the system comes to equilibrium at 800°C? $K_c$ for the reaction is 0.0132.
   CaCO$_{3(s)}$ → CaO$_{(s)}$ + CO$_{2(g)}$
   (b) If the volume of the reaction vessel were 2.00 L, what would be the concentration of carbon dioxide gas?

34. An industrial process results in the formation of 60 mol/L of product each minute at 25°C. What could you do to double the rate of this process without changing the concentrations of the reactants or the temperature of the reaction?

35. For the following reaction, predict the change, if any, that will occur as a result of each identified stress.
   4NH$_{3(g)}$ + 5O$_{2(g)}$ → 4NO$_{(g)}$ + 6H$_2$O$_{(g)}$ + heat
   (a) removing O$_2$
   (b) adding NO
   (c) adding H$_2$O$_{(g)}$
   (d) adding a catalyst
   (e) decreasing the volume
   (f) increasing the pressure
   (g) adding NH$_3$

## Communication

36. The decomposition of aqueous hydrogen peroxide can be catalyzed by different substances. One such substance is aqueous sodium iodide. Sketch potential energy diagrams to compare the catalyzed and uncatalyzed reactions.

37. Interpret, in as much detail as possible, the following graph.

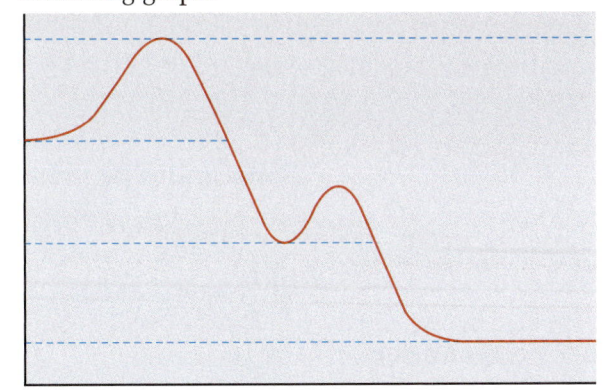

38. Compare the effect of a catalyst on the reaction rate of a reaction that is proceeding toward equilibrium.

39. A chemistry student wrote the following sentences to remind herself of important concepts. Write a paragraph to expand on each concept. Use examples and diagrams as appropriate to illustrate your answer.
    (a) A catalyst speeds up a reaction, but it does not affect the overall change in energy of the reaction.
    (b) Le Chatelier's principle states that a dynamic equilibrium tends to respond in a way that relieves the effect of any change in the conditions that affect the equilibrium.
    (c) The size of $K_c$ indicates the extent of a reaction.

40. Hydrogen and carbon dioxide react at a high temperature to give water and carbon monoxide.
    (a) Laboratory measurements at 986°C show that there is 0.11 mol each of CO and water vapor and 0.087 mol each of $H_2$ and $CO_2$ at equilibrium in a 1.0 L container. Calculate the equilibrium constant for the reaction at 986°C.
    (b) If there is 0.050 mol each of $H_2$ and $CO_2$ in a 2.0 L container at equilibrium at 986°C, what amounts of $CO_{(g)}$ and $H_2O_{(g)}$, in moles, are present?

41. For the reaction shown below, more $CO_2$ is added at constant volume. Eventually, the reaction reaches a new equilibrium state.
    $2CO_{(g)} + O_{2(g)} \rightarrow 2CO_{2(g)} +$ heat
    Agree or disagree with the following statement. For the new equilibrium, the rate of the forward reaction has decreased and the rate of the reverse reaction has increased.

42. Explain the significance of the reaction quotient for a reaction being greater than the equilibrium constant for the same reaction.

43. (a) Predict the effect of adding carbon monoxide to this reaction at equilibrium:
    $2CO_{(g)} + O_{2(g)} \rightleftharpoons 2CO_{2(g)}$
    (b) Explain your answer to part (a) in terms of reaction rates.
    (c) What would happen instead if you added $CO_2$ to this reaction?

44. The following paragraph contains a variety of inaccurate statements. Rewrite the paragraph so that it is completely accurate.
    In equilibrium constant expressions, the rate terms are added and subtracted, not multiplied and divided. The amounts in the numerator are those of the substances on the left side of the chemical equation, and those in the denominator are for the substances on the right side. The coefficients in the equilibrium expression correspond to the exponents in the chemical equation.

45. Explain why pure solids and liquids do not appear in the equilibrium constant expression.

## Making Connections

46. Scientists have noticed various shifts in Earth's equilibrium in recent years. Since Earth is in a dynamic equilibrium, some shifts occur naturally. There is considerable evidence, however, that shifts such as the following have been caused by, or are strongly influenced by, human activities:
    - an increase in atmospheric concentrations of carbon dioxide and methane
    - an increase in the amounts and concentrations of pollutants
    - an increase in average global temperatures
    - a decrease in plant and animal species, both by observable causes and by unknown causes
    - a decrease in arable land and a corresponding increase in desert land
    - an increase in extreme weather conditions such as hurricanes and tornadoes

    Choose two of these questions to answer.
    (a) Where on Earth do you think this shift is most visible? Explain why you think so.
    (b) What are some possible causes of this shift?
    (c) What are some possible consequences of this shift?

# UNIT 6

# Acids and Bases

## UNIT 6 CONTENTS

**CHAPTER 14**
Properties of Acids and Bases

**CHAPTER 15**
Acid-Base Equilibria and Reactions

## UNIT 6 OVERALL OUTCOMES

- In what ways can you define and describe acids and bases?
- What predictions can you make about the chemical composition and the reactivity of acids and bases?
- How does your understanding of chemical equilibrium help you to compare, qualitatively and quantitatively, strong acids and strong bases and weak acids and weak bases?

People who own or maintain swimming pools rely on acids and bases to ensure that the pool water remains safe for fun and for sport.

Chlorine compounds are commonly added to pool water to kill harmful micro-organisms. The most effective chlorine compound is hypochlorous acid, $HClO_{(aq)}$. Bubbling chlorine gas through water produces $HClO_{(aq)}$. Since chlorine is corrosive and toxic, sodium hypochlorite, a salt of hypochlorous acid is used instead. It is safer to handle and reacts with water to form $HClO_{(aq)}$.

Hypochlorous acid dissociates in water to form hypochlorite ions, $OCl^-$. The concentrations of HOCl and $OCl^-$ change with the pH of the pool water. As the pH gets lower, the water becomes too acidic. As pH gets higher, the water becomes too basic. Both conditions can cause skin irritation, sore throats, and stinging eyes. Pool materials and equipment may also be damaged. As well, higher pH levels reduce the effectiveness of the HClO.

Pool managers continually monitor pH. They add either acids or bases to maintain pool pH within a narrow range that is only slightly greater than the pH of pure, neutral water.

What is pH? Why do pH changes affect HClO and $OCl^-$? It is only since the early 1900s that chemists can offer adequate answers to questions such as these. In this unit, you will learn theories that will enable you to describe acids and bases, and to explain and make predictions about their behaviour.

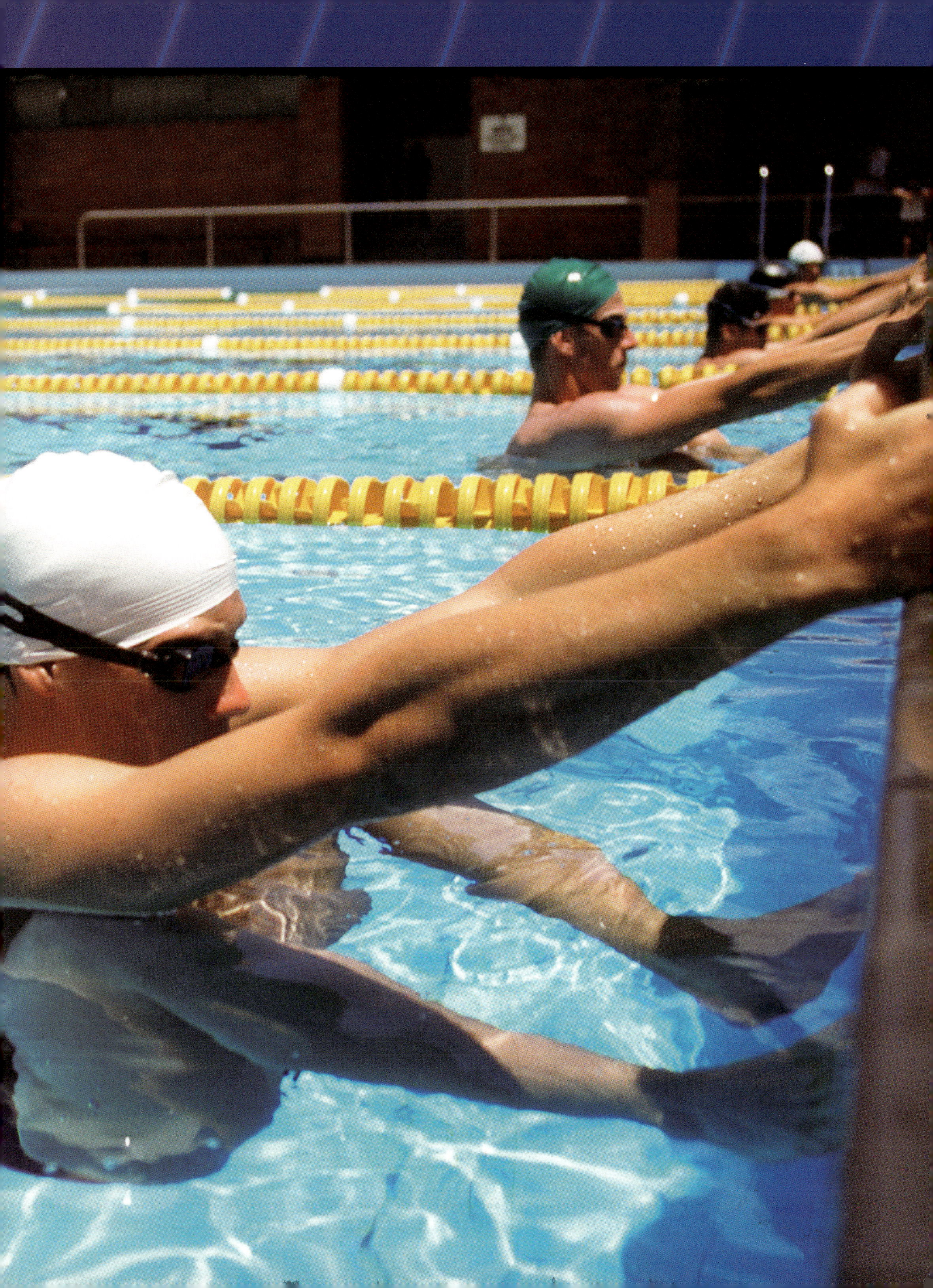

# CHAPTER 14
# Properties of Acids and Bases

**Chapter Preview**

14.1 Defining Acids and Bases

14.2 Strong and Weak Acids and Bases

For many people, the word "acid" evokes the image of a fuming, highly corrosive, dangerous liquid. This image is fairly accurate for concentrated hydrochloric acid, which chemists classify as a strong acid. Most acids, however, are not as corrosive as hydrochloric acid, although they may still be very hazardous. For example, hydrofluoric acid can cause deep, slow-healing tissue burns if it is handled carelessly. It is used by artists and artisans who etch glass. It reacts with the silica in glass to form a compound that dissolves, leaving contoured patterns in the glass surface. Hydrofluoric acid is highly corrosive. Even a 1% solution is considered to be hazardous. Yet chemists classify hydrofluoric acid as a weak acid.

In this chapter, you will examine the properties of acids and bases, starting with those you observe directly. Afterwards, you will learn how chemists developed theories to explain the properties of acids and bases, as well as the properties of the solutions that result when they are mixed together. You will learn how to compare the strengths of acids and bases qualitatively. As well, you will use another equilibrium constant and the pH of an aqueous solution to compare the strengths of acids and bases quantitatively.

**Prerequisite Concepts and Skills**

Before you begin this chapter, review the following concepts and skills:

- calculating molar concentrations (Chapter 7)
- writing net ionic equations (Chapter 8)
- solving equilibrium problems (Chapter 13)

Some acids can etch glass, corrode metal, and burn skin. Other acids are safe to eat. Which safe acids do you come in contact with frequently? What makes them safe?

# Defining Acids and Bases  14.1

Acids and bases are as much a part of people's lives today as they were thousands of years ago. (See Figure 14.1.) For example, vinegar is an acidic solution that is common in many food and cleaning products. It was discovered long before people invented the skill of writing to record its use. Today, acids are also used in the manufacturing of fertilizers, explosives, medicines, plastics, motor vehicles, and computer circuit boards.

Bases also have numerous uses in the home and in chemical industries. Nearly 5000 years ago, in the Middle East, the Babylonians made soap using the bases in wood ash. Today, one of Canada's most important industries, the pulp and paper industry, uses huge quantities of a base called sodium hydroxide—one of the bases in wood ash. Sodium hydroxide is also used to make soap, detergents, drain cleaner, dyes, medicines, and many other products. Table 14.1 lists some common bases and acids you can find in and around the home. You will work with some of these in Investigation 14-A, as you compare the properties of acids and bases.

**Figure 14.1** It is easy to identify some products as acids, because the word "acid" appears in the list of ingredients. Identifying bases is more difficult. What acids and bases do you have in and around your home?

### Section Preview/Outcomes

In this section, you will

- **describe** and **compare** theories of acids and bases
- **identify** conjugate acid-base pairs
- **write** chemical equations to show the amphoteric nature of water
- **communicate** your understanding of the following terms: *Arrhenius theory of acids and bases, hydronium ion, Brønsted-Lowry theory of acids and bases, conjugate acid-base pair, conjugate base, conjugate acid, amphoteric*

### Language LINK

The word "acid" comes from the Latin *acidus*, meaning "sour tasting." As you will learn in this chapter, bases are the "base" (the foundation) from which many other compounds form. A base that is soluble in water is called an alkali. The word "alkali" comes from an Arabic word meaning "ashes of a plant." In the ancient Middle East, people rinsed plant ashes with hot water to obtain a basic solution. The basic solution was then reacted with animal fats to make soap.

**Table 14.1** Common Acids and Bases in the Home

| Acids | |
|---|---|
| **Product** | **Acid(s) contained in the product** |
| citrus fruits (such as lemons, limes, oranges and tomatoes) | citric acid and ascorbic acid |
| dairy products (such as cheese, milk, and yogurt) | lactic acid |
| vinegar | acetic acid |
| soft drinks | carbonic acid; may also contain phosphoric acid and citric acid |
| underarm odour | 3-methyl-2-hexenoic acid |

| Bases | |
|---|---|
| **Product** | **Base contained in the product** |
| oven cleaner | sodium hydroxide |
| baking soda | sodium hydrogen carbonate |
| washing soda | sodium carbonate |
| glass cleaner (some brands) | ammonia |

Chapter 14 Properties of Acids and Bases • MHR  545

# Investigation 14-A

**SKILL FOCUS**
Performing and recording
Analyzing and interpreting

## Observing Properties of Acids and Bases

In this investigation, you will conduct tests in order to develop your own set of identifying properties and definitions for acids and bases. Two of the tests will involve the use of an indicator—a chemical that changes colour in the presence of acids or bases. One of these indicators, litmus paper, is made from a compound that is extracted from lichens, a plant-like member of the fungi kingdom. Litmus paper is made by dipping paper in a solution made with this compound, litmus.

### Question

How can the use of chemical tests enable you to develop a definition for acids and bases?

### Safety Precautions

- Hydrochloric acid, sulfuric acid, sodium hydroxide, and ammonia are toxic and/or corrosive. Wash any spills on skin or clothing with plenty of cool water. Inform your teacher immediately.
- If you are using a conductivity tester with two separate electrodes, be extremely careful to keep the electrodes well separated while you perform your tests.
- Phenolphthalein is flammable. Keep it well away from sparks or flames.

### Materials

1.0 mol/L solutions of the following:
  hydrochloric acid, $HCl_{(aq)}$, sulfuric acid, $H_2SO_{4(aq)}$, acetic acid, $HC_3COOH_{(aq)}$, sodium hydroxide, $NaOH_{(aq)}$, ammonia, $NH_{3(aq)}$
saturated limewater solution
baking soda (sodium hydrogen carbonate, $NaHCO_{3(s)}$)
phenolphthalein solution
magnesium ribbon, finely cut to 1 mm–2 mm lengths
6 test tubes
25 mL graduated cylinder
glass stirring rod
well plate or glass plate
red litmus paper
blue litmus paper
rubber stopper
wooden splint
evaporating dish or 50 mL beaker
medicine dropper or dropper pipette
scoopula
conductivity apparatus

### Procedure

1. Read through the entire Procedure, and design a table to record your observations.

2. Add 10 mL of hydrochloric acid solution to a test tube and place it in the test tube rack. Test the conductivity of the solution.

3. Place a strip of red litmus paper on a glass cover plate, and use a stirring rod to put one drop of the acid on the litmus paper. Do the same with another glass plate and a strip of blue litmus paper. Clean the stirring rod, and record your observations.

4. Read this step completely before proceeding. Add limewater to a test tube until it is half full, and place it in the rack. Obtain a third test tube, and pour half of the hydrochloric acid from the first test tube into it. Still holding this test tube, use the scoopula to gently add about 0.25 mL of baking soda to the acid. Working quickly but carefully, tip the tube to pour the gas into the limewater. Take care not to pour any liquid into the limewater. Place a stopper firmly in the test tube and shake it for several seconds. Then put it in the rack and record your observations.

546 MHR • Unit 6 Acids and Bases

5. Add six to eight small pieces of magnesium ribbon to the acid in the remaining (second) test tube. Record your observations. Then carefully lower a burning wooden splint into the test tube. Record your observations.

6. Wash the test tubes so they are completely clean. Ensure that no unreacted magnesium ribbon is placed down the sink.

7. Repeat steps 2 to 6 with the sulfuric acid and acetic acid solutions.

8. Add 5 mL of sodium hydroxide solution to a clean test tube. Use the clean stirring rod to put a drop of the sodium hydroxide solution on the strip of red litmus paper. Do the same with a strip of blue litmus paper. Clean the stirring rod, and record your observations.

9. Add six to eight small pieces of magnesium ribbon to the sodium hydroxide solution, and perform the splint test as you did for the acids in step 4. Record your observations.

10. Repeat steps 8 and 9 with the ammonia solution and the limewater.

11. Add 10 mL of sodium hydroxide solution to a clean 50 mL beaker or evaporating dish. With a stirring rod, add two drops of phenolphthalein. Then stir and record your observations. Continue stirring and use a dropper pipette or medicine dropper to add hydrochloric acid solution, slowly, one drop at a time. Stop adding acid when the pink colour you are observing disappears.

12. Test the clear, colourless solution that results from step 10 with blue and red litmus paper. Also test its conductivity. Record your observations.

## Analysis

1. Examine the chemical formulas for the acids and bases you tested in this investigation. Identify anything you notice that they have in common.

2. What is the effect of acids and bases on red and blue litmus paper?

3. What reaction, if any, occurs when magnesium is mixed with an acid and with a base? Explain your answer in detail.

4. Describe the reaction that occurs when baking soda is mixed with an acid, and identify the gas that is produced.

5. (a) The reaction that occurs when an acid is mixed with a base is called a neutralization reaction. Write a balanced chemical equation to describe the neutralization reaction that you observed.

   (b) What evidence from the procedure supports your balanced equation?

## Conclusion

6. (a) Design a chart or concept map to compare the properties of acids and bases, based on your observations.

   (b) Collaborate with the class to write a definition for acids and bases.

## Properties of Acids and Bases

One way to distinguish acids from bases is to describe their observable properties. For example, acids taste sour, and they change colour when mixed with coloured dyes called indicators. Bases taste bitter and feel slippery. They also change colour when mixed with indicators.

> **CAUTION** You should never taste or touch acids, bases, or any other chemicals. Early chemists used their senses of taste and touch to observe the properties of many chemicals. This dangerous practice often led to serious injury, and sometimes death.

Another property that can be used to distinguish acids from bases is their conductivity in solution. As you can see in Figure 14.2, aqueous solutions of acids and bases conduct electricity. This is evidence that ions are present in acidic and basic solutions. Some of these solutions, such as hydrochloric acid and sodium hydroxide (a base), cause the bulb to glow brightly. Most acidic and basic solutions, however, cause the bulb to glow dimly.

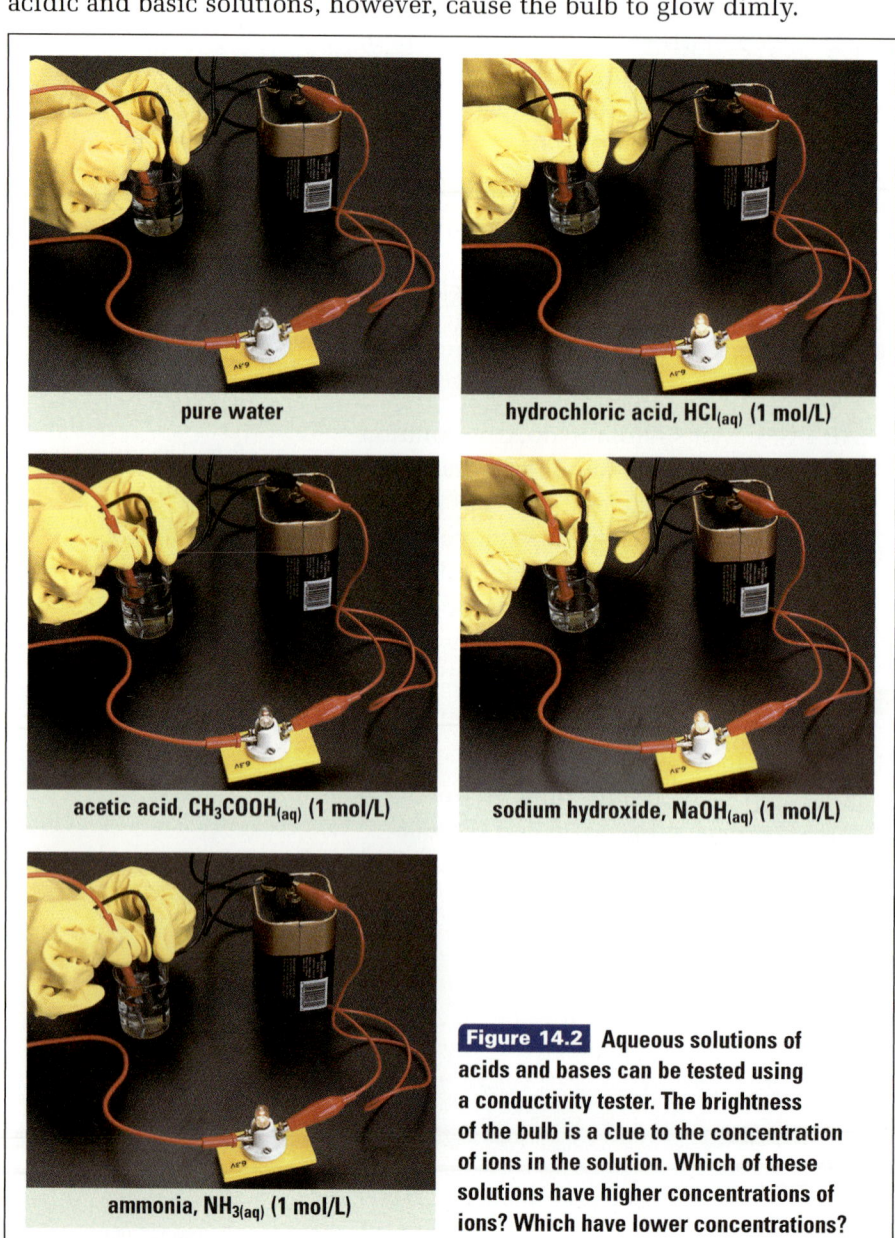

pure water

hydrochloric acid, $HCl_{(aq)}$ (1 mol/L)

acetic acid, $CH_3COOH_{(aq)}$ (1 mol/L)

sodium hydroxide, $NaOH_{(aq)}$ (1 mol/L)

ammonia, $NH_{3(aq)}$ (1 mol/L)

**Figure 14.2** Aqueous solutions of acids and bases can be tested using a conductivity tester. The brightness of the bulb is a clue to the concentration of ions in the solution. Which of these solutions have higher concentrations of ions? Which have lower concentrations?

Table 14.2 on the next page summarizes some other observable properties of acids and bases. These properties include their physical characteristics and their chemical behaviour. Taken together, the observable properties of acids and bases constitute their operational definitions. For example, an operational definition of an acid might be, "An acid is a sour-tasting compound that turns blue litmus paper red and reacts with bases to form a salt and water." An operational definition of a base might be, "A base is a bitter-tasting, slippery-feeling compound that turns red litmus paper blue and reacts with acids to form a salt and water." Operational definitions are useful for classifying substances and describing their macroscopic (observable) characteristics and behaviour. However, an operational definition cannot describe or explain the microscopic (unobservable) *causes* of those characteristics and behaviour. A more comprehensive definition of acids and bases requires a theory—something that chemists can use to explain established observations and properties and to make predictions in new situations. You will consider several theories of acids and bases in the remainder of this section.

## The Arrhenius Theory of Acids and Bases

In Figure 14.2, you saw evidence that ions are present in solutions of acids and bases. When hydrogen chloride dissolves in water, for example, it dissociates (breaks apart) into hydrogen ions and chloride ions.

$$HCl_{(aq)} \rightarrow H^+_{(aq)} + Cl^-_{(aq)}$$

When sodium hydroxide dissolves in water, it dissociates to form sodium ions and hydroxide ions.

$$NaOH_{(aq)} \rightarrow Na^+_{(aq)} + OH^-_{(aq)}$$

The dissociations of other acids and bases in water reveal a pattern. This pattern was first noticed in the late nineteenth century by a Swedish chemist named Svanté Arrhenius. (See Figure 14.3.)

$$HBr_{(aq)} \rightarrow \mathbf{H^+}_{(aq)} + Br^-_{(aq)}$$
$$H_2SO_{4(aq)} \rightarrow \mathbf{H^+}_{(aq)} + HSO_4^-_{(aq)}$$
$$HClO_{4(aq)} \rightarrow \mathbf{H^+}_{(aq)} + ClO_4^-_{(aq)}$$

*acids dissociating in water, and their resulting ions*

$$LiOH_{(aq)} \rightarrow Li^+_{(aq)} + \mathbf{OH^-}_{(aq)}$$
$$KOH_{(aq)} \rightarrow K^+_{(aq)} + \mathbf{OH^-}_{(aq)}$$
$$Ba(OH)_{2(aq)} \rightarrow Ba^+_{(aq)} + 2\mathbf{OH^-}_{(aq)}$$

*bases dissociating in water, and their resulting ions*

In 1887, Arrhenius published a theory to explain the nature of acids and bases. It is called the Arrhenius theory of acids and bases.

### The Arrhenius Theory of Acids and Bases
- An acid is a substance that dissociates in water to produce one or more hydrogen ions, $H^+$.
- A base is a substance that dissociates in water to form one or more hydroxide ions, $OH^-$.

According to the Arrhenius theory, acids increase the concentration of $H^+$ in aqueous solutions. Thus, *an Arrhenius acid must contain hydrogen as the source of* $H^+$. You can see this in the preceding dissociation reactions for acids.

**Figure 14.3** Svanté Arrhenius (1859–1927)–contributed to chemists understanding of electrolytes and reaction rates, as well as acids and bases.

**Web LINK**

www.mcgrawhill.ca/links/atlchemistry

Scientists did not embrace the Arrhenius theory when they first heard about it during the 1880s. Why were they unimpressed with this theory? What was necessary to convince them? Arrhenius is featured on several web sites on the Internet. To link with these web sites, go to the web site above and click on **Web Links**. Write a brief report to answer the above questions.

**Table 14.2** Some Observable Properties of Acids and Bases

| | Property | | |
|---|---|---|---|
| | Taste | Electrical conductivity in solution | Feel of solution |
| ACIDS | taste sour | conduct electricity | have no characteristic feel |
| BASES | taste bitter | conduct electricity | feel slippery |

| | Property | | |
|---|---|---|---|
| | Reaction with litmus paper | Reaction with active metals | Reaction with carbonate compounds |
| ACIDS | Acids turns blue litmus red | produce hydrogen gas | produce carbon dioxide gas |
| BASES | Bases turn red litmus blue | do not react | do not react |

| | Property |
|---|---|
| | Reaction with each other |
| ACIDS | Acids neutralize basic solutions, forming a salt and water. |
| BASES | Bases neutralize acidic solutions, forming a salt and water. |

550  MHR • Unit 6  Acids and Bases

Bases, on the other hand, increase the concentration of OH⁻ in aqueous solutions. *An Arrhenius base must contain the hydroxyl group, —OH.* You can see this in the preceding dissociation reactions for bases.

### Limitations of the Arrhenius Theory

The Arrhenius theory is useful if you are interested in the ions that result when an acid or a base dissociates in water. It also helps explain what happens when an acid and a base undergo a neutralization reaction. In such a reaction, an acid combines with a base to form an ionic compound and water. Examine the following reactions:

$$HCl_{(aq)} + NaOH_{(aq)} \rightarrow NaCl_{(aq)} + H_2O_{(\ell)}$$

The net ionic equation for this reaction shows the principal ions in the Arrhenius theory.

$$H^+_{(aq)} + OH^-_{(aq)} \rightarrow H_2O_{(\ell)}$$

Since acids and bases produce hydrogen ions and hydroxide ions, water is an inevitable result of acid-base reactions.

Problems arise with the Arrhenius theory, however. One problem involves the ion that is responsible for acidity: H⁺. Look again at the equation for the dissociation of hydrochloric acid.

$$HCl_{(aq)} \rightarrow H^+_{(aq)} + Cl^-_{(aq)}$$

This dissociation occurs in aqueous solution, but chemists often leave out H₂O as a component of the reaction. They simply assume that it is there. What happens if you put H₂O into the equation?

$$HCl_{(aq)} + H_2O_{(\ell)} \rightarrow H^+_{(aq)} + Cl^-_{(aq)} + H_2O_{(\ell)}$$

Notice that the water is unchanged when the reaction is represented this way. However, you learned earlier that water is a polar molecule. The O atom has a partial negative charge, and the H atoms have partial positive charges. Thus, H₂O must interact in some way with the ions H⁺ and Cl⁻. In fact, chemists made a discovery in the early twentieth century. They realized that protons do not exist in isolation in aqueous solution. (The hydrogen ion is simply a proton. It is a positively charged nuclear particle.) Instead, protons are always hydrated: they are attached to water molecules. A hydrated proton is called a **hydronium ion**, H₃O⁺₍aq₎. (See Figure 14.4.)

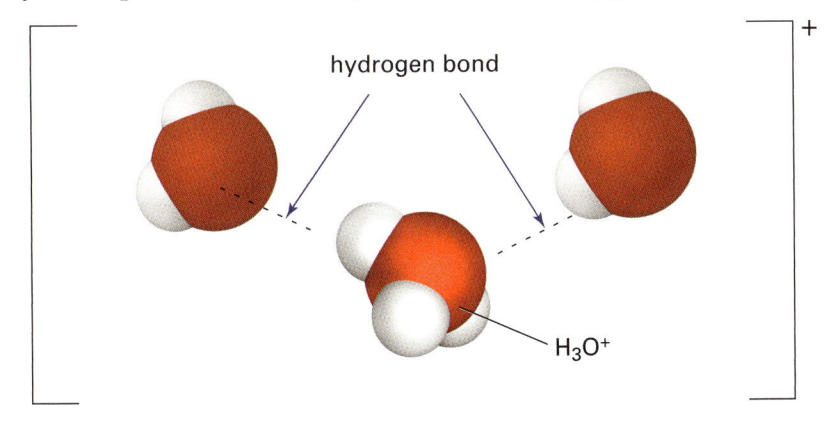

**Figure 14.4** For convenience, chemists often use H⁺₍aq₎ as a shorthand notation for the hydronium ion, H₃O⁺₍aq₎. Hydronium ions do not exist independently. Instead, they form hydrogen bonds with other water molecules. Thus, a more correct formula is H⁺(H₂O)$_n$, where $n$ is usually 4 or 5.

There is another problem with the Arrhenius theory. Consider the reaction of ammonia, NH₃, with water.

$$NH_{3(g)} + H_2O_{(\ell)} \rightleftharpoons NH_4^+_{(aq)} + OH^-_{(aq)}$$

Ammonia is one of several substances that produce basic solutions in water. As you can see, ammonia does not contain hydroxide ions. However, it does produce these ions when it reacts with water. Ammonia also undergoes a neutralization reaction with acids. The Arrhenius theory cannot explain the basic properties of ammonia.

**Modernizing the Arrhenius Theory**

One of the strengths of the Arrhenius theory is that chemists can modernize it to account for some of its limitations. For example, here again is the dissociation equation for $HCl_{(aq)}$ according to Arrhenius theory.

$$HCl_{(aq)} \rightarrow H^+_{(aq)} + Cl^-_{(aq)}$$

A modernized Arrhenius theory acknowledges the role of water and the production of hydronium ions in this reaction. Thus:

$$HCl_{(aq)} + H_2O_{(\ell)} \rightarrow H_3O^+_{(aq)} + Cl^-_{(aq)}$$

Similarly, the dissociation of sulfuric acid may be reinterpreted in light of a modernized Arrhenius theory.

$$H_2SO_{4(aq)} + H_2O_{(\ell)} \rightarrow H_3O^+_{(aq)} + HSO_4^-_{(aq)}$$

You have read that another limitation of the Arrhenius theory is that it cannot account for the basic properties of aqueous ammonia, because ammonia does not contain hydroxide ions. Thus, the dissociation of ammonia in water would occur by way of a two-step process:

$$NH_{3(g)} + H_2O_{(\ell)} \rightarrow NH_4OH_{(aq)}$$
$$NH_4OH_{(aq)} \rightarrow NH_4^+_{(aq)} + OH^-_{(aq)}$$

The compound inserted into this process is ammonium chloride. Its ionic formula, $NH_4OH$, is written correctly. However, there is *no* experimental evidence to suggest that this hypothesized compound actually exists. Chemists' attempt to "rescue" the Arrhenius theory in this case fails.

There is a further limitation with the Arrhenius theory—one that chemists cannot overcome. The theory assumes that all acid-base reactions occur in a single solvent: water. However, many acid-base reactions may occur in other solvents.

Yet another limitation of the Arrhenius theory is its inability to predict successfully whether a given aqueous compound is acidic or basic. For example, the hydrogen phosphate ion, $HPO_4^{2-}$, appears to contain $H^+$ ions. You would expect that solutions containing this ion would be acidic. And yet, when tested with litmus paper, solutions that contain this ion are found to be *basic*. Similarly, salts that contain carbonate ions also have basic properties when dissolved in water. The Arrhenius theory cannot predict or explain these facts.

Despite its shortcomings, the Arrhenius theory provides chemists—both past and present—with a way of understanding the properties and behaviour of certain compounds when they react with water and with each other. An underlying strength of the Arrhenius theory is that it provides a simple, effective way to understand neutralization reactions. You can see this clearly by examining the ionic and net ionic equations for a typical neutralization reaction.

Balanced equation: $HCl_{(aq)} + NaOH_{(aq)} \rightarrow NaCl_{(aq)} + H_2O_{(\ell)}$

Ionic equation: $H^+_{(aq)} + Cl^-_{(aq)} + Na^+_{(aq)} + OH^-_{(aq)} \rightarrow Na^+_{(aq)} + Cl^-_{(aq)} + H_2O_{(\ell)}$

Net ionic equation: $H^+_{(aq)} + OH^-_{(aq)} \rightarrow H_2O_{(\ell)}$

The net ionic equation clearly shows that hydrogen ions and hydroxide ions combine in a neutralization reaction to form water.

On the basis of its limitations, however, both the original and the modernized Arrhenius definitions for acids and bases are incomplete. Chemists needed a more comprehensive theory to account for a much broader range of observations and properties.

## The Brønsted-Lowry Theory of Acids and Bases

In 1923, two chemists working independently of each other proposed a new theory of acids and bases. (See Figure 14.5.) Johannes Brønsted in Copenhagen, Denmark, and Thomas Lowry in London, England, proposed what is called the **Brønsted-Lowry theory of acids and bases.** This theory overcame the limitations of the Arrhenius theory.

> **The Brønsted-Lowry Theory of Acids and Bases**
> - An acid is a substance from which a proton ($H^+$ ion) can be removed.
> - A base is a substance that can remove a proton ($H^+$ ion) from an acid.

**Figure 14.5** Johannes Brønsted (1879–1947), left, and Thomas Lowry (1874–1936), right. Brønsted published many more articles about ions in solution than Lowry did. Thus, some chemistry resources refer to the "Brønsted theory of acids and bases."

Like an Arrhenius acid, a Brønsted-Lowry acid must contain H in its formula. This means that all Arrhenius acids are also Brønsted-Lowry acids. However, any negative ion (not just $OH^-$) can be a Brønsted-Lowry base. In addition, water is not the only solvent that can be used.

According to the Brønsted-Lowry theory, there is only one requirement for an acid-base reaction. One substance must provide a proton, and another substance must receive the same proton. In other words, *an acid-base reaction involves the transfer of a proton.* Note that the word "proton" refers to the nucleus of a hydrogen atom—an $H^+$ ion that has been removed from the acid molecule. It does not refer to a proton removed from the nucleus of another atom, such as oxygen or sulfur, that may be present in the acid molecule.

The idea of proton transfer has major implications for understanding the nature of acids and bases. According to the Brønsted-Lowry theory, any substance can behave as an acid, but only if another substance behaves as a base at the same time. Similarly, any substance can behave as a base, but only if another substance behaves as an acid at the same time.

For example, consider the reaction between hydrochloric acid and water shown in Figure 14.6. In this reaction, hydrochloric acid is an acid

### CHEM FACT

In many chemistry references, Brønsted-Lowry acids are called "proton donors." Brønsted-Lowry bases are called "proton acceptors." Although these terms are common, they create a false impression about the energy that is involved in acid-base reactions. Breaking bonds always requires energy. For example, removing a proton from a hydrochloric acid molecule requires $1.4 \times 10^3$ kJ/mol. This is far more energy than the word "donor" implies.

Chapter 14 Properties of Acids and Bases • MHR  **553**

because it provides a proton (H⁺) to the water. The water molecule receives the proton. Therefore, according to the Brønsted-Lowry theory, water is a base in this reaction. When the water receives the proton, it becomes a hydronium ion ($H_3O^+$). Notice the hydronium ion on the right side of the equation.

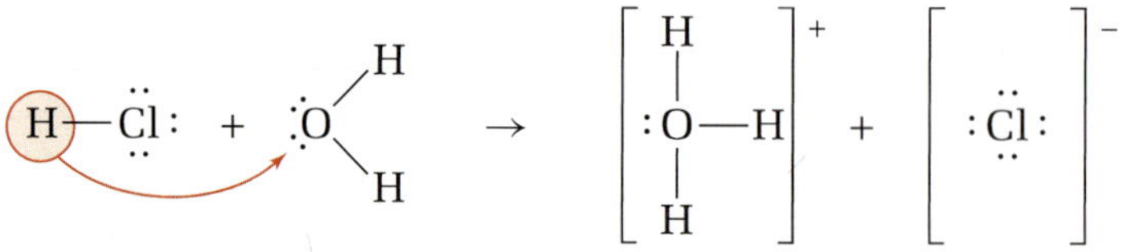

**Figure 14.6** The reaction between hydrochloric acid and water, according to the Brønsted-Lowry theory.

Two molecules or ions that are related by the transfer of a proton are called a **conjugate acid-base pair.** (Conjugate means "linked together.") The **conjugate base** of an acid is the particle that remains when a proton is removed from the acid. The **conjugate acid** of a base is the particle that results when the base receives the proton from the acid. In the reaction between hydrochloric acid and water, the hydronium ion is the conjugate acid of the base, water. The chloride ion is the conjugate base of the acid, hydrochloric acid.

According to the Brønsted-Lowry theory, every acid has a conjugate base, and every base has a conjugate acid. The conjugate base and conjugate acid of an acid-base pair are linked by the transfer of a proton. The conjugate base of the acid-base pair has one less hydrogen than the acid. It also has one more negative charge than the acid. The conjugate acid of the acid-base pair has one more hydrogen than the base. It also has one less negative charge than the base. See how this works with another reaction involving acetic acid.

The dissociation of acetic acid in water is represented in Figure 14.7. Acetic acid is a Brønsted-Lowry acid, because it provides a proton (H⁺) to the water. In receiving the proton, the water molecule is the base in this reaction, and becomes a hydronium ion. Notice, however, that this reaction is an equilibrium reaction; it proceeds in both directions. When acetic acid reacts with water, only a few ions dissociate. (You will learn about the significance of this fact shortly.) The position of equilibrium lies to the left, and the reverse reaction is favoured. In the reverse reaction, the hydronium ion gives up a proton to the acetate ion. Thus, in the reverse reaction, the hydronium ion is a Brønsted-Lowry acid and the acetate ion is a Brønsted-Lowry base. The acid on the left ($CH_3COOH$) and the base on the right ($CH_3COO^-$) differ by one proton, so they are a conjugate acid-base pair. Similarly, $H_2O$ and $H_3O^+$ are a conjugate acid-base pair, because they, too, differ by one proton. This relationship is shown in Figure 14.8.

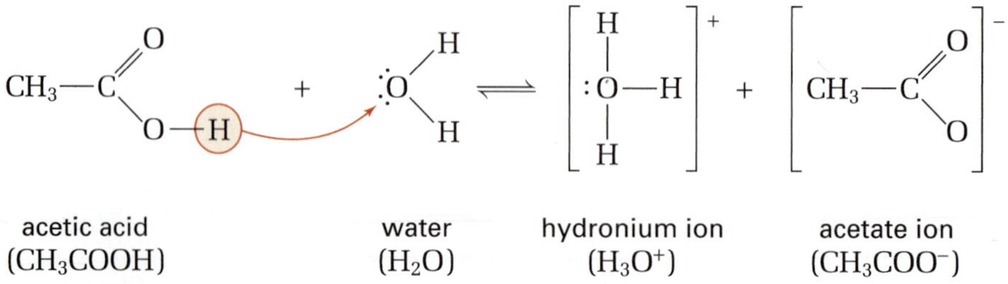

acetic acid ($CH_3COOH$)  water ($H_2O$)  hydronium ion ($H_3O^+$)  acetate ion ($CH_3COO^-$)

**Figure 14.7** The dissociation of acetic acid in water

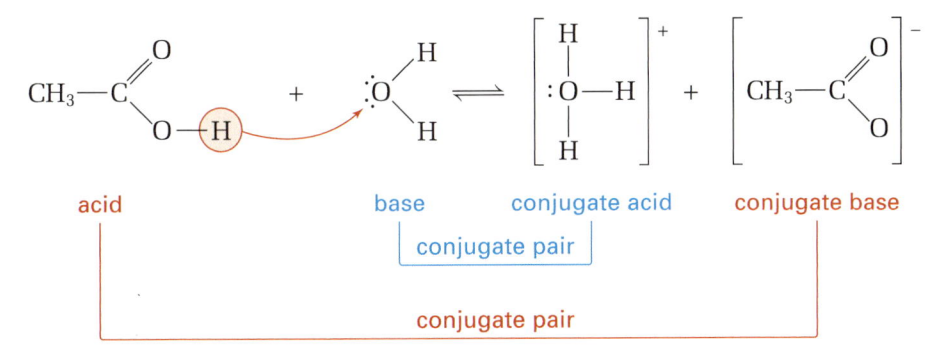

**Figure 14.8** Conjugate acid-base pairs in the dissociation of acetic acid in water

These ideas about acid-base reactions and conjugate acid-base pairs will become clearer as you study the following Sample Problems and Practice Problems.

## Sample Problem

### Conjugate Acid-Base Pairs

**Problem**

Hydrogen bromide is a gas at room temperature. It is soluble in water, forming hydrobromic acid. Identify the conjugate acid-base pairs.

**What Is Required?**

You need to identify two sets of conjugate acid-base pairs.

**What Is Given?**

You know that hydrogen bromide forms hydrobromic acid in aqueous solution.

**Plan Your Strategy**

- Write a balanced chemical equation.
- On the left side of the equation, identify the acid as the molecule that provides the proton. Identify the base as the molecule that accepts the proton.
- On the right side of the equation, identify the particle that has one proton less than the acid on the left side as the conjugate base of the acid. Identify the particle on the right side that has one proton more than the base on the left side as the conjugate acid of the base.

**Act on Your Strategy**

Hydrogen bromide provides the proton, so it is the Brønsted-Lowry acid in the reaction. Water receives the proton, so it is the Brønsted-Lowry base. The conjugate acid-base pairs are $HBr/Br^-$ and $H_2O/H_3O^+$.

*Continued...*

*Continued...*

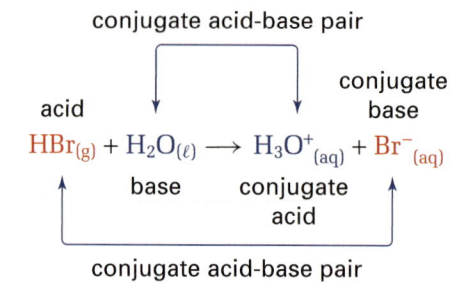

$$\underset{\text{acid}}{\text{HBr}_{(g)}} + \underset{\text{base}}{\text{H}_2\text{O}_{(\ell)}} \longrightarrow \underset{\text{conjugate acid}}{\text{H}_3\text{O}^+_{(aq)}} + \underset{\text{conjugate base}}{\text{Br}^-_{(aq)}}$$

conjugate acid-base pair (HBr / Br⁻)
conjugate acid-base pair (H₂O / H₃O⁺)

**Check Your Solution**

The formulas of the conjugate pairs differ by one proton, H⁺, as expected.

## Sample Problem

### More Conjugate Acid-Base Pairs

**Problem**

Ammonia is a pungent gas at room temperature. Its main use is in the production of fertilizers and explosives. It is very soluble in water. It forms a basic solution that is used in common products, such as glass cleaners. Identify the conjugate acid-base pairs in the reaction between aqueous ammonia and water.

$$\text{NH}_{3(g)} + \text{H}_2\text{O}_{(\ell)} \rightleftharpoons \text{NH}_4^+_{(aq)} + \text{OH}^-_{(aq)}$$

**What Is Required?**

You need to identify the conjugate acid-base pairs.

**What Is Given?**

The chemical equation is given.

**Plan Your Strategy**

- Identify the proton-provider on the left side of the equation as the acid. Identify the proton-remover (or proton-receiver) as the base.
- Identify the conjugate acid and base on the right side of the equation by the difference of a single proton from the acid and base on the left side.

**Act on Your Strategy**

The conjugate acid base pairs are NH₄⁺/NH₃ and H₂O/OH⁻.

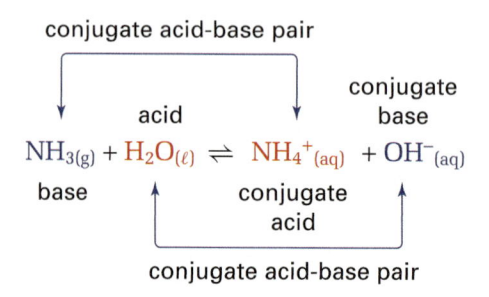

$$\underset{\text{base}}{\text{NH}_{3(g)}} + \underset{\text{acid}}{\text{H}_2\text{O}_{(\ell)}} \rightleftharpoons \underset{\text{conjugate acid}}{\text{NH}_4^+_{(aq)}} + \underset{\text{conjugate base}}{\text{OH}^-_{(aq)}}$$

conjugate acid-base pair
conjugate acid-base pair

**Check Your Solution**

The formulas of the conjugate pairs differ by a proton, as expected.

> **PROBLEM TIP**
>
> In the previous Sample Problem, water acted as a base. In this Sample Problem, water acts as an acid. Water can act as a proton-provider (an acid) in some reactions and as a proton-receiver (a base) in others. You will learn more about this dual-behaviour of certain substances following the Practice Problems.

556 MHR • Unit 6 Acids and Bases

## Practice Problems

1. When perchloric acid dissolves in water, the following reaction occurs:
$$HClO_{4(aq)} + H_2O_{(\ell)} \rightarrow H_3O^+_{(aq)} + ClO_4^-_{(aq)}$$
Identify the conjugate acid-base pairs.

2. Sodium acetate is a good electrolyte. In water, the acetate ion reacts as follows:
$$CH_3COO^-_{(aq)} + H_2O_{(\ell)} \rightleftharpoons CH_3COOH_{(aq)} + OH^-_{(aq)}$$
Identify the conjugate acid-base pairs.

3. Name and write the formula of the conjugate base of each molecule or ion.
   (a) $HCl$   (b) $HCO_3^-$   (c) $H_2SO_4$   (d) $N_2H_5^+$

4. Name and write the formula of the conjugate acid of each molecule or ion.
   (a) $NO_3^-$   (b) $OH^-$   (c) $H_2O$   (d) $HCO_3^-$

## Acting Like an Acid *or* a Base: Amphoteric Substances

In the Sample Problem on page 556, you saw that $H_2O$ gives up a proton, and thus plays the role of an acid in the dissociation reaction involving ammonia. In Figure 14.7, you saw that $H_2O$ removes a proton, and thus plays the role of a base in the dissociation of acetic acid. Substances that can act as an acid in one reaction and as a base in a different reaction are said to be **amphoteric**.

Amphoteric substances may be molecules, as in the case of water, or anions with an available hydrogen that can dissociate. For example, the hydrogen carbonate ion, $HCO_3^-$, is amphoteric. When baking soda is mixed with water, $HCO_3^-$ may remove a proton from water, thus acting as a base. Conversely, $HCO_3^-$ may lose a proton to water, thus acting as an acid.

$$HCO_3^-_{(aq)} + OH^-_{(aq)} \rightleftharpoons CO_3^{2-}_{(aq)} + H_2O_{(\ell)}$$
$$HCO_3^-_{(aq)} + H_3O^+_{(aq)} \rightleftharpoons H_2CO_{3(aq)} + H_2O_{(\ell)}$$

In the following Practice Problems, remember that a Brønsted-Lowry acid is a substance that can donate a proton to another substance. A Brønsted-Lowry base is a substance that can accept a proton from an acid. Therefore, look carefully for hydrogen-containing compounds that could provide or accept a proton.

## Practice Problems

5. Write equations to show how the hydrogen sulfide ion, $HS^-$, can be classified as amphoteric. First show the ion acting as an acid. Then show the ion acting as a base.

6. Identify the conjugate acid-base pair in each reaction.
   (a) $H_3PO_{4(aq)} + H_2O_{(\ell)} \rightleftharpoons H_2PO_4^-_{(aq)} + H_3O^+_{(aq)}$
   (b) $O^{2-}_{(aq)} + H_2O_{(\ell)} \rightleftharpoons 2OH^-_{(aq)}$

7. For each reaction in question 6, identify the amphoteric chemical species, and identify its role as either an acid or a base.

*Continued...*

> Continued...
>
> 8. Identify the conjugate acid-base pair in each reaction.
>    (a) $NH_{3(aq)} + H_2O_{(\ell)} \rightleftharpoons NH_4^+{}_{(aq)} + OH^-{}_{(aq)}$
>    (b) $CH_3COOH_{(aq)} + H_2O_{(\ell)} \rightleftharpoons CH_3COO^-{}_{(aq)} + H_3O^+{}_{(aq)}$
>
> 9. For each reaction in question 8, identify the amphoteric chemical species, and identify its role as either an acid or a base.

## Section Wrap-up

The theories that you have considered in this section attempt to explain the chemical nature of acids and bases. Table 14.3 summarizes the key points of these theories.

In the Arrhenius theory, the modern Arrhenius theory, and the Brønsted-Lowry theory, acids and bases form ions in solution. Many characteristics of acid-base behaviour are linked to the number of ions that form from a particular acid or base. One of these characteristics is strength.

In the next section, you will learn why a dilute solution of vinegar is safe to ingest, while the same molar concentration of hydrochloric acid would be extremely poisonous.

**Table 14.3** Comparing the Arrhenius Theory, Modern Arrhenius Theory, and the Brønsted-Lowry Theory.

| Theory | Arrhenius | Modern Arrhenius | Brønstead-Lowry |
|---|---|---|---|
| Acid | any substance that dissociates to form $H^+$ in aqueous solution | any substance that dissociates to form $H_3O^+$ in aqueous solution | any substance that provides a proton to another substance (or any substance from which a proton may be removed) |
| Base | any substance that dissociates to form $OH^-$ in aqueous solution | any substance that dissociates to form $OH^-$ in aqueous solution | any substance that receives a proton from an acid (or any substance that removes a proton from an acid) |
| Example | $HCl_{(aq)} \rightarrow H^+{}_{(aq)} + Cl^-{}_{(aq)}$ | $HCl_{(aq)} + H_2O_{(\ell)} \rightarrow H_3O^+{}_{(aq)} + Cl^-{}_{(aq)}$ | $NH_{3(g)} + H_2O_{(\ell)} \rightleftharpoons NH_4^+{}_{(aq)} + OH^-{}_{(aq)}$ |

## Section Review

| Solution | Results of Conductivity Test |
|---|---|
| A | the bulb glows dimly |
| B | the bulb glows strongly |
| C | the bulb does not glow |
| D | the bulb glows strongly |

**1** Suppose that you have four unknown solutions, labelled A, B, C, and D. You use a conductivity apparatus to test their conductivity, and obtain the results shown below. Use these results to answer the questions that follow.

(a) Which of these solutions has a high concentration of dissolved ions? What is your evidence?

(b) Which of these solutions has a low concentration of dissolved ions? What is your evidence?

(c) Which of the four unknowns are probably aqueous solutions of acids or bases?

(d) Based on these tests alone, can you distinguish the acidic solution(s) from the basic solutions(s)? Why or why not?

(e) Suggest one way that you could distinguish the acidic solution(s) from the basic solution(s).

**2** (a) Define an acid and a base according to the Arrhenius theory.

(b) Give two examples of observations that the Arrhenius theory can explain.

(c) Give two examples of observations that the Arrhenius theory can not explain.

**3** Sodium bicarbonate, $NaHCO_{3(aq)}$ turns red litmus paper blue and undergoes a neutralization reaction with acids.

(a) Use Arrhenius theory to write a chemical equation showing the dissociation of $NaHCO_{3(aq)}$ in water.

(b) Use a modernized Arrhenius theory to show this same dissociation.

(c) Does the chemical behaviour of $NaHCO_{3(aq)}$ support or refute either forms of the Arrhenius theory? Give reasons to justify your answer.

**4** (a) Define an acid and a base according to the Brønsted-Lowry theory.

(b) What does the Brønsted-Lowry theory have in common with the Arrhenius theory? In what ways is it different?

(c) Which of the two acid-base theories is more comprehensive? (In other words, which explains a broader body of observations?)

**5** (a) What is the conjugate acid of a base?

(b) What is the conjugate base of an acid?

(c) Use an example to illustrate your answers to parts (a) and (b) above.

**6** Write the formula for the conjugate acid of the following:

(a) the hydroxide ion, $OH^-$

(b) the carbonate ion, $CO_3^{2-}$

**7** Write the formula for the conjugate base of the following:

(a) nitric acid, $HNO_3$

(b) the hydrogen sulfate ion, $HSO_4^-$

**8** Which of the following compounds is an acid according to the Arrhenius theory?

(a) $H_2O$  (b) $Ca(OH)_2$
(c) $H_3PO_3$  (d) HF

**9** Which of the following compounds is a base according to the Arrhenius theory?

(a) KOH  (b) $Ba(OH)_2$
(c) HClO  (d) $H_3PO_4$

**10** Hydrofluoric acid dissociates in water to form fluoride ions.

(a) Write a balanced chemical equation for this reaction.

(b) Identify the conjugate acid-base pairs.

(c) Explain how you know whether or not you have correctly identified the conjugate acid-base pairs.

**11** Identify the conjugate acid-base pairs in the following reactions:

(a) $H_2PO_4^-{}_{(aq)} + CO^{2-}{}_{3(aq)} \rightleftharpoons HPO_4^{2-}{}_{(aq)} + HCO_3^-{}_{(aq)}$

(b) $HCOOH_{(aq)} + CN^-{}_{(aq)} \rightleftharpoons HCOO^-{}_{(aq)} + HCN_{(aq)}$

(c) $H_2PO_4^-{}_{(aq)} + OH^-{}_{(aq)} \rightleftharpoons HPO_4^{2-}{}_{(aq)} + H_2O_{(\ell)}$

## 14.2 Strong and Weak Acids and Bases

**Section Preview/Outcomes**

In this section, you will

- **explain**, in terms of the degree to which they dissociate, the difference between strong and weak acids and bases
- **compare**, qualitatively, the relative strengths of acids and bases
- **predict** whether reactants or products are favoured in an acid-base reaction
- **define** $K_w$, and relate its value to $[H_3O^+]$ and $[OH^-]$
- **define** and **calculate** values associated with pH, pOH, $[H_3O^+]$, and $[OH^-]$
- **communicate** your understanding of the following terms: *strong acid, weak acid, strong base, weak base, ion product constant for water ($K_w$), pH, pOH*

You have learned that substances such as water and the hydrogen carbonate ion are amphoteric—that is, they can act either as a Brønsted-Lowry acid or a Brønsted-Lowry base in a particular reaction. For example, the hydrogen carbonate ion behaves as an acid in the first of the two chemical equations below, but as a base in the second.

$$HCO_3^-{}_{(aq)} + HPO_4^{2-}{}_{(aq)} \rightleftharpoons CO_3^{2-}{}_{(aq)} + H_2PO_4^-{}_{(aq)}$$

($HCO_3^-{}_{(aq)}$ behaving as an acid)

$$HCO_3^-{}_{(aq)} + HF_{(aq)} \rightleftharpoons H_2CO_{3(aq)} + F^-{}_{(aq)}$$

($HCO_3^-{}_{(aq)}$ behaving as a base)

How could you predict which role the ion plays in each reaction? The answer to this question involves the strength of the ion in relation to the other substances present in each reaction. Keep this idea in mind as you learn about strong and weak acids and bases. You will return to it later in this section.

### Strong and Weak Acids

In terms of acid-base reactions, strength refers to the extent to which a substance dissociates in its solvent. An acid that dissociates completely is termed a **strong acid**. As you can see in Figure 14.9, hydrochloric acid is a strong acid. *All* the molecules of hydrochloric acid in an aqueous solution dissociate into $H^+$ and $Cl^-$ ions. Table 14.4 lists the strong acids. Note that *the concentration of hydronium ions in a dilute solution of a strong acid is equal to the concentration of the acid*. Therefore, a 1.0 mol/L solution of hydrochloric acid contains 1.0 mol/L of hydronium ions and 1.0 mol/L of chloride ions.

**Table 14.4**
Common Strong Acids

| |
|---|
| hydrochloric acid, HCl |
| hydrobromic acid, HBr |
| hydroiodic acid, HI |
| nitric acid, HNO₃ |
| sulfuric acid, H₂SO₄ |
| perchloric acid, HClO₄ |

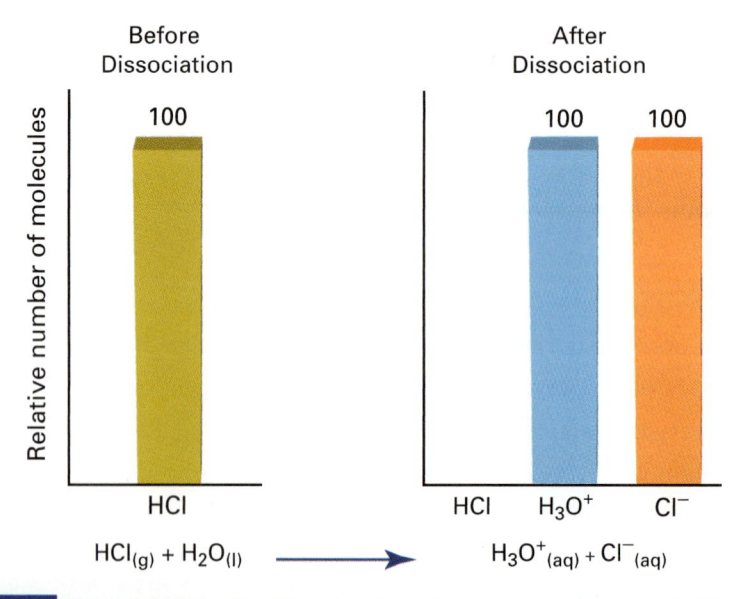

$HCl_{(g)} + H_2O_{(l)} \longrightarrow H_3O^+{}_{(aq)} + Cl^-{}_{(aq)}$

**Figure 14.9** When hydrogen chloride molecules enter an aqueous solution, 100% of the hydrogen chloride molecules dissociate. In other words, the percent dissociation of HCl in water is 100%. As a result, the solution contains the same percent of $H^+$ ions (in the form of $H_3O^+$) and $Cl^-$ ions: 100%.

560 MHR • Unit 6 Acids and Bases

A **weak acid** is an acid that dissociates only slightly in a water solution. Thus, only a small percentage of the acid molecules dissociate. Most of the acid molecules remain intact. The majority of acids are weak acids. For example, acetic acid is a weak acid. The percent dissociation of acetic acid molecules is only about 1% in a 0.1 mol/L solution. (The number of acid molecules that dissociate depends on the concentration and temperature of the solution.) Note that the concentration of hydronium ions in a solution of a weak acid is always less than the concentration of the dissolved acid. (See Figure 14.10.)

> **CONCEPT CHECK**
>
> Notice that the chemical equation for the dissociation of hydrochloric acid—a strong acid—proceeds in a single direction, to the right. The chemical equation for the dissociation of acetic acid—a weak acid—shows an equilibrium reaction. Explain why this makes sense. Look for other instances of single- and double-arrows up to this point in the chapter, and explain what you see.

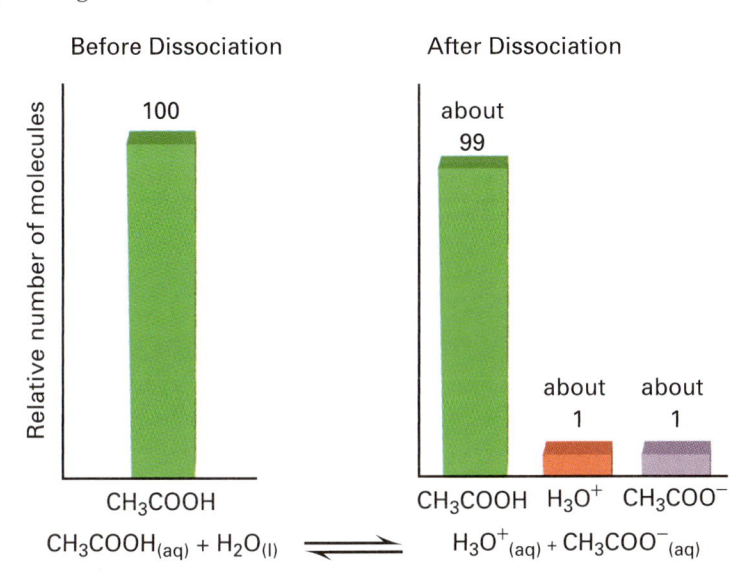

**Figure 14.10** When acetic acid molecules enter an aqueous solution, only about 1% of them dissociate. Thus, the number of acetic acid molecules in solution is far greater than the number of hydronium ions and acetate ions.

A few acids contain only a single hydrogen ion that can dissociate. These acids are called *monoprotic acids*. (The prefix mono- means "one." The root -protic refers to "proton.") Hydrochloric acid, hydrobromic acid, and hydroiodic acid are strong monoprotic acids. Hydrofluoric acid, HF, is weak monoprotic acid.

Many acids contain two or more hydrogen ions that can dissociate. For example, sulfuric acid, $H_2SO_{4(aq)}$, has two hydrogen ions that can dissociate. Sulfuric acid is a strong acid, but *only* for its first dissociation.

$$H_2SO_{4(aq)} \rightarrow H^+_{(aq)} + HSO_4^-_{(aq)}$$

The resulting aqueous hydrogen sulfate ion, $HSO_4^-$, is a weak acid. It dissociates to form the sulfate ion in the following equilibrium reaction:

$$HSO_4^-_{(aq)} \rightleftharpoons H^+_{(aq)} + SO_4^{2-}_{(aq)}$$

Thus, acids that contain two hydrogen ions dissociate to form two anions. These acids are sometimes called *diprotic acids*. (The prefix di-, as you know, means "two.") The acid that is formed by the first dissociation is stronger than the acid that is formed by the second dissociation.

Acids that contain three hydrogen ions are called *triprotic acids*. Phosphoric acid, $H_3PO_{4(aq)}$, is a triprotic acid. It gives rise to three anions, as follows:

$$H_3PO_{4(aq)} + H_2O_{(\ell)} \rightleftharpoons H^+_{(aq)} + H_2PO_4^-_{(aq)}$$
$$H_2PO_4^-_{(aq)} + H_2O_{(\ell)} \rightleftharpoons H^+_{(aq)} + HPO_4^{2-}_{(aq)}$$
$$HPO_4^{2-}_{(aq)} + H_2O_{(\ell)} \rightleftharpoons H^+_{(aq)} + PO_4^{3-}_{(aq)}$$

Here again, the acid that is formed by the first dissociation is stronger than the acid that is formed by the second dissociation. This acid is stronger than the acid that is formed by the third dissociation. Keep in mind, however, that all three of these acids are weak, because only a small proportion of them dissociates.

### Strong and Weak Bases

Like a strong acid, a **strong base** dissociates completely into ions in water. All oxides and hydroxides of the alkali metals—Group 1 (IA)—are strong bases. The oxides and hydroxides of the alkaline earth metals—Group 2 (IIA)—below beryllium are also strong bases.

Recall that the concentration of hydronium ions in a dilute solution of a strong acid is equal to the concentration of the acid. Similarly, the concentration of hydroxide ions in a dilute solution of a strong base is equal to the concentration of the base. For example, a 1.0 mol/L solution of sodium hydroxide (a strong base) contains 1.0 mol/L of hydroxide ions.

Table 14.5 lists some common strong bases. Barium hydroxide, $Ba(OH)_2$, and strontium hydroxide, $Sr(OH)_2$, are strong bases that are soluble in water. Magnesium oxide, MgO, and magnesium hydroxide, $Mg(OH)_2$, are also strong bases, but they are considered to be insoluble. A small amount of these compounds does dissolve in water, however. Virtually all of this small amount dissociates completely.

Most bases are weak. A **weak base** dissociates very slightly in a water solution. The most common weak base is aqueous ammonia. In a 0.1 mol/L solution, only about 1% of the ammonia molecules react with water to form hydroxide ions. This equilibrium is represented in Figure 14.11.

**Table 14.5**
Common Strong Bases

| sodium hydroxide, NaOH |
| potassium hydroxide, KOH |
| calcium hydroxide, $Ca(OH)_2$ |
| strontium hydroxide, $Sr(OH)_2$ |
| barium hydroxide, $Ba(OH)_2$ |

$$NH_3 + H_2O \rightleftharpoons NH_4^+ + OH^-$$

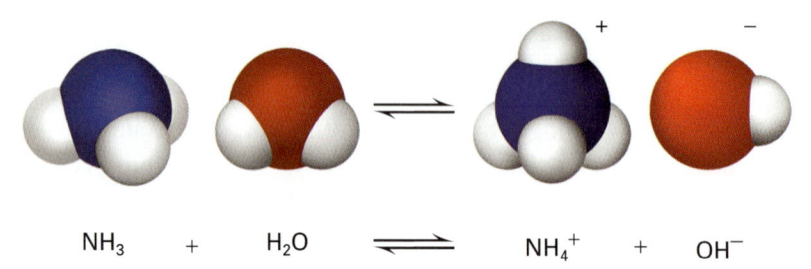

**Figure 14.11** Ammonia does not contain hydroxide ions, so it is not an Arrhenius base. As you can see, however, an ammonia molecule can remove a proton from water, leaving a hydroxide ion behind. Thus, ammonia is a Brønsted-Lowry weak base.

### Comparing the Relative Strengths of Acids and Bases

Earlier, you saw that the hydrogen carbonate ion may behave as an acid or a base in the following reactions.

$$HCO_3^-_{(aq)} + HPO_4^{2-}_{(aq)} \rightleftharpoons CO_3^{2-}_{(aq)} + H_2PO_4^-_{(aq)}$$
($HCO_3^-_{(aq)}$ behaving as an acid)

$$HCO_3^-_{(aq)} + HF_{(aq)} \rightleftharpoons H_2CO_{3(aq)} + F^-_{(aq)}$$
($HCO_3^-_{(aq)}$ behaving as a base)

You are now in a position to reconsider the question: How could you predict the role that each ion plays?

Over the centuries, chemists have performed countless experiments involving acids and bases. The data from these experiments have enabled chemists to rank acids and bases according to their strengths in relation to one another, as shown in Figure 14.12. Using these data, you can easily compare the relative strengths of acids and bases. For example, the chart shows that $HCO_3^-$ is above $HPO_4^{2-}$. Since the acids are listed in order of increasing strength, you know that $HCO_3^-$ is a stronger acid than $HPO_4^{2-}$. Therefore, you can (if the remainder of the balanced equation were not given to you) correctly predict the products of the reaction. Why? Because you know that the proton from the acid must go to its conjugate base, $H_2PO_4^{2-}$.

| Acid | Base |
|---|---|
| HCl | Cl$^-$ |
| $H_2SO_4$ | $HSO_4^-$ |
| $HNO_3$ | $NO_3^-$ |
| $H_3O^+$ | $H_2O$ |
| $HSO_4^-$ | $SO_4^{2-}$ |
| $H_2SO_3$ | $HSO_3^-$ |
| $H_3PO_4$ | $H_2PO_4^-$ |
| HF | F$^-$ |
| $CH_3COOH$ | $CH_3COO^-$ |
| $H_2CO_3$ | $HCO_3^-$ |
| $H_2S$ | HS$^-$ |
| $HSO_3^-$ | $SO_3^{2-}$ |
| $H_2PO_4^-$ | $HPO_4^{2-}$ |
| $NH_4^+$ | $NH_3$ |
| $HCO_3^-$ | $CO_3^{2-}$ |
| $HPO_4^{2-}$ | $PO_4^{3-}$ |
| $H_2O$ | OH$^-$ |
| HS$^-$ | S$^{2-}$ |
| OH$^-$ | O$^{2-}$ |

ACID STRENGTH ↑     BASE STRENGTH ↓

**Figure 14.12** Relative strengths of selected acids and bases. Notice that the strongest acids appear at the top of the list on the left, and the strongest bases appear at the bottom of the list on the right. Notice also that a stronger acid has a weaker conjugate base, and a stronger base has a weaker conjugate acid.

Figure 14.12 is useful in another way. You can use it to predict the direction in which an acid-base reaction will proceed. Or, to put it another way, you can use the chart to help you determine whether products or reactants are favoured in an acid-base reaction. The direction of an acid-base reaction usually proceeds from a stronger acid and a stronger base to a weaker acid and a weaker base. If the reaction proceeds to the right (that is, if the stronger acid and stronger base are on the left side of the equation), products are favoured. If the reaction goes to the left (if the stronger acid and stronger base are on the right side of the equation), reactants are favoured. Use the next Sample Problem and Practice Problems to help you understand these ideas better.

### Sample Problem

### Predicting the Direction of an Acid-Base Reaction

**Problem**

Predict the direction in which the following reaction will proceed. State whether reactants or products are favoured at equilibrium, and briefly defend your reasoning.

$$SO_4^{2-}{}_{(aq)} + CH_3COOH_{(aq)} \rightleftharpoons HSO_4^-{}_{(aq)} + CH_2COO^-{}_{(aq)}$$

**Solution**

Use Figure 14.12 to assess the relative strengths of the acids and bases. You can see that $HSO_4^-$ is above $CH_3COOH$, so $HSO_4^-$ is a stronger acid than $CH_3COOH$. Similarly, $CH_2COO^-$ is lower than $SO_4^-$, so $CH_2COO^-$ is a stronger base. The stronger acid and stronger base form the weaker acid and the weaker base, so the direction proceeds in that direction—that is, to the left.

$$\underset{\text{weaker acid}}{SO_4^{2-}{}_{(aq)}} + \underset{\text{weaker base}}{CH_3COOH_{(aq)}} \rightleftharpoons \underset{\text{stronger acid}}{HSO_4^-{}_{(aq)}} + \underset{\text{stronger base}}{CH_2COO^-{}_{(aq)}}$$

the reactants are favoured in this reaction.

### Practice Problems

10. Predict the direction for the following equations. State whether reactants or products are favoured, and give reasons to support your decision.

    (a) $NH_4^+{}_{(aq)} + H_2PO_4^-{}_{(aq)} \rightleftharpoons NH_{3(aq)} + H_3PO_{4(aq)}$

    (b) $H_2O_{(\ell)} + HS^-{}_{(aq)} \rightleftharpoons OH^-{}_{(aq)} + H_2S_{(aq)}$

    (c) $HF_{(aq)} + SO_4^{2-}{}_{(aq)} \rightleftharpoons F^-{}_{(aq)} + HSO_4^-{}_{(aq)}$

11. In which direction will the following reactions proceed? Explain why in each case.

    (a) $HPO_4^{2-}{}_{(aq)} + NH_4^+{}_{(aq)} \rightleftharpoons H_2PO_4^- + NH_{3(aq)}$

    (b) $H_2SO_{4(aq)} + H_2O_{(\ell)} \rightleftharpoons HSO_4^-{}_{(aq)} + OH^-{}_{(aq)}$

    (c) $H_2S + NH_{3(aq)} \rightleftharpoons HS^-{}_{(aq)} + NH_4^+{}_{(aq)}$

## pH: One Method for Describing Acids and Bases Quantitatively

You are probably familiar with the term "pH" from a variety of sources. Advertisers talk about the "pH balance" of products such as soaps, shampoos, and skin creams. People who own aquariums and swimming pools must monitor the pH of the water. Gardeners and farmers use simple tests to determine the pH of the soil. They know that plants and food crops grow best within a narrow range of pH. Similarly, the pH of your blood must remain within narrow limits for you to stay healthy.

What exactly is pH? How is it measured? To answer these questions, you must look again at the chemical equilibrium of water.

### The Ion Product Constant for Water

As you know, all aqueous solutions contain ions. Even pure water contains a few ions that are produced by the dissociation of water molecules.

$$H_2O_{(\ell)} + H_2O_{(\ell)} \rightleftharpoons H_3O^+{}_{(aq)} + OH^-{}_{(aq)}$$

At 25°C, only about two water molecules in one billion dissociate. This is why pure water is such a poor conductor of electricity. Chemists have determined that the concentration of hydronium ions in pure water at 25°C is $1.0 \times 10^{-7}$ mol/L. The dissociation of water also produces the same very small number of hydroxide ions: $1.0 \times 10^{-7}$ mol/L. These concentrations must be the same, because the dissociation of water produces equal number of hydronium and hydroxide ions. That is, the ions are produced in a 1:1 ratio.

Because the dissociation of water is an equilibrium, you can write an equilibrium constant expression for it. However, recall that the concentration of a liquid is itself a constant, so it may be factored out of the expression for $K$. The resulting constant is called the **ion product constant for water**, and is given the symbol $K_w$.

$$K_w = [H_3O^+][OH^-]$$

The equilibrium value of the concentration ion product, $[H_3O^+][OH^-]$, at 25°C is as follows:

$$K_w = 1.0 \times 10^{-7} \text{ mol/L} \times 1.0 \times 10^{-7} \text{ mol/L}$$
$$= 1.0 \times 10^{-14}$$

The concentration of $H_3O^+$ in the solution of a strong acid is equal to the concentration of the dissolved acid, unless the solution is very dilute. Consider $[H_3O^+]$ in a solution of 0.1 mol/L hydrochloric acid. All the molecules of HCl dissociate in water, forming a hydronium ion concentration that equals 0.1 mol/L. The increased $[H_3O^+]$ pushes the dissociation reaction between water molecules to the left, in accordance with Le Châtelier's principle. Therefore, the concentration of hydronium ions that results from the dissociation of water is even less than $1 \times 10^{-7}$ mol/L. This $[H_3O^+]$ is negligible compared with the 0.1 mol/L concentration of the hydrochloric acid. Unless the solution is very dilute (about $1 \times 10^{-7}$ mol/L), the dissociation of water molecules can be ignored when determining $[H_3O^+]$ of a strong acid.

Similarly, you can determine the concentration of hydroxide ions from the concentration of the dissolved base. If the solution is a strong base, you can ignore the dissociation of water molecules when determining $[OH^-]$, unless the solution is very dilute. When either $[H_3O^+]$ or $[OH^-]$ is known, you can use the ion product constant for water, $K_w$, to determine the concentration of the other ion. Although the value of $K_w$ for water is $1.0 \times 10^{-14}$ at 25°C only, you can use this value unless another value is given for a different temperature.

### CONCEPT CHECK

At 24°C or at any temperature other than 25°C, the concentration of hydronium (or hydroxide) ions in water is *not* $1.0 \times 10^{-7}$ mol/L. Explain in detail why this is the case.

**$[H_3O^+]$ and $[OH^-]$ in Aqueous Solutions at 25°C**

In an acidic solution, $[H_3O^+]$ is greater than $1.0 \times 10^{-7}$ mol/L and $[OH^-]$ is less than $1.0 \times 10^{-7}$ mol/L.

In a neutral solution, both $[H_3O^+]$ and $[OH^-]$ are equal to $1.0 \times 10^{-7}$ mol/L.

In a basic solution, $[H_3O^+]$ is less than $1.0 \times 10^{-7}$ mol/L and $[OH^-]$ is greater than $1.0 \times 10^{-7}$ mol/L.

## Sample Problem

### Determining [H₃O⁺] and [OH⁻]

**Problem**

Find [H₃O⁺] and [OH⁻] in each solution.

(a) 2.5 mol/L nitric acid

(b) 0.16 mol/L barium hydroxide

**Solution**

You know that nitric acid is a strong acid and barium hydroxide is a strong base. Since both dissociate completely in aqueous solutions, you can use their molar concentrations to determine [H₃O⁺] or [OH⁻]. You can find the concentration of the other ion using $K_w$:

$K_w = 1.0 \times 10^{-14}$

$\quad = [H_3O^+][OH^-]$

(a) [HNO₃] = 2.5 mol/L, so [H₃O⁺] = 2.5 mol/L

$$[OH^-] = \frac{1.0 \times 10^{-14} \text{ mol/L}}{2.5}$$

$$\quad\quad\quad = 4.0 \times 10^{-15} \text{ mol/L}$$

(b) $Ba(OH)_2 \xrightarrow{H_2O} Ba^{2+}_{(aq)} + 2OH^-_{(aq)}$

Each mole of Ba(OH)₂ in solution forms two moles of OH⁻ ions.

∴ [OH⁻] = 2 × 0.16 = 0.32 mol/L

$$[H_3O^+] = \frac{1.0 \times 10^{-14} \text{ mol/L}}{0.32}$$

$$\quad\quad\quad = 3.1 \times 10^{-14} \text{ mol/L}$$

**Check Your Solution**

For a solution of a strong acid, as in part (a), [H₃O⁺] should be greater than $1.0 \times 10^{-14}$ and [OH⁻] should be less than $1.0 \times 10^{-14}$. For a solution of strong base, [OH⁻] should be greater than, and [H₃O⁺] should be less than, $1.0 \times 10^{-14}$.

---

**PROBLEM TIP**

In the solution for part (b), be sure you understand why [OH⁻] = 0.32 mol/L, and 0.16 mol/L. If you are unsure, review Chapter 8, page 300. As well, try this problem: What is the molar concentration of [H₃O⁺] and [OH⁻] in each of the following aqueous solutions of strong acids and strong bases:

- $1.0 \times 10^{-6}$ mol/L KOH
- 0.100 mol/L Ba(OH)₂
- 1.50 mol/L HBr
- $1.0 \times 10^{-2}$ mol/L HNO₃

---

## Practice Problems

12. Determine [H₃O⁺] and [OH⁻] in each solution.

    (a) 0.45 mol/L hydrochloric acid

    (b) 1.1 mol/L sodium hydroxide

13. Determine [H₃O⁺] and [OH⁻] in each solution.

    (a) 0.95 mol/L hydroiodic acid

    (b) 0.012 mol/L calcium hydroxide

14. [OH⁻] is $5.6 \times 10^{-14}$ mol/L in a solution of hydrochloric acid. What is the molar concentration of the HCl$_{(aq)}$?

15. [H₃O⁺] is $1.7 \times 10^{-14}$ in a solution of calcium hydroxide. What is the molar concentration of the Ca(OH)$_{2(aq)}$?

## The pH Scale: Measuring by Powers of Ten

The concentration of hydronium ions ranges from about 10 mol/L for a concentrated strong acid to about $10^{-15}$ mol/L for a concentrated strong base. This wide range of concentrations, and the negative powers of 10, are not very convenient to work with. In 1909, a Danish biochemist, Søren Sørensen, suggested a method for converting concentrations to positive numbers. His method involved using the numerical system of logarithms.

The logarithm of a number is the power to which you must raise 10 to equal that number. For example, the logarithm of 10 is 1, because $10^1 = 10$. The logarithm of 100 is 2, because $10^2 = 100$. (See Appendix E for more information about exponents and logarithms.)

Sørensen defined **pH** as $-\log[H^+]$. Since Sørensen did not know about hydronium ions, his definition of pH is based on Arrhenius' hydrogen ion. Many chemistry references reinterpret the H so that it refers to the Brønsted-Lowry hydronium ion, $H_3O^+$, instead. This textbook adopts the hydronium ion usage. Thus, the definition for pH becomes $pH = -\log[H_3O^+]$. Recall, though, that chemists use $[H^+]$ as a shorthand notation for $[H_3O^+]$. As a result, both equations give the same product.

As you can see in Figure 14.13, the "p" in pH stands for the word "power." The power referred to is exponential power: the power of 10. The "H" stands for the concentration of hydrogen ions (or $H_3O^+$ ions), measured in mol/L.

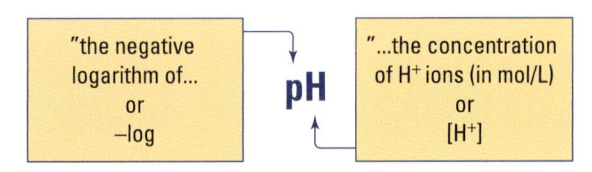

**Figure 14.13** The concept of pH makes working with very small values, such as 0.000 000 000 000 01, much easier.

The concept of pH allows hydronium (or hydrogen) ion concentrations to be expressed as positive numbers, rather than negative exponents. For example, recall that $[H_3O^+]$ of neutral water at 25°C is $1.0 \times 10^{-7}$.

$$\begin{aligned}
\therefore pH &= -\log[H_3O^+] \\
&= -\log(1.0 \times 10^{-7}) \\
&= -(-7.00) \\
&= 7.00
\end{aligned}$$

$[H_3O^+]$ in acidic solutions is greater than $[H_3O^+]$ in neutral water. For example, if $[H_3O^+]$ in an acid is $1.0 \times 10^{-4}$ mol/L, this is 1000 times greater than $[H_3O^+]$ in neutral water. Use Table 14.6 to make sure that you understand why. The pH of the acid is 4.00. All acidic solutions have a pH that is less than 7.

### Science LINK

Using logarithms is a convenient way to count a wide range of values by powers of 10. Chemists are not the only scientists who use such logarithms, however. Audiologists (scientists who study human hearing) use logarithms, too. Research the decibel scale to find out how it works. Present your findings in the medium of your choice.

**Table 14.6** Understanding pH

| Range of acidity and basicity | [H$_3$O$^+$] (mol/L) | Exponential notation (mol/L) | log | pH (−log [H$_3$O$^+$]) |
|---|---|---|---|---|
| strong acid | 1 | $1 \times 10^0$ | 0 | 0 |
| | 0.1 | $1 \times 10^{-1}$ | −1 | 1 |
| | 0.01 | $1 \times 10^{-2}$ | −2 | 2 |
| | 0.001 | $1 \times 10^{-3}$ | −3 | 3 |
| | 0.000 1 | $1 \times 10^{-4}$ | −4 | 4 |
| | 0.000 01 | $1 \times 10^{-5}$ | −5 | 5 |
| | 0.000 001 | $1 \times 10^{-6}$ | −6 | 6 |
| neutral [H$_3$O$^+$] = [OH$^-$] = $1.0 \times 10^{-7}$ | 0.000 000 1 | $1 \times 10^{-7}$ | −7 | 7 |
| | 0.000 000 01 | $1 \times 10^{-8}$ | −8 | 8 |
| | 0.000 000 001 | $1 \times 10^{-9}$ | −9 | 9 |
| | 0.000 000 000 1 | $1 \times 10^{-10}$ | −10 | 10 |
| | 0.000 000 000 01 | $1 \times 10^{-11}$ | −11 | 11 |
| | 0.000 000 000 001 | $1 \times 10^{-12}$ | −12 | 12 |
| | 0.000 000 000 000 1 | $1 \times 10^{-13}$ | −13 | 13 |
| strong base | 0.000 000 000 000 01 | $1 \times 10^{-14}$ | −14 | 14 |

[H$_3$O$^+$] in basic solutions is less than [H$_3$O$^+$] in pure water. For example, if [H$_3$O$^+$] in a base is $1.0 \times 10^{-11}$ mol/L, this is 10 000 times less than [H$_3$O$^+$] in neutral water. The pH of the base is 11.00. All basic solutions have a pH that is greater than 7.

How do you determine the number of significant digits in a pH? You count only the digits to the right of the decimal point. For example, suppose that the concentration of hydronium ions in a sample of orange juice is $2.5 \times 10^{-4}$ mol/L. This number has two significant digits: the 2 and the 5. The power of 10 only tells us where to place the decimal: 0.000 25. The pH of the sample is −log ($2.5 \times 10^{-4}$) = 3.602 059. The digit to the left of the decimal (the 3) is derived from the power of 10. Therefore, it is not considered to be a significant digit. Only the two digits to the right of the decimal are significant. Thus, the pH value is rounded off to 3.60.

The relationship among pH, [H$_3$O$^+$], and the strength of acids and bases is summarized in Table 14.7. Use the following Sample Problem and Practice Problems to assess your understanding of this relationship.

**Table 14.7** The Relation of pH, [H$_3$O$^+$], and Acid-Base Strength

| Type of solution | [H$_3$O+] (mol/L) | Concentration of hydronium and hydroxide ions | pH at 25°C |
|---|---|---|---|
| acidic solution | greater than $1 \times 10^{-7}$ | [H$_3$O$^+$] > [OH$^-$] | < 7.00 |
| neutral solution | $1 \times 10^{-7}$ | [H$_3$O$^+$] = [OH$^-$] | 7.00 |
| basic solution | less than $1 \times 10^{-7}$ | [H$_3$O$^+$] < [OH$^-$] | > 7.00 |

## Sample Problem

### Calculating the pH of a Solution

**Problem**

Calculate the pH of a solution with [H$_3$O$^+$] = 3.8 × 10$^{-3}$ mol/L.

**What Is Required?**

You need to calculate the pH, given [H$_3$O$^+$].

**What Is Given?**

You know that [H$_3$O$^+$] is 3.8 × 10$^{-3}$ mol/L.

**Plan Your Strategy**

Use the equation pH = −log [H$_3$O$^+$] to solve for the unknown.

**Act on Your Strategy**

pH = −log (3.8 × 10$^{-3}$)
 = 2.42

**Check Your Solution**

[H$_3$O$^+$] is greater than 1.0 × 10$^{-7}$ mol/L. Therefore, the pH should be less than 7.00. The solution is acidic, as you would expect.

> **PROBLEM TIP**
>
> Appendix D, "Math and Chemistry", explains how you can do these calculations with a calculator.

## Practice Problems

16. Calculate the pH of each solution, given the hydronium ion concentration.
    (a) [H$_3$O$^+$] = 0.0027 mol/L
    (b) [H$_3$O$^+$] = 7.28 × 10$^{-8}$ mol/L
    (c) [H$_3$O$^+$] = 9.7 × 10$^{-5}$ mol/L
    (d) [H$_3$O$^+$] = 8.27 × 10$^{-12}$

17. [H$_3$O$^+$] in a cola drink is about 5.0 × 10$^{-3}$ mol/L. Calculate the pH of the drink. State whether the drink is acidic or basic.

18. A glass of orange juice has [H$_3$O$^+$] of 2.9 × 10$^{-4}$ mol/L. Calculate the pH of the juice. State whether the result is acidic or basic.

19. (a) [H$_3$O$^+$] in a dilute solution of nitric acid, HNO$_3$, is 6.3 × 10$^{-3}$ mol/L. Calculate the pH of the solution.

    (b) [H$_3$O$^+$] of a solution of sodium hydroxide is 6.59 × 10$^{-10}$ mol/L. Calculate the pH of the solution.

> **Web LINK**
>
> www.mcgrawhill.ca/links/atlchemistry
>
> Many people, for both personal and professional reasons, rely on pH meters to provide quick, reliable pH measurements. Use the Internet to find out how a pH meter works, and what jobs or tasks it is used for. To start your research, go to the web site above and click on **Web Links**. Prepare a brief report, a web page, or a brochure to present your findings.

Just as pH refers to the exponential power of the hydronium ion concentration in a solution, **pOH** refers to the power of hydroxide ion concentration. You can calculate the **pOH** of a solution from the [OH$^-$]. Notice the relationship between pH and pOH shown below, and in the Concept Organizer.

$$pOH = -\log[OH^-]$$
$$K_w = [H_3O^+][OH^-] = 1.0 \times 10^{-14} \text{ at } 25°C$$
$$\therefore pH + pOH = 14$$

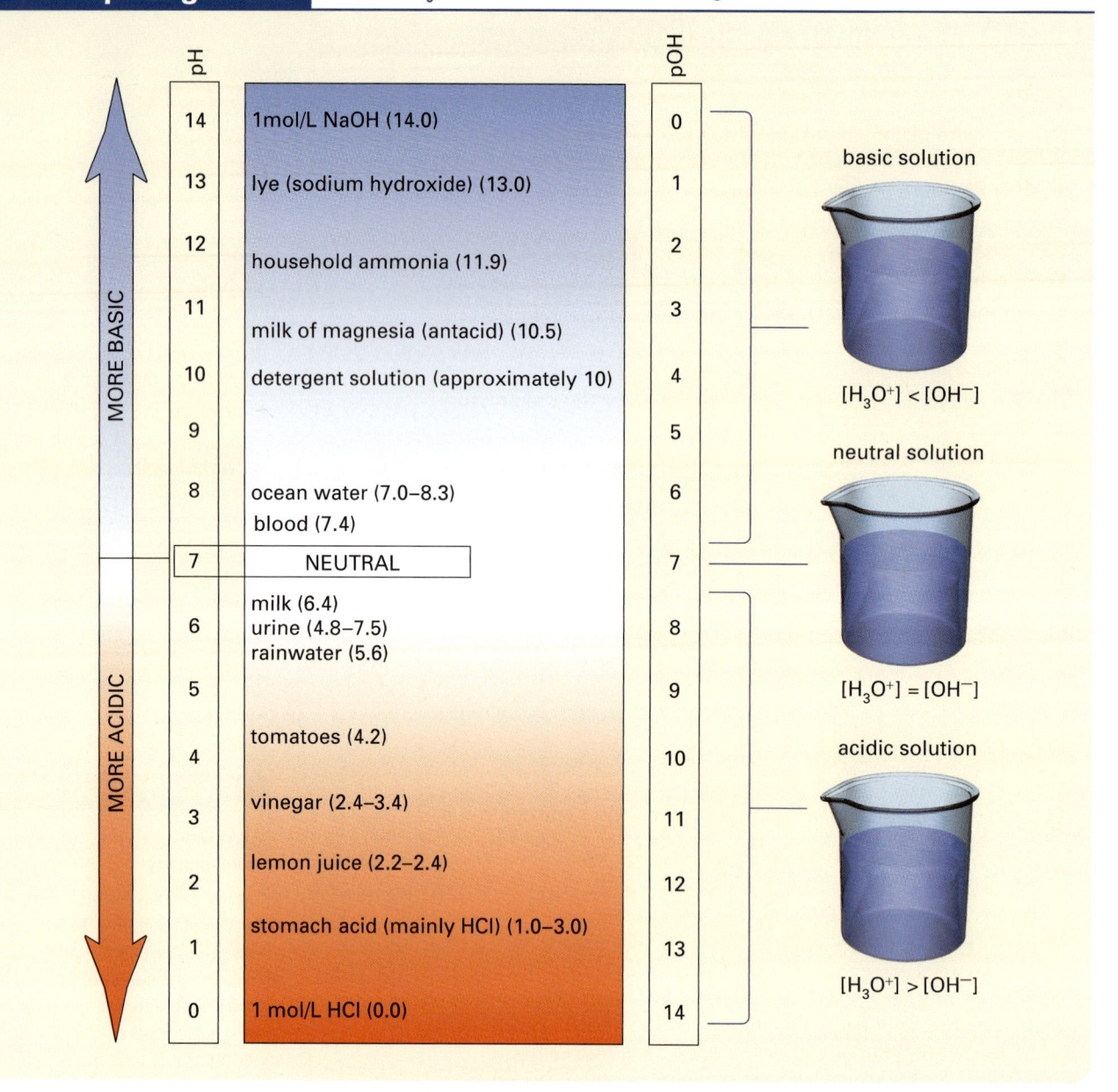

## Concept Organizer: pH, pOH, [H₃O⁺], [OH⁻], and Acid-Base Strength

**pH scale (MORE BASIC ↑ / MORE ACIDIC ↓):**

| pH | Example | pOH |
|---|---|---|
| 14 | 1 mol/L NaOH (14.0) | 0 |
| 13 | lye (sodium hydroxide) (13.0) | 1 |
| 12 |  | 2 |
| 11 | household ammonia (11.9) | 3 |
| 10 | milk of magnesia (antacid) (10.5) | 4 |
| 10 | detergent solution (approximately 10) | 4 |
| 9 |  | 5 |
| 8 | ocean water (7.0–8.3) | 6 |
|  | blood (7.4) |  |
| 7 | NEUTRAL | 7 |
| 6 | milk (6.4) | 8 |
|  | urine (4.8–7.5) |  |
|  | rainwater (5.6) |  |
| 5 |  | 9 |
| 4 | tomatoes (4.2) | 10 |
| 3 | vinegar (2.4–3.4) | 11 |
| 2 | lemon juice (2.2–2.4) | 12 |
| 1 | stomach acid (mainly HCl) (1.0–3.0) | 13 |
| 0 | 1 mol/L HCl (0.0) | 14 |

- basic solution: $[H_3O^+] < [OH^-]$
- neutral solution: $[H_3O^+] = [OH^-]$
- acidic solution: $[H_3O^+] > [OH^-]$

### Math LINK

Prove the relationship pH + pOH = 14 as follows. Record the ion product equation and its value at 25°C. Take the logarithm of both sides. Then reverse the sign of each term. What is your result?

### Sample Problem

#### Calculating pH and pOH

**Problem**

A liquid shampoo has a hydroxide ion concentration of $6.8 \times 10^{-5}$ mol/L at 25°C.

(a) Is the shampoo acidic, basic, or neutral?
(b) Calculate the hydronium ion concentration.
(c) What is the pH and the pOH of the shampoo?

### Solution

(a) Compare [OH⁻] in the shampoo with [OH⁻] in neutral water at 25°C.
    $[OH^-] = 6.8 \times 10^{-5}$ mol/L, which is greater than $1 \times 10^{-7}$ mol/L. Therefore, the shampoo is basic.

(b) Use the equation $[H_3O^+] = \dfrac{1.0 \times 10^{-14}}{[OH^-]}$ to find the hydronium ion concentration.
$$[H_3O^+] = \dfrac{1.0 \times 10^{-14}}{6.8 \times 10^{-5}}$$
$$= 1.5 \times 10^{-10} \text{ mol/L}$$

(c) Substitute known values into the equations $pH = -\log[H_3O^+]$ and $pOH = -\log[OH^-]$.
$$pH = -\log(1.5 \times 10^{-10})$$
$$= 9.83$$
$$pOH = -\log(6.8 \times 10^{-5})$$
$$= 4.17$$

### Check Your Solution
pH + pOH = 14

> **PROBLEM TIP**
>
> When you work with logarithms, *the number of significant digits in a number must equal the number of digits after the decimal in the number's logarithm.* Here $1.5 \times 10^{-10}$ has two significant digits. Therefore, the calculated pH, **9.83**, must have two significant digits after the decimal.

### Another Way to Find [H₃O⁺] and [OH⁻]

You can calculate $[H_3O^+]$ or $[OH^-]$ by finding the *antilog* of the pH or pOH.
$$[H_3O^+] = 10^{-pH}$$
$$[OH^-] = 10^{-pOH}$$

If you are using a calculator, you can use it to find the antilog of a number in one of two ways. If the logarithm is entered in the calculator, you can press the two keys [INV] and [LOG] in sequence. (Some calculators may have a [10ˣ] button instead.) Alternatively, since $[H_3O^+] = 10^{-pH}$ and $[OH^-] = 10^{-pOH}$, you can enter 10, press the [yˣ] button, enter the negative value of pH (or pOH), and then press [=].

## Sample Problem
### Finding pOH, [H₃O⁺], and [OH⁻]

**Problem**

If the pH of urine is outside the normal range of values, this can indicate medical problems. Suppose that the pH of a urine sample was measured to be 5.53 at 25°C. Calculate pOH, $[H_3O^+]$, and $[OH^-]$ for the sample.

**Solution**

You use the known value, pH = 5.53, to calculate the required values.
$$pOH = 14.00 - 5.53$$
$$= 8.47$$

$$[H_3O^+] = 10^{-5.53}$$
$$= 3.0 \times 10^{-6} \text{ mol/L}$$

$$[OH^-] = 10^{-8.47}$$
$$= 3.4 \times 10^{-9} \text{ mol/L}$$

*Continued ...*

### Check Your Solution

In this problem, the ion product constant is a useful check:
$[H_3O^+][OH^-] = (3.0 \times 10^{-6}) \times (3.4 \times 10^{-9})$
$= 1.0 \times 10^{-14}$
This value equals the expected value for $K_w$ at 25°C.

### Practice Problems

20. $[H_3O^+]$ of a sample of milk is found to be $3.98 \times 10^{-7}$ mol/L. Is the milk acidic, neutral, or basic? Calculate the pH and $[OH^-]$ of the sample.

21. A sample of household ammonia has a pH of 11.9. What is the pOH and $[OH^-]$ of the sample?

22. Phenol, $C_6H_5OH$, is used as a disinfectant. An aqueous solution of phenol was found to have a pH of 4.72. Is phenol acidic, neutral, or basic? Calculate $[H_3O^+]$, $[OH^-]$, and pOH of the solution.

23. At normal body temperature, 37°C, the value of $K_w$ for water is $2.5 \times 10^{-14}$. Calculate $[H_3O^+]$ and $[OH^-]$ at this temperature. Is pure water at 37°C acidic, neutral, or basic?

24. A sample of baking soda was dissolved in water and the pOH of the solution was found to be 5.81 at 25°C. Is the solution acidic, basic, or neutral? Calculate the pH, $[H_3O^+]$, and $[OH^-]$ of the solution.

25. A chemist dissolved some Aspirin™ in water. The chemist then measured the pH of the solution and found it to be 2.73 at 25°C. What are $[H_3O^+]$ and $[OH^-]$ of the solution?

## Dilution Calculations involving Acids and Bases

When chemists go to use an acid in the lab, they commonly use a stock solution of known concentration and dilute it to the concentration they need. A chemist may want to dilute the stock solution either to a specific $[H_3O^+]$, a specific pH, or a specific $[OH^-]$ before use. Thus, calculations involving dilutions of acids and bases are very common in a practical lab setting. You have already studied the basis of these calculations in Unit 3. Where diluting acids and bases is concerned, the main idea is that *the number of moles of acid (or base) remains constant, and thus the number of moles (n = CV) before dilution equals the number of moles after dilution.*

This means that acid (or base) dilution problems may be summarized as shown in Figure 14.14. Note that the given or wanted parameters may be any of pH, pOH, $[H^+]$ or $[OH^-]$. This is possible because any given pH or pOH is directly related to the concentration of its respective hydronium or hydroxide ion.

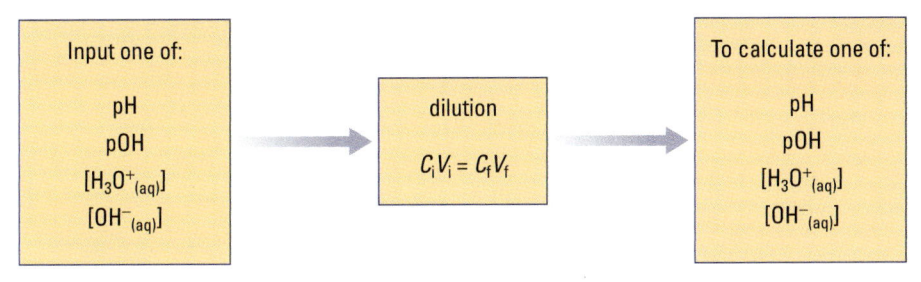

```
Input one of:          dilution           To calculate one of:
   pH                                           pH
   pOH               $C_iV_i = C_fV_f$          pOH
   [H_3O^+_{(aq)}]                              [H_3O^+_{(aq)}]
   [OH^-_{(aq)}]                                [OH^-_{(aq)}]
```

**Figure 14.14** A summary for making dilution calculations involving strengths and concentrations of acids and bases

Note also that if one of the volumes ($V_i$ or $V_f$) and all other concentration-related terms are given, the missing volume may be the variable that you must calculate.

After reading these paragraphs, you may feel that the number of possible problems that can be developed are daunting, yet if you remember that it is the number of moles of solute that is constant and that pH/pOH are simply "disguised" versions of [H$^+$]/[OH$^-$], then any version of the above problem can be tackled with confidence. Use the Sample Problem and Practice Problems below to develop your calculation skills. You will use these skills in a hands-on setting in Investigation 14-B, which follows.

## Sample Problem

### Calculating Unknown Quantities Involving Dilution of Acids or Bases

#### Problem
In preparing a solution, a chemist takes a stock hydrochloric acid solution of pH = 1.50 from the chemical storeroom. If the chemist uses 125 mL of the stock solution and dilutes it to 495 mL, what is the final hydronium ion concentration?

#### What Is Required?
You need to find the final [H$_3$O$^+$] of the solution after it has been diluted (that is, [H$_3$O$^+$]$_f$), to determine the final pH (that is pH$_f$).

#### What Is Given?
Since this is a dilution problem, you may separate the given quantities as either initial or final quantities in the dilution process.

Initial: pH$_i$ = 1.50
$V_i$ = 125 mL
Final: $V_f$ = 495 mL

#### Plan Your Strategy
You must first recognize the initial pH as a "disguised form" of the initial hydronium ion concentration, and then calculate the [H$_3$O$^+$]$_i$ directly from this given initial pH. Then relate the initial given data to the final given data by remembering that $n = CV$, and that during a dilution, the moles of solute remain constant ($n_i = n_f$).

*Continued...*

**PROBLEM TIP**

You may prefer to substitute the values directly into the dilution formula rather than rearrange the formula. This procedure will also give the correct result.

**PROBLEM TIP**

The conversion from mL to L here serves as a reminder that the units for both $V_i$ and $V_f$ are consistent and, thus, cancel. Converting units in litres avoids possible problems with cancellation.

*Continued...*

### Act on Your Strategy

$[H_3O^+]_i = 10^{-pH_i} = 10^{-1.50} = 0.032$ mol/L $H_3O^+$

Since $n_i = n_f$,
therefore,
$$[H_3O^+]_i V_i = [H_3O^+]_f V_f$$

Rearranging for $[H_3O^+]_f$:

$$[H_3O^+] = \frac{[H_3O^+]_i}{V_f}$$
$$= \frac{(0.032 \text{ mol/L})(0.125 \text{ L})}{(0.495 \text{ L})}$$
$$= 0.0081 \text{ mol/L}$$

Finally, convert $[H_3O^+]_f$ to the final pH:

$$pH_f = -\log[H_3O^+]_f$$
$$= -\log(0.00799)$$
$$= 2.09$$

### Check Your Solution

For dilution problems, you can check the initial and final concentrations by looking at the ratio of the final volume to the initial volume. From the given volumes, the final volume is larger by a factor of almost four (495 mL/125 mL). The final solution also has a concentration that is smaller (that is, diluted) by a factor of almost four (0.0316 mol/L/0.00799 mol/L). Remember also that the pH is a logarithmic scale, so there will not be a direct relationship between the factor of the dilution and the pH; you can only check your final answer against concentration and volume values.

### Practice Problems

26. Calculate the pH of a $HNO_{3(aq)}$ solution which is formed by diluting 45 mL of 0.0115 mol/L $HNO_{3(aq)}$ to a final volume of 2.00 L.

27. (a) A solution of lithium hydroxide is diluted from $3.25 \times 10^{-3}$ mol/L to $3.25 \times 10^{-6}$ mol/L. If the initial volume was 36.0 mL, calculate the final volume of the solution.
    (b) Calculate the pOH and pH of the solution.

28. (a) Calculate the volume of concentrated hydrochloric acid (12.4 mol/L) required to prepare 950.0 mL of a solution that has a pH of 1.50.
    (b) What is the pOH and $[OH^-]$ of the solution?

29. (a) Calculate the pOH of a solution that forms when 150.0 mL of 0.0000223 mol/L $Ca(OH)_{2(aq)}$ is diluted to 15.0 L.
    (b) Calculate the pH of the final solution.
    (c) Calculate the hydronium ion concentration of the final solution.
    (d) Is the solution acidic or basic? Briefly explain your choice.

574 MHR • Unit 6 Acids and Bases

# Investigation 14-B

**SKILL FOCUS**
Performing and recording
Analyzing and interpreting

## The Effect of Dilution on the pH of an Acid

In this investigation, you will compare the effects of diluting a strong acid and a weak acid.

In Part 1, you will measure the pH of a strong acid. Then you will perform a series of ten-fold dilutions. That is, each solution will be one-tenth as dilute as the previous solution. You will measure and compare the pH after each dilution.

In Part 2, you will measure the pH of a weak acid with the same initial concentration as the strong acid. Then you will perform a series of ten-fold dilutions with the weak acid. Again, you will measure and compare the pH after each dilution.

### Problem
How does the pH of dilutions of a strong acid compare with the pH of dilutions of a weak acid?

### Prediction
Predict each pH, and explain your reasoning.
(a) the pH of 0.10 mol/L hydrochloric acid
(b) the pH of the hydrochloric acid after one ten-fold dilution
(c) the pH of the hydrochloric acid after each of six more ten-fold dilutions
(d) the pH of 0.10 mol/L acetic acid, compared with the pH of 0.10 mol/L hydrochloric acid
(e) the pH of the acetic acid after one ten-fold dilution

### Safety Precautions

Hydrochloric acid is corrosive. Wash any spills on skin or clothing with plenty of cool water. Inform your teacher immediately.

### Materials
100 mL graduated cylinder
100 mL beaker
2 beakers (250 mL)
universal indicator paper and glass rod
pH meter
0.10 mol/L hydrochloric acid (for Part 1)
0.10 mol/L acetic acid (for Part 2)
distilled water

### Procedure

**Part 1 The pH of Solutions of a Strong Acid**

1. Copy the table below into your notebook. Record the pH you predicted for each dilution.

2. Pour about 40 mL of 0.10 mol/L hydrochloric acid into a clean, dry 100 mL beaker. Use the end of a glass rod to transfer a drop of solution to a piece of universal pH paper into the acid. Compare the colour against the colour chart to determine the pH. Record the pH. Then measure and record the pH of the acid using a pH meter. Rinse the electrode with distilled water afterward.

**Data Table for Part 1**

| [HCl$_{(aq)}$] mol/L | Predicted pH | pH measured with universal indicator | pH measured with pH meter |
|---|---|---|---|
| $1 \times 10^{-1}$ | | | |
| $1 \times 10^{-2}$ | | | |
| $1 \times 10^{-3}$ | | | |
| $1 \times 10^{-4}$ | | | |
| $1 \times 10^{-5}$ | | | |
| $1 \times 10^{-6}$ | | | |
| $1 \times 10^{-7}$ | | | |
| $1 \times 10^{-8}$ | | | |

*continued...*

continued

3. Measure 90 mL of distilled water in a 100 mL graduated cylinder. Add 10 mL of the acid from step 2. The resulting 100 mL of solution is one-tenth as concentrated as the acid from step 2. Pour the dilute solution into a clean, dry 250 mL beaker. Use universal pH paper and a pH meter to measure the pH. Record your results.

4. Repeat step 3. Pour the new dilute solution into a second clean, dry beaker. Dispose of the more concentrated acid solution as directed by your teacher. Rinse and dry the beaker so you can use it for the next dilution.

5. Make further dilutions and pH measurements until the hydrochloric acid solution is $1.0 \times 10^{-8}$ mol/L

### Part 2  The pH of Solutions of a Weak Acid

1. Design a table to record your predictions and measurements for 0.10 mol/L and 0.010 mol/L concentrations of acetic acid.

2. Use the same procedure that you used in Part 1 to measure and record the pH of a 0.10 mol/L sample of acetic acid. Then dilute the solution to 0.010 mol/L. Measure the pH again.

### Analysis

1. Which do you think gave the more accurate pH: the universal indicator paper or the pH meter? Explain.

2. For the strong acid, compare the pH values you predicted with the measurements you made. How can you explain any differences for the first few dilutions?

3. What was the pH of the solution that had a concentration of $1.0 \times 10^{-8}$ mol/L? Explain the pH you obtained.

4. Compare the pH of 0.10 mol/L acetic acid with the pH of 0.10 mol/L hydrochloric acid. Why do you think the pH values are different, even though the concentrations of the acids were the same?

5. What effect does a ten-fold dilution of a strong acid (hydrochloric acid) have on the pH of the acid? What effect does the same dilution of a weak acid (acetic acid) have on its pH? Compare the effects for a strong acid and a weak acid. Account for any differences.

### Conclusion

6. Use evidence from your investigation to support the conclusion that a weak acid ionizes less than a strong acid of identical concentration.

7. Why is the method for calculating the pH of a strong acid (if it is not too dilute) not appropriate for a weak acid?

### Applications

8. Nicotinic acid is a B vitamin. The pH of a 0.050 mol/L solution of this acid is measured to be 3.08. Is it a strong acid or a weak acid? Explain. What would be the pH of a solution of nitric acid having the same concentration?

9. Would you expect to be able to predict the pH of a weak base, given its concentration? Explain. Design an experiment you could perform to check your answer.

# Chemistry Bulletin

**Science** · **Technology** · **Science** · **Environment**

## The Chemistry of Oven Cleaning

Oven cleaning is not a job that most people enjoy. Removing baked-on grease from inside an oven requires serious scrubbing. Any chemical oven cleaners that help to make the job easier are usually welcome. Like all chemicals, however, the most effective oven cleaners require attention to safety.

Cleaners that contain strong bases are the most effective for dissolving grease and grime. Bases are effective because they produce soaps when they react with the fatty acids in grease. When a strong base (such as sodium hydroxide, NaOH, or potassium hydroxide, KOH) is used on a dirty oven, the fat molecules that make up the grease are split into smaller molecules. Anions from the base then bond with some of these molecules to form soap.

One end of a soap molecule is non-polar (uncharged), so it is soluble in dirt and grease, which are also non-polar. The other end of a soap molecule is polar (charged), so it is soluble in water. Because of its two different properties, soap acts like a "bridge" between the grease and the water. Soap enables grease to dissolve in water and be washed away, thus allowing the cleaner to remove the grease from the oven surface.

Cleaners that contain sodium hydroxide and potassium hydroxide are very effective. They are also caustic and potentially very dangerous. For example, sodium hydroxide, in the concentrations that are used in oven cleaners, can irritate the skin and cause blindness if it gets in the eyes. As well, it is damaging to paints and fabrics.

There are alternatives to sodium hydroxide and other strong base cleaners. One alternative involves using ammonia, $NH_3$, which is a weak base. If a bowl of dilute ammonia solution is placed in an oven and left for several hours, most of the grease and grime can be wiped off.

Ammonia does not completely ionize in water. Only a small portion dissociates. Although an ammonia solution is less caustic than sodium hydroxide, it can be toxic if inhaled directly. As well, ammonia vapours can cause eye, lung, and skin irritations. At higher concentrations, ammonia can be extremely toxic.

Baking soda is a non-toxic alternative, but it is much less effective. Therefore, it requires even more scrubbing. An abrasive paste can be made by mixing baking soda and water. The basic properties of baking soda also have a small effect on grease and grime if it is applied to the oven and left for several hours.

### Making Connections

1. Survey the cleaners in your home or school. Which cleaners contain bases and which contain acids? What cleaning jobs can an acid cleaner perform well? How do most acid cleaners work?

2. Some companies claim to make environmentally sensitive cleaners. Investigate these cleaners. What chemicals do they contain? See if you can infer how they work. You might like to design a controlled experiment to test the effectiveness of several oven cleaners. **CAUTION** Obtain permission from your teacher before performing such an experiment.

## Section Wrap-up

In this section, you considered the relationship among the strength of acids and bases, the concentration of hydronium and hydroxide ions, and their relation to pH and pOH. Earlier in the section, pH was introduced to you as "one method" for quantitatively describing acids and bases. There is another method, which involves equilibrium constants. In the second chapter of this unit, you will learn about these equilibrium constants, and how they apply to acid-base reactions involving weak acids and weak bases. As well, you will look more closely at what happens, both macroscopically and chemically, during neutralization reactions.

## Section Review

**1** Distinguish, in terms of degree of dissociation, between a strong acid and a weak acid, and a strong base and a weak base.

**2** Give one example of the following:
 (a) a weak acid
 (b) a strong acid
 (c) a strong base
 (d) a weak base

**3** Explain the meaning of pH, both in terms of hydrogen ions and hydronium ions.

**4** Arrange the following foods in order of increasing acidity: beets, pH = 5.0; camembert cheese, pH = 7.4; egg white, pH = 8.0; sauerkraut, pH = 3.5; yogurt, pH = 4.5.

**5** Calculate the pH of each body fluid, given the concentration of hydronium ions.
 (a) tears, $[H_3O^+] = 4.0 \times 10^{-8}$ mol/L
 (b) stomach acid, $[H_3O^+] = 4.0 \times 10^{-2}$ mol/L

**6** Calculate the pH of the solution that is formed by diluting 50 mL of 0.025 mol/L hydrochloric acid to a final volume of 1.0 L.

**7** What is $[H_3O^+]$ in a solution with pH = 0? Why do chemists not usually use pH to describe $[H_3O^+]$ when the pH value would be a negative number?

**8** Complete the following table by calculating the missing values and indicating whether each solution is acidic or basic.

| $[H_3O^+]$ (mol/L) | pH | $[OH^-]$ (mol/L) | pOH | Acidic or basic? |
|---|---|---|---|---|
| $3.7 \times 10^{-5}$ | (a) | (b) | (c) | (d) |
| (e) | 10.41 | (f) | (g) | (h) |
| (i) | (j) | $7.0 \times 10^{-2}$ | (k) | (l) |
| (m) | (n) | (o) | 8.9 | (p) |

# CHAPTER 14 Review

## Reflecting on Chapter 14
- Summarize this chapter in the format of your choice. Here are a few ideas to use as guidelines:
- Compare the properties and theories of acids and bases.
- Identify conjugate acid-base pairs for selected acid-base reactions, and compare their strengths.
- Distinguish strong and weak acids and bases on the basis of their dissociation in water.
- Outline the relationship among $[H_3O^+]$, pH, $[OH^-]$, and pOH.
- Dilute an acid and describe the effect on its pH.

## Reviewing Key Terms
For each of the following terms, write a sentence that shows your understanding of its meaning.

Arrhenius theory of acids and bases
hydronium ion
Brønsted-Lowry theory of acids and bases
conjugate acid-base pair
conjugate base
conjugate acid
amphoteric
strong acid
weak acid
strong base
weak base
ion product constant for water ($K_w$)
pH
pOH

## Knowledge/Understanding
1. Use the Arrhenius theory, the modernized Arrhenius theory, and the Brønsted-Lowry theory to describe the following concepts. If one or more of the theories do not apply, state that this is the case.
   (a) composition of an acid and a base
   (b) conductivity of an acidic or basic solution
   (c) interaction between an acid and water
   (d) interaction between a base and water
   (e) conjugate acid-base pairs
   (f) an aqueous solution of ammonia
   (g) strong and weak acids and bases
   (h) the pH of a solution
2. How does diluting an acidic or basic solution affect the pH of the solution?
3. Codeine is a compound that is extracted from opium. It is used for pain relief. The pH of a 0.020 mol/L solution of codeine is 10.26. Is codeine an acid or a base? Is it strong or weak? Explain how you decided.
4. Sodium hydrogen carbonate, $NaHCO_3$ (commonly called sodium bicarbonate, or bicarbonate of soda), is commonly used in baked goods. It dissolves in water to form an alkaline solution.
   (a) Is the pH of $NaHCO_{3(aq)}$ greater or less than 7.00?
   (b) Write the name and formula of an acid and a base that react together to form this compound. Identify each as strong or weak.
5. Classify each compound as a strong acid, strong base, weak acid, or weak base.
   (a) phosphoric acid, $H_3PO_4$ (used in cola beverages and rust-proofing products)
   (b) chromic acid, $H_2CrO_4$ (used in the production of wood preservatives)
   (c) barium hydroxide, $Ba(OH)_2$, a white, toxic base (can be used to de-acidify paper)
   (d) $CH_3NH_2$, commonly called methylamine (is responsible for the characteristic smell of fish that are no longer fresh)
6. Write a chemical formula for each acid or base.
   (a) the conjugate base of $OH^-$
   (b) the conjugate acid of ammonia, $NH_3$
   (c) the conjugate acid of $HCO_3^-$
   (d) the conjugate base of $HCO_3^-$
7. Decide whether each statement is true or false, and explain your reasoning.
   (a) HBr is a stronger acid than HI.
   (b) $HBrO_2$ is a stronger acid than HBrO.
   (c) $H_2SO_3$ is a stronger acid than $HSO_3^-$.
8. In each pair of bases, which is the stronger base?
   (a) $HSO_4^-{}_{(aq)}$ or $SO_4^{2-}{}_{(aq)}$
   (b) $S^{2-}{}_{(aq)}$ or $HS^-{}_{(aq)}$
   (c) $HPO_4^{2-}{}_{(aq)}$ or $H_2PO_4^-{}_{(aq)}$
   (d) $HCO_3^-{}_{(aq)}$ or $CO_3^{2-}{}_{(aq)}$
9. Sodium hydrogen carbonate can be used to neutralize an acid. The hydrogen carbonate ion is the conjugate base of which acid?
10. Explain the significance of the following statement, and give an example to illustrate its meaning: Water is amphoteric.

Answers to questions highlighted in red type are provided in Appendix A.

11. In different reactions in aqueous solution, the hydrogen carbonate ion can act as an acid or a base. Write the chemical formula of the conjugate acid and the conjugate base of the hydrogen carbonate ion, $HCO_3^-{}_{(aq)}$. Then complete the following equations. State whether the ion is a Brønsted-Lowry acid or a base.
    (a) $HCO_3^-{}_{(aq)} + H_3O^+{}_{(aq)}$
    (b) $HCO_3^-{}_{(aq)} + OH^-{}_{(aq)}$

12. Which of the following are conjugate acid-base pairs? For those pairs that are not conjugates, write the correct conjugate acid or base for each compound or ion.
    (a) $HNO_3/OH^-$
    (b) $NH_4^+/NH_3$
    (c) $HSO_4^-/SO_4^{2-}$
    (d) $H_3PO_4/PO_4^{3-}$

13. How is $K_w$ related to the pH of acids and bases?

14. How are pH and pOH related to each other?

## Inquiry

15. In the laboratory, you have samples of three different acids of equal concentration: a 1.0 mol/L solution of acetic acid, a 1.0 mol/L solution of hydrochloric acid, and a 1.0 mol/L solution of sulfuric acid.
    (a) How would the pH of each acid solution compare? Explain.
    (b) If samples of each acid were used in separate titration experiments with 0.50 mol/L sodium hydroxide solution, how would the volume of acid required for neutralization compare? State your reasoning.

16. Write balanced chemical equations for the following reactions:
    (a) calcium oxide with hydrochloric acid
    (b) magnesium with sulfuric acid
    (c) sodium carbonate with nitric acid
    (d) Are products or reactants favoured in each of the above three reactions? Explain how you know.

17. Domestic bleach is typically a 5% solution of sodium hypochlorite, $NaOCl_{(aq)}$. It is made by bubbling chlorine gas through a solution of sodium hydroxide.
    (a) Write a balanced chemical equation showing the reaction that takes place.
    (b) In aqueous solution, the hypochlorite ion combines with $H^+{}_{(aq)}$ present in water to form hypochlorous acid. Write the equation for this reaction. Is the hypochlorite ion acting as an acid or a base?

18. In this chapter, you are told that $[H_3O^+]$ in pure water is $1.0 \times 10^{-7}$ mol/L at 25°C. Thus, two out of every one billion water molecules have dissociated. Check these data by answering the following questions.
    (a) What is the mass (in g) of 1.0 L of water?
    (b) Calculate the amount (in mol) of water in 1.0 L. This is the concentration of water in mol/L.
    (c) Divide the concentration of hydronium ions by the concentration of water. Your answer should be about 2 ppb.

19. 80.0 mL of 4.00 mol/L, $H_2SO_4$ are diluted to 400.0 mL by adding water. What is the molar concentration of the sulfuric acid after dilution?

20. (a) Calculate the pOH of the solution that forms when 375 mL of a 0.000102 mol/L $Ba(OH)_{2(aq)}$ is diluted to 1.50 L.
    (b) Calculate the pH of the final solution.
    (c) Calculate the hydronium ion concentration of the final solution.

21. How is a 1.0 mol/L solution of hydrochloric acid different from a 1.0 mol/L solution of acetic acid? Suppose that you added a strip of magnesium metal to each acid. Would you observe any differences in the reactions? Explain your answer so that grade 9 students could understand it.

## Communicate

22. Predict the products of the following aqueous reactions.
    (a) $NO_2^-{}_{(aq)} + H_2O_{(\ell)}$
    (b) $H_2PO_4^-{}_{(aq)} + H_2O_{(\ell)}$

23. Rank the following in order of decreasing acidity: HF, $H_2O$, $OH^-$, HCl, $HSO_4^-$.

24. Rank the following in order of increasing basicity: $F^-$, $Cl^-$, $HSO_3^-$, $S^{2-}$, $NH_3$.

25. 40.0 g of sodium hydroxide is dissolved in enough water to make 1.00 L of solution. Afterward, 2.50 mL of the solution is diluted to 250 mL. What is the pH of the final, diluted solution?

26. You want to prepare 2.0 L of a solution that is 0.50 mol/L KOH. How many moles of KOH do you need? Explain how you arrived at your answer.

27. Harmful microorganisms grow on cheese unless the pH of the cheese is maintained in a range between 5.5 and 5.9. You test a sample of cheese and determine that $[OH^-] = 1.0 \times 10^{-8}$. Is the cheese safe to eat? Why or why not?

28. Predict whether the product of each of the following is acidic or basic.
    (a) $NO_2 + H_2O$
    (b) $SO_3 + H_2O$
    (c) $K_2O + H_2O$
    (d) $MgO + H_2O$

## Making Connections

29. Citric acid can be added to candy to give a sour taste. The structure of citric acid is shown below.

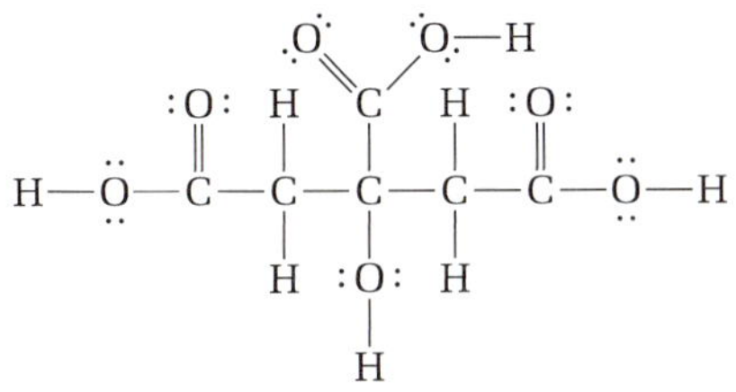

    (a) Identify the acidic hydrogen atoms that are removed by water in aqueous solution. Why do water molecules pull these hydrogen atoms away, rather than other hydrogen atoms in citric acid?
    (b) Why does citric acid not form $OH^-$ ions in aqueous solution, and act as a base?

30. (a) Imagine that you have collected a sample of rainwater in your community. The pH of your sample is 4.52. Unpolluted rainwater has a pH of about 5.6. How many more hydronium ions are present in your sample, compared with normal rainwater? Calculate the ratio of the concentration of hydronium ions in your sample to the concentration of hydronium ions in unpolluted rainwater.
    (b) You have been invited to a community meeting to explain your findings to local residents. No one at the meeting has a background in chemistry. In a paragraph, write what you would say at this meeting.
    (c) Suggest at least two possible factors that could be responsible for the pH you measured. What observations would you want to make, and what data would you want to collect, to help you gain confidence that one of these factors is responsible?

### Answers to Practice Problems and Short Answers to Section Review Questions:

**Practice Problems: 1.** $HClO_{(aq)}/ClO_4^-{}_{(aq)}$; $H_3O^+{}_{(aq)}/H_2O_{(\ell)}$ **2.** $CH_3COOH_{(aq)}/CH_3COO^-{}_{(aq)}$; $H_2O_{(\ell)}/OH^-{}_{(aq)}$ **3.(a)** chloride ion, $Cl^-$ **(b)** carbonate ion, $CO_3^{2-}$ **(c)** hydrogen sulfate ion, $HSO_4^-$ **(d)** hydrazine, $N_2H_4$ **4.(a)** nitric acid, $HNO_3$ **(b)** water, $H_2O$ **(c)** hydronium ion, $H_3O^+$ **(d)** carbonic acid, $H_2CO_3$ **5.** as an acid: $HS^-{}_{(aq)} + H_2O_{(\ell)} \rightleftharpoons S^{2-}{}_{(aq)} + H_3O^+{}_{(aq)}$ ; as a base: $HS^-{}_{(aq)} + H_2O_{(\ell)} \rightleftharpoons H_2S_{(aq)} + OH^-{}_{(aq)}$ **6.(a)** $H_3PO_4/H_2PO_4^-$; $H_2O/H_3O^+$ **(b)** $O^{2-}/OH^-$; $H_2O/OH^-$ **7.** In (a) water is the amphoteric species, acting as an acid. In (b), water is the amphoteric species, acting as an acid. **8.(a)** $H_2O/OH^-$; $NH_3/NH_4^+$ **(b)** $CH_3COOH/CH_3COO^-$; $H_2O/H_3O^+$ **9.** Water is the amphoteric chemical, acting as an acid with ammonia and as a base with acetic acid. **10.(a)** direction left, favouring reactants **(b)** direction left, favouring reactants **(c)** direction left, favouring reactants **11.(a)** left, favouring reactants **(b)** right, favouring products **(c)** right, favouring products **12.(a)** $[H_3O^+] = 0.45$ mol/L; $[OH^-] = 2.2 \times 10^{-14}$ mol/L **(b)** $[OH^-] = 1.1$ mol/L; $[H_3O^+] = 9.1 \times 10^{-15}$ mol/L **13.(a)** $[H_3O^+] = 0.95$ mol/L; $[OH^-] = 1.1 \times 10^{-14}$ mol/L **(b)** $[OH^-] = 0.024$ mol/L; $[H_3O^+] = 4.2 \times 10^{-13}$ mol/L **14.** $[HCl] = 0.18$ mol/L **15.** $[Ca(OH)_2] = 0.29$ mol/L **16.(a)** 2.57 **(b)** 7.138 **(c)** 4.01 **(d)** 11.082 **17.** 2.30; acidic **18.** 3.54; acidic **19.(a)** 2.20 **(b)** 9.181; basic **20.** acidic; pH = 6.400; $[OH^-] = 2.51 \times 10^{-8}$ mol/L **21.** pOH = 2.1; $[OH^-] = 8 \times 10^{-3}$ mol/L **22.** acidic; $[H_3O^+] = 1.9 \times 10^{-5}$ mol/L; $[OH^-] = 5.2 \times 10^{-10}$; pOH = 9.28 **23.** $[H_3O^+] = [OH^-] = 1.6 \times 10^{-7}$; neutral **24.** basic; pH = 8.19; $[H_3O^+] = 6.5 \times 10^{-9}$ mol/L; $[OH^-] = 1.5 \times 10^{-6}$ mol/L **25.** $[H_3O] = 1.9 \times 10^{-3}$ mol/L; $[OH^-] = 5.4 \times 10^{-12}$ mol/L **26.** pH = 3.59 **27.(a)** 36.0 L **(b)** pOH = 5.49; pH = 8.51 **28.(a)** 2.42 mL **(b)** pOH = 12.50; $[OH^-] = 3.16 \times 10^{-13}$ mol/L **29.(a)** 6.350 **(b)** 7.650 **(c)** $2.24 \times 10^{-8}$ mol/L **(d)** pH > 7; basic **Section Review 14.1: 1.(a)** B and D **(b)** A (C, although unintended, would also be correct) **(c)** A, B, D **(d)** no **(e)** litmus test, for example **6.(a)** $H_2O$ **(b)** $HCO_3^-$ **7.(a)** $NO_3^-{}_{(aq)}$ **(b)** $SO_4^{2-}{}_{(aq)}$ **8.** (c) and (d) **9.** (a) and (b) **10.(a)** $HF_{(aq)} + H_2O_{(\ell)} \rightleftharpoons H_3O^+{}_{(aq)} + F^-{}_{(aq)}$ **(b)** $HF/F^-$; $H_2O/H_3O^+$ **11.(a)** $H_2PO_4^-/HPO_4^{2-}$ and $CO_3^{2-}/HCO_3^-$ **(b)** $HCOOH/HCOO^-$ and $CN^-/HCN$ **(c)** $H_2PO_4^-/HPO_4^{2-}$ and $OH^-/H_2O$ **14.2: 4.** egg white, camembert cheese, beets, yogurt, sauerkraut **5.(a)** 7.4 **(b)** 1.4 **6.** 2.90 **7.** 1.0 mol/L **8.(a)** 4.43 **(b)** $2.70 \times 10^{-10}$ **(c)** 9.57 **(d)** acidic **(e)** $3.9 \times 10^{-11}$ **(f)** $2.6 \times 10^{-4}$ **(g)** 3.59 **(h)** basic **(i)** $1.4 \times 10^{-13}$ **(j)** 12.85 **(k)** 1.15 **(l)** basic **(m)** $8 \times 10^{-6}$ **(n)** 5.1 **(o)** $1 \times 10^{-9}$ **(p)** acidic

# CHAPTER 15
# Acid-Base Equilibria and Reactions

### Chapter Preview
- **15.1** Revisiting Acid-Base Strength
- **15.2** Acid-Base Reactions and Titration Curves

### Prerequisite Concepts and Skills

Before you begin this chapter, review the following concepts and skills:

- solving equilibrium problems (Chapter 13)
- identifying conjugate acid-base pairs and comparing their relative strengths (Chapter 14)
- calculating molar concentrations (Chapter 7)
- define the relationship among [$H_3O^+$], [$OH^-$], pH, and pOH (Chapter 14)

Normal rain, with a pH ranging from 5.5 to 6.2, is slightly acidic from the reaction between atmospheric carbon dioxide and water vapour to form carbonic acid. Industrial activity and motor-vehicle exhaust produce oxides of sulfur ($SO_2$ and $SO_3$) and nitrogen (usually NO and $NO_2$), which also react with water vapour to form sulfuric and nitric acids (strong acids) and sulfurous and nitrous acids (weak acids). The net effect of these additional acids in the atmosphere is a lowering of precipitation pH. For example, the pH of precipitation in most of eastern Canada and the United States is between 4 and 5. During times of excessive industrial activity, the pH has been as low as 3.

Because it is so soluble in water, acid precipitation poses a serious threat to bodies of water, especially lakes and ponds. For example, as the pH approaches 6, insects and other aquatic animals begin to die. As the pH approaches 5, many plants and plant-like microorganisms die, reducing the food source of remaining aquatic animals. Below a pH of 5, all life in a lake or pond is gone, and the water appears ominously crystal clear.

You could drink this same water safely, despite the fact that it contains acids that are strong enough to dissolve metal and burn organic matter such as skin. On the other hand, a concentrated solution of "harmless" vinegar could damage your skin or eyes. In this chapter, you will look more closely at the equilibria of strong and weak acids and bases—especially weak acids and bases. You will learn how to calculate their concentrations. You will also chart pH changes as acids and bases of different strengths undergo neutralization reactions.

Why is acid rain a bigger problem in eastern Canada than in western Canada?

# Revisiting Acid-Base Strength

## 15.1

In this section, you will examine equilibria involving weak acids and weak bases. To support and enhance your understanding, you will first review what you know about strong acids and strong bases. This time, however, you will consider these substances at a "deeper" level—a molecular level.

As you know, strong acids and strong bases dissociate completely in water. For example, hydrobromic acid is a strong acid and potassium hydroxide is a strong base.

$$HBr_{(g)} \rightarrow H^+_{(aq)} + Br^-_{(aq)}$$
$$KOH \rightarrow K^+_{(aq)} + OH^-_{(aq)}$$

Notice the use of single arrows to represent these reactions. Because the equilibrium in strong acid and strong base dissociations lies so far to the right ($K_c$ is very large), you can view the reactions as going to completion. Thus, in a dilute aqueous solution of a strong, monoprotic acid, the molar concentration of $H^+$ is essentially equal to the molar concentration of the acid. Similarly, in a dilute aqueous solution of a strong base, the molar concentration of hydroxide ions is essentially equal to the molar concentration of the base.

While there are many acids and bases, most are weak. Thus, the number of strong acids and strong bases is fairly small.

### Section Preview/Outcomes

In this section, you will

- **solve problems** that involve strong acids and strong bases
- **solve problems** involving weak acids and weak bases, using the equilibrium constants $K_a$ and $K_b$
- **communicate** your understanding of the following terms: *monoprotic acids, polyprotic acids, acid dissociation constant* ($K_a$), *percent dissociation, base dissociation constant* ($K_b$), *buffer solution, buffer capacity*

### Strong Acids

- *binary acids* that have the general formula $HX_{(aq)}$, where X = Cl, Br, and I (but not F): for example, hydrochloric acid, HCl, and hydrobromic acid, HBr (HCl and HBr are *hydrohalic acids*: acids that have hydrogen bonded to atoms of the halogen elements.)
- *oxoacids* (acids containing oxygen atoms) in which the number of oxygen atoms exceeds, by two or more, the number of protons that can be dissociated: for example, nitric acid, $HNO_3$, sulfuric acid, $H_2SO_4$, perchloric acid, $HClO_4$, and chloric acid, $HClO_3$

**Figure 15.1** The binary acids show periodic trends, which are related to electronegativity and bond strength.

The binary acids of non-metals exhibit periodic trends in their acid strength, as shown in Figure 15.1. Two factors are responsible for this trend: the electronegativity of the atom that is bonded to hydrogen, and the strength of the bond.

Across a period, electronegativity is the most important factor. The acid strength increases as the electronegativity of the element bonded to the H atom increases. This happens because an electronegative atom draws electrons away from the hydrogen atom, making it relatively positive. As well, when the compound dissolves in water, the negative pole of a water molecule then strongly attracts the hydrogen atom and pulls it away.

Down a group, bond strength is the most important factor. Acid strength increases as bond strength decreases. A weaker bond means that the hydrogen atom is more easily pulled

increasing acid strength →

increasing electronegativity

| | 13 (IIIA) | 14 (IVA) | 15 (VA) | 16 (VIA) | 17 (VIIA) | 18 (VIIIA) |
|---|---|---|---|---|---|---|
| | | $CH_4$ | $NH_3$ | $H_2O$ | HF | |
| | | | | $H_2S$ | HCl | |
| | | | | $H_2Se$ | HBr | |
| | | | | $H_2Te$ | HI | |
| | | | | | | |

↓ decreasing bond strength    ↓ increasing acid strength

**CHEM FACT**

You might wonder how HCl$_{(aq)}$, HBr$_{(aq)}$, and HI$_{(aq)}$ can be described as increasing in strength, since each acid dissociates completely in water. This trend becomes apparent if you add equal concentrations of the acids to a solvent that is less basic than water, such as pure acetic acid. You will find that the acids dissociate to different extents.

away from the atom to which it is attached. For example, hydrofluoric acid is a stronger acid than water, but HF is the weakest of the hydrohalic acids because the H-F bond is relatively strong.

Oxoacids increase in strength with increasing numbers of oxygen atoms, as shown in Figure 15.2. The hydrogen atoms that dissociate in water are always attached to oxygen atoms. Oxygen is more electronegative than hydrogen, so oxygen atoms draw electrons away from hydrogen atoms. The more oxygen atoms there are in a molecule, the greater is the polarity of the bond between each hydrogen atom and the oxygen atom it is attached to, and the more easily the water molecule can tear the hydrogen atom away.

increasing acid strength →

| hypochlorous acid | chlorous acid | chloric acid | perchloric acid |

**Figure 15.2** The relative strength of oxoacids increases with the number of oxygen atoms.

Acids such as HCl, CH$_3$COOH, and HF are **monoprotic acids**. They have only a single hydrogen atom that dissociates in water. Some acids have more than one hydrogen atom that dissociates. These acids are called **polyprotic acids**. For example, sulfuric acid has two hydrogen atoms that can dissociate.

$$H_2SO_{4(aq)} + H_2O_{(\ell)} \rightarrow H_3O^+_{(aq)} + HSO_4^-_{(aq)}$$
$$HSO_4^-_{(aq)} + H_2O_{(\ell)} \rightleftharpoons H_3O^+_{(aq)} + SO_4^{2-}_{(aq)}$$

Sulfuric acid is a far stronger acid than the hydrogen sulfate ion, because much more energy is required to remove a proton from a negatively charged ion. *The strength of a polyprotic acid decreases as the number of hydrogen atoms that have dissociated increases.*

Strong bases are confined to the oxides and hydroxides from Groups 1 (IA) and 2 (IIA).

**Strong Bases**
- all oxides and hydroxides of the alkali metals: for example, sodium hydroxide, NaOH, and potassium hydroxide, KOH
- alkaline earth (Group 2 (IIA)) metal oxides and hydroxides below beryllium: for example, calcium hydroxide, Ca(OH)$_2$, and barium hydroxide, Ba(OH)$_2$

The strong basic oxides have metal atoms with low electronegativity. Thus, the bond to oxygen is ionic and is relatively easily broken by the attraction of polar water molecules. The oxide ion always reacts with water molecules to produce hydroxide ions.

$$O^{2-}_{(aq)} + H_2O_{(\ell)} \rightarrow 2OH^-_{(aq)}$$

Magnesium oxide and magnesium hydroxide are not very soluble. They are strong bases, however, because the small amount that does dissolve dissociates almost completely into ions. Beryllium oxide is a weak base. (It is the exception in Group 2 (IIA).) It is a relatively small atom, so the bond to oxygen is strong and not easily broken by water molecules.

## Calculations That Involve Strong Acids and Bases

When a strong, monoprotic acid dissociates in water, the concentration of $H_3O^+_{(aq)}$ is equal to the concentration of the strong acid. Similarly, when a strong base dissociates completely in water, the concentration of $OH^-_{(aq)}$ is equal to the concentration of the strong base.

### Sample Problem

### Calculating Ion Concentrations in Acidic and Basic Solutions

#### Problem
During an experiment, a student pours 25.0 mL of 1.40 mol/L nitric acid into a beaker that contains 15.0 mL of 2.00 mol/L sodium hydroxide solution. Is the resulting solution acidic or basic? What is the concentration of the ion that causes the solution to be acidic or basic? What is the final pH of the solution?

#### What Is Required?
You must determine the ion in excess, its concentration, and the final pH.

#### What Is Given?
You have the following data:
Volume of nitric acid = 25.0 mL
$[HNO_3]$ = 1.40 mol/L
Volume of sodium hydroxide = 15.0 mL
$[NaOH]$ = 2.00 mol/L

#### Plan Your Strategy

**Step 1** Write the chemical equation for the reaction.

**Step 2** Calculate the amount of each reactant using the following equation.
Amount (in mol) = Concentration (in mol/L) × Volume (in L)

**Step 3** Determine the limiting reactant.

**Step 4** The reactant in excess is a strong acid or base. Thus, the excess amount results in the same amount of $H_3O^+$ or $OH^-$.

**Step 5** Calculate the concentration of the excess ion by using the amount in excess and the total volume of the solution. Calculate the final pH.

#### Act on Your Strategy

**Step 1** $HNO_{3(aq)} + NaOH_{(aq)} \rightarrow NaNO_{3(aq)} + H_2O_{(\ell)}$

**Step 2** Amount of $HNO_3$ = 1.40 mol/L × 0.0250 L
= 0.0350 mol
Amount of NaOH = 2.00 mol/L × 0.0150 L
= 0.0300 mol

**Step 3** The reactants combine in a 1:1 ratio. The amount of NaOH is less, so this reactant must be the limiting reactant.

*Continued...*

> **Step 4** Amount of excess $HNO_{3(aq)}$ = 0.0350 mol − 0.0300 mol
> = 0.005 0 mol
>
> Therefore, the amount of $H_3O^+_{(aq)}$ is $5.0 \times 10^{-3}$ mol.
>
> **Step 5** Total volume of solution = 25.0 mL + 15.0 mL = 40.0 mL
>
> $$[H_3O^+] = \frac{5.0 \times 10^{-3} \text{ mol}}{0.0400 \text{ L}}$$
> $$= 0.12 \text{ mol/L}$$
>
> The final pH = $-\log[H_3O^+]$
> = 0.92
>
> The solution is acidic, $[H_3O^+]$ is 0.12 mol/L, and final pH is 0.92.
>
> **Check Your Solution**
>
> The chemical equation has a 1:1 ratio between reactants. The amount of acid is greater than the amount of base. Therefore, the resulting solution should be acidic, which it is.

## Practice Problems

1. Calculate the concentration of hydronium ions in each solution.
   - **(a)** 4.5 mol/L $HCl_{(aq)}$
   - **(b)** 30.0 mL of 4.50 mol/L $HBr_{(aq)}$ diluted to 100.0 mL
   - **(c)** 18.6 mL of 2.60 mol/L $HClO_{4(aq)}$ added to 24.8 mL of 1.92 mol/L $NaOH_{(aq)}$
   - **(d)** 17.9 mL of 0.175 mol/L $HNO_{3(aq)}$ added to 35.4 mL of 0.0160 mol/L $Ca(OH)_{2(aq)}$

2. Calculate the concentration of hydroxide ions in each solution.
   - **(a)** 3.1 mol/L $KOH_{(aq)}$
   - **(b)** 21.0 mL of 3.1 mol/L KOH diluted to 75.0 mL
   - **(c)** 23.2 mL of 1.58 mol/L $HCl_{(aq)}$ added to 18.9 mL of 3.50 mol/L $NaOH_{(aq)}$
   - **(d)** 16.5 mL of 1.50 mol/L $H_2SO_{4(aq)}$ added to 12.7 mL of 5.50 mol/L $NaOH_{(aq)}$

3. Determine whether reacting each pair of solutions results in an acidic solution or a basic solution. Then calculate the concentration of the ion that causes the solution to be acidic or basic. (Assume that the volumes in part (a) are additive. Assume that the volumes in part (b) stay the same.) Determine the final pH in each case.
   - **(a)** 31.9 mL of 2.75 mol/L $HCl_{(aq)}$ added to 125 mL of 0.0500 mol/L $Mg(OH)_{2(aq)}$
   - **(b)** 4.87 g of $NaOH_{(s)}$ added to 80.0 mL of 3.50 mol/L $HBr_{(aq)}$

4. 2.75 g of $MgO_{(s)}$ is added to 70.0 mL of 2.40 mol/L $HNO_{3(aq)}$. Is the solution that results from the reaction acidic or basic? What is the concentration of the ion that is responsible for the character of the solution? What is the pH of the solution?

## The Acid Dissociation Constant, $K_a$

Many common foods (such as citrus fruits), pharmaceuticals (such as Aspirin™), and some vitamins (such as niacin, vitamin B3) are weak acids. When a weak acid dissolves in water, it does not completely dissociate. The concentration of the hydronium ions, and the concentration of the conjugate base of the acid that is formed in solution, depend on the initial concentration of the acid and the amount of acid that dissociates.

You can represent any weak monoprotic acid with the general formula HA. Thus you can express the equilibrium of a weak monoprotic acid in aqueous solution as follows:

$$HA_{(aq)} + H_2O_{(\ell)} \rightleftharpoons H_3O^+_{(aq)} + A^-_{(aq)}$$

The equilibrium constant expression for this reaction is

$$K_c = \frac{[H_3O^+][A^-]}{[HA][H_2O]}$$

In dilute solutions, the concentration of water is almost constant. Multiplying both sides of the equilibrium expression by $[H_2O]$ gives the product of two constants on the left side. This new constant is called the **acid dissociation constant**, $K_a$. (Some chemists refer to the acid dissociation constant as the *acid ionization constant*. With either name, the symbol is $K_a$.)

$$K_c[H_2O] = K_a = \frac{[H_3O^+][A^-]}{[HA]}$$

## pH and $K_a$ of a Weak Acid

Table 15.1 lists the acid dissociation constants for selected acids at 25°C. Notice that weak acids have $K_a$ values that are between 1 and about $1 \times 10^{-16}$. Very weak acids have $K_a$ values that are less than $1 \times 10^{-16}$. The smaller the value of $K_a$, the less the acid dissociates in aqueous solution.

**Table 15.1** Some Acid Dissociation Constants for Weak Acids at 25°C

| Acid | Formula | Acid dissociation constant, $K_a$ |
|---|---|---|
| acetic acid | $CH_3COOH$ | $1.8 \times 10^{-5}$ |
| chlorous acid | $HClO_2$ | $1.1 \times 10^{-2}$ |
| formic acid | $HCOOH$ | $1.8 \times 10^{-4}$ |
| hydrocyanic acid | $HCN$ | $6.2 \times 10^{-10}$ |
| hydrofluoric acid | $HF$ | $6.6 \times 10^{-4}$ |
| hydrogen oxide (water) | $H_2O$ | $1.0 \times 10^{-14}$ |
| lactic acid | $CH_3CHOHCOOH$ | $1.4 \times 10^{-4}$ |
| nitrous acid | $HNO_2$ | $7.2 \times 10^{-4}$ |
| phenol | $C_6H_5OH$ | $1.3 \times 10^{-10}$ |

### CHEM FACT

Niacin is found in many foods, including corn. The niacin in corn, however, cannot be absorbed in the intestinal tract. In regions of the world where corn is a major part of the diet, niacin deficiency can occur. If you add calcium oxide or wood ash to the water in which you boil corn, the resulting basic solution allows the niacin to be absorbed. The flour for making corn tortillas is commonly prepared using this method.

For example, 10.6% of 1 mol/L chlorous acid ($K_a = 1.1 \times 10^{-2}$) dissociates into ions. Chlorous acid has a relatively high $K_a$. Compare this with 1 mol/L acetic acid ($K_a = 1.8 \times 10^{-5}$), which dissociates 0.42%, and 1 mol/L hydrocyanic acid ($K_a = 6.2 \times 10^{-10}$), which dissociates only 0.0025%. (A more extensive table of $K_a$ values for weak acids may be found in Appendix E.)

Problems that involve the concentrations of ions formed in aqueous solutions are considered to be equilibrium problems. The steps for solving acid and base equilibrium problems are similar to the steps you learned in Chapter 13 for solving equilibrium problems.

> **Solving Equilibrium Problems That Involve Acids and Bases**
>
> The steps that you will use to solve acid and base equilibrium problems will vary depending on the problem. Below are a few general steps to guide you.
>
> - Write the chemical equation. Use the chemical equation to set up an ICE table for the reacting substances whenever initial acid concentrations are involved. Enter any values that are given in the problem. It is often convenient to represent the concentration of dissociated hydronium ion and conjugate base with a value of $x$. (**Note:** For the problems in this textbook, you can assume that the concentrations of hydronium ions and hydroxide ions in pure water are negligible compared with the concentrations of these ions when a weak acid or weak base is dissolved in water.)
>
> - For problems that give the initial concentration of the acid, [HA], compare the initial concentration of the acid with the acid dissociation constant, $K_a$.
>
> - If $\dfrac{[HA]}{K_a} > 500$, the change in the initial concentration, $x$, is negligible and can be ignored.
>
> - If $\dfrac{[HA]}{K_a} < 500$, the change in the initial concentration, $x$, may not be negligible. Solving the equilibrium constant expression for $x$ will be more complex, possibly requiring the solution of a quadratic equation.

## Percent Dissociation

**CONCEPT CHECK**

Using appropriate calculations, demonstrate how acetic acid (a weak acid) can have a pH lower than hydrochloric acid (a strong acid). Explain your answer using the concept of percent dissociation.

The **percent dissociation** (also called percent ionization) of a weak acid is the fraction of acid molecules that dissociate compared with the initial concentration of the acid, expressed as a percent. The percent dissociation depends on the value of $K_a$ for the acid, as well as the initial concentration of the weak acid. The following Sample Problems show how to solve problems that involve percent dissociation.

Before you begin, use the Concept Check to make sure that you are clear on the distinction between the concentration and strength of an acid.

# Sample Problem

## Determining $K_a$ and Percent Dissociation

### Problem
Propanoic acid, $CH_3CH_2COOH$, is a weak monoprotic acid that is used to inhibit mould formation in bread. A student prepared a 0.10 mol/L solution of propanoic acid and found that the pH was 2.96. What is the acid dissociation constant for propanoic acid? What percent of its molecules were dissociated in the solution?

### What Is Required?
You need to find $K_a$ and the percent dissociation for propanoic acid.

### What Is Given?
You have the following data:
Initial $[CH_3CH_2COOH]$ = 0.10 mol/L
pH = 2.96

### Plan Your Strategy

**Step 1** Write the equation for the dissociation equilibrium of propanoic acid in water. Then set up an ICE table.

**Step 2** Write the equation for the acid dissociation constant. Substitute equilibrium terms into the equation.

**Step 3** Calculate $[H_3O^+]$ using $[H_3O^+] = 10^{-pH}$

**Step 4** Use the stoichiometry of the equation and $[H_3O^+]$ to substitute for the unknown term, $x$, and calculate $K_a$.

**Step 5** Calculate the percent dissociation by expressing the fraction of molecules that dissociate out of 100.

### Act on Your Strategy

**Step 1** Use the equation for the dissociation equilibrium of propanoic acid in water to set up an ICE table.

| Concentration (mol/L) | $CH_3CH_2COOH_{(aq)}$ + $H_2O_{(\ell)}$ ⇌ | $CH_3CH_2COO^-_{(aq)}$ | + $H_3O^+_{(aq)}$ |
|---|---|---|---|
| Initial | 0.10 | 0 | ~0 |
| Change | $-x$ | $+x$ | $+x$ |
| Equilibrium | $0.10 - x$ | $+x$ | $+x$ |

**Step 2** $K_a = \dfrac{[CH_3CH_2COO^-][H_3O^+]}{[CH_3CH_2COOH]}$

$= \dfrac{(x)(x)}{(0.10 - x)}$

**Step 3** The value of $x$ is equal to $[H_3O^+]$ and $[CH_3CH_2COO^-]$.

$[H_3O^+] = 10^{-2.96}$
$= 1.1 \times 10^{-3}$ mol/L

**Step 4** $K_a = \dfrac{(1.1 \times 10^{-3})^2}{0.10 - (1.1 \times 10^{-3})}$

$= 1.2 \times 10^{-5}$

*Continued...*

> **Step 5** Percent dissociation = $\dfrac{1.1 \times 10^{-3} \text{ mol/L}}{0.10 \text{ mol/L}} \times 100$
> = 1.1%
>
> **Check Your Solution**
>
> The value of $K_a$ and the percent dissociation are reasonable for a weak acid.

## Sample Problem

### Calculating pH

**Problem**

Formic acid, HCOOH, is present in the sting of certain ants. What is the pH of a 0.025 mol/L solution of formic acid?

**What Is Required?**

You need to calculate the pH of the solution.

**What Is Given?**

You know the concentration of formic acid:
[HCOOH] = 0.025 mol/L

The acid dissociation constant for formic acid is listed in Table 15.1:
$K_a = 1.8 \times 10^{-4}$

**Plan Your Strategy**

**Step 1** Write the equation for the dissociation equilibrium of formic acid in water. Then set up an ICE table.

**Step 2** Write the equation for the acid dissociation constant. Substitute equilibrium terms into the equation.

**Step 3** Check the value of $\dfrac{[\text{HCOOH}]}{K_a}$ to see whether or not the amount that dissociates is negligible compared with the initial concentration of the acid.

**Step 4** Solve the equation for $x$. If the amount that dissociates is not negligible compared with the initial concentration of acid, you will need to use a quadratic equation.

**Step 5** pH = $-\log [\text{H}_3\text{O}^+]$

**Act on Your Strategy**

**Step 1**

| Concentration (mol/L) | HCOOH$_{(aq)}$ + H$_2$O$_{(\ell)}$ ⇌ | HCOO$^-_{(aq)}$ | + H$_3$O$^+_{(aq)}$ |
|---|---|---|---|
| Initial | 0.025 | 0 | ~0 |
| Change | $-x$ | $+x$ | $+x$ |
| Equilibrium | 0.025 $-x$ | $+x$ | $+x$ |

**Step 2** $K_a = \dfrac{[\text{HCOO}^-][\text{H}_3\text{O}^+]}{[\text{HCOOH}]}$

$= \dfrac{(x)(x)}{(0.025 - x)}$

$= 1.8 \times 10^{-4}$

**Step 3** $\dfrac{[HCOOH]}{K_a} = \dfrac{0.025}{1.8 \times 10^{-4}}$

$\qquad\qquad = 139$

Since this value is less than 500, the amount that dissociates is not negligible compared with the initial concentration of the acid.

**Step 4** Rearrange the equation into a quadratic equation.

$$\dfrac{x^2}{(0.025 - x)} = 1.8 \times 10^{-4}$$

$$x^2 + (1.8 \times 10^{-4})x - (4.5 \times 10^{-6}) = 0$$

$$x = \dfrac{-b \pm \sqrt{b^2 - 4ac}}{2a}$$

$$= \dfrac{-(1.8 \times 10^{-4}) \pm \sqrt{(1.8 \times 10^{-4})^2 - 4 \times 1 \times (-4.5 \times 10^{-6})}}{2 \times 1}$$

$x = 0.0020$ or $x = -0.002$

The negative value is not reasonable, since a concentration term cannot be negative.

$\therefore x = 0.0020$ mol/L $= [H_3O^+]$

**Step 5** pH $= -\log 0.0020$

$\qquad = 2.70$

The pH of a solution of 0.025 mol/L formic acid is 2.70.

### Check Your Solution

The pH indicates an acidic solution, as expected. Data that was given in the problem has two significant digits, and the pH has two digits following the decimal place. It is easy to make a mistake when solving a quadratic equation. You can estimate a solution to this problem, by assuming that $(0.025 - x)$ is approximately equal to 0.025.

$\dfrac{x^2}{0.025} = 1.8 \times 10^{-4}$

$\quad x^2 = 4.5 \times 10^{-6}$

$\quad\; x = 2.1 \times 10^{-3}$ mol/L

This answer is very close to the answer obtained by solving the quadratic equation. Therefore, the solution is probably correct.

### Practice Problems

**5.** Calculate the pH of a sample of vinegar that contains 0.83 mol/L acetic acid. What is the percent dissociation of the vinegar?

**6.** In low doses, barbiturates act as sedatives. Barbiturates are made from barbituric acid, a weak monoprotic acid that was first prepared by the German chemist Adolph von Baeyer in 1864. The formula of barbituric acid is $C_4H_4N_2O_3$. A chemist prepares a 0.10 mol/L solution of barbituric acid. The chemist finds the pH of the solution to be 2.50. What is the acid dissociation constant for barbituric acid? What percent of its molecules dissociate?

**7.** A solution of hydrofluoric acid has a molar concentration of 0.0100 mol/L. What is the pH of this solution?

*Continued ...*

8. Hypochlorous acid, HOCl, is used as a bleach and a germ-killer. A chemist finds that 0.027% of hypochlorous acid molecules are dissociated in a 0.40 mol/L solution of the acid. What is the value of $K_a$ for the acid?

9. The word "butter" comes from the Greek *butyros*. Butanoic acid (common name: butyric acid) gives rancid butter its distinctive odour. Calculate the pH of a $1.0 \times 10^{-2}$ mol/L solution of butanoic acid ($K_a = 1.51 \times 10^{-5}$).

10. Caproic acid, $C_5H_{11}COOH$, occurs naturally in coconut and palm oil. It is a weak monoprotic acid, with $K_a = 1.3 \times 10^{-5}$. A certain aqueous solution of caproic acid has a pH of 2.94. How much acid was dissolved to make 100 mL of this solution?

## The Base Dissociation Constant, $K_b$

Many compounds that are present in plants are weak bases. Caffeine in coffee and piperidine in black pepper are two examples. A weak base, represented by B, reacts with water to form an equilibrium solution of ions.

$$B_{(aq)} + H_2O_{(\ell)} \rightleftharpoons HB^+_{(aq)} + OH^-_{(aq)}$$

The equilibrium expression for this general reaction is given as follows:

$$K_c = \frac{[HB^+][OH^-]}{[B][H_2O]}$$

The concentration of water is almost constant in dilute solutions. Multiplying both sides of the equilibrium expression by $[H_2O]$ gives the product of two constants on the left side. The new constant is called the **base dissociation constant**, $K_b$.

$$K_c[H_2O] = \frac{[HB^+][OH^-]}{[B]} = K_b$$

Table 15.2 lists the base dissociation constants for several weak bases at 25°C. Nitrogen-containing compounds are Brønsted-Lowry bases, because the lone pair of electrons on a nitrogen atom can bond with $H^+$ from water. The steps for solving problems that involve weak bases are similar to the steps you learned for solving problems that involve weak acids.

**Table 15.2** Some Base Dissociation Constants at 25°C

| Base | Formula | Base dissociation constant, $K_b$ |
| --- | --- | --- |
| ethylenediamine | $NH_2CH_2CH_2NH_2$ | $5.2 \times 10^{-4}$ |
| dimethylamine | $(CH_3)_2NH$ | $5.1 \times 10^{-4}$ |
| methylamine | $CH_3NH_2$ | $4.4 \times 10^{-4}$ |
| trimethylamine | $(CH_3)_3N$ | $6.5 \times 10^{-5}$ |
| ammonia | $NH_3$ | $1.8 \times 10^{-5}$ |
| hydrazine | $N_2H_4$ | $1.7 \times 10^{-6}$ |
| pyridine | $C_5H_5N$ | $1.4 \times 10^{-9}$ |
| aniline | $C_6H_5NH_2$ | $4.2 \times 10^{-10}$ |
| urea | $NH_2CONH_2$ | $1.5 \times 10^{-14}$ |

## Sample Problem

### Solving Problems Involving $K_b$

**Problem**

The characteristic taste of tonic water is due to the addition of quinine. Quinine is a naturally occurring compound that is also used to treat malaria. The base dissociation constant, $K_b$, for quinine is $3.3 \times 10^{-6}$. Calculate [OH⁻] and the pH of a $1.7 \times 10^{-3}$ mol/L solution of quinine.

**What Is Required?**

You need to find [OH⁻] and pH.

**What Is Given?**

$K_b = 3.3 \times 10^{-6}$
Concentration of quinine = $1.7 \times 10^{-3}$ mol/L

**Plan Your Strategy**

**Step 1** Let Q represent the formula of quinine. Write the equation for the equilibrium reaction of quinine in water. Then set up an ICE table.

**Step 2** Write the equation for the base dissociation constant. Substitute equilibrium terms into the equation.

**Step 3** Calculate the value of $\dfrac{[Q]}{K_b}$ to determine whether or not the amount of quinine that dissociates is negligible compared with the initial concentration.

**Step 4** Solve the equation for $x$. If the amount that dissociates is not negligible compared with the initial concentration of the base, you will need to use a quadratic equation.

**Step 5** pOH = −log [OH⁻]
pH = 14.00 − pOH

**Act on Your Strategy**

**Step 1**

| Concentration (mol/L) | $Q_{(aq)}$ + $H_2O_{(\ell)}$ ⇌ | $HQ^+_{(aq)}$ | + $OH^-_{(aq)}$ |
|---|---|---|---|
| Initial | $1.7 \times 10^{-3}$ | 0 | ~0 |
| Change | −x | +x | +x |
| Equilibrium | $(1.7 \times 10^{-3}) - x$ | x | x |

**Step 2** $K_b = \dfrac{[HQ^+][OH^-]}{[Q]}$

$3.3 \times 10^{-6} = \dfrac{(x)(x)}{(1.7 \times 10^{-3}) - x}$

**Step 3** $\dfrac{[Q]}{K_b} = \dfrac{1.7 \times 10^{-3}}{3.3 \times 10^{-6}}$

$= 515$

Since this value is greater than 500, the amount that dissociates is probably negligible, compared with the initial concentration of the base.

*Continued...*

*Continued...*

**Step 4** $3.3 \times 10^{-6} = \dfrac{x^2}{1.7 \times 10^{-3}}$

$x = \pm 7.5 \times 10^{-5}$

The negative root is not reasonable.
$\therefore x = 7.5 \times 10^{-5}$ mol/L = [OH$^-$]

**Step 5** pOH = $-\log 7.5 \times 10^{-5}$

= 4.13

pH = 14.00 − pOH

$\therefore$ pH = 9.87

### Check Your Solution

The pH of the solution is greater than 7, as expected for a basic solution.

## Sample Problem

### Calculating $K_b$

#### Problem

Pyridine, $C_5H_5N$, is used to manufacture medications and vitamins. Calculate the base dissociation constant for pyridine if a 0.125 mol/L aqueous solution has a pH of 9.10.

#### What Is Required?

You need to find $K_b$.

#### What Is Given?

[$C_5H_5N$] = 0.125 mol/L
pH = 9.10

#### Plan Your Strategy

**Step 1** Write the equation for the equilibrium reaction of pyridine in water. Then set up an ICE table.

**Step 2** Write the equation for the base dissociation constant. Substitute equilibrium terms into the equation.

**Step 3** pOH = 14.0 − pH

**Step 4** [OH$^-$] = $10^{-pOH}$

**Step 5** Substitute for $x$ into the equilibrium equation. Calculate the value of $K_b$.

#### Act on Your Strategy

**Step 1**

| Concentration (mol/L) | HCOOH$_{(aq)}$ + H$_2$O$_{(\ell)}$ ⇌ HCOO$^-_{(aq)}$ + H$_3$O$^+_{(aq)}$ |
|---|---|
| Initial | 0.025      0      ~0 |
| Change | −$x$      +$x$      +$x$ |
| Equilibrium | 0.025 − $x$      +$x$      +$x$ |

**Step 2** $K_b = \dfrac{[C_5H_5NH^+][OH^-]}{[C_5H_5N]}$

$= \dfrac{(x)(x)}{(0.125 - x)}$

**Step 3** pOH = 14.00 − 9.10 = 4.90

**Step 4** $[OH^-] = 10^{-4.90}$
$= 1.3 \times 10^{-5}$ mol/L

**Step 5** $0.125 - (1.3 \times 10^{-5}) = 0.125$

$$K_b = \dfrac{(1.3 \times 10^{-5})^2}{0.125}$$
$$= 1.4 \times 10^{-9}$$

### Check Your Solution

The value of $K_b$ is reasonable for a weak organic base. (See Table 15.2.) The final answer has two significant digits, consistent with the two decimal places in the given pH.

### Practice Problems

11. An aqueous solution of household ammonia has a molar concentration of 0.105 mol/L. Calculate the pH of the solution.

12. Hydrazine, $N_2H_4$, has been used as a rocket fuel. The concentration of an aqueous solution of hydrazine is $5.9 \times 10^{-2}$ mol/L. Calculate the pH of the solution.

13. Morphine, $C_{17}H_{19}NO_3$, is a naturally occurring base that is used to control pain. A $4.5 \times 10^{-3}$ mol/L solution has a pH of 9.93. Calculate $K_b$ for morphine.

14. Methylamine, $CH_3NH_2$, is a fishy-smelling gas at room temperature. It is used to manufacture several prescription drugs, including methamphetamine. Calculate $[OH^-]$ and pOH of a 0.25 mol/L aqueous solution of methylamine.

15. At room temperature, trimethylamine, $(CH_3)_3N$, is a gas with a strong ammonia-like odour. Calculate $[OH^-]$ and the percent of trimethylamine molecules that react with water in a 0.22 mol/L aqueous solution.

16. An aqueous solution of ammonia has a pH of 10.85. What is the concentration of the solution?

## Buffer Solutions

A solution that contains a weak acid/conjugate base mixture or a weak base/conjugate acid mixture is called a **buffer solution**. A buffer solution resists changes in pH when a moderate amount of an acid or a base is added to it. (See Figure 15.3.) For example, adding 10 mL of 1.0 mol/L hydrochloric acid to 1 L of water changes the pH from 7 to about 3, a difference of 4 units. Adding the same amount of acid to 1 L of buffered solution might change the pH by only 0.1 unit.

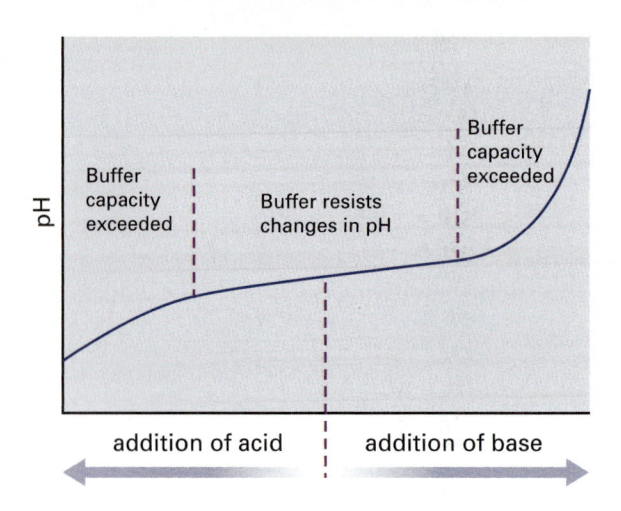

**Figure 15.3** Adding a moderate amount of an acid or a base to a buffer solution causes little change in pH.

Buffer solutions can be made in two different ways:

1. by using a weak acid and one of its salts: for example, by mixing acetic acid and sodium acetate

2. by using a weak base and one of its salts: for example, by mixing ammonia and ammonium chloride

How does a buffer solution resist changes in pH when an acid or a base is added? Consider a buffer solution that is made using acetic acid and sodium acetate. Acetic acid is weak, so most of its molecules are not dissociated and [$CH_3COOH$] is high. Sodium acetate is soluble and a good electrolyte, so [$CH_3COO^-$] is also high. Adding an acid or a base has little effect because the added $H_3O^+$ or $OH^-$ ions are removed by one of the components in the buffer solution. The equilibrium of the reactions between the ions in solution shifts, as predicted by Le Châtelier's principle and described below.

- Adding an acid to a buffer: Acetate ions react with the hydronium ions added to the solution.

$$CH_3COO^-_{(aq)} + H_3O^+_{(aq)} \rightleftharpoons CH_3COOH_{(aq)} + H_2O_{(\ell)}$$

The position of equilibrium shifts to the right. Here hydronium ions are removed, by acetate ions, from the sodium acetate component.

- Adding a base to a buffer: Hydroxide ions react with the hydronium ions that are formed by the dissociation of acetic acid.

$$H_3O^+_{(aq)} + OH^-_{(aq)} \rightleftharpoons 2H_2O_{(\ell)}$$

The position of this water equilibrium shifts to the right, replacing hydronium ions.

Buffer solutions have two important characteristics. One of these characteristics is the pH of the solution. The other is its **buffer capacity**: the amount of acid or base that can be added before considerable change occurs to the pH. The buffer capacity depends on the concentration of the acid/conjugate base (or the base/conjugate acid) in the buffer solution. When the ratio of the concentration of the buffer components is close to 1, the buffer capacity has reached its maximum. As well, a buffer that is more concentrated resists changes to pH more than than a buffer that is more dilute. This idea is illustrated in Figure 15.4, with buffer solutions of acetic acid and acetate of different concentrations.

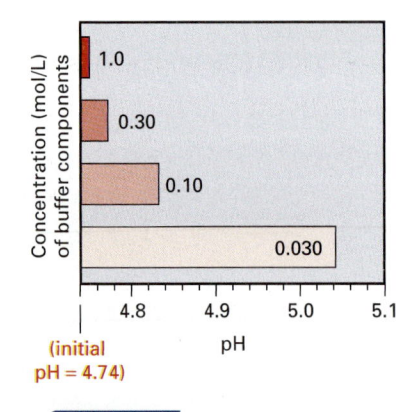

**Figure 15.4** These four buffer solutions have the same initial pH but different concentrations (shown by the numbers beside or on the bars). The pH increases with the addition of a certain amount of strong base. The more concentrated the buffer solution is (that is, the higher its buffer capacity), the smaller the change in pH is.

596 MHR • Unit 6 Acids and Bases

## Buffers in the Blood

Buffers are extremely important in biological systems. The pH of arterial blood is about 7.4. The pH of the blood in your veins is just slightly less. If the pH of blood drops to 7.0, or rises above 7.5, life-threatening problems develop. To maintain its pH within a narrow range, blood contains a number of buffer systems. The most important buffer system in the blood depends on an equilibrium between hydrogen carbonate ions and carbonate ions. Dissolved carbon dioxide reacts with water to form hydrogen carbonate ions.

$$CO_{2(aq)} + 2H_2O_{(\ell)} \rightleftharpoons HCO_{3(aq)}^- + H_3O^+_{(aq)}$$

The $HCO_3^-$ dissociates in water to form $CO_3^{2-}$.

$$HCO_{3(aq)}^- + H_2O_{(\ell)} \rightleftharpoons CO_{3(aq)}^{2-} + H_3O^+_{(aq)}$$

If metabolic changes add $H_3O^+$ ions to the blood, the excess $H_3O^+$ ions are removed by combining with $HCO_3^-$ ions.

$$H_3O^+_{(aq)} + HCO_{3(aq)}^- \rightleftharpoons CO_{2(aq)} + 2H_2O_{(\ell)}$$

If excess $OH^-$ ions enter the blood, they are removed by reacting with the hydrogen carbonate, $HCO_3^-$, ions.

$$OH^-_{(aq)} + HCO_{3(aq)}^- \rightleftharpoons CO_{3(aq)}^{2-} + H_2O_{(\ell)}$$

> **Web LINK**
>
> www.mcgrawhill.ca/links/atlchemistry
>
> Some aspirin products are sold in buffered form. Infer the reasoning behind this practice. Is there clinical evidence to support it? To find out, go to the web site above and click on **Web Links**. Conduct further research to investigate whether there is more recent evidence either in support of or disproving the effectiveness of buffering aspirin.

## Section Summary

In this section, you learned about the relationship between pH and the concentrations of the ions that form in aqueous solutions of weak acids and weak bases. You also saw that solutions containing a mixture of a weak acid and a salt of its conjugate base have properties with important biochemical and industrial applications. In the next section, you will focus your attention on neutralization reactions. In doing so, you will use a laboratory technique called titration to help you examine and compare pH changes that occur when solutions of acids and bases with different strengths react.

## Section Review

**❶** Distinguish between a concentrated solution of a weak base and a dilute solution of a strong base. Give an example of each.

**❷** Lactic acid, $CH_3CHOHCOOH$, is a monoprotic acid that is produced by muscle activity. It is also produced from milk by the action of certain kinds of bacteria. What is the pH of a 0.12 mol/L solution of lactic acid?

**❸** A 0.10 mol/L solution of a weak acid was found to be 5.0% dissociated. Calculate $K_a$.

**❹** Phenol, $C_6H_5OH$, is an aromatic alcohol with weak basic properties. It is used as a disinfectant and cleanser. Calculate the molar concentration of $OH^-$ ions in a 0.75 mol/L solution of phenolate, $C_6H_5O^-$, ions ($K_b = 7.7 \times 10^{-5}$). What is the pH of the solution?

**❺** Potassium sorbate is a common additive in foods. It is used to inhibit the formation of mould. A solution contains 1.82 g of sorbate, $C_6H_7O_2^-$, ions ($K_a = 1.7 \times 10^{-5}$) dissolved in 250 mL of water. What is the pH of the solution?

**6** Write the chemical formula for the conjugate base of hypobromous acid, HOBr. Calculate $K_b$ for this ion.

**7** Describe how a buffer solution differs from an aqueous acidic or basic solution.

**8** Explain the function and importance of buffers in blood.

**9** Explain why an aqueous mixture of NaCl and HCl does not act as a buffer, but an aqueous mixture of $NH_3$ and $NH_4Cl$ does.

**10** In Photo A below, the pH of a 100 mL sample of dilute $HCl_{(aq)}$ reads 5.00. In Photo B, the pH of a 100 mL sample of solution containing a mixture of 1 mol/L acetic acid and 1 mol/L sodium acetate also reads 5.00. The left side of Photo B shows the pH of the dilute $HCl_{(aq)}$ solution after 1 mL of 1 mol/L hydrochloric acid was added. The right side of the photo shows the pH after 1 mL of 1 mol/L sodium hydroxide was added. In Photo D, 1 mL of 1 mol/L hydrochloric acid (left) and 1 mL of 1 mol/L sodium hydroxide (right) have been added to the acetic acid-sodium acetate mixture. Examine the pH readings in Photos B and D and explain the results.

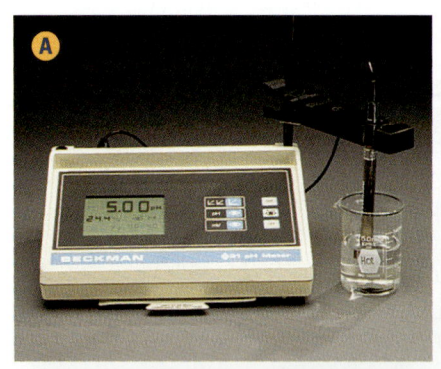

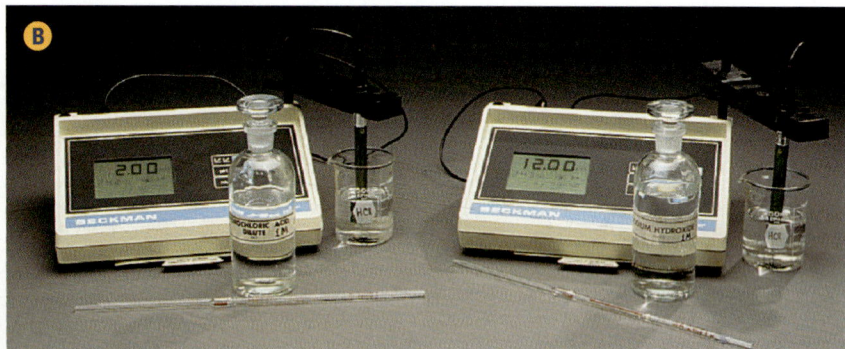

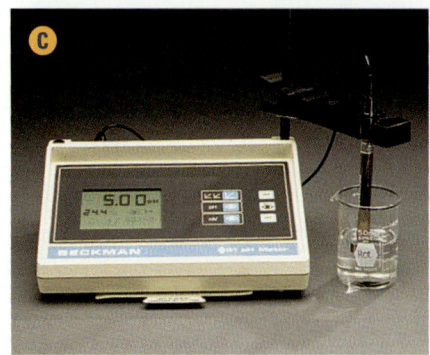

# Acid-Base Reactions and Titration Curves

## 15.2

Adding a base to an acid neutralizes the acid's acidic properties. This type of reaction is called a **neutralization reaction**. There are many different acids and bases. Being able to predict the results of reactions between them is important. Bakers, for example, depend on neutralization reactions to create light, fluffy baked goods. Gardeners and farmers depend on these reactions to modify the characteristics of the soil. Industrial chemists rely on these reactions to produce the raw materials that are used to make a wide variety of chemicals and chemical products.

### Neutralization Reactions

The quantitative reaction between an acid and a base produces an ionic compound (a salt) and water.

$$\text{acid} + \text{base} \rightarrow \text{a salt} + \text{water}$$

A **salt** is any ionic compound that is composed of the anion from an acid and a cation from a base. For example, sodium nitrate is a salt that is found in many kitchens. It is often added to processed meat to preserve the colour and to slow the rate of spoiling by inhibiting bacterial growth. Sodium nitrate can be prepared in a laboratory by reacting nitric acid with sodium hydroxide, as shown below.

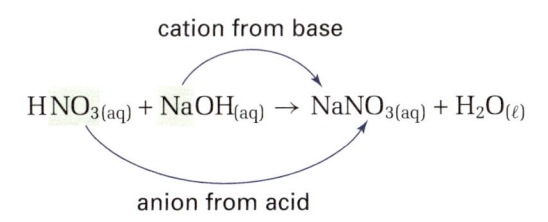

$$\underset{\text{anion from acid}}{H\overbrace{NO_3}_{(aq)}} + \underset{}{\overbrace{Na}^{\text{cation from base}}OH_{(aq)}} \rightarrow NaNO_{3(aq)} + H_2O_{(\ell)}$$

The balanced chemical equation for this reaction shows that 1 mol of nitric acid reacts with 1 mol of sodium hydroxide. If equal molar quantities of nitric acid and sodium hydroxide are used, the result is a neutral (pH 7) aqueous solution of sodium nitrate. In fact, *when any strong acid reacts with any strong base in the mole ratio from the balanced chemical equation, a neutral aqueous solution of a salt is formed*. Reactions between acids and bases of different strengths usually do not result in neutral solutions.

For most neutralization reactions, there are no visible signs that a reaction is occurring. How can you determine that a neutralization reaction is taking place? One way is to use an **acid-base indicator**. *Operationally*, an indicator is a substance that changes colour over a given pH range. *Theoretically*, an indicator is a weak monoprotic acid—usually one with a complex organic structure. The equilibrium below combines both definitions to give an overall view of an acid-base indicator.

$$\underset{\text{colour 2}}{\underset{\text{colour 1}}{H(\text{indicator})_{(aq)}}} \rightleftharpoons H^+ + (\text{indicator})^-_{(aq)}$$

### Section Preview/Outcomes

In this section, you will

- **perform** calculations involving neutralization reactions
- **determine** the concentration of an acid in solution by conducting a titration
- **interpret** acid-base titration curves and the pH at the equivalence point
- **communicate** your understanding of the following terms: *neutralization reaction, salt, acid-base indicator, titration, equivalence point, primary standard end point, acid-base titration curve*

### CHEM FACT

If a small quantity of an acid or a base is spilled in a laboratory, you can use a neutralization reaction to minimize the hazard. To neutralize a basic solution spill, you can add solid sodium hydrogen sulfate or citric acid. For an acidic solution spill, you can use sodium hydrogen carbonate (baking soda). Note that you cannot use a strong acid or base to clean up a spill. This would result in another hazardous spill. As well, the neutralization reaction would generate a lot of heat, and thus produce a very hot solution.

## CONCEPT CHECK

For the indicator equilibrium on the previous page, use Le Châtelier's principle to explain why colour 1 persists in an acidic solution, while colour 2 persists in a basic solution.

In an acidic solution, the indicator does not dissociate very much. It appears as colour 1. In a basic solution, the indicator dissociates much more. It appears as colour 2. Often a single drop of indicator causes a dramatic change in colour. For example, phenolphthalein is an indicator that chemists often use for reactions between a strong acid and a strong base. It is colourless between pH 0 and pH 8. It turns pink between pH 8 and pH 10. (See Figure 15.5.) Consult Appendix E for a selection of other indicators and their effective pH ranges.

**Figure 15.5** A good indicator, such as the phenolphthalein shown here, must give a vivid colour change.

### Calculations Involving Neutralization Reactions

Suppose that a solution of an acid reacts with a solution of a base. You can determine the concentration of one solution if you know the concentration of the other. (This assumes that the volumes of both are accurately measured.) Use the concentration and volume of one solution to determine the amount (in moles) of reactant that it contains. The balanced chemical equation for the reaction describes the mole ratio in which the compounds combine. In the following Sample Problems and Practice Problems, you will see how to do these calculations.

### CHEM FACT

An old remedy to relieve the prickly sting of a nettle plant is to rub the area with the leaf of a dock plant. The sting contains an acid. This acid is neutralized by a base that is present in the dock leaf. Bees and ants also have an acidic sting. You can wash the sting with soap, because soap is basic. You can also apply baking soda (a base) to the skin for more effective relief. If you are stung by a wasp, however, you should apply vinegar. The sting of a wasp contains a base.

### Sample Problem

#### Finding Concentration

**Problem**

13.84 mL of hydrochloric acid, $HCl_{(aq)}$, just neutralizes 25.00 mL of a 0.1000 mol/L solution of sodium hydroxide, $NaOH_{(aq)}$. What is the concentration of the hydrochloric acid?

**What Is Required?**

You need to find the concentration of the hydrochloric acid.

**What Is Given?**

Volume of hydrochloric acid, HCl = 13.84 mL
Volume of sodium hydroxide, NaOH = 25.00 mL
Concentration of sodium hydroxide, NaOH = 0.1000 mol/L

**Plan Your Strategy**

**Step 1** Write the balanced chemical equation for the reaction.
**Step 2** Calculate the amount (in mol) of sodium hydroxide added, based on the volume and concentration of the sodium hydroxide solution.

**600** MHR • Unit 6 Acids and Bases

**Step 3** Determine the amount (in mol) of hydrochloric acid needed to neutralize the sodium hydroxide.

**Step 4** Find [HCl$_{(aq)}$], based on the amount and volume of hydrochloric acid solution needed.

### Act on Your Strategy

**Step 1** The balanced chemical equation is
$$HCl_{(aq)} + NaOH_{(aq)} \rightarrow NaCl_{(aq)} + H_2O_{(\ell)}$$

**Step 2** Amount (in mol) = Concentration (in mol/L) × Volume (in L)

Amount NaOH (in mol) added = 0.1000 mol/L × 0.02500 L
= 2.500 × 10$^{-3}$ mol

**Step 3** HCl reacts with NaOH in a 1:1 ratio; therefore,

Amount (in mol) of HCl = amount (in mol) of NaOH × mole ratio

$$= 2.500 \times 10^{-3} \text{ mol NaOH} \left( \frac{1 \text{ mol HCl}}{1 \text{ mol NaOH}} \right)$$

= 2.500 × 10$^{-3}$ mol HCl

**Step 4** Concentration (in mol/L) = $\frac{\text{Amount (in mol)}}{\text{Volume (in L)}}$

$$[HCl_{(aq)}] = \frac{2.500 \times 10^3 \text{ mol}}{0.01384 \text{ L}}$$

= 0.1806 mol/L

Therefore, the concentration of hydrochloric acid is 0.1806 mol/L.

### Check Your Solution

[HCl$_{(aq)}$] is greater than [NaOH$_{(aq)}$]. This is reasonable because a smaller volume of hydrochloric acid was required. As well, the balanced equation shows a 1:1 mole ratio between these reactants.

## Sample Problem
### Finding Volume

#### Problem
What volume of 0.250 mol/L sulfuric acid, H$_2$SO$_{4(aq)}$, is needed to react completely with 37.2 mL of 0.650 mol/L potassium hydroxide, KOH$_{(aq)}$?

#### What Is Required?
You need to find the volume of sulfuric acid.

#### What Is Given?
Concentration of sulfuric acid, H$_2$SO$_4$ = 0.250 mol/L
Concentration of potassium hydroxide, KOH = 0.650 mol/L
Volume of potassium hydroxide, KOH = 37.2 mL.

#### Plan Your Strategy
**Step 1** Write the balanced chemical equation for the reaction.
**Step 2** Calculate the amount (in mol) of potassium hydroxide, based on the volume and concentration of the potassium hydroxide solution.
**Step 3** Determine the amount (in mol) of sulfuric acid that is needed to neutralize the potassium hydroxide.

*Continued...*

---

### PROBLEM TIP

1. Make sure that the values you use in your calculations refer to the same reactant. For example, you can use the concentration and volume of sodium hydroxide to find the amount of sodium hydroxide in this problem. You cannot use the concentration of sodium hydroxide and the volume of hydrochloric acid.

2. In the solution, the volumes are converted to litres. If all the volumes are expressed in the same unit, the conversion step is not necessary.

3. Do not drop significant digits, even zeros, during your calculations.

### Web LINK

www.mcgrawhill.ca/links/atlchemistry

Limestone (calcium carbonate) reacts with acids in a neutralization reaction. You might think, then, that adding limestone to water systems damaged by acid precipitation would solve the problem of acidified lakes. This process, called "liming", works on a smaller scale to modify the alkalinity of soil in gardens and farmers' fields. On the larger scale of a lake, however, liming presents several difficulties. Go to the web site above and click on **Web Links**. Conduct research to examine the problems associated with liming, and to suggest alternatives to the much larger problem of environmental damage associated with acid precipitation.

**Step 4** Find the volume of the sulfuric acid, based on the amount and concentration of sulfuric acid needed.

### Act on Your Strategy

**Step 1** The balanced chemical equation is
$$H_2SO_{4(aq)} + 2KOH_{(aq)} \rightarrow K_2SO_{4(aq)} + 2H_2O_{(\ell)}$$

**Step 2** Amount (in mol) of KOH = 0.650 mol/L × 0.0372 L
= 0.02418 mol KOH

**Step 3** $H_2SO_4$ reacts with KOH in a 1:2 mole ratio. Therefore,

Amount (in mol) of $H_2SO_4$ = amount (in mol) of KOH × mole ratio

$$= 0.02418 \text{ mol KOH} \left( \frac{1 \text{ mol } H_2SO_4}{2 \text{ mol KOH}} \right)$$

$$= 0.01209 \text{ mol } H_2SO_4$$

**Step 4** Amount (in mol) $H_2SO_4$
= 0.01209 mol
= 0.250 mol/L × Volume $H_2SO_{4(aq)}$ (in L)

$$\text{Volume } H_2SO_{4(aq)} = \frac{0.01209 \text{ mol}}{0.250 \text{ mol/L}}$$

= 0.04836 L

Therefore, the volume of sulfuric acid that is needed is 48.4 mL.

### Check Your Solution

The balanced chemical equation shows that half the amount of sulfuric acid will neutralize a given amount of potassium hydroxide. The concentration of sulfuric acid, however, is less than half the concentration of potassium hydroxide. Therefore, the volume of sulfuric acid should be greater than the volume of potassium hydroxide.

## Practice Problems

17. 17.85 mL of nitric acid neutralizes 25.00 mL of 0.150 mol/L $NaOH_{(aq)}$. What is the concentration of the nitric acid?

18. What volume of 1.015 mol/L magnesium hydroxide is needed to neutralize 40.0 mL of 1.60 mol/L hydrochloric acid?

19. What volume of 0.150 mol/L hydrochloric acid is needed to neutralize each solution below?
    (a) 25.0 mL of 0.135 mol/L sodium hydroxide
    (b) 20.0 mL of 0.185 mol/L ammonia solution
    (c) 80 mL of 0.0045 mol/L calcium hydroxide

20. What concentration of sodium hydroxide solution is needed for each neutralization reaction?
    (a) 37.82 mL of sodium hydroxide neutralizes 15.00 mL of 0.250 mol/L hydrofluoric acid.
    (b) 21.56 mL of sodium hydroxide neutralizes 20.00 mL of 0.145 mol/L sulfuric acid.
    (c) 14.27 mL of sodium hydroxide neutralizes 25.00 mL of 0.105 mol/L phosphoric acid.

## Acid-Base Titration

In the previous Sample Problems and Practice Problems, you were given the concentrations and volumes you needed to solve the problems. What if you did not have some of this information? Chemists often need to know the concentration of an acidic or basic solution. To acquire this information, they use an experimental procedure called a titration. In a **titration**, the concentration of one solution is determined by quantitatively observing its reaction with a solution of known concentration. The solution of known concentration is called a *standard solution*. Standard solutions are prepared in one of two ways. It may be made from a **primary standard**—a substance of known purity and molecular mass that may be used to prepare a solution with a precisely determined concentration. Standard solutions can also be obtained by titrating a non-standard solution with a primary standard solution. This is called a standardization titration, and is the most common method of obtaining a standard solution for a compound that is not a primary standard. Once it is ready for titration, the standard solution is also commonly called the titrant. The aim of a titration is to find the point at which the number of moles of the standard solution is stoichiometrically equal to the original number of moles of the unknown solution. This point is referred to as the **equivalence point**. At the equivalence point, all the moles of hydrogen ions that were present in the original volume of one solution have reacted with an equal number of moles of hydroxide ions from the other solution.

Precise and accurate volume measurements are needed when you perform a titration. Chemists use special glass apparatus to collect these measurements. (See Figure 15.16.) As well, an acid-base indicator is needed to monitor changes in pH during the titration.

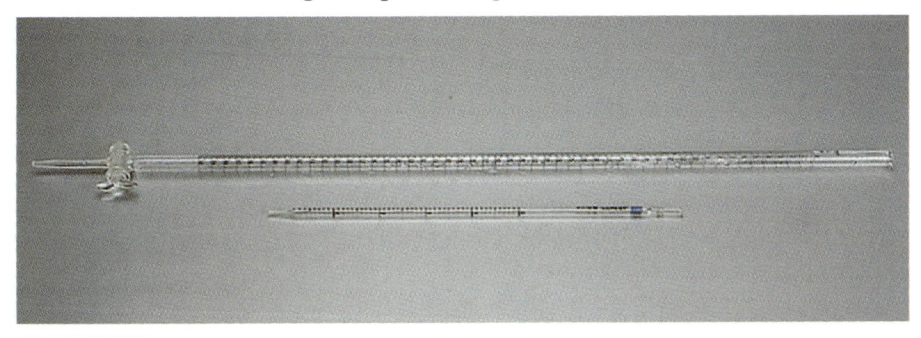

**Figure 15.6** A transfer pipette (bottom) measures a fixed volume of liquid, such as 10.00 mL, 25.00 mL, or 50.0 mL. A burette (top) measures a variable volume of liquid.

In a titration, a pipette is used to measure an accurate volume of standard solution into a flask. The flask sits under a burette that contains the solution of unknown concentration. After adding a few drops of indicator, you take an initial burette reading. Then you start adding the known solution, slowly, to the flask. The **end point** of the titration occurs when the indicator changes colour. The indicator is chosen so that it changes colour over a pH range that includes the end point pH value.

## Titration Step by Step

The following pages outline the steps that you need to follow to prepare for a titration. Review these steps carefully, and observe as your teacher demonstrates them for you. In Investigation 15-A, you will perform your own titration of a common substance: vinegar.

---

**Web LINK**

www.mcgrawhill.ca/links/atlchemistry

Sometimes a moving picture is worth a thousand words. To enhance your understanding, your teacher will demonstrate the titration procedure described in this textbook. In addition, some web sites provide downloadable or real-time titration movies to help students visualize the procedure and its techniques. Go to the web site above, and click on **Web Links**. Compare the different demonstrations you can find and observe, including your teacher's. Prepare your own set of "Titration Tips" to help you recall important details.

**PROBEWARE**

www.mcgrawhill.ca/links/atlchemistry

If you have access to probeware, try the lab, Titrating an Unknown. Go to the web site above and click on **Probeware**.

**TITRATION TIP**

Never use your mouth in place of a suction bulb to draw a liquid into a pipette. The liquid could be corrosive or poisonous. As well, you will contaminate the glass stem.

## Rinsing the Pipette

A pipette is used to measure and transfer an accurate volume of liquid. You rinse a pipette with the solution whose volume you are measuring. This ensures that any drops that remain inside the pipette will form part of the measured volume.

1. Pour a sample of standard solution into a clean, dry beaker.
2. Place the pipette tip in a beaker of distilled water. Squeeze the suction bulb. Maintain your grip while placing it over the stem of the pipette. (If your suction bulbs have valves, your teacher will show you how to use them.)
3. Relax your grip on the bulb to draw up a small volume of distilled water.
4. Remove the bulb and discard the water by letting it drain out.
5. Rinse the pipette by drawing several millilitres of solution from the beaker into it. Rotate and rock the pipette to coat the inner surface with solution. Discard the rinse. Rinse the pipette twice in this way. It is now ready to fill with standard solution.

## Filling the Pipette

6. Place the tip of the pipette below the surface of the solution.
7. Hold the suction bulb loosely on the end of the glass stem. Use the suction bulb to draw liquid up just past the etched volume mark. (See Figure 15.7.)
8. As quickly and smoothly as you can, slide the bulb off and place your index finger over the end of the glass stem.
9. Gently roll your finger slightly away from end of the stem to let solution drain slowly out.
10. When the bottom of the meniscus aligns with the etched mark, as in Figure 15.8, press your finger back over the end of the stem. This will prevent more solution from draining out.
11. Touch the tip of the pipette to the side of the beaker to remove any clinging drop. See Figure 15.9. The measured volume inside the pipette is now ready to transfer to an Erlenmeyer flask or a volumetric flask.

## Transferring the Solution

12. Place the tip of the pipette against the inside glass wall of the flask. Let the solution drain slowly, by removing your finger from the stem.
13. After the solution drains, wait several seconds, then touch the tip to the inside wall of the flask to remove any drop on the end. Note: You may notice a small amount of liquid remaining in the tip. The pipette was calibrated to retain this amount. Do not try to remove it.

**Figure 15.7** Draw a bit more liquid than you need into the pipette. It is easier to reduce this volume than it is to add more solution to the pipettes.

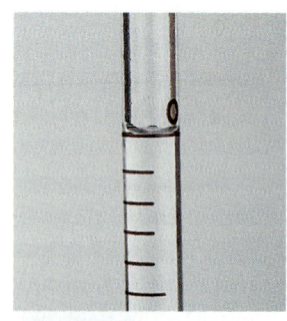

**Figure 15.8** The bottom of the meniscus must align exactly with the etched mark.

**TITRATION TIP**

Practice removing the bulb and replacing it with your index finger. You need to be able to perform this action quickly and smoothly.

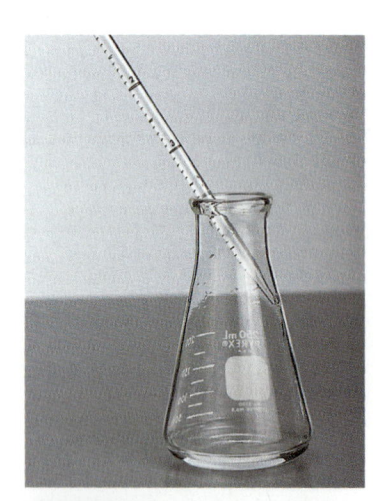

**Figure 15.9** You can prevent a "stubborn" drop from clinging to the pipette tip by touching the tip to the inside of the glass surface.

### Adding the Indicator

14. Add two or three drops of indicator to the flask and its contents. Do not add too much indicator. Using more does not make the colour change easier to see. Also, indicators are usually weak acids. Too much can change the amount of base needed for neutralization. You are now ready to prepare the apparatus for the titration.

### Rinsing the Burette

A burette is used to accurately measure the volume of liquid added during a titration experiment. It is a graduated glass tube with a tap at one end.

15. To rinse the burette, close the tap and add about 10 mL of distilled water from a wash bottle.
16. Tip the burette to one side and roll it gently back and forth so that the water comes in contact with all inner surfaces.
17. Hold the burette over a sink. Open the tap, and let the water drain out. While you do this, check that the tap does not leak. Make sure that it turns smoothly and easily.
18. Rinse the burette with 5 mL to 10 mL of the solution that will be measured. Remember to open the tap to rinse the lower portion of the burette. Rinse the burette twice, discarding the liquid each time.

### Filling the Burette

19. Assemble a retort stand and burette clamp to hold the burette. Place a funnel in the top of the burette.
20. With the tap closed, add solution until the liquid is above the zero mark. Remove the funnel. Carefully open the tap. Drain the liquid into a beaker until the meniscus is just below the zero mark.
21. Touch the tip of the burette against the beaker to remove any clinging drop. Check that the portion of the burette that is below the tap is filled with liquid and contains no air bubbles.
22. Record the initial burette reading in your notebook. Do not try to get an initial reading of 0.00 mL
23. Replace the beaker with the Erlenmeyer flask that you prepared earlier. Place a sheet of white paper under the Erlenmeyer to help you see the indicator colour change that will occur near the end point.

### Reading the Burette

24. A meniscus reader is a small white card with a thick black line on it. Hold the card behind the burette, with the black line just under the meniscus, as in Figure 15.10. Record the volume added from the burette to the nearest 0.05 mL.

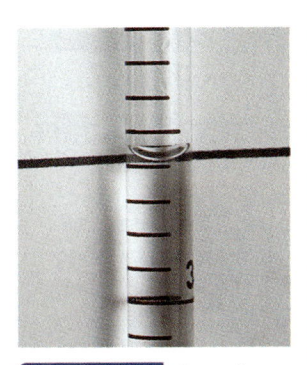

**Figure 15.10** A meniscus reader helps you read the volume of liquid in the burette more easily.

**TITRATION TIP**

If you are right-handed, the tap should be on your right as you face the burette. Use your left hand to operate the tap. Use your right hand to swirl the liquid in the Erlenmeyer flask. If you are left-handed, reverse this arrangement.

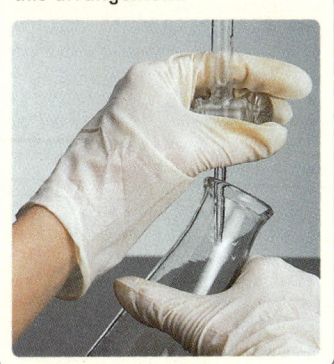

**TITRATION TIP**

Near the end point, when you see the indicator change colour as liquid enters the flask from the burette, slow the addition of liquid. The end point can occur very quickly.

**TITRATION TIP**

Observe the level of solution in the burette so that your eye is level with the bottom of the meniscus.

# Investigation 15-A

**SKILL FOCUS**
Predicting
Performing and recording
Analyzing and interpreting

## The Concentration of Acetic Acid in Vinegar

Vinegar is a dilute solution of acetic acid, CH₃COOH. Only the hydrogen atom that is attached to an oxygen atom is acidic. Thus, acetic acid is monoprotic. As a consumer, you can buy vinegar with different concentrations. For example, the concentration of table vinegar is different from the concentration of the vinegar that is used for pickling foods. To maintain consistency and quality, manufacturers of vinegar need to determine the percent concentration of acetic acid in the vinegar. In this investigation, you will determine the concentration of acetic acid in a sample of vinegar.

### Prediction

Which do you predict has the greater concentration of acetic acid: table vinegar or pickling vinegar? Give reasons for your prediction.

### Materials

pipette
suction bulb
retort stand
burette
burette clamp
3 beakers (250 mL)
3 Erlenmeyer flasks (250 mL)
labels
meniscus reader
sheet of white paper
funnel
table vinegar
pickling vinegar
sodium hydroxide solution
distilled water
dropper bottle containing phenolphthalein

### Safety Precautions

Both vinegar and sodium hydroxide solutions are corrosive. Wash any spills on skin or clothing with plenty of water. Inform your teacher immediately.

### Procedure

1. Record the following information in your notebook. Your teacher will tell you the concentration of the sodium hydroxide solution.
   - concentration of NaOH$_{(aq)}$ (in mol/L)
   - type of vinegar solution
   - volume of pipette (in mL)

2. Copy the table below into your notebook, to record your observations.

   **Burette Readings for the Titration of Acetic Acid**

   | Reading (mL) | Trial 1 | Trial 2 | Trial 3 | Trial 4 |
   |---|---|---|---|---|
   | final reading | | | | |
   | initial reading | | | | |
   | volume added | | | | |

3. Label a clean, dry beaker for each liquid: NaOH$_{(aq)}$, vinegar, and distilled water. Obtain each liquid. Record the type of vinegar you will be testing.

4. Obtain a pipette and a suction bulb. Record the volume of the pipette for trial 1. Rinse it with distilled water, and then with vinegar.

5. Pipette some vinegar into the first Erlenmeyer flask. Record this amount. Add approximately 50 mL of water. Also add two or three drops of phenolphthalein indicator.

606    MHR • Unit 6 Acids and Bases

6. Set up a retort stand, burette clamp, burette, and funnel. Rinse the burette first with distilled water. Then rinse it with sodium hydroxide solution. Make sure that there are no air bubbles in the burette. Also make sure that the liquid fills the tube below the glass tap. Remove the funnel before beginning the titration.

7. Place a sheet of white paper under the Erlenmeyer flask. Titrate sodium hydroxide into the Erlenmeyer flask while swirling the contents. The end point of the titration is reached when a permanent pale pink colour appears. If you are not sure whether you have reached the end point, take the burette reading. Add one drop of sodium hydroxide, or part of a drop. Observe the colour of the solution. If you go past the end point, the solution will become quite pink.

8. Repeat the titration at least twice more until two trials agree within ±0.1 mL for the volume added. Record your results for each of these trials.

9. When you have finished all three trials, dispose of the chemicals as directed by your teacher. Rinse the pipette and burette with distilled water. Leave the burette tap open.

## Analysis

1. For two trials that agree within ±0.1 mL, find the average volume added. If you have more than two trials within ±0.1 mL, include these readings in your average, too.

2. Write the chemical equation for the reaction of acetic acid with sodium hydroxide.

3. Calculate the concentration of acetic acid in your vinegar sample. Use the average volume and concentration of sodium hydroxide, and the volume of vinegar.

4. Find the molar mass of acetic acid. Then calculate the mass of acid in the volume of vinegar you used.

5. The density of vinegar is 1.01 g/mL. (The density of the more concentrated vinegar solution is greater than the density of the less concentrated solution. You can ignore the difference, however.) Calculate the mass of the vinegar sample. Find the percent by mass of acetic acid in the sample.

## Conclusions

6. Compare your results with the results of other students who used the same type of vinegar. Then compare the concentration of acetic acid in table vinegar with the concentration in pickling vinegar. How did your results compare with your prediction?

7. List several possible sources of error in this investigation.

## Application

8. Most shampoos are basic. Why do some people rinse their hair with vinegar after washing it?

## Acid-Base Titration Curves

At the end of this section, you will perform another titration involving acetic acid. In that lab, Investigation 15-B, the concentration of acid will be unknown, and one goal will be to calculate its value. To do so, you will graph the data you collect by plotting pH against the volume of titrant that you add. A graph of the pH of an acid (or base) against the volume of an added base (or acid) is called an **acid-base titration curve**.

Titrations are common analytical procedures that chemists perform, with the usual goal of determining the concentration of one of the reactants. As you know, the equivalence point is the point in a titration when the acid and base that are present completely react with each other. If chemists know the volumes of both solutions at the equivalence point, and the concentration of one of them, they can calculate the unknown concentration.

As you can see in Figure 15.11, the middle of the steep rise that occurs in a titration curve is the equivalence point. The endpoint of a titration occurs when the indicator changes colour, which happens over a range of about 2 pH units. The pH changes rapidly near the equivalence point. Therefore, the change in colour usually takes place in a fraction of a millilitre, with the addition of a single drop of solution. Chemists have access to a variety of indicators that change colour at different pH values. Refer to Appendix E.

When an indicator is chosen for a titration, the end point pH (which is essentially the equivalence point pH) must be within the pH range over which the indicator changes colour. Some representative acid-base titration curves, shown in Figures 15.11, 15.12, and 15.13, will illustrate this point. (You may notice the absence of a curve for the reaction of a weak acid with a weak base. A weak acid-weak base titration curve is difficult to describe quantitatively, because it has competing equilibria. You may learn about this curve in future chemistry courses.)

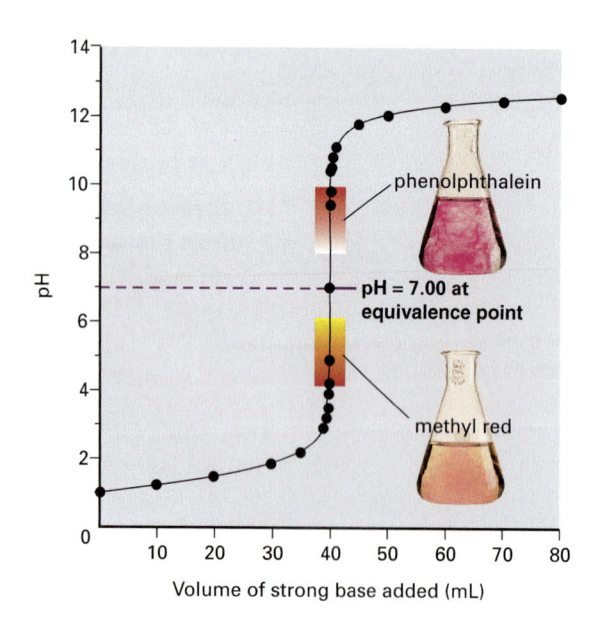

### Titration Curve for a Strong Acid With a Strong Base

These titrations have a pH of 7 at equivalence. Indicators such as phenolphthalein, methyl red, and bromocresol green can be used, because their endpoints are close to the equivalence point. Many chemists prefer phenolphthalein because the change from colourless to pink is easy to see.

**Figure 15.11** The curve for a strong acid-strong base titration

## Titration Curve for a Weak Acid With a Strong Base

These titrations have pH values that are greater than 7 at equivalence. In the titration shown in Figure 15.12, the equivalence point occurs at a pH of 8.80. Therefore, phenolphthalein is a good indicator for this titration. Methyl red is not, because its endpoint is too far from the equivalence point.

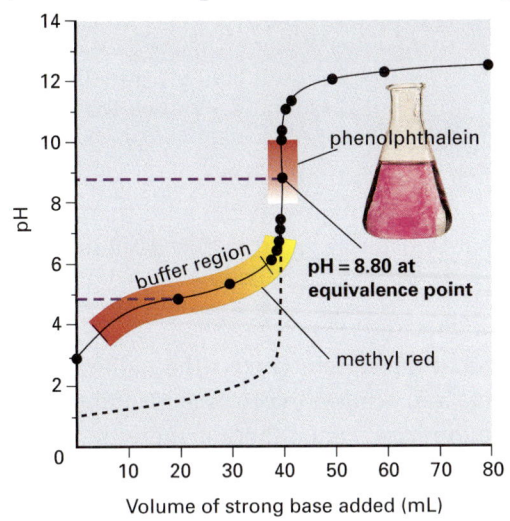

**Figure 15.12** The curve for a weak acid-strong base titration: The weak acid here is propanoic acid, $CH_3CH_2COOH$.

### CHEM FACT

Notice in Figures 15.12 and 15.13 the portions of the titration curves labelled "buffer region." The gently rising portion of the curve in this region represents the slow pH changes during this part of the titration. At this stage in Figure 15.12, some amount of the conjugate base of the weak acid has formed. In Figure 15.13, some amount of the conjugate acid of the weak base has formed. In both titrations, the solution at this stage is a buffer solution.

## Titration Curve for a Weak Base With a Strong Acid

These titrations have pH values that are less than 7 at the equivalence point. The equivalence point in the titration shown in Figure 15.13, involving ammonia and hydrochloric acid, occurs at a pH of 5.27. Either methyl red or bromocresol green could be used as an indicator, but not phenolphthalein.

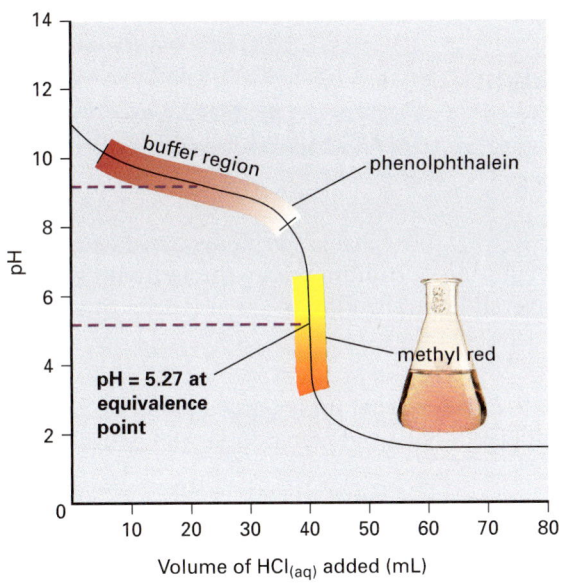

**Figure 15.13** The curve for a weak base-strong acid titration

## Neutralization Reactions for Polyprotic and Polybasic Species

The titration curves you have considered so far have a single defined equivalence point, because the acids are monoprotic. They have a single proton ($H^+$) that may dissociate. Polyprotic acids have more than one proton that dissociates. For example, oxalic acid ($H_2C_2O_{4(aq)}$) is *di*protic. It has two protons that can dissociate. (The *di-* prefix means two.) Phosphoric acid ($H_3PO_{4(aq)}$) is *tri*protic. It has three protons that can dissociate. (The *tri-* prefix means three.)

Polyprotic acids do not donate all their H⁺ protons at the same time when they dissociate. The dissociation of a polyprotic acid occurs as a series of separate steps. For example:

$$H_3PO_{4(aq)} + H_2O_{(\ell)} \rightleftharpoons H_2PO_4^{2-}{}_{(aq)} + H_3O^+{}_{(aq)} \qquad K_a = 6.9 \times 10^{-3}$$

$$H_2PO_4^-{}_{(aq)} + H_2O_{(\ell)} \rightleftharpoons HPO_4^-{}_{(aq)} + H_3O^+{}_{(aq)} \qquad K_a = 6.2 \times 10^{-8}$$

$$HPO_4^{2-}{}_{(aq)} + H_2O_{(\ell)} \rightleftharpoons PO_4^{3-}{}_{(aq)} + H_3O^+{}_{(aq)} \qquad K_a = 4.8 \times 10^{-13}$$

Notice that each of these dissociation steps is an equilibrium with its own $K_a$ value. (You can find acid dissociation values for other polyprotic acids in Appendix E.) Notice also how these acid dissociation values change with each successive dissociation—$H_3PO_{4(aq)}$ is a much stronger acid than $H_2PO_4^-{}_{(aq)}$, which is a much stronger acid than $HPO_4^{2-}{}_{(aq)}$. (Don't forget, however, that phosphoric acid is a *weak* acid.) In other words, for phosphoric acid:

$$K_{a1} > K_{a2} > K_{a3}$$

Finally, remember that the stoichiometry (the mole ratio) for the neutralization reactions of this and other polyprotic acids depends on the number of protons donated to the base. When there is more than one proton that may dissociate, the proton will be donated one proton at a time.

Basic compounds and ions that can accept two or more protons are *polybasic*. For example, the sulfide ion, $S^{2-}{}_{(aq)}$ is dibasic, because it can accept up to two protons. The phosphate ion, $PO_4^{3-}{}_{(aq)}$ is tribasic, because it can accept up to three protons. In the following Sample Problem, you will see how to write stepwise and overall equations to describe the dissociation of a polybasic ion. The Practice Problems that follow provide opportunities for you to apply this skill to several other polybasic and polyprotic species.

### Sample Problem

#### Writing Equations for a Polybasic Ion

**Problem**

The borate ion, $BO_3^{3-}{}_{(aq)}$, is tribasic. Write balanced, stepwise net ionic equations for a solution of sodium borate titrated with hydrochloric acid, and write the overall neutralization reaction for the titration.

**What Is Required?**

You need to write balanced, net ionic equations for each step of the process, and an overall balanced equation for the entire reaction.

**What Is Given?**

You know the word equation for the reaction, as well as the fact that the borate ion (being tribasic) can accept three protons.

**Plan Your Strategy**

To write the stepwise net ionic equations for this reaction, you first take into account that the sodium ion is a spectator, while the hydrochloric acid (a strong acid) is essentially a quantitative source of hydronium ions.

**Step 1** For all bases, remember that the metal ion is a spectator. Therefore, from a solution of sodium borate, only the borate ion is included in the first stepwise equation. The other reactant is a strong monoprotic acid. You know that any strong acid will dissociate completely in water, so you may write it as $H_3O^+$.

**Step 2** Transfer a proton (an "$H^+$") from the acid, $H_3O^+$, to the base, and write the conjugate products. This constitutes the first stepwise equation. Continue to write stepwise equations between the titrant and the other substance until the conjugate product cannot accept any more protons. This usually is the point at which you obtain a neutral product from the ion that you started to titrate.

**Step 3** Add and/or cancel common terms from the stepwise equations.

### Act on Your Strategy

**Step 1 and Step 2** Omitting all spectator ions, the two reactants are $BO_3^{3-}$ and $H_3O^+$. Start writing stepwise equations, stopping when you reach a neutral species based on the borate ion.

**Stepwise Equation 1** $BO_3^{3-}{}_{(aq)} + H_3O^+{}_{(aq)} \rightarrow HBO_3^{2-}{}_{(aq)} + H_2O_{(\ell)}$

**Stepwise Equation 2** $HBO_3^{2-}{}_{(aq)} + H_3O^+{}_{(aq)} \rightarrow H_2BO_3^-{}_{(aq)} + H_2O_{(\ell)}$

**Stepwise Equation 3** $H_2BO_3^-{}_{(aq)} + H_3O^+{}_{(aq)} \rightarrow H_3BO_{3(aq)} + H_2O_{(\ell)}$

**Step 3** To write the overall equation, notice that there are three hydronium ions on the left side of the equation; these add. There are three water molecules on the right side; these also add. All borate-related conjugate bases cancel, leaving only the $BO_3^{3-}$ reactant and the $H_3BO_3$ product:

**Stepwise Equation 1** $BO_3^{3-}{}_{(aq)} + (1)H_3O^+{}_{(aq)} \rightarrow \cancel{HBO_3^{2-}{}_{(aq)}} + (1)H_2O_{(\ell)}$

**Stepwise Equation 2** $\cancel{HBO_3^{2-}{}_{(aq)}} + (1)H_3O^+{}_{(aq)} \rightarrow \cancel{H_2BO_3^-{}_{(aq)}} + (1)H_2O_{(\ell)}$

**Stepwise Equation 3** $\cancel{H_2BO_3^-{}_{(aq)}} + (1)H_3O^+{}_{(aq)} \rightarrow H_3BO_{3(aq)} + (1)H_2O_{(\ell)}$

**Overall Equation** $BO_3^{3-}{}_{(aq)} + 3H_3O^+{}_{(aq)} \rightarrow H_3BO_{3(aq)} + 3H_2O_{(\ell)}$

### Check Your Solution

The product, $H_3BO_{3(aq)}$ is a neutral substance to which no more protons may be transferred. The borate ion is tribasic, and you have a three-step neutralization, providing further confidence in your answer.

## Practice Problems

21. Write balanced stepwise net ionic equations and the overall equation for the reaction between the following substances in aqueous solution:
    (a) sodium thiosulfate and excess hydronium ions
    (b) carbonic acid and excess hydroxide ions

22. Sodium carbonate solution, $Na_2CO_{3(aq)}$, is a common primary standard for acid titrants. Write reactions showing the quantitative transfer of protons from the hydronium ions of a strong acid to the sodium carbonate, and the overall reaction for the titration.

23. Write the stepwise reactions that occur for the proton transfers between sulfurous acid and sodium hydroxide, and write the overall neutralization reaction.

### CHEM FACT

With one exception, all common polyprotic acids are weak. The exception is sulfuric acid. It is the only common strong polyprotic acid, but it is strong only for the first dissociation ($K_{a1}$ is very large). Like the second dissociation of other polyprotic acids, the second dissociation of sulfuric acid is weak ($K_{a2} = 1.0 \times 10^{-2}$).

## Titration Curves for Polyprotic and Polybasic Species

Sulfurous acid is a weak diprotic acid ($K_{a1} = 1.4 \times 10^{-2}$; $K_{a2} = 6.3 \times 10^{-8}$). Figure 15.14 shows the titration curve for sulfurous acid with a strong base. Notice that separate equivalence points appear with the removal of each mole of H$^+$. You can think of the acid as neutralizing in two steps, one for each dissociation. For polyprotic acids with three protons that may dissociate, such as phosphoric acid, the quality of their titration curves deteriorates with each successive equivalence point. Except in extremely dilute solutions, the third dissociation of most triprotic acids is so weak that precise data is difficult to obtain.)

The general shape of a titration curve for a polybasic substance is shown in Figure 15.15. As you might expect, the quality of the titration curves for polybasic species also deteriorates with each successive proton added. Thus, the third proton addition for a tribasic titration curve is rarely evident unless the titrated solution is very dilute.

**Web LINK**

www.mcgrawhill.ca/links/atlchemistry

Atlantic Canada's fogs blanket some coastal areas of the region for over 2000 hours per year. Studies on fog chemistry show that these fogs can be a major source of acidification, having concentrations of nitrates and sulfates greater than those measured in normal precipitation. In the early 1990s, the median pH of fog at Cape Forchu, Nova Scotia, and Cape Race, Newfoundland and Labrador, was 3.86 and 3.71 respectively. Go to the web site above and follow the links to the Environment Canada web site. What are the main sources of acid precipitation in Atlantic Canada? Has the problem increased or decreased in recent years? Why?

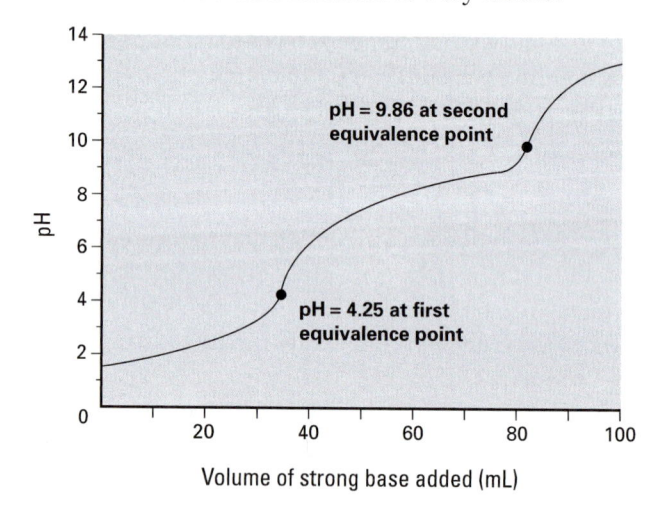

**Figure 15.14** A titration curve obtained by titrating 40.00 mL of 0.1000 mol/L H$_2$SO$_3$ with 0.1000 mol/L NaOH. This titration curve has characteristics that are typical of titration curves for other polyprotic acids. Notice two equivalence points—one for each H$^+$ dissociation.

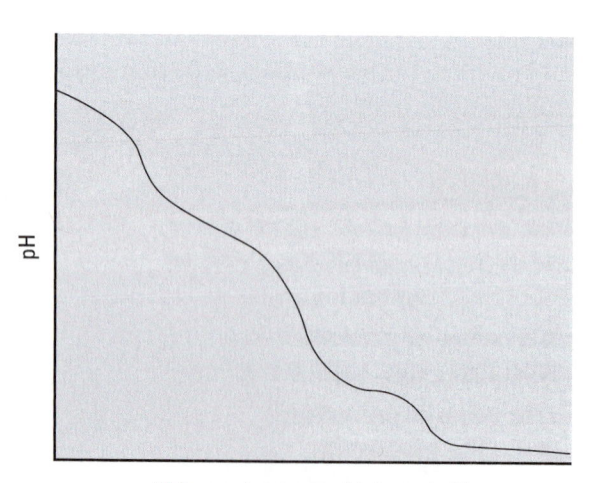

**Figure 15.15** How does the titration curve for a polybasic species compare with the titration curve for a monobasic substance such as ammonia, shown in Figure 15.13?

# Investigation 15-B

**SKILL FOCUS**
Predicting
Performing and recording
Analyzing and interpreting

## $K_a$ of Acetic Acid

In Investigation 15-A, you learned how to perform a titration. You determined the molar concentration of an acid by adding a basic solution of known concentration and measuring the volume of the basic solution required to reach the end point.

In this investigation, you will be given a sample of acetic acid with an unknown concentration. Instead of measuring the volume of the basic solution required to reach the end point, however, you will measure the pH. Then you will graph the data you collected and use the graph to calculate the molar concentration of the acetic acid and its $K_a$.

### Question

In a solution of acetic acid, how does the concentration of hydronium ions compare with the concentration of acetic acid?

### Materials

25 mL pipette and pipette bulb
retort stand
burette and burette clamp
2 beakers (150 mL)
Erlenmeyer flask (150 mL)
labels
meniscus reader
sheet of white paper
funnel
acetic acid, $CH_3COOH$, solution
sodium hydroxide, NaOH, solution
dropper bottle containing phenolphthalein
pH meter or pH paper

### Safety Precautions

Both $CH_3COOH$ and NaOH are corrosive. Wash any spills on your skin or clothing with plenty of cool water. Inform your teacher immediately.

### Procedure

1. Your teacher will give you the molar concentration of the NaOH solution. Record this concentration in your notebook, as well as the volume of the pipette (in mL).

2. Copy the table below into your notebook, to record your observations. Leave plenty of space. You will collect 15 to 30 sets of data, depending on the concentration of the acid.

| Volume of NaOH added (mL) | pH |
|---|---|
| 0.00 | |
| | |
| | |

3. Label a clean, dry beaker for each solution. Obtain about 40 mL of acetic acid and about 70 mL of NaOH solution.

4. Rinse a clean burette twice with about 10 mL of NaOH solution. On the second rinse, drain all NaOH except for a very small amount just above the tap. Then set up a retort stand, burette clamp, meniscus reader, and funnel. Fill the burette with NaOH solution. Make sure that the solution fills the tube below the tap with no air bubbles. Remove the funnel.

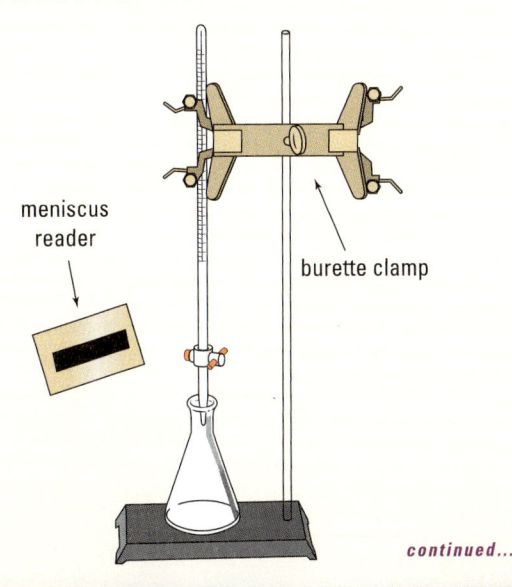

*continued...*

Chapter 15 Acid-Base Equilibria and Reactions • MHR 613

continued

5. Obtain a clean 25 mL pipette and a suction bulb. Rinse the pipette twice with 5 mL to 10 mL of $CH_3COOH$. It is not necessary to rinse the pipette. Pipette 25.00 mL of $CH_3COOH$ into an Erlenmeyer flask. Add two or three drops of phenolphthalein indicator.

6. Record the initial pH of the solution. Make sure that the glass electrode is immersed deeply enough to get an accurate reading. If necessary, tip the flask to one side. Place a sheet of white paper under the flask.

7. Add 2 mL of NaOH from the burette. Record the volume carefully, correct to two decimal places. Swirl the contents of the Erlenmeyer flask, then measure the pH of the solution.

8. Repeat step 7 until the pH rises above 2.0. Add 1 mL amounts until the pH reaches 5.0.

9. Above pH = 5.0, add NaOH in 0.2 mL or 0.1 mL portions. Continue to swirl the contents of the flask and take pH readings. Record the volume at which the phenolphthalein changes from colourless to pink.

10. Above pH = 11, add 1 mL portions until the pH reaches at least 12.

11. Wash the liquids down the sink with plenty of water. Rinse the pipette and burette with distilled water. Leave the burette tap open.

## Analysis

1. Write the chemical equation for the neutralization reaction you observed.

2. Plot a graph of your data, with pH on the vertical axis and volume of NaOH on the horizontal axis. Your graph should show a steep rise in pH as the volume of NaOH becomes enough to neutralize all the $CH_3COOH$. Take the midpoint on the graph (where the graph rises steeply) and read off the volume of NaOH. This is the volume of NaOH that was needed to neutralize all the $CH_3COOH$. Compare the volume on your graph with the volume you recorded when the phenolphthalein indicator first turned pink.

3. Calculate the molar concentration of the $CH_3COOH$. Use the ratio in which the acid and base react, determined from the chemical equation. You can use the following equation to find the amount of a chemical in solution.

   Amount (in mol) = Concentration (in mol/L) × Volume (in L)

   Determine the amount of NaOH added, using its concentration (given by your teacher) and the volume on your graph (from question 2).

4. Write the expression for $K_a$ for the dissociation of $CH_3COOH$ in water.

5. Use the initial pH of the $CH_3COOH$ (before you added any base) to find the initial $[H_3O^+]$. What was the initial $[CH_3COO^-]$?

6. Assume that the amount of $CH_3COOH$ that dissociates is small compared with the initial concentration of the acid. If this is true, the equilibrium value of $[CH_3COOH]$ is equal to the initial concentration of the acid. Use your values of $[H_3O^+]$, $[CH_3COO^-]$, and $[CH_3COOH]$ to calculate $K_a$ for acetic acid. **Hint:** $[H_3O^+] = [CH_3COO^-]$

7. Refer to the volume of NaOH on your graph (from question 2). Calculate half this volume. On your graph, find the pH when the solution was half-neutralized.

8. Calculate $[H_3O^+]$ when the $CH_3COOH$ was half-neutralized. How does this value compare with your value of $K_a$ for $CH_3COOH$?

## Conclusion

9. Calculate the percent difference between your value for $K_a$ and the accepted value. State two sources of error that might account for any differences.

## Application

10. Do the values you calculated for $[H_3O^+]$ and $[CH_3COOH]$ prove that $CH_3COOH$ is a weak acid? Explain.

# Careers in Chemistry

## Dangerous Goods Inspection

**Joanne Seviour**

You get a phone call in the night. A tanker truck has overturned and spilled its chemical cargo on a busy highway. Your job is to find out exactly what the chemical is and how to deal with it so that it does not injure people or harm the environment.

Dangerous materials are transported throughout the Canada every day—by road, rail, water and even air. Although accidents do happen, spills and leaks are rare, thanks to strict federal codes regulating how materials must be packaged and moved. One of the people responsible for developing these codes, monitoring the movements of dangerous goods, and supervising the inspectors who go out on calls to spills is Joanne Seviour. Joanne works for Transport Canada as the Regional Manager of Transportation of Dangerous Goods for the Atlantic Region. Before that, she was employed for two years as a dangerous goods inspector herself.

"I did a lot of travelling around Nova Scotia as an inspector," says Joanne. "Realistically, I was in more danger of being involved in a car accident than getting burned, blown up, or poisoned," she laughs. Inspectors are well trained, equipped, and prepared to deal with hazardous materials. Joanne studied chemistry at Memorial University of Newfoundland and worked for six years as an environmental officer for the Department of National Defence in the Atlantic Region before joining the Dangerous Goods Division of Transport Canada.

What made Joanne interested in chemistry? "I recall when I first saw the chemical equation for photosynthesis in grade four," she says. "I was amazed that you could abbreviate such complex words with symbols. I was not very good at spelling at the time and I thought: Hey! I like these symbols. On a more serious note, chemistry is the study of the fundamental buildings blocks of our world and universe. If you can understand chemistry you can understand how many, many things work."

A recent project managed by Joanne was to compile statistics on the movement of dangerous goods in Atlantic Canada. Student researchers hired during the summer visited companies throughout the region to collect information on what dangerous goods were regularly transported, in what quantities, and by what routes. This was combined with records of accidents, and the resulting database is being used to plan inspections and develop a risk assessment strategy, helping focus inspections on commodities which pose the greatest risk to the public.

Dangerous goods are classified in such categories as explosive, flammable (liquids, gases, and solids), corrosive, radioactive, and poisonous. For each type of hazard, a specific remedy is needed. For example, you must not add water to an acid to dilute it as that may cause splashing and bubbling and spread the acid further. You need to know not only which neutralizing agent to add but also in what quantity (which varies depending on the strength of the acid.)

A major responsibility for staff in Joanne's department is to help enforce laws designed for public safety. Inspectors have the authority to enter businesses to inspect dangerous goods and ensure they are being packaged and handled as required by government regulations. An inspector who finds violations can lay charges and will help prepare the legal case for prosecution. This part of the job requires not only a knowledge of chemistry but also skill and tact in dealing with people.

Joanne recently completed a master's degree in environmental studies at Dalhousie University. Her knowledge and skills in the field of dangerous materials not only helps prevent injury and death but also helps reduce the risk of environmental pollution.

## Section Summary

In this chapter, as well as in all of Unit 6, you have learned about equilibria involving acids and bases of different strengths. You can have used a variety of constants to help you determine concentrations, pH, and pOH, and you have learned an important analytical technique that chemists use to determine the amounts of substances as they react. In the next unit, you will revisit aspects of your earlier study of kinetics and equilibrium to study the energy changes that occur when substances react.

## Section Review

**1** Write a generalized word equation to describe what happens during a neutralization reaction.

**2** Write a chemical equation for each neutralization reaction.
- (a) KOH with $HNO_3$
- (b) HBr with $Ca(OH)_2$
- (c) $H_3PO_4$ with NaOH
- (d) $Mg(OH)_2$ with HCl

**3** (a) Distinguish between the equivalence point and the end point for a titration.

(b) When choosing an indicator, do the pH values of the two points need to coincide exactly? Explain.

**4** Sketch the pH curve for the titration of a weak acid with a strong base. Show the equivalence point on your sketch. Suggest an indicator that might be used, and explain your selection.

**5** Suggest an indicator that could be used for the titration of potassium hydroxide with nitrous acid. Explain your suggestion.

**6** A 25.0 mL sample of sulfuric acid is completely neutralized by adding 32.8 mL of 0.116 mol/L ammonia solution. Ammonium sulfate, $(NH_4)_2SO_4$, and water are formed. What is the concentration of the sulfuric acid?

**7** The following data were collected during a titration. Calculate the concentration of the sodium hydroxide solution.

**Titration Data**

| | |
|---|---|
| Volume of $HCl_{(aq)}$ | 10.00 mL |
| Final volume of $NaOH_{(aq)}$ | 23.08 mL |
| Initial volume of $NaOH_{(aq)}$ | 1.06 mL |
| Concentration of $HCl_{(aq)}$ | 0.235 mol/L |

**8** Estimate the pH of a solution in which bromocresol green is blue, and methyl red is orange.

**9** Classify these substances in aqueous solution as monoprotic, diprotic, triprotic, monobasic, dibasic, or tribasic.
- (a) $H_2S$
- (b) $HNO_3$
- (c) $CrO_4^{2-}$
- (d) $H_3P$
- (e) $CH_3COOH$
- (f) HOOCCOOH

**10** Sketch the curve for a titration of a polyprotic substance titrated with a strong base. How does it compare with the curve for a polybasic substance titrated with a strong acid?

# CHAPTER 15 Review

## Reflecting on Chapter 15
Summarize this chapter in the format of your choice. Here are a few ideas to use as guidelines.
- Differentiate strong and weak acids and bases on the basis of their dissociation.
- Describe how to conduct a titration.
- Explain the use of $K_a$ and $K_b$ and percent dissociation in describing the strength of acids and bases.
- Calculate pH from the initial concentration of a weak acid or a weak base and $K_a$ or $K_b$.
- Describe buffer solutions in your daily life and explain how they function.
- Sketch and interpret acid-base titration curves.

## Reviewing Key Terms
For each of the following terms, write a sentence that shows your understanding of its meaning.

monoprotic acids
acid dissociation constant ($K_a$)
percent dissociation
base dissociation constant ($K_b$)
buffer solution
buffer capacity
neutralization reaction
salt
acid-base indicator
titration
equivalence point
primary standard
end point
acid-base titration curve

## Knowledge/Understanding
1. (a) Use Appendix E to find the values of $K_a$ for hydrogen sulfate ion, $HSO_4^-{}_{(aq)}$, and hydrogen sulfite ion, $HSO_3^-{}_{(aq)}$.
   (b) Write equations for the base dissociation constants of $HSO_4^-{}_{(aq)}$ and $HSO_3^-{}_{(aq)}$.
   (c) Calculate the value of $K_b$ for each ion.
   (d) Which is the stronger base, $HSO_4^-{}_{(aq)}$ or $HSO_3^-{}_{(aq)}$? Explain.

2. While the pH of blood must be maintained within strict limits, the pH of urine can vary. The sulfur in foods, such as eggs, is oxidized in the body and excreted in the urine. Does the presence of sulfide ions in urine tend to increase or decrease the pH? Explain.

3. Sodium methanoate, NaHCOO, and methanoic acid, HCOOH, can be used to make a buffer solution. Explain how this combination resists changes in pH when small amounts of acid or base are added.

4. Oxoacids contain an atom that is bonded to one or more oxygen atoms. One or more of these oxygen atoms may also be bonded to hydrogen. Consider the following oxoacids: $HBrO_{3(aq)}$, $HClO_{3(aq)}$, $HClO_{4(aq)}$, and $H_2SO_{4(aq)}$.
   (a) What factors are used to predict the strengths of oxoacids?
   (b) Arrange the oxoacids above in the order of increasing acid strength.

5. In a titration, a basic solution is added to an acidic solution, and measurements of pH are taken. Compare a strong acid-strong base titration and a strong acid-weak base titration in terms of
   (a) the initial pH
   (b) the quantity of $OH^-{}_{(aq)}$ that is needed to reach the equivalence point
   (c) the pH at the equivalence point
   Assume that the concentrations of the two solutions are identical.

## Inquiry
6. Suppose that 15.0 mL of sulfuric acid just neutralized 18.0 mL of 0.500 mol/L sodium hydroxide solution. What is the concentration of the sulfuric acid?

7. A student dissolved 5.0 g of vitamin C in 250 mL of water. The molar mass of ascorbic acid is 176 g/mol, and its $K_a$ is $8.0 \times 10^{-5}$. Calculate the pH of the solution. **Note:** Abbreviate the formula of ascorbic acid to $H_{Asc}$.

8. Benzoic acid is a weak, monoprotic acid ($K_a = 6.3 \times 10^{-5}$). Its structure is shown below. Calculate the pH and the percent dissociation of each of the following solutions of benzoic acid. Then use Le Châtelier's principle to explain the trend in percent dissociation of the acid as the solution becomes more dilute.
   (a) 1.0 mol/L   (b) 0.10 mol/L   (c) 0.01 mol/L

9. Hypochlorous acid, HOCl, is a weak acid that is found in household bleach. It is made by dissolving chlorine gas in water.

Answers to questions highlighted in red type are provided in Appendix A.

$Cl_{2(g)} + 2H_2O_{(\ell)} \rightleftharpoons H_3O^+_{(aq)} + Cl^-_{(aq)} + HOCl_{(aq)}$

(a) Calculate the pH and the percent dissociation of a 0.065 mol/L solution of hypochlorous acid.

(b) What is the conjugate base of hypochlorous acid? What is its value for $K_b$?

10. Calculate the pH of a 1.0 mol/L aqueous solution of sodium benzoate. **Note:** Only the benzoate ion affects the pH of the solution.

11. Calculate the pH of a 0.10 mol/L aqueous solution of sodium nitrite, $NaNO_2$. **Note:** Only the nitrite ion affects the pH of the solution.

12. A student prepared a saturated solution of salicylic acid and measured the pH of the solution. The student then carefully evaporated 100 mL of the solution and collected the solid. If the pH of the solution was 2.43, and 0.22 g was collected after evaporating 100 mL of the solution, what is the acid dissociation constant for salicylic acid?

(Structure: benzene ring with COOH and OH groups)

## Communication

13. List the oxoacids of bromine (HOBr, $HBrO_2$, $HBrO_3$, and $HBrO_4$) in order of increasing strength. What is the order of increasing strength for the conjugate bases of these acids?

14. Consider the following acid-base reactions.
$HBrO_{2(aq)} + CH_3COO^-_{(aq)} \rightleftharpoons CH_3COOH_{(aq)} + BrO_2^-_{(aq)}$
$H_2S_{(aq)} + OH^-_{(aq)} \rightleftharpoons HS^-_{(aq)} + H_2O_{(\ell)}$
$HS^-_{(aq)} + CH_3COOH_{(aq)} \rightleftharpoons H_2S_{(aq)} + CH_3COO^-_{(aq)}$

If each equilibrium lies to the right, arrange the following compounds in order of increasing acid strength: $HBrO_2$, $CH_3COOH$, $H_2S$, $H_2O$.

15. Discuss the factors that can be used to predict the relative strength of different oxoacids.

16. The formula of methyl red indicator can be abbreviated to HMr. Like most indicators, methyl red is a weak acid.
$HMr_{(aq)} + H_2O_{(\ell)} \rightleftharpoons H_3O^+_{(aq)} + Mr^-_{(aq)}$
The change between colours (when the indicator colour is orange) occurs at a pH of 5.4. What is the equilibrium constant for the reaction?

17. Gallic acid is the common name for 3,4,5-trihydroxybenzoic acid.
(a) Draw the structure of gallic acid.
(b) $K_a$ for gallic acid is $3.9 \times 10^{-5}$. Calculate $K_b$ for the conjugate base of gallic acid. Then write the formula of the ion.

18. (a) Sketch the pH curves you would expect if you titrated
- a strong monoprotic acid with a strong base
- a strong monoprotic acid with a weak base
- a weak monoprotic acid with a strong base
- a weak polyprotic acid with a strong base
- a weak polybasic species with a strong acid

(b) Congo red changes colour over a pH range of 3.0 to 5.0. For which of the above titrations would Congo red be a good indicator to use?

19. In a titration experiment, 25.0 mL of an aqueous solution of sodium hydroxide was required to neutralize 50.0 mL of 0.010 mol/L hydrochloric acid. What is the molar concentration of the sodium hydroxide solution?

20. A burette delivers 20 drops of solution per 1.0 mL. What amount (mol) of $H^+_{(aq)}$ is present in one drop of a 0.20 mol/L HCl solution?

21. When hydrochloric acid is titrated with magnesium silicate solution, pH measurements indicate that two proton transfers occur. Write reactions for each proton transfer, as well as the overall reaction for the titration. (**Hint:** Remember that hydrochloric acid is a strong acid.)

22. If a 10.00 mL volume of magnesium silicate is pipetted in the titration described in question 21, and if 23.42 mL of 0.1000 mol/L hydrochloric acid is needed to complete the titration, calculate the concentration of the magnesium silicate solution. (**Hint:** Which of the three reactions from your answer to question 21 will you use to solve this problem?)

## Making Connections

23. Commercial processors of potatoes remove the skin by using a 10–20% by mass solution of sodium hydroxide. The potatoes are soaked in the solution for a few minutes at 60–70°C, after

which the peel can be sprayed off using fresh water. You work in the laboratory at a large food processor and must analyse a batch of sodium hydroxide solution. You pipette 25.00 mL of NaOH$_{(aq)}$, and find it has a mass of 25.75 g. Then you titrate the basic solution against 1.986 mol/L HCl, and find it requires 30.21 mL of acid to reach an end point.

(a) Inform your supervisor what the molar concentration of the sodium hydroxide is.

(b) The mass percent of NaOH present must be a minimum of 10% for the solution to be used. Advise your supervisor whether or not the solution can be used to process more potatoes, and explain your reasoning.

24. On several occasions during the past few years, you have studied the environmental issue of acid rain. Now that you have further developed your understanding of acids and bases in this chapter, reflect on your earlier understandings.

(a) List two facts about acid rain that you now understand in a more comprehensive way. Explain what is different between your previous and your current understanding in each case.

(b) Identify three questions that your teacher could assign as a research project on acid rain. The emphasis of the research must be on how an understanding of chemistry can contribute clarifying the questions and possible solutions involved in this issue. Develop a rubric that would be used to assess any student who is assigned this research project.

25. Research the use of hypochlorous acid in the management of swimming pools and write a report on your findings. Include a discussion on the importance of controlling pool water.

**Answers to Practice Problems and Short Answers to Section Review Questions**

**Practice Problems:** 1.(a) 4.5 mol/L (b) 1.35 mol/L (c) 0.0171 mol/L (d) 0.0375 mol/L 2.(a) 3.1 mol/L (b) 0.87 mol/L (c) 0.701 mol/L (d) 0.697 mol/L 3.(a) acidic solution; [H$_3$O$^+$] = 0.479 mol/L; pH= 0.320 (b) acidic solution; [H$_3$O$^+$] = 1.98 mol/L; pH= 0.297 4. acidic solution; [H$_3$O$^+$] = 0.46 mol/L; pH= 0.34 5. pH = 2.41; 0.47% dissociation 6. $K_a$ = 1.0 × 10$^{-4}$; percent dissociation = 3.2% 7. pH = 2.65 8. $K_a$ = 2.9 × 10$^{-8}$ 9. pH = 3.411 10. 1.2 g 11. pH = 11.14 12. pH = 10.50 13. $K_b$ = 1.6 × 10$^{-6}$ 14. [OH$^-$] = 1.0 × 10$^{-2}$ mol/L; pOH = 1.98 15. [OH$^-$] = 3.8 × 10$^{-3}$ mol/L; percent dissociation = 1.7% 16. [NH$_3$] = 2.8 × 10$^{-2}$ mol/L 17. 0.210 mol/L 18. 31.5 mL 19.(a) 22.5 mL (b) 24.7 mL (c) 4.8 mL 20.(a) 0.0992 mol/L (b) 0.269 mol/L (c) 0.552 mol/L 21.(a) Na$_2$S$_2$O$_3$ is dibasic, dissociates completely, and Na$^+$ ions are spectators, so do not appear in the net or overall equations. First stepwise reaction, S$_2$O$_3^{2-}$$_{(aq)}$ gains a proton to become HS$_2$O$_3^-$. Second stepwise reaction, HS$_2$O$_3^-$ gains a proton to become H$_2$S$_2$O$_3$. Overall: S$_2$O$_3^{2-}$$_{(aq)}$ + 2H$_3$O$^+$$_{(aq)}$ → H$_2$S$_2$O$_3$ + 2H$_2$O$_{(\ell)}$ (b) Carbonic acid is diprotic, losing one hydrogen in the first stepwise reaction to become HCO$_3^-$, and losing its second hydrogen in the second stepwise reaction to become CO$_3^{2-}$. Overall: H$_2$CO$_{3(aq)}$ + 2OH$^-$$_{(aq)}$ → CO$_3^{2-}$$_{(aq)}$ + 2H$_2$O$_{(\ell)}$ 22. Again, Na$^+$ ions are spectators. Thus, CO$_3^{2-}$ gains a proton to become HCO$_3^-$, and this ion gains a proton to become H$_2$CO$_3$. Overall: CO$_3^{2-}$$_{(aq)}$ + 2H$_3$O$^+$$_{(aq)}$ → H$_2$CO$_3$ + 2H$_2$O$_{(\ell)}$ 23. H$_2$SO$_{3(aq)}$ is diprotic, and sodium ions are spectators. First stepwise reaction, H$_2$SO$_3$ loses one proton to the hydroxide ion to become HSO$_3^-$. Second stepwise reaction, HSO$_3^-$ loses its proton to become SO$_3^{2-}$. Overall: H$_2$SO$_{3(aq)}$ + 2OH$^-$$_{(aq)}$ → SO$_3^{2-}$$_{(aq)}$ + 2H$_2$O$_{(\ell)}$

**Section Review: 15.1:** 2. pH = 2.39 3. $K_a$ = 2.5 × 10$^{-4}$ 4. [OH$^-$] = 7.6 × 10$^{-3}$ mol/L; pH = 11.88 5. pH = 8.79 6. OBr$^-$; $K_b$ = 3.6 × 10$^{-6}$ **15.2:** 1. acid plus base yields salt plus water 2.(a) KOH$_{(aq)}$ + HNO$_{3(aq)}$ → KNO$_{3(aq)}$ + H$_2$O$_{(\ell)}$ (b) 2HBr$_{(aq)}$ + Ca(OH$_2$)$_{(aq)}$ → CaBr$_{2(aq)}$ + 2H$_2$O$_{(\ell)}$ (c) H$_3$PO$_{4(aq)}$ + 3NaOH$_{(aq)}$ → Na$_3$PO$_{4(aq)}$ + 3H$_2$O$_{(\ell)}$ (d) 2HCl$_{(aq)}$ + MgO$_{\ell(aq)}$ → Mg$_{\ell(aq)}$ + 2H$_2$O$_{(\ell)}$ 5. phenolphthalein 6. 7.6 × 10$^{-2}$ mol/L 7. 0.107 mol/L 8. ~5.0 9.(a) diprotic (b) monoprotic (c) dibasic (d) tribasic (e) monoprotic (f) diprotic

# UNIT 6 Review

## Knowledge/Understanding

**Multiple Choice**

In your notebook, write the letter for the best answer to each question.

1. A sample of lemon juice has a pH of 2. A sample of an ammonia cleaner has a pH of 10. If the two samples are combined, what ratio by volume of lemon juice to ammonia cleaner is needed to yield a neutral solution?
   (a) 1:10
   (b) 1:8
   (c) 1:100
   (d) 1:1000
   (e) 10:1

2. A solution of a weak acid is titrated with a solution of a strong base. What is the pH at the end point?
   (a) 7
   (b) >7
   (c) <7
   (d) Either (b) or (c) may be correct, depending on the $K_a$ of the acid.
   (e) More information about the number of hydrogen ions is needed.

3. In which reaction is water acting as an acid?
   (a) $H_2O_{(\ell)} + NH_{3(aq)} \rightleftharpoons OH^-_{(aq)} + NH_4^+_{(aq)}$
   (b) $H_2O_{(\ell)} + H_3PO_{4(aq)} \rightleftharpoons H_3O^+_{(aq)} + H_2PO_4^-_{(aq)}$
   (c) $H_2O_{(\ell)} \rightleftharpoons H_{2(g)} + \frac{1}{2}O_{2(g)}$
   (d) $2H_2O_{(\ell)} + BaCl_{2(s)} \rightleftharpoons BaCl_2 + 2H_2O_{(g)}$
   (e) $2Na_2O_{2(s)} + 2H_2O_{(\ell)} \rightleftharpoons 4NaOH_{(aq)} + O_{2(g)}$

4. Which of the following statements is appropriate as an operational definition for an acid?
   (a) Acids are electrolytes.
   (b) Acids taste sour.
   (c) Brønsted-Lowry acids have one or more protons that may be donated to a base.
   (d) statements (a) and (b)
   (e) statements (b) and (c)

5. If 4.0 mL of 6.00 mol/L sulfuric acid is diluted to 120 mL by the addition of water, what is the molar concentration of the sulfuric acid after dilution?
   (a) $5.00 \times 10^{-2}$ mol/L
   (b) $7.50 \times 10^{-2}$ mol/L
   (c) 1.00 mol/L
   (d) $2.0 \times 10^{-1}$ mol/L
   (e) 4.0 mol/L

6. 10 mL of 1.0 mol/L HCl is diluted by adding 990 mL of water. What is the pH of the new solution?
   (a) decrease by 2 pH units?
   (b) decrease by 0.5 pH units?
   (c) increase by 0.5 pH units?
   (d) increase by 2 pH units?
   (e) increase by a factor of 2 pH units?

7. Select the correct statement about the following equilibrium.
   $HBO_3^{2-}_{(aq)} + HSiO_3^-_{(aq)} \rightleftharpoons SiO_3^{2-}_{(aq)} + H_2BO_3^-_{(aq)}$
   (a) $HBO_3^{2-}_{(aq)} + HSiO_3^-_{(aq)}$ are a conjugate acid-base pair.
   (b) $HSiO_3^-_{(aq)}$ and $SiO_3^{2-}_{(aq)}$ are both acting as acids.
   (c) $HBO_3^{2-}_{(aq)}$ and $SiO_3^{2-}_{(aq)}$ are both bases.
   (d) $HSiO_3^-_{(aq)}$ and $H_2BO_3^-_{(aq)}$ are a conjugate acid-base pair.
   (e) $SiO_3^{2-}_{(aq)}$ and $H_2BO_3^-_{(aq)}$ are a conjugate acid-base pair.

8. Which of the following statements is true?
   (a) All Brønsted-Lowry acids are Arrhenius acids.
   (b) All Brønsted-Lowry bases are Arrhenius bases.
   (c) All Arrhenius bases are Brønsted-Lowry bases.
   (d) All Arrhenius acids are Brønsted-Lowry acids.
   (e) All Brønsted-Lowry acids are modernized Arrhenius acids.

9. The following represent acid-base reactions. In which reaction must the conjugate acid-base pair be incorrect?
   (a) $HF_{(aq)} + H_2O_{(\ell)} \rightleftharpoons F^-_{(aq)} + H_3O^+_{(aq)}$
   (b) $HCOOH_{(aq)} + CN^-_{(aq)} \rightleftharpoons HCOO^-_{(aq)} + HCN$
   (c) $NH_4^+_{(aq)} + CO_3^{2-}_{(aq)} \rightleftharpoons NH_{3(aq)} + HCO_{3(aq)}$
   (d) $H_2PO_4^-_{(aq)} + OH^-_{(aq)} \rightleftharpoons HPO_4^{2-}_{(aq)} + H_2O_{(\ell)}$
   (e) $HPO_4^{2-}_{(aq)} + SO_3^{2-}_{(aq)} \rightleftharpoons PO_4^{3-}_{(aq)} + HSO_3^-_{(aq)}$

10. The following statements are related to question 9. Which statement is false?
    (a) Each reaction has an acid and a base as reactants and an acid and a base as products.
    (b) Arrhenius theory cannot explain all these acid-base reactions.
    (c) The reactions shown in question 9 (a), (d), and (e) include amphoteric species.
    (d) Acids and bases can be neutral substances.
    (e) The modernized Arrhenius theory can explain all these acid-base reactions.

Answers to questions highlighted in red type are provided in Appendix A.

## Short Answer

In your notebook, write a sentence or a short paragraph to answer each question.

11. Solutions of $NaHCO_{3(aq)}$ and $CaHPO_{4(aq)}$ are mixed.
    (a) Write the net ionic equation that represents the equilibrium for this mixture.
    (b) What are the conjugate acid-base pairs?
    (c) Are reactants or products favoured in this reaction? Give a reason for your answer.

12. What is the pH of a 100 mL sample of 0.002 mol/L $H_2SO_4$?

13. What is $[H^+_{(aq)}]$, $[OH^-_{(aq)}]$, pH, and pOH for a 0.048 mol/L solution of benzoic acid, $C_6H_5COOH$?

14. If you did not know that acids and bases are electrolytes, could you reasonably infer this fact? Why or why not?

15. (a) Name one Arrhenius acid and Arrhenius base.
    (b) Name one Brønsted-Lowry acid and Brønsted-Lowry base that cannot be an Arrhenius acid or base. Explain why they cannot be.

16. Agree or disagree with the following statement, and justify your opinion: a solution with a pH of 4 has a greater concentration of hydronium ions than a solution with a pH of 3.

17. State the pH of pure water, and explain, in terms of ion concentration, why this is the case.

18. (a) Acids and bases may be described as strong and weak. Explain what this means.
    (b) Acids and bases may be described as dilute and concentrated. Explain how these terms are different from those used in question (a).

19. Write the expression for $K_a$ for the following.
    (a) HClO
    (b) $H_2S$

20. Sort the following in order of decreasing acid strength: $H_3PO_4$, HI, $CH_3COOH$, HF.

21. Sort the conjugate bases for the acids in question 20 in order of decreasing strength.

22. Which of these solutions has a higher pH, and explain why.
    (a) 0.1 mol/L solution of a weak acid; 0.01 mol/L solution of the same acid
    (b) 0.1 mol/L solution of an acid; 0.1 mol/L solution of a base
    (c) a solution with a pOH of 7.0; a solution with a pOH of 8.0
    (d) 0.1 mol/L solution of an acid with $K_a = 1 \times 10^{-4}$; 0.1 mol/L solution of an acid with $K_a = 4 \times 10^{-5}$

23. (a) Determine the pH of $5.04 \times 10^{-3}$ mol/L HI.
    (b) Is the solution in (a) acidic, basic, or neutral?

24. Explain what the term amphoteric means. Use an example (including balanced chemical equations) to illustrate your explanation.

## Inquiry

25. Cyanide ion, $CN^-$, reacts with $Fe^{3+}$ to form the blue dye that is used in blueprint paper. Hydrocyanic acid, $HCN_{(aq)}$, is a weak acid, with $K_a = 6.2 \times 10^{-10}$.
    (a) Calculate $K_b$ for the conjugate base of $HCN_{(aq)}$.
    (b) Calculate the pH of 0.120 mol/L $HCN_{(aq)}$.

26. The following solutions have been prepared at a concentration of 0.1 mol/L: $NH_{3(aq)}$, $NaOH_{(aq)}$, $CH_3COONa_{(aq)}$, $CH_3COOH_{(aq)}$. Arrange these solutions in order of decreasing pH. For each solution, write the balanced equation that supports your answer.

27. A 10.0 g sample of a commercial washing powder contains ammonium sulfate, $(NH_4)_2SO_4$. The sample is treated with an excess of NaOH.
    $(NH_4)_2SO_{4(aq)} + 2NaOH_{(aq)} \rightarrow$
    $2NH_{3(g)} + Na_2SO_{4(aq)} + H_2O_{(\ell)}$
    The ammonia gas that forms is dissolved in 50.0 mL of 0.250 mol/L $H_2SO_4$. The following reaction occurs.
    $2NH_{3(g)} + H_2SO_{4(aq)} \rightarrow (NH_4)_2SO_{4(aq)}$
    The $H_2SO_{4(aq)}$ that is not used up in this reaction is titrated to the end point with 27.9 mL of 0.230 mol/L NaOH. Use this information to calculate the percent of $(NH_4)_2SO_4$ in the 10.0 g sample of washing powder.

28. Use sketches to compare titrations curves for the following:
    (a) a strong acid with a strong base
    (b) a weak acid with a strong base
    (c) a strong base with a weak acid

29. How does a titration curve for a polyprotic acid compare with the titration curves you drew in question 28?

30. Will the concentration of acid, before and after dissociation, be about the same or different in each of the following? Give reasons in each case.
    (a) a concentrated solution of a strong acid
    (b) a dilute solution of a weak acid
    (c) a concentrated solution of a weak acid
    (d) a dilute solution of a weak acid

31. A weak acid with a molar concentration of 0.25 mol/L is 3.0% dissociated. Calculate $[H_3O^+]$, $[OH^-]$, pH, pOH, and $K_a$ for this acid.

32. Use the chart of indicators in Appendix E to answer the following questions.
    (a) If a vinegar solution has a pH of 5, what colour would you expect the following indicators to show if placed into separate samples of the vinegar?
      - thymol blue
      - bromophenol blue
      - phenolphthalein
    (b) An aqueous solution of sodium acetate, used in photographic development, makes phenol red indicator red and phenolphthalein pink. What is the pH of this sodium acetate solution?

33. Calculate the percent dissociation of 0.25 mol/L benzoic acid ($C_6H_5COOH$). ($K_a = 6.3 \times 10^{-5}$)

34. A solution with a volume of 655 mL forms when a 0.250 mol sample of acid HX is dissolved in a certain amount of water. If the pH of the solution is 3.44, what is the $K_a$ of HX?

35. HZ is a weak acid with a $K_a$ of $1.55 \times 10^{-4}$. What is the pH of 0.075 mol/L HZ? What is the pOH of 0.045 mol/L HZ?

## Communication

36. You have a sample of acetic acid with a concentration of 0.1 mol/L and a sample of hydrochloric acid with a concentration of 0.0001 mol/L. Both samples have nearly the same concentration of hydronium ions. Are these two particular samples equally strong? Justify your answer.

37. Use equations to demonstrate how a buffer system, such as $HNO_2 : NO_2^-$ reacts with $H_3O^+$ and $OH^-$.

38. (a) Explain how you would select an indicator to use in a titration involving a strong acid with a strong base.
    (b) Would your selection of an indicator be different if you were titrating a weak base with a strong base? Explain why or why not.

39. Explain the difference between the end point and the equivalence point of a titration. Do you always reach the equivalence point first? Explain.

40. Phenolphthalein is colourless in an acidic solution and red in a basic solution. Two drops of phenolphthalein are added to 100 mL of an acid solution that is colourless. Drops of basic solution are added slowly until the acid solution is neutralized. Describe the change in colour you would expect to observe during this process.

41. What is the concentration of a solution of sulfuric acid when 25.00 mL of solution is neutralized by 38.93 mL of $4.500 \times 10^{-3}$ mol/L sodium hydroxide solution?

42. What is the concentration of each ion in the solutions below?
    (a) 4.12 mol/L $NH_4Cl$
    (b) 0.275 mol/L $Ba(OH)_2$
    (c) 0.543 mol/L $(NH_4)_3PO_4$
    (d) 0.704 mol/L $KClO_4$

43. The equilibrium constant for the following reaction is $2.8 \times 10^2$ at 727°C.
    $$2SO_{2(g)} + O_{2(g)} \rightleftharpoons 2SO_{3(g)}$$
    Is the percent of sulfur dioxide that reacts with oxygen at this temperature relatively large or relatively small? How does the percent of $SO_2$ reacting compare with the percent $O_2$ that reacts?

44. Nitrogen dioxide and dinitrogen tetroxide form an equilibrium mixture:
    $$2NO_{2(g)} \rightleftharpoons N_2O_{4(g)}$$
    At 25°C, 0.100 mol of nitrogen dioxide was placed into a 4.00 L container and equilibrium was established. At equilibrium, 74% of the nitrogen dioxide had reacted. What is the equilibrium constant for this reaction at 25°C?

## Making Connections

**45.** Bones and teeth consist mostly of a compound called hydroxyapatite, $Ca_{10}(PO_4)_6(OH)_2$. This compound contains $PO_4^{3-}$ and $OH^-$ ions.
  (a) Do you expect hydroxyapatite will be an acid or a base?
  (b) Foods that contain sucrose form lactic acid in the mouth and the pH drops. As a result, eating candy promotes a reaction between hydroxyapatite and $H^+_{(aq)}$. Balance the skeleton reaction:
  $Ca_{10}(PO_4)_6(OH)_{2(s)} + H^+_{(aq)} \rightleftharpoons$
  $CaHPO_{4(s)} + H_2O_{(\ell)} + Ca^{2+}_{(aq)}$
  (c) At lower pH values, the $CaHPO_{4(s)}$ also reacts with $H^+_{(aq)}$:
  $CaHPO_{4(s)} + H^+_{(aq)} \rightleftharpoons Ca^{2+}_{(aq)} + H_2PO_4^-_{(aq)}$
  Dentists and toothpaste manufacturers warn that eating candy promotes tooth decay. What chemical evidence have you seen to support this advice?

**46.** Putrescine, $NH_2CH_2CH_2CH_2CH_2NH_2$, is a base that gives rotting animal tissue its distinctive odour. Putrescine also plays a role in the growth of animal cells. Calculate $K_b$ for a 0.10 mol/L aqueous solution of putrescine, in which $[OH^-] = 2.1 \times 10^{-3}$.

**47.** Acids and bases are electrolytes and their strengths in both "roles" are related. Explain whether the electrical conductivity of 0.1 mol/L HCl is higher, lower, or the same as that of 0.1 mol/L $CH_3COOH$. Is the conductivity of $1 \times 10^{-7}$ mol/L HCl higher, lower, or the same as that of $1 \times 10^{-7}$ mol/L $CH_3COOH$? Defend your answer.

**48.** One drop of a concentrated strong acid such as sulfuric acid will burn a hole through a piece of cloth (or your skin!). If you add one drop of the same acid to a large bucket filled with water, a cloth (or your hand) can rest in the solution with no change—other than being wet. Explain the difference.

**49.** Phosphoric acid, $H_3PO_{4(aq)}$ is triprotic.
  (a) Explain what this means.
  (b) Write stepwise equations to show the dissociation of each proton.
  (c) Show that $H_2PO_4^-_{(aq)}$ can act as either an acid or a base.
  (d) Which is the stronger acid: $H_3PO_{4(aq)}$ or $H_2PO_4^-_{(aq)}$? Explain your answer.

**50.** Are there any circumstances in which pure water could have a pH greater than 7.0? Explain.

**51.** Sodium bicarbonate, $NaHCO_{3(aq)}$ turns red litmus paper blue and undergoes a neutralization reaction with acids.
  (a) Use Arrhenius theory to write a chemical equation showing the dissociation of $NaHCO_{3(aq)}$ in water.
  (b) A solution of $NaHCO_3$ causes red litmus to turn blue in an instant. Does this support or contradict your dissociation equation from part (a)?
  (c) Use a modernized Arrhenius theory to show the dissociation for $NaHCO_3$ Does this dissociation equation support or contradict the above litmus test? Give reasons to justify your answer.

# UNIT 7

# Thermochemistry

**UNIT 7 CONTENTS**

**CHAPTER 16**
Theories of Energy Changes

**CHAPTER 17**
Measuring and Using Energy Changes

**UNIT 7 OVERALL OUTCOMES**

- What happens to the particles of a substance when energy is added?
- What skills are involved in determining energy changes for physical and chemical processes
- How do chemical technologies and processes depend on the energetics of chemical reactions?

In the nineteenth century, railway tunnels were blasted through the Rocky Mountains to connect British Columbia with the rest of Canada. Workers used nitroglycerin to blast through the rock. This compound is so unstable, however, that accidents were frequent and many workers died. Alfred B. Nobel found a way to stabilize nitroglycerin, and make it safer to use, when he invented dynamite.

What makes nitroglycerin such a dangerous substance? Nitroglycerin, $C_3H_5(NO_3)_{3(\ell)}$, gives off a large amount of energy when it decomposes. In fact, about 1500 kJ of energy is released for every mole of nitroglycerin that reacts. The fast, exothermic decomposition of nitroglycerin is accompanied by a tremendous shock wave, which is caused by the expansion of the gaseous products: nitrogen, oxygen, water, and carbon dioxide. Due to the great instability of nitroglycerin, simply jarring or shaking a container of nitroglycerin can initiate the decomposition reaction.

Energy changes accompany every chemical and physical process, from the decomposition of nitroglycerin to the vaporization of water. What causes these energy changes? How can you determine the energy change associated with a given chemical reaction or physical change? In this unit, you will explore the energy changes that accompany various chemical and physical processes.

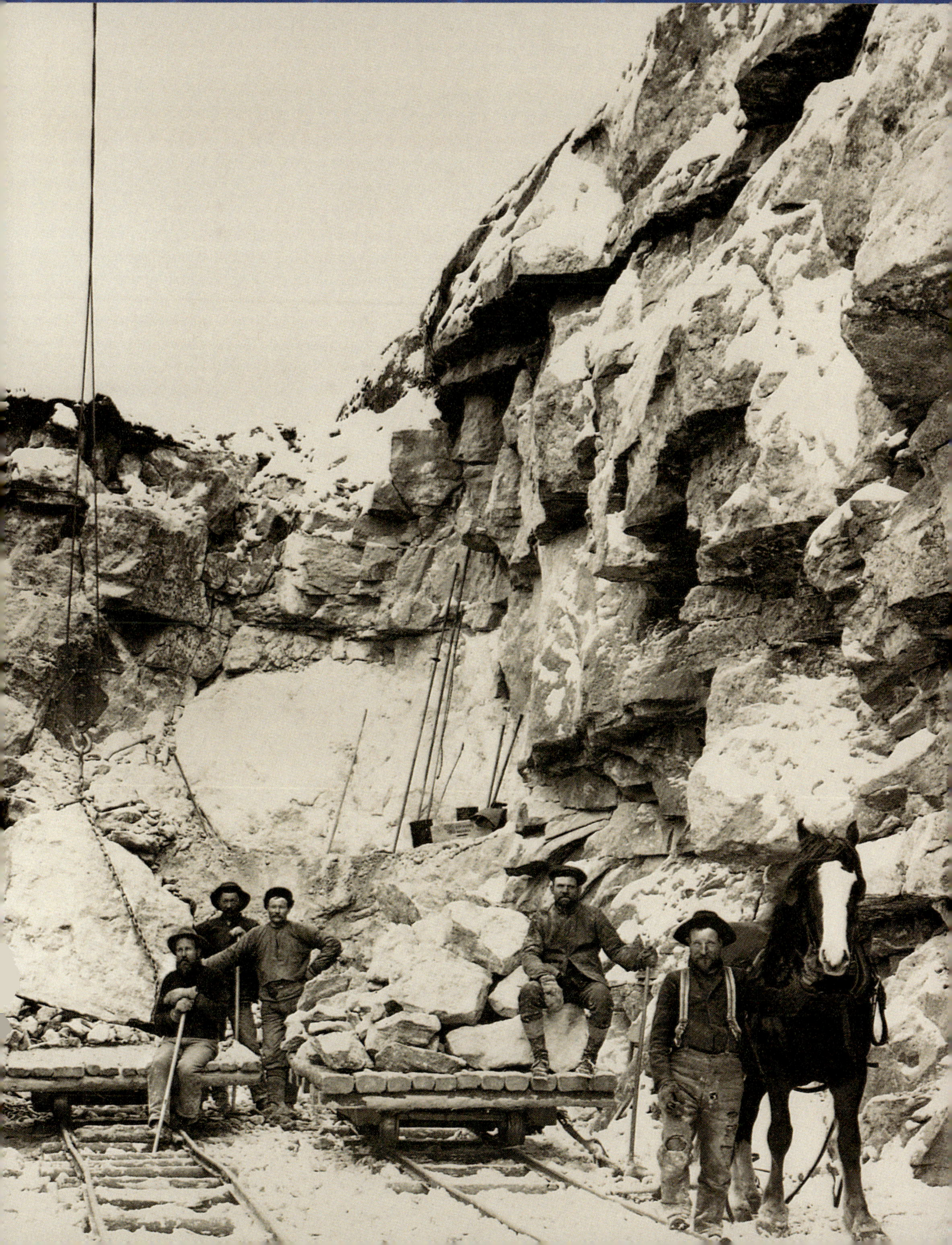

# CHAPTER 16
# Theories of Energy Changes

**Chapter Preview**

- 16.1 Temperature Change and Heat
- 16.2 Enthalpy Changes
- 16.3 Heating and Cooling Curves

Think about a prehistoric family group building a fire. It may seem as though this fire does not have much in common with a coal-burning power plant. Both the fire and the power plant, however, are technologies that harness energy-producing processes.

Humans continually devise new technologies that use chemical reactions to produce materials with useful properties. Since the invention of fire, humans have also worked to devise technologies that harness energy. These technologies depend on the fact that every chemical and physical process is accompanied by a characteristic energy change.

You take advantage of such energy changes every day, whether you are using ice cubes to cool a drink or a gas stove to cook a meal. Canadian society depends on the energy changes that are associated with physical and chemical changes.

In this chapter, you will learn about the energy changes associated with heating and cooling. Next, you will study the energy changes that accompany physical changes and chemical reactions. You will learn how to calculate the heat that is absorbed or released during physical or chemical changes. Finally, you will use your skills to analyze what happens as a substance is heated or cooled, changing its state once or more.

**Prerequisite Concepts and Skills**

Before you begin this chapter, review the following concepts and skills:

- writing balanced chemical equations (Chapter 1)
- performing stoichiometric calculations (Chapter 4)

This chef is using the heat released by the combustion of natural gas to cook food. What kinds of energy changes are involved in this process?

# Temperature Change and Heat

## 16.1

All physical changes and chemical reactions are accompanied by changes in energy. These energy changes are crucial to life on Earth. For example, chemical reactions in your body generate the heat that helps to regulate your body temperature. Physical changes, such as evaporation, help to keep your body cool. On a much larger scale, there would be no life on Earth without the energy from the nuclear reactions that take place in the Sun.

The study of energy and energy transfer is known as **thermodynamics**. Chemists are interested in the branch of thermodynamics known as **thermochemistry**: the study of energy involved in chemical reactions. In order to discuss energy and its interconversions, thermochemists have agreed on a number of terms and definitions. You will learn about these terms and definitions over the next few pages. Then you will examine the energy changes that accompany temperature changes, chemical reactions, and physical changes.

### Studying Energy Changes

The **law of conservation of energy** states that the total energy of the universe is constant. In other words, energy can be neither destroyed nor created. This idea can be expressed by the following equation:

$$\Delta E_{universe} = 0$$

Energy can, however, be transferred from one substance to another. It can also be converted into various forms. In order to interpret energy changes, scientists must clearly define what part of the universe they are dealing with. The **system** is defined as the part of the universe that is being studied and observed. In a chemical reaction, the system is usually made up of the reactants and products. By contrast, the **surroundings** are everything else in the universe. The two equations below show the relationship between the universe, a system, and the system's surroundings.

$$\text{Universe} = \text{System} + \text{Surroundings}$$

$$\Delta E_{universe} = \Delta E_{system} + \Delta E_{surroundings} = 0$$

This relationship is known as the **first law of thermodynamics**.

According to the first law of thermodynamics, any change in the energy of a system is accompanied by an equal and opposite change in the energy of the surroundings.

$$\Delta E_{system} = -\Delta E_{surroundings}$$

Consider the chemical reaction that is taking place in the flask in Figure 16.1. A chemist would probably define the system as the contents of the flask—the reactants and products. Technically, the rest of the universe is the surroundings. In reality, however, the entire universe changes very little when the system changes. Therefore, the surroundings are usually considered to be only the part of the universe that is likely to be affected by the energy changes of the system. In Figure 16.1, the flask, the lab bench, the air in the room, and the student all make up the surroundings. The system is more likely to significantly influence its immediate surroundings than it is to influence a mountaintop in Japan (also, technically, part of the surroundings).

### Section Preview/Outcomes

In this section, you will

- **identify** and **describe** the changes to particle movement that accompany a change in temperature
- **describe** heat as a transfer of kinetic energy from a system of higher temperature to a system of lower temperature
- **perform** calculations involving heat capacity, specific heat capacity, and mass
- **communicate** your understanding of the following terms: *thermodynamics, thermochemistry, system, surroundings, first law of thermodynamics, open system, closed system, isolated system, kinetic energy, potential energy, joule (J), temperature (T), heat (q), specific heat capacity (c), heat capacity (C)*

**Figure 16.1** The solution in the flask is the system. The flask, the laboratory, and the student are the surroundings.

Depending on how they are separated from their surroundings, systems are defined in three different ways.

- An **open system**, as its name implies, is open to its surroundings. Both energy and matter may be exchanged between an open system and its surroundings. A reaction in an open beaker is an open system.
- In a **closed system**, matter cannot move between the system and surroundings. Energy, however, can be transferred between a closed system and its surroundings. A reaction in a stoppered Erlenmeyer flask is a closed system.
- An **isolated system** is completely insulated from the surroundings. Neither matter nor energy is exchanged between an isolated system and its surroundings. You will learn more about the importance of isolated systems in Chapter 17.

## Types of Energy

You may recall from earlier science courses that energy is classified into two fundamental types. These types of energy are:

- **kinetic energy** — the energy of motion
- **potential energy** — energy that is stored

If you pick up a rock and lift it several metres above the ground, the rock gains potential energy. Once you let go of the rock, the rock falls to the ground as the potential energy is converted to kinetic energy. As you will learn, energy changes involved in chemical and physical processes fit into one or both of these two categories.

The SI unit for both kinetic and potential energy is the **joule** (symbol **J**). The joule is derived from other SI units. One joule is equal to $1 \text{ kg}\left(\frac{\text{m}^2}{\text{s}^2}\right)$.

## Temperature Change and Heat

**Temperature**, *T*, is a measure of the average kinetic energy of the particles that make up a substance or system. You can think of temperature as a way of quantifying how hot or cold a substance is, relative to another substance.

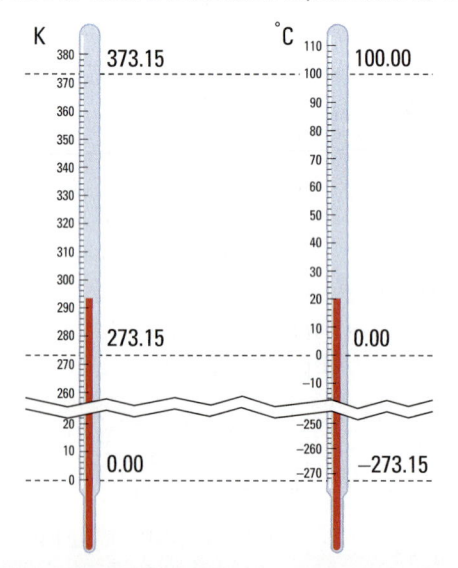

**Figure 16.2** Celsius degrees and Kelvin degrees are the same size. The Kelvin scale begins at absolute zero. This is the temperature at which the particles in a substance have no kinetic energy. Therefore, Kelvin temperatures are never negative. By contrast, 0°C is set at the melting point of water. Celsius temperatures can be positive or negative.

Temperature is measured in either Celsius degrees (°C) or kelvins (K). The Celsius scale is a relative scale. It was designed so that the boiling point of water is at 100°C and the melting point of water is at 0°C. The Kelvin scale, on the other hand, is an absolute scale. It was designed so that 0 K is the temperature at which a substance possesses no kinetic energy. The relationship between the Kelvin and Celsius scales is shown in Figure 16.2 on the previous page, and by the following equation.

Temperature in Kelvin degrees = Temperature in Celsius degrees + 273.15

Temperature change is an indication of a change in kinetic energy. The temperature variable that is used is the change in temperature, symbolized $\Delta T$. The change in temperature is always equal to the final temperature subtracted from the initial temperature, as shown by the following equation.

$$\Delta T = T_f - T_i$$

A positive value for $\Delta T$ indicates an increase in temperature. A negative value for $\Delta T$ indicates a decrease in temperature.

## Transfer of Kinetic Energy

**Heat**, $q$, refers to the transfer of kinetic energy between objects with different temperatures. Heat, therefore, has the same units as energy—joules (J).

According to the particle model of matter, matter is made up of particles in constant motion. When a substance absorbs heat, the average speed of the particles in the object increases. Therefore the temperature of the substance increases. Figure 16.3 models what happens when hot chocolate is heated on a stove.

**A** hot chocolate at 25°C

**B** hot chocolate at 75°C

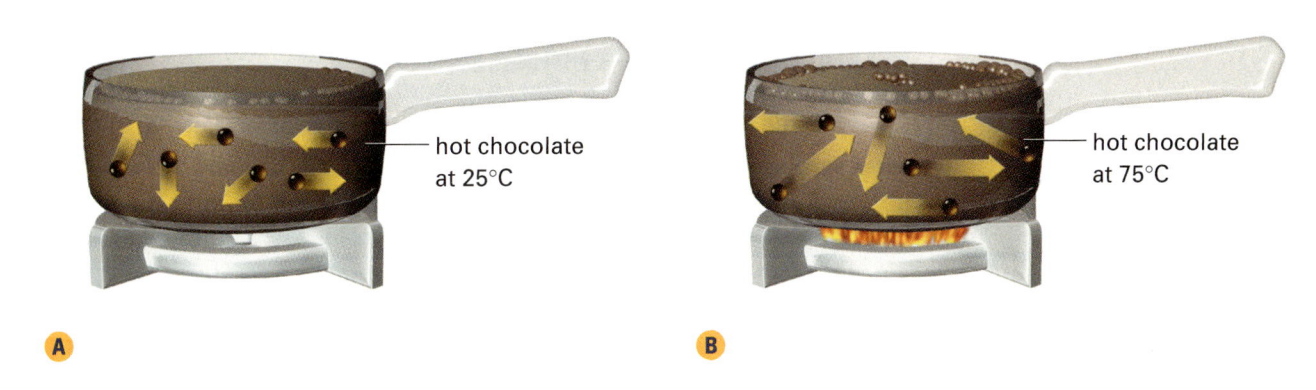

**Figure 16.3** The length of each arrow represents the speed of the particle. As the hot chocolate absorbs heat, the average speed of the particles increases.

According to the first law of thermodynamics, any energy released by a system is gained by the surroundings, and vice versa. When the energy is transferred as heat, you can express the relationship as follows.

$$q_{system} = -q_{surroundings}$$

When substances with different temperatures come into contact, kinetic energy is transferred as heat from the particles of the warmer substance to the particles of the cooler substance.

For example, hot chocolate at 75°C has a higher temperature than your body, which is about 37°C. You can feel the difference as warmth when you drink the hot chocolate, as shown in Figure 16.4.

As you drink, the particles in the hot chocolate transfer kinetic energy as heat to particles in your mouth. The particles in the hot chocolate slow down, and the particles in your mouth speed up.

The moving particles in the hot chocolate also transfer kinetic energy as heat to the particles in the mug. The average speed of the particles of the mug increases and the temperature of the mug increases as a result. You feel the increased kinetic energy of the particles of the mug as warmth. The particles in the mug transfer kinetic energy as heat to the particles in your fingertips.

**Figure 16.4** Kinetic energy is transferred as heat from the particles in the hot chocolate to the particles in the mug to the particles in this person's fingertips.

When you blow on hot chocolate to cool it, as shown in Figure 16.5, what happens? You cause high-pressure air to move over the surface of the liquid. As the air leaves your mouth, it begins to expand and cool. As the air moves across the surface of the hot chocolate, the particles in the hot chocolate transfer some kinetic energy to the particles of the cool air. The hot chocolate cools down.

**Figure 16.5** Particles in the hot chocolate transfer kinetic energy as heat to particles in the cool, moving air. The particles in the hot chocolate slow down, and the temperature of the hot chocolate decreases.

## mind STRETCH

Would you rather drink hot chocolate out of a cup made of glass, a ceramic mug, or a mug insulated with a vacuum layer, such as a Thermos™? Explain your reasoning.

Temperature is an important factor in the transfer of kinetic energy as heat. What other factors are important? Work through the following ThoughtLab to find out.

# ThoughtLab: Factors in Heat Transfer

Two students performed an experiment to determine what factors need to be considered when determining the quantity of thermal energy lost or gained by a substance undergoing an energy change. They set up their experiment as follows.

### Part A

The students placed two different masses of water, at the same initial temperature, in separate beakers. They placed an equal mass of ice (from the same freezer) in each beaker. Then they monitored the temperature of each beaker. Their results are listed in the table below.

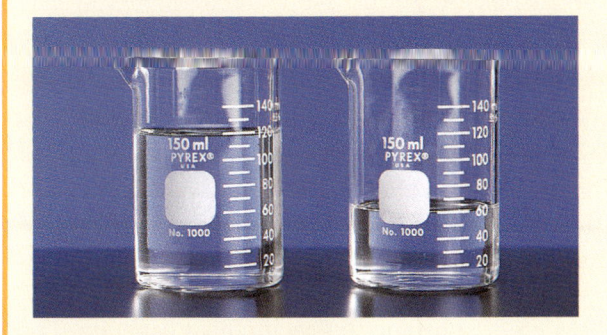

**Different Masses of Water**

| Beaker | 1 | 2 |
|---|---|---|
| Mass of water (g) | 60.0 | 120.0 |
| Initial temperature of water (°C) | 26.5 | 26.5 |
| Mass of ice added (g) | 10.0 | 10.0 |
| Final temperature of mixture (°C) | 9.7 | 17.4 |
| Temperature change (°C) | 16.8 | 9.1 |

### Part B

The students placed equal masses of canola oil and water, at the same initial temperature, in separate beakers. They placed equal masses of ice (from the same freezer) in the two beakers. Then they monitored the temperature of each beaker. Their results are listed in the following table.

**Different Liquids**

| Beaker | 1 (canola oil) | 2 (water) |
|---|---|---|
| Mass of liquid (g) | 60.0 | 60.0 |
| Initial temperature of liquid (°C) | 35.0 | 35.0 |
| Mass of ice added (g) | 10.0 | 10.0 |
| Final temperature of mixture (°C) | 5.2 | 16.9 |
| Temperature change (°C) | 29.8 | 18.1 |

### Procedure

1. For each part of the experiment, identify
   (a) the variable that was changed by the students (the manipulated variable)
   (b) the variable that changed as a result of changing the manipulated variable (the responding variable)
   (c) the variables that were kept constant to ensure a fair test (the controlled variables)
2. Interpret the students' results by answering the following questions.
   (a) If ice is added to two different masses of water, how does the temperature change?
   (b) If ice is added to two different liquids, how does the temperature change?

### Analysis

3. Think about your interpretation of the students' experiment and the discussion prior to this ThoughtLab. What are three important factors to consider when measuring the thermal energy change of a substance?

Chapter 16 Theories of Energy Changes • MHR **631**

## Mass and Energy Transfer

The ThoughtLab on page 631 gave you some insight into the factors that are important when measuring energy changes. How can you use these factors to calculate the amount of kinetic energy that is transferred as heat?

First you must examine each factor and determine its relationship to heat. You have already learned that a change in temperature indicates a change in energy. For a given object or sample, the larger the temperature change, the larger the energy change. How does the mass of a substance affect energy transfer? To answer that question, think about water.

About 70% of Earth's surface is covered with water. This enormous mass of water absorbs and releases tremendous amounts of energy. Water makes our climate more moderate by absorbing heat in hot weather and releasing heat in cold weather. The greater the mass of the water, the greater the amount of heat it can absorb and release. Mass is directly related to heat transfer. Mass is therefore a variable in the calculation of heat energy. It is symbolized by a lower-case $m$.

## Type of Substance and Energy Transfer

In the ThoughtLab, you probably noticed that the quantity of heat being transferred depends on the type of substance. When you added equal masses of ice to the same mass of oil and water, the temperature change of the oil was almost double the temperature change of the water. "Type of substance" cannot be used as a variable, however, when calculating energy changes. Instead, we use a variable that reflects the individual nature of different substances: specific heat capacity. The **specific heat capacity** of a substance is the quantity of energy, in joules (J), that is required to change one gram (g) of the substance by one degree Celsius (°C). The specific heat capacity of a substance reflects how well the substance can store energy. A substance with a large specific heat capacity can absorb and release more energy than a substance with a smaller specific heat capacity. The symbol that is used for specific heat capacity is a lower-case $c$. The units are J/g·°C.

The specific heat capacity of water is relatively large: 4.184 J/g·°C. This value helps to explain how water can absorb and release enough energy to moderate Earth's temperature. Examine the values in Table 16.1. Notice that the specific heat capacities of most substances shown are much lower than the specific heat capacity of water.

Areas without much water, such as Canada's prairie provinces, experience huge variations in temperature from summer to winter. Provinces near the Atlantic and Pacific Oceans, however, such as Newfoundland and Labrador or British Columbia, experience much smaller variations in year-round temperatures. These more moderate temperatures are due to the ability of water to absorb and release large quantities of heat.

**Table 16.1** Specific Heat Capacities of Various Substances

| Substance | Specific heat capacity (J/g·°C at SATP) |
|---|---|
| **Elements** | |
| aluminum | 0.900 |
| carbon (graphite) | 0.711 |
| copper | 0.385 |
| gold | 0.129 |
| hydrogen | 14.267 |
| iron | 0.444 |
| **Compounds** | |
| ammonia (liquid) | 4.70 |
| ethanol | 2.46 |
| water (solid) | 2.01 |
| water (liquid) | 4.184 |
| water (gas) | 2.01 |
| **Other materials** | |
| air | 1.02 |
| concrete | 0.88 |
| glass | 0.84 |
| granite | 0.79 |
| wood | 1.76 |

## Calculating Heat Transfer

You have just considered three variables: change in temperature (Δ$T$), mass ($m$), and type of substance, which is characterized by specific heat capacity ($c$). The formula in Figure 16.6 shows how to combine these variables to calculate the heat involved.

$$q = m \cdot c \cdot \Delta T$$

where:
- $q$ = heat (J)
- $m$ = mass (g)
- $c$ = specific heat capacity (J/g·°C), depends on type of substance and state of substance
- $\Delta T$ = change in temperature (°C or K)

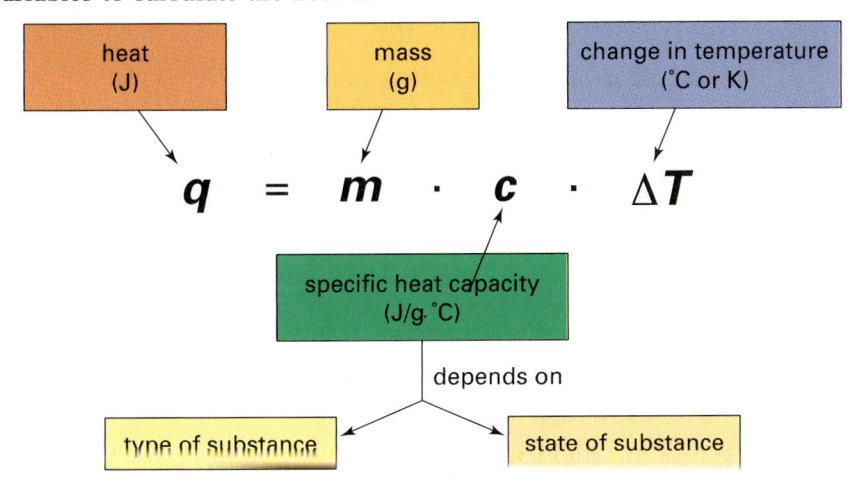

**Figure 16.6** Use this formula to calculate heat ($q$) transfer.

How do you solve heat problems using $q = mc\Delta T$? Go back to the ThoughtLab. Some of the data in this ThoughtLab can be used to illustrate the calculation of heat transfer, as shown below.

### Sample Problem

#### Heat Transferred From Water to Ice

**Problem**

In the ThoughtLab, 10.0 g of ice was added to 60.0 g of water. The initial temperature of the water was 26.5°C. The final temperature of the mixture was 9.7°C. How much heat was lost by the water?

**What Is Required?**

You need to calculate the quantity of heat ($q$) that was lost by the water.

**What Is Given?**

You know the mass of the water. You also know the initial and final temperatures of the water.
Mass of water ($m$) = 60.0 g
Initial temperature ($T_i$) = 26.5°C
Final temperature ($T_f$) = 9.7°C

**Plan Your Strategy**

You have enough information to solve this problem using $q = mc\Delta T$. Use the initial and final temperatures to calculate $\Delta T$. You need the specific heat capacity ($c$) of liquid water. This is given in Table 16.1 (4.184 J/g·°C). Because you are concerned only with the water, you will not use the mass of the ice.

*Continued...*

### mind STRETCH

Why does water have such a high specific heat capacity? Do some research to find out.
**Hint:** Water's specific heat capacity has something to do with bonding.

Chapter 16 Theories of Energy Changes • MHR **633**

### Act on Your Strategy

Substitute the values into the following heat formula, and solve.
Remember that $\Delta T = T_f - T_i$

$$q = mc\Delta T$$
$$= (60.0 \text{ g})(4.184 \tfrac{J}{g \cdot ^\circ C})(9.7^\circ C - 26.5^\circ C)$$
$$= -4.22 \times 10^3 \ (\cancel{g})(\tfrac{J}{\cancel{g \cdot ^\circ C}})(\cancel{^\circ C})$$
$$= -4.22 \times 10^3 \text{ J (or } -4.22 \text{ kJ)}$$

The water lost $4.22 \times 10^3$ J of heat.

### Check Your Solution

The water lost heat, so the heat value should be negative.
Heat is measured in joules or kilojoules. Make sure that the units cancel out to give the appropriate unit for your answer.

## Practice Problems

1. 100 g of ethanol at 25°C is heated until it reaches 50°C. How much heat does the ethanol gain? **Hint:** Find the specific heat capacity of ethanol in Table 16.1.

2. In Part A of the ThoughtLab on page 631, the students added ice to 120.0 g of water in beaker 2. Calculate the heat lost by the water. Use the information given for beaker 2, as well as specific heat capacities in Table 16.1.

3. Beaker A contains 50 g of liquid at room temperature. The beaker is heated until the liquid gains 10°C. Beaker B contains 100 g of the same liquid at room temperature. This beaker is also heated until the liquid gains 10°C. In which beaker does the liquid absorb the most heat? Explain.

4. As the diagram below illustrates, the sign of the heat value tells you whether a substance has lost or gained heat energy. Consider the following descriptions. Write each heat value, and give it the appropriate sign to indicate whether heat was lost from or gained by the system.

    (a) In Part A of the ThoughtLab on page 631, the ice absorbed the heat that was lost by the water. When ice was added to 60.0 g of water, it absorbed 4.22 kJ of heat. When ice was added to 120.0 g of water, it absorbed 4.6 kJ of heat.

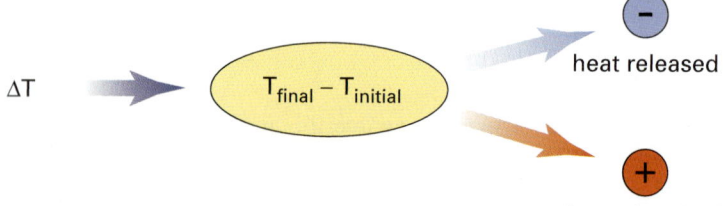

    $\Delta T \rightarrow T_{final} - T_{initial} \rightarrow$ − heat released / + heat absorbed

    (b) When 2.0 L of water was heated over a campfire, the water gained 487 kJ of energy.

    (c) A student baked a cherry pie and put it outside on a cold winter day. There was a change of 290 kJ of heat energy in the pie.

In the Sample Problem, heat was *lost* by the water. Therefore the value of $q$ was *negative*. If the value of $q$ is positive, this indicates that heat is *gained* by a substance.

The heat equation $q = mc\Delta T$ can be rearranged to solve for any of the variables. For example, in Part B of the ThoughtLab on page 631, ice was added to both canola oil and water. How can you use the information given in Part B to calculate the specific heat capacity of the canola oil?

### Math LINK

$$q = m \cdot c \cdot \Delta T$$

$$\text{Units:} \rightarrow \text{kJ} = \text{kg} \cdot \frac{4.184 \text{ kJ}}{\text{kg} \cdot \text{°C}} \cdot \text{°C}$$

mass—must be kg

Specific heat capacity
- must have kJ (top) and kg (bottom)
- Since "k" is on top and bottom, the number stays the same

## Sample Problem
### Calculating Specific Heat Capacity

**Problem**

Calculate the specific heat capacity of canola oil, using the information given in Part B of the ThoughtLab on page 631. Note that the ice gained $4.0 \times 10^3$ J of energy when it came in contact with the canola oil.

**What Is Required?**

You need to calculate the specific heat capacity ($c$) of the canola oil.

**What Is Given?**

From the ThoughtLab, you know the mass ($m$) and the initial and final temperatures of the canola oil.

Mass of oil ($m$) = 60.0 g

Initial temperature ($T_i$) = 35.0°C

Final temperature ($T_f$) = 5.2°C

You also know the quantity of heat gained by the ice. This must be the same as the heat lost by the oil.
Heat gained by the ice = Heat lost by the canola oil = $4.0 \times 10^3$ J

**Plan Your Strategy**

Rearrange the equation $q = mc\Delta T$ to solve for c. Then substitute the values for $q$, $m$, and $\Delta T$ ($T_f - T_i$) into the equation.

**Act on Your Strategy**

$$c = \frac{q}{m\Delta T}$$

$$= \frac{-4.0 \times 10^3 \text{ J}}{(60.0 \text{ g})(5.2°C - 35.0°C)}$$

$$= 2.2437 \frac{\text{J}}{\text{g} \cdot °C}$$

$$= 2.24 \frac{\text{J}}{\text{g} \cdot °C}$$

**Check Your Solution**

The specific heat capacity should be positive, and it is. It should have the units $\frac{\text{J}}{\text{g} \cdot °C}$.

Heat values are often very large. Therefore it is convenient to use kilojoules (kJ) to calculate heat. How does this affect the units of the other variables in the heat equation? Does the specific heat capacity have to change? The above diagram shows how units must be modified in order to end up with kilojoules.

**Figure 16.7** Canola oil is a vegetable oil that is used in salads and cooking.

### Practice Problems

5. Solve the equation $q = mc\Delta T$ for the following quantities.
   (a) $m$
   (b) $c$
   (c) $\Delta T$

6. You know that $\Delta T = T_f - T_i$. Combine this equation with the heat equation, $q = mc\Delta T$, to solve for the following quantities.
   (a) $T_i$ (in terms of $q$, $m$, $c$, and $T_f$)
   (b) $T_f$ (in terms of $q$, $m$, $c$, and $T_i$)

7. How much heat is required to raise the temperature of 789 g of liquid ammonia, from 25.0°C to 82.7°C?

8. A solid substance has a mass of 250.00 g. It is cooled by 25.00°C and loses 4937.50 J of heat. What is its specific heat capacity? Look at Table 16.1 to identify the substance.

9. A piece of metal with a mass of 14.9 g is heated to 98.0°C. When the metal is placed in 75.0 g of water at 20.0°C, the temperature of the water rises by 28.5°C. What is the specific heat capacity of the metal? Assume all heat released by metal is absorbed by the water.

10. A piece of gold ($c = 0.129$ J/g°C) with mass of 45.5 g and a temperature of 80.5°C is dropped into 192 g of water at 15.0°C. What is the final temperature of the water? (**Hint:** Use the equation $q_w = -q_g$.)

## Heat Capacity

Consider the bathtub and teacup shown in Figure 16.8. Both are filled with water at 20°C. The water in the bathtub and the water in the teacup have the same heat capacity, but the water in the bathtub has a much greater mass. Therefore, it would take a great deal more energy to heat the water in the bathtub to 60°C than it would to heat the water in the teacup to 60°C. Although all water has the same *specific* heat capacity, different samples of water have different masses. Different samples of water have different *heat capacities* if they have different masses.

**Figure 16.8** The water in the bathtub has a higher heat capacity than the water in the teacup.

**Heat capacity,** *C*, relates the heat of a given sample, object, or system to its change in temperature. Heat capacity is usually expressed in units of kJ/°C. When you know the specific heat capacity and mass of a sample, you can determine the heat capacity using the following equation.

$$C = m \times c$$

where $C$ = heat capacity (kJ/°C)
$m$ = mass (kg)
$c$ = specific heat capacity (kJ/kg · °C)

Given the heat capacity of a sample, object, or system, you can determine the heat associated with a given temperature change, using the following equation.

$$q = C \Delta T$$

Practice using the equations above in the following Practice Problems. You will need to refer to Table 16.1.

### Math LINK

Use the relationships $q = mc\Delta T$ and $C = c \times m$ to show why $q = C\Delta T$.

## Practice Problems

11. A bathtub contains 100.0 kg of water.
    (a) What is the heat capacity, $C$, of the water in the bathtub?
    (b) How much heat is transferred to the surroundings as the water in the bathtub cools from 60.0°C to 20.0°C? Use your answer from (a) in your calculations.
    (c) Calculate the heat transferred in (b) using a different method.

12. A teacup contains 0.100 kg of water.
    (a) What is the heat capacity, $C$, of the water in the teacup?
    (b) How much heat is transferred to the surroundings as the water in the teacup cools from 60.0°C to 20.0°C? Use your answer from (a) in your calculations.
    (c) Calculate the heat transferred in (b) using a different method.

13. A ring of pure gold with a mass of 18.8 g is tossed into a fire, and then removed with a pair of tongs. The initial temperature of the ring was 23.2°C and the final temperature of the ring is 55.8°C.
    (a) What is the heat capacity of the ring?
    (b) How much heat did the ring absorb from the fire?

14. Because humans are about 85% water, the specific heat capacity of a human is sometimes approximated as 0.85 times the specific heat capacity of water.
    (a) Using the above approximation, what is the heat capacity of a human that has a mass of 68.0 kg?
    (b) How much heat is required to raise the temperature of a 68.0 kg human by 1.00°C?

### mind STRETCH

In Practice Problem 14, you estimated the heat capacity of a human. Your body temperature is kept at a constant 37°C. If your body did not have mechanisms for cooling down and keeping warm, you would freeze on cold days and overheat on warm days. List several ways in which your body warms itself and several ways in which your body cools itself. You will learn more about how your body regulates its temperature in sections 16.2 and 16.3.

## Section Summary

In this section, you have learned about the factors involved in temperature change and kinetic energy transfer as heat. So far, you have focussed on changes in kinetic energy. In the next section, you will learn about physical and chemical processes, which involve a change in potential energy.

## Section Review

**1** In your own words, state the first law of thermodynamics. Then express the first law of thermodynamics as an equation.

**2** Define the term "heat."

**3** What are three important factors to consider when measuring heat transfer?

**4** In Part B of the ThoughtLab on page 631, 60.0 g of water was in beaker 2. The initial temperature of the water was 35.0°C, and the final temperature was 16.9°C.

(a) Calculate the heat that was lost by the water in beaker 2.

(b) Where did the heat go?

**5** When iron nails are hammered into wood, friction causes the nails to heat up.

(a) Calculate the heat that is gained by a 5.2 g iron nail as it changes from 22.0°C to 38.5°C. (See Table 16.1.)

(b) Calculate the heat that is gained by a 10.4 g iron nail as it changes from 22.0°C to 38.5°C.

(c) Calculate the heat that is gained by the 5.2 g nail if its temperature changes from 22.0°C to 55.0°C.

**6** (a) A 23.9 g silver spoon is put in a cup of hot chocolate. It takes 0.343 kJ of energy to change the temperature of the spoon from 24.5°C to 85.0°C. What is the specific heat capacity of solid silver?

(b) What is the heat capacity, $C$, of the silver spoon?

**7** The specific heat capacity of aluminum is 0.902 J/g°C. The specific heat capacity of copper is 0.389 J/g°C. The same amount of heat is transferred to equal masses of these two metals. Which metal increases more in temperature? Explain your answer.

**8** You have prepared some hot chocolate to take with you on a hike. You are about to fill your Thermos™ bottle with the hot chocolate when your friend stops you. Your friend suggests you rinse the inside of the bottle with hot water before filling the bottle with hot chocolate.

(a) Explain in detail why rinsing the bottle with hot water will help keep your hot chocolate hot.

(b) A Thermos™ bottle has a vacuum between its inner layer and its outer layer. Explain why the vacuum insulation helps to keep cold drinks cold and hot drinks hot.

(c) What kind of system (open, closed, or insulated) is hot chocolate inside a sealed Thermos™ bottle? Explain your answer.

# Enthalpy Changes

## 16.2

In the previous section, you explored what happens when the average kinetic energy of the particles in a substance changes. You saw that a change in average kinetic energy can be observed by monitoring temperature change. For example, you saw how the energy change of water relates to the temperature change, the mass of the water, and its specific heat capacity.

What happens during a phase change? What happens during a chemical reaction? These changes involve changes in the potential energy of a system, not the kinetic energy.

Chemists use the term **enthalpy change ($\Delta H$)** to refer to the potential energy change of a system during a process such as a chemical reaction or a physical change. Enthalpy changes are measured at constant pressure. The units of enthalpy change are kJ/mol.

### Enthalpy Changes in Chemical Changes

The enthalpy change of a chemical reaction represents the difference between the potential energy of the products and the potential energy of the reactants. In chemical reactions, potential energy changes result from chemical bonds being broken and formed. Chemical bonds are sources of stored energy (potential energy). *Breaking a bond is a process that requires energy. Creating a bond is a process that releases energy.* Figure 16.9 represents these ideas.

Chemists define the total internal energy of a substance at a constant pressure as its enthalpy, H. Chemists do not work with the *absolute* enthalpy of the reactants and products in a physical or chemical process. Instead, they study the enthalpy change, that accompanies a process. That is, they study the *relative* enthalpy of the reactants and products in a system. This is like saying that the distance between your home and your school is 2 km. You do not usually talk about the *absolute* position of your home and school in terms of their latitude, longitude, and elevation. You talk about their *relative* position, in relation to each other.

**Section Preview/Outcomes**

In this section, you will

- **use** and **interpret** enthalpy notation for communicating energy changes
- **explain** that chemical changes and changes of state involve changes in potential energy only
- **explain** the energy changes involved in bond breaking and bond formation
- **express** the enthalpy involved in chemical and physical changes using thermochemical equations and enthalpy diagrams
- **write** thermochemical equations for combustion reactions of alkanes
- **calculate** the heat released or absorbed by a system during a chemical or physical change
- **communicate** your understanding of the following terms: *enthalpy change ($\Delta H$), endothermic reaction, exothermic reaction, enthalpy of reaction ($\Delta H_{rxn}$), standard enthalpy of reaction, ($\Delta H^0_{rxn}$), thermochemical equation, standard molar enthalpy of formation ($\Delta H^0_f$)*

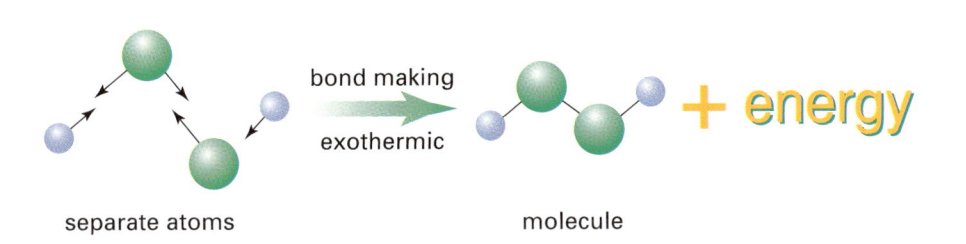

**Figure 16.9** This illustration shows bonds being broken and made during a chemical reaction. If the bonds are strong, there is a large change in energy. If the bonds are weak, there is a small change in energy.

Chapter 16 Theories of Energy Changes • MHR **639**

A chemical bond is caused by the attraction between the electrons and nuclei of two atoms. Energy is needed to break a chemical bond, just like energy is needed to break a link in a chain. On the other hand, making a chemical bond releases energy. The strength of a bond depends on how much energy is needed to break the bond.

Consider the combustion reaction that takes place when nitrogen reacts with oxygen.

$$N_{2(g)} + O_{2(g)} \rightarrow 2NO_{(g)}$$

In this reaction, one mole of nitrogen-nitrogen triple bonds and one mole of oxygen-oxygen double bonds are broken. Two moles of nitrogen-oxygen bonds are formed. This reaction absorbs energy. In other words, more energy is released to form two nitrogen-oxygen bonds than is used to break one nitrogen-nitrogen bond and one oxygen-oxygen bond. When a reaction results in a net *absorption* of energy, it is called an **endothermic reaction**.

On the other hand, when a reaction results in a net *release* of energy, it is called an **exothermic reaction**. In an exothermic reaction, more energy is released to form bonds than is used to break bonds. Therefore, energy is released. Figure 16.10 shows the relationship between bond breaking, bond formation, and endothermic and exothermic reactions.

> **CHEM FACT**
>
> The biggest explosion created by humans before the development of the atomic bomb was caused by exothermic reactions that occurred on December 6, 1917, in Halifax, Nova Scotia. The reactions occurred in the harbour when two ships collided. One of the ships was carrying 300 tons of picric acid (2,4,6-trinitrophenol), 200 tons of TNT (trinitrotoluene), 10 tons of gun cotton (nitrocellulose) and 35 tons of benzol. The blast killed over 1900 people. Its effects were later studied by J. Robert Oppenheimer, one of the scientists involved in developing the bombs dropped on Hiroshima and Nagasaki.

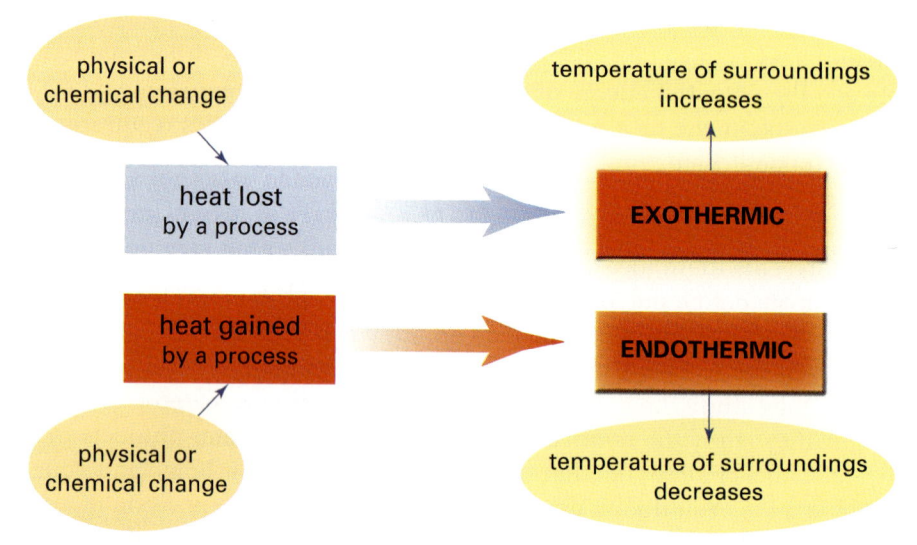

**Figure 16.10** The overall change in potential energy when bonds break and form determines whether a reaction is exothermic or endothermic.

## Representing Enthalpy Changes

The enthalpy change of a chemical reaction is known as the **enthalpy of reaction**, $\Delta H_{rxn}$. The enthalpy of reaction is dependent on conditions such as temperature and pressure. Therefore, chemists often talk about the **standard enthalpy of reaction**, $\Delta H^0_{rxn}$: the enthalpy change of a chemical reaction that occurs at SATP (25°C and 100 kPa). Often, $\Delta H^0_{rxn}$ is written simply as $\Delta H^0$. The $^0$ symbol is called "nought." It refers to a property of a substance at a standard state or under standard conditions. You may see the enthalpy of reaction referred to as the *heat of reaction* in other chemistry books.

> **CHEM FACT**
>
> Chemists use different subscripts to represent enthalpy changes for specific kinds of reactions. For example, $\Delta H_{comb}$ represents the enthalpy change of a combustion reaction.

## Representing Exothermic Reactions

There are three different ways to represent the enthalpy change of an exothermic reaction. The simplest way is to use a **thermochemical equation**: a balanced chemical equation that indicates the amount of heat that is absorbed or released by the reaction it represents. For example, consider the exothermic reaction of one mole of hydrogen gas with half a mole of oxygen gas to produce liquid water. For each mole of hydrogen gas that reacts, 285.8 kJ of heat is produced. Notice that the heat term is included with the products because heat is produced.

$$H_{2(g)} + \frac{1}{2}O_{2(g)} \rightarrow H_2O_{(\ell)} + 285.8 \text{ kJ}$$

You can also indicate the enthalpy of reaction as a separate expression beside the chemical equation. For exothermic reactions, $\Delta H°$ is always negative.

$$H_{2(g)} + \frac{1}{2}O_{2(g)} \rightarrow H_2O_{(\ell)} \quad \Delta H°_{rxn} = -285.8 \text{ kJ/mol}$$

A third way to represent the enthalpy of reaction is to use an enthalpy diagram. Examine Figure 16.11 to see how this is done.

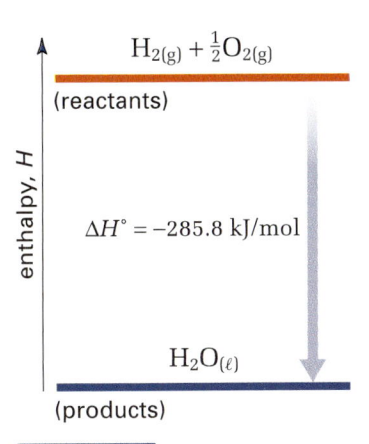

**Figure 16.11** In an exothermic reaction, the enthalpy of the system decreases as energy is released to the surroundings.

## Representing Endothermic Reactions

The endothermic decomposition of solid magnesium carbonate produces solid magnesium oxide and carbon dioxide gas. For each mole of magnesium carbonate that decomposes, 117.3 kJ of energy is absorbed. As for an exothermic reaction, there are three different ways to represent the enthalpy change of an endothermic reaction.

You can include the enthalpy of reaction as a heat term in the chemical equation. Because heat is absorbed in an endothermic reaction, the heat term is included on the reactant side of the equation.

$$117.3 \text{ kJ} + MgCO_{3(s)} \rightarrow MgO_{(s)} + CO_{2(g)}$$

You can also indicate the enthalpy of reaction as a separate expression beside the chemical reaction. For endothermic reactions, the enthalpy of reaction is always positive.

$$MgCO_{3(s)} \rightarrow MgO_{(s)} + CO_{2(g)} \quad \Delta H°_{rxn} = 117.3 \text{ kJ/mol}$$

Finally, you can use a diagram to show the enthalpy of reaction. Figure 16.12 shows how the decomposition of solid magnesium carbonate can be represented graphically.

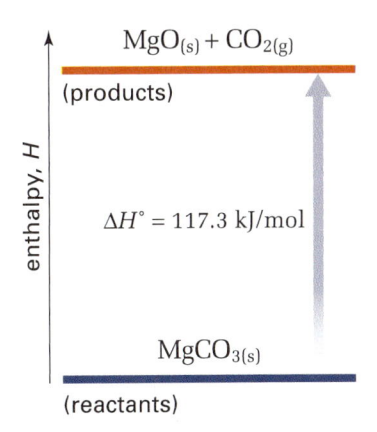

**Figure 16.12** In an endothermic reaction, the enthalpy of the system increases as heat energy is absorbed from the surroundings.

## Enthalpy of Reaction

The enthalpy change associated with a reaction depends on the amount of reactants involved. For example, the thermochemical equation for the decomposition of magnesium carbonate, shown above, indicates that 117.3 kJ of energy is absorbed when one mole, or 84.32 g, of magnesium carbonate decomposes. The decomposition of two moles of magnesium carbonate absorbs twice as much energy, or 234.6 kJ.

$$117.3 \text{ kJ} + MgCO_{3(s)} \rightarrow MgO_{(s)} + CO_{2(g)}$$
$$234.6 \text{ kJ} + 2MgCO_{3(s)} \rightarrow 2MgO_{(s)} + 2CO_{2(g)}$$

Enthalpy of reaction is *linearly dependent* on the amount of substances that react. That is, if the amount of reactants doubles, the enthalpy change also doubles.

In other words, when you multiply the stoichiometric coefficients of a thermochemical equation by any factor, you must multiply the heat term or enthalpy expression by the same factor.

## Enthalpy and Catalysts

In Chapter 12, you learned that catalysts change the rate of a reaction by providing an alternative pathway. The catalyst itself is the same at the beginning and at the end of the reaction.

Does using a catalyst affect the enthalpy change of the reaction? No. For a given process, the reactants and products are the same, regardless of whether a catalyst is used or not. The catalyst itself is regenerated unchanged after the reaction. The change in potential energy between the reactants and products is the same, and the catalyst itself has no net change in potential energy. The enthalpy change is therefore the same whether a catalyst is used or not.

## Standard Molar Enthalpy of Formation

In a formation reaction, a substance is formed from elements in their standard states. The enthalpy change of a formation reaction is called the **standard molar enthalpy of formation**, $\Delta H^0_f$. *The standard molar enthalpy of formation is the quantity of energy that is absorbed or released when one mole of a compound is formed directly from its elements in their standard states.* A more comprehensive list is provided in Appendix E.

Some standard molar enthalpies of formation are listed in Table 16.2. The standard state of an element is its most stable or common form at SATP (25°C and 100 kPa).

**Table 16.2** Selected Standard Molar Enthalpies of Formation

| Compound | $\Delta H^°_f$ kJ/mol | Formation equations |
|---|---|---|
| $CO_{(g)}$ | −110.5 | $C_{(s)} + \frac{1}{2}O_{2(g)} \rightarrow CO_{(g)}$ |
| $CO_{2(g)}$ | −393.5 | $C_{(s)} + O_{2(g)} \rightarrow CO_{2(g)}$ |
| $CH_{4(g)}$ | −74.6 | $C_{(s)} + 2H_{2(g)} \rightarrow CH_{4(g)}$ |
| $CH_3OH_{(\ell)}$ | −238.6 | $C_{(s)} + 2H_{2(g)} + \frac{1}{2}O_{2(g)} \rightarrow CH_3OH_{(\ell)}$ |
| $C_2H_5OH_{(\ell)}$ | −277.6 | $2C_{(s)} + 3H_{2(g)} + \frac{1}{2}O_{2(g)} \rightarrow C_2H_5OH_{(\ell)}$ |
| $C_6H_{6(\ell)}$ | +49.0 | $6C_{(s)} + 3H_{2(g)} \rightarrow C_6H_{6(\ell)}$ |
| $C_6H_{12}O_{6(s)}$ | −1274.5 | $6C_{(s)} + 6H_{2(g)} + 3O_{2(g)} \rightarrow C_6H_{12}O_{6(s)}$ |
| $H_2O_{(\ell)}$ | −285.8 | $H_{2(s)} + \frac{1}{2}O_{2(g)} \rightarrow H_2O_{(\ell)}$ |
| $H_2O_{(g)}$ | −241.8 | $H_{2(s)} + \frac{1}{2}O_{2(g)} \rightarrow H_2O_{(g)}$ |
| $CaCl_{2(s)}$ | −795.4 | $Ca_{(s)} + Cl_{2(g)} \rightarrow CaCl_{2(s)}$ |
| $CaCO_{3(s)}$ | −1206.9 | $Ca_{(s)} + C_{(s)} + \frac{3}{2}O_{2(g)} \rightarrow CaCO_{3(s)}$ |
| $NaCl_{(s)}$ | −411.1 | $Na_{(s)} + \frac{1}{2}Cl_{2(g)} \rightarrow NaCl_{(g)}$ |
| $HCl_{(g)}$ | −92.3 | $\frac{1}{2}H_{2(s)} + \frac{1}{2}Cl_{2(g)} \rightarrow HCl_{(g)}$ |
| $HCl_{(aq)}$ | −167.5 | $\frac{1}{2}H_{2(s)} + \frac{1}{2}Cl_{2(g)} \rightarrow HCl_{(aq)}$ |

When writing a formation equation, always write the elements in their standard states. For example, examine the equation for the formation of water directly from its elements under standard conditions.

$$H_{2(g)} + \frac{1}{2}O_{2(g)} \rightarrow H_2O_{(\ell)} \quad \Delta H^0_f = -285.8 \text{ kJ}$$

A formation equation should show the formation of exactly one mole of the compound of interest. The following equation shows the formation of benzene, $C_6H_6$ under standard conditions.

$$6C_{(graphite)} + 3H_{2(g)} \rightarrow C_6H_{6(\ell)} \quad \Delta H^0_f = 49.1 \text{ kJ}$$

### Standard Molar Enthalpy of Combustion

The standard molar enthalpy of combustion, $\Delta H^0_{comb}$, is the enthalpy associated with the combustion of *1 mol* of a given substance. The change in enthalpy is measured for the products and reactants in their standard states. For example, for methane, $\Delta H^0_{comb} = -965.1$ kJ/mol. You can represent the standard molar enthalpy of combustion using a thermochemical equation or using an enthalpy diagram, as shown in Figure 16.13.

$$CH_{4(g)} + 2O_{2(g)} \rightarrow CO_{2(g)} + 2H_2O_{(\ell)} + 965.1 \text{ kJ}$$

Notice that water is shown in the liquid form. Although water is formed as vapour during a combustion reaction, the enthalpy change for a standard molar enthalpy of combustion is measured with the energy change required for products to cool to SATP taken into consideration.

Table 16.3 lists selected standard molar enthalpies of combustion for the first eight straight-chain alkanes.

Practice representing formation and combustion reactions by working through the following problems.

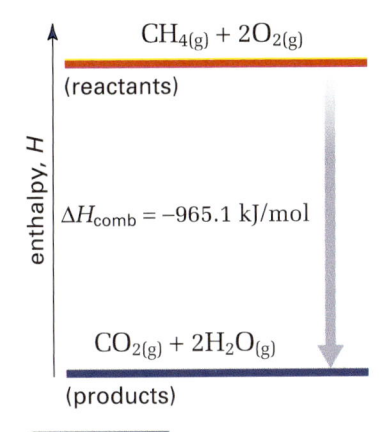

**Figure 16.13** This enthalpy diagram shows the standard molar enthalpy of combustion of methane.

### Practice Problems

**15.** Write a balanced thermochemical equation to represent the standard molar enthalpy of formation of each of the following substances. Include the heat term within the equation.

(a) $H_2O_{(g)}$

(b) $CaCl_{2(s)}$

(c) $CH_{4(g)}$

(d) $C_6H_{6(\ell)}$

(e) Repeat (c) and (d), showing the standard molar enthalpy of formation in a different way.

**16.** Draw an enthalpy diagram to represent the standard molar enthalpy formation of sodium chloride.

**17.** Write a balanced thermochemical equation to represent the standard molar enthalpy of combustion of each of the following alkanes (see Table 16.3). Remember: the products and reactants must be in their standard states.

(a) ethane

(b) propane

(c) butane

(d) pentane

**18.** Draw an enthalpy diagram to represent the standard molar enthalpy of the combustion of heptane (see Table 16.3). Remember: the products and reactants must be in their standard states.

**Table 16.3** Standard Molar Enthalpies of Combustion for Alkanes

| Alkane name | $\Delta H_{comb}$ (kJ/mol) |
|---|---|
| methane | −965.1 |
| ethane | −1250.9 |
| propane | −2323.7 |
| butane | −3003.0 |
| pentane | −3682.3 |
| hexane | −4361.6 |
| heptane | −5040.9 |
| octane | −5720.2 |

### Calculating Enthalpy Changes

The energy released or absorbed during a chemical reaction depends on the reactants involved. For example, the reaction of hydrogen and oxygen to form water releases heat to the surroundings, while the reaction of magnesium carbonate to form magnesium oxide and carbon dioxide absorbs heat from the surroundings. For each chemical reaction at SATP, there is a specific enthalpy change.

$$q = n \cdot \Delta H$$

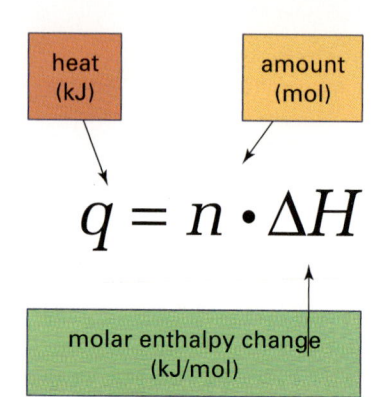

where heat (kJ), amount (mol), and molar enthalpy change (kJ/mol).

**Figure 16.14** Use this formula to calculate the heat absorbed or released by a chemical reaction.

What other factors affect the enthalpy change of a given chemical reaction? Compare the warmth you feel from a burning wooden match and the warmth you feel from a roaring bonfire. Both involve the same chemical reactions (the combustion of wood, primarily cellulose), but the heat released in each case is clearly different. There is a great deal more cellulose and oxygen involved in the reactions in the bonfire compared to the burning matchstick. Therefore, the *amount* of reactants present plays a role in the energy change. As you know, the symbol for amount is $n$, and the unit is the mole (mol).

The heat released or absorbed by a system during a chemical change can be calculated using the relationship shown in Figure 16.14.

Work through the following Sample Problem to learn how to use the formula in Figure 16.14 for formation and combustion reactions. Then try the Practice Problems that follow.

### Sample Problem
### Heat of Formation and Combustion Reactions

**Problem**

Methane is the main component of natural gas. Natural gas undergoes combustion to provide energy for heating homes and cooking food.

(a) How much heat is released when 50.00 g of methane forms from its elements?

(b) How much heat is released when 50.00 g of methane undergoes complete combustion?

**What Is Required?**

(a) You need to determine $q$ for the formation reaction that produces 50.00 g of methane.

(b) You need to determine $q$ for the combustion reaction that consumes 50.00 g of methane.

**What Is Given?**

$m$ = 50.00 g

From Table 16.2, you can see that for methane, $\Delta H°_f = -74.6$ kJ/mol.

From Table 16.3, you can see that for methane, $\Delta H°_{comb} = -965.1$ kJ/mol.

**Plan Your Strategy**

**Step 1** Determine molar mass, $M$, for methane.

**Step 2** Determine $n$ based on the formula $n = \dfrac{m}{M}$.

**Step 3** (a) Determine $q$ for the formation reaction using $q = n\Delta H°_f$.
(b) Determine $q$ for the combustion reaction using $q = n\Delta H°_{comb}$.

**Act on Your Strategy**

**Step 1** For methane, $M$ = 12.01 g/mol + (4 × 1.01 g/mol) = 16.05 g/mol

**Step 2** $n = \dfrac{m}{M}$

$= \dfrac{50.00 \text{ g}}{16.05 \text{ g/mol}}$

= 3.115 mol

**Step 3 (a)** $q = n\Delta H^0{}_f$
   $= (3.115 \text{ mol})(-74.6 \text{ kJ/mol})$
   $= -232 \text{ kJ}$

Therefore, 232 kJ of heat is released when 50.0 g of methane forms from its elements at SATP.

**(b)** $q = n\Delta H^0{}_{comb}$
   $= (3.115 \text{ mol})(-965.1 \text{ kJ/mol})$
   $= -3.006 \times 10^3 \text{ kJ}$

Therefore, $3.006 \times 10^3$ kJ of heat is released to the surroundings when 50.0 g of methane undergoes complete combustion.

**Check Your Solution**

The units are correct. The sign of the answer is negative, meaning heat is released by the reactions to the surroundings. As a quick check, use approximate numbers.

(a) There are about 3 mol of methane present, and the enthalpy of formation is about −75 kJ/mol. (3)(−75) = −225, which is close to the actual solution.

(b) The enthalpy of combustion is about −1000 kJ. (3)(−1000) = 3000, which is close to the actual solution.

### Practice Problems

**19. (a)** Hydrogen gas and oxygen gas react to form 0.534 g of *liquid* water. How much heat is released to the surroundings?

**(b)** Hydrogen gas and oxygen gas react to form 0.534 g of *gaseous* water. How much heat is released to the surroundings?

**20.** Carbon and oxygen react to form carbon dioxide. At STP, the carbon dioxide has a volume of 78.2 L. How much heat was released to the surroundings during the reaction?

**21.** Determine the heat released by the combustion of each of the following samples of hydrocarbons.
   **(a)** 56.78 g pentane, $C_5H_{12(\ell)}$
   **(b)** 1.36 kg octane, $C_8H_{18(\ell)}$
   **(c)** $2.344 \times 10^4$ g hexane, $C_6H_{14(\ell)}$

**22.** How much heat is released by the combustion of a sample of methane, $CH_{4(g)}$, that has a volume of 5.34 mL at STP?

**23.** What mass of methanol, $CH_3OH_{(\ell)}$, is formed from its elements if $2.34 \times 10^4$ kJ of energy is released during the process?

## Enthalpy Changes and Changes of State

Like chemical reactions, changes of state involve changes in the potential energy of a system only. The temperature of the system undergoing the state change remains constant. Because energy is absorbed or released as heat, however, the temperature of the surroundings often changes.

The energy changes associated with changes of state are important in regulating body temperature. When you sweat, for example, the water absorbs heat from your skin as the water vaporizes. Cats lick their fur, and the vaporizing liquid helps keep them cool. Dogs cool off in a similar way, as shown in Figure 16.15. Again, the evaporating water absorbs heat from their mouths and helps keep them cool.

In general, the energy change associated with a physical change is smaller than the energy change associated with a chemical change. In the case of molecular substances, for example, changes of state involve the breaking of intermolecular forces. Intermolecular forces are generally much weaker than chemical bonds. The energy released or absorbed when intermolecular bonds form or break is much less than the energy released or absorbed when chemical bonds form or break.

Figure 16.16 summarizes the terminology and enthalpy changes associated with changes of state. Note that in general, enthalpy changes for changes of state between liquid and gas are greater than changes of state between liquid and solid. For example, when a solid melts to form a liquid the attractive forces between particles are not completely broken. The particles remain close together. When a liquid changes to a gas, however, the attractive forces are completely broken as relatively great distances separate the particles from each other. The difference in potential energy between a liquid and a gas is much greater than the difference in potential energy between a solid and a liquid.

**Figure 16.15** Dogs let their tongues hang out and pant when they are hot. The evaporating water absorbs heat from the dog's tongue.

### Web LINK

www.mcgrawhill.ca/links/atlchemistry

Frozen fish and plastic buckets both exhibit a phenomenon known as the glass transition, which is unique to polymers. Below its glass transition temperature, a polymer is hard and brittle, like glass (or like a frozen fish or a plastic bucket left outside in winter). Above the glass transition phase, the polymer is soft and flexible. This change is not the same as melting, which occurs when polymer chains lose their structured order as they transform from a crystal to a liquid state. Research the subject of glass transition and describe some of its applications in everyday life. To begin your research, go to the web site above and click on **Web Links**.

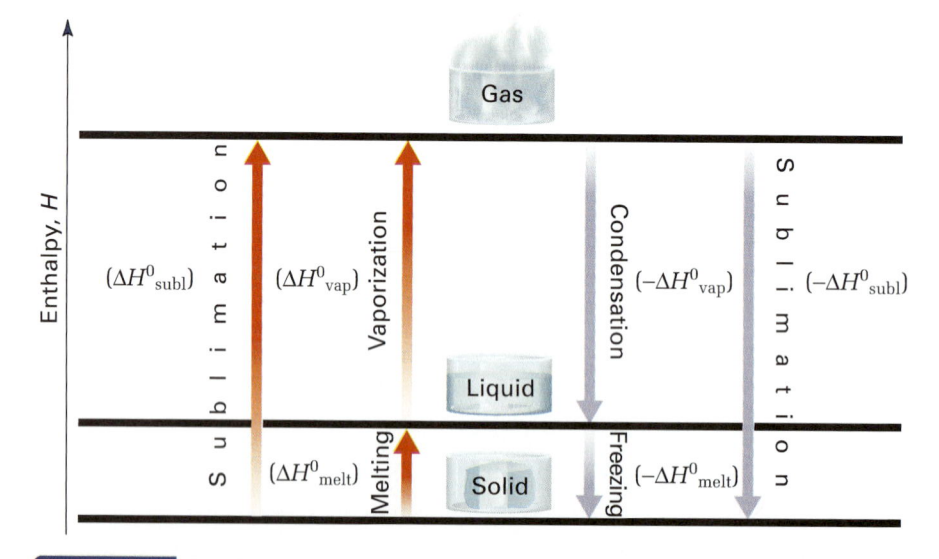

**Figure 16.16** This figure shows the enthalpy changes associated with changes of state.

You can represent the enthalpy change that accompanies a change of state—from liquid to solid, for example—just like you represented the enthalpy change of a chemical reaction. You can include a heat term in the equation, or you can use a separate expression of enthalpy change.

For example, when one mole of water melts, it absorbs 6.02 kJ of energy.

$$H_2O_{(s)} + 6.02 \text{ kJ} \rightarrow H_2O_{(\ell)}$$
$$H_2O_{(s)} \rightarrow H_2O_{(\ell)} \quad \Delta H = 6.02 \text{ kJ/mol}$$

Normally, however, chemists represent enthalpy changes associated with phase changes using modified $\Delta H$ symbols. These symbols are described below.

- *molar enthalpy of vaporization*, $\Delta H_{vap}$: the enthalpy change for the state change of one mole from liquid to gas
- *molar enthalpy of condensation*, $\Delta H_{cond}$: the enthalpy change for the state change of one mole of a substance from gas to liquid
- *molar enthalpy of melting*, $\Delta H_{melt}$: the enthalpy change for the state change of one mole of a substance from solid to liquid
- *molar enthalpy of freezing*, $\Delta H_{fre}$: the enthalpy change for the state change of one mole of a substance from liquid to solid

> **CHEM FACT**
>
> The process of melting is also known as *fusion*. Therefore, you will sometimes see the enthalpy of melting referred to as the *enthalpy of fusion*, $\Delta H_{fus}$.

Vaporization and condensation are opposite processes. Thus, the enthalpy changes for these processes have the same value but opposite signs. For example, 6.02 kJ of heat is needed to vaporize one mole of water. Therefore, 6.02 kJ of heat is released when one mole of water freezes.

$$\Delta H_{vap} = -\Delta H_{cond}$$

Similarly, melting and freezing are opposite processes.

$$\Delta H_{melt} = -\Delta H_{fre}$$

Several molar enthalpies of melting and vaporization are shown in Table 16.4. Notice that the same units (kJ/mol) are used for the enthalpies of melting, vaporization, condensation, and freezing. Also notice that energy changes associated with changes of state can vary widely.

**Table 16.4** Enthalpies of Melting and Vaporization for Several Substances

| Substance | Enthalpy of melting, $\Delta H_{melt}$ (kJ/mol) | Enthalpy of vaporization, $\Delta H_{vap}$ (kJ/mol) |
|---|---|---|
| argon | 1.3 | 6.3 |
| diethyl ether | 7.3 | 29 |
| ethanol | 5.0 | 40.5 |
| mercury | 23.4 | 59 |
| methane | 8.9 | 0.94 |
| sodium chloride | 27.2 | 207 |
| water | 6.02 | 40.7 |

## Enthalpy of Solution

Another type of physical change that involves a heat transfer is dissolution. When 1 mol of a solute dissolves in a solvent, the enthalpy change that occurs is called the *molar enthalpy of solution*, $\Delta H_{soln}$. Dissolution can be either endothermic or exothermic.

Manufacturers take advantage of endothermic dissolution to produce cold packs that athletes can use to treat injuries. One type of cold pack contains water and a salt, such as ammonium nitrate, in separate compartments. When you crush the pack, the membrane that divides the compartments breaks, and the salt dissolves. This dissolution process is endothermic. It absorbs heat for a short time, so the cold pack feels cold. Figure 16.7 on the next page shows how a cold pack works.

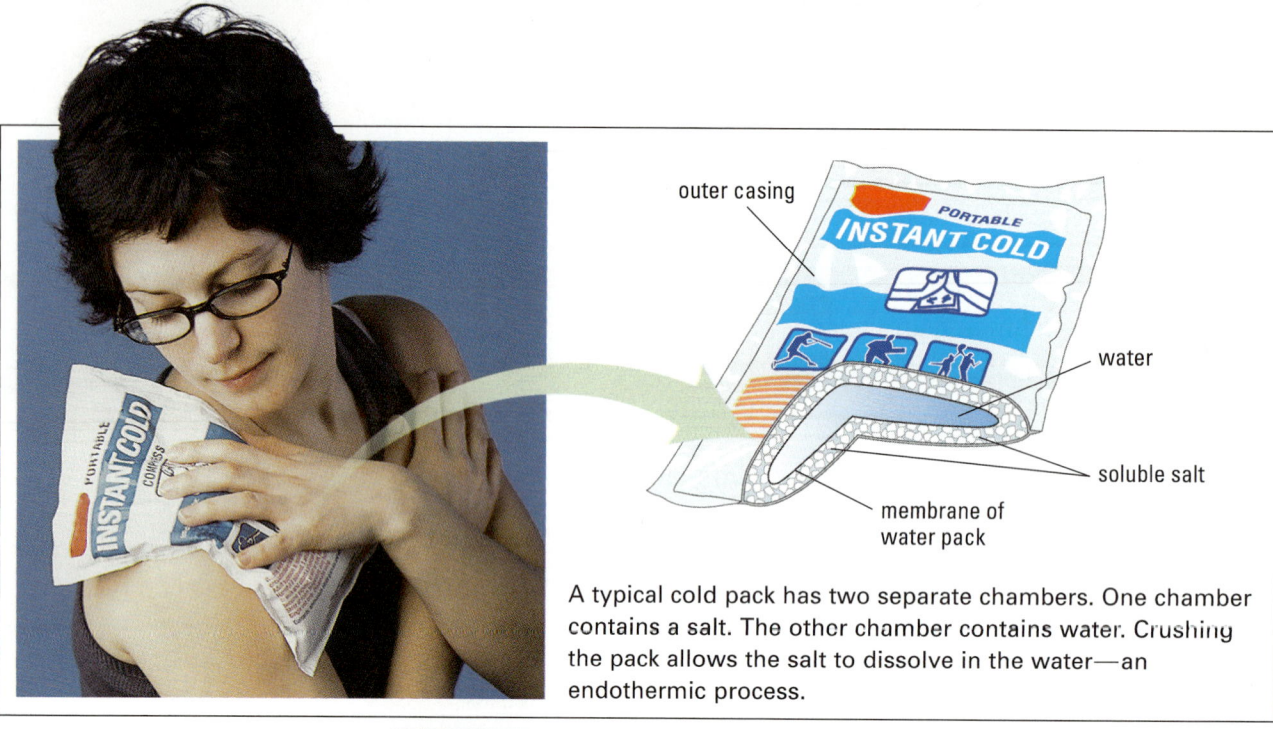

A typical cold pack has two separate chambers. One chamber contains a salt. The other chamber contains water. Crushing the pack allows the salt to dissolve in the water—an endothermic process.

**Figure 16.17** This person's shoulder was injured. Using a cold pack helps to reduce the inflammation of the joint.

Some types of hot packs are constructed in much the same way as the cold packs described above. They have two compartments. One compartment contains a salt, such as calcium chloride. The other compartment contains water. In hot packs, however, the dissolution process is exothermic. The process releases heat to the surroundings.

The molar enthalpy of solution, $\Delta H_{soln}$, of ammonium nitrate is 25.7 kJ/mol.

$$NH_4NO_{3(s)} + 25.7 \text{ kJ} \rightarrow NH_4NO_{3(aq)}$$

The molar enthalpy of solution, $\Delta H_{soln}$, of calcium chloride is −82.8 kJ/mol.

$$CaCl_{2(s)} \rightarrow CaCl_{2(aq)} + 82.8 \text{ kJ}$$

## Heat Absorbed or Released by a Physical Change

You can determine the heat absorbed or released by a state change or dissolution using the equation $q = n\Delta H$, just as you did with chemical reactions. Work through the following problems to practice determining $q$ for physical changes. Refer to Table 16.4 for enthalpies of melting and vaporization.

### Practice Problems

24. An ice cube with a mass of 8.2 g is placed in some lemonade. The ice cube melts completely. How much heat does the ice cube absorb from the lemonade as it melts?

25. A teacup contains 0.100 kg of water at its freezing point. The water freezes solid.
    (a) How much heat is released to the surroundings?
    (b) How much heat would be required to melt the ice in the teacup?

26. A sample of liquid mercury vaporizes. The mercury is at its boiling point and has a mass of 0.325 g.
   (a) Is heat absorbed from the surroundings or released to the surroundings?
   (b) How much heat is absorbed or released to the surroundings by the process?

27. The molar enthalpy of solution for sodium chloride, NaCl, is 3.9 kJ/mol.
   (a) Write a thermochemical equation to represent the dissolution of sodium chloride.
   (b) Suppose you dissolve 25.3 g of sodium chloride in a glass of water at room temperature. How much heat is absorbed or released by the process?
   (c) Do you expect the glass containing the salt solution to feel warm or cool? Explain your answer.

28. What mass of diethyl ether, $C_4H_{10}O$, can be vaporized by adding 80.7 kJ of heat?

29. $3.97 \times 10^4$ J of heat is required to vaporize 100 g of benzene, $C_6H_6$. What is the molar enthalpy of vaporization of benzene?

## Section Summary

In this section, you learned that chemical reactions and physical changes involve changes in potential energy. In the next section, you will incorporate what you know about changes in kinetic energy and potential energy. You will analyze what happens when you heat a solid until it melts, then continue to heat the liquid until it vaporizes. You will track the temperature change and comment on whether a change in kinetic energy or potential energy is taking place.

## Section Review

**1** In your own words, explain why exothermic reactions have $\Delta H < 0$.

**2** Label each thermochemical equation with the most specific form(s) of $\Delta H$. Remember to pay attention to the *sign* of $\Delta H$.
   (a) $Ag_{(s)} + \frac{1}{2}Cl_{2(g)} \rightarrow AgCl_{(s)} + 127.0$ kJ (at 25°C and 100 kPa)
   (b) $44.0$ kJ $+ H_2O_{(\ell)} \rightarrow H_2O_{(g)}$ (at 25°C and 100 kPa)
   (c) $C_2H_{4(g)} + 3O_{2(g)} \rightarrow 2CO_{2(g)} + 2H_2O_{(g)} +$ energy

**3** A pot of water boils on a stove. The temperature of the liquid water remains at 100°C.
   (a) What type of energy change is taking place within the water—a change in kinetic energy or a change in potential energy?
   (b) Explain why the temperature of the water does not change as the water boils, even though heat is continuously absorbed by the water from the stove element.
   (c) How much heat is needed to boil 2.32 kg of water at 100°C?

**4** A group of campers light a bonfire by touching the flame from a butane lighter to some kindling. They roast marshmallows over the blaze. Describe the energy transfers involved using the terms potential energy, kinetic energy, temperature, system, surroundings, endothermic, exothermic, and combustion reaction.

**5** Acetylene, $C_2H_2$, undergoes complete combustion in oxygen. Carbon dioxide and water are formed. The standard molar enthalpy of the complete combustion of acetylene is $1.3 \times 10^3$ kJ/mol.

(a) Write a thermochemical equation for this reaction.

(b) Draw a diagram to represent the thermochemical equation.

(c) How much heat is released during the complete combustion of 2.17 g of acetylene?

**6** Write an equation to represent each phase change in Table 16.4. Include the enthalpy change as a heat term in the equation.

**7** When one mole of gaseous water forms from its elements, 241.8 kJ of energy is released. In other words, when hydrogen burns in oxygen or air, it produces a great deal of energy. Since the nineteenth century, scientists have been researching the potential of hydrogen as a fuel. One way in which the energy of the combustion of hydrogen has been successfully harnessed is as rocket fuel for aircraft.

(a) Write a thermodynamic equation for the combustion of hydrogen.

(b) Suggest three reasons why hydrogen gas is a desirable rocket fuel.

(c) Suggest challenges that engineers might have had to overcome in order to make hydrogen a workable rocket fuel for aircraft.

**8** Calcium oxide, CaO, reacts with carbon in the form of graphite. Calcium carbide, $CaC_2$, and carbon monoxide, CO, are produced in an endothermic reaction.

$CaO_{(s)} + 3C_{(s)} + 462.3 \text{ kJ} \rightarrow CaC_{2(s)} + CO_{(g)}$

(a) 246.7 kJ of heat is absorbed by the system. What mass of calcium carbide is produced?

(b) How much heat is absorbed by the system if 46.7 g of graphite reacts with excess calcium oxide?

(c) $1.38 \times 10^{24}$ formula units of calcium oxide react with excess graphite. How much energy is needed?

# Heating and Cooling Curves

## 16.3

Water is a remarkable substance in many ways. For example, water is one of very few substances that is found naturally on Earth in all three states: gas, liquid, and solid. You observe water changing state all the time in your daily life. Water condenses in the air and falls as rain; when the sun comes out, the puddles evaporate. Icebergs melt as they travel through the ocean. Ice cubes form from liquid water in a freezer.

Chemists use heating curves and cooling curves to represent temperature and phase changes in substances such as water. A **cooling curve** shows how the temperature of a substance changes as heat is removed from it. A **heating curve** shows how the temperature of a substance changes as heat is added to it. Heating curves and cooling curves show temperature on the y-axis, and either heat transfer or time on the x-axis. Figure 16.18 shows a cooling curve that represents how the temperature of water changes as it cools from a gas at 130°C to ice at −40°C.

### Section Preview/Outcomes

In this section, you will

- **use** heating and cooling curves to represent and explain changes in potential energy and kinetic energy of a system
- **calculate** the total heat for a multi-step process that includes a temperature change and change in state
- **communicate** your understanding of the following terms: *cooling curve, heating curve*

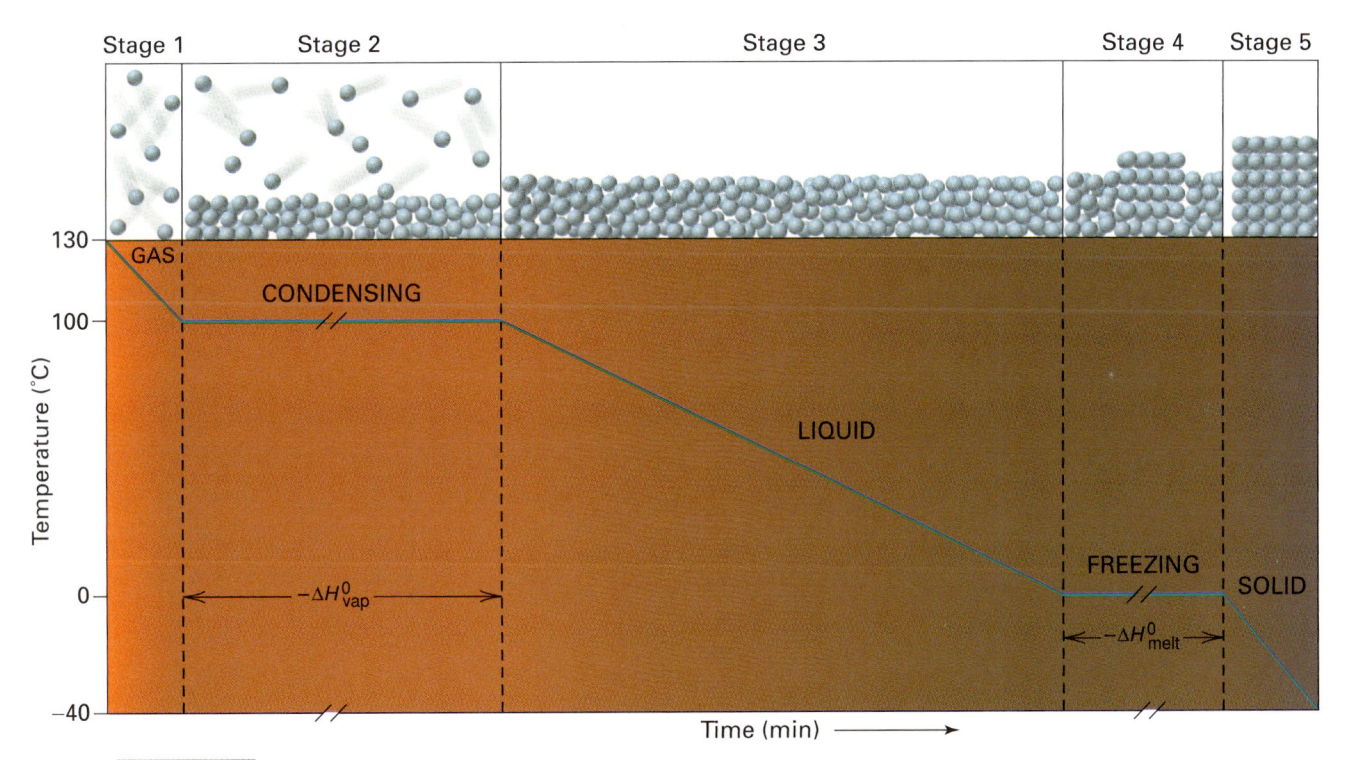

**Figure 16.18** This cooling curve shows the conversion of water vapour to ice as heat is removed from the system.

Refer to Figure 16.18 as you read about Stages 1–5.

**Stage 1** The gaseous water cools until it reaches the boiling point of water (100°C). The average kinetic energy of the water particles is decreasing, therefore the temperature decreases.

**Stage 2** The gaseous water condenses, forming liquid water. Only potential energy is changing. Because the average kinetic energy of the water particles does not change, the temperature does not change.

Chapter 16 Theories of Energy Changes • MHR  **651**

**Stage 3** The liquid water cools until it reaches the melting point of water (0°C). The average kinetic energy of the water particles is decreasing, therefore the temperature decreases.

**Stage 4** The liquid water freezes, forming ice. Only potential energy is changing. The average kinetic energy of the particles does not change, therefore the temperature does not change.

**Stage 5** The ice cools further. The average kinetic energy of the water particles is decreasing, therefore the temperature decreases.

Heating and cooling curves do not need to show all possible changes of state. For example, the cooling curve in Figure 16.19 shows the conversion of liquid mercury to solid mercury. As you now know, the flat portions of a heating or cooling curve occur at either the melting point or the boiling point of the substance. In this case, since the title of the curve indicates that the mercury started out as a liquid, you know that the flat portion represents the melting point (which is the same as the freezing point). Based on the graph, therefore, you can tell that the melting point of mercury is −39°C.

The heating curve in Figure 16.20 shows the conversion of liquid ethanol to gaseous ethanol. Since the title of the curve indicates that the ethanol started out as a liquid, you know that as heat is added, the liquid will eventually vaporize. While the liquid is vaporizing, the temperature does not change. Based on the graph, therefore, you can tell that the boiling point of ethanol is 78.5°C.

It is often helpful to draw heating curves and cooling curves when solving problems involving temperature change and state change. In the laboratory, you can determine melting points or boiling points by tracking temperature changes and constructing heating or cooling curves.

In the following ExpressLab, you will construct a heating curve and a cooling curve for lauric acid.

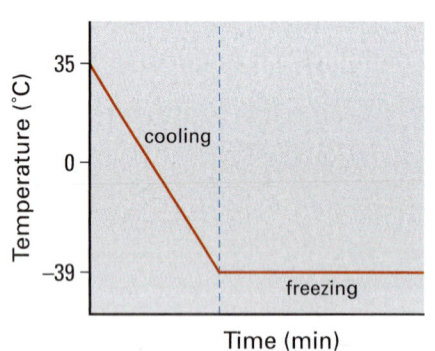

**Figure 16.19** A cooling curve for mercury. The graph is not drawn to scale.

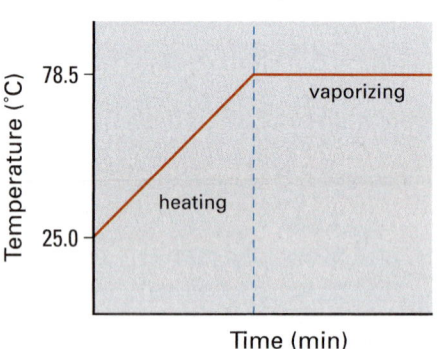

**Figure 16.20** A heating curve for ethanol. The graph is not drawn to scale.

# ExpressLab: Construct a Heating Curve and a Cooling Curve

## Safety

## Materials
- 10 g lauric acid
- test tube
- test tube tongs
- test tube holder
- hot plate
- thermometer
- stopwatch
- 250 mL beakers

## Procedure

1. Read the procedure carefully. Prepare a data table as shown below to record your data.

| Time (s) | Temperature (°C) for cooling curve | Temperature (°C) for heating curve |
|----------|-----------------------------------|-----------------------------------|
| 0        |                                   |                                   |
| 30       |                                   |                                   |

2. Fill a 250 mL beaker with about 200 mL of warm water. Using the hot plate, heat the water to about 60°C. Adjust the hot plate settings so that you maintain the temperature at 60°C.

3. Place your sample of lauric acid in the 60°C water on the hot plate. When the lauric acid starts to melt, place the thermometer in the test tube. Do not allow the thermometer to touch the bottom of the test tube.

4. When the temperature has reached about 55°C, use the test tube holder to transfer the test tube to the test-tube holder.

5. Immediately record the temperature of the lauric acid. Every 30 seconds, record the temperature again. Stir the lauric acid gently with the thermometer.

6. Continue to record data until the lauric acid solidifies. Do not stop taking data until the temperature is well below the freezing point.

7. Check that your 60°C bath is still at 60°C. Return the test tube to the 60°C bath.

8. Immediately record the temperature of the lauric acid. Every 30 seconds, record the temperature again. Stir the lauric acid gently with the thermometer once the lauric acid has melted somewhat. Stop recording data once the temperature reaches 55°C.

9. Remove the thermometer and wipe it clean. Clean up as directed by your teacher.

10. Use your data to create two graphs: one to show cooling data and one to show heating data. Label your x-axis "time (s)" and your y-axis "temperature (°C)". Name your graphs appropriately.

## Analysis

1. What is the freezing point of lauric acid? What is the melting point of lauric acid?

2. Your graphs will probably have both "flat" and "sloped" portions. Answer the following questions for each flat and sloped portion of each graph.
   (a) What type of energy change (kinetic or potential) is taking place? How do you know?
   (b) What is happening to the average speed of the lauric acid molecules?

3. Label the graph appropriately with the following labels: heating, cooling, melting, freezing, melting point, freezing point. (You will need to use some labels multiple times.)

## Total Heat Absorbed or Released by a System

As you discovered in the ExpressLab, substances undergo a sequence of temperature changes and state changes as they absorb or release heat. Suppose you wanted to determine how much heat is absorbed by a system that experiences both a temperature change and a change of state?

For example, suppose you are preparing to boil some water for pasta. You place 5 L of water at 10°C on the stove. Unfortunately, you forget about the water. When you return, the water has completely boiled away. You want to know how much heat was required to heat the liquid water to 100°C, and then completely vaporize it. How would you carry out this calculation?

You can calculate the heat required to heat and boil the water using the two equations you learned in sections 16.1 and 16.2.

$$q = mc\Delta T \quad \text{(equation 1)}$$
$$q = n\Delta H \quad \text{(equation 2)}$$

Recall that equation 1 is used to determine the heat absorbed or released by a system in which only the kinetic energy is changing. Conversely, equation 2 is used to determine the heat absorbed or released by a system in which only the potential energy is changing. You must use both equations, because heating the water involves a change in kinetic energy, while boiling the water involves a change in potential energy. Work through the following Sample Problem to see how to use the equations to determine the total heat absorbed by the water.

### Sample Problem

### Determining the Overall Heat Transfer of a System

**Problem**

How much heat is absorbed by 5.00 kg of water at 10.0°C if it is heated until it is converted completely into water vapour?

**What Is Required?**

You need to determine $q$ for 5.00 kg of water heated from 10.0°C to the boiling point of water, plus $q$ required for the change of state from water to steam.

**What Is Given?**

$T_i = 10.0°C$

$T_f = 100.0°C$ (the boiling point of water)

$m = 5.00$ kg $= 5.00 \times 10^3$ g

$c = 4.184$ J/g°C (from Table 16.1)

$\Delta H_{vap} = 40.7$ kJ/mol

**Plan Your Strategy**

**Step 1** Use $q_1 = mc\Delta T$ to determine the heat absorbed by the water as it heats from 10°C to 100°C.

**Step 2** Use $q_2 = n\Delta H$ to determine the heat absorbed by the water as it vaporizes.

**Step 3** Add your answers together. Be sure to express both answers in kJ before adding them.

**Act on Your Strategy**

**Step 1** $q_1 = mc\Delta T$

$= (5.00 \times 10^3 \text{ g})(4.184 \text{ J/g} \cdot °C)(100.0°C - 10.0°C)$

$= (5.00 \times 10^3 \text{ g})(4.184 \text{ J/g} \cdot °C)(90.0°C)$

$= 1.88 \times 10^6$ J

The heat absorbed by the water as it heats to the boiling point is $1.88 \times 10^6$ J.

**Step 2** $M = (2 \times 1.01 \text{ g/mol}) + 16.00 \text{ g/mol} = 18.02 \text{ g/mol}$

$n = \dfrac{m}{M}$

$= \dfrac{5.00 \times 10^3 \text{ g}}{18.02 \text{ g/mol}}$

$= 277 \text{ mol}$

$q_2 = n\Delta H°_{vap}$

$= (277)(40.7 \text{ kJ/mol})$

$= 1.13 \times 10^4 \text{ kJ}$

The water absorbs $1.13 \times 10^4$ kJ of heat as it vaporizes.

**Step 3** Convert the answer from Step 1 to kJ before adding the answers together.

$q_1 = (1.88 \times 10^6 \text{ J})\left(\dfrac{1 \text{ kJ}}{1000 \text{ J}}\right) = 1.88 \times 10^3 \text{ kJ}$

$q_{total} = q_1 + q_2$

$= 1.88 \times 10^3 \text{ kJ} + 1.13 \times 10^4 \text{ kJ}$

$= 1.32 \times 10^4 \text{ kJ}$

The total heat absorbed by the water is $1.32 \times 10^4$ kJ.

### Check Your Solution

The solution has the correct number of significant digits. The units are correct. Values for $q_1$ and $q_2$ are positive, which makes sense since the water absorbed heat as its temperature increased and as it changed state from liquid to gas.

## Practice Problems

**30.** 1.451 kg of water at 25.2°C is added to a kettle and the water is completely vaporized.
  (a) Draw a heating curve for this process.
  (b) How much heat is required for the vaporization of all the water?

**31.** A metal bucket containing 532.1 g of ice at 0.00°C is placed on a wood-burning stove. The ice is melted and then heated to 45.21°C.
  (a) Draw a heating curve for this process.
  (b) How much heat is required to melt the ice and warm the water to 45.21°C?

**32.** A sample of liquid mercury having a mass of 0.0145 g cools from 35.1°C and forms solid mercury at −38.8°C. The melting point of mercury is −38.8°C and its specific heat capacity is 0.140 J/g°C.
  (a) Draw a cooling curve for this process.
  (b) How much heat is released during this process?

**33.** A sample of 36.8 g of ethanol gas at 300.0°C is cooled to liquid ethanol at 25.5°C. The specific heat capacity of ethanol gas is 1.43 J/g°C, and the specific heat capacity of liquid ethanol is 2.45 J/g°C. The boiling point of ethanol is 78.5°C.
  (a) Draw a cooling curve for this process.
  (b) Calculate the heat released during this process.

**34.** Calculate the total heat released when 200 g of water vapour at 300.0°C is cooled until it reaches −20.0°C. The specific heat capacity of ice is 2.02 J/g°C. The specific heat capacity of water vapour is 1.99 J/g°C.

> **PROBLEM TIP**
> You will need to refer to Table 16.4 to solve the Practice Problems.

## Section Summary

In this section, you learned how to calculate the total heat associated with changes involving both kinetic energy and potential energy. In Chapter 17, you will find out how to use changes in kinetic energy to determine potential energy changes associated with chemical reactions.

## Section Review

**1** The portions of heating curves and cooling curves that represent changes of state are horizontal (flat). Explain why this is the case.

**2** Iron melts at 1535°C. Draw a cooling curve to show what happens when molten iron at 1600°C cools to solid iron at 50°C. Label the axes and sections of your cooling curve appropriately.

**3** A pure gold chain with a mass of 10.23 g is heated until it forms liquid gold at the melting point of 1064.2°C. The temperature of the chain is initially 21.0°C. The enthalpy of freezing for gold is 12.5 kJ/mol.

  **(a)** Draw a heating curve to represent what happens to the gold.

  **(b)** What is the total heat absorbed by the gold as it warms and then melts?

**4** An ice cube with a mass of 3.375 g sits in an empty glass of water on a hot day. The ice cube is initially at −5.2°C. It melts and the liquid water warms to 27.3°C.

  **(a)** Draw a heating curve to represent what happens to the ice cube.

  **(b)** What is the total heat absorbed by the water as it warms from ice at −5.2°C to liquid water at 27.3°C? The specific heat capacity of ice is 2.02 J/g°C.

**5** 1.55 kg of water vapour at 125.4°C cools to eventually form ice at −5.5°C. The specific heat capacity of ice is 2.02 J/g°C. The specific heat capacity of water vapour is 1.99 J/°C.

  **(a)** Draw a cooling curve to represent the process.

  **(b)** How much heat is released during this process?

**6** During cold weather, fruit farmers spray their fruit with water to prevent frost damage. Explain why this practice works.

**7** Water vapour at 100°C causes more severe burns than liquid water at 100°C. Explain why this is the case.

**8** If you became lost in a forest in winter, why would it be better to drink water from a stream rather than eat snow to quench your thirst?

**9** Examine the heating curve for lead shown in Figure 16.21.

  **(a)** What is the boiling point of lead?

  **(b)** What is the melting point of lead?

  **(c)** A 500 g sample of solid lead at −25.0°C is heated until it melts completely. How much heat is absorbed? The specific heat capacity of solid lead is 0.127 J/g°C. The enthalpy of melting of lead is 5.08 kJ/mol.

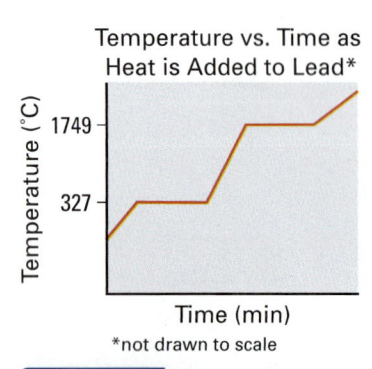

**Figure 16.21** Use this heating curve to determine the melting point and the boiling point of lead.

# CHAPTER 16 Review

## Reflecting on Chapter 16

Summarize this chapter in the format of your choice. Here are a few ideas to use as guidelines:
- Explain what happens to the particles of a system during heat transfer.
- Describe heat in terms of transfer of kinetic energy.
- Explain how the temperature change of a substance depends on the nature of the substance, the heat absorbed or released, and the mass of the substance.
- Show how to calculate the heat absorbed or released by a system during a temperature change.
- Demonstrate various ways to communicate the enthalpy change associated with a process.
- Explain the relationship between enthalpy change and potential energy.
- Explain the energy changes related to bond breaking and bond making.
- List various processes that involve a change in enthalpy.
- Show how to calculate the heat absorbed or released by a system during a chemical or physical change.
- Explain what is happening to the particles of a substance during the different stages shown on a heating curve or a cooling curve.
- Show how to calculate the total heat absorbed or released during a multi-stage process that involves changes in both kinetic energy and potential energy.

## Reviewing Key Terms

closed system
endothermic reaction
enthalpy change ($\Delta H$)
exothermic reaction
heat capacity ($C$)
isolated system
kinetic energy
potential energy
standard enthalpy of reaction
system
thermochemical equation
thermodynamics
cooling curve
enthalpy ($H$)
enthalpy of reaction
heat ($q$)
heating curve
Joule (J)
open system
specific heat capacity ($c$)
surroundings
temperature ($T$)
thermochemistry

## Knowledge/Understanding

1. In your own words, define and distinguish between closed, open, and isolated systems.
2. The vaporization of liquid carbon disulfide, $CS_2$, requires an energy input of 29 kJ/mol.
   (a) Is this reaction exothermic or endothermic?
   (b) What is the molar enthalpy of vaporization of carbon disulfide?
   (d) Write a thermochemical equation to represent the vaporization of carbon disulfide. Include 29 kJ as a heat term in the equation.
   (e) Draw and label an enthalpy diagram for the vaporization of liquid carbon disulfide.
3. Describe the relationship between the amount of heat released by water and the following factors:
   (a) the mass of the water
   (b) the temperature change of the water
4. (a) What are the products of the complete combustion of a hydrocarbon?
   (b) What products can form if the combustion is incomplete?
   (c) Why can incomplete combustion be dangerous if it occurs in your home?
   (d) Under what circumstances does incomplete combustion take place?
5. Define and distinguish between kinetic energy and potential energy and give one example of each of the following types of changes.
   (a) a process that involves changes in kinetic energy only
   (b) a process that involves changes in potential energy only
   (c) a process that involves changes in both potential and kinetic energy
6. State the first law of thermodynamics and express the law as an equation.
7. Explain briefly why areas of land that are not close to water tend to experience large variations in temperature, as compared to areas of land near lakes or oceans.
8. Define and distinguish between heat capacity and specific heat capacity. Use examples in your answer.
9. Explain how sweating helps your body to cool off on a hot day.

Answers to questions highlighted in red type are provided in Appendix A.

10. An equal amount of heat is absorbed by a 25 g sample of aluminum ($c = 0.900$ J/g°C) and a 25 g sample of nickel ($c = 0.444$ J/g°C). Which metal will show a greater increase in temperature? Explain your answer.

## Inquiry

11. To make four cups of tea, 1.00 kg of water is heated from 22.0°C to 99.0°C. How much heat is added?

12. A sample of graphite with a mass of 2.35 kg experiences a temperature increase of 3.45°C. How much heat did the graphite absorb? Refer to Table 16.1 for the specific heat capacity of graphite.

13. In an ice cube tray in a freezer, 98.4 g of water at 0.00°C freezes solid. How much heat is released when the water freezes?

14. Hydrogen sulfide gas, $H_2S_{(g)}$, has a distinct, powerful smell of rotten eggs. The gas undergoes a combustion reaction with oxygen to produce gaseous sulfur dioxide and gaseous water. The molar enthalpy of combustion of hydrogen sulfide gas is −519 kJ/mol.
    (a) Write a balanced thermochemical equation to represent the combustion of hydrogen sulfide gas.
    (b) How much heat is released when 15.0 g of hydrogen sulfide gas undergoes combustion?
    (c) A sample of hydrogen sulfide gas undergoes combustion. 47.2 kJ of heat are released. What volume of sulfur dioxide at STP is produced by the reaction?

15. A 10.0 g sample of pure acetic acid, $CH_3COOH$, is completely burned in oxygen. 144.77 kJ of heat are released. What is the molar enthalpy of combustion of acetic acid?

16. The molar enthalpy of combustion of methanol, $CH_3OH_{(\ell)}$, is −727 kJ/mol.
    (a) Write a balanced thermochemical equation to show the complete combustion of methanol to form water and carbon dioxide.
    (b) 44.3 g of methanol undergoes complete combustion. How much heat is released?
    (c) A sample of methanol undergoes complete combustion to form 56.2 L of carbon dioxide at STP. How much heat is released?

17. The standard molar enthalpy of formation of poisonous hydrogen cyanide gas, $HCN_{(g)}$, is 135 kJ/mol.
    (a) Write a balanced thermochemical equation to show the formation of 1 mol of hydrogen cyanide gas from its elements.
    (b) How much heat would be required to form 50.0 L of $HCN_{(g)}$ at STP from its elements?
    (c) How much heat would be released if 25.3 g of $HCN_{(g)}$ were decomposed to form its elements?

18. A certain type of hot pack uses the enthalpy of solution of calcium chloride to release heat. The molar enthalpy of solution of calcium chloride is −82.8 kJ/mol.
    (a) Write a balanced thermochemical equation to show the enthalpy of solution of calcium chloride.
    (b) A hot pack contains 15.0 g of calcium chloride. How much heat is released when the 15.0 g of calcium chloride dissolves in water?
    (c) Assume that all of the heat released in question (b) is transferred to 215.0 g of calcium chloride solution. The initial temperature is 25°C. What is the final temperature of the solution? (Assume the solution has the same specific heat capacity as water.)

19. 2.5 kg of ice at −5.3°C is heated until it becomes water vapour at 250.0°C. The specific heat capacity of ice is 2.02 J/g°C. The specific heat capacity of water vapour is 1.99 J/g°C.
    (a) Calculate the total heat absorbed by the process.
    (b) Draw a heating curve to show the process.

20. Examine the heating curve on the following page to answer the following questions.
    (a) What is the melting point of methanol?

(b) What is the boiling point of methanol?
(c) The molar enthalpy of vaporization of methanol is 38 kJ/mol. 1.50 kg of liquid methanol at 25.0°C is heated to form methanol vapour at 65.0°C. How much heat is absorbed in the process? The specific heat capacity of liquid methanol is 2.350 J/g°C.

**Temperature vs. Time as Heat is Added to Methanol***

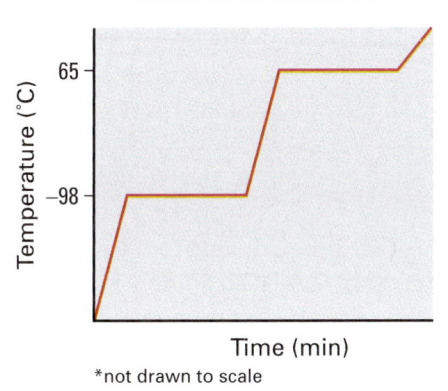

*not drawn to scale

## Communication

21. Aqueous hydrogen peroxide, $H_2O_{2(aq)}$, decomposes slowly to form water and oxygen. When 1 mol of aqueous hydrogen peroxide decomposes, 94.64 kJ of heat is released. The rate of the reaction is increased by the addition of a catalyst such as manganese(IV) oxide, $MnO_2$.
    (a) Write a thermochemical equation to show the decomposition of aqueous hydrogen peroxide.
    (b) Does the heat released by the decomposition of hydrogen peroxide change if a catalyst is added to speed up the reaction? Explain your answer in detail.

22. You are heating soup on the stove, and leave a metal spoon in the pot of hot soup. You accidentally touch the hot spoon with your hand. Immediately you turn on the cold water tap and hold your hand under the cold water. Describe in detail how heat is transferred in this scenario, using your knowledge of kinetic energy and the particle model of matter.

23. Explain in detail how the energy changes involved in bond breaking and bond making determine whether a chemical reaction is endothermic or exothermic. Use a diagram to illustrate your explanation.

## Making Connections

24. When walking briskly, you use about 20 kJ of energy per minute. Eating one serving of a whole wheat cereal (37.5 g) provides about 527.18 kJ of energy.
    (a) How long could you walk after eating one serving of cereal?
    (b) What mass of cereal would provide enough energy for you to walk for 4.0 h?
    (c) An average apple provides 283 kJ of energy. How many apples would provide enough energy for you to walk for 4.0 h?

**Answers to Practice Problems and Short Answers to Section Review Questions**

**Practice Problems** 1. $6.2 \times 10^3$ J  2. $-4.6 \times 10^3$ J
3. Beaker B  4.(a) +4.22 J; +4.6J (b) +487 kJ (c) −290 kJ
5.(a) $m = q/c\Delta T$ (b) $c = q/m\Delta T$ (c) $\Delta T = q/mc$
6.(a) $T_i = T_f - q/mc$ (b) $T_f = T_i + q/mc$  7. $2.14 \times 10^5$ J
8. 0.7900 J/g°C; granite  9. 1.21 J/g°C  10. 15.5°C
11.(a) 418.4 kJ/°C (b) $1.67 \times 10^4$ kJ (c) $1.67 \times 10^4$ kJ
12.(a) 0.4184 kJ/°C (b) 16.7 kJ (c) 16.7 kJ
13.(a) $2.43 \times 10^{-3}$ kJ/°C (b) 0.0792 kJ  14.(a) 242 kJ/°C (b) 242 kJ
15.(a) $H_{2(g)} + 1/2O_{2(g)} \rightarrow H_2O_{(g)} + 241.8$ kJ
(b) $Ca_{(s)} + Cl_{2(g)} \rightarrow CaCl_{2(s)} + 795.4$ kJ
(c) $C_{(s)} + 2H_{2(g)} \rightarrow CH_{4(g)} + 74.6$ kJ
(d) $49 kJ + 6C_{(s)} + 3H_{2(g)} \rightarrow C_6H_{6(\ell)}$
(e) $C_{(s)} + 2H_{2(g)} \rightarrow CH_{4(g)} \Delta H^0_f = -74.6$ kJ/mol;
$6C_{(s)} + 3H_{2(g)} \rightarrow C_6H_{6(\ell)} \Delta H^0_f = +49$ kJ/mol  16. products $(NaCl_{(s)})$ should be shown below reactants $(Na_{(s)} + 1/2Cl_{2(g)})$; $\Delta H^0_f = -411.1$ kJ/mol
17.(a) $C_2H_{6(g)} + 7/2O_{2(g)} \rightarrow 3H_2O_{(\ell)} + 2CO_{2(g)} + 1250.9$ kJ
(b) $C_3H_{8(g)} + 5O_{2(g)} \rightarrow 4H_2O_{(\ell)} + 3CO_{2(g)} + 2323.7$ kJ
(c) $C_4H_{10(g)} + 13/2O_{2(g)} \rightarrow 5H_2O_{(\ell)} + 4CO_{2(g)} + 3003.0$ kJ
(d) $C_5H_{12(\ell)} + 8O_{2(g)} \rightarrow 6H_2O_{(\ell)} + 5CO_{2(g)} + 3682.3$ kJ
18. products $(8H_2O_{(\ell)} + 7CO_{2(g)})$ should be shown below reactants $(C_7H_{16(\ell)} + 11O_{2(g)})$; $\Delta H^0_f = -5040.9$ kJ/mol
19.(a) 8.46 kJ (b) 7.16 kJ  20. $1.37 \times 10^3$ kJ
21.(a) $2.896 \times 10^3$ kJ (b) $6.81 \times 10^4$ kJ (c) $1.186 \times 10^6$ kJ
22. 0.230 kJ  23. $3.14 \times 10^3$ g  24. 2.74 kJ  25.(a) 33.4 kJ
(b) 33.4 kJ  26.(a) absorbed (b) 0.0956 kJ
27.(a) $NaCl_{(s)} + 3.9$ kJ/mol $\rightarrow NaCl_{(aq)}$ (b) 1.69 kJ (c) cool; heat absorbed from water  28. 206 g  29. $3.10 \times 10^4$ kJ/mol
30.(b) $3.73 \times 10^3$ kJ  31.(b) 279 kJ  32.(b) $-1.84 \times 10^{-3}$ kJ
33.(b) −48.8 kJ  34. −690 kJ
**Section Review  16.1:** 4.(a) $-4.54 \times 10^3$ J  5.(a) 38 J
(b) 76 J (c) 76 J  6.(a) 0.237 J/g°C (b) 5.66 J/°C
7. copper  **16.2:** 3.(c) $5.24 \times 10^3$ kJ
5.(a) $C_2H_{2(g)} + 5/2O_{2(g)} \rightarrow H_2O_{(\ell)} + 2CO_{2(g)} + 1.3 \times 10^3$ kJ
(c) 104 kJ  8.(a) 34.21 g (b) 599 kJ (c) $1.06 \times 10^3$ kJ
**16.3:** 3.(b) 2.03 kJ  4.(b) 1.55 kJ  5.(b) $4.76 \times 10^3$ kJ  9.(a) 1749°C
(b) 327.5°C (c) 34.7 kJ

# CHAPTER 17
# Measuring and Using Energy Changes

**Chapter Preview**

17.1 The Technology of Heat Measurement

17.2 Hess's Law of Heat Summation

17.3 Fuelling Society

**W**hy is it important to know how to determine the energy changes associated with chemical and physical changes?

Engineers need to know how much energy is released from different fuels when they design an engine and decide between different fuels. Firefighters need to know how much heat can be given off by the combustion of different materials so they can decide on the best way to fight a specific fire. Manufacturers of hot packs need to know how much heat is released by a given exothermic process so that their pack will warm but not harm the user.

How do you determine the heat absorbed or released by chemical and physical processes? In this section, you will learn some ways to determine the enthalpy changes of various processes by experiment, based on the heat they release or absorb. You will apply what you have learned by performing your own heat experiments. You will also learn how to use tabulated values to determine enthalpies of physical and chemical processes. Finally, you will examine the efficiency and environmental impact of traditional and alternative energy sources.

**Prerequisite Concepts and Skills**

Before you begin this chapter, review the following concepts and skills:

- writing balanced chemical equations (Chapter 1)
- performing stoichiometric calculations and conversions (Chapter 2)

Different substances release different amounts of heat when they burn. What are some ways to determine the enthalpy of combustion of a substance?

# The Technology of Heat Measurement

## 17.1

In the ThoughtLab in section 16.1, two students used beakers with no lids when they measured change in temperature. The students assumed that energy was being exchanged only between the ice and the water. In fact, energy was also being exchanged with the beaker, the lab bench, and the surrounding air. As a result, the data that the students obtained had a large experimental error. How could the students have prevented this error?

Much of the technology in our lives is designed to stop the flow of kinetic energy as heat. Your home is insulated to prevent heat loss in the winter and heat gain in the summer. If you take hot soup to school for your lunch, you probably use a Thermos™ to prevent heat loss to the environment. Whenever there is a temperature difference between two objects, kinetic energy is transferred as heat from the hotter object to the colder object. You measure the heat being transferred in a reaction or other process by monitoring temperature change. Therefore, you must minimize any heat transfer between the system and portions of the surroundings whose temperature change you are *not* measuring.

### Section Preview/Outcomes

In this section, you will

- **identify** the basic instrument for measuring heat transfer
- **determine** the enthalpy changes associated with processes such as neutralization, dissolution, and combustion
- **draw** and interpret enthalpy diagrams based on experimental data for chemical changes
- **communicate** your understanding of the following terms: *calorimeter, calorimetry, coffee-cup calorimeter, thermal equilibrium, bomb calorimeter*

## Calorimetry

To measure the heat flow in a process, you need an isolated system, such as a Thermos™. As you learned in section 16.1, an isolated system stops matter and energy from flowing into or out of the system. You also need a known amount of a substance, usually water. The water absorbs the heat that is released by the process, or the water releases heat if the process is endothermic. To determine the heat flow, you can measure the temperature change of the water. With its large specific heat capacity (4.184 J/ g·°C) and its broad temperature range (0°C to 100°C), liquid water can absorb and release a lot of heat.

Water, a thermometer, and an isolated system are the basic components of a calorimeter. A **calorimeter** is a device that is used to measure changes in kinetic energy. The technological process of measuring changes in kinetic energy is called **calorimetry**.

## Using a Calorimeter

In a **coffee-cup calorimeter**, shown in Figure 17.1, a known mass of water is inside the coffee cup. The water surrounds, and is in direct contact with, the process that produces the energy change. The initial temperature of the water is measured. Then the process takes place and the final temperature of the water is measured. The water is stirred to maintain even energy distribution, and the system is kept at a constant pressure. This type of calorimeter can measure heat changes during processes such as dissolving, neutralization, heating, and cooling.

The law of conservation of energy states that energy can be changed into different forms, but it cannot be created or destroyed. This law allows you to calculate the energy change in a calorimetry experiment. However, you need to make the assumptions listed on the following page.

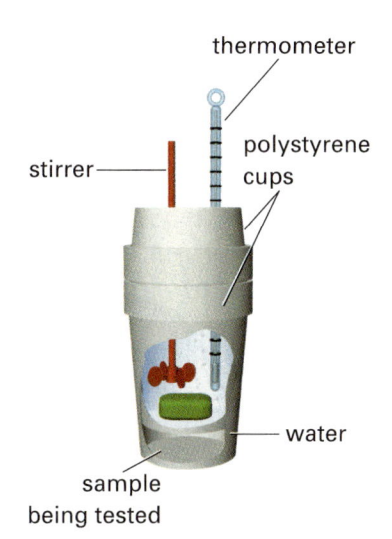

**Figure 17.1** A coffee-cup calorimeter usually consists of two nested polystyrene cups with a polystyrene lid, to provide insulation from the surroundings.

- The system is isolated. (No heat is exchanged with the surroundings outside the calorimeter.)
- The amount of heat that is exchanged with the calorimeter itself is small enough to be ignored.
- If something dissolves or reacts in the calorimeter water, the solution still retains the properties of water. (For example, density and specific heat capacity remain the same.)

Once you make those assumptions, the following equation applies:

$$q_{system} = -q_{surroundings}$$

The system is the chemical or physical process you are studying. The surroundings consist of the water or solution in the calorimeter. When a process causes an energy change in a calorimeter, the change in temperature is measured by a thermometer in the water. If you know the mass of the water and its specific heat capacity, you can calculate the change in kinetic energy caused by the process using the equation $q = m \cdot c \cdot \Delta T$. See Figures 17.2 and 17.3 on the following page for examples.

## Tools & Techniques

### The First Ice Calorimeter

A calorimeter measures the thermal energy that is absorbed or released by a material. Today we measure heat using joules (J) or calories (cal). Early scientists accepted one unit of heat as the amount of heat required to melt 1 kg of ice. Thus two units of heat could melt 2 kg of ice.

The earliest measurements of heat energy were taken around 1760, by a Scottish chemist named Joseph Black. He hollowed out a chamber in a block of ice. Then he wiped the chamber dry and placed a piece of platinum, heated to 38°C, inside. He used another slab of ice as a lid. As the platinum cooled, it gave up its heat to the ice. The ice melted, and water collected in the chamber. When the platinum reached the temperature of the ice, Black removed the water and weighed it to find out how much ice had melted. In this way, he measured the quantity of heat that was released by the platinum.

In 1780, two French scientists, Antoine Lavoisier and Pierre Laplace, developed the first apparatus formally called a calorimeter. Like Black, they used the amount of melted ice to measure the heat released by a material. Their calorimeter consisted of three concentric chambers. The object to be tested was placed in the innermost chamber. Broken chunks of ice were placed in the middle chamber. Ice was also placed in the outer chamber to prevent any heat reaching the apparatus from outside. As the object in the inner chamber released heat, the ice in the middle chamber melted. Water was drawn from the middle chamber by a tube, and then measured.

Lavoisier made many important contributions to the science of chemistry. Unfortunately, his involvement with a company that collected taxes for the government led to his arrest during the French Revolution. He was beheaded after a trial that lasted less than a day.

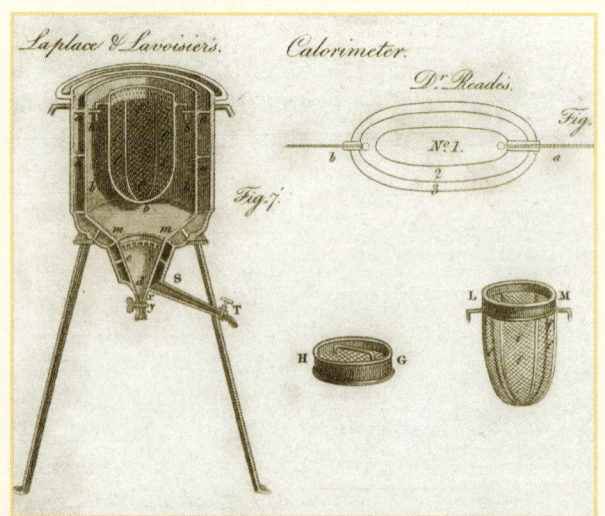

**The original calorimeter used by Lavoisier and Laplace**

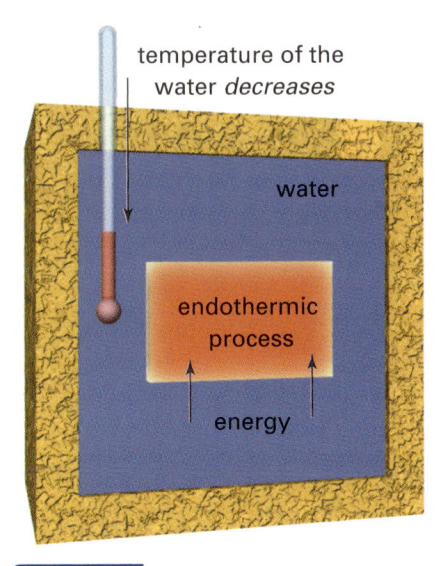

**Figure 17.2** An endothermic process, such as ice melting

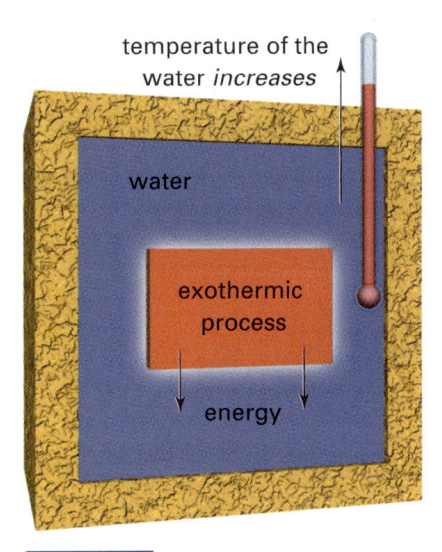

**Figure 17.3** An exothermic process, such as a neutralization reaction

In the next Sample Problem, you will use what you have just learned to calculate the specific heat capacity of a metal.

Notice that all the materials in the calorimeter in the following Sample Problem have the same final temperature. A system is said to be at **thermal equilibrium** when all its components have the same temperature.

## Sample Problem
### Determining a Metal's Specific Heat Capacity

**Problem**

A 70.0 g sample of a metal was heated to 95.0°C in a hot water bath. Then it was quickly transferred to a coffee-cup calorimeter. The calorimeter contained 100.0 g of water at an initial temperature of 19.8°C. The final temperature of the contents of the calorimeter was 22.6°C. What is the specific heat capacity of the metal?

**What Is Required?**

You need to calculate the specific heat capacity of the metal.

**What Is Given?**

You know the mass of the metal, and its initial and final temperatures.

Mass of metal ($m_m$) = 70.0 g

Initial temperature of metal ($T_i$) = 95.0°C

Final temperature of metal ($T_f$) = 22.6°C

You also know the mass of the water, and its initial and final temperatures.

Mass of water ($m_w$) = 100.0 g

Initial temperature of water ($T_i$) = 19.8°C

Final temperature of water ($T_f$) = 22.6°C

As well, you know the specific heat capacity of water: 4.184 J/g·°C.

*Continued...*

### Plan Your Strategy

Assume that the following relationship holds:
$q_{system} = -q_{surroundings}$
In this case, the system is the metal and the water in the calorimeter is the surroundings.
Therefore,
$q_m = -q_w$
Both $q_m$ and $q_w$ involve a temperature change only, not a change in potential energy.
Therefore,
$q_m = m_m c_m \Delta T_m$
and
$q_w = m_w c_w \Delta T_w$
Since $q_m = -q_w$, then
$m_m c_m \Delta T_m = -m_w c_w \Delta T_w$
Rearrange to solve for $c_m$.
$$c_m = \frac{-m_w c_w \Delta T_w}{m_m \Delta T_m}$$
It is very important that you do not mix up the given information. For example, when solving for $\Delta T_w$, the temperature change of the water, make sure that you only use variables for the water. You must use the initial temperature of the water, 19.8°C, *not* the initial temperature of the metal, 95.0°C. Also, remember that $\Delta T = T_f - T_i$.

### Act on Your Strategy

Solve for $c_m$.
$$c_m = \frac{-m_w c_w \Delta T_w}{m_m \Delta T_m}$$
$$= \frac{-(100.0 \text{ g})(4.184 \text{ J/g°C})(22.6°C - 19.8°C)}{(70.0 \text{ g})(22.6°C - 95.0°C)}$$
$$= -0.23 \text{ J/g°C}$$
The specific heat capacity of the metal is 0.23 J/g • °C.

### Check Your Solution

The specific heat capacity of the metal is positive, and it has the correct units. The specific heat capacity of the metal is much smaller than the specific heat capacity of water. This makes sense, since the metal released the same amount of heat as the water absorbed, but the temperature change of the metal was much greater.

## Practice Problems

1. A 92.0 g sample of a substance, with a temperature of 55°C, is placed in a large-scale polystyrene calorimeter. The calorimeter contains 1.00 kg of water at 20.0°C. The final temperature of the system is 25.2°C.

    (a) How much heat did the substance release? How much heat did the water absorb?

    (b) What is the specific heat capacity of the substance?

2. A coffee-cup calorimeter contains 125.3 g of water at 20.3°C. To this water is added 25.3 g of water. The water is mixed. The final temperature is 32.4°C.
   (a) How much heat did the 125.3 g sample of water absorb? How much heat did the 25.3 g sample of water release?
   (b) What was the initial temperature of the water that was added to the calorimeter?

3. A 2.4 g diamond is heated to 85.30°C and placed in a coffee-cup calorimeter containing some water. The initial temperature of the water is 25.50°C, and the final temperature of the contents of the calorimeter is 26.20°C. The specific heat capacity of diamond is 0.519 J/g°C. What mass of water was in the calorimeter?

4. A sample of iron is heated to 98.0°C in a hot water bath. The iron is then transferred to a coffee-cup calorimeter, which contains 125.2 g of water at 22.3°C. The iron and water are allowed to come to thermal equilibrium. The final temperature of the contents of the calorimeter is 24.3°C. The specific heat capacity of iron is 0.444 J/g°C.
   (a) What was the mass of the sample of iron?
   (b) Suppose the process above was repeated with a 60.4 g sample of iron. All other initial conditions were the same as above. What is the final temperature of the contents of the calorimeter?

## Using a Calorimeter to Determine the Enthalpy of a Reaction

A coffee-cup calorimeter is well-suited to determining the enthalpy changes of reactions in dilute aqueous solutions. The water in the calorimeter absorbs (or provides) the energy that is released (or absorbed) by a chemical reaction. When carrying out an experiment in a dilute solution, *the solution itself* absorbs or releases the energy. You can calculate the amount of energy that is absorbed or released by the solution using $q = m \cdot c \cdot \Delta T$. The mass, *m*, is the mass of the *solution*. The following Sample Problem shows how calorimetry can be used to determine the enthalpy change of a chemical reaction.

### Sample Problem
### Determining the Enthalpy of a Chemical Reaction

#### Problem
Copper(II) sulfate, $CuSO_4$, reacts with sodium hydroxide, NaOH, in a double displacement reaction. A precipitate of copper(II) hydroxide, $Cu(OH)_2$, and aqueous sodium sulfate, $Na_2SO_4$, is produced.

$$CuSO_{4(aq)} + 2NaOH_{(aq)} \rightarrow Cu(OH)_{2(s)} + Na_2SO_{4(aq)}$$

50.0 mL of 0.300 mol/L $CuSO_4$ solution is mixed with an equal volume of 0.600 mol/L NaOH. The initial temperature of both solutions is 21.4°C. After mixing the solutions in the coffee-cup calorimeter, the highest temperature that is reached is 24.6°C. Determine the enthalpy change of the reaction. Then write the thermochemical equation.

#### What Is Required?
You need to calculate $\Delta H$ of the given reaction.

*Continued...*

*Continued...*

### What Is Given?

You know the volume of each solution. You also know the initial temperature of each solution and the final temperature of the reaction mixture.

Volume of $CuSO_4$ solution, $V_{CuSO_4}$ = 50.0 mL
Volume of NaOH solution, $V_{NaOH}$ = 50.0 mL
Initial temperature, $T_i$ = 21.4°C
Final temperature, $T_f$ = 24.6°C

### Plan Your Strategy

**Step 1** Determine the total volume by adding the volumes of the two solutions. Determine the total mass of the reaction mixture, assuming a density of 1.00 g/mL (the density of water).

**Step 2** Determine the number of moles of $CuSO_4$ and NaOH that reacted. If necessary, determine the limiting reactant.

**Step 3** Assume that $q_{system} = -q_{surroundings}$ or $q_{rxn} = -q_{solution}$.

$$q_{rxn} \cdot \Delta H_{rxn}$$
$$q_{solution} = m_{solution} \cdot c_{solution} \cdot \Delta T_{solution}$$

Since $q_{rxn} = -q_{solution}$ you can use the equations above to solve for $\Delta H_{rxn}$ as follows.

$$n \cdot \Delta H_{rxn} = -m_{solutions} \cdot c_{solution} \cdot \Delta T_{solution}$$
$$\Delta H_{rxn} = \frac{-m_{solution} \cdot c_{solution} \cdot \Delta T_{solution}}{n}$$

**Step 4** Use your $\Delta H$ to write the thermochemical equation for the reaction.

### Act on Your Strategy

**Step 1** The total volume of the reaction mixture is

50.0 mL + 50.0 mL = 100.0 mL

The mass of the reaction mixture, assuming a density of 1.00 g/mL, is

$m = DV$
$\quad = (1.00 \text{ g/mL})(100.0 \text{ mL})$
$\quad = 1.00 \times 10^2 \text{ g}$

**Step 2** Calculate the number of moles of $CuSO_4$ as follows.

$n = c \cdot V$
$\quad = (0.300 \text{ mol/L})(50.0 \times 10^{-3} \text{ L})$
$\quad = 0.0150 \text{ mol}$

Calculate the number of moles of NaOH.

$n \text{ mol NaOH} = (0.600 \text{ mol/L})(50.0 \times 10^{-3} \text{ L})$
$\qquad\qquad\qquad = 0.0300 \text{ mol}$

The reactants are present in stoichiometric amounts. (There is no limiting reactant.) Use 0.0150 mol in Step 3, since the stoichiometric coefficient of $CuSO_4$ in the equation is 1.

---

**PROBLEM TIP**

Because the solutions are dilute, you can make the assumption that the density of the reaction mixture is the same as the density of water.

$$\Delta H_{\text{rxn}} = \frac{-m_{\text{solution}} \cdot C_{\text{solution}} \cdot \Delta T_{\text{solution}}}{n}$$

$$= \frac{-(100\text{ g})(4.184\text{ J/g}°\text{C})(24.6°\text{C} - 21.4°\text{C})}{0.0150\text{ mol}}$$

$$= -8.9 \times 10^4 \text{ J/mol}$$

$$= -89 \text{ kJ/mol}$$

The enthalpy change of the reaction is −89 kJ/mol CuSO4.

**Step 4** The thermochemical equation is

$$CuSO_{4(aq)} + 2NaOH_{(aq)} \rightarrow Cu(OH)_{2(s)} + Na_2SO_{4(aq)} + 89\text{ kJ}$$

### Check Your Solution

The solution has the correct number of significant digits. The units are correct. You know that the reaction was exothermic, because the temperature of the solution increased. The calculated value for $\Delta H$ is negative, which is correct for an exothermic reaction.

## Practice Problems

5. A chemist wants to determine the enthalpy of neutralization for the following reaction.

   $HCl_{(aq)} + NaOH_{(aq)} \rightarrow NaCl_{(aq)} + H_2O_{(\ell)}$

   The chemist uses a coffee-cup calorimeter to neutralize completely 61.1 mL of 0.543 mol/L $HCl_{(aq)}$ with 42.6 mL of sufficiently concentrated $NaOH_{(aq)}$. The initial temperature of both solutions is 17.8°C. After neutralization, the highest recorded temperature is 21.6°C. Calculate the enthalpy of neutralization, in kJ/mol of HCl. Assume that the density of both solutions is 1.00 g/mL.

6. A chemist wants to determine empirically the enthalpy change for the following reaction.

   $Mg_{(s)} + 2HCl_{(aq)} \rightarrow MgCl_{2(aq)} + H_{2(g)}$

   The chemist uses a coffee-cup calorimeter to react 0.50 g of Mg ribbon with 100 mL of 1.00 mol/L $HCl_{(aq)}$. The initial temperature of the $HCl_{(aq)}$ is 20.4°C. After neutralization, the highest recorded temperature is 40.7°C.

   **(a)** Calculate the enthalpy change, in kJ/mol of Mg, for the reaction.

   **(b)** State any assumptions that you made in order to determine the enthalpy change.

7. Nitric acid is neutralized with potassium hydroxide in the following reaction.

   $HNO_{3(aq)} + KOH_{(aq)} \rightarrow KNO_{3(aq)} + H_2O_{(\ell)}$   $\Delta H_{\text{rxn}} = -53.4$ kJ/mol

   55.0 mL of 1.30 mol/L solutions of both reactants, at 21.4°C, are mixed in a calorimeter. What is the final temperature of the mixture? Assume that the density of both solutions is 1.00 g/mL. Also assume that the specific heat capacity of both solutions is the same as the specific heat capacity of water. No heat is lost to the calorimeter itself.

In the following investigation, you will construct a coffee-cup calorimeter and use it to determine the enthalpy of a neutralization reaction.

# Investigation 17-A

**SKILL FOCUS**
Predicting
Performing and recording
Analyzing and interpreting

## Determining the Enthalpy of a Neutralization Reaction

The neutralization of hydrochloric acid with sodium hydroxide solution is represented by the following equation.

$HCl_{(aq)} + NaOH_{(aq)} \rightarrow NaCl_{(aq)} + H_2O_{(\ell)}$

Using a coffee-cup calorimeter, you will determine the enthalpy change for this reaction.

### Question

What is the enthalpy of neutralization for hydrochloric acid and sodium hydroxide solution?

### Prediction

Will the neutralization reaction be endothermic or exothermic? Record your prediction, and give reasons.

### Safety Precautions

If you get any hydrochloric acid or sodium hydroxide solution on your skin, flush your skin with plenty of cold water.

### Materials

100 mL graduated cylinder
400 mL beaker
2 polystyrene cups that are the same size
polystyrene lid
thermometer
stirring rod
1.00 mol/L $HCl_{(aq)}$
1.00 mol/L $NaOH_{(aq)}$

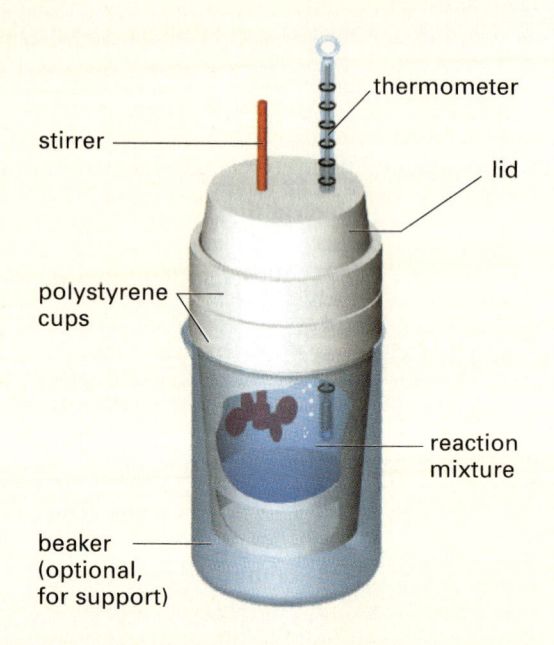

### Procedure

1. Your teacher will allow the hydrochloric acid and sodium hydroxide solution to come to room temperature overnight.

2. Read the rest of this Procedure carefully before you continue. Set up a graph to record your temperature observations.

3. Build a coffee-cup calorimeter, using the diagram above as a guide. You will need to make two holes in the lid—one for the thermometer and one for the stirring rod. The holes should be as small as possible to minimize heat loss to the surroundings.

4. Rinse the graduated cylinder with a small quantity of 1.00 mol/L $NaOH_{(aq)}$. Use the cylinder to add 50.0 mL of 1.00 mol/L $NaOH_{(aq)}$ to the calorimeter. Record the initial temperature of the $NaOH_{(aq)}$. (This will also represent the initial temperature of the $HCl_{(aq)}$.) **CAUTION** The $NaOH_{(aq)}$ can burn your skin.

**668** MHR • Unit 7 Thermochemistry

5. Rinse the graduated cylinder with tap water. Then rinse it with a small quantity of 1.00 mol/L HCl$_{(aq)}$. Quickly and carefully, add 50.0 mL of 1.00 mol/L HCl$_{(aq)}$ to the NaOH$_{(aq)}$ in the calorimeter. **CAUTION** The HCl$_{(aq)}$ can burn your skin.

6. Cover the calorimeter. Record the temperature every 30 s, stirring gently and continuously.

7. When the temperature levels off, record the final temperature, $T_f$.

8. If time permits, repeat steps 4 to 7.

## Analysis

1. Determine the amount of heat that is absorbed by the solution in the calorimeter.

2. Use the following equation to determine the amount of heat that is released by the reaction:

   $-q_{reaction} = q_{solution}$

3. Determine the number of moles of HCl$_{(aq)}$ and NaOH$_{(aq)}$ that were involved in the reaction.

4. Use your knowledge of solutions to explain what happens during a neutralization reaction. Use equations in your answer. Was heat released or absorbed during the neutralization reaction? Explain your answer.

## Conclusion

5. Use your results to determine the enthalpy change of the neutralization reaction, in kJ/mol of NaOH. Write the thermochemical equation for the neutralization reaction.

## Applications

6. When an acid gets on your skin, why must you flush the area with plenty of water rather than neutralizing the acid with a base?

7. Suppose that you had added solid sodium hydroxide pellets to hydrochloric acid, instead of adding hydrochloric acid to sodium hydroxide solution?

    (a) Do you think you would have obtained a different enthalpy change?

    (b) Would the enthalpy change have been higher or lower?

    (c) How can you test your answer? Design an investigation, and carry it out with the permission of your teacher.

    (d) What change do you need to make to the thermochemical equation if you perform the investigation using solid sodium hydroxide?

8. In this investigation, you assumed that the heat capacity of your calorimeter was 0 J/°C.

    (a) Design an investigation to determine the actual heat capacity of your coffee-cup calorimeter, $C_{calorimeter}$. Include equations for any calculations you will need to do. If time permits, have your teacher approve your procedure and carry out the investigation. **Hint:** If you mix hot and cold water together and no heat is absorbed by the calorimeter itself, then the amount of heat absorbed by the cold water should equal the amount of heat released by the hot water. If more heat is released by the hot water than is absorbed by the cold water, the difference must be absorbed by the calorimeter.

    (b) Include the heat capacity of your calorimeter in your calculations for $\Delta H_{neutralization}$. Use the following equation:

    $-q_{reaction} = (m_{solution} \cdot c_{solution} \cdot \Delta T) + (C_{calorimeter} \cdot \Delta T)$

---

**PROBEWARE**

www.mcgrawhill.ca/links/atlchemistry

If you have access to probeware, go to the website above for a probeware investigation related to this investigation.

## Determining Enthalpy of Solution

In section 16.2, you learned that some substances dissolve exothermically, and some substances dissolve endothermically. In the following ExpressLab, use a coffee-cup calorimeter to determine the molar enthalpy of solution for two different substances.

### ExpressLab: The Energy of Dissolving

In this lab you will measure the heat of solution of two solids.

**Safety Precautions**

- NaOH and KOH can burn skin. If you accidentally spill NaOH or KOH on your skin, wash immediately with copious amounts of cold water.

**Materials**

balance and beakers or weigh boats
polystyrene calorimeter
thermometer and stirring rod
distilled water
2 pairs of solid compounds:
- ammonium nitrate and potassium hydroxide
- potassium nitrate and sodium hydroxide

**Procedure**

1. Choose *one* pair of chemicals from the list.
2. For each of the two chemicals, calculate the mass required to make 100.0 mL of a 1.00 mol/L aqueous solution.
3. Measure the required mass of one of the chemicals in a beaker or a weigh boat.
4. Measure exactly 100 g of distilled water directly into your calorimeter.
5. Measure the initial temperature of the water.
6. Pour one of the chemicals into the calorimeter. Put the lid on the calorimeter.
7. Stir the solution. Record the temperature until there is a maximum temperature change.
8. Dispose of the chemical as directed by your teacher. Clean your apparatus.
9. Repeat steps 3 to 8, using the other chemical.

**Analysis**

1. For each chemical you used, calculate the molar enthalpy of solution.
2. For each chemical you used, write a thermochemical equation and draw an enthalpy diagram to represent the dissolution process.
3. Which chemical dissolved endothermically? Which chemical dissolved exothermically?
4. As you learned in section 16.2, one type of cold pack contains a compartment of powder and a compartment of water. When the barrier between the two compartments is broken, the solid dissolves in the water and causes an energy change. Which of the two chemicals you tested could be used in this type of cold pack? Why?

## Determining Enthalpy of Combustion

In chapter 16, you compared the molar enthalpies of combustion for short, straight-chain alkanes. In your everyday life, you may have encountered another type of hydrocarbon: paraffins. Paraffins are long-chain hydrocarbons. They are semisolid or solid at room temperature. One type of paraffin has been a household item for centuries—paraffin wax, $C_{25}H_{52(s)}$, better known as candle wax. (See Figure 17.4)

Like other hydrocarbons, the paraffin wax in candles undergoes combustion when burned. It releases thermal energy in the process. In the following investigation, you will determine the molar enthalpy of combustion of paraffin.

**Figure 17.4** Paraffin wax candles have been an important light source for hundreds of years.

# Investigation 17-B

**SKILL FOCUS**
Predicting
Performing and recording
Analyzing and interpreting

## The Enthalpy of Combustion of a Candle

You have probably gazed into the flame of a candle without thinking about chemistry! Now, however, you will use the combustion of candle wax to gain insight into the measurement of heat changes. You will also evaluate the design of this investigation and make suggestions for improvement.

### Question

What is the molar enthalpy of combustion of candle wax?

### Prediction

Will the enthalpy of combustion of candle wax be greater or less than the enthalpy of combustion of other fuels, such as propane and butane? Record your prediction, and give reasons.

### Safety Precautions

- Tie back long hair and secure any loose clothing. Before you light the candle, check that there are no flammable solvents nearby.

### Materials

balance
calorimeter apparatus (see the diagram to the right)
thermometer
stirring rod
matches
water
candle

### Procedure

1. Burn the candle to melt some wax. Use the wax to attach the candle to the smaller can lid. Blow out the candle.

2. Set up the apparatus as shown in the diagram, but do not include the large can yet. Adjust the ring stand so that the small can is about 5 cm above the wick of the candle. The tip of the flame should just touch the bottom of the small can.

3. Measure the mass of the candle and the lid.

4. Measure the mass of the small can. Measure the mass of the hanger.

5. Place the candle inside the large can on the retort stand.

6. Fill the small can about two-thirds full of cold water (10°C to 15°C). You will measure the mass of the water later.

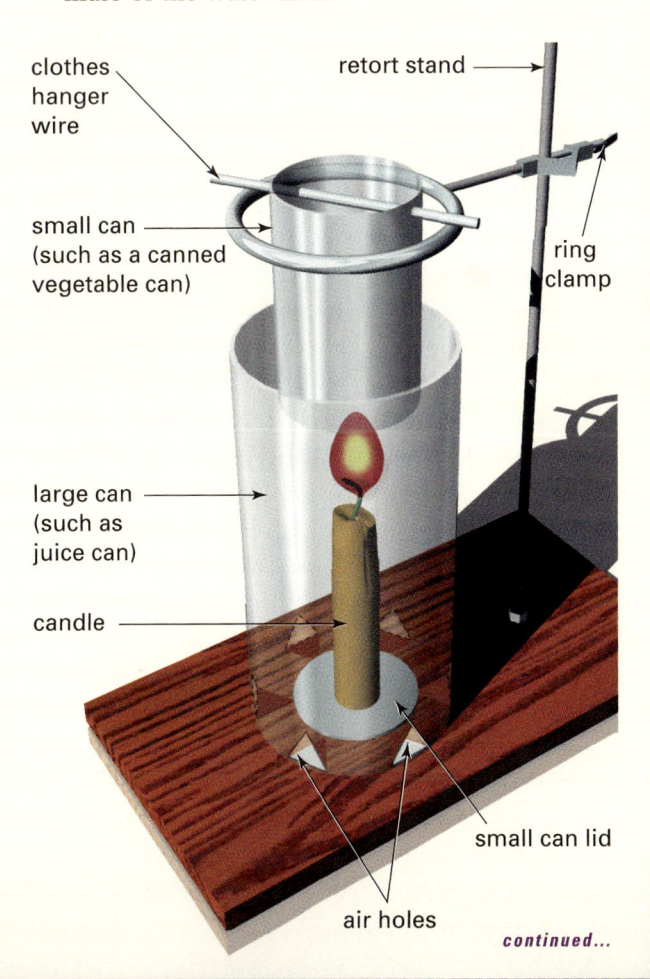

*continued...*

**continued**

7. Stir the water in the can. Measure the temperature of the water.

8. Light the candle. Quickly place the small can in position over the candle. **CAUTION** Be careful of the open flame.

9. Continue stirring. Monitor the temperature of the water until it has reached 10°C to 15°C above room temperature.

10. Blow out the candle. Continue to stir. Monitor the temperature until you observe no further change.

11. Record the final temperature of the water. Examine the bottom of the small can, and record your observations.

12. Measure the mass of the small can and the water.

13. Measure the mass of the candle, lid, and any drops of candle wax.

## Analysis

1. (a) Calculate the mass of the water.
   (b) Calculate the mass of candle wax that burned.

2. Calculate the heat that was absorbed by the water.

## Conclusions

3. (a) Assume that the candle wax is pure paraffin wax, $C_{25}H_{52(s)}$. Calculate the molar enthalpy of combustion of paraffin wax.
   (b) Write a balanced thermochemical equation for the complete combustion of paraffin wax.
   (c) Draw an enthalpy diagram showing the complete combustion of paraffin wax.

4. (a) List some possible sources of error that may have affected the results you obtained.
   (b) Evaluate the design and the procedure of this investigation. Consider the apparatus, the combustion, and anything else you can think of. Make suggestions for possible improvements.

5. What if soot (unburned carbon) accumulated on the bottom of the small can? Would this produce a greater or a lower heat value than the value you expected? Explain.

6. The aluminum can absorbs some of the heat from the combustion reaction.
   (a) Repeat your determination of the molar enthalpy of combustion of paraffin as shown:
   $q_{system} = -(q_{water} + q_{aluminum})$
   $q_{system} = n \cdot \Delta H_{comb}$
   (b) Is your result from 6. (a) more accurate than your result from 3. (a)? Is it more precise? Explain your answer.

## Bomb Calorimetry

In Investigation 17-B, you constructed a calorimeter to determine the enthalpy of combustion of a paraffin wax. Your calorimeter was more flame-resistant than a coffee-cup calorimeter, but some heat was transferred to the air and the metal containers. To measure precisely and accurately the enthalpy changes of combustion reactions, chemists use a calorimeter called a **bomb calorimeter**, shown in Figures 17.5. and 17.6. A bomb calorimeter measures enthalpy changes during combustion reactions at a constant volume.

The bomb calorimeter works on the same general principle as the polystyrene calorimeter. The reaction, however, takes place inside an inner metal chamber, called a "bomb." This "bomb" contains pure oxygen. The reactants are ignited using an electric coil. A known quantity of water surrounds the bomb and absorbs the energy that is released by the reaction.

**Figure 17.5** A bomb calorimeter is more sophisticated than a polystyrene calorimeter.

**Figure 17.6** Figure 17.6 shows a chemist preparing a sample for testing in a bomb calorimeter.

A bomb calorimeter has many more parts than a polystyrene calorimeter. All of these parts can absorb or release small quantities of energy. Therefore, you cannot assume that the heat lost to the calorimeter is small enough to be negligible. To obtain precise heat measurements, you must know or find out the heat capacity of the bomb calorimeter. The heat capacity of a calorimeter takes into account the heat that *all* parts of the calorimeter can lose or gain. (See Figure 17.7)

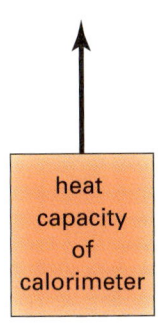

$$C_{total} = C_{water} + C_{thermometer} + C_{stirrer} + C_{container}$$

**Figure 17.7** The heat capacity of the calorimeter incorporates the heat capacity of all its components.

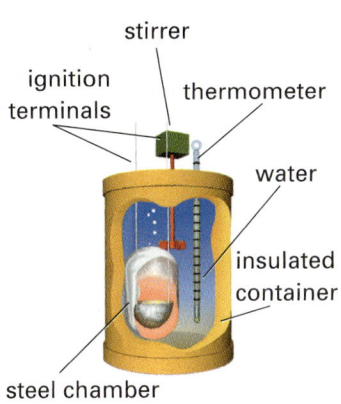

A bomb calorimeter is calibrated for a constant mass of water. Since the mass of the other parts remains constant, there is no need for mass units in the heat capacity value. The manufacturer usually includes the heat capacity value in the instructions for the calorimeter.

Heat calculations must be done differently when the heat capacity of a calorimeter is included. The next Sample Problem illustrates how to use the heat capacity of a calorimeter in your calculations.

### Sample Problem
#### Calculating Heat Change in a Bomb Calorimeter

**Problem**

A laboratory decided to test the energy content of peanut butter. A technician placed a 16.0 g sample of peanut butter in the steel bomb of a calorimeter, along with sufficient oxygen to burn the sample completely. She ignited the mixture and took heat measurements. The heat capacity of the calorimeter was calibrated at 8.28 kJ/°C. During the experiment, the temperature increased by 50.5°C.

**(a)** What was the thermal energy released by the sample of peanut butter?

**(b)** What is the enthalpy of combustion of the peanut butter per gram of sample?

**What Is Required?**

**(a)** You need to calculate the heat ($q_{sample}$) lost by the peanut butter.

**(b)** You need to calculate the enthalpy change per gram of peanut butter.

**What Is Given?**

You know the mass of the peanut butter, the heat capacity of the calorimeter, and the change in temperature of the system.

Mass of peanut butter ($m$) = 16.0 g
Heat capacity of calorimeter ($C$) = 8.28 kJ/°C
Change in temperature ($\Delta T$) = 50.5°C

**Plan Your Strategy**

**(a)** The heat capacity of the calorimeter takes into account the specific heat capacities and masses of all the parts of the calorimeter. Calculate the heat change of the calorimeter, $q_{cal}$, using the equation

$$q_{cal} = C\Delta T$$

**Note:** $C$ is the heat capacity of the calorimeter in J/°C or kJ/°C. It replaces the $m$ and $c$ in other calculations involving specific heat capacity.

First calculate the heat gained by the calorimeter. When the peanut butter burns, the heat released by the peanut butter sample equals the heat absorbed by the calorimeter.

$$q_{sample} = -q_{cal}$$

**(b)** To find the heat of combustion per gram, divide the heat by the mass of the sample.

### Act on Your Strategy

(a) $q_{cal} = C\Delta T$
   $= (8.28 \text{kJ/°C})(50.5\text{°C})$
   $= 418.14 (\text{kJ/°C})(\text{°C})$
   $= 418 \text{ kJ}$

The calorimeter gained 418 kJ of thermal energy.

$$q_{sample} = -q_{cal}$$
$$= -418 \text{ kJ}$$

The sample of peanut butter released 418 kJ of thermal energy.

(b) Heat of combustion per gram $= \dfrac{q_{sample}}{m}$
   $= \dfrac{-418 \text{ kJ}}{16.0 \text{ g}}$
   $= -26.2 \text{ kJ/g}$

The heat of combustion per gram of peanut butter is −26.2 kJ/g.

### Check Your Solution

Heat was released by the peanut butter, so the heat value is negative.

> **CHEM FACT**
>
> In the sample problem, you will notice that the enthalpy of combustion of peanut butter is expressed in kJ/g, not kJ/mol. Molar enthalpy of combustion is used for pure substances. Peanut butter, however, like most food, is not a pure substance. It is a mixture of different organic compounds such as fats and sugars. Therefore, there is no such thing as "a mole of peanut butter." Instead, you describe energy changes per gram of peanut butter.

### Practice Problems

**8.** Use the heat equation for a calibrated calorimeter, $q_{cal} = C\Delta T$. Recall that $\Delta T = T_f - T_i$. Solve for the following quantities.

(a) $C$
(b) $\Delta T$
(c) $T_f$ (in terms of C, $\Delta T$, and $T_i$)
(d) $T_i$ (in terms of C, $\Delta T$, and $T_f$)

**9.** A lab technician places a 5.00 g food sample into a bomb calorimeter that is calibrated at 9.23 kJ/°C. The initial temperature of the calorimeter system is 21.0°C. After burning the food, the final temperature of the system is 32.0°C. What is the heat of combustion of the food in kJ/g?

**10.** A scientist places a small block of ice in an uncalibrated bomb calorimeter. The ice melts, gaining 10.5 kJ ($10.5 \times 10^3$ J) of heat. The liquid water undergoes a temperature change of 25.0°C. The calorimeter undergoes a temperature change of 1.2°C.

(a) What mass of ice was added to the calorimeter? (Use the heat capacity of liquid water.)
(b) What is the calibration of the bomb calorimeter in kJ/°C?

### Section Summary

In this section, you measured the enthalpy changes of several processes using constant-pressure calorimeters. You learned how to carry out calculations involving data obtained from a bomb calorimeter. You determined the enthalpy associated with a neutralization reaction, a dissolution, and the combustion of a hydrocarbon. In section 17.2, you will learn several methods for calculating enthalpies of reaction based on tabulated thermochemical data. This skill will allow you to determine enthalpies of reaction without carrying out experiments.

## Section Review

**1** List two characteristics of a calorimeter that are necessary for successful heat measurement.

**2** A calorimeter is calibrated at 7.61 kJ/°C. When a sample of coal is burned in the calorimeter, the temperature increases by 5.23°C. How much heat was lost by the coal?

**3** A reaction in a calorimeter causes 150 g of water to decrease in temperature by 5.0°C. What is the kinetic energy change of the water?

**4** What properties of polystyrene make it a suitable material for a constant-pressure calorimeter?

**5** Suppose that you use concentrated reactant solutions in an experiment with a coffee-cup calorimeter. Should you make the same assumptions that you did when you used dilute solutions? Explain.

**6** Concentrated sulfuric acid can be diluted by adding it to water. The reaction is extremely exothermic. In this question, you will design an experiment to measure the enthalpy change (in kJ/mol) for the dilution of concentrated sulfuric acid. Assume that you have access to any equipment in your school's chemistry laboratory. Do not carry out this experiment.

(a) State the equipment and chemicals that you need.

(b) Write a step-by-step procedure.

(c) Set up an appropriate data table.

(d) State any information that you need.

(e) State any simplifying assumptions that you will make.

**7** A chemist mixes 100.0 mL of 0.050 mol/L aqueous potassium hydroxide with 100.0 mL of 0.050 mol/L nitric acid in a coffee-cup calorimeter. The temperature of the reactants is 21.01°C. The temperature of the products is 21.34°C.

(a) Determine the molar enthalpy of neutralization of $KOH_{(aq)}$ with $HNO_{3(aq)}$

(b) Write a thermochemical equation and draw an enthalpy diagram for the reaction.

(c) If you performed this investigation, how would you change the procedure? Explain your answer.

**8** From experience, you know that you produce significantly more heat when you are exercising than when you are resting. Scientists can study the heat that is produced by human metabolism reactions using a "human calorimeter." Based on what you know about calorimetry, how would you design a human calorimeter? What variables would you control and study in an investigation using your calorimeter? Write a brief proposal outlining the design of your human calorimeter and the experimental approach you would take.

# Hess's Law of Heat Summation

## 17.2

In section 17.1, you learned how to use a coffee-cup calorimeter to determine the heat that was released or absorbed in a chemical reaction. Coffee-cup calorimeters are generally used only for dilute aqueous solutions. There are many non-aqueous chemical reactions, however. There are also many reactions that release so much energy they are not safe to perform using a coffee-cup calorimeter. Imagine trying to determine the enthalpy of reaction for the detonation of nitroglycerin, an unstable and powerfully explosive compound. Furthermore, there are reactions that occur too slowly for the calorimetric method to be practical.

Chemists can determine the enthalpy change of any reaction using an important law, known as **Hess's law of heat summation**. This law states that *the enthalpy change of a physical or chemical process depends only on the beginning conditions (reactants) and the end conditions (products). The enthalpy change is independent of the pathway of the process and the number of intermediate steps in the process. It is the sum of the enthalpy changes of all the individual steps that make up the process.*

For example, carbon and oxygen can form carbon dioxide via two pathways.

1. Carbon can react with oxygen to form carbon monoxide. The carbon monoxide then reacts with oxygen to produce carbon dioxide. The two equations below represent this pathway.

$$C_{(s)} + \tfrac{1}{2}O_{2(g)} \rightarrow CO_{(g)} \quad \Delta H° = -110.5 \text{ kJ/mol}$$
$$CO_{(g)} + \tfrac{1}{2}O_{2(g)} \rightarrow CO_{2(g)} \quad \Delta H° = -283.0 \text{ kJ/mol}$$

2. Carbon can also react with oxygen to produce carbon dioxide directly.

$$C_{(s)} + O_{2(g)} \rightarrow CO_{2(g)} \quad \Delta H° = -393.5 \text{ kJ/mol}$$

In both cases, the net result is that one mole of carbon reacts with one mole of oxygen to produce one mole of carbon dioxide. (In the first pathway, all the carbon monoxide that is produced reacts with oxygen to form carbon dioxide.) Notice that the sum of the enthalpy changes for the first pathway is the same as the enthalpy change for the second pathway.

Examine Figure 17.8 to see how to represent the two pathways using one enthalpy diagram.

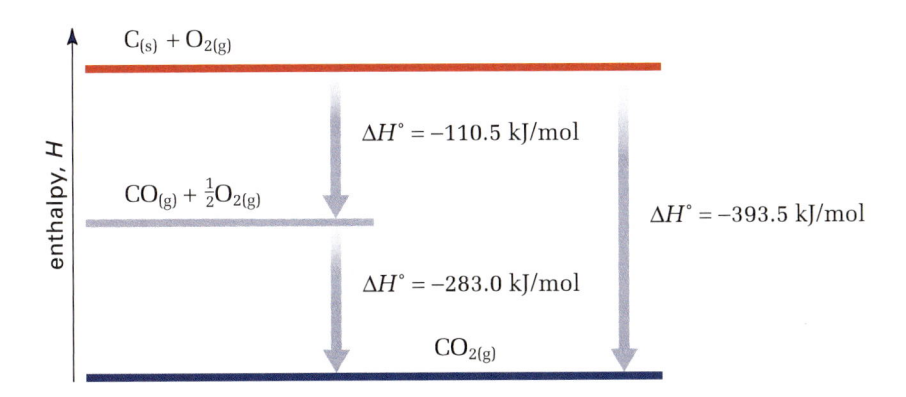

### Section Preview/Outcomes

In this section, you will

- **determine** enthalpy of reaction for a chemical change using a table of standard molar enthalpies of formation
- **determine** the enthalpy change of an overall process using the method of addition of chemical equations and corresponding enthalpy changes
- **apply** Hess's Law to determine enthalpy of a reaction by experiment
- **communicate** your understanding of the following terms: Hess's law of heat summation, state function

**Figure 17.8** Carbon dioxide can be formed by the reaction of oxygen with carbon to form carbon monoxide, followed by the reaction of carbon monoxide with oxygen. Carbon dioxide can also be formed directly from carbon and oxygen. No matter which pathway is used, the enthalpy change of the reaction is the same.

Chapter 17 Measuring and Using Energy Changes • MHR **677**

One way to think about Hess's law is to compare the energy changes that occur in a chemical reaction with the changes in the potential energy of a cyclist on hilly terrain. This comparison is shown in Figure 17.9.

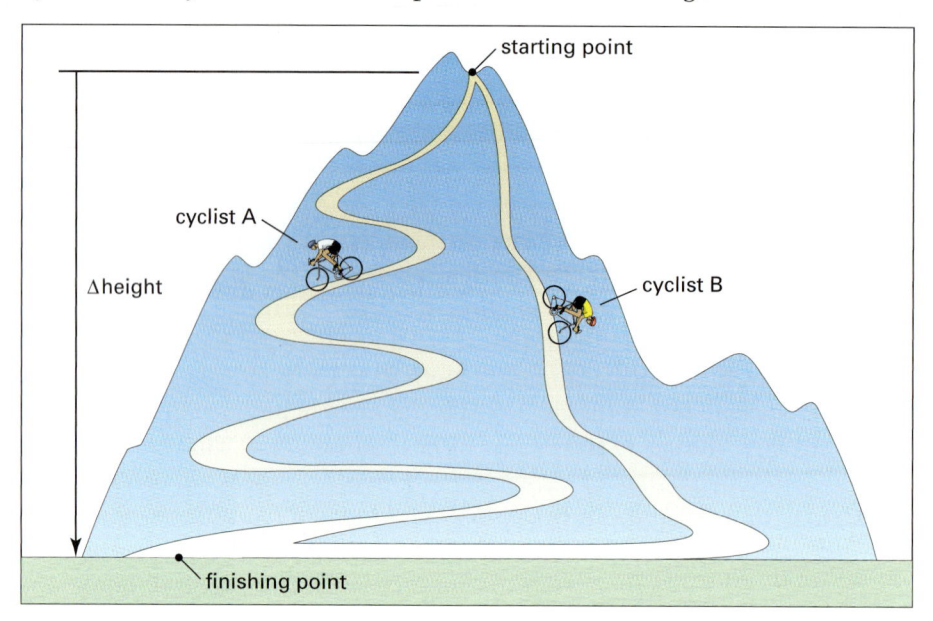

**Figure 17.9** The routes that cyclists take to get from the starting point to the finishing point has no effect on the *net* change in the cyclists' gravitational potential energy.

Hess's Law is valid because enthalpy is considered to be a state function. A **state function** is a property of a system that is determined only by the current conditions of the system. It is not dependent on the path taken by the system to reach those conditions. As you have seen, there are several different ways for a given change to occur. If the beginning and end conditions (e.g., substances present, temperature, pressure) of a system are the same, however, the enthalpy change is the same, regardless of the steps taken in between.

Hess's law allows you to determine the energy of a chemical reaction without directly measuring it. In this section, you will examine two ways in which you can use Hess's law to calculate the enthalpy change of a chemical reaction:

1. by combining chemical equations algebraically
2. by using the enthalpy of formation

You will also learn how to use bond energies and Hess's law to *estimate* the enthalpy change of a chemical reaction.

## Combining Chemical Equations Algebraically

According to Hess's law, the pathway that is taken in a chemical reaction has no effect on the enthalpy change of the reaction. How can you use Hess's law to calculate the enthalpy change of a reaction? One way is to add equations for reactions with known enthalpy changes, so that their net result is the reaction you are interested in.

For example, you can combine thermochemical equations (1) and (2) below to find the enthalpy change for the decomposition of hydrogen peroxide, equation (3).

(1) $H_2O_{2(\ell)} \rightarrow H_{2(g)} + O_{2(g)}$ $\Delta H° = +188$ kJ/mol

(2) $H_{2(g)} + \frac{1}{2}O_{2(g)} \rightarrow H_2O_{(\ell)}$ $\Delta H° = -286$ kJ/mol

(3) $H_2O_{2(\ell)} \rightarrow H_2O_{(\ell)} + \frac{1}{2}O_{2(g)}$ $\Delta H° = ?$

Carefully examine equation (3), the *target* equation. Notice that $H_2O_2$ is on the left (reactant) side, while $H_2O$ and $\frac{1}{2}O_2$ are on the right (product) side. Now examine equations (1) and (2). Notice which sides of the equations $H_2O_2$ and $H_2O$ are on. They are on the correct sides, based on equation (3). Also notice that hydrogen does not appear in equation (3). Therefore, hydrogen must cancel out when equations (1) and (2) are added. Since there is one mole of $H_{2(g)}$ on the product side of equation (1) and one mole of $H_{2(g)}$ on the reactant side of equation (2), these two terms cancel. Set up equations (1) and (2) as shown below. Add the products and the reactants. Then cancel any substances that appear on opposite sides.

(1) $\qquad H_2O_{2(\ell)} \rightarrow H_{2(g)} + O_{2(g)} \qquad \Delta H° = +188$ kJ/mol

(2) $\qquad H_{2(g)} + \frac{1}{2}O_{2(g)} \rightarrow H_2O_{(\ell)} \qquad \Delta H° = -286$ kJ/mol

$\qquad H_2O_{2(\ell)} + \cancel{H_{2(g)}} + \frac{1}{2}O_{2(g)} \rightarrow H_2O_{(\ell)} + O_{2(g)} + \cancel{H_{2(g)}} \qquad \Delta H° = ?$

or

(3) $\qquad H_2O_{2(\ell)} \rightarrow H_2O_{(\ell)} + \frac{1}{2}O_{2(g)} \qquad \Delta H° = ?$

Equations (1) and (2) add to give equation (3). Therefore, you know that the enthalpy change for equation (3) is the sum of the enthalpy changes of equations (1) and (2).

$H_2O_{2(\ell)} \rightarrow H_2O_{(\ell)} + \frac{1}{2}O_{2(g)} \quad \Delta H° = 188$ kJ/mol $- 286$ kJ/mol $= -98$ kJ/mol

Figure 17.10 illustrates this combination of chemical equations in an enthalpy diagram.

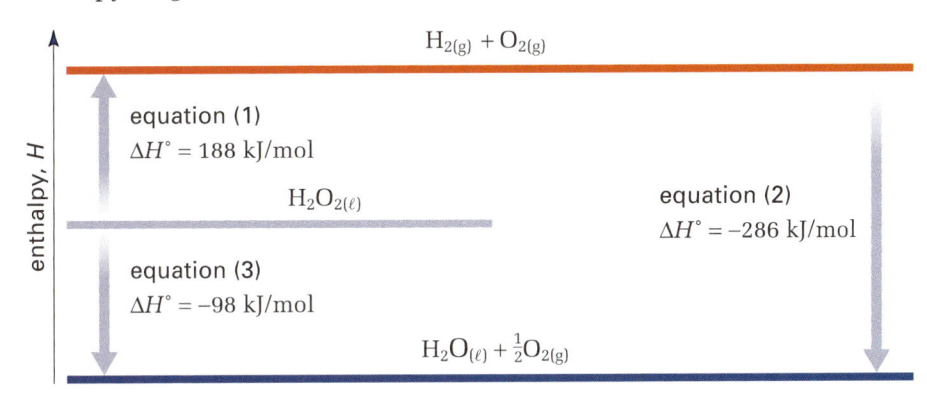

**Figure 17.10** The algebraic combination of chemical reactions can be represented in an enthalpy diagram.

In the previous example, you did not need to manipulate the two equations with known enthalpy changes. They added to the target equation as they were written. In many cases, however, you *will* need to manipulate the equations before adding them. There are two key ways in which you can manipulate an equation:

1. *Reverse an equation* so that the products become reactants and the reactants become products. When you reverse an equation, you need to change the sign of $\Delta H°$ (multiply by $-1$).

2. *Multiply each coefficient in an equation* by the same integer or fraction. When you multiply an equation, you need to multiply $\Delta H°$ by the same number.

Examine the following Sample Problem to see how to manipulate equations so that they add to the target equation. Try the problems that follow to practise finding the enthalpy change by adding equations.

### Sample Problem

## Using Hess's Law to Determine Enthalpy Change

**Problem**

One of the methods that the steel industry uses to obtain metallic iron is to react iron(III) oxide, $Fe_2O_3$, with carbon monoxide, CO.

$Fe_2O_{3(s)} + 3CO_{(g)} \rightarrow 3CO_{2(g)} + 2Fe_{(s)}$

Determine the enthalpy change of this reaction, given the following equations and their enthalpy changes.

(1) $CO_{(g)} + \frac{1}{2}O_{2(g)} \rightarrow CO_{2(g)}$    $\Delta H° = -283.0$ kJ/mol

(2) $2Fe_{(s)} + \frac{3}{2}O_{2(g)} \rightarrow Fe_2O_3$    $\Delta H° = -822.3$ kJ/mol

**What Is Required?**

You need to find $\Delta H°$ of the target reaction.

$Fe_2O_{3(s)} + 3CO_{(g)} \rightarrow 3CO_{2(g)} + 2Fe_{(s)}$

**What Is Given?**

You know the chemical equations for reactions (1) and (2), and their corresponding enthalpy changes.

**Plan Your Strategy**

**Step 1** Examine equations (1) and (2) to see how they compare with the target equation. Decide how you need to manipulate equations (1) and (2) so that they add to the target equation. (Reverse the equation, multiply the equation, do both, or do neither). Remember to adjust $\Delta H°$ accordingly for each equation.

**Step 2** Write the manipulated equations so that their equation arrows line up. Add the reactants and products on each side, and cancel substances that appear on both sides.

**Step 3** Ensure that you have obtained the target equation. Add $\Delta H°$ for the combined equations.

**Act on Your Strategy**

**Step 1** Equation (1) has CO as a reactant and $CO_2$ as a product, as does the target reaction. The stoichiometric coefficients do not match the coefficients in the target equation, however. To achieve the same coefficients, you must multiply equation (1) by 3.
Equation (2) has the required stoichiometric coefficients, but Fe and $Fe_2O_3$ are on the wrong sides of the equation. You need to reverse equation (2) and change the sign of $\Delta H°$.

**Step 2** Multiply each equation as required, and add them.

$3 \times (1)$   $3CO_{(g)} + \frac{3}{2}O_{2(g)} \rightarrow 3CO_{2(g)}$   $\Delta H° = 3(-283.0$ kJ/mol$)$

$-1 \times (2)$   $Fe_2O_{3(s)} \rightarrow 2Fe_{(s)} + \frac{3}{2}O_{2(g)}$   $\Delta H° = -1(-824.2$ kJ/mol$)$

$Fe_2O_{3(s)} + \frac{3}{2}O_{2(g)} + 3CO_{(g)} \rightarrow 3CO_{2(g)} + 2Fe_{(s)} + \frac{3}{2}O_{2(g)}$

or

$Fe_2O_{3(s)} + 3CO_{(g)} \rightarrow 3CO_{2(g)} + 2Fe_{(s)}$

---

**PROBLEM TIP**

Before adding chemical equations, be sure to line up the equation arrows.

---

**Web LINK**

www.mcgrawhill.ca/links/atlchemistry

For more information about aspects of material covered in this section of the chapter, go to the web site above and click on **Electronic Learning Partner**.

**Step 3** The desired equation is achieved. Therefore, you can calculate the enthalpy change of the target reaction by adding the heats of reaction for the manipulated equations.

$$\Delta H° = 3(-283.0 \text{ kJ/mol}) + 824.2 \text{ kJ/mol} = -24.8 \text{ kJ/mol}$$
$$\therefore Fe_2O_{3(s)} + 3CO_{(g)} \rightarrow 3CO_{2(g)} + 2Fe_{(s)} \quad \Delta H° = -24.8 \text{ kJ/mol}$$

**Check Your Solution**

The equations added correctly to the target equation. Check to ensure that you adjusted $\Delta H°$ accordingly for each equation.

## Practice Problems

**11.** Ethene, $C_2H_4$, reacts with water to form ethanol, $CH_3CH_2OH_{(\ell)}$.

$$C_2H_{4(g)} + H_2O_{(\ell)} \rightarrow CH_3CH_2OH_{(\ell)}$$

Determine the enthalpy change of this reaction, given the following thermochemical equations.

(1) $CH_3CH_2OH_{(\ell)} + 3O_{2(g)} \rightarrow 3H_2O_{(\ell)} + 2CO_{2(g)} \quad \Delta H° = -1367 \text{ kJ/mol}$
(2) $C_2H_{4(g)} + 3O_{2(g)} \rightarrow 2H_2O_{(\ell)} + 2CO_{2(g)} \quad \Delta H° = -1411 \text{ kJ/mol}$

**12.** A typical automobile engine uses a lead-acid battery. During discharge, the following chemical reaction takes place.

$$Pb_{(s)} + PbO_{2(s)} + 2H_2SO_{4(\ell)} \rightarrow 2PbSO_{4(aq)} + 2H_2O_{(\ell)}$$

Determine the enthalpy change of this reaction, given the following equations.

(1) $Pb_{(s)} + PbO_{2(s)} + 2SO_{3(g)} \rightarrow 2PbSO_{4(s)} \quad \Delta H° = -775 \text{ kJ/mol}$
(2) $SO_{3(g)} + H_2O_{(\ell)} \rightarrow H_2SO_{4(\ell)} \quad \Delta H° = -133 \text{ kJ/mol}$

**13.** Mixing household cleansers can result in the production of hydrogen chloride gas, $HCl_{(g)}$. Not only is this gas dangerous in its own right, but it also reacts with oxygen to form chlorine gas and water vapour.

$$4HCl_{(g)} + O_{2(g)} \rightarrow 2Cl_{2(g)} + 2H_2O_{(g)}$$

Determine the enthalpy change of this reaction, given the following equations.

(1) $H_{2(g)} + Cl_{2(g)} \rightarrow 2HCl_{(g)} \quad \Delta H° = -185 \text{ kJ/mol}$
(2) $H_{2(g)} + \frac{1}{2}O_{2(g)} \rightarrow H_2O_{(\ell)} \quad \Delta H° = -285.8 \text{ kJ/mol}$
(3) $H_2O_{(g)} \rightarrow H_2O_{(\ell)} \quad \Delta H° = -40.7 \text{ kJ/mol}$

**14.** Calculate the enthalpy change of the following reaction between nitrogen gas and oxygen gas, given thermochemical equations (1), (2), and (3).

$$2N_{2(g)} + 5O_{2(g)} \rightarrow 2N_2O_{5(g)}$$

(1) $2H_{2(g)} + O_{2(g)} \rightarrow 2H_2O_{(\ell)} \quad \Delta H° = -572 \text{ kJ/mol}$
(2) $N_2O_{5(g)} + H_2O_{(\ell)} \rightarrow 2HNO_{3(\ell)} \quad \Delta H° = -77 \text{ kJ/mol}$
(3) $\frac{1}{2}N_{2(g)} + \frac{3}{2}O_{2(g)} + \frac{1}{2}H_{2(g)} \rightarrow HNO_{3(\ell)} \quad \Delta H° = -174 \text{ kJ/mol}$

Sometimes it is impractical to use a coffee-cup calorimeter to find the enthalpy change of a reaction. You can, however, use the calorimeter to find the enthalpy changes of other reactions, which you can combine to arrive at the desired reaction. In the following investigation, you will apply Hess's law to determine the enthalpy change of a reaction.

# Investigation 17-C

**SKILL FOCUS**
Performing and recording
Analyzing and interpreting
Communicating results

## Hess's Law and the Enthalpy of Combustion of Magnesium

Magnesium ribbon burns in air in a highly exothermic combustion reaction. (See equation (1).) A very bright flame accompanies the production of magnesium oxide, as shown in the photograph below. It is impractical and dangerous to use a coffee-cup calorimeter to determine the enthalpy change for this reaction.

(1) $Mg_{(s)} + \frac{1}{2}O_{2(g)} \rightarrow MgO_{(s)}$

Instead, you will determine the enthalpy changes for two other reactions (equations (2) and (3) below). You will use these enthalpy changes, along with the known enthalpy change for another reaction (equation (4) below), to determine the enthalpy change for the combustion of magnesium.

(2) $MgO_{(s)} + 2HCl_{(aq)} \rightarrow MgCl_{2(aq)} + H_2O_{(\ell)}$

(3) $Mg_{(s)} + 2HCl_{(aq)} \rightarrow MgCl_{2(aq)} + H_{2(g)}$

(4) $H_{2(g)} + \frac{1}{2}O_{2(g)} \rightarrow H_2O_{(\ell)} + 285.8$ kJ

Notice that equations (2) and (3) occur in aqueous solution. You can use a coffee-cup calorimeter to determine the enthalpy changes for these reactions. Equation (4) represents the formation of water directly from its elements in their standard state.

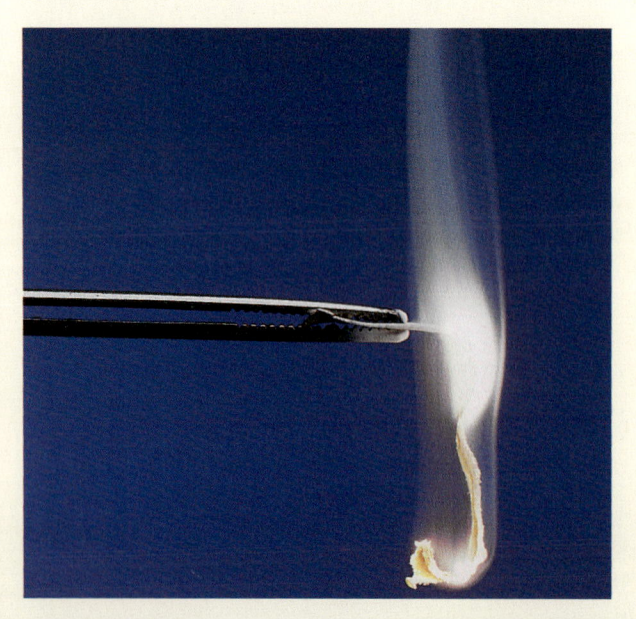

### Question

How can you use equations (2), (3), and (4) to determine the enthalpy change of equation (1)?

### Prediction

Predict whether reactions (2) and (3) will be exothermic or endothermic.

### Materials

coffee cup calorimeter (2 nested coffee cups sitting in a 250 mL beaker)
thermometer
100 mL graduated cylinder
scoopula
electronic balance
MgO powder
Mg ribbon (or Mg turnings)
sandpaper or emery paper
1.00 mol/L $HCl_{(aq)}$

### Safety Precautions

- Hydrochloric acid is corrosive. Use care when handling it.
- Be careful not to inhale the magnesium oxide powder.

### Procedure

**Part 1 Determining ΔH of Equation (2)**

1. Read the Procedure for Part 1. Prepare a fully-labelled set of axes to graph your temperature observations.

2. Set up the coffee-cup calorimeter. (Refer to Investigation 5-A) Using a graduated cylinder, add 100 mL of 1.00 mol/L $HCl_{(aq)}$ to the calorimeter. **CAUTION** $HCl_{(aq)}$ can burn your skin.

3. Record the initial temperature, $T_i$, of the $HCl_{(aq)}$, to the nearest tenth of a degree.

4. Find the mass of no more than 0.80 g of MgO. Record the exact mass.

5. Add the MgO powder to the calorimeter containing the $HCl_{(aq)}$. Swirl the solution gently, recording the temperature every 30 s until the highest temperature, $T_f$, is reached.

6. Dispose of the reaction solution as directed by your teacher.

**Part 2 Determining ΔH of Equation (3)**

1. Read the Procedure for Part 2. Prepare a fully-labelled set of axes to graph your temperature observations.

2. Using a graduated cylinder, add 100 mL of 1.00 mol/L $HCl_{(aq)}$ to the calorimeter.

3. Record the initial temperature, $T_i$, of the $HCl_{(aq)}$, to the nearest tenth of a degree.

4. If you are using magnesium ribbon (as opposed to turnings), sand the ribbon. Accurately determine the mass of no more than 0.50 g of magnesium. Record the exact mass.

5. Add the Mg to the calorimeter containing the $HCl_{(aq)}$. Swirl the solution gently, recording the temperature every 30 s until the highest temperature, $T_f$, is reached.

6. Dispose of the solution as directed by your teacher.

**Analysis**

1. Use the equation $q = m \cdot c \cdot \Delta T$ to determine the amount of heat that is released or absorbed by reactions (2) and (3). List any assumptions you make.

2. Convert the mass of MgO and Mg to moles. Calculate ΔH of each reaction in units of kJ/mol of MgO or Mg. Remember to put the proper sign (+ or −) in front of each ΔH value.

3. Algebraically combine equations (2), (3), and (4), and their corresponding ΔH values, to get equation (1) and ΔH of the combustion of magnesium.

4. (a) Your teacher will tell you the accepted value of ΔH of the combustion of magnesium. Based on the accepted value, calculate your percent error.

   (b) Suggest some sources of error in the investigation. In what ways could you improve the procedure?

5. What assumption did you make about the amount of heat that was lost to the calorimeter? Do you think that this is a fair assumption? Explain.

6. Why was it fair to assume that the hydrochloric acid solution has the same density and specific heat capacity as water?

**Conclusion**

7. Explain how you used Hess's law of heat summation to determine ΔH of the combustion of magnesium. State the result you obtained for the thermochemical equation that corresponds to chemical equation (1).

**Extension**

8. Design an investigation to verify Hess's law, using the following equations.

   (1) $NaOH_{(s)} \rightarrow Na^+_{(aq)} + OH^-_{(aq)}$
   (2) $NaOH_{(s)} + H^+_{(aq)} + Cl^-_{(aq)} \rightarrow Na^+_{(aq)} + Cl^-_{(aq)} + H_2O_{(\ell)}$
   (3) $Na^+_{(aq)} + OH^-_{(aq)} + H^+_{(aq)} + Cl^-_{(aq)} \rightarrow Na^+_{(aq)} + Cl^-_{(aq)} + H_2O_{(\ell)}$

   Assume that you have a coffee-cup calorimeter, solid NaOH, 1.00 mol/L $HCl_{(aq)}$, 1.00 mol/L $NaOH_{(aq)}$, and standard laboratory equipment. Write a step-by-step procedure for the investigation. Then outline a plan for analyzing your data. Be sure to include appropriate safety precautions. If time permits, obtain your teacher's approval and carry out the investigation.

**Table 17.1** Selected Standard Molar Enthalpies of Formation

| Compound | $\Delta H°_f$ |
|---|---|
| $CO_{(g)}$ | −110.5 |
| $CO_{2(g)}$ | −393.5 |
| $CH_{4(g)}$ | −74.6 |
| $CH_3OH_{(\ell)}$ | −238.6 |
| $C_2H_5OH_{(\ell)}$ | −277.6 |
| $C_6H_{6(\ell)}$ | +49.0 |
| $C_6H_{12}O_{6(s)}$ | −1274.5 |
| $H_2O_{(\ell)}$ | −285.8 |
| $H_2O_{(g)}$ | −241.8 |
| $CaCl_{2(s)}$ | −795.4 |
| $CaCO_{3(s)}$ | −1206.9 |
| $NaCl_{(s)}$ | −411.1 |
| $HCl_{(g)}$ | −92.3 |
| $HCl_{(aq)}$ | −167.5 |

**Figure 17.11** Carbon can exist as graphite or diamond under standard conditions. It can, however, have only one standard state. Carbon's standard state is graphite.

## Using Standard Molar Enthalpies of Formation

You have learned how to add equations with known enthalpy changes to obtain the enthalpy change for another equation. This method can be time-consuming and difficult, however, because you need to find reactions with known enthalpy changes that will add to give your target equation. There is another way to use Hess's law to find the enthalpy of an equation.

### Formation Reactions

In section 16.2, you learned about molar enthalpies of formation. Recall that in a formation reaction, a substance is formed from elements in their standard states. Table 17.1 shows some standard molar enthalpies of formation for easy reference. You will find additional standard molar enthalpies of formation listed in Appendix E.

*By definition, the enthalpy of formation of an element in its standard state is zero.* The standard state of an element is usually its most stable form under standard conditions. Standard conditions are 25°C and 100 kPa (close to room temperature and pressure). Therefore, the standard state of nitrogen is $N_{2(g)}$. The standard state of magnesium is $Mg_{(s)}$.

Some elements exist in more than one form under standard conditions. For example, carbon can exist as either graphite or diamond, as shown in Figure 17.11. Graphite is defined as the standard state of carbon. Therefore, the standard molar enthalpy of formation of graphite carbon is 0 kJ/mol. The standard molar enthalpy of formation of diamond is 1.9 kJ/mol. Another example is oxygen, $O_{2(g)}$. Oxygen also exists in the form of ozone, $O_{3(g)}$, under standard conditions. The diatomic molecule is defined as the standard state of oxygen, however, because it is far more stable than ozone. Therefore, the standard molar enthalpy of formation of oxygen gas, $O_{2(g)}$, is 0 kJ/mol. The standard molar enthalpy of formation of ozone is 143 kJ/mol.

$$C_{(graphite)} \longrightarrow C_{(diamond)} \quad \Delta H_f = 1.9 \text{ kJ/mol}$$

Recall from section 16.2 that a formation equation shows the formation of one mole of a substance from its elements in their standard states. For example, the following equation shows the formation of liquid water from its elements under standard conditions.

$$H_{2(g)} + \frac{1}{2}O_{2(g)} \rightarrow H_2O_{(\ell)} \quad \Delta H°_f = -285.8 \text{ kJ/mol}$$

Try the following problems to practice writing formation equations.

## Practice Problems

**15.** Write a thermochemical equation for the formation of each substance. Be sure to include the physical state of all the elements and compounds in the equation. You can find the standard enthalpy of formation of each substance in Appendix E.
   (a) $CH_4$
   (b) NaCl
   (c) MgO
   (d) $CaCO_3$

**16.** Liquid sulfuric acid has a very large negative standard enthalpy of formation (−814.0 kJ/mol). Write an equation to show the formation of liquid sulfuric acid. The standard state of sulfur is rhombic sulfur ($S_{(s)}$).

**17.** Write a thermochemical equation for the formation of gaseous cesium. The standard enthalpy of formation of $Cs_{(g)}$ is 76.7 kJ/mol.

**18.** Solid phosphorus is found in two forms: white phosphorus ($P_4$) and red phosphorus (P). White phosphorus is the standard state.
   (a) The enthalpy of formation of red phosphorus is −17.6 kJ/mol. Write a thermochemical equation for the formation of red phosphorus.
   (b) 32.6 g of white phosphorus forms red phosphorus. What is the enthalpy change?

## Calculating Enthalpy Changes

You can calculate the enthalpy change of a chemical reaction by adding the heats of formation of the products and subtracting the heats of formation of the reactants. The following equation can be used to determine the enthalpy change of a chemical reaction.

$$\Delta H°_{rxn} = \Sigma(n\Delta H°_f \text{ products}) - \Sigma(n\Delta H°_f \text{ reactants})$$

In this equation, $n$ represents the molar coefficient of each compound in the balanced chemical equation and $\Sigma$ means "the sum of."

As usual, you need to begin with a balanced chemical equation. If a given reactant or product has a molar coefficient that is not 1, you need to multiply its $\Delta H°_f$ by the same molar coefficient. This makes sense because the units of $\Delta H°_f$ are kJ/mol. Consider, for example, the complete combustion of methane, $CH_{4(g)}$.

$$CH_{4(g)} + 2O_{2(g)} \rightarrow CO_{2(g)} + 2H_2O_{(g)}$$

Using the equation for the enthalpy change, and the standard enthalpies of formation in Appendix E, you can calculate the enthalpy change of this reaction.

$$\Delta H°_{rxn} = [(\Delta H°_f \text{ of } CO_{2(g)}) + 2(\Delta H°_f \text{ of } H_2O_{(g)})] - [(\Delta H°_f \text{ of } CH_{4(g)}) + 2(\Delta H°_f \text{ of } O_{2(g)})]$$

Substitute the standard enthalpies of formation from Appendix E to get the following calculation.

$$\Delta H°_{rxn} = [(-393.5 \text{ kJ/mol}) + 2(-241.8 \text{ kJ/mol})] - [(-74.8 \text{ kJ/mol}) + 2(0 \text{ kJ/mol})]$$
$$= -802.3 \text{ kJ/mol of } CH_4$$

How does this method of adding heats of formation relate to Hess's law? Consider the equations for the formation of each compound that is involved in the reaction of methane with oxygen.

(1) $H_{2(g)} + \frac{1}{2}O_{2(g)} \rightarrow H_2O_{(g)}$ $\quad \Delta H°_f = -241.8$ kJ

(2) $C_{(s)} + O_{2(g)} \rightarrow CO_{2(g)}$ $\quad \Delta H°_f = -393.5$ kJ

(3) $C_{(s)} + 2H_{2(g)} \rightarrow CH_{4(g)}$ $\quad \Delta H°_f = -74.6$ kJ

There is no equation for the formation of oxygen, because oxygen is an element in its standard state.

By adding the formation equations, you can obtain the target equation. Notice that you need to reverse equation (3) and multiply equation (1) by 2.

2 × (1) $\quad 2H_{2(g)} + O_{2(g)} \rightarrow 2H_2O_{(g)}$ $\quad \Delta H°_f = 2(-241.8)$ kJ

(2) $\quad C_{(s)} + O_{2(g)} \rightarrow CO_{2(g)}$ $\quad \Delta H°_f = -393.5$ kJ

−1 × (3) $\quad CH_{4(g)} \rightarrow C_{(s)} + 2H_{2(g)}$ $\quad \Delta H°_f = -1(-74.6)$ kJ

$CH_{4(g)} + 2O_{2(g)} + \cancel{C_{(s)}} + \cancel{2H_{2(g)}} \rightarrow 2H_2O_{(g)} + CO_{2(g)} + \cancel{C_{(s)}} + \cancel{2H_{2(g)}}$

or

$CH_{4(g)} + 2O_{2(g)} \rightarrow 2H_2O_{(g)} + CO_{2(g)}$

Add the manipulated $\Delta H°_f$ values:

$\Delta H°_{rxn} = 2(-241.8)$ kJ $- 393.5$ kJ $+ 74.6$ kJ.
$= -802.3$ kJ

This value of $\Delta H°$ is the same as the value you obtained using $\Delta H°_f$ data. When you used the addition method, you performed the same operations on the enthalpies of formation before adding them. Therefore, using enthalpies of formation to determine the enthalpy of a reaction is consistent with Hess's law. Figure 17.12 shows the general process for determining the enthalpy of a reaction from enthalpies of formation.

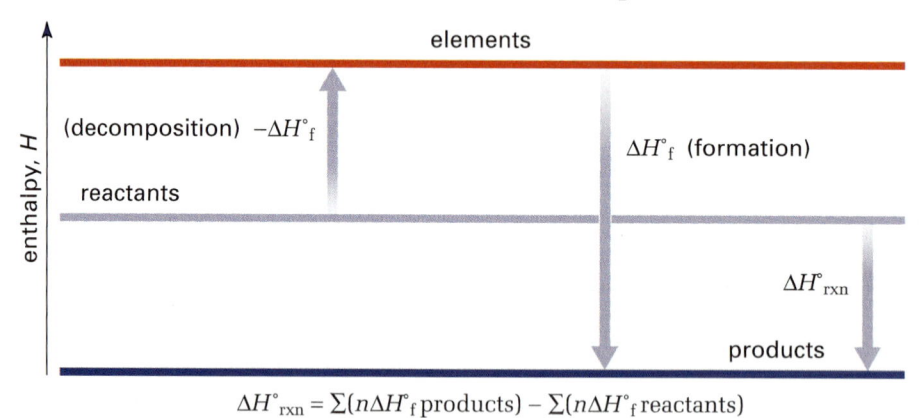

**CONCEPT CHECK**

"Using enthalpies of formation is like a shortcut for adding equations to obtain $\Delta H°$." Do you agree with this statement? Explain your answer.

**Figure 17.12** The overall enthalpy change of any reaction is the sum of the enthalpy change of the decomposition of the reactants to their elements and the enthalpy change of the formation of the products from their elements.

$\Delta H°_{rxn} = \Sigma(n\Delta H°_f \text{ products}) - \Sigma(n\Delta H°_f \text{ reactants})$

It is important to realize that, in most reactions, *the reactants do not actually break down into their elements and then react to form products.* Since there is extensive data about enthalpies of formation, however, it is useful to calculate the overall enthalpy change this way. Moreover, according to Hess's law, the enthalpy change is the same, regardless of the pathway. Examine the following Sample Problem to see how to use enthalpies of formation to determine the enthalpy change of a reaction. Then try the Practice Problems that follow.

## Sample Problem

### Using Enthalpies of Formation

**Problem**

Iron(III) oxide reacts with carbon monoxide to produce elemental iron and carbon dioxide. Determine the enthalpy change of this reaction, using known enthalpies of formation.

$Fe_2O_{3(s)} + 3CO_{(g)} \rightarrow 3CO_{2(g)} + 2Fe_{(s)}$

**What Is Required?**

You need to find $\Delta H°$ of the given chemical equation, using $\Delta H°_f$ data.

**What Is Given?**

From Appendix E, you can obtain the enthalpies of formation.

$\Delta H°_f$ of $Fe_2O_{3(s)} = -824.2$ kJ/mol

$\Delta H°_f$ of $CO_{(g)} = -110.5$ kJ/mol

$\Delta H°_f$ of $CO_{2(g)} = -393.5$ kJ/mol

$\Delta H°_f$ of $Fe_{(s)} = 0$ kJ/mol (by definition)

**Plan Your Strategy**

Multiply each $\Delta H°_f$ value by its molar coefficient from the balanced chemical equation. Substitute into the following equation, and then solve.

$\Delta H° = \Sigma(n\Delta H°_f \text{ products}) - \Sigma(n\Delta H°_f \text{ reactants})$

**Act on Your Strategy**

$\Delta H° = \Sigma(n\Delta H°_f \text{ products}) - \Sigma(n\Delta H°_f \text{ reactants})$

$= [3(\Delta H°_f\ CO_{2(g)}) + 2(\Delta H°_f\ Fe_{(s)})] - [(\Delta H°_f\ Fe_2O_{3(s)}) + 3(\Delta H°_f\ CO_{(g)})]$

$= [(-393.5 \text{ kJ/mol}) + 2(0 \text{ kJ/mol})] - [(-824.2 \text{ kJ/mol}) + 3(-110.5 \text{ kJ/mol})]$

$= -24.8$ kJ/mol

$\therefore Fe_2O_{3(s)} + 3CO_{(g)} \rightarrow 3CO_{2(g)} + 2Fe_{(s)} \quad \Delta H° = -24.8$ kJ/mol

**Check Your Solution**

A balanced chemical equation was used in the calculation. The number of significant digits is correct. The units are also correct.

> **CONCEPT CHECK**
>
> You saw the reaction between iron(III) oxide and carbon monoxide in the Sample Problem on page 680. Which method for determining the enthalpy of reaction do you prefer? Explain your answer.

### Practice Problems

19. Hydrogen can be added to ethene, $C_2H_4$, to obtain ethane, $C_2H_6$.

    $C_2H_{4(g)} + H_{2(g)} \rightarrow C_2H_{6(g)}$

    Show that the equations for the formation of ethene and ethane from their elements can be algebraically combined to obtain the equation for the addition of hydrogen to ethene.

20. Zinc sulfide reacts with oxygen gas to produce zinc oxide and sulfur dioxide.

    $2ZnS_{(s)} + 3O_{2(g)} \rightarrow 2ZnO_{(s)} + 2SO_{2(g)}$

    Write the chemical equation for the formation of the indicated number of moles of each compound from its elements. Algebraically combine these equations to obtain the given equation.

*Continued...*

> **Continued...**
>
> 21. Small amounts of oxygen gas can be produced in a laboratory by heating potassium chlorate, KClO₃.
>
>     2KClO₃₍s₎ → 2KCl₍s₎ + 3O₂₍g₎
>
>     Calculate the enthalpy change of this reaction, using enthalpies of formation from Appendix E.
>
> 22. Use the following equation to answer the questions below.
>
>     CH₃OH₍ℓ₎ + 1.5O₂₍g₎ → CO₂₍g₎ + 2H₂O₍g₎
>
>     **(a)** Calculate the enthalpy change of the complete combustion of one mole of methanol, using enthalpies of formation.
>
>     **(b)** How much energy is released when 125 g of methanol undergoes complete combustion.

## Using Bond Energies

As you learned in Chapter 5, a chemical bond is caused by the attractions among the electrons and nuclei of two atoms. Breaking bonds is an exothermic process, while making bonds is an endothermic process.

A specific amount of energy is needed to break each type of bond. The energy that is required to break a bond is called bond energy. Bond energy is usually measured in kJ/mol. Table 17.2 reproduces the bond energy table you saw in Chapter 5, for easy reference. You will find a more comprehensive list of bond energies in Appendix E.

Every chemical reaction involves bond breaking and bond formation. Since there are different types of bonds in the reactants and products, the total energy required for bond breaking and the total energy released when new bonds form are different. The difference represents the energy change for the reaction.

To use bond energies to estimate the enthalpy change for a reaction, add together the total bond energies of the reactants. Bond breakage requires energy, so the sign will be positive. From the total, subtract the total bond energies of the products. If the reaction is endothermic, the result will be positive. If the reaction is exothermic, the result will be negative. The following equation summarizes the process for using bond energies to estimate the enthalpy of a reaction.

$$\Delta H_{rxn} = \sum \text{bond energies (reactants)} - \sum \text{bond energies (products)}$$

Since bond energies are only a way of *estimating* the energy of a reaction, your answer will not agree exactly with the recognized enthalpy of reaction. Bond energies are not exact; rather, they are averages.

For example, the table shows that 413 kJ of energy is needed to break one mole of C—H bonds. In fact, breaking a C—H bond in a methane molecule (CH₄) would require a slightly different quantity of energy than breaking a C—H bond in, say, a molecule of dichloromethane (CH₂Cl₂). The value 413 kJ/mol is a value that reflects the average of known bond energies for C—H bonds in different molecules.

The Sample Problem on the following page shows you how to estimate enthalpy changes of reactions using average bond energies. Try the Practice Problems that follow to compare how this method of estimating enthalpy change compares with the other methods you have learned in this chapter.

**Table 17.2** Average Bond Energies

| Bond | Average bond energy (kJ/mol) |
| --- | --- |
| C—C | 346 |
| C=C | 610 |
| C≡C | 835 |
| Si—Si | 226 |
| C—H | 413 |
| Si—H | 318 |
| H—H | 432 |
| C—O | 358 |
| C=O | 745 |
| O—H | 467 |
| O=O | 498 |

## Sample Problem
### Using Bond Energies

**Problem**

Consider the following familiar combustion reaction.

$CH_{4(g)} + 2O_{2(g)} \rightarrow CO_{2(g)} + 2H_2O_{(g)}$

**(a)** Use bond energies to estimate the enthalpy change for the combustion of methane.

**(b)** Compare your answer to the accepted value for the combustion of methane.

**What Is Required?**

**(a)** You need to determine the difference in energy between bonds formed in the products and bonds formed in the products.

**(b)** You need to find the accepted value for the combustion of methane.

**What Is Given?**

You know the equation for the combustion reaction.
From Table 17.2, you know the following bond energies:

C—H (413 kJ/mol)
O=O (498 kJ/mol)
C=O (745 kJ/mol)
O—H (467 kJ/mol)

**Plan Your Strategy**

**(a)** Draw Lewis dot diagrams or structural diagrams to determine the number and type of bonds broken in the reactants and the number and type of bonds formed in the products.

Then use the following equation to estimate the enthalpy change for the reaction.

$\Delta H_{rxn} = \sum$ bond energies (reactants) $- \sum$ bond energies (products)

**(b)** Look up the accepted value for the standard molar enthalpy of combustion of methane in Table 16.3.

**Act on Your Strategy**

reactants: bonds broken    products: bonds formed

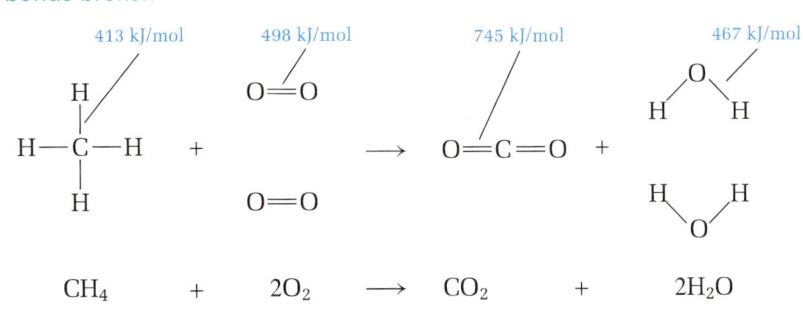

**Bonds broken**

$\sum$ bond energies (reactants) $= (4 \times C - H) + (2 \times O = O)$
$= (4 \times 413 \text{ kJ/mol}) + (2 \times 498 \text{ kJ/mol})$
$= 2648 \text{ kJ/mol}$

*Continued...*

**Bonds formed**

$$\sum \text{bond energies (products)} = (2 \times C=O) + (4 \times O-H)$$
$$= (2 \times 745 \text{ kJ/mol}) + (4 \times 467 \text{ kJ/mol})$$
$$= 3358 \text{ kJ/mol}$$

**Estimated enthalpy change**

$$\Delta H_{\text{rxn}} = \sum \text{bond energies (reactants)} - \sum \text{bond energies (products)}$$
$$= 2648 \text{ kJ/mol} - 3358 \text{kJ/mol}$$
$$= -710 \text{ kJ/mol}$$

(b) From Table 16.3, the standard molar enthalpy of combustion of methane is −965.1 kJ/mol.

**Check Your Solution**

The sign of your answer is negative, which makes sense for a combustion reaction. You know the combustion of methane is exothermic. Your answer is reasonably close to the accepted value. It is not surprising that the values are somewhat different, since you are using *average* bond energies. Also, the accepted value you used assumes that the water product is liquid, which accounts for some of the deviation.

## Practice Problems

23. Consider the following equation for the combustion of butane.

    $$C_4H_{10(g)} + \tfrac{13}{2}O_{2(g)} \rightarrow 4CO_{2(g)} + 5H_2O_{(g)}$$

    (a) Use bond energies to estimate the enthalpy change for the reaction as written.

    (b) Compare your answer to the accepted value using the standard molar enthalpy of combustion of butane.

24. Consider the following equation for the formation of ammonia from its elements.

    $$N_{2(g)} + 3H_{2(g)} \rightarrow 2NH_{3(g)}$$

    (a) Use bond energies to estimate the enthalpy change for the reaction as written (see Appendix E).

    (b) Compare your answer to the accepted value using the standard molar enthalpy of formation of ammonia (see Appendix E).

25. (a) Use bond energies to estimate the standard molar enthalpy of formation of gaseous water.

    (b) Compare your answer to the accepted value.

26. Consider the following equation for the reaction of methane with chlorine.

    $$CH_{4(g)} + 3Cl_{2(g)} \rightarrow CHCl_{3(g)} + 3HCl_{(g)}$$

    (a) Use bond energies to estimate the enthalpy change for the reaction (see Appendix E).

    (b) Use standard molar enthalpies to determine the enthalpy change for the reaction. (For gaseous $CHCl_3$, $\Delta H°_f = -103.18$ kJ/mol.) Compare your results from (a) and (b).

## Section Summary

In this section, you learned how to calculate the enthalpy change of a chemical reaction using Hess's law of heat summation. In the next section, you will see how the use of energy affects your lifestyle and the environment.

## Section Review

**1** Explain why you need to reverse the sign of $\Delta H°$ when you reverse an equation. Use an example in your answer.

**2** In section 5.3, you learned two methods for calculating enthalpy changes using Hess's law. If you had only this textbook as a reference, which method would allow you to calculate enthalpy changes for the largest number of reactions? Explain your answer.

**3** In the early 1960s, Neil Bartlett, at the University of British Columbia, was the first person to synthesize compounds of the noble gas xenon. A number of noble gas compounds (such as $XeF_2$, $XeF_4$, $XeF_6$, and $XeO_3$) have since been synthesized. Consider the reaction of xenon difluoride with fluorine gas to produce xenon tetrafluoride.

$XeF_{2(g)} + F_{2(g)} \rightarrow XeF_{4(s)}$

Use the standard molar enthalpies of formation on the right to calculate the enthalpy change for this reaction.

| Compound | $\Delta H°_f$ (kJ/mol) |
|---|---|
| $XeF_{2(g)}$ | −108 |
| $XeF_{4(s)}$ | −251 |

**4** Calculate the enthalpy change of the following reaction, given equations (1), (2), and (3).

$2H_3BO_{3(aq)} \rightarrow B_2O_{3(s)} + 3H_2O_{(\ell)}$

(1) $H_3BO_{3(aq)} \rightarrow HBO_{2(aq)} + H_2O_{(\ell)}$ $\Delta H° = -0.02$ kJ/mol

(2) $H_2B_4O_{7(s)} + H_2O_{(\ell)} \rightarrow 4HBO_{2(aq)}$ $\Delta H° = -11.3$ kJ/mol

(3) $H_2B_4O_{7(s)} \rightarrow 2B_2O_{3(s)} + H_2O_{(\ell)}$ $\Delta H° = 17.5$ kJ/mol

**5** The standard molar enthalpy of formation of calcium carbonate is −1207.6 kJ/mol. Calculate the standard molar enthalpy of formation of calcium oxide, given the following equation.

$CaO_{(g)} + CO_{2(g)} \rightarrow CaCO_{3(s)}$ $\Delta H° = -178.1$ kJ/mol

**6** A classmate is having difficulty understanding Hess's law. Write a few paragraphs to explain the law. Include examples, diagrams, and an original analogy.

**7** The combustion of acetylene gas, $C_2H_{2(g)}$, is highly exothermic, as shown in Figure 17.13. The following thermochemical equation represents the complete combustion of acetylene.

$C_2H_2(g) + \frac{5}{2}O_{2(g)} \rightarrow 2CO_{2(g)} + H_2O_{(g)} + 1259$ kJ

(a) Draw structural diagrams to represent each of the molecules involved in the reaction.

(b) Use the information provided and Table 17.2 to estimate the bond energy of the carbon-carbon triple bond in acetylene.

(c) Compare your result to the average bond energy for a carbon-carbon triple bond listed in Table 17.2.

**Figure 17.13** The combustion of acetylene (ethyne) in an oxyacetylene torch produces the highest temperature (about 3300°C) of any known mixture of combustible gases. Metal workers can use the heat from this combustion to cut through most metal alloys.

# 17.3 Fuelling Society

**Section Preview/Outcomes**

In this section, you will

- **compare** physical, chemical, and nuclear changes in terms of the species and the magnitude of energy involved
- **identify** and **describe** sources of energy including present sources and possible new ones
- **communicate** your understanding of the following terms: *non-renewable, renewable, greenhouse gases, global warming, sustainable development, risk, benefit, risk-benefit analysis*

Compared to many other countries, Canada has huge energy requirements per capita. This energy demand is due in part to Canada's vast size compared to its population, and the energy required for transportation of goods and people as a result. Also, Canada's northern climate means that for many months of the year, Canadians rely on natural gas, oil, or electricity to heat their homes.

Where does the energy come from? As you can see in Figure 17.14, most of Canada's energy comes from chemical processes such as the combustion of petroleum, coal, or natural gas. A significant portion of energy is also derived from nuclear processes. Although the energy changes associated with physical processes are not used directly to provide energy for Canadian society on a large scale, changes of state play an important role in generating electricity. Most power plants, whether nuclear or chemical, use the heat generated by exothermic processes to convert water into steam. The steam moves a turbine, which generates electrical energy.

**Primary Energy by Source, Canada, 1871 to 1996**
**(Percent of Energy Consumption)**

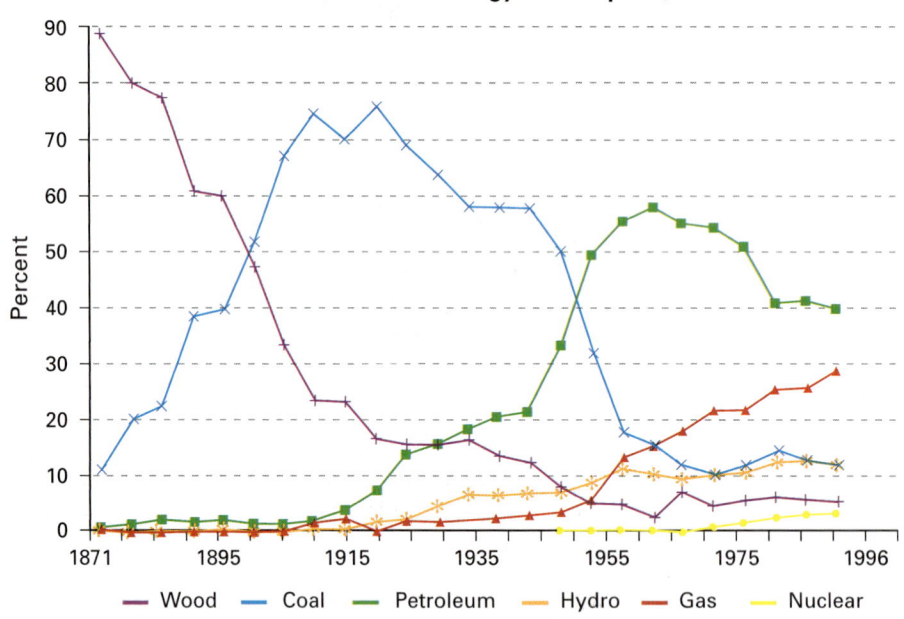

Wood — Coal — Petroleum — Hydro — Gas — Nuclear

**Figure 17.14** The energy that Canadians use comes from a variety of sources. What factors account for the changes you can see in this graph? How do you think energy use has changed since 1996?

## Energy from Physical, Chemical, and Nuclear Processes

How does the energy from physical, chemical, and nuclear processes compare?

- Changes of state involve the breaking of intermolecular bonds. Physical changes such as changes of state usually involve tens of kJ/mol. When water vapour condenses, 40.7 kJ/mol of heat is released.

692 MHR • Unit 7 Thermochemistry

- Chemical changes involve the breaking of chemical bonds, which are stronger than intermolecular bonds. Chemical changes usually involve hundreds or thousands of kJ/mol. For example, the molar enthalpy of combustion of coal is about 3900 kJ/mol.
- Nuclear changes involve changes within the nuclei of atoms. These types of changes involve enormous changes in energy. Nuclear changes involve millions and billions of kJ/mol. For example, nuclear power plants derive their energy from the fusion (splitting) of uranium-235. The fusion of 1 mol of uranium-235 releases about $2.1 \times 10^{10}$ kJ, or 21 billion kilojoules!

Figure 17.15 shows the energy changes associated with various different processes.

## Energy and Efficiency

When you think about energy efficiency, what comes to mind? You may think about taking the stairs instead of the elevator, choosing to drive a small car instead of a sport utility vehicle, or turning off lights when you are not using them. What, however, does efficiency really mean? How do you quantify it?

There are several ways to define efficiency. One general definition says that energy efficiency is the ability to produce a desired effect with minimum energy expenditure. For example, suppose that you want to bake a potato. You can use a microwave oven or a conventional oven. Both options achieve the same effect (baking the potato), but the first option uses less energy. According to the general definition above, using the microwave oven is more energy-efficient than using the conventional oven. The general definition is useful, but it is not quantitative.

Another definition of efficiency suggests that it is *the ratio of useful energy produced to energy used in its production, expressed as a percent.* This definition quantitatively compares input and output of energy. When you use it, however, you need to be clear about what you mean by "energy used." Figure 17.16 shows factors to consider when calculating efficiency or analyzing efficiency data.

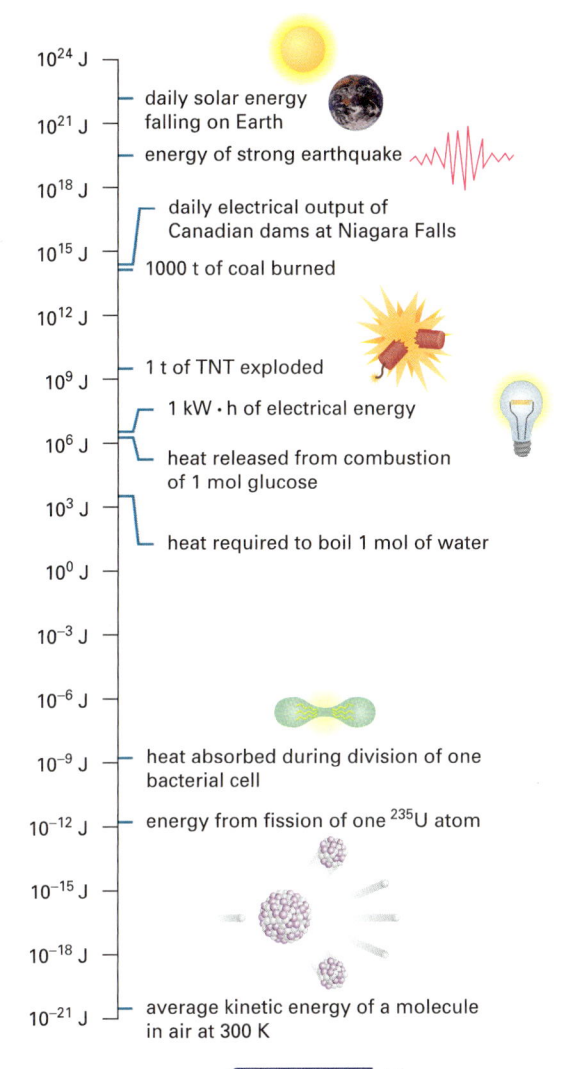

$10^{24}$ J — daily solar energy falling on Earth
$10^{21}$ J — energy of strong earthquake
$10^{18}$ J — daily electrical output of Canadian dams at Niagara Falls
$10^{15}$ J — 1000 t of coal burned
$10^{12}$ J
$10^{9}$ J — 1 t of TNT exploded
— 1 kW · h of electrical energy
$10^{6}$ J — heat released from combustion of 1 mol glucose
$10^{3}$ J — heat required to boil 1 mol of water
$10^{0}$ J
$10^{-3}$ J
$10^{-6}$ J
$10^{-9}$ J — heat absorbed during division of one bacterial cell
$10^{-12}$ J — energy from fission of one $^{235}$U atom
$10^{-15}$ J
$10^{-18}$ J
$10^{-21}$ J — average kinetic energy of a molecule in air at 300 K

**Figure 17.15** The energy changes of physical, nuclear, and chemical processes vary widely. Solar energy, necessary for life on earth, comes from nuclear reactions that take place in the sun.

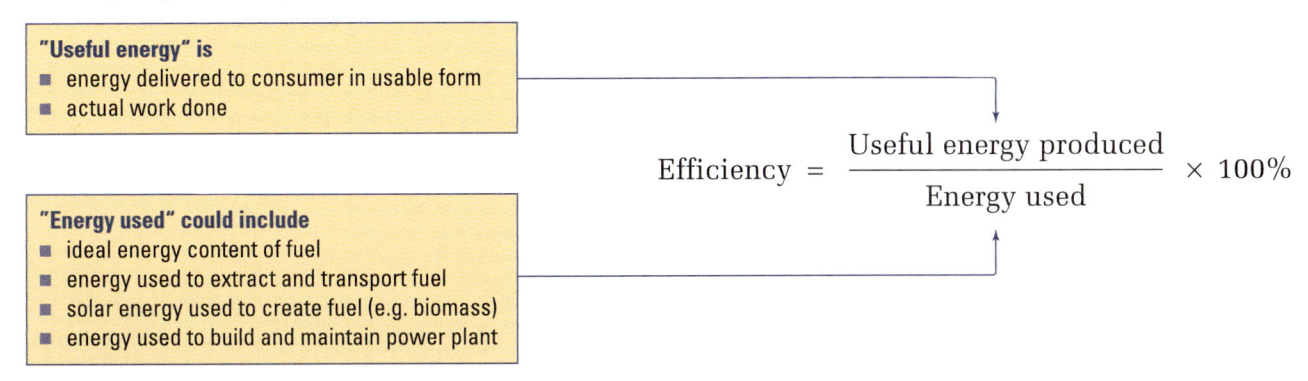

"Useful energy" is
- energy delivered to consumer in usable form
- actual work done

"Energy used" could include
- ideal energy content of fuel
- energy used to extract and transport fuel
- solar energy used to create fuel (e.g. biomass)
- energy used to build and maintain power plant

$$\text{Efficiency} = \frac{\text{Useful energy produced}}{\text{Energy used}} \times 100\%$$

**Figure 17.16** Efficiency is expressed as a percent. Always specify what is included in the "energy used" part of the ratio.

It is often difficult to determine how much energy is used to produce useful energy. Often an efficiency percent only takes into account the "ideal" energy output of a system, based on the energy content of the fuel.

### Efficiency and Natural Gas

When discussing the efficiency of a fuel, you need to specify how that fuel is being used. Consider, for example, natural gas. Natural gas is primarily methane. Therefore, you can estimate an ideal value for energy production using the enthalpy of combustion of methane.

$$CH_{4(g)} + 2O_{2(g)} \rightarrow CO_{2(g)} + 2H_2O_{(g)} \quad \Delta H° = -802 \text{ kJ}$$

In other words, 16 g of methane produces 802 kJ of heat (under constant pressure conditions).

When natural gas is used directly in cooking devices, its efficiency can be as high as 90%. Thus, for every 16 g of gas burned, you get about 720 kJ (0.90 × 802 kJ) of usable energy as heat for cooking. This is a much higher fuel efficiency than you can get with appliances that use electrical energy produced in a power plant that runs on a fuel such as coal.

If natural gas is used to produce electricity in a power plant, however, the efficiency is much lower—around 37%. Why? The heat from the burning natural gas is used to boil water. The kinetic energy of the resulting steam is transformed to mechanical energy for turning a turbine. The turbine generates the electrical energy. Each of these steps has an associated efficiency that is less than 100%. Thus, at each step, the overall efficiency of the fuel decreases.

### Thinking About the Environment

Efficiency is not the only criterion for selecting an energy source. Since the 1970s, society has become increasingly conscious of the impact of energy technologies on the environment.

Suppose that you want to analyze the environmental impact of an energy source. You can ask the following questions:

- Are any waste products or by-products of the energy production process harmful to the environment? For example, any process in which a hydrocarbon is burned produces carbon dioxide, a compound known to contribute to global warming. Any combustion process provides the heat required to form oxides of nitrogen from nitrogen gas. Nitrogen oxides contribute to acid precipitation.

- Is obtaining or harnessing the fuel harmful to the environment? For example, oil wells and strip coal mines destroy habitat. Natural gas pipelines, shown in Figure 17.16, are visually unappealing. They also split up habitat, which harms the ecosystem.

- Will using the energy source permanently remove the fuel from the environment? A **non-renewable** energy source (such as coal, oil, or natural gas) is effectively gone once we have used it up. Non-renewable energy sources take millions of years to form. We use them up at a much faster rate than they can be replenished. An energy source that is clearly **renewable** is solar energy. The Sun will continue to radiate energy toward Earth over its lifetime—many millions of years. A somewhat renewable energy source is wood. Trees can be grown to replace those cut down. It takes trees a long time to grow, however, and habitat is often destroyed in the meantime.

**Figure 17.16** This gas pipeline harms the ecosystem by splitting up habitat.

# Chemistry Bulletin

**Science** · **Technology** · **Science** · **Environment**

## Hot Ice

When engineers first began extending natural gas pipelines through regions of bitter cold, they noticed that their lines plugged with a dangerous slush of ice and gas. The intense pressure of the lines, combined with the cold, led to the formation of *methane hydrates*, a kind of gas-permeated ice. More than a mere nuisance, methane hydrate plugs were a potential threat to pipelines. The build-up of gas pressure behind a methane hydrate plug could lead to an explosion. Now, however, this same substance may hold the key to a vast fuel supply.

Methane hydrates form when methane molecules become trapped within an ice lattice as water freezes. They can form in very cold conditions or under high-pressure conditions. Both of these conditions are met in deep oceans and in permafrost. In Canada, hydrates have already been found in large quantities in the Canadian Arctic. Methane hydrate has a number of remarkable properties. For example, when brought into an oxygen atmosphere, the methane fumes can be ignited, making it appear that the ice is burning!

Methane releases 25% less carbon dioxide per gram than coal, and it emits none of the oxides of nitrogen and sulfur that contribute to acid precipitation. Therefore, using methane in place of other fossil fuels is very desirable. Methane hydrates seem to be an ideal and plentiful "pre-packaged" source of natural gas. Estimates of the exact amount of methane stored in hydrates suggest there could be enough to serve our energy needs anywhere from 350 years to 3500 years, based on current levels of energy consumption. This would constitute a significant source of fossil fuels, if we can find a way to extract the gas safely and economically.

Unfortunately, hydrates become unstable when the pressure or temperature changes. Even small changes in these conditions can cause hydrates to degrade rapidly. Methane hydrates are stable at ocean depths greater than 300 m, but offshore drilling at these depths has been known to disturb the hydrate formations, causing large, uncontrolled releases of flammable methane gas. Also, methane hydrates often hold sediment layers together. Therefore, in addition to the danger of a gas explosion, there is the danger of the sea floor collapsing where drilling occurs.

Methane is a significant greenhouse gas. A massive release of methane could cause catastrophic global climate change. Some researchers believe that the drastic climate change that occurred during the Pleistocene era was due to methane hydrate destabilization and widespread methane release.

Nonetheless, Canada, Japan, the United States, and Russia all have active research and exploration programs in this area. As global oil supplies dwindle, using methane hydrates might increasingly be seen as worth the risk and cost.

### Making Connections

1. Compare using methane from natural gas with using methane from methane hydrates in terms of environmental impact and efficiency. You will need to do some research to find out extraction methods for each source of methane.

2. On the Internet, research one possible structure of methane hydrate. Create a physical model or a three-dimensional computer model to represent it. Use your model to explain why methane hydrates are unstable at temperatures that are warmer than 0°C.

**Web LINK**

www.mcgrawhill.ca/links/atlchemistry

To learn more about greenhouse gases, global warming, and acid rain, go to the web site above and click on **Web Links**.

## Hydrocarbon Fuels and the Environment

Hydrocarbon fuels have changed the way we live. Our dependence on them, however, has affected the world around us. The greenhouse effect, global warming, acid rain, and pollution are familiar topics on the news today. Our use of petroleum products, such as oil and gasoline, is linked directly to these problems.

### The Greenhouse Effect and Global Warming

Roads, expressways, service stations, and parking lots occupy almost 40% of Toronto. They are the result of our demand for fast and efficient transportation. Every day, Toronto's vehicles produce nearly 16 000 t of carbon dioxide by the combustion of fossil fuels. Carbon dioxide is an example of a greenhouse gas. **Greenhouse gases** trap heat in Earth's atmosphere and prevent the heat from escaping into outer space. Scientists think that a build-up of carbon dioxide in the atmosphere may lead to an increase in global temperature, known as **global warming**. The diagram below shows how these concepts are connected to fossil fuels.

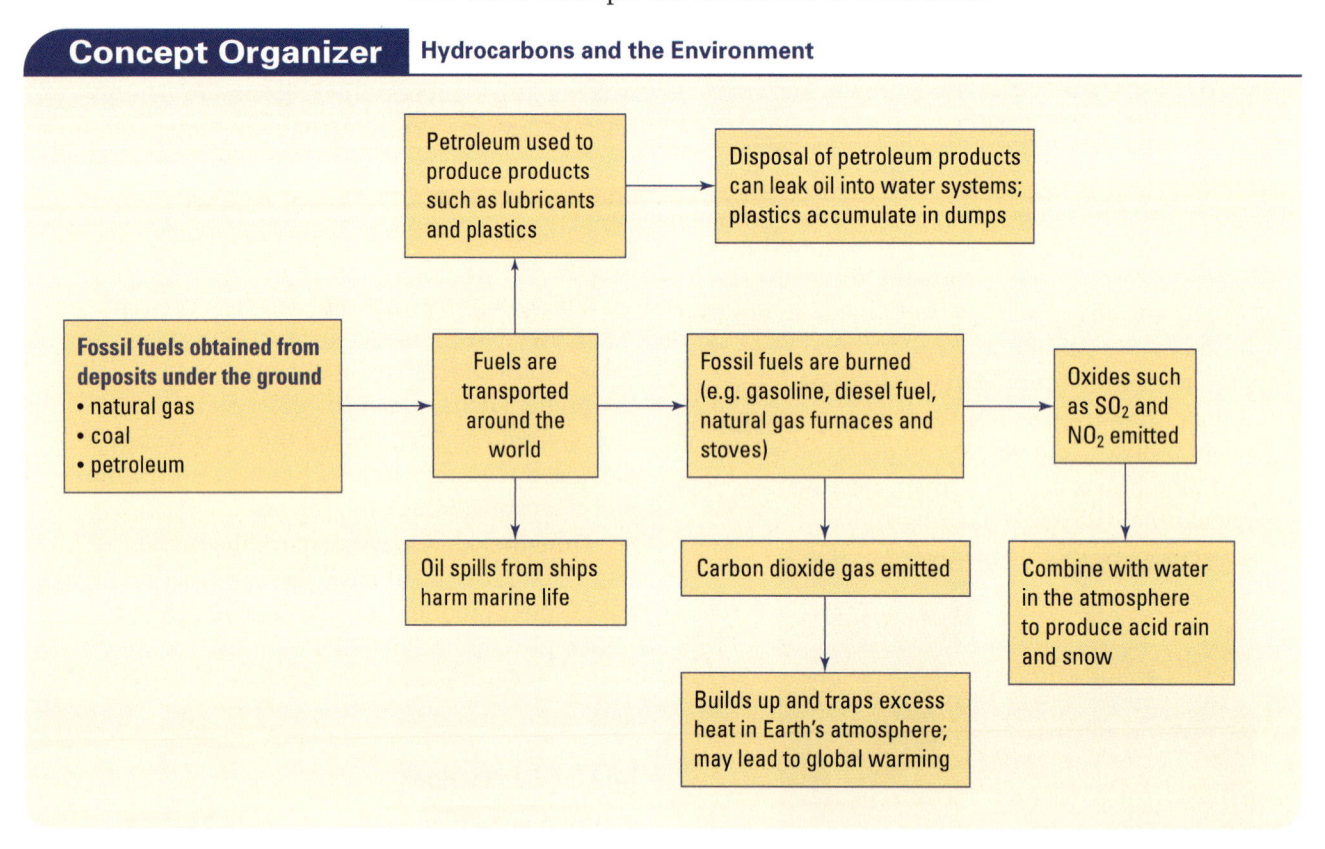

**Concept Organizer** — Hydrocarbons and the Environment

- Fossil fuels obtained from deposits under the ground
  - natural gas
  - coal
  - petroleum
- → Fuels are transported around the world
  - → Petroleum used to produce products such as lubricants and plastics → Disposal of petroleum products can leak oil into water systems; plastics accumulate in dumps
  - → Oil spills from ships harm marine life
  - → Fossil fuels are burned (e.g. gasoline, diesel fuel, natural gas furnaces and stoves)
    - → Carbon dioxide gas emitted → Builds up and traps excess heat in Earth's atmosphere; may lead to global warming
    - → Oxides such as $SO_2$ and $NO_2$ emitted → Combine with water in the atmosphere to produce acid rain and snow

### Acid Rain

The combustion of fossil fuels releases sulfur and nitrogen oxides. These oxides react with water vapour in the atmosphere to produce acid rain. Some lakes in northern Canada are "dead" because acid rain has killed the plants, algae, and fish that used to live in them. Forests in Québec and other parts of Canada have also been harmed by acid rain.

### Oil Spill Pollution

Our society demands a regular supply of fossil fuels. Petroleum is transported from oil-rich countries to the rest of the world. If an oil tanker carrying petroleum has an accident, the resulting oil spill can be disastrous to the environment.

# Careers in Chemistry

## Oil Spill Advisor

Developed nations, such as Canada, depend heavily on petroleum. Our dependence affects the environment in many ways. Oil spills are a dramatic example of environmental harm caused by petrochemicals. In the news, you may have seen oceans on fire and wildlife choked with tar. What can we do?

The best thing to do is to prevent oil spills from taking place. Stricter regulations and periodic inspections of oil storage companies help to prevent oil leakage. Once an oil spill has occurred, however, *biological*, *mechanical*, and *chemical* technologies can help to minimize harm to the environment.

Biological methods involve helpful microorganisms that break down, or *biodegrade*, the excess oil. Mechanical methods depend on machines that physically separate spilled oil from the environment. For example, barriers and booms are used to contain an oil spill and prevent it from spreading. Materials such as sawdust are sprinkled on a spill to soak up the oil.

Two main chemical strategies are also used to clean up oil spills. In the first strategy, *gelling agents* are added to react with the oil. The reaction results in a bulky product that is easier to collect using mechanical methods. In the second strategy, *dispersing agents* break up oil into small droplets that mix with the water. This prevents the oil from reaching nearby shorelines. Dispersing agents work in much the same way as a bar of soap!

The scientific advisor for an oil spill response unit assesses a spill and determines the appropriate clean-up methods. She or he acts as part of a team of advisors. Most advisors have an M.Sc. or Ph.D. in an area of expertise such as organic chemistry, physical chemistry, environmental chemistry, biology, oceanography, computer modelling, or chemical engineering.

Oil spill response is handled by private and public organizations. All these organizations look for people with a background in chemistry. In fact, much of what you are learning about hydrocarbons can be related to oil spill response. Hydrocarbon chemistry can lead you directly to an important career, helping to protect the environment.

### Make Career Connections

Create a technology scrapbook. Go through the business and employment sections in a newspaper. Cut out articles about clean-up technologies. What kinds of companies are doing this work? What can you learn about jobs in this field? What qualifications does a candidate need to apply for this type of job?

## Everyday Oil Pollution

The biggest source of oil pollution comes from the everyday use of oil by ordinary people. Oil that is dumped into water in urban areas adds to oil pollution from ships and tankers. In total, *three million tonnes* of oil reach the ocean each year. This is equivalent to having an oil spill disaster every day!

A high school student did a home experiment to discover how much oil remains in "empty" motor oil containers that are thrown out. He collected 100 empty oil containers from a local gas station. Then he measured the amount of oil that was left in each container. He found an average of 36 mL per container. Over 130 million oil containers are sold and thrown out in Canada each year. Using these figures, he calculated that nearly five million litres of oil are dumped into landfill sites every year, just in "empty" oil containers!

Once oil reaches the environment, it is almost impossible to clean up. Oil leaking from a landfill site can contaminate drinking water in the area. Because oil can dissolve similar substances, pollutants such as chlorine and pesticides, and other organic toxins, mix with the oil. They are carried with it into the water system, increasing the problem.

### Solutions to Environmental Problems Caused by Fossil Fuels

All of the problems described above hinge on our use of fossil fuels. Thus, cutting back on our use of fossil fuels will help to reduce environmental damage. Cutting back on fossil fuels, however, depends on the consumers who buy petrochemicals and use fossil fuels. In other words, it depends on you and the people you know.

Corporations that are looking for profit have little incentive to change their use of fossil fuels. For example, the technology is available to build cars that can drive about 32 km on a single litre of fuel. Because this technology is not financially profitable, cars are still being produced that drive about 8 km per litre of fuel. If consumers demand and purchase more fuel-efficient cars, however, car manufacturers will have an incentive to produce such cars. Tougher government standards may also help to push the vehicle industry towards greater fuel efficiency.

Governments can also bring about change by endorsing the principle of sustainable development. This principle was introduced at the 1992 Earth Summit Conference. **Sustainable development** takes into account *the environment, the economy, and the health and needs of society.* (See Figure 17.18.)

### Web LINK

www.mcgrawhill.ca/links/atlchemistry

The Kyoto Protocol is a set of terms for the reduction of greenhouse gas emissions. This set of terms was developed at a meeting that took place in Kyoto, Japan in 1997. Over 160 countries signed the protocol. Go to the web site above and click on **Web Links**. Follow the links to find answers to the following questions:

- What are some of the specific goals of the Kyoto Protocol?
- How is Canada participating in the Kyoto Protocol?
- What are some arguments against the Kyoto Protocol?

Write a brief essay either defending or arguing against the Kyoto Protocol.

**Figure 17.18** Canada and other members of the United Nations endorse the principle of sustainable development. This principle states that the world must find ways to meet our current needs, without compromising the needs of future generations.

Hydrocarbon fuels and products can benefit our society if they are managed well. They can cause great environmental damage, however, if they are managed irresponsibly. With enough knowledge, you can learn to make informed decisions on these important issues. Here are some suggestions of ways you can reduce your consumption of petroleum products. Brainstorm with your classmates to think of other ways to reduce consumption.

- Contact your local government and local power companies. Suggest using alternative fuels, such as solar energy and wind power.
- Ride a bicycle or walk more.
- Express your concerns by writing letters to the government or to newspapers.
- Become more informed by researching issues that concern you.
- Fix oil leaks in vehicles, and avoid dumping oil down the sink.
- If you are cold at home, put on an extra sweater instead of turning up the heat.

# Chemistry Bulletin

## Lamp Oil and the Petroleum Age

Abraham Gesner was born in 1797 near Cornwallis, Nova Scotia. Although Gesner became a medical doctor, he was much more interested in fossils. Gesner was fascinated by hydrocarbon substances, such as coal, asphaltum (asphalt), and bitumen. These substances were formed long ago from fossilized plants, algae, fish, and animals.

When Gesner was a young man, the main light sources available were fire, candles, and whale oil lamps. Gesner had made several trips to Trinidad. He began to experiment with asphaltum, a semisolid hydrocarbon from Trinidad's famous "pitch lake." In 1846, while giving a lecture in Prince Edward Island, he startled his audience by lighting a lamp that was filled with a fuel he had distilled from asphaltum. Gesner's lamp fuel gave more light and produced less smoke than any other lamp fuel the audience had ever seen used.

Gesner needed a more easily obtainable raw material to make his new lamp fuel. He tried a solid, black, coal-like bitumen from Albert County, New Brunswick. This substance, called albertite, worked better than any other substance that Gesner had tested.

### Making Kerosene

One residue from Gesner's distillation process was a type of wax. Therefore, he called his lamp fuel *kerosolain*, from the Greek word for "wax oil." He soon shortened the name to *kerosene*. To produce kerosene, Gesner heated chunks of albertite in a retort (a distilling vessel with a long downward-bending neck). As the albertite was heated, it gave off vapours. The vapours passed into the neck of the retort, condensed into liquids, and trickled down into a holding tank. Once Gesner had finished the first distillation, he let the tank's contents stand for several hours. This allowed water and solid to settle to the bottom. Then he drew off the oil that remained on top.

Gesner distilled this oil again, and then treated it with sulfuric acid and calcium oxide. Finally he distilled the oil once more.

By 1853, Gesner had perfected his process. In New York, he helped to start the North American Kerosene Gas Light Company. Gesner distinguished between three grades of kerosene: grades A, B, and C. Grade C, he said, was the best lamp oil. Grades A and B could also be burned in lamps, but they were dangerous because they could cause explosions and fires.

Although Gesner never knew, his grades A and B kerosene became even more useful than the purer grade C. These grades were later produced from crude oil, or petroleum, and given a new name: gasoline!

Gesner laid the groundwork for the entire petroleum industry. All the basics of later petroleum refining can be found in his technology.

### Making Connections

1. In the early nineteenth century, whales were hunted extensively for their oil, which was used mainly as lamp fuel. When kerosene became widely available, the demand for whale oil decreased. Find out what effect this had on whalers and whales.

2. How do you think the introduction of kerosene as a lamp oil changed people's lives at the time? What conclusions can you draw about the possible impact of technology?

**CONCEPT CHECK**

Hydrocarbons, such as fossil fuels, carry both risks and benefits. In a group, brainstorm to identify some risks and benefits.

## Analyzing Risks and Benefits of Energy Sources

A **risk** is a chance of possible negative or dangerous results. Riding a bicycle carries the risk of falling off. Driving a car carries the risk of an accident. Almost everything you do has some kind of risk attached. Fortunately most risks are relatively small, and they may never happen. Many of the activities that carry risks also carry benefits. A **benefit** is an advantage, or positive result. For example, riding a bicycle provides the benefits of exercise, transportation, and enjoyment. When deciding to do an activity, it may be a good idea to compare the risks and benefits involved. (See Figure 17.19.)

**Figure 17.19** Would you like to have a coal-burning power plant near your home? Some people might be upset by this idea because there is a health risk caused by pollution from the plant. Other people might think that a coal-burning power plant poses very little threat, and the benefits are worth the risk. Who is right? How do you decide?

### Risk-Benefit Analysis

Knowing more about an issue helps you assess its risks and benefits more accurately. How can you make the most informed decision possible? Follow these steps to do your own assessment of risks and benefits, called a **risk-benefit analysis**.

**Step 1** Identify possible risks and benefits of the activity. Decide how to research these risks and benefits.

**Step 2** Research the risks and benefits. You need information from reliable sources to make an accurate analysis.

**Step 3** Weigh the effects of the risks and benefits. You may find that the risks are too great and decide not to do the activity. On the other hand, you may find that the benefits are greater than the risks.

**Step 4** Compare your method for doing the activity with other possible methods. Do you use the safest method to do the activity? One method may be much safer than another.

Both efficiency and environmental impact are important factors to consider when comparing the risks and benefits of various energy sources. In the following ThoughtLab, you will conduct a risk-benefit analysis of an alternative or a conventional energy source.

## ThoughtLab — Comparing Energy Sources

In this ThoughtLab, you will work as a class to compare two different energy sources.

### Procedure

1. On your own, or with a group, choose an energy source from the following list. Other energy sources may be discussed and added in class.

   - solar (radiant) energy
   - petroleum
   - hydrogen fuel cell
   - natural gas fuel cell
   - wind energy
   - hydroelectric power
   - geothermal energy
   - wood
   - biomass
   - nuclear fission
   - natural gas
   - coal
   - tar sands

2. Before beginning your research, record your current ideas about the risks and benefits of your chosen energy source.

3. Conduct a risk-benefit analysis of your energy source. Focus especially on the efficiency and environmental impact. If possible, determine what the efficiency data means. For example, suppose that a source tells you that natural gas is 90% efficient. Is the source referring to natural gas burned directly for heat or for cooking? Is the energy being converted from heat to electricity in a power plant? Be as specific as possible.

   - Ensure that you use a variety of sources to find your data. Be aware of any bias that might be present in your sources.
   - Trace the energy source as far back as you can. For example, you can trace the energy in fossil fuels back to solar energy. Write a brief outline of your findings.

4. Your teacher will pair you (or your group) with another student (or group) that has researched a different energy source. Work together to compare the risks and benefits of the two energy sources, based on your research.

5. Write a conclusion that summarizes the benefits and risks of both energy sources.

6. Present your findings to the class.

### Analysis

1. Discuss the presentations as a class.
   (a) Decide which energy sources are most efficient. Also decide which energy sources are least damaging to the environment.
   (b) Decide which energy source is best overall in terms of both efficiency and environmental impact.

2. Could the "best overall" energy source be used to provide a significant portion of Canada's energy needs? What obstacles would need to be overcome for this to happen?

3. Besides efficiency and environmental impact, what other factors are involved in developing and delivering an energy source to consumers?

## Emerging Energy Sources

In the ThoughtLab above, you probably noticed that all energy sources have drawbacks as well as benefits. Scientists and engineers are striving to find and develop new and better energy sources. One energy source that engineers are trying to harness is nuclear fusion. Nuclear fusion provides a great deal of energy from readily available fuel (isotopes of hydrogen). In addition, nuclear fusion produces a more benign waste product than nuclear fission. Unfortunately, fusion is not yet practical and controllable on a large scale because of the enormous temperatures involved.

Chemists are also striving to find new sources for existing fuels that work well. The Chemistry Bulletin on page 695, for example, discussed a new potential source of methane.

## Section Summary

In this section, you learned how efficiency can be defined in different ways for different purposes. You used your understanding of processes that produce energy to investigate the efficiency and environmental impact of different energy sources.

## Section Review

**1** Compare the energy changes associated with physical, chemical, and nuclear processes. Give one example of each process.

**2** Your friend tells you about an energy source that is supposed to be 46% efficient. What questions do you need to ask your friend in order to clarify this claim?

**3** Design an experiment to determine the efficiency of a laboratory burner. You will first need to decide how to define the efficiency, and you will also need to find out what fuel your burner uses. Include a complete procedure and safety precautions.

**4** Some high-efficiency gas furnaces can heat with an efficiency of up to 97%. These gas furnaces work by allowing the water vapour produced during combustion to condense. Condensation is an exothermic reaction that releases further energy for heating. Explain why allowing the water to condense increases the energy output of the furnace.

**5** The label on an electric kettle claims that the kettle is 95% efficient.

(a) What definition of efficiency is the manufacturer using?

(b) Write an expression that shows how the manufacturer might have arrived at an efficiency of 95% for the kettle.

(c) Design a detailed experiment to test the manufacturer's claim. Include safety precautions.

**6** Read the Chemistry Bulletin on page 695. How does the efficiency of using methane as a fuel source compare to using methane hydrates? Justify your answer.

**7** Hydrogen is a very appealing fuel, in part because burning it produces only water. One of the challenges that researchers face in making hydrogen fuel a reality is how to produce hydrogen economically. Researchers are investigating methods of producing hydrogen indirectly. The following series of equations represent one such method.

$$3FeCl_{2(s)} + 4H_2O_{(g)} \rightarrow Fe_3O_{4(s)} + 6HCl_{(g)} + H_{2(g)} \qquad \Delta H° = 318 \text{ kJ}$$

$$Fe_3O_{4(s)} + \tfrac{3}{2}Cl_{2(g)} + 6HCl_{(g)} \rightarrow 3FeCl_{3(s)} + 3H_2O_{(g)} + \tfrac{1}{2}O_{2(g)} \qquad \Delta H° = -249 \text{ kJ}$$

$$3FeCl_{3(s)} \rightarrow 3FeCl_{2(s)} + \tfrac{3}{2}Cl_{2(g)} \qquad \Delta H° = 173 \text{ kJ}$$

(a) Show that the net result of the three reactions is the decomposition of water to produce hydrogen and oxygen.

(b) Use Hess's law and the enthalpy changes for the reactions to determine the enthalpy change for the decomposition of one mole of water. Check your answer, using the enthalpy of formation of water.

# CHAPTER 17 Review

## Reflecting on Chapter 17

Summarize this chapter in the format of your choice. Here are a few ideas to use as guidelines:
- Explain what a calorimeter is and how it is used to determine the energy changes associated with physical and chemical processes.
- Compare a coffee-cup calorimeter and a bomb calorimeter.
- Use examples and analogies to explain Hess's Law.
- Show how to calculate enthalpy of reaction using known enthalpies of reaction, and explain how this calculation relates to Hess's law.
- Show how to calculate enthalpy of reaction using known enthalpies of formation, and explain how this calculation relates to Hess's law.
- Describe how to use bond energies to estimate enthalpy of reaction. Explain why the result is an estimate only.
- Explain the concept of efficiency, and discuss the efficiency and environmental impact of conventional and alternative energy sources.

## Reviewing Key Terms

benefit
bomb calorimeter
calorimeter
calorimetry
coffee-cup calorimeter
global warming
greenhouse gases
Hess's law of heat summation
non-renewable
renewable
risk
risk-benefit analysis
state function
sustainable development
thermal equilibrium

## Knowledge/Understanding

1. List three assumptions that you make when using a coffee-cup calorimeter.
2. A given chemical equation is tripled and then reversed. What effect, if any, will there be on the enthalpy change of the reaction?
3. Explain why two nested polystyrene coffee cups, with a lid, make a good constant-pressure calorimeter.
4. Write the balanced equation for the formation of each substance.
   (a) $LiCl_{(s)}$
   (b) $C_2H_5OH_{(\ell)}$
   (c) $NH_4NO_{3(s)}$
5. If the enthalpy of formation of an element in its standard state is equal to zero, explain why the enthalpy of formation of iodine gas, $I_{2(g)}$, is 21 kJ/mol.

## Inquiry

6. In an oxygen-rich atmosphere, carbon burns to produce carbon dioxide, $CO_2$. Both carbon monoxide, CO, and carbon dioxide are produced when carbon is burned in an oxygen-deficient atmosphere. This makes the direct measurement of the enthalpy of formation of CO difficult. CO, however, also burns in oxygen, $O_2$, to produce pure carbon dioxide. Explain how you would experimentally determine the enthalpy of formation of carbon monoxide.

7. Two 30.0 g pieces of aluminium, Al, are placed in a calorimeter.
   (a) One piece of Al has an initial temperature of 100.0°C. The other piece has an initial temperature of 20.0°C. What is the temperature of the contents of the calorimeter after the system has reached thermal equilibrium?
   (b) Repeat the calculation in part (a) with the following change: The piece of Al at 20.0°C has a mass of 50.0 g.

8. The complete combustion of 1.00 mol of sucrose, $C_{12}H_{22}O_{11}$, releases −5641 kJ of energy (at 25°C and 100 kPa).
   $C_{12}H_{22}O_{11(s)} + 12O_{2(g)} \rightarrow 12CO_{2(g)} + 11H_2O_{(\ell)}$
   (a) Use the enthalpy change of this reaction, and enthalpies of formation from Appendix E, to determine the enthalpy of formation of sucrose.
   (b) Draw and label an enthalpy diagram for this reaction.

*Answers to questions highlighted in red type are provided in Appendix A.*

9. A horseshoe can be shaped from an iron bar when the iron is heated to temperatures near 1500°C. The hot iron is then dropped into a bucket of water and cooled. An iron bar is heated to 1500°C and then cooled in 1000 g of water that was initially at 20.0°C. How much heat does the water absorb if its final temperature is 65.0°C?

10. A chemist wants to calibrate a new bomb calorimeter. He completely burns a mass of 0.930 g of carbon in a calorimeter. The temperature of the calorimeter changes from 25.00°C to 28.15°C. If the molar enthalpy of combustion of carbon is 394 kJ/mol, what is the heat capacity of the new calorimeter? What evidence shows that the reaction was exothermic?

11. 200 g of iron at 350°C is added to 225 g of water at 10.0°C. What is the final temperature of the iron-water mixture?

12. Fats have long hydrocarbon sections in their molecular structure. Therefore, they have many C—C and C—H bonds. Sugars have fewer C—C and C—H bonds but more C—O bonds. Explain why you can obtain more energy from burning a fat than from burning a sugar.

13. Use equations (1), (2), and (3) to find the enthalpy change of the formation of methane, $CH_4$, from chloroform, $CHCl_3$.
$CHCl_{3(\ell)} + 3HCl_{(g)} \rightarrow CH_{4(g)} + 3Cl_{2(g)}$
(1) $\frac{1}{2}H_{2(g)} + \frac{1}{2}Cl_{2(g)} \rightarrow HCl_{(g)}$   $\Delta H° = -92.3$ kJ
(2) $C_{(s)} + 2H_{2(g)} \rightarrow CH_{4(g)}$   $\Delta H° = -74.8$ kJ
(3) $C_{(s)} + \frac{1}{2}H_{2(g)} + \frac{3}{2}Cl_{2(g)} \rightarrow CHCl_{3(\ell)}$   $\Delta H° = -134.5$ kJ

14. The following equation represents the combustion of ethylene glycol, $(CH_2OH)_2$.
$(CH_2OH)_{2(\ell)} + \frac{5}{2}O_{2(g)} \rightarrow 2CO_{2(g)} + 3H_2O_{(\ell)}$
$\Delta H° = -1178$ kJ
Use known enthalpies of formation and the given enthalpy change to determine the enthalpy of formation of ethylene glycol.

15. Hydrogen peroxide, $H_2O_2$, is a strong oxidizing agent. It is used as an antiseptic in a 3.0% aqueous solution. Some chlorine-free bleaches contain 6.0% hydrogen peroxide.
(a) Write the balanced chemical equation for the formation of one mole of $H_2O_{2(\ell)}$.
(b) Using the following equations, determine the enthalpy of formation of $H_2O_2$.
(1) $2H_2O_{2(\ell)} \rightarrow 2H_2O_{(\ell)} + O_{2(g)}$   $\Delta H° = -196$ kJ
(2) $H_{2(g)} + \frac{1}{2}O_{2(g)} \rightarrow H_2O_{(\ell)}$   $\Delta H° = -286$ kJ

16. Hydrogen cyanide is a highly poisonous gas. It is produced from methane and ammonia.
$CH_{4(g)} + NH_{3(g)} \rightarrow HCN_{(g)} + 3H_{2(g)}$
Find the enthalpy change of this reaction, using the following thermochemical equations.
(1) $H_{2(g)} + 2C_{(graphite)} + N_{2(g)} \rightarrow 2HCN_{(g)}$
$\Delta H° = 270$ kJ
(2) $N_{2(g)} + 3H_{2(g)} \rightarrow 2NH_{3(g)}$   $\Delta H° = -92$ kJ
(3) $C_{(graphite)} + 2H_{2(g)} \rightarrow CH_{4(g)}$   $\Delta H° = -75$ kJ

17. The following equation represents the complete combustion of butane, $C_4H_{10}$.
$C_4H_{10(g)} + 6.5O_{2(g)} \rightarrow 4CO_{2(g)} + 5H_2O_{(g)}$
(a) Using known enthalpies of formation, calculate the enthalpy change of the complete combustion of $C_4H_{10}$. (The enthalpy of formation of $C_4H_{10}$ is −126 kJ/mol.)
(b) Using known enthalpies of formation, calculate the enthalpy change of the complete combustion of ethane, $C_2H_6$, to produce carbon dioxide and water vapour. Express your answer in units of kJ/mol and kJ/g.
(c) A 10.0 g sample that is 30% $C_2H_6$ and 70% $C_4H_{10}$, by mass, is burned in excess oxygen. How much heat is released?

18. Design an investigation to determine the enthalpy change of the combustion of ethanol using a wick-type burner, similar to that in a kerosene lamp.
(a) Draw and label a diagram of the apparatus.
(b) Write a step-by-step procedure.
(c) Prepare a table to record your data and other observations.
(d) State any assumptions that you will make when carrying out the calculations.

## Communication

19. Suppose that you need to determine the enthalpy change of a chemical reaction. Unfortunately, you are unable to carry out the reaction in your school laboratory. Does this mean that you cannot determine the enthalpy change of the reaction? Explain.

20. Acetylene, $C_2H_2$, and ethylene, $C_2H_4$, are both used as fuels. They combine with oxygen gas to produce carbon dioxide and water in an exothermic reaction. Acetylene also reacts with hydrogen to produce ethylene, as shown.
$C_2H_{2(g)} + H_{2(g)} \rightarrow C_2H_{4(g)}$    $\Delta H° = -175.1$ kJ
(a) Without referring to any tables or doing any calculations, explain how you know that $C_2H_2$ has a more positive enthalpy of formation than $C_2H_4$.
(b) Do you think $C_2H_2$ or $C_2H_4$ is a more energetic fuel? Explain.

## Making Connections

21. When energy is "wasted" during an industrial process, what actually happens to this energy?
22. On an episode of *The Nature of Things*, Dr. David Suzuki made the following comment: "As a society and as individuals, we're hooked on it [oil]." Discuss his comment. Explain how our society has benefitted from hydrocarbons. Describe some of the problems that are associated with the use of hydrocarbons. Also describe some possible alternatives for the future.
23. Consider methane, $CH_4$, and hydrogen, $H_2$, as possible fuel sources.
    (a) Write the chemical equation for the complete combustion of each fuel. Then find the enthalpy of combustion, $\Delta H_{comb}$, of each fuel. Express your answers in kJ/mol and kJ/g. Assume that water *vapour*, rather than liquid water, is formed in both reactions.
    (b) Which is the more energetic fuel, per unit mass?
    (c) Consider a fixed mass of each fuel. Which fuel would allow you to drive a greater distance? Explain briefly.
    (d) Describe how methane and hydrogen could be obtained. Which of these methods do you think is less expensive? Explain.
    (e) Which fuel do you think is more environmentally friendly? Explain.

**Answers to Practice Problem and Short Section Review Questions:**
**Practice Problems** 1.(a) $-2.18 \times 10^4$ J, $2.18 \times 10^4$ J (b) 8.01 J/g°C
2.(a) $6.34 \times 10^3$ J, $-6.34 \times 10^3$ J (b) 92.3°C  3. 25 g  4.(a) 32.0 g
(b) 26.0°C  5. $-51.2$ kJ/mol  6.(a) $-4.0 \times 10^2$ kJ/mol (b) The density and specific heat capacity of the solutions are assumed to be the same as the density and specific heat capacity of water. It is assumed that no heat is lost to the calorimeter.
7. 29.7°C  8.(a) $C = q/\Delta T$ (b) $\Delta T = q/C$ (c) $T_f = T_i + q/C$
(d) $T_i = T_f - q/C$  9. $-20.3$ kJ/g  10.(a) $1.00 \times 10^2$ g
(b) 8.75 kJ/°C  11. $-44$ kJ  12. $-509$ kJ  13. $-120$ kJ
14. $+30$ kJ  15.(a) $C_{(s)} + 2H_{2(g)} \rightarrow H_{4(g)} + 74.6$ kJ
(b) $Na_{(s)} + \frac{1}{2}Cl_{2(g)} \rightarrow NaCl_{(s)} + 411.2$ kJ
(c) $Mg_{(s)} + \frac{1}{2}O_{2(g)} \rightarrow MgO_{(s)} 601.2$ kJ
(d) $Ca_{(s)} + C_{(s)} + \frac{3}{2}O_{2(g)} \rightarrow CaCO_{3(s)} + 1207.6$ kJ
16. $H_{2(g)} + \frac{1}{8}S_{8(s)} + 2O_{2(g)} \rightarrow H_2SO_{4(\ell)} + 814.0$ kJ
17. $Cs_{(s)} + 76.7$ kJ $\rightarrow Cs_{(g)}$  18.(a) $1/4P_{4(s)} \rightarrow P_{(s)} + 17.6$ kJ (b) $-18.5$ kJ
19. add the formation equation of ethane to the reverse of the formation equation of ethene  20. add the formation equation of sulfur dioxide, times two, to the formation equation of zinc oxide, times two, to the reverse of the formation equation of zinc sulfide, times two  21. $-77.6$ kJ/mol
22.(a) $-637.9$ kJ (b) $2.49 \times 10^3$ kJ  23.(a) $-2225$ kJ/mol
(b) $-3003$ kJ/mol  24.(a) 219 kJ (b) $-91.8$ kJ  25.(a) $-249$ kJ/mol
(b) $-242$ kJ/mol  26.(a) $-519$ J (b) $-305$ kJ
**Section Review 17.1** 2. $-39.8$ kJ  3. $-3.1 \times 10^3$ J  7.(a) $-55$ kJ/mol
(b) $KOH_{(aq)} + HNO_{3(aq)} \rightarrow H_2O_{(\ell)} + KNO_{3(aq)} + 55$ kJ  **17.2** 3. $-143$ kJ
4. 14.36 kJ  5. $-636.0$ kJ/mol  7.(b) 584 kJ/mol (c) 835 kJ/mol
**17.3** 7.(b) 242 kJ

# UNIT 7 Review

## Knowledge/Understanding

### Multiple Choice

In your notebook, write the letter for the best answer to each question.

1. Which situation describes an exothermic reaction?
   (a) The energy that is released to form the product bonds is greater than the energy that is used to break the reactant bonds.
   (b) The energy that is released to form the product bonds is less than the energy that is used to break the reactant bonds.
   (c) The energy that is released to form the product bonds is equal to the energy that is used to break the reactant bonds.
   (d) The energy that is used to break the product bonds is greater than the energy that is released to form the reactant bonds.
   (e) The energy that is used to break the product bonds is less than the energy that is released to form the reactant bonds.

2. Beaker A contains 500 mL of a liquid. Beaker B contains 1000 mL of the same liquid. What happens when 200 kJ of heat is absorbed by the liquid in each beaker?
   (a) The temperature of the liquid increases the same amount in both beakers.
   (b) The temperature of the liquid decreases the same amount in both beakers.
   (c) The temperature of the liquid remains the same in both beakers.
   (d) The temperature change of the liquid in Beaker B is twice as large as the temperature change of the liquid in Beaker A.
   (e) The temperature change of the liquid in Beaker B is one half as large as the temperature change of the liquid in Beaker A.

3. Which of the following processes are exothermic?
   I. boiling water
   II. freezing water
   III. condensing steam
   IV. melting ice
   (a) I and II only
   (b) II and III only
   (c) I and IV only
   (d) II, III, and IV only
   (e) II and IV only

4. The $\Delta H°_f$ of an element in its standard state is defined to be
   (a) 0 kJ/mol
   (b) 10 kJ/mol
   (c) −10 kJ/mol
   (d) greater than 0 kJ/mol
   (e) a unique value for each element

5. 10.9 kJ of heat is needed to vaporize 60.0 g of liquid $Br_2$ vapour at 60°C. What is the molar enthalpy of vaporization of $Br_2$ at 60°C?
   (a) 3.64 kJ/mol
   (b) 7.27 kJ/mol
   (c) 14.6 kJ/mol
   (d) 29.1 kJ/mol
   (e) 10.9 kJ/mol

6. What is the molar enthalpy of vaporization of water, given the following thermochemical equations?
   $H_{2(g)} + \frac{1}{2}O_{2(g)} \rightarrow H_2O_{(g)} + 241.8 \text{ kJ}$
   $H_{2(g)} + \frac{1}{2}O_{2(g)} \rightarrow H_2O_{(\ell)} + 285.8 \text{ kJ}$
   (a) 44.0 kJ/mol
   (b) −527.6 kJ/mol
   (c) −44.0 kJ/mol
   (d) −527.6 kJ/mol
   (e) 241.8 kJ/mol

7. Which substance has a standard enthalpy of formation, $\Delta H°_f$, equal to zero?
   (a) gold, $Au_{(s)}$
   (b) water, $H_2O_{(\ell)}$
   (c) carbon monoxide, $CO_{(s)}$
   (d) zinc, Zn
   (e) water, $H_2O_{(g)}$

8. Which of the following statements are true?
   I. The temperature of the surroundings decreases when an endothermic reaction occurs.
   II. An endothermic reaction has a negative value of $\Delta H$.
   III. Heat is liberated when an exothermic reaction occurs.
   (a) I and II
   (b) I, II, and III
   (c) I and III only
   (d) II and III only
   (e) none of them

### Short Answer

In your notebook, write a sentence or a short paragraph to answer each question.

9. Distinguish between a closed system and an insulated system.

10. In your own words, define the terms "system" and "surroundings." Use an example to illustrate your definition.

Answers to questions highlighted in red type are provided in Appendix A.

11. In a chemical reaction, bonds are formed and broken.
    (a) How would you characterize the enthalpy change of bond breaking?
    (b) How would you characterize the enthalpy change of bond formation?
    (c) State the relationship between the enthalpy change of the overall reaction (exothermic or endothermic) and bond breakage and formation.

12. "The reactants have more potential energy than the products." What kind of reaction does this statement describe? Justify your answer.

13. Compare and contrast enthalpy of vaporization and enthalpy of condensation.

14. Four fuels are listed in the table below. Equal masses of these fuels are burned completely.

| Fuel | Heat of combustion at SATP (kJ/mol) |
|---|---|
| octane, $C_8H_{18(\ell)}$ | 5513 |
| methane, $CH_{4(g)}$ | 890 |
| ethanol, $C_2H_5OH_{(\ell)}$ | 1367 |
| hydrogen $H_{2(g)}$ | 285 |

   (a) Put the fuels in order, from greatest to least amount of energy provided.
   (b) Write a balanced chemical equation for the complete combustion of each fuel.
   (c) Suggest one benefit and one risk of using hydrocarbon fuels.

15. You are given a 70 g sample of each of the following metals, all at 25°C. You heat each metal under identical conditions. Which metal will be first to reach 30°C? Which will be last? Explain your reasoning.

| Metal | Specific Heat Capacity (J/g•°C) |
|---|---|
| Platinum | 0.133 |
| Titanium | 0.528 |
| Zinc | 0.388 |

16. Describe the information that is included in the following thermochemical equation:

    $$N_{2(g)} + 3H_{2(g)} \rightarrow 2HN_{3(g)} + 92.2 \text{ kJ}$$

17. Define calorimetry. Describe two commonly used calorimeters.

18. When taking a calorimetric measurement, why do you need to know the heat capacity of the calorimeter?

19. Describe two exothermic processes and two endothermic processes.

20. Compare the following terms: specific heat capacity and heat capacity.
    (a) Write their symbols and their units.
    (b) Write a mathematical formula in which each term would be used.

## Inquiry

21. When kerosene, $C_{12}H_{26(\ell)}$, is burned in a device such a space heater, energy is released. Design an investigation to determine the amount of energy that is released, per gram of kerosene. Discuss potential problems with using a kerosene heater in a confined area, such as a camper trailer, where the supply of air may be limited. Support your discussion with balanced chemical equations.

22. A group of students tested two white, crystalline solids, A and B, to determine their enthalpies of solution. The students dissolved 10.00 g of each solid in 100.0 mL of water in a polystyrene calorimeter and collected the temperature data. They obtained the following data:

| Time | Temperature (A) | Temperature (B) |
|---|---|---|
| 0.0 | 15 | 25 |
| 0.5 | 20.1 | 18.8 |
| 1.0 | 25 | 16.7 |
| 1.5 | 29.8 | 15.8 |
| 2.0 | 31.9 | 15.2 |
| 2.5 | 32.8 | 15 |
| 3.0 | 33 | 15 |
| 3.5 | 33 | 15.2 |
| 4.0 | 32.8 | 15.5 |
| 4.5 | 32.5 | 15.8 |
| 5.0 | 32.2 | 16.1 |
| 5.5 | 31.9 | 16.4 |

   (a) Graph the data in the table above, placing time on the x-axis and temperature on the y-axis. Use your graph to answer (b) to (d).
   (b) Classify the enthalpy of solution of each solid as exothermic or endothermic.

Unit 7 Review • MHR **707**

(c) From the data given, calculate the enthalpy of solution for each solid. Your answer should be in kJ/g.

(d) Is there any evidence from the data that the students' calorimeter could have been more efficient? Explain your answer.

23. Butane, $C_4H_{10}$, is the fuel that is used in disposable lighters. Consider the following equation for the complete combustion of butane.
$C_4H_{10(g)} + 6.5O_{2(g)} \rightarrow 4CO_{2(g)} + 5H_2O_{(\ell)}$
(a) Write a separate balanced chemical equation for the formation of $C_4H_{10}$, the formation of $CO_2$, and the formation of $H_2O$, directly from the elements in their standard states.
(b) Algebraically combine these equations to get the balanced chemical equation for the complete combustion of $C_4H_{10}$.

24. A piece of aluminium metal at 160.0°C is added to a calorimeter containing 80.0 g of ice at 0.0°C. Thermal equilibrium is reached at 15.0°C.
(a) What further pieces of data do you need to determine the mass of the aluminum?
(b) Locate the missing data in your textbook, and determine the mass of the aluminum.

25. 543 g of ice is removed from a freezer that has a temperature of −15.3°C. The ice is dropped into a bowl containing 4.00 L of water.
(a) If thermal equilibrium is reached at 38.6°C, calculate the initial temperature of the 4.00 L of water. The specific heat capacity of ice is 2.01 J/g°C.
(b) State the assumptions you made in order to answer the question.

26. A student wants to determine the enthalpy change associated with dissolving solid sodium hydroxide, NaOH, in water. The student dissolves 1.96 g of NaOH in 100.0 mL of water in a coffee-cup calorimeter. The initial temperature of the water is 23.4°C. After the NaOH dissolves, the temperature of the water rises to 28.7°C.
(a) Use these data to determine the enthalpy of dissolution of sodium hydroxide, in kJ/mol NaOH. Assume that the heat capacity of the calorimeter is negligible.
(b) Suppose that the heat capacity of the calorimeter was not negligible. Explain how the value of ΔH that you calculated in part (a) would compare with the *actual* ΔH.

(c) Draw and label an enthalpy diagram for this reaction.

27. Some solid ammonium nitrate, $NH_4NO_3$, is added to a coffee-cup calorimeter that contains water at room temperature. After the $NH_4NO_3$ has dissolved, the temperature of the solution drops to near 0°C. Explain this observation.

28. Consider the following chemical equations and their enthalpy changes.
$CH_{4(g)} + 2O_{2(g)} \rightarrow$
$\qquad CO_{2(g)} + 2H_2O_{(g)}$  ΔH = −8.0 × 10² kJ
$CaO_{(s)} + H_2O_{(\ell)} \rightarrow Ca(OH)_{2(aq)}$  ΔH = −65 kJ
What volume of methane, at STP, would have to be combusted in order to release the same amount of energy as the reaction of 1.0 × 10² g of CaO with sufficient water?

29. The molar enthalpy of combustion of sucrose (table sugar), $C_{12}H_{22}O_{11}$, is 5.65 × 10³ kJ.
(a) Write a balanced thermochemical equation for the complete combustion of sucrose.
(b) Calculate the amount of energy that is released when 5.00 g of sucrose (about one teaspoon) is combusted.

30. Carbon monoxide reacts with hydrogen gas to produce a mixture of methane, carbon dioxide, and water. (This mixture is known as substitute natural gas.)
$4CO_{(g)} + 8H_{2(g)} \rightarrow 3CH_{4(g)} + CO_{2(g)} + 2H_2O_{(\ell)}$
Use the following thermochemical equations to determine the enthalpy change of the reaction.
$C_{(graphite)} + 2H_{2(g)} \rightarrow CH_{4(g)} + 74.8$ kJ
$CO_{(g)} + \frac{1}{2}O_{2(g)} \rightarrow CO_{2(g)} + 283.1$ kJ
$H_{2(g)} + \frac{1}{2}O_{2(g)} \rightarrow H_2O_{(g)} + 241.8$ kJ
$C_{(graphite)} + \frac{1}{2}O_{2(g)} \rightarrow CO_{(g)} + 110.5$ kJ
$H_2O_{(\ell)} + 44.0$ kJ $\rightarrow H_2O_{(g)}$

31. The decomposition of aqueous hydrogen peroxide, $H_2O_2$, can be catalyzed by different catalysts, such as aqueous sodium iodide, NaI, or aqueous iron(II) nitrate, $Fe(NO_3)_2$.
(a) The enthalpy change, in kJ/mol of $H_2O_2$, would be the same for this reaction, regardless of the catalyst. Explain why, with the help of a potential energy diagram.
(b) Design an investigation to verify your explanation in part (a). Do not attempt to carry out the investigation without the supervision of your teacher.

708 MHR • Unit 7 Thermochemistry

## Communication

**32.** If a solution of acid accidentally comes in contact with your skin, you are told to run the affected area under cold water for several minutes. Explain why it is not advisable to simply neutralize the acid with a basic solution.

**33.** A classmate is having difficulty understanding how the concepts of system, insulated system, and surroundings are related to exothermic and endothermic reactions. Write a note to explain to your classmate how the concepts are related. Use diagrams to help clarify your explanation.

**34.** Consider the following data for the complete combustion of the $C_1$ to $C_8$ straight-chain alkanes.

| Name | Formula | $\Delta H_{comb}$ (kJ/mol of alkane) |
|---|---|---|
| methane | $CH_4$ | $-8.90 \times 10^2$ |
| ethane | $C_2H_6$ | $-1.56 \times 10^3$ |
| propane | $C_3H_8$ | $-2.22 \times 10^3$ |
| butane | $C_4H_{10}$ | $-2.88 \times 10^3$ |
| pentane | $C_5H_{12}$ | $-3.54 \times 10^3$ |
| hexane | $C_6H_{14}$ | $-4.16 \times 10^3$ |
| heptane | $C_7H_{16}$ | $-4.81 \times 10^3$ |
| octane | $C_8H_{18}$ | $-5.45 \times 10^3$ |

(a) Using either graph paper or spreadsheet software, plot a graph of $\Delta H_{comb}$ (y-axis) versus the number of C atoms in the fuel (x-axis).

(b) Extrapolate your graph to predict $\Delta H_{comb}$ of decane, $C_{10}H_{22}$.

(c) From your graph, develop an equation to determine $\Delta H_{comb}$ of a straight-chain alkane, given the number of carbons. Your equation should be of the form $\Delta H = \ldots$.

(d) Use the equation you developed in part (c) to determine $\Delta H_{comb}$ of $C_{10}H_{22}$. How does this value compare with the value of $\Delta H$ you determined by extrapolation from the graph? Explain why.

(e) Methane, ethane, propane, and butane are all gases at room temperature. You know that equal volumes of different gases contain the same number of moles under identical conditions of temperature and pressure. Which of these gases do you think would make the best fuel? Explain your answer.

## Making Connections

**35.** Suppose that you are having a new home built in a rural area, where natural gas is not available. You have two choices for fuelling your furnace:
- propane, $C_3H_8$, delivered as a liquid under pressure and stored in a tank
- home heating oil, delivered as a liquid (not under pressure) and stored in a tank

What factors do you need to consider in order to decide on the best fuel? What assumptions do you need to make?

**36.** Suppose that you read the following statement in a magazine: 0.95 thousand cubic feet of natural gas is equal to a gigajoule, GJ, of energy. Being a media-literate student, you are sceptical of this claim and wish to verify it. The following assumptions/information may be useful.
- Natural gas is pure methane.
- Methane undergoes complete combustion.
- $H_2O_{(\ell)}$ is formed, rather than $H_2O_{(g)}$.
- 1.00 mol of any gas occupies 24 L at 20°C and 100 kPa.
- 1 foot = 12 inches; 1 inch = 2.54 cm; 1 L = 1 dm$^3$

# UNIT 8

# Electrochemistry

## UNIT 8 CONTENTS

**CHAPTER 18**
Oxidation-Reduction Reactions

**CHAPTER 19**
Cells and Batteries

## UNIT 8 OVERALL OUTCOMES

- What are oxidation-reduction reactions? How are they involved in the interconversion of chemical and electrical energy?

- How are galvanic and electrolytic cells built, and how do they function? What equations are used to describe these types of cells? How can you solve quantitative problems related to electrolysis?

- What are the uses of batteries and fuel cells? How is electrochemical technology used to produce and protect metals? How can you assess the environmental and safety issues associated with these technologies?

Canadian engineer Dr. John Hopps, working with a team of medical researchers in the late 1940s, developed one of the most significant medical inventions of the twentieth century: the pacemaker. The photograph on the right shows a pacemaker embedded in the body of a heart patient. A modern pacemaker is essentially a tiny computer that monitors a person's heartbeat and corrects irregularities as needed. Pacemakers are particularly useful in correcting a heartbeat that is too slow.

The pacemaker device is surgically placed in a "pocket" of tissue near the patient's collarbone. One or more wires, called "leads," are connected to the pacemaker and threaded down through a major vein to the patient's heart. By sending electrical impulses along the leads to the heart, the pacemaker can induce a heartbeat.

A pacemaker obtains electrical energy from a tiny battery that lasts for about seven years before it must be replaced. But how do batteries supply electrical energy? The answer lies in a branch of chemistry known as electrochemistry. In this unit, you will learn about the connection between chemical reactions and electricity. You will also learn about the chemical reactions that take place inside batteries.

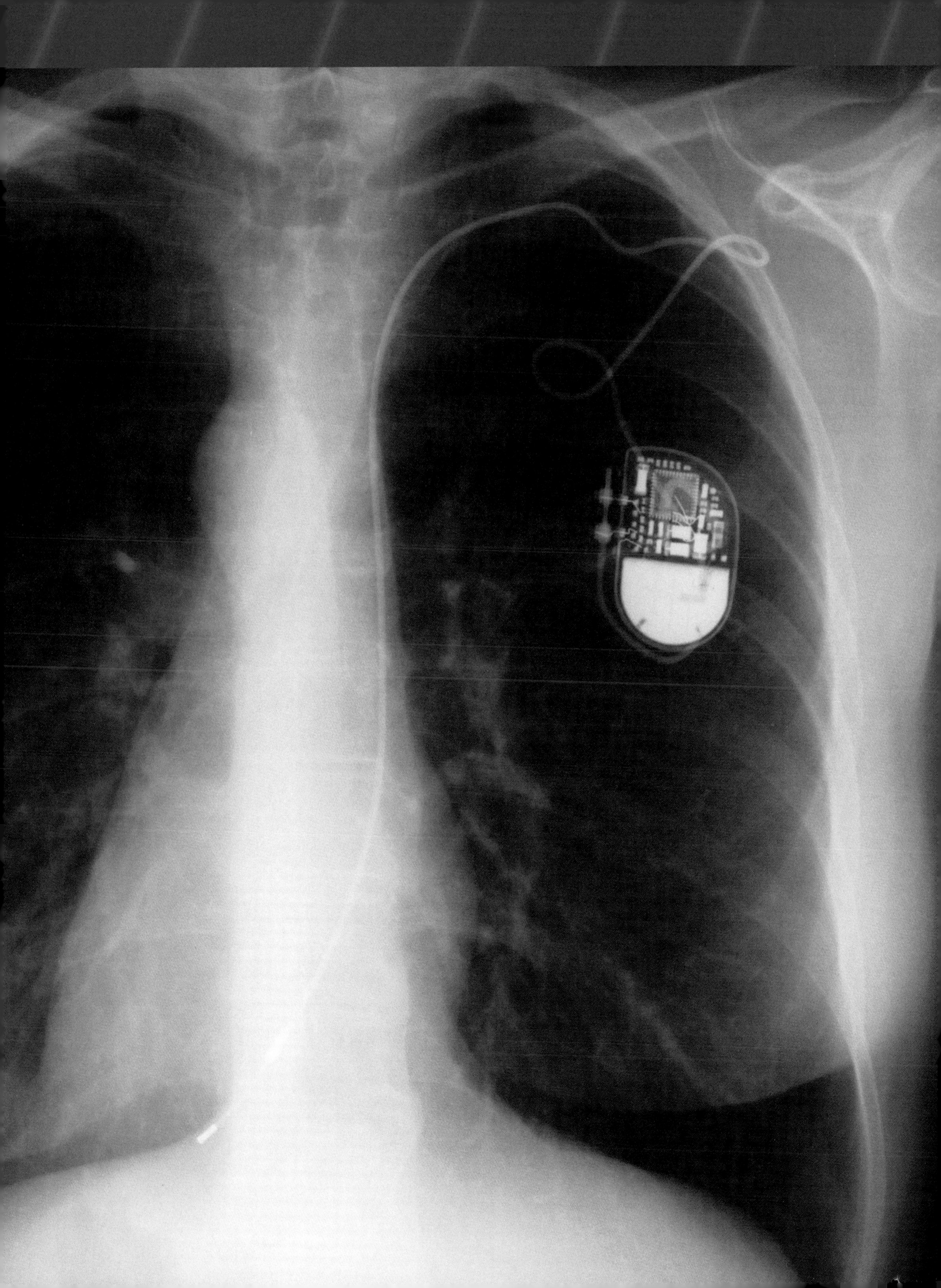

# CHAPTER 18
# Oxidation-Reduction Reactions

## Chapter Preview

- 18.1 Defining Oxidation and Reduction
- 18.2 Oxidation Numbers
- 18.3 The Half-Reaction Method for Balancing Equations
- 18.4 The Oxidation Number Method or Balancing Equations

## Prerequisite Concepts and Skills

Before you begin this chapter, review the following concepts and skills:

- balancing chemical, total ionic, and net ionic equations (Chapter 1, Chapter 8)
- reaction types, including synthesis, decomposition, single displacement, and double displacement reactions (Chapter 1)
- the common ionic charges of metal ions and non-metal ions, and the formulas of common polyatomic ions
- drawing Lewis structures (Chapter 5)
- electronegativities and bond polarities (Chapter 5)

Kitchen chemistry is an important part of daily life. Cooks use chemistry all the time to prepare and preserve food. Even the simplest things you do in the kitchen can involve chemical reactions. For example, you have probably seen a sliced apple turn brown. The same thing happens to pears, bananas, avocados, and several other fruits. Slicing the fruit exposes the flesh to oxygen in the air. Compounds in the fruit react with oxygen to form brown products. An enzyme in the fruit acts as a catalyst, speeding up this reaction. How can you stop fruit from turning brown after it is sliced?

A Waldorf salad uses a simple method to prevent fruit from browning. This type of salad usually consists of diced apples, celery, and walnuts, covered with a mayonnaise dressing. The dressing keeps the air away from the food ingredients. Without air, the fruit does not turn brown.

Another way to solve this problem is to prevent the enzyme in the fruit from acting as a catalyst. Enzymes are sensitive to pH. Therefore, adding an acid such as lemon juice or vinegar to fruit can prevent the enzyme from acting. You may have noticed that avocado salad recipes often include lemon juice. In addition to hindering the enzyme, lemon juice contains vitamin C, which is very reactive toward oxygen. The vitamin C reacts with oxygen before the sliced fruit can do so.

In this chapter, you will be introduced to oxidation-reduction reactions, also called redox reactions. You will discover how to identify this type of reaction. You will also find out how to balance equations for a redox reaction.

A redox reaction causes fruit to go brown. How can you recognize other redox reactions?

# Defining Oxidation and Reduction 18.1

The term *oxidation* can be used to describe the process in which certain fruits turn brown by reacting with oxygen. The original, historical definition of this term was "to combine with oxygen." Thus, oxidation occurred when iron rusted, and when magnesium was burned in oxygen gas. The term *reduction* was used historically to describe the opposite of oxidation, that is, the formation of a metal from its compounds. An **ore** is a naturally occurring solid compound or mixture of compounds from which a metal can be extracted. Thus, the process of obtaining a metal from an ore was known as a reduction. Copper ore was reduced to yield copper, and iron ore was reduced to yield iron.

As you will learn in this chapter, the modern definitions for oxidation and reduction are much broader. The current definitions are based on the idea of electron transfers, and can now be applied to numerous chemical reactions.

When a piece of zinc is placed in an aqueous solution of copper(II) sulfate, the zinc displaces the copper in a single displacement reaction. This reaction is shown in Figure 18.1. As the zinc dissolves, the zinc strip gets smaller. A dark red-brown layer of solid copper forms on the zinc strip, and some copper is deposited on the bottom of the beaker. The blue colour of the solution fades, as blue copper(II) ions are replaced by colourless zinc ions.

### Section Preview/Outcomes

In this section, you will

- **describe** oxidation and reduction in terms of the loss and the gain of electrons
- **write** half-reactions from balanced chemical equations for oxidation-reduction systems
- **investigate** oxidation-reduction reactions by comparing the reactivities of some metals
- **communicate** your understanding of the terms *ore, oxidation, reduction, oxidation-reduction reaction, redox reaction, oxidizing agent, reducing agent, half-reaction, disproportionation*

**Figure 18.1** A solid zinc strip reacts with a solution that contains blue copper(II) ions.

**CONCEPT CHECK**

From your earlier work, you will recognize the sulfate ion, $SO_4^{2-}$, as a polyatomic ion. To review the names and formulas of common polyatomic ions, refer to Appendix E, Table E.9.

The reaction in Figure 18.1 is represented by the following equation.

$$Zn_{(s)} + CuSO_{4(aq)} \rightarrow Cu_{(s)} + ZnSO_{4(aq)}$$

This equation can be written as a total ionic equation.

$$Zn_{(s)} + Cu^{2+}_{(aq)} + SO_4^{2-}_{(aq)} \rightarrow Cu_{(s)} + Zn^{2+}_{(aq)} + SO_4^{2-}_{(aq)}$$

The sulfate ions are *spectator ions*, meaning ions that are not involved in the chemical reaction. By omitting the spectator ions, you obtain the following net ionic equation.

$$Zn_{(s)} + Cu^{2+}_{(aq)} \rightarrow Cu_{(s)} + Zn^{2+}_{(aq)}$$

Notice what happens to the reactants in this equation. The zinc atoms *lose* electrons to form zinc ions. The copper ions *gain* electrons to form copper atoms.

$$Zn_{(s)} + Cu^{2+}_{(aq)} \rightarrow Cu_{(s)} + Zn^{2+}_{(aq)}$$
(gains 2e⁻ / loses 2e⁻)

The following chemical definitions describe these changes.

- **Oxidation** is the loss of electrons.
- **Reduction** is the gain of electrons.

In the reaction of zinc atoms with copper(II) ions, the zinc atoms lose electrons and undergo oxidation. In other words, the zinc atoms are *oxidized*. The copper(II) ions gain electrons and undergo reduction. In other words, the copper(II) ions are *reduced*. Because oxidation and reduction both occur in the reaction, it is known as an **oxidation-reduction reaction** or **redox reaction**.

Notice that electrons are transferred from zinc atoms to copper(II) ions. The copper(II) ions are responsible for the oxidation of the zinc atoms. A reactant that oxidizes another reactant is called an **oxidizing agent**. The oxidizing agent accepts electrons in a redox reaction. In this reaction, copper(II) is the oxidizing agent. The zinc atoms are responsible for the reduction of the copper(II) ions. A reactant that reduces another reactant is called a **reducing agent**. The reducing agent gives or donates electrons in a redox reaction. In this reaction, zinc is the reducing agent.

A redox reaction can also be defined as a reaction between an oxidizing agent and a reducing agent, as illustrated in Figure 18.2.

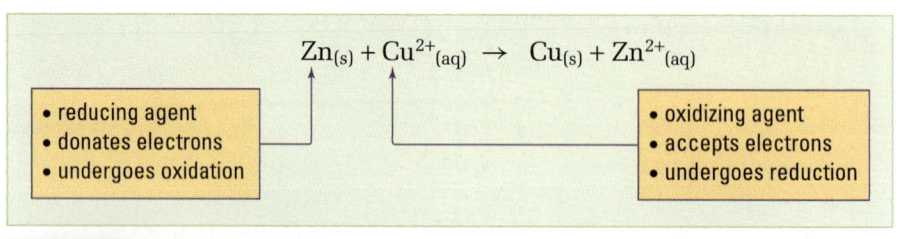

**CHEM FACT**

Try using a mnemonic to remember the definitions for oxidation and reduction. For example, in "LEO the lion says GER," LEO stands for "Loss of Electrons is Oxidation." GER stands for "Gain of Electrons is Reduction." The mnemonic "OIL RIG" stands for "Oxidation Is Loss. Reduction Is Gain." Make up your own mnemonic to help you remember these definitions.

$$Zn_{(s)} + Cu^{2+}_{(aq)} \rightarrow Cu_{(s)} + Zn^{2+}_{(aq)}$$

- reducing agent
- donates electrons
- undergoes oxidation

- oxidizing agent
- accepts electrons
- undergoes reduction

**Figure 18.2** In a redox reaction, the reducing agent is oxidized, and the oxidizing agent is reduced. Note that the oxidizing agent *does not* undergo oxidation, and that the reducing agent *does not* undergo reduction.

Try the following practice problems to review your understanding of net ionic equations, and to work with the new concepts of oxidation and reduction.

> **Practice Problems**
>
> 1. Write a balanced net ionic equation for the reaction of zinc with aqueous iron(II) chloride. Include the physical states of the reactants and products.
>
> 2. Write a balanced net ionic equation for each reaction, including physical states.
>    (a) magnesium with aqueous aluminum sulfate
>    (b) a solution of silver nitrate with metallic cadmium
>
> 3. Identify the reactant oxidized and the reactant reduced in each reaction in question 2.
>
> 4. Identify the oxidizing agent and the reducing agent in each reaction in question 2.

## Half-Reactions

To monitor the transfer of electrons in a redox reaction, you can represent the oxidation and reduction separately. A **half-reaction** is a balanced equation that shows the number of electrons involved in either oxidation or reduction. Because a redox reaction involves both oxidation and reduction, two half-reactions are needed to represent a redox reaction. One half-reaction shows oxidation, and the other half-reaction shows reduction.

As you saw earlier, the reaction of zinc with aqueous copper(II) sulfate can be represented by the following net ionic equation.

$$Zn_{(s)} + Cu^{2+}_{(aq)} \rightarrow Cu_{(s)} + Zn^{2+}_{(aq)}$$

Each neutral Zn atom is oxidized to form a $Zn^{2+}$ ion. Thus, each Zn atom must lose two electrons. You can write an oxidation half-reaction to show this change.

$$Zn_{(s)} \rightarrow Zn^{2+}_{(aq)} + 2e^-$$

Each $Cu^{2+}$ ion is reduced to form a neutral Cu atom. Thus, each $Cu^{2+}$ ion must gain two electrons. You can write a reduction half-reaction to show this change.

$$Cu^{2+}_{(aq)} + 2e^- \rightarrow Cu_{(s)}$$

If you look again at each half-reaction above, you will notice that the atoms and the charges are balanced. Like other types of balanced equations, half-reactions are balanced using the smallest possible whole-number coefficients. In the following equation, the atoms and charges are balanced, but the coefficients can all be divided by 2 to give the usual form of the half-reaction.

$$2Cu^{2+}_{(aq)} + 4e^- \rightarrow 2Cu_{(s)}$$

**CONCEPT CHECK**

You can write separate oxidation and reduction half-reactions to represent a redox reaction, but one half-reaction cannot occur on its own. Explain why this statement must be true.

In most redox reactions, one substance is oxidized and a different substance is reduced. In a **disproportionation** reaction, however, a single element undergoes both oxidation and reduction in the same reaction. For example, a copper(I) solution undergoes disproportionation in the following reaction.

$$2Cu^+_{(aq)} \rightarrow Cu_{(s)} + Cu^{2+}_{(aq)}$$

In this reaction, some copper(I) ions gain electrons, while other copper(I) ions lose electrons.

$$Cu^+_{(aq)} + Cu^+_{(aq)} \rightarrow Cu_{(s)} + Cu^{2+}_{(aq)}$$

(gains 1e⁻ / loses 1e⁻)

The two half-reactions are as follows.

Oxidation: $Cu^+_{(aq)} \rightarrow Cu^{2+}_{(aq)} + 1e^-$

Reduction: $Cu^+_{(aq)} + 1e^- \rightarrow Cu_{(s)}$

You have learned that half-reactions can be used to represent oxidation and reduction separately. Half-reactions always come in pairs: an oxidation half-reaction is always accompanied by a reduction half-reaction, and vice versa. Try writing and balancing half-reactions using the following practice problems.

### Practice Problems

5. Write balanced half-reactions from the net ionic equation for the reaction between solid aluminum and aqueous iron(III) sulfate. The sulfate ions are spectator ions, and are not included.

    $Al_{(s)} + Fe^{3+}_{(aq)} \rightarrow Al^{3+}_{(aq)} + Fe_{(s)}$

6. Write balanced half-reactions from the following net ionic equations.

    (a) $Fe_{(s)} + Cu^{2+}_{(aq)} \rightarrow Fe^{2+}_{(aq)} + Cu_{(s)}$
    (b) $Cd_{(s)} + 2Ag^+_{(aq)} \rightarrow Cd^{2+}_{(aq)} + 2Ag_{(s)}$

7. Write balanced half-reactions for each of the following reactions.

    (a) $Sn_{(s)} + PbCl_{2(aq)} \rightarrow SnCl_{2(aq)} + Pb_{(s)}$
    (b) $Au(NO_3)_{3(aq)} + 3Ag_{(s)} \rightarrow 3AgNO_{3(aq)} + Au_{(s)}$
    (c) $3Zn_{(s)} + Fe_2(SO_4)_{3(aq)} \rightarrow 3ZnSO_{4(aq)} + 2Fe_{(s)}$

8. Write the net ionic equation and the half-reactions for the disproportionation of mercury(I) ions in aqueous solution to give liquid mercury and aqueous mercury(II) ions. Assume that mercury(I) ions exist in solution as $Hg_2^{2+}$.

You may already know that some metals are more reactive than others. For instance, magnesium combusts readily, while gold does not. The arrangement of metals according to their reactivity is called a *metal activity series*. In Investigation 18-A, you will discover how this series is related to oxidation and reduction. You will write chemical equations, ionic equations, and half-reactions for the single displacement reactions of several metals.

# Chemistry Bulletin

**Science** **Technology** **Science** **Environment**

## Aging: Is Oxidation a Factor?

Why do we grow old? Despite advances in molecular biology and medical research, the reasons for aging remain mysterious. One theory suggests that aging may be influenced by oxidizing agents, also known as *oxidants*.

Oxidants are present in the environment and in foods. Nitrogen oxides are oxidants present in cigarette smoke and urban smog. Other oxidants include the copper and iron salts in meat and some plants. Inhaling and ingesting oxidants such as these can increase the level of oxidants in our bodies.

Oxidants are also naturally present in the body, where they participate in important redox reactions. For example, mitochondria consume oxygen during aerobic respiration, and cells ingest and destroy bacteria. Both these processes involve oxidation and reduction.

As you have just seen, redox reactions are an essential part of your body's processes. However, these reactions can produce *free radicals*, which are highly reactive atoms or molecules with one or more unpaired electrons. Because they are so reactive, free radicals can oxidize surrounding molecules by robbing them of electrons. This process can damage DNA, proteins, and other macromolecules. Such damage may contribute to aging, and to diseases that are common among the aging, such as cancer, cardiovascular disease, and cataracts.

The study of oxidative damage has sparked a debate about the role that antioxidants might play in illness and aging. *Antioxidants* are reducing agents. They donate electrons to substances that have been oxidized, decreasing the damage caused by free radicals. Dietary antioxidants include vitamins C and E, beta-carotene, and carotenoids.

Most medical researchers agree that people with diets rich in fruits and vegetables have a lower incidence of cardiovascular disease, certain cancers, and cataracts. Although fruits and vegetables are high in antioxidants, they also contain fibre and many different vitamins

**Carotenoids are pigments found in some fruits and vegetables, including spinach.**

and plant chemicals. It is hard to disentangle the effects of antioxidants from the beneficial effects of these other substances.

As a result, the benefits of antioxidant dietary supplements are under debate. According to one study, vitamin E supplements may lower the risk of heart disease. Another study, however, concludes that taking beta-carotene supplements does *not* reduce the risk of certain cancers.

We can be sure that a balanced diet including fruits and vegetables is beneficial to human health. Whether antioxidants confer these benefits, and whether these benefits include longevity, remain to be seen.

### Making Connections

1. Research vitamins C, E, alpha- and beta-carotenes, and folic acid. How do they affect our health? What fruits and vegetables contain these vitamins?

2. Lycopene is a carotenoid that has been linked to a decreased risk of pancreatic, cervical, and prostate cancer. Find out what fruits and vegetables contain lycopene. What colour are these fruits and vegetables?

# Investigation 18-A

MICROSCALE

**SKILL FOCUS**
Predicting
Performing and recording
Analyzing and interpreting

## Single Displacement Reactions

The metal activity series is shown in the table below. The more reactive metals are near the top of the series, and the less reactive metals are near the bottom. In this investigation, you will relate the activity series to the ease with which metals are oxidized and metal ions are reduced.

### Activity Series of Metals

| Metal | |
|---|---|
| lithium | Most Reactive |
| potassium | |
| barium | |
| calcium | |
| sodium | |
| magnesium | |
| aluminum | |
| zinc | |
| chromium | |
| iron | |
| cadmium | |
| cobalt | |
| nickel | |
| tin | |
| lead | |
| copper | |
| mercury | |
| silver | |
| platinum | |
| gold | Least Reactive |

### Question

How is the order of the metals in the activity series related to the ease with which metals are oxidized and metal ions are reduced?

### Predictions

Predict the relative ease with which the metals aluminum, copper, iron, magnesium, and zinc can be oxidized. Predict the relative ease with which the ions of these same metals can be reduced. Explain your reasoning in both cases.

### Materials

well plate
test tube rack
4 small test tubes
4 small pieces of each of these metals:
   aluminum foil, thin copper wire or tiny copper beads, iron filings, magnesium, and zinc
dropper bottles containing dilute solutions of
   aluminum sulfate, copper(II) sulfate, iron(II) sulfate, magnesium sulfate, and zinc nitrate

### Safety Precautions

- Wear goggles, gloves, and an apron for all parts of this investigation.

### Procedure

1. Place the well plate on a white piece of paper. Label it to match the table on the next page.

2. In each well plate, place a small piece of the appropriate metal, about the size of a grain of rice. Cover each piece with a few drops of the appropriate solution. Wait 3–5 min to observe if a reaction occurs.

3. Look for evidence of a chemical reaction in each mixture. Record the results, using "y" for a reaction, or "n" for no reaction. If you are unsure, repeat the process on a larger scale in a small test tube.

| Metal \ Compound | Al$_2$(SO$_4$)$_3$ | CuSO$_4$ | FeSO$_4$ | MgSO$_4$ | Zn(NO$_3$)$_2$ |
|---|---|---|---|---|---|
| Al | | | | | |
| Cu | | | | | |
| Fe | | | | | |
| Mg | | | | | |
| Zn | | | | | |

4. Discard the mixtures in the waste beaker supplied by your teacher. Do not pour anything down the drain.

## Analysis

1. For each single displacement reaction you observed, write
   - (a) a balanced chemical equation
   - (b) a total ionic equation
   - (c) a net ionic equation

2. Write an oxidation half-reaction and a reduction half-reaction for each net ionic equation you wrote in question 1. Use the smallest possible whole-number coefficients in each half-reaction.

3. Look at each balanced net ionic equation. Compare the total number of electrons lost by the reducing agent with the total number of electrons gained by the oxidizing agent.

4. List the different oxidation half-reactions. Start with the half-reaction for the most easily oxidized metal, and end with the half-reaction for the least easily oxidized metal. Explain your reasoning. Compare your list with your first prediction from the beginning of this investigation.

5. List the different reduction half-reactions. Start with the half-reaction for the most easily reduced metal ion, and end with the half-reaction for the least easily reduced metal ion. Explain your reasoning. Compare your list with your second prediction from the beginning of this investigation.

## Conclusions

6. Which list from questions 4 and 5 puts the metals in the same order as they appear in the activity series?

7. How is the order of the metals in the activity series related to the ease with which metals are oxidized and metal ions are reduced?

## Applications

8. Use the activity series to choose a reducing agent that will reduce aqueous nickel(II) ions to metallic nickel. Explain your reasoning.

9. Use the activity series to choose an oxidizing agent that will oxidize metallic cobalt to form aqueous cobalt(II) ions. Explain your reasoning.

### Section Summary

In this section, you learned to define and recognize redox reactions, and to write oxidation and reduction half-reactions. In Investigation 18-A, you observed the connection between the metal activity series and redox reactions. However, thus far, you have only worked with redox reactions that involve atoms and ions as reactants or products. In the next section, you will learn about redox reactions that involve covalent reactants or products.

## Section Review

**1** Predict whether each of the following single displacement reactions will occur. If so, write a balanced chemical equation, a balanced net ionic equation, and two balanced half-reactions. Include the physical states of the reactants and products in each case.

(a) aqueous silver nitrate and metallic cadmium

(b) gold and aqueous copper(II) sulfate

(c) aluminum and aqueous mercury(II) chloride

**2** (a) On which side of an oxidation half-reaction are the electrons? Why?

(b) On which side of a reduction half-reaction are the electrons? Why?

**3** Explain why, in a redox reaction, the oxidizing agent undergoes reduction.

**4** In a combination reaction, does metallic lithium act as an oxidizing agent or a reducing agent? Explain.

**5** Write a net ionic equation for a reaction in which

(a) $Fe^{2+}$ acts as an oxidizing agent

(b) Al acts as a reducing agent

(c) $Au^{3+}$ acts as an oxidizing agent

(d) Cu acts as a reducing agent

(e) $Sn^{2+}$ acts as an oxidizing agent and as a reducing agent

**6** The element potassium is made industrially by the single displacement reaction of molten sodium with molten potassium chloride.

(a) Write a net ionic equation for the reaction, assuming that all reactants and products are in the liquid state.

(b) Identify the oxidizing agent and the reducing agent in the reaction.

(c) Explain why the reaction is carried out in the liquid state and not in aqueous solution.

# Oxidation Numbers

## 18.2

Redox reactions are very common. Some of them produce light in a process known as *chemiluminescence*. In living things, the production of light in redox reactions is known as *bioluminescence*. You can actually see the light from redox reactions occurring in some organisms, such as glowworms and fireflies, as shown in Figure 18.3.

**Figure 18.3** Fireflies use flashes of light produced by redox reactions to attract a mate.

Not all redox reactions give off light, however. How can you recognize a redox reaction, and how can you identify the oxidizing and reducing agents? In section 18.1, you saw net ionic equations with monatomic elements, such as Cu and Zn, and with ions containing a single element, such as $Cu^{2+}$ and $Zn^{2+}$. In these cases, you could use ionic charges to describe the transfer of electrons. However, many redox reactions involve reactants or products with covalent bonds, including elements that exist as covalent molecules, such as oxygen, $O_2$; covalent compounds, such as water, $H_2O$; or polyatomic ions that are not spectator ions, such as permanganate, $MnO_4^-$. For reactions involving covalent reactants and products, you cannot use ionic charges to describe the transfer of electrons.

**Oxidation numbers** are actual or hypothetical charges, assigned using a set of rules. They are used to describe redox reactions with covalent reactants or products. They are also used to identify redox reactions, and to identify oxidizing and reducing agents. In this section, you will see how oxidation numbers were developed from Lewis structures, and then learn the rules to assign oxidation numbers.

## Oxidation Numbers from Lewis Structures

You are probably familiar with the Lewis structure of water, shown in Figure 18.4A. From the electronegativities on the periodic table in Figure 18.5, on the next page, you can see that oxygen (electronegativity 3.44) is more electronegative than hydrogen (electronegativity 2.20). The electronegativity difference is less than 1.7, so the two hydrogen-oxygen bonds are polar covalent, not ionic. In each bond, the electrons are more strongly attracted to the oxygen atom than to the hydrogen atom.

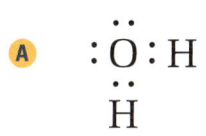

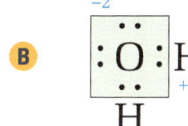

### Section Preview/Outcomes

In this section, you will

- **describe** oxidation and reduction in terms of changes in oxidation number
- **assign** oxidation numbers to elements in covalent molecules and polyatomic ions
- **identify** the species oxidized, the species reduced, the oxidizing agent, and the reducing agent in simple redox reactions
- **compare** redox reactions with other kinds of reactions
- **communicate** your understanding of the terms *oxidation numbers, oxidation, reduction*

### CHEM FACT

Oxidation numbers are just a bookkeeping method used to keep track of electron transfers. In a covalent molecule or a polyatomic ion, the oxidation number of each element does *not* represent an ionic charge, because the elements are not present as ions. However, to assign oxidation numbers to the elements in a covalent molecule or polyatomic ion, you can *pretend* the bonds are ionic.

**Figure 18.4** (A) The Lewis structure of water; (B) The formal counting of electrons with the more electronegative element assigned a negative charge

To assign oxidation numbers to the atoms in a water molecule, you can consider all the bonding electrons to be "owned" by the more electronegative oxygen atom, as shown in Figure 18.4B. Thus, each hydrogen atom in a water molecule is considered to have no electrons, as hydrogen would in a hydrogen ion, $H^+$. Therefore, the element hydrogen is assigned an oxidation number of +1 in water. On the other hand, the oxygen atom in a water molecule is considered to have a filled octet of electrons, as oxygen would in an oxide ion, $O^{2-}$. Therefore, the element oxygen is assigned an oxidation number of −2 in water. (**Note:** These are *not* ionic charges, since water is a covalent molecule. Also, note that the plus or minus sign in an oxidation number, such as −2, is written *before* the number. The plus or minus sign in an ionic charge, such as 2−, is written *after* the number.)

| | | | | | | | | | | | | | | | | | |
|---|---|---|---|---|---|---|---|---|---|---|---|---|---|---|---|---|---|
| H 2.20 | | | | | | | | | | | | | | | | | He - |
| Li 0.98 | Be 1.57 | | | | | | | | | | | B 2.04 | C 2.55 | N 3.04 | O 3.44 | F 3.98 | Ne - |
| Na 0.93 | Mg 1.31 | | | | | | | | | | | Al 1.61 | Si 1.90 | P 2.19 | S 2.58 | Cl 3.16 | Ar - |
| K 0.82 | Ca 1.00 | Sc 1.36 | Ti 1.54 | V 1.63 | Cr 1.66 | Mn 1.55 | Fe 1.83 | Co 1.88 | Ni 1.91 | Cu 1.90 | Zn 1.65 | Ga 1.81 | Ge 2.01 | As 2.18 | Se 2.55 | Br 2.96 | Kr - |
| Rb 0.82 | Sr 0.95 | Y 1.22 | Zr 1.33 | Nb 1.6 | Mo 2.16 | Tc 2.10 | Ru 2.2 | Rh 2.28 | Pd 2.20 | Ag 1.93 | Cd 1.69 | In 1.78 | Sn 1.96 | Sb 2.05 | Te 2.1 | I 2.66 | Xe - |
| Cs 0.79 | Ba 0.89 | Lu 1.0 | Hf 1.3 | Ta 1.5 | W 1.7 | Re 1.9 | Os 2.2 | Ir 2.2 | Pt 2.2 | Au 2.4 | Hg 1.9 | Tl 1.8 | Pb 1.8 | Bi 1.9 | Po 2.0 | At 2.2 | Rn - |
| Fr 0.7 | Ra 0.9 | Lr - | Rf - | Db - | Sg - | Bh - | Hs - | Mt - | Uun - | Uuu - | Uub - | - | Uuq - | - | Uuh - | | |

| | | | | | | | | | | | | | |
|---|---|---|---|---|---|---|---|---|---|---|---|---|---|
| La 1.10 | Ce 1.12 | Pr 1.13 | Nd 1.14 | Pm - | Sm 1.17 | Eu - | Gd 1.20 | Tb - | Dy 1.22 | Ho 1.23 | Er 1.24 | Tm 1.25 | Yb - |
| Ac 1.1 | Th 1.3 | Pa 1.5 | U 1.7 | Np 1.3 | Pu 1.3 | Am - | Cm - | Bk - | Cf - | Es - | Fm - | Md - | No - |

**Figure 18.5** **The periodic table, showing electronegativity values**

In a chlorine molecule, $Cl_2$, each atom has the same electronegativity, so the bond is non-polar covalent. Because the electrons are equally shared, you can consider each chlorine atom to "own" one of the shared electrons, as shown in Figure 18.6. Thus, each chlorine atom in the molecule is considered to have the same number of electrons as a neutral chlorine atom. Each chlorine atom is therefore assigned an oxidation number of 0.

**Figure 18.6** **(A) The Lewis structure of a chlorine molecule; (B) The formal counting of electrons in a chlorine molecule for oxidation number purposes**

Figure 18.7 shows how oxidation numbers are assigned for the polyatomic cyanide ion, CN⁻. The electronegativity of nitrogen (3.04) is greater than the electronegativity of carbon (2.55). Thus, the three shared pairs of electrons are all considered to belong to the nitrogen atom. As a result, the carbon atom is considered to have two valence electrons, which is two electrons less than the four valence electrons of a neutral carbon atom. Therefore, the carbon atom in CN⁻ is assigned an oxidation number of +2. The nitrogen atom is considered to have eight valence electrons, which is three electrons more than the five valence electrons of a neutral nitrogen atom. Therefore, the nitrogen atom in CN⁻ is assigned an oxidation number of −3.

**A**

$[:C:::N:]^-$

**B**

$[:C:::N:]^-$
 +2   −3

**Figure 18.7** (A) The Lewis structure of a cyanide ion; (B) The formal counting of electrons in a cyanide ion for oxidation number purposes

You have seen examples of how Lewis structures can be used to assign oxidation numbers for polar molecules such as water, non-polar molecules such as chlorine, and polar polyatomic ions such as the cyanide ion. In the following ThoughtLab, you will use Lewis structures to assign oxidation number values, and then look for patterns in your results.

## ThoughtLab — Finding Rules for Oxidation Numbers

### Procedure

1. Use Lewis structures to assign an oxidation number to each element in the following covalent molecules.
   (a) HI   (b) $O_2$   (c) $PCl_5$   (d) $BBr_3$

2. Use Lewis structures to assign an oxidation number to each element in the following polyatomic ions.
   (a) $OH^-$   (b) $NH_4^+$   (c) $CO_3^{2-}$

3. Assign an oxidation number to each of the following atoms or monatomic ions. Explain your reasoning.
   (a) Ne   (b) K   (c) $I^-$   (d) $Mg^{2+}$

### Analysis

1. For each molecule in question 1 of the procedure, find the sum of the oxidation numbers of all the atoms present. What do you notice? Explain why the observed sum must be true for a neutral molecule.

2. For each polyatomic ion in question 2 of the procedure, find the sum of the oxidation numbers of all the atoms present. Describe and explain any pattern you see.

### Extension

3. Predict the sum of the oxidation numbers of the atoms in the hypochlorite ion, $OCl^-$.

4. Test your prediction from question 3.

## Using Rules to Find Oxidation Numbers

Drawing Lewis structures to assign oxidation numbers can be a very time-consuming process for large molecules or large polyatomic ions. Instead, the results from Lewis structures have been summarized to produce a more convenient set of rules, which can be applied more quickly. Table 18.1 summarizes the rules used to assign oxidation numbers. You may have discovered some of these rules for yourself in the ThoughtLab you just completed.

**Table 18.1** Oxidation Number Rules

| Rules | Examples |
|---|---|
| 1. A pure element has an oxidation number of 0. | Na in $Na_{(s)}$, Br in $Br_{2(\ell)}$, and P in $P_{4(s)}$ all have an oxidation number of 0. |
| 2. The oxidation number of an element in a monatomic ion equals the charge of the ion. | The oxidation number of Al in $Al^{3+}$ is +3. The oxidation number of Se in $Se^{2-}$ is −2. |
| 3. The oxidation number of hydrogen in its compounds is +1, except in metal hydrides, where the oxidation number of hydrogen is −1. | The oxidation number of H in $H_2S$ or $CH_4$ is +1. The oxidation number of H in NaH or in $CaH_2$ is −1. |
| 4. The oxidation number of oxygen in its compounds is usually −2, but there are exceptions. These include peroxides, such as $H_2O_2$, and the compound $OF_2$. | The oxidation number of O in $Li_2O$ or in $KNO_3$ is −2. |
| 5. In covalent compounds that do not contain hydrogen or oxygen, the more electronegative element is assigned an oxidation number that equals the negative charge it usually has in its ionic compounds. | The oxidation number of Cl in $PCl_3$ is −1. The oxidation number of S in $CS_2$ is −2. |
| 6. The sum of the oxidation numbers of all the elements in a compound is 0. | In $CF_4$, the oxidation number of F is −1, and the oxidation number of C is +4. (+4) + 4(−1) = 0 |
| 7. The sum of the oxidation numbers of all the elements in a polyatomic ion equals the charge on the ion. | In $NO_2^-$, the oxidation number of O is −2, and the oxidation number of N is +3. (+3) + 2(−2) = −1 |

Some oxidation numbers found using these rules are not integers. For example, an important iron ore called magnetite has the formula $Fe_3O_4$. Using the oxidation number rules, you can assign oxygen an oxidation number of −2, and calculate an oxidation number of $+\frac{8}{3}$ for iron. However, magnetite contains no iron atoms with this oxidation number. It actually contains iron(III) ions and iron(II) ions in a 2:1 ratio. The formula of magnetite is sometimes written as $Fe_2O_3 \cdot FeO$ to indicate that there are two different oxidation numbers. The value $+\frac{8}{3}$ for the oxidation number of iron is an average value.

$$\frac{2(+3) + (+2)}{3} = +\frac{8}{3}$$

Even though some oxidation numbers found using these rules are averages, the rules are still useful for monitoring electron transfers in redox reactions.

In the following Sample Problem, you will find out how to apply these rules to covalent molecules and polyatomic ions.

## Sample Problem

### Assigning Oxidation Numbers

**Problem**

Assign an oxidation number to each element.

(a) SiBr$_4$    (b) HClO$_4$    (c) Cr$_2$O$_7^{2-}$

**Solution**

(a) • Because the compound SiBr$_4$ does not contain hydrogen or oxygen, rule 5 applies. Because SiBr$_4$ is a compound, rule 6 also applies.
- Silicon has an electronegativity of 1.90. Bromine has an electronegativity of 2.96. From rule 5, therefore, you can assign bromine an oxidation number of −1.
- The oxidation number of silicon is unknown, so let it be $x$. You know from rule 6 that the sum of the oxidation numbers is 0. Then,

$$x + 4(-1) = 0$$
$$x - 4 = 0$$
$$x = 4$$

The oxidation number of silicon is +4. The oxidation number of bromine is −1.

(b) • Because the compound HClO$_4$ contains hydrogen and oxygen, rules 3 and 4 apply. Because HClO$_4$ is a compound, rule 6 also applies.
- Hydrogen has its usual oxidation number of +1. Oxygen has its usual oxidation number of −2. The oxidation number of chlorine is unknown, so let it be $x$. You know from rule 6 that the sum of the oxidation numbers is 0. Then,

$$(+1) + x + 4(-2) = 0$$
$$x - 7 = 0$$
$$x = 7$$

The oxidation number of hydrogen is +1. The oxidation number of chlorine is +7. The oxidation number of oxygen is −2.

(c) • Because the polyatomic ion Cr$_2$O$_7^{2-}$ contains oxygen, rule 4 applies. Because Cr$_2$O$_7^{2-}$ is a polyatomic ion, rule 7 also applies.
- Oxygen has its usual oxidation number of −2.
- The oxidation number of chromium is unknown, so let it be $x$. You know from rule 7 that the sum of the oxidation numbers is −2. Then,

$$2x + 7(-2) = -2$$
$$2x - 14 = -2$$
$$2x = 12$$
$$x = 6$$

The oxidation number of chromium is +6. The oxidation number of oxygen is −2.

> **PROBLEM TIP**
>
> When finding the oxidation numbers of elements in ionic compounds, you can work with the ions separately. For example, Na$_2$Cr$_2$O$_7$ contains two Na$^+$ ions, and so sodium has an oxidation number of +1. The oxidation numbers of Cr and O can then be calculated as shown in part (c) of the Sample Problem.

### Practice Problems

**9.** Determine the oxidation number of the specified element in each of the following.

(a) N in $NF_3$  
(b) S in $S_8$  
(c) Cr in $CrO_4^{2-}$  
(d) P in $P_2O_5$  
(e) C in $C_{12}H_{22}O_{11}$  
(f) C in $CHCl_3$

**10.** Determine the oxidation number of each element in each of the following.

(a) $H_2SO_3$  
(b) $OH^-$  
(c) $HPO_4^{2-}$

**11.** As stated in rule 4, oxygen does not always have its usual oxidation number of –2. Determine the oxidation number of oxygen in each of the following.

(a) the compound oxygen difluoride, $OF_2$  
(b) the peroxide ion, $O_2^{2-}$

**12.** Determine the oxidation number of each element in each of the following ionic compounds by considering the ions separately.
**Hint:** One formula unit of the compound in part (c) contains two identical monatomic ions and one polyatomic ion.

(a) $Al(HCO_3)_3$  
(b) $(NH_4)_3PO_4$  
(c) $K_2H_3IO_6$

## Applying Oxidation Numbers to Redox Reactions

You have seen that the single displacement reaction of zinc with copper(II) sulfate is a redox reaction, represented by the following chemical equation and net ionic equation.

$$Zn_{(s)} + CuSO_{4(aq)} \rightarrow Cu_{(s)} + ZnSO_{4(aq)}$$

$$Zn_{(s)} + Cu^{2+}_{(aq)} \rightarrow Cu_{(s)} + Zn^{2+}_{(aq)}$$

Each atom or ion shown in the net ionic equation can be assigned an oxidation number. Zn has an oxidation number of 0; $Cu^{2+}$ has an oxidation number of +2; Cu has an oxidation number of 0; and $Zn^{2+}$ has an oxidation number of +2. Thus, there are changes in oxidation numbers in this reaction. The oxidation number of zinc increases, while the oxidation number of copper decreases.

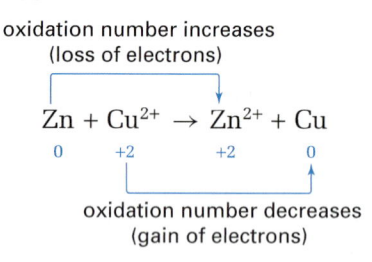

oxidation number increases
(loss of electrons)

$$Zn + Cu^{2+} \rightarrow Zn^{2+} + Cu$$
$$\phantom{Zn + }0 \phantom{ + }+2 \phantom{\rightarrow }+2 \phantom{ + }0$$

oxidation number decreases
(gain of electrons)

In the oxidation half-reaction, the element zinc undergoes an increase in its oxidation number from 0 to +2.

$$Zn \rightarrow Zn^{2+} + 2e^-$$
$$0 \phantom{\rightarrow }+2$$

In the reduction half-reaction, the element copper undergoes a decrease in its oxidation number from +2 to 0.

$$Cu^{2+} + 2e^- \rightarrow Cu$$
$$+2 \phantom{ + 2e^- \rightarrow }0$$

Therefore, you can describe oxidation and reduction as follows. (Also see Figure 18.8.)

- **Oxidation** is an increase in oxidation number.
- **Reduction** is a decrease in oxidation number.

You can also monitor changes in oxidation numbers in reactions that involve covalent molecules. For example, oxidation number changes occur in the reaction of hydrogen and oxygen to form water.

$$2H_{2(g)} + O_{2(g)} \rightarrow 2H_2O_{(\ell)}$$
$$\phantom{2H_{2(g)}}\ 0 \phantom{\ + \ } 0 \phantom{\rightarrow \ }\ +1\ -2$$

Because hydrogen combines with oxygen in this reaction, hydrogen undergoes oxidation, according to the historical definition given at the beginning of section 10.1. Hydrogen also undergoes oxidation according to the modern definition, because the oxidation number of hydrogen increases from 0 to +1. Hydrogen is the reducing agent in this reaction. The oxygen undergoes reduction, because its oxidation number decreases from 0 to −2. Oxygen is the oxidizing agent in this reaction.

The following Sample Problem illustrates how to use oxidation numbers to identify redox reactions, oxidizing agents, and reducing agents.

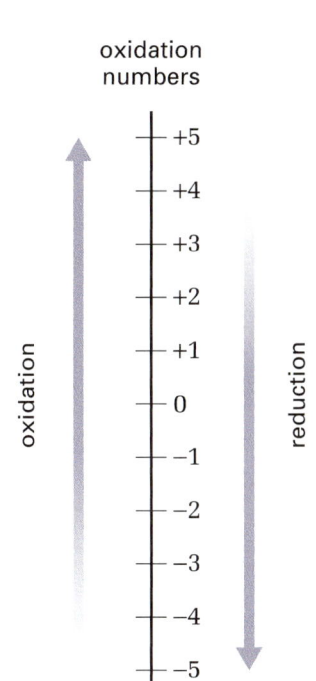

**Figure 18.8** Oxidation and reduction are directly related to changes in oxidation numbers.

## Sample Problem

### Identifying Redox Reactions

#### Problem
Determine whether each of the following reactions is a redox reaction. If so, identify the oxidizing agent and the reducing agent.

(a) $CH_{4(g)} + Cl_{2(g)} \rightarrow CH_3Cl_{(g)} + HCl_{(g)}$

(b) $CaCO_{3(s)} + 2HCl_{(aq)} \rightarrow CaCl_{2(aq)} + H_2O_{(\ell)} + CO_{2(g)}$

#### Solution
Find the oxidation number of each element in the reactants and products. Identify any elements that undergo an increase or a decrease in oxidation number during the reaction.

(a) The oxidation number of each element in the reactants and products is as shown.

$$CH_{4(g)} + Cl_{2(g)} \rightarrow CH_3Cl_{(g)} + HCl_{(g)}$$
$$-4\ +1 \phantom{\ +\ }\ 0 \phantom{\rightarrow}\ -2\ +1\ -1\phantom{\ +\ }\ +1\ -1$$

- The oxidation number of hydrogen is +1 on both sides of the equation, so hydrogen is neither oxidized nor reduced.
- Both carbon and chlorine undergo changes in oxidation number, so the reaction is a redox reaction.
- The oxidation number of carbon increases from −4 to −2. The carbon atoms on the reactant side exist in methane molecules, $CH_{4(g)}$, so methane is oxidized. Therefore, methane is the reducing agent.
- The oxidation number of chlorine decreases from 0 to −1, so elemental chlorine, $Cl_{2(g)}$, is reduced. Therefore, elemental chlorine is the oxidizing agent.

(b) Because this reaction involves ions, write the equation in its total ionic form.

$$CaCO_{3(s)} + 2H^+_{(aq)} + 2Cl^-_{(aq)} \rightarrow Ca^{2+}_{(aq)} + 2Cl^-_{(aq)} + H_2O_{(\ell)} + CO_{2(g)}$$

*Continued ...*

### PROBLEM TIP
- Use the fact that the sum of the oxidation numbers in a molecule is zero to check the assignment of the oxidation numbers.
- Make sure that a reaction does not include only a reduction or only an oxidation. Oxidation and reduction must occur together in a redox reaction.

### CONCEPT CHECK
In part (b) of the Sample Problem, you can assign oxidation numbers to each element in the given chemical equation *or* in the net ionic equation. What are the advantages and the disadvantages of each method?

## CONCEPT CHECK

In previous courses, you classified reactions into four main types: synthesis, decomposition, single displacement, and double displacement. You also learned to recognize combustion reactions and neutralization reactions. You have now learned to classify redox reactions. In addition, you have also learned about a special type of redox reaction known as a disproportionation reaction.

1. Classify each reaction in two ways.
   (a) magnesium reacting with a solution of iron(II) nitrate
   $Mg + Fe(NO_3)_2 \rightarrow Fe + Mg(NO_3)_2$
   (b) hydrogen sulfide burning in oxygen
   $2H_2S + 3O_2 \rightarrow 2SO_2 + 2H_2O$
   (c) calcium reacting with chlorine
   $Ca + Cl_2 \rightarrow CaCl_2$
2. Classify the formation of water and oxygen from hydrogen peroxide in three ways.
   $2H_2O_2 \rightarrow 2H_2O + O_2$

---

*Continued ...*

The chloride ions are spectator ions, which do not undergo oxidation or reduction. The net ionic equation is as follows.

$CaCO_{3(s)} + 2H^+_{(aq)} \rightarrow Ca^{2+}_{(aq)} + H_2O_{(\ell)} + CO_{2(g)}$

For the net ionic equation, the oxidation number of each element in the reactants and products is as shown.

$CaCO_{3(s)} + 2H^+_{(aq)} \rightarrow Ca^{2+}_{(aq)} + H_2O_{(\ell)} + CO_{2(g)}$
+2 +4 −2          +1                    +2             +1 −2         +4 −2

No elements undergo changes in oxidation numbers, so the reaction is not a redox reaction.

## Practice Problems

13. Determine whether each reaction is a redox reaction.
    (a) $H_2O_2 + 2Fe(OH)_2 \rightarrow 2Fe(OH)_3$
    (b) $PCl_3 + 3H_2O \rightarrow H_3PO_3 + 3HCl$

14. Identify the oxidizing agent and the reducing agent for the redox reaction(s) in the previous question.

15. For the following balanced net ionic equation, identify the reactant that undergoes oxidation and the reactant that undergoes reduction.
    $Br_2 + 2ClO_2^- \rightarrow 2Br^- + 2ClO_2$

16. Nickel and copper are two metals that have played a role in the economy of Newfoundland and Labrador. Nickel and copper ores usually contain the metals as sulfides, such as NiS and $Cu_2S$. Do the extractions of these pure elemental metals from their ores involve redox reactions? Explain your reasoning.

## Section Summary

In this section, you extended your knowledge of redox reactions to include covalent reactants and products. You did this by learning how to assign oxidation numbers and how to use them to recognize redox reactions, oxidizing agents, and reducing agents. In the next section, you will extend your knowledge further by learning how to write balanced equations that represent redox reactions.

## Section Review

**1** At the beginning of section 18.1, it was stated that oxidation originally meant "to combine with oxygen." Explain why a metal that combines with the element oxygen undergoes oxidation as we now define it. What happens to the oxygen in this reaction? Write a balanced chemical equation for a reaction that illustrates your answer.

**2** Determine whether each of the following reactions is a redox reaction.
   (a) $H_2 + I_2 \rightarrow 2HI$
   (b) $2NaHCO_3 \rightarrow Na_2CO_3 + H_2O + CO_2$
   (c) $2HBr + Ca(OH)_2 \rightarrow CaBr_2 + 2H_2O$
   (d) $PCl_5 \rightarrow PCl_3 + Cl_2$

**3** Write three different definitions for a redox reaction.

**4** Explain why fluorine has an oxidation number of –1 in all its compounds.

**5** When an element combines with another element, is the reaction a redox reaction? Explain your answer.

**6 (a)** Use the oxidation number rules to find the oxidation number of sulfur in a thiosulfate ion, $S_2O_3^{2-}$.

**(b)** The Lewis structure of a thiosulfate ion is given here. Use the Lewis structure to find the oxidation number of each sulfur atom.

$$\left[ \begin{array}{c} :\ddot{O}: \\ :\ddot{O}:S:S: \\ :\ddot{O}: \end{array} \right]^{2-}$$

**(c)** Compare your results from parts (a) and (b) and explain any differences.

**(d)** What are the advantages and disadvantages of using Lewis structures to assign oxidation numbers?

**(e)** What are the advantages and disadvantages of using the oxidation number rules to assign oxidation numbers?

**7 (a)** The Haber Process for the production of ammonia from nitrogen and hydrogen is a very important industrial process. Write a balanced chemical equation for the reaction. Use oxidation numbers to identify the oxidizing agent and the reducing agent.

**(b)** When ammonia is reacted with nitric acid to make the common fertilizer ammonium nitrate, is the reaction a redox reaction? Explain. (**Hint:** Consider the two polyatomic ions in the product separately.)

**8** Historically, the extraction of a metal from its ore was known as reduction. One way to reduce iron ore on an industrial scale is to use a huge reaction vessel, 30 m to 40 m high, called a blast furnace. The reactants in a blast furnace are an impure iron ore, such as $Fe_2O_3$, mixed with limestone, $CaCO_3$, and coke, C, which is made from coal. The solid mixture is fed into the top of the blast furnace. A blast of very hot air, at about 900°C, is blown in near the bottom of the furnace. The following reactions occur.

$2C + O_2 \rightarrow 2CO$

$Fe_2O_3 + 3CO \rightarrow 2Fe + 3CO_2$

The limestone is present to convert sand or quartz, $SiO_2$, which is present as an impurity in the ore, to calcium silicate, $CaSiO_3$.

$CaCO_3 \rightarrow CaO + CO_2$

$CaO + SiO_2 \rightarrow CaSiO_3$

**(a)** Which of the four reactions above are redox reactions?

**(b)** For each redox reaction that you identified in part (a), name the oxidizing agent and the reducing agent.

---

**CHEM FACT**

Redox reactions are involved in some very important industrial processes, such as iron and steel production. However, the widespread use of metals has occupied a relatively small part of human history. In the Stone Age, humans relied on stone, wood, and bone to make tools and weapons. The Stone Age ended in many parts of the world with the start of the Bronze Age, which was marked by the use of copper and then bronze (an alloy of copper and tin). In the Iron Age, bronze was replaced by the use of iron. The dates of the Bronze Age and the Iron Age vary for different parts of the world.

# 18.3 The Half-Reaction Method for Balancing Equations

**Section Preview/Outcomes**

In this section, you will

- **investigate** oxidation-reduction reactions by reacting metals with acids and by combusting hydrocarbons
- **identify** a redox reaction as the sum of the oxidation half-reaction and the reduction half-reaction
- **write** balanced equations for redox reactions using the half-reaction method

Did you know that redox reactions are an important part of CD manufacturing? The CDs you buy at a music store are made of Lexan®, the same plastic used for riot shields and bulletproof windows. The CDs are coated with a thin aluminum film. They are copies of a single master disc, which is made of glass coated with silver, as seen in Figure 18.9. Silver is deposited on a glass disc by the reduction of silver ions with methanal, HCHO, also known as formaldehyde. In the same reaction, formaldehyde is oxidized to methanoic acid, HCOOH, also known as formic acid. The redox reaction occurs under acidic conditions.

**Figure 18.9** The production of CDs depends on a redox reaction used to coat the master disc with silver.

You have seen many balanced chemical equations and net ionic equations that represent redox reactions. There are specific techniques for balancing these equations. These techniques are especially useful for reactions that take place under acidic or basic conditions, such as the acidic conditions used in coating a master CD with silver.

In section 18.1, you learned to divide the balanced equations for some redox reactions into separate oxidation and reduction half-reactions. You will now use the reverse approach, and discover how to write a balanced equation by combining two half-reactions. To do this, you must first understand how to write a wide range of half-reactions.

730  MHR • Unit 8 Electrochemistry

## Balancing Half-Reactions

In the synthesis of potassium chloride from its elements, metallic potassium is oxidized to form potassium ions, and gaseous chlorine is reduced to form chloride ions. This reaction is shown in Figure 18.10. Each half-reaction can be balanced by writing the correct formulas for the reactant and product, balancing the numbers of atoms, and then adding the correct number of electrons to balance the charges. For the oxidation half-reaction,

$$K \rightarrow K^+ + e^-$$

The atoms are balanced. The net charge on each side is 0. For the reduction half-reaction,

$$Cl_2 + 2e^- \rightarrow 2Cl^-$$

The atoms are balanced. The net charge on each side is −2.

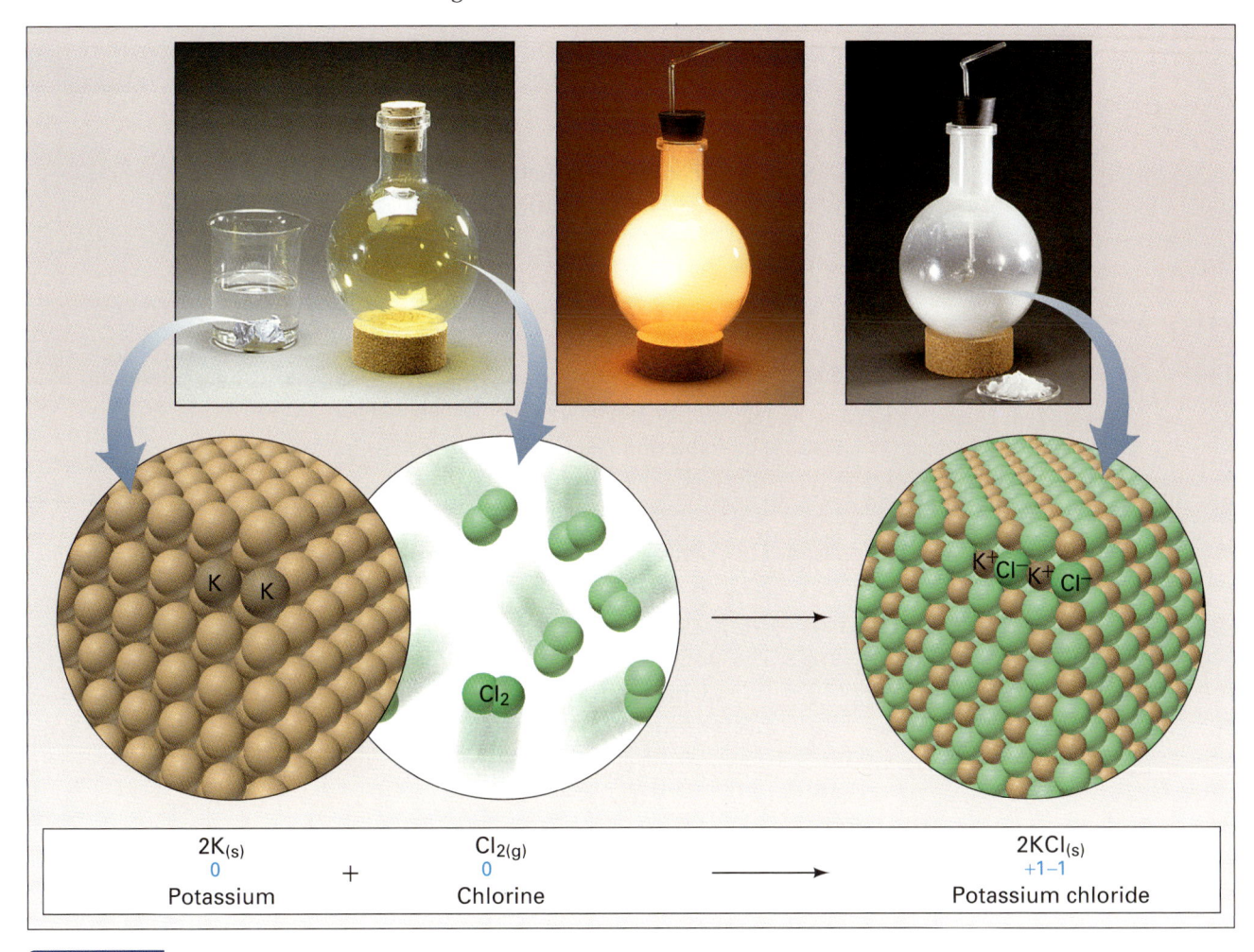

**Figure 18.10** Grey potassium metal, which is stored under oil, reacts very vigorously with greenish-yellow chlorine gas to form white potassium chloride. The changes in oxidation numbers show that this synthesis reaction is also a redox reaction.

Redox reactions do not always take place under neutral conditions. Balancing half-reactions is more complicated for reactions that take place in acidic or basic solutions. When an acid or base is present, H$^+$ or OH$^-$ ions must also be considered. However, the overall approach is similar. This approach involves writing the correct formulas for the reactants and products, balancing the atoms, and adding the appropriate number of electrons to one side of the half-reaction to balance the charges.

## Balancing Half-Reactions for Acidic Solutions

The following steps are used to balance a half-reaction for an acidic solution. The Sample Problem that follows applies these steps.

**Step 1** Write an unbalanced half-reaction that shows the formulas of the given reactant(s) and product(s).
**Step 2** Balance any atoms other than oxygen and hydrogen first.
**Step 3** Balance any oxygen atoms by adding water molecules.
**Step 4** Balance any hydrogen atoms by adding hydrogen ions.
**Step 5** Balance the charges by adding electrons.

### Sample Problem
### Balancing a Half-Reaction in Acidic Solution

**Problem**
Write a balanced half-reaction that shows the reduction of permanganate ions, $MnO_4^-$, to manganese(II) ions in an acidic solution.

**Solution**
**Step 1** Represent the given reactant and product with correct formulas.
$$MnO_4^- \rightarrow Mn^{2+}$$
**Step 2** Balance the atoms, starting with the manganese atoms. Here, the manganese atoms are already balanced.
**Step 3** The reduction occurs in aqueous solution, so add water molecules to balance the oxygen atoms.
$$MnO_4^- \rightarrow Mn^{2+} + 4H_2O$$
**Step 4** The reaction occurs in acidic solution, so add hydrogen ions to balance the hydrogen atoms.
$$MnO_4^- + 8H^+ \rightarrow Mn^{2+} + 4H_2O$$
**Step 5** The atoms are now balanced, but the net charge on the left side is 7+, whereas the net charge on the right side is 2+. Add five electrons to the left side to balance the charges.
$$MnO_4^- + 8H^+ + 5e^- \rightarrow Mn^{2+} + 4H_2O$$

### Practice Problems

**CONCEPT CHECK**
The ability to balance a single half-reaction as a bookkeeping exercise does not mean that a single half-reaction can occur on its own. In a redox reaction, oxidation and reduction must both occur.

17. Write a balanced half-reaction for the reduction of cerium(IV) ions to cerium(III) ions.

18. Write a balanced half-reaction for the oxidation of bromide ions to bromine.

19. Balance each of the following half-reactions under acidic conditions.
    (a) $O_2 \rightarrow H_2O_2$   (b) $H_2O \rightarrow O_2$   (c) $NO_3^- \rightarrow N_2$

20. Balance each of the following half-reactions under acidic conditions.
    (a) $ClO_3^- \rightarrow Cl^-$   (b) $NO \rightarrow NO_3^-$   (c) $Cr_2O_7^{2-} \rightarrow Cr^{3+}$

## Balancing Half-Reactions for Basic Solutions

The following steps are used to balance a half-reaction for a basic solution. The Sample Problem that follows applies these steps.

> **Step 1** Write an unbalanced half-reaction that shows the formulas of the given reactant(s) and product(s).
>
> **Step 2** Balance any atoms other than oxygen and hydrogen first.
>
> **Step 3** Balance any oxygen and hydrogen atoms as if the conditions are acidic.
>
> **Step 4** Adjust for basic conditions by adding to both sides the same number of hydroxide ions as the number of hydrogen ions already present.
>
> **Step 5** Simplify the half-reaction by combining the hydrogen ions and hydroxide ions on the same side of the equation into water molecules.
>
> **Step 6** Remove any water molecules present on both sides of the half-reaction.
>
> **Step 7** Balance the charges by adding electrons.

### Sample Problem
### Balancing a Half-Reaction in Basic Solution

**Problem**

Write a balanced half-reaction that shows the oxidation of thiosulfate ions, $S_2O_3^{2-}$, to sulfite ions, $SO_3^{2-}$, in a basic solution.

**Solution**

**Step 1** Represent the given reactant and product with correct formulas.
$$S_2O_3^{2-} \rightarrow SO_3^{2-}$$

**Step 2** Balance the atoms, beginning with the sulfur atoms.
$$S_2O_3^{2-} \rightarrow 2SO_3^{2-}$$

**Step 3** Balance the oxygen and hydrogen atoms as if the solution is acidic.
$$S_2O_3^{2-} + 3H_2O \rightarrow 2SO_3^{2-}$$
$$S_2O_3^{2-} + 3H_2O \rightarrow 2SO_3^{2-} + 6H^+$$

**Step 4** There are six hydrogen ions present, so adjust for basic conditions by adding six hydroxide ions to each side.
$$S_2O_3^{2-} + 3H_2O + 6OH^- \rightarrow 2SO_3^{2-} + 6H^+ + 6OH^-$$

**Step 5** Combine the hydrogen ions and hydroxide ions on the right side into water molecules.
$$S_2O_3^{2-} + 3H_2O + 6OH^- \rightarrow 2SO_3^{2-} + 6H_2O$$

**Step 6** Remove three water molecules from each side.
$$S_2O_3^{2-} + 6OH^- \rightarrow 2SO_3^{2-} + 3H_2O$$

**Step 7** The atoms are now balanced, but the net charge on the left side is 8–, whereas the net charge on the right side is 4–. Add four electrons to the right side to balance the charges.
$$S_2O_3^{2-} + 6OH^- \rightarrow 2SO_3^{2-} + 3H_2O + 4e^-$$

## Practice Problems

**21.** Write a balanced half-reaction for the oxidation of chromium(II) ions to chromium(III) ions.

**22.** Write a balanced half-reaction for the reduction of oxygen to oxide ions.

**23.** Balance each of the following half-reactions under basic conditions.
   **(a)** $Al \rightarrow Al(OH)_4^-$
   **(b)** $CN^- \rightarrow CNO^-$
   **(c)** $MnO_4^- \rightarrow MnO_2$
   **(d)** $CrO_4^{2-} \rightarrow Cr(OH)_3$
   **(e)** $CO_3^{2-} \rightarrow C_2O_4^{2-}$

**24.** Balance each of the following half-reactions.
   **(a)** $FeO_4^{2-} \rightarrow Fe^{3+}$ (acidic conditions)
   **(b)** $ClO_2^- \rightarrow Cl^-$ (basic conditions)

## Half-Reaction Method for Balancing Redox Reactions

Recall that, if you consider a redox reaction as two half-reactions, electrons are lost in the oxidation half-reaction, and electrons are gained in the reduction half-reaction. For example, you know the reaction of zinc with aqueous copper(II) sulfate.

$$Zn_{(s)} + CuSO_{4(aq)} \rightarrow Cu_{(s)} + ZnSO_{4(aq)}$$

Removing the spectator ions leaves the following net ionic equation.

$$Zn_{(s)} + Cu^{2+}_{(aq)} \rightarrow Cu_{(s)} + Zn^{2+}_{(aq)}$$

- You can break the net ionic equation into two half-reactions:
  Oxidation half-reaction: $Zn \rightarrow Zn^{2+} + 2e^-$
  Reduction half-reaction: $Cu^{2+} + 2e^- \rightarrow Cu$

- You can also start with the half-reactions and use them to produce a net ionic equation. If you add the two half-reactions, the result is as follows.

$$Zn + Cu^{2+} + 2e^- \rightarrow Cu + Zn^{2+} + 2e^-$$

Removing the two electrons from each side results in the original net ionic equation.

As shown above, you can use half-reactions to write balanced net ionic equations for redox reactions. In doing so, you use the fact that *no electrons are created or destroyed in a redox reaction*. Electrons are transferred from one reactant (the reducing agent) to another (the oxidizing agent).

### Balancing a Net Ionic Equation

You know from Investigation 18-A that magnesium metal, $Mg_{(s)}$, displaces aluminum from an aqueous solution of one of its compounds, such as aluminum nitrate, $Al(NO_3)_{3(aq)}$. To obtain a balanced net ionic equation for this reaction, you can start by looking at the half-reactions. Magnesium atoms undergo oxidation to form magnesium ions, which have a 2+ charge. The oxidation half-reaction is as follows.

$$Mg \rightarrow Mg^{2+} + 2e^-$$

Aluminum ions, which have a 3+ charge, undergo reduction to form aluminum atoms. The reduction half-reaction is as follows.

$$Al^{3+} + 3e^- \rightarrow Al$$

To balance the net ionic equation for this redox reaction, you can combine the two half-reactions in such a way that the number of electrons lost through oxidation equals the number of electrons gained through reduction. In other words, you can model the transfer of a certain number of electrons from the reducing agent to the oxidizing agent.

For the reaction of magnesium metal with aluminum ions, the two balanced half-reactions include different numbers of electrons, 2 and 3. The least common multiple of 2 and 3 is 6. To combine the half-reactions and give a balanced net ionic equation, multiply the balanced half-reactions by different numbers so that the results both include six electrons, as shown below.

- Multiply the oxidation half-reaction by 3. Multiply the reduction half-reaction by 2.

$$3Mg \rightarrow 3Mg^{2+} + 6e^-$$
$$2Al^{3+} + 6e^- \rightarrow 2Al$$

- Add the results.

$$3Mg + 2Al^{3+} + 6e^- \rightarrow 3Mg^{2+} + 2Al + 6e^-$$

- Remove $6e^-$ from each side to obtain the balanced net ionic equation.

$$3Mg + 2Al^{3+} \rightarrow 3Mg^{2+} + 2Al$$

To produce the balanced chemical equation, you can include the spectator ions, which are nitrate ions in this example. Include the states, if necessary. The balanced chemical equation is:

$$3Mg_{(s)} + 2Al(NO_3)_{3(aq)} \rightarrow 3Mg(NO_3)_{2(aq)} + 2Al_{(s)}$$

### Steps for Balancing by the Half-Reaction Method

You could balance the chemical equation for the reaction of magnesium with aluminum nitrate by inspection, instead of writing half-reactions. However, many redox equations are difficult to balance by the inspection method. In general, you can balance the net ionic equation for a redox reaction by a process known as the half-reaction method. The preceding example of the reaction of magnesium with aluminum nitrate illustrates this method. Specific steps for following the half-reaction method are given below.

**Step 1** Write an unbalanced net ionic equation, if it is not already given.

**Step 2** Divide the unbalanced net ionic equation into an oxidation half-reaction and a reduction half-reaction. To do this, you may need to assign oxidation numbers to all the elements in the net ionic equation to determine what is oxidized and what is reduced.

**Step 3** Balance the oxidation half-reaction and the reduction half-reaction independently.

**Step 4** Determine the least common multiple (LCM) of the numbers of electrons in the oxidation half-reaction and the reduction half-reaction.

*Continued on the next page*

---

**Math LINK**

The lowest or least common multiple (LCM) of two numbers is the smallest multiple of each number. For example, the LCM of 2 and 1 is 2; the LCM of 3 and 6 is 6; and the LCM of 2 and 5 is 10. One way to find the LCM of two numbers is to list the multiples of each number and to find the smallest number that appears in both lists. For the numbers 6 and 8,

- the multiples of 6 are: 6, 12, 18, **24**, 30,...
- the multiples of 8 are: 8, 16, **24**, 32, 40,...

Thus, the LCM of 6 and 8 is 24. What is the LCM of the numbers 4 and 12? What is the LCM of the numbers 7 and 3?

**Step 5** Use coefficients to write each half-reaction so that it includes the LCM of the numbers of electrons.

**Step 6** Add the balanced half-reactions that include the equal numbers of electrons.

**Step 7** Remove the electrons from both sides of the equation.

**Step 8** Remove any identical molecules or ions that are present on both sides of the equation.

**Step 9** If you require a balanced chemical equation, include any spectator ions in the chemical formulas.

**Step 10** If necessary, include the states.

When using the half-reaction method, keep in mind that, in a redox reaction, *the number of electrons lost through oxidation must equal the number of electrons gained through reduction.* Figure 10.11 provides another example.

**Figure 18.11** Lithium displaces hydrogen from water to form lithium hydroxide.

**Oxidation half-reaction:**

$Li \rightarrow Li^+ + e^-$

**Reduction half-reaction:**

$2H_2O + 2e^- \rightarrow H_2 + 2OH^-$

Multiply the oxidation half-reaction by 2, add the half-reactions, and simplify the result to obtain the balanced net ionic equation.

$2Li + 2H_2O \rightarrow 2Li^+ + 2OH^- + H_2$

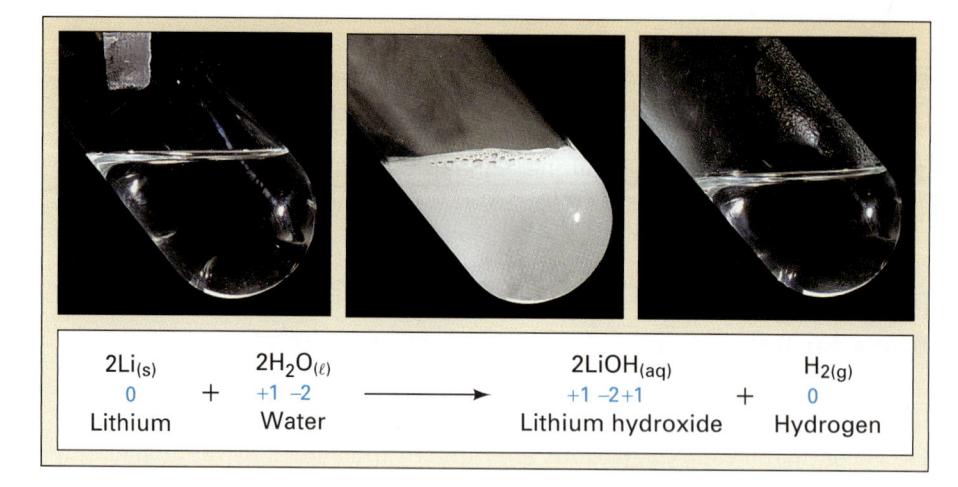

$$2Li_{(s)} + 2H_2O_{(\ell)} \rightarrow 2LiOH_{(aq)} + H_{2(g)}$$
$$0 \quad\quad +1\ -2 \quad\quad\quad +1\ -2\ +1 \quad\quad 0$$
Lithium + Water → Lithium hydroxide + Hydrogen

## Balancing Redox Reactions in Acidic and Basic Solutions

The half-reaction method of balancing equations can be more complicated for reactions that take place under acidic or basic conditions. The overall approach, however, is the same. You need to balance the two half-reactions, find the LCM of the numbers of electrons, and then multiply by coefficients to equate the number of electrons lost and gained. Finally, add the half-reactions and simplify to give a balanced net ionic equation for the reaction. The ten steps listed above show this process in more detail.

The Sample Problem on the next page illustrates the use of these steps for an acidic solution. To balance a net ionic equation for basic conditions by the half-reaction method, balance each half-reaction for acidic conditions, adjust for basic conditions, and then combine the half-reactions to obtain the balanced net ionic equation. The following Concept Organizer summarizes how to use the half-reaction method in both acidic and basic conditions.

**CONCEPT CHECK**

Explain why a balanced chemical equation or net ionic equation for a redox reaction does not include any electrons.

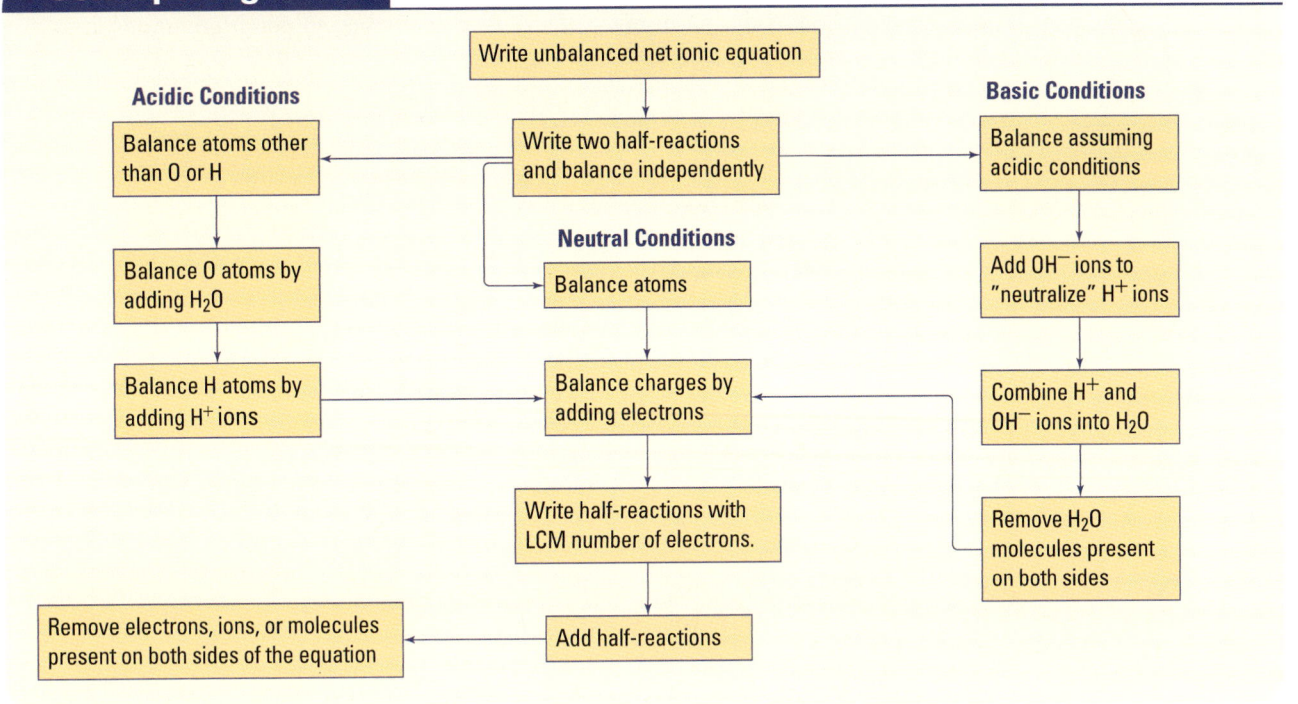

# Concept Organizer: The Half-Reaction Method for Balancing Redox Equations

**Acidic Conditions**

- Balance atoms other than O or H
- Balance O atoms by adding H₂O
- Balance H atoms by adding H⁺ ions

Write unbalanced net ionic equation → Write two half-reactions and balance independently

**Neutral Conditions**
- Balance atoms
- Balance charges by adding electrons
- Write half-reactions with LCM number of electrons.
- Add half-reactions
- Remove electrons, ions, or molecules present on both sides of the equation

**Basic Conditions**
- Balance assuming acidic conditions
- Add OH⁻ ions to "neutralize" H⁺ ions
- Combine H⁺ and OH⁻ ions into H₂O
- Remove H₂O molecules present on both sides

## Sample Problem

### Balancing a Redox Equation in Acidic Solution

**Problem**

Write a balanced net ionic equation to show the reaction of perchlorate ions, $ClO_4^-$, and nitrogen dioxide in acidic solution to produce chloride ions and nitrate ions.

**What Is Required?**

You need to write a balanced net ionic equation for the given reaction.

**What Is Given?**

You know the identities of two reactants and two products, and that the reaction takes place in acidic solution.

**Plan Your Strategy**

- Write an unbalanced ionic equation.
- Determine whether the reaction is a redox reaction.
- If it is not a redox reaction, balance by inspection.
- If it is a redox reaction, follow the steps for balancing by the half-reaction method.

**Act on Your Strategy**

- The unbalanced ionic equation is: $ClO_4^- + NO_2 \rightarrow Cl^- + NO_3^-$
- Assign oxidation numbers to all the elements to determine which reactant, if any, is oxidized or reduced.

$ClO_4^- + NO_2 \rightarrow Cl^- + NO_3^-$
  +7 −2    +4 −2    −1    +5 −2

*Continued ...*

Chapter 18 Oxidation-Reduction Reactions • MHR  **737**

The oxidation number of chlorine decreases, so perchlorate ions are reduced to chloride ions.

The oxidation number of nitrogen increases, so nitrogen dioxide is oxidized to nitrate ions.

- This is a redox reaction. Use the half-reaction method to balance the equation.

**Step 1** The unbalanced net ionic equation is already written.
$ClO_4^- + NO_2 \rightarrow Cl^- + NO_3^-$

**Step 2** Write two unbalanced half-reactions.
Oxidation: $NO_2 \rightarrow NO_3^-$
Reduction: $ClO_4^- \rightarrow Cl^-$

**Step 3** Balance the two half-reactions for acidic conditions.

**Oxidation**
$NO_2 \rightarrow NO_3^-$
$NO_2 + H_2O \rightarrow NO_3^-$
$NO_2 + H_2O \rightarrow NO_3^- + 2H^+$
$NO_2 + H_2O \rightarrow NO_3^- + 2H^+ + e^-$

**Reduction**
$ClO_4^- \rightarrow Cl^-$
$ClO_4^- \rightarrow Cl^- + 4H_2O$
$ClO_4^- + 8H^+ \rightarrow Cl^- + 4H_2O$
$ClO_4^- + 8H^+ + 8e^- \rightarrow Cl^- + 4H_2O$

**Step 4** The LCM of 1 and 8 is 8.

**Step 5** Multiply the oxidation half-reaction by 8, so that equal numbers of electrons are lost and gained.
$8NO_2 + 8H_2O \rightarrow 8NO_3^- + 16H^+ + 8e^-$

**Step 6** Add the half reactions.
$8NO_2 + 8H_2O \rightarrow 8NO_3^- + 16H^+ + 8e^-$
$\underline{ClO_4^- + 8H^+ + 8e^- \rightarrow Cl^- + 4H_2O}$
$8NO_2 + 8H_2O + ClO_4^- + 8H^+ + 8e^- \rightarrow 8NO_3^- + 16H^+ + 8e^- + Cl^- + 4H_2O$

**Step 7** Simplify by removing 8 electrons from both sides.
$8NO_2 + 8H_2O + ClO_4^- + 8H^+ \rightarrow 8NO_3^- + 16H^+ + Cl^- + 4H_2O$

**Step 8** Simplify by removing 4 water molecules, and 8 hydrogen ions from each side.
$8NO_2 + ClO_4^- + 4H_2O \rightarrow 8NO_3^- + 8H^+ + Cl^-$

(Steps 9 and 10 are not required for this problem.)

### Check Your Solution

- The atoms are balanced.
- The charges are balanced.

## Practice Problems

**25.** Balance each of the following redox equations by inspection. Write the balanced half-reactions in each case.

(a) $Na + F_2 \rightarrow NaF$

(b) $Mg + N_2 \rightarrow Mg_3N_2$

(c) $HgO \rightarrow Hg + O_2$

**26.** Balance the following equation by the half-reaction method.
$Cu^{2+} + I^- \rightarrow CuI + I_3^-$

**27.** Balance each of the following ionic equations for acidic conditions. Identify the oxidizing agent and the reducing agent in each case.

(a) $MnO_4^- + Ag \rightarrow Mn^{2+} + Ag^+$

(b) $Hg + NO_3^- + Cl^- \rightarrow HgCl_4^{2-} + NO_2$

(c) $AsH_3 + Zn^{2+} \rightarrow H_3AsO_4 + Zn$

(d) $I_3^- \rightarrow I^- + IO_3^-$

**28.** Balance each of the following ionic equations for basic conditions. Identify the oxidizing agent and the reducing agent in each case.

(a) $CN^- + MnO_4^- \rightarrow CNO^- + MnO_2$

(b) $H_2O_2 + ClO_2 \rightarrow ClO_2^- + O_2$

(c) $ClO^- + CrO_2^- \rightarrow CrO_4^{2-} + Cl_2$

(d) $Al + NO_2^- \rightarrow NH_3 + AlO_2^-$

In the next investigation, you will carry out several redox reactions, including reactions of acids with metals, and the combustion of hydrocarbons.

# Tools & Techniques

## The Breathalyzer Test: A Redox Reaction

The police may pull over a driver weaving erratically on the highway on suspicion of drunk driving. A police officer must confirm this suspicion by assessing whether the driver has a blood alcohol concentration over the "legal limit." The "Breathalyzer" test checks a person's breath using a redox reaction to determine blood alcohol concentration. This test was invented in 1953 by Robert Borkenstein, a former member of the Indiana State Police, and a professor of forensic studies.

What does a person's breath have to do with the alcohol in his or her blood? In fact, there is a direct correlation between the concentration of alcohol in an exhaled breath and the concentration of alcohol in the blood.

As blood moves through the lungs, it comes in close contact with inhaled gases. If the blood contains alcohol, the concentration of alcohol in the blood quickly reaches equilibrium with the concentration of alcohol in each inhaled breath. Thus, the alcohol content in an exhaled breath is a measure of the alcohol concentration in the blood itself. For example, if a person has been drinking alcohol, every 2100 mL of air exhaled contains about the same amount of alcohol as 1 mL of blood.

In the Breathalyzer test, the subject blows into a tube connected to a vial. The exhaled air collects in the vial, which already contains a mixture of sulfuric acid, potassium dichromate, water, and the catalyst silver nitrate. The alcohol reacts with the dichromate ion in the following redox reaction.

$16H^+_{(aq)} + 2Cr_2O_7^{2-}_{(aq)} + 3C_2H_5OH_{(\ell)} \rightarrow$
(orange)    $4Cr^{3+}_{(aq)} + 3C_2H_4O_{2(aq)} + 11H_2O_{(\ell)}$
(green)

This reaction is accompanied by a visible colour change, as orange dichromate ions become green chromium(III) ions. The concentration of alcohol in the blood is determined by measuring the intensity of the final colour.

A recent modification of the Breathalyzer test prevents drivers from starting their cars if they have been drinking. Alcohol ignition locks involve a type of Breathalyzer test that is linked to the car's ignition system. Until the driver passes the test, the car will not start. This test is useful in regulating the driving habits of people who have been previously convicted of drinking and driving.

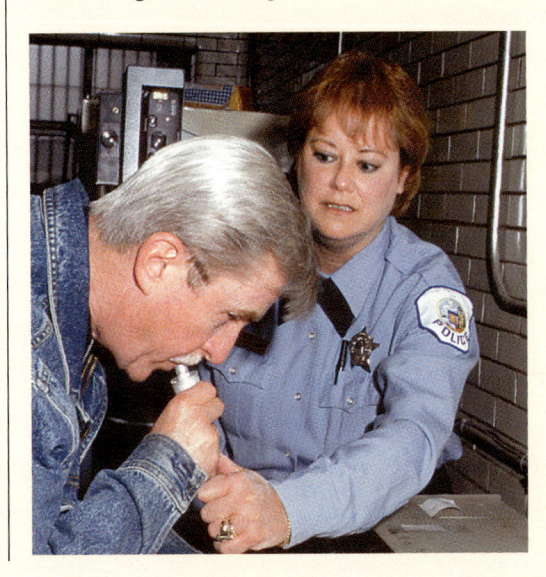

# Investigation 18-B

**SKILL FOCUS**
Predicting
Performing and recording
Analyzing and interpreting

## Redox Reactions and Balanced Equations

A redox reaction involves the transfer of electrons between reactants. A reactant that loses electrons is oxidized and acts as a reducing agent. A reactant that gains electrons is reduced and acts as an oxidizing agent. Redox reactions can be represented by balanced equations.

### Questions

How can you tell if a redox reaction occurs when reactants are mixed? Can you observe the transfer of electrons in the mixture?

### Predictions

- Predict which of the metals magnesium, zinc, copper, and aluminum can be oxidized by aqueous hydrogen ions. Explain your reasoning.

- Predict whether metals that cannot be oxidized by hydrogen ions can dissolve in acids. Explain your reasoning.

- Predict whether the combustion of a hydrocarbon is a redox reaction. What assumptions have you made about the products?

### Materials

well plate
4 small test tubes
test tube rack
small pieces of each of the metals magnesium, zinc, copper, and aluminum
dilute hydrochloric acid (1 mol/L)
dilute sulfuric acid (1 mol/L)
Bunsen burner
candle

### Safety Precautions

- The acid solutions are corrosive. Handle them with care.

- If you accidentally spill a solution on your skin, wash the area immediately with copious amounts of cool water. If you get any acid in your eyes, wash at the eye wash station. Inform your teacher.

- Before lighting a Bunsen burner or candle, make sure that there are no flammable liquids nearby. Also, tie back long hair, and confine any loose clothing.

### Procedure

#### Part 1 Reactions of Acids

1. Place a small piece of each metal on the well plate. Add a few drops of hydrochloric acid to each metal. Record your observations. If you are unsure of your observations, repeat the procedure on a larger scale in a small test tube.

2. Place another small piece of each metal on clean sections of the well plate. Add a few drops of sulfuric acid to each metal. Record your observations. If you are unsure of your observations, repeat the procedure on a larger scale in a small test tube.

3. Dispose of the mixtures in the beaker supplied by your teacher.

#### Part 2 Combustion of Hydrocarbons

4. Observe the combustion of natural gas in a Bunsen burner. Adjust the colour of the flame by varying the quantity of oxygen admitted to the burner. How does the colour depend on the quantity of oxygen?

5. Observe the combustion of a candle. Compare the colour of the flame with the colour of the Bunsen burner flame. Which adjustment of the burner makes the colours of the two flames most similar?

## Analysis

### Part 1 Reactions of Acids

1. Write a balanced chemical equation for each of the reactions of an acid with a metal.
2. Write each equation from question 1 in net ionic form.
3. Determine which of the reactions from question 1 are redox reactions.
4. Write each redox reaction from question 3 as two half-reactions.
5. Explain any similarities in your answers to question 4.
6. In the reactions you observed, are the hydrogen ions acting as an oxidizing agent, a reducing agent, or neither?
7. In the neutralization reaction of hydrochloric acid and sodium hydroxide, do the hydrogen ions behave in the same way as you found in question 6? Explain.
8. Your teacher may demonstrate the reaction of copper with concentrated nitric acid to produce copper(II) ions and brown, toxic nitrogen dioxide gas. Write a balanced net ionic equation for this reaction. Do the hydrogen ions behave in the same way as you found in question 6? Identify the oxidizing agent and the reducing agent in this reaction.
9. From your observations of copper with hydrochloric acid and nitric acid, can you tell whether hydrogen ions or nitrate ions are the better oxidizing agent? Explain.

### Part 2 Combustion of Hydrocarbons

10. The main component of natural gas is methane, $CH_4$. The products of the combustion of this gas in a Bunsen burner depend on how the burner is adjusted. A blue flame indicates complete combustion. What are the products in this case? Write a balanced chemical equation for this reaction.
11. A yellow or orange flame from a Bunsen burner indicates incomplete combustion and the presence of carbon in the flame. Write a balanced chemical equation for this reaction.
12. Name another possible carbon-containing product from the incomplete combustion of methane. Write a balanced chemical equation for this reaction.
13. The fuel in a burning candle is paraffin wax, $C_{25}H_{52}$. Write a balanced chemical equation for the complete combustion of paraffin wax.
14. Write two balanced equations that represent the incomplete combustion of paraffin wax.
15. How do you know that at least one of the incomplete combustion reactions is taking place when a candle burns?
16. Are combustion reactions also redox reactions? Does your answer depend on whether the combustion is complete or incomplete? Explain.

## Conclusion

17. How could you tell if a redox reaction occurred when reactants were mixed? Could you observe the transfer of electrons in the mixture?

## Applications

18. Gold is very unreactive and does not dissolve in most acids. However, it does dissolve in *aqua regia* (Latin for "royal water"), which is a mixture of concentrated hydrochloric and nitric acids. The unbalanced ionic equation for the reaction is as follows.

    $$Au + NO_3^- + Cl^- \rightarrow AuCl_4^- + NO_2$$

    Balance the equation, and identify the oxidizing agent and reducing agent.
19. Natural gas is burned in gas furnaces. Give at least three reasons why this combustion reaction should be as complete as possible. How would you try to ensure complete combustion?

## Stoichiometry and Redox Titrations

Redox titrations are an important application of redox chemistry and stoichiometry. In an acid-base titration, a base is used to find the concentration of an acid, or vice versa. Similarly, in a redox titration, a known concentration of an oxidizing agent can be used to find the unknown concentration of a reducing agent, or vice versa. Redox titrations are used in a wide range of situations, including measuring the iron content in drinking water and the vitamin C content in foods or vitamin supplements.

The permanganate ion, $MnO_4^-$, is commonly used as an oxidizing agent in redox titrations. It is particularly useful because it has a strong purple colour, meaning that no additional indicator is needed to determine the endpoint of the titration. Figure 18.X shows the titration of a solution of sodium oxalate, $Na_2C_2O_{4(aq)}$, with a solution of potassium permanganate, $KMnO_{4(aq)}$.

As long as there are oxalate ions present in solution, they reduce the manganese with oxidation number +7 present in purple permanganate ions to nearly colourless manganese(II) ions, $Mn^{2+}$, where the oxidation number is +2). Once all oxalate ions have been oxidized, the next drop of potassium permanganate solution turns the solution faint purple. This purple colour signals the endpoint of the titration.

**Figure 18.12** The photo on the right shows the endpoint of the titration of a solution containing oxalate ions with a solution containing permanganate ions. In the balanced equation, the ratio of $MnO_4^-$ to $C_2O_4^{2-}$ is 2:5. This ratio means that at the endpoint, 2 mol of $MnO_4^-$ have been added for every 5 mol of $C_2O_4^{2-}$ that were present initially.

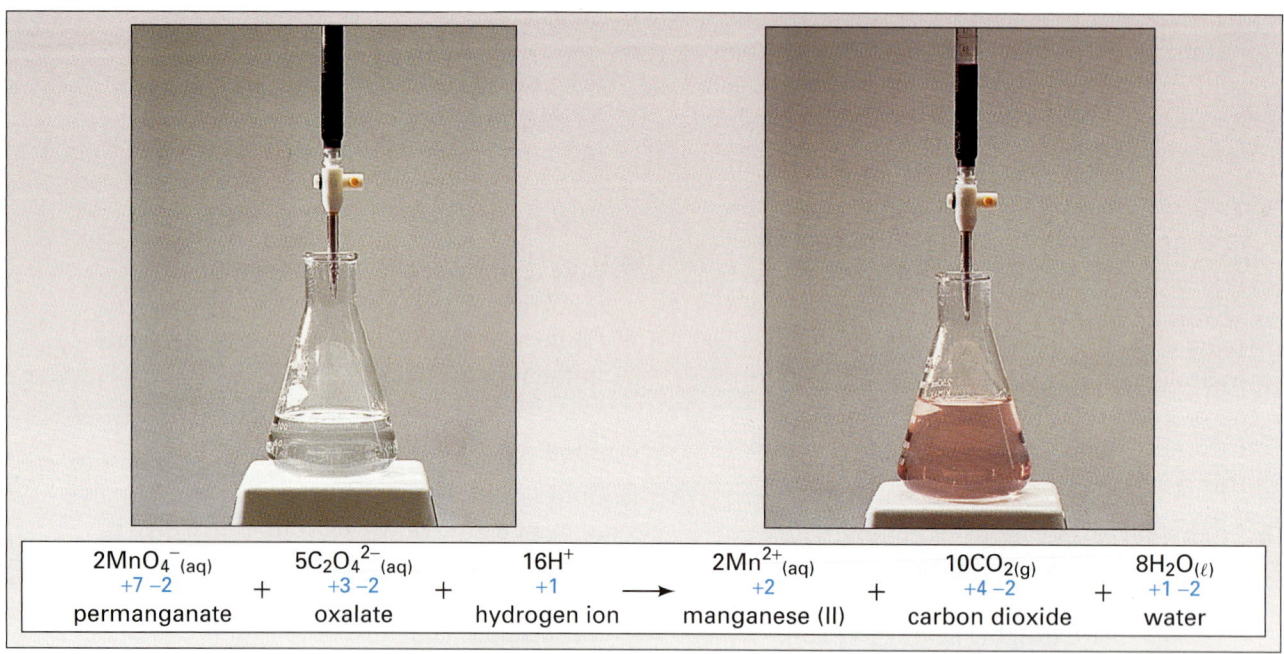

| $2MnO_4^-{}_{(aq)}$ +7 −2 permanganate | + | $5C_2O_4^{2-}{}_{(aq)}$ +3 −2 oxalate | + | $16H^+$ +1 hydrogen ion | → | $2Mn^{2+}{}_{(aq)}$ +2 manganese (II) | + | $10CO_{2(g)}$ +4 −2 carbon dioxide | + | $8H_2O_{(\ell)}$ +1 −2 water |

The permanganate ion may also be used to oxidize hydrogen peroxide, $H_2O_2$. The following equation shows the redox reaction in acidic conditions.

$$5H_2O_{2(aq)} + 2MnO_4^-{}_{(aq)} + 6H^+{}_{(aq)} \rightarrow 5O_{2(g)} + 2Mn^{2+}{}_{(aq)} + 8H_2O_{(\ell)}$$

Aqueous solutions of hydrogen peroxide sold in pharmacies are often about 3% $H_2O_2$ by mass. In solution, however, hydrogen peroxide decomposes steadily to form water and oxygen.

Suppose that you need to use hydrogen peroxide solution with a concentration of at least 2.5% by mass for a certain experiment. Your 3% $H_2O_2$ is not fresh, so it may have decomposed significantly. How do you find out if your $H_2O_2$ solution is concentrated enough? The following Sample Problem shows how to solve this problem using data from the titration of the hydrogen peroxide solution with a solution of potassium permanganate.

## Sample Problem
## Redox Titrations

### Problem
You are using 0.01143 mol/L $KMnO_{4(aq)}$ to determine the percentage by mass of an aqueous solution of $H_2O_2$. You know that the solution is about 3% $H_2O_2$ by mass.

You prepare the sample by adding 1.423 g of the hydrogen peroxide solution to an Erlenmeyer flask. You add about 75 mL of water to dilute the solution. You also add some dilute sulfuric acid to acidify the solution.

You reach the purple-coloured endpoint of the titration when you have added 40.22 mL of the $KMnO_{4(aq)}$ solution.

### What is Required?
You need to determine the mass of $H_2O_2$ in the sample. You need to express your result as a mass percent.

### What is Given?
Concentration of $KMnO_{4(aq)}$ = 0.01143 mol/L
Volume of $KMnO_{4(aq)}$ = 40.22 mL
Mass of 3% $H_2O_2$ solution = 1.423 g

### Plan Your Strategy
**Step 1** Write the balanced chemical equation for the reaction.

**Step 2** Calculate the amount (in mol) of permanganate ion added, based on the volume and concentration of the potassium permanganate solution.

**Step 3** Determine the amount (in mol) of hydrogen peroxide needed to reduce the permanganate ions.

**Step 4** Determine the mass of hydrogen peroxide, based on the molar mass of hydrogen peroxide. Finally, express your answer as a mass percent of the hydrogen peroxide solution, as the question directs.

### Act on Your Strategy
**Step 1** The redox equation was provided on the previous page. It is already balanced.

$$5H_2O_2 + 2MnO_4^- + 6H^+ \rightarrow 5O_2 + 2Mn^{2+} + 8H_2O$$

**Step 2** The concentration of $MnO_4^-{}_{(aq)}$ is the same as the concentration of $KMnO4_{(aq)}$.

$n = C \times V$
$= 0.01143 \text{ mol/L} \times 0.04022 \text{ L}$
$= 4.597 \times 10^{-4}$ mol

**Step 3** Permanganate ion reacts with hydrogen peroxide in a 2:5 ratio.

Amount (in mol)$H_2O_2 = \frac{5 \text{ mol } H_2O_2}{2 \text{ mol } MnO_4^-} \times 4.597 \times 10^{-4} \text{ mol } MnO_4^-$
$= 1.149 \times 10^{-3}$ mol $H_2O_2$

**Step 4**
$M_{H_2O_2} = 2(1.01 \text{ g/mol}) + 2(16.00 \text{ g/mol})$
$= 34.02$ g/mol

Mass (in g) $H_2O_2 = 1.149 \times 10^{-3}$ mol $\times 34.02$ g/mol
$= 0.03909$ g

*Continued...*

Mass percent H₂O₂ in solution = 0.03909 g/1.423 g
= 2.747%

**Check Your Solution**

The units are correct. The value for the mass of pure H₂O₂ that you obtained is less than the mass of the H₂O₂₍aq₎ sample solution, as you would expect. The mass percent you obtained for the solution is close to the expected value. It makes sense that the value is somewhat less than 3%, since H₂O₂ decomposes in solution, forming water and oxygen.

## Practice Problems

**29.** An analyst uses 0.02045 mol/L KMnO₄ to titrate a sample solution of H₂O₂. The analyst knows that the sample solution is about 6% H₂O₂ by mass. The analyst places 1.284 g of H₂O₂ solution in a flask, dilutes it with water, and adds a small amount of sulfuric acid to acidify it. It takes 38.95 mL of KMnO₄ solution to reach the endpoint. What mass of pure H₂O₂ was present? What is the mass percent of pure H₂O₂ in the original sample solution?

**30.** A forensic chemist wants to determine the level of alcohol in a sample of blood plasma. The chemist titrates the plasma with a solution of potassium dichromate. The balanced equation is:

$$16H^+ + 2Cr_2O_7^{2-} + C_2H_5OH \rightarrow 4Cr^{3+} + 2CO_2 + 11H_2O$$

If 32.35 mL of 0.05023 mol/L $Cr_2O_7^{2-}$ is required to titrate 27.00 g plasma, what is the mass percent of alcohol in the plasma?

**31.** An analyst titrates an acidified solution containing 0.153 g of purified sodium oxalate, $Na_2C_2O_4$, with a potassium permanganate solution, $KMnO_{4(aq)}$. The purple endpoint is reached when the chemist has added 41.45 mL of potassium permanganate solution. What is the molar concentration of the potassium permanganate solution? The balanced equation is:

$$2MnO_4^- + 5H_2C_2O_4 + 6H^+ \rightarrow 2Mn^{2+} + 10CO_2 + 8H_2O$$

**32.** 25.00 mL of a solution containing iron(II) ions was titrated with a 0.02043 mol/L potassium dichromate solution. The endpoint was reached when 35.55 mL of potassium dichromate solution had been added. What was the molar concentration of iron(II) ions in the original, acidic solution? The unbalanced equation is:

$$Cr_2O_7^{2-} + Fe^{2+} \rightarrow Cr^{3+} + Fe^{3+}$$

**PROBLEM TIP**

In question 32, you will need to balance the redox equation before solving the problem.

## Section Summary

In this section, you learned the half-reaction method for balancing equations for redox reactions. You investigated the redox reactions of metals with acids, and the combustion of two hydrocarbons. You also learned how known concentrations of oxidizing agents can be used to find the unknown concentrations of reducing agents, or vice versa, in redox titrations.

## Section Review

**1** Balance each half-reaction. Identify it as an oxidation or reduction half-reaction.

(a) $C \rightarrow C_2^{2-}$

(b) $S_2O_3^{2-} \rightarrow S_4O_6^{2-}$

(c) $AsO_4^{3-} \rightarrow As_4O_6$ (acidic conditions)

(d) $Br_2 \rightarrow BrO_3^-$ (basic conditions)

**2** Balance each equation.

(a) $Co^{3+} + Au \rightarrow Co^{2+} + Au^{3+}$

(b) $Cu + NO_3^- \rightarrow Cu^{2+} + NO$ (acidic conditions)

(c) $NO_3^- + Al \rightarrow NH_3 + AlO_2^-$ (basic conditions)

**3** This section began with a description of the use of a redox reaction to make the master disc in the production of CDs. Write a balanced net ionic equation for the reaction of silver ions with methanal under acidic conditions to form metallic silver and methanoic acid.

**4** In basic solution, ammonia, $NH_3$, can be oxidized to dinitrogen monoxide, $N_2O$.

(a) Try to balance the half-reaction by adding water molecules, hydroxide ions, and electrons, without first assuming acidic conditions. Describe any difficulties you encounter.

(b) Balance the half-reaction by first assuming acidic conditions and then adjusting to introduce the hydroxide ions. Compare your findings with those from part (a).

**5** A mixture of liquid hydrazine, $N_2H_4$, and liquid dinitrogen tetroxide can be used as a rocket fuel. The products of the reaction are nitrogen gas and water vapour.

(a) Write a balanced chemical equation for the reaction by inspection.

(b) Identify the oxidizing agent and the reducing agent.

(c) Hydrazine is made in the Raschig Process. In this process, ammonia reacts with hypochlorite ions in a basic solution to form hydrazine and chloride ions. Write the balanced net ionic equation.

**6** Ben and Larissa were working together to balance the following equation for a redox reaction.

$$Zn + SO_4^{2-} + H^+ \rightarrow Zn^{2+} + S + H_2O$$

Ben suggested balancing by inspection, with the following result.

$$Zn + SO_4^{2-} + 8H^+ \rightarrow Zn^{2+} + S + 4H_2O$$

Larissa said: "That's not balanced."

(a) Was Larissa right? Explain.

(b) How would you balance the equation, and what would the result be?

**7** Oxygen and other oxidizing agents can react harmfully with your body. Vitamin C acts as an antioxidant. It is very easily oxidized, so it reacts with oxidizing agents, preventing them from reacting with other important molecules in the body. Vitamin C is relatively stable when oxidized. Therefore, it does not propagate a harmful series of oxidation reactions. One way to determine the vitamin C content of a sample is to titrate it with an iodine/iodine solution. The diagram below shows the reaction involved.

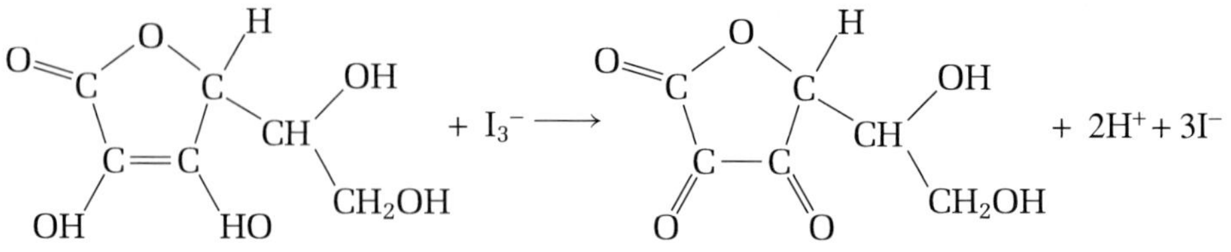

ascorbic acid (vitamin C) + $I_3^-$ ⟶ dehydroascorbic acid + $2H^+ + 3I^-$

In the iodine solution, iodine, $I_2$, and iodide, $I^-$, ions are in equilibrium with triiodide, as shown below.

$$I_2 + I^- \rightleftharpoons I_3^-$$

Molecular iodine is a deep violet-red colour. Iodine ions are colourless. Thus, when an antioxidant reduces the iodine molecules in a solution, the iodine colour disappears completely.

**(a)** In storage, the concentration of iodine in solution decreases fairly quickly over time. Why do you think this happens?

**(b)** Because the iodine solution's concentration is not stable, it should be standardized frequently. To standardize an iodine solution, use it to titrate a solution that contains a known quantity of vitamin C. Explain how you would standardize a solution of iodine using vitamin C tablets from a pharmacy.

**(c)** To standardize iodine using vitamin C tablets, you should use fresh tablets. Explain why.

**8** If you titrate orange juice that has been exposed to the air for a week, will the vitamin C concentration be different from the vitamin C concentration in fresh juice? If so, will it decrease or increase? Explain your prediction, in terms of redox reactions.

**9** A chemist adds a few drops of deep violet-red iodine to a vitamin C tablet. The iodine solution quickly becomes colourless. Then the chemist adds a solution that contains chlorine, $Cl_2$. The chemist observes that the violet-red colour of the iodine reappears. Explain the chemist's observations, in terms of redox reactions.

# The Oxidation Number Method for Balancing Equations

## 18.4

**Section Preview/Outcomes**

In this section, you will

- **write** balanced equations for redox reactions using the oxidation number method

In section 18.2, you learned that a redox reaction involves changes in oxidation numbers. If an element undergoes oxidation, its oxidation number increases. If an element undergoes reduction, its oxidation number decreases. When balancing equations by the half-reaction method in section 18.3, you sometimes used oxidation numbers to determine the reactant(s) and product(s) in each half-reaction.

In fact, you can use oxidation numbers to balance a chemical equation by a new method. The oxidation number method is a method of balancing redox equations by ensuring that *the total increase in the oxidation numbers of the oxidized element(s) equals the total decrease in the oxidation numbers of the reduced element(s)*.

For example, the combustion of ammonia in oxygen produces nitrogen dioxide and water.

$$NH_3 + O_2 \rightarrow NO_2 + H_2O$$
$$\phantom{NH_3}_{-3\ +1} \phantom{+}\ _{0} \phantom{\rightarrow}\ _{+4\ -2} \phantom{+}\ _{+1\ -2}$$

The oxidation number of nitrogen increases from −3 to +4, an increase of 7. The oxidation number of oxygen decreases from 0 to −2, a decrease of 2. The least common multiple of 7 and 2 is 14. In this case, two nitrogen atoms must react for every seven oxygen atoms so that the total increase and decrease in oxidation numbers both equal 14.

$$2NH_3 + \frac{7}{2}O_2 \rightarrow NO_2 + H_2O$$

Complete the equation by inspection. If necessary, eliminate the fraction.

$$2NH_3 + \frac{7}{2}O_2 \rightarrow 2NO_2 + 3H_2O$$

$$4NH_3 + 7O_2 \rightarrow 4NO_2 + 6H_2O$$

A summary of the steps of the oxidation number method is given below. The following Sample Problem shows how these steps are applied.

---

**Step 1** Write an unbalanced equation, if it is not given.

**Step 2** Determine whether the reaction is a redox reaction by assigning an oxidation number to each element wherever it appears in the equation.

**Step 3** If the reaction is a redox reaction, identify the element(s) that undergo an increase in oxidation number and the element(s) that undergo a decrease in oxidation number.

**Step 4** Find the numerical values of the increase and the decrease in oxidation numbers.

**Step 5** Determine the smallest whole-number ratio of the oxidized and reduced elements so that the total increase in oxidation numbers equals the total decrease in oxidation numbers.

**Step 6** Use the smallest whole-number ratio to balance the numbers of atoms of the element(s) oxidized and the element(s) reduced.

*continued on the next page*

**Step 7** Balance the other elements by inspection, if possible.

**Step 8** For reactions that occur in acidic or basic solutions, include water molecules, hydrogen ions, or hydroxide ions as needed to balance the equation.

## Sample Problem

### Balancing a Redox Equation in Basic Solution

**Problem**

Write a balanced net ionic equation to show the formation of iodine by bubbling oxygen gas through a basic solution that contains iodide ions.

**Solution**

**Step 1** Write an unbalanced equation from the given information.
$$O_2 + I^- \rightarrow I_2$$

**Step 2** Assign oxidation numbers to see if it is a redox reaction.
$$O_2 + I^- \rightarrow I_2$$
$$\phantom{O_2 +\ } 0 \phantom{\ +\ } -1 \phantom{\rightarrow\ } 0$$

Because iodide is oxidized to iodine, the reaction is a redox reaction. Though the product that contains oxygen is unknown at this stage, oxygen must be reduced.

**Step 3** Iodine is the element that undergoes an increase in oxidation number. Oxygen is the element that undergoes a decrease in oxidation number.

**Step 4** Iodine undergoes an increase in its oxidation number from −1 to 0, an increase of 1. Assume that the oxidation number of oxygen after reduction is its normal value, that is, −2. Thus, oxygen undergoes a decrease in its oxidation number from 0 to −2, a decrease of 2.

**Step 5** A 2:1 ratio of iodine atoms to oxygen atoms ensures that the total increase in oxidation numbers and the total decrease in oxidation numbers are both equal to 2. This is the smallest whole-number ratio.

**Step 6** Use the ratio to balance the numbers of atoms of iodine and oxygen. Make sure there are two iodine atoms for every oxygen atom.
$$O_2 + 4I^- \rightarrow 2I_2$$

**Step 7** No other reactants or products can be balanced by inspection.

**Step 8** The reaction occurs in basic solution. As you learned in section 10.3, for basic conditions, start by assuming that the conditions are acidic. Add water molecules and hydrogen ions as necessary to balance the atoms.
$$O_2 + 4I^- \rightarrow 2I_2 + 2H_2O$$
$$O_2 + 4I^- + 4H^+ \rightarrow 2I_2 + 2H_2O$$

Add hydroxide ions to adjust for basic conditions. Simplify the resulting equation.
$$O_2 + 4I^- + 4H^+ + 4OH^- \rightarrow 2I_2 + 2H_2O + 4OH^-$$
$$O_2 + 4I^- + 4H_2O \rightarrow 2I_2 + 2H_2O + 4OH^-$$
$$O_2 + 4I^- + 2H_2O \rightarrow 2I_2 + 4OH^-$$

## CONCEPT CHECK

Explain why none of the steps in the oxidation number method result in equations that include electrons, $e^-$.

## Practice Problems

**33.** Use the oxidation number method to balance the following equation for the combustion of carbon disulfide.

$CS_2 + O_2 \rightarrow CO_2 + SO_2$

**34.** Use the oxidation number method to balance the following equations.

(a) $B_2O_3 + Mg \rightarrow MgO + Mg_3B_2$

(b) $H_2S + H_2O_2 \rightarrow S_8 + H_2O$

**35.** Use the oxidation number method to balance each ionic equation in acidic solution.

(a) $Cr_2O_7^{2-} + Fe^{2+} \rightarrow Cr^{3+} + Fe^{3+}$

(b) $I_2 + NO_3^- \rightarrow IO_3^- + NO_2$

(c) $PbSO_4 \rightarrow Pb + PbO_2 + SO_4^{2-}$

**36.** Use the oxidation number method to balance each ionic equation in basic solution.

(a) $Cl^- + CrO_4^{2-} \rightarrow ClO^- + CrO_2^-$

(b) $Ni + MnO_4^- \rightarrow NiO + MnO_2$

(c) $I^- + Ce^{4+} \rightarrow IO_3^- + Ce^{3+}$

## Concept Organizer: The Oxidation Number Method for Balancing Redox Equations

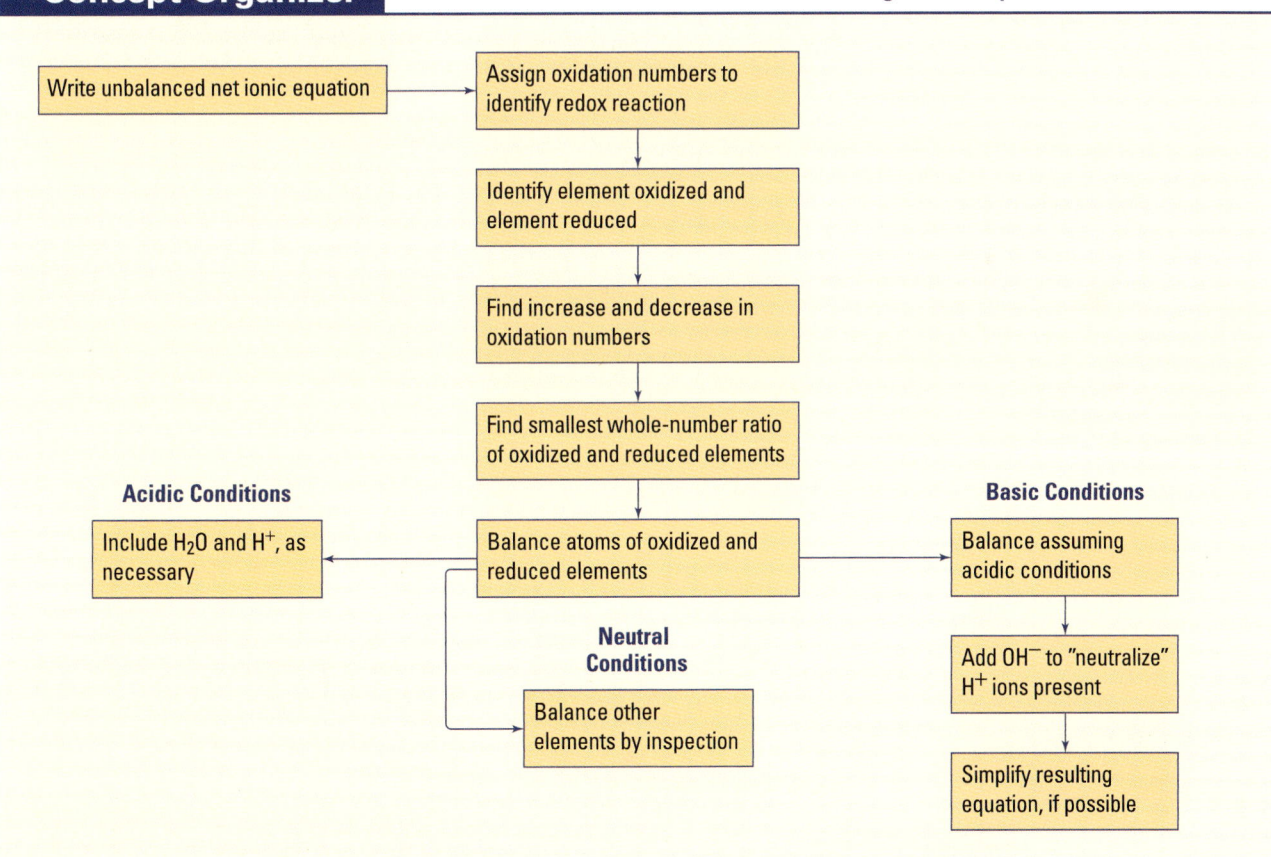

- Write unbalanced net ionic equation
- Assign oxidation numbers to identify redox reaction
- Identify element oxidized and element reduced
- Find increase and decrease in oxidation numbers
- Find smallest whole-number ratio of oxidized and reduced elements
- Balance atoms of oxidized and reduced elements

**Acidic Conditions**
- Include $H_2O$ and $H^+$, as necessary

**Neutral Conditions**
- Balance other elements by inspection

**Basic Conditions**
- Balance assuming acidic conditions
- Add $OH^-$ to "neutralize" $H^+$ ions present
- Simplify resulting equation, if possible

Chapter 18 Oxidation-Reduction Reactions • MHR **749**

## Section Summary

In this section, you learned how to use the oxidation number method to balance redox equations. You now know various techniques for recognizing and representing redox reactions. In Chapter 19, you will use these techniques to examine specific applications of redox reactions in the business world and in your daily life.

## Section Review

**1** Is it possible to use the half-reaction method or the oxidation number method to balance the following equation? Explain your answer.

$Al_2S_3 + H_2O \rightarrow Al(OH)_3 + H_2S$

**2** Balance each equation by the method of your choice. Explain your choice of method in each case.

(a) $CH_3COOH + O_2 \rightarrow CO_2 + H_2O$

(b) $O_2 + H_2SO_3 \rightarrow HSO_4^-$ (acidic conditions)

**3** Use the oxidation number method to balance the following equations.

(a) $NH_3 + Cl_2 \rightarrow NH_4Cl + N_2$

(b) $Mn_3O_4 + Al \rightarrow Al_2O_3 + Mn$

**4** Explain why, in redox reactions, the total increase in the oxidation numbers of the oxidized elements must equal the total decrease in the oxidation numbers of the reduced elements.

**5** The combustion of ammonia in oxygen to form nitrogen dioxide and water vapour involves covalent molecules in the gas phase. The oxidation number method for balancing the equation was shown in an example in this section. Devise a half-reaction method for balancing the equation. Describe the assumptions you made in order to balance the equation. Also, describe why these assumptions did not affect the final result.

# CHAPTER 18 Review

## Reflecting on Chapter 18

Summarize this chapter in the format of your choice. Here are a few ideas to use as guidelines:
- Give two different definitions for the term oxidation.
- Give two different definitions for the term reduction.
- Define a half-reaction. Give an example of an oxidation half-reaction and a reduction half-reaction.
- Practise balancing equations using the half-reaction method.
- Write an example of a balanced chemical equation for a redox reaction. Assign oxidation numbers to each element in the equation, then explain how you know it is a redox reaction.

## Reviewing Key Terms

For each of the following terms, write a sentence that shows your understanding of its meaning.

ore
reduction
redox reaction
oxidizing agent
half-reaction
oxidation numbers
oxidation
oxidation-reduction reaction
reducing agent
disproportionation

## Knowledge/Understanding

1. For each reaction below, write a balanced chemical equation by inspection.
   (a) zinc metal with aqueous silver nitrate
   (b) aqueous cobalt(II) bromide with aluminum metal
   (c) metallic cadmium with aqueous tin(II) chloride

2. For each reaction in question 1, write the total ionic and net ionic equations.

3. For each reaction in question 1, identify the oxidizing agent and reducing agent.

4. For each reaction in question 1, write the two half-reactions.

5. When a metallic element reacts with a non-metallic element, which reactant is
   (a) oxidized?
   (b) reduced?
   (c) the oxidizing agent?
   (d) the reducing agent?

6. Use a Lewis structure to assign an oxidation number to each element in the following compounds.
   (a) $BaCl_2$
   (b) $CS_2$
   (c) $XeF_4$

7. Determine the oxidation number of each element present in the following substances.
   (a) $BaH_2$
   (b) $Al_4C_3$
   (c) $KCN$
   (d) $LiNO_2$
   (e) $(NH_4)_2C_2O_4$
   (f) $S_8$
   (g) $AsO_3^{3-}$
   (h) $VO_2^+$
   (i) $XeO_3F^-$
   (j) $S_4O_6^{2-}$

8. Identify a polyatomic ion in which chlorine has an oxidation number of +3.

9. Determine which of the following balanced chemical equations represent redox reactions. For each redox reaction, identify the oxidizing agent and the reducing agent.
   (a) $2C_6H_6 + 15O_2 \rightarrow 12CO_2 + 6H_2O$
   (b) $CaO + SO_2 \rightarrow CaSO_3$
   (c) $H_2 + I_2 \rightarrow 2HI$
   (d) $KMnO_4 + 5CuCl + 8HCl \rightarrow KCl + MnCl_2 + 5CuCl_2 + 4H_2O$

10. Determine which of the following balanced net ionic equations represent redox reactions. For each redox reaction, identify the reactant that undergoes oxidation and the reactant that undergoes reduction.
    (a) $2Ag^+_{(aq)} + Cu_{(s)} \rightarrow 2Ag_{(s)} + Cu^{2+}_{(aq)}$
    (b) $Pb^{2+}_{(aq)} + S^{2-}_{(aq)} \rightarrow PbS_{(s)}$
    (c) $2Mn^{2+} + 5BiO_3^- + 14H^+ \rightarrow 2MnO_4^- + 5Bi^{3+} + 7H_2O$

Answers to questions highlighted in red type are provided in Appendix A.

11. (a) Examples of molecules and ions composed only of vanadium and oxygen are listed below. In this list, identify molecules and ions in which the oxidation number of vanadium is the same.
    $V_2O_5$
    $V_2O_3$
    $VO_2$
    $VO$
    $VO_2^+$
    $VO^{2+}$
    $VO_3^-$
    $VO_4^{3-}$
    $V_3O_9^{3-}$
    (b) Is the following reaction a redox reaction?
    $2NH_4VO_3 \rightarrow V_2O_5 + 2NH_3 + H_2O$

12. The method used to manufacture nitric acid involves the following three steps.
    **Step 1** $4NH_{3(g)} + 5O_{2(g)} \rightarrow 4NO_{(g)} + 6H_2O_{(g)}$
    **Step 2** $2NO_{(g)} + O_{2(g)} \rightarrow 2NO_{2(g)}$
    **Step 3** $3NO_{2(g)} + H_2O_{(\ell)} \rightarrow 2HNO_{3(aq)} + NO_{(g)}$
    (a) Which of these steps are redox reactions?
    (b) Identify the oxidizing agent and the reducing agent in each redox reaction.

13. In a synthesis reaction involving elements A and B, the oxidation number of element A increases. What happens to the oxidation number of element B? How do you know?

14. Balance each of the following half-reactions.
    (a) The reduction of iodine to iodide ions
    (b) The oxidation of lead to lead(IV) ions
    (c) The reduction of tetrachlorogold(III) ions, $AuCl_4^-$, to chloride ions and metallic gold
    (d) $C_2H_5OH \rightarrow CH_3COOH$ (acidic conditions)
    (e) $S_8 \rightarrow H_2S$ (acidic conditions)
    (f) $AsO_2^- \rightarrow AsO_4^{3-}$ (basic conditions)

15. Use the half-reaction method to balance each of the following equations.
    (a) $MnO_2 + Cl^- \rightarrow Mn^{2+} + Cl_2$ (acidic conditions)
    (b) $NO + Sn \rightarrow NH_2OH + Sn^{2+}$ (acidic conditions)
    (c) $Cd^{2+} + V^{2+} \rightarrow Cd + VO_3^-$ (acidic conditions)
    (d) $Cr \rightarrow Cr(OH)_4^- + H_2$ (basic conditions)
    (e) $S_2O_3^{2-} + NiO_2 \rightarrow Ni(OH)_2 + SO_3^{2-}$ (basic conditions)
    (f) $Sn^{2+} + O_2 \rightarrow Sn^{4+}$ (basic conditions)

16. Use the oxidation number method to balance each of the following equations.
    (a) $SiCl_4 + Al \rightarrow Si + AlCl_3$
    (b) $PH_3 + O_2 \rightarrow P_4O_{10} + H_2O$
    (c) $I_2O_5 + CO \rightarrow I_2 + CO_2$
    (d) $SO_3^{2-} + O_2 \rightarrow SO_4^{2-}$

17. Complete and balance a net ionic equation for each of the following disproportionation reactions.
    (a) $NO_2 \rightarrow NO_2^- + NO_3^-$ (acidic conditions), which is one of the reactions involved in acid rain formation
    (b) $Cl_2 \rightarrow ClO^- + Cl^-$ (basic conditions), which is one of the reactions involved in the bleaching action of chlorine in basic solution

18. Balance each of the following net ionic equations. Then include the named spectator ions to write a balanced chemical equation. Include the states.
    (a) $Co^{3+} + Cd \rightarrow Co^{2+} + Cd^{2+}$ (spectator ions $NO_3^-$)
    (b) $Ag^+ + SO_2 \rightarrow Ag + SO_4^{2-}$ (acidic conditions; spectator ions $NO_3^-$)
    (c) $Al + CrO_4^{2-} \rightarrow Al(OH)_3 + Cr(OH)_3$ (basic conditions; spectator ions $Na^+$)

19. If possible, give an example for each.
    (a) a synthesis reaction that is a redox reaction
    (b) a synthesis reaction that is not a redox reaction
    (c) a decomposition reaction that is a redox reaction
    (d) a decomposition reaction that is not a redox reaction
    (e) a double displacement reaction that is a redox reaction
    (f) a double displacement reaction that is not a redox reaction

20. Give an example of a reaction in which sulfur behaves as
    (a) an oxidizing agent
    (b) a reducing agent

21. Write a balanced equation for a synthesis reaction in which elemental oxygen acts as a reducing agent.

22. Phosphorus, $P_{4(s)}$, reacts with hot water to form phosphine, $PH_{3(g)}$, and phosphoric acid.
    (a) Write a balanced chemical equation for this reaction.
    (b) Is the phosphorus oxidized or reduced? Explain your answer.

23. The thermite reaction, which is highly exothermic, can be used to weld metals. In the thermite reaction, aluminum reacts with iron(III) oxide to form iron and aluminum oxide. The temperature becomes so high that the iron is formed as a liquid.
    (a) Write a balanced chemical equation for the reaction.
    (b) Is the reaction a redox reaction? If so, identify the oxidizing agent and the reducing agent.

## Inquiry

24. Iodine reacts with concentrated nitric acid to form iodic acid, gaseous nitrogen dioxide, and water.
    (a) Write the balanced chemical equation.
    (b) Calculate the mass of iodine needed to produce 28.0 L of nitrogen dioxide at STP.

25. Describe a laboratory investigation you could perform to decide whether tin or nickel is the better reducing agent. Include in your description all the materials and equipment you would need, and the procedure you would follow.

26. The following table shows the average composition, by volume, of the air we inhale and exhale, as part of a biochemical process called respiration. (The values are rounded.)

| Gas | Inhaled Air (% by volume) | Exhaled Air (% by volume) |
|---|---|---|
| Oxygen | 21 | 16 |
| Carbon dioxide | 0.04 | 4 |
| Nitrogen and other gases | 79 | 80 |

How do the data indicate that at least one redox reaction is involved in respiration?

27. Highly toxic phosphine gas, $PH_3$, is used in industry to produce flame retardants. One way to make phosphine on a large scale is by heating elemental phosphorus with a strong base.
    (a) Balance the following net ionic equation for the reaction under basic conditions.
    $P_4 \rightarrow H_2PO_2^- + PH_3$
    (b) Show that the reaction in part (a) is a disproportionation reaction.
    (c) Calculate the mass of phosphine that can theoretically be made from 10.0 kg of phosphorus by this method.

## Communication

28. Explain why, in a redox reaction, the reducing agent undergoes oxidation.

29. Explain why you would not expect sulfide ions to act as an oxidizing agent.

30. Why can't the oxidation number of an element in a compound be greater than the number of valence electrons in one atom of that element?

31. Explain why the historical use of the word "reduction," that is, the production of a metal from its ore, is consistent with the modern definitions of reduction.

32. Organic chemists sometimes describe redox reactions in terms of the loss or gain of pairs of hydrogen atoms. Examples include the addition of hydrogen to ethene to form ethane, and the elimination of hydrogen from ethanol to form ethanal.
    (a) Write a balanced equation for each reaction.
    (b) Determine whether the organic reactant is oxidized or reduced in each reaction.
    (c) Write a definition of oxidation and a definition of reduction based on an organic reactant losing or gaining hydrogen.
    (d) Would your definitions be valid for the synthesis and decomposition of a metal hydride? Explain your answer.

## Making Connections

**33.** The compound NaAl(OH)$_2$CO$_3$ is a component of some common stomach acid remedies.
   **(a)** Determine the oxidation number of each element in the compound.
   **(b)** Predict the products of the reaction of the compound with stomach acid (hydrochloric acid), and write a balanced chemical equation for the reaction.
   **(c)** Were the oxidation numbers from part (a) useful in part (b)? Explain your answer.
   **(d)** What type of reaction is this?
   **(e)** Check your medicine cabinet at home for stomach acid remedies. If possible, identify the active ingredient in each remedy.

**34.** Two of the substances on the head of a safety match are potassium chlorate and sulfur. When the match is struck, the potassium chlorate decomposes to give potassium chloride and oxygen. The sulfur then burns in the oxygen and ignites the wood of the match.
   **(a)** Write balanced chemical equations for the decomposition of potassium chlorate and for the burning of sulfur in oxygen.
   **(b)** Identify the oxidizing agent and the reducing agent in each reaction in part (a).
   **(c)** Does any element in potassium chlorate undergo disproportionation in the reaction? Explain your answer.
   **(d)** Research the history of the safety match to determine when it was invented, why it was invented, and what it replaced.

**35.** Ammonium ions, from fertilizers or animal waste, are oxidized by atmospheric oxygen. The reaction results in the acidification of soil on farms and the pollution of ground water with nitrate ions.
   **(a)** Write a balanced net ionic equation for this reaction.
   **(b)** Why do farmers use fertilizers? What alternative farming methods have you heard of? Which farming method(s) do you support, and why?

**36.** One of the most important discoveries in the history of the Canadian chemical industry was accidental. Thomas "Carbide" Willson (1860–1915) was trying to make the element calcium from lime, CaO, by heating the lime with coal tar. Instead, he made the compound calcium carbide, CaC$_2$. This compound reacts with water to form a precipitate of calcium hydroxide and gaseous ethyne (acetylene). Willson's discovery led to the large-scale use of ethyne in numerous applications.
   **(a)** Was Willson trying to perform a redox reaction? How do you know? Why do you not need to know the substances in coal tar to answer this question?
   **(b)** Write a balanced chemical equation for the reaction of calcium carbide with water. Is this reaction a redox reaction?
   **(c)** An early use of Willson's discovery was in car headlights. Inside a headlight, the reaction of calcium carbide and water produced ethyne, which was burned to produce light and heat. Write a balanced chemical equation for the complete combustion of ethyne. Is this reaction a redox reaction?
   **(d)** Research the impact of Willson's discovery on society, from his lifetime to the present day.

### Answers to Practice Problems and Short Answers to Section Review Questions

**Practice Problems: 1.** $Zn_{(s)} + Fe^{2+}_{(aq)} \rightarrow Zn^{2+}_{(aq)} + Fe_{(s)}$
**2.(a)** $3Mg_{(s)} + 2Al^{3+}_{(aq)} \rightarrow 3Mg^{2+}_{(aq)} + 2Al_{(s)}$
**(b)** $2Ag^{+}_{(aq)} + Cd_{(s)} \rightarrow 2Ag_{(s)} + Cd^{2+}_{(aq)}$
**3.(a)** Mg oxidized, Al$^{3+}$ reduced
**(b)** Cd oxidized, Ag$^+$ reduced
**4.(a)** Al$^{3+}$ oxidizing agent, Mg reducing agent
**(b)** Ag$^+$ oxidizing agent, Cd reducing agent
**5.** $Al_{(s)} \rightarrow Al^{3+}_{(aq)} + 3e^-$, $Fe^{3+}_{(aq)} + 3e^- \rightarrow Fe_{(s)}$
**6.(a)** $Fe_{(s)} \rightarrow Fe^{2+}_{(aq)} + 2e^-$, $Cu^{2+}_{(aq)} + 2e^- \rightarrow Cu_{(s)}$
**(b)** $Cd_{(s)} \rightarrow Cd^{2+}_{(aq)} + 2e^-$, $Ag^{+}_{(aq)} + 1e^- \rightarrow Ag_{(s)}$
**7.(a)** $Sn_{(s)} \rightarrow Sn^{2+}_{(aq)} + 2e^-$, $Pb^{2+}_{(aq)} + 2e^- \rightarrow Pb_{(s)}$
**(b)** $Ag_{(s)} \rightarrow Ag^+ + e^-$, $Au^{3+}_{(aq)} + 3e^- \rightarrow Au_{(s)}$
**(c)** $Zn_{(s)} \rightarrow Zn^{2+}_{(aq)} + 2e^-$, $Fe^{3+}_{(aq)} + 3e^- \rightarrow Fe_{(s)}$
**8.** $Hg_2^{2+}{}_{(aq)} \rightarrow Hg_{(\ell)} + Hg^{2+}_{(aq)}$, $Hg_2^{2+}{}_{(aq)} + 2e^- \rightarrow 2Hg_{(\ell)}$, $Hg_2^{2+}{}_{(aq)} \rightarrow 2Hg^{2+}_{(aq)} + 2e^-$
**9.(a)** +3 **(b)** 0 **(c)** +6 **(d)** +5 **(e)** 0 **(f)** +2

10.(a) H, +1; S, +4; O, −2
(b) H, +1; O, −2 (c) H, +1; P, +5; O, −2
11.(a) +2 (b) −1  12.(a) Al, +3; H, +1; C, +4; O, −2
(b) N, −3; H, +1; P, +5; O, −2 (c) K, +1; H, +1; I, +7; O, −2
13.(a) yes (b) no
14.(a) $H_2O_2$ oxidizing agent, $Fe^{2+}$ reducing agent
15. $ClO_2^-$ oxidized, $Br_2$ reduced  16. yes
17. $Ce^{4+} + e^- \rightarrow Ce^{3+}$  18. $2Br^- \rightarrow Br_2 + 2e^-$
19.(a) $O_2 + 2H^+ + 2e^- \rightarrow H_2O_2$ (b) $2H_2O \rightarrow O_2 + 4H^+ + 4e^-$
(c) $2NO_3^- + 12H^+ + 10e^- \rightarrow N_2 + 6H_2O$
20.(a) $ClO_3^- + 6H^+ + 6e^- \rightarrow Cl^- + 3H_2O$
(b) $NO + 2H_2O \rightarrow NO_3^- + 4H^+ + 3e^-$
(c) $Cr_2O_7^{2-} + 14H^+ + 6e^- \rightarrow 2Cr^{3+} + 7H_2O$
21. $Cr^{2+} \rightarrow Cr^{3+} + e^-$  22. $O_2 + 4e^- \rightarrow 2O^{2-}$
23.(a) $Al + 4OH^- \rightarrow Al(OH)_4^- + 3e^-$
(b) $CN^- + 2OH^- \rightarrow CNO^- + H_2O + 2e^-$
(c) $MnO_4^- + 2H_2O + 3e^- \rightarrow MnO_2 + 4OH^-$
(d) $CrO_4^{2-} + 4H_2O + 3e^- \rightarrow Cr(OH)_3 + 5OH^-$
(e) $2CO_3^{2-} + 2H_2O + 2e^- \rightarrow C_2O_4^{2-} + 4OH^-$
24.(a) $FeO_4^{2-} + 8H^+ + 3e^- \rightarrow Fe^{3+} + 4H_2O$
(b) $ClO_2^- + 2H_2O + 4e^- \rightarrow Cl^- + 4OH^-$
25.(a) $2Na + F_2 \rightarrow 2NaF$
ox: $Na \rightarrow Na^+ + e^-$
red: $F_2 + 2e^- \rightarrow 2F^-$
(b) $3Mg + N_2 \rightarrow Mg_3N_2$
ox: $Mg \rightarrow Mg^{2+} + 2e^-$
red: $N_2 + 6e^- \rightarrow 2N^{3-}$
(c) $2HgO \rightarrow 2Hg + O_2$
ox: $2O^{2-} \rightarrow O_2 + 4e^-$
red: $Hg^{2+} + 2e^- \rightarrow Hg$
26. $2Cu^{2+} + 5I^- \rightarrow 2CuI + I_3^-$
27.(a) $MnO_4^- + 5Ag + 8H^+ \rightarrow Mn^{2+} + 5Ag^+ + 4H_2O$
oxidizing agent, $MnO_4^-$; reducing agent, Ag
(b) $Hg + 2NO_3^- + 4Cl^- + 4H^+ \rightarrow HgCl_4^{2-} + 2NO_2 + 2H_2O$
oxidizing agent, $NO_3^-$; reducing agent, Hg
(c) $AsH_3 + 4Zn^{2+} + 4H_2O \rightarrow H_3AsO_4 + 4Zn + 8H^+$
oxidizing agent, $Zn^{2+}$; reducing agent, $AsH_3$
(d) $3I_3^- + 3H_2O \rightarrow 8I^- + IO_3^- + 6H^+$
oxidizing agent, $I_3^-$; reducing agent, $I_3^-$
28.(a) $3CN^- + 2MnO_4^- + H_2O \rightarrow 3CNO^- + 2MnO_2 + 2OH^-$
oxidizing agent, $MnO_4^-$; reducing agent, $CN^-$
(b) $H_2O_2 + 2ClO_2 + 2OH^- \rightarrow 2ClO_2^- + O_2 + 2H_2O$
oxidizing agent, $ClO_2$; reducing agent, $H_2O_2$
(c) $6ClO^- + 2CrO_2^- + 2H_2O \rightarrow 3Cl_2 + 2CrO_4^{2-} + 4OH^-$
oxidizing agent, $ClO^-$; reducing agent, $CrO_2^-$
(d) $2Al + NO_2^- + H_2O + OH^- \rightarrow NH_3 + 2AlO_2^-$
oxidizing agent, $NO_2^-$; reducing agent, Al
29. 0.06774 g; 5.276%
30. 0.1387%

31. 0.0110 mol/L
32. $Cr_2O_7^{2-} + 6Fe^{2+} + 14H^+ \leftrightarrow 2Cr^{3+} + 6Fe^{3+} + 7H_2O$; 0.1743 mol/L
33. $CS_2 + 3O_2 \rightarrow CO_2 + 2SO_2$
34.(a) $B_2O_3 + 6Mg \rightarrow 3MgO + Mg_3B_2$
(b) $8H_2S + 8H_2O_2 \rightarrow S_8 + 16H_2O$
35.(a) $Cr_2O_7^{2-} + 6Fe^{2+} + 14H^+ \rightarrow 2Cr^{3+} + 6Fe^{3+} + 7H_2O$
(b) $I_2 + 10NO_3^- + 8H^+ \rightarrow 2IO_3^- + 10NO_2 + 4H_2O$
(c) $2PbSO_4 + 2H_2O \rightarrow Pb + PbO_2 + 2SO_4^{2-} + 4H^+$
36.(a) $3Cl^- + 2CrO_4^{2-} + H_2O \rightarrow 3ClO^- + 2CrO_2^- + 2OH^-$
(b) $3Ni + 2MnO_4^- + H_2O \rightarrow 3NiO + 2MnO_2 + 2OH^-$
(c) $I^- + 6Ce^{4+} + 6OH^- \rightarrow IO_3^- + 6Ce^{3+} + 3H_2O$

**Section Review 18.1:** 1.(a) yes;
$2AgNO_{3(aq)} + Cd_{(s)} \rightarrow 2Ag_{(s)} + Cd(NO_3)_{2(aq)}$;
$2Ag^+_{(aq)} + Cd_{(s)} \rightarrow 2Ag_{(s)} + Cd^{2+}_{(aq)}$;
$Ag^+_{(aq)} + e^- \rightarrow Ag_{(s)}$; $Cd_{(s)} \rightarrow Cd^{2+}_{(aq)} + 2e^-$
(b) no (c) yes; $2Al_{(s)} + 3HgCl_{2(aq)} \rightarrow 2AlCl_{3(aq)} + 3Hg_{(\ell)}$;
$2Al_{(s)} + 3Hg^{2+}_{(aq)} \rightarrow 2Al^{3+}_{(aq)} + 3Hg_{(\ell)}$;
$Al_{(s)} \rightarrow Al^{3+}_{(aq)} + 3e^-$; $Hg^{2+}_{(aq)} + 2e^- \rightarrow Hg_{(\ell)}$
2.(a) right side (b) left side  4. reducing agent
6.(a) $K^+ + Na_{(\ell)} \rightarrow Na^+ + K_{(\ell)}$
(b) oxidizing agent, $K^+$; reducing agent, $Na_{(\ell)}$
**18.2:** 2.(a) yes (b) no (c) no (d) yes
6.(a) +2 (b) +5 and −1
7.(a) $N_2 + 3H_2 \rightarrow 2NH_3$ (0, 0, −3, +1); $N_2$ is oxidizing agent, $H_2$ is reducing agent (b) no
8.(a) $2C + O_2 \rightarrow 2CO$, $Fe_2O_3 + 3CO \rightarrow 2Fe + 3CO_2$
(b) carbon—reducing agent, oxygen—oxidizing agent; carbon monoxide—reducing agent, iron(III) oxide—oxidizing agent
**18.3:** 1.(a) $2C + 2e^- \rightarrow C_2^{2-}$, reduction
(b) $2S_2O_3^{2-} \rightarrow S_4O_6^{2-} + 2e^-$, oxidation
(c) $4AsO_4^{3-} + 20H^+ + 8e^- \rightarrow As_4O_6 + 10H_2O$, reduction
(d) $Br_2 + 12OH^- \rightarrow 2BrO_3^- + 6H_2O + 10e^-$, oxidation
2.(a) $3Co^{3+} + Au \rightarrow 3Co^{2+} + Au^{3+}$
(b) $3Cu + 2NO_3^- + 8H^+ \rightarrow 3Cu^{2+} + 2NO + 4H_2O$
(c) $3NO_3^- + 8Al + 5OH^- + 2H_2O \rightarrow 3NH_3 + 8AlO_2^-$
3. $2Ag^+ + HCHO + H_2O \rightarrow 2Ag + HCOOH + 2H^+$
4.(b) $2NH_3 + 8OH^- \rightarrow N_2O + 7H_2O + 8e^-$
5.(a) $2N_2H_4 + N_2O_4 \rightarrow 3N_2 + 4H_2O$
(b) $N_2O_4$ is oxidizing agent, $N_2H_4$ is reducing agent
(c) $2NH_3 + ClO^- \rightarrow N_2H_4 + Cl^- + H_2O$
6.(a) Larissa was right
(b) $3Zn + SO_4^{2-} + 8H^+ \rightarrow 3Zn^{2+} + S + 4H_2O$
**18.4:** 1. no, not a redox reaction
2.(a) $CH_3COOH + 2O_2 \rightarrow 2CO_2 + 2H_2O$
(b) $O_2 + 2H_2SO_3 \rightarrow 2HSO_4^- + 2H^+$
3.(a) $8NH_3 + 3Cl_2 \rightarrow 6NH_4Cl + N_2$
(b) $3Mn_3O_4 + 8Al \rightarrow 4Al_2O_3 + 9Mn$

# CHAPTER 19 Cells and Batteries

## Chapter Preview

- 19.1 Galvanic Cells
- 19.2 Standard Cell Potentials
- 19.3 Electrolytic Cells
- 19.4 Faraday's Law
- 19.5 Issues Involving Electrochemistry

## Prerequisite Concepts and Skills

Before you begin this chapter, review the following concepts and skills:

- recognizing oxidation, reduction, and redox reactions (Chapter 18)
- writing balanced half-reactions (Chapter 18)
- using half-reactions to write balanced net ionic equations (Chapter 18)
- using stoichiometry to calculate quantities of reactants and products in a chemical reaction (Chapter 4)

Batteries come in a wide range of sizes and shapes, from the tiny button battery in a watch, to a large and heavy car battery. Although you probably use batteries every day, they may still surprise you. For example, did you know that you could make a battery from a lemon and two pieces of metal? To make a lemon battery, you could insert two different electrodes, such as copper and zinc strips, into a lemon. The battery can provide electricity for a practical use. For example, the battery can power a small light bulb, or turn a small motor.

For obvious reasons, lemon batteries are not a convenient way to power a portable device, such as a cell phone. Scientists and inventors have worked to develop a variety of batteries that are inexpensive, compact, and easy to store and to carry. Our society uses vast numbers of batteries.

What exactly is a battery, and how does it work? What is the relationship between batteries and the redox reactions studied in Chapter 18? You will find out the answers to these questions in this chapter.

**Can other fruits or vegetables be used to make batteries?**

# Galvanic Cells

## 19.1

You know that redox reactions involve the transfer of electrons from one reactant to another. You may also recall that an **electric current** is a flow of electrons in a circuit. These two concepts form the basis of **electrochemistry**, which is the study of the processes involved in converting chemical energy to electrical energy, and in converting electrical energy to chemical energy.

As you learned in Chapter 18, a zinc strip reacts with a solution containing copper(II) ions, forming zinc ions and metallic copper. The reaction is spontaneous (a **spontaneous reaction** is a reaction that occurs by itself; that is, without an ongoing input of energy). It releases energy in the form of heat; in other words, this reaction is exothermic.

$$Zn_{(s)} + Cu^{2+}_{(aq)} \rightarrow Zn^{2+}_{(aq)} + Cu_{(s)}$$

This reaction occurs on the surface of the zinc strip, where electrons are transferred from zinc atoms to copper(II) ions when these atoms and ions are in direct contact. A common technological invention called a galvanic cell uses redox reactions, such as the one described above, to release energy in the form of electricity.

### The Galvanic Cell

A **galvanic cell**, also called a **voltaic cell**, is a device that converts chemical energy to electrical energy. The key to this invention is to prevent the reactants in a redox reaction from coming into direct contact with each other. Instead, electrons flow from one reactant to the other through an **external circuit**, which is a circuit outside the reaction vessel. This flow of electrons through the external circuit is an electric current.

#### An Example of a Galvanic Cell: The Daniell Cell

Figure 19.1 shows one example of a galvanic cell, called the Daniell cell. One half of the cell consists of a piece of zinc placed in a zinc sulfate solution. The other half of the cell consists of a piece of copper placed in a copper(II) sulfate solution. A *porous barrier*, sometimes called a *semi-permeable membrane*, separates these two half-cells. It stops the copper(II) ions from coming into direct contact with the zinc electrode.

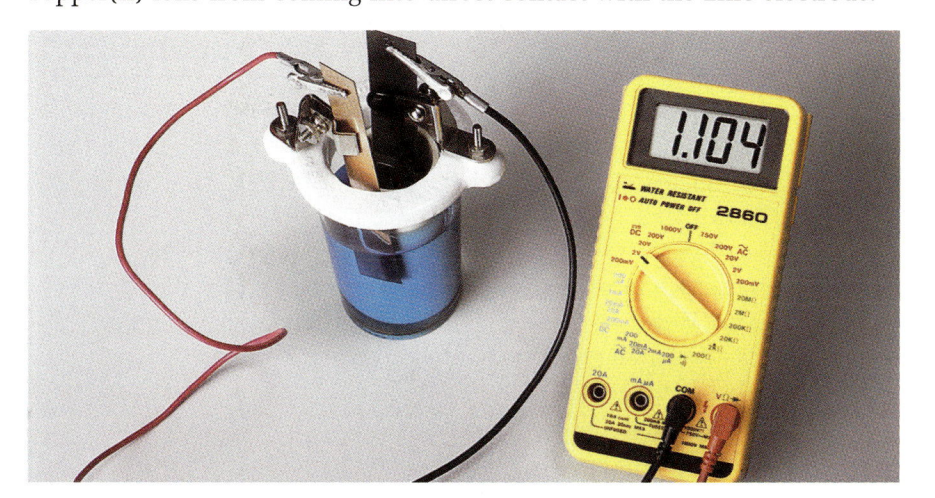

**Section Preview/Outcomes**

In this section, you will

- **identify** the components in galvanic cells and **describe** how they work
- **describe** the oxidation and reduction half-cells for some galvanic cells
- **determine** half-cell reactions, the direction of current flow, electrode polarity, cell potential, and ion movement in some galvanic cells
- **build** galvanic cells in the laboratory and **investigate** galvanic cell potentials
- **develop** a table of redox half-reactions from experimental results
- **communicate** your understanding of the following terms: *electric current, electrochemistry, spontaneous reaction, galvanic cell, voltaic cell, external circuit, electrodes, electrolytes, anode, cathode, salt bridge, inert electrode, electric potential, cell voltage, cell potential, dry cell, battery, primary battery, secondary battery*

**Figure 19.1** The Daniell cell is named after its inventor, the English chemist John Frederic Daniell (1790–1845). In the photograph shown here, the zinc sulfate solution is placed inside a porous cup, which is placed in a larger container of copper sulfate solution. The cup acts as the porous barrier.

> **Web LINK**
>
> www.mcgrawhill.ca/links/atlchemistry
>
> Galvanic cells are named after the Italian doctor Luigi Galvani (1737–1798), who generated electricity using two metals. These cells are also called voltaic cells, after the Italian physicist Count Alessandro Volta (1745–1827), who built the first chemical batteries. To learn more about scientists who made important discoveries in electrochemistry, such as Galvani, Volta, and Faraday, go to the web site above. Click on **Web Links** to find out where to go next.

In a Daniell cell, the pieces of metallic zinc and copper act as electrical conductors. The conductors that carry electrons into and out of a cell are named **electrodes**. The zinc sulfate and copper(II) sulfate act as electrolytes. **Electrolytes** are substances that conduct electricity when dissolved in water. (The fact that a solution of an electrolyte conducts electricity does not mean that free electrons travel through the solution. An electrolyte solution conducts electricity because of ion movements, and the loss and gain of electrons at the electrodes.) The terms *electrode* and *electrolyte* were invented by the leading pioneer of electrochemistry, Michael Faraday (1791–1867).

The redox reaction takes place in a galvanic cell when an external circuit, such as a metal wire, connects the electrodes. The oxidation half-reaction occurs in one half-cell, and the reduction half-reaction occurs in the other half-cell. For the Daniell cell:

Oxidation (loss of electrons): $Zn_{(s)} \rightarrow Zn^{2+}_{(aq)} + 2e^-$
Reduction (gain of electrons): $Cu^{2+}_{(aq)} + 2e^- \rightarrow Cu_{(s)}$

The electrode at which oxidation occurs is named the **anode**. In this example, zinc atoms undergo oxidation at the zinc electrode. Thus, the zinc electrode is the anode of the Daniell cell. The electrode at which reduction occurs is named the **cathode**. Here, copper(II) ions undergo reduction at the copper electrode. Thus, the copper electrode is the cathode of the Daniell cell.

Free electrons cannot travel through the solution. Instead, *the external circuit conducts electrons from the anode to the cathode of a galvanic cell.* Figure 19.2 gives a diagram of a typical galvanic cell.

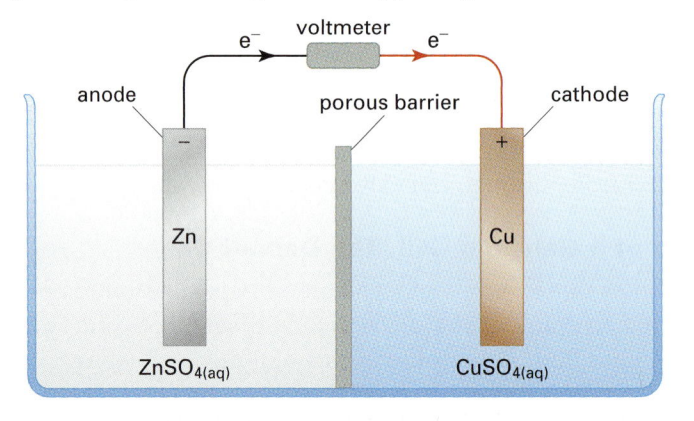

**Figure 19.2** A typical galvanic cell, such as the Daniell cell shown here, includes two electrodes, electrolyte solutions, a porous barrier, and an external circuit. Electrons flow through the external circuit from the negative anode to the positive cathode.

At the anode of a galvanic cell, electrons are released by oxidation. For example, at the zinc anode of the Daniell cell, zinc atoms release electrons to become positive zinc ions. Thus, the anode of a galvanic cell is negatively charged. Relative to the anode, the cathode of a galvanic cell is positively charged. In galvanic cells, electrons flow through the external circuit from the negative electrode to the positive electrode. These electrode polarities may already be familiar to you. An example is shown in Figure 19.3.

Each half-cell contains a solution of a neutral compound. In a Daniell cell, these solutions are aqueous zinc sulfate and aqueous copper(II) sulfate. How can these electrolyte solutions remain neutral when electrons are leaving the anode of one half-cell and arriving at the cathode of the other half-cell? To maintain electrical neutrality in each half-cell, some ions migrate through the porous barrier, as shown in Figure 11.4, on the next page. Negative ions (anions) migrate toward the anode, and positive ions (cations) migrate toward the cathode.

**Figure 19.3** Batteries contain galvanic cells. The + mark labels the positive cathode. If there is a − mark, it labels the negative anode.

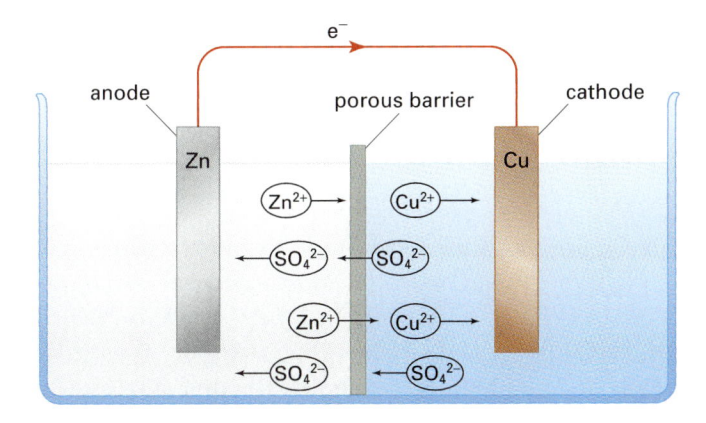

**Figure 19.4** Ion migration in a Daniell cell. Some sulfate ions migrate from the copper(II) sulfate solution into the zinc sulfate solution. Some zinc ions migrate from the zinc sulfate solution into the copper(II) sulfate solution.

The separator between the half-cells does not need to be a porous barrier. Figure 19.5 shows an alternative device. This device, called a **salt bridge**, contains an electrolyte solution that does not interfere in the reaction. The open ends of the salt bridge are plugged with a porous material, such as glass wool, to stop the electrolyte from leaking out quickly. The plugs allow ion migration to maintain electrical neutrality.

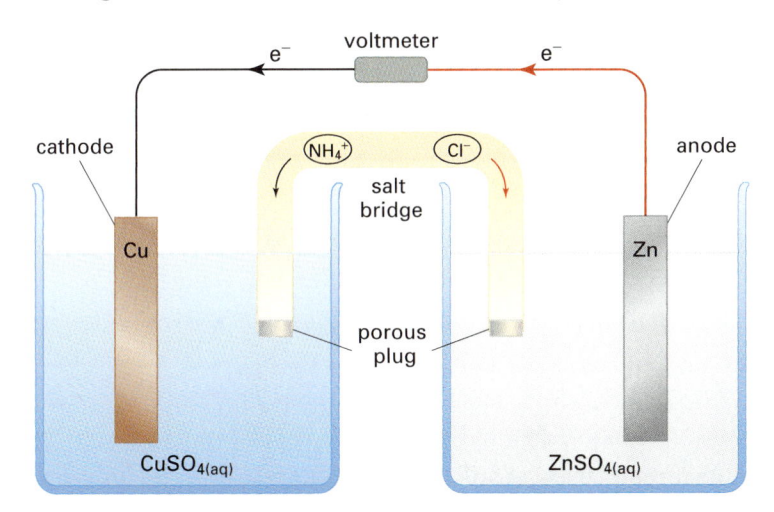

**Figure 19.5** This Daniell cell includes a salt bridge instead of a porous barrier. In a diagram, the anode may appear on the left or on the right.

Suppose the salt bridge of a Daniell cell contains ammonium chloride solution, $NH_4Cl_{(aq)}$. As positive zinc ions are produced at the anode, negative chloride ions migrate from the salt bridge into the half-cell that contains the anode. As positive copper(II) ions are removed from solution at the cathode, positive ammonium ions migrate from the salt bridge into the half-cell that contains the cathode.

Other electrolytes, such as sodium sulfate or potassium nitrate, could be chosen for the salt bridge. Neither of these electrolytes interferes in the cell reaction. Silver nitrate, $AgNO_{3(aq)}$, would be a poor choice for the salt bridge, however. Positive silver ions would migrate into the half-cell that contains the cathode. Zinc displaces both copper and silver from solution, so both copper(II) ions and silver ions would be reduced at the cathode. The copper produced would be contaminated with silver.

## Galvanic Cell Notation

A convenient shorthand method exists for representing galvanic cells. The shorthand representation of a Daniell cell is as follows.

$$Zn \mid Zn^{2+} \parallel Cu^{2+} \mid Cu$$

The phases or states may be included.

$$Zn_{(s)} \mid Zn^{2+}_{(aq)} \parallel Cu^{2+}_{(aq)} \mid Cu_{(s)}$$

As you saw in Figure 11.4 and Figure 11.5, the anode may appear on the left or on the right of a diagram. *In the shorthand representation, however, the anode is always shown on the left and the cathode on the right.* Each single vertical line, |, represents a phase boundary between the electrode and the solution in a half-cell. For example, the first single vertical line shows that the solid zinc and aqueous zinc ions are in different phases or states. The double vertical line, ||, represents the porous barrier or salt bridge between the half-cells. Spectator ions are usually omitted.

## Inert Electrodes

The zinc anode and copper cathode of a Daniell cell are both metals, and can act as electrical conductors. However, some redox reactions involve substances that cannot act as electrodes, such as gases or dissolved electrolytes. Galvanic cells that involve such redox reactions use inert electrodes. An **inert electrode** is an electrode made from a material that is neither a reactant nor a product of the cell reaction. Figure 19.6 shows a cell that contains one inert electrode. The chemical equation, net ionic equation, and half-reactions for this cell are given below.

Chemical equation: $Pb_{(s)} + 2FeCl_{3(aq)} \rightarrow 2FeCl_{2(aq)} + PbCl_{2(aq)}$

Net ionic equation: $Pb_{(s)} + 2Fe^{3+}_{(aq)} \rightarrow 2Fe^{2+}_{(aq)} + Pb^{2+}_{(aq)}$

Oxidation half-reaction: $Pb_{(s)} \rightarrow Pb^{2+}_{(aq)} + 2e^-$

Reduction half-reaction: $Fe^{3+}_{(aq)} + e^- \rightarrow Fe^{2+}_{(aq)}$

The reduction half-reaction does not include a solid conductor of electrons, so an inert platinum electrode is used in this half-cell. The platinum electrode is chemically unchanged, so it does not appear in the chemical equation or half-reactions. However, it is included in the shorthand representation of the cell.

$$Pb \mid Pb^{2+} \parallel Fe^{3+}, Fe^{2+} \mid Pt$$

A comma separates the formulas $Fe^{3+}$ and $Fe^{2+}$ for the ions involved in the reduction half-reaction. The formulas are not separated by a vertical line, because there is no phase boundary between these ions. The $Fe^{3+}$ and $Fe^{2+}$ ions exist in the same aqueous solution.

**Figure 19.6** This cell uses an inert electrode to conduct electrons. Why do you think that platinum is often chosen as an inert electrode? Another common choice is graphite.

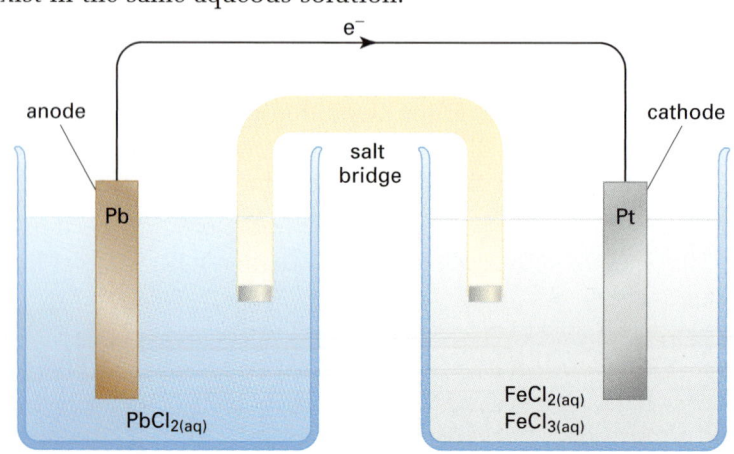

### Practice Problems

1. **(a)** If the reaction of zinc with copper(II) ions is carried out in a test tube, what is the oxidizing agent and what is the reducing agent?
   **(b)** In a Daniell cell, what is the oxidizing agent and what is the reducing agent? Explain your answer.

2. Write the oxidation half-reaction, the reduction half-reaction, and the overall cell reaction for each of the following galvanic cells. Identify the anode and the cathode in each case. In part (b), platinum is present as an inert electrode.
   **(a)** $Sn_{(s)} | Sn^{2+}_{(aq)} || Tl^{+}_{(aq)} | Tl_{(s)}$
   **(b)** $Cd_{(s)} | Cd^{2+}_{(aq)} || H^{+}_{(aq)} | H_{2(g)} | Pt_{(s)}$

3. A galvanic cell involves the overall reaction of iodide ions with acidified permanganate ions to form manganese(II) ions and iodine. The salt bridge contains potassium nitrate.
   **(a)** Write the half-reactions, and the overall cell reaction.
   **(b)** Identify the oxidizing agent and the reducing agent.
   **(c)** The inert anode and cathode are both made of graphite. Solid iodine forms on one of them. Which one?

4. As you saw earlier, pushing a zinc electrode and a copper electrode into a lemon makes a "lemon cell". In the following representation of the cell, $C_6H_8O_7$ is the formula of citric acid. Explain why the representation does not include a double vertical line.
   $Zn_{(s)} | C_6H_8O_{7(aq)} | Cu_{(s)}$

## Introducing Cell Potentials

You know that water spontaneously flows from a higher position to a lower position. In other words, water flows from a state of higher gravitational potential energy to a state of lower gravitational potential energy. As water flows downhill, it can do work, such as turning a water wheel or a turbine. The chemical changes that take place in galvanic cells are also accompanied by changes in potential energy. Electrons spontaneously flow from a position of higher potential energy at the anode to a position of lower potential energy at the cathode. The moving electrons can do work, such as lighting a bulb or turning a motor.

The difference between the potential energy at the anode and the potential energy at the cathode is the **electric potential**, $E$, of a cell. The unit used to measure electric potential is called the *volt*, with symbol V. Because of the name of this unit, electric potential is more commonly known as **cell voltage**. Another name for it is **cell potential**. A cell potential can be measured using an electrical device called a voltmeter.

A cell potential of 0 V means that the cell has no electric potential, and no electrons will flow. You know that you can generate electricity by connecting a zinc electrode and a copper electrode that have been inserted into a lemon. However, you cannot generate electricity by connecting two copper electrodes that have been inserted into the lemon. The two copper electrodes are the same and are in contact with the same electrolyte. There is no potential difference between the two electrodes.

Electric potentials vary from one cell to another, depending on various factors. You will examine some of these factors in the next investigation.

> **CONCEPT CHECK**
>
> The cell voltage is sometimes called the *electromotive force*, abbreviated *emf*. However, this term can be misleading. A cell voltage is a potential difference, not a force. The unit of cell voltage, the volt, is not a unit of force.

## Investigation 19-A

**SKILL FOCUS**
Predicting
Performing and recording
Analyzing and interpreting
Communicating results

# Measuring Cell Potentials of Galvanic Cells

In this investigation, you will build some galvanic cells and measure their cell potentials.

## Question
What factors affect the cell potential of a galvanic cell?

## Prediction
Predict whether the cell potentials of galvanic cells depend on the electrodes and electrolytes in the half-cells. Give reasons for your prediction.

## Materials
25 cm clear aquarium rubber tubing (Tygon®), internal diameter 4–6 mm
1 Styrofoam or clear plastic egg carton with 12 wells
5 cm strip of Mg ribbon
1 cm × 5 cm strips of Cu, Al, Ni, Zn, Sn, Fe, and Ag
5 cm of thick graphite pencil lead or a graphite rod
5 mL of 0.1 mol/L solutions of each of the following: $Mg(NO_3)_2$, $Cu(NO_3)_2$, $Al(NO_3)_3$, $Ni(NO_3)_2$, $Zn(NO_3)_2$, $SnSO_4$, $Fe(NO_3)_3$, $AgNO_3$, $HNO_3$
15 mL of 1.0 mol/L $KNO_3$
5 mL of saturated NaCl solution
disposable pipette
cotton batting
sandpaper
black and red electrical leads with alligator clips
voltmeter set to a scale of 0 V to 20 V
paper towel

## Safety Precautions
Handle the nitric acid solution with care. It is an irritant. Wash any spills on your skin with copious amounts of water, and inform your teacher.

## Procedure
1. Use tape or a permanent marker to label the outside of nine wells of your egg carton with the nine different half-cells. Each well should correspond to one of the eight different metal/metal ion pairs: $Mg/Mg^{2+}$, $Cu/Cu^{2+}$, $Al/Al^{3+}$, $Ni/Ni^{2+}$, $Zn/Zn^{2+}$, $Sn/Sn^{2+}$, $Fe/Fe^{3+}$, and $Ag/Ag^+$. Label the ninth well $H^+/H_2$.

2. Prepare a 9 × 9 grid in your notebook. Label the nine columns to match the nine half-cells. Label the nine rows in the same way. You will use this chart to mark the positive cell potentials you obtain when you connect two half-cells to build a galvanic cell. You will also record the anode and the cathode for each galvanic cell you build. (You may not need to fill out the entire chart.)

3. Sand each of the metals to remove any oxides.

4. Pour 5 mL of each metal salt solution into the appropriate well of the egg carton. Pour 5 mL of the nitric acid into the well labelled $H^+/H_2$.

5. Prepare your salt bridge as follows.
   (a) Roll a small piece of cotton batting so that it forms a plug about the size of a grain of rice. Place the plug in one end of your aquarium tubing, but leave a small amount of the cotton hanging out, so you can remove the plug later.
   (b) Fill a disposable pipette as full as possible with the 1 mol/L $KNO_3$ electrolyte solution. Fit the tip of the pipette firmly into the open end of the tubing. Slowly inject the electrolyte solution into the tubing. Fill the tubing completely, so that the cotton on the other side becomes wet.
   (c) With the tubing completely full, insert another cotton plug into the other end. There should be no air bubbles. (You may have to repeat this step from the beginning if you have air bubbles.)

6. Insert each metal strip into the corresponding well. Place the graphite rod in the well with the nitric acid. The metal strips and the graphite rod are your electrodes.

7. Attach the alligator clip on the red lead to the red probe of the voltmeter. Attach the black lead to the black probe.

8. Choose two wells to test. Insert one end of the salt bridge into the solution in the first well. Insert the other end of the salt bridge into the solution in the second well. Attach a free alligator clip to the electrode in each well. (**Note:** The graphite electrode is very fragile. Be gentle when using it.) You have built a galvanic cell.

9. If you get a negative reading, switch the alligator clips. Once you obtain a positive value, record it in your chart. The black lead should be attached to the anode (electrons flowing into the voltmeter). Record which metal is acting as the anode and which is acting as the cathode in this galvanic cell.

10. Remove the salt bridge and wipe any excess salt solution off the outside of the tubing. Remove the alligator clips from the electrodes.

11. Repeat steps 8 to 10 for all other combinations of electrodes. Record your results.

12. Reattach the leads to the silver and magnesium electrodes, and insert your salt bridge back into the two appropriate wells. While observing the reading on the voltmeter, slowly add 5 mL of saturated NaCl solution to the Ag/Ag$^+$ well to precipitate AgCl. Record any changes in the voltmeter reading. Observe the Ag/Ag$^+$ well.

13. Rinse off the metals and the graphite rod with water. Dispose of the salt solutions into the heavy metal salts container your teacher has set aside. Rinse out your egg carton. Remove and discard the plugs of the salt bridge, and dispose of the KNO$_3$ solution as directed by your teacher. Return all your materials to their appropriate locations.

## Analysis

1. For each cell in which you measured a cell potential, identify
   (a) the anode and the cathode
   (b) the positive and negative electrodes

2. For each cell in which you measured a cell potential, write a balanced equation for the reduction half-reaction, the oxidation half-reaction, and the overall cell reaction.

3. For any one cell in which you measured a cell potential, describe
   (a) the direction in which electrons flow through the external circuit
   (b) the movements of ions in the cell

4. Use your observations to decide which of the metals used as electrodes is the most effective reducing agent. Explain your reasoning.

5. List all the reduction half-reactions you wrote in question 2 so that the metallic elements in the half-reactions appear in order of their ability as reducing agents. Put the least effective reducing agent at the top of the list and the most effective reducing agent at the bottom.

6. In which part of your list from question 5 are the metal ions that are the best oxidizing agents? Explain.

7. (a) When saturated sodium chloride solution was added to the silver nitrate solution, what reaction took place? Explain.
   (b) Does the concentration of an electrolyte affect the cell potential of a galvanic cell? How do you know?

## Conclusion

8. Identify factors that affect the cell potential of a galvanic cell.

## Applications

9. Predict any other factors that you think might affect the voltage of a galvanic cell. Describe an investigation you could complete to test your prediction.

## Disposable Batteries

The Daniell cell is fairly large and full of liquid. Realistically, you could not use this type of cell to power a wristwatch, a remote control, or a flashlight. Galvanic cells have been modified, however, to make them more useful.

### The Dry Cell Battery

A **dry cell** is a galvanic cell with the electrolyte contained in a paste thickened with starch. This cell is much more portable than the Daniell cell. The first dry cell, invented by the French chemist Georges Leclanché in 1866, was called the Leclanché cell.

Modern dry cells are closely modelled on the Leclanché cell, and also contain electrolyte pastes. You have probably used dry cells in all kinds of applications, such as lighting a flashlight, powering a remote control, or ringing a doorbell. Dry cells are inexpensive. The cheapest AAA-, AA-, C-, and D-size 1.5-V batteries are dry cells.

A **battery** is defined as a set of galvanic cells connected in series. The negative electrode of one cell is connected to the positive electrode of the next cell in the set. *The voltage of a set of cells connected in series is the sum of the voltages of the individual cells.* Thus, a 9-V battery contains six 1.5-V dry cells connected in series. Often, the term "battery" is also used to describe a single cell. For example, a 1.5-V dry cell battery contains only a single cell.

A dry cell battery stops producing electricity when the reactants are used up. This type of battery is disposable after it has run down completely. A disposable battery is known as a **primary battery**. Some other batteries are rechargeable. A rechargeable battery is known as a **secondary battery**. The rest of this section will deal with primary batteries. You will learn about secondary batteries in section 11.3.

A dry cell contains a zinc anode and an inert graphite cathode, as shown in Figure 19.7. The electrolyte is a moist paste of manganese(IV) oxide, $MnO_2$, zinc chloride, $ZnCl_2$, ammonium chloride, $NH_4Cl$, and "carbon black," $C_{(s)}$, also known as soot.

The oxidation half-reaction at the zinc anode is already familiar to you.

$$Zn_{(s)} \rightarrow Zn^{2+}_{(aq)} + 2e^-$$

The reduction half-reaction at the cathode is more complicated. An approximation is given here.

$$2MnO_{2(s)} + H_2O_{(\ell)} + 2e^- \rightarrow Mn_2O_{3(s)} + 2OH^-_{(aq)}$$

Therefore, an approximation of the overall cell reaction is:

$$2MnO_{2(s)} + Zn_{(s)} + H_2O_{(\ell)} \rightarrow Mn_2O_{3(s)} + Zn^{2+}_{(aq)} + 2OH^-_{(aq)}$$

**Figure 19.7** The D-size dry cell battery is shown whole, and cut in two. The anode is the zinc container, located just inside the outer paper, steel, or plastic case. The graphite cathode runs through the centre of the cylinder.

### The Alkaline Cell Battery

The more expensive alkaline cell, shown in Figure 19.8, is an improved, longer-lasting version of the dry cell.

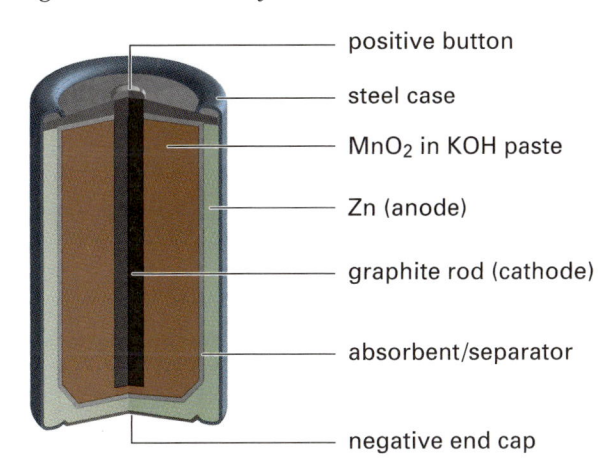

**Figure 19.8** The structure of an alkaline cell is similar to the structure of a dry cell. Each type has a voltage of 1.5 V.

Billions of alkaline batteries, each containing a single alkaline cell, are made every year. The ammonium chloride and zinc chloride used in a dry cell are replaced by strongly alkaline (basic) potassium hydroxide, KOH. The half-reactions and the overall reaction in an alkaline cell are given here.

Oxidation (at the anode): $Zn_{(s)} + 2OH^-_{(aq)} \rightarrow ZnO_{(s)} + H_2O_{(\ell)} + 2e^-$

Reduction (at the cathode): $MnO_{2(s)} + 2H_2O_{(\ell)} + 2e^- \rightarrow Mn(OH)_{2(s)} + 2OH^-_{(aq)}$

Overall cell reaction: $Zn_{(s)} + MnO_{2(s)} + H_2O_{(\ell)} \rightarrow ZnO_{(s)} + Mn(OH)_{2(s)}$

### The Button Cell Battery

A button battery is much smaller than an alkaline battery. Button batteries are commonly used in watches, as shown in Figure 19.9. Because of its small size, the button battery is also used for hearing aids, pacemakers, and some calculators and cameras. The development of smaller batteries has had an enormous impact on portable devices, as shown in Figure 19.10.

Two common types of button batteries both use a zinc container, which acts as the anode, and an inert stainless steel cathode, as shown in Figure 11.11 on the next page. In the mercury button battery, the alkaline electrolyte paste contains mercury(II) oxide, HgO. In the silver button battery, the electrolyte paste contains silver oxide, $Ag_2O$. The batteries have similar voltages: about 1.3 V for the mercury cell, and about 1.6 V for the silver cell.

The reaction products in a mercury button battery are solid zinc oxide and liquid mercury. The two half-reactions and the overall equation are as follows.

Oxidation half-reaction: $Zn_{(s)} + 2OH^-_{(aq)} \rightarrow ZnO_{(s)} + H_2O_{(\ell)} + 2e^-$

Reduction half-reaction: $HgO_{(s)} + H_2O_{(\ell)} + 2e^- \rightarrow Hg_{(\ell)} + 2OH^-_{(aq)}$

Overall reaction: $Zn_{(s)} + HgO_{(s)} \rightarrow ZnO_{(s)} + Hg_{(\ell)}$

**Figure 19.9** Button batteries are small and long-lasting.

**Figure 19.10** With small, long-lasting batteries, a pacemaker can now be implanted in a heart patient's chest. Early pacemakers, such as this one developed at the University of Toronto in the 1950s, were so big and heavy that patients had to wheel them around on a cart.

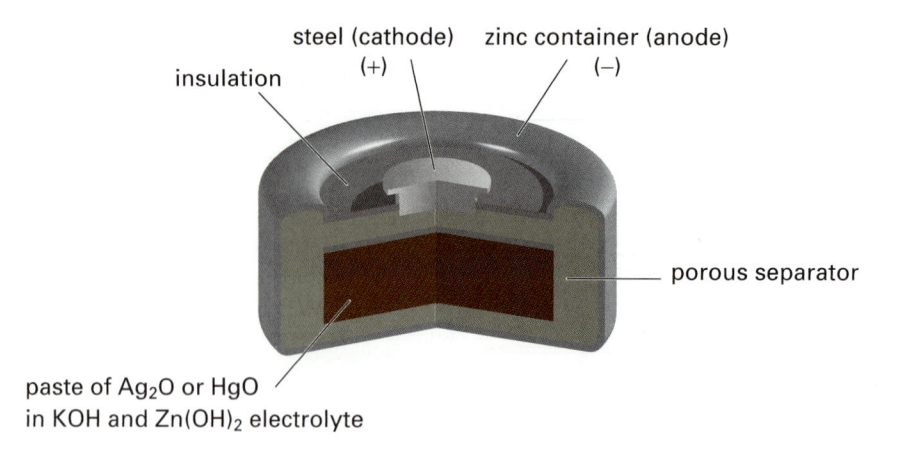

**Figure 19.11** A common type of button battery, shown here, contains silver oxide or mercury(II) oxide. Mercury is cheaper than silver, but discarded mercury batteries release toxic mercury metal into the environment.

- insulation
- steel (cathode) (+)
- zinc container (anode) (−)
- porous separator
- paste of Ag$_2$O or HgO in KOH and Zn(OH)$_2$ electrolyte

## Careers in Chemistry

### Explosives Chemist

Fortunato Villamagna works as the vice-president of technology for an Australian-owned company with offices worldwide, including Canada. Villamagna's job has given him the opportunity to invent new products, build chemical plants, and conduct projects in Africa and Australia.

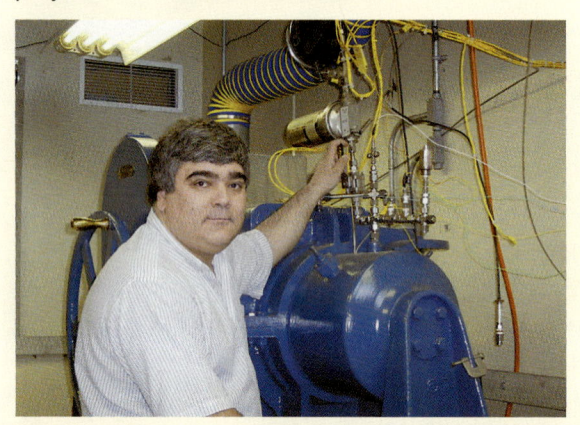

Born in Italy, Villamagna moved to Canada when he was eight. He grew up in Montréal, where his interest in chemistry was sparked by a Grade 10 teacher.

At university, Villamagna gained a greater understanding of the value of chemistry research. "Research creates new technologies and concepts, results in new products and services, creates jobs and prosperity, and in the end improves people's lives," he says. Villamagna decided to pursue graduate studies. He obtained a Masters of Science in physical chemistry at Concordia University, and a Ph.D. in physical chemistry at McGill University. Today, Villamagna leads a team of researchers responsible for developing new products and techniques involving explosives.

Redox reactions play an important role in industrial safety. Explosives are used in controlled ways in the mining, highway, and construction industries. The use of explosives allows modern workers to break up bedrock and carry out necessary demolitions from a safe distance. Chemists are involved in the development and production of explosives. They are also involved in making recommendations for the safe handling and disposal of explosives.

Many explosives are based on redox reactions. For example, the decomposition of nitroglycerin into nitrogen, carbon dioxide, water vapour, and oxygen is a redox reaction that results in a powerful explosion. The three nitrate groups of a nitroglycerin molecule act as powerful oxidizing agents, and the glycerol portion of the compound acts as a fuel. Fuels are very easily oxidized.

Nitroglycerin is highly unstable and can explode very easily. Therefore, it is difficult to manufacture and transport safely. Ammonium nitrate, an explosive that can act as both an oxidizing agent and a reducing agent, is often used to modify other explosives such as nitroglycerin. Ammonium nitrate is one of the products made by Villamagna's company.

### Making Career Connections

- The chemical formula of nitroglycerin is $C_3O_9N_3H_5$. Write the balanced chemical equation for the decomposition of nitroglycerin, as described in this feature.
- Find out which Canadian companies employ chemists who specialize in safe applications of explosives. Contact those companies for more information.

## Section Summary

In this section, you learned how to identify the different components of a galvanic cell. Also, you found out how galvanic cells convert chemical energy into electrical energy. You were introduced to several common primary batteries that contain galvanic cells. In the next section, you will learn more about the cell potentials of galvanic cells.

**PROBEWARE**

www.mcgrawhill.ca/links/atlchemistry

If you have access to probeware, do the Galvanic Cells investigation, or a similar investigation from a probeware company.

## Section Review

**1** Identify the oxidizing agent and the reducing agent in a dry cell.

**2** Explain why the top of a commercial 1.5-V dry cell battery is always marked with a plus sign.

**3** The reaction products in a silver button battery are solid zinc oxide and solid silver.
   **(a)** Write the two half-reactions and the equation for the overall reaction in the battery.
   **(b)** Name the materials used to make the anode and the cathode.

**4** If two 1.5-V D-size batteries power a flashlight, at what voltage is the flashlight operating? Explain.

**5** How many dry cells are needed to make a 6-V dry cell battery? Explain.

**6** Research the environmental impact of mercury pollution. Describe the main sources of mercury in the environment, the effects of mercury on human health, and at least one incident in which humans were harmed by mercury pollution.

**7** When a dry cell produces electricity, what happens to the container? Explain.

**8** Use the following shorthand representation to sketch a possible design of the cell. Include as much information as you can. Identify the anode and cathode, and write the half-reactions and the overall cell reaction.

$$Fe_{(s)} \mid Fe^{2+}_{(aq)} \parallel Ag^{+}_{(aq)} \mid Ag_{(s)}$$

## 19.2 Standard Cell Potentials

**Section Preview/Outcomes**

In this section, you will

- **describe** the use of the hydrogen half-cell as a reference
- **predict** cell voltage and whether redox reactions are spontaneous, using a table of reduction potentials
- **calculate** overall cell potentials under standard conditions
- **communicate** your understanding of the following terms: *reduction potential*, *oxidation potential*

In section 19.1, you learned that a cell potential is the difference between the potential energies at the anode and the cathode of a cell. In other words, a cell potential is the difference between the potentials of two half-cells. You cannot measure the potential of one half-cell, because a single half-reaction cannot occur alone. However, you can use measured cell potentials to construct tables of half-cell potentials. A table of standard half-cell potentials allows you to calculate cell potentials, rather than building the cells and measuring their potentials. Table 19.1 includes a few standard half-cell potentials. A larger table of standard half-cell potentials is given in Appendix E.

**Table 19.1** Standard Half-Cell Potentials (298 K)

| Half-reaction | $E°$ (V) |
|---|---|
| $F_{2(g)} + 2e^- \rightleftharpoons 2F^-_{(aq)}$ | 2.866 |
| $Br_{2(\ell)} + 2e^- \rightleftharpoons 2Br^-_{(aq)}$ | 1.066 |
| $I_{2(s)} + 2e^- \rightleftharpoons 2I^-_{(aq)}$ | 0.536 |
| $Cu^{2+}_{(aq)} + 2e^- \rightleftharpoons Cu_{(s)}$ | 0.342 |
| $2H^+_{(aq)} + 2e^- \rightleftharpoons H_{2(g)}$ | 0.000 |
| $Fe^{2+}_{(aq)} + 2e^- \rightleftharpoons Fe_{(s)}$ | −0.447 |
| $Zn^{2+}_{(aq)} + 2e^- \rightleftharpoons Zn_{(s)}$ | −0.762 |
| $Al^{3+}_{(aq)} + 3e^- \rightleftharpoons Al_{(s)}$ | −1.662 |
| $Na^+_{(aq)} + e^- \rightleftharpoons Na_{(s)}$ | −2.711 |

Table 19.1 and the larger table in Appendix E are based on the following conventions.

- Each half-reaction is written as a reduction. The half-cell potential for a reduction half-reaction is called a **reduction potential**. Look at the molecules and ions on the left side of each half-reaction. The most easily reduced molecules and ions (best oxidizing agents), such as $F_2$, $MnO_4^-$, and $O_2$, are near the top of the list. The least easily reduced molecules and ions (worst oxidizing agents), such as $Na^+$, $Ca^{2+}$, and $H_2O$, are near the bottom of the list.

- The numerical values of cell potentials and half-cell potentials depend on various conditions, so tables of *standard* reduction potentials are true when ions and molecules are in their *standard states*. These standard states are the same as for tables of standard enthalpy changes. Aqueous molecules and ions have a standard concentration of 1 mol/L. Gases have a standard pressure of 101.3 kPa or 1 atm. The standard temperature is 25°C or 298 K. Standard reduction potentials are designated by the symbol $E°$, where the superscript ° indicates standard states.

- Because you can measure potential differences, but not individual reduction potentials, all values in the table are relative. Each half-cell reduction potential is given relative to the reduction potential of the standard hydrogen electrode, which has been assigned a value of zero. The design of this electrode is shown in Figure 19.12.

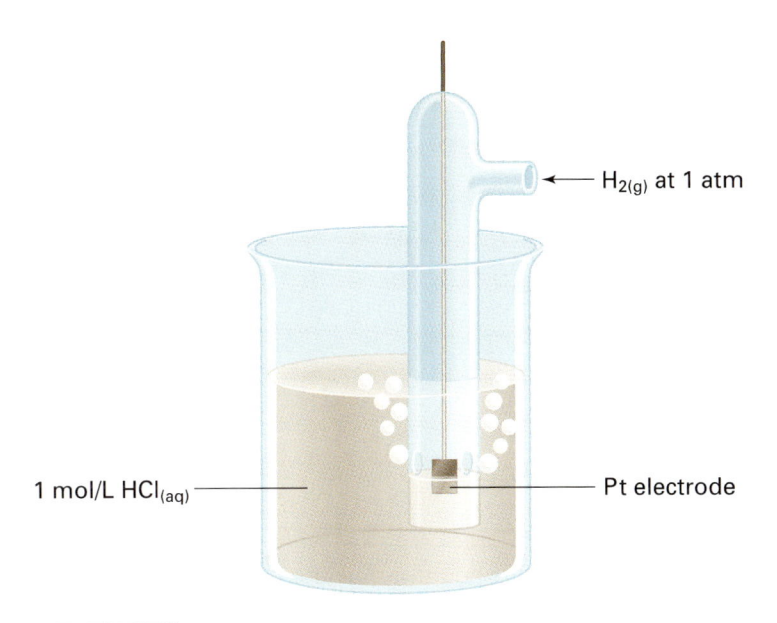

**Figure 19.12** In a standard hydrogen electrode, which is open to the atmosphere, hydrogen gas at 1 atm pressure bubbles over an inert platinum electrode. The electrode is immersed in a solution containing 1 mol/L H⁺ ions.

## Calculating Standard Cell Potentials

You can use Table 19.1 to calculate the standard cell potential of the familiar Daniell cell. This cell has its standard potential when the solution concentrations are 1 mol/L, as shown in the shorthand representation below.

$$Zn \mid Zn^{2+} (1 \text{ mol/L}) \parallel Cu^{2+} (1 \text{ mol/L}) \mid Cu$$

One method to calculate the standard cell potential is to subtract the standard reduction potential of the anode from the standard reduction potential of the cathode.

Method 1: $E°_{cell} = E°_{cathode} - E°_{anode}$

For a Daniell cell, you know that copper is the cathode and zinc is the anode. The relevant half-reactions and standard reduction potentials from Table 19.1 are as follows.

$$Cu^{2+}_{(aq)} + 2e^- \rightleftharpoons Cu_{(s)} \quad E° = 0.342 \text{ V}$$
$$Zn^{2+}_{(aq)} + 2e^- \rightleftharpoons Zn_{(s)} \quad E° = -0.762 \text{ V}$$

Use these values to calculate the cell potential.

$$\begin{aligned} E°_{cell} &= E°_{cathode} - E°_{anode} \\ &= 0.342 \text{ V} - (-0.762 \text{ V}) \\ &= 0.342 \text{ V} + 0.762 \text{ V} \\ &= 1.104 \text{ V} \end{aligned}$$

Thus, the standard cell potential for a Daniell cell is 1.104 V.
*The standard cell potentials of all galvanic cells have positive values*, as explained in Figure 19.13, on the following page. Figure 19.13 shows a "potential ladder" diagram. A "potential ladder" diagram models the potential difference. The rungs on the ladder correspond to the values of the reduction potentials.

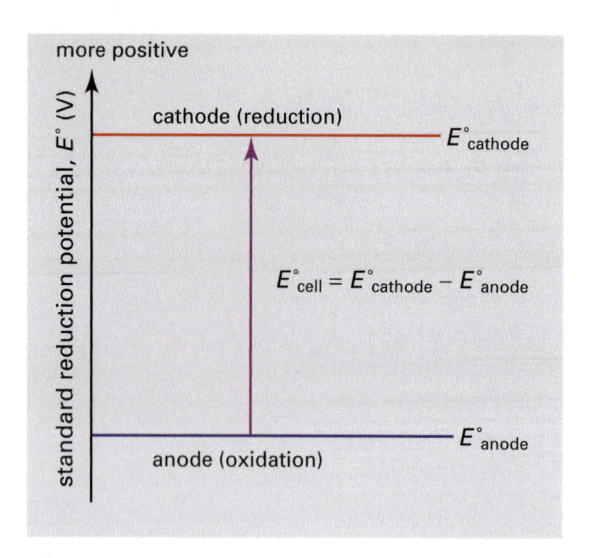

**Figure 19.13** For a galvanic cell, $E°_{cathode}$ is more positive (or less negative) than $E°_{anode}$. Thus, $E°_{cell}$ is always positive.

This calculation of the standard cell potential for the Daniell cell used the mathematical concept that the subtraction of a negative number is equivalent to the addition of its positive value. You saw that

$$0.342 \text{ V} - (-0.762 \text{ V}) = 0.342 \text{ V} + 0.762 \text{ V}$$

In other words, the *subtraction* of the reduction potential for a half-reaction is equivalent to the *addition* of the potential for the reverse half-reaction. The reverse half-reaction of a reduction is an oxidation. The half-cell potential for an oxidation half-reaction is called an **oxidation potential**. If the reduction half-reaction is as follows,

$$Zn^{2+}_{(aq)} + 2e^- \rightleftharpoons Zn_{(s)} \quad E° = -0.762 \text{ V}$$

then the oxidation half-reaction is

$$Zn_{(s)} \rightleftharpoons Zn^{2+}_{(aq)} + 2e^- \quad E°_{ox} = +0.762 \text{ V}$$

To summarize, the standard cell potential can also be calculated as the sum of a standard reduction potential and a standard oxidation potential.

Method 2: $E°_{cell} = E°_{red} + E°_{ox}$

As shown above, you can obtain the standard oxidation potential from a table of standard reduction potentials by reversing the reduction half-reaction, and changing the sign of the relevant potential. The reduction and oxidation half-reactions for the previous example are as follows.

$$Cu^{2+}_{(aq)} + 2e^- \rightleftharpoons Cu_{(s)} \quad E°_{red} = 0.342 \text{ V}$$
$$Zn_{(s)} \rightleftharpoons Zn^{2+}_{(aq)} + 2e^- \quad E°_{ox} = +0.762 \text{ V}$$

The calculation of the standard cell potential using these standard half-reaction potentials is as follows.

$$E°_{cell} = E°_{red} + E°_{ox}$$
$$= 0.342 \text{ V} + 0.762 \text{ V}$$
$$= 1.104 \text{ V}$$

Finding the difference between two reduction potentials, and finding the sum of a reduction potential and an oxidation potential are exactly equivalent methods for finding a cell potential. Use whichever method you prefer. The first Sample Problem includes both methods for finding cell potentials. The second Sample Problem uses only the subtraction of two reduction potentials. Practice problems are included after the second Sample Problem.

### Web LINK

www.mcgrawhill.ca/links/atlchemistry

For a movie describing the operation of a galvanic cell, and the calculation of its cell potential, go to the website above and click on **Electronic Learning Partner**.

## Sample Problem

### Calculating a Standard Cell Potential, Given a Net Ionic Equation

**Problem**

Calculate the standard cell potential for the galvanic cell in which the following reaction occurs.

$2I^-_{(aq)} + Br_{2(\ell)} \rightarrow I_{2(s)} + 2Br^-_{(aq)}$

**What Is Required?**

You need to find the standard cell potential for the given reaction.

**What Is Given?**

You have the balanced net ionic equation and a table of standard reduction potentials.

**Plan Your Strategy**

**Method 1: Subtracting Two Reduction Potentials**

**Step 1** Write the oxidation and reduction half-reactions.

**Step 2** Locate the relevant reduction potentials in a table of standard reduction potentials.

**Step 3** Subtract the reduction potentials to find the cell potential, using $E°_{cell} = E°_{cathode} - E°_{anode}$

**Method 2: Adding an Oxidation Potential and a Reduction Potential**

**Step 1** Write the oxidation and reduction half-reactions.

**Step 2** Locate the relevant reduction potentials in a table of standard reduction potentials.

**Step 3** Change the sign of the reduction potential for the oxidation half-reaction to find the oxidation potential.

**Step 4** Add the reduction potential and the oxidation potential, using $E°_{cell} = E°_{red} + E°_{ox}$

> **PROBLEM TIP**
>
> Think of a *red cat* to remember that *red*uction occurs at the *cat*hode. Think of *an ox* to remember that the *an*ode is where *ox*idation occurs.

**Act on Your Strategy**

**Method 1: Subtracting Two Reduction Potentials**

**Step 1** The oxidation and reduction half-reactions are as follows.

Oxidation half-reaction (occurs at the anode): $2I^-_{(aq)} \rightarrow I_{2(s)} + 2e^-$

Reduction half-reaction (occurs at the cathode): $Br_{2(\ell)} + 2e^- \rightarrow 2Br^-_{(aq)}$

**Step 2** The relevant reduction potentials in the table of standard reduction potentials are:

$I_{2(s)} + 2e^- \rightleftharpoons 2I^-_{(aq)}$    $E°_{anode} = 0.536$ V

$Br_{2(\ell)} + 2e^- \rightleftharpoons 2Br^-_{(aq)}$    $E°_{cathode} = 1.066$ V

**Step 3** Calculate the cell potential by subtraction.

$E°_{cell} = E°_{cathode} - E°_{anode}$
$= 1.066 \text{ V} - 0.536 \text{ V}$
$= 0.530 \text{ V}$

*Continued ...*

**Method 2: Adding an Oxidation Potential and a Reduction Potential**

**Step 1** The oxidation and reduction half-reactions are as follows.

Oxidation half-reaction (occurs at the anode): $2I^-_{(aq)} \rightarrow I_{2(s)} + 2e^-$

Reduction half-reaction (occurs at the cathode): $Br_{2(\ell)} + 2e^- \rightarrow 2Br^-_{(aq)}$

**Step 2** The relevant reduction potentials in the table of standard reduction potentials are:

$I_{2(s)} + 2e^- \rightleftharpoons 2I^-_{(aq)}$   $E°_{anode} = 0.536$ V

$Br_{2(\ell)} + 2e^- \rightleftharpoons 2Br^-_{(aq)}$   $E°_{cathode} = 1.066$ V

**Step 3** The standard electrode potential for the reduction half-reaction is $E°_{red} = 1.066$ V. Changing the sign of the potential for the oxidation half-reaction gives

$2I^-_{(aq)} \rightleftharpoons I_{2(s)} + 2e^-$   $E°_{ox} = -0.536$ V

**Step 4** Calculate the cell potential by addition.

$E°_{cell} = E°_{red} + E°_{ox}$
$= 1.066$ V $+ (-0.536$ V$)$
$= 0.530$ V

**Check Your Solution**

Both methods give the same answer. The cell potential is positive, as expected for a galvanic cell.

---

A standard cell potential depends only on the identities of the reactants and products in their standard states. As you will see in the next Sample Problem, *you do not need to consider the amounts of reactants or products present, or the reaction stoichiometry, when calculating a standard cell potential.* Since you have just completed a similar Sample Problem, only a brief solution using the subtraction method is given here. Check that you can solve this problem by adding a reduction potential and an oxidation potential.

### Sample Problem

### Calculating a Standard Cell Potential, Given a Chemical Equation

**Problem**

Calculate the standard cell potential for the galvanic cell in which the following reaction occurs.

$2Na_{(s)} + 2H_2O_{(\ell)} \rightarrow 2NaOH_{(aq)} + H_{2(g)}$

**Solution**

**Step 1** Write the equation in ionic form to identify the half-reactions.

$2Na_{(s)} + 2H_2O_{(\ell)} \rightarrow 2Na^+_{(aq)} + 2OH^-_{(aq)} + H_{2(g)}$

Write the oxidation and reduction half-reactions.

Oxidation half-reaction (occurs at the anode): $Na_{(s)} \rightarrow Na^+_{(aq)} + e^-$

Reduction half-reaction (occurs at the cathode):
$2H_2O_{(\ell)} + 2e^- \rightarrow 2OH^-_{(aq)} + H_{2(g)}$

**Step 2** Locate the relevant reduction potentials in a table of standard reduction potentials.

$Na_{(s)} + e^- \rightleftharpoons Na^+_{(aq)}$     $E°_{anode} = -2.711$ V

$2H_2O_{(\ell)} + 2e^- \rightleftharpoons H_{2(g)} + 2OH^-_{(aq)}$     $E°_{cathode} = -0.828$ V

**Step 3** Subtract the standard reduction potentials to calculate the cell potential.

$E°_{cell} = E°_{cathode} - E°_{anode}$
$= -0.828 \text{ V} - (-2.711 \text{ V})$
$= 1.883 \text{ V}$

The standard cell potential is 1.883 V. The Problem Tip on this page illustrates this calculation.

---

### PROBLEM TIP

A "potential ladder" diagram models the potential difference. The rungs on the ladder correspond to the values of the reduction potentials. For a galvanic cell, the half-reaction at the cathode is always on the upper rung, and the subtraction $E°_{cathode} - E°_{anode}$ always gives a positive cell potential.

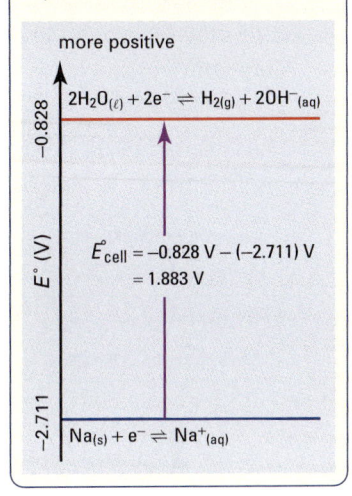

more positive

$-0.828$ : $2H_2O_{(\ell)} + 2e^- \rightleftharpoons H_{2(g)} + 2OH^-_{(aq)}$

$E°_{cell} = -0.828 \text{ V} - (-2.711) \text{ V}$
$= 1.883 \text{ V}$

$-2.711$ : $Na_{(s)} + e^- \rightleftharpoons Na^+_{(aq)}$

---

### Practice Problems

(**Note:** Obtain the necessary standard reduction potential values from the table in Appendix E.)

5. Write the two half-reactions for the following redox reaction. Subtract the two reduction potentials to find the standard cell potential for a galvanic cell in which this reaction occurs.

   $Cl_{2(g)} + 2Br^-_{(aq)} \rightarrow 2Cl^-_{(aq)} + Br_{2(\ell)}$

6. Write the two half-reactions for the following redox reaction. Add the reduction potential and the oxidation potential to find the standard cell potential for a galvanic cell in which this reaction occurs.

   $2Cu^+_{(aq)} + 2H^+_{(aq)} + O_{2(g)} \rightarrow 2Cu^{2+}_{(aq)} + H_2O_{2(aq)}$

7. Write the two half-reactions for the following redox reaction. Subtract the two standard reduction potentials to find the standard cell potential for the reaction.

   $Sn_{(s)} + 2HBr_{(aq)} \rightarrow SnBr_{2(aq)} + H_{2(g)}$

8. Write the two half-reactions for the following redox reaction. Add the standard reduction potential and the standard oxidation potential to find the standard cell potential for the reaction.

   $Cr_{(s)} + 3AgCl_{(s)} \rightarrow CrCl_{3(aq)} + 3Ag_{(s)}$

---

You have learned that the standard hydrogen electrode has an assigned standard reduction potential of exactly 0 V, and is the reference for all half-cell standard reduction potentials. What would happen to cell potentials if a different reference were used? You will address this question in the following ThoughtLab.

### Web LINK

www.mcgrawhill.ca/links/atlchemistry

Your **Electronic Learning Partner** has more information about aspects of material covered in this section of the chapter.

## ThoughtLab: Assigning Reference Values

Many scales of measurement have zero values that are arbitrary. For example, on Earth, average sea level is often assigned as the zero of altitude. In this ThoughtLab, you will investigate what happens to calculated cell potentials when the reference half-cell is changed.

### Procedure

1. Copy the following table of reduction potentials into your notebook. Change the zero on the scale by adding 1.662 V to each value to create new, adjusted reduction potentials.

| Reduction half-reaction | $E°$ (V) | $E° + 1.662$ (V) |
|---|---|---|
| $F_{2(g)} + 2e^- \rightleftharpoons 2F^-_{(aq)}$ | 2.866 | |
| $Fe^{3+}_{(aq)} + e^- \rightleftharpoons Fe^{2+}_{(aq)}$ | 0.771 | |
| $2H^+_{(aq)} + 2e^- \rightleftharpoons H_{2(g)}$ | 0.000 | |
| $Al^{3+}_{(aq)} + 3e^- \rightleftharpoons Al_{(s)}$ | −1.662 | |
| $Li^+_{(aq)} + e^- \rightleftharpoons Li_{(s)}$ | −3.040 | |

2. Use the given standard reduction potentials to calculate the standard cell potentials for the following redox reactions.

   (a) $2Li_{(s)} + 2H^+_{(aq)} \rightarrow 2Li^+_{(aq)} + H_{2(g)}$

   (b) $2Al_{(s)} + 3F_{2(g)} \rightarrow 2Al^{3+}_{(aq)} + 6F^-_{(aq)}$

   (c) $2FeCl_{3(aq)} + H_{2(g)} \rightarrow 2FeCl_{2(aq)} + 2HCl_{(aq)}$

   (d) $Al(NO_3)_{3(aq)} + 3Li_{(s)} \rightarrow 3LiNO_{3(aq)} + Al_{(s)}$

3. Repeat your calculations using the new, adjusted reduction potentials.

### Analysis

1. Compare your calculations from questions 2 and 3 of the procedure. What effect does changing the zero on the scale of reduction potentials have on

   (a) reduction potentials?

   (b) cell potentials?

### Applications

2. Find the difference between the temperatures at which water boils and freezes on the following scales. (Assume that a difference is positive, rather than negative.)

   (a) the Celsius temperature scale

   (b) the Kelvin temperature scale

3. What do your answers for the previous question tell you about these two temperature scales?

4. The zero on a scale of masses is not arbitrary. Why not?

## Section Summary

In this section, you learned that you can calculate cell potentials by using tables of half-cell potentials. The half-cell potential for a reduction half-reaction is called a reduction potential. The half-cell potential for an oxidation half-reaction is called an oxidation potential. Standard half-cell potentials are written as reduction potentials. The values of standard reduction potentials for half-reactions are relative to the reduction potential of the standard hydrogen electrode. You used standard reduction potentials to calculate standard cell potentials for galvanic cells. You learned two methods of calculating standard cell potentials. One method is to subtract the standard reduction potential of the anode from the standard reduction potential of the cathode. The other method is to add the standard reduction potential of the cathode and the standard oxidation potential of the anode. In the next section, you will learn about a different type of cell, called an electrolytic cell.

## Section Review

**1** Determine the standard cell potential for each of the following redox reactions.

(a) $3Mg_{(s)} + 2Al^{3+}_{(aq)} \rightarrow 3Mg^{2+}_{(aq)} + 2Al_{(s)}$

(b) $2K_{(s)} + F_{2(g)} \rightarrow 2K^+_{(aq)} + 2F^-_{(aq)}$

(c) $Cr_2O_7^{2-}{}_{(aq)} + 14H^+_{(aq)} + 6Ag_{(s)} \rightarrow 2Cr^{3+}_{(aq)} + 6Ag^+_{(aq)} + 7H_2O_{(\ell)}$

**2** Determine the standard cell potential for each of the following redox reactions.

(a) $CuSO_{4(aq)} + Ni_{(s)} \rightarrow NiSO_{4(aq)} + Cu_{(s)}$

(b) $4Au(OH)_{3(aq)} \rightarrow 4Au_{(s)} + 6H_2O_{(\ell)} + 3O_{2(g)}$

(c) $Fe_{(s)} + 4HNO_{3(aq)} \rightarrow Fe(NO_3)_{3(aq)} + NO_{(g)} + 2H_2O_{(\ell)}$

**3** For which half-cell are the values of the standard reduction potential and the standard oxidation potential equal?

**4** Look at the half-cells in the table of standard reduction potentials in Appendix E. Could you use two of the standard half-cells to build a galvanic cell with a standard cell potential of 7 V? Explain your answer.

**5** Compare the positions of metals in the metal activity series with their positions in the table of standard reduction potentials. Describe the similarities and differences.

**6** The cell potential for the following galvanic cell is given.

Zn | Zn²⁺ (1 mol/L) ‖ Pd²⁺ (1 mol/L) | Pd   $E°_{cell}$ = 1.750 V

Determine the standard reduction potential for the following half-reaction.

$Pd^{2+}_{(aq)} + 2e^- \rightleftharpoons Pd_{(s)}$

---

### Web LINK

www.mcgrawhill.ca/links/altchemistry

Battery makers have been challenged to make batteries smaller, lighter, longer-lasting, and more powerful. In March, 2003, Dalhousie University established a Research Chair in Battery and Fuel Cell Materials. The university's new laboratory is one of the few in the world equipped to use a new mode of research developed in 1995, called Combinatorial Materials Synthesis (CMS). CMS rapidly accelerates the time needed to test new combinations of materials by uncovering thousands of distinct compositions in a single experiment. Chief researcher Jeffery Dahn is applying CMS methods to improve the safety of lithium-ion batteries and produce cells large enough to power electric vehicles.

To research the history and latest developments in the field of battery technology, go to the web site above and click on **Web Links**.

# 19.3 Electrolytic Cells

**Section Preview/Outcomes**

In this section, you will
- **identify** the components of an electrolytic cell, and **describe** how they work
- **describe** electrolytic cells using oxidation and reduction half-cells
- **determine** oxidation and reduction half-cell reactions, direction of current flow, electrode polarity, cell potential, and ion movement in some electrolytic cells
- **build** and **investigate** an electrolytic cell in the laboratory
- **predict** whether or not redox reactions are spontaneous, using standard cell potentials
- **compare** galvanic and electrolytic cells in terms of energy efficiency, electron flow, and chemical change
- **describe** some common rechargeable batteries, and **evaluate** their impact on the environment and on society
- **communicate** your understanding of the following terms: *electrolytic cell, electrolysis, overvoltage, electroplating*

For a galvanic cell, you have learned that the overall reaction is spontaneous, and that the cell potential has a positive value. A galvanic cell converts chemical energy to electrical energy. Electrons flow from a higher potential energy to a lower potential energy. As described earlier, the flow of electrons in the external circuit of a galvanic cell can be compared to water flowing downhill.

Although water flows downhill spontaneously, you can also pump water uphill. This process requires energy because it moves water from a position of lower potential energy to a position of higher potential energy. You will now learn about a type of cell that uses energy to move electrons from lower potential energy to higher potential energy. This type of cell, called an **electrolytic cell**, is a device that converts electrical energy to chemical energy. The process that takes place in an electrolytic cell is called **electrolysis**. The overall reaction in an electrolytic cell is non-spontaneous, and requires energy to occur. This type of reaction is the reverse of a spontaneous reaction, which generates energy when it occurs.

Like a galvanic cell, an electrolytic cell includes electrodes, at least one electrolyte, and an external circuit. Unlike galvanic cells, electrolytic cells require an external source of electricity, sometimes called the *external voltage*. This is included in the external circuit. Except for the external source of electricity, an electrolytic cell may look just like a galvanic cell. Some electrolytic cells include a porous barrier or salt bridge. In other electrolytic cells, the two half-reactions are not separated, and take place in the same container.

## Electrolysis of Molten Salts

The electrolytic cell shown in Figure 19.14 decomposes sodium chloride into its elements. The cell consists of a single container with two inert electrodes dipping into liquid sodium chloride. To melt the sodium chloride, the temperature must be above its melting point of about 800°C. As in an aqueous solution of sodium chloride, the ions in molten sodium chloride have some freedom of movement. In other words, molten sodium chloride is the electrolyte of this cell.

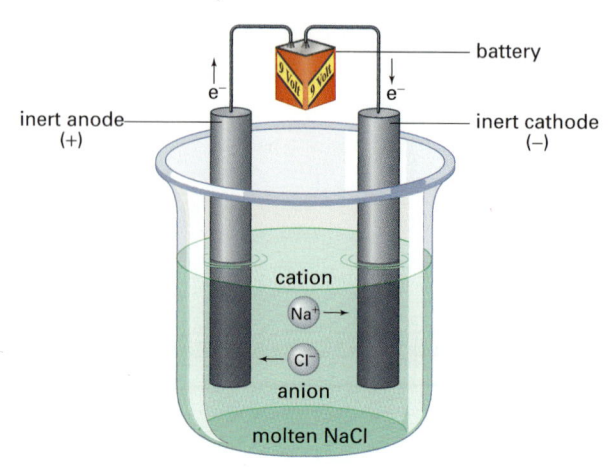

**Figure 19.14** Molten sodium chloride decomposes into sodium and chlorine in this electrolytic cell. The sodium chloride is said to undergo electrolysis, or to be *electrolyzed*.

The external source of electricity forces electrons onto one electrode. As a result, this electrode becomes negative relative to the other electrode. The positive sodium ions move toward the negative electrode, where they gain electrons and are reduced to the element sodium. At this temperature, sodium metal is produced as a liquid. The negative chloride ions move toward the positive electrode, where they lose electrons and are oxidized to the element chlorine, a gas. *As in a galvanic cell, reduction occurs at the cathode, and oxidation occurs at the anode of an electrolytic cell.* The half-reactions for this electrolytic cell are as follows.

Reduction half-reaction (occurs at the cathode): $Na^+_{(\ell)} + e^- \rightarrow Na_{(\ell)}$

Oxidation half-reaction (occurs at the anode): $2Cl^-_{(\ell)} \rightarrow Cl_{2(g)} + 2e^-$

Because of the external voltage of the electrolytic cell, the electrodes do not have the same polarities in electrolytic and galvanic cells. In a galvanic cell, the cathode is positive and the anode is negative. In an electrolytic cell, the anode is positive and the cathode is negative.

The electrolysis of molten sodium chloride is an important industrial reaction. Figure 19.15 shows the large electrolytic cell used in the industrial production of sodium and chlorine. You will meet other industrial electrolytic processes later in this chapter.

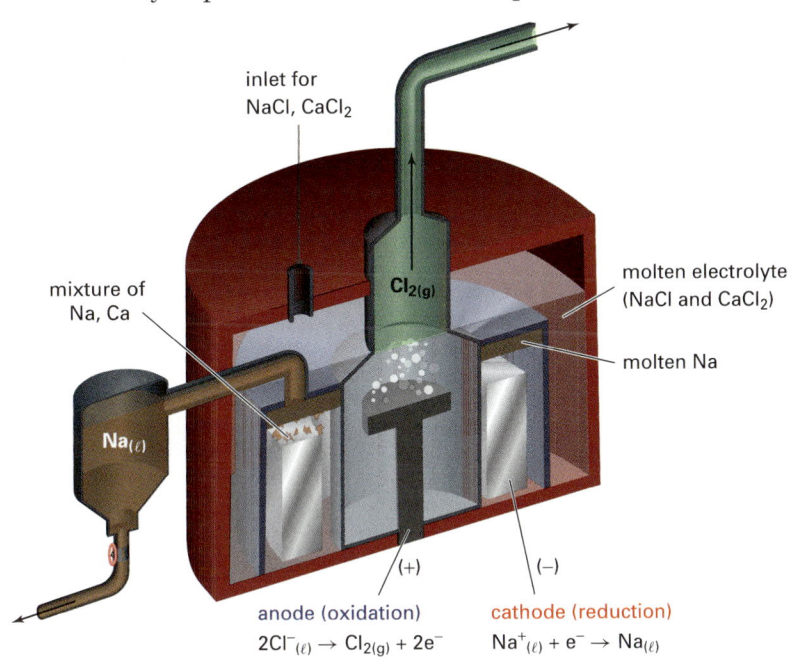

> **CONCEPT CHECK**
>
> "Electrochemical cell" is a common term in electrochemistry. Some scientists include both galvanic cells and electrolytic cells as types of electrochemical cells. Other scientists consider galvanic cells, but not electrolytic cells, as electrochemical cells. If you meet the term "electrochemical cell," always check its exact meaning.

**Figure 19.15** The large cell used for the electrolysis of sodium chloride in industry is known as a *Downs cell*. To decrease heating costs, calcium chloride is added to lower the melting point of sodium chloride from about 800°C to about 600°C. The reaction produces sodium and calcium by reduction at the cathode, and chlorine by oxidation at the anode.

anode (oxidation)
$2Cl^-_{(\ell)} \rightarrow Cl_{2(g)} + 2e^-$

cathode (reduction)
$Na^+_{(\ell)} + e^- \rightarrow Na_{(\ell)}$

Check your understanding of the introduction to electrolytic cells by completing the following practice problems.

### Practice Problems

9. The electrolysis of molten calcium chloride produces calcium and chlorine. Write
   (a) the half-reaction that occurs at the anode
   (b) the half-reaction that occurs at the cathode
   (c) the chemical equation for the overall cell reaction

*Continued...*

> **Continued...**
>
> **10.** For the electrolysis of molten lithium bromide, write
>   **(a)** the half-reaction that occurs at the negative electrode
>   **(b)** the half-reaction that occurs at the positive electrode
>   **(c)** the net ionic equation for the overall cell reaction
>
> **11.** A galvanic cell produces direct current, which flows in one direction. The mains supply at your home is a source of alternating current, which changes direction every fraction of a second. Explain why the external electrical supply for an electrolytic cell must be a source of direct current, rather than alternating current.
>
> **12.** Suppose a battery is used as the external electrical supply for an electrolytic cell. Explain why the negative terminal of the battery must be connected to the cathode of the cell.

### Electrolysis of Water

The electrolysis of aqueous solutions may not yield the desired products. Sir Humphry Davy (1778–1829) discovered the elements sodium and potassium by electrolyzing their molten salts. Before this discovery, Davy had electrolyzed aqueous solutions of sodium and potassium salts. He had not succeeded in reducing the metal ions to the pure metals at the cathode. Instead, his first experiments had produced hydrogen gas. Where did the hydrogen gas come from?

When electrolyzing an aqueous solution, there are two compounds present: water, and the dissolved electrolyte. Water may be electrolyzed as well as, or instead of, the electrolyte. The electrolysis of water produces oxygen gas and hydrogen gas, as shown in Figure 19.16.

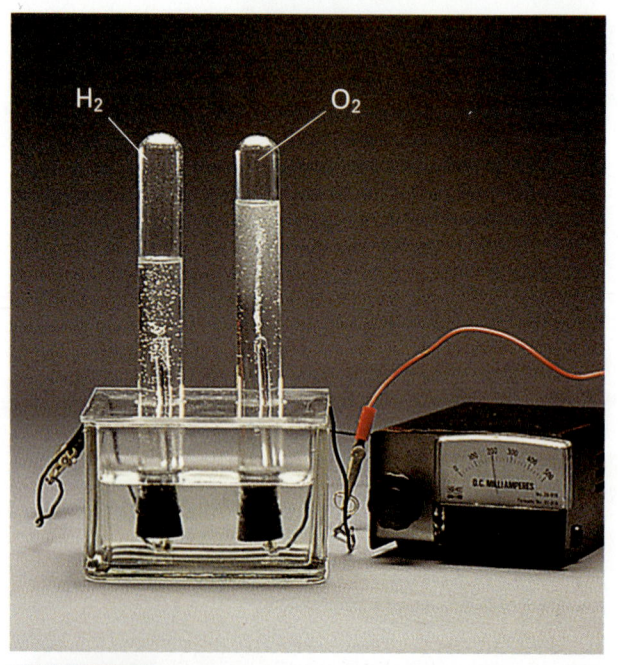

**Figure 19.16** The electrolysis of water produces hydrogen gas at the cathode and oxygen gas at the anode. Explain why the volume of hydrogen gas is twice the volume of oxygen gas.

The half-reactions for the electrolysis of water are given below.

Oxidation half-reaction (occurs at the anode):
$2H_2O_{(\ell)} \rightarrow O_{2(g)} + 4H^+_{(aq)} + 4e^-$

Reduction half-reaction (occurs at the cathode):
$2H_2O_{(\ell)} + 2e^- \rightarrow H_{2(g)} + 2OH^-_{(aq)}$

Because the number of electrons lost and gained must be equal, multiply the reduction half-reaction by 2. Then add and simplify to obtain the overall cell reaction.

Overall cell reaction: $2H_2O_{(\ell)} \rightarrow 2H_{2(g)} + O_{2(g)}$

> **CONCEPT CHECK**
>
> Check that you recall how to combine the two half-reactions to obtain the overall cell reaction. You learned how to do this in section 18.3.

The standard reduction potentials are as follows.

$O_{2(g)} + 4H^+_{(aq)} + 4e^- \rightleftharpoons 2H_2O_{(\ell)}$   $E° = 1.229$ V

$2H_2O_{(\ell)} + 2e^- \rightleftharpoons H_{2(g)} + 2OH^-_{(aq)}$   $E° = -0.828$ V

You can use these values to calculate the $E°_{cell}$ value for the decomposition of water.

$$E°_{cell} = E°_{cathode} - E°_{anode}$$
$$= -0.828 \text{ V} - 1.229 \text{ V}$$
$$= -2.057 \text{ V}$$

Therefore,

$2H_2O_{(\ell)} \rightarrow 2H_{2(g)} + O_{2(g)}$   $E°_{cell} = -2.057$ V

The negative cell potential shows that the reaction is not spontaneous. Electrolytic cells are used for non-spontaneous redox reactions, so *all electrolytic cells have negative cell potentials*.

The standard reduction potentials used to calculate $E°_{cell}$ for the decomposition of water apply only to reactants and products in their standard states. However, in pure water at 25°C, the hydrogen ions and hydroxide ions each have concentrations of $1 \times 10^{-7}$ mol/L. This is not the standard state value of 1 mol/L. The reduction potential values for the non-standard conditions in pure water are given below. The superscript zero is now omitted from the $E$ symbol, because the values are no longer standard.

$O_{2(g)} + 4H^+_{(aq)} + 4e^- \rightleftharpoons 2H_2O_{(\ell)}$   $E = 0.815$ V

$2H_2O_{(\ell)} + 2e^- \rightleftharpoons H_{2(g)} + 2OH^-_{(aq)}$   $E = -0.414$ V

Using these new half-cell potentials, $E_{cell}$ for the decomposition of pure water at 25°C by electrolysis has a calculated value of −1.229 V. Therefore, the calculated value of the external voltage needed is 1.229.

In practice, the external voltage needed for an electrolytic cell is always greater than the calculated value, especially for reactions involving gases. Therefore, the actual voltage needed to electrolyze pure water is *greater than* 1.229 V. The excess voltage required above the calculated value is called the **overvoltage**. Overvoltages depend on the gases involved and on the materials in the electrodes.

When electrolyzing water, there is another practical difficulty to consider. Pure water is a very poor electrical conductor. To increase the conductivity, an electrolyte that does not interfere in the reaction is added to the water.

> **CHEM FACT**
>
> Galvanic cells release electrical energy. In contrast, the redox reactions that take place in electrolytic cells require electrical energy in order to proceed. However, more energy (overvoltage) is required to cause a non-spontaneous reaction to occur in an electrolytic cell compared to the energy released by the reverse, spontaneous, reaction in a galvanic cell. For example, it takes more than 12 V to recharge a 12-V car battery. As a result of overvoltage, electrolytic cells are less efficient than galvanic cells.

## Electrolysis of Aqueous Solutions

As stated previously, an electrolytic cell may have the same design as a galvanic cell, except for the external source of electricity. Consider, for example, the familiar Daniell cell. (This cell was described in section 19.1 and shown in Figure 19.5.) By adding an external electrical supply, with a voltage greater than the voltage of the Daniell cell, you can push electrons in the opposite direction. By pushing electrons in the opposite direction, you reverse the chemical reaction. Figure 19.17 shows both cells, while their properties are compared in Table 19.2.

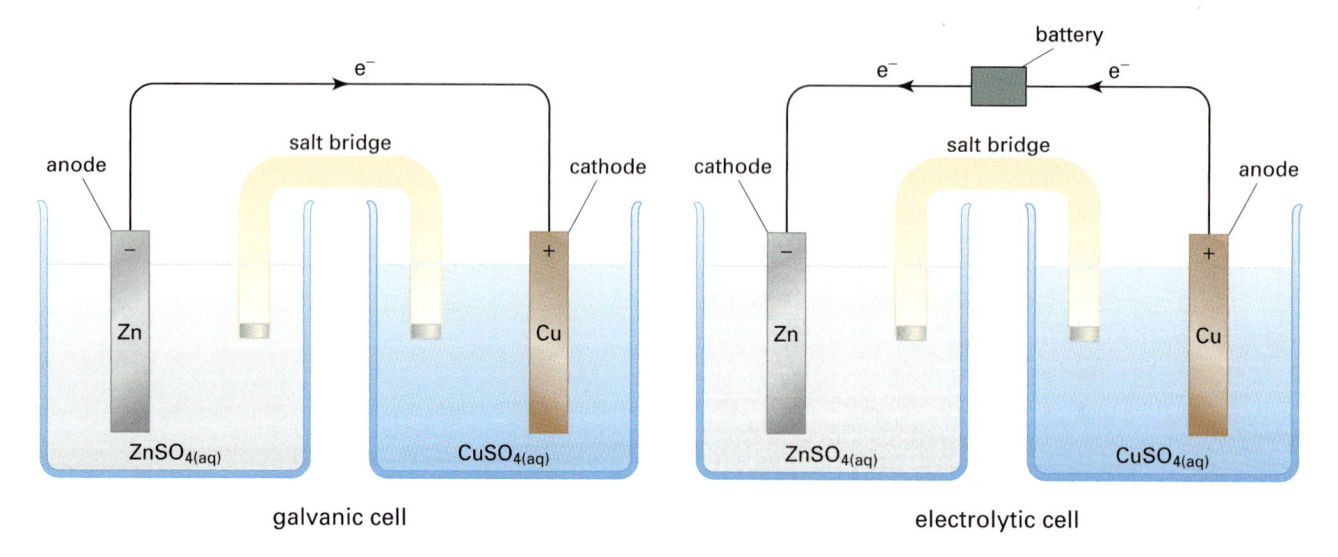

**Figure 19.17** Adding an external voltage to reverse the electron flow converts a Daniell cell from a galvanic cell into an electrolytic cell. The result is to switch the anode and cathode.

**Table 19.2** Cell Comparison

| Galvanic Cell | Electrolytic Cell |
|---|---|
| Spontaneous reaction | Non-spontaneous reaction |
| Converts chemical energy to electrical energy | Converts electrical energy to chemical energy |
| Anode (negative): Zinc | Anode (positive): Copper |
| Cathode (positive): Copper | Cathode (negative): Zinc |
| Oxidation (at anode): $Zn_{(s)} \rightarrow Zn^{2+}_{(aq)} + 2e^-$ | Oxidation (at anode): $Cu_{(s)} \rightarrow Cu^{2+}_{(aq)} + 2e^-$ |
| Reduction (at cathode): $Cu^{2+}_{(aq)} + 2e^- \rightarrow Cu_{(s)}$ | Reduction (at cathode): $Zn^{2+}_{(aq)} + 2e^- \rightarrow Zn_{(s)}$ |
| Cell reaction: $Zn_{(s)} + Cu^{2+}_{(aq)} \rightarrow Zn^{2+}_{(aq)} + Cu_{(s)}$ | Cell reaction: $Cu_{(s)} + Zn^{2+}_{(aq)} \rightarrow Cu^{2+}_{(aq)} + Zn_{(s)}$ |

In the galvanic cell, the zinc anode gradually dissolves. The copper cathode grows as more copper is deposited onto it. In the electrolytic cell, the copper anode gradually dissolves. The zinc cathode grows as more zinc is deposited onto it. The process in which a metal is deposited, or plated, onto the cathode in an electrolytic cell is known as **electroplating**. Electroplating is very important in industry, as you will learn later in this chapter.

### Predicting the Products of Electrolysis for an Aqueous Solution

The comparison of the Daniell cell with the electrolytic version of the cell appears straightforward. One reaction is the reverse of the other. However, you have just learned that the electrolysis of an aqueous solution may involve the electrolysis of water. How can you predict the actual products for this type of electrolysis reaction?

To predict the products of an electrolysis involving an aqueous solution, you must examine all possible half-reactions and their reduction potentials. Then, you must find the overall reaction that requires the *lowest* external voltage. That is, you must find the overall cell reaction with a negative cell potential that is closest to zero. The next Sample Problem shows you how to predict the products of the electrolysis of an aqueous solution.

In practice, reaction products are sometimes different from the products predicted, using the method described here. Predictions are least reliable when the reduction potentials are close together, especially when gaseous products are expected. However, there are many cases in which the predictions are correct.

## Sample Problem
### Electrolysis of an Aqueous Solution

**Problem**

Predict the products of the electrolysis of 1 mol/L $LiBr_{(aq)}$.

**What Is Required?**

You need to predict the products of the electrolysis of 1 mol/L $LiBr_{(aq)}$.

**What Is Given?**

This is an aqueous solution. You are given the formula and concentration of the electrolyte. You have a table of standard reduction potentials, and you know the non-standard reduction potentials for water.

**Plan Your Strategy**

**Step 1** List the four relevant half-reactions and their reduction potentials.

**Step 2** Predict the products by finding the cell reaction that requires the lowest external voltage.

**Act on Your Strategy**

**Step 1** The $Li^+$ and $Br^-$ concentrations are 1 mol/L, so use the standard reduction potentials for the half-reactions that involve these ions. Use the non-standard values for water.

$Br_{2(\ell)} + 2e^- \rightleftharpoons 2Br^-_{(aq)}$    $E° = 1.066$ V

$O_{2(g)} + 4H^+_{(aq)} + 4e^- \rightleftharpoons 2H_2O_{(\ell)}$    $E = 0.815$ V

$2H_2O_{(\ell)} + 2e^- \rightleftharpoons H_{2(g)} + 2OH^-_{(aq)}$    $E = -0.414$ V

$Li^+_{(aq)} + e^- \rightleftharpoons Li_{(s)}$    $E° = -3.040$ V

*Continued...*

There are two possible oxidation half-reactions at the anode: the oxidation of bromide ion in the electrolyte, or the oxidation of water.

$$2Br^-_{(aq)} \rightarrow Br_{2(\ell)} + 2e^-$$

$$2H_2O_{(\ell)} \rightarrow O_{2(g)} + 4H^+_{(aq)} + 4e^-$$

There are two possible reduction half-reactions at the cathode: the reduction of lithium ions in the electrolyte, or the reduction of water.

$$Li^+_{(aq)} + e^- \rightarrow Li_{(s)}$$

$$2H_2O_{(\ell)} + 2e^- \rightarrow H_{2(g)} + 2OH^-_{(aq)}$$

**Step 2** Combine pairs of half-reactions to produce four possible overall reactions. (You learned how to do this in Chapter 10.)

Reaction 1: the production of lithium and bromine

$$2Li^+_{(aq)} + 2Br^-_{(aq)} \rightarrow 2Li_{(s)} + Br_{2(\ell)}$$

$$E°_{cell} = E°_{cathode} - E°_{anode}$$
$$= -3.040 \text{ V} - 1.066 \text{ V}$$
$$= -4.106 \text{ V}$$

Reaction 2: the production of hydrogen and oxygen

$$2H_2O_{(\ell)} \rightarrow 2H_{2(g)} + O_{2(g)}$$

$$E_{cell} = E_{cathode} - E_{anode}$$
$$= -0.414 \text{ V} - 0.815 \text{ V}$$
$$= -1.229 \text{ V}$$

Reaction 3: the production of lithium and oxygen

$$4Li^+_{(aq)} + 2H_2O_{(\ell)} \rightarrow 4Li_{(s)} + O_{2(g)} + 4H^+_{(aq)}$$

$$E_{cell} = E°_{cathode} - E'_{anode}$$
$$= -3.040 \text{ V} - 0.815 \text{ V}$$
$$= -3.855 \text{ V}$$

Reaction 4: the production of hydrogen and bromine

$$2H_2O_{(\ell)} + 2Br^-_{(aq)} \rightarrow H_{2(g)} + 2OH^-_{(aq)} + Br_{2(\ell)}$$

$$E_{cell} = E_{cathode} - E°_{anode}$$
$$= -0.414 \text{ V} - 1.066 \text{ V}$$
$$= -1.480 \text{ V}$$

The electrolysis of water requires the lowest external voltage. Therefore, the predicted products of this electrolysis are hydrogen and oxygen.

### Check Your Solution

Use a potential ladder diagram, such as the one on the next page, part A, to visualize the cell potentials. For an electrolytic cell, the half-reaction at the anode is always on the upper rung, and the subtraction $E°_{cathode} - E°_{anode}$ always gives a negative cell potential, as shown in part B.

> **PROBLEM TIP**
>
> - Remember that spectator ions do not appear in net ionic equations. In Reaction 3, the bromide ions are spectator ions. In Reaction 4, lithium ions are spectator ions.
>
> - As for a galvanic cell, the cell potential for an electrolytic cell is the sum of a reduction potential and an oxidation potential. Using $E_{cell} = E_{red} + E_{ox}$ gives the same predicted products as using $E_{cell} = E_{cathode} - E_{anode}$

**A**

more positive

$E$ (V)

1.066 — $Br_{2(aq)} + 2e^- \rightleftharpoons 2Br^-_{(aq)}$
0.815 — $O_{2(g)} + 4H^+_{(aq)} + 4e^- \rightleftharpoons 2H_2O_{(\ell)}$

$E_{cell} = -1.229$ V

−0.414 — $2H_2O_{(\ell)} + 2e^- \rightleftharpoons H_{2(g)} + 2OH^-_{(aq)}$

$E_{cell} = -1.480$ V

$E°_{cell} = -4.106$ V       $E_{cell} = -3.855$ V

−3.040 — $Li^+_{(aq)} + e^- \rightleftharpoons Li_{(s)}$

**B**

more positive

$E$ (V)

anode (oxidation) —— $E_{anode}$

$E_{cell} = E°_{cathode} - E_{anode}$

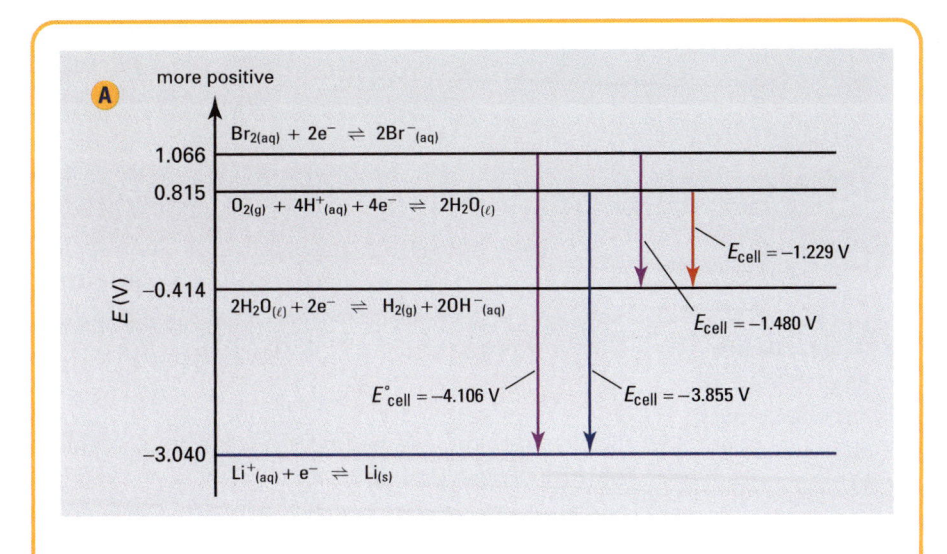

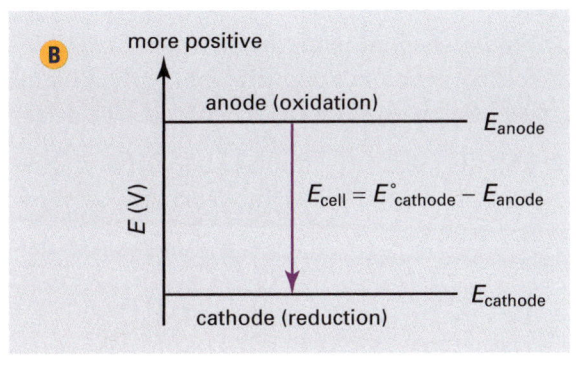

cathode (reduction) —— $E_{cathode}$

## Practice Problems

**13.** Predict the products of the electrolysis of a 1 mol/L solution of sodium chloride.

**14.** Explain why calcium can be produced by the electrolysis of molten calcium chloride, but not by the electrolysis of aqueous calcium chloride.

**15.** One half-cell of a galvanic cell has a nickel electrode in a 1 mol/L nickel(II) chloride solution. The other half-cell has a cadmium electrode in a 1 mol/L cadmium chloride solution.
  **(a)** Find the cell potential.
  **(b)** Identify the anode and the cathode.
  **(c)** Write the oxidation half-reaction, the reduction half-reaction, and the overall cell reaction.

**16.** An external voltage is applied to change the galvanic cell in question 15 into an electrolytic cell. Repeat parts (a) to (c) for the electrolytic cell.

In Investigation 19-B, you will build an electrolytic cell for the electrolysis of an aqueous solution of potassium iodide. You will predict the products of the electrolysis, and compare the observed products with your predictions.

# Investigation 19-B

**SKILL FOCUS**
Predicting
Performing and recording
Analyzing and interpreting

## Electrolysis of Aqueous Potassium Iodide

When an aqueous solution is electrolyzed, the electrolyte or water can undergo electrolysis. In this investigation, you will build an electrolytic cell, carry out the electrolysis of an aqueous solution, and identify the products.

### Questions

What are the products from the electrolysis of a 1 mol/L aqueous solution of potassium iodide? Are the observed products the ones predicted using reduction potentials?

### Predictions

Use the relevant standard reduction potentials from the table in Appendix E, and the non-standard reduction potentials you used previously for water, to predict the electrolysis products. Predict which product(s) are formed at the anode and which product(s) are formed at the cathode.

### Materials

25 cm clear aquarium rubber tubing (Tygon®), internal diameter 4–6 mm
1 graphite pencil lead, 2 cm long
2 wire leads (black and red) with alligator clips
600 mL or 400 mL beaker
sheet of white paper
1 elastic band
3 toothpicks
3 disposable pipettes
2 cm piece of copper wire (20 gauge)
1 drop 1% starch solution
10 mL 1 mol/L KI
1 drop 1% phenolphthalein
9-V battery or variable power source set to 9 V

### Safety Precautions

Make sure your lab bench is dry before carrying out this investigation.

### Procedure

1. Fold a sheet of paper lengthwise. Curl the folded paper so that it fits inside the 600 mL beaker. Invert the beaker on your lab bench.

2. Use the elastic to strap the aquarium tubing to the side of the beaker in a U shape, as shown in the diagram.

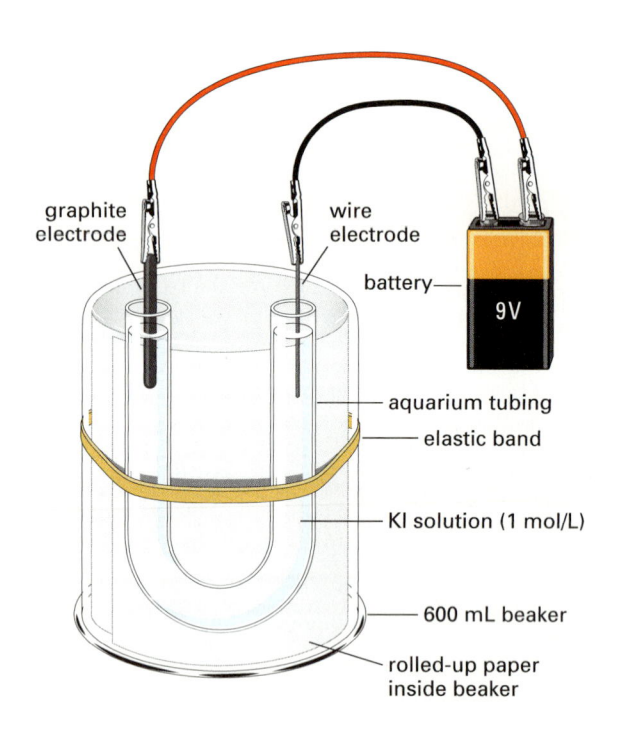

784   MHR • Unit 8 Electrochemistry

3. Fill a pipette as completely as possible with 1 mol/L KI solution. Insert the tip of the pipette firmly into one end of the aquarium tubing. Slowly inject the solution into the U-tube until the level of the solution is within 1 cm to 2 cm from the top of both ends. If air bubbles are present, try to remove them by poking them with a toothpick. You may need to repeat this step from the beginning.

4. Attach the black lead to the 2 cm piece of wire. Insert the wire into one end of the U-tube. Attach the red electrical lead to the graphite. Insert the graphite into the other end of the U-tube.

5. Attach the leads to the 9-V battery or to a variable power source set to 9 V. Attach the black lead to the negative terminal, and the red lead to the positive terminal.

6. Let the reaction proceed for three minutes, while you examine the U-tube. Record your observations. Shut off the power source and remove the electrodes. Determine the product formed around the anode by adding a drop of starch solution to the end of the U-tube that contains the anode. Push the starch solution down with a toothpick if there is an air lock. Determine one of the products around the cathode by adding a drop of phenolphthalein to the appropriate end of the U-tube.

7. Dispose of your reactants and products as instructed by your teacher. Take your apparatus apart, rinse out the tubing, and rinse off the electrodes. Return your equipment to its appropriate location.

## Analysis

1. Sketch the cell you made in this investigation. On your sketch, show
   (a) the direction of the electron flow in the external circuit
   (b) the anode and the cathode
   (c) the positive electrode and the negative electrode
   (d) the movement of ions in the cell

2. Use your observations to identify the product(s) formed at the anode and the product(s) formed at the cathode.

3. Write a balanced equation for the half-reaction that occurs at the anode.

4. Write a balanced equation for the half-reaction that occurs at the cathode.

5. Write a balanced equation for the overall cell reaction.

6. Calculate the external voltage required to carry out the electrolysis. Why was the external voltage used in the investigation significantly higher than the calculated value?

## Conclusion

7. What are the products from the electrolysis of a 1 mol/L aqueous solution of potassium iodide? Are the observed products the same as the products predicted using reduction potentials?

## Applications

8. If you repeated the electrolysis using aqueous sodium iodide instead of aqueous potassium iodide, would your observations change? Explain your answer.

9. To make potassium by electrolyzing potassium iodide, would you need to modify the procedure? Explain your answer.

## Spontaneity of Reactions

You know that galvanic cells have positive standard cell potentials, and that these cells use spontaneous chemical reactions to produce electricity. You also know that electrolytic cells have negative standard cell potentials, and that these cells use electricity to perform non-spontaneous chemical reactions. Thus, you can use the sign of the standard cell potential to predict whether a reaction is spontaneous or not under standard conditions.

### Sample Problem
### Predicting Spontaneity

**Problem**

Predict whether each reaction is spontaneous or non-spontaneous under standard conditions.

**(a)** $Cd_{(s)} + Cu^{2+}_{(aq)} \rightarrow Cd^{2+}_{(aq)} + Cu_{(s)}$  **(b)** $I_{2(s)} + 2Cl^-_{(aq)} \rightarrow 2I^-_{(aq)} + Cl_{2(g)}$

**Solution**

**(a)** The two half-reactions are as follows.

Oxidation (occurs at the anode): $Cd_{(s)} \rightarrow Cd^{2+}_{(aq)} + 2e^-$

Reduction (occurs at the cathode): $Cu^{2+}_{(aq)} + 2e^- \rightarrow Cu_{(s)}$

The relevant standard reduction potentials are:

$Cu^{2+}_{(aq)} + 2e^- \rightleftharpoons Cu_{(s)}$    $E° = 0.342$ V

$Cd^{2+}_{(aq)} + 2e^- \rightleftharpoons Cd_{(s)}$    $E° = -0.403$ V

$E°_{cell} = E°_{cathode} - E°_{anode}$

$= 0.342$ V $- (-0.403$ V$)$

$= 0.745$ V

The standard cell potential is positive, so the reaction is spontaneous under standard conditions.

**(b)** The two half-reactions are as follows.

Oxidation (occurs at the anode): $2Cl^-_{(aq)} \rightarrow Cl_{2(g)} + 2e^-$

Reduction (occurs at the cathode): $I_{2(s)} + 2e^- \rightarrow 2I^-_{(aq)}$

The relevant standard reduction potentials are:

$Cl_{2(g)} + 2e^- \rightleftharpoons 2Cl^-_{(aq)}$    $E° = 1.358$ V

$I_{2(s)} + 2e^- \rightleftharpoons 2I^-_{(aq)}$    $E° = 0.536$ V

$E°_{cell} = E°_{cathode} - E°_{anode}$

$= 0.536$ V $- 1.358$ V

$= -0.822$ V

The standard cell potential is negative, so the reaction is non-spontaneous under standard conditions.

### Practice Problems

17. Look up the standard reduction potentials of the following half-reactions. Predict whether acidified nitrate ions will oxidize manganese(II) ions to manganese(IV) oxide under standard conditions.

    $MnO_{2(s)} + 4H^+_{(aq)} + 2e^- \rightarrow Mn^{2+}_{(aq)} + 2H_2O_{(\ell)}$

    $NO_3^-_{(aq)} + 4H^+_{(aq)} + 3e^- \rightarrow NO_{(g)} + 2H_2O_{(\ell)}$

18. Predict whether each reaction is spontaneous or non-spontaneous under standard conditions.

    (a) $2Cr_{(s)} + 3Cl_{2(g)} \rightarrow 2Cr^{3+}_{(aq)} + 6Cl^-_{(aq)}$
    (b) $Zn^{2+}_{(aq)} + Fe_{(s)} \rightarrow Zn_{(s)} + Fe^{2+}_{(aq)}$
    (c) $5Ag_{(s)} + MnO_4^-_{(aq)} + 8H^+_{(aq)} \rightarrow 5Ag^+_{(aq)} + Mn^{2+}_{(aq)} + 4H_2O_{(\ell)}$

19. Explain why an aqueous copper(I) compound disproportionates to form copper metal and an aqueous copper(II) compound under standard conditions. (You learned about disproportionation in Chapter 10.)

20. Predict whether each reaction is spontaneous or non-spontaneous under standard conditions in an acidic solution.

    (a) $H_2O_{2(aq)} \rightarrow H_{2(g)} + O_{2(g)}$
    (b) $3H_{2(g)} + Cr_2O_7^{2-}_{(aq)} + 8H^+_{(aq)} \rightarrow 2Cr^{3+}_{(aq)} + 7H_2O_{(\ell)}$

## Rechargeable Batteries

In section 19.1, you learned about several primary (disposable) batteries that contain galvanic cells. One of the most common secondary (rechargeable) batteries is found in car engines. Most cars contain a lead-acid battery, shown in Figure 19.18. When you turn the ignition, a surge of electricity from the battery starts the motor.

When in use, a lead-acid battery partially discharges. In other words, the cells in the battery operate as galvanic cells, and produce electricity. The reaction in each cell proceeds spontaneously in one direction. To recharge the battery, a generator driven by the car engine supplies electricity to the battery. The external voltage of the generator reverses the reaction in the cells. The reaction in each cell now proceeds non-spontaneously, and the cells operate as electrolytic cells. All secondary batteries, including the lead-acid battery, operate some of the time as galvanic cells, and some of the time as electrolytic cells.

As the name suggests, the materials used in a lead-acid battery include lead and an acid. Figure 19.19 shows that the electrodes in each cell are constructed using lead grids. One electrode consists of powdered lead packed into one grid. The other electrode consists of powdered lead(IV) oxide packed into the other grid. The electrolyte solution is fairly concentrated sulfuric acid, at about 4.5 mol/L.

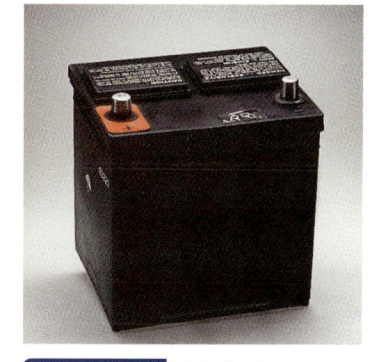

**Figure 19.18** A typical car battery consists of six 2-V cells. The cells are connected in series to give a total potential of 12 V.

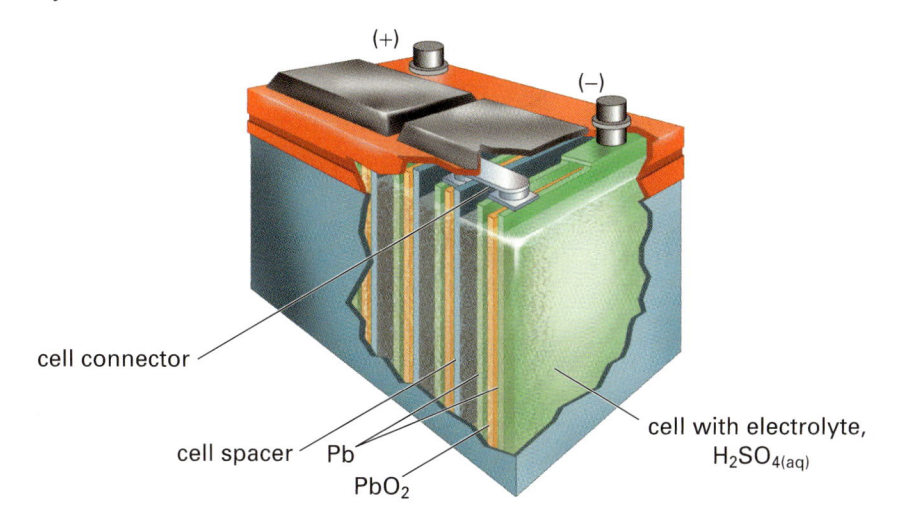

**Figure 19.19** Each cell of a lead-acid battery is a single compartment, with no porous barrier or salt bridge. Fibreglass or wooden sheets are placed between the electrodes to prevent them from touching.

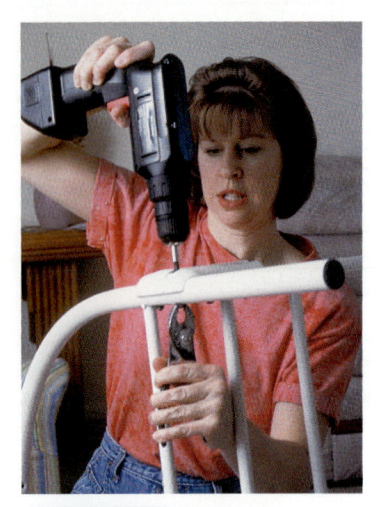

When the battery supplies electricity, the half-reactions and overall cell reaction are as follows.

Oxidation (at the Pb anode): $Pb_{(s)} + SO_4^{2-}{}_{(aq)} \rightarrow PbSO_{4(s)} + 2e^-$

Reduction (at the PbO$_2$ cathode):
$PbO_{2(s)} + 4H^+{}_{(aq)} + SO_4^{2-}{}_{(aq)} + 2e^- \rightarrow PbSO_{4(s)} + 2H_2O_{(\ell)}$

Overall cell reaction:
$Pb_{(s)} + PbO_{2(s)} + 4H^+{}_{(aq)} + 2SO_4^{2-}{}_{(aq)} \rightarrow 2PbSO_{4(s)} + 2H_2O_{(\ell)}$

You can see that the reaction consumes some of the lead in the anode, some of the lead(IV) oxide in the cathode, and some of the sulfuric acid. A precipitate of lead(II) sulfate forms.

When the battery is recharged, the half-reactions and the overall cell reaction are reversed. In this reverse reaction, lead and lead(IV) oxide are redeposited in their original locations, and sulfuric acid is re-formed.

Reduction (at the Pb cathode): $PbSO_{4(s)} + 2e^- \rightarrow Pb_{(s)} + SO_4^{2-}{}_{(aq)}$

Oxidation (at the PbO$_2$ anode):
$PbSO_{4(s)} + 2H_2O_{(\ell)} \rightarrow PbO_{2(s)} + 4H^+{}_{(aq)} + SO_4^{2-}{}_{(aq)} + 2e^-$

Overall cell reaction:
$2PbSO_{4(s)} + 2H_2O_{(\ell)} \rightarrow Pb_{(s)} + PbO_{2(s)} + 4H^+{}_{(aq)} + 2SO_4^{2-}{}_{(aq)}$

In practice, this reversibility is not perfect. However, the battery can go through many charge/discharge cycles before it eventually wears out.

Many types of rechargeable batteries are much more portable than a car battery. For example, there is now a rechargeable version of the alkaline battery. Another example, shown in Figure 19.20, is the rechargeable nickel-cadmium (nicad) battery. Figure 19.21 shows a nickel-cadmium cell, which has a potential of about 1.4 V. A typical nicad battery contains three cells in series to produce a suitable voltage for electronic devices. When the cells in a nicad battery operate as galvanic cells, the half-reactions and the overall cell reaction are as follows.

Oxidation (at the Cd anode): $Cd_{(s)} + 2OH^-{}_{(aq)} \rightarrow Cd(OH)_{2(s)} + 2e^-$

Reduction (at the NiO(OH) cathode):
$NiO(OH)_{(s)} + H_2O_{(\ell)} + e^- \rightarrow Ni(OH)_{2(s)} + OH^-{}_{(aq)}$

Overall cell reaction:
$Cd_{(s)} + 2NiO(OH)_{(s)} + 2H_2O_{(\ell)} \rightarrow Cd(OH)_{2(s)} + 2Ni(OH)_{2(s)}$

Like many technological innovations, nickel-cadmium batteries carry risks as well as benefits. After being discharged repeatedly, they eventually wear out. In theory, worn-out nicad batteries should be recycled. In practice, however, many end up in garbage dumps. Over time, discarded nicad batteries release toxic cadmium. The toxicity of this substance makes it hazardous to the environment, as cadmium can enter the food chain. Long-term exposure to low levels of cadmium can have serious medical effects on humans, such as high blood pressure and heart disease.

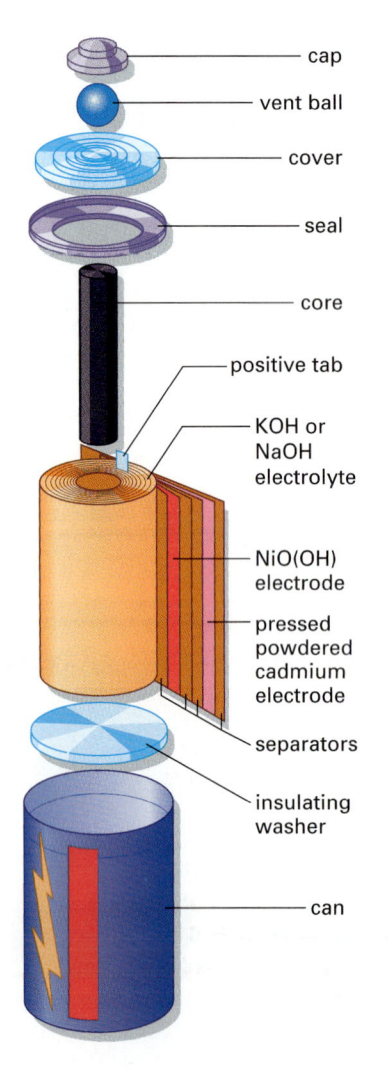

**Figure 19.20** Billions of rechargeable nicad batteries are produced every year. They are used in portable devices such as cordless razors and cordless power tools.

**Figure 19.21** A nicad cell has a cadmium electrode and another electrode that contains nickel(III) oxyhydroxide, NiO(OH). When the cell is discharging, cadmium is the anode. When the cell is recharging, cadmium is the cathode. The electrolyte is a base, sodium hydroxide or potassium hydroxide.

## Section Summary

In this section, you learned about electrolytic cells, which convert electrical energy into chemical energy. You compared the spontaneous reactions in galvanic cells, which have positive cell potentials, with the non-spontaneous reactions in electrolytic cells, which have negative cell potentials. You then considered cells that act as both galvanic cells and electrolytic cells in some common rechargeable batteries. These batteries are an important application of electrochemistry. In the next two sections, you will learn about many more electrochemical applications.

## Section Review

**1** Predict the products of the electrolysis of a 1 mol/L aqueous solution of copper(I) bromide.

**2** In this section, you learned that an external electrical supply reverses the cell reaction in a Daniell cell so that the products are zinc atoms and copper(II) ions.
  (a) What are the predicted products of this electrolysis reaction?
  (b) Explain the observed products.

**3** Predict whether each reaction is spontaneous or non-spontaneous under standard conditions.
  (a) $2FeI_{3(aq)} \rightarrow 2Fe_{(s)} + 3I_{2(s)}$
  (b) $2Ag^+_{(aq)} + H_2SO_{3(aq)} + H_2O_{(\ell)} \rightarrow 2Ag_{(s)} + SO_4^{2-}_{(aq)} + 4H^+_{(aq)}$

**4** Write the two half-reactions and the overall cell reaction for the process that occurs when a nicad battery is being recharged.

**5** What external voltage is required to recharge a lead-acid car battery?

**6** The equation for the overall reaction in an electrolytic cell does not include any electrons. Why is an external source of electrons needed for the reaction to proceed?

**7** (a) Predict whether aluminum will displace hydrogen from water.
  (b) Water boiling in an aluminum saucepan does not react with the aluminum. Give possible reasons why.

**8** Research the impact of lead pollution on the environment. Do lead-acid batteries contribute significantly to lead pollution?

**9** Lithium batteries are increasingly common. The lithium anode undergoes oxidation when the battery discharges. Various cathodes and electrolytes are used to make lithium batteries with different characteristics. Research lithium batteries. Prepare a report describing the designs, cell reactions, and uses of lithium batteries. Include a description of the advantages and disadvantages of these batteries.

## 19.4 Faraday's Law

**Section Preview/Outcomes**

In this section, you will

- **describe** the relationship between time, current, and the amount of substance produced or consumed in an electrolytic process
- **solve** problems using Faraday's law
- **investigate** Faraday's law by performing an electroplating process in the laboratory
- **explain** how electrolytic processes are used to refine metals
- **research** and **assess** some environmental, health, and safety issues involving electrochemistry
- **communicate** your understanding of the following terms: *quantity of electricity, electric charge, Faraday's law, extraction, refining*

As mentioned earlier in this chapter, Michael Faraday (1791–1867) was the leading pioneer of electrochemistry. One of Faraday's major contributions was to connect the concepts of stoichiometry and electrochemistry.

**Figure 19.22** A depiction of Faraday's laboratory.

You know that a balanced equation represents relationships between the quantities of reactants and products. For a reaction that takes place in a cell, stoichiometric calculations can also include the quantity of electricity produced or consumed. Stoichiometric calculations in electrochemistry make use of a familiar unit—the mole.

As a first step, you need information about measurements in electricity. You know that the flow of electrons through an external circuit is called the *electric current*. It is measured in a unit called the *ampere* (symbol A), named after the French physicist André Ampère (1775–1836). The **quantity of electricity**, also known as the **electric charge**, is the product of the current flowing through a circuit and the time for which it flows. The quantity of electricity is measured in a unit called the *coulomb* (symbol C). This unit is named after another French physicist, Charles Coulomb (1736–1806). The ampere and the coulomb are related, in that *one coulomb is the quantity of electricity that flows through a circuit in one second if the current is one ampere*. This relationship can be written mathematically.

$$\text{charge (in coulombs)} = \text{current (in amperes)} \times \text{time (in seconds)}$$

For example, suppose a current of 2.00 A flows for 5.00 min. You can use this information to find the quantity of electricity, in coulombs.

$$5.00 \text{ min} = 300 \text{ s}$$

$$2.00 \text{ A} \times 300 \text{ s} = 600 \text{ C, or } 6.00 \times 10^2 \text{ C}$$

For stoichiometric calculations, you also need to know the electric charge on a mole of electrons. This charge can be calculated by multiplying the charge on one electron and the number of electrons in one mole (Avogadro's number). The charge on a mole of electrons is known as one faraday (1 F), named after Michael Faraday.

$$\text{Charge on one mole of electrons} = \frac{1.602 \times 10^{-19} \text{ C}}{1 e^-} \times \frac{6.022 \times 10^{23} e^-}{1 \text{ mol}}$$
$$= 9.647 \times 10^4 \text{ C/mol}$$

A rounded value of 96 500 C/mol is often used in calculations. Note that this rounded value has three significant digits.

The information you have just learned permits a very precise control of electrolysis. For example, suppose you modify a Daniell cell to operate as an electrolytic cell. You want to plate 0.1 mol of zinc onto the zinc electrode. The coefficients in the half-reaction for the reduction represent stoichiometric relationships. Figure 19.23 shows that two moles of electrons are needed for each mole of zinc deposited. Therefore, to deposit 0.1 mol of zinc, you need to use 0.2 mol of electrons.

$$0.1 \text{ mol Zn} \times \frac{2 \text{ mol } e^-}{1 \text{ mol Zn}} = 0.2 \text{ mol } e^-$$

| $Zn^{2+}$ | + | $2e^-$ | → | Zn |
|---|---|---|---|---|
| 1 ion | | 2 electrons | | 1 atom |
| $1 \times 6.02 \times 10^{23}$ ions | | $2 \times 6.02 \times 10^{23}$ electrons | | $1 \times 6.02 \times 10^{23}$ atoms |
| 1 mol of ions | | 2 mol of electrons | | 1 mol of atoms |

**Figure 19.23** A balanced half-reaction shows relationships between the amounts of reactants and products and the amount of electrons transferred.

In the next Sample Problem, you will learn to apply the relationship between the amount of electrons and the amount of an electrolysis product.

### Sample Problem
### Calculating the Mass of an Electrolysis Product

**Problem**

Calculate the mass of aluminum produced by the electrolysis of molten aluminum chloride, if a current of 500 mA passes for 1.50 h.

**What Is Required?**

You need to calculate the mass of aluminum produced.

**What Is Given?**

You know the name of the electrolyte, the current, and the time.
electrolyte: $AlCl_{3(\ell)}$
current: 500 mA
time: 1.50 h

From the previous calculation, you know the charge on one mole of electrons is 96 500 C/mol.

*Continued ...*

### History LINK

Considered by many the greatest experimental chemist ever, Michael Faraday did not receive any formal education beyond the primary grades. At the age of 14, Faraday worked as an apprentice at a book bindery in London, where he educated himself by reading many of the books brought there for binding, including the section on electricity in the *Encyclopaedia Britannica*. A client of the bookbindery gave Faraday tickets to lectures at the Royal Institution given by Sir Humphry Davy. Faraday eagerly attended the lectures, and afterwards presented his detailed and precise notes on them to Davy. Impressed by the young Faraday's diligence, Davy hired him as his laboratory assistant in 1813, saying "his disposition is active and cheerful, his manner intelligent." In 1825, Faraday took over from Davy directing the laboratory at the Royal Institution, and went on to contribute even more to the study of electricity and its applications than Davy, himself an eminent figure in the field, did.

Continued ...

### Plan Your Strategy

**Step 1** Use the current and the time to find the quantity of electricity used.

**Step 2** From the quantity of electricity, find the amount of electrons that passed through the circuit.

**Step 3** Use the stoichiometry of the relevant half-reaction to relate the amount of electrons to the amount of aluminum produced.

**Step 4** Use the molar mass of aluminum to convert the amount of aluminum to a mass.

### Act on Your Strategy

**Step 1** To calculate the quantity of electricity in coulombs, work in amperes and seconds.

$$1000 \text{ mA} = 1 \text{ A}$$

$$\begin{aligned} 500 \text{ mA} &= 500 \cancel{\text{mA}} \times \frac{1 \text{ A}}{1000 \cancel{\text{mA}}} \\ &= 0.500 \text{ A} \end{aligned}$$

$$\begin{aligned} 1.50 \text{ h} &= 1.50 \cancel{\text{h}} \times \frac{60 \cancel{\text{min}}}{1 \cancel{\text{h}}} \times \frac{60 \text{ s}}{1 \cancel{\text{min}}} \\ &= 5400 \text{ s, or } 5.40 \times 10^3 \text{ s} \end{aligned}$$

$$\begin{aligned} \text{Quantity of electricity} &= 0.500 \text{ A} \times 5400 \text{ s} \\ &= 2700 \text{ C, or } 2.70 \times 10^3 \text{ C} \end{aligned}$$

**Step 2** Find the amount of electrons. One mole of electrons has a charge of 96 500 C.

$$\begin{aligned} \text{Amount of electrons} &= 2700 \cancel{\text{C}} \times \frac{1 \text{ mol e}^-}{96\,500 \cancel{\text{C}}} \\ &= 0.0280 \text{ mol e}^- \end{aligned}$$

**Step 3** The half-reaction for the reduction of aluminum ions to aluminum is $Al^{3+} + 3e^- \rightarrow Al$.

$$\begin{aligned} \text{Amount of aluminum formed} &= 0.0280 \cancel{\text{mol e}^-} \times \frac{1 \text{ mol Al}}{3 \cancel{\text{mol e}^-}} \\ &= 0.00933 \text{ mol Al} \end{aligned}$$

**Step 4** Convert the amount of aluminum to a mass.

$$\begin{aligned} \text{Mass of Al formed} &= 0.00933 \cancel{\text{mol}} \text{ Al} \times \frac{27.0 \text{ g Al}}{1 \cancel{\text{mol}} \text{ Al}} \\ &= 0.252 \text{ g} \end{aligned}$$

### Check Your Solution

The answer is expressed in units of mass. To check your answer, use estimation. If the current were 1 A, then 1 mol of electrons would pass in 96 500 s. In this example, the current is less than 1 A, and the time is much less than 96 500 s. Therefore, much less than 1 mol of electrons would be used, and much less than 1 mol (27 g) of aluminum would be formed.

## Practice Problems

**21.** Calculate the mass of zinc plated onto the cathode of an electrolytic cell by a current of 750 mA in 3.25 h.

**22.** How many minutes does it take to plate 0.925 g of silver onto the cathode of an electrolytic cell using a current of 1.55 A?

**23.** The nickel anode in an electrolytic cell decreases in mass by 1.20 g in 35.5 min. The oxidation half-reaction converts nickel atoms to nickel(II) ions. What is the constant current?

**24.** The following two half-reactions take place in an electrolytic cell with an iron anode and a chromium cathode.

Oxidation: $Fe_{(s)} \rightarrow Fe^{2+}_{(aq)} + 2e^-$

Reduction: $Cr^{3+}_{(aq)} + 3e^- \rightarrow Cr_{(s)}$

During the process, the mass of the iron anode decreases by 1.75 g.

**(a)** Find the change in mass of the chromium cathode.

**(b)** Explain why you do not need to know the electric current or the time to complete part (a).

---

The preceding Sample Problem gave an example of the mathematical use of Faraday's law. **Faraday's law** states that *the amount of a substance produced or consumed in an electrolysis reaction is directly proportional to the quantity of electricity that flows through the circuit.*

To illustrate this statement, think about changing the quantity of electricity used in the Sample Problem. Suppose this quantity were doubled by using the same current, 500 mA, for twice the time, 3 h. As a result, the amount of electrons passing into the cell would also be doubled.

$$500 \text{ mA} = 0.500 \text{ A}$$
$$3\text{h} = 2 \times 1.5 \text{ h}$$
$$= 2 \times 5400 \text{ s}$$

$$\text{Quantity of electricity} = 0.500 \text{ A} \times (2 \times 5400 \text{ s})$$
$$= 2 \times 2700 \text{ C}$$
$$= 5400 \text{ C, or } 5.40 \times 10^3 \text{ C}$$

$$\text{Amount of electrons} = 5400 \text{ C} \times \frac{1 \text{ mol e}^-}{96\ 500 \text{ C}}$$
$$= 0.0560 \text{ mol e}^-$$

Then, as you can see from the relevant half-reaction, the mass of aluminum produced would be doubled. The mass of aluminum produced is clearly proportional to the quantity of electricity used.

In Investigation 19-C, you will apply Faraday's law to an electrolytic cell that you construct.

# Investigation 19-C

## Electroplating

**SKILL FOCUS**
Predicting
Performing and recording
Analyzing and interpreting
Communicating results

You have learned that electroplating is a process in which a metal is deposited, or plated, onto the cathode of an electrolytic cell. In this investigation, you will build an electrolytic cell and electrolyze a copper(II) sulfate solution to plate copper onto the cathode. You will use Faraday's law to relate the mass of metal deposited to the quantity of electricity used.

### Question
Does the measured mass of copper plated onto the cathode of an electrolytic cell agree with the mass calculated from Faraday's law?

### Prediction
Predict whether the measured mass of copper plated onto the cathode of an electrolytic cell will be greater than, equal to, or less than the mass calculated using Faraday's law.

### Materials
150 mL 1.0 mol/L $HNO_3$ in a 250 mL beaker
120 mL acidified 0.50 mol/L $CuSO_4$ solution
   (with 5 mL of 6 mol/L $H_2SO_4$ and 3 mL of
   0.1 mol/L HCl added)
drying oven, or acetone in a wash bottle
3 cm × 12 cm × 1 mm Cu strip
50 cm 16-gauge bare solid copper wire
250 mL beaker
adjustable D.C. power supply with ammeter
deionized water in a wash bottle
fine sandpaper
2 electrical leads with alligator clips
electronic balance

### Safety Precautions

- Nitric acid is corrosive. Also, note that the $CuSO_4$ solution contains sulfuric acid and hydrochloric acid. Wash any spills on your skin with plenty of cold water. Inform your teacher immediately.
- Avoid touching the parts of the electrodes that have been washed with nitric acid.
- Acetone is flammable. Use acetone in the fume hood.
- Make sure your hands and your lab bench are dry before handling any electrical equipment.

### Procedure

1. Clean off any tarnish on the copper strip by sanding it gently. Dip the bottom of the copper strip in the nitric acid for a few seconds, and then rinse off the strip carefully with deionized water. Avoid touching the section that has been cleaned by the acid.

2. Place the copper strip in the beaker, with the clean part of the strip at the bottom. Bend the top of the strip over the rim of the beaker so that the copper strip is secured in a vertical position. This copper strip will serve as the anode.

3. Wrap the copper wire around a pencil to make a closely spaced coil. Leave 10 cm of the wire unwrapped. Measure and record the mass of the wire. Dip the coil in the nitric acid, and rinse the coil with water. Use the 10 cm of uncoiled wire to secure the coil on the opposite side of the beaker from the anode, as shown in the diagram. This copper wire will serve as the cathode.

## Analysis

1. Write a balanced equation for the half-reaction that occurs at the cathode.

2. Use the measured current and the time for which the current passed to calculate the quantity of electricity used.

3. Use your answers to questions 1 and 2 to calculate the mass of copper plated onto the cathode.

4. Compare the calculated mass from question 3 with the measured increase in mass of the cathode. Give possible reasons for any difference between the two values.

## Conclusion

5. How did the mass of copper electroplated onto the cathode of the electrolytic cell compare with the mass calculated using Faraday's law? Compare your answer with your prediction from the beginning of this investigation.

## Applications

6. Suppose you repeated this investigation using iron electrodes, and 0.5 mol/L iron(II) sulfate solution as the electrolyte. If you used the same current for the same time, would you expect the increase in mass of the cathode to be greater than, less than, or equal to the increase in mass that you measured? Explain your answer.

7. Suppose you repeated the investigation with the copper(II) sulfate solution, but you passed the current for only half as long as before. How would the masses of copper plated onto the cathode compare in the two investigations? Explain your answer.

8. Could you build a galvanic cell without changing the electrodes or the electrolyte solution you used in this investigation? Explain your answer.

---

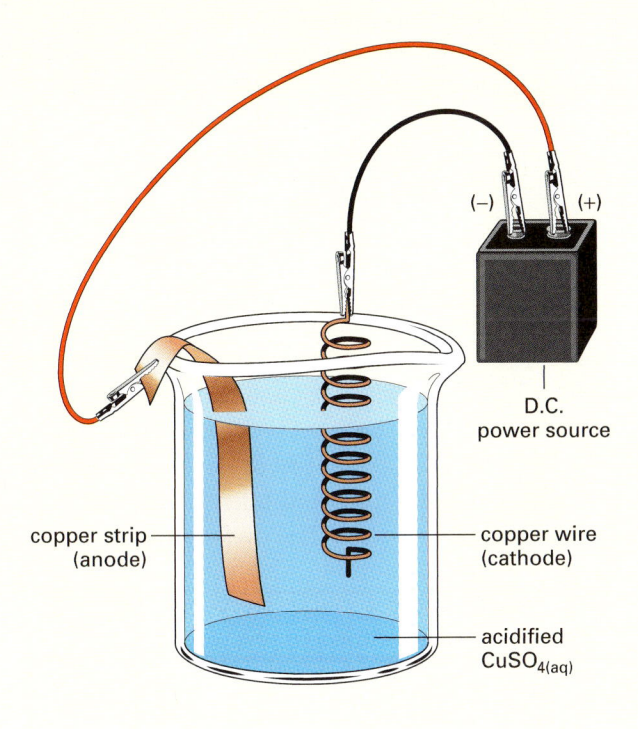

4. Pour 120 mL of the acidified $CuSO_4$ solution into the beaker. Attach the lead from the negative terminal of the power supply to the cathode. Attach the positive terminal to the anode.

5. Turn on the power supply and set the current to 1 A. Maintain this current for 20 min by adjusting the variable current knob as needed.

6. After 20 min, turn off the power. Remove the cathode and rinse it very gently with deionized water. Place the cathode in a drying oven for 20 min. Alternatively, rinse the cathode gently with acetone, and let the acetone evaporate in the fume hood for 5 min.

7. Measure and record the new mass of the cathode.

8. Dispose of all materials as instructed by your teacher.

## Industrial Extraction and Refining of Metals

Many metals, and their alloys, are widely used in modern society. The enormous variety of metal objects ranges from large vehicles, such as cars and aircraft, to small items, such as the pop cans shown in Figure 19.24.

**Extraction** is a process by which a metal is obtained from an ore. Some metals are extracted in electrolytic cells. In section 19.3, you saw the extraction of sodium from molten sodium chloride in a Downs cell. Other reactive metals, including lithium, beryllium, magnesium, calcium, and radium, are also extracted industrially by the electrolysis of their molten chlorides.

One of the most important electrolytic processes is the extraction of aluminum from an ore called bauxite. This ore is mainly composed of hydrated aluminum oxide, $Al_2O_3 \cdot xH_2O$. (The "$x$" in the formula indicates that the number of water molecules per formula unit is variable.) In industry, the scale of production of metals is huge. The electrolytic production of aluminum is over two million tonnes per year in Canada alone. As you know from Faraday's law, the amount of a metal produced by electrolysis is directly proportional to the quantity of electricity used. Therefore, the industrial extraction of aluminum and other metals by electrolysis requires vast quantities of electricity. The availability and cost of electricity greatly influence the location of industrial plants.

In industry, the process of purifying a material is known as **refining**. After the extraction stage, some metals are refined in electrolytic cells. For example, copper is about 99% pure after extraction. This copper is pure enough for some uses, such as the manufacture of copper pipes for plumbing. However, the copper is not pure enough for one of its principal uses, electrical wiring. Therefore, some of the impure copper is refined electrolytically, as shown in Figure 19.25. Nickel can be refined electrolytically in a similar way. You refined copper on a small scale in Investigation 19-C.

**Figure 19.24** The alloy used to make pop cans contains about 97% aluminum, by mass. The other elements in the alloy are magnesium, manganese, iron, silicon, and copper.

**Figure 19.25** This electrolytic cell is used to refine copper. The anode is impure copper, and the cathode is pure copper. During electrolysis, the impure copper anode dissolves, and pure copper is plated onto the cathode. The resulting cathode is 99.99% pure metal. Most impurities that were present in the anode either remain in solution or fall to the bottom of the cell as a sludge.

## Section Summary

In this section, you learned how stoichiometry relates the quantities of reactants and products to the quantity of electricity consumed in an electrolytic cell. You used Faraday's law to solve problems relating to electrolysis. You also learned that the extraction and refining of some metals are carried out electrolytically. In the next section, you will see several other important applications of electrochemistry to modern society.

## Section Review

**1** In Section 18.3, you learned about a redox reaction used in the production of compact discs. In another step of this production process, nickel is electroplated onto the silver-coated master disc. The nickel layer is removed and used to make pressings of the CD onto plastic discs. The plastic pressings are then coated with aluminum to make the finished CDs.

  **(a)** When nickel is plated onto the silver master disc, is the master disc the anode or the cathode of the cell? Explain.

  **(b)** Calculate the quantity of electricity needed to plate each gram of nickel onto the master disc. Assume that the plating process involves the reduction of nickel(II) ions.

**2** Most industrial reactions take place on a much larger scale than reactions in a laboratory or classroom. The voltage used in a Downs cell for the industrial electrolysis of molten sodium chloride is not very high, about 7 V to 8 V. However, the current used is 25 000 A to 40 000 A. Assuming a current of $3.0 \times 10^4$ A, determine the mass of sodium and the mass of chlorine made in 24 h in one Downs cell. Express your answers in kilograms.

**3** An industrial cell that purifies copper by electrolysis operates at $2.00 \times 10^2$ A. Calculate the mass, in tonnes, of pure copper produced if the cell is supplied with raw materials whenever necessary, and if it works continuously for a year that is not a leap year.

**4** Canada is a major producer of aluminum by the electrolysis of bauxite. However, there are no bauxite mines in Canada, and all the ore must be imported. Explain why aluminum is produced in Canada.

**5** Research the extraction of aluminum by the electrolysis of bauxite. Write a report on your findings. Include a description of the electrolytic cell and how it operates. Indicate where aluminum is produced in Canada. Also include any environmental concerns associated with aluminum production by electrolysis.

**6** Nickel and copper are both very important to the Ontario economy. Before they can be refined by electrolysis, they must be extracted from their ores. Both metals can be extracted from a sulfide ore, NiS or $Cu_2S$. The sulfide is roasted to form an oxide, and then the oxide is reduced to the metal. Research the extraction processes for both nickel and copper, and write balanced equations for the redox reactions involved. One product of each extraction process is sulfur dioxide. Research the environmental effects of this compound. Describe any steps taken to decrease these effects.

# 19.5 Issues Involving Electrochemistry

**Section Preview/Outcomes**

In this section, you will

- **explain** corrosion as an electrochemical process, and **describe** some techniques used to prevent corrosion
- **evaluate** the environmental and social impact of some common cells, including the hydrogen fuel cell in electric cars
- **explain** how electrolytic processes are used in the industrial production of chlorine, and how this element is used in the purification of water
- **research** and **assess** some environmental, health, and safety issues connected to electrochemistry
- **communicate** your understanding of the following terms: *corrosion, galvanizing, sacrificial anode, cathodic protection, fuel cell, chlor-alkali process*

You have probably seen many examples of rusty objects, such as the one in Figure 19.26. However, you may not realize that rusting costs many billions of dollars per year in prevention, maintenance, and replacement costs. You will now learn more about rusting and about other issues involving electrochemistry.

**Figure 19.26** Rust is a common sight on iron objects.

## Corrosion

Rusting is an example of **corrosion**, which is a spontaneous redox reaction of materials with substances in their environment. Figure 19.27 shows an example of the hazards that result from corrosion.

Many metals are fairly easily oxidized. The atmosphere contains a powerful oxidizing agent: oxygen. Because metals are constantly in contact with oxygen, they are vulnerable to corrosion. In fact, the term "corrosion" is sometimes defined as the oxidation of *metals* exposed to the environment. In North America, about 20% to 25% of iron and steel production is used to replace objects that have been damaged or destroyed by corrosion. However, not all corrosion is harmful. For example, the green layer formed by the corrosion of a copper roof is considered attractive by many people.

**Figure 19.27** On April 28, 1988, this Aloha Airlines aircraft was flying at an altitude of 7300 m when a large part of the upper fuselage ripped off. This accident was caused by undetected corrosion damage. The pilot showed tremendous skill in landing the plane.

798 MHR • Unit 8 Electrochemistry

Rust is a hydrated iron(III) oxide, $Fe_2O_3 \cdot xH_2O$. The electrochemical formation of rust occurs in small galvanic cells on the surface of a piece of iron, as shown in Figure 19.28. In each small cell, iron acts as the anode. The cathode is inert, and may be an impurity that exists in the iron or is deposited onto it. For example, the cathode could be a piece of soot that has been deposited onto the iron surface from the air.

Water, in the form of rain, is needed for rusting to occur. Carbon dioxide in the air dissolves in water to form carbonic acid, $H_2CO_{3(aq)}$. This weak acid partially dissociates into ions. Thus, the carbonic acid is an electrolyte for the corrosion process. Other electrolytes, such as road salt, may also be involved. The circuit is completed by the iron itself, which conducts electrons from the anode to the cathode.

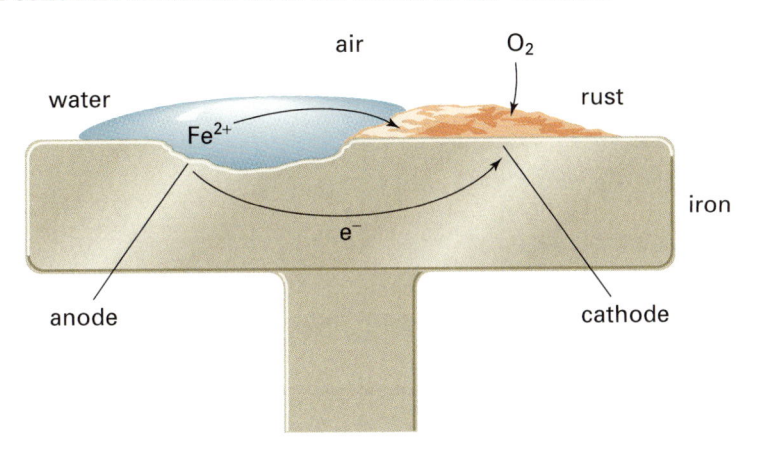

**Figure 19.28** The rusting of iron involves the reaction of iron, oxygen, and water in a naturally occurring galvanic cell on the exposed surface of the metal. There may be many of these small cells on the surface of the same piece of iron.

The rusting process is complex, and the equations may be written in various ways. A simplified description of the half-reactions and the overall cell reaction is given here.

Oxidation half-reaction (occurs at the anode):
$$Fe_{(s)} \rightarrow Fe^{2+}_{(aq)} + 2e^-$$

Reduction half-reaction (occurs at the cathode):
$$O_{2(g)} + 2H_2O_{(\ell)} + 4e^- \rightarrow 4OH^-_{(aq)}$$

Multiply the oxidation half-reaction by two and add the half-reactions to obtain the overall cell reaction.
$$2Fe_{(s)} + O_{2(g)} + 2H_2O_{(\ell)} \rightarrow 2Fe^{2+}_{(aq)} + 4OH^-_{(aq)}$$

There is no barrier in the cell, so nothing stops the dissolved $Fe^{2+}$ and $OH^-$ ions from mixing. The iron(II) ions produced at the anode and the hydroxide ions produced at the cathode react to form a precipitate of iron(II) hydroxide, $Fe(OH)_2$. Therefore, the overall cell reaction could be written as follows.
$$2Fe_{(s)} + O_{2(g)} + 2H_2O_{(\ell)} \rightarrow 2Fe(OH)_{2(s)}$$

The iron(II) hydroxide undergoes further oxidation by reaction with the oxygen in the air to form iron(III) hydroxide.
$$4Fe(OH)_{2(s)} + O_{2(g)} + 2H_2O_{(\ell)} \rightarrow 4Fe(OH)_{3(s)}$$

Iron(III) hydroxide readily breaks down to form hydrated iron(III) oxide, $Fe_2O_3 \cdot xH_2O$, more commonly known as rust. As noted earlier, the "$x$" signifies a variable number of water molecules per formula unit.

$$2Fe(OH)_{3(s)} \rightarrow Fe_2O_3 \cdot 3H_2O_{(s)}$$

Both $Fe(OH)_3$ and $Fe_2O_3 \cdot xH_2O$ are reddish-brown, or "rust coloured." A rust deposit may contain a mixture of these compounds.

Fortunately, not all metals corrode to the same extent as iron. Many metals do corrode in air to form a surface coating of metal oxide. However, in many cases, the oxide layer adheres, or sticks firmly, to the metal surface. This layer protects the metal from further corrosion. For example, aluminum, chromium, and magnesium are readily oxidized in air to form their oxides, $Al_2O_3$, $Cr_2O_3$, and $MgO$. Unless the oxide layer is broken by a cut or a scratch, the layer prevents further corrosion. In contrast, rust easily flakes off from the surface of an iron object and provides little protection against further corrosion.

**Corrosion Prevention**

Corrosion, and especially the corrosion of iron, can be very destructive. For this reason, a great deal of effort goes into corrosion prevention. The simplest method of preventing corrosion is to paint an iron object. The protective coating of paint prevents air and water from reaching the metal surface. Other effective protective layers include grease, oil, plastic, or a metal that is more resistant to corrosion than iron. For example, a layer of chromium protects bumpers and metal trim on cars. An enamel coating is often used to protect metal plates, pots, and pans. *Enamel* is a shiny, hard, and very unreactive type of glass that can be melted onto a metal surface. A protective layer is effective as long as it completely covers the iron object. If a hole or scratch breaks the layer, the metal underneath can corrode.

It is also possible to protect iron against corrosion by forming an alloy with a different metal. *Stainless steel* is an alloy of iron that contains at least 10% chromium, by mass, in addition to small quantities of carbon and occasionally metals such as nickel. Stainless steel is much more resistant to corrosion than pure iron. Therefore, stainless steel is often used for cutlery, taps, and various other applications where rust-resistance is important. However, chromium is much more expensive than iron. As a result, stainless steel is too expensive for use in large-scale applications, such as building bridges.

**Galvanizing** is a process in which iron is covered with a protective layer of zinc. Galvanized iron is often used to make metal buckets and chain-link fences. Galvanizing protects iron in two ways. First, the zinc acts as a protective layer. If this layer is broken, the iron is exposed to air and water. When this happens, however, the iron is still protected. Zinc is more easily oxidized than iron. Therefore, zinc, not iron, becomes the anode in the galvanic cell. The zinc metal is oxidized to zinc ions. In this situation, zinc is known as a **sacrificial anode**, because it is destroyed (sacrificed) to protect the iron. Iron acts as the cathode when zinc is present. Thus, iron does not undergo oxidation until all the zinc has reacted.

**Cathodic protection** is another method of preventing rusting, as shown in Figure 19.29. As in galvanizing, a more reactive metal is attached to the iron object. This reactive metal acts as a sacrificial anode, and the iron becomes the cathode of a galvanic cell. Unlike galvanizing, the metal used in cathodic protection does not completely cover the iron. Because the sacrificial anode is slowly destroyed by oxidation, it must be replaced periodically.

If iron is covered with a protective layer of a metal that is *less* reactive than iron, there can be unfortunate results. A "tin" can is actually a steel can coated with a thin layer of tin. While the tin layer remains intact, it provides effective protection against rusting. If the tin layer is broken or scratched, however, the iron in the steel corrodes *faster* in contact with the tin than the iron would on its own. Since tin is less reactive than iron, tin acts as a cathode in each galvanic cell on the surface of the can. Therefore, the tin provides a large area of available cathodes for the small galvanic cells involved in the rusting process. Iron acts as the anode of each cell, which is its normal role when rusting.

Sometimes, the rusting of iron is promoted accidentally. For example, by connecting an iron pipe to a copper pipe in a plumbing system, an inexperienced plumber could accidentally speed up the corrosion of the iron pipe. Copper is less reactive than iron. Therefore, copper acts as the cathode and iron as the anode in numerous small galvanic cells at the intersection of the two pipes.

Build on your understanding of corrosion by completing the following practice problems.

**Figure 19.29** Magnesium, zinc, or aluminum blocks are attached to ships' hulls, oil and gas pipelines, underground iron pipes, and gasoline storage tanks. These reactive metals provide cathodic protection by acting as a sacrificial anode.

### Practice Problems

**25. (a)** Use the two half-reactions for the rusting process, and a table of standard reduction potentials. Determine the standard cell potential for this reaction.

**(b)** Do you think that your calculated value is the actual cell potential for each of the small galvanic cells on the surface of a rusting iron object? Explain.

**26.** Explain why aluminum provides cathodic protection to an iron object.

**27.** In the year 2000, Transport Canada reported that thousands of cars sold in the Atlantic Provinces between 1989 and 1999 had corroded engine cradle mounts. Failure of these mounts can cause the steering shaft to separate from the car. The manufacturer recalled the cars so that repairs could be made, where necessary. The same cars were sold across the country. Why do you think that the corrosion problems showed up in the Atlantic Provinces?

**28. (a)** Use a table of standard reduction potentials to determine whether elemental oxygen, $O_{2(g)}$, is a better oxidizing agent under acidic conditions or basic conditions.

**(b)** From your answer to part (a), do you think that acid rain promotes or helps prevent the rusting of iron?

## Automobile Engines

The internal combustion engine found in most automobiles uses gasoline as a fuel. Unfortunately, this type of engine produces pollutants, such as carbon dioxide ($CO_2$), nitrogen oxides ($NO_x$), and volatile organic compounds (VOCs). These pollutants contribute to health and environmental problems, such as smog and the greenhouse effect. In addition, the internal combustion engine is very inefficient. It converts only about 25% of the chemical energy of the fuel into the kinetic energy of the car. Electric cars, such as those shown in Figure 19.30, may provide a more efficient and less harmful alternative.

**Figure 19.30** Many electric cars are currently under development. In fact, the idea of electric cars is not new. In the early days of the automobile, electric cars were more common than gasoline-powered cars. The production of electric cars peaked in 1912, but then completely stopped in the 1930s.

Manufacturers and researchers have attempted to power electric cars with rechargeable batteries, such as modified lead-acid and nickel-cadmium batteries. However, rechargeable batteries run down fairly quickly. The distance driven before recharging a battery may be 250 km or less. The battery must then be recharged from an external electrical source. Recharging the lead-acid battery of an electric car takes several hours. Cars based on a version of the nickel-cadmium battery can be recharged in only fifteen minutes. However, recharging the batteries of an electric car is still inconvenient.

A new type of power supply for electric cars eliminates the need for recharging. A **fuel cell** is a battery that produces electricity while reactants are supplied continuously from an external source. Because reactants continuously flow into the cell, a fuel cell is also known as a *flow battery*. Unlike the fuel supply of a more conventional battery, the fuel supply in a fuel cell is unlimited. As in the combustion of gasoline in a conventional engine, the overall reaction in a fuel cell is the oxidation of a fuel by oxygen.

The space shuttle uses a fuel cell as a source of energy. This cell depends on the oxidation of hydrogen by oxygen to form water. The fuel cell operates under basic conditions, so it is sometimes referred to as an *alkaline fuel cell*. Figure 19.31, on the next page, shows the design of the cell. The half-reactions and the overall reaction are as follows.

Reduction (occurs at the cathode): $O_{2(g)} + 2H_2O_{(\ell)} + 4e^- \rightarrow 4OH^-_{(aq)}$

Oxidation (occurs at the anode): $H_{2(g)} + 2OH^-_{(aq)} \rightarrow 2H_2O_{(\ell)} + 2e^-$

Multiply the oxidation half-reaction by 2, add the two half-reactions, and simplify to obtain the overall cell reaction.

$$2H_{2(g)} + O_{2(g)} \rightarrow 2H_2O_{(\ell)}$$

Notice that the overall equation is the same as the equation for the combustion of hydrogen. The combustion of hydrogen is an exothermic reaction. In the fuel cell, this reaction produces energy in the form of electricity, rather than heat.

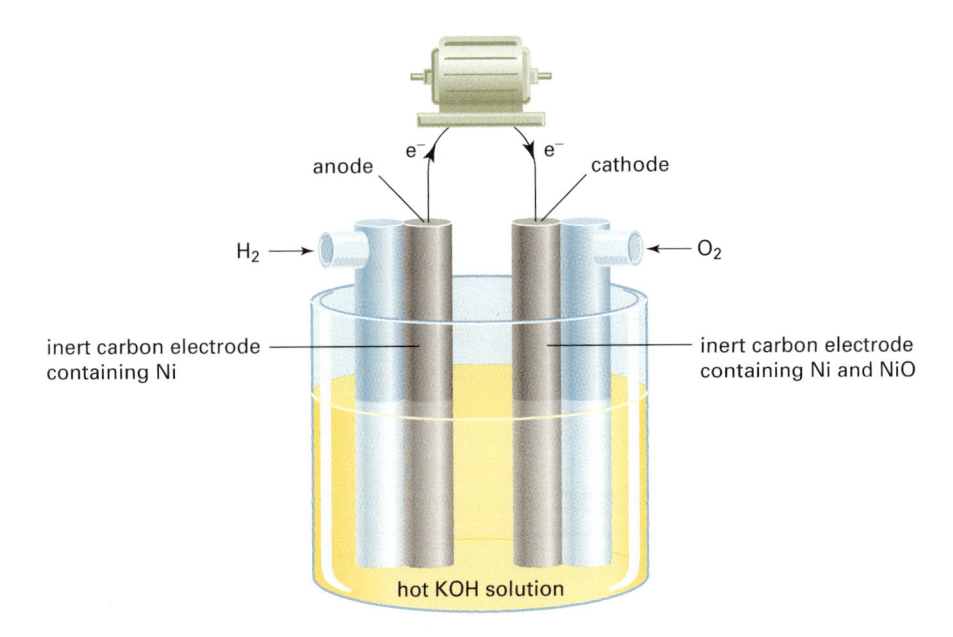

**Figure 19.31** In an alkaline fuel cell, the half-reactions do not include solid conductors of electrons. Therefore, the cell has two inert electrodes.

The hydrogen fuel cell produces water vapour, which does not contribute to smog formation or to the greenhouse effect. This product makes the hydrogen fuel cell an attractive energy source for cars. Also, the hydrogen fuel cell is much more efficient than the internal combustion engine. A hydrogen fuel cell converts about 80% of the chemical energy of the fuel into the kinetic energy of the car.

There is a possible problem with the hydrogen fuel cell. The cell requires hydrogen fuel. Unfortunately, uncombined hydrogen is not found naturally on Earth. Most hydrogen is produced from hydrocarbon fuels, such as petroleum or methane. These manufacturing processes may contribute significantly to pollution problems. However, hydrogen can also be produced by the electrolysis of water. If a source such as solar energy or hydroelectricity is used to power the electrolysis, the overall quantity of pollution is low.

The following practice problems will allow you to test your understanding of fuel cells.

### Practice Problems

29. Calculate $E°_{cell}$ for a hydrogen fuel cell.

30. In one type of fuel cell, methane is oxidized by oxygen to form carbon dioxide and water.
    (a) Write the equation for the overall cell reaction.
    (b) Write the two half-reactions, assuming acidic conditions.

31. Reactions that occur in fuel cells can be thought of as being "flameless combustion reactions." Explain why.

32. If a hydrogen fuel cell produces an electric current of 0.600 A for 120 min, what mass of hydrogen is consumed by the cell?

# Canadians in Chemistry

## Dr. Viola Birss

As a science student, Viola Birss decided to focus on chemistry. "I have always had a concern for the environment," she says. "I am particularly interested in identifying new, non-polluting ways of converting, storing, and using energy." Dr. Birss's interest in non-polluting energy narrowed her field of interest to electrochemistry.

Today, Dr. Birss is a chemistry professor at the University of Calgary. Her research focuses on developing films to coat metal surfaces. Among other uses, these films can serve as protective barriers against corrosion, and as catalysts in fuel cells.

Magnesium alloys are very lightweight, and are being used in the aerospace industry. Because they are very reactive, these alloys need to be protected from corrosion. Dr. Birss holds a patent on a new approach to the electrochemical formation of protective oxide films on magnesium alloys. Dr. Birss also works on developing new catalysts for fuel cells, and studies the factors that lead to the breakdown of fuel cells.

After finishing an undergraduate degree at the University of Calgary, Dr. Birss went on to complete her doctoral degree as a Commonwealth Scholar at the University of Auckland in New Zealand. During her postdoctoral studies at the University of Ottawa, she worked with Dr. Brian Conway, a famous Canadian electrochemist. Dr. Conway's work in electrochemistry has led to progress in a range of electrochemical devices including fuel cells, advanced batteries, and electrolytic cells.

Dr. Birss takes pride in her team of undergraduate, graduate, and post-doctoral students. She works hard to provide a creative and inspiring environment for them. Together, they go on wilderness hiking and cross country ski trips. In addition to creating a sense of community within her team, Birss feels that sports and nature help to recharge her internal "battery." Dr. Birss comments, "These activities seem to provide me with mental and physical rest, so that my creativity and energy are catalyzed."

## Water Treatment and the Chlor-Alkali Process

In many countries, water is not safe to drink. Untreated water is sometimes polluted with toxic chemicals. It may also carry numerous water-borne diseases, including typhoid fever, cholera, and dysentery. In Canada, the water that comes through your tap has been through an elaborate purification process. This process is designed to remove solid particles and toxic chemicals, and to reduce the number of bacteria to safe levels. Adding chlorine to water is the most common way to destroy bacteria.

You have already seen that chlorine gas can be made by the electrolysis of molten sodium chloride. In industry, some chlorine is produced in this way using the Downs cell described earlier. However, more chlorine is produced in Canada using a different method, called the **chlor-alkali process**. In this process, brine is electrolyzed in a cell like the one shown in Figure 19.32. *Brine* is a saturated solution of sodium chloride.

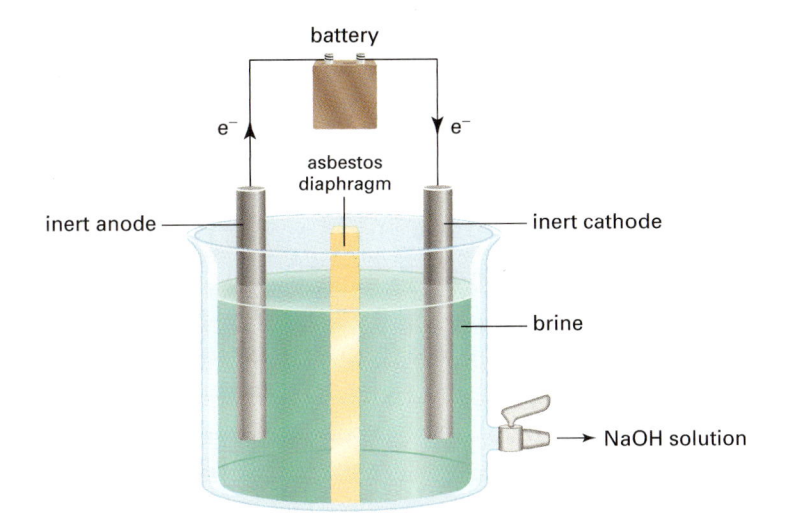

**Figure 19.32** The chlor-alkali cell in this diagram electrolyzes an aqueous solution of sodium chloride to produce chlorine gas, hydrogen gas, and aqueous sodium hydroxide. The asbestos diaphragm stops the chlorine gas produced at the anode from mixing with the hydrogen gas produced at the cathode. Sodium hydroxide solution is removed from the cell periodically, and fresh brine is added to the cell.

The half-reactions and the overall cell reaction in the chlor-alkali process are as follows.

Oxidation: $2Cl^-_{(aq)} \rightarrow Cl_{2(g)} + 2e^-$

Reduction: $2H_2O_{(\ell)} + 2e^- \rightarrow H_{2(g)} + 2OH^-_{(aq)}$

Overall: $2Cl^-_{(aq)} + 2H_2O_{(\ell)} \rightarrow Cl_{2(g)} + H_{2(g)} + 2OH^-_{(aq)}$

Note that the sodium ion is a spectator ion, and does not take part in this reaction. However, it combines with OH⁻ to produce sodium hydroxide, as shown by the balanced chemical equation below.

$$2NaCl_{(aq)} + 2H_2O_{(\ell)} \rightarrow Cl_{2(g)} + H_{2(g)} + 2NaOH_{(aq)}$$

The products of the chlor-alkali process are all useful. Sodium hydroxide is used to make soaps and detergents. It is widely used as a base in many other industrial chemical reactions, as well. The hydrogen produced by the chlor-alkali process is used as a fuel. Chlorine has many uses besides water treatment. For example, chlorine is used as a bleach in the pulp and paper industry. Chlorine is also used in the manufacture of chlorinated organic compounds, such as the common plastic polyvinyl chloride (PVC).

The chlorination of water is usually carried out by adding chlorine gas, sodium hypochlorite, or calcium hypochlorite to the water in low concentrations. The active antibacterial agent in each case is hypochlorous acid, $HClO_{(aq)}$. For example, when chlorine gas is added to water, hypochlorous acid is formed by the following reaction.

$$Cl_{2(g)} + H_2O_{(\ell)} \rightarrow HClO_{(aq)} + HCl_{(aq)}$$

Some people object to the chlorination of water, and prefer to drink bottled spring water. There is controversy over the level of risk associated with chlorination, and over the possible benefits of spring water. For example, hypochlorous acid reacts with traces of organic materials in the water supply. These reactions can produce toxic substances, such as chloroform. Supporters of chlorination believe that these substances are present at very low, safe levels, but opponents of chlorination disagree. Complete the following practice problems to help you decide on your own opinion of chlorination.

### Practice Problems

33. Show that the reaction of chlorine gas with water is a disproportionation reaction.

34. Would you predict the products of the chlor-alkali process to be hydrogen and chlorine? Explain.

35. Research and assess the most recent information you can find on the health and safety aspects of the chlorination of water. Are you in favour of chlorination, or opposed to it? Explain your answer.

36. Some municipalities use ozone gas rather than chlorine to kill bacteria in water. Research the advantages and disadvantages of using ozone in place of chlorine.

**Web LINK**

www.mcgrawhill.ca/links/atlchemistry

For more information about aspects of material covered in this section of the chapter, go to the website above, and click on **Electronic Learning Partner**.

### Section Summary

In this section, you learned about some important electrochemical processes. You had the opportunity to weigh some positive and negative effects of electrochemical technologies. The questions that follow in the section review and chapter review will encourage you to think further about the science of electrochemistry, and about its impact on society.

### Section Review

**1** Why does the use of road salt cause cars to rust faster than they otherwise would?

**2** Aluminum is a more reactive metal than any of the metals present in steel. However, discarded steel cans disintegrate much more quickly than discarded aluminum cans when both are left open to the environment in the same location. Give an explanation.

**3** Explain why zinc acts as a sacrificial anode in contact with iron.

**4** (a) Identify two metals that do not corrode easily in the presence of oxygen and water. Explain why they do not corrode.

  (b) How are these metals useful? How do the uses of these metals depend on their resistance to corrosion?

**5** A silver utensil is said to *tarnish* when its surface corrodes to form a brown or black layer of silver sulfide. Research and describe a chemical procedure that can be used to remove this layer. Write balanced half-reactions and a chemical equation for the process.

**6** In a chlor-alkali cell, the current is very high. A typical current would be about 100 000 A. Calculate the mass of sodium hydroxide, in kilograms, that a cell using this current can produce in one minute.

**7** Research the advances made in the development of fuel cells since this book was written. Describe how any new types of fuel cells operate. Evaluate their advantages and disadvantages, as compared to the internal combustion engine and other fuel cells.

# CHAPTER 19 Review

## Reflecting on Chapter 19

Summarize this chapter in the format of your choice. Here are a few ideas to use as guidelines:
- Represent one example of a galvanic cell, and one example of an electrolytic cell, using chemical equations, half-reactions, and diagrams.
- Calculate a standard cell potential using a table of standard reduction potentials.
- Compare primary and secondary batteries.
- Predict the products of the electrolysis of molten salts and aqueous solutions.
- Perform a sample stoichiometric calculation involving the quantity of electricity used in electrolytic processes.
- Describe some electrolytic processes involved in extracting and refining metals.
- Describe the process of corrosion. Explain some methods used to prevent it.
- Describe the design of fuel cells, and their potential use in automobiles.
- Describe the industrial production of chlorine, and its use in the purification of water.
- Give some examples of how electrochemistry affects the environment, human health, and safety.

## Reviewing Key Terms

For each of the following terms, write a sentence that shows your understanding of its meaning.

electric current
galvanic cell
spontaneous reaction
external circuit
electrolytes
cathode
inert electrode
cell voltage
dry cell
primary battery
reduction potential
electrolytic cell
overvoltage
quantity of electricity
Faraday's law
refining
galvanizing
cathodic protection
fuel cell
electrochemistry
voltaic cell
electrodes
anode
salt bridge
electric potential
cell potential
battery
secondary battery
oxidation potential
electrolysis
electroplating
electric charge
extraction
corrosion
sacrificial anode
chlor-alkali process

## Knowledge/Understanding

1. Explain the function of the following parts of an electrolytic cell.
   (a) electrodes  (c) external voltage
   (b) electrolyte

2. In a galvanic cell, one half-cell has a cadmium electrode in a 1 mol/L solution of cadmium nitrate. The other half-cell has a magnesium electrode in a 1 mol/L solution of magnesium nitrate. Write the shorthand representation.

3. Write the oxidation half-reaction, the reduction half-reaction, and the overall cell reaction for the following galvanic cell.
   $Pt \mid NO_{(g)} \mid NO_3^-{}_{(aq)}, H^+{}_{(aq)} \parallel I^-{}_{(aq)} \mid I_{2(s)}, Pt$

4. What is the importance of the hydrogen electrode?

5. Lithium, sodium, beryllium, magnesium, calcium, and radium are all made industrially by the electrolysis of their molten chlorides. These salts are all soluble in water, but aqueous solutions are not used for the electrolytic process. Explain why.

6. Use the following two half-reactions to write balanced net ionic equations for one spontaneous reaction and one non-spontaneous reaction. State the standard cell potential for each reaction.
   $N_2O_{(g)} + 2H^+{}_{(aq)} + 2e^- \rightleftharpoons N_{2(g)} + H_2O_{(\ell)}$   $E° = 1.770$ V
   $CuI_{(s)} + e^- \rightleftharpoons Cu_{(s)} + I^-{}_{(aq)}$   $E° = -0.185$ V

7. Identify the oxidizing agent and the reducing agent in a lead-acid battery that is
   (a) discharging  (b) recharging

8. Rank the following in order from most effective to least effective oxidizing agents under standard conditions.
   $Zn^{2+}{}_{(aq)}, Co^{3+}{}_{(aq)}, Br_{2(\ell)}, H^+{}_{(aq)}$

9. Rank the following in order from most effective to least effective reducing agents under standard conditions.
   $H_{2(g)}, Cl^-{}_{(aq)}, Al_{(s)}, Ag_{(s)}$

10. The ions $Fe^{2+}{}_{(aq)}$, $Ag^+{}_{(aq)}$, and $Cu^{2+}{}_{(aq)}$ are present in the half-cell that contains the cathode of an electrolytic cell. The concentration of each of these ions is 1 mol/L. If the external voltage is very slowly increased from zero, in what order will the three metals Fe, Ag, and Cu be plated onto the cathode? Explain your answer.

Answers to questions highlighted in red type are provided in Appendix A.

## Inquiry

**11.** Write the half-reactions and calculate the standard cell potential for each reaction. Identify each reaction as spontaneous or non-spontaneous.
(a) $Zn_{(s)} + Fe^{2+}_{(aq)} \rightarrow Zn^{2+}_{(aq)} + Fe_{(s)}$
(b) $Cr_{(s)} + AlCl_{3(aq)} \rightarrow CrCl_{3(aq)} + Al_{(s)}$
(c) $2AgNO_{3(aq)} + H_2O_{2(aq)} \rightarrow 2Ag_{(s)} + 2HNO_{3(aq)} + O_{2(g)}$

**12.** Calculate the mass of magnesium that can be plated onto the cathode by the electrolysis of molten magnesium chloride, using a current of 3.65 A for 55.0 min.

**13.** (a) Describe a method you could use to measure the standard cell potential of the following galvanic cell.
$Sn \mid Sn^{2+}$ (1 mol/L) $\parallel Pb^{2+}$ (1 mol/L) $\mid Pb$
(b) Why is this cell unlikely to find many practical uses?

**14.** The two half-cells in a galvanic cell consist of one iron electrode in a 1 mol/L iron(II) sulfate solution, and a silver electrode in a 1 mol/L silver nitrate solution.
(a) Assume the cell is operating as a galvanic cell. State the cell potential, the oxidation half-reaction, the reduction half-reaction, and the overall cell reaction.
(b) Repeat part (a), but this time assume that the cell is operating as an electrolytic cell.
(c) For the galvanic cell in part (a), do the mass of the anode, the mass of the cathode, and the total mass of the two electrodes increase, decrease, or stay the same while the cell is operating?
(d) Repeat part (c) for the electrolytic cell in part (b).

**15.** (a) Describe an experiment you could perform to determine the products from the electrolysis of aqueous zinc bromide. How would you identify the electrolysis products?
(b) Zinc and bromine are the observed products from the electrolysis of aqueous zinc bromide solution under standard conditions. They are also the observed products from the electrolysis of molten zinc bromide. Explain why the first observation is surprising.

**16.** Use the half-cells shown in a table of standard reduction potentials. Could you build a battery with a potential of 8 V? If your answer is yes, give an example.

**17.** Suppose you produce a kilogram of sodium and a kilogram of aluminum by electrolysis. Compare your electricity costs for these two processes. Assume that electricity is used for electrolysis only, and not for heating.

## Communication

**18.** Research the following information. Prepare a short presentation or booklet on the early history of electrochemistry.
(a) the contributions of Galvani and Volta to the development of electrochemistry
(b) how Humphry Davy and Michael Faraday explained the operation of galvanic and electrolytic cells. (Note that these scientists could not describe them in terms of electron transfers, because the electron was not discovered until 1897.)

**19.** How rapidly do you think that iron would corrode on the surface of the moon? Explain your answer.

**20.** Reactions that are the reverse of each other have standard cell potentials that are equal in size but opposite in sign. Explain why.

**21.** Use a labelled diagram to represent each of the following.
(a) a galvanic cell in which the hydrogen electrode is the anode
(b) a galvanic cell in which the hydrogen electrode is the cathode

## Making Connections

**22.** A D-size dry cell flashlight battery is much bigger than a AAA-size dry cell calculator battery. However, both have cell potentials of 1.5 V. Do they supply the same quantity of electricity? Explain your answer.

**23.** (a) Would you use aluminum nails to attach an iron gutter to a house? Explain your answer.
(b) Would you use iron nails to attach aluminum siding to a house? Explain your answer

24. Research the aluminum-air battery, and the sodium-sulfur battery. Both are rechargeable batteries that have been used to power electric cars. In each case, describe the design of the battery, the half-reactions that occur at the electrodes, and the overall cell reaction. Also, describe the advantages and disadvantages of using the battery as a power source for a car.

25. Explain why the recycling of aluminum is more economically viable than the recycling of many other metals.

26. Suppose you live in a small town with high unemployment. A company plans to build a smelter there to produce copper and nickel by roasting their sulfide ores and reducing the oxides formed. Would you be in favour of the plant being built, or opposed to it? Explain and justify your views.

27. Many metal objects are vulnerable to damage from corrosion. A famous example is the Statue of Liberty. Research the history of the effects of corrosion on the Statue of Liberty. Give a chemical explanation for the processes involved. Describe the steps taken to solve the problem and the chemical reasons for these steps.

28. (a) Estimate the number of used batteries you discard in a year. Survey the class to determine an average number. Now estimate the number of used batteries discarded by all the high school students in your province in a year.
    (b) Prepare an action plan suggesting ways of decreasing the number of batteries discarded each year.

**Answers to Practice Problems and Short Answers to Section Review Questions**

**Practice Problems:** 1.(a) oxidizing agent, Cu(II); reducing agent, Zn (b) same as previous 2.(a) ox: $Sn_{(s)} \rightarrow Sn^{2+}_{(aq)} + 2e^-$, red: $Tl^+_{(aq)} + e^- \rightarrow Tl_{(s)}$, overall: $Sn_{(s)} + 2Tl^+_{(aq)} \rightarrow Sn^{2+}_{(aq)} + 2Tl_{(s)}$, tin anode, thallium cathode (b) ox: $Cd_{(s)} \rightarrow Cd^{2+}_{(aq)} + 2e^-$, red: $2H^+_{(aq)} + 2e^- \rightarrow H_{2(g)}$, overall: $Cd_{(s)} + 2H^+_{(aq)} \rightarrow Cd^{2+}_{(aq)} + H_{2(g)}$, cadmium anode, platinum cathode 3.(a) ox: $2I^-_{(aq)} \rightarrow I_{2(s)} + 2e^-$, red: $MnO_4^-_{(aq)} + 8H^+_{(aq)} + 5e^- \rightarrow Mn^{2+}_{(aq)} + 4H_2O_{(\ell)}$, overall: $10I^-_{(aq)} + 2MnO_4^-_{(aq)} + 16H^+_{(aq)} \rightarrow 5I_{2(s)} + 2Mn^{2+}_{(aq)} + 8H_2O_{(\ell)}$ (b) oxidizing agent, $MnO_4^-$, reducing agent, $I^-$ (c) the anode 4. one electrolyte, no barrier 5. ox: $2Br^-_{(aq)} \rightarrow Br_{2(\ell)} + 2e^-$, red: $Cl_{2(g)} + 2e^- \rightarrow 2Cl^-_{(aq)}$, $E°_{cell} = 0.292$ V 6. ox: $Cu^+_{(aq)} \rightarrow Cu^{2+}_{(aq)} + e^-$, red: $2H^+_{(aq)} + O_{2(g)} + 2e^- \rightarrow H_2O_{2(aq)}$, $E°_{cell} = 0.542$ V 7. ox: $Sn_{(s)} \rightarrow Sn^{2+}_{(aq)} + 2e^-$, red: $2H^+_{(aq)} + 2e^- \rightarrow H_{2(g)}$, $E°_{cell} = 0.138$ V 8. ox: $Cr_{(s)} \rightarrow Cr^{3+}_{(aq)} + 3e^-$, red: $AgCl_{(s)} + e^- \rightarrow Ag_{(s)} + Cl^-_{(aq)}$, $E°_{cell} = 0.966$ V 9.(a) $2Cl^- \rightarrow Cl_2 + 2e^-$ (b) $Ca^{2+} + 2e^- \rightarrow Ca$ (c) $CaCl_2 \rightarrow Ca + Cl_2$ 10.(a) cathode: $Li^+ + e^- \rightarrow Li$ (b) anode: $2Br^- \rightarrow Br_2 + 2e^-$ (c) $2Li^+ + 2Br^- \rightarrow 2Li + Br_2$ 11. direct current: reaction proceeds steadily in one direction 12. reduction (gain of electrons) at cathode of electrolytic cell; electrons come from negative electrode (anode) of battery 13. hydrogen and oxygen 14. hydrogen, not calcium, produced at cathode 15.(a) 0.146 V (b) Cd anode, Ni cathode (c) ox: $Cd \rightarrow Cd^{2+} + 2e^-$, red: $Ni^{2+} + 2e^- \rightarrow Ni$, overall: $Cd + Ni^{2+} \rightarrow Cd^{2+} + Ni$ 16.(a) −0.146 V (b) Ni anode, Cd cathode (c) ox: $Ni \rightarrow Ni^{2+} + 2e^-$, red: $Cd^{2+} + 2e^- \rightarrow Cd$, overall: $Cd^{2+} + Ni \rightarrow Cd + Ni^{2+}$ 17. No 18.(a) spontaneous (b) non-spontaneous (c) spontaneous 19. $2Cu^+_{(aq)} \rightarrow Cu^{2+}_{(aq)} + Cu_{(s)}$, $E°_{cell} = 0.368$ V 20.(a) non-spontaneous (b) spontaneous 21. 2.98 g 22. 8.90 min 23. 1.85 A 24.(a) increases by 1.09 g (b) You can use the stoichiometry of the equations. 25.(a) 0.848 V (b) no; conditions are not standard 26. aluminum is more easily oxidized than iron 27. higher levels of salt (an electrolyte) and moisture 28.(a) acidic conditions (b) promotes it 29. 1.229 V 30.(a) overall: $CH_{4(g)} + 2O_{2(g)} \rightarrow CO_{2(g)} + 2H_2O_{(\ell)}$ (b) ox: $CH_{4(g)} + 2H_2O_{(\ell)} \rightarrow CO_{2(g)} + 8H^+_{(aq)} + 8e^-$, red: $O_{2(g)} + 4H^+_{(aq)} + 4e^- \rightarrow 2H_2O_{(\ell)}$, 31. same equation as combustion, but fuel does not burn 32. 0.0452 g 33. oxidation number of Cl increases from 0 to +1 in forming HClO, decreases from 0 to −1 in forming HCl; chlorine undergoes both oxidation and reduction 34. hydrogen and oxygen, but conditions are far from standard

**Section Review: 19.1:** 1. ox: manganese(IV) oxide, red: zinc 3.(a) ox: $Zn_{(s)} + 2OH^-_{(aq)} \rightarrow ZnO_{(s)} + H_2O_{(\ell)} + 2e^-$, red: $Ag_2O_{(s)} + H_2O_{(\ell)} + 2e^- \rightarrow 2Ag_{(s)} + 2OH^-_{(aq)}$, overall: $Zn_{(s)} + Ag_2O_{(s)} \rightarrow ZnO_{(s)} + 2Ag_{(s)}$ (b) zinc anode, stainless steel cathode 4. 3 V, two cells connected in series 5. four dry cells (4 × 1.5 V = 6 V) 7. it dissolves; inner container is zinc anode 8. Fe anode, Ag cathode, ox: $Fe_{(s)} \rightarrow Fe^{2+}_{(aq)} + 2e^-$, red: $Ag^+_{(aq)} + e^- \rightarrow Ag_{(s)}$, overall: $Fe_{(s)} + 2Ag^+_{(aq)} \rightarrow Fe^{2+}_{(aq)} + 2Ag_{(s)}$ **19.2:** 1.(a) 0.710 V (b) 5.797 V (c) 0.432 V 2.(a) 0.599 V (b) 1.899 V (c) 0.994 V 3. standard hydrogen electrode 4. no; biggest difference in $E°_{red}$ is less than 7 V 6. 0.987 V **19.3:** 1. copper, oxygen, $H^+_{(aq)}$ 2.(a) copper(II) ions, hydrogen (b) hydrogen gas production requires overvoltage, so zinc forms at cathode 3.(a) non-spontaneous (b) spontaneous 4. ox: $Ni(OH)_{2(s)} + OH^-_{(aq)} \rightarrow NiO(OH)_{(s)} + H_2O_{(\ell)} + e^-$, red: $Cd(OH)_{2(s)} + 2e^- \rightarrow Cd_{(s)} + 2OH^-_{(aq)}$, overall: $Cd(OH)_{2(s)} + 2Ni(OH)_{2(s)} \rightarrow Cd_{(s)} + 2NiO(OH)_{(s)} + 2H_2O_{(\ell)}$ 5. over 12 V 7.(a) yes **19.4:** 1.(a) cathode (b) $3.29 \times 10^3$ C 2. 616 kg Na, 950 kg $Cl_2$ 3. 2.08 t 4. plentiful and inexpensive electricity **19.5:** 1. salt acts as electrolyte 2. aluminum protected by oxide layer 4.(a) e.g., gold, platinum 6. 2.49 kg

# UNIT 8 Review

## Knowledge/Understanding

### Multiple Choice

In your notebook, write the letter for the best answer to each question.

1. The oxidation number of carbon in a compound cannot be
   (a) +4
   (b) +2
   (c) 0
   (d) +6
   (e) −4

2. Rusting is
   (a) a decomposition reaction
   (b) a combustion reaction
   (c) a redox reaction
   (d) a neutralization reaction
   (e) a double displacement reaction

3. Of the following, the most effective reducing agent is
   (a) Na
   (b) Pb
   (c) Cl$^-$
   (d) H$_2$
   (e) H$_2$O

4. Primary batteries contain
   (a) rechargeable cells
   (b) no cells
   (c) fuel cells
   (d) electrolytic cells
   (e) galvanic cells

5. In an electroplating experiment, the mass of metal plated onto the cathode
   (a) doubles, if the external voltage is doubled
   (b) doubles, if the current is doubled
   (c) depends on the material used to make the cathode
   (d) depends only on the time for which the current flows
   (e) equals the decrease in mass of the electrolyte

6. A fuel cell
   (a) does not form any products
   (b) has a negative cell potential
   (c) uses a spontaneous reaction to generate electricity
   (d) is not very efficient
   (e) produces its own fuel

7. A metal cathode in a cell always
   (a) decreases in mass while the cell is in operation
   (b) conducts electrons out of the cell
   (c) has a positive polarity
   (d) undergoes reduction
   (e) consists of a different metal than the metal anode in the same cell

8. In both a galvanic cell and an electrolytic cell,
   (a) reduction occurs at the cathode
   (b) the cell potential is positive
   (c) the overall reaction is spontaneous
   (d) an external voltage is required
   (e) chemical energy is converted to electrical energy

9. A type of reaction that is always a redox reaction is
   (a) synthesis
   (b) decomposition
   (c) single displacement
   (d) double displacement
   (e) neutralization

10. The sum of the oxidation numbers of all the elements in a compound
    (a) depends on the formula of the compound
    (b) is positive
    (c) is negative
    (d) is zero
    (e) depends on the type(s) of chemical bonds in the compound

11. While a lead-acid battery is being recharged,
    (a) the mass of lead(II) sulfate in the battery in increasing
    (b) the number of electrons in the battery is increasing
    (c) chemical energy is being converted to electrical energy
    (d) no chemical reaction is taking place
    (e) the cell potential is increasing

12. The oxidation number of oxygen
    (a) is always negative
    (b) is usually negative
    (c) is never zero
    (d) is never positive
    (e) is never −1

Answers to questions highlighted in red type are provided in Appendix A.

## Short Answer

In your notebook, write a sentence or a short paragraph to answer each question.

13. Determine the oxidation number of
    (a) N in $N_2O_3$
    (b) P in $H_4P_2O_7$
    (c) Si in $SiF_6^{2-}$
    (d) each element in $(NH_4)_2SO_4$

14. Does the fact that you can assign oxidation numbers of +1 to hydrogen and −2 to oxygen in water mean that water is an ionic substance? Explain.

15. (a) Limewater, $Ca(OH)_{2(aq)}$, turns cloudy in the presence of carbon dioxide. Write a balanced chemical equation for the reaction. Include the states of the substances.
    (b) Is this reaction a redox reaction?

16. Balance each of the following half-reactions.
    (a) $Hg_2^{2+} \rightarrow Hg$
    (b) $TiO_2 \rightarrow Ti^{2+}$ (acidic conditions)
    (c) $I_2 \rightarrow H_3IO_6^{3-}$ (basic conditions)

17. Explain why you cannot use the table of standard reduction potentials in Appendix E to calculate the external voltage required to electrolyze molten sodium chloride.

18. Why is the density of the electrolyte solution in a lead-acid battery greatest when the battery is fully recharged?

19. Some metals can have different oxidation numbers in different compounds. In the following reactions of iron with concentrated nitric acid, assume that one of the products in each case is gaseous nitrogen monoxide. Include the states of all the reactants and products in the equations.
    (a) Write a balanced chemical equation for the reaction of iron with concentrated nitric acid to form iron(II) nitrate.
    (b) Write a balanced chemical equation for the reaction of iron(II) nitrate with concentrated nitric acid to form iron(III) nitrate.
    (c) Write a balanced chemical equation for the reaction of iron with concentrated nitric acid to form iron(III) nitrate.
    (d) How are the equations in parts (a), (b), and (c) related?

20. For the following galvanic cell,
    $C_{(s)}, I_{2(s)} | I^-_{(aq)} || Ag^+_{(aq)} | Ag_{(s)}$
    (a) identify the anode, the cathode, the positive electrode, and the negative electrode
    (b) write the two half-reactions and the overall cell reaction
    (c) identify the oxidizing agent and the reducing agent
    (d) determine the standard cell potential

21. State whether the reaction shown in each of the following unbalanced equations is a redox reaction. If so, identify the oxidizing agent and the reducing agent. Balance each equation.
    (a) $Cl_2O_7 + H_2O \rightarrow HClO_4$
    (b) $I_2 + ClO_3^- \rightarrow IO_3^- + Cl^-$ (acidic conditions)
    (c) $S^{2-} + Br_2 \rightarrow SO_4^{2-} + Br^-$ (basic conditions)
    (d) $HNO_3 + H_2S \rightarrow NO + S + H_2O$

22. Determine the standard cell potential for each of the following reactions. State whether each reaction is spontaneous or non-spontaneous.
    (a) $2Fe^{2+}_{(aq)} + I_{2(s)} \rightarrow 2Fe^{3+}_{(aq)} + 2I^-_{(aq)}$
    (b) $Au(NO_3)_{3(aq)} + 3Ag_{(s)} \rightarrow 3AgNO_{3(aq)} + Au_{(s)}$
    (c) $H_2O_{2(aq)} + 2HCl_{(aq)} \rightarrow Cl_{2(g)} + 2H_2O_{(\ell)}$

## Inquiry

23. Predict the products from the electrolysis of a 1 mol/L solution of hydrochloric acid.

24. Suppose you decide to protect a piece of iron from rusting, by covering it with a layer of lead.
    (a) Would the iron rust if the lead layer completely covered the iron? Explain.
    (b) Would the iron rust if the lead layer partially covered the iron? Explain.

25. When aqueous solutions of potassium permanganate, $KMnO_4$, and sodium oxalate, $Na_2C_2O_4$, react in acidic solution, the intense purple colour of the permanganate ion fades and is replaced by the very pale pink colour of manganese(II) ions. Gas bubbles are observed as the oxalate ions are converted to carbon dioxide. If the redox reaction is carried out as a titration, with the permanganate being added to the oxalate, the permanganate acts as both reactant and indicator. The persistence of the purple colour in the solution with the addition of one drop of permanganate at the end-point shows that the reaction is complete.

(a) Complete and balance the equation for acidic conditions.
$MnO_4^- + C_2O_4^{2-} \rightarrow Mn^{2+} + CO_2$

(b) If 14.28 mL of a 0.1575 mol/L potassium permanganate solution reacts completely with 25.00 mL of a sodium oxalate solution, what is the concentration of the sodium oxalate solution?

26. Redox titrations can be used for chemical analysis in industry. For example, the percent by mass of tin in an alloy can be found by dissolving a sample of the alloy in an acid to form aqueous tin(II) ions. Titrating with cerium(IV) ions produces aqueous tin(IV) and cerium(III) ions. A 1.475 g sample of an alloy was dissolved in an acid and reacted completely with 24.38 mL of a 0.2113 mol/L cerium(IV) nitrate solution. Calculate the percent by mass of tin in the alloy.

27. Design and describe a procedure you could use to galvanize an iron nail. Explain your choice of materials.

28. Suppose you build four different half-cells under standard conditions in the laboratory.
    (a) What is the greatest number of different galvanic cells you could make from the four half-cells?
    (b) What is the smallest number of standard cell potentials you would need to measure to rank the half-cells from greatest to least standard reduction potentials? Explain.

29. In a galvanic cell, the mass of the magnesium anode decreased by 3.38 g while the cell produced electricity.
    (a) Calculate the quantity of electricity produced by the cell.
    (b) If the constant current flowing through the external circuit was 100 mA, for how many hours did the cell produce electricity?

30. Calculate the mass of aluminum plated onto the cathode by a current of 2.92 A that is supplied for 71.0 min to an electrolytic cell containing aqueous aluminum nitrate as the electrolyte.

31. An industrial method for manufacturing fluorine gas is the electrolysis of liquid hydrogen fluoride.
$2HF_{(\ell)} \rightarrow H_{2(g)} + F_{2(g)}$
If the current supplied to the electrolytic cell is 5000 A, what mass of fluorine, in tonnes, is produced by one cell in one week?

32. Describe how you could build a galvanic cell and an electrolytic cell in which the two electrodes are made of lead and silver. Include a list of the materials you would require.

33. It is possible to use the standard reduction potentials for the reduction of hydrogen ions and the reduction of water molecules to show that the dissociation of water molecules into hydrogen ions and hydroxide ions is non-spontaneous under standard conditions. Describe how you would do this. How is this result consistent with the observed concentrations of hydrogen ions and hydroxide ions in pure water?

## Communication

34. List all the information you can obtain from a balanced half-reaction. Give two examples to illustrate your answer.

35. (a) If one metal (metal A) displaces another metal (metal B) in a single displacement reaction, which metal is the more effective reducing agent? Explain. Give an example to illustrate your answer.
    (b) If one non-metal (non-metal A) displaces another non-metal (non-metal B) in a single displacement reaction, which non-metal is the more effective oxidizing agent? Explain. Give an example to illustrate your answer.

36. (a) Sketch a cell that has a standard cell potential of 0 V.
    (b) Can the cell be operated as an electrolytic cell? Explain.

37. If a redox reaction cannot create or destroy electrons, how is it possible that the balanced half-reactions for a redox reaction may include different numbers of electrons?

38. **(a)** A friend in your class has difficulty recognizing disproportionation reactions. Write a clear explanation for your friend on how to recognize a disproportionation reaction.
    **(b)** Write a balanced chemical equation that represents a disproportionation reaction.

39. **(a)** An electrolytic cell contains a standard hydrogen electrode as the anode and another standard half-cell. Is the standard reduction potential for the half-reaction that occurs in the second half-cell greater than or less than 0 V? Explain how you know.
    **(b)** Will your answer for part (a) change if the hydrogen electrode is the cathode of the electrolytic cell? Explain.

40. Write a descriptive paragraph to compare the reactions that occur in a Downs cell and a chlor-alkali cell. Describe the similarities and differences.

41. One of your classmates is having trouble understanding some of the main concepts in this unit. Use your own words to write an explanation for each of the following concepts. Include any diagrams or examples that will help to make the concepts clear to your classmate.
    **(a)** oxidation and reduction
    **(b)** galvanic and electrolytic cells

## Making Connections

42. Black-and-white photographic film contains silver bromide. When exposed to light, silver bromide decomposes in a redox reaction. Silver ions are reduced to silver metal, and bromide ions are oxidized to bromine atoms. The brighter the light that hits the film, the greater is the decomposition. When the film is developed, the parts of the film that contain the most silver metal produce the darkest regions on the negative. In other words, the brightest parts of the photographed scene give the darkest parts of the image on the negative. High-speed black-and-white film uses silver iodide in place of silver bromide. Do you think that silver iodide is more sensitive or less sensitive to light than silver bromide? Explain.

43. The two rocket booster engines used in a space shuttle launch contain a solid mixture of aluminum and ammonium perchlorate. The products of the reaction after ignition are aluminum oxide and ammonium chloride.
    **(a)** Write a balanced chemical equation for this reaction, and identify the oxidizing agent and reducing agent.
    **(b)** The other four engines used in a space shuttle launch use the redox reaction of liquid hydrogen and liquid oxygen to form water vapour. Write a balanced chemical equation for this reaction.
    **(c)** Which of the reactions described in parts (a) and (b) do you think is the more environmentally friendly reaction? Explain.

44. Every year, corrosion is responsible for the failure of many thousands of water mains. These are the pipes that transport water to Canadian homes and businesses. Research and describe the economic, environmental, health, and safety issues associated with rupture and repair of water mains. Describe the methods that are being used to improve the situation.

45. Since the corrosion of iron is such an expensive problem, why do you think that iron is still used for so many purposes?

46. Industrial plants that make use of electrochemical processes are often located close to sources of hydroelectricity. Research and identify an example in Ontario. Write a brief description of what the plant produces and how it produces it.

# Appendix A

## Answers to Selected and Numerical Chapter and Unit Review Questions

### Chapter 1

6. 32 neutrons, 27 electrons
7. (a) 7   (b) 7   (c) 10
   (d) 3−   (e) $^{79}_{34}$Se   (f) 2−
   (g) Cr   (h) 24   (i) 28
   (j) 21   (k) 3+   (l) 19
   (m) 9   (n) 9   (o) 9
8. (a) SnF$_2$   (b) BaSO$_4$   (c) Mg(OH)$_2$
   (d) CsBr   (e) NH$_4$HCO$_3$   (f) K$_3$PO$_4$
   (g) Fe$_2$(SO$_4$)$_3$
9. (a) sodium hydrogen carbonate
   (b) iron(II) oxide
   (c) copper(II) chloride
   (d) PbO$_2$
   (e) tin(IV) chloride
   (f) diphosphorus pentoxide
   (g) aluminum sulfate
12. (a) PdCl$_{2(aq)}$ + 2HNO$_{3(aq)}$ → Pd(NO$_3$)$_{2(aq)}$ + 2HCl$_{(aq)}$
    (b) Cr$_{(s)}$ + 2HCl$_{(aq)}$ → CrCl$_{2(aq)}$ + H$_{2(g)}$
    (c) 2K$_{(s)}$ + 2H$_2$O$_{(\ell)}$ → 2KOH$_{(aq)}$ + H$_{2(g)}$
13. (a) single displacement;
       H$_{2(g)}$ + CuO$_{(s)}$ → Cu$_{(s)}$ + H$_2$O$_{(g)}$
    (b) synthesis; 16Ag$_{(s)}$ + S$_{8(s)}$ → 8Ag$_2$S$_{(s)}$
    (c) hydrocarbon combustion;
       C$_4$H$_{8(g)}$ + 6O$_{2(g)}$ → 4CO$_{2(g)}$ + 4H$_2$O$_{(g)}$
    (d) synthesis; NH$_{3(aq)}$ + HCl$_{(aq)}$ → NH$_4$Cl$_{(aq)}$
    (e) synthesis; 2Mg$_{(s)}$ + O$_{2(g)}$ → 2MgO$_{(s)}$
14. (a) NaOH$_{(aq)}$ + Fe(NO$_3$)$_{3(aq)}$ → NaNO$_{3(aq)}$ + Fe(OH)$_{3(s)}$
    (b) 2Sb$_{(s)}$ + 3Cl$_{2(g)}$ → 2SbCl$_{3(s)}$
    (c) 2Hg$_{(\ell)}$ + O$_{2(g)}$ → 2HgO$_{(s)}$
    (d) NH$_4$NO$_{3(s)}$ → N$_{2(g)}$ + H$_2$O$_{(\ell)}$
    (e) 2Al$_{(s)}$ + 3Zn(NO$_3$)$_{2(aq)}$ → 2Al(NO$_3$)$_{3(aq)}$ + 3Zn$_{(s)}$
15. (a) synthesis
    (b) Pb$_{(s)}$ + Cr$_{(s)}$ + 2O$_{2(g)}$ → PbCrO$_{4(s)}$
16. double displacement;
    HCl$_{(aq)}$ + CaCO$_{3(s)}$ → CaCl$_{2(aq)}$ + H$_2$CO$_{3(aq)}$
    (the H$_2$CO$_3$ then decomposes to form H$_2$O$_{(\ell)}$ and CO$_{2(g)}$)

## UNIT 1

### Chapter 2

9. 137   10. 77   11. 6.94 u
12. 95% K-39, 5% K-41
13. (a) $2.84 \times 10^{-3}$ mol   (b) 0.517 mol
    (c) $8.15 \times 10^{-5}$ mol   (d) 0.483 mol
    (e) 0.126 mol
14. (a) $7.09 \times 10^{-3}$ mol   (b) 1.23 mol
    (c) $8.94 \times 10^{-2}$ mol   (d) 0.483 mol
    (e) 2.69 mol   (f) $9.27 \times 10^{-4}$ mol

15. (a) 17.0 g/mol   (b) 1.46 mol
    (c) $8.79 \times 10^{23}$ molecules   (d) $3.52 \times 10^{24}$
    (e) 18.0 g/mol   (f) 1.58 g
    (g) $8.77 \times 10^2$ mol   (h) $1.58 \times 10^{23}$
    (i) 157.9 g/mol   (j) 10.5 g
    (k) $6.64 \times 10^{-2}$ mol   (l) $4.00 \times 10^{22}$
    (m) 194.2 g/mol   (n) $4.98 \times 10^{-3}$ mol
    (o) $3.00 \times 10^{21}$   (p) $2.10 \times 10^{22}$ atoms
    (q) 152.2 g/mol   (r) $1.99 \times 10^3$
    (s) 13.1 mol   (t) $1.50 \times 10^{26}$
    (u) 78.0 g/mol   (v) $6.66 \times 10^4$ g
    (w) $5.14 \times 10^{26}$   (x) $3.60 \times 10^{27}$
16. (a) 354.88 g/mol   (b) 74.09 g/mol
    (c) 142.05 g/mol   (d) 252.10 g/mol
    (e) 310.18 g/mol   (f) 182.90 g/mol
17. (a) 66.7 g   (b) 335 g
    (c) $3.75 \times 10^3$ g   (d) 1.45 g
    (e) 57.4 g   (f) $2.05 \times 10^{-2}$ g
18. $21.1 \times 10^{24}$ atoms
19. $4.22 \times 10^{24}$
20. (a) 131.29 u   (b) 131.29 g
    (c) $2.18 \times 10^{-22}$ g   (d) $7.90 \times 10^{25}$ u/mol
    (e) $6.02 \times 10^{23}$ u/g
21. $3.01 \times 10^{24}$
22. $6.54 \times 10^{24}$
23. $4.53 \times 10^{23}$
24. 192 g
25. NaCl + AgNO$_3$ → AgCl + NaNO$_3$
26. (a) 131 g   (b) 380 g
    (c) 4.7 g   (d) $4.0 \times 10^{-5}$
    (e) $6.4 \times 10^{-3}$ g   (f) 78.5 g
27. $M$ = 83.8 g/mol, krypton, Kr$_{(g)}$
32. (a) $2.65 \times 10^{-4}$ mol   (b) 118.2 mg
33. (a) 101 mg

### Chapter 3

5. 2.64 g
6. (a) 9.93% C; 58.6% Cl; 31.4% F
   (b) 80.1% Pb; 16.5% O; 0.3% H; 3.10% C
7. (a) 6.86 g   (b) 1.74 g
8. (a) 63.5%   (b) $1.27 \times 10^2$ kg
9. 58.8%   10. 62.8%   11. C$_4$H$_8$O$_4$
12. C$_{12}$H$_4$O$_2$Cl$_4$
13. (c) C$_3$H$_8$
14. C$_{21}$O$_2$H$_{30}$   15. Na$_2$Cr$_2$O$_7$   16. HgSO$_4$
17. (a) Ca$_3$P$_2$O$_8$   (b) Ca$_3$(PO$_4$)$_2$
18. C$_{18}$H$_{26}$O$_3$N   19. vanadium   20. C$_3$H$_6$O
21. FeSO$_4$ · 7H$_2$O

**22. (a)** 37.50% C; 4.20% H; 58.30% O
   **(b)** $C_6H_8O_7$   **(c)** $C_6H_8O_7$
**23.** 1.37 g $CO_2$; 1.12 g $H_2O$
**28. (a)** 0.87 g/L   **(b)** 0.56 g
**29. (a)** sodium carbonate heptahydrate
   **(b)** 54%   **(c)** $5.25 \times 10^4$ g
**30. (b)** $C_9H_8O_4$

## Chapter 4
**6.** 9.60 g   **7.** $8.0 \times 10^{22}$   **8.** 292 g
**9. (a)** Zn   **(b)** 8.95 g   **(c)** 0.391 g
**10.** 36.0 g   **11.** 2.03 g   **12.** 22.7 g
**13.** 1.655 g   **14.** 2.58 g
**15. (a)** $2CH_3OH_{(\ell)} + 3O_{2(g)} \rightarrow 2CO_{2(g)} + 4H_2O_{(g)}$
   **(b)** 6.67 L   **(c)** 9.52 g
**16.** 4.35 g   **17.** 2.80 L
**18.** $1.396 \times 10^3$ L   **19.** 1.54 L
**20. (a)** 7.32 g   **(b)** 34.2%   **(c)** 7.23 g
**21.** 53.7%
**22. (a)** 15.1 g   **(b)** 0.414 g
   **(c)** 14.2 g   **(d)** 94.0%
**25. (d)** 0.0721 kg
**26. (a)** 380 L   **(b)** 19 L

## Unit 1 Review
**29. (a)** $6.02 \times 10^{23}$ $N_2$ molecules, $1.20 \times 10^{24}$ N atoms
   **(b)** $3.01 \times 10^{24}$   **(c)** $2.26 \times 10^{24}$
**33. (a)** 0.167 mol   **(b)** $1.00 \times 10^{23}$   **(c)** $3.00 \times 10^{23}$
**34.** $1.94 \times 10^{23}$ P atoms   **35.** $1.20 \times 10^{22}$
**36.** $C_{32}H_{64}O_2$; 13.4% H; $C_{32}H_{64}O_2$: 13.4% H
**37.** 68.1% C; 13.7% H; 18.1% O
**38.** 3.74 L
**39. (a)** $P_4S_{3(s)} + 8O_{2(g)} \rightarrow P_4O_{10(g)} + 3SO_{2(g)}$
   **(b)** 2.0 L
   **(c)** 6.5 g
**40.** $C_4H_{10}O$
**44. (a)** $1.48 \times 10^{-2}$ mm   **(b)** 0.1275 nm   **(c)** 0158 nm

# UNIT 2

## Chapter 5
**3.** letters—nucleus and all except valance electrons; dots—valance electrons
**4. (a)** KBr   **(b)** $CaF_2$   **(c)** MgO   **(d)** $Li_2O$
**5.** unpaired electrons will most likely be involved in bonding, electron pairs will most likely not be involved in bonding
**6. (a)** :O::Si::O:   **(b)** :Br:O:Br:
   **(c)** :Cl:F:   **(d)** :F:N:F:
           :F:

**9.** The ratio of potassium to sulfur in the compound is 2:1 **15.** the attraction of the nucleus for electrons in a bond
**18. (a)** 1.79   **(b)** 1.11   **(c)** 1.28   **(d)** 0.40
**19. (a)** ionic   **(b)** polar covalent
   **(c)** polar covalent   **(d)** covalent
**24. (a)** Mn and F, 2.43; Mn and O, 1.89; Mn and N, 1.49
   **(b)** Be and F, 2.41; Be and Cl, 1.59; Be and Br, 1.39
   **(c)** Ti and Cl, 1.62; Fe and Cl, 1.33; Cu and Cl = Hg and Cl, 1.26; Ag and Cl, 1.23
**27.** exothermic—more energy is released when the products form than is required to break the bonds of the reactants.

## Chapter 6
**2.** Diatomic molecules that consist of a single element are non-polar and small. The intermolecular forces are very weak and break easily at room temperature.
**3.** No—ionic compounds are not made of molecules therefore cannot have "intermolecular" forces.
**7. (a)** dipole-dipole   **(b)** ionic
   **(c)** metallic   **(d)** dispersion
**8.** $H_2O$, O–H bonds are more polar than N–H bonds
**9.** $C_3H_8$, $C_2H_5OH$, $SiO_2$; dispersion forces < dipole-dipole forces < network covalent forces
**10.** $CHCl_3 > CH_3Cl > CCl_4$
**12.** If the vectors representing the polar bonds add to zero, the molecules will be non-polar. That is, the effects of the polar bonds cancel each other.
**13. (a)** $NH_2Cl$   **(b)** SiC   **(c)** Xe
   **(d)** $CH_3OH$   **(e)** $CH_3F$   **(f)** $AlCl_3$
   **(g)** $C_4H_{10}$   **(h)** $NH_3$   **(i)** $C_4H_9F$
   **(j)** ammonia   **(k)** silicon dioxide
   **(l)** krypton   **(m)** KCl   **(n)** KBr
   **(o)** $NH_2Cl$   **(p)** $NH_4Cl$   **(q)** $C_2H_5F$
   **(r)** $CH_3NH_2$   **(s)** $C_2HCl$   **(t)** $H_2O$
   **(u)** $SiO_2$   **(v)** NaCl
**22.** $CO_2$ has covalent bonding while $SiO_2$ is a covalent network solid.

## Unit 2 Review
**14.** If the compound consists of a metal and a nonmetal, it is probably an ionic compound.
**18.** The diatomic molecules have a different number of valance electrons so they have to share a different number of electrons to achieve a noble gas configuration.
**23.** short
**26. (a)** linear
   **(b)** trigonal planar
   **(c)** bent
**33.** number of electrons in the molecule and shape of the molecule

35. when a line connecting the two electronegative atoms goes directly through the hydrogen atom
41. LiBr
42. (a) N ← H  (b) F ← N  (c) I → Cl
43. (a) Cl–Cl, Br → Cl, Cl → F
    (b) Si → Cl, P → Cl, S → Cl, Si–Si
46. ammonium ion is tetrahedral and symmetrical, polarity of bonds cancels; ammonia pyramidal and is polar

# UNIT 3

## Chapter 7
11. 5.62 g
12. (a) 25 g  (b) 225 g  (c) 2.5 mol/L
13. 96 mL
14. 1.2 mg NaF; 1.2 ppm
15. 5.67 mol/L
16. (a) 0.427 mol/L  (b) 6.3 mol/L
17. (a) 9.89 g  (b) 83 g
18. (a) 1.7 mol/L  (b) 1.44 mol/L
19. (a) 0.381 mol/L  (b) 0.25 mol/L
20. 25.0 g
23. (b) 85 g/100 g water  (c) 140 g/100 g water
    (d) 77°C
24. 390 g
25. 40°C
29. 0.25 ppm; 250 ppb
32. 55.5 mol/L

## Chapter 8
11. 0.015 mol/L
12. (a) 0.172 mol/L  (b) 0.578 mol/L  (c) 0.694 mol/L
13. 0.104 mol/L
14. $K^+_{(aq)}$ = 0.0667 mol/L; $Ca^{2+}_{(aq)}$ = 0.0667 mol/L; $NO_3^-_{(aq)}$ = 0.200 mol/L
16. (a) $Ca^{2+}_{(aq)}$ = 0.045 mol/L; $Mg^{2+}_{(aq)}$ = 0.024 mol/L
    (b) $Ca^{2+}$ = 1.8 × $10^3$ ppm; $Mg^{2+}$ = 5.8 × $10^2$ ppm
17. 0.014 mol
18. 12.2 g $Pb_3(PO_4)_2$
20. (b) 0.01057 mol  (c) 120.4 g/mol
22. (b) 0.09600%
23. (a) 1.3 L  (b) 2.0 g  (c) 2.0 mL

## Unit 3 Review
16. 3.6 × $10^{-3}$ g
27. 3.89 × $10^{-2}$ g
28. 1.94 × $10^{-3}$
29. 2.43 × $10^{-2}$ mol/L
30. (b) 7.41 × $10^{-2}$ mol/L
31. (b) 0.364 g
32. (a) 119 g
33. (a) 0.031 mol  (b) 3.40 × $10^{-3}$ mol
34. 2.295 × $10^{-3}$ mol/L
38. (a) 0.291 mol/L  (b) 0.436 mol/L  (c) 0.257 mol/L
41. 5.65 mol
42. (a) 0.420 mol/L  (b) 0.840 mol/L
43. $[H^+]$ = 0.705 mol/L; $[OH^-]$ negligible; $[Ba^{2+}_{(aq)}]$ = 0.0640 mol/L; $[Cl^-_{(aq)}]$ = 1.66 mol/L
44. (a) 1.8 mol  (b) 1.8 mol  (c) 0.90 mol/L
45. (a) 0.212 mol  (b) 12.4 g

# UNIT 4

## Chapter 9
1. (a) fossil fuels
   (b) wood, plant fermentation products, fossil fuels
   (c) petroleum
2. bacterial activity, heat, pressure
5. increases
11. (a) $C_nH_{2n+2}$  (b) $C_nH_{2n}$  (c) $C_nH_{2n-2}$
    (d) $C_nH_{2n}$  (e) $C_nH_{2n-2}$
12. (a) alkene  (b) alkane
    (c) alkyne  (d) alkane
13. (a) aliphatic  (b) aromatic  (c) aliphatic
    (d) aliphatic  (e) aromatic
14. (a) alkyne  (b) alkene  (c) alkyne
    (d) alkane  (e) alkene (cycloalkene)
15. (b) (a) unsaturated, (b) unsaturated, (c) unsaturated, (d) saturated, (e) unsaturated
16. (a) butane
    (b) ethane
    (c) 2,3,4-trimethylpentane
    (d) 4-ethyl-3-methylhexane
17. (a) 2-methylpropene
    (b) 1-ethyl-2-methylcyclohexane
    (c) 3-methyl-1-pentyne
    (d) trans-3-methyl-2-pentene
    (e) 3-ethyl-3,4-dimethylhexane
    (f) ethylbenzene
18. trans-2-heptene and cis-2-heptene
19. (A)—(d); (B)—(f); (C)—(c); (D)—(a)
20. no; 3-methyl-1-cyclobutene
21. oil floats on water, does not dissolve
22. the methyl group would be part of the main chain
23. (a) and (c), (b) and (d) are the same
24. (a) ethylbenzene
    (b) 1-propyl-3-ethylbenzene
    (c) 1-ethyl-2-methylbenzene
    (d) 1,4-dimethylbenzene

29. (a) ethane  (b) Br$_2$  (c) benzene
    (d) $\frac{9}{2}$O$_2$  (e) Cl$_2$  (e) 2-pentyne
30. (a) complete  (b) incomplete  (c) incomplete
31. (a) 2-bromobutane + HBr, substitution
    (b) 3-methyl-2,2-dichloropentane, addition
32. (b) cracking
43. (b), (c), (d), (a)

## Chapter 10

1. (a) tetrahedral, trigonal planar, trigonal planar
   (b) C=O
   (c) non-polar, polar, non-polar
2. (a) polar  (b) non-polar
   (c) polar  (d) non-polar
   (e) polar
4. (a) R—OH  (b) R—NR'$_2$
   (c) R—COO—R'  (d) R—CON—R'$_2$
6. (a) amine  (b) alkane
   (c) alcohol  (d) ester
   (e) amide  (f) carboxylic acid
7. (a) hydroxyl
   (b) double bond, amine
   (c) double bond, ester
   (d) triple bond, carbonyl, ether
   (e) chlorine, amide
   (f) carboxyl group, carbonyl group
8. (a) amine  (b) aldehyde
   (c) ester  (d) carboxylic acid
   (e) ketone
9. (a) alcohol
10. (a) alkane
11. (a) 1-propanol  (b) 3-octanol
    (c) 2-hexanamine  (d) 1-methoxybutane
    (e) 2-propanone  (f) methyl pentanoate
    (g) 3-methyl-3-hexanol  (h) butanal
    (i) 2-bromopropane  (j) cyclopentanol
    (k) pentanoic acid  (l) ethyl propanoate
    (m) 5-ethyl-5,6-dimethyl-3-heptanone
    (n) *trans*-1-bromo-2-iodo-ethene
12. (a) ethoxyethane, ether
    (b) octanal, aldehyde
    (c) hexanamide, amide
    (d) 3-ethyl-2-hexanol
    (e) 2-methylpentanoic acid
    (f) 1-butanamine, amine
    (g) N-propylpropanamide, amide
13. (a) ethyl alcohol  (b) methyl alcohol
    (c) formaldehyde  (d) acetic acid
    (e) acetone  (f) isopropyl alcohol
15. (a) elimination  (b) substitution
    (c) addition  (d) substitution
16. (a) A, ethanamine; B, ethane; C, methoxy methane; D, ethanol; E, ethanoic acid
17. (a) 1-bromopentane + water
    (b) propene + water
    (c) methyl propanoate + water
    (d) ethanoic acid + ethanol
19. (a) propanoic acid + 1-butanol
    (b) cyclobutanol
25. (a) either 2-methylhexanal or pentanal
    (b) N-ethyl-N-methylpentanamide
    (c) 2-methylpropanoic acid
26. 1-butanol, 2-butanol, 2-methyl-1-propanol, 2-methyl-1-propanol, 1-methoxypropane, 1-ethoxyethane, 2-methoxypropane
27. butanoic acid, 2-methylpropanic acid, methyl propanoate, ethyl ethanoate, 1-propyl methanoate, 2-propyl methanoate (**Note:** Other isomers are possible that contain hydroxyl and carbonyl groups, or alkoxy and carbonyl groups, but these are beyond the scope of this course.)

## Chapter 11

2. Polymers are made of repeating units called monomers.
6. (a) propene, H$_2$C=CHCH$_3$
   (b) addition
   (c) polypropylene
8. polyamide; Amino acids are joined by amide linkages.
14. (a) monosaccharide or simple sugar
    (b) glucose
16. (a) ethene, CH$_2$=CH$_2$
    (b) CH$_3$CHOHCH$_2$COOH
    (c) CH$_2$=CHCH$_2$CH$_3$
    (d) NH$_2$CH$_2$CH(CH$_3$)NH$_2$ + HOOCCH$_2$COOH
    (e) HOCH$_2$(benzene)CH$_2$OH + HOOCCH$_2$COOH

## Unit 4 Review

20. alkenes, alkynes and aromatics do
21. higher  23. grain alcohol  24. ethanoic acid
26. (a) butanoic acid  (b) methanol
    (c) methyl pentanoate  (d) 2-butanone
    (e) N-methylethanamine  (f) 4,5-dimethylheptanal
27. (a) methyl alcohol or wood alcohol
    (b) isopropyl alcohol
    (c) ether
28. (a) 3-ethyl-4-methyl-1-hexene
    (b) methoxypropane
    (c) 3-methyl-1-pentanamine
    (d) 5-ethyl-3,7,9-trimethyl-2-decanol
    (e) butyl propanoate
29. (a) 5-methyloctanoic acid
    (b) N-ethyl-1-butanamine
    (c) 6-ethyl-5-methyl-2-octyne
    (d) butanamide
    (e) 6-ethyl-7-methyl-4-decanone
    (f) 2-ethyl-3-methylcylcohexanol

(g) 4,4,6,6-tetramethylheptanal
(h) 4-ethyl-3,5-dimethylcyclopentene
(i) N-ethyl-N-methylethanamide
(j) 3,6,7-trimethylcyclononanone

30. (a) 4-methyl-2-heptanone
(b) 3,3-dimethylhexane
(c) 2,5-octanediol

34. (a) 2,3-diiodopentane
(b) 1-chloropropane
(c) propyl methanoate + water
(d) 1-heptene + water

35. (a) 2-propanol
(b) decanoic acid + 1-propanol
(c) 3-methyl-1-butene
(d) 1,3 dichlorobutane

36. (a) ethyne
(b) 2-methylpentene
(c) heptanoic acid + 1-butanol

# UNIT 5

## Chapter 13

8. $[H_2] = 5.6 \times 10^{-4}$ mol/L; $[HI] = 1.12 \times 10^{-3}$ mol/L
9. $[SO_3] = 0.13$ mol/L
10. 37.4 g
11. $[CO] = [H_2O] = 0.013$ mol/L; $[H_2] = [CO_2] = 0.037$ mol/L
12. $K_c = 2.7 \times 10^{-4}$
13. (b) $1.7 \times 10^{-2}$
16. 0.26
19. $1.2 \times 10^{-17}$ mol/L

## Unit 5 Review

18. $6.9 \times 10^{-2}$
20. 0.16
21. $\dfrac{[N_{2(g)}][H_{2(g)}]^3}{[NH_{3(g)}]^2}$
22. $6 \times 10^3$
24. 3.24
25. $1.86 \times 10^{-2}$
26. $1.3 \times 10^{-2}$
27. (b) $[PCl_5] = 0.060$ mol/L; $[PCl_3] = 0.38$ mol/L; $[Cl_2] = 0.16$ mol/L
31. (c) $[XO_2] = 0.00985$ mol/L; $[O_2] = 0.992$ mol/L
33. (a) $[CO_2] = 0.0132$ mol/L
(b) $[CO_2] = 0.0132$ mol/L
40. (a) $K_c = 1.6$
(b) 0.032 (both CO and $H_2O$) or 0.064 mol/2 L (both CO and $H_2O$)

# UNIT 6

## Chapter 14

18. (a) $1.0 \times 10^3$ g
(b) 55 mol
(c) $1.8 \times 10^{-9}$ mol/L or about 2 ppb
19. 0.800 mol/L
20. (a) pOH = 4.292
(b) pH = 9.708
(c) $[H_3O^+] = 1.96 \times 10^{-10}$ mol/L
25. pH = 12
26. 1 mol
30. (a) $2.5 \times 10^{-6}$ mol/L; 12:1

## Chapter 15

1. (a) $K_a$ for $HSO_4^- = 1.0 \times 10^{-2}$; $K_a$ for $HSO_3^- = 6.3 \times 10^{-8}$
(b) $K_b$ for $HSO_4^- = 1.0 \times 10^{-12}$; $K_b$ for $HSO_3^- = 1.6 \times 10^{-7}$
6. $1.4 \times 10^{-1}$ mol/L
7. 2.52
8. (a) pH = 2.10; 0.79% dissociation
(b) pH = 2.60; 2.5% dissociation
(c) pH = 3.10; 7.9% dissociation
9. (a) pH = 4.29; 0.078%
(b) $OCl^-$, $2.5 \times 10^{-7}$
10. pH = 9.10
11. pH = 8.11
12. $K_a = 1.1 \times 10^{-3}$
16. $4.0 \times 10^{-6}$
17. (b) $2.6 \times 10^{-10}$
19. 0.20 mol/L
20. $1.0 \times 10^{-5}$ mol
22. 0.1171 mol/L
23. (a) 2.399 mol/L

## Unit 6 Review

12. 2.4
13. $[H_3O^+] = 1.7 \times 10^{-3}$ mol/L; pH = 2.77; pOH = 11.23
19. (a) $\dfrac{[H_3O^+][ClO^-]}{[HClO]}$ (b) $\dfrac{[H_3O^+][HSO_4^-]}{[H_2SO_4]}$
23. (a) 2.298
25. (a) $1.6 \times 10^{-5}$ (b) 11.14
27. 12.3%
31. $[H_3O^+] = 0.0075$ mol/L; $K_a = 2.2 \times 10^{-4}$; pH = 2.12; pOH = 11.88; $[OH^-] = 1.3 \times 10^{-12}$
32. (b) about 10
33. 1.6%
34. $3.4 \times 10^{-7}$

35. **(a)** 2.47
    **(b)** 11.42
41. $3.504 \times 10^{-3}$
42. **(a)** 4.12 mol/L $NH_4^+$; 4.12 mol/L $Cl^-$
    **(b)** 0.275 mol/L $Ba^{2+}$; 0.550 mol/L $OH^-$
    **(c)** 1.63 mol/L $NH_4^+$; 0.543 mol/L $PO_4^{3-}$
    **(d)** 0.704 $K^+$; 0.704 $ClO_4^-$
44. $2.2 \times 10^2$
46. $4.5 \times 10^{-5}$

## UNIT 7

### Chapter 16

11. 322 kJ
12. 5.76 kJ
13. −32.9 kJ
14. **(a)** $H_2S_{(g)} + \frac{3}{2}O_{2(g)} \rightarrow SO_{2(g)} + H_2O_{(g)} + 519$ kJ
    **(b)** 228 kJ
    **(c)** 2.04 L
15. 869 kJ/mol
16. **(a)** $CH_3OH_{(\ell)} + \frac{3}{2}O_{2(g)} \rightarrow CO_{2(g)} + 2H_2O_{(g)} + 727$ kJ
    **(b)** $-1.00 \times 10^3$ kJ
    **(c)** $-1.82 \times 10^3$ kJ
17. **(a)** $\frac{1}{2}H_{2(g)} + C_{(s)} + \frac{1}{2}N_{2(g)} + 135$ kJ $\rightarrow HCN_{(g)}$
    **(b)** 301 kJ
    **(c)** −126 kJ
18. **(a)** $CaCl_{2(s)} \rightarrow CaCl_{2(aq)} + 82.8$ kJ
    **(b)** −11.2 kJ
    **(c)** 37°C
19. **(a)** $8.3 \times 10^3$ kJ
20. **(a)** −98°C
    **(b)** 65°C
    **(c)** $1.9 \times 10^3$ kJ
24. **(a)** 26 min
    **(b)** 0.34 kg
    **(c)** 17

### Chapter 17

7. **(a)** 60.0°C
   **(b)** 50.0°C
8. **(a)** −2225 kJ/mol
9. 188 kJ
10. 9.68 kJ/°C
11. 39.3°C
13. 336.6 kJ/mol
14. −466 kJ/mol
15. **(a)** $H_{2(g)} + O_{2(g)} \rightarrow H_2O_{2(\ell)}$
    **(b)** −188 kJ/mol
16. 256 kJ/mol
17. **(a)** −2657 kJ
    **(b)** −1428.4 kJ/mol; −47.5 kJ/g
    **(c)** $4.6 \times 10^2$ kJ
23. **(a)** $CH_{4(g)} + 2O_{2(g)} \rightarrow CO_{2(g)} + 2H_2O_{(g)}$;
    $\Delta H_{comb} = -802.5$ kJ/mol, or −50.00 kJ/g $CH_4$
    $H_{2(g)} + \frac{1}{2}O_{2(g)} \rightarrow H_2O_{(g)}$; $\Delta H_{comb} = -241.8$ kJ/mol, or −120 kJ/g $H_2$

### Unit 7 Review

22. **(c)** −0.828 kJ/g, 0.460 kJ/g
24. **(b)** 243 g
25. **(a)** 55.7°C
26. **(a)** −46 kJ/mol
    **(b)** $|\text{calculated } \Delta H| < |\text{actual } \Delta H|$
28. 3.4 L
29. **(a)** $C_{12}H_{22}O_{11(s)} + 35/2 O_{2(g)} \rightarrow 12CO_{2(g)} + 11H_2O_{(\ell)} + 5.65 \times 10^3$ kJ
    **(b)** −82.5 kJ
30. −747.6 kJ/mol
34. **(b)** about −6800 kJ/mol

## UNIT 8

### Chapter 18

1. **(a)** $Zn_{(s)} + 2AgNO_{3(aq)} \rightarrow Zn(NO_3)_{2(aq)} + 2Ag_{(s)}$
   **(b)** $3CoBr_{2(aq)} + 2Al_{(s)} \rightarrow 3Co_{(s)} + 2AlBr_{3(aq)}$
   **(c)** $Cd_{(s)} + SnCl_{2(aq)} \rightarrow CdCl_{2(aq)} + Sn_{(s)}$
2. **(a)** $Zn_{(s)} + 2Ag^+_{(aq)} + 2NO_3^-_{(aq)} \rightarrow Zn^{2+}_{(aq)} + 2NO_3^-_{(aq)} + 2Ag_{(s)}$
   $Zn_{(s)} + 2Ag^+_{(aq)} \rightarrow Zn^{2+}_{(aq)} + 2Ag_{(s)}$
   **(b)** $3Co^{2+}_{(aq)} + 6Br^-_{(aq)} + 2Al_{(s)} \rightarrow 3Co_{(s)} + 2Al^{3+}_{(aq)} + 6Br^-_{(aq)}$
   $3Co^{2+}_{(aq)} + 2Al_{(s)} \rightarrow 3Co_{(s)} + 2Al^{3+}_{(aq)}$
   **(c)** $Cd_{(s)} + Sn^{2+}_{(aq)} + 2Cl^-_{(aq)} \rightarrow Cd^{2+}_{(aq)} + 2Cl^-_{(aq)} + Sn_{(s)}$
   $Cd_{(s)} + Sn^{2+}_{(aq)} \rightarrow Cd^{2+}_{(aq)} + Sn_{(s)}$
3. **(a)** oxidizing agent: $Ag^+$; reducing agent: Zn
   **(b)** oxidizing agent: $Co^{2+}$; reducing agent: Al
   **(c)** oxidizing agent: $Sn^{2+}$; reducing agent: Cd
4. **(a)** oxidation half-reaction: $Zn \rightarrow Zn^{2+} + 2e^-$
   reduction half-reaction: $Ag^+ + e^- \rightarrow Ag$
   **(b)** oxidation half-reaction: $Al \rightarrow Al^{3+} + 3e^-$
   reduction half-reaction: $Co^{2+} + 2e^- \rightarrow Co$
   **(c)** oxidation half-reaction: $Cd \rightarrow Cd^{2+} + 2e^-$
   reduction half-reaction: $Sn^{2+} + 2e^- \rightarrow Sn$
5. **(a)** metallic element
   **(b)** non-metallic element
   **(c)** non-metallic element
   **(d)** metallic element
6. **(a)** Ba = +2; Cl = −1
   **(b)** C = +4; S = −2
   **(c)** Xe = +4; F = −1

7. (a) Ba +2; H −1  (f) S 0
   (b) Al +3; C −4  (g) As +3; O −2
   (c) K +1; C +2; N −3  (h) V +5; O −2
   (d) Li +1; N +3; O −2  (i) Xe +6; O −2; F −1
   (e) N −3; H +1; C +3; O −2  (j) S +2.5; O −2

8. $ClO_2^-$

9. (a) yes; oxidizing agent: $O_2$; reducing agent: $C_6H_6$
   (b) no
   (c) yes; oxidizing agent: $I_2$; reducing agent: $H_2$
   (d) yes; oxidizing agent: $KMnO_4$; reducing agent: CuCl.

10. (a) yes; Cu undergoes oxidation; $Ag^+$ undergoes reduction.
    (b) no
    (c) yes; $Mn^{2+}$ undergoes oxidation; $BiO_3^-$ undergoes reduction.

11. (a) +5 in $V_2O_5$, $VO_2^+$, $VO_3^-$, $VO_4^{3-}$, $V_3O_9^{3-}$; +4 in $VO_2$, $VO^{2+}$
    (b) no

12. (a) all three steps
    (b) Step 1: oxidizing agent: $O_2$; reducing agent: $NH_3$
    Step 2: oxidizing agent: $O_2$; reducing agent: NO
    Step 3: oxidizing agent: $NO_2$; reducing agent: $NO_2$

14. (a) $I_2 + 2e^- \rightarrow 2I^-$
    (b) $Pb \rightarrow Pb^{4+} + 4e^-$
    (c) $AuCl_4^- + 3e^- \rightarrow Au + 4Cl^-$
    (d) $C_2H_5OH + H_2O \rightarrow CH_3COOH + 4H^+ + 4e^-$
    (e) $S_8 + 16H^+ + 16e^- \rightarrow 8H_2S$
    (f) $AsO_2^- + 4OH^- \rightarrow AsO_4^{3-} + 2H_2O + 2e^-$

15. (a) $MnO_2 + 2Cl^- + 4H^+ \rightarrow Mn^{2+} + Cl_2 + 2H_2O$
    (b) $2NO + 3Sn + 6H^+ \rightarrow 2NH_2OH + 3Sn^{2+}$
    (c) $3Cd^{2+} + 2V^{2+} + 6H_2O \rightarrow 3Cd + 2VO_3^- + 12H^+$
    (d) $2Cr + 6H_2O + 2OH^- \rightarrow 2Cr(OH)_4^- + 3H_2$
    (e) $S_2O_3^{2-} + 2NiO_2 + H_2O + 2OH^- \rightarrow 2Ni(OH)_2 + 2SO_3^{2-}$
    (f) $2Sn^{2+} + O_2 + 2H_2O \rightarrow 2Sn^{4+} + 4OH^-$

16. (a) $3SiCl_4 + 4Al \rightarrow 3Si + 4AlCl_3$
    (b) $4PH_3 + 8O_2 \rightarrow P_4O_{10} + 6H_2O$
    (c) $I_2O_5 + 5CO \rightarrow I_2 + 5CO_2$
    (d) $2SO_3^{2-} + O_2 \rightarrow 2SO_4^{2-}$

17. (a) $2NO_2 + H_2O \rightarrow NO_2^- + NO_3^- + 2H^+$
    (b) $Cl_2 + 2OH^- \rightarrow ClO^- + Cl^- + H_2O$

18. (a) $2Co^{3+} + Cd \rightarrow 2Co^{2+} + Cd^{2+}$
    $2Co(NO_3)_{3(aq)} + Cd_{(s)} \rightarrow 2Co(NO_3)_{2(aq)} + Cd(NO_3)_{2(aq)}$
    (b) $2Ag^+ + SO_2 + 2H_2O \rightarrow 2Ag + SO_4^{2-} + 4H^+$
    $2AgNO_{3(aq)} + SO_{2(g)} + 2H_2O_{(\ell)} \rightarrow 2Ag_{(s)} + H_2SO_{4(aq)} + 2HNO_{3(aq)}$
    (c) $Al + CrO_4^{2-} + 4H_2O \rightarrow Al(OH)_3 + Cr(OH)_3 + 2OH^-$
    $Al_{(s)} + Na_2CrO_{4(aq)} + 4H_2O_{(\ell)} \rightarrow Al(OH)_{3(s)} + Cr(OH)_{3(s)} + 2NaOH_{(aq)}$

21. $O_2 + 2F_2 \rightarrow 2OF_2$

22. (a) $2P_4 + 12H_2O \rightarrow 5PH_3 + 3H_3PO_4$
    (b) reduced

23. (a) $2Al + Fe_2O_3 \rightarrow 2Fe + Al_2O_3$
    (b) yes; oxidizing agent: $Fe_2O_3$; reducing agent: Al

24. (a) $I_2 + 10HNO_3 \rightarrow 2HIO_3 + 10NO_2 + 4H_2O$
    (b) 31.7 g

27. (a) $P_4 + 3H_2O + 3OH^- \rightarrow 3H_2PO_2^- + PH_3$
    (c) 2.74 kg

32. (a) $C_2H_4 + H_2 \rightarrow C_2H_6$
    $C_2H_5OH \rightarrow CH_3CHO + H_2$

33. (a) Na +1; Al +3; O −2; H +1; C +4
    (b) $NaAl(OH)_2CO_3 + 4HCl \rightarrow NaCl + AlCl_3 + 3H_2O + CO_2$

34. (a) $2KClO_3 \rightarrow 2KCl + 3O_2$
    $S + O_2 \rightarrow SO_2$
    (b) oxidizing agent and reducing agent: $KClO_3$; oxidizing agent: $O_2$; reducing agent: S

35. $NH_4^+ + 2O_2 \rightarrow NO_3^- + H_2O + 2H^+$

36. (b) $CaC_2 + 2H_2O \rightarrow Ca(OH)_2 + C_2H_2$; not redox
    (c) $2C_2H_2 + 5O_2 \rightarrow 4CO_2 + 2H_2O$; redox

## Chapter 19

2. $Mg_{(s)} | Mg^{2+}_{(aq)}$ (1 mol/L) $|| Cd^{2+}_{(aq)}$ (1 mol/L) $| Cd_{(s)}$

3. oxidation: $NO_{(g)} + 2H_2O_{(\ell)} \rightarrow NO_3^-_{(aq)} + 4H^+_{(aq)} + 3e^-$
   reduction: $I_{2(s)} + 2e^- \rightarrow 2I^-_{(aq)}$
   overall: $2NO_{(g)} + 4H_2O_{(\ell)} + 3I_{2(s)} \rightarrow 2NO_3^-_{(aq)} + 8H^+_{(aq)} + 6I^-_{(aq)}$

6. spontaneous: $N_2O_{(g)} + 2H^+_{(aq)} + 2Cu_{(s)} + 2I^-_{(aq)} \rightarrow N_{2(g)} + H_2O_{(\ell)} + 2CuI_{(s)}$ $E°_{cell}$ = 1.955 V
   non-spontaneous: $N_{2(g)} + H_2O_{(\ell)} + 2CuI_{(s)} \rightarrow N_2O_{(g)} + 2H^+_{(aq)} + 2Cu_{(s)} + 2I^-_{(aq)}$ $E°_{cell}$ = −1.955 V

7. (a) oxidizing agent: $PbO_2$; reducing agent: Pb
   (b) oxidizing agent: $PbSO_4$; reducing agent: $PbSO_4$

8. $Co^{3+}_{(aq)}, Br_{2(aq)}, H^+_{(aq)}, Zn^{2+}_{(aq)}$

9. $Al_{(s)}, H_{2(g)}, Ag_{(s)}, Cl^-_{(aq)}$

11. (a) oxidation: $Zn_{(s)} \rightarrow Zn^{2+}_{(aq)} + 2e^-$
    reduction: $Fe^{2+}_{(aq)} + 2e^- \rightarrow Fe_{(s)}$
    cell potential: 0.315 V; spontaneous
    (b) oxidation: $Cr_{(s)} \rightarrow Cr^{3+}_{(aq)} + 3e^-$
    reduction: $Al^{3+}_{(aq)} + 3e^- \rightarrow Al_{(s)}$
    cell potential: −0.918 V; non-spontaneous
    (c) oxidation: $H_2O_{2(aq)} \rightarrow 2H^+_{(aq)} + O_{2(g)} + 2e^-$
    reduction: $Ag^+_{(aq)} + e^- \rightarrow Ag_{(s)}$
    cell potential: 0.105 V; spontaneous

12. 1.52 g

14. (a) $E°_{cell}$ = 1.247 V
    oxidation: $Fe_{(s)} \rightarrow Fe^{2+}_{(aq)} + 2e^-$
    reduction: $Ag^+_{(aq)} + e^- \rightarrow Ag_{(s)}$
    overall: $Fe_{(s)} + 2Ag^+_{(aq)} \rightarrow Fe^{2+}_{(aq)} + 2Ag_{(s)}$
    (b) $E°_{cell}$ = −1.247 V
    oxidation: $Ag_{(s)} \rightarrow Ag^+_{(aq)} + e^-$
    reduction: $Fe^{2+}_{(aq)} + 2e^- \rightarrow Fe_{(s)}$
    overall: $Fe^{2+}_{(aq)} + 2Ag_{(s)} \rightarrow 2Ag^+_{(aq)} + Fe_{(s)}$

17. The cost for Al would be 2.56 times the cost for Na.

## Unit 8 Review

**13.** (a) +3
(b) +5
(c) +4
(d) N −3; H +1; S +6; O −2

**14.** no

**15.** (a) $Ca(OH)_{2(aq)} + CO_{2(g)} \rightarrow CaCO_{3(s)} + H_2O_{(\ell)}$

**16.** (a) $Hg_2^{2+} + 2e^- \rightarrow 2Hg$
(b) $TiO_2 + 4H^+ + 2e^- \rightarrow Ti^{2+} + 2H_2O$
(c) $I_2 + 18OH^- \rightarrow 2H_3IO_6^{3-} + 6H_2O + 12e^-$

**19.** (a) $3Fe_{(s)} + 8HNO_{3(aq)} \rightarrow$
$3Fe(NO_3)_{2(aq)} + 2NO_{(g)} + 4H_2O_{(\ell)}$
(b) $3Fe(NO_3)_{2(aq)} + 4HNO_{3(aq)} \rightarrow$
$3Fe(NO_3)_{3(aq)} + NO_{(g)} + 2H_2O_{(\ell)}$
(c) $3Fe_{(s)} + 12HNO_{3(aq)} \rightarrow$
$3Fe(NO_3)_{3(aq)} + 3NO_{(g)} + 6H_2O_{(\ell)}$

**20.** (b) oxidation: $2I^- \rightarrow I_2 + 2e^-$
reduction: $Ag^+ + e^- \rightarrow Ag$
overall: $2I^- + 2Ag^+ \rightarrow I_2 + 2Ag$
(d) 0.264 V

**21.** (a) no
$Cl_2O_7 + H_2O \rightarrow 2HClO_4$
(b) yes; oxidizing agent: $ClO_3^-$; reducing agent: $I_2$
$3I_2 + 5ClO_3^- + 3H_2O \rightarrow 6IO_3^- + 5Cl^- + 6H^+$
(c) yes; oxidizing agent: $Br_2$; reducing agent: $S^{2-}$
$S^{2-} + 4Br_2 + 8OH^- \rightarrow SO_4^{2-} + 8Br^- + 4H_2O$
(d) yes; oxidizing agent: $HNO_3$; reducing agent: $H_2S$
$2HNO_3 + 3H_2S \rightarrow 2NO + 3S + 4H_2O$

**22.** (a) −0.235 V; non-spontaneous
(b) 0.698 V; spontaneous
(c) 0.418 V; spontaneous

**25.** (a) $2MnO_4^- + 5C_2O_4^{2-} + 16H^+ \rightarrow$
$2Mn^{2+} + 10CO_2 + 8H_2O$
(b) 0.2249 mol/L

**26.** 20.72%

**28.** (a) 6
(b) 3

**29.** (a) $2.68 \times 10^4$ C
(b) 74.4 h

**30.** 1.16 g

**31.** 0.595 t

**43.** (a) $8Al + 3NH_4ClO_4 \rightarrow 4Al_2O_3 + 3NH_4Cl$
oxidizing agent: $NH_4ClO_4$; reducing agent: Al
(b) $2H_{2(\ell)} + O_{2(\ell)} \rightarrow 2H_2O_{(g)}$

# Appendix B

## Supplemental Practice Problems

### Chapter 1

1. Label each as either a physical or a chemical property.
   (a) The boiling point of water is 100°C.
   (b) Chlorine gas reacts violently with sodium metal.
   (c) Bromine has a brown colour.
   (d) Sulfuric acid causes burns when it comes in contact with skin.

2. How many significant digits are in the following quantities?
   (a) 624 students
   (b) 22.40 mL of water
   (c) 0.00786 g of platinum

3. Characterize each of the following occurrences as a physical or as a chemical change.
   (a) sugar is heated over a flame and caramelises (turns black)
   (b) blood clots
   (c) a rubber band is stretched until it snaps
   (d) a match burns
   (e) a grape is crushed
   (f) salt is put on the roads in the winter, melting the ice.

4. Name each of the following ionic compounds.
   (a) $MgCl_2$
   (b) $Na_2O$
   (c) $FeCl_3$
   (d) $CuO$
   (e) $AlBr_3$

5. Write the chemical formula for each of the following compounds.
   (a) aluminum bromide
   (b) magnesium oxide
   (c) sodium sulfide
   (d) iron(II) oxide
   (e) copper(II) chloride

6. Write the formula for each of the following.
   (a) sodium hydrogen carbonate
   (b) potassium dichromate
   (c) sodium hypochlorite
   (d) lithium hydroxide
   (e) potassium permanganate

7. Name each of the following compounds.
   (a) $K_2CrO_4$
   (b) $NH_4NO_3$
   (c) $Na_2SO_4$
   (d) $KH_2PO_4$
   (e) $Sr_3(PO_4)_2$

8. Name each of the following covalent compounds.
   (a) $Cl_2O_7$
   (b) $H_2O$
   (c) $BF_3$
   (d) $N_2O_4$
   (e) $N_2O$

9. Write the formula for each of the following compounds.
   (a) tetraphosphorus decoxide
   (b) nitrogen trichloride
   (c) sulfur tetrafluoride
   (d) xenon hexafluoride

10. Balance each of the following skeleton equations. Classify each chemical reaction.
    (a) $Fe + Cl_2 \rightarrow FeCl_2$
    (b) $FeCl_2 + Cl_2 \rightarrow FeCl_3$
    (c) $C_4H_{10}O + O_2 \rightarrow CO_2 + H_2O$
    (d) $Al + H_2SO_4 \rightarrow Al_2(SO_4)_3 + H_2$
    (e) $N_2O_5 + H_2O \rightarrow HNO_3$
    (f) $(NH_4)_2CO_3 \rightarrow NH_3 + CO_2 + H_2O$

11. Balance the following chemical equation.
    $BiCl_3 + NH_3 + H_2O \rightarrow Bi(OH)_3 + NH_4Cl$

12. Balance the equation.
    $NiSO_4 + NH_3 + H_2O \rightarrow Ni(NH_3)_6(OH)_2 + (NH_4)_2SO_4$

## Unit 1

### Chapter 2

13. Gallium exists as two isotopes, Ga-69 and Ga-71.
    (a) How many protons and neutrons are in each isotope?
    (b) If Ga-69 exists in 60.0% relative abundance, estimate the average atomic mass of gallium using the mass numbers of the isotopes.

14. Rubidium exists as two isotopes: Rb-85 has a mass of 84.9117 u and Rb-87 has a mass of 86.9085 u. If the average atomic mass of rubidium is 85.4678, determine the relative abundance of each isotope.

15. You have 10 mL of isotopically labelled water, $^3H_2O$. That is, the water is made with the radioactive isotope of hydrogen, tritium, $^3H$. You pour the 10 mL of tritium-labelled water into an ocean and allow it to thoroughly mix with all the bodies of water on the earth. After the tritium-labelled water mixes thoroughly with the earth's ocean water, you remove 100 mL of ocean water. Estimate how many molecules of $^3H_2O$ will be in this 100 mL sample. (Assume that the average depth of the ocean is 5 km. The earth's surface is covered roughly two-thirds with water. The radius of the earth is about 6400 km.)

16. Calculate the molar mass of each of the following compounds.
    (a) $Al_2(CrO_4)_3$
    (b) $C_4H_9SiCl_3$ (n-butyltrichlorosilane, an intermediate in the synthesis of silicones)
    (c) $Cd(ClO_3)_2 \cdot 2H_2O$ (cadmium chlorate dihydrate, an oxidizing agent)

17. How many atoms are contained in 3.49 moles of manganese?

18. How many atoms are there in 8.56 g of sodium?

19. What is the mass of $5.67 \times 10^{23}$ molecules of pentane, $C_5H_{12}$?

Answers to Supplemental Practice Problems are provided on the *Teachers' Resource* CD.

20. Consider a 23.9 g sample of ammonium carbonate, $(NH_4)_2CO_3$.
    (a) How many moles are in this sample?
    (b) How many formula units are in this sample?
    (c) How many atoms are in this sample?

21. One litre of a certain gas has a mass of 2.05 g at STP. What is the molar mass of this gas?

## Chapter 3

22. Pyridine, $C_5H_5N$, is a slightly yellow liquid with a nauseating odour. It is flammable and toxic by ingestion and inhalation. Pyridine is used in the synthesis of vitamins and drugs, and has many other uses in industrial chemistry. Determine the percentage composition of pyridine.

23. Bromine azide is an explosive compound that is composed of bromine and nitrogen. A sample of bromine azide was found to contain 2.35 g Br and 1.24 g N.
    (a) Calculate the percentage by mass of Br and N in bromine azide.
    (b) Calculate the empirical formula of bromine azide.
    (c) The molar mass of bromine azide is 122 g/mol. Determine its molecular formula.

24. Progesterone is a female hormone. It is 80.2% C, 9.62% H and 10.2% O by mass.
    (a) Determine the empirical formula of progesterone.
    (b) From the given data, is it possible to determine the molecular formula of progesterone? Explain your answer.

25. Potassium tartrate is a colourless, crystalline solid. It is 34.6% K, 21.1% C, 1.78% H, 42.4% O by mass.
    (a) Calculate the empirical formula of potassium tartrate.
    (b) If the molar mass of potassium tartrate is 226 g/mol, what is the molecular formula of potassium tartrate?

26. Menthol is a compound that contains C, H and O. It is derived from peppermint oil and is used in cough drops and chest rubs. When 0.2393 g of menthol is subjected to carbon-hydrogen combustion analysis, 0.6735 g of $CO_2$ and 0.2760 g of $H_2O$ are obtained.
    (a) Determine the empirical formula of menthol.
    (b) If each menthol molecule contains one oxygen atom, what is the molecular formula of menthol?

27. Glycerol, $C_3H_8O_3$, also known as glycerin, is used in products that claim to protect and soften skin. Glycerol can be purchased at the drug store. If 0.784 g of glycerol is placed in a carbon-hydrogen combustion analyzer, what mass of $CO_2$ and $H_2O$ will be expected?

28. Calculate the percentage by mass of water in potassium sulfite dihydrate, $K_2SO_3 \cdot 2H_2O$.

29. What mass of water is present in 24.7 g of cobaltous nitrate hexahydrate, $Co(NO_3)_2 \cdot 6H_2O$?

30. A chemist requires 1.28 g of sodium hypochlorite, NaOCl, to carry out an experiment, but only has sodium hypochlorite pentahydrate, $NaOCl \cdot 5H_2O$ in the lab. How many grams of the hydrate should the chemist use?

31. A compound was found to contain 54.5% carbon, 9.10% hydrogen, and 36.4% oxygen. The vapour from 0.082 g of the compound occupied 20.9 mL at STP.
    (a) Calculate the empirical formula of the compound.
    (b) Calculate the molecular mass of the compound.
    (c) Calculate the molecular formula of the compound.

## Chapter 4

32. Consider the equation corresponding to the decomposition of mercuric oxide.

    $2HgO_{(s)} \rightarrow 2Hg_{(\ell)} + O_{2(g)}$

    What mass of liquid mercury is produced when 5.79 g of mercuric oxide decomposes?

33. Examine the following equation.

    $C_3H_{8(g)} + 5\ O_{2(g)} \rightarrow 3\ CO_{2(g)} + 4\ H_2O_{(g)}$

    (a) What mass of propane, $C_3H_8$, reacting with excess oxygen, is required to produce 26.7 g of carbon dioxide gas?
    (b) How many oxygen molecules are required to react with 26.7 g of propane?

34. Metal hydrides, such as strontium hydride, $SrH_2$, react with water to form hydrogen gas and the corresponding metal hydroxide.

    $SrH_{2(s)} + 2H_2O_{(\ell)} \rightarrow Sr(OH)_{2(s)} + 2H_{2(g)}$

    (a) When 2.50 g of $SrH_2$ is reacted with $8.03 \times 10^{22}$ molecules of water, what is the limiting reagent?
    (b) What mass of strontium hydroxide will be produced?

35. Consider the following successive reactions.
    reaction (1): A → B
    reaction (2): B → C
    If reaction (1) proceeds with a 45% yield and reaction (2) has a 70% yield, what is the overall yield for the reactions that convert A to C?

36. Disposable cigarette lighters contain liquid butane, $C_4H_{10}$. Butane undergoes complete combustion to carbon dioxide gas and water vapour according to the skeleton equation below:

    $C_4H_{10(\ell)} + O_{2(g)} \rightarrow CO_{2(g)} + H_2O_{(\ell)}$

    A particular lighter contains 5.00 mL of butane, which has a density of 0.579 g/mL.
    (a) How many grams of $O_2$ are required to combust all of the butane?
    (b) How many molecules of water will be produced?
    (c) Air contains 21.0% $O_2$ by volume. What mass of air is required to combust 5.00 mL of butane?

37. If the following reaction proceeds with a 75% yield, how much diborane, $B_2H_6$, will be produced when 23.5 g of sodium borohydride, $NaBH_4$ reacts with 50.0 g of boron trifluoride, $BF_3$?

    $3NaBH_{4(s)} + BF_{3(g)} \rightarrow 2B_2H_{6(g)} + 3NaF_{(s)}$

38. Drinking a solution of baking soda (sodium hydrogen carbonate, $NaHCO_3$) can neutralize excess hydrochloric acid in the stomach in water. A student stirred 5.0 g of baking soda in water and drank the solution, then calculated the size of "burp" expected from the carbon dioxide generated in the following reaction.

    $NaHCO_{3(aq)} + HCl_{(aq)} \rightarrow NaCl_{(aq)} + H_2O_{(\ell)} + CO_{2(g)}$

    What volume of carbon dioxide will be generated at STP?

39. Examine the reaction below and answer the following questions.

    $C_7H_{16(g)} + 11O_{2(g)} \rightarrow 7CO_{2(g)} + 8H_2O_{(g)}$

    (a) if 10.0 L of $C_7H_{16(g)}$ at STP are burned, what volume of oxygen gas at STP is required?
    (b) if 200 g of $CO_2$ are formed, what mass of $C_7H_{16(g)}$ was burned?
    (c) if 200 L of $CO_2$ are formed, measured at STP, what mass of oxygen was consumed?

# Unit 2

## Chapter 5

40. Draw Lewis structures to represent each of the following atoms.
    (a) Mg (b) B (c) K
    (d) C (e) Ne (f) Al

41. Draw Lewis structures to represent each of the following ions.
    (a) $H^+$ (b) $K^+$ (c) $Br^-$
    (d) $S^{2-}$ (e) $Al^{3+}$ (f) $Mg^{2+}$

42. Draw Lewis structures to represent each of the following ionic compounds.
    (a) KBr (b) $MgCl_2$ (c) CaO
    (d) $Mg_3N_2$ (e) $Na_2O$ (f) $AlBr_3$

43. Draw Lewis structures to represent each of the following molecular compounds.
    (a) $I_2$ (b) $CO_2$
    (c) NO (d) $N_2$

44. Use Lewis structures to predict the correct formulas for the following.
    (a) potassium iodide
    (b) magnesium sulfide
    (c) calcium chloride
    (d) lithium fluoride
    (e) barium chloride
    (f) cesium bromide

45. Determine the $\Delta EN$ for each of the following pairs or atoms.
    (a) carbon and nitrogen
    (b) silver and iodine
    (c) manganese and oxygen
    (d) hydrogen and iodine
    (e) copper and oxygen
    (f) hydrogen and sulfur

46. For each of the pairs of atoms in question 6, state whether a bond between them will be mostly ionic, polar covalent, slightly polar covalent, or non-polar covalent.

47. For each of the following bonds, indicate which will be polar or non-polar. For the polar bonds, use arrows above the symbols to indicate the direction of the polarity.
    (a) B—F (b) Na—I (c) Si—O
    (d) N—H (e) O—O (f) Cl—F

48. List six pairs of atoms that will have polar covalent bonds between them.

49. Arrange the following bonds in order of decreasing polarity:

    hydrogen—nitrogen, calcium—chlorine, nitrogen—nitrogen, magnesium—oxygen, hydrogen—sulfur

50. List the following bonds in order of increasing bond energy.

    C—O, C=C, H—H, C≡C, Si—H

51. List the bonds in question 50 in order of decreasing length.

52. How do the orders you listed in questions 50 and 51 compare?

## Chapter 6

53. Draw Lewis structures of the following compounds.
    (a) $CH_3COOH$ (b) $N_2O$ (c) $SO_2Cl_2$
    (d) $XeO_2F_2$ (e) $NO_2^-$ (f) $C_2H_2$

54. Use VSEPR theory to determine the shape of each molecule in question 53. For part (a), state the shape around the central carbon atom.

55. Which compound(s) in questions 53 has resonance structures?

56. Which of the compounds in question 53 are polar?

57. When the iodide ion is in a solution of iodine, a triiodide ion forms. Draw the Lewis structure for $I_3^-$.

58. Determine the shape of the sulfate, $SO_4^-$, and sulfite, $SO_3^-$, ions.

59. The amide linkage that holds amino acids together in proteins creates a rigid structure. Construct a Lewis diagram and determine the shape around the carbon atom and the nitrogen atom in an amide linkage, $H_2C(O)NH_2$.

60. Draw Lewis structures for CO, $NO^+$, $CN^-$, $N_2$. State any similarities among these compounds.

61. Draw a Lewis structure for urea, $H_2NC(O)NH_2$. State whether it is a polar molecule.

62. Compare the molecules $CH_3F$ and $NH_2F$ with respect to their
    (a) Lewis structures
    (b) dispersion forces
    (c) hydrogen bonding

63. Determine the shape of the phosphate, $PO_4^{3-}$, and hydrogen phosphate, $HPO_4^{2-}$, ions. Which, if any, of the ions is polar?

# Unit 3

## Chapter 7

64. How many litres of solution are needed to accommodate 2.11 mol of 0.988 mol/L solute?

65. What volume of 2.00 mol/L hydrochloric acid is needed to react with 36.5 mL of 1.85 mol/L sodium hydroxide?

66. 2.68 L of 2.11 mol/L HCl reacts with 3.17 L of 2.28 mol/L NaOH. How many moles of NaCl are produced by this reaction?

67. Determine the final concentration when 1.50 L of 2.50 mol/L solution is diluted with 3.00 L of solvent. How is this concentration different if 1.50 L of 2.50 mol/L solution is diluted to 3.00 L with solvent?

68. What is the molar concentration of the solution made by dissolving 1.00 g of solid sodium nitrate, $NaNO_3$, in enough water to make 315 mL of solution?

69. What volume of $4.00 \times 10^{-2}$ mol/L calcium nitrate solution, $Ca(NO_3)_{2(aq)}$ will contain $5.0 \times 10^{-2}$ mol of nitrate ions?

70. By the addition of water, 80.0 mL of 4.00 mol/L sulfuric acid, $H_2SO_4$, is diluted to 400.0 mL. What is the molar concentration of the sulfuric acid after dilution?

71. How many moles of NaOH are in 100.0 mL of 0.00100 mol/L NaOH solution?

72. If a burette delivers 20 drops of solution per 1.0 mL, how many moles of $HCl_{(aq)}$ are in one drop of a 0.20 mol/L HCl solution?

73. What is the mass percent concentration of nicotine in the body of a 70 kg person who smokes a pack of cigarettes (20 cigarettes) in one day? Assume that there is 1.0 mg of nicotine per cigarette, and that all the nicotine is absorbed into the person's body.

74. Ozone is a highly irritating gas that reduces the lung capability of healthy people in concentrations as low as 0.12 ppm. Older photocopy machines could generate ozone gas and they were often placed in closed rooms with little air circulation. Calculate the volume of ozone gas that would result in a concentration of 0.12 ppm in a room with dimensions of 5.0 m × 4.0 m × 3.0 m.

75. Human blood serum contains about 3.4 g/L of sodium ions. What is the molar concentration of $Na^+$ in blood serum?

## Chapter 8

76. A student makes a solution by adding 55.0 mL of 1.77 mol/L sodium chloride to 105 mL of 0.446 mol/L calcium chloride, and diluting the solution to 200.0 mL. Determine the concentration of each ion in the solution.

77. What is the total concentration of all the ions in each of the solutions listed below?
    (a) 2.600 mol/L NaCl
    (b) 1.20 mol/L $Mg(ClO_3)_2$
    (c) 10.3 g of $(NH_4)_2SO_3$ in 420 mL of solution

78. Complete the following reaction by writing a balanced equation. Then write the total ionic equation and net ionic equation for the reaction, and identify the spectator ions.
    $CaCl_{2(aq)} + Cs_3PO_{4(aq)}$

79. Complete the following reaction by writing a balanced equation. Then write the total ionic equation and net ionic equation for the reaction, and identify the spectator ions.
    $Na_2S_{(aq)} + ZnSO_{4(aq)}$

80. When 25.0 mL of silver nitrate solution reacts with excess potassium chloride solution, a precipitate with a mass of 0.842 g results. Determine the molar concentration of silver ion in the original silver nitrate solution.

81. For each of the following, predict whether a reaction will occur. If it does, write a balanced chemical equation, the total ionic equation, and the net ionic equation, and identify the spectator ions. Assume all the reactions occur in aqueous solution.
    (a) ammonium perchlorate + sodium bromide
    (b) sodium hydroxide and cadmium nitrate

82. Solid iron reacts with aqueous iron(III) chlorate to produce aqueous iron(II) chlorate. Write the net ionic equation for this reaction.

83. When ammonium sulfide reacts with each of the following substances in aqueous solution, an insoluble sulfide precipitate results. Write a net ionic equation for each.
    (a) zinc chlorate
    (b) copper(II) chlorate
    (c) manganese(II) chlorate

84. Consider the following net ionic equation:
    $2H^+_{(aq)} + CO_3^{2-}{}_{(aq)} \rightarrow H_2O_{(\ell)} + CO_{2(g)}$
    Write five balanced chemical equations that result in this net ionic equation.

85. Write the net ionic equation for the reaction between aqueous solutions of barium chloride and sodium sulfate. Be sure to include the state of each reactant and product.

86. Write the net ionic equation for the reaction between aqueous sodium hydroxide and aqueous nitric acid. Be sure to include the state of each reactant and product.
87. What are the spectator ions when solutions of $Na_2SO_4$ and $Pb(NO_3)_2$ are mixed?
88. Iron(II) sulfate reacts with potassium hydroxide in aqueous solution to form a precipitate.
    (a) What is the net ionic equation for this reaction?
    (b) Which ions are spectator ions?
89. Write the balanced molecular and net ionic equations for the following reactions:
    (a) $Na_3PO_{4(aq)} + Ca(OH)_{2(aq)} \rightarrow NaOH_{(aq)} + Ca_3(PO_4)_{2(s)}$
    (b) $Zn_{(s)} + Fe_2(SO_4)_{3(aq)} \rightarrow ZnSO_{4(aq)} + Fe_{(s)}$

# Unit 4

## Chapter 9

90. Name each of the following hydrocarbons.
    (a) 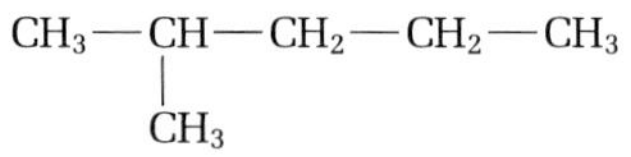 CH₃—CH—CH₂—CH₂—CH₃
              |
              CH₃
    (b) 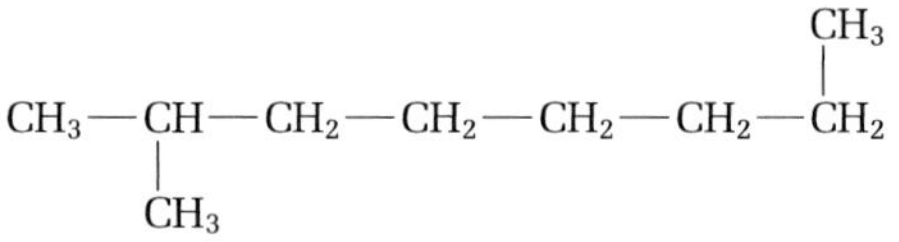 CH₃—CH—CH₂—CH₂—CH₂—CH₂—CH₂
              |                    |
              CH₃                  CH₃
    (c) 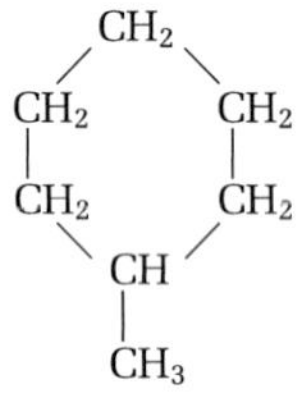 [cyclohexane ring with CH₃ substituent]
    (d) 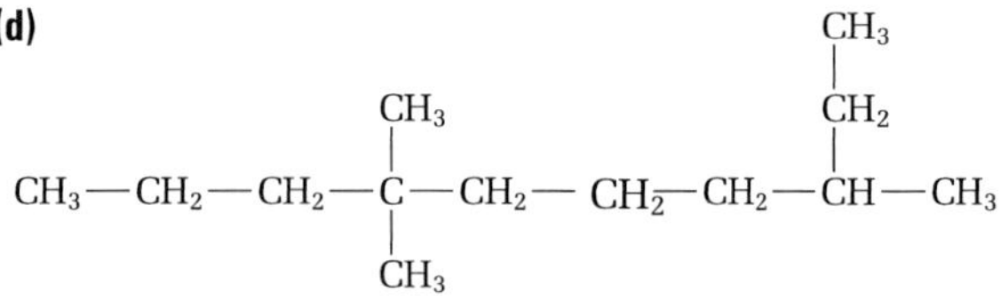 CH₃—CH₂—CH₂—C(CH₃)(CH₃)—CH₂—CH₂—CH₂—CH(CH₃)—CH₃
    (e) 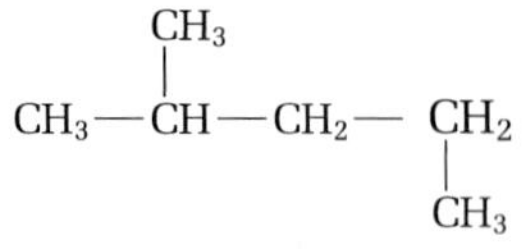 CH₃—CH(CH₃)—CH₂—CH₂—CH₃
    (f) 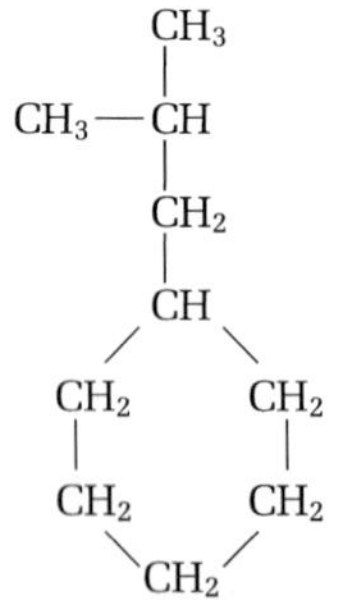 CH₃—CH(CH₃)—CH₂—CH [attached to cyclohexane ring]

91. Draw condensed structural diagrams for each of the following compounds.
    (a) 2,5-dimethylhexane
    (b) 2-ethyl-5-methylhexane
    (c) 2,3,-trimethylpentane
    (d) 2,2,3-trimethyl-4-propyloctane
    (e) 3-ethyl-2,4,6-trimethyl-5-propylnonane

92. Use each *incorrect* name to draw the corresponding hydrocarbon. Examine your drawing and rename the hydrocarbon correctly.
    (a) 2-ethyl-3,3-dimethyl-4-propylpentane
    (b) 2,3-dimethyl-3-butylpropane
    (c) 1-ethyl-4-methylbutane
    (d) 2,4-dibutylpentane

93. Write a balanced equation for the complete combustion of 3-ethyl-2,2,5-trimethylheptane.

94. Write two possible, correct equations for the following reaction:
    $CH_3CH(CH_3)CH_3 + Cl_2 \xrightarrow{UV}$

95. Name the following compounds.
    (a) $CH_2=CH-CH_3$
    (b) 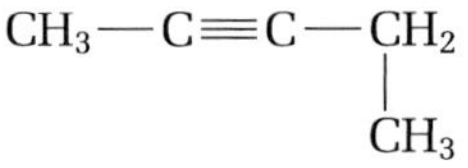 $CH_3-C\equiv C-CH_2$
                        |
                        $CH_3$
    (c) [cyclohexene ring with CH₃ substituent]
    (d) $CH_3-CH=CH-CH_2-CH_2-CH_2-CH_2-CH_3$
    (e) 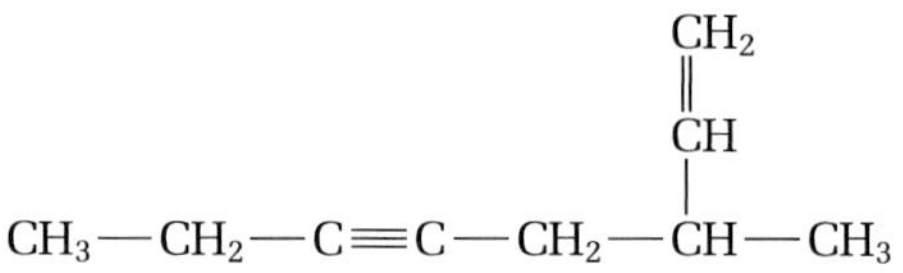 $CH_3-CH_2-C\equiv C-CH_2-CH-CH_3$
                                    |
                                    $CH$
                                    ‖
                                    $CH_2$
    (f) 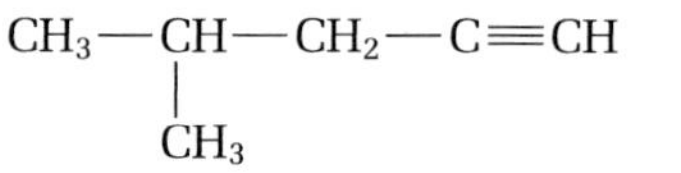 $CH_3-CH-CH_2-C\equiv CH$
             |
             $CH_3$
    (g) 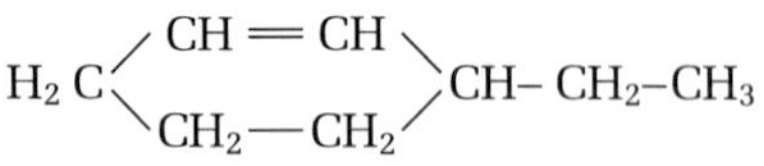 [cyclic structure with CH=CH, H₂C, CH₂—CH₂, CH—CH₂—CH₃]
    (h) 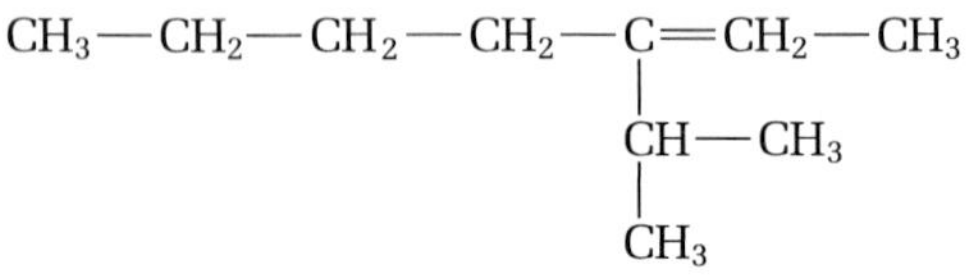 $CH_3-CH_2-CH_2-CH_2-C=CH_2-CH_3$
                                    |
                                    $CH-CH_3$
                                    |
                                    $CH_3$

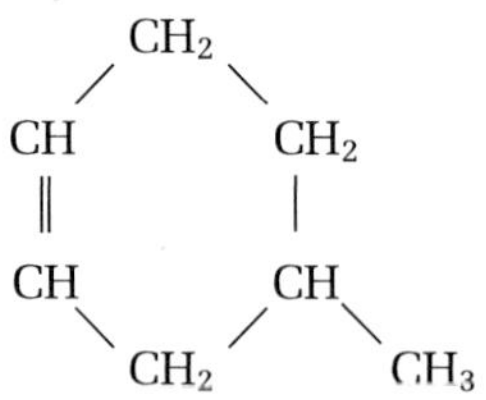

96. Examine the following compounds. Correct any flaws that you see in the structural diagrams.
    (a) Is this compound 4-ethyl-2-methylpentane?
        $CH_3-CH-CH_2-CH-CH_2-CH_3$
              |           |
              $CH_3$       $CH_2$
                          |
                          $CH_3$
                          (above CH in middle)

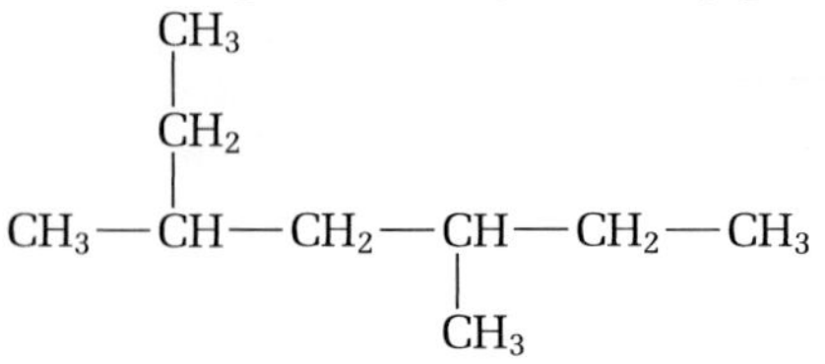

(b) Is this compound 4,5-dimethylhexane?

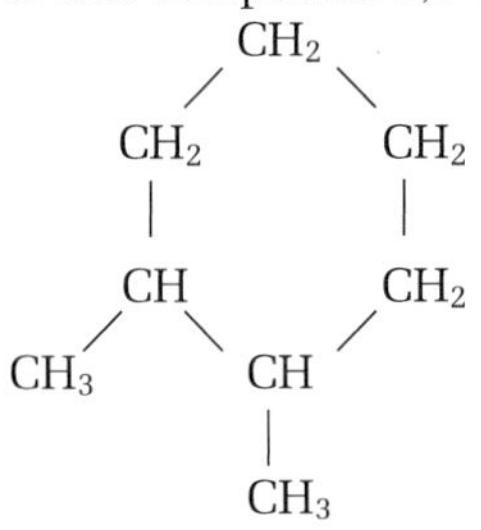

(c) Is this compound 2-methyl-3-ethylpentane?

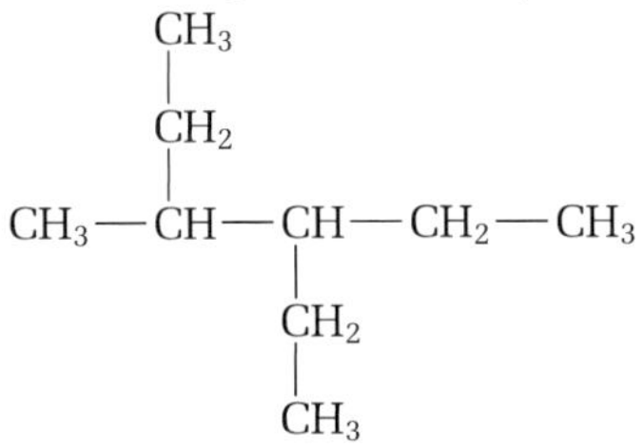

(d) Is this compound 1-methyl-3-cyclobutene?

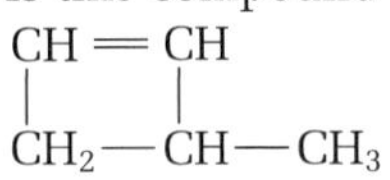

**97.** Draw condensed structural diagrams for each of the following compounds.
(a) 2-ethyl-1-pentene
(b) 2,6-dimethyl-3-nonene
(c) 2,5-dimethyl-3-hexyne
(d) 2-butyl-3-ethyl-1-cyclobutene
(e) 1-butyl-3-propyl-cyclooctene
(f) 1,3,5-triethyl-2-cyclohexane

**98.** Name each compound.
(a) 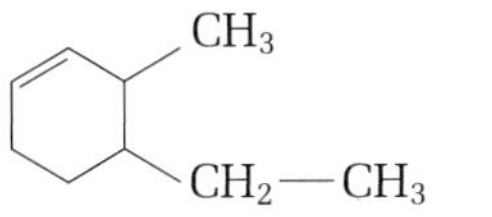
(b) $CH_2-CH_2-CH_2-CH_3$
(c) $CH_3 \quad CH_2-CH_3$
(d) $CH_3-CH_3-CH-CH_2-CH_3$
(e)

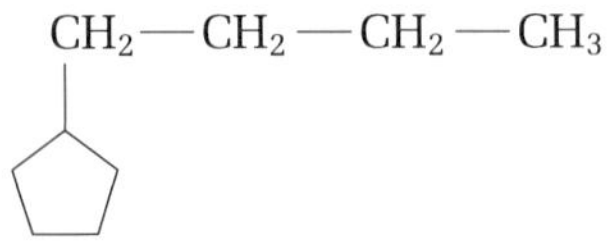

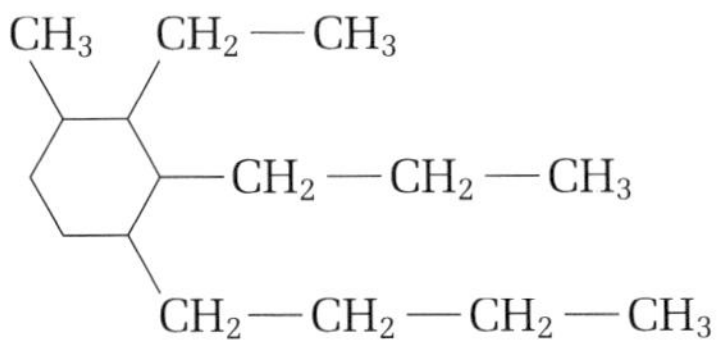

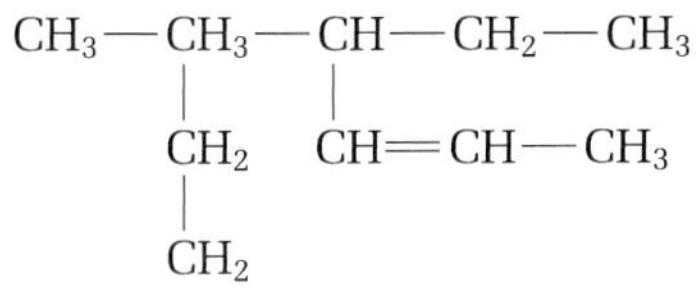

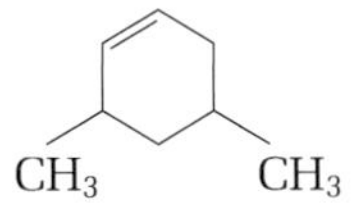

**99.** Draw and name the products of each reaction.
(a) $CH_3-C{\equiv}C-CH_2-CH_3 + Br_2 \rightarrow$
(b) 1-methylcyclopentene + $H_2 \rightarrow$
(c) $CH_3(CH_2)_4CH = CH_2 + Cl_2 \rightarrow$
(d) 1,3-dimethylcyclobutene + HCl $\rightarrow$

**100.** Name the following compounds.
(a) 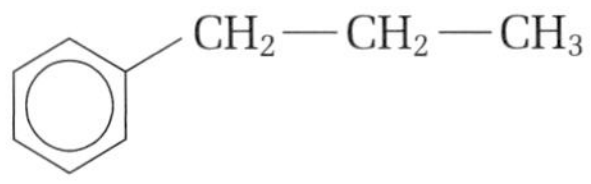
(b) 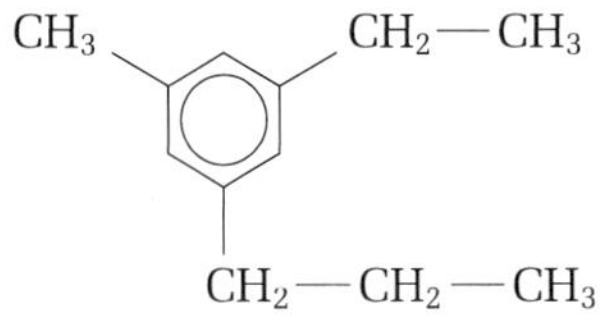
(c) 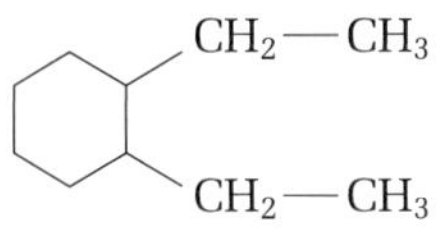
(d)

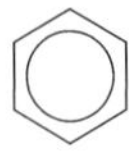

**101.** Draw and name the reactants for the following reactions.
(a) ? + ? $\xrightarrow{FeBr_3}$ (Cl-benzene) + HCl
(b) ? + ? $\longrightarrow$ (NO_2-benzene) + $H_2$
(c) ? + ? $\longrightarrow$ (1,2-dichlorocyclohexane)
(d) ? + ? $\longrightarrow$ (cyclopentanol)
(e) ? + ? $\longrightarrow$ (bromocyclobutane)

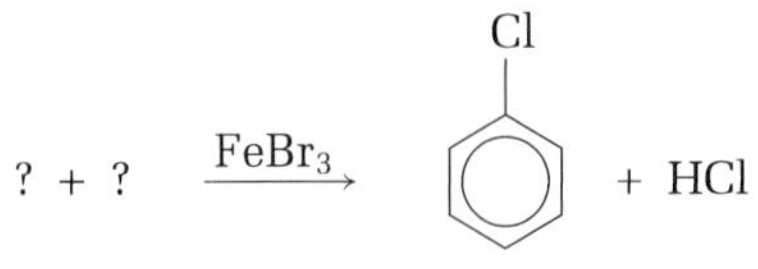

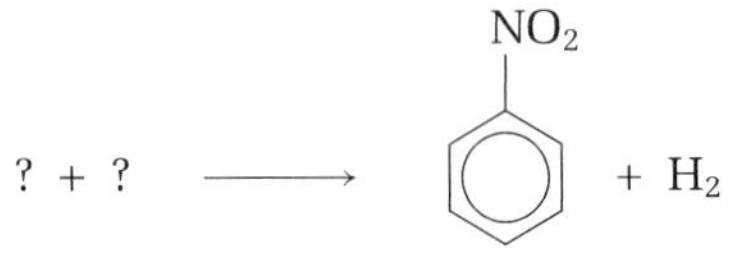

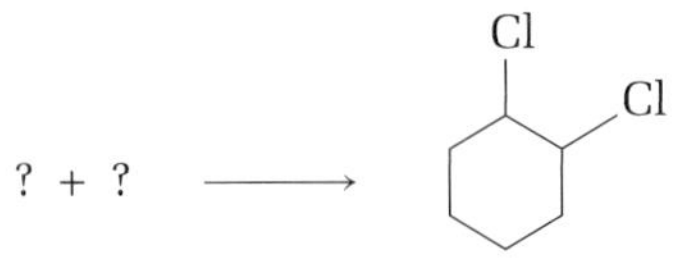

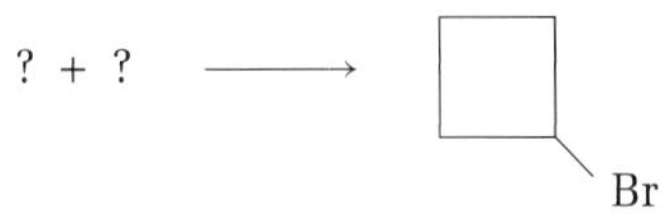

## Chapter 10

**102.** Identify the type of reaction.
(a) $CH_3CH_2OH \rightarrow CH_2 = CH_2 + H_2O$
(b) $CH_3CH_2CH_2CH_2OH + HBr \rightarrow CH_3CH_2CH_2CH_2Br + HOH$
(c) $CH_3CH_2CH(OH)CH_3 \rightarrow CH_3CH_2CH = CH_2 + H_2O$
(d) $CH_3CH_2CH(Br)CH_3 + OH^- \rightarrow CH_3CH_2CH = CH_2 + H_2O + Br^-$

103. Name the compounds.
   (a) 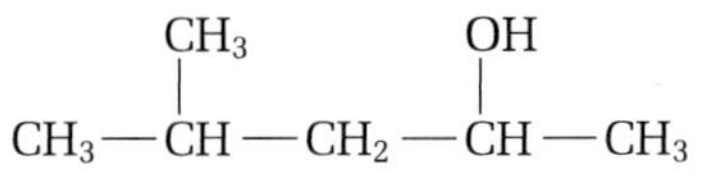 CH₃—CH(CH₃)—CH₂—CH(OH)—CH₃
   (b) 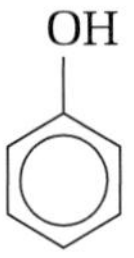 OH on benzene ring (phenol)
   (c) 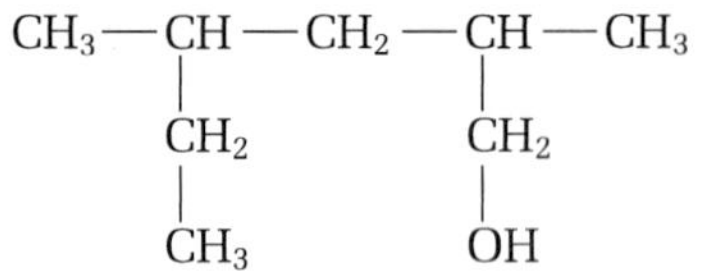 CH₃—CH(CH₂CH₃)—CH₂—CH(CH₂OH)—CH₃
   (d) 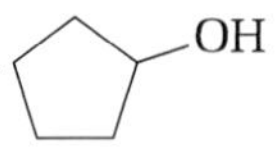 cyclopentane with OH

104. Draw a condensed structural diagram for each compound.
   (a) 1,3-difluoro-4-methyl-1-pentyne
   (b) *trans*-7,7-dimethyl-1-octene
   (c) 2,3,4-trichloropentane
   (d) 3-methylcyclopentanol
   (e) ethoxy pentane
   (f) dichloromethane

105. Draw and name the products of the following reactions.
   (a) 3-methyl-5-chlorononane + OH⁻ →
   (b) 4-ethyl-3-octanol + HBr →
   (c) 3-nonanol

106. Name the compounds.
   (a) 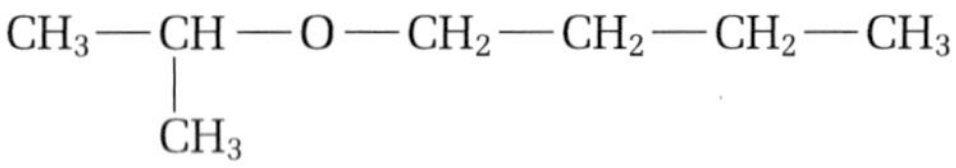 CH₃—CH(CH₃)—O—CH₂—CH₂—CH₂—CH₃
   (b) 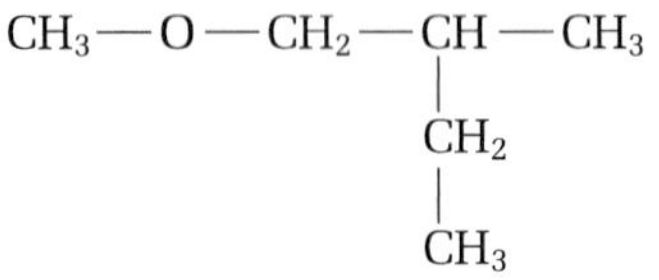 CH₃—O—CH₂—CH(CH₂CH₃)—CH₃
   (c) 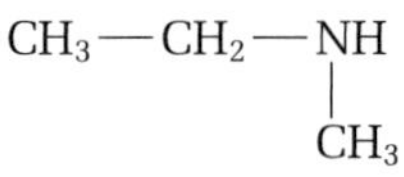 CH₃—CH₂—NH(CH₃)
   (d) 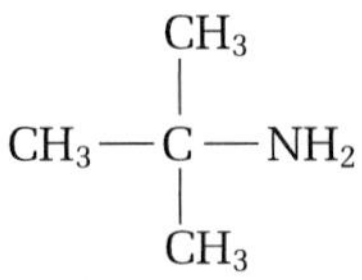 CH₃—C(CH₃)(CH₃)—NH₂

107. Write a condensed structural diagram for each compound.
   (a) cyclopentanamine
   (b) N-butyl-N-ethyl-3-octanamine
   (c) 2-ethoxy-2-heptane
   (d) methoxy methane
   (e) N,N-diethyl-1-hexanamine
   (f) 2-butanamine

108. Classify the amines in #18 as primary, secondary, or tertiary.

109. Name the compounds.
   (a) 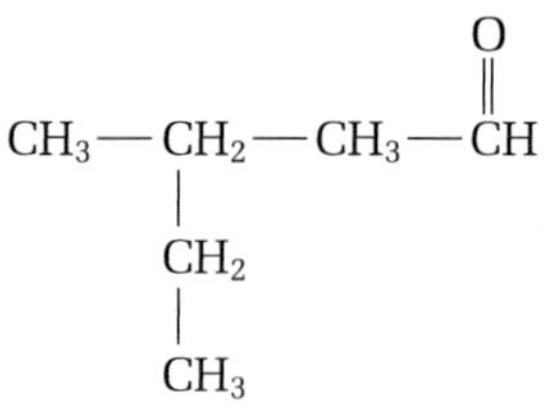 CH₃—CH₂—CH₃—CH(=O) with CH₂—CH₃ branch
   (b) 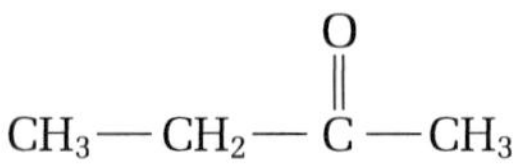 CH₃—CH₂—C(=O)—CH₃
   (c) HC(=O)—CH(CH₃)—CH₂—CH₃
   (d) 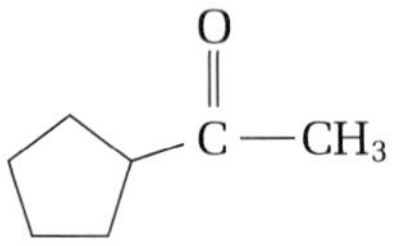 cyclopentyl—C(=O)—CH₃

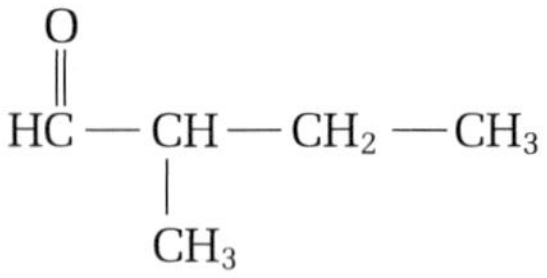

110. Explain why the following compounds could not exist.
   (a) 3-methyl-3-pentanone
   (b) 2-methyl-1-butyne
   (c) 3,3-diethyl-2-pentene
   (d) 1-methyl propanal

111. Draw a condensed structural diagram of the compounds.
   (a) 2,4,5-trimethyl octanal
   (b) 3-methyl-2-butanone
   (c) 2-ethyl-4,4-dimethyl-3-hexanone
   (d) methanal

112. Name the compounds.
   (a) 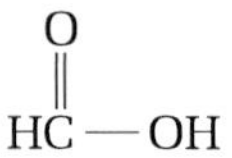 HC(=O)—OH
   (b) 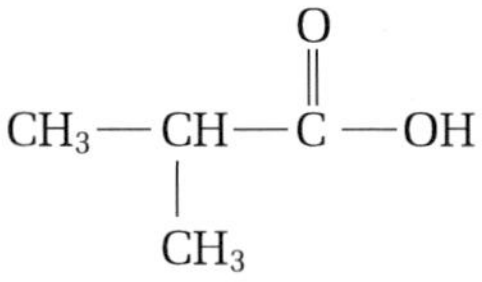 CH₃—CH(CH₃)—C(=O)—OH
   (c) CH₃—CH₂—CH₂—CH₂—O—CH(=O)
   (d) 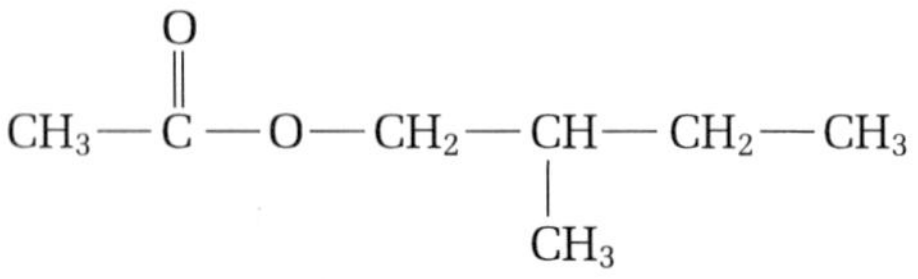 CH₃—C(=O)—O—CH₂—CH(CH₃)—CH₂—CH₃

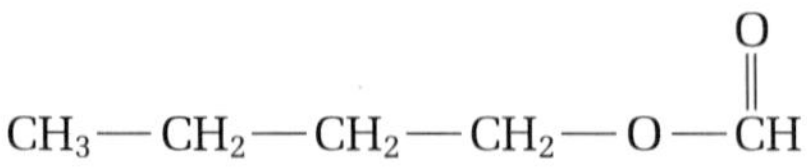

113. Write reactions for the following.
   (a) synthesis of methyl pentanoate
   (b) esterification of ethanol and propanoic acid
   (c) hydrolysis of butyl butanoate
   (d) esterification of 2-propanol and butyric acid

114. Draw a condensed structural diagram for each compound.
    (a) N-ethyl-nonanamide
    (b) hexanamide
    (c) N-methyl propanamide
    (d) N,N-dimethyl butanamide

115. Write the reaction, using names and condensed structural diagrams for the following reactions.
    (a) acid hydrolysis of N-ethyl hexanamide
    (b) base hydrolysis of N,N-dimethyl butanamide
    (c) acid hydrolysis of N-propyl 2-methyl-pentanamide
    (d) base hydrolysis of 3,3-dimethyl hexanamide

## Chapter 11

116. Draw the monomers that were used to make the polymer shown here.

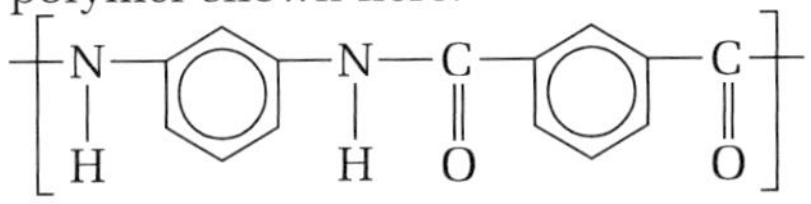

117. What is the name of the class of the polymer shown in #116?

118. Was the polymer in #116 an addition or condensation monomer?

119. Sketch three units of the polymer made from the structure shown here.

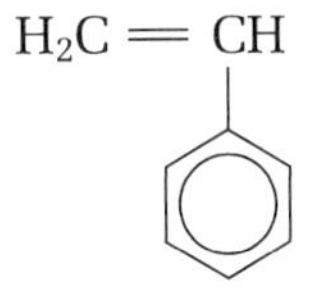

    $H_2C = CH$

120. Was the polymer you sketched in #119 an addition or condensation polymer?

121. Sketch two units of the polymer formed by a reaction between the following.

    $nHO-CH_2-CH_2-OH \ + \ nHO-\overset{O}{\overset{\|}{C}}-CH_2-CH_2-\overset{O}{\overset{\|}{C}}-OH$

122. What class of polymer did you sketch in #121?

123. Draw the monomer units that were used to make the polymer shown here.

124. What class of polymer is shown in #121?

125. Sketch monomer units that could be used to make a polyester. What are the functional groups on the monomer units?

126. Sketch the type of bond that joins amino acids to make proteins.

## Unit 5

## Chapter 12

127. (a) Write the overall equation for the following two-step mechanism:
    Step 1  $NO_{2(g)} + NO_{2(g)} \rightarrow NO_{(g)} + NO_{3(g)}$
    Step 2  $NO_{3(g)} + CO_{(g)} \rightarrow NO_{2(g)} + CO_{2(g)}$

    (b) If Step 1 is slow and Step 2 is fast, which is the rate-determining step?

128. For the equation $2H_{2(g)} + 2NO_{(g)} \rightarrow N_{2(g)} + 2H_2O_{(g)}$, a chemist proposes a three-step mechanism. Write Step 2 of this mechanism.
    Step 1  $H_{2(g)} + NO_{(g)} \rightarrow H_2O_{(g)} + N_{(g)}$
    Step 2  ?
    Step 3  $H_{2(g)} + O_{(g)} \rightarrow H_2O_{(g)}$

129. (a) Write the overall equation for the two-step mechanism below.
    Step 1  $O_{3(g)} + NO_{(g)} \rightarrow NO_{2(g)} + O_{2(g)}$
    Step 2  $NO_{2(g)} + O_{(g)} \rightarrow NO_{(g)} + O_{2(g)}$

    (b) Identify the reaction intermediates and the catalyst, if any.

130. A chemist proposes the following two-step mechanism:
    Step 1  $NO_2Cl_{(g)} \rightarrow NO_{2(g)} + Cl_{(g)}$
    Step 2  $NO_2Cl_{(g)} + Cl_{(g)} \rightarrow NO_{2(g)} + Cl_{2(g)}$

    (a) Write the overall equation for this mechanism.
    (b) Identify the rate-determining step if Step 1 is slow and Step 2 is fast.

131. A chemist proposes a three-step mechanism for the reaction
    $Br_{2(aq)} + OCl_{2(aq)} \rightarrow BrOCl_{(aq)} + BrCl_{(aq)}$
    Given Steps 2 and 3 below, write the mechanism for Step 1.
    Step 1  ?
    Step 2  $Br_{(aq)} + OCl_{2(aq)} \rightarrow BrOCl_{(aq)} + Cl_{(aq)}$
    Step 3  $Br_{(aq)} + Cl_{(aq)} \rightarrow BrCl_{(aq)}$

132. (a) Write the overall balanced chemical equation for the following mechanism.
    Step 1  $OCl^-_{(aq)} + H_2O_{(\ell)} \rightarrow HOCl_{(aq)} + OH^-_{(aq)}$
    Step 2  $HOCl_{(aq)} + I^-_{(aq)} \rightarrow HOI_{(aq)} + Cl^-_{(aq)}$
    Step 3  $HOI_{(aq)} + OH^-_{(aq)} \rightarrow H_2O_{(\ell)} + OI^-_{(aq)}$

    (b) List the reaction intermediates and the catalyst, if any.

133. In the upper atmosphere, ozone decomposes in the presence of ultraviolet radiation. Nitric oxide, NO, is also a part of this process. A chemist proposes the following mechanism for this reaction:
    Step 1  $O_{3(g)} + \text{ultraviolet radiation} \rightarrow O_{2(g)} + O_{(g)}$
    Step 2  $O_{3(g)} + NO_{(g)} \rightarrow NO_{2(g)} + O_{2(g)}$
    Step 3  $NO_{2(g)} + O_{(g)} \rightarrow NO_{(g)} + O_{2(g)}$

    (a) What is the overall equation for this reaction? Identify the reaction intermediates and the catalyst, if any.
    (b) Identify the role played by ultraviolet radiation in this reaction.

134. Write the overall equation for the following mechanism:
    $O_{3(g)} \rightarrow O_{2(g)} + O$
    $O_{(g)} + O_{3(g)} \rightarrow 2O_{2(g)}$

135. Write the missing step in this mechanism for the reaction $2NO_{2(g)} + F_{2(g)} \rightarrow 2NO_2F_{(g)}$.
    Step 1  $NO_{2(g)} + F_{2(g)} \rightarrow NO_2F_{(g)} + F_{(g)}$
    Step 2  ?

136. Sketch a potential energy diagram, given the following criteria for a reaction: potential energy of the reactants = 200 kJ; potential energy of the activated complex = 300 kJ; potential energy of the products = 100 kJ. State whether the reaction is exothermic or endothermic.

137. Show how your potential energy diagram for the previous question would be different with the addition of a catalyst, and state the effect on the reactants, the activated complex, the products, and the reaction rate.

138. For the three reactions below, identify the reaction that has the fastest rate and the reaction that has the slowest rate. Assume all reactions take place at the same temperature.
    (a) $2HCl_{(aq)} + CaCO_{3(s)} \rightarrow CaCl_{2(aq)} + CO_{2(g)} + H_2O_{(\ell)}$
    (b) $H_{2(g)} + I_{2(g)} \rightarrow 2HI_{(g)}$
    (c) $Na^+_{(aq)} + Cl^-_{(aq)} \rightarrow NaCl_{(s)}$

## Chapter 13

139. For the following reaction at 963 K, $K_c = 10.0$:
    $CO_{(g)} + H_2O_{(g)} \rightleftharpoons CO_{2(g)} + H_{2(g)}$
    State whether or not this system is at equilibrium when reactants and products have the following concentrations: $[CO_{(g)}] = 2.0$ mol/L, $[H_2O_{(g)}] = 0.10$ mol/L, $[H_2] = 1.0$ mol/L, $[CO_2] = 2.0$ mol/L.

140. A chemist places 0.300 mol of $Br_2$ and 0.600 mol of $Cl_2$ into a 3.00 L reaction vessel. At equilibrium at high temperature, 0.045 mol/L of $Br_2$ remain. Calculate $K_c$ for this reaction.
    $Br_{2(g)} + Cl_{2(g)} \rightleftharpoons 2BrCl_{(g)}$

141. For the equation below, predict the direction in which the equilibrium will shift in response to the changes specified.
    $CO_{(g)} + H_2O_{(g)} + \text{energy} \rightleftharpoons CO_{2(g)} + H_{2(g)}$
    (a) add water and decrease the temperature
    (b) add a catalyst and remove carbon monoxide
    (c) add hydrogen and water
    (d) add carbon monoxide and increase the temperature

142. For the system $A + 2B \rightleftharpoons C + D$, 2.00 mol of A and 2.00 mol of B are placed into a 2.00 L reaction vessel. When the system reaches equilibrium at a certain temperature, $K_c = 4.00 \times 10^{-9}$. What is [C] at equilibrium?

143. For the reaction $FeSCN^{2+}_{(aq)} \rightleftharpoons Fe^{3+}_{(aq)} + SCN^-_{(aq)}$ at a certain temperature, $K_c = 9.1 \times 10^{-4}$. Determine the concentrations of each species in a solution in which the initial concentration of $FeSCN^{2+}$ is 1.5 mol/L.

144. At 400°C, $K_c = 50.0$ for the following reaction:
    $H_{2(g)} + I_{2(g)} \rightleftharpoons 2HI_{(g)}$
    A chemist places 0.0800 mol of $H_2$ and 0.0800 mol of $I_2$ in a 1.00 L reaction vessel. Calculate the concentration of $I_{2(g)}$ at equilibrium.

145. Determine $[CO_2]$ at equilibrium when 2.50 mol of $CO_{(g)}$ and 2.50 mol of $H_2O_{(g)}$ react in a 1.00 L vessel at a certain high temperature. $K_c = 0.63$ for this reaction.
    $CO_{(g)} + H_2O_{(g)} \rightleftharpoons CO_{2(g)} + H_{2(g)}$

146. Consider the equilibrium below.
    $S_{8(s)} + 8O_{2(g)} \rightleftharpoons 8SO_{2(g)} + \text{heat}$
    What is the effect on the concentration of each substance when the equilibrium is altered as follows.
    (a) More $S_{8(s)}$ is added.
    (b) The temperature is lowered.
    (c) The volume of the container is increased.
    (d) More $SO_2$ is injected into the system.

147. At 2000 K, the concentration of the components in the following equilibrium system are $[CO_{2(g)}] = 0.30$ mol/L, $[H_{2(g)}] = 0.20$ mol/L, and $[H_2O_{(g)}] = [CO_{(g)}] = 0.55$ mol/L.
    $CO_{2(g)} + H_{2(g)} \rightleftharpoons H_2O_{(g)} + CO_{(g)}$
    (a) What is the value of the equilibrium constant?
    (b) When the temperature is lowered, 20.0% of the $CO_{(g)}$ is converted back to $CO_{2(g)}$. Calculate the equilibrium constant at the lower temperature.
    (c) Rewrite the equilibrium equation, and indicate on which side of the equation the heat term should be placed.

148. After 5.00 g of $SO_2Cl_2$ are placed in a 2.00 L flask, the following equilibrium is established.
    $SO_2Cl_{2(g)} \rightleftharpoons SO_{2(g)} + Cl_{2(g)}$
    $K_c$ for this equilibrium is 0.0410. Determine the mass of $SO_2Cl_{2(g)}$ that is present at equilibrium.

149. For the equilibrium below, $K_c = 6.00 \times 10^{-2}$.
    $N_{2(g)} + 3H_{2(g)} \rightleftharpoons 2NH_{3(g)}$
    Explain why this value of $K_c$ does not apply when the equation is written as follows.
    $\frac{1}{2}N_{2(g)} + \frac{3}{2}H_{2(g)} \rightleftharpoons NH_{3(g)}$

150. A 11.5 g sample of $I_{2(g)}$ is sealed in a 250 mL flask. An equilibrium, shown below, is established as this molecular form of $I_2$ dissociates into iodine atoms.
    $I_{2(g)} \rightleftharpoons 2I_{(g)}$
    $K_c$ for the equilibrium is $3.80 \times 10^{-5}$. Calculate the equilibrium concentration of both forms of the iodine.

# Unit 6

## Chapter 14

151. A sample of lemon juice was found to have a pH of 2.50. What is the concentration of hydronium ions and hydroxide ions in the lemon juice?

What are the concentrations of hydronium and hydroxide ions in a solution that has a pH of 5?

**153.** What is the pH of a $1.0 \times 10^{-5}$ mol/L $Ca(OH)_{2(aq)}$?

**154.** What is the pH of a solution in which $2.0 \times 10^{-4}$ mol of HCl are dissolved in enough distilled water to make 300 mL of solution?

**155.** What is the pH of a solution containing 2.5 g of NaOH dissolved in 100 mL of water?

**156.** For each of the following reactions, identify the acid, the base, the conjugate base, and the conjugate acid:
 (a) $HF_{(aq)} + NH_{3(aq)} \rightleftharpoons NH_4^+{}_{(aq)} + F^-{}_{(aq)}$
 (b) $Fe(H_2O)_6^{3+}{}_{(aq)} + H_2O_{(\ell)} \rightleftharpoons Fe(H_2O)_5(OH)_2^+{}_{(aq)} + H_3O^+{}_{(aq)}$
 (c) $NH_4^+{}_{(aq)} + CN^-{}_{(aq)} \rightleftharpoons HCN_{(aq)} + NH_{3(aq)}$
 (d) $(CH_3)_3N_{(aq)} + H_2O_{(\ell)} \rightleftharpoons (CH_3)_3NH^+{}_{(aq)} + OH^-{}_{(aq)}$

**157.** A solution was prepared by mixing 70.0 mL of 4.00 mol/L $HCl_{(aq)}$ and 30.0 mL of 8.00 mol/L $HNO_{3(aq)}$. Water was then added until the final volume was 500 mL. Calculate [H⁺] and the final pH.

**158.** Name the conjugate base and the conjugate acid of $HPO_4^{2-}$. What determines if $HPO_4^{2-}$ will act as an acid or a base?

**159.** Rewrite each equation below, identify each species as an acid or a base, and draw lines to connect the conjugate acid-base pairs.
 (a) $H_2SO_{4(aq)} + OH^-{}_{(aq)} \rightleftharpoons HSO_4^-{}_{(aq)} + H_2O_{(\ell)}$
 (b) $HSO_4^-{}_{(aq)} + OH^- \rightleftharpoons SO_4^{2-}{}_{(aq)} + H_2O_{(\ell)}$

**160.** Calculate pOH for the following aqueous solutions.
 (a) $3.2 \times 10^{-3}$ mol/L HCl
 (b) 0.0040025 mol/L $HNO_3$
 (c) $2.6 \times 10^{-4}$ mol/L $Ca(OH)_2$

**161.** A solution of acetic acid has a pH of 2.4. Is the pH of this solution higher than, lower than, or the same as the pH of a solution of 0.0005 mol/L sulfuric acid?

**162.** Identify each of the following as on Arrhenius acid/base, a Brønsted-Lowry acid/base, or both.
 (a) $Ba(OH)_2$
 (b) HOH
 (c) KOH
 (d) $H_3AsO_4$
 (e) $CH_3COOH$
 (f) HClO

## Chapter 15

**163.** How many milliliters of sodium hydroxide solution are required to neutralize 20 mL of 1.0 mol/L acetic acid if 32 mL of the same sodium hydroxide solution neutralized 20 mL of 1.0 mol/L hydrochloric acid?

**164.** How many moles of calcium hydroxide will be neutralized by one mole of hydrochloric acid, according to the following equation?
$Ca(OH)_{2(aq)} + 2HCl_{(aq)} \rightarrow CaCl_{2(aq)} + 2H_2O_{(\ell)}$

**165.** In an experiment, 50.0 mL of 0.0800 mol/L NaOH is titrated by the addition of 0.0500 mol/L $HNO_3$. What is the hydroxide ion concentration after 30.0 mL $HNO_3$ solution has been added.

**166.** A 100 mL volume of 0.200 mol/L HCl was placed in a flask. What volume of 0.400 mol/L NaOH solution must be added to bring the solution to a pH of 7.0?

**167.** A blood sample has a pH of 7.32. What is the hydronium ion concentration? How can hydronium ions exist in this basic solution?

**168.** A solution is made by combining 200.0 mL of 0.23 mol/L $H_2SO_4$, 600.0 mL of 0.16 mol/L KOH, and 200.0 mL of water. What is the pH of this solution?

**169.** $K_a$ for $CH_3COOH$ is $1.8 \times 10^{-5}$, and $K_a$ for $HNO_2$ is $7.2 \times 10^{-4}$. When the conjugate bases of these two acids are compared, which has the larger $K_b$? What is the value of this larger $K_b$?

**170.** $K_a$ for hydrazoic acid, $HN_3$, is $2.80 \times 10^{-5}$.
 (a) Write the equation for this dissociation of this acid in water.
 (b) What is the $K_b$ for the conjugate base of this acid?
 (c) Compare the pH of 0.100 mol/L $HN_3$ with the pH of a sample of the same volume in which 0.600 g of sodium azide, $NaN_3$, has been dissolved?

**171.** 10.0 mL of $NH_{3(aq)}$, with a concentration of $5.70 \times 10^{-2}$ mol/L, is titrated to the end point with 2.85 mL of HBr solution. What is the concentration of the HBr solution?

**172.** 0.250 mol of acetic acid and 0.100 mol of sodium hydroxide are dissolved in enough water to produce 1.00 L of solution. Determine the concentration of the acetate ion, the acetic acid, and the hydronium ion in the solution.

**173.** The hydroxide concentration of a 0.350 mol/L solution of B, a weak base, is $7.11 \times 10^{-5}$ mol/L. Find $K_b$ for B. (The dissociation equation for B is $B_{(aq)} + H_2O_{(\ell)} \rightleftharpoons BH^+{}_{(aq)} + OH^-{}_{(aq)}$.)

**174.** $K_a$ for formic acid, $HCHO_2$, is $1.7 \times 10^{-4}$. Determine the hydronium ion concentration of 0.400 mol/L formic acid.

# Unit 7

## Chapter 16

**175.** Using the compound ethanol, $C_2H_5OH$, write equations to distinguish between $\Delta H°_f$ and $\Delta H_{comb}$.

**176.** Glass has a specific heat capacity of 0.84 J/g°C. A certain metal has a specific heat capacity of 0.500 J/g°C. A metal tray and a glass tray have the same mass. They are placed in an oven at 80°C.
 (a) After 1 h, how does the temperature of the two trays compare?
 (b) How does the quantity of heat absorbed by the two trays compare?
 (c) Does either tray feel hotter to the touch?

**177.** $\Delta H°_f$ of $HI_{(g)}$ is +25.9 kJ/mol.
  (a) Write the equation that represents this reaction.
  (b) Which has more enthalpy, the elements $H_{2(g)}$ and $I_{2(g)}$, or the product $HI_{(g)}$?
  (c) What does the positive value of $\Delta H°_f$ indicate about the energy that is needed to break bonds in $H_{2(g)}$ and $I_{2(g)}$, compared with the energy released when H—I bonds form?

**178.** A 120.0 g sample of water at 30.0°C is placed in the freezer compartment of a refrigerator. How much heat has the sample lost when it changes completely to ice at 0°C?

**179.** $\Delta H°$ is +106.9 kJ for the reaction $C_{(s)} + PbO_{(s)} \rightarrow Pb_{(s)} + CO_{(g)}$. How much heat is needed to convert 50.0 g of $PbO_{(s)}$ to $Pb_{(s)}$?

**180.** 100 g of ethanol at 25°C is heated until it reaches 75°C. How much heat did the ethanol absorb?

**181.** A beaker containing 25 g of liquid at room temperature is heated until it gains 5°C. A second beaker containing 50 g of the same liquid at room temperature is heated until it also gains 5°C. Which beaker has gained the most thermal energy? Explain.

**182.** An 80.0 g lump of iron at room temperature (23°C) was placed in an insulated container that held 100.0 mL of water at 85°C. What was the final temperature of the iron and water?

**183.** A student claimed that a certain ring was pure gold. To test this claim, a classmate first determined that the mass of the ring was 12 g. The classmate then placed the ring in a beaker of hot water until the ring reached a temperature of 62°C. The classmate then removed the ring from the water with tongs and placed into an insulated cup containing 25 mL of water at 20°C. The final temperature of the ring and water was 16.7°C. Was the ring pure gold? Explain your reasoning.

**184.** Your hot water heater is not working and you want to take a hot bath. You decide to boil water in a teapot on the stove and pour into the tub. First, you fill the tub with 100 kg of water at room temperature (23°C). You start boiling water on the stove. By the time you get the teapot to the tub to pour the hot water into the tub, the water in the teapot is at 96°C. How much hot water would you have to pour into the tub to bring the water to a temperature of 35°C? (Assume that a negligible amount of heat was lost to the air.)

## Chapter 17

**185.** A 20.00 g sample of metal is warmed to 165°C in an oil bath. The sample is then transferred to a coffee-cup calorimeter that contains 125.0 g of water at 5.0°C. The final temperature of the water is 8.8°C.
  (a) Calculate the specific heat capacity of the metal.
  (b) What are three sources of experimental error that occur in this experiment?

**186.** Use $\Delta H°_f$ data to determine the heat of reaction for each reaction.
  (a) $4NH_{3(g)} + 3O_{2(g)} \rightarrow 2N_{2(g)} + 6H_2O_{(\ell)}$
  (b) $2H_2O_{(g)} + CS_{2(g)} \rightarrow 2H_2S_{(g)} + CO_{2(g)}$
  (c) $4NO_{(g)} + 6H_2O_{(g)} \rightarrow 4NH_{3(g)} + 5O_{2(g)}$

**187.** Use the enthalpies of combustion for the burning of $CO_{(g)}$, $H_{2(g)}$, and $C_{(s)}$ to determine $\Delta H°$ for the reaction $C_{(s)} + H_2O_{(g)} \rightarrow H_{2(g)} + CO_{(g)}$.

$CO_{(g)} + \frac{1}{2}O_{2(g)} \rightarrow CO_{2(g)}$     $\Delta H°_{comb} = -238$ kJ/mol
$H_{2(g)} + \frac{1}{2}O_{2(g)} \rightarrow H_2O_{(g)}$     $\Delta H°_{comb} = -241$ kJ/mol
$C_{(s)} + O_{2(g)} \rightarrow CO_{2(g)}$     $\Delta H°_{comb} = -393$ kJ/mol

**188.** $\Delta H_{comb}$ for ethene is –337 kJ/mol.
  (a) Write the balanced equation for the complete combustion of ethene in air. Include the heat term in your balanced equation.
  (b) The heat that is produced by burning 1.00 kg of ethene warms a quantity of water from 15.0°C to 85.0°C. What is the mass of the water if the heat transfer is 60.0% efficient?

**189.** Determine $\Delta H°$ for
$Ca^{2+}_{(aq)} + 2OH^-_{(aq)} + CO_{2(g)} \rightarrow CaCO_{3(s)} + H_2O_{(\ell)}$, given the following information.

$CaO_{(s)} + H_2O_{(\ell)} \rightarrow Ca(OH)_{2(s)}$     $\Delta H° = -65.2$ kJ/mol
$CaCO_{3(s)} \rightarrow CaO_{(s)} + CO_{2(g)}$     $\Delta H° = +178.1$ kJ/mol
$Ca(OH)_{2(s)} \rightarrow Ca^{2+}_{(aq)} + 2OH^-_{(aq)}$     $\Delta H° = -16.2$ kJ/mol

**190.** Calculate the enthalpy of formation of manganese(IV) oxide, based on the following information.
$4Al_{(s)} + 3MnO_{2(s)} \rightarrow 3Mn_{(s)} + 2Al_2O_{3(s)} + 1790$ kJ
$2Al_{(s)} + \frac{3}{2}O_{2(g)} \rightarrow Al_2O_{3(s)} + 1676$ kJ

**191.** $\Delta H_{comb}$ for toluene, $C_7H_{8(\ell)}$, is –3904 kJ/mol.
  (a) Write the equation for the complete combustion of toluene.
  (b) Use the combustion equation and $\Delta H°_f$ values to determine the enthalpy of formation of toluene.

**192.** An impure sample of zinc has a mass of 7.35 g. The sample reacts with 150.0 g of dilute hydrochloric acid solution inside a calorimeter. The calorimeter has a mass of 520.57 g and a specific heat capacity of 0.400 J/g°C. $\Delta H°$ for the following reaction is –153.9 kJ/mol.
$Zn_{(s)} + 2HCl_{(aq)} \rightarrow ZnCl_{2(aq)} + H_{2(g)}$
When the reaction occurs, the temperature of the solution rises from 14.5°C to 29.7°C. What is the percentage purity of the sample? Assume that the specific heat capacity of the hydrochloric acid is 4.184 J/g°C. Also assume that all of the zinc in the impure sample reacts.

193. An unknown solid was dissolved in the water of a calorimeter in order to find its heat of solution. The following data was recorded:
mass of solid = 5.5 g
mass of calorimeter water = 120.0 g
initial temperature of water = 21.7°C
final temperature of solution = 32.6°C
(a) Calculate the heat change of the water.
(b) Calculate the heat change caused by the solid dissolving.
(c) What is the heat of solution per gram of solid dissolved?

194. A 100-g sample of food is placed in a bomb calorimeter calibrated at 7.23 kJ/°C. When the food is burned, the calorimeter gains 512 kJ of heat. If the initial temperature of the calorimeter was 19°C, what is the final temperature of the calorimeter and its contents?

# Unit 8

## Chapter 18

195. Identify the redox reactions.
(a) $CCl_4 + HF \rightarrow CFCl_3 + HCl$
(b) $Al_2O_3 + 3H_2SO_4 \rightarrow Al_2(SO_4)_3 + 3H_2O$
(c) $CH_4 + 2O_2 \rightarrow CO_2 + 2H_2O$
(d) $P_4 + 3OH^- + 3H_2O \rightarrow PH_3 + 3H_2PO_2^-$

196. Consider the following reaction.
$S_8 + 8Na_2SO_3 \rightarrow 8Na_2S_2O_3$
(a) Assign oxidation numbers to all the elements.
(b) Identify the reactant that undergoes reduction.
(c) Identify the reactant that is the reducing agent.

197. Find the oxidation number for each element.
(a) sulfur in $HS^-, S_8, SO_3^{2-}, S_2O_3^{2-}$, and $S_4O_6^{2-}$
(b) boron in $B_4O_7^{2-}, BO_3^-, BO_2^-, B_2H_6$, and $B_2O_3$

198. List the following oxides of nitrogen in order of decreasing oxidation number of nitrogen:
$NO_2, N_2O_5, NO, N_2O_3, N_2O, N_2O_4$.

199. For the following redox reaction, indicate which statements, if any, are true.
$C + D \rightarrow E + F$
(a) If C is the oxidizing agent, then it loses electrons.
(b) If D is the reducing agent, then it is reduced.
(c) If C is the reducing agent, and if it is an element, then its oxidation number will increase.
(d) If D is oxidized, then C must be a reducing agent.
(e) If C is reduced, then D must lose electrons.

200. Consider the following reaction.
$3SF_4 + 4BCl_3 \rightarrow 4BF_3 + 3SCl_2 + 3Cl_2$
(a) Why is this reaction classified as a redox reaction?
(b) What is the oxidizing agent in this reaction?

201. The metals $Ga_{(s)}, In_{(s)}, Mn_{(s)}$, and $Np_{(s)}$, and their salts, react as follows:
$3Mn^{2+}_{(aq)} + 2Np_{(s)} \rightarrow 3Mn_{(s)} + 2Np^{3+}_{(aq)}$
$In^{3+}_{(aq)} + Ga_{(s)} \rightarrow In_{(s)} + Ga^{3+}_{(aq)}$
$Mn^{2+} + Ga_{(s)} \rightarrow$ no reaction
Analyze this information, and list the reducing agents from the worst to the best.

202. The following redox reactions occur in basic solution. Balance the equations using the oxidation number method.
(a) $Ti^{3+} + RuCl_5^{2-} \rightarrow Ru + TiO^{2+} + Cl^-$
(b) $ClO_2 \rightarrow ClO_2^- + ClO_3^-$

203. The following redox reactions occur in basic solution. Use the half-reaction method to balance the equations.
(a) $NO_2 \rightarrow NO_2^- + NO_3^-$
(b) $CrO_4^- + HSnO_2^- \rightarrow CrO_2^- + HSnO_3^-$
(c) $Al + NO_3^- \rightarrow Al(OH)_4^- + NH_3$

204. The following redox reactions occur in acidic solution. Balance the equations using the half-reaction method.
(a) $ClO_3^- + I_2 \rightarrow IO_3^- + Cl^-$
(b) $C_2H_4 + MnO_4^- \rightarrow CO_2 + Mn^{2+}$
(c) $Cu + SO_4^{2-} \rightarrow Cu^{2+} + SO_2$

205. The following redox reactions occur in acidic solution. Use oxidation numbers to balance the equations.
(a) $Se + NO_3^- \rightarrow SeO_2 + NO$
(b) $Ag + Cr_2O_7^{2-} \rightarrow Ag^+ + Cr^{3+}$

206. Balance each equation for a redox reaction.
(a) $P_4 + NO_3^- \rightarrow H_3PO_4 + NO$ (acidic)
(b) $MnO_2 + NO_2^- \rightarrow NO_3^- + Mn^{2+}$ (acidic)
(c) $TeO_3^{2-} + N_2O_4 \rightarrow Te + NO_3^-$ (basic)
(d) $MnO_4^- + N_2H_4 \rightarrow MnO_2 + N_2$ (basic)
(e) $S_2O_3^{2-} + OCl^- \rightarrow SO_4^{2-} + Cl^-$ (basic)
(f) $Br_2 + SO_2 \rightarrow Br^- + SO_4^{2-}$ (acidic)
(g) $PbO_2 + Cl^- \rightarrow PbCl_2 + Cl_2$ (acidic)
(h) $MnO_4^- + H_2O_2 \rightarrow Mn^{2+} + O_2$ (acidic)

207. The following reaction takes place in acidic solution.
$CH_3OH_{(aq)} + MnO_4^-_{(aq)} \rightarrow HCOOH_{(aq)} + Mn^{2+}_{(aq)}$
(a) Balance the equation.
(b) How many electrons are gained by the oxidizing agent when ten molecules of methanol are oxidized?
(c) What volume of 0.150 mol/L $MnO_4^-$ solution is needed to react completely with 20.0 g of methanol?

## Chapter 19

208. Consider the galvanic cell represented as
$Al|Al^{3+}||Ce^{4+}|Ce^{3+}|Pt$.
(a) What is the cathode?
(b) In which direction do electrons flow through the external circuit?
(c) What is the oxidizing agent?
(d) Will the aluminum electrode increase or decrease in mass as the cell operates?

Appendix B • MHR  833

**209. (a)** Which is a better oxidizing agent in acidic solution, $MnO_4^-$ or $Cr_2O_7^{2-}$? From what information did you determine your answer?

**(b)** Suppose that you have the metals Ni, Cu, Fe, and Ag, as well as 1.0 mol/L aqueous solutions of the nitrates of these metals. Which metals should be paired in a galvanic cell to produce the highest standard cell potential? (Use the most common ion for each metal.)

**210.** $E°_{cell}$ for the cell No|No$^{3+}$||Cu$^{2+}$|Cu is 2.842 V. Use this information, as well as the standard reduction potential for the Cu$^{2+}$/Cu half-reaction given in tables, to calculate the standard reduction potential for the No/No$^{3+}$ half-cell.

**211.** A galvanic cell is set up using tin in a 1.0 mol/L Sn$^{2+}$ solution and iron in a 1.0 mol/L Fe$^{2+}$ solution.
**(a)** Write the equation for the overall reaction that occurs in this cell.
**(b)** What is the standard cell potential?
**(c)** Which electrode is positive in this cell?
**(d)** What change in mass will occur at the anode when the cathode undergoes a change in mass of 1.50 g?

**212.** Predict the products that would be expected from the electrolysis of 1.0 mol/L NaI. Use the non-standard $E$ values for water.

**213.** Given the half reactions below, determine if the thiosulfate ion, $S_2O_3^{2-}$, can exist in an acidic solution under standard conditions.
$S_2O_3^{2-} + H_2O \rightarrow 2SO_2 + 2H^+ + 4e^-$  $E° = 0.40$ V
$2S + H_2O \rightarrow S_2O_3^{2-} + 6H^+ + 4e^-$  $E° = -0.50$ V

**214.** Refer to the table of half-cell potentials to determine if $MnO_2$ can oxidize Br$^-$ to Br$_2$ in acidic solution under standard conditions.

**215.** A galvanic cell contains 50.0 mL of 0.150 mol/L CuSO$_4$. If the Cu$^{2+}$ ions are completely used up, what is the maximum quantity of electricity that the cell can generate?

**216.** Determine the standard cell potential for each redox reaction.
**(a)** $IO_3^- + 6ClO_2^- + 6H^+ \rightarrow I^- + 3H_2O + 6ClO_2$
**(b)** $2Cu^{2+} + Hg_2^{2+} \rightarrow 2Hg^{2+} + 2Cu^+$
**(c)** $Ba^{2+} + Pb \rightarrow Pb^{2+} + Ba$
**(d)** $Ni + I_2 \rightarrow Ni^{2+} + 2I^-$

**217.** A current of 3.0 A flows for 1.0 h during an electrolysis of copper(II) sulfate. What mass of copper is deposited?

**218.** To recover aluminum metal, $Al_2O_3$ is first converted to $AlCl_3$. Then an electrolysis of molten $AlCl_3$ is performed using inert carbon electrodes.
**(a)** Write the half-reaction that occurs at the anode and at the cathode.
**(b)** Can the standard reduction potentials be used to calculate the external voltage needed for this process? Explain your answer.

**219.** Summarize the differences between a hydrogen fuel cell and a dry cell.

# Appendix C

# Alphabetical List of Elements

| Element | Symbol | Atomic Number | Element | Symbol | Atomic Number |
|---|---|---|---|---|---|
| Actinium | Ac | 89 | Neodymium | Nd | 60 |
| Aluminum | Al | 13 | Neon | Ne | 10 |
| Americium | Am | 95 | Neptunium | Np | 93 |
| Antimony | Sb | 51 | Nickel | Ni | 28 |
| Argon | Ar | 18 | Niobium | Nb | 41 |
| Arsenic | As | 33 | Nitrogen | N | 7 |
| Astatine | At | 85 | Nobelium | No | 102 |
| Barium | Ba | 56 | Osmium | Os | 76 |
| Berkelium | Bk | 97 | Oxygen | O | 8 |
| Beryllium | Be | 4 | Palladium | Pd | 46 |
| Bismuth | Bi | 83 | Phosphorus | P | 15 |
| Bohrium | Bh | 107 | Platinum | Pt | 78 |
| Boron | B | 5 | Plutonium | Pu | 94 |
| Bromine | Br | 35 | Polonium | Po | 84 |
| Cadmium | Cd | 48 | Potassium | K | 19 |
| Calcium | Ca | 20 | Praseodymium | Pr | 59 |
| Californium | Cf | 98 | Promethium | Pm | 61 |
| Carbon | C | 6 | Protactinium | Pa | 91 |
| Cerium | Ce | 58 | Radium | Ra | 88 |
| Cesium | Cs | 55 | Radon | Rn | 86 |
| Chlorine | Cl | 17 | Rhenium | Re | 75 |
| Chromium | Cr | 24 | Rhodium | Rh | 45 |
| Cobalt | Co | 27 | Rubidium | Rb | 37 |
| Copper | Cu | 29 | Ruthenium | Ru | 44 |
| Curium | Cm | 96 | Rutherfordium | Rf | 104 |
| Dubnium | Db | 105 | Samarium | Sm | 62 |
| Dysprosium | Dy | 66 | Scandium | Sc | 21 |
| Einsteinium | Es | 99 | Seaborgium | Sg | 106 |
| Erbium | Er | 68 | Selenium | Se | 34 |
| Europium | Eu | 63 | Silicon | Si | 14 |
| Fermium | Fm | 100 | Silver | Ag | 47 |
| Fluorine | F | 9 | Sodium | Na | 11 |
| Francium | Fr | 87 | Strontium | Sr | 38 |
| Gadolinium | Gd | 64 | Sulfur | S | 16 |
| Gallium | Ga | 31 | Tantalum | Ta | 73 |
| Germanium | Ge | 32 | Technetium | Tc | 43 |
| Gold | Au | 79 | Tellurium | Te | 52 |
| Hafnium | Hf | 72 | Terbium | Tb | 65 |
| Hassium | Hs | 108 | Thallium | Tl | 81 |
| Helium | He | 2 | Thorium | Th | 90 |
| Holmium | Ho | 67 | Thulium | Tm | 69 |
| Hydrogen | H | 1 | Tin | Sn | 50 |
| Indium | In | 49 | Titanium | Ti | 22 |
| Iodine | I | 53 | Tungsten | W | 74 |
| Iridium | Ir | 77 | Ununbium | Uub | 112 |
| Iron | Fe | 26 | Ununhexium | Uuh | 116 |
| Krypton | Kr | 36 | Ununnilium | Uun | 110** |
| Lanthanum | La | 57 | Ununquadium | Uuq | 114 |
| Lawrencium | Lr | 103 | Unununium | Uuu | 111 |
| Lead | Pb | 82 | Uranium | U | 92 |
| Lithium | Li | 3 | Vanadium | V | 23 |
| Lutetium | Lu | 71 | Xenon | Xe | 54 |
| Magnesium | Mg | 12 | Ytterbium | Yb | 70 |
| Manganese | Mn | 25 | Yttrium | Y | 39 |
| Meitnerium | Mt | 109 | Zinc | Zn | 30 |
| Mendelevium | Md | 101 | Zirconium | Zr | 40 |
| Mercury | Hg | 80 | | | |
| Molybdenum | Mo | 42 | | | |

**The names and symbols for elements 110 through 118 have not yet been chosen

## Periodic Table of the Elements

**MAIN-GROUP ELEMENTS**

Legend (sample element):
- Atomic number: 6
- Electronegativity: 2.5
- First ionization energy (kJ/mol): 1086
- Melting point (K): 3800
- Boiling point (K): 4300
- Average atomic mass*: 12.01
- Common oxidation number: ±4
- Other oxidation numbers: ±2
- Symbol: C (carbon)

Color key:
- Gases (red)
- Liquids (blue)
- Synthetics (orange)
- metals (main group)
- metals (transition)
- metals (inner transition)
- metalloids
- nonmetals

### Group 1 (IA)

**Period 1:** H — 1, 1.01, 2.20, 1312, 13.81, 20.28, +1, −1, hydrogen

### Group 2 (IIA)

**Period 2:**
- Li — 3, 6.94, 0.98, 520, 453.7, 1615, +1, lithium
- Be — 4, 9.01, 1.57, 899, 1560, 2744, +2, beryllium

**Period 3:**
- Na — 11, 22.99, 0.93, 496, 371, 1156, +1, sodium
- Mg — 12, 24.31, 1.31, 738, 923.2, 1363, +2, magnesium

### TRANSITION ELEMENTS

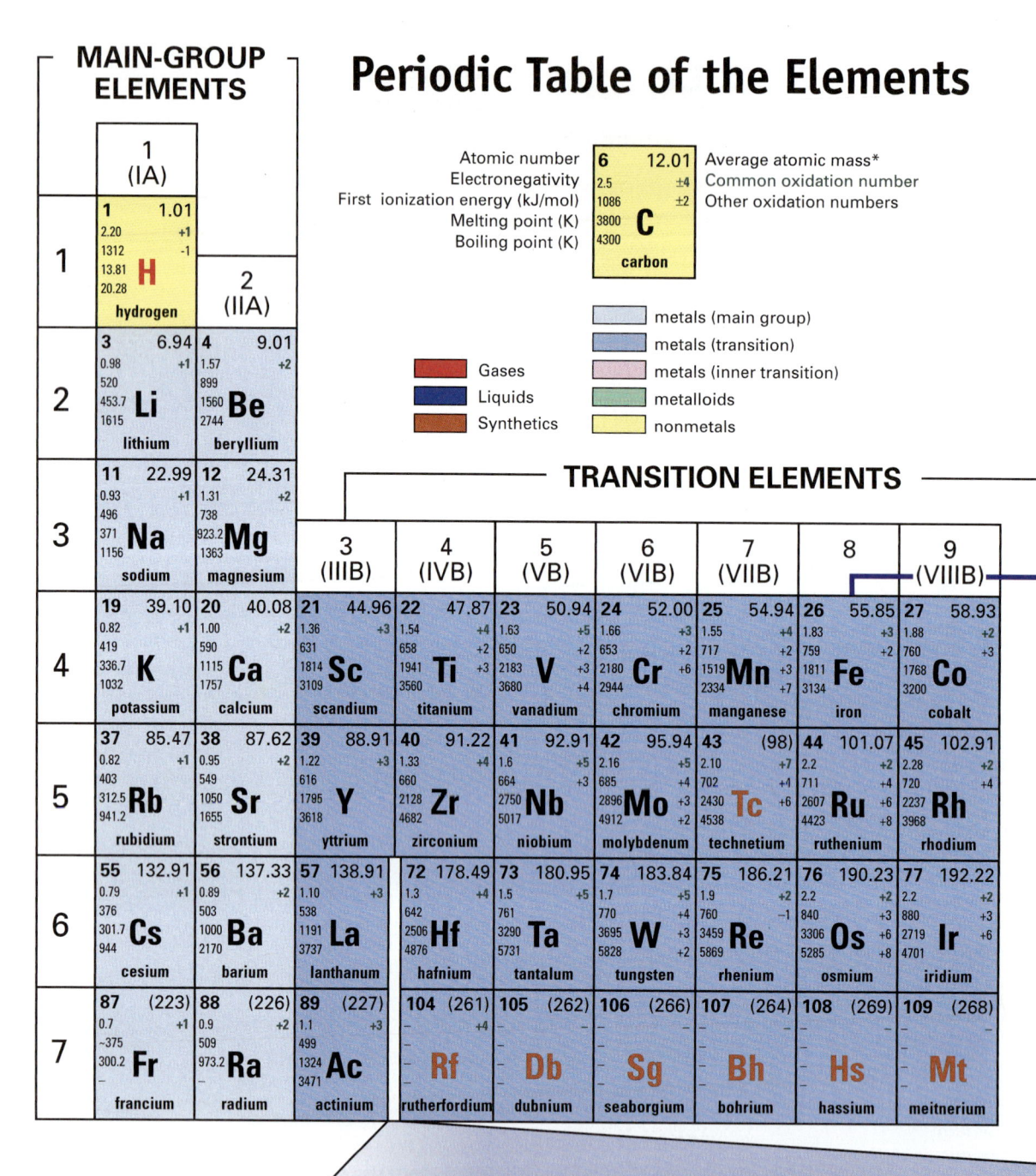

Groups: 3 (IIIB), 4 (IVB), 5 (VB), 6 (VIB), 7 (VIIB), 8 (VIIIB), 9 (VIIIB)

**Period 4:**
- K — 19, 39.10, 0.82, 419, 336.7, 1032, +1, potassium
- Ca — 20, 40.08, 1.00, 590, 1115, 1757, +2, calcium
- Sc — 21, 44.96, 1.36, 631, 1814, 3109, +3, scandium
- Ti — 22, 47.87, 1.54, 658, 1941, 3560, +4, +2, +3, titanium
- V — 23, 50.94, 1.63, 650, 2183, 3680, +5, +2, +3, +4, vanadium
- Cr — 24, 52.00, 1.66, 653, 2180, 2944, +3, +2, +6, chromium
- Mn — 25, 54.94, 1.55, 717, 1519, 2334, +2, +3, +7, manganese
- Fe — 26, 55.85, 1.83, 759, 1811, 3134, +3, +2, iron
- Co — 27, 58.93, 1.88, 760, 1768, 3200, +2, +3, cobalt

**Period 5:**
- Rb — 37, 85.47, 0.82, 403, 312.5, 941.2, +1, rubidium
- Sr — 38, 87.62, 0.95, 549, 1050, 1655, +2, strontium
- Y — 39, 88.91, 1.22, 616, 1795, 3618, +3, yttrium
- Zr — 40, 91.22, 1.33, 660, 2128, 4682, +4, zirconium
- Nb — 41, 92.91, 1.6, 664, 2750, 5017, +5, +3, niobium
- Mo — 42, 95.94, 2.16, 685, 2896, 4912, +6, +4, +3, molybdenum
- Tc — 43, (98), 2.10, 702, 2430, 4538, +7, technetium
- Ru — 44, 101.07, 2.2, 711, 2607, 4423, +2, +4, +6, ruthenium
- Rh — 45, 102.91, 2.28, 720, 2237, 3968, +2, +4, rhodium

**Period 6:**
- Cs — 55, 132.91, 0.79, 376, 301.7, 944, +1, cesium
- Ba — 56, 137.33, 0.89, 503, 1000, 2170, +2, barium
- La — 57, 138.91, 1.10, 538, 1191, 3737, +3, lanthanum
- Hf — 72, 178.49, 1.3, 642, 2506, 4876, +4, hafnium
- Ta — 73, 180.95, 1.5, 761, 3290, 5731, +5, tantalum
- W — 74, 183.84, 1.7, 770, 3695, 5828, +5, +4, +3, +2, tungsten
- Re — 75, 186.21, 1.9, 760, 3459, 5869, +2, −1, rhenium
- Os — 76, 190.23, 2.2, 840, 3306, 5285, +3, +6, +8, osmium
- Ir — 77, 192.22, 2.2, 880, 2719, 4701, +2, +3, +6, iridium

**Period 7:**
- Fr — 87, (223), 0.7, ~375, 300.2, —, +1, francium
- Ra — 88, (226), 0.9, 509, 973.2, —, +2, radium
- Ac — 89, (227), 1.1, 499, 1324, 3471, +3, actinium
- Rf — 104, (261), —, —, —, —, +4, rutherfordium
- Db — 105, (262), —, —, —, —, dubnium
- Sg — 106, (266), —, —, —, —, seaborgium
- Bh — 107, (264), —, —, —, —, bohrium
- Hs — 108, (269), —, —, —, —, hassium
- Mt — 109, (268), —, —, —, —, meitnerium

### INNER TRANSITION ELEMENTS

**6 Lanthanoids:**
- Ce — 58, 140.12, 1.12, 527, 1071, 3716, +3, +4, cerium
- Pr — 59, 140.91, 1.13, 523, 1204, 3793, +3, praseodymium
- Nd — 60, 144.24, 1.14, 530, 1294, 3347, +3, neodymium
- Pm — 61, (145), —, 536, 1315, 3273, +3, promethium
- Sm — 62, 150.36, 1.17, 543, 1347, 2067, +3, +2, samarium
- Eu — 63, 151.96, —, 547, 1095, 1802, +3, +2, europium
- Gd — 64, 157.25, 1.20, 593, 1586, 3546, +3, gadolinium

**7 Actinoids:**
- Th — 90, 232.04, 1.3, 587, 2023, 5061, +4, thorium
- Pa — 91, 231.04, 1.5, 568, 1845, —, +5, +4, protactinium
- U — 92, 238.03, 1.7, 584, 1408, 4404, +6, +3, +4, +5, uranium
- Np — 93, 237.05, 1.3, 597, 917, —, +5, +3, +4, +6, neptunium
- Pu — 94, (244), 1.3, 585, 913.2, 3501, +4, +3, +5, +6, plutonium
- Am — 95, (243), —, 578, 1449, 2284, +3, +4, +5, +6, americium
- Cm — 96, (247), —, 581, 1618, 3373, +3, curium

*Average atomic mass data in brackets indicate atomic mass of most stable isotope of the element.

## MAIN-GROUP ELEMENTS

| 13 (IIIA) | 14 (IVA) | 15 (VA) | 16 (VIA) | 17 (VIIA) | 18 (VIIIA) |
|---|---|---|---|---|---|
| | | | | | 2  4.00<br>—<br>2372<br>5.19<br>5.02<br>**He**<br>helium |
| 5  10.81<br>2.04  +3<br>800<br>2348<br>4273<br>**B**<br>boron | 6  12.01<br>2.55  ±4<br>1086  ±2<br>3800<br>4300<br>**C**<br>carbon | 7  14.01<br>3.04  −3<br>1402<br>63.15<br>77.36<br>**N**<br>nitrogen | 8  16.00<br>3.44  −2<br>1314<br>54.36<br>90.2<br>**O**<br>oxygen | 9  19.00<br>3.98  −1<br>1681<br>53.48<br>84.88<br>**F**<br>fluorine | 10  20.18<br>—<br>2080<br>24.56<br>27.07<br>**Ne**<br>neon |
| 13  26.98<br>1.61  +3<br>577<br>933.5<br>2792<br>**Al**<br>aluminum | 14  28.09<br>1.90  +4<br>786<br>1687<br>3538<br>**Si**<br>silicon | 15  30.97<br>2.19  ±5<br>1012  ±4<br>317.3  ±3<br>553.7<br>**P**<br>phosphorus | 16  32.07<br>2.58  ±6<br>999  ±2<br>392.8  ±4<br>717.8<br>**S**<br>sulfur | 17  35.45<br>3.16  −1<br>1256  +1<br>171.7  ±3<br>239.1  ±5<br>**Cl**  ±7<br>chlorine | 18  39.95<br>—<br>1520<br>83.8<br>87.3<br>**Ar**<br>argon |

| 10 | 11 (IB) | 12 (IIB) | | | | | | |
|---|---|---|---|---|---|---|---|---|
| 28  58.69<br>1.91  +2<br>737  +3<br>1728<br>3186<br>**Ni**<br>nickel | 29  63.55<br>1.90  +2<br>745  +1<br>1358<br>2835<br>**Cu**<br>copper | 30  65.39<br>1.65  +2<br>906<br>692.7<br>1180<br>**Zn**<br>zinc | 31  69.72<br>1.81  +3<br>579<br>302.9<br>2477<br>**Ga**<br>gallium | 32  72.61<br>2.01  +4<br>761<br>1211<br>3106<br>**Ge**<br>germanium | 33  74.92<br>2.18  ±3<br>947  ±5<br>1090<br>876.2<br>**As**<br>arsenic | 34  78.96<br>2.55  −4<br>941  −2<br>493.7  +4<br>958.2  −6<br>**Se**<br>selenium | 35  79.90<br>2.96  −1<br>1143  +1<br>266  ±5<br>332<br>**Br**<br>bromine | 36  83.80<br>—<br>1351<br>115.8<br>119.9<br>**Kr**<br>krypton |
| 46  106.42<br>2.20  +2<br>805  +4<br>1828<br>3236<br>**Pd**<br>palladium | 47  107.87<br>1.93  +1<br>731<br>1235<br>2435<br>**Ag**<br>silver | 48  112.41<br>1.69  +2<br>868<br>594.2<br>1040<br>**Cd**<br>cadmium | 49  114.82<br>1.78  +3<br>558<br>429.8<br>3345<br>**In**<br>indium | 50  118.71<br>1.96  +4<br>708  +2<br>505<br>2875<br>**Sn**<br>tin | 51  121.76<br>2.05  ±3<br>834  +5<br>903.8  −3<br>1860<br>**Sb**<br>antimony | 52  127.60<br>2.1  −4<br>869  −2<br>722.7  +6<br>1261<br>**Te**<br>tellurium | 53  126.90<br>2.66  −1<br>1009  +1<br>386.9  ±5<br>457.4  ±7<br>**I**<br>iodine | 54  131.29<br>—<br>1170<br>161.4<br>165<br>**Xe**<br>xenon |
| 78  195.08<br>2.2  +4<br>870  +2<br>2042<br>4098<br>**Pt**<br>platinum | 79  196.97<br>2.4  +3<br>890  +1<br>1337<br>3129<br>**Au**<br>gold | 80  200.59<br>1.9  +2<br>1107  +1<br>234.3<br>629.9<br>**Hg**<br>mercury | 81  204.38<br>1.8  +1<br>589  +3<br>577.2<br>1746<br>**Tl**<br>thallium | 82  207.20<br>1.8  +2<br>715  +4<br>600.6<br>2022<br>**Pb**<br>lead | 83  208.98<br>1.9  +3<br>703  +5<br>544.6<br>1837<br>**Bi**<br>bismuth | 84  (209)<br>2.0  +4<br>813  +2<br>527.2<br>1235<br>**Po**<br>polonium | 85  (210)<br>2.2  −1<br>(926)  +1<br>575  ±3<br>±5<br>**At**  ±7<br>astatine | 86  (222)<br>—<br>1037<br>202.2<br>211.5<br>**Rn**<br>radon |
| 110  (271)<br>—<br>**Uun**<br>ununnilium | 111  (272)<br>—<br>**Uuu**<br>unununium | 112  (277)<br>—<br>**Uub**<br>ununbium | 113 | 114  (285)<br>—<br>**Uuq**<br>ununquadium | 115 | 116  (289)<br>—<br>**Uuh**<br>ununhexium | | |

| 65  158.93<br>—  +3<br>565<br>1629<br>3503<br>**Tb**<br>terbium | 66  162.50<br>1.22  +3<br>572<br>1685<br>2840<br>**Dy**<br>dysprosium | 67  164.93<br>1.23  +3<br>581<br>1747<br>2973<br>**Ho**<br>holmium | 68  167.26<br>1.24  +3<br>589<br>1802<br>3141<br>**Er**<br>erbium | 69  168.93<br>1.25  +3<br>597<br>1818<br>2223<br>**Tm**<br>thulium | 70  173.04<br>—  +3<br>603  2+<br>1092<br>1469<br>**Yb**<br>ytterbium | 71  174.97<br>1.0  +3<br>524<br>1936<br>3675<br>**Lu**<br>lutetium |
|---|---|---|---|---|---|---|
| 97  (247)<br>—  +3<br>601  +4<br>1323<br>**Bk**<br>berkelium | 98  (251)<br>—  +3<br>608<br>1173<br>**Cf**<br>californium | 99  (252)<br>—  +3<br>619<br>1133<br>**Es**<br>einsteinium | 100  (257)<br>—  +3<br>627<br>1800<br>**Fm**<br>fermium | 101  (258)<br>—  +3<br>635  +2<br>1100<br>**Md**<br>mendelevium | 102  (259)<br>—  +3<br>642  +2<br>1100<br>**No**<br>nobelium | 103  (262)<br>—  +3<br>1900<br>**Lr**<br>lawrencium |

Appendix C • MHR  837

# Appendix D

# Math and Chemistry

## Precision, Error, and Accuracy

A major component of the scientific inquiry process is the comparison of experimental results with predicted or accepted theoretical values. In conducting experiments, realize that all measurements have a maximum degree of certainty, beyond which there is uncertainty. The uncertainty, often referred to as "error," is not a result of a mistake, but rather it is caused by the limitations of the equipment or the experimenter. The best scientist, using all possible care, could not measure the height of a doorway to a fraction of a millimetre accuracy using a metre stick. The uncertainty introduced through measurement must be communicated using specific vocabulary.

Experimental results can be characterized by both their accuracy and their precision.

**Precision** describes the exactness and repeatabilty of a value or set of values. A set of data could be grouped very tightly, demonstrating good precision, but not necessarily be accurate. The darts in illustration (A) missed the bull's-eye and yet are tightly grouped, demonstrating precision without accuracy.

**Differentiating between accuracy and precision**

**Accuracy** describes the degree to which the result of an experiment or calculation approximates the true value. The darts in illustration (B) missed the bull's-eye in different directions, but are all relatively the same distance away from the centre. The darts demonstrate three throws that share approximately the same accuracy, with limited precision.

The darts in illustration (C) demonstrate accuracy and precision.

## Random Error

- Random error results from small variations in measurements due to randomly changing conditions (weather, humidity, quality of equipment, level of care, etc.).
- Repeating trials will reduce but never eliminate the effects of random error.
- Random error is unbiased.
- Random error affects precision, and, usually, accuracy.

## Systematic Error

- Systematic error results from consistent bias in observation.
- Repeating trials will not reduce systematic error.
- Three sources of systematic error are natural error, instrument-calibration error, and personal error.
- Systematic error affects accuracy.

## Error Analysis

Error exists in every measured or experimentally obtained value. The error could deal with extremely tiny values, such as wavelengths of light, or with large values, such as the distances between stars. A practical way to illustrate the error is to compare it to the specific data as a percentage.

## Relative Uncertainty

Relative uncertainty calculations are used to determine the error introduced by the natural limitations of the equipment used to collect the data. For instance, measuring the width of your textbook will have a certain degree of error due to the quality of the equipment used. This error, called "estimated uncertainty," has been deemed by the scientific community to be half of the smallest division of the measuring device. A metre stick with only centimetres marked would have an error of ±0.5 cm. A ruler that includes millimetre divisions would have a smaller error of ±0.5 mm (0.05 cm or ten-fold decrease in error). The measure should be recorded showing the estimated uncertainty, such as 22.0 ±0.5 cm. Use the relative uncertainty equation to convert the estimated uncertainty into a percentage of the actual measured value.

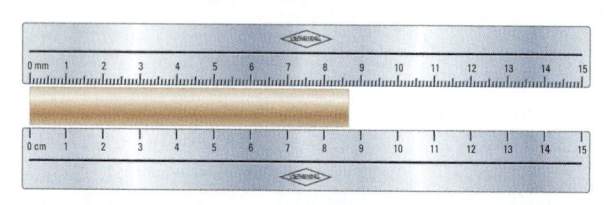

**Estimated uncertainty is accepted to be half of the smallest visible division. In this case, the estimated uncertainty is ±0.5 mm for the top ruler and ±0.5 cm for the bottom ruler.**

$$\text{relative uncertainty} = \frac{\text{estimated uncertainty}}{\text{actual measurement}} \times 100\%$$

**Example:**
Converting the error represented by 22.0 ±0.5 cm to a percentage

$$\text{relative uncertainty} = \frac{0.5 \text{ cm}}{22.00 \text{ cm}} \times 100\%$$

$$\text{relative uncertainty} = 2\%$$

## Percent Deviation

In conducting experiments, it frequently is unreasonable to expect that accepted theoretical values can be verified, because of the limitations of available equipment. In such cases, percent deviation calculations are made. For instance, the standard value for acceleration due to gravity on Earth is 9.81 m/s² toward the centre of Earth in a vacuum. Conducting a crude experiment to verify this value might yield a value of 9.6 m/s². This result deviates from the accepted standard value. It is not necessarily due to error. The deviation, as with most high school experiments, might be due to physical differences in the actual lab (for example, the experiment might not have been conducted in a vacuum). Therefore, deviation is not necessarily due to error, but could be the result of experimental conditions that should be explained as part of the error analysis. Use the percent deviation equation to determine how close the experimental results are to the accepted or theoretical value.

percent deviation = $\left|\dfrac{\text{experimental value} - \text{theoretical value}}{\text{theoretical value}}\right| \times 100\%$

**Example:**

percent deviation = $\dfrac{|9.6 \frac{m}{s^2} - 9.8 \frac{m}{s^2}|}{9.8 \frac{m}{s^2}} \times 100\%$

percent deviation = 2%

## Percent Difference

Experimental inquiry does not always involve an attempt at verifying a theoretical value. For instance, measurements made in determining the width of your textbook do not have a theoretical value based on a scientific theory. You still might want to know, however, how precise your measurements were. Suppose you measured the width 100 times and found that the smallest width measurement was 21.6 cm, the largest was 22.4 cm, and the average measurement of all 100 trials was 22.0 cm. The error contained in your ability to measure the width of the textbook can be estimated using the percent difference equation.

percent difference = $\dfrac{\text{maximum difference in measurements}}{\text{average measurement}} \times 100\%$

**Example:**

percent difference = $\dfrac{(22.4 \text{ cm} - 21.6 \text{ cm})}{22.0 \text{ cm}} \times 100\%$

percent difference = 4%

### Practice Problems

1. In Sèvres, France, a platinum–iridium cylinder is kept in a vacuum under lock and key. It is the standard kilogram with mass 1.0000 kg. Imagine you were granted the opportunity to experiment with this special mass, and obtained the following data: 1.32 kg, 1.33 kg, and 1.31 kg. Describe your results in terms of precision and accuracy.

2. You found that an improperly zeroed triple-beam balance affected the results obtained in question 1. If you used this balance for each measure, what type of error did it introduce?

3. Describe a fictitious experiment with obvious random error.

4. Describe a fictitious experiment with obvious systematic error.

5. (a) Using common scientific practice, find the estimated uncertainty of a stopwatch that displays up to a hundredth of a second.

   (b) If you were to use the stopwatch in part (a) to time repeated events that lasted less than 2.0 s, could you argue that the estimated uncertainty from part (a) is not sufficient? Explain.

## Significant Digits

All measurements involve uncertainty. One source of this uncertainty is the measuring device itself. Another source is your ability to perceive and interpret a reading. In fact, you cannot measure anything with complete certainty. The last (farthest right) digit in any measurement is always an estimate.

The digits that you record when you measure something are called *significant digits*. Significant digits include the digits that you are certain about, *and* a final, uncertain digit that you estimate. Follow the rules below to identify the number of significant digits in a measurement.

### Rules for Determining Significant Digits

**Rule 1** All non-zero numbers are significant.
- 7.886 has four significant digits.
- 19.4 has three significant digits.
- 527.266 992 has nine significant digits.

**Rule 2** All zeros that are located between two non-zero numbers are significant.
- 408 has three significant digits.
- 25 074 has five significant digits.

**Rule 3** Zeros that are located to the left of a measurement are *not* significant.
- 0.0907 has three significant digits: the 9, the third 0 to the right, and the 7.

**Rule 4** Zeros that are located to the right of a measurement may or may not be significant.
- 22 700 may have three significant digits, if the measurement is approximate.
- 22 700 may have five significant digits, if the measurement is taken carefully.

When you take measurements and use them to calculate other quantities, you must be careful to keep track of which digits in your calculations and results are significant. Why? Your results should not imply more certainty than your measured quantities justify. This is especially important when you use a calculator. Calculators usually report results with far more digits than your data warrant. Always remember that calculators do not make decisions about certainty. You do. Follow the rules given below to report significant digits in a calculated answer.

## Rules for Reporting Significant Digits in Calculations

**Rule 1 Multiplying and Dividing**
The value with the fewest number of significant digits, going into a calculation, determines the number of significant digits that you should report in your answer.

**Rule 2 Adding and Subtracting**
The value with the fewest number of decimal places, going into a calculation, determines the number of decimal places that you should report in your answer.

**Rule 3 Rounding**
To get the appropriate number of significant digits (rule 1) or decimal places (rule 2), you may need to round your answer.

- If your answer ends in a number that is greater than 5, increase the preceding digit by 1. For example, 2.346 can be rounded to 2.35.
- If your answer ends with a number that is less than 5, leave the preceding number unchanged. For example, 5.73 can be rounded to 5.7.
- If your answer ends with 5, increase the preceding number by 1 if it is odd. Leave the preceding number unchanged if it is even. For example, 18.35 can be rounded to 18.4, but 18.25 is rounded to 18.2.

### Sample Problem
**Using Significant Digits**

**Problem**
Suppose that you measure the masses of four objects as 12.5 g, 145.67 g, 79.0 g, and 38.438 g. What is the total mass?

**What Is Required?**
You need to calculate the total mass of the objects.

**What Is Given?**
You know the mass of each object.

**Plan Your Strategy**
- Add the masses together, aligning them at the decimal point.
- Underline the estimated (farthest right) digit in each value. This is a technique you can use to help you keep track of the number of estimated digits in your final answer.
- In the question, two values have the fewest decimal places: 12.5 and 79.0. You need to round your answer so that it has only one decimal place.

**Act on Your Strategy**

12.5
145.67
79.0
+ 38.438
―――――
275.608

Total mass = 275.608 g
Therefore, the total mass of the objects is 275.6 g.

**Check Your Solution**
- Your answer is in grams. This is a unit of mass.
- Your answer has one decimal place. This is the same as the values in the question with the fewest decimal places.

### Practice Problems

**Significant Digits**

1. Express each answer using the correct number of significant digits.
   a) 55.671 g + 45.78 g
   b) 1.9 mm + 0.62 mm
   c) 87.9478 L − 86.25 L
   d) 0.350 mL + 1.70 mL + 1.019 mL
   e) 5.841 cm × 6.03 cm
   f) $\dfrac{17.51 \text{ g}}{2.2 \text{ cm}^3}$

## Scientific Notation

One mole of water, $H_2O$, contains 602 214 199 000 000 000 000 000 molecules. Each molecule has a mass of 0.000 000 000 000 000 000 000 029 9 g. As you can see, it would be very awkward to calculate the mass of one mole of water using these values. To simplify large numbers (and clarify the number of significant digits), when reporting them and doing calculations, you can use scientific notation.

**Step 1** Move the decimal point so that only one non-zero digit is in front of the decimal point. (Note that this number is now between 1.0 and 9.99999999.) Count the number of places that the decimal point moves to the left or to the right.

**Step 2** Multiply the value by a power of 10. Use the number of places that the decimal point moved as the exponent for the power of 10. If the decimal point moved to the right, exponent is negative. If the decimal point moved to the left, the exponent is positive.

6.02 000 000 000 000 000 000 000.
  23   21   18   15   12   9   6   3

$6.02 \times 10^{23}$

**Figure D.1** The decimal point moves to the left.

0.000 000 000 000 000 000 000 02.9 9 g

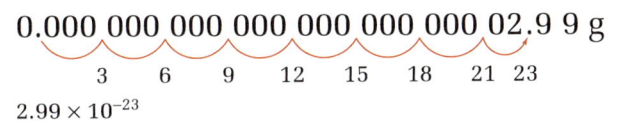

$2.99 \times 10^{-23}$

**Figure D.2** The decimal point moves to the right.

Figure D.3 shows you how to calculate the mass of one mole of water using a scientific calculator. When you enter an exponent on a scientific calculator, you do not have to enter ($\times 10$).

| Keystrokes | Display |
|---|---|
| 6 . 0 2 EXP 2 3 | 6.02  23 |
| × 2 . 9 9 EXP 2 3 ± | 2.99 −23 |
| = | 17.998 |

Round to three significant digits and express in scientific notation: $1.80 \times 10^1$ g/mol

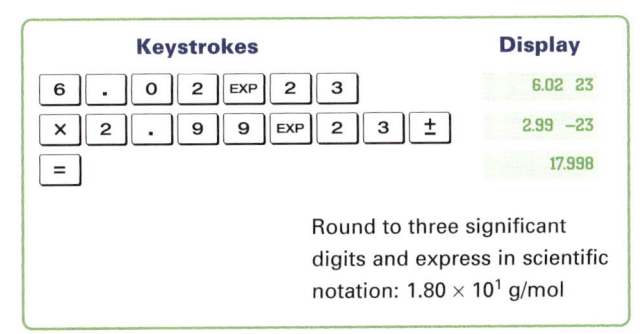

**Figure D.3** On some scientific calculators, the EXP key is labelled EE. Key in negative exponents by entering the exponent, then striking the ± key.

### Rules for Scientific Notation

**Rule 1** To multiply two numbers in scientific notation, add the exponents.

$(7.32 \times 10^{-3}) \times (8.91 \times 10^{-2})$
$= (7.32 \times 8.91) \times 10^{(-3 + -2)}$
$= 65.2212 \times 10^{-5}$
$\rightarrow 6.52 \times 10^{-4}$

**Rule 2** To divide two numbers in scientific notation, subtract the exponents.

$(1.842 \times 10^6 \text{ g}) \div (1.0787 \times 10^2 \text{ g/mol})$
$= (1.842 \div 1.0787) \times 10^{(6-2)}$
$= 1.707611 \times 10^4 \text{ g}$
$\rightarrow 1.708 \times 10^4 \text{ g}$

**Rule 3** To add or subtract numbers in scientific notation, first convert the numbers so they have the same exponent. Each number should have the same exponent as the number with the greatest power of 10. Once the numbers are all expressed to the same power of 10, the power of 10 is neither added nor subtracted in the calculation.

$(3.42 \times 10^6 \text{ cm}) + (8.53 \times 10^3 \text{ cm})$
$= (3.42 \times 10^6 \text{ cm}) + (0.00853 \times 10^6 \text{ cm})$
$= 3.42853 \times 10^6 \text{ cm}$
$\rightarrow 3.43 \times 10^6 \text{ cm}$

$(9.93 \times 10^1 \text{ L}) - (7.86 \times 10^{-1} \text{ L})$
$= (9.93 \times 10^1 \text{ L}) - (0.0786 \times 10^1 \text{ L})$
$= 9.8514 \times 10^1 \text{ L}$
$\rightarrow 9.85 \times 10^1 \text{ L}$

Practice problems are given on the following page.

### Practice Problems

**Scientific Notation**

1. Convert each value into correct scientific notation.
   (a) 0.000 934
   (b) 7 983 000 000
   (c) 0.000 000 000 820 57
   (d) $496 \times 10^6$
   (e) $0.000\ 06 \times 10^1$
   (f) $309\ 72 \times 10^{-8}$

2. Add, subtract, multiply, or divide. Round off your answer, and express it in scientific notation to the correct number of significant digits.
   (a) $(3.21 \times 10^{-3}) + (9.2 \times 10^2)$
   (b) $(8.1 \times 10^3) + (9.21 \times 10^2)$
   (c) $(1.010\ 1 \times 10^1) - (4.823 \times 10^{-2})$
   (d) $(1.209 \times 10^6) \times (8.4 \times 10^7)$
   (e) $(4.89 \times 10^{-4}) \div (3.20 \times 10^{-2})$

## Logarithms

Logarithms are a convenient method for communicating large and small numbers. The *logarithm*, or "log," of a number is the value of the exponent that 10 would have to be raised to, in order to equal this number. Every positive number has a logarithm. Numbers that are greater than 1 have a positive logarithm. Numbers that are between 0 and 1 have a negative logarithm. Table D1 gives some examples of the logarithm values of numbers.

**Table D.1** Some Numbers and Their Logarithms

| Number | Scientific notation | As a power of 10 | Logarithm |
|---|---|---|---|
| 1 000 000 | $1 \times 10^6$ | $10^6$ | 6 |
| 7 895 900 | $7.859 \times 10^5$ | $10^{5.8954}$ | 5.8954 |
| 1 | $1 \times 10^0$ | $10^0$ | 0 |
| 0.000 001 | $1 \times 10^{-6}$ | $10^{-6}$ | −6 |
| 0.004 276 | $4.276 \times 10^{-3}$ | $10^{-2.3690}$ | −2.3690 |

Logarithms are especially useful for expressing values that span a range of powers of 10. The Richter scale for earthquakes, the decibel scale for sound, and the pH scale for acids and bases all use logarithmic scales.

### Logarithms and pH

The pH of an acid solution is defined as $-\log[H_3O^+]$. (The square brackets mean "concentration.") For example, suppose that the hydronium ion concentration in a solution is 0.0001 mol/L ($10^{-4}$ mol/L). The pH is $-\log(0.0001)$. To calculate this, enter 0.0001 into your calculator. Then press the [LOG] key. Press the [±] key. The answer in the display is 4. Therefore, the pH of the solution is 4.

There are logarithms for all numbers, not just whole multiples of 10. What is the pH of a solution if $[H_3O^+] = 0.004\ 76$ mol/L? Enter 0.00476. Press the [LOG] key and then the [±] key. The answer is 2.322. This result has three significant digits—the same number of significant digits as the concentration.

### CONCEPT CHECK

For logarithmic values, only the digits to the right of the decimal point count as significant digits. The digit to the left of the decimal point fixes the location of the decimal point of the original value. For example:

$[H_3O^+] = 0.0476$ mol/L; the pH = 1.322
$[H_3O^+] = 0.00476$ mol/L; the pH = 2.322
$[H_3O^+] = 0.00476$ mol/L; the pH = 3.322

Notice only the decimal place in the $[H_3O^+]$ changes; therefore, only the first digit in the pH changes.

What if you want to find $[H_3O^+]$ from the pH? You would need to find $10^{-pH}$. For example, what is $[H_3O^+]$ if the pH is 5.78? Enter 5.78, and press the [±] key. Then use the [10$^x$] function. The answer is $10^{-5.78}$. Therefore, $[H_3O^+]$ is $1.7 \times 10^{-6}$ mol/L.

Remember that the pH scale is a negative log scale. Thus, a decrease in pH from pH 7 to pH 4 is an increase of $10^3$, or 1000, in the acidity of a solution. An increase from pH 3 to pH 6 is a decrease of $10^3$, or 1000, in acidity.

### Practice Problems

#### Logarithms

1. Calculate the logarithm of each number. Note the trend in your answers.
   (a) 1
   (b) 5
   (c) 10
   (d) 50
   (e) 100
   (f) 500
   (g) 50 000
   (h) 100 000

2. Calculate the antilogarithm of each number.
   (a) 0
   (b) 1
   (c) −1
   (d) 2
   (e) −2
   (f) 3
   (g) −3

3. (a) How are your answers for question 2, parts (b) and (c), related?
   (b) How are your answers for question 2, parts (d) and (e), related?
   (c) How are your answers for question 2, parts (f) and (g), related?
   (d) Calculate the antilogarithm of 3.5.
   (e) Calculate the antilogarithm of −3.5.
   (f) Take the reciprocal of your answer for part (d).
   (g) How are your answers for parts (e) and (f) related?

4. (a) Calculate log 76 and log 55.
   (b) Add your answers for part (a).
   (c) Find the antilogarithm of your answer for part (b).
   (d) Multiply 76 and 55.
   (e) How are your answers for parts (c) and (d) related?

### The Unit Analysis Method of Problem Solving

The unit analysis method of problem solving is extremely versatile. You can use it to convert between units or to solve some formula problems. If you forget a formula, you may still be able to solve the problem using unit analysis.

The unit analysis method involves analyzing the units and setting up conversion factors. You match and arrange the units so that they divide out to give the desired unit in the answer. Then you multiply and divide the numbers that correspond to the units.

### Steps for Solving Problems Using Unit Analysis

**Step 1** Determine which data you have and which conversion factors you need to use. (A conversion factor is usually a ratio of two numbers with units, such as 1000 g/1 kg. You multiply the given data by the conversion factor to get the desired units for the answer.) It is often convenient to use the following three categories to set up your solution: Have, Need, and Conversion factor.

**Step 2** Arrange the data and conversion factors so that you can cross out the undesired units. Decide whether you need any additional conversion factors to get the desired units for the answer.

**Step 3** Multiply all the numbers on the top of the ratio. Then multiply all the numbers on the bottom of the ratio. Divide the top result by the bottom result.

**Step 4** Check that you have cancelled the units correctly. Also check that the answer seems reasonable, and that the significant digits are correct.

### CONCEPT CHECK

Remember that counting numbers for exact quantities are considered to have infinite significant digits. For example, if you have 3 apples, the number is exact, and has an infinite number of significant digits. Conversion factors for unit analysis are a form of counting or record keeping. Therefore, you do not need to consider the number of significant digits in conversion factors, such as 1000 mL/1 L, when deciding on the number of significant digits in the answer.

## Sample Problem
### Active ASA

**Problem**

In the past, pharmacists measured the active ingredients in many medications in a unit called grains (gr). A grain is equal to 64.8 mg. If one headache tablet contains 5.0 gr of active acetylsalicylic acid (ASA), how many grams of ASA are in two tablets?

**What Is Required?**

You need to find the mass in grams of ASA in two tablets.

**What Is Given?**

There are 5.0 gr of ASA in one tablet. A conversion factor for grains to milligrams is given.

**Plan Your Strategy**

Multiply the given quantity by conversion factors until all the unwanted units cancel out and only the desired units remain.

| Have | Need | Conversion factors |
|------|------|-------------------|
| 5.0 gr | ? g | 64.8 mg/1 gr and 1 g/1000 mg |

**Act on Your Strategy**

$$\frac{5.0 \text{ gr}}{1 \text{ tablet}} \times \frac{64.8 \text{ mg}}{1 \text{ gr}} \times \frac{1 \text{ g}}{1000 \text{ mg}} \times 2 \text{ tablets}$$

$$= \frac{5.0 \times 64.8 \times 1 \times 2 \text{ g}}{1000}$$

$$= 0.648 \text{ g}$$

$$= 0.65 \text{ g}$$

There are 0.65 g of active ASA in two headache tablets.

**Check Your Solution**

There are two significant digits in the answer. This is the least number of significant digits in the given data.

Notice how conversion factors are multiplied until all the unwanted units are cancelled out, leaving only the desired unit in the answer.

The next sample problem will show you how to solve a stoichiometric problem.

## Sample Problem
### Stoichiometry and Unit Analysis

**Problem**

What mass of oxygen, $O_2$, can be obtained by the decomposition of 5.0 g of potassium chlorate, $KClO_3$? The balanced equation is given below.

$$2KClO_3 \rightarrow 2KCl + 3O_2$$

**What Is Required?**

You need to calculate the mass of oxygen, in grams, that is produced by the decomposition of 5.0 g of potassium chlorate.

**What Is Given?**

You know the mass of potassium chlorate that decomposes.

$$\text{Mass} = 5.0 \text{ g}$$

From the balanced equation, you can obtain the molar ratio of the reactant and the product.

$$\frac{3 \text{ mol } O_2}{2 \text{ mol } KClO_3}$$

**Plan Your Strategy**

Calculate the molar masses of potassium chlorate and oxygen. Use the molar mass of potassium chlorate to find the number of moles in the sample.

Use the molar ratio to find the number of moles of oxygen produced. Use the molar mass of oxygen to convert this value to grams.

**Act on Your Strategy**

The molar mass of potassium chlorate is

$1 \times M_K = 39.10$
$1 \times M_{Cl} = 35.45$
$3 \times M_O = 48.00$
$\overline{\phantom{xxxxxxxx} 122.55 \text{ g/mol}}$

The molar mass of oxygen is

$$2 \times M_O = 32.00 \text{ g/mol}$$

Find the number of moles of potassium chlorate.

$$\text{mol } KClO_3 = 5.0 \text{ g} \times \left(\frac{1 \text{ mol}}{122.55 \text{ g } KClO_3}\right)$$

$$= 0.0408 \text{ mol}$$

Find the number of moles of oxygen produced.

$$\frac{\text{mol } O_2}{0.0408 \text{ mol } KClO_3} = \frac{3 \text{ mol } O_2}{2 \text{ mol } KClO_3}$$

$$\text{mol } O_2 = 0.0408 \text{ mol } KClO_3 \times \frac{3 \text{ mol } O_2}{2 \text{ mol } KClO_3}$$

$$= 0.0612 \text{ mol}$$

Convert this value to grams.

$$\text{mass } O_2 = 0.0612 \text{ mol} \times \frac{32.00 \text{ g}}{1 \text{ mol } O_2}$$

$$= 1.96 \text{ g}$$

$$= 2.0 \text{ g}$$

Therefore, 2.0 g of oxygen are produced by the decomposition of 5.0 g of potassium chlorate. As you become more familiar with this type of question, you will be able to complete more than one step at once. Below, you can see how the conversion factors we used in each step above can be combined. Set these conversion ratios so that the units cancel out correctly.

$$\text{mass } O_2 = 5.0 \text{ g KClO}_3 \times \left(\frac{1 \text{ mol}}{122.6 \text{ g KClO}_3}\right) \times$$
$$\left(\frac{3 \text{ mol } O_2}{2 \text{ mol KClO}_3}\right) \times \left(\frac{32.0 \text{ g}}{1 \text{ mol } O_2}\right)$$
$$= 1.96 \text{ g}$$
$$= 2.0 \text{ g}$$

**Check Your Solution**

The oxygen makes up only part of the potassium chlorate. Thus, we would expect less than 5.0 g of oxygen, as was calculated.

The smallest number of significant digits in the question is two. Thus, the answer must also have two significant digits.

**Practice Problems**

**Unit Analysis**

Use the unit analysis method to solve each problem.

1. The molar mass of cupric chloride is 134.45 g/mol. What is the mass, in grams, of $8.19 \times 10^{-3}$ mol of this compound?

2. To make a salt solution, 0.82 mol of $CaCl_2$ are dissolved in 1650 mL of water. What is the concentration, in g/L, of the solution?

3. The density of solid sulfur is 2.07 g/cm$^3$. What is the mass, in kg, of a 1.8 dm$^3$ sample?

4. How many grams of dissolved sodium bromide are in 689 mL of a 1.32 mol/L solution?

# Appendix E

# Chemistry Data Tables

**Table E.1** Useful Math Relationships

$D = \dfrac{m}{V}$

$P = \dfrac{F}{A}$

$\pi = 3.1416$

Volume of sphere $V = \dfrac{4}{3}\pi r^3$

Volume of cylinder $= \pi r^2 h$

**Table E.2** Fundamental Physical Constants (to six significant digits)

| | |
|---|---|
| acceleration due to gravity ($g$) | 9.806 65 m/s$^2$ |
| Avogadro constant ($N_a$) | 6.022 14 × 10$^{23}$/mol |
| charge on one mole of electrons (Faraday constant) | 96 485.3 C/mol |
| mass of electron ($m_e$) | 9.109 38 × 10$^{-31}$ kg |
| mass of neutron ($m_n$) | 1.674 93 × 10$^{-27}$ kg |
| mass of proton ($m_p$) | 1.672 62 × 10$^{-27}$ kg |
| molar gas constant ($R$) | 8.314 47 J/mol·K |
| molar volume of gas at STP | 22.414 0 L/mol |
| speed of light in vacuo ($c$) | 2.997 92 × 10$^8$ m/s |
| unified atomic mass ($u$) | 1.660 54 × 10$^{-27}$ kg |

**Table E.3** Common SI Prefixes

| | |
|---|---|
| tera (T) | 10$^{12}$ |
| giga (G) | 10$^9$ |
| mega (M) | 10$^6$ |
| kilo (k) | 10$^3$ |
| deci (d) | 10$^{-1}$ |
| centi (c) | 10$^{-2}$ |
| milli (m) | 10$^{-3}$ |
| micro (μ) | 10$^{-6}$ |
| nano (n) | 10$^{-9}$ |
| pico (p) | 10$^{-12}$ |

**Table E.4** Conversion Factors

| Quantity | Relationships between units |
|---|---|
| length | 1 m = 10$^{-3}$ km <br> = 10$^3$ mm <br> = 10$^2$ cm |
| | 1 pm = 10$^{-12}$ m |
| mass | 1 kg = 10$^3$ g <br> = 10$^{-3}$ t |
| | 1 u = 1.66 × 10$^{-27}$ kg |
| temperature | 0 K = −273.15°C |
| | $T$ (K) = $T$ (°C) + 273.15 <br> $T$ (°C) = $T$ (K) − 273.15 |
| | mp of H$_2$O = 273.15 K (0°C) <br> bp of H$_2$O = 373.15 K (100°C) |
| volume | 1 L = 1 dm$^3$ <br> = 10$^{-3}$ m$^3$ <br> = 10$^3$ mL |
| | 1 mL = 1 cm$^3$ |
| pressure | 101 325 Pa = 101.325 kPa <br> = 760 mm Hg <br> = 760 torr <br> = 1 atm |
| density | 1 kg/m$^3$ = 10$^3$ g/m$^3$ <br> = 10$^{-3}$ g/mL <br> = 1 g/L |
| energy | 1 J = 6.24 × 10$^{18}$ eV |

### Table E.5 Summary of Naming Rules for Ions

| Type of ion | Prefix or suffix | Example |
|---|---|---|
| **Polyatomic Ions** | | |
| if the ion is the most common oxoanion | -ate | chlorate, $ClO_3^-$ |
| if the ion has one O atom less than the most common oxoanion | -ite | chlorite, $ClO_2^-$ |
| if the ion has two O atoms less than the most common oxoanion | hypo-___-ite | hypochlorite, $ClO^-$ |
| if the ion has 1 O atom more than the most common oxoanion | per-___-ate | perchlorate, $ClO_4^-$ |
| if the ion has 1 H atom added to the most common oxoanion | bi- | bicarbonate, $HCO_3^-$ |
| if the ion has 1 O atom less and 1 S atom more than the most common oxoanion | thio- | thiosulphate, $S_2O_3^{2-}$ |
| **Metallic Ions** | | |
| if the ion has the higher possible charge | -ic | titanic, $Ti^{4+}$ |
| if the ion has the lower possible charge | -ous | cuprous, $Cu^+$ |
| Note: According to the Stock system, metallic ions are named using Roman numerals. | The Roman numeral shows the charge on the metal ion | titanium(IV) $Ti^{4+}$ copper(I), $Cu^+$ manganese(VII), $Mn^{7+}$ |

### Table E.6 Common Metal Ions with More Than One Ionic Charge

| Formula | Stock Name | Classical Name |
|---|---|---|
| $Cu^+$ | copper(I) ion | cuprous ion |
| $Cu^{2+}$ | copper(II) ion | cupric ion |
| $Fe^{2+}$ | iron(II) ion | ferrous ion |
| $Fe^{3+}$ | iron(III) ion | ferric ion |
| $Hg_2^{2+}(Hg^+)$ | mercury(I) ion | mercurous ion |
| $Hg^{2+}$ | mercury(II) ion | mercuric ion |
| $Pb^{2+}$ | lead(II) ion | plumbous ion |
| $Pb^{4+}$ | lead(IV) ion | plumbic ion |
| $Sn^{2+}$ | tin(II) ion | stannous ion |
| $Sn^{4+}$ | tin(IV) ion | stannic ion |
| $Cr^{2+}$ | chromium(II) ion | chromous ion |
| $Cr^{3+}$ | chromium(III) ion | chromic ion |
| $Mn^{2+}$ | manganese(II) ion | |
| $Mn^{3+}$ | manganese(III) ion | |
| $Mn^{4+}$ | manganese(IV) ion | |
| $Co^{2+}$ | cobalt(II) ion | cobaltous ion |
| $Co^{3+}$ | cobalt(III) ion | cobaltic ion |

### Table E.7 Ionic Charges of Representative Elements

| IA 1 | IIA 2 | IIIA 13 | IVA 14 | VA 15 | VIA 16 | VIIA 17 | VIIIA 18 |
|---|---|---|---|---|---|---|---|
| $H^+$ | | | | | | $H^-$ | noble |
| $Li^+$ | $Be^{2+}$ | | | $N^{3-}$ | $O^{2-}$ | $F^-$ | gases |
| $Na^+$ | $Mg^{2+}$ | $Al^{3+}$ | | $P^{3-}$ | $S^{2-}$ | $Cl^-$ | do not |
| $K^+$ | $Ca^{2+}$ | | | | $Se^{2-}$ | $Br^-$ | ionize |
| $Rb^+$ | $Sr^{2+}$ | | | | | $I^-$ | |
| $Cs^+$ | $Ba^{2+}$ | | | | | | |

### Table E.8 Charges of Some Transition Metal Ions

| 1+ | 2+ | 3+ |
|---|---|---|
| silver, $Ag^+$ | cadmium, $Cd^{2+}$ nickel, $Ni^{2+}$ zinc, $Zn^{2+}$ | scandium, $Sc^{3+}$ |

### Table E.9 Alphabetical Listing of Common Polyatomic Ions

| Most common ion | | Common related ions | |
|---|---|---|---|
| acetate | $CH_3COO^-$ | | |
| ammonium | $NH_4^+$ | | |
| arsenate | $AsO_4^{3-}$ | arsenite | $AsO_3^{3-}$ |
| benzoate | $C_6H_5COO^-$ | | |
| borate | $BO_3^{3-}$ | tetraborate | $B_4O_7^{2-}$ |
| bromate | $BrO_3^-$ | | |
| carbonate | $CO_3^{2-}$ | bicarbonate (hydrogen carbonate) | $HCO_3^-$ |
| chlorate | $ClO_3^-$ | perchlorate chlorite hypochlorite | $ClO_4^-$ $ClO_2^-$ $ClO^-$ |
| chromate | $CrO_4^{2-}$ | dichromate | $Cr_2O_7^{2-}$ |
| cyanide | $CN^-$ | cyanate thiocyanate | $OCN^-$ $SCN^-$ |
| glutamate | $C_5H_8NO_4^-$ | | |
| hydroxide | $OH^-$ | peroxide | $O_2^{2-}$ |
| iodate | $IO_3^-$ | iodide | $I^-$ |
| nitrate | $NO_3^-$ | nitrite | $NO_2^-$ |
| oxalate | $OOCCOO^{2-}$ | | |
| permanganate | $MnO_4^-$ | | |
| phosphate | $PO_4^{3-}$ | phosphite tripolyphosphate | $PO_3^{3-}$ $P_3O_{10}^{5-}$ |
| silicate | $SiO_3^{2-}$ | orthosilicate | $SiO_4^{4-}$ |
| stearate | $C_{17}H_{35}COO^-$ | | |
| sulfate | $SO_4^{2-}$ | bisulfate (hydrogen sulfate) sulfite bisulfite (hydrogen sulfite) thiosulfate | $HSO_4^-$ $SO_3^{2-}$ $HSO_3^-$ $S_2O_3^{2-}$ |
| sulfide | $S^{2-}$ | bisulfide (hydrogen sulfide) | $HS^-$ |

### Table E.10 Specific Heat Capacities of Various Substances

| Substance | Specific heat capacity (J/g·°C at 25°C) |
|---|---|
| **Element** | |
| aluminum | 0.900 |
| carbon (graphite) | 0.711 |
| copper | 0.385 |
| gold | 0.129 |
| hydrogen | 14.267 |
| iron | 0.444 |
| **Compound** | |
| ammonia (liquid) | 4.70 |
| ethanol | 2.46 |
| water (solid) | 2.01 |
| water (liquid) | 4.184 |
| water (gas) | 2.01 |
| **Other material** | |
| air | 1.02 |
| concrete | 0.88 |
| glass | 0.84 |
| granite | 0.79 |
| wood | 1.76 |

### Table E.11 Average Bond Energies

| Bond | Energy (kJ/mol) | Bond | Energy (kJ/mol) | Bond | Energy (kJ/mol) | Bond | Energy (kJ/mol) |
|---|---|---|---|---|---|---|---|
| **Hydrogen** | | **Carbon** | | **Nitrogen** | | **Phosphorus and sulfur** | |
| H—H | 436 | C—C | 347 | N—N | 160 | P—P | 210 |
| H—C | 338 | C—N | 305 | N—O | 201 | P—S | 444 |
| H—N | 339 | C—O | 358 | N—F | 272 | P—F | 490 |
| H—O | 460 | C—F | 552 | N—Si | 330 | P—Cl | 331 |
| H—F | 570 | C—Si | 305 | N—P | 209 | P—Br | 272 |
| H—Si | 299 | C—P | 264 | N—S | 464 | P—I | 184 |
| H—P | 297 | C—S | 259 | N—Cl | 200 | S—S | 266 |
| H—S | 344 | C—Cl | 397 | N—Br | 276 | S—F | 343 |
| H—Cl | 432 | C—Br | 280 | N—I | 159 | S—Cl | 277 |
| H—Br | 366 | C—I | 209 | | | S—Br | 218 |
| H—I | 298 | | | | | S—I | 170 |
| H—Mg | 126 | | | | | | |
| **Oxygen** | | **Silicon** | | **Halogens** | | **Multiple bonds** | |
| O—O | 204 | Si—Si | 226 | F—Cl | 256 | C=C | 607 |
| O—F | 222 | Si—P | 364 | F—Br | 280 | C=N | 615 |
| O—Si | 368 | Si—S | 226 | F—I | 272 | C=O | 745 |
| O—P | 351 | Si—F | 553 | Cl—Br | 217 | N=N | 418 |
| O—S | 265 | Si—Cl | 381 | Cl—I | 211 | N=O | 631 |
| O—Cl | 269 | Si—Br | 368 | Br—I | 179 | O=O | 498 |
| O—Br | 235 | Si—I | 293 | F—F | 159 | C≡C | 839 |

*continued...*

| Bond | Energy (kJ/mol) | Bond | Energy (kJ/mol) | Bond | Energy (kJ/mol) | Bond | Energy (kJ/mol) |
|---|---|---|---|---|---|---|---|
| **Oxygen** | | **Silicon** | | **Halogens** | | **Multiple bonds** | |
| O—I | 249 | Si=O | 640 | Cl—Cl | 243 | C≡N | 891 |
| | | | | Br—Br | 193 | C≡O | 1077 |
| | | | | I—I | 151 | N≡N | 945 |

**Note:** The values in this table represent average values for the dissociation of bonds between the pairs of atoms listed. The true values may vary for different molecules.

### Table E.12 Average Bond Lengths

| Bond | Length (pm) | Bond | Length (pm) | Bond | Length (pm) | Bond | Length (pm) |
|---|---|---|---|---|---|---|---|
| **Hydrogen** | | **Carbon** | | **Nitrogen** | | **Phosphorus and sulfur** | |
| H—H | 74 | C—C | 154 | N—N | 146 | P—P | 221 |
| H—C | 109 | C—N | 147 | N—O | 144 | P—S | 210 |
| H—N | 101 | C—O | 143 | N—F | 139 | P—F | 156 |
| H—O | 96 | C—F | 133 | N—Si | 172 | P—Cl | 204 |
| H—F | 92 | C—Si | 186 | N—P | 177 | P—Br | 222 |
| H—Si | 148 | C—P | 187 | N—S | 168 | P—I | 243 |
| H—P | 142 | C—S | 181 | N—Cl | 191 | S—S | 204 |
| H—S | 134 | C—Cl | 177 | N—Br | 214 | S—F | 158 |
| H—Cl | 127 | C—Br | 194 | N—I | 222 | S—Cl | 201 |
| H—Br | 141 | C—I | 213 | | | S—Br | 225 |
| H—I | 161 | | | | | S—I | 234 |
| H—Mg | 173 | | | | | | |
| **Oxygen** | | **Silicon** | | **Halogens** | | **Multiple bonds** | |
| O—O | 148 | Si—Si | 234 | F—Cl | 166 | C=C | 134 |
| O—F | 142 | Si—P | 227 | F—Br | 178 | C=N | 127 |
| O—Si | 161 | Si—S | 210 | F—I | 187 | C=O | 123 |
| O—P | 160 | Si—F | 156 | Cl—Br | 214 | N=N | 122 |
| O—S | 151 | Si—Cl | 204 | Cl—I | 243 | N=O | 120 |
| O—Cl | 164 | Si—Br | 216 | Br—I | 248 | O=O | 121 |
| O—Br | 172 | Si—I | 240 | F—F | 143 | C≡C | 121 |
| O—I | 194 | | | Cl—Cl | 199 | C≡N | 115 |
| | | | | Br—Br | 228 | C≡O | 113 |
| | | | | I—I | 266 | N≡N | 110 |

**Note:** The values in this table are average values. The length of a bond may be slightly different in different molecules, depending on the intramolecular forces within the molecules.

## Table E.13 Standard Molar Enthalpies of Formation

| Substance | ΔH°f (kJ/mol) | Substance | ΔH°f (kJ/mol) | Substance | ΔH°f (kJ/mol) |
|---|---|---|---|---|---|
| $Al_2O_{3(s)}$ | −1675.7 | $HBr_{(g)}$ | −36.3 | $NH_{3(g)}$ | −45.9 |
| $CaCO_{3(s)}$ | −1207.6 | $HCl_{(g)}$ | −92.3 | $N_2H_{4(\ell)}$ | +50.6 |
| $CaCl_{2(s)}$ | −795.4 | $HF_{(g)}$ | −273.3 | $NH_4Cl_{(s)}$ | −314.4 |
| $Ca(OH)_{2(s)}$ | −985.2 | $HCN_{(g)}$ | +135.1 | $NH_4NO_{3(s)}$ | −365.6 |
| $CCl_{4(\ell)}$ | −128.2 | $H_2O_{(\ell)}$ | −285.8 | $NO_{(g)}$ | +91.3 |
| $CCl_{4(g)}$ | −95.7 | $H_2O_{(g)}$ | −241.8 | $NO_{2(g)}$ | +33.2 |
| $CHCl_{3(\ell)}$ | −134.1 | $H_2O_{2(\ell)}$ | −187.8 | $N_2O_{(g)}$ | +81.6 |
| $CH_{4(g)}$ | −74.6 | $HNO_{3(\ell)}$ | −174.1 | $N_2O_{4(g)}$ | +11.1 |
| $C_2H_{2(g)}$ | +227.4 | $H_3PO_{4(s)}$ | −1284.4 | $PH_{3(g)}$ | +5.4 |
| $C_2H_{4(g)}$ | +52.4 | $H_2S_{(g)}$ | −20.6 | $PCl_{3(g)}$ | −287.0 |
| $C_2H_{6(g)}$ | −84.0 | $H_2SO_{4(\ell)}$ | −814.0 | $P_4O_{6(s)}$ | −2144.3 |
| $C_3H_{8(g)}$ | −103.8 | $FeO_{(s)}$ | −272.0 | $P_4O_{10(s)}$ | −2984.0 |
| $C_6H_{6(\ell)}$ | +49.1 | $Fe_2O_{3(s)}$ | −824.2 | $KBr_{(s)}$ | −393.8 |
| $CH_3OH_{(\ell)}$ | −239.2 | $Fe_3O_{4(s)}$ | −1118.4 | $KCl_{(s)}$ | −436.5 |
| $C_2H_5OH_{(\ell)}$ | −277.6 | $FeCl_{2(s)}$ | −341.8 | $KClO_{3(s)}$ | −397.7 |
| $CH_3COOH_{(\ell)}$ | −484.3 | $FeCl_{3(s)}$ | −399.5 | $KOH_{(s)}$ | −424.6 |
| $CO_{(g)}$ | −110.5 | $FeS_{2(s)}$ | −178.2 | $Ag_2CO_{3(s)}$ | −505.8 |
| $CO_{2(g)}$ | −393.5 | $PbCl_{2(s)}$ | −359.4 | $AgCl_{(s)}$ | −127.0 |
| $COCl_{2(g)}$ | −219.1 | $MgCl_{2(s)}$ | −641.3 | $AgNO_{3(s)}$ | −124.4 |
| $CS_{2(\ell)}$ | +89.0 | $MgO_{(s)}$ | −601.6 | $Ag_2S_{(s)}$ | −32.6 |
| $CS_{2(g)}$ | +116.7 | $Mg(OH)_{2(s)}$ | −924.5 | $SF_{6(g)}$ | −1220.5 |
| $CrCl_{3(s)}$ | −556.5 | $HgS_{(s)}$ | −58.2 | $SO_{2(g)}$ | −296.8 |
| $Cu(NO_3)_{2(s)}$ | −302.9 | $NaCl_{(s)}$ | −411.2 | $SO_{3(g)}$ | −395.7 |
| $CuO_{(s)}$ | −157.3 | $NaOH_{(s)}$ | −425.6 | $SnCl_{2(s)}$ | −325.1 |
| $CuCl_{(s)}$ | −137.2 | $Na_2CO_{3(s)}$ | −1130.7 | $SnCl_{4(\ell)}$ | −511.3 |
| $CuCl_{2(s)}$ | −220.1 | | | | |

**Note:** The enthalpy of formation of an element in its standard state is defined as zero.

## Table E.14 Standard Reduction Potentials

| Reduction half reaction | E°(V) |
|---|---|
| $F_{2(g)} + 2e^- \rightleftharpoons 2F^-_{(aq)}$ | 2.866 |
| $Co^{3+}_{(aq)} + e^- \rightleftharpoons Co^{2+}_{(aq)}$ | 1.92 |
| $H_2O_{2(aq)} + 2H^+_{(aq)} + 2e^- \rightleftharpoons 2H_2O_{(\ell)}$ | 1.776 |
| $Ce^{4+}_{(aq)} + e^- \rightleftharpoons Ce^{3+}_{(aq)}$ | 1.72 |
| $PbO_{2(s)} + 4H^+_{(aq)} + SO_4^{2-}_{(aq)} + 2e^- \rightleftharpoons PbSO_{4(s)} + H_2O_{(\ell)}$ | 1.691 |
| $MnO_4^-_{(aq)} + 8H^+_{(aq)} + 5e^- \rightleftharpoons Mn^{2+}_{(aq)} + 4H_2O_{(\ell)}$ | 1.507 |
| $Au^{3+}_{(aq)} + 3e^- \rightleftharpoons Au_{(s)}$ | 1.498 |
| $PbO_{2(s)} + 4H^+_{(aq)} + 2e^- \rightleftharpoons Pb^{2+}_{(aq)} + 2H_2O_{(\ell)}$ | 1.455 |
| $Cl_{2(g)} + 2e^- \rightleftharpoons 2Cl^-_{(aq)}$ | 1.358 |
| $Cr_2O_7^{2-}_{(aq)} + 14H^+_{(aq)} + 6e^- \rightleftharpoons 2Cr^{3+}_{(aq)} + 7H_2O_{(\ell)}$ | 1.232 |
| $O_{2(g)} + 4H^+_{(aq)} + 4e^- \rightleftharpoons 2H_2O_{(\ell)}$ | 1.229 |
| $MnO_{2(s)} + 4H^+_{(aq)} + 2e^- \rightleftharpoons Mn^{2+}_{(aq)} + 2H_2O_{(\ell)}$ | 1.224 |
| $IO_3^-_{(aq)} + 6H^+_{(aq)} + 6e^- \rightleftharpoons I^-_{(aq)} + 3H_2O_{(\ell)}$ | 1.085 |
| $Br_{2(\ell)} + 2e^- \rightleftharpoons 2Br^-_{(aq)}$ | 1.066 |

| Reduction half reaction | E°(V) |
|---|---|
| $AuCl_4^-_{(aq)} + 3e^- \rightleftharpoons Au_{(s)} + 4Cl^-_{(aq)}$ | 1.002 |
| $NO_3^-_{(aq)} + 4H^+_{(aq)} + 3e^- \rightleftharpoons NO_{(g)} + 2H_2O_{(\ell)}$ | 0.957 |
| $2Hg^{2+}_{(aq)} + 2e^- \rightleftharpoons Hg_2^{2+}_{(aq)}$ | 0.920 |
| $Ag^+_{(aq)} + e^- \rightleftharpoons Ag_{(s)}$ | 0.800 |
| $Hg_2^{2+}_{(aq)} + 2e^- \rightleftharpoons 2Hg_{(\ell)}$ | 0.797 |
| $Fe^{3+}_{(aq)} + e^- \rightleftharpoons Fe^{2+}_{(aq)}$ | 0.771 |
| $O_{2(g)} + 2H^+_{(aq)} + 2e^- \rightleftharpoons H_2O_{2(aq)}$ | 0.695 |
| $I_{2(s)} + 2e^- \rightleftharpoons 2I^-_{(aq)}$ | 0.536 |
| $Cu^+_{(aq)} + e^- \rightleftharpoons Cu_{(s)}$ | 0.521 |
| $O_{2(g)} + 2H_2O_{(\ell)} + 4e^- \rightleftharpoons 4OH^-_{(aq)}$ | 0.401 |
| $Cu^{2+}_{(aq)} + 2e^- \rightleftharpoons Cu_{(s)}$ | 0.342 |
| $AgCl_{(s)} + e^- \rightleftharpoons Ag_{(s)} + Cl^-_{(aq)}$ | 0.222 |
| $4H^+_{(aq)} + SO_4^{2-}_{(aq)} + 2e^- \rightleftharpoons H_2SO_{3(aq)} + H_2O_{(\ell)}$ | 0.172 |
| $Cu^{2+}_{(aq)} + e^- \rightleftharpoons Cu^+_{(aq)}$ | 0.153 |
| $2H^+_{(aq)} + 2e^- \rightleftharpoons H_{2(g)}$ | 0.000 |
| $Fe^{3+}_{(aq)} + 3e^- \rightleftharpoons Fe_{(s)}$ | −0.037 |
| $Pb^{2+}_{(aq)} + 2e^- \rightleftharpoons Pb_{(s)}$ | −0.126 |
| $Sn^{2+}_{(aq)} + 2e^- \rightleftharpoons Sn_{(s)}$ | −0.138 |
| $Ni^{2+}_{(aq)} + 2e^- \rightleftharpoons Ni_{(s)}$ | −0.257 |
| $Cd^{2+}_{(aq)} + 2e^- \rightleftharpoons Cd_{(s)}$ | −0.403 |
| $Cr^{3+}_{(aq)} + e^- \rightleftharpoons Cr^{2+}_{(aq)}$ | −0.407 |
| $Fe^{2+}_{(aq)} + 2e^- \rightleftharpoons Fe_{(s)}$ | −0.447 |
| $Cr^{3+}_{(aq)} + 3e^- \rightleftharpoons Cr_{(s)}$ | −0.744 |
| $Zn^{2+}_{(aq)} + 2e^- \rightleftharpoons Zn_{(s)}$ | −0.762 |
| $2H_2O_{(\ell)} + 2e^- \rightleftharpoons H_{2(g)} + 2OH^-_{(aq)}$ | −0.828 |
| $Al^{3+}_{(aq)} + 3e^- \rightleftharpoons Al_{(s)}$ | −1.662 |
| $Mg^{2+}_{(aq)} + 2e^- \rightleftharpoons Mg_{(s)}$ | −2.372 |
| $La^{3+}_{(aq)} + 3e^- \rightleftharpoons La_{(s)}$ | −2.379 |
| $Na^+_{(aq)} + e^- \rightleftharpoons Na_{(s)}$ | −2.711 |
| $Ca^{2+}_{(aq)} + 2e^- \rightleftharpoons Ca_{(s)}$ | −2.868 |
| $Ba^{2+}_{(aq)} + 2e^- \rightleftharpoons Ba_{(s)}$ | −2.912 |
| $K^+_{(aq)} + e^- \rightleftharpoons K_{(s)}$ | −2.931 |
| $Li^+_{(aq)} + e^- \rightleftharpoons Li_{(s)}$ | −3.040 |

## Table E.15 Summary of Naming Rules for Acids

| Modern name | Classical acid name | Example |
|---|---|---|
| aqueous hydrogen ___ide | hydro___ic acid | HCl, aqueous hydrogen chloride or hydrochloric acid |
| aqueous hydrogen ___ate | ___ic acid | $H_2CO_3$, aqueous hydrogen carbonate or carbonic acid |
| aqueous hydrogen ___ite | ___ous acid | $HNO_2$, aqueous hydrogen nitrite or nitrous acid |

*continued...*

**Table E.16** The Colour of Some Common Ions in Aqueous Solutions

| | Ions | Symbol | Colour |
|---|---|---|---|
| **Cations** | chromium (II)<br>copper(II) | $Cr^{2+}$<br>$Cu^{2+}$ | blue |
| | chromium(III)<br>copper(I)<br>iron(II)<br>nickel(II) | $Cr^{3+}$<br>$Cu^{+}$<br>$Fe^{2+}$<br>$Ni^{2+}$ | green |
| | iron(III) | $Fe^{3+}$ | pale yellow |
| | cobalt(II)<br>manganese(II) | $Co^{2+}$<br>$Mn^{2+}$ | pink |
| **Anions** | chromate | $CrO_4^{2-}$ | yellow |
| | dichromate | $Cr_2O_7^{2-}$ | orange |
| | permanganate | $MnO_4^{-}$ | purple |

**Table E.17** The Flame Colour of Selected Metallic Ions

| Ion | Symbol | Colour |
|---|---|---|
| lithium | $Li^{+}$ | red |
| sodium | $Na^{+}$ | yellow |
| potassium | $K^{+}$ | violet |
| cesium | $Cs^{+}$ | violet |
| calcium | $Ca^{2+}$ | red |
| strontium | $Sr^{2+}$ | red |
| barium | $Ba^{2+}$ | yellowish-green |
| copper | $Cu^{2+}$ | bluish-green |
| boron | $B^{2+}$ | green |
| lead | $Pb^{2+}$ | bluish-white |

**Table E.18** General Solubility Guidelines for Some Common Ionic Compounds in Water at 25°C

| Ion | Group 1 (IA)<br>$NH_4^+$, $H^+$<br>$(H_3O^+)$ | $ClO_3^-$<br>$NO_3^-$<br>$ClO_4^-$ | $Cl^-$<br>$Br^-$<br>$I^-$ | $CH_3COO^-$ | $SO_4^{2-}$ | $S^{2-}$ | $OH^-$ | $PO_4^{3-}$<br>$SO_3^{2-}$<br>$CO_3^{2-}$ |
|---|---|---|---|---|---|---|---|---|
| Very soluble (solubility greater than or equal to 0.1 mol/L) | all | all | most | most | most | Group 1 (IA)<br>Group (2 IIA)<br>$NH_4^+$ | Group 1 (IA)<br>$NH_4^+$<br>$Sr^{2+}$<br>$Ba^{2+}$<br>$Tl^+$ | Group 1 (IA)<br>$NH_4^+$ |
| Slightly soluble (solubility less than 0.1 mol/L) | none | none | $Ag^+$<br>$Hg^+$<br>$Tl^+$ | $Ag^+$<br>$Pb^{2+}$<br>$Hg^+$<br>$Cu^+$ | $Ca^{2+}$<br>$Sr^{2+}$<br>$Ba^{2+}$<br>$Ra^{2+}$<br>$Pb^{2+}$<br>$Ag^+$ | most | most | most |

**Table E.19** End-point Indicators

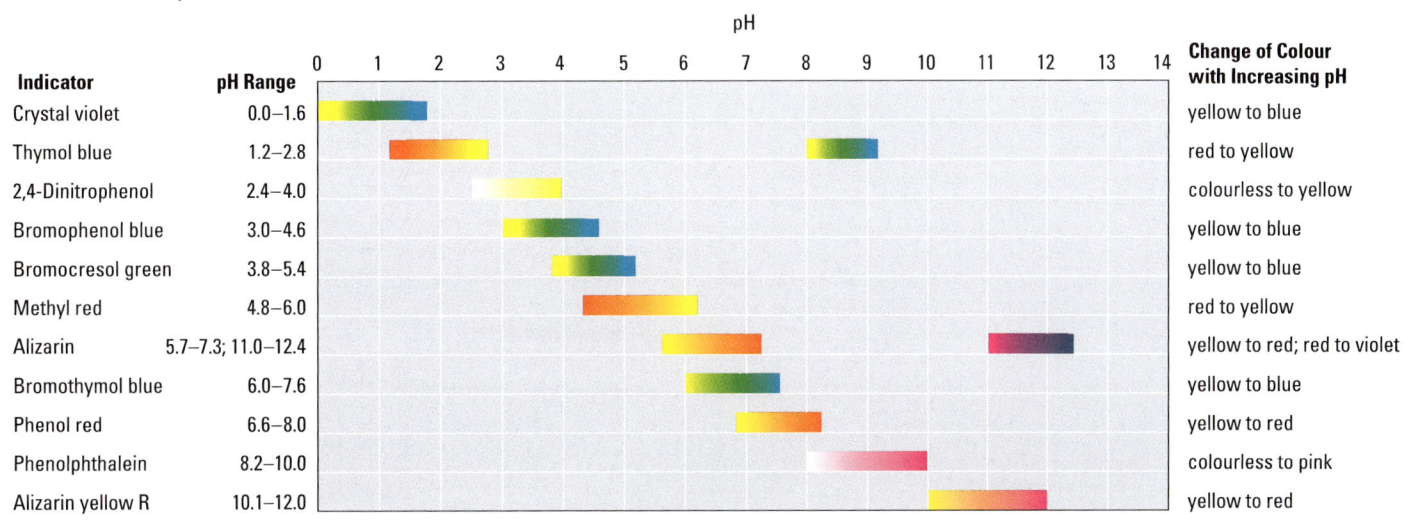

| Indicator | pH Range | Change of Colour with Increasing pH |
|---|---|---|
| Crystal violet | 0.0–1.6 | yellow to blue |
| Thymol blue | 1.2–2.8 | red to yellow |
| 2,4-Dinitrophenol | 2.4–4.0 | colourless to yellow |
| Bromophenol blue | 3.0–4.6 | yellow to blue |
| Bromocresol green | 3.8–5.4 | yellow to blue |
| Methyl red | 4.8–6.0 | red to yellow |
| Alizarin | 5.7–7.3; 11.0–12.4 | yellow to red; red to violet |
| Bromothymol blue | 6.0–7.6 | yellow to blue |
| Phenol red | 6.6–8.0 | yellow to red |
| Phenolphthalein | 8.2–10.0 | colourless to pink |
| Alizarin yellow R | 10.1–12.0 | yellow to red |

**Table E.20** Relative Strengths of Acids and Bases (concentration = 0.10 mol/L) at 25°C

| Acid Name | Acid Formula | Formula of Conjugate Base | $K_a$ |
|---|---|---|---|
| perchloric acid | $HClO_{4(aq)}$ | $ClO_4^-{}_{(aq)}$ | very large |
| hydroiodic acid | $HI_{(aq)}$ | $I^-{}_{(aq)}$ | very large |
| hydrobromic acid | $HBr_{(aq)}$ | $Br^-{}_{(aq)}$ | very large |
| hydrochloric acid | $HCl_{(aq)}$ | $Cl^-{}_{(aq)}$ | very large |
| sulfuric acid | $H_2SO_{4(aq)}$ | $HSO_4^-{}_{(aq)}$ | very large |
| nitric acid | $HNO_{3(aq)}$ | $NO_3^-{}_{(aq)}$ | very large |
| hydronium ion | $H_3O^+{}_{(aq)}$ | $H_2O_{(\ell)}$ | 1 |
| oxalic acid | $HOOCCOOH_{(aq)}$ | $HOOCCOO^-{}_{(aq)}$ | $5.6 \times 10^{-2}$ |
| sulfurous acid ($SO_2 + H_2O$) | $H_2SO_{3(aq)}$ | $HSO_3^-{}_{(aq)}$ | $1.4 \times 10^{-2}$ |
| hydrogen sulfate ion | $HSO_4^-{}_{(aq)}$ | $SO_4^{2-}{}_{(aq)}$ | $1.0 \times 10^{-2}$ |
| phosphoric acid | $H_3PO_{4(aq)}$ | $H_2PO_4^-{}_{(aq)}$ | $6.9 \times 10^{-3}$ |
| nitrous acid | $HNO_{2(aq)}$ | $NO_2^-{}_{(aq)}$ | $5.6 \times 10^{-3}$ |
| citric acid | $H_3C_6H_5O_{7(aq)}$ | $H_2C_6H_5O_7^-{}_{(aq)}$ | $7.4 \times 10^{-4}$ |
| hydrofluoric acid | $HF_{(aq)}$ | $F^-{}_{(aq)}$ | $6.3 \times 10^{-4}$ |
| methanoic acid | $HCOOH_{(aq)}$ | $HCOO^-{}_{(aq)}$ | $1.8 \times 10^{-4}$ |
| hydrogen oxalate ion | $HOOCCOO^-{}_{(aq)}$ | $OOCCOO^{2-}{}_{(aq)}$ | $1.5 \times 10^{-4}$ |
| ascorbic acid | $C_6H_8O_{6(aq)}$ | $C_6H_7O_6^-{}_{(aq)}$ | $9.1 \times 10^{-5}$ |
| benzoic acid | $C_6H_5COOH_{(aq)}$ | $C_6H_5COO^-{}_{(aq)}$ | $6.3 \times 10^{-5}$ |
| ethanoic (acetic) acid | $CH_3COOH_{(aq)}$ | $CH_3COO^-{}_{(aq)}$ | $1.8 \times 10^{-5}$ |
| dihydrogen citrate ion | $H_2C_6H_5O_7^-{}_{(aq)}$ | $HC_6H_5O_7^{2-}{}_{(aq)}$ | $1.7 \times 10^{-5}$ |
| carbonic acid ($CO_2 + H_2O$) | $H_2CHO_{3(aq)}$ | $HCO_3^-{}_{(aq)}$ | $4.5 \times 10^{-7}$ |
| hydrogen citrate ion | $HC_6H_5O_7^{2-}{}_{(aq)}$ | $C_6H_5O_7^{3-}{}_{(aq)}$ | $4.0 \times 10^{-7}$ |
| hydrosulfuric acid | $H_2S_{(aq)}$ | $HS^-{}_{(aq)}$ | $8.9 \times 10^{-8}$ |
| hydrogen sulfite ion | $HSO_3^-{}_{(aq)}$ | $SO_3^{2-}{}_{(aq)}$ | $6.3 \times 10^{-8}$ |
| dihydrogen phosphate ion | $H_2PO_4^-{}_{(aq)}$ | $HPO_4^{2-}{}_{(aq)}$ | $6.2 \times 10^{-8}$ |
| hypochlorous acid | $HOCl_{(aq)}$ | $OCl^-{}_{(aq)}$ | $4.0 \times 10^{-8}$ |
| hydrocyanic acid | $HCN_{(aq)}$ | $CN^-{}_{(aq)}$ | $6.2 \times 10^{-10}$ |
| ammonium ion | $NH_4^+{}_{(aq)}$ | $NH_{3(aq)}$ | $5.6 \times 10^{-10}$ |
| hydrogen carbonate ion | $HCO_3^-{}_{(aq)}$ | $CO_3^{2-}{}_{(aq)}$ | $4.7 \times 10^{-11}$ |
| hydrogen ascorbate ion | $C_6H_7O_6^-{}_{(aq)}$ | $C_6H_6O_6^{2-}{}_{(aq)}$ | $2.0 \times 10^{-12}$ |
| hydrogen phosphate ion | $HPO_4^{2-}{}_{(aq)}$ | $PO_4^{3-}{}_{(aq)}$ | $4.8 \times 10^{-13}$ |
| water (55.5 mol/L) | $H_2O_{(\ell)}$ | $HO^-{}_{(aq)}$ | $1.0 \times 10^{-14}$ |

Increasing Acid Strength ↑

Increasing Base Strength ↓

# Glossary

## A

**acid dissociation constant ($K_a$):** the equilibrium constant for the dissociation of an acid in water; the product of the concentrations of hydronium ion and conjugate base divided by the acid concentration (15.1)

**acid-base indicator:** a substance, usually a weak, monoprotic acid, that changes colour in acidic and basic solutions (15.2)

**acid-base titration curve:** a graph for an acid-base titration in which the pH of an acid (or base) is plotted against the volume of the base (or acid) added (15.2)

**activated complex:** a highly unstable transition species that has partial bonds and is neither product nor reactant (12.2)

**activation energy ($E_a$):** the minimum energy that is required for a successful reaction between colliding molecules (12.2)

**actual yield:** the measured quantity of product obtained in a chemical reaction (4.3)

**addition reaction:** an organic reaction in which atoms are added to a multiple carbon-carbon bond (9.3)

**additional polymerization:** a reaction in which monomers with double bonds are joined together, through multiple addition reactions, to form a polymer (11.1)

**alcohol:** an organic compound that contains the —OH group bonded to a carbon atom (10.2)

**aldehyde:** an organic compound that has a double-bonded oxygen atom and a hydrogen atom bonded to the last carbon of a carbon chain (10.3)

**aliphatic hydrocarbons:** compounds containing only carbon and hydrogen in which carbon atoms form chains and/or non-aromatic rings; alkanes, alkenes, and alkynes are aliphatic hydrocarbons (9.3)

**alkane:** a hydrocarbon molecule in which the carbon atoms are joined by single covalent bonds (9.3)

**alkene:** a hydrocarbon that contains one or more carbon-carbon double bonds (9.3)

**alkyl group:** an alkane branch that is obtained by removing one hydrogen atom from an alkane (9.3)

**alkyl halide (haloalkane):** an alkane in which one or more hydrogen atoms have been replaced with halogen atoms; written as R—X, where R stands for an alkyl group and X stands for a halogen atom (10.2)

**alkyne:** a hydrocarbon molecule that contains one or more carbon-carbon triple bonds (9.3)

**allotropes:** different crystalline or molecular forms of the same element that differ in physical and chemical properties (6.1)

**alloy:** a solid solution of two or more metals (7.1)

**amide:** an organic compound that has a carbon atom double-bonded to an oxygen atom and single-bonded to a nitrogen atom (10.3)

**amine:** an organic compound that has the functional group —NR$_2$, where N stands for a nitrogen atom and R stands for a hydrogen atom or an alkyl group (10.2)

**amino acid:** an organic compound that contains a carboxylic acid group and an amino group attached to a central carbon atom; the monomer from which proteins are made (11.2)

**amphoteric:** the ability of a substance to act as a proton donor (an acid) in one reaction and a proton acceptor (a base) in a different reaction (14.1)

**anhydrous:** a term used to describe a compound that does not have any water molecules bonded to it; applies to compounds that can be hydrated (3.4)

**anode:** the electrode at which oxidation occurs in a cell (19.1)

**aqueous solution:** a solution in which water is the solvent (7.1)

**aromatic hydrocarbon:** a hydrocarbon that is based on the aromatic benzene group; benzene has the molecular formula $C_6H_6$ (9.3)

**Arrhenius theory of acids and bases:** the theory explaining the nature of acids and bases in terms of their structure and the ions produced when they dissolve in water; defines an acid as a substance that dissociates in water to produce one or more hydrogen ions, H$^+$, and a base as a substance that dissociates in water to produce one or more hydroxide ions, OH$^-$ (14.1)

**atom:** the smallest particle of an element that still retains the element's properties (1.2)

**atomic mass unit (u):** a unit of mass that is 1/12 of the mass of a carbon-12 atom; equal to $1.66 \times 10^{-24}$ g (1.2)

**atomic number ($Z$):** the unique number of protons in the nucleus of the atom of a particular element (1.2)

**atomic symbol:** a one- or two-letter abbreviation of the name of an element; also called *element symbol* (1.2)

**average atomic mass:** the average of the masses of all naturally occurring isotopes of an element weighted according to their abundances (2.1)

**Avogadro constant ($N_A$):** the experimentally determined number of particles in 1 mol of a substance; currently accepted value is $6.022\ 141\ 99 \times 10^{23}$ mol$^{-1}$ (2.2)

**Avogadro's hypothesis:** states that equal volumes of ideal gases, at the same temperature and pressure, contain the same number of particles (2.4)

## B

**base dissociation constant ($K_b$):** the equilibrium constant for the reaction of a weak base with water; the product of the concentrations of hydroxide ion and conjugate acid divided by the base concentration (15.1)

**battery:** a set of galvanic cells connected in series (19.1)

**benefit:** an advantage, or positive result, of a course of action (17.3)

**bent:** an angular molecular shape that results when a central atom has one or two lone pairs and two electron groupings around it (6.1)

**binary compound:** any compound that is composed of two elements (1.3)

**binary ionic compound:** a compound that is composed of ions of one metal element and ions of one non-metal element, grouped together in a lattice structure (1.3)

**binary molecular compound:** a compound consisting of only two elements, joined by covalent bonds (1.3)

**biochemistry:** the study of the organic compounds and reactions that occur in living things (11.2)

**bomb calorimeter:** a device that measures enthalpy changes during combustion reactions at a constant volume (17.1)

**bond angle:** the angle formed by two bonds around a central atom (6.1)

**bond dipole:** the charge in a bond, being polarized into a positive area and a negative area (5.3, 10.1)

**bond energy:** the amount of energy required to break a specific bond in one mole of molecules, given in kJ/mol (5.3)

**bonding electron pair:** a pair of electrons that is shared between two atoms in covalent bonding (5.2)

**Brønsted-Lowry theory of acids and bases:** the theory recognizing acid-base reactions as a chemical equilibrium; defines an acid as a substance from which a proton (H$^+$ ion) can be removed and a base as a substance that can remove a proton from an acid (14.1)

**buffer capacity:** the amount of acid or base that can be added to a solution before considerable change occurs to the pH of the solution (15.1)

**buffer solution:** a solution that resists changes in pH when a moderate amount of acid or base is added; contains a weak acid/conjugate base mixture or a weak base/conjugate acid mixture (15.1)

## C

**calorimeter:** a device that is used to measure the heat released or absorbed during a chemical or physical process occurring within it (17.1)

**calorimetry:** a technological process of measuring the heat released or absorbed during a chemical or physical process (17.1)

**carbohydrate (saccharide):** a biological molecule that, in its linear form, contains either an aldehyde group or a ketone group, along with two or more hydroxyl groups (11.2)

**carbonyl group:** the functional group of aldehydes and ketones, composed of a carbon atom double-bonded to an oxygen atom (10.3)

**carbon-hydrogen combustion analyzer:** a device that uses the combustion-enabling properties of $O_2$ to determine the composition of compounds containing carbon, hydrogen, and oxygen (3.4)

**carboxyl group:** the functional group of carboxylic acids, composed of a carbon atom double-bonded to an oxygen atom and single-bonded to an —OH group; written as —COOH (10.3)

**carboxylic acid:** an organic compound that contains the carboxyl functional group, composed of a carbon atom double-bonded to an oxygen atom and single-bonded to an —OH group (10.3)

**catalyst:** a substance that increases the rate of a chemical reaction without being consumed by the reaction (12.2)

**cathode:** the electrode at which reduction occurs in a cell (19.1)

**cathodic protection:** a method of preventing the rusting of an iron object, using a more reactive metal that acts as the sacrificial anode (19.5)

**cell potential (cell voltage):** the electric potential difference between the anode and cathode of an electrochemical cell when no current is flowing (19.1)

**chemical change:** a change that alters the composition of matter (1.1)

**chemical property:** a property of a substance that can only be observed as the substance changes into another substance (1.1)

**chemistry:** the study of matter, its composition, and its interactions (1.1)

**chlor-alkali process:** an industrial process in which brine, a saturated solution of sodium chloride, is electrolyzed in a cell to produce chlorine and sodium hydroxide (19.5)

**cis-trans isomers (geometric isomers):** two compounds that are identical except for the arrangement of groups across a carbon-carbon double bond (9.3)

**closed system:** a system in which energy, but not matter, may be exchanged between the system and its surroundings (16.1)

**coffee-cup calorimeter:** a calorimeter that consists of two nested polystyrene cups (coffee cups) with a known mass of water inside; used to measure heat changes during processes such as dissolving, neutralization, heating, and cooling (17.1)

**collision theory:** reacting particles (atoms, molecules, or ions) must collide with one another successfully for a reaction to occur (12.2)

**competing reaction:** a reaction that occurs at the same time as a principal reaction and consumes some of the reactants and/or products of the principal reaction (4.3)

**complete structural diagram:** a symbolic representation of all the atoms in a molecule and the way they are bonded to one another (9.2)

**compound:** a pure substance that is composed of two or more elements chemically combined in fixed proportions (1.1)

**concentrated solution:** has a higher proportion of solute to solvent than a dilute solution (7.1)

**concentration:** the amount of solute per quantity of solvent in a solution (7.3)

**condensation polymerization:** a reaction in which monomers with two functional groups, usually one at each end, are joined together by the formation of ester bonds or amide bonds to produce a polymer (11.1)

**condensation reaction:** an organic reaction in which two molecules combine to form a larger molecule, producing a small molecule, usually water, as a second product (10.1)

**condensed structural diagram:** a symbolic representation of an organic compound showing the atoms present and the bonds between carbon atoms (9.2)

**conjugate acid:** of a base, is the particle that remains when the base receives a proton from an acid (14.1)

**conjugate acid-base pair:** an acid and a base that differ by one proton (14.1)

**conjugate base:** of an acid, is the particle that remains when a proton is removed from the acid (14.1)

**corrosion:** a spontaneous redox reaction causing unwanted oxidation of a metal (19.5)

**covalent bond:** the electrostatic attraction between the nuclei of two adjacent atoms and a pair of shared bonding electrons (5.2)

**co-ordinate covalent bond:** a covalent bond in which one atom contributes both electrons to the shared electron pair (6.1)

**cracking:** a process that uses heat, in the absence of air, to break large hydrocarbon molecules into smaller gasoline molecules (9.4)

**crystal lattice:** a three-dimensional array of alternating positive and negative ions closely packed together in a crystal (5.2)

**cyclic hydrocarbon:** a hydrocarbon ring structure formed when the two ends of a hydrocarbon chain join together; can be a cycloalkane, cycloalkene, or cycloalkyne (9.3)

## D

**derivative:** an organic compound that is structurally based on, or derived from, another organic compound (10.3)

**dilute solution:** has a lower proportion of solute to solvent than a concentrated solution (7.1)

**dipole:** a distribution of molecular charge consisting of two opposite charges that are separated by a short distance (7.2)

**dipole-dipole attraction:** the intermolecular force between oppositely charged ends of two polar molecules (molecules with dipoles) (7.2)

**dipole-dipole forces:** electrostatic attractions between the oppositely charged ends of polar molecules (6.2)

**dipole-dipole interactions:** the attractive forces between polar molecules (10.1)

**dipole-induced dipole force:** an intermolecular force in which a polar molecule distorts the electron density of a non-polar molecule to produce a temporary dipole; similar to an ion-induced dipole force (6.2)

**disaccharide:** a biological molecule (carbohydrate) that is composed of two saccharide units; sucrose (table sugar) is a disaccharide (11.2)

**dispersion (London) force:** a weak intermolecular force of attraction that is present between all molecules due to temporary dipoles (6.2, 10.1)

**disproportionation reaction:** a reaction in which a single substance undergoes both oxidation and reduction (18.1)

**dissociation equation:** a balanced chemical equation showing all ions produced when an ionic compound dissolves (8.3)

**DNA (2-deoxyribonucleic acid):** the biological molecules that carry the genetic information that determines the structure and function of a living organism (11.2)

**dry cell:** a galvanic cell with the electrolyte contained in a paste thickened with starch (19.1)

**dynamic equilibrium:** an equilibrium in which opposing changes to a closed chemical system occur at the same time and at the same rate (13.1)

Glossary • MHR **853**

## E

**electric charge:** the product of the current flowing through a circuit and the time for which it flows (19.4)

**electric current:** net movement of electrical charge; often the flow of electrons in a circuit (19.1)

**electric potential ($E$):** the difference in the electric potential energy of a unit charge between the potential energy at the anode and the potential energy at the cathode, given in volts (V) (19.1)

**electrochemistry:** the study of the processes involved in converting chemical energy to electrical energy, and in converting electrical energy to chemical energy (19.1)

**electrode:** a conductor that carries electrons into and out of a cell (19.1)

**electrolysis:** the splitting (lysing) of a substance caused by electrical energy (19.3)

**electrolyte:** a solute that conducts an electric current in an aqueous solution (7.2, 19.1)

**electrolytic cell:** an electrochemical cell that uses electical energy to drive a non-spontaneous chemical reaction (19.3)

**electronegativity ($EN$):** a relative measure of an atom's ability to attract the shared electrons in a chemical bond (5.3, 10.1)

**electroplating:** the process in which a metal is deposited, or plated, onto the cathode in an electrolytic cell (19.3)

**element:** a pure substance that cannot be separated chemically into any simpler substances (e.g. copper, oxygen, hydrogen, carbon) (1.1)

**elementary reaction:** a reaction that involves a single molecular event, such as a simple collision between atoms, molecules, or ions; can involve the formation of different molecules or ions; may involve a change in the energy or geometry of the starting molecules; cannot be broken down into simpler steps (12.2)

**elimination reaction:** a reaction in which atoms are removed from a molecule to form a multiple bond (such as C=C bond) (10.1)

**empirical formula:** shows the lowest whole number ratio of atoms of each element in a compound (3.2)

**end point:** the point in a titration at which the acid-base indicator changes colour (15.2)

**endothermic reaction:** a reaction that results in a net absorption of energy (16.2)

**energy level:** fixed, three-dimension volume in which electrons travel around the nucleus (1.2)

**enthalpy change ($\Delta H$):** the difference between the potential energy of the reactants and the potential energy of the products in a process (16.2)

**enthalpy of reaction ($\Delta H_{rxn}$):** the enthalpy change of a chemical reaction (16.2)

**equilibrium constant ($K_c$):** the ratio of the concentrations of the products over the concentrations of the reactants, with all concentration terms raised to the power of the coefficients in the chemical equation (13.2)

**equivalence point:** the point in a titration at which the number of moles of the unknown solution is stoichiometrically equal to the number of moles of the standard solution (15.2)

**esterification reaction:** the reaction of a carboxylic acid with an alcohol to form an ester; a specific type of condensation reaction (10.3)

**ether:** an organic compound that has two alkyl groups joined by an oxygen atom; written as R—O—R, where R stands for an alkyl group (10.2)

**excess reactant:** a reactant that remains after a chemical reaction is over (4.2)

**exothermic reaction:** a reaction that results in a net release of energy (16.2)

**expanded molecular formula:** a symbolic representation that shows the groupings of atoms in a molecule; often gives an idea of the molecular structure; an example is $CH_3CH_2CH_3$ for propane (9.2)

**external circuit:** in a galvanic cell, a circuit outside the reaction vessel (19.1)

**extraction:** a process by which a pure compound is obtained from a mixture (19.4)

## F

**Faraday's law:** the law stating that the amount of a substance produced or consumed in an electrolysis reaction is directly proportional to the quantity of electric current that flows through the circuit (19.4)

**fat:** a lipid that contains a glycerol molecule bonded by ester linkages to three long-chain carboxylic acids (11.2)

**first law of thermodynamics:** the law stating that energy can be neither created nor destroyed; any change in the energy system is accompanied by an equal and opposite change in the energy of the surroundings (16.1)

**forensic scientist:** a scientist who uses specialized knowledge to analyze evidence in criminal and legal cases (3.3)

**fractional distillation:** a process that uses the specific boiling points of substances to separate (refine) a mixture into its components; used in oil refineries to separate petroleum into its hydrocarbon components (9.4)

**free-electron model:** a model in which a metal is thought of as a densely packed core of metallic kernels (nuclei and inner electrons) embedded in a region of shared, mobile valence electrons (5.2)

**fuel cell:** a battery that produces electricity while reactants are supplied continuously from an external source (19.5)

**functional group:** a reactive group of bonded atoms that appears in all the members of a chemical family (10.1)

### G

**galvanic cell (voltaic cell):** a device that converts chemical energy to electrical energy (19.1)

**galvanizing:** a process in which iron is covered with a protective layer of zinc (19.5)

**gas stoichiometry:** stoichiometric analyses involving the volume of gases (4.1)

**general formula:** a formula that represents a family of simple organic compounds; written as R + functional group, where R stands for an alkyl group (10.1)

**global warming:** a gradual increase in the average global temperature due to the build-up of greenhouse gases in the atmosphere (17.3)

**gravimetric stoichiometry:** stoichiometric analyses involving mass (4.1)

**greenhouse gas:** a gas that traps heat in Earth's atmosphere and prevents the heat from escaping into outer space (17.3)

### H

**half-reaction:** a balanced equation that shows the number of electrons involved in either oxidation or reduction (18.1)

**haloalkane (alkyl halide):** an alkane in which one or more hydrogen atoms have been replaced with halogen atoms; written as R—X, where R stands for an alkyl group and X stands for a halogen atom (10.2)

**heat ($q$):** the transfer of molecular kinetic energy between objects of different temperatures; measured in joules (J) (16.1)

**heat capacity ($C$):** of a substance is the quantity of energy, in joules (J), required to change the temperature of the substance by one degree Celsius (°C) (16.1)

**Hess's law of heat summation:** the law stating that the enthalpy change of a physical or chemical process depends only on the beginning conditions (reactants) and the end conditions (products) and is independent of the pathway of the process or the number of intermediate steps in the process (17.2)

**heterogeneous catalyst:** a catalyst that exists in a phase that is different from the phase of the reaction it catalyzes (12.2)

**heterogeneous equilibrium:** an equilibrium in which reactants and products in the chemical system are in different phases; an example is a solution containing a crystal of the solute (13.1)

**homogeneous catalyst:** a catalyst that exists in the same phase as the reactants (12.2)

**homogeneous equilibrium:** an equilibrium in which all reactants and products in the chemical system are in the same phase (13.1)

**homologous series:** a series of molecules in which each member differs from the next by an additional specific structural unit —$CH_2$— (9.3)

**hydrate:** a compound that has a specific number of water molecules bonded to each formula unit (3.4)

**hydrated:** a term used to describe ions in aqueous solutions, surrounded by and attached to water molecules (7.2)

**hydrocarbon:** a molecular compound that contains only carbon and hydrogen atoms (1.3, 9.1)

**hydrogen bonding:** the strong intermolecular attraction between the nucleus of a hydrogen atom, bonded to an atom of a highly electronegative element such as oxygen (as in water), and the negative end of a dipole nearby (6.1, 7.2, 10.1)

**hydrolysis reaction:** a reaction in which a molecule is split in two by the addition of a water molecule; the reverse of a condensation reaction (10.1)

**hydronium ion:** an aqueous hydrated proton; written as $H_3O^+_{(aq)}$ (14.1)

**hydroxyl group:** the functional group of the alcohol family of organic compounds; written as —OH group (10.1)

### I

**ICE table:** a problem-solving table that records the initial conditions, the change, and the equilibrium conditions of a chemical system (13.2)

**ideal gas:** a hypothetical gas with particles that have mass but no volume or attractive forces between them (2.4)

**immiscible:** a term used to describe substances that are not able to combine with each other in a solution (7.1)

**inert electrode:** an electrode made from a material that is neither a reactant nor a product of the cell reaction (19.1)

**intermolecular forces:** forces that act *between* molecules or ions to influence the physical properties of compounds (6.2, 10.1)

**intramolecular forces:** forces that are exerted *within* a molecule or polyatomic ion (6.2)

**ion:** a positively or negatively charged particle that results from a neutral atom or group of atoms losing or gaining electrons (1.2)

**ion product constant for water ($K_w$):** the product of the concentration of hydronium ions and the concentration of hydroxide ions in a solution ($[H_3O^+][OH^-]$); always equal to $1.0 \times 10^{-14}$ at 25°C (14.2)

**ion-dipole attractions:** the intermolecular forces between ions and polar molecules (7.2)

**ion-dipole force:** the force of attraction between an ion and a polar molecule (6.2)

**ion-induced dipole force:** an intermolecular force in which an ion distorts the electron density of a non-polar molecule to produce a temporary dipole (6.2)

**isolated system:** a system that is completely isolated from its surroundings; neither matter nor energy is exchanged between the system and its surroundings (16.1)

**isomers:** compounds that have the same chemical formula, but different structural arrangements (9.2)

**isotopic abundance:** the relative amount of an isotope of an element; expressed as a percent or a decimal fraction (2.1)

## J

**joule (J):** the SI unit for kinetic energy and potential energy (16.1)

## K

**ketone:** an organic compound that has an oxygen atom double-bonded to a carbon within the carbon chain (10.3)

**kilopascal (kPa):** a unit of pressure; equal to 1000 Pa (2.4)

**kinetic energy:** the energy of motion (16.1)

**kinetic molecular theory (KMT):** a theory that explains ideal gas behaviour on the basis of random motion of gas particles (12.2)

## L

**law of chemical equilibrium:** the law stating that there is a constant ratio, at equilibrium, between the concentrations of products and reactants in any change (13.2)

**law of combining volumes:** the law stating that when gases react, the volumes of the gaseous reactants and products, at constant temperatures and pressures, are always in whole number ratios (2.4)

**law of conservation of energy:** the law stating that the total energy of the universe is constant (16.1)

**law of conservation of mass:** the law stating that matter can be neither created or destroyed; in any chemical reaction, the mass of the products is always equal to the mass of the reactants (4.1)

**law of definite proportions:** the law stating that the elements in a given chemical compound are always present in the same proportions by mass (3.1)

**Le Châtelier's principle:** the principle stating that a dynamic equilibrium tends to respond so as to relieve the effect of any change in the conditions that affect the equilibrium (13.3)

**Lewis structure:** a symbolic representation of the arrangement of the valence electrons of an element or compound (5.1)

**limiting reactant:** the reactant that is completely consumed during a chemical reaction, limiting the amount of product produced (4.2)

**line structural diagram:** a graphical representation of the bonds between carbon atoms in a hydrocarbon or the hydrocarbon portion of an organic compound; hydrogen atoms and carbon-hydrogen bonds are omitted (9.2)

**linear shape:** a molecular shape in which the angle formed by two bonds around a central atom is 180° (6.1)

**lipid:** a biological molecule that is not soluble in water but is soluble in non-polar solvents, such as benzene and hexanes (11.2)

**lone pair:** a pair of electrons in an atom's valence shell that is not involved in covalent bonding (5.2)

## M

**Markovnikov's rule:** the rule stating that the halogen atom or OH group in an addition reaction is usually added to the more substituted carbon atom—the carbon atom that is bonded to the largest number of other carbon atoms (9.3)

**mass number (A):** the sum of the protons and neutrons in the nucleus of one atom of a particular element (1.2)

**mass percent:** the mass of an element in a compound, expressed as a percent of the compound's total mass (3.1)

**mass spectrometer:** an instrument that uses magnetic fields to separate atoms and/or molecules according to their mass and determines the exact mass and abundance of each species present; e.g., can determine the mass and abundance of isotopes (2.1)

**mass/mass percent:** the mass of a solute divided by the mass of the solution, expressed as a percent (7.3)

**mass/volume percent:** the mass of a solute divided by the volume of the solution, expressed as a percent (7.3)

**matter:** anything that has mass and occupies space (1.1)

**metallic bond:** the electrostatic attraction between the positively charged metal ions and the pool of valence electrons that moves freely among them (5.2)

**miscible:** a term used to describe substances that are able to combine with each other in any proportion (7.1)

**mixture:** a physical combination of two or more kinds of matter, in which each component retains its own characteristics (1.1)

**molar concentration (C):** a unit of concentration expressed as the number of moles of solute present in one litre of solution; also called *molarity* (7.3)

**molar mass (M):** the mass of one mole of a substance, numerically equal to the element's average atomic mass; expressed in g/mol (2.3)

**molar volume:** the amount of space that is occupied by 1 mol of a substance; equal to 22.4 L for a gas at standard temperature and pressure (STP) (2.4)

**mole (mol):** the amount of substance that contains the same number of atoms, molecules, or formula units as exactly 12 g of carbon-12 (2.2)

**mole ratio:** a ratio that compares the number of moles of different substances in a balanced chemical equation (4.1)

**molecular dipole:** a molecule in which the charge is polarized into a positive area and a negative area across the molecule (6.1)

**molecular formula:** a formula that gives the actual number of atoms of each element in a molecule (3.2)

**molecular polarity:** the distribution of charge on a molecule; if a molecule is polar, it has some molecular polarity (10.1)

**molecularity:** the number of reactant particles (molecules, atoms, or ions) that are involved in an elementary reaction (12.2)

**monomers:** small molecules that are combined in long chains to produce very large molecules called polymers (11.1)

**monoprotic acid:** an acid that has only a single hydrogen atom that dissociates in water (15.1)

**monosaccharide:** the smallest molecule possible for a carbohydrate; composed of one saccharide unit (11.2)

### N

**net ionic equation:** a representation of a chemical reaction in a solution that shows only the ions involved in the chemical change (8.2)

**network solids:** solids in which atoms are bonded covalently into continuous two-dimensional or three-dimensional arrays with a wide range of properties (6.1)

**neutralization reaction:** a reaction between an acid and a base that produces an ionic compound (salt) and water (15.2)

**non-electrolyte:** a solute that does not conduct an electric current in an aqueous solution (7.2)

**non-polar:** describes molecules that have no polar bonds or molecules in which polar bonds cancel each other (6.1)

**non-polar covalent bond:** a bond in which electrons are equally shared between two atoms (5.3)

**non-renewable energy source:** resource that will run out (such as coal, oil, or natural gas); once the supply is used up, there will be no more (17.3)

**nucleotide:** the monomer from which DNA and RNA are composed; consists of three parts: a sugar, a phosphate group, and a base; a strand of DNA may be made of more than one million nucleotides (11.2)

**nylon (polyamide):** a condensation polymer that contains amide bonds (11.1)

### O

**octet rule:** the rule stating that when bonds form, atoms gain, lose, or share electrons in such a way as to achieve an octet or filled outer energy level (5.2)

**oil:** a fat that is liquid at room temperature (11.2)

**open system:** a system that is completely open to its surroundings; both matter and energy are exchanged between the system and its surroundings (16.1)

**ore:** a solid, naturally occurring compound, or mixture of compounds, from which a metal can be extracted (18.1)

**organic compound:** a molecular compound that is based on carbon (9.1)

**oxidation:** a change with the loss of electrons (18.1); a change with an increase in oxidation number (18.2)

**oxidation number:** an actual or a hypothetical charge that is used to describe an element in a redox reaction with covalent reactants or products; assigned using a set of rules (18.2)

**oxidation potential:** the half-cell potential for a reduction half-reaction (19.2)

**oxidation-reduction reaction (redox reaction):** a reaction in which one reactant loses electrons (oxidation) and another reactant gains electrons (reduction) (18.1)

**oxidizing agent:** a reactant that oxidizes another reactant and gains electrons itself (is reduced) in the process (18.1)

**overvoltage:** the excess voltage required above the calculated value for electrolysis to take place in a cell (19.3)

### P

**parent alkane:** an alkane that contains the same number of carbons and the same basic structure as a more complex organic compound; a concept used primarily for nomenclature (10.2)

**parts per million/parts per billion:** units of concentration used to express very small quantities of solute (7.3)

**pascal (Pa):** the SI unit of pressure; equal to 1 N/m$^2$ (2.4)

**percent dissociation (percent ionization):** of a weak acid, is the fraction of acid molecules that dissociate compared with the initial concentration of the acid, expressed as a percent (15.1)

**percentage composition:** the relative mass of each element in a compound (3.1)

**percentage purity:** describes what proportion, by mass, of a sample is composed of a specific compound or element (4.3)

**percentage yield:** the actual yield of a reaction, expressed as a percent of the theoretical yield (4.3)

**periodic law:** the law stating that the chemical and physical properties of the elements repeat in a regular, periodic pattern when they are arranged according to their atomic numbers (1.2)

**periodic trends:** patterns that are evident when elements are organized by their atomic numbers (1.2)

**petrochemicals:** products derived from petroleum; basic hydrocarbons, such as ethene and propene, that are converted into plastics and other synthetic materials (9.4)

**petroleum:** a complex mixture of solid, liquid, and gaseous hydrocarbons; sometimes referred to as crude oil (9.1)

**pH:** the negative of the logarithm of the hydronium ion concentration (14.2)

**physical change:** a change, such as change of state, that alters the appearance but not the composition of matter (1.1)

**physical property:** a property that can be observed without the substance changing into or interacting with another substance (1.1)

**plastics:** synthetic polymers that can be heated and moulded into specific shapes and forms (11.1)

**pOH:** the negative of the logarithm of the hydroxide ion concentration (14.2)

**polar:** having an uneven distribution of charge, forming a positive end and a negative end (6.1, 10.1)

**polar covalent bond:** a bond in which electrons are unequally shared between two atoms (5.3)

**polyester:** a condensation polymer that contains ester bonds (11.1)

**polymer:** a large long-chain molecule with repeating units; made by linking together many small molecules called monomers (11.1)

**polyprotic acid:** an acid that has more than one hydrogen atom that dissociates in water (15.1)

**polysaccharide:** a polymer that contains many saccharide units (11.2)

**potential energy:** energy that is stored (16.1)

**potential energy diagram:** a diagram that charts the potential energy of a reaction against the progress of the reaction (12.2)

**precipitate:** an insoluble solid that is formed by a chemical reaction between two soluble compounds (8.1)

**pressure:** the force that is exerted on an object, per unit of surface area (2.4)

**primary battery:** a disposable battery (19.1)

**primary standard:** a substance of known purity and molecular mass that can be used to prepare a solution with a precisely determined concentration (15.2)

**properties:** characteristics that distinguish different types of matter (e.g., colour, melting point, boiling point, conductivity, density) (1.1)

**protein:** a natural polymer that is present in living systems; composed of monomers called amino acids (11.2)

**pure substance:** a material that is composed of only one type of particle (e.g., iron, water, sodium chloride) (1.1)

**pyramidal:** a molecular shape that results when a central atom has one lone pair and three single bonds (6.1)

## Q

**qualitative analysis:** the process of separating and identifying ions in an aqueous solution (8.2)

## R

**rate of dissolving:** the speed at which a solute dissolves in a solvent (7.2)

**rate of reaction (reaction rate):** the rate at which a reaction occurs; measured in terms of the amount of reactant used or product formed per unit time (12.1)

**rate-determining step:** the slowest reaction step; determines the rate of the reaction (12.2)

**reaction intermediates:** molecules, atoms, or ions that are formed in an elementary reaction and consumed in a subsequent elementary reaction, but are not present in the overall reaction (12.2)

**reaction mechanism:** a series of steps that make up an overall reaction (12.2)

**reaction quotient ($Q_c$):** an expression that is identical to the equilibrium constant expression, but its value is calculated using concentrations that are not necessarily those at equilibrium (13.3)

**reaction rate (rate of reaction):** the rate at which a reaction occurs; measured in terms of the amount of reactant used or product formed per unit time (12.1)

**redox reaction (oxidation-reduction reaction):** a reaction in which one reactant loses electrons (oxidation) and another reactant gains electrons (reduction) (18.1)

**reducing agent:** a reactant that reduces another reactant and loses electrons itself (is oxidized) in the process (18.1)

**reduction:** a change with the gain of electrons (18.1); a change with a decrease in oxidation number (18.2)

**reduction potential:** the half-cell potential for a reduction half-reaction (19.2)

**refining:** the process of purifying a material (19.4)

**reforming:** a technique that uses heat, pressure, and catalysts to convert naturally occurring hydrocarbons into a more useful form; usually involves the removal of hydrogen atoms (9.4)

**renewable energy source:** resource that exists in infinite supply (such as solar energy) (17.3)

**resonance structures:** models that give the same relative position of atoms as in Lewis structures, but show different placing of the bonding pairs and lone pairs (6.1)

**risk:** a chance of possible negative or dangerous results (17.3)

**risk-benefit analysis:** a thoughtful assessment of both the positive and negative results that may follow upon a particular course of action (17.3)

**RNA (ribonucleic acid):** a nucleic acid that is present in the body's cells; works closely with DNA to produce the proteins in the body (11.2)

### S

**sacrificial anode:** an anode made of material that is more easily oxidized than iron and is destroyed to protect the iron during the galvanizing process (19.5)

**salt:** an ionic compound formed in a neutralization reaction; composed of the anion from an acid and the cation from a base (15.2)

**salt bridge:** a device containing a salt that completes a circuit between electrochemical half cells; mantains electrical neutrality without interfering with the cell reaction (19.1)

**saturated hydrocarbons:** hydrocarbons in which each carbon atom is bonded to the maximum possible number of atoms (either hydrogen or carbon atoms) (9.3)

**saturated solution:** a solution in which no more of a particular solute can be dissolved at a specific temperature (7.1)

**secondary battery:** a rechargeable battery (19.1)

**solubility:** the mass of solute that dissolves in a given quanity of solvent at a specific temperature (7.1)

**solute:** a substance that is dissolved in a solution (7.1)

**solution:** a homogeneous mixture of a solvent and one or more solutes (7.1)

**solution stoichiometry:** the calculation of quantities of substances in chemical reactions that take place in solutions (8.3)

**solvent:** a substance that has other substances dissolved in it (7.1)

**specific heat capacity (*c*):** of a substance, is the quantity of energy, in joules (J), required to change the temperature of one gram of the substance by one degree Celsius (°C) (16.1)

**spectator ions:** ions that are present in a solution but are not involved in the chemical reaction (8.2)

**standard enthalpy of reaction ($\Delta H^0_{rxn}$):** the enthalpy change of a chemical reaction that occurs at SATP (25°C and 100 kPa) (16.2)

**standard molar enthalpy of formation ($\Delta H^0_f$):** the quantity of energy that is absorbed or released when one mole of a compound is formed directly from its elements in their standard states (16.2)

**standard solution:** a solution of known concentration (7.4)

**standard temperature and pressure (STP):** 0°C and 101.325 kPa (2.4)

**starch:** a glucose polysaccharide that is used by plants to store energy (11.2)

**state function:** a property of a system that is determined only by the current conditions of the system and is independent of the path taken by the system to reach these conditions (17.2)

**stoichiometric amount:** the exact molar amount of a reactant or product, as predicted by a balanced chemical equation (4.2)

**stoichiometric coefficient:** a number that is placed in front of the formula of a product or a reactant of a chemical equation to indicate the mole ratio of the reactants and products in a reaction (4.2)

**stoichiometry:** the study of the relative quantities of reactants and products in chemical reactions (4.1)

**strong acid:** an acid that dissociates completely into ions in a solution (14.2)

**strong base:** a base that dissociates completely into ions in a solution (14.2)

**structural diagram (structural formula):** a two-dimensional representation of the structure of a compound; can be complete, condensed, or line diagrams (5.2, 9.2)

**substitution reaction:** a reaction in which a hydrogen atom or a group of atoms is replaced by a different atom or group of atoms (9.3)

**surroundings:** everything in the universe outside the system (16.1)

**sustainable development:** the use of resources in a way that meets our current needs, without jeopardizing the ability of future generations to meet their needs (17.3)

**system:** the part of the universe that is being studied and observed (16.1)

**systematic name:** a name that is based on the IUPAC rules for naming compounds (1.3)

## T

**temperature (*T*):** a measure of the average kinetic energy of the particles that make up a substance or system; a way of quantifying how hot or cold a substance is, relative to another substance; measured in either Celsius degrees (°C) or kelvins (K) (16.1)

**tetrahedral:** the most stable shape for a compound that contains four atoms bonded to a central atom with no lone pairs; the atoms are positioned at the four corners of an imaginary tetrahedron, and the angles between the bonds are approximately 109.5° (6.1, 9.2)

**theoretical yield:** the amount of product that is produced by a chemical reaction as predicted by stoichiometry (4.3)

**thermal equilibrium:** the state achieved when all components in a system have the same final temperature (17.1)

**thermochemical equation:** a balanced chemical equation that indicates the amount of heat that is absorbed or released by the reaction the equation represents (16.2)

**thermochemistry:** the study of energy involved in chemical reactions (16.1)

**thermodynamics:** the study of energy and energy transfer (16.1)

**titration:** a laboratory process to determine the concentration of a compound in solution by monitoring its reaction with a solution of known concentration (15.2)

**total ionic equation:** a form of chemical equation that shows dissociated ions of soluble ionic compounds (8.2)

**transition state:** the condition of the reactants and products of a reaction in the activated complex (12.2)

**transition state theory:** the theory explaining what happens once colliding particles react (12.2)

**trigonal planar:** a molecular shape in which three bonding groups surround a central atom; the three bonded atoms are all in the same plane as the central atom, at the corners of an invisible triangle (6.1, 9.2)

## U

**unsaturated hydrocarbons:** hydrocarbons that contain carbon-carbon double or triple bonds, whose carbon atoms can potentially bond to additional atoms (9.3)

**unsaturated solution:** a solution in which more of a particular solute can be dissolved at a specific temperature (7.1)

## V

**valence electron:** an electron that occupies the outermost energy level, or shell, of an atom (1.2, 5.1)

**Valence-Shell Electron-Pair Repulsion (VSEPR) theory:** the theory stating that the magnitude of repulsion between lone pairs (LP) and bond pairs (BP) of electrons is in the order:
$$LP-LP > LP-BP > BP-BP \quad (6.1)$$

**variable composition:** a term used to describe a solution; capable of having different ratios of solutes to solvent (7.1)

**voltaic cell (galvanic cell):** a device that converts chemical energy to electrical energy (19.1)

**volumetric flask:** a flat-bottomed, tapered glass vessel that is used to prepare standard solutions (7.4)

**volume/volume percent:** the volume of a liquid solute divided by the volume of the solution, expressed as a percent (7.3)

## W

**wax:** a biological molecule that is an ester of a long-chain alcohol or a long-chain carboxylic acid (11.2)

**weak acid:** an acid that dissociates only slightly into ions in a solution (14.2)

**weak base:** a base that dissociates very slightly into ions in a solution (14.2)

## Z

**zero sum rule:** the rule stating that for chemical formulas of neutral compounds involving ions, the sum of positive valences and negative valences must equal zero (1.3)

# Index

The page numbers in **boldface** type indicate the pages where the terms are defined. Terms that occur in Sample Problems (*SP*), Investigations (*inv*), ExpressLabs (*EL*), and Thoughtlabs (*TL*) are also indicated.

Abiogenic theory, 323
Acetaminophen, 446
Acetic acid, 548, 554, 561, 596
   acid dissociation constant, 588
   acid ionization constant, 603–604*inv*
   concentration, 606–607*inv*
Acetone, 448
Acetylene, 87
Acetylsalicylic acid (ASA), 320, 446
Acid dissociation constant, **587**
Acid ionization constant, 587
   acetic acid, 603–604*inv*
Acid rain, 696
Acid-base indicator, **599**
Acid-base reactions/equilibria, 582
   predicting direction of, 564*SP*
   titration curve, 599
Acid-base titration, 603
Acid-base titration curve, **608**
Acidic solutions
   half-reaction method, 732*SP*
   redox reactions, 736, 737–738*SP*
Acids, 544, 545
   Arrhenius theory of, 549–552
   bond strength, 583
   Brønsted-Lowry theory of, 553
   diluting, 276
   dilution calculations, 572, 573–574*SP*
   effect of dilution, 575–576*inv*
   electronegativity, 583
   ion concentration, 585–586*SP*
   properties of, 546–547*inv*,
     548–549, 550
Activated complex, **474**
Activation energy ($E_a$), **471**, 472
Actual yield, 137
   predicting, 140*SP*
Addition polymerization/polymer,
   **428**, 431
Addition reactions, **349**, 355–356, 381
Adenine, 441
Adhesive forces, 210
AIM Theory, 194
Air bag, 462
Airy, George, 525
Alcohol, 378, **386**, 387
   additional characteristics, 389
   combustion, 389
   elimination reaction, 390
   naming, 387–388*SP*
   parent alkane, 387
   physical properties, 389
   reactions, 389
   reactions, predicting, 392*SP*
   substitution reaction, 389

Aldehyde, 378, **492**
   characteristics, additional, 404
   drawing and naming, 402, 403*SP*
   physical properties, 404
Aliphatic compounds, physical
   properties/structures of, 359*inv*
Aliphatic hydrocarbons, 331
Alkali, 545
Alkali metals, 18, 281
Alkaline cell battery, 765
Alkaline earth metals, 18
Alkaline fuel cell, 802
Alkanes, **331**, 378
   boiling points, 339
   branched-chain, 333
   combustion, 340
   condensed structural diagrams, 337
   drawing, 337, 338*SP*
   naming, 332–333, 334, 335–336*SP*
   physical properties, 339
   reactions of/reactivity, 340,
     350–351*inv*
   standard molar enthalpy of
     combustion, 643
   straight-chain, 333
   substitution reactions, 344
   unbranched, 333
Alkenes, **344**, 345, 378
   addition reactions, 349
   asymmetrical addition reaction, 352
   drawing, 347
   naming, 345, 346*SP*
   physical properties, 348
   reactions of/reactivity, 349,
     350–351*inv*
   symmetrical addition reaction, 352
Alkyl group, **334**
Alkyl halide, **390**
   naming, 390*SP*
   reactions, 391, 392*SP*
Alkynes, **354**, 378
   addition reactions, 355–356
   drawing/naming, 354
   physical properties, 355
Allotropes, **199**
Alloys, **239**, 259
Aluminum, 796
Amalgam, 239, 259
Amides, 378, **413**
   characteristics, additional, 415
   naming, 413
   physical properties, 415
   reactions, 415
   secondary, naming, 414*SP*
Amines, 378, **396**
   characteristics, additional, 399
   naming, 397

   physical properties, 399
   primary, naming, 397*SP*
   reactions, 399
   secondary, naming, 398*SP*
Amino acid, 438
Ammonia, 29, 548
   base dissociation constant, 592
Ampere (A), 790
Ampère, André, 790
Amphoteric, **557**
Amylopectin, 439
Amylose, 439
Analytical chemistry, 96
Anesthesiology, 534
Aniline, 87
   base dissociation constant, 592
Anode, **758**
Antioxidants, 717
Aqueous ionic reactions, 292
Aqueous solutions, **239**, 544
   electrolysis, 780, 781–783*SP*
   equilibrium and ions, 526
   ion identification, 295
   limiting reactants, 304–307
   matter, 31
   pH, 544
   reactions, 288
Argon, 165
Aromatic compounds, 360
   reactions of/substitution reaction, 362
Aromatic hydrocarbon, **360**, 361
Arrhenius theory of acids
   and bases, 549–552
Arrhenius, Svanté, 549
Aspartame, 447
Asymmetrical addition reaction
   and alkenes, 352
Atomic mass units (u), **14**, **43**
Atomic number (Z), **15**, 16
Atomic symbol, **15**
Atomic theory, 14
Atoms, **14**, 162
   carbon, 324
   moles, 51*SP*
Average atomic mass, **44**, 45*SP*
Avogadro constant/number, 15, **48**, 49,
   50, 55, 57, 791
   molar mass, 62
   mole, 47–48
Avogadro's hypothesis, **67**, 68
Avogadro's law, 69
Avogadro, Amedeo, 53, 67
Ayotte, Dr. Christine, 96

Bader, Richard, 194
Balanced chemical equation, 299

Balanced equations, 552
    half-reaction method, 730–731
    redox reactions, 740–741*inv*
Balancing equations and oxidation
    numbers, 747
Ball-and-stick models, 327
Barium hydroxide, 562
Bartlett, Neil, 166
Base dissociation constant, **592**
    calculating, 594–595*SP*
    solving problems involving,
        593–594*SP*
Bases, 544, 545
    Arrhenius theory of, 549–552
    Brønsted-Lowry theory of, 553
    dilution calculations, 572, 573–574*SP*
    electonegativity, 584
    ion concentration, 585–586*SP*
    properties of, 546–547*inv*,
        548–549, 550
Basic solutions
    half-reaction method, 733*SP*
    redox reaction, 736, 748*SP*
Battery, 756, **764**
Bauxite, 796
Bayer, Friedrich, 321
Beauchamp, Dr. Stephen, 123
Benefit, **700**
Bent, **191**
Benzene, 87
Bernstein, Dr. Richard, 213
Berzelius, Jons Jakob, 321
Bimolecular, 478
Binary acids, 583
Binary compounds, **24**
Binary ionic compound, **26**
    writing formulas for, 27
Binary molecular compounds, **24**
Biochemistry, **437**
Biodegradable plastic, 435
Biogenic theory, 323
Biological molecules and bonding, 211
Bioluminescence, 721
Biomolecules, 437
Black, Joseph, 662
Blood, 597
Bohr, Niels, 20
Bohr-Rutherford diagram, 20
Boiling, 10
Boiling point, 7, 8, 216–218
    alkanes, 339
    bonding, 217
    ionic bonds, 217
    metallic bonds, 217
    solution, 237
    water, 216
Bomb calorimeter, **673**
    heat change, calculating, 674–675*SP*
Bond angle, **190**
Bond dipole, **178**

Bond energy, **179**
    enthalpy, 688
    using, 688, 689–690*SP*
Bond formation, 162
Bond length, 179
Bond properties, 162
Bond strength in acids, 583
Bonding
    boiling point, 217
    melting point, 217
    structure, 325
    types of, 209
Bonding electron pairs, **169**
Bosch, Carl, 531
Brass, 259
Breathalyzer and redox reactions, 739
Brine, 804
Briss, Dr. Viola, 804
Brittleness, 7
Brønsted, Johannes, 553
Brønsted-Lowry base, 558
Brønsted-Lowry theory of acids
    and bases, **553**
Bronze, 239, 259
Bronze Age, 729
Buckminsterfullerene, 200
Buckyballs, 200
Buffer capacity, **596**
Buffer solution, **595**, 596
Burette, 603–605
Button cell battery, 765

Calcium hydroxide, 562
Calorimeter, **661**, 662, 665
Calorimetry, **661**
Cancer, 376
Carbaryl, 447
Carbohydrates, 97, **438**, 442
Carbon, 159
    atom, 324
    bond formation, 165
Carbon-hydrogen combustion
    analyzer, 99
    calculations, 100–101*SP*
Carbonate, 29
Carbonyl group, **492**
Carboxyl acid, **405**
    characteristics, additional, 407
    derivatives of, 407, 408
    naming, 405
    physical properties, 406
    preparation of derivative, 408–409*inv*
    reactions, 407–408
Carboxyl group, **405**
Carboxylic acid, 378
Catalysts, 477, **481**
    enthalpy, 642
    equilibrium constant, 527–528
    reaction mechanism, 477
Cathode, **758**
Cathodic protection, **801**

Cell potential, 761, 761
    measuring potential, 762–763*inv*
Cell voltage, **761**
Cells, 756
Cellular respiration, 439
Cellulose, 426, 437, 439
Celsius, 628–629
Centre for Forensic Science, 95
CFCs, 449
Change of state and enthalpy, 645–647
Chemical bonding, 158
Chemical changes, **10**
Chemical engineer, 144
Chemical equations
    algebraic combination, 678–679
    balanced/balancing, 31, 32*SP*, 111,
        112–113
    mass, 118
    moles, 114–115
    standard cell potential, 772–773*SP*
Chemical formula
    compound, 27*SP*
    determining formula of, 104–105*inv*
    experiment, 99–100
    percentage composition, 83,
        84*SP*, 85*SP*
    polyatomic ions, 29–30SP
Chemical property, **7**
Chemical reactions, 33
    classifying, 33–35
    enthalpy, 665–667*SP*
    moles, 112*EL*
    percentage yield, determining,
        142–143*inv*
    quantities in, 110
    rates of, 462
Chemical species, 474
Chemical vapour deposition, 200
Chemiluminescence, 721
Chemistry, **5**
Chlor-alkali process, **804**
Chloric acid, 583
Chlorination of water supply, 309,
    804, 805
Chlorofluorocarbons, 370
Chlorous acid and acid dissociation
    constant, 588
Cholesterol, 356
Cinnabar, 84
Circle of life, 488
Cis isomer, 348
Cis-trans isomer, **348**
Classical system, 26
Closed system, **628**
Co-ordinate covalent bond, **187**
    Lewis structure, 187–188*SP*
Coffee-cup calorimeter, **661**
Cohesive forces, 210

Collision theory, **470**
    concentration, 470
    reactants, 471
    surface area, 470
    temperature, 471
Colour and reaction rates, 466
Combustibility, 7
Combustion
    alcohol, 389
    alkanes, 340
    balancing equations, 341
    complete, of butane, 341–342*SP*
    enthalpy, 670, 671–672*inv*, 682–683*inv*
    Hess's law of heat summation, 682–683*inv*
    incomplete, 342–343*SP*
Combustion reactions, 35, 382
    heat, 644–645*SP*
Common ion effect, 526
Competing reaction, **137**
Complete combustion, 35, **340**
Complete structural diagram, **325**
Compounds, **11**, **12**, 159
    chemical formula, 27*SP*
    chemical proportions, 78
    law of definite proportions, 79
    molar mass, 56*SP*
    molecular formula, 95
    percentage composition, 79
    physical properties, 159–160
Concentrated solution, **237**
Concentration, 236, **255**
    acetic acid, 606–607*inv*
    collision theory, 470
    estimating, 274–275*inv*
    finding, 600–601*SP*
    reaction rates, 464*inv*
    solution, 255–269, 269*inv*
Condensation, 10
Condensation polymer, 431
Condensation polymerization, **429**
Condensation reaction, 382
Condensed structural diagram, **325**, 337
Conductivity and reaction rates, 466
Conjugate acid, **554**
Conjugate acid-base pair, **554**, 555–557*SP*
Conjugate base, **554**
Cooling curve, **651**
    constructing, 65*EL*
Copper sulfate pentahydrate, 490
Corrosion, **798**, 799
    prevention, 800–801
Coulomb (C), 790
Coulomb, Charles, 790
Covalent bonding, 168–169, 209
Covalent bonds
    Lewis structures, 186
    molecules, 185–187
    structures, 185

Covalent compounds and solubility, 247
Covalent pairs, **169**
Cracking, 367
Crystal lattice, **167**, 185
Cupronickel, 259
Cyclic hydrocarbons, 356, **357**
    drawing/naming, 357
    physical properties, 358
    reactions, 358
Cytosine, 441

D-limonene, 265
Dacron, 430
Dalton, John, 14, 67, 87
Dangerous goods in*sp*ection, 605
Daniell cell, 757, 758, 759, 764
    standard cell potential, 769–770
Daniell, John Frederic, 757
Davy, Sir Humphrey, 778, 791
DDT, 450
Decomposition reaction, 7, 34, 492
DEET, 191
Degradable plastics, 434, 435
Delocalized electrons, 360
Density, 7, 8
Derivative, **410**
Diatomic, 25
Dickson, Alison, 173
Diluted solution, **237**
Dipole, **245**
Dipole interactions in hydrocarbons, 379
Dipole-dipole attraction, **245**
Dipole-dipole bonding, 209
Dipole-dipole forces, **202**
Dipole-induced dipole bonding, 209
Dipole-induced dipole force, **204**
Diprotic acids, 561
Disaccharide, **439**
Dispersion bonding, 209
Dispersion force, **204**, 380
Disproportionation reaction, **716**
Dissociation, 240, 292
Dissociation equation, **299**
Distillation, 237
DNA (2-deoxyribonucleic acid), **440**
    bonding, 211
    structure, 441
Doping Control Laboratory, 96
Double covalent bond, **169**
Double displacement reactions, 34
    gases, 291
    precipitates, 289
    water, 292
Double helix, 211, 442
Downs cell, 777
Dry cell, **764**
Dry ice, 5
Ductility, 7, 219
Duralumin, 259

Dynamic equilibrium, 240, 490, **489**
    applicable conditions, 492
    modelling, 491*EL*

E. coli, 309
Electonegativity, 584
Electric current, **757**, 790
Electric potential, **761**
Electrical conductivity, 7, 220, 221
Electrochemical cell, 777
Electrochemistry, **757**
    stoichiometry, 790
Electrodes, **758**
Electrolysis, 10, **776**
    aqueous potassium iodide, 784–785*inv*
    aqueous solution, 780, 781–783*SP*
    mass, 791–792*SP*
    molten salts, 776–777
    water, 778–779
Electrolytes, 246, **758**
Electrolytic cell, **776**
Electromotive force, 761
Electron, 15
Electron density maps, 194
Electron pair, **163**
Electronegativity (*EN*), **174**, 175, 176, 583
    acids, 583
    bond type, 176–178
    oxidation numbers, 723
    solubility, 246
Electrons, 14
    atomic number, 16
    energy levels, 19
    isotopes, 43
    periodic table, 19
    valence. *See* Valence electrons
Electroplating, **780**, 794–795*inv*
Element symbol, 15
Elementary reaction, **477**, 478–479
Elements, **11**, **12**, 84, 159, 162
    electronegativity, 174–176
    inner transition, 18
    main group, 18
    physical properties, 159–160
    transition, 18
Elimination reaction, **381**
Empirical formula, **87**
    determining, 88, 91, 92–93*inv*
    experiment, 91, 99–100
    percentage composition, 88–89*SP*, 90–91*SP*
    tips for solving problems, 89–90
End point, **603**
Endothermic, 180, 526
Endothermic reaction, **640**, 641
Endrin, 447

Energy, 626, 628
  benefit, 700
  efficiency, 693–694
  nuclear, 692–693
  risk, 700
  sources of, 692
  sources, comparing, 701*TL*
  sources, emerging, 702
  transfer and mass, 632
  transfer and type of substance, 632
  types of, 628
Energy levels, **19**, 20
  electron arrangements, 21
  electrons, 19
  group-related pattern, 21
  period-related pattern, 21
  valence electrons, 21
Enthalpy
  bond energy, 688
  calculating changes, 643–644, 685
  calorimeter, 665
  catalysts, 642
  change of state, 645–647
  chemical reaction, 665–667*SP*
  combustion, 670, 671–672*inv*, 682–683*inv*
  formation, 687*SP*
  Hess's law of heat summation, 682–683*inv*
  neutralization reaction, 668–669*inv*
  reaction, 665–667*SP*
  solution, 670
  using, 680–681*SP*
Enthalpy change, **639**
Enthalpy of reaction, **640**, 641
Environment, 694, 696, 698
Enzymes, 438, 483
Equilibrium
  direction of reaction, 517
  modelling, 491*EL*
Equilibrium amounts, finding, 505–506*SP*
Equilibrium concentration, measuring, 499–500
Equilibrium constant, 495, **496**, 544
  approximation method, using, 513–515*SP*
  calculating, 498*SP*
  catalyst, 527–528
  determining direction of shift, 518*SP*
  determining from percent reaction, 509–510*SP*
  determining qualitatively, 511–512
  interpreting small value, 513
  Le Châtelier's Principle, 519–520, 528–529*SP*
  measuring, 501–502*inv*
  percent reaction, 509, 512*SP*
  perturbing, 521–524*inv*
  pressure, 527

reaction quotient, 517
temperature, 497, 512*SP*, 526
volume, 527
Equilibrium expression/reaction, 382
  quadratic equation, 506–507*SP*
  writing, 496*SP*
Equivalence point, **603**, **608**
Ester, **378**, **410**
  characteristics, additional, 412
  drawing and naming, 411*SP*
  physical properties, 412
  reactions, 412
Esterification reaction, **410**
Estrone, 356
Ethene, 427
Ethers, **394**
  characteristics, additional, 396
  common name, 395
  IUPAC name, 395
  naming, 395*SP*
  physical properties, 396
Ethylene, 428
Ethylene glycol, 448
Ethyne, 87
Excess reactant, **129**
  predicting, 132*inv*
Exothermic reaction, 180, 526, **640**
  representing, 641
Expanded molecular formula, **326**
Experiment, chemical/empirical formula, 99–100
Explosives chemist, 766
External circuit, **757**
External voltage, 776
Extraction, **796**

Fahrenheit degrees, 11
Faraday's Law, 790, 793, 796
Faraday, Michael, 758, 790, 791
Fats, 97, **442**
Fatty acids, 442
Fennerty, Charles, 437
Fermi problems, 50
Fermi, Enrico, 50
Filipovic, Dusanka, 370
First law of thermodynamics, **627**
Flammability, 7, 8
Flow battery, 802
Fool's gold, 145, 159
Forensic scientists, **95**
Formic acid, dissociation constant, 588
Fossil fuels, 322
  environment, 698
Fractional distillation, **366**
Free radicals, 717
Free-electron model, **172**
Freezing, 10
Fructose, 97, 438
Fuel cell, **802**
Fullerenes, 200

Functional groups, **377**
  C–O bond, 402
  single-bonded, 386
Fusion, 10
Fuzzy fibre, 426

Galactose, 97
Galvani, Luigi, 758
Galvanic cells, **757**, 758
  electrical energy, 779
  measuring potential, 762–763*inv*
  notation, 760
Galvanizing, **800**
Gas stoichiometry, **119**
Gases, 18
  double displacement reactions, 291
  matter, 31
  molar volume, 69*TL*
  moles, 71–73*SP*
  pressure, 66, 252
  solubility, 251, 252
  solution, 238
  temperature and solubility, 251*EL*
  volume, 70–71*SP*
Gasoline, 450
Gay-Lussac, Joseph Louis, 67, 68
General formula, **377**
General solubility guidelines, **285**
Geometric isomer, **348**
Gesner, Abraham, 699
Gillespie, Ronald, 189
Gimli Glider, 9
Global warming, **696**
Glucose, 87, 97, 438
Glyceryl trioleate, 442
Glycogen, 439
Gold, 158, 159, 171
  bond formation, 165
Gravimetric stoichiometry, **119**, 303
Greenhouse effect, 696, 802
Greenhouse gases, **696**
Guanine, 441
Guldberg, Cato, 494, 495

Haber Process, 113, 530
Haber, Fritz, 530
Haber-Bosch process, 532
Half bond, 189
Half-cell potentials, 768
Half-reaction, **715**
  acidic solutions, 732*SP*
  balanced equations, 730–731
  basic solutions, 733*SP*
  redox reactions, 734*SP*
Haloalkane, **390**
Halogens, 18
Hardness, 7

Heat, 628, **629**
　combustion reactions, 644–645*SP*
　factors in transfer, 631*TL*
　physical change, 648
　system, 653, 654–655*SP*
　transfer, calculating, 633
Heat capacity, 636, **637**
Heat measurement, technology of, 661
Heating curve, **651**
　constructing, 653*EL*
Helium, 165
Henri, Victor, 483
Hess's law of heat of summation, **677**
　combustion, 682–683*inv*
　enthalpy, 682–683*inv*
　using, 680–681*SP*
Heterogeneous catalyst, **482**
Heterogeneous equilibrium, **492**
Heterogeneous mixtures, 11, 12
High-performance liquid chromatography (HPLC), 61
Homogeneous catalyst, **482**
Homogeneous equilibrium, **492**
　measuring, 501–504*inv*
Homogeneous mixtures, 11, 12
Homologous series, **331**
Hormones, 438
Hot ice, 695
Hydrate, **101**
　determining formula of, 102–103*SP*, 104–105*inv*
Hydrated, **246**
Hydrazine, base dissociation constant, 592
Hydrobromic acid, 560, 561, 583
Hydrocarbon derivatives, 376
Hydrocarbon fuel, 696
　environment, 696
Hydrocarbons, **35**, **320**
　applications, 365
　classifying, 331
　cracking and reforming, 367
　dipole interactions, 379
　dispersion forces, 380
　hydrogen bonds, 379
　origins of, 322–323
　physical properties, 379
　refining, 365
　representing, 324
　sources of, 323
　substituted, 379
　using, 365
Hydrochloric acid, 544, 548, 560, 561, 583
Hydrocyanic acid, dissociation constant, 588
Hydrofluoric acid, 544, 561, 584
　acid dissociation constant, 588
Hydrogen and bond formation, 165
Hydrogen binding, **245**

Hydrogen bonds, **205**, 206, 209
　hydrocarbons, 379
Hydrogen carbonate, 29
Hydrogen compounds, 30–31
Hydrogen fuel cell, 802–803
Hydrogen peroxide, 7, 87
Hydrogen sulfate, 29
Hydrohalic acids, 583
Hydroiodic acid, 560, 561
Hydrolysis reaction, **382**
Hydronium ion, **551**, 560, 596
Hydrophobic, 173
Hydroxide, 29
Hydroxyl group, **377**, 551

ICE table, **505**
Ideal gas, **70**
Immiscible, **239**
Impurities, 144
Incomplete combustion, **35**, **340**
Induced intermolecular forces, 203
Industrial process, analyzing, 36*inv*
Inert electrode, **760**
Inner transition elements, 18
Insaturated solution, **239**
Insoluble, 240
Insulin, 438
Intermolecular, **202**
Intermolecular forces
　applications, 214
　organic compounds, 380*TL*
International System of Unts (SI), 53
International Union of Pure and Applied Chemistry (IUPAC), 24, 331
Invertase, 483
Ion, 16
Ion concentration, 301–302*SP*
Ion product constant for water, **565**
Ion size and solubility, 282
Ion-dipole attractions, **246**
Ion-dipole bonding, 209
Ion-dipole force, **203**
Ion-induced dipole bonding, 209
Ion-induced dipole force, **204**
Ionic bonding, 165–168, 209
　boiling point, 217
　crystal lattice, 185
　melting point, 217
Ionic charge and solubility, 281
Ionic compounds, 161*TL*
　classifying, 241*TL*
　general solubility guidelines, 285
　hydrated, 101
　properties, 160
　solubility, 246, 281, 283–284*inv*
Ionic equation, 552
Ionic liquids, 223
Iron Age, 729
Isolated system, **628**

Isomers, 326, 327
　modelling, 328*inv*
Isotopes, 43, 44
Isotopic abundance, **44**

Joule (J), **628**

Karwa, Dr. Laila Rafik, 534
Kekulé, August, 360
Kelvin, 628–629
Kerosene, 699
Ketone, 378, **492**
　characteristics, additional, 404
　drawing and naming, 403*SP*
　physical properties, 404
Kevlar, 214, 451
Kilopascal (kPa), **66**
Kinetic energy, **628**, 629–630
Kinetic molecular theory, **469**
Krypton, 165
Kwolek, Stephanie, 214, 451

Lactic acid, dissociation constant, 588
Laplace, Pierre, 662
Laterite ores, 145
Lavoisier, Antoine, 118, 662
Lavoisier, Marie-Anne, 118
Law of chemical equilibrium, **494**
Law of combining volumes, **67**
Law of conservation of energy, **627**, 661
Law of conservation of mass, **118**
Law of definite proportions, **79**
Law of mass action, 496
Law of multiple proportions, 80
Le Châtelier's Principle, **519–520**, 526, 596
　application to ecosystems, 525
　equilibrium constant, 528–529*SP*
　manufacturing ammonia, 530–532
　perturbing equilibrium, 521–524*inv*
Le Châtelier, Henri, 520
Le Roy, Dr. R. J., 213
Lead-acid battery, 787–788
Least common multiple (LCM), 735
Leclanché cell, 764
Leclanché, Georges, 764
Lemieux, Dr. Raymond, 443
Lenz, Heinrich, 525
Lewis structures, **162**, 163, 167–169, 171
　co-ordinate covalent bond, 187–188*SP*
　covalent bonds, 186
　drawing, 170*SP*
　methane, 324
　molecule, 186–187*SP*
　oxidation numbers, 721–724
　resonance structures, 188–189
Lidocaine, 446
Like dissolves like, 247
Liming, 601

Limiting reactant, 128, **129**
  analogy, 129*TL*
  aqueous solutions, 304–307
  determining, 129–130*TL*
  identifying, 130*SP*
  predicting, 132*inv*
  stoichiometry, 133–134*SP*
Line structural diagram, **325**
Linear, **190**, **329**
Lipids, **442**, 443
Liquid
  matter, 31
  properties of, 210TL
  solution, 238
Litmus paper, 549
Logarithms, 567
London dispersion forces/bonding, 203, 206, 209, 211, 212, 219
London, Fritz, 203
Lone pairs, **169**
Lonsdale, Kathleen Yardley, 361
Lowry, Thomas, 553

Magnesium, industrial process, 36–37*inv*
Magnesium hydroxide, 562
Magnesium oxide, determining, 92–93*inv*
Main group elements, 18
Malleability, 7, 219
Mannose, 97
Markovnikov's rule, **352**, 353*SP*
Mass
  chemical equations, 118
  electrolysis, 791–792*SP*
  moles, 55, 58*SP*, 59, 60*SP*, 62*SP*
  particles, 63–64*SP*
  percentage composition, 81–82*SP*, 83*TL*
  reaction rates, 466
  stoichiometry, 124–125*SP*
  subatomic particles, 14
  volume, 6
Mass number ($A$), **15**, **43**
Mass percent, **80**
  See also Mass/mass percent
Mass spectrometer, **44**, 96
Mass/mass percent, 258–261bbb
  solving for, 260–261*SP*
Mass/volume percent, **255**
  solving for, 256–258*SP*
Material Safety Data Sheet, 255
Matter, **6**, 12
  aqueous solution, 31
  chemical changes, 10
  classification of, 10, 11
  describing, 7
  gases, 31
  liquid, 31
  physical changes, 10
  properties, 7
  solid, 31
  states of, 31

Mechanical mixture, 12
Melting, 10
Melting point, 7, 216–218
  bonding, 217
  ionic bonds, 217
  metallic bonds, 217
  solution, 237
  water, 216
Mendeleev, Dmitri, 17, 19
Menten, Dr. Maud L., 483
Menthol, 447
Metal activity series, 716, 718–719*inv*
Metallic bonding, 171, 172, 209
Metallic bonds, **172**, 217
Metalloids, 18, 19
Metallurgist, 173
Metals, 18, 19
Methane, 169
  Lewis structure, 324
Methane hydrates, 695
Methylamine, base dissociation constant, 592
Michaelis, Leonor, 483
Michaelis-Menten equation, 483
Micronutrients, 61
Miscible, **239**
Mixture, **11**, 12
  heterogeneous, 11
  homogeneous, 11
  pure substance, 12*TL*
Moissan, Ferdinand, 232
Mol, **48**
Molar absorption coefficient, 213
Molar concentration, **266**
  calculating, 266–267*SP*
  mass, finding, 267–268*SP*
Molar enthalpy of condensation/ freezing/melting/solution/ vaporization, 647
Molar mass ($M$), **55**, 96
  Avogadro constant, 62
  compounds, 56*SP*
  moles, 57
Molar solubility, 239
Molar volume, **69**
  gases, 69*TL*
  standard, 70
Molarity. See Molar concentration
Mole, 42
  atoms, 51*SP*
  Avogadro constant, 47–48
  chemical equations, 114–115
  chemical reaction, 112*EL*
  converting, 50, 52
  definition, 48
  gases, 71–73*SP*
  mass, 55, 58*SP*, 59, 60*SP*, 62*SP*
  molar mass, 57
  molecules, 53*SP*

Mole ratios, **114**
  reactants, 116–117*SP*
Molecular compounds, **24**, 161*TL*
  properties, 160
  writing formulas for, 25
Molecular dipoles, **195**
Molecular formula, **87**
  compound, 95
  determining, 95–96, 97*SP*
Molecular polarity, 195–196
Molecularity, **478**, 479
Molecule, 167, **191**
  covalent bonds, 185–187
  Lewis structure, 186–187*SP*
  modelling, 197–198*inv*
  polar bonds, 194–196
  properties, 184
  shape, 184, 190–191, 192, 193*SP*, 329*EL*
  structure, 184
Molecule size and solubility, 248
Molecules and covalent bonds, 186
Monomers, **427**
Monoprotic acids, 561, 584
Monosaccharide, **439**
Monosodium glutamate, 447
Morphine, 446

Nanotubes, 200
National Institute of Standards and Technology (NIST), 56
Neon, 165
Net ionic equation, **293**, 552
  balancing, 734–736
  guidelines for writing, 293
  standard cell potential, 771–772*SP*
  writing, 293–294*SP*
Network solids, **199**
Neutralization, calculations involving, 600
Neutralization reaction, **599**
  enthalpy, 668–669*inv*
  polybasic, 609–610
  polyprotic, 609–610
Neutrons, 14, 43
Nicad cell, 787–788
Nitrate, 29, 488
Nitric acid, 560, 583
Nitrogen, 165, 488
Nitrogen cycle, 488
Nitrous acid, dissociation constant, 588
Noble gases, 18, 165, 166
Non-electrolytes, **247**
Non-metals, 18, 19
Non-polar, **195**
Non-polar compound
  classifying, 241*TL*
  solubility, 244
Non-polar molecules, oxidation numbers, 723

Non-renewable, **694**
Nonpolar covalent bonds, **176**, **177**
Nucleic acids, 440
Nucleotides, **440**
Nucleus, 14
Nyholm, Ronald, 189
Nylon, 427, **429**, 430

Octane, 450
Octet rule, **165**
Oil pollution, 697
Oil refining, 369
Oil spill advisor, 697
Oil spill pollution, 696
Oils, **442**
Open system, **628**
Oppenheimer, Robert, 640
Ore, 713
Organic compounds, **320**, 426
 functional group.
  *See* Functional group
 identifying reactions, 382–383*SP*
 intermolecular forces, 380*TL*
 modelling, 328*inv*
 natural, 321
 physical properties, comparing, 418*inv*
 predicting products, 416*SP*
 problem solving with, 451*TL*
 risk-benefit analysis, 445–448
 synthetic, 321
Oven cleaning, 577
Overvoltage, **779**
Oxidants, 717
Oxidation, 713, **714**, 717, **727**, 758
Oxidation numbers, **721**
 assigning, 725*SP*
 balancing equations, 747
 electronegativity, 723
 finding rules, 723*TL*
 Lewis structures, 721–724
 non-polar molecules, 723
 polar molecules, 723
 polar polyatomic ions, 723
 redox reactions, 726–727
 using rules to find, 724
Oxidation potential, **770**, 771
Oxidation-reduction reaction, 712, **714**
Oxides, 282
Oxidizing agent, **714**
Oxoacids, 583
Oxygen and bond formation, 165
Ozin, Dr. Geoffrey, 220

Parathion, 447
Parent alkane, **387**
Parts per billion (ppb), **263**, 264*SP*
Parts per million (ppm), **263**
Pascal (Pa), **66**
Percent (m/m). *See* Mass/mass percent

Percent (v/v). *See* Volume/volume percent
Percent by volume. *See* Volume/volume percent
Percent dissociation, **588**, 589–590*SP*
Percent ionization, 588
Percent reaction and equilibrium constant, 509, 512*SP*
Percentage composition, **81**
 chemical formula, 83, 84*SP*, 85*SP*
 compounds, 79
 empirical formula, 88–89*SP*, 90–91*SP*
 mass, 81–82*SP*, 83*TL*
Percentage purity, **146**
 finding, 146–147*SP*
Percentage yield, 137, **138**
 applications, 141
 calculating, 138–139*SP*
 determining, 142–143*inv*
 stoichiometry, 137
Perchloric acid, 560, 583
Periodic law, **17**
Periodic table, **17**
 average atomic mass, 44
 electrons, 19
Periodic trends, **20**
Petrochemicals, 365
Petroleum, **323**
Petroleum Age, 699
Pewter, 259
pH, 544, 564, **567**, 568, 596
 buffer solution, 596
 calculating, 569*SP*, 570–571*SP*, 590–591*SP*
 effect of dilution, 575–576*inv*
 reaction rates, 466
Phenol and acid dissociation constant, 588
Phosphate, 29, 281
Phosphoric acid, 561
Photodegradable plastic, 435
Photosynthesis, 5
Physical changes, **10**
Physical properties, **7**
 compounds, 159–160
 elements, 159–160
 solute, 237
 solvent, 237
Pipette, 603–605
Plastic, **427**, 434
Platinum, 159, 165, 171
pOH, 569–570
 calulating, 570–571*SP*
 finding, 571–572*SP*
Polar, **195**
Polar bonds in molecule, 194–196
Polar compound
 classifying, 241*TL*
 solubility, 244
Polar covalent bonds, **177**

Polar molecules and oxidation numbers, 723
Polar polyatomic ions and oxidation numbers, 723
Polyacrylonitrile, 429
Polyamide, **429**, 430
Polyatomic ions, 29
 chemical formula, 27*SP*, 29–30*SP*
Polyester, 427, 430
Polyethene, 427, 428, 429
Polyethylene terephthalate, 434
Polymer chemist, 451
Polymerization reaction, classifying, 430*SP*
Polymers, 426, **427**, 432–433*inv*
Polypropylene, 428
Polyprotic acids, 584
Polysaccharides, **439**
Polystyrene, 429
Polyvinylchloride, 428, 429, 434
Porous barrier, 757
Potassium hydroxide, 562
Potential energy, **628**
Potential energy diagram, **473**
 drawing, 475–476*SP*
 tracing a reaction, 474–475
Potential ladder diagram, 769, 773, 782
Pratt, John, 525
Precious metals, 159
Precipitate, 292
 double displacement reactions, 289
 finding mass of, 304–306*SP*
 finding minimum volume, 303*SP*
 predicting formation of, 289–290*SP*
Pressure, **66**
 equilibrium constant, 527
 gases, 252
 reaction rates, 466
 solubility, 252
 temperature, 67
 volume, 67
Pressure acid leaching (PAL) process, 145
Primary battery, **764**
Primary solution, **603**
Primary standard, **603**
Primary structure, 211
Product development chemist, 265
Products, calculations for, 121–122*SP*
Properties, **7**
Propylene, 428
Proteins, 97, 211, **438**
Proton acceptors, 553
Proton donors, 553
Protons, 14, 15
 atomic number, 16
 isotopes, 43
Proust, Joseph Louis, 79
Pure substance, **11**, 12
 mixture, 12*TL*
Pyramidal, **191**

Index • MHR **867**

Pyrethrin, 447
Pyridine, base dissociation constant, 592

Qualitative analysis, **295**, 296–297*inv*
Qualitative properties, 8
Quality control, 61
Quantitative properties, 8

Radon, 165
Rate of dissolving, **243**
    factors affecting, 243–254
Rate of reaction, **463**
Rate-determining step, **479**
Rayon, 427
Reactants, 128
    calculations for, 120*SP*, 121–122*SP*
    collision theory, 471
    limiting, 128, 129
    mole ratios, 116–117*SP*
    ratios, 116
    reaction rates, 465*inv*
Reaction intermediates, **477**
    determining, 477–478*S*
Reaction mechanism, **477**
Reaction quotient, **517**
Reaction rates, **463**
    colour, 466
    concentration, 464*inv*
    conductivity, 466
    factors that affect, 463, 464–465*inv*, 467
    mass, 466
    methods of measuring, 466
    pH, 466
    pressure, 466
    reactants, 465*inv*
    surface area, 465*inv*
    temperature, 464*inv*, 465*inv*
    theories of, 469–483
    volume, 466
Reactions
    activation energy, 472
    aqueous solutions, 288
    enthalpy, 665–667*SP*
    orientation of reactants, 472
Reactivity, 7
Rechargeable batteries, 787–788
Recommended Nutrient Intake, 61
Red #40, 447
Redox reactions, **714**
    acidic solutions, 736, 737–738*SP*
    balanced equations, 740–741*inv*
    basic solution, 736, 748*SP*
    breathalyzer, 739
    half-reaction method, 734*SP*
    identifying, 727–728*SP*
    oxidation numbers, 726–727
Redox titrations, 742, 743–744*SP*
Reduce, reuse, and recycle, 434
Reducing agent, **714**
Reduction, 713, **714**, **727**, 758

Reduction potential, **768**, 771
Reference values, assigning, 774*TL*
Refining, **796**
Reforming, **368**
Renewable, **694**
Resmethrin, 447
Resonance hybrid, 189
Resonance structures, **188**, 189
Reversible changes, 488
Richter, Jeremias Benjamin, 119
Risk, **700**
Risk-benefit analysis and organic compounds, 445–448
RNA (ribonucleic acid), **440**, 441
Rusting, 798, 799

Saccharide, **438**
Sacrificial anode, **800**
Salbutalmol, 446
Salicin, 320
Salt, **599**
Salt bridge, **759**
Saturated hydrocarbons, **331**
Saturated solution, **239**
Saturation, 239–240
Secondary battery, **764**
Secondary structure, 212
Semi-permeable membrane, 757
Seviour, Joanne, 605
Silver, 159, 165, 171
Single covalent bond, **169**
Single displacement reactions, 34, 718–719*inv*
Slag, 173
Slightly soluble, 240
Smog, 802
Sodium hydroxide, 545, 548, 562
Solids
    matter, 31
    mechanical properties, 219
    properties of, 218, 222*inv*
    solution, 238
Solubility, 7, 239–240
    covalent compounds, 247
    curve, 248
    curve, plotting, 249–250*inv*
    electronegativity, 246
    factors affecting, 243–254, 281
    gases, 251, 252
    heat pollution problem, 251
    intermolecular forces, 245
    ion size, 282
    ionic charge, 281
    ionic compounds, 246, 281, 283–284*inv*
    molecule size, 248
    non-polar compounds, 244
    particle attractions, 244
    polar compounds, 244
    predictions about, 281, 282
    pressure, 252
    temperature, 248, 251

Soluble, 240
Solutes, **237**
Solution, 12, 223, 236, **237**
    boiling point, 237
    concentration, 255–269, 269*inv*
    enthalpy, 670
    gases, 238
    heat of, measuring, 670*EL*
    liquid, 238
    melting point, 237
    molar enthalpy, 647
    solid, 238
    stoichiometry, 280
    types of, 237
Solution stoichiometry, **299**, 303
Solvents, **237**
    coffee, 253
    identifying, 240
    matching with solutes, 241*TL*
    matching with solvents, 241*TL*
    physical properties, 237
    suitable, 240
Sorbose, 97
Sørensen, Søren, 567
Space-filling models, 327
Sparingly soluble, 240
Specific heat capacity, **632**
    calculating, 635*SP*
    determining, 663–4*SP*
Spectator ions, **292**, 391
Spectrophotometer, 479
Spontaneity of reactions, 786*SP*
Stainless steel, 259, 800
Standard cell potentials, 768
    calculating, 769–770
    chemical equation, 772–773*SP*
    net ionic equation, 771–772*SP*
Standard enthalpy of reaction, **640**
Standard molar enthalpy of combustion, 643
Standard molar enthalpy of formation, **642**, 684
Standard solution, **271**, 603
    diluting, 272–273*SP*
Standard states, 768
Standard temperature and pressure (STP), **67**
Starch, 159, **439**
State function, **678**
Sterling silver, 259
Steroids, 356, 357
Stock system, 26
Stock, Alfred, 26
Stoichiometric amounts, **128**
Stoichiometric coefficients, **128**

Stoichiometry, 111, **119**
   actual yield, 137
   electrochemistry, 790
   general process for solving, 124
   mass, 124–125*SP*
   percentage yield, 137
   redox titrations, 742
   solution, 280
   theoretical yield, 137
Stone Age, 729
Strong acids, **560**, 583
Strong bases, **562**, 584
Strontium hydroxide, 562
Structural diagram, **325**
Structural formulas, **171**
Structural models, 327
Structure
   bonding, 325
   covalent bonds, 185
   properties, 216
Styrofoam, 427
Subatomic particles, 14
Sublimation, 10
Substitution reaction, **362**, 381
   alcohol, 389
   alkanes, 344
   aromatic compounds, 362
Sucrose, 443
Sulfate, 29
Sulfides, 282
Sulfuric acid, 560, 561, 583
Supersaturated solution, 490
Surface area
   collision theory, 470
   reaction rates, 465*inv*
Surroundings, **627**
Sustainable development, **698**
Swiss Water Process, 253
Symmetrical addition reaction
   in alkenes, 352
Synthesis reaction, 33, 34
System, **627**
   heat, 653, 654–655*SP*
Systematic names, **24**, 331

TAXOL™, 376
Technology, 6
Temperature, **628**
   collision theory, 471
   equilibrium constant, 497, 512*SP*, 526
   pressure, 67
   reaction rates, 464*inv*, 465*inv*
   solubility, 248, 251
   volume, 67
Termolecular reaction, 479
Tertiary structure, 212
Testosterone, 356
Tetra-ethyl lead, 450
Tetrachloroethene, 448
Tetrahedral, **191**, **329**

Theoretical yield, **137**
Thermal conductivity, 7, 221
Thermal equilibrium, **663**
Thermochemical equation, **641**
Thermochemistry, **627**
Thermodynamics, **627**
Thermoplastics/thermosets, 428
Three-dimensional structural diagrams, 330
Thymine, 441
Titration, **603**
Titration curve
   acid-base reactions, 599
   polybasic, 602
   polyprotic, 602
Total ionic equation, **292**
Toxicity, 7
Trans isomer, 348
Transition elements, 18
Transition state, **474**
Transition state theory, **473**
Trigonal planar, **190**, **329**
Triple covalent bond, **169**
Triprotic acids, 561
Typhoid, 309

Unimolecular elementary reaction, 479
Unpaired electrons, **163**
Unsaturated hydrocarbons, **348**
Uracil, 441
Urea and base dissociation constant, 592

Valence electrons, **21**, **162**, 163, 168–169, 172
Valence-shell electron-pair repulsion (VSEPR) theory, 189–191
van der Waals, Johannes, 202
Vaporization, 10
Villamagna, Fortunato, 766
Vitamin supplements, 61
Vitamins, 443
Voisey's Bay, 145
Volta, Count Alessandro, 758
Voltaic cells, **757**
Volume
   equilibrium constant, 527
   finding, 601–602*SP*
   gases, 70–71*SP*
   mass, 6
   pressure, 67
   reaction rates, 466
   temperature, 67
Volume percent. *See* Volume/volume percent
Volume percent concentration. *See* Volume/volume percent
Volume/volume percent, **261**, 262*SP*
Volumetric flask, **271**

Waage, Peter, 494, 495
Water, 87, 236, 280, 564
   boiling point, 216
   chlorination, 804
   double di*sp*lacement reactions, 292
   electrolysis, 778–779
   ion product constant for, 564
   melting point, 216
   properties of, 207–208*inv*
   treatment, 804
Water quality, 309
Waxes, **442**
Weak acid, **561**
Weak base, **562**
Westray Mine explosion, 471
Wet chemical techniques, 295
Wohler, Friedrich, 321

Xenon, 165

Zero sum rule, **27**

## Photo Credits

**2** (bottom centre), Artbase Inc.; **3** (centre left) Ian Crysler; **3** (centre right) Artbase Inc.; **3** (bottom right) Ian Crysler; **5** (top right) Ian Crysler; **6** (top right) Ian Crysler; **7** (top left) Artbase Inc.; **8** (bottom left); from Chemistry 11, © 2001, McGraw-Hill Ryerson, a subsidiary of The McGraw-Hill Companies; **9** (centre right) Ian Crysler; **15** (centre) © Stephen Frish Photography; **15** (centre) © Stephen Frish Photography; **15** (centre) © Stephen Frish Photography; **15** (centre) © Stephen Frish Photography; **15** (centre) © Stephen Frish Photography; **32** (centre) © Charles D. Winters/Photo Researchers; **32** (centre) Jerry Mason/Science Photo Library/Photo Researchers; **32** (bottom centre) Ian Crysler; **32** (top centre) Ian Crysler; **42** (bottom centre) Ian Crysler; **44** (centre left) Ian Crysler; **47** (centre) Ian Crysler; **50** (bottom left) Ian Crysler; **66** (centre left) Tim Wright/CORBIS/MAGMA; **67** (top right) Jeff Hunter/ Photographers Choice/Getty Images; **74** (bottom right) Norman Owen Tomalin/Bruce Coleman Inc.; **78** (bottom) Ian Crysler; **80** (centre) Ian Crysler; **86** (centre left) Ian Crysler; **86** (bottom left) Ian Crysler; **86** (bottom centre) Ian Crysler; **86** (bottom right) Ian Crysler; **87** (bottom right) Ian Crysler; **102** (centre left) Ian Crysler; **104** (bottom left) Ian Crysler; **104** (bottom centre) Ian Crysler; **106** (centre left) Ian Crysler; **116** (centre left) Ian Crysler; **136** (top left) Ian Crysler; **145** (bottom left) Mike Dobel/Masterfile; **166** (centre left) Neil Bartlett/B. C. Jennings Photographer; **184** (bottom centre) Ian Crysler; **185** (centre left) Mark A. Schneider/Photo Researchers Inc.; **194** (centre left) Photo Courtesy of Dr. Richard Bader; **195** (top left) © Stephen Dalton/Photo Researchers, Inc.; **197** (centre left) Ian Crysler; **213** (top left) Photo courtesy the Photo Imaging Department at the University of Waterloo; **216** (top right) Dale Wilson/Masterfile; **219** (bottom right) John Elk III; **236** (bottom centre) Courtesy of Parks and Natural Areas division files Government of Newfoundland and Labrador; **280** (bottom centre) Vision/Photo Researchers Inc.; **281** (centre left) Ian Crysler; **285** (top left) Ian Crysler; **288** (bottom left) Ian Crysler; **291** (top right) Ian Crysler; **298** (bottom centre) Ian Crysler; **324** (top right) (diamond), © 2000, Rosemary Weller/Stone; (top left) (pencil) Ian Crysler; **365** (centre left) © Mike Dobel, MCMXCI/Masterfile; **369** (top left) Barrett & MacKay Photography Inc ; **370** (top left) Blue-Zone Technologies; **386** (centre left) Archive Holdings Inc./The Image Bank/Getty Images; **394** (top centre) Paul Grebliunas/Getty Images; **398** (bottom left) Plus Pix/Firstlight.ca; **402** (bottom left) (nail polish) Artbase Inc.; **404** (centre left) (cinnamon sticks) Paul Eekhoff/Masterfile **404** (bottom right) (rose) Dick Keen/Visuals Unlimited; **405** (top right) (salad dressing) © Wally Eberhart/Visuals Unlimited **426** (bottom centre) Gilbert Grant/Photo Researchers Inc.; **428** (top left) Neal Preston/CORBIS/MAGMA; **434** (bottom right) Grambo Photography/Firstlight.ca; **437** (bottom left) J. Litray/Visuals Unlimited; **462** (bottom centre) Photo Courtesy Hibernia Management and Development Company Ltd.; **463** (centre right) Stu Forster/Allsport; **466** (bottom left) Ian Crysler; **469** (bottom left, bottom right, and bottom centre) from Chemistry 11, © 2001 McGraw-Hill Ryerson Limited, a subsidiary of The McGraw-Hill Companies; **482** (bottom centre) © Marcos Welsh/Firstlight.ca; **482** (top left) Ian Crysler; **482** (top left) Ian Crysler; **482** (centre) Ian Crysler; **482** (centre) Ian Crysler; **489** (bottom left) © Pat Ivy/Ivy Images; **489** (bottom centre) Ian Crysler; **490** (top left) Ian Crysler; **490** (top right) Ian Crysler; **490** (bottom left) Ian Crysler; **490** (bottom right) Ian Crysler; **501** (bottom left) Ian Crysler; **501** (bottom right) Ian Crysler; **520** (centre left) Science Photo Library/Photo Researchers Inc.; **523** (bottom left) Ian Crysler; **523** (bottom centre) Ian Crysler; **523** (bottom right) Ian Crysler; **524** (bottom left) Ian Crysler; **524** (bottom centre) Ian Crysler; **524** (bottom right) Ian Crysler; **525** (top left) © Leonard Rue III/Visuals Unlimited Inc.; **534** (top right) Adrian Holmes Photography Ltd.; **615** (top left) Photo Courtesy of Transport Canada; **626** (bottom right) Beneluxpress/Firstlight.ca; **646** (top left) Andrew Wenzel/Masterfile; **684** (bottom right) (diamond ring) © Thom Lang/ The Stock Market/Firstlight.ca; (bottom centre) (pencil) **712** (centre) Artbase Inc.; **713** (top left) Stephen Frisch; **713** (top right) Stephen Frisch; **717** (top right) Josh Mitchell/Index Stock Imagery/Picture Quest; **721** (top left) © Jeff Daly/Stock Boston; **730** (centre right) © 1997 Brownie Harris/The Stock Market/Firstlight.ca; **736** (centre left) Stephen Frisch; **736** (centre) Stephen Frisch; **736** (centre right) Stephen Frisch; **739** (bottom right) © Phil Martin/Photo Edit; **156–157** Artbase Inc. (Gord Pronk); **318–319** Paul McCormaick/Image Bank; **460–461** Malcome Hanes/Bruce Coleman Inc.; **542–543** Photobank Yokohama/Firstlight.ca; **710–711** Artbase Inc.